Properties of Water (English)

Temperature (°F)	*Density* (slugs/ft^3)	*Specific weight* (lb/ft^3)	*Viscosity* (lb-sec/ft^2)	*Kinematic viscosity* (ft^2/sec)	*Bulk modulus* (lb/in^2)	*Surface tension* (lb/ft)	*Vapor pressure* (lb/in^2)
32	1.94	62.4	3.75×10^{-5}	1.93×10^{-5}	293,000	0.518×10^{-2}	0.089
40	1.94	62.4	3.23	1.66	294,000	0.514	0.12
50	1.94	62.4	2.74	1.41	305,000	0.509	0.178
60	1.94	62.4	2.36	1.22	311,000	0.504	0.256
70	1.94	62.4	2.05	1.06	320,000	0.500	0.340
80	1.93	62.1	1.80	0.93	322,000	0.492	0.507
90	1.93	62.1	1.60	0.826	323,000	0.486	0.698
100	1.93	62.1	1.42	0.739	327,000	0.480	0.949
120	1.92	61.8	1.17	0.609	333,000	0.465	1.69
140	1.91	61.5	0.98	0.514	330,000	0.454	2.89
160	1.90	61.1	0.84	0.442	326,000	0.441	4.74
180	1.88	60.5	0.73	0.385	318,000	0.426	7.51
200	1.87	60.2	0.64	0.341	308,000	0.412	11.53
212	1.86	59.8	0.59×10^{-5}	0.319×10^{-5}	300,000	0.404×10^{-2}	14.7

Properties of Air at Atmospheric Pressure (English)

Temperature (°F)	*Density* (slugs/ft^3)	*Viscosity* (lb-sec/ft^2)	*Kinematic viscosity* (ft^2/sec)	*Speed of sound* (ft/sec)
−20	0.00280	3.34×10^{-7}	11.9×10^{-5}	1028
0	0.00268	3.38	12.6	1051
20	0.00257	3.50	13.6	1074
40	0.00247	3.62	14.6	1096
60	0.00237	3.74	15.8	1117
68	0.00233	3.81	16.0	1125
80	0.00228	3.85	16.9	1138
100	0.00220	3.96	18.0	1159
120	0.00213	4.07	18.9	1180
160	0.00199	4.23	21.3	1220
200	0.00187	4.50	24.1	1258
300	0.00162	4.98	30.7	1348
400	0.00144	5.26	36.7	1431
1000	0.000844	7.87×10^{-7}	93.2×10^{-5}	1839

Thermal Sciences

An Introduction to Thermodynamics, Fluid Mechanics, and Heat Transfer

Thermal Sciences

An Introduction to Thermodynamics, Fluid Mechanics, and Heat Transfer

Merle C. Potter
Michigan State University

Elaine P. Scott
Virginia Polytechnic Institute and State University

CENGAGE Learning™

Australia • Brazil • Japan • Korea • Mexico • Singapore • Spain • United Kingdom • United States

CENGAGE
Learning™

Thermal Sciences: An Introduction to Thermodynamics, Fluid Mechanics, and Heat Transfer
Merle C. Potter and Elaine P. Scott

Publisher: Bill Stenquist

Editorial Assistants: Valerie Boyajian, Julie Ruggiero

Technology Project Manager: Burke Taft

Executive Marketing Manager: Tom Ziolkowski

Marketing Communications: Margaret Parks

Editorial Production Project Managers: Mary Vezilich, Kelsey McGee

Composition Buyer: Micky Lawler

Print/Media Buyer: Rebecca Cross

Permissions Editors: Sue Ewing, Kiely Sexton

Production Service: RPK Editorial Services

Copy Editor: Rose Kernan

Text Designer: Roy R. Neuhaus

Cover Designer: Roger Knox

Cover Image: Ryan McVay

Compositor: Asco Typesetters

Library of Congress Control Number: 2003102725

ISBN-13: 978-0-534-38521-7

ISBN-10: 0-534-38521-4

Cengage Learning
5191 Natorp Boulevard
Mason, OH 45040
USA

Printed in the United States of America
3 4 5 6 15 14

Contents

Preface

The suggestion to write this book was offered by a good friend, Professor Ward Frea, of Michigan Technological University. Sadly, he died while cross-country skiing (he was a champion skier) before he was able to contribute to the text. But his idea of combining the thermal sciences into one text was a good one and we have done just that. There are often three courses, Thermodynamics, Fluid Mechanics, and Heat Transfer, which make up the introductory series in thermal sciences. In some curricula there may be only two required courses, and in some only one thermal science course that introduces all three subjects. This text is written so it can be used in either the two-semester sequence of thermodynamics and fluid mechanics or in the course that introduces all three subjects. Possibly, in a second edition, the heat transfer material, which is introduced in several sections at appropriate locations in the text, will be expanded to provide for a full semester course.

Thermodynamics and fluid mechanics texts have increased in length over the years so that now they each may contain 1000 pages. Much of that material is never used in the classroom and much of it tends to confuse the students with material that's not significant to the subject at hand. We have attempted to eliminate much of that material, especially the material that's most often reserved for an advanced course.

By combining the thermal science material into one text, we have maintained the same nomenclature as much as possible throughout the book. This eliminates the confusion that often results when using texts written by different authors. For example, we use Q to denote heat transfer and Q_f to denote flow rate. We use q'' to denote energy emitted per unit area due to radiation, rather than the more conventional but unusual G. This makes for a more unified approach to thermal science and presents thermal science as one subject rather than three separate subjects. There are situations, however, where the desire to use the same variable gives way to a more logical choice: we use v for specific volume in thermodynamics but in fluid mechanics it represents a velocity component. The context eliminates any possible confusion in these cases.

We have maintained the mathematical level such that the solution of differential equations is not required except in a few relatively simple applications. We emphasize the physical aspects with applications to numerous everyday situations. It is important that students understand the phenomena being studied and not be burdened with unnecessary math. They will then be better able to apply the basic laws to solve problems encountered in their engineering professions.

The Thermodynamics Part includes more material than can be covered in a one-semester course; this allows for the introductory course in thermodynamics to be structured to fit the desire of the faculty at a particular school. Selected material could come from the chapters on vapor and gas power and refrigeration cycles, psychrometrics, and combustion. The Fluid Mechanics Part also contains more material than can be covered in a one-semester course; material on potential flows, boundary layers, and compressible flows can be selected for a particular course. The heat transfer material that is included in various chapters can be inserted, if desired, as it is encountered in the text. A one-semester service course for non-mechanical engineers may be organized with selected sections from both the Thermodynamics Part and the Fluid Mechanics Part including the sections on heat transfer.

Thermodynamics is presented in chapters 1 through 9, fluid mechanics in Chapters 10 through 17, and the introductory material of heat transfer is included in Sections 3.6, 4.11, and 16.6.6. All the material is presented so that students can follow the derivations with relative ease; reference is made to figures and previous equations using an easy-to-follow style of presentation. Numerous examples then illustrate all the basic principles of the text. Problems at the end of each chapter then allow for application of those principles to numerous situations encountered in real life.

The problems at the end of each chapter begin with a set of questions that are typical of the questions encountered on the Fundamentals of Engineering Exam (the exam usually taken at the end of the senior year to begin the process of licensure) and the Graduate Record Exam/Engineering. Those questions are followed with problems, often grouped according to topics and ordered by level of difficulty, which illustrate the principles presented in the text material. Answers to selected problems are included at the end of the text.

When presenting the solution to an example we have, in general, included only the numerical values with the assumption that the units will follow providing basic units are used in the equations. Units are included in those cases where confusion may result or when they are first used. We do not desire that students work two problems every time: a problem using the numbers and a problem using the units. If basic units are used, the answer will also result with the basic unit. That's the way physical laws work and we do not wish to complicate the students' solutions of problems. For example, when substituting into Newton's second law to calculate a force, if we use the mass in kilograms and the acceleration in m/s^2, the force will be in newtons. We don't have to make sure it's in newtons by checking the units every time. That's the way nature works and we can be assured that it works that way every time!

We have used SI units primarily but have presented about 20% of the examples and home problems using English units since a significant number of industries in the United States continue to use the English units; we feel that it is important for students to have some exposure to such units. In the Thermodynamics Part we have used the pound force and pound mass along with the Btu, the conventional English units used in thermodynamics. In the Fluid Mechanics Part we have used the pound force and the slug along with the foot-pound, the usual English units used in fluids. The tables present properties using both sets of units. In addition to the tables, a CD is included that provides properties of solids, liquids, and gases.

We have also included the use of TK Solver, a powerful calculation tool, to determine the properties of steam, R134a, and gases. We have not used its calculation abilities but only its property determination abilities. Some faculty and students may desire to use its more advanced abilities and the CD in the back of the book will allow that possibility. There are short tutorials included on the CD.

A solutions manual is available for all instructors from the publisher. It is written so the instructors do not have to spend precious time solving all the problems assigned to the students and to allow the instructors to check their solutions. It is not intended to provide a detailed solution for the students thereby replacing the important student-instructor relationship. So, our solutions manual is not an encyclopedic production but a concise presentation of the solution to every problem in the text. We have included sample exams in the manual to aid the instructor in the creation of exams, a time-consuming part of the educational process.

We wish to thank those who have put up with us while we've spent the many hours needed to put this book together. Professor Ward Frea deserves special thanks even though he's not here to see the final product. We know he would be pleased with the result. Professor JB Jones, of Virginia Polytechnic Institute, also deserves much thanks for answering many questions about the use of TK Solver and adapting it to this text. And all of you who use this text are encouraged to send us your comments so that we can improve possible future editions for both the students and the instructors. We will develop a web site that will list any errors that we discover and other related information.

To my friend,
Professor Ward Free

To my boys,
Charlie and Jacob

Merle C. Potter, East Lansing, Michigan
Elaine P. Scott, Blacksburg, Virginia

Thermal sciences include heat transfer, phase change of substances, and the flow of fluids.

Introduction to Thermal Sciences

Many physical systems involve energy storage, transformation, transfer, and transport; *Thermal Sciences* presents an introductory analysis of these phenomena. To completely analyze each of these phenomena, several courses in the engineering curricula are required. In fact, in addition to the introductory courses, numerous graduate courses are needed to understand the more complicated applications. In the study of Engineering Thermodynamics, we present a more global analysis and do not investigate the details of how energy is transferred or transported. The details of how energy is transported by a fluid, inside a pipe or around a body, are considered in Fluid Mechanics. Heat Transfer presents the details of how energy is transferred from a boundary, such as a pipe, to another body in contact with the boundary, or to a body some distance away, or to the fluid next to the boundary.

As an example, consider a power plant that produces electricity. In Thermodynamics, we will present the overall performance of the power plant and each of its primary components: the boiler, the turbine, the condenser, and the circulating pump. The influence of efficiencies will also be considered. The pressure drop through the many tubes and the design of the turbine blades for maximum efficiency would be considered in Fluid Mechanics. The transfer of heat in the boiler from the coal fire to the tubes and in the condenser from the tubes to the cooling water would be calculated in a Heat Transfer course. The subjects are usually presented in the above order since Engineering Thermodynamics does not require a knowledge of either fluid mechanics or heat transfer.

In this text, we will present Thermodynamics in the first Part and Fluid Mechanics in the second Part. There is sufficient material to present a one-semester course in each of these two subjects or topics from both Parts can be selected to form a one-semester course. The subject of Heat Transfer has been incorporated into the Thermodynamics and Fluid Mechanics Parts at the appropriate locations. Heat Transfer that occurs due to conduction and radiation does not require a knowledge of Fluid Mechanics so it will be found in the Thermodynamics Part, and Heat Transfer that occurs due to convection will be found in the Fluid Mechanics Part. The more complicated heat transfer modes are not considered in this text but can be found in a Heat Transfer text.

Part 1

Thermodynamics

Space vehicles orbit the earth at a radius where the force due to gravity and the centripetal force due to the vehicle's speed just balance each other. (Image courtesy of Getty Images.)

1

Concepts, Definitions and Basic Principles

Outline

Chapter Objectives

The objectives of this chapter are to:

- ▲ Introduce the basic concepts and definitions used in our study of thermodynamics.
- ▲ Discuss many of the constraints that limit the applications that are presented.
- ▲ Introduce the units used in our study.
- ▲ Define pressure and introduce temperature.
- ▲ Present numerous examples and practice problems that illustrate the concepts discussed.

Nomenclature

The following variables are introduced in this chapter:

a	Acceleration	v	Specific volume
A	Area	V	Velocity
F	Force	$\forall$	Volume
g	Gravity	W	Weight
h	Vertical downward	z	Vertical direction
KE	Kinetic energy		
m	Mass	ε	A small volume
PE	Potential energy	ϕ	A general property
p	Pressure	γ	Specific weight
T	Temperature	ρ	Density
U	Internal energy		

1.1 INTRODUCTION

Key Concept: *Energy is stored, transformed and transferred.*

Thermodynamics is a science in which the storage, transformation, and transfer of energy are studied. Energy is *stored* as internal energy (associated with temperature), kinetic energy (due to motion), potential energy (due to elevation) and chemical energy (due to chemical composition); it is *transformed* from one of these forms to another; and it is *transferred* across a boundary as either heat or work. In thermodynamics, we will derive equations that relate the transformations and transfers of energy to properties such as temperature, pressure, and density. Substances and their properties, thus, become very important in thermodynamics. Many of our equations will be based on experimental observations that have been organized into mathematical statements or *laws*; the first and second laws of thermodynamics are the most widely used.

Key Concept: *Experimental observations have been organized into laws.*

The engineer's objective in studying thermodynamics is most often the analysis or design of a large-scale system—anything from an air conditioner to a nuclear power plant. The flow of a fluid through such a system may be regarded as a continuum in which the activity of the constituent molecules is averaged into measurable quantities such as pressure, temperature, and velocity. This book, then, will be restricted to *macroscopic* or *engineering thermodynamics.* If the behavior of individual molecules is important, a text in *statistical thermodynamics* must be consulted.

1.2 THERMODYNAMIC SYSTEMS AND CONTROL VOLUMES

System: *A fixed quantity of matter contained within some enclosure.*

A thermodynamic **system** is a fixed quantity of matter contained within some enclosure. The surface is usually an obvious one (like that surrounding the gas in the cylinder of Fig. 1.1). However, it may be an imagined boundary (like the

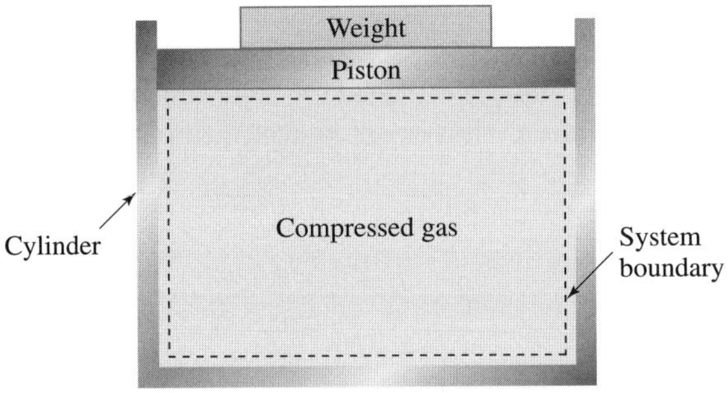

FIGURE 1.1 A system.

deforming boundary of a certain amount of mass as it flows through a pump). In Fig. 1.1, the system is the compressed gas (the *working fluid*), and the dashed line shows the *system boundary*.

All matter and space external to a system is collectively called its **surroundings**. Thermodynamics is concerned with the interactions of a system and its surroundings, or one system interacting with another. A system interacts with its surroundings by transferring energy across its boundary. No material crosses the boundary of a system. If the system does not exchange energy with the surroundings, it is an *isolated* system.

Surroundings: *All matter and space external to a system.*

In many cases, an analysis is simplified if attention is focused on a particular volume in space into which, or from which, a substance flows. Such a volume is a **control volume**. A pump, a turbine, and an inflating or deflating balloon are examples of control volumes. The surface that completely surrounds the control volume is called a **control surface**. An example is sketched in Fig. 1.2.

Control volume: *A volume in space into which, or from which, a substance flows.*

Thus, we must choose, in a particular problem, whether a system is to be considered or whether a control volume is more useful. If there is mass flux across a boundary, then a control volume is required; otherwise, a system is identified. We will present the analysis of systems first and follow that with a study using control volumes.

Control surface: *The surface that completely surrounds the volume control.*

Key Concept: *If there is a mass flux across a boundary, then a control volume is required.*

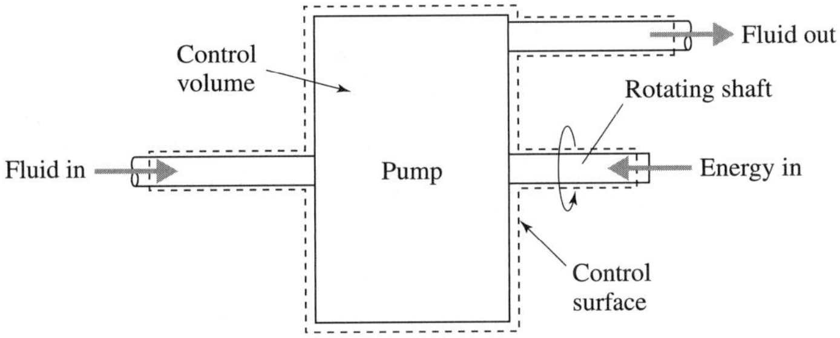

FIGURE 1.2 A control volume.

1.3 MACROSCOPIC DESCRIPTION

Continuum: *The material that is continuously distributed throughout the region.*

In engineering thermodynamics, we postulate that the material in our system or control volume is a **continuum**; that is, it is continuously distributed throughout the region of interest. Such a postulate allows us to describe a system or control volume using only a few measurable properties.

Consider the definition of *density* given by

$$\rho = \lim_{\Delta V \to 0} \frac{\Delta m}{\Delta V} \tag{1.1}$$

where Δm is the mass contained in the volume ΔV, shown in Fig. 1.3. Physically, ΔV cannot be allowed to shrink to zero since, if ΔV became extremely small, Δm would vary discontinuously, depending on the number of molecules in ΔV. So, the zero in the definition of ρ should be replaced by some quantity ε, small, but large enough to eliminate molecular effects. Noting that there are about 3×10^{16} molecules in a cubic millimeter of air at standard conditions, ε need not be very large to contain billions and billions of molecules. For most engineering applications, ε is sufficiently small that we can let it be zero, as in Eq. 1.1.

There are, however, situations where the continuum assumption is not valid (i.e., the re-entry of satellites). At an elevation of 100 km, the *mean free path* (the average distance a molecule travels before it collides with another molecule) is about 30 mm; the macroscopic approach is already questionable. At 150 km, the mean free path exceeds 3 m, which is comparable to the dimensions of the satellite! Under these conditions, statistical methods based on molecular activity must be used.

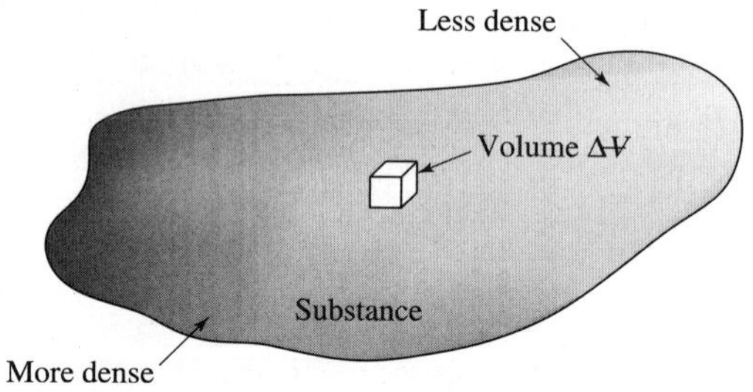

FIGURE 1.3 Mass as a continuum.

1.4 PROPERTIES AND STATE OF A SYSTEM

Phase: *A quantity of matter that has the same chemical composition throughout.*

The matter in a system may exist in several phases: as a solid, a liquid, or a gas. A **phase** is a quantity of matter that has the same chemical composition throughout, that is, it is *homogeneous*. It is all solid, all liquid, or all gas. Phase boundaries separate the phases in what, when taken as a whole, is called a *mixture*. Gases can be mixed in any ratio to form a single phase. Two liquids that are miscible form a mixture when mixed, but liquids that are not miscible, such as water and oil, form two phases.

A *pure substance* is uniform in chemical composition. It may exist in more than one phase (e.g., ice, liquid water, and vapor) in which each phase would have the same composition. A uniform mixture of gases is a pure substance as long as it does not react chemically (as in combustion) or liquefy, in which case the composition would change.

A **property** is any quantity that serves to describe a system. The **state** of a system is its condition as described by giving values to its properties at a particular instant. The common properties are pressure, temperature, volume, velocity, and position; but others must occasionally be considered. *Shape* is important when surface effects are significant; *color* is important when radiation heat transfer is being investigated.

Property: *Any quantity that serves to describe a system.*

State: *A condition described by giving values to its properties.*

The essential feature of a property is that it has a unique value when a system is in a particular state, and this value does not depend on the previous states that the system passed through; that is, it is not a path function. Since a property is not dependent on the path, any change depends only on the initial and final states of the system. Using the symbol ϕ to represent a property, the mathematical statement is

$$\int_{\phi_2}^{\phi_1} d\phi = \phi_2 - \phi_1$$

This requires that $d\phi$ be an exact differential; $\phi_2 - \phi_1$ represents the change in the property as the system changes from state 1 to state 2. There are several quantities that we will encounter, such as work, that are path functions for which an exact differential does not exist.

Key Concept: *There are several quantities, such as work, that are path functions for which an exact differential does not exist.*

A relatively small number of *independent properties* suffice to fix all other properties and, thus, the state of the system. If the system is composed of a single phase, free from magnetic, electrical and surface effects, the state is fixed when any two properties are fixed; this **simple system** receives the most attention in engineering thermodynamics.

Simply system: *The state is fixed when any two properties are fixed.*

Thermodynamic properties are divided into two general types, intensive and extensive. An **intensive property** is one that does not depend on the mass of the system; temperature, pressure, density and velocity are examples, since they are the same for the entire system or for parts of the system. If we bring two systems together, intensive properties are not summed.

Intensive property: *One that does not depend on the mass of the system.*

An **extensive property** is one that depends on the mass of the system; mass, volume, momentum, and kinetic energy are examples. If two systems are brought together, the extensive property of the new system is the sum of the extensive properties of the original two systems.

Extensive property: *One that depends on the mass of the system.*

If we divide an extensive property by the mass, a **specific property** results. The *specific volume* is thus defined to be

Specific property: *Divide an extensive property by the mass.*

$$v = \frac{V\!\!\!/}{m} \tag{1.2}$$

We will generally use an uppercase letter to represent an extensive property [exception: m for mass] and a lowercase letter to denote the associated intensive property.

1.5 EQUILIBRIUM, PROCESSES, AND CYCLES

Thermodynamic equilibrium: *When the properties are constant from point to point and there is no tendency for change with time.*

When the temperature of a system is referred to, it is assumed that all points of the system have the same, or essentially the same, temperature. When the properties are constant from point to point and when there is no tendency for change with time, a condition of **thermodynamic equilibrium** exists. If the temperature, say, is suddenly increased at some part of the system boundary, spontaneous redistribution is assumed to occur until all parts of the system are at the same temperature. If the effects of gravity were negligible, pressure, as well as specific volume, would also be uniform throughout the system, as occurs with gases. If gravity effects are significant, pressure could vary with elevation, such as occurs behind a dam.

If a system would undergo a large change in its properties when subjected to some small disturbance, it is said to be in *metastable equilibrium.* A mixture of gasoline and air and a bowling ball on top of a pyramid are examples.

Process: *When a system changes from one equilibrium state to another, it is the path of successive states.*

Key Concept: *Quasi-equilibrium processes can approximate many processes, such as the compression and expansion of gases in an internal combustion engine.*

When a system changes from one equilibrium state to another, the path of successive states through which the system passes is called a **process**. If, in the passing from one state to the next, the deviation from equilibrium is infinitesimal, a *quasi-equilibrium* process occurs, and each state in the process may be idealized as an equilibrium state. Quasi-equilibrium processes can approximate many processes, such as the compression and expansion of gases in an internal combustion engine, with no significant loss of accuracy. If a system undergoes a quasi-equilibrium process (such as the compression of air in a cylinder), it may be sketched on appropriate coordinates by using a solid line, as shown for the volume and pressure in Fig. 1.4a. If the system, however, goes from one equilibrium state to another through a series of nonequilibrium states (as in combustion), a *nonequilibrium process* occurs. In Fig. 1.4b, the dashed curve may represent such a process; between $(\mathcal{V}_1, p_1)$ and $(\mathcal{V}_2, p_2)$ properties are not uniform throughout the system; thus, the state of the system is not known at each state between the two end states.

Whether a particular process may be considered quasi-equilibrium or nonequilibrium depends on how the process is carried out. Let us add the weight W to the piston of Fig. 1.5, and explain how W can be added in a nonequilibrium manner or in an equilibrium manner. If the weight is added suddenly as one large

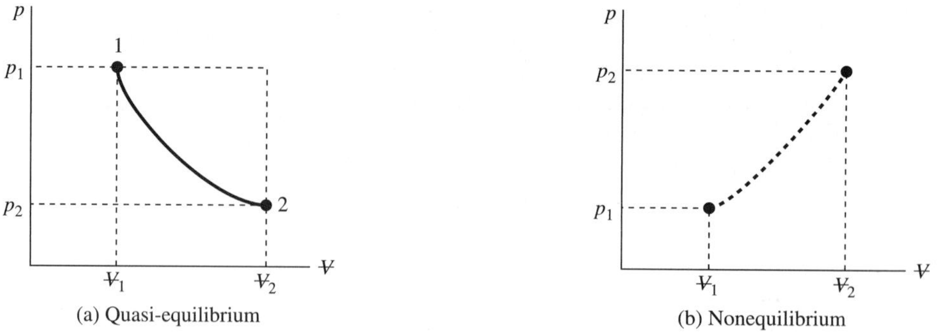

FIGURE 1.4 A process.

FIGURE 1.5 Equilibrium and nonequilibrium additions of weight.

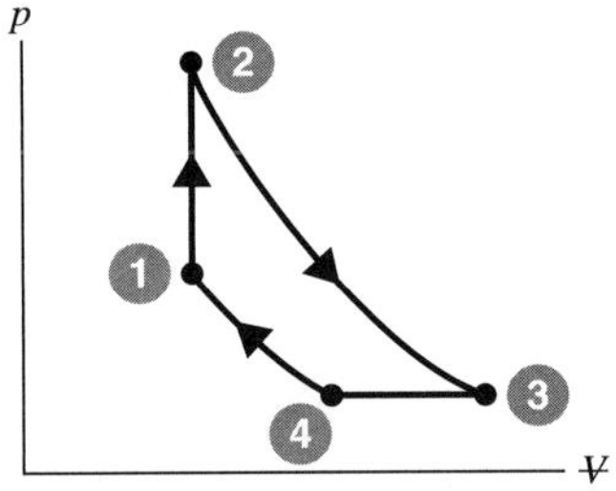

FIGURE 1.6 Four processes that make up a cycle.

weight, as in part (a), a nonequilibrium process will occur in the gas (the system). If we divide the weight into a large number of small weights and add them one at a time, as in part (b), a quasi-equilibrium process will occur.

Note that the surroundings play no part in the notion of equilibrium. It is possible that the surroundings do work on the system via friction; for quasi-equilibrium, it is only required that the properties of the system be uniform at any instant during a process.

Key Concept: *For quasi-equilibrium, it is only required that the properties of the system be uniform at any instant during a process.*

When a system in a given initial state experiences a series of processes and returns to the initial state, the system undergoes a **cycle**. At the end of the cycle, the properties of the system have the same values they had at the beginning (see Fig. 1.6).

The prefix *iso-* is attached to the name of any property that remains unchanged in a process. An **isothermal** process is one in which the temperature is held constant; in an **isobaric** process, the pressure remains constant; an **isometric** process is a constant-volume process. Note the isobaric and the isometric legs in Fig. 1.6.

Cycle: *When a system in a given initial state experiences a series of processes and returns to the initial state.*

Isothermal: *Temperature is held constant.*

Isobaric: *Pressure is held constant.*

Isometric: *Volume is held constant.*

1.6 UNITS

While the student is undoubtedly most at home with SI units, much of the data gathered in the United States is in English units. Therefore, a certain number of examples and problems will be presented in English units. Table 1.1 lists units of the principal thermodynamic quantities. A table of unit conversions is presented in Appendix A.2. Observe the use of $\mathcal{V}$ for volume and V for velocity.

TABLE 1.1 Conversion Factors

Quantity	Symbol	SI units	English units	To convert from English to SI units multiply by
Length	L	m	ft	0.3048
Mass	m	kg	lbm or slug	0.4536 or 14.59
Time	t	s	sec	1
Area	A	m^2	ft^2	0.09290
Volume	$\forall$	m^3	ft^3	0.02832
Velocity	V	m/s	ft/sec	0.3048
Acceleration	a	m/s^2	ft/sec^2	0.3048
Angular velocity	Ω	rad/s	sec^{-1}	1
Force, Weight	F, W	N	lbf	4.448
Density	ρ	kg/m^3	lbm/ft^3	16.02
Specific weight	γ	N/m^3	lbf/ft^3	157.1
Pressure, Stress	p, τ	kPa	lbf/ft^2	0.04788
Work, Energy	W, E, U	J	ft-lbf	1.356
Internal energy	E, U	J	Btu	1055
Heat transfer	Q	J	Btu	1055
Power	$\dot{W}$	W	ft-lbf/sec or hp	1.356 or 745.7
Heat flux	$\dot{Q}$	W or J/s	Btu/sec	1055
Mass flux	$\dot{m}$	kg/s	lbm/sec	0.4536
Flow rate	Q_f	m^3/s	ft^3/sec	0.02832
Specific heat	c	kJ/kg·K	Btu/lbm-°R	4.187
Specific enthalpy	h	kJ/kg	Btu/lbm	2.326
Specific entropy	s	kJ/kg·K	Btu/lbm-°R	4.187
Specific volume	v	m^3/kg	ft^3/lbm	0.06242

When expressing a quantity in SI units, certain letter prefixes shown in Table 1.2 may be used to represent multiplication by a power of 10. So, rather than writing 30 000 000 W or 30×10^6 W, we may simply write 30 MW.

The units of various quantities are interrelated via the physical laws obeyed by the quantities. It follows that, in either system, all units may be expressed as

TABLE 1.2 Prefixes for SI units

Multiplication factor	Prefix	Symbol
10^{12}	tera	T
10^9	giga	G
10^6	mega	M
10^3	kilo	k
10^{-2}	centi*	c
10^{-3}	mili	m
10^{-6}	micro	μ
10^{-9}	nano	n
10^{-12}	pico	p

*Discouraged except in cm, cm^2, or cm^3.

algebraic combinations of a selected set of *base units.* There are seven base units in the SI system: m, kg, s, K, mol (mole), A (ampere), cd (candela). The last one is rarely encountered in engineering thermodynamics. The SI system of units is used primarily in the United States; however, various other systems are also in use. The English Engineering System will be the other system used in this text. It uses ft, lbf, lbm, sec, and °R as base units. To better understand this English system, consider the following.

We often use Newton's second law,

$$F = ma \tag{1.3a}$$

If we measure F in newtons, m in kg, and a in m/s^2, no additional constant is needed. If we measure F in kN, m in kg, and a in m/s^2, we must insert a constant to make the units consistent: $1000F = ma$, where the constant 1000 converts kN to N. The same applies when using English units; if we measure F in lbf, m in slugs, and a in ft/sec^2, no additional constant is needed. If, however, we choose to measure m in lbm, we must insert a constant: $F = (m/32.2)a$, where the constant 32.2 converts lbm to slugs. In thermodynamics, we traditionally choose to measure mass in lbm when using the English system; so the constant 32.2 lbm/slug, or equivalently, 32.2 lbm-ft/(lbf-sec^2), often appears in the solution of problems. It merely converts the units so that the equation is dimensionally acceptable.

Weight is the force of gravity; by Newton's second law,

$$W = mg \tag{1.3b}$$

Since mass remains constant, the variation of W with elevation is due to the change in the acceleration of gravity g (from about 9.77 m/s^2 on the highest mountain to 9.83 m/s^2 in the deepest ocean trench, only about a 0.3% variation from 9.80 m/s^2). We will use the standard sea-level value of 9.81 m/s^2 (32.2 ft/sec^2), unless otherwise stated.

Finally, a note on significant figures. In engineering calculations, we often do not have confidence in a calculation beyond three significant digits since the information given in the problem statement is often not known to more than three significant digits; in fact, density and other material properties may not be known to even three significant digits. The diameter of a pipe may be stated as 2 cm; this would, in general, not be as precise as 2.000 cm would imply. If information used in the solution of a problem is known to only two significant digits, it is incorrect to express a result to more than two significant digits. In the examples and problems, we will assume that all information given is known to three or four significant digits, and the results will be expressed accordingly. If the numeral 1 begins a number, it is not counted in the number of significant digits (i.e., the number 1.210 has three significant digits).

Key Concept: *We will assume that all information given is known to three significant digits.*

Example 1.1

Newton's second law, $F = ma$, relates a net force acting on a body to its mass and acceleration. If a force of one newton accelerates a mass of one kilogram at one m/s^2 and a force of one lbf accelerates 32.2 lbm (1 slug) at a rate of one ft/sec^2, how are the units related?

Solution

The units are related by referring to Newton's second law, $F = ma$,

$$1 \text{ N} = 1 \text{ kg·m/s}^2 \qquad \text{or} \qquad 1 \text{ lbf} = 32.2 \text{ lbm-ft/sec}^2$$

Example 1.2

Express the kinetic energy $mV^2/2$ in acceptable terms if (a) $m = 10$ kg, $V = 5$ m/s, and (b) $m = 10$ lbm, $V = 5$ ft/sec.

Solution

(a) Using the SI system, the unit conversion of Example 1.1 is

$$\frac{1}{2}mV^2 = \frac{1}{2} \times 10 \times 5^2 = 125 \text{ kg·m/s}^2 = 125 \text{ (N·s}^2\text{/m)m}^2\text{/s}^2 = 125 \text{ N·m} = 125 \text{ J}$$

(b) Using the English system, the unit conversion of Example 1.1 is

$$\frac{1}{2}mV^2 = \frac{1}{2} \times \frac{10}{32.2} \times 5^2 = 3.88 \frac{\text{lbm}}{\text{lbm-ft/(lbf-sec}^2)} \frac{\text{ft}^2}{\text{sec}^2} = 3.88 \text{ ft-lbf}$$

1.7 DENSITY, SPECIFIC VOLUME, AND SPECIFIC WEIGHT

By Eq. 1.1, density is mass per unit volume; by Eq. 1.2, specific volume is volume per unit mass. By comparing their definitions, we see that

$$v = \frac{1}{\rho} \tag{1.4}$$

Associated with (mass) density is *specific weight* γ where

$$\gamma = \frac{W}{V\!\!\!/} \tag{1.5}$$

with units N/m^3 (lbf/ft^3). (Note that γ is volume-specific, not mass-specific.) Specific weight is related to density through $W = mg$:

$$\gamma = \rho g \tag{1.6}$$

For water, nominal values of ρ and γ are, respectively, 1000 kg/m^3 (62.4 lbm/ft^3) and 9810 N/m^3 (62.4 lbf/ft^3). For air, the nominal values at sea level are 1.21 kg/m^3 (0.0755 lbm/ft^3) and 11.86 N/m^3 (0.0755 lbf/ft^3).

Example 1.3

The mass of air in a room 3 × 5 × 20 m is known to be 350 kg. Determine the density, specific volume, and specific weight.

Solution

Equations 1.1, 1.4, and 1.6 are used:

$$\rho = \frac{m}{V\!\!\!\!/} = \frac{350}{3 \times 5 \times 20} = 1.167 \text{ kg/m}^3$$

$$v = \frac{1}{\rho} = \frac{1}{1.167} = 0.857 \text{ m}^3/\text{kg}$$

$$\gamma = \rho g = 1.167 \times 9.81 = 11.45 \text{ N/m}^3$$

1.8 PRESSURE

In gases and liquids, it is common to call the effect of a normal force acting on an area the *pressure.* If a force ΔF acts at an angle to an area ΔA (Fig. 1.7), only the normal component ΔF_n enters into the definition of pressure:

$$p = \lim_{\Delta A \to 0} \frac{\Delta F_n}{\Delta A} \tag{1.7}$$

The SI unit of pressure is the pascal (Pa), where 1 Pa = 1 N/m^2. The pascal is a relatively small unit so pressure is usually measured in kPa. The corresponding English unit is lbf/ft^2 (psf), although lbf/in^2 (psi) is commonly used. By consid-

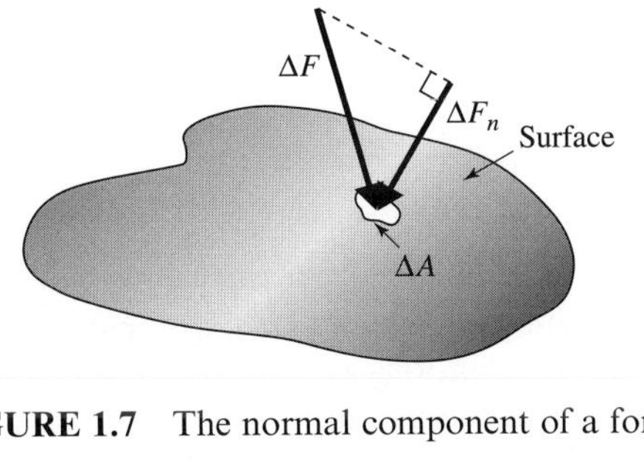

FIGURE 1.7 The normal component of a force.

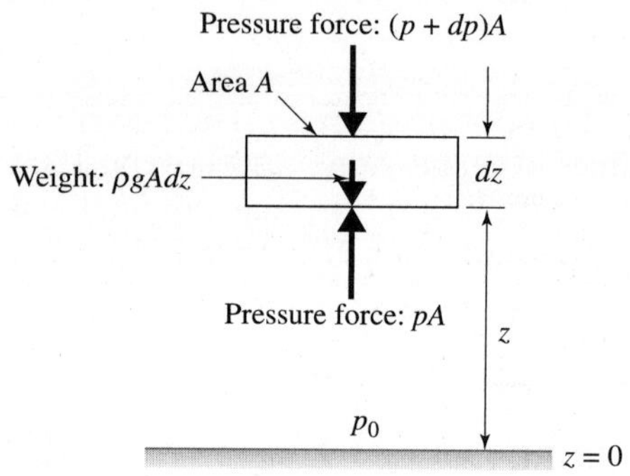

FIGURE 1.8 Forces acting on an element of air.

ering the pressure forces acting on a triangular fluid element of constant depth, we can show that the pressure at a point in a fluid in equilibrium is the same in all directions; it is a scalar quantity. For gases and liquids in relative motion, the pressure may vary from point to point, even at the same elevation, but it does not vary with direction at a given point.

Key Concept: *Pressure may vary from point to point, but it does not vary with direction at a given point.*

Pressure Variation with Elevation. In the atmosphere, pressure varies with elevation. This variation can be expressed mathematically by considering the equilibrium of the element of air shown in Fig. 1.8. Summing forces on the element in the vertical direction (up is positive) gives

$$\sum F_z = pA - (p + dp)A - \rho g A\, dz = 0 \qquad \text{or} \qquad dp = -\rho g\, dz \tag{1.8}$$

Now, if ρ is a known function of z, the above equation can be integrated to give $p(z)$:

$$p(z) - p_0 = -\int_0^z \gamma\, dz \tag{1.9}$$

where we used $\rho g = \gamma$. For a liquid, γ is constant. If we write Eq. 1.8 using $dh = -dz$, we have

$$dp = \gamma\, dh \tag{1.10}$$

where h is measured positive downward. Integrating this equation, starting at a liquid surface where $p = 0$, results in

$$p = \gamma h \tag{1.11}$$

This equation can be used to convert a pressure to pascals when that pressure is measured in, for example, meters of water or millimeters of mercury.

In most thermodynamic relations *absolute pressure* must be used. Absolute pressure is *gage* pressure plus the local atmospheric pressure:

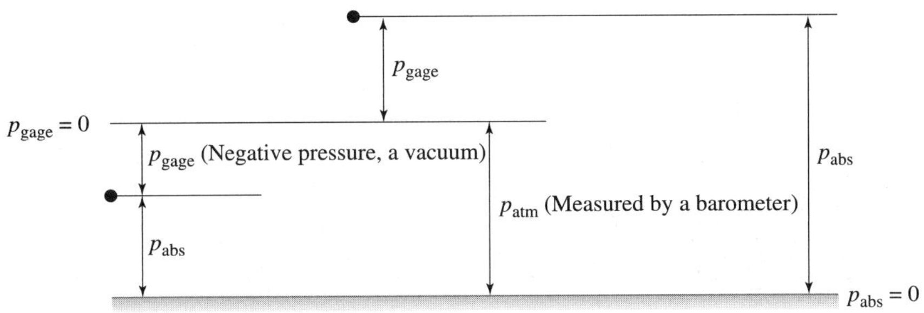

FIGURE 1.9 Absolute and gage pressure.

$$p_{abs} = p_{gage} + p_{atm} \tag{1.12}$$

A negative gage pressure is often called a *vacuum*, and gages capable of reading negative pressures are *vacuum gages.* A gage pressure of −50 kPa would be referred to as a vacuum of 50 kPa, with the sign omitted. Figure 1.9 shows the relationships between absolute and gage pressure.

The word "gage" is generally used in statements of gage pressure (e.g., $p = 200$ kPa gage). If "gage" is not present, the pressure will, in general, be an absolute pressure. Atmospheric pressure is an absolute pressure, and will be taken as 100 kPa (at sea level), unless otherwise stated. It is more accurately 101.3 kPa at standard conditions. It should be noted that atmospheric pressure is highly dependent on elevation; in Denver, Colorado, it is about 84 kPa; and in a mountain city with elevation 3000 m, it is only 70 kPa. In Appendix Table B.1, the variation of atmospheric pressure with elevation is listed.

Key Concept: *Pressure will, in general, be an absolute pressure.*

Example 1.4

Express a pressure gage reading of 35 psi in absolute pascals at an elevation of 2000 m.

Solution

We convert the pressure reading into pascals. We have

$$\left(35\frac{\text{lbf}}{\text{in}^2}\right)\left(144\frac{\text{in}^2}{\text{ft}^2}\right)\left(0.04788\frac{\text{kPa}}{\text{lbf/ft}^2}\right) = 241 \text{ kPa gage}$$

To find the absolute pressure, we simply add the atmospheric pressure to the above value. Using $p_{atm} = 0.7846 \times 101.3 = 79.5$ kPa as found in the Appendix Table B.1, we obtain

$$p = p_{gage} + p_{atm} = 241 + 79.5 = 320.5 \text{ kPa}$$

Example 1.5

The manometer shown in Fig. E1.5 is used to measure the pressure in the water pipe. Determine the water pressure if the manometer reading is 0.6 m. Mercury is 13.6 times heavier than water.

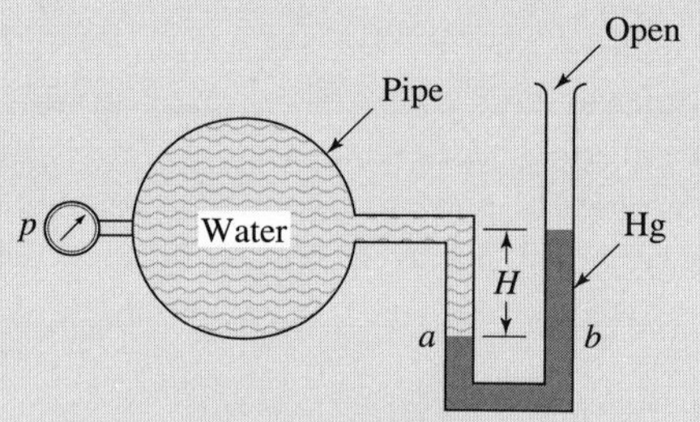

FIGURE E1.5

Solution

To solve the manometer problem, we use the fact that $p_a = p_b$, since a and b are at the same elevation in the same fluid. The pressure p_a is simply the pressure p in the water pipe plus the pressure due to the 0.6 m of water; the pressure p_b is the pressure due to 0.6 m of mercury. Thus,

$$p + \gamma_w h = \gamma_{Hg} h$$

$$p + (9810\ \text{N/m}^3)(0.6\ \text{m}) = (13.6 \times 9810\ \text{N/m}^3)(0.6\ \text{m})$$

$$\therefore p = 74\,200\ \text{Pa or } 74.2\ \text{kPa gage}$$

Example 1.6

Calculate the force due to the pressure acting on the 1-m-diameter horizontal hatch of a submarine submerged 600 m below the surface.

Solution

The pressure acting on the hatch at a depth of 600 m is found from Eq. 1.11 (let's use ρgh) as

$$p = \rho gh = (1000\ \text{kg/m}^3)(9.81\ \text{m/s}^2)(600\ \text{m}) = 5.89 \times 10^6\ \text{kg/(m·s}^2) \text{ or } 5.89\ \text{MPa}$$

Check with Example 1.1 to observe that the units check.

The pressure is constant over the area; hence, the force due to the pressure is given by

$$F = pA = (5.89 \times 10^6\ \text{N/m}^2)\left[\frac{\pi(1)^2}{4}\text{m}^2\right] = 4.62 \times 10^6\ \text{N}$$

1.9 TEMPERATURE

Temperature is, in reality, a measure of molecular activity. However, in classical thermodynamics, the quantities of interest are defined in terms of macroscopic observations only, and a definition of temperature using molecular measurements is not useful. Thus, we must proceed without actually defining temperature. What we shall do instead is discuss *equality of temperatures.*

Equality of Temperatures. Let two bodies be isolated from the surroundings but placed in contact with each other. If one is hotter than the other, the hotter body will become cooler and the cooler body will become hotter; both bodies will undergo change until all properties (e.g., electrical resistance) of the bodies cease to change. When this occurs, *thermal equilibrium* is said to have been established between the two bodies. Hence, we state that two systems have equal temperatures if no change occurs in any of their properties when the systems are brought into contact with each other. In other words, if two systems are in thermal equilibrium, their temperatures are postulated to be equal.

A rather obvious observation is referred to as the *zeroth law of thermodynamics*: if two systems are equal in temperature to a third, they are equal in temperature to each other.

Relative Temperature Scale. To establish a temperature scale, we choose the number of subdivisions, called degrees, between two fixed, easily duplicated points, the ice point and the steam point. The *ice point* exists when ice and water are in equilibrium at a pressure of 101 kPa; the *steam point* exists when liquid water and its vapor are in a state of equilibrium at a pressure of 101 kPa. On the Fahrenheit scale there are 180° between these two points; on the Celsius scale, 100°. On the Fahrenheit scale, the ice point is assigned the value of 32, and on the Celsius scale, it is assigned the value 0. These selections allow us to write:

$$T_F = \frac{9}{5}T_C + 32 \tag{1.13}$$

$$T_C = \frac{5}{9}(T_F - 32) \tag{1.14}$$

Absolute Temperature Scale. The second law of thermodynamics will allow us to define an absolute temperature scale; however, since we do not have the second law at this point and we have immediate use for absolute temperature, an empirical absolute temperature scale will be presented.

The relations between absolute and relative temperatures are

$$T_K = T_C + 273.15 \tag{1.15}$$

$$T_R = T_F + 459.67 \tag{1.16}$$

(The values 273 and 460 are used where precise accuracy is not required, the case for most engineering situations.)

The absolute temperature on the Fahrenheit scale is given in degrees Rankine (°R), and on the Celsius scale it is given in kelvins (K). *Note:* We do not use the degree symbol when writing kelvins; also, we do not say "degrees kelvin."

Example 1.7

The temperature of a body is 50°F. Find its temperature in °C, K, and °R.

Solution

Using the conversion equations,

$$T_C = \frac{5}{9}(50 - 32) = 10°C \qquad T_K = 10 + 273 = 283 \text{ K} \qquad T_R = 50 + 460 = 510°R$$

1.10 ENERGY

A system may possess several different forms of energy. Assuming uniform properties throughout the system, the *kinetic energy* is given by

$$KE = \frac{1}{2}mV^2 \tag{1.17}$$

where V is the velocity of each lump of substance, assumed constant over the entire system. If the velocity is not constant for each lump, then the kinetic energy is found by integrating over the system. The energy that a system possesses due to its elevation h above some arbitrarily selected datum is its *potential energy*; it is determined from the equation

$$PE = mgh \tag{1.18}$$

Other forms of energy include the energy stored in a battery, energy stored in an electrical condenser, electrostatic potential energy, and surface energy. In addition, there is the energy associated with the translation, rotation, and vibration of the molecules, electrons, protons, and neutrons, and the chemical energy due to bonding between atoms and between subatomic particles. These molecular and atomic forms of energy will be referred to as *internal energy* and designated by the letter U. In combustion, energy is released when the chemical bonds between atoms are rearranged; nuclear reactions result when changes occur between the subatomic particles. In thermodynamics, our attention will be focused initially on the internal energy associated with the motion of molecules that is influenced by various macroscopic properties such as pressure, temperature, and specific volume. In Chapter 9, the combustion process is studied in some detail.

Internal energy, like pressure and temperature, is a property of fundamental importance. A substance always has internal energy; if there is molecular activity, there is internal energy. We need not know, however, the absolute value of internal energy, since we will be interested only in its increase or decrease.

Key Concept: *We need not know the absolute value of internal energy, since we will be interested only in its increase or decrease.*

We now come to an important law, which is often of use when considering isolated systems. The law of *conservation of energy* states that the energy of an isolated system remains constant. Energy cannot be created or destroyed in an isolated system; it can only be transformed from one form to another. This is expressed as

$$KE + PE + U = \text{const} \qquad \text{or} \qquad \frac{1}{2}mV^2 + mgh + U = \text{const} \tag{1.19}$$

Consider a system composed of two automobiles that hit head on and come to rest. Because the energy of the system is the same before and after the collision, the initial total kinetic energy KE must simply have been transformed into another kind of energy, in this case, internal energy U, stored primarily in the deformed metal.

Example 1.8

A 2200-kg automobile traveling at 90 kph (25 m/s) hits the rear of a stationary, 1000-kg automobile. After the collision, the large automobile slows to 50 kph (13.89 m/s), and the smaller vehicle has a speed of 88 kph (24.44 m/s). What has been the increase in internal energy, taking both vehicles as the system?

Solution

The kinetic energy before the collision is ($V = 25$ m/s)

$$KE_1 = \frac{1}{2}m_a V_a^2 = \frac{1}{2} \times 2200 \times 25^2 = 687\,500 \text{ J}$$

where the subscript a refers to the first automobile; the subscript b will refer to the second one. After the collision, the kinetic energy is

$$KE_2 = \frac{1}{2}m_a V_a^2 + \frac{1}{2}m_b V_b^2 = \frac{1}{2} \times 2200 \times 13.89^2 + \frac{1}{2} \times 1000 \times 24.44^2 = 510\,900 \text{ J}$$

The conservation of energy requires that

$$E_1 = E_2 \qquad \text{or} \qquad KE_1 + U_1 = KE_2 + U_2$$

Thus,

$$U_2 - U_1 = KE_1 - KE_2 = 687\,500 - 510\,900 = 176\,600 \text{ J or } 176.6 \text{ kJ}$$

1.11 SUMMARY

In this chapter, we defined thermodynamics: The science in which the storage, the transformation, and the transfer of energy are studied. We also defined a number of important terms:

System: A fixed quantity of matter contained within some enclosure.
Surroundings: All matter and space external to the system.
An isolated system: A system that does not exchange energy with its surroundings.
Control volume: A volume into which and/or from which a substance flows.
Control surface: The surface that surrounds the control volume.
Phase: Matter that has the same chemical composition throughout.
Property: Any quantity that describes a system.
Simple system: A system that is defined by only two properties.
Extensive property: A property that depends on the mass of the system.
Quasi-equilibrium: When the properties are constant from point to point in the system.
Isothermal process: A constant temperature process.
Isobaric process: A constant pressure process.
Isometric process: A constant volume process.
Weight: The force of gravity on a system.

There were several equations introduced:

Newton's second law:

$$F = ma \tag{1.20}$$

Pressure: Normal force F_n divided by area:

$$p = \frac{F_n}{A} \tag{1.21}$$

Pressure variation: The pressure varies in the vertical direction according to:

$$p = \gamma h \tag{1.22}$$

where h is positive downward and $p = 0$ at $h = 0$ (e.g., the surface of a manometer or a lake).

Absolute pressure: Pressures in thermodynamics are absolute unless otherwise stated:

$$p = p_{\text{gage}} + p_{\text{atm}} \tag{1.23}$$

where p_{atm} is assumed to be 100 kPa if not given.

Temperature: The temperature in the SI system is degrees Celsius and kelvins:

$$T_C = \frac{5}{9}(T_F - 32) \qquad T_K = T_C + 273 \tag{1.24}$$

Conservation of energy: The energy of an isolated system remains constant:

$$KE + PE + U = \text{const} \quad \text{or} \quad \frac{1}{2}mV^2 + mgh + U = \text{const} \tag{1.25}$$

REVIEW PROBLEMS FOR THE FE/EIT EXAMINATION

1.1FE Engineering thermodynamics does not include energy
A. Storage **B.** Utilization
C. Transfer **D.** Transformation

1.2FE In a quasi-equilibrium process, the pressure in a system
A. Remains constant
B. Varies with temperature
C. Is everywhere constant at an instant
D. Force results in a positive work input

1.3FE Which of the following is not an extensive property?
A. Momentum **B.** Kinetic energy
C. Density **D.** Mass

1.4FE Which of the following would be identified as a control volume?
A. Compression of the air-fuel mixture in a cylinder
B. Filling a tire with air at a service station
C. Expansion of the gases immediately after combustion in a cylinder
D. The Goodyear blimp during flight

1.5FE Which of the following is not an intensive property?
A. Velocity **B.** Pressure
C. Temperature **D.** Volume

1.6FE Which of the following is a quasi-equilibrium process?
A. Mixing a fluid
B. Compression of the air-fuel mixture in a cylinder
C. Combustion
D. A balloon bursting

1.7FE Which of the following is not an acceptable SI unit?
A. Distance measured in centimeters
B. Pressure measured in newtons per square meter
C. Density measured in grams per cubic centimeter
D. Volume measured in cubic centimeters

1.8FE The units of joules can be converted to which of the following?
A. $Pa{\cdot}m^2$ **B.** $N{\cdot}kg$
C. Pa/m^2 **D.** $Pa{\cdot}m^3$

1.9FE The standard atmosphere in meters of ammonia ($\rho = 600\ kg/m^3$) is:
A. 17 m **B.** 19 m
C. 23 m **D.** 31 m

1.10FE Convert 200 kPa gage of pressure to absolute millimeters of mercury ($\rho_{Hg} = 13.6\rho_{water}$).
A. 1500 mm **B.** 1750 mm
C. 2050 mm **D.** 2250 mm

1.11FE A gage pressure of 400 kPa acting on a 4-cm-diameter piston is resisted by a spring with a spring constant of 800 N/m. How much is the spring compressed? Neglect the piston weight.
A. 0.63 m **B.** 0.95 m
C. 1.32 m **D.** 1.98 m

1.12FE Calculate the pressure in the 200-mm-diameter cylinder shown in Fig. P1.12FE. The spring is compressed 40 cm.

A. 138 kPa **B.** 125 kPa
C. 110 kPa **D.** 76 kPa

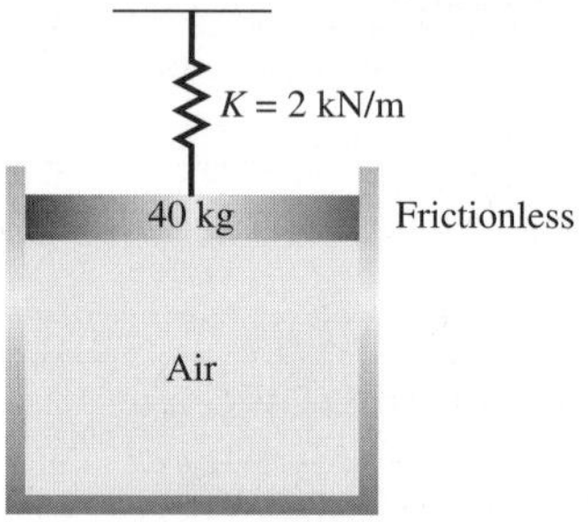

FIGURE P1.12FE

1.13FE What pressure force exists on an 80-cm-diameter horizontal area at a depth of 300 meters below the surface of a lake?

A. 150 kN **B.** 840 kN
C. 1480 kN **D.** 5910 kN

1.14FE In thermodynamics, our attention is focused on which form of energy?

A. Kinetic energy **B.** Internal energy
C. Potential energy **D.** Surface energy

PROBLEMS

1.1 Draw a sketch of the following situations identifying the system or control volume and the boundary of the system or the control surface. **(a)** The combustion gases in a cylinder during the power stroke. **(b)** The combustion gases in a cylinder during the exhaust stroke. **(c)** A balloon exhausting air. **(d)** An automobile tire being heated while driving. **(e)** A pressure cooker during operation.

1.2 Identify which of the following are extensive properties and which are intensive properties: **(a)** a 10-m^3 volume, **(b)** 30 J of kinetic energy, **(c)** a pressure of 90 kPa, **(d)** a stress of 1000 kPa, **(e)** a mass of 75 kg, and **(f)** a velocity of 60 m/s. Convert all extensive properties to intensive properties assuming $m = 75$ kg.

1.3 Which of the following processes can be approximated by a quasi-equilibrium process? **(a)** The expansion of combustion gases in the cylinder of an automobile engine. **(b)** The rupturing of a membrane separating a high and low pressure region in a tube. **(c)** The heating of the air in a room with a baseboard heater.

1.4 The gas in a cubical volume with sides at different temperatures is suddenly isolated with reference to transfer of mass and energy. Is this system in thermodynamic equilibrium? Why or why not?

1.5 A supercooled liquid is a liquid which is cooled to a temperature below that at which it ordinarily solidifies. Is this system in thermodynamic equilibrium? Why or why not?

1.6 Express the following quantities in terms of base SI units (kg, m, and s): **(a)** power, **(b)** heat flux, and **(c)** specific weight.

1.7 Convert the following to SI units: **(a)** 6 ft, **(b)** 4 in^3, **(c)** 2 slugs, **(d)** 40 ft-lbf, **(e)** 2000 ft-lbf/sec, **(f)** 150 hp, **(g)** 10 ft^3/sec.

Newton's Second Law

1.8 Determine the weight of a mass of 10 kg at a location where the acceleration of gravity is 9.77 m/s^2.

1.9 Determine the force necessary to accelerate a mass of 20 lbm at a rate of 60 ft/sec^2 vertically upward.

1.10 The weight of a 10-lb mass is measured at a location where $g = 32.1$ ft/sec^2 on a spring scale originally calibrated in a region where $g = 32.3$ ft/sec^2. What will be the reading?

1.11 A cubic meter of water at room temperature has a weight of 9800 N at a location where $g = 9.80$ m/s^2. What is its specific weight and its density at a location where $g = 9.77$ m/s^2?

1.12 The acceleration of gravity is given as a function of elevation above sea level by the relation $g = 9.81 - 3.32 \times 10^{-6}h$ m/s^2, with h measured in meters. What is the weight of an airplane at 10 km elevation when its weight at sea level is 40 kN?

1.13 Calculate the force necessary to accelerate a 20 000-lbm rocket vertically upward at the rate of 100 ft/sec^2. Assume $g = 32.2$ ft/sec^2.

1.14 Assume the acceleration of gravity on a celestial body to be given as a function of altitude by the expression $g = 4 - 1.6 \times 10^{-6}h$ m/s^2, where h is in meters above the surface of the planet. A space probe weighed 100 kN on earth at sea level. Determine: **(a)** The mass of the probe. **(b)** Its weight on the surface of the planet. **(c)** Its weight at an elevation of 200 km above the surface of the planet.

1.15 Determine the deceleration of **(a)** a 2200-kg car and **(b)** a 1100-kg car, if the brakes are suddenly applied so that all four tires slide. The coefficient of friction $\eta = 0.6$ on the dry asphalt. ($\eta = F/N$ where N is the normal force and F is the frictional force.)

1.16 When a body is accelerated under water, some of the water is also accelerated. This makes the body appear to have a larger mass than it actually has. For a sphere at rest, this added mass is equal to the mass of one half of the displaced water. Calculate the force necessary to accelerate a 10-kg, 300-mm-diameter sphere that is at rest under water at the rate of 10 m/s^2 in the horizontal direction. Use $\rho_{\text{water}} =$ 1000 kg/m^3.

1.17 The force of attraction between two masses m_1 and m_2, having dimensions that are small compared with their separation distance R, is given by Newton's third law, $F = km_1m_2/R^2$, where $k = 6.67 \times 10^{-11}$ N·m^2/kg^2. What is the total gravitational force that the sun (1.97×10^{30} kg) and the earth (5.95×10^{24} kg) exert on the moon (7.37×10^{22} kg) at an instant when the earth, moon, and sun form a 90° angle? The earth–moon and sun–moon distances are 380×10^3 km and 150×10^6 km, respectively.

1.18 The mass that enters into Newton's third law of gravitation (Problem 1.17) is the same as the mass defined by Newton's second law of motion. **(a)** Show that if g is the gravitational acceleration, then $g = km_e/R^2$, where m_e is the mass of the earth and R is the radius of the earth. **(b)** The radius of the earth is 6370 km. Calculate its mass if the acceleration of gravity is 9.81 m/s^2.

1.19 **(a)** A satellite is orbiting the earth at 500 km above the surface with only the attraction of the earth acting on it. Estimate the speed of the satellite. [*Hint:* The acceleration in the radial direction of a body moving with velocity V in a circular path of radius r is V^2/r; this must be equal to the gravitational acceleration (see Problem 1.12).]
(b) The first earth satellite was reported to have circled the earth at 27 000 km/h and its maximum height above the earth's surface was given as 900 km. Assuming the orbit to be circular and taking the mean diameter of the earth to be 12 700 km, determine the gravitational acceleration at this height using **(i)** the force of attraction between two bodies, and **(ii)** the radial acceleration of a moving object.

1.20 Complete the following if $g = 9.81$ m/s^2 and $\mathcal{V} =$ 10 m^3.

	v (m^3/kg)	ρ (kg/m^3)	γ (N/m^3)	m (kg)	W (N)
(a)	20				
(b)		2			
(c)			4		
(d)				100	
(e)					100

1.21 Calculate the density, specific weight, mass, and weight of a body that occupies 200 ft^3 if its specific volume is 10 ft^3/lbm.

Pressure and Temperature

1.22 Complete the following if $p_{atm} = 100$ kPa ($\gamma_{Hg} = 13.6\gamma_{water}$).

	kPa gage	kPa absolute	mm Hg abs	m H_2O abs
(a)	5			
(b)		150		
(c)			30	
(d)				30

1.23 The pressure at a given point is 50-mm Hg absolute. Express this pressure in kPa, kPa gage, and m of H_2O abs if $p_{atm} = 80$ kPa. Use the fact that mercury is 13.6 times heavier than water.

1.24 A manometer tube that contains mercury (Fig. P1.24) is used to measure the pressure p_A in the air pipe. Determine the gage pressure p_{air}. Use $\gamma_{Hg} = 13.6\gamma_{water}$.

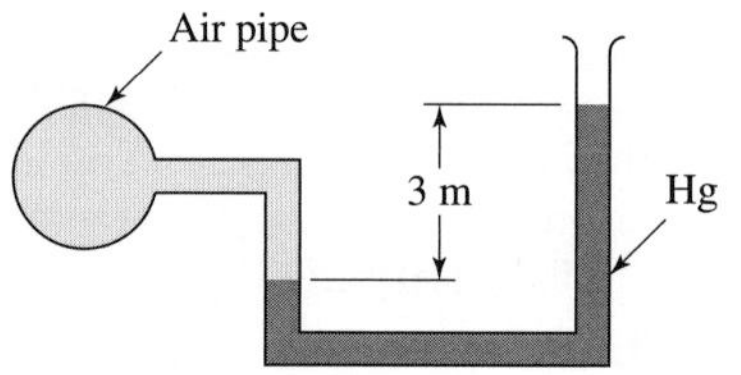

FIGURE P1.24

1.25 Determine the pressure difference between the water pipe and the oil pipe (Fig. P1.25). The top of the mercury column is at the same elevation as the middle of the oil pipe.

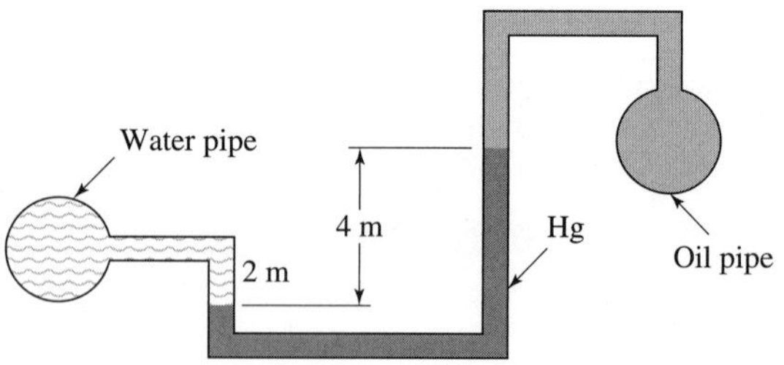

FIGURE P1.25

1.26 A large chamber is separated into compartments 1 and 2, as shown in Fig. P1.26, that are kept at different pressures. Pressure gage *A* reads 300 kPa and pressure gage *B* reads 120 kPa. If the local barometer reads 720 mm Hg, determine the absolute pressures existing in the compartments and the reading of gage *C*.

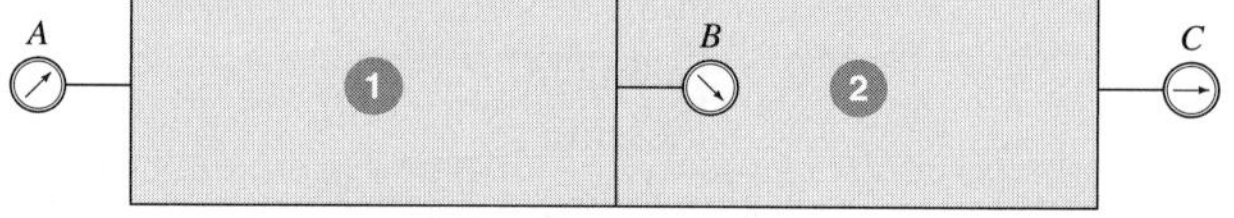

FIGURE P1.26

1.27 A tube can be inserted into the top of a pipe transporting liquids, providing the pressure is relatively low, so that the liquid fills the tube a height *h*. Determine the pressure in a water pipe if the water seeks a level at height $h = 6$ ft above the center of the pipe.

1.28 A bell jar 250 mm in diameter sits on a flat plate and is evacuated until a vacuum of 700-mm Hg exists. The local barometer reads 760-mm Hg. Find the absolute pressure inside the jar, and determine the force required to lift the jar off the plate. Neglect the weight of the jar.

1.29 A horizontal 2-m-diameter gate is located in the bottom of a water tank as shown in Fig. P1.29. Determine the force *F* required to just open the gate.

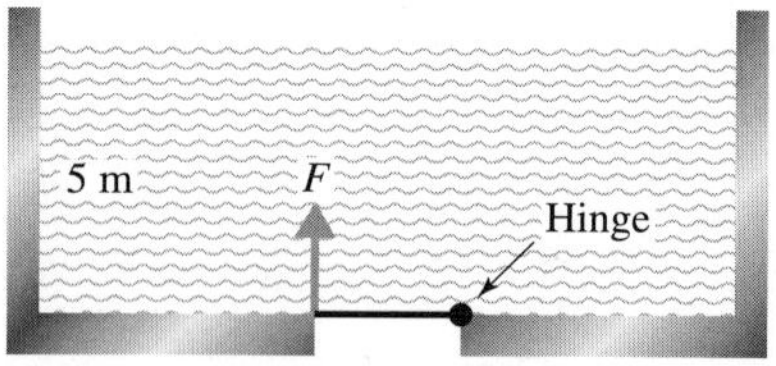

FIGURE P1.29

1.30 A temperature of a body is measured to be 26°C. Determine the temperature in °R, K, and °F.

Energy

1.31 A 10-kg body falls from rest, with negligible interaction with its surroundings (no friction). Determine its velocity after it falls 5 m.

1.32 The potential energy stored in a spring is given by $\frac{1}{2}Kx^2$, where K is the spring constant and x is the distance the spring is compressed. Two springs are designed to absorb the kinetic energy of a 2000-kg vehicle. Determine the spring constant necessary if the maximum compression is to be 100 mm for a vehicle speed of 10 m/s.

1.33 A 0.8-lbm object traveling at 200 ft/sec enters a viscous liquid and is essentially brought to rest before it strikes the bottom. What is the increase in internal energy, taking the object and the liquid as the system? Neglect the potential energy change.

1.34 A 1500-kg vehicle traveling at 60 km/h collides head-on with a 1000 kg vehicle traveling at 90 km/h. If they come to rest immediately after impact, determine the increase in internal energy, taking both vehicles as the system.

1.35 Gravity is given by $g = 9.81 - 3.32 \times 10^{-6}h$ m/s^2, where h is the height above sea level. An airplane is traveling at 900 km/h at an elevation of 10 km. If its weight at sea level is 40 kN, determine **(a)** its kinetic energy and **(b)** its potential energy relative to sea level.

Cooling towers at power plants exhaust steam high off the ground. (Image courtesy of Getty Images.)

2

Properties of Pure Substances

Outline

Chapter Objectives

The objectives of the chapter are to:

- ▲ Introduce the tables needed to obtain properties of substances that change phase.
- ▲ Introduce TK Solver for property determination.
- ▲ Review the equations of state for an ideal gas.
- ▲ Present equations of state for nonideal gases.
- ▲ Present numerous examples and practice problems that illustrate the concepts discussed.

Nomenclature

The following is the nomenclature introduced in this chapter:

a	A constant	p_R	Reduced pressure
b	A constant	R	The gas constant
f	Subscript to denote a liquid	$\bar{R}$	Universal gas constant
g	Subscript to denote a vapor	T_c	Critical-point temperature
K	Spring constant	T_R	Reduced temperature
M	Molar mass	x	Quality
p_c	Critical-point pressure	Z	Compressibility factor

2.1 INTRODUCTION

In this chapter, the relationships between pressure, specific volume, and temperature will be presented for a pure substance. A pure substance is homogeneous. It may exist in more than one phase, but each phase must have the same chemical composition. Water is a pure substance. The various combinations of its three phases have the same chemical composition. Air, in the gas phase, is a pure substance, but liquid air has a different chemical composition. Air is not a pure substance if it exists in more than one phase. In addition, only a *simple compressible substance* will be considered (i.e., a substance that is essentially free of magnetic, electrical, or surface tension effects). We will find the pure, simple, compressible substance of much use in our study of thermodynamics.

2.2 THE *p-v-T* SURFACE

Key Concept: *During the phase change from liquid to vapor, the temperature remains constant as heat is added at constant pressure.*

It is well known that a substance can exist in three different phases: solid, liquid, and gas. Consider an experiment in which a solid (ice) is contained in a piston-cylinder arrangement such that the pressure is maintained at a constant value. Heat is added to the cylinder causing the substance to experience all three phases. Our experiment is shown at various stages in Fig. 2.1. We will record the temperature and specific volume during the experiment. Start with the solid at some low temperature such as point A in Fig. 2.2; then add heat until it just begins to melt at point B. Additional heat will completely melt the solid, with the temperature remaining constant until point C is reached. After all the solid is melted, the temperature of the liquid again rises until vapor just begins to form at point D, the *saturated liquid* state. Again, during the phase change from liquid to vapor, called *vaporization*, the temperature remains constant as heat is added. Finally, all the liquid is vaporized and the state of *saturated vapor* exists at point E, after which the temperature again rises with heat addition. Each specific volume of the solid and liquid is much less than the specific volume of vapor. The scale is exaggerated in this figure so that the differences are apparent.

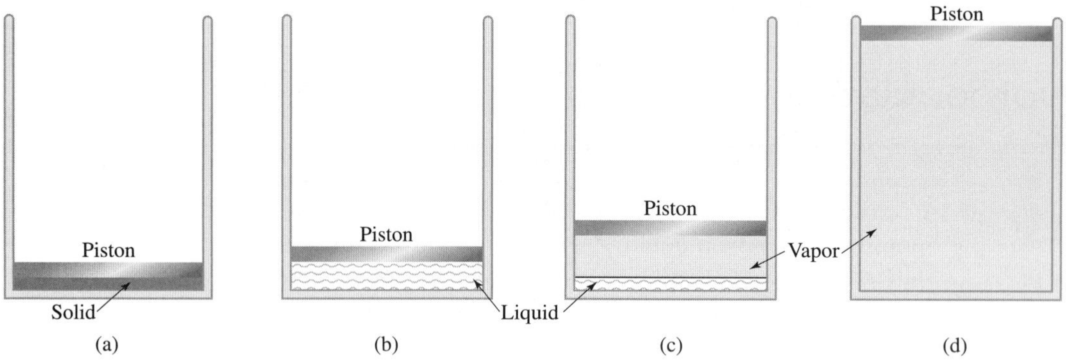

FIGURE 2.1 The solid, liquid, and vapor phases of a substance.

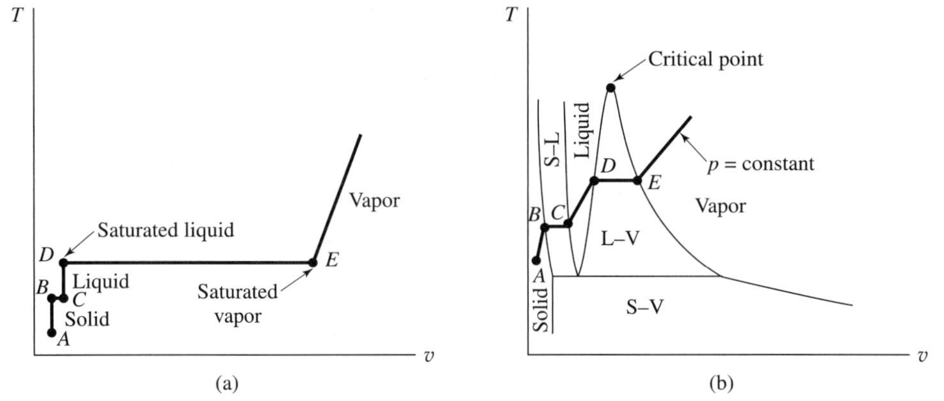

FIGURE 2.2 The *T-v* diagram.

If the experiment is repeated a number of times using different pressures, a *T-v* diagram results, shown in Fig. 2.2b. At pressures that exceed the pressure of the *critical point*, the liquid simply changes to a vapor without a constant-temperature vaporization process. Property values of the critical point for various substances are included in Appendix Table B.3.

The data obtained in an actual experiment could be presented as a three-dimensional surface with $p = p(v, T)$. Figure 2.3 shows a qualitative rendering of a substance that contracts on freezing. For a substance that expands on freezing, the solid–liquid surface would be at a smaller specific volume than for the solid surface. The regions where only one phase exists are labeled solid, liquid, and vapor. Where two phases exist simultaneously the regions are labeled solid–liquid

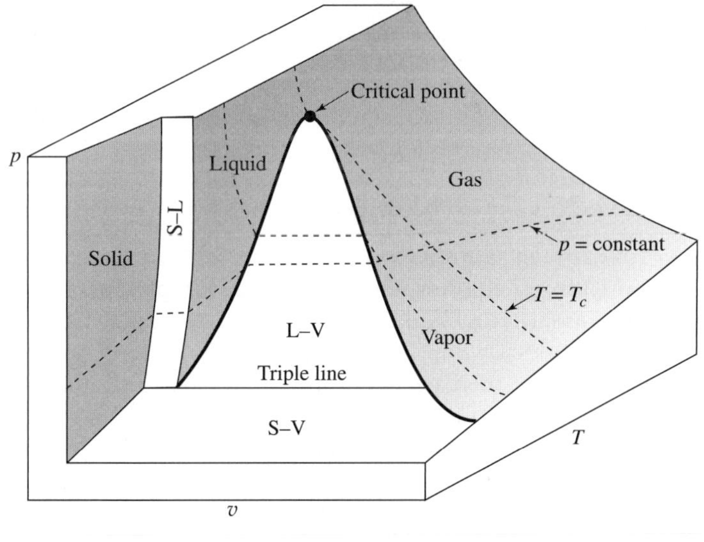

FIGURE 2.3 The *p-v-T* rendering of a substance that contracts on freezing.

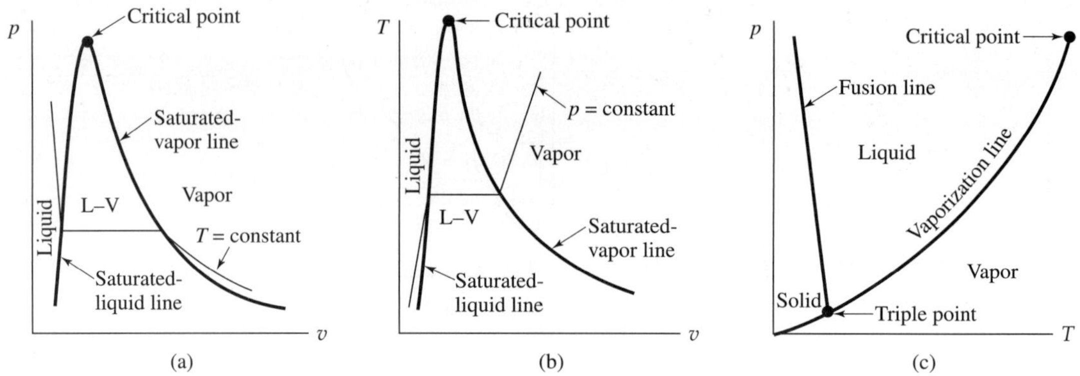

FIGURE 2.4 The *p*-*v*, *T*-*v*, and *p*-*T* diagrams.

(S–L), solid–vapor (S–V), and liquid–vapor (L–V). Along the triple line, a line of constant temperature and pressure, all three phases coexist.

The *p*-*v*-*T* surface may be projected unto the *p*-*v* plane, the *T*-*v* plane, and the *p*-*T* plane, thus, obtaining the *p*-*v*, *T*-*v*, and *p*-*T* diagrams shown in Fig. 2.4. The process of melting, vaporization, and sublimation (the transformation of a solid directly to a vapor) are shown in part (c). Again, distortions are made so that the various regions are displayed. Note that when the triple line of Fig. 2.3 is viewed parallel to the *v* axis, it appears to be a point, hence the name *triple point*. A constant pressure line is shown on the *T*-*v* diagram and a constant temperature line on the *p*-*v* diagram, the two diagrams often sketched in problems involving phase change. Become very familiar with these two diagrams; they are very helpful when analyzing problems of this nature.

Primary practical interest is in situations involving the liquid, liquid–vapor, and vapor regions. A saturated vapor lies on the *saturated-vapor line* and a saturated liquid on the *saturated-liquid line*. The region to the right of the saturated-vapor line is the **superheated region**; the region to the left of the saturated liquid line is the **compressed-liquid region** (also called the *subcooled-liquid region*). A *supercritical state* is encountered when the pressure and temperature are greater than the critical values.

Superheated region: *The region to the right of the saturated-vapor line.*

Compressed-liquid region: *The region to the left of the saturated-liquid line.*

2.3 THE LIQUID–VAPOR REGION

At any state (T, v) between saturated points f and g, shown in Fig. 2.5, liquid and vapor exist as a mixture in equilibrium. Let v_f and v_g represent, respectively, the specific volumes of the saturated liquid and the saturated vapor. Let m be the total mass of a system (such as shown in Fig. 2.1c), m_f the amount of mass in the liquid phase, and m_g the amount of mass in the vapor phase. Then, for a state of the system represented by any (T, v), such as state 1, the total volume of the mixture is the sum of the volume occupied by the liquid and that occupied by the vapor[1], or

[1]A liquid was referred to as a fluid and a vapor as a gas, hence the subscripts "f" and "g".

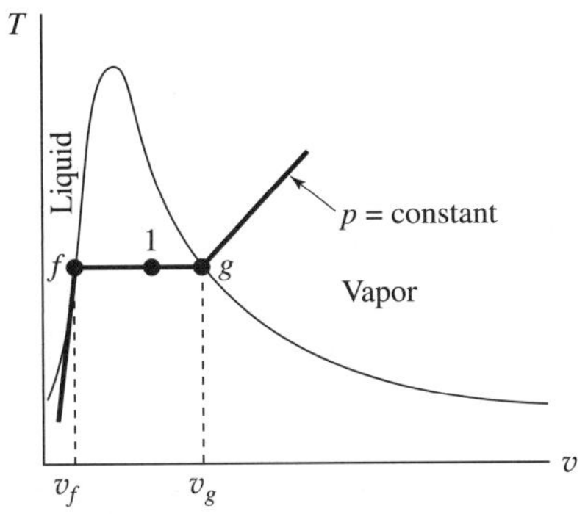

FIGURE 2.5 A T-v diagram showing the saturated-liquid and saturated-vapor points.

$$\forall = \forall_f + \forall_g \qquad \text{or} \qquad mv = m_f v_f + m_g v_g \tag{2.1}$$

The ratio of the mass of saturated vapor to the total mass is called the **quality** of the mixture, designated by the symbol x; it is

Quality: *The ratio of the mass of saturated vapor to the total mass.*

$$x = \frac{m_g}{m} \tag{2.2}$$

We often refer to the region under the saturation lines as the *quality region*, the *mixture region*, or the *wet region*; it is the only region where quality x has a meaning.

Recognizing that $m_f = m - m_g$, we may write Eq. 2.1, using our definition of quality, as

$$v = v_f + x(v_g - v_f) \tag{2.3}$$

Because the difference in saturated vapor and saturated liquid values frequently appears in calculations, we often let the subscript "fg" denote this difference; that is,

$$v_{fg} = v_g - v_f \tag{2.4}$$

Thus, Eq. 2.3 is

$$v = v_f + xv_{fg} \tag{2.5}$$

Note that the percentage liquid by mass in a mixture is $100(1 - x)$ and the percentage vapor is $100x$.

2.4 PROPERTIES OF STEAM

2.4.1 Steam Tables

Tabulations have been made for many substances of the thermodynamic properties p, v, and T, and additional properties to be identified in subsequent chapters. Values are presented in the Appendix in both tabular and graphical form. Appendix Table C.1 gives the saturation properties of water as a function of saturation temperature, and Table C.2 gives these properties as a function of saturation pressure. The information contained in the two tables is essentially the same, the choice being a matter of convenience. We should note, however, that in the mixture region, pressure and temperature are dependent. Thus, to establish the state of a mixture, if we specify the pressure, we need to specify a property other than temperature. Conversely, if we specify temperature, we must specify a property other than pressure.

Key Concept: *To establish the state of a mixture, if we specify the pressure, we need to specify a property other than temperature.*

Appendix Table C.3 lists the properties of superheated water vapor. To establish the state of a simple substance in the superheated region, it is necessary to specify two properties. While any two may be used, the most common procedure is to use pressure and temperature since these are easily measured. Thus, properties such as v are given in terms of the set of independent properties p and T.

Key Concept: *It is common in calculations to assume that the specific volume of a compressed liquid is equal to the specific volume of the saturated liquid at the same temperature.*

Appendix Table C.4 lists data pertaining to compressed liquid. At a given temperature, the specific volume of a liquid is essentially independent of the pressure. For example, for a temperature of 100°C in Appendix Table C.1, the specific volume v_f of liquid is 0.001044 m^3/kg at a pressure of 100 kPa, whereas at a pressure of 10 MPa the specific volume is 0.001038 m^3/kg, less than a 1 percent decrease in specific volume. Thus, it is common in calculations to assume that the specific volume (and other properties, as well) of a compressed liquid is equal to the specific volume of the saturated liquid at the same temperature. Note, however, that the specific volume of saturated liquid increases significantly with temperature, especially at higher temperatures.

Appendix Table C.5 gives the properties of a saturated solid and a saturated vapor for an equilibrium condition. Note that the value of the specific volume of ice is relatively insensitive to temperature and pressure for the saturated-solid line. Also, it has a greater value (almost 10 percent greater) than the minimum value on the saturated-liquid line.

Tables D in the Appendix provide the properties of refrigerant R134a, the refrigerant now being used in most air conditioners and commercial coolers. The refrigerant R12 (Freon) is no longer used because of its harmful effects on the environment.

2.4.2 TK Solver

Rather than use tables to find properties of steam, the CD inside the back cover of this text may be used. It contains the powerful TK Solver which we will use for property determination. It can also be used for extensive calculations, but that feature will not be demonstrated in this text. Only property evaluation will be utilized here, however, the student is encouraged to explore its calculation power.

To demonstrate its use in property evaluation, repeat the following steps. On the C: drive of your computer select the following:

1. The "Program Files" folder
2. The "Uts" folder
3. Engineering Thermodynamics on TK
4. TKW4.exe
5. Applications (on the very top tool bar)
6. Thermodynamics on TK
7. Data (double click)
8. Steam (double click)
9. SI units, 1 state (double click)
10. Click "OK" at the very bottom of the screen.
11. You may have to click "OK" on the next screen.
12. Two screens may appear, the "Rules" screen where equations are input and the "Variables" screen where we input the data (if you lose the "Variables" screen, select "Window" from the top tool bar, then select "Select Sheet" and then "Variables").
13. Select the "Variables" screen, and make sure the "Input" column to the left is showing.
14. Input two variables and press "Enter."
15. Click on the light bulb in the tool bar or select F9.
16. The values of the remaining variables should appear.

Let's determine the specific volume of steam for the following states:

(a) $T = 200°\text{C}$, $x = 0.8$
(b) $p = 200$ kPa, $x = 40\%$
(c) $T = 400°\text{C}$, $p = 2$ MPa
(d) $T = 423°\text{F}$, $p = 826$ psia

Following the steps outlined above, we find:

(a) $v = 0.10209\ \text{m}^3/\text{kg}$
(b) Use $x = 0.4$: $v = 0.35496\ \text{m}^3/\text{kg}$
(c) Use $p = 2000$ kPa: $v = 0.15119\ \text{m}^3/\text{kg}$
(d) At Step 9 select "English Units": $v = 0.018987\ \text{ft}^3/\text{lbm}$

There are several additional properties listed in the program. They will be introduced in subsequent sections. There are also other routes using TK Solver to find properties, but it is best to use the method outlined above.

It is good to create a shortcut to go directly to TK Solver. At Step 4 above, click the right side of the mouse and select "Create Shortcut." That will place a symbol on the desktop that will take you directly to TK Solver so that you can start at Step 5 above.

Example 2.1

Determine the volume change when 10 kg of saturated water is completely vaporized at a pressure of (a) 1 kPa, (b) 100 kPa, (c) 260 kPa, and (d) 10 000 kPa.

Solution

Appendix Table C.2 provides the necessary values. The quantity being sought is $\Delta V = mv_{fg}$ where $v_{fg} = v_g - v_f$. Note that p is given in MPa in the table and $m = 10$ kg.

(a) 1 kPa = 0.001 MPa. Thus, $v_{fg} = 129.2 - 0.001 = 129.2\ \text{m}^3/\text{kg}$. $\therefore \Delta V = mv_{fg} = 1292\ \text{m}^3$.

(b) 100 kPa = 0.1 MPa. Again, $v_{fg} = 1.694 - 0.001 = 1.693\ \text{m}^3/\text{kg}$. $\therefore \Delta V = mv_{fg} = 16.93\ \text{m}^3$.

(c) 260 kPa = 0.26 MPa. This requires us to interpolate if we use the tables. The tabulated values are used at 0.2 MPa and 0.3 MPa:

$$v_g = \frac{0.26 - 0.2}{0.3 - 0.2}(0.6058 - 0.8857) + 0.8857 = 0.718\ \text{m}^3/\text{kg}$$

The value for v_f is, to four decimal places, 0.0011 m³/kg at 0.2 MPa and at 0.3 MPa; hence no need to interpolate for v_f. We then have

$$v_{fg} = 0.718 - 0.0011 = 0.717\ \text{m}^3/\text{kg}. \qquad \therefore \Delta V = 7.17\ \text{m}^3$$

(d) 10 000 kPa = 10 MPa. Then, $v_{fg} = 0.01803 - 0.00145 = 0.01658\ \text{m}^3/\text{kg}$ and $\Delta V = 0.1658\ \text{m}^3$.

Notice the large change in specific volume at low pressures compared with the small change as the critical point is approached. This underscores the distortion of the p-v diagram in Fig. 2.4.

Example 2.2

Four kg of water is placed in an enclosed volume of 1 m³. Heat is added until the temperature is 150°C. Find (a) the pressure, (b) the mass of vapor, and (c) the volume of the vapor.

Solution

Appendix Table C.1 is used. The volume of 4 kg of saturated vapor at 150°C is $(0.3928)(4) = 1.5712\ \text{m}^3$. Since the given volume is less than this, we assume the state to be in the quality region.

(a) In the quality region, the pressure is given as $p = 475.8$ kPa, the value next to the temperature.

(b) To find the mass of the vapor, we must determine the quality. It is found from Eq. 2.3, using $v = \frac{1}{4}\ \text{m}^3/\text{kg}$, as

$$v = v_f + x(v_g - v_f)$$

$$0.25 = 0.00109 + x(0.3928 - 0.00109)$$

(continued)

Thus, $x = 0.6354$. Using Eq. 2.2, the vapor mass is

$$m_g = mx = (4)(0.6354) = 2.542 \text{ kg}$$

(c) Finally, the volume of the vapor is found from

$$\forall_g = v_g m_g = (0.3928)(2.542) = 0.9985 \text{ m}^3$$

Note that, in mixtures where the quality is not very close to zero, the vapor phase occupies most of the volume. In this example, with a quality of 63.54 percent, it occupies 99.85 percent of the volume.

Example 2.3

Four kg of water is heated at a pressure of 220 kPa to produce a mixture with quality $x = 0.8$. Determine the final volume occupied by the mixture using (a) the steam tables and (b) TK Solver.

Solution

(a) Use Appendix Table C.2. To determine the appropriate numbers at 220 kPa, we linearly interpolate[1] between 0.2 and 0.3 MPa. This provides, at 220 kPa,

$$v_g = \frac{220 - 200}{300 - 200}(0.6058 - 0.8857) + 0.8857 = 0.8297 \text{ m}^3/\text{kg}, \qquad v_f = 0.00106 \text{ m}^3/\text{kg}$$

Note that no interpolation is necessary for v_f, since for both pressures v_f is the same to four decimal places. Using Eq. 2.3, we now find

$$v = v_f + x(v_g - v_f) = 0.00106 + (0.8)(0.8297 - 0.00106) = 0.664 \text{ m}^3/\text{kg}$$

The total volume occupied by 4 kg is

$$\forall = mv = (4 \text{ kg})(0.664 \text{ m}^3/\text{kg}) = 2.66 \text{ m}^3$$

(b) Follow the steps outlined in Section 2.4.2 and find $v = 0.6481 \text{ m}^3/\text{kg}$. Then

$$m = \frac{\forall}{\text{v1}} = \frac{0.6}{0.2804} = 2.140 \text{ kg} \qquad \forall = mv = 4 \times 0.6481 = 2.5924 \text{ m}^3$$

The 2.7% error in part (a) is due to the straight line interpolation. It is certainly within acceptable engineering accuracy.

[1] Make sure you can quickly interpolate. It will be required of you often when using the tables.

Example 2.4

Two pounds of water is contained in a constant-pressure container held at 540 psia. Heat is added until the temperature reaches 700°F. Determine the final volume of the container using (a) the steam tables and (b) TK Solver.

(continued)

Solution

(a) Use Appendix Table C.3E. Since 540 psia lies between 500 psia and 600 psia, the specific volume is interpolated to be

$$v = 1.3040 + (0.4)(1.0727 - 1.3040) = 1.211\ \text{ft}^3/\text{lbm}$$

The final volume is then

$$\mathcal{V} = mv = (2)(1.211) = 2.422\ \text{ft}^3$$

(b) Follow the steps outlined in Section 2.4.2 and find $v = 1.1993\ \text{ft}^3/\text{lbm}$. The volume is then

$$\mathcal{V} = mv = 2 \times 1.1993 = 2.399\ \text{ft}^3$$

The 1% error in part (a) is due to the straight-line interpolation.

2.5 EQUATIONS OF STATE

When the vapor of a substance has relatively low density, the pressure, specific volume, and temperature are related by the simple equation

$$pv = RT \tag{2.6}$$

Equation of state: *Relates the state properties p, v, and T.*

Key Concept: *When using the ideal-gas equation, the pressure and temperature must be expressed as absolute quantities.*

where R is a constant for a particular gas and is called the *gas constant*. This equation is an **equation of state** in that it relates the state properties p, v, and T; any gas for which this equation is valid is called an *ideal gas* or sometimes a *perfect gas*. Hence, Eq. 2.6 is called the ideal-gas equation of state. Note that, when using the ideal-gas equation, the pressure and temperature must be expressed as absolute quantities.

The gas constant R is related to a *universal gas constant* $\bar{R}$, which has the same value for all gases, by the relationship

$$R = \frac{\bar{R}}{M} \tag{2.7}$$

where M is the *molar mass*, values of which are tabulated in Appendix Tables B.2 and B.3. The *mole* is that quantity of a substance (i.e., that number of atoms or molecules) having a mass which is numerically equal to the atomic or molecular weight of the substance. In the SI system, it is convenient to use the kilomole (kmol), which amounts to x kilograms of a substance of molecular weight x. For instance, 1 kmol of carbon is a mass of 12 kg; 1 kmol of molecular oxygen is 32 kg. Stated otherwise, $M = 12$ kg/kmol for C, and $M = 32$ kg/kmol for O_2. In the English system, the pound-mole (lbmol) is used for O_2, where $M = 32$ lbm/lbmol.

The value of $\bar{R}$ is

$$\bar{R} = 8.314\ \text{kJ}/(\text{kmol·K}) = 1545\ \text{ft-lbf}/(\text{lbmol-°R}) \tag{2.8}$$

For air, M is 28.97 kg/kmol (28.97 lbm/lbmol); so that for air, R is 0.287 kJ/kg·K (53.3 ft-lbf/lbm-°R), a value used extensively in calculations involving air.

Other forms of the ideal-gas equation are

$$p\mathcal{V} = mRT \qquad p = \rho RT \qquad p\mathcal{V} = n\bar{R}T \tag{2.9}$$

where n is the number of moles.

Care must be taken in using this simple, convenient equation of state. A low density can result from either a low pressure or a high temperature. For air, the ideal-gas equation is surprisingly accurate for a wide range of temperatures and pressures; less than 1 percent error is encountered for pressures as high as 3000 kPa at room temperature, or for temperatures as low as −130°C at atmospheric pressure.

The *compressibility factor* Z helps us determine whether or not the ideal-gas equation can be used. It is defined as

$$Z = \frac{pv}{RT} \tag{2.10}$$

and is displayed in Fig. 2.6 for nitrogen. Since air is composed mainly of nitrogen, this figure is acceptable for air also. If $Z = 1$, or very nearly 1, the ideal-gas equation can be used. If Z is not approximately 1, then Eq. 2.10 may be used.

The compressibility factor can be determined for any gas by using a generalized compressibility chart presented in Fig. H.1 in the Appendix. In the generalized

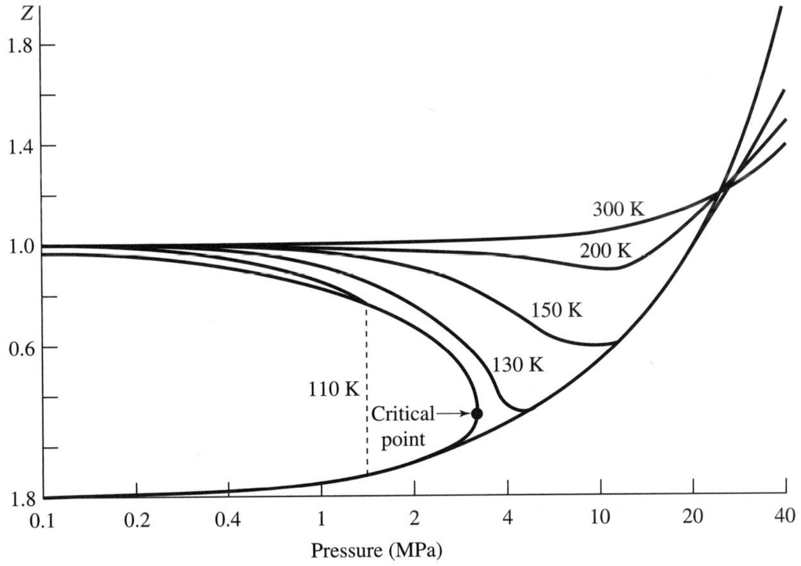

FIGURE 2.6 The compressibility factor for nitrogen.

chart, the *reduced pressure* p_R and *reduced temperature* T_R must be used. They are calculated from

$$p_R = \frac{p}{p_c} \qquad T_R = \frac{T}{T_c} \tag{2.11}$$

where p_c and T_c are the critical-point pressure and temperature, respectively, of Appendix Table B.3.

Example 2.5

An automobile tire with a volume of 0.6 m^3 is inflated to a gage pressure of 200 kPa. Calculate the mass of air in the tire if the temperature is 20°C using (a) the ideal-gas equation of state and (b) TK Solver.

Solution

(a) Air is assumed to be an ideal gas at the conditions of this example. In the ideal-gas equation, $p\mathcal{V} = mRT$, we use absolute pressure and absolute temperature. Thus, using $p_{atm} = 100$ kPa (to use a pressure of 101 kPa is unnecessary; the difference of 1% is not significant in most engineering problems),

$$p_{abs} = p_{gage} + p_{atm} = 200 + 100 = 300 \text{ kPa} \quad \text{and} \quad T = T_C + 273 = 20 + 273 = 293 \text{ K}$$

The mass is then calculated to be

$$m = \frac{p\mathcal{V}}{RT} = \frac{(300\,000 \text{ N/m}^2)(0.6 \text{ m}^3)}{(287 \text{ N·m/kg·K})(293 \text{ K})} = 2.14 \text{ kg}$$

Become familiar with the units in the above equation.

(b) Refer to the steps outlined in Section 2.4.2 and select "Equations of State" and then "Ideal Gases." Click "OK" on the bottom bar, and on the "Variables" screen under the "Input" column insert 'Air as the gas, 300 for p1, and 20 for T1. Click on the light bulb and find v1 = 0.2804 m^3/kg. Then

$$m = \frac{\mathcal{V}}{\text{v1}} = \frac{0.6}{0.2804} = 2.140 \text{ kg}$$

Example 2.6

The temperature in the atmosphere near the surface of the earth (up to an elevation of 10 000 m) can be approximated by $T(z) = 15 - 0.0065z$ °C. Determine the pressure at an elevation of 3000 m, if at $z = 0$, $p = 101$ kPa.

Solution

Equation 1.8 relates the pressure change to the elevation change. We can put the ideal-gas equation for air in the form:

$$p = \rho RT = \rho g \frac{R}{g} T = \rho g \frac{287}{9.81} T = 29.3 \rho g T$$

(continued)

Hence, Eq. 1.8 can be written as

$$dp = -\frac{p}{29.3T}\,dz$$

Using the given equation for $T(z)$, we have

$$dp = -\frac{p}{29.3(288 - 0.0065z)}\,dz$$

where we have added 273 to express the temperature in kelvins. To integrate the above equation, we must separate variables as

$$\frac{dp}{p} = -\frac{dz}{29.3(288 - 0.0065z)}$$

Now integrate between the appropriate limits:

$$\int_{101}^{p}\frac{dp}{p} = -\int_0^{3000}\frac{dz}{29.3(288 - 0.0065z)} = \frac{1}{29.3}\frac{1}{0.0065}\int_0^{3000}\frac{-0.0065\,dz}{288 - 0.0065z}$$

$$\ln\frac{p}{101} = [5.25\ln(288 - 0.0065z)]_0^{3000} = -0.368 \qquad \therefore p = 101e^{-0.368} = 69.9\text{ kPa}$$

This compares favorably with Appendix Table B.1 where we find $p = 69.9$ kPa.

2.6 EQUATIONS OF STATE FOR A NONIDEAL GAS

There are many equations of state that have been recommended for use to account for nonideal-gas behavior. Such behavior occurs where the pressure is relatively high (>2 MPa for many gases) or when the temperature is near the saturation temperature. There are no acceptable criteria that can be used to determine if the ideal-gas equation can be used or if the nonideal-gas equations of this section must be used. Usually, a problem is stated in such a way that it is obvious that nonideal-gas effects must be included; otherwise a problem is solved assuming an ideal gas.

The *van der Waals equation of state* is intended to account for the volume occupied by the gas molecules and for the attractive forces between molecules. It is

$$p = \frac{RT}{v - b} - \frac{a}{v^2} \tag{2.12}$$

where the constants a and b are related to the critical-point data of Appendix Table B.3 by

$$a = \frac{27R^2T_c^2}{64p_c} \qquad b = \frac{RT_c}{8p_c} \tag{2.13}$$

These constants are also presented in Appendix Table B.8 to simplify calculations.

An improved equation is the *Redlich–Kwong equation of state*:

$$p = \frac{RT}{v-b} - \frac{a}{v(v+b)\sqrt{T}} \tag{2.14}$$

where the constants are given by

$$a = 0.4275\frac{R^2T_c^{2.5}}{p_c} \qquad b = 0.0867\frac{RT_c}{p_c} \tag{2.15}$$

and are also included in Appendix Table B.8.

A *virial equation of state* presents the product pv as a series expansion. The most common expansion is

$$pv = RT + \frac{B(T)}{v} + \frac{C(T)}{v^2} + \cdots \tag{2.16}$$

where interest is focused on $B(T)$ since it represents the first-order correction to the ideal-gas law. The functions $B(T)$, $C(T)$, etc., must be specified for the particular gas.

Example 2.7

Calculate the pressure of steam at a temperature of 500°C and a density of 24 kg/m^3 using (a) the ideal-gas equation, (b) the van der Waals equation, (c) the Redlich–Kwong equation, (d) the compressibility factor, and (e) the steam table. Also, for parts (a), (b), (c), and (e), find the pressure using TK Solver.

Solution

(a) Using the ideal-gas equation,

$$p = \rho RT = (24)(0.462)(773) = 8570 \text{ kPa}$$

where the gas constant for steam is found in Appendix Table B.2.

(b) Using values for a and b from Appendix Table B.8, the van der Waals equation provides

$$p = \frac{RT}{v-b} - \frac{a}{v^2} = \frac{0.462 \times 773}{\frac{1}{24} - 0.00169} - \frac{1.703}{\frac{1}{24^2}} = 7950 \text{ kPa}$$

(c) Using values for a and b from Appendix Table B.8, the Redlich–Kwong equation gives

$$p = \frac{RT}{v-b} - \frac{a}{v(v+b)\sqrt{T}} = \frac{0.462 \times 773}{\frac{1}{24} - 0.00117} - \frac{43.9}{\frac{1}{24}\left(\frac{1}{24} + 0.00117\right)\sqrt{773}} = 7930 \text{ kPa}$$

(d) The compressibility factor is found from the generalized compressibility chart of Fig. H.1 in the Appendix. To use the chart, we must know the reduced temperature and pressure:

$$T_R = \frac{T}{T_c} = \frac{773}{647.4} = 1.19 \qquad p_R = \frac{p}{p_c} = \frac{8570}{22\,100} = 0.39$$

(continued)

where we have used the ideal-gas pressure from part (a). Using the compressibility chart (it is fairly insensitive to the precise values of T_R and p_R, so estimates of these values are quite acceptable) and Eq. 2.10, we find

$$p = \frac{ZRT}{v} = \frac{0.93 \times 0.462 \times 773}{1/24} = 7970 \text{ kPa}$$

(e) The steam table provides the most precise value for the pressure. Using $T = 500°\text{C}$ and $v = 1/24 = 0.0417 \text{ m}^3/\text{kg}$, we find $p = 8000$ kPa.

Note that the ideal-gas law has an error of 7.1 percent, and the error of each of the other three equations is less than 1 percent.

To use TK Solver, follow the steps outlined in Section 2.4.2 but select "Equation of State."

1. Select "Ideal Gas" and find pl = 8572 kPa
2. Select "van der Waals" and find pl = 7942 kPa
3. Select "Redlich–Kwong" and find pl = 7925 kPa
4. Select "Steam" under "Data" and find pl = 8013 kPa

These values are more accurate, but those obtained from the tables and equations are of acceptable engineering accuracy.

2.7 SUMMARY

In this chapter, we defined the following:

Saturated liquid: The state at which all solid is melted and a vapor just begins to form.
Saturated vapor: The state at which all liquid is just vaporized.
Saturated-liquid line: The line made up of the saturated liquid states.
Saturated-vapor line: The line made up of the saturated vapor states.
Critical point: The point that connects the saturated liquid line and the saturated vapor line.
Triple point: The state where all phases—solid, liquid, gas—are in equilibrium.
Compressed-liquid region: The region to the left of the saturated liquid line.
Superheated region: The region to the right of the saturated vapor line.
Quality (mixture or wet) region: The region under the saturation lines.
Equation of state: An equation that relates p, v, and T.

The equations introduced include the definition of quality:

$$x = \frac{m_g}{m_f + m_g} \tag{2.17}$$

where m_g is the mass of the vapor, and m_f is the mass of the liquid. The specific volume, in the quality region, is

$$\begin{aligned} v &= v_f + x(v_g - v_f) \\ &= v_f + x v_{fg} \end{aligned} \tag{2.18}$$

The ideal gas law takes several forms:

$$pv = RT, \qquad p\mathcal{V} = mRT, \qquad p = \rho RT, \qquad p\mathcal{V} = n\bar{R}T \tag{2.19}$$

where $\bar{R} = 8.314$ kJ/(kmol·K) = 1545 ft-lbf/(lbmol-°R) and n is the number of moles. Also, $R = \bar{R}/M$, where M is the molar mass given in Appendix Table B.2.

For a nonideal gas, the compressibility factor Z may be used:

$$pv = ZRT \tag{2.20}$$

The factor Z is found in Appendix H. The van der Waals equation, the Redlich–Kwong equation, or a virial equation of state may also be used for the nonideal gas. Such nonideal-gas equations of state must be used when the pressure is relatively high (greater than about 2000 kPa for many gases) and when the temperature is relatively low (near the critical point temperature).

REVIEW PROBLEMS FOR THE FE/EIT EXAMINATION

2.1FE The phase change from a liquid to a vapor is referred to as:
A. Vaporization **B.** Condensation
C. Sublimation **D.** Melting

2.2FE The point that connects the saturated-liquid line to the saturated-vapor line is called the:
A. Triple point
B. Critical point
C. Superheated point
D. Compressed-liquid point

2.3FE Find the volume occupied by 4 kg of 200°C steam at a quality of 80%.
A. 0.004 m^3 **B.** 0.104 m^3
C. 0.410 m^3 **D.** 4.10 m^3

2.4FE Estimate the volume occupied by 10 kg of water at 200°C and 2 MPa.
A. 0.099 m^3 **B.** 0.012 m^3
C. 9.4 L **D.** 11.8 L

2.5FE Saturated steam is heated in a rigid tank from 50°C to 250°C. Estimate the final pressure.
A. 20 kPa **B.** 27 kPa
C. 33 kPa **D.** 44 kPa

2.6FE A 10-L tank holds 0.1 L of liquid water and 9.9 L of steam at 300°C. What is the quality?
A. 43% **B.** 61%
C. 76% **D.** 86%

2.7FE Water at an initial quality of 50% is heated at constant volume from $p_1 = 100$ kPa to $T_2 = 200$°C. What is p_2?
A. 162 kPa **B.** 216 kPa
C. 286 kPa **D.** 485 kPa

2.8FE A vertical circular cylinder holds 1 cm of liquid water and 100 cm of vapor. If $p = 200$ kPa, the quality is nearest:
A. 0.0102
B. 0.107
C. 0.406
D. 0.803

2.9FE Find the quality of steam at 120°C if the vapor occupies 1200 L and the liquid occupies 2 L.
A. 42% **B.** 28%
C. 18% **D.** 2%

2.10FE Select the correct T-v diagram (Fig. P2.10FE) if steam at $v_1 = 0.005$ m^3/kg is heated to $v_2 = 0.5$ m^3/kg while maintaining $p = 500$ kPa. The dots are states 1 and 2 with 1 being on the left.

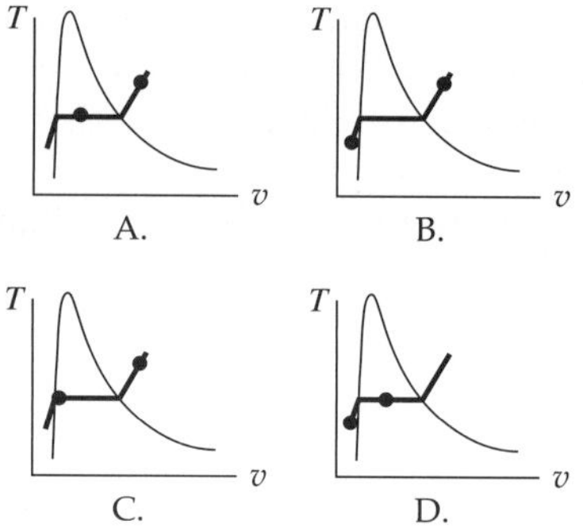

FIGURE P2.10FE

2.11FE Steam exists in a volume at 288°C and 2 MPa. Estimate the specific volume.
A. 0.1221 m^3/kg **B.** 0.1148 m^3/kg
C. 0.1192 m^3/kg **D.** 0.1207 m^3/kg

2.12FE Calculate the mass of air in a tire with a volume of 0.2 m^3 if it is filled to a gage pressure of 280 kPa at 25°C.
A. 7.8 kg **B.** 0.888 kg
C. 0.732 kg **D.** 0.655 kg

2.13FE The mass of nitrogen at 150 K and 2 MPa contained in 0.2 m^3 is nearest:
A. 10.2 kg **B.** 8.89 kg
C. 6.23 kg **D.** 2.13 kg

2.14FE Estimate the temperature of 2 kg of air contained in a 40 L volume at 2 MPa.
A. 120 K **B.** 140 K
C. 160 K **D.** 180 K

2.15FE Find the mass of air in the volume shown in Fig. P2.15FE if $T = 20$°C and the frictionless piston has a mass of 75 kg.
A. 0.0512 kg **B.** 0.0256 kg
C. 0.0064 kg **D.** 0.0016 kg

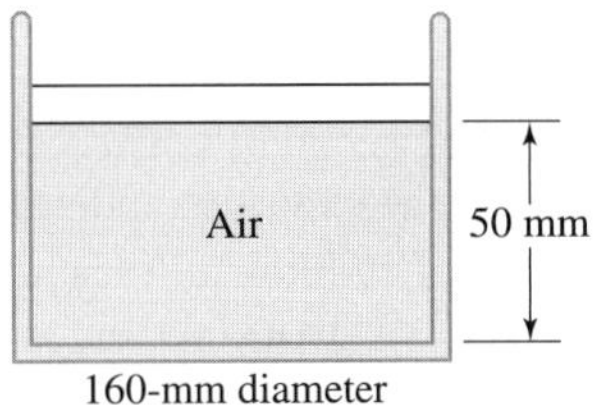

FIGURE P2.15FE

2.16FE Estimate the difference in density between the inside and outside of a house in the winter when $p = 100$ kPa and $T_{\text{inside}} = 20$°C and $T_{\text{outside}} = -20$°C. (Would you expect any tendency for air exchange due to this difference?)
A. 0.188 kg/m^3 **B.** 0.165 kg/m^3
C. 0.151 kg/m^3 **D.** 0.092 kg/m^3

PROBLEMS

In the following problem set, work each problem using the appropriate equations and/or Appendix tables. Use TK Solver for property evaluation when requested.

2.1 Using the steam tables C.1 and C.2 in the Appendix, plot to scale the **(a)** p-v and **(b)** p-T diagrams. Choose a linear-linear plot. Note the distortions of the various figures in Sections 2.2 and 2.3. Such distortions are necessary if the various regions are to be displayed.

2.2 For a specific volume of 0.2 m^3/kg, find the quality of steam if the absolute pressure is **(a)** 40 kPa and **(b)** 630 kPa. What is the temperature of each case? Use the steam tables. **(c)** Answer parts **(a)** and **(b)** using TK Solver.

2.3 Calculate the specific volume for the following situations: **(a)** water at 200°C, 80% quality and **(b)** R134a at −40°C, 90% quality.

2.4 The quality of each of the following substances is 80%. Using the steam tables, calculate the specific volume of **(a)** water at 500 psia; **(b)** R134a at 80 psia. **(c)** Answer parts **(a)** and **(b)** using TK Solver.

2.5 Calculate the specific volume of water at **(a)** 160°C and **(b)** 221°C if the quality is 85 percent.

2.6 Five kg of steam occupies a volume of 10 m^3. Find the quality and the pressure if the temperature is measured at **(a)** 40°C and **(b)** 86°C.

2.7 Ten lb of steam is contained in a volume of 50 ft^3. Find the quality and the pressure if the temperature is 263°F using **(a)** the steam tables and **(b)** TK Solver.

2.8 Determine the final volume of a mixture of water and steam if 3 kg of water is heated at a constant pressure until the quality is 60 percent. The pressure is **(a)** 25 kPa and **(b)** 270 kPa.

2.9 Saturated water occupies a volume of 1.2 m^3. Heat is added until it is completely vaporized. If the pressure is held constant at 600 kPa, calculate the final volume using **(a)** the steam tables and **(b)** TK Solver.

2.10 Two kg of saturated water at 125 kPa are completely vaporized. Calculate the volume **(a)** before and **(b)** after.

2.11 Water is contained in a rigid vessel of 5 m^3 at a quality of 0.8 and a pressure of 2 MPa. If the pressure is reduced to 400 kPa by cooling the vessel, find the final mass of vapor m_g and mass of liquid m_f.

2.12 The temperature of 10 lb of water is held constant at 205°F. The pressure is reduced from a very high value until vaporization is complete. Determine the final volume of the steam using **(a)** the steam tables and **(b)** TK Solver.

2.13 Water exists at the critical point in a rigid container. The container and water are cooled until a pressure of 10 psia is reached. Calculate the final quality.

2.14 A rigid vessel with a volume of 10 m^3 contains a water-vapor mixture at 400 kPa. If the quality is 60 percent, find the mass. The pressure is lowered to 300 kPa by cooling the vessel; find m_g and m_f.

2.15 Steam with a quality of 0.85 is contained in a rigid vessel at a pressure of 200 kPa. Heat is then added until the temperature reaches **(a)** 400°C and **(b)** 140°C. Determine the final pressures using the steam tables and **(c)** TK Solver.

2.16 Two kg of R134a is contained in a piston-cylinder arrangement, as sketched in Fig. 2.1. The 20-mm-dia, 48-kg piston is allowed to rise freely until the temperature reaches 160°C. Calculate the final volume.

2.17 A mass of 0.01 kg of steam at a quality of 0.9 is contained in the cylinder shown in Fig. P2.17. The spring just touches the top of the piston. Heat is added until the spring is compressed 15.7 cm. Calculate the final temperature.

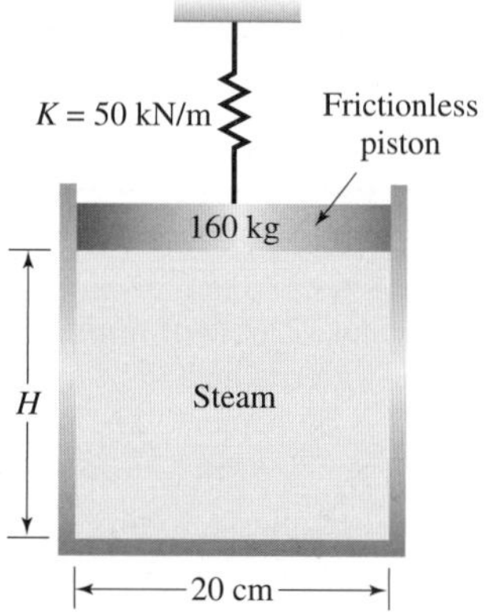

FIGURE P2.17

2.18 A rigid vessel contains water at 400°F. Heat is to be added so that the water passes through the critical point. What should the quality be at the temperature of 400°F?

2.19 R134a is contained in a sealed glass container at 50°C. As it is cooled, vapor droplets are noted condensing on the sidewalls at 20°C. Find the original pressure in the container using **(a)** the steam tables and **(b)** TK Solver.

2.20 Two kg of water is contained in a piston-cylinder arrangement by a 16 000-kg, 2-m-diameter, frictionless piston. See Fig. 2.1. Heat is added until the temperature reaches **(a)** 140°C, **(b)** 400°C, and **(c)** 650°C. Calculate the final volume.

2.21 Two kg of steam at a quality of 0.80 are contained in the volume shown (Fig. P2.21). A spring is then brought in contact with the top of the piston, and heat is added until the temperature reaches 500°C. Determine the final pressure. *Note:* the force in the spring is Kx, where x is the displacement of the spring. This results in a trial-and-error solution.

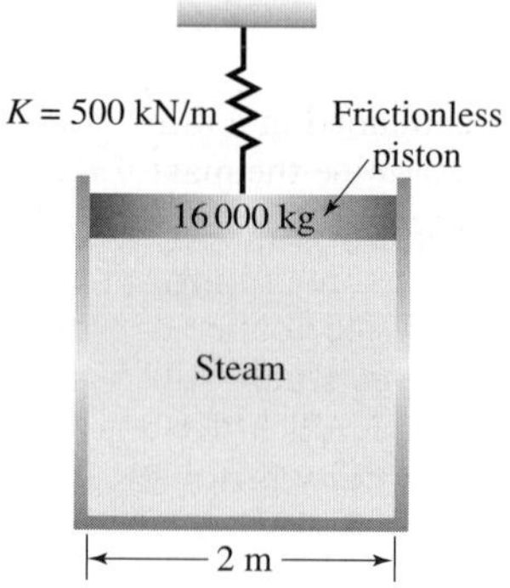

FIGURE P2.21

2.22 Determine the volume occupied by 10 kg of water at a pressure of 10 MPa and a temperature of **(a)** 5°C, **(b)** 200°C, **(c)** 400°C, **(d)** 800°C, **(e)** 1500°C, and **(f)** −10°C using the steam tables, and **(g)** solve all parts using TK Solver.

Ideal Gas

2.23 For air at 100 psia and 60°F, calculate **(a)** the density, **(b)** the specific volume, **(c)** the specific weight if $g = 32.1$ ft/sec^2, and **(d)** the mass contained in 200 ft^3.

2.24 Estimate the difference between the weight of air in a room that measures 20 ft × 100 ft × 10 ft in the summer when $T = 90$°F and the winter when $T = 10$°F. Use $p = 14$ psia.

2.25 Provide the missing information for air at an elevation where $g = 9.82$ m/s^2.

	p (kPa)	T (°C)	v (m^3/kg)	ρ (kg/m^3)	γ (N/m^3)
(a)	100	20			
(b)		100	2		
(c)	500		0.1		
(d)		400			20
(e)	200			2	

2.26 A pressurized can contains air at a gage pressure of 40 psi when the temperature is 70°F. The can will burst when the gage pressure reaches 200 psi. At what temperature will the can burst?

2.27 Assuming the atmosphere to be isothermal at an average temperature of −20°C, determine the pressure at elevations of **(a)** 3000 m and **(b)** 10 000 m. Let $p = 101$ kPa at the earth's surface. Compare with tabulated values in Appendix Table B.1 of 70.1 kPa and 26.5 kPa, respectively, by calculating the percent error.

2.28 **(a)** Assuming the temperature in the atmosphere to be given by $T = 15 - 0.00651z$ °C, determine the pressure at an elevation of 10 km. Let $p = 101$ kPa at sea level. **(b)** Compare the result of **(a)** with a tabulated value in Appendix Table B.1 of 26.5 kPa by calculating the percent error.

2.29 The gage pressure reading on an automobile tire is 35 psi when the temperature is 0°F. The automobile is driven to a warmer climate and the temperature increases to 120°F. Estimate the increased pressure in the tire using **(a)** the equations and **(b)** TK Solver.

Non-ideal Gas

2.30 Nitrogen is contained in a 4-m^3 vessel at a pressure of 4200 kPa. Determine the mass if the temperature is **(a)** 30°C and **(b)** −120°C.

2.31 Estimate the pressure of nitrogen at a temperature of 220 K and a specific volume of 0.04 m^3/kg using **(a)** the ideal-gas equation, **(b)** the van der Waals equation, **(c)** the Redlich–Kwong equation, and **(d)** the compressibility factor. **(e)** Answer parts **(a)**, **(b)**, and **(c)** using TK Solver.

2.32 Ten kg of 600°C steam is contained in a 182-liter tank. Find the pressure using **(a)** the ideal-gas equation, **(b)** the van der Waals equation, **(c)** the Redlich–Kwong equation, **(d)** the compressibility factor, and **(e)** the steam tables.

2.33 Propane at 200°F has a density of 2.0 lbm/ft^3. Find the pressure using **(a)** the ideal-gas equation, **(b)** the van der Waals equation, **(c)** the Redlich–Kwong equation, and **(d)** the compressibility factor.

Heat transferred to a large volume makes a pressure cooker operate at a relatively high temperature. (Image courtesy of Getty Images.)

3

Work and Heat

Outline

Chapter Objectives

The objectives of this chapter are to:

- Introduce the basic concept of work.
- Discuss quasi-equilibrium work and nonequilibrium work modes.
- Introduce the basic concepts of heat transfer including conduction, convection, and radiation.
- Present numerous examples and practice problems that illustrate the concepts discussed.

Nomenclature

The following is the nomenclature introduced in this chapter:

dl	Differential distance	q	Heat transfer per unit mass, electrical charge
E_R	Emissive power	q'	Heat transfer per unit length
G_α	Absorbed radiation	q''	Heat flux
G_ρ	Reflected radiation	q''_ε	Emissive power
G_R	Incident radiation	$q''_{\varepsilon b}$	Black body radiation
h_c	Convective heat transfer coefficient	q'''	Volumetric heat generation
i	Electrical current	Q	Heat transfer
k	Thermal conductivity		

$\dot{Q}$	Rate of heat transfer
r	Radius
R	Radius
R_{mat}	Thermal resitivity
t	Time
T	Torque
V	Voltage
w	Work per unit mass
W	Work
$\dot{W}$	Rate of doing work
x	Distance
α_R	Absorbtivity
δQ	Differential heat transfer
δW	Differential work
ε	Emissivity
ρ_R	Reflectivity
σ	Stefan–Boltzmann constant
τ	Shearing stress
τ_R	Transmissitivity
ω	Angular velocity

3.1 INTRODUCTION

In this chapter, we will discuss the two quantities that result from energy transfer across the boundary of a system: work and heat. This will lead into a presentation of the first law of thermodynamics. Work will be discussed in detail and will be calculated for several common situations. Heat transfer, often simply called "heat," however, is a quantity that requires substantial analysis for its calculation. The study of heat transfer—conduction, convection, and radiation—is included at the appropriate places where it arises in our studies of thermodynamics and fluid mechanics.

3.2 DEFINITION OF WORK

The term "work" is so broad that we must be very particular in a technical definition. It must include, for example, the work done by expanding exhaust gases after combustion occurs in the cylinder of an automobile engine, as shown in Fig. 3.1. The energy released during the combustion process is transferred to the crankshaft by means of the connecting rod in the form of work. Thus, in thermodynamics, *work* can be thought of as energy being transferred across the boundary of a system, where, in this example, the system is the gas in the cylinder.

Work, designated W, is often defined as the product of a force and the distance moved in the direction of the force. This is a mechanical definition of work.

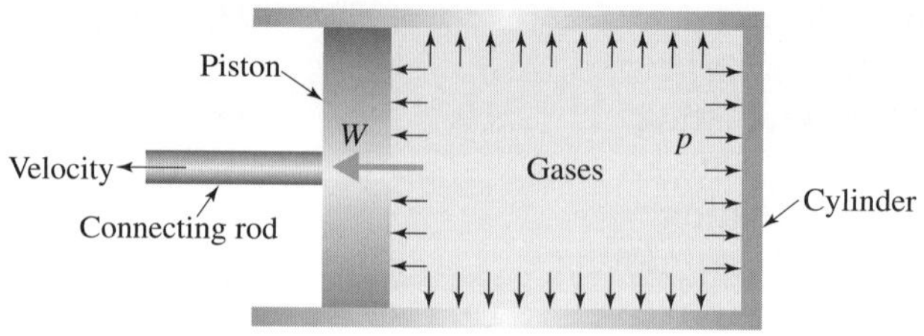

FIGURE 3.1 Work being done by expanding gases in a cylinder.

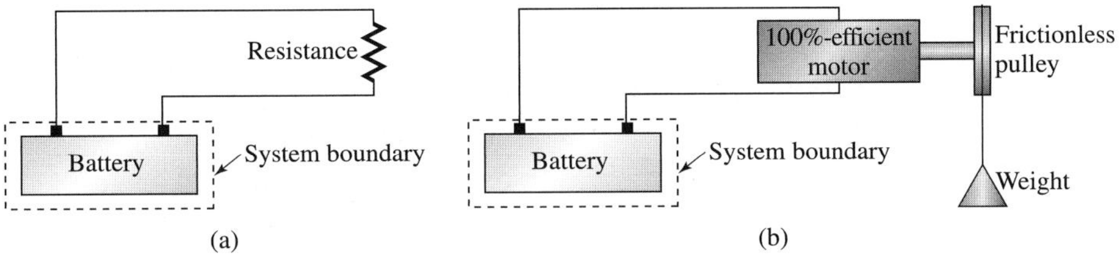

FIGURE 3.2 Work being done by electrical means.

A more general definition of work is the thermodynamic definition: **Work**, an interaction between a system and its surroundings, is done by a system if the sole external effect on the surroundings could be the raising of a weight. The magnitude of the work is the product of that weight and the distance it could be lifted. The schematic of Fig. 3.2a qualifies as work in the thermodynamic sense since it could lift the weight of Fig. 3.2b.

Work: *An interaction between a system and its surroundings if the sole external effect on the surrounding could be the raising of a weight.*

The convention chosen for positive work is that if the system performs work on the surroundings it is positive.[1] A piston compressing a fluid is doing negative work on the system, whereas a fluid expanding against a piston is doing positive work. The units of work are quickly observed from the units of a force multiplied by a distance: in the SI system, newton-meters (N·m) or joules (J); in the English system, the units are ft-lbf or Btu (778 ft-lbf = 1 Btu).

Key Concept: *If the system performs work on the surroundings, it is positive.*

The rate of doing work, designated $\dot{W}$, is called **power**. In the SI system, power has units of joules per second (J/s), or watts (W); in the English system, the units are ft-lbf/sec. We will find occasion to use the unit of horsepower because of its widespread use in rating engines. To convert, we simply use 1 hp = 0.746 kW = 550 ft-lbf/sec.

Power: *The rate of doing work.*

The work associated with a unit mass will be designated w:

$$w = \frac{W}{m} \tag{3.1}$$

A final general comment concerning work relates to the choice of the system. Note that if the system in Fig. 3.2 included the entire battery-resistor setup in part (a) or the entire battery-motor-pulley-weight setup in part (b), no energy would cross the system boundary, with the result that no work would be done. The identification of the system is very important in determining work.

3.3 QUASI-EQUILIBRIUM WORK DUE TO A MOVING BOUNDARY

There are a number of work modes that occur in various engineering situations. These include the work needed to stretch a wire, to rotate a shaft, to move against

[1]This may be defined as negative work by other authors.

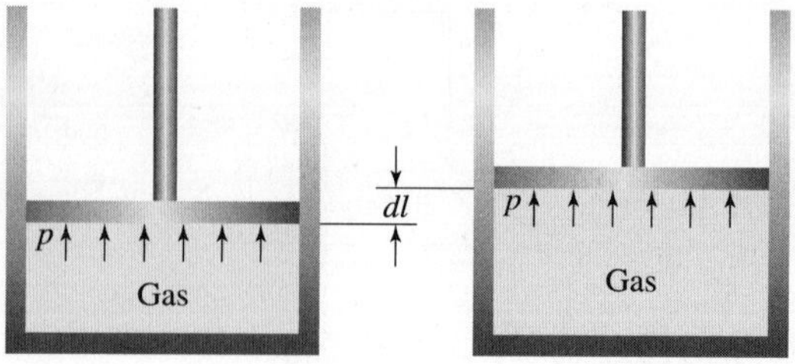

FIGURE 3.3 Work due to a moving boundary.

friction, to cause a current to flow through a resistor, and to charge a capacitor. Many of these work modes are presented in other courses. In this book, we are primarily concerned with the work required to move a boundary against a pressure force.

Key Concept: *The gas in a system is assumed to pass through a series of equilibrium states.*

Consider the piston-cylinder arrangement shown in Fig. 3.3. There is a seal to contain the gas in the cylinder; the pressure is uniform throughout the cylinder, and there are no gravity, magnetic, or electrical effects. This assures us of a quasi-equilibrium process, one in which the gas is assumed to pass through a series of equilibrium states. Now, allow an expansion of the gas to occur by moving the piston upward a small distance dl. The total force acting on the piston is the pressure times the area of the piston. This pressure is expressed as absolute pressure since pressure is a result of molecular activity; any molecular activity will yield a pressure that will result in work being done when the boundary moves. The infinitesimal work that the system (the gas) does on the surroundings (the piston) is then the force multiplied by the distance:

$$\delta W = pA\,dl \tag{3.2}$$

The symbol δW will be discussed shortly. The quantity $A\,dl$ is simply $d\forall$, the differential volume, allowing Eq. 3.2 to be written in the form

$$\delta W = p\,d\forall \tag{3.3}$$

As the piston moves from some position 1 to another position 2, the above expression can be integrated to give

$$W_{1-2} = \int_{\forall_1}^{\forall_2} p\,d\forall \tag{3.4}$$

where we assume the pressure is known for every position as the piston moves from volume $\forall_1$ to volume $\forall_2$. Typical pressure-volume diagrams are shown in Fig. 3.4. The work W_{1-2} is the area under the p-$\forall$ curve.

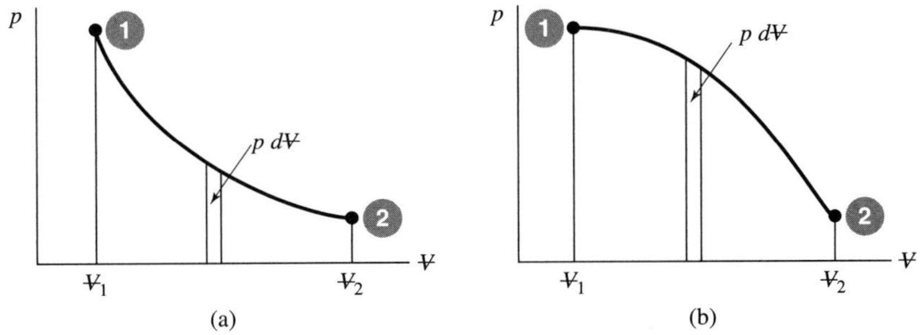

FIGURE 3.4 Work depends on the path between two states.

Consideration of the integration process highlights two very important features in Eq. 3.4. First, as we proceed from state 1 to state 2, the area representing the work is very dependent on the path that we follow. That is, states 1 and 2 in Fig. 3.4a and b are identical; yet the areas under the p-V curves are very different. In addition to being dependent on the end points, work depends on the actual path that connects the two end points. Thus, work is a path function, as contrasted to a point function that is dependent only on the end points. The differential of a path function is called an **inexact differential**, whereas the differential of a point function is an *exact differential*. An inexact differential will be denoted with the symbol δ. The integral of δW is W_{1-2}, where the subscript emphasizes that the work is associated with the path as the process passes from state 1 to state 2. The subscript may be omitted, however, and work done written simply as W. We would never write W_1 or W_2, since work is not associated with a state but with a process. Work is not a property. The integral of an exact differential, for example dT, would be

Key Concept: *Work is a path function.*

Inexact differential: *The differential of a path function.*

$$\int_{T_1}^{T_2} dT = T_2 - T_1 \tag{3.5}$$

where T_1 is the temperature at state 1 and T_2 is the temperature at state 2.

The second observation to be made from Eq. 3.4 is that the pressure is assumed to be constant throughout the volume at each intermediate position. The system passes through each equilibrium state shown in the p-V diagrams of Fig. 3.4. An equilibrium state can usually be assumed, even though the variables may appear to be changing quite rapidly. Combustion is a very rapid process that cannot be modeled as a quasi-equilibrium process. The other processes in the internal combustion engine—expansion, exhaust, intake, and compression—can be assumed to be quasi-equilibrium processes; they occur at a relatively slow rate, thermodynamically.

Key Concept: *Processes, except combustion, in the internal combustion engine can be assumed to be quasi-equilibrium processes.*

As a final comment regarding work, we may now discuss what is meant by a simple system, as defined in Chapter 1. For a system free of surface, magnetic, and electrical effects, the only work mode is that due to pressure acting on a moving boundary. For such simple systems, only two independent variables are necessary to establish an equilibrium state of the system composed of a homogeneous substance. If other work modes are present, such as a work mode due to an electric field, then additional independent variables would be necessary, such as the electric field intensity.

Example 3.1

One kg of steam with a quality of 20 percent is heated at a constant pressure of 200 kPa until the temperature reaches 400°C. Calculate the work done by the steam using (a) the steam tables and (b) TK Solver.

Solution

The work is given by

$$W = \int_{V_1}^{V_2} p\,dV = p(V_2 - V_1) = mp(v_2 - v_1)$$

(a) To evaluate the work, we must determine v_1 and v_2. Using Table C.2, we find

$$\begin{aligned} v_1 &= v_f + x(v_g - v_f) \\ &= 0.001061 + 0.2(0.8857 - 0.001061) = 0.1780\ \text{m}^3/\text{kg} \end{aligned}$$

From the superheat table, we locate state 2 at $T_2 = 400$°C and $p_2 = 0.2$ MPa:

$$v_2 = 1.549\ \text{m}^3/\text{kg}$$

The work is then

$$W = 1 \times 200 \times (1.549 - 0.1780) = 274.2\ \text{kJ}$$

Since the pressure has units of kPa, the result is in kJ.

(b) Using TK Solver, we find $v_1 = 0.1780\ \text{m}^3/\text{kg}$ and $v_2 = 1.549\ \text{m}^3/\text{kg}$. Therefore,

$$W = mp(v_2 - v_1) = 1 \times 200 \times (1.549 - 0.1780) = 274.2\ \text{kJ}$$

the same as in part (a).

Example 3.2

A 110-mm-diameter cylinder contains 100 cm^3 of water at 60°C. A 50-kg piston sits on top of the water. If heat is added until the temperature is 200°C, find the work done using (a) the steam tables and (b) TK Solver.

Solution

The pressure in the cylinder is due to the weight of the piston and remains constant. Assuming a frictionless seal (this is always done unless information is given to the contrary), a force balance provides

$$mg = pA - p_{\text{atm}}A \qquad 50 \times 9.81 = (p - 100\,000)\frac{\pi \times 0.110^2}{4} \qquad \therefore p = 151\,600\ \text{Pa}$$

(continued)

The atmospheric pressure is included so that absolute pressure results. The volume at the initial state 1 is given as

$$V_1 = 100 \times 10^{-6} = 10^{-4} \text{ m}^3$$

(a) Using v_1 at 60°C from Table C.1, the mass is calculated to be

$$m = \frac{V_1}{v_1} = \frac{10^{-4}}{0.001017} = 0.09833 \text{ kg}$$

At state 2, the temperature is 200°C and the pressure is 0.1516 MPa. Interpolation in Table C.3 results in $v_2 = 1.609 \text{ m}^3/\text{kg}$. The volume is then

$$V_2 = mv_2 = 0.09833 \times 1.609 = 0.1582 \text{ m}^3$$

Finally, the work is calculated to be

$$W = p(V_2 - V_1) = 151.6(0.1582 - 0.0001) = 24.0 \text{ kJ}$$

It has units of kJ, since the pressure has units of kPa.

(b) Using TK Solver, we find $v_1 = 0.0010176 \text{ m}^3/\text{kg}$ so that $m = 10^{-4}/0.0010176 = 0.09827$ kg. Also, $v_2 =$ at 200°C and 151.6 kPa is 1.4276 m^3/kg. Then we have

$$W = mp(v_2 - v_1) = 0.09827 \times 151.6 \times (1.4276 - 0.0010176) = 21.3 \text{ kJ}$$

The straight-line interpolation in part (a) resulted in 13% error.

Example 3.3

Energy is added to a piston-cylinder arrangement, and the piston is withdrawn in such a way that the quantity pV remains constant. The initial pressure and volume are 200 kPa and 2 m^3, respectively. If the final pressure is 100 kPa, calculate the work done by the gas on the piston.

Solution

The work, using Eq. 3.4, can be expressed as

$$W_{1\text{-}2} = \int_{V_1}^{V_2} p\,dV = \int_{V_1}^{V_2} \frac{C}{V} dV$$

where we have used $pV = C$. To calculate the work we must find C and V_2. The constant C is found from

$$C = p_1 V_1 = 200 \times 2 = 400 \text{ kJ}$$

To find V_2, we use $p_2 V_2 = p_1 V_1$, which is, of course, the equation that would result from an isothermal process (constant temperature) involving an ideal gas. This can be written as

$$V_2 = \frac{p_1 V_1}{p_2} = \frac{200 \times 2}{100} = 4 \text{ m}^3$$

(continued)

Finally,

$$W_{1-2} = \int_2^4 \frac{400}{V} dV = 400 \ln \frac{4}{2} = 277 \text{ kJ}$$

This is positive, since work is done during the expansion process by the system (the gas contained in the cylinder).

Example 3.4

Determine the horsepower required to overcome the wind drag on a streamlined car traveling 90 km/h if the drag coefficient C_D is 0.2. The drag force is given by $F_D = \frac{1}{2}\rho V^2 A C_D$, where A is the projected area of the car and V is the velocity. The density ρ of air is 1.23 kg/m^3. Use $A = 2.3$ m^2.

Solution

To find the drag force on a car, we must express the velocity in m/s: $V = (90 \times 1000/3600) = 25$ m/s. The drag force is then

$$F_D = \frac{1}{2}\rho V^2 A C_D$$

$$= 0.5 \times 1.23 \times 25^2 \times 2.3 \times 0.3 = 265 \text{ N}$$

To move this drag force at 25 m/s, the engine must do work at the rate

$$\dot{W} = F_D \times V = 265 \times 25 = 6630 \text{ W}$$

The horsepower is then

$$\text{Horsepower} = \frac{6630 \text{ W}}{746 \text{ W/hp}} = 8.89 \text{ hp}$$

Note that it doesn't require much power to move an automobile at 90 km/h on a flat highway. Our high-powered engines simply allow us to accelerate rapidly.

3.4 NONEQUILIBRIUM WORK

Key Concept: *For nonequilibrium processes, the work cannot be calculated using $\int p\, dV$.*

It must be emphasized that the area on a p-V diagram represents the work for a quasi-equilibrium process only. For nonequilibrium processes, the work cannot be calculated using $\int p\, dV$; either it must be given for the particular process, or it must be determined by some other means. Two examples will be given. Consider a system to be formed by the gas in Fig. 3.5. In part (a) work is obviously crossing the boundary of the system by means of the rotating shaft, yet the volume does not change. We could calculate the work input by multiplying the weight by the distance it dropped, neglecting friction in the pulley system. This would not, however, be equal to $\int p\, dV$, which is zero. The paddle wheel provides us with a nonequilibrium work mode.

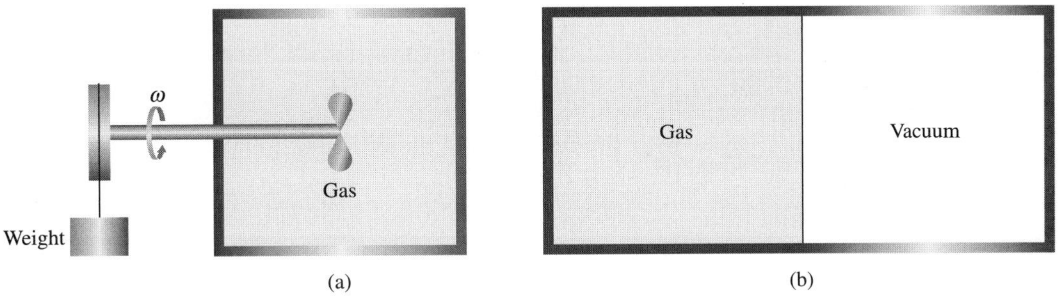

FIGURE 3.5 Nonequilibrium work.

Suppose the membrane in Fig. 3.5b ruptures, allowing the gas to expand and fill the evacuated volume. There is no resistance to the expansion of the gas at the moving boundary as the gas fills the volume; hence, there is no work done, yet there is a change in volume. The sudden expansion is a nonequilibrium process, and, again, we cannot use $\int p\,dV$ to calculate the work for this nonequilibrium work mode.

Key Concept: *A paddle wheel provides us with a nonequilibrium work mode.*

Example 3.5

A 100-kg mass drops 3 m, resulting in an increased volume in the cylinder of 0.002 m^3 (Fig. E3.5). The weight W and the piston maintain a constant gage pressure of 100 kPa. Determine the net work done by the gas on the surroundings. Neglect all friction.

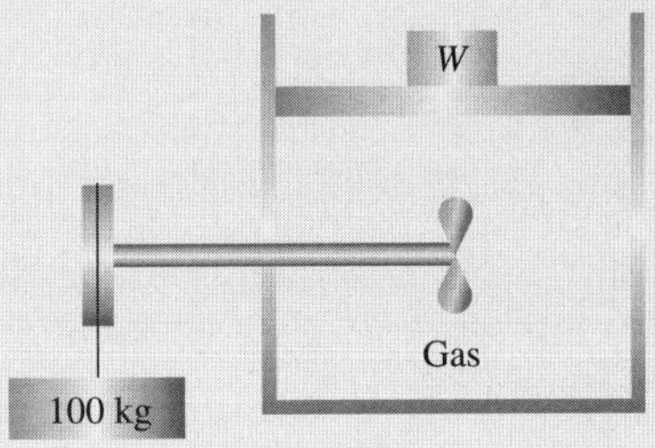

FIGURE E3.5

Solution

The paddle wheel does work on the system (the gas) due to the 100-kg mass dropping 3 m. That work is negative and is

$$W = -F \times d = -(100 \times 9.81) \times 3 = -2940 \text{ N·m}$$

The work done by the system on this frictionless piston is positive, since the system is doing the work. It is

$$W = pA \times h = pV = 200\,000 \times 0.002 = 400 \text{ N·m}$$

where absolute pressure has been used. The net work done is thus,

$$W_{\text{net}} = W_{\text{paddle wheel}} + W_{\text{boundary}} = -2940 + 400 = -2540 \text{ J}$$

3.5 OTHER WORK MODES

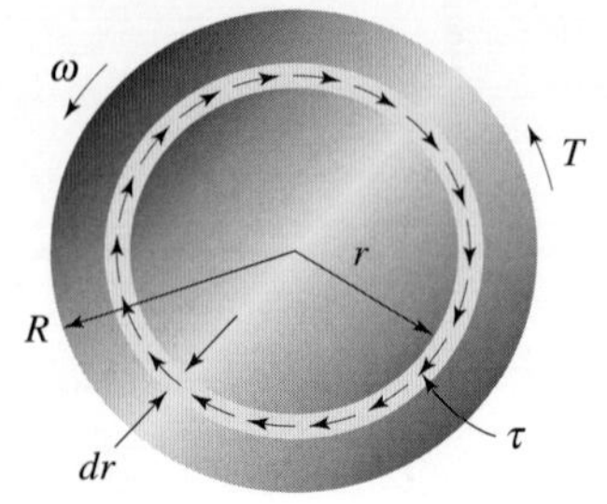

FIGURE 3.6 Work due to a rotating shaft transmitting a torque.

Work transferred by a rotating shaft (Fig. 3.6) is a common occurrence in mechanical systems. The work results from the shearing forces due to the shearing stress τ, which varies with the radius over the cross-sectional area, moving with angular velocity ω as the shaft rotates. The differential shearing force due to the shearing stress is

$$dF = \tau\, dA = \tau(2\pi r\, dr) \tag{3.6}$$

The linear velocity with which this force moves is $r\omega$. Hence, the rate of doing work (force times velocity) is

$$\dot{W} = \int_A r\omega\, dF = \int_0^R r\omega \times \tau \times 2\pi r\, dr = 2\pi\omega \int_0^R \tau \times r^2\, dr \tag{3.7}$$

where R is the radius of the shaft. The torque T is found from the shearing stresses by integrating over the area:

$$T = \int_A r\, dF = 2\pi \int_0^R \tau \times r^2\, dr \tag{3.8}$$

Combining this with Eq. 3.7 above, we have

$$\dot{W} = T\omega \tag{3.9}$$

To find the work transferred in a given time Δt, we simply multiply Eq. 3.9 by the number of seconds:

$$W = T\omega\Delta t \tag{3.10}$$

The angular velocity must be expressed in rad/s. If it is expressed in rev/min (rpm), we must multiply by $2\pi/60$.

The work necessary to stretch a linear spring (Fig. 3.7) with spring constant K from a length x_1 to x_2 can be found by using the relation for the force:

$$F = Kx \tag{3.11}$$

where x is the distance the spring is stretched from the unstretched position. Note that the force is dependent on the variable x. Hence, we must integrate the force over the distance the spring is stretched; this results in

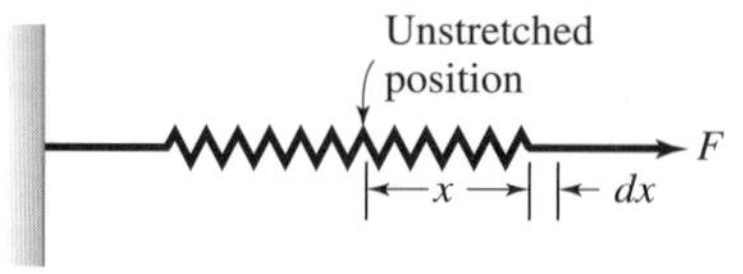

FIGURE 3.7 Work needed to stretch a spring.

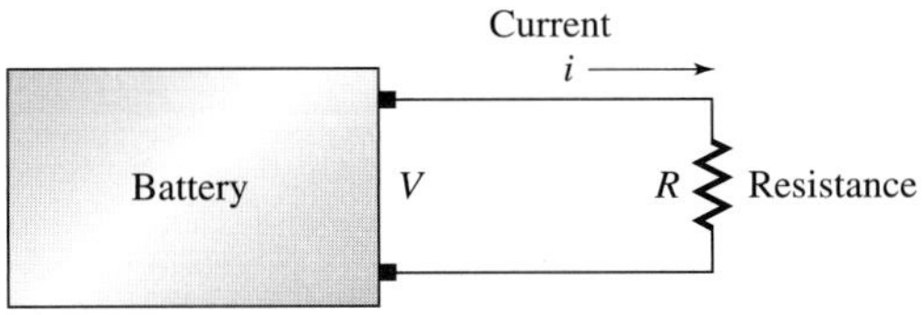

FIGURE 3.8 Work due to a current flowing through a resistor.

$$W = \int_{x_1}^{x_2} F\,dx = \int_{x_1}^{x_2} Kx\,dx = \frac{1}{2}K(x_2^2 - x_1^2) \tag{3.12}$$

As a final type, let us discuss an electrical work mode, illustrated in Fig. 3.8. The potential difference V across the battery terminals is the "force" that drives the charge q through the resistor during the time increment Δt. The current i is related to the charge by

$$i = \frac{dq}{dt} \tag{3.13}$$

For a constant current, the charge is

$$q = i\Delta t \tag{3.14}$$

The work is then given by the expression

$$W = Vi\Delta t \tag{3.15}$$

The power would be the rate of doing work, or

$$\dot{W} = Vi \tag{3.16}$$

This relationship is actually used to define the *electric potential* (the voltage V), since the ampere is a base unit and the watt has already been defined. One volt is one watt divided by one ampere.

Example 3.6

The drive shaft in an automobile delivers 100 N·m of torque as it rotates at 3000 rpm. Calculate the horsepower transmitted.

Solution

The power is found by using $\dot{W} = T\omega$. This requires ω to be expressed in rad/s:

$$\omega = 3000 \times \frac{2\pi}{60} = 314.2 \text{ rad/s}$$

Hence

$$\dot{W} = T\omega = 100 \times 314.2 = 31\,420 \text{ W} \quad \text{or} \quad \frac{31\,420}{746} = 42.1 \text{ hp}$$

Example 3.7

The air in a circular cylinder (Fig. E3.7) is heated until the spring is compressed 50 mm. Find the work done by the air on the frictionless piston. The spring is initially unstretched, as shown.

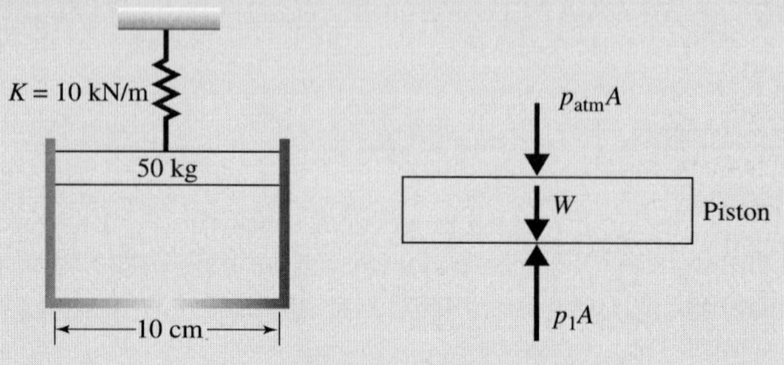

FIGURE E3.7

Solution

The pressure in the cylinder is initially found from a force balance as shown:

$$p_1 A = p_{atm} A + W$$

$$p_1 \pi \times 0.05^2 = 100\,000 \times \pi \times 0.05^2 + 50 \times 9.81 \qquad \therefore p_1 = 162\,500 \text{ Pa}$$

To raise the piston a distance of 50 mm without the spring, the pressure would be constant and the work required would be force times distance:

$$W = pA \times d = 162\,500 \times (\pi \times 0.05^2) \times 0.05 = 63.81 \text{ J}$$

Using Eq. 3.12, the work required to compress the spring is calculated to be

$$W = \frac{1}{2} K(x_2^2 - x_1^2) = \frac{1}{2} \times 10\,000 \times 0.05^2 = 12.5 \text{ J}$$

The total work is then found by summing the above two values:

$$W_{total} = 63.81 + 12.5 = 76.31 \text{ J}$$

3.6 HEAT TRANSFER

In the preceding sections, we considered several work modes by which energy is transferred macroscopically to or from a system. Energy can also be transferred microscopically to, or from, a system by means of interactions between the molecules that form the surface of the system and those that form the surface of the surroundings. If the molecules of the system boundary are more active than those of the surroundings, they will transfer energy from the system to the surroundings, with the faster molecules transferring energy to the slower molecules. On this microscopic scale, the energy is transferred by a work mode: collisions between particles. A force occurs over an extremely short time span, with work transferring energy from the faster molecules to the slower ones. Our problem is that this microscopic transfer of energy is not observable macroscopically as any of the work modes; we must devise a macroscopic quantity to account for this microscopic transfer of energy.

We have noted that temperature is a property that increases with increased molecular activity. Thus, it is not surprising that we can relate microscopic energy transfer to the macroscopic property temperature. This macroscopic transfer of energy that we cannot account for by any of the macroscopic work modes is called **heat transfer**. We can define heat transfer as energy transferred across the boundary of a system due to a difference in temperature between the system and the surroundings of the system, and *heat* as the amount of energy that is transferred. A process in which there is zero heat transfer is called an **adiabatic process**. Such a process is approximated experimentally by sufficiently insulating the system so that the heat transfer is negligible. An insulated system in thermodynamics will always be assumed to have zero heat transfer.

Heat transfer: *Energy transferred across the boundary of a system due to a difference in temperature.*

Adiabatic process: *A process in which there is zero heat transfer.*

Key Concept: *An insulated system will always be assumed to have zero heat transfer.*

To illustrate these concepts, consider a system composed of a hot block and a cold block of equal mass. The hot block contains more energy than the cold block due to its greater molecular activity, that is, its higher temperature. When the blocks are brought into contact with each other, energy flows from the hot block to the cold one by means of heat transfer until the blocks attain thermal equilibrium, with both blocks arriving at the same temperature. Once in thermal equilibrium, there are no longer any temperature differences; therefore, there is no longer any heat transfer. Some energy in the hot block has simply been transferred to the cold block; so that if the system is insulated, the cold block gains the amount of energy lost by the hot block.

Key Concept: *Heat and work are qualitatively equivalent and are expressed in units of energy.*

It should be noted that the energy contained in a system might be transferred to the surroundings either by work done by the system or by heat transferred from the system. In either case, heat and work are qualitatively equivalent and are expressed in units of energy. An equivalent reduction in energy is accomplished if 100 J of heat is transferred from a system or if 100 J of work is performed by a system. In Fig. 3.9, the burner illustrates heat being added to the system, and the rotating shaft illustrates work being done on the system. In either case, energy is being added to the system.

Key Concept: *Positive heat transfer adds energy to a system. Positive work subtracts energy from a system.*

Although heat and work are similar in that they both represent energy crossing a boundary, they differ in that, unlike work, convention prescribes that heat transferred to a system is considered positive, and heat transferred from a system is considered negative. Thus, positive heat transfer adds energy to a system; whereas positive work subtracts energy from a system.

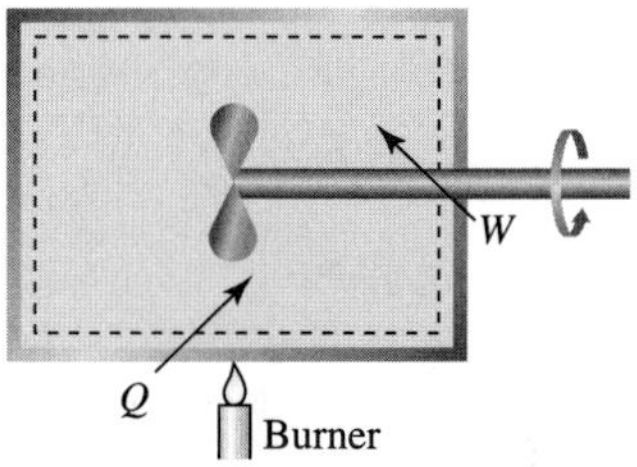

FIGURE 3.9 Energy added to a system.

Because a system does not contain heat, heat is not a property. Thus, its differential is inexact and is written as δQ, where Q is the heat transfer. For a particular process between state 1 and state 2, the heat transfer could be written as Q_{1-2}, but it will generally be denoted simply as Q. It is sometimes convenient to refer to the heat transfer per unit mass q defined as

$$q = \frac{Q}{m} \tag{3.17}$$

Often, we are interested in the *rate of heat transfer*, denoted by $\dot{Q}$, with units of $J/s = W$, or in the *heat flux*, which is defined as

$$q'' = \frac{\dot{Q}}{A} \tag{3.18}$$

and has units of W/m^2. Similarly, it is sometimes convenient to define the *heat transfer rate per unit length*:

$$q' = \frac{\dot{Q}}{L} \tag{3.19}$$

with units of W/m. Finally, some heat transfer analyses involve materials that generate heat, such as certain components on a circuit board or the drying of concrete. For these, we define *volumetric heat generation* as the rate of heat transfer per volume. It is common in heat transfer analysis to define this as $\dot{q}$; however, since we use the dot to indicate rate, we will define volumetric heat generation as

$$q''' = \frac{\dot{Q}}{\forall} \tag{3.20}$$

with units of W/m^3.

We can define three heat transfer modes: conduction, convection, and radiation. In general, each of these three heat transfer modes refers to a separate rate of energy crossing a specific plane, or surface. Often, engineering applications involving heat transfer must consider all three modes in an analysis. In each case, we can write relationships between the rate of heat transfer and temperature.

Example 3.8

A paddle wheel adds work to a rigid container by rotations caused by dropping a 50-kg weight a distance of 2 m from a pulley. How much heat must be transferred to result in an equivalent effect?

Solution

For this nonquasi-equilibrium process the work is given by

$$W = mg \times d = 50 \times 9.8 \times 2 = 980 \text{ J}$$

The heat Q that must be transferred equals the work. Hence,

$$Q = W = 980 \text{ J}$$

3.6.1 Conduction

Conduction: *Heat transfer due to the presence of temperature differences within a material.*

Fourier's law: *The relationship between the rate of heat transfer and temperature for conduction.*

Conduction heat transfer exists in a material due to the presence of temperature differences within the material. It represents heat transfer due to random molecular motion or diffusion in the absence of velocity gradients. Thus, conduction can exist in solids, liquids, and gases, but it is most often associated with solids. The relationship between the rate of heat transfer and temperature for conduction is known as **Fourier's law** *of heat transfer*. For a one-dimensional plane wall, as shown in Fig. 3.10, the relationship between temperature and the heat flux is

$$q''_x = -k\frac{dT}{dx} \tag{3.21}$$

where k is a transport property called **thermal conductivity** and has units of W/m·K, and the subscript indicates heat transfer in the x-direction. For steady-state one-dimensional heat conduction through a planar wall of thickness L, the temperature distribution is linear, as shown in Fig. 3.10. Therefore, for this case, the heat flux can be expressed in terms of the temperatures T_1 and T_2 at each boundary:

$$q''_x = -k\frac{T_2 - T_1}{L} = k\frac{T_1 - T_2}{L} \tag{3.22}$$

Thermal conductivity: *A property which indicates the rate in which a given material can transport energy.*

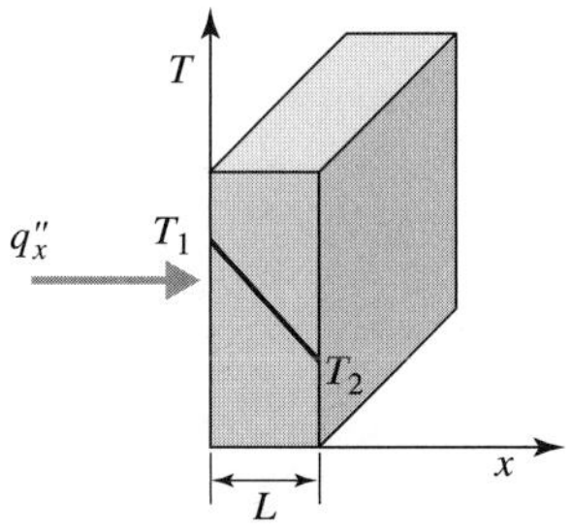

FIGURE 3.10 One-dimensional steady-state heat transfer through a plane wall.

Thermal conductivity is a material dependent property, which indicates the rate in which a given material can transport energy. Since it is most often associated with solids, it is useful to understand the general mechanisms of the property for solid materials. Here, thermal conductivity can be divided into two components: one is associated with the flow of free electrons, and the other is associated with lattice vibrational waves. The first component is proportional to the inverse of the electrical resistivity; hence, for metallic materials with low electrical resistivity, this is the dominant effect. The second factor is a function of the regularity of the lattice arrangement; thus, this component is dominant in highly crystalline materials. Since the effect of the flow of free electrons generally has a stronger influence than the lattice structure, the thermal conductivities of pure metals are generally higher than those of nonmetallic materials (with the exception of extremely highly ordered crystalline structures, such as diamond), and the thermal conductivity of crystalline structures are generally larger than those of amorphous materials. Due to the higher intermolecular spacing of gases and liquids, the thermal conductivities of these materials are typically much smaller than those for solids, as shown in Fig. 3.11.

One important group of materials is generally known as insulation materials. If one quickly made the assumption that these materials are nonmetallic materials, we would expect from Fig. 3.11 the thermal conductivities of these materials to be between 0.5 and 50 W/m·K. However, the thermal conductivities of these materials can be ten times lower than this, because insulation materials are generally a combination of a nonmetallic material and air. The nonmetallic material is carefully processed into such forms as foams and fibers, which can trap air in small spaces throughout the material. Thus, the thermal conductivity of these materials is an effective value that includes the properties of both the air and the solid

Key Concept: *Insulation material is processed into foams and fibers, which can trap air in smal spaces.*

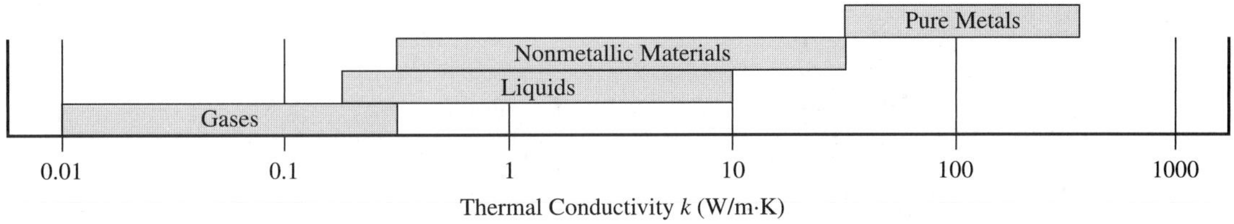

FIGURE 3.11 General ranges of thermal conductivites for various materials at atmospheric conditions.

material. It is commonly expressed in terms of a *thermal resistivity* R_{mat}, which represents resistivity rather than conductivity for a given thickness of material. Hence, the higher R_{mat}, the lower the heat transfer. Using Eq. 3.22 for a one-dimensional planar wall under steady-state conditions, we can express R_{mat} using the equation

$$q''_x = k\frac{T_1 - T_2}{L} = \frac{T_1 - T_2}{R_{mat}} \tag{3.23}$$

where R_{mat} has units of m^2·°C/W (ft^2-°F-sec/Btu). The resistivity is related to the conductivity by

$$R_{mat} = \frac{L}{k} \tag{3.24}$$

Key Concept: *Although the thermal conductivity of gases increases with temperature, the thermal conductivity of liquids can increase or decrease.*

In addition to the effects of the state of the material, thermal conductivity is also a function of temperature. For many solids, however, it can be assumed constant over small temperature ranges. For example, the thermal conductivity of pure copper changes less than 2 percent from 300 K to 400 K. Note that this is not true for all materials, as seen by the thermal conductivity of quartz, which changes approximately 25 percent over the same temperature range. The thermal conductivity of gases generally increases with temperature and decreases with molar mass. As a result, helium, with a molar mass of 4 kg/kmol, has a thermal conductivity of 0.15 W/m·K at 300 K, while carbon dioxide, with a molar mass of 44 kg/kmol, has a thermal conductivity of 0.165 W/m·K at the same temperature. In either case, the temperature dependence is significant, particularly with higher molar masses. To illustrate, the thermal conductivity of helium increases about 20 percent from 300 K to 400 K, while that of carbon dioxide increases about 45 percent over the same temperature range. Although the thermal conductivity of gases generally increases with temperature, the thermal conductivity for liquids can increase, decrease, or as in the case of water, increase and then decrease as the temperature increases.

Example 3.9

One side of a 10-cm-thick large, planar wall is exposed to a heat flux of 100 W/m^2. The difference across the wall is measured to be 10°C. What is the thermal conductivity of the wall?

Solution

Using Eq. 3.22, the thermal conductivity is

$$k = \frac{q''_x \cdot L}{T_1 - T_2} = \frac{100 \times 10/100}{10} = 1 \text{ W/m·K}$$

Make sure you check the units.

Example 3.10

Consider Example 3.9. If a different material with a thermal conductivity of 0.1 W/m·K were chosen for the planar wall, but the wall thickness and the temperature difference across the wall remained the same, what would be the resulting heat flux?

Solution

From Eq. 3.22,

$$q''_{x,\text{old}} = k_{\text{old}} \frac{T_1 - T_2}{L} \quad \text{and} \quad q''_{x,\text{new}} = k_{\text{new}} \frac{T_1 - T_2}{L}$$

Therefore,

$$\frac{T_1 - T_2}{L} = \frac{q''_{x,\text{old}}}{k_{\text{old}}} = \frac{q''_{x,\text{new}}}{k_{\text{new}}}$$

so that

$$q''_{x,\text{new}} = \frac{k_{\text{new}}}{k_{\text{old}}} \cdot q''_{x,\text{old}} = \frac{0.1}{1} \times 100 = 10 \text{ W/m}^2$$

3.6.2 Convection

Energy transferred from a solid surface to a fluid in motion is called **convection**. It is a combination of energy transfer by both random molecular motion (*conduction*) and the bulk motion of the fluid (*advection*). Therefore, if there is no bulk motion of the fluid, energy is transferred by conduction only. Convection can be expressed in terms of the temperature difference between the bulk fluid temperature T_∞ and the temperature at the surface of the solid T_s. *Newton's law of cooling* describes the flux as

Convection: *Energy transferred from a solid surface to a fluid in motion.*

$$q'' = h_c(T_s - T_\infty) \tag{3.25}$$

where h_c is called the *convective heat transfer coefficient* with units of W/m^2·K, and is a function of the flow conditions, the transport properties of the fluid, and the geometry of the wall. Usually, h_c is determined from correlating equations that are based on generalized experimental and/or analytical results. It must be noted that the convective heat transfer coefficient h_c is not itself a fluid property, because its value does not depend only on fluid properties. For example, changes in the fluid velocities and/or wall geometry can drastically change h_c, even though fluid properties may have remained unchanged.

Key Concept: *The convective heat transfer coefficient is not a fluid property.*

Convection can be classified as forced or free (natural). Free convection differs from forced convection in that the motion of the fluid in free convection is primarily due to temperature dependence and the density of a fluid. Thus, as a fluid is heated, it's density decreases and it rises above the cooler, more dense fluid. Since the fluid flow in this case is generally small, the heat transfer coefficients associated with forced convection are generally much greater than those associated with free convection. For example, typical values for free convection in an enclosed room are from 5 to 15 W/m^2·K, while values associated with forced convection with air can be as high as 250 W/m^2·K. Fans used for cooling illustrate this quite well. Furthermore, due to the higher conductive properties of liquids, the convective heat transfer coefficients of liquids are typically higher than those associated with gases under similar flow conditions.

Example 3.11

Heat is transferred by convection from a wall to an air stream at 25°C. If the heat flux remains constant at 100 W/m^2, find the surface temperature of the wall if (a) $h_c =$ 10 W/m^2·K, (b) $h_c = 100$ W/m^2·K, and (c) $h_c = 1000$ W/m^2·K.

Solution

Equation (3.25) can be written as

$$T_s = T_\infty + \frac{q''}{h_c}$$

Let $T_\infty = 25$°C and $q'' = 100$ W/m^2:

(a) for $h_c = 10$ W/m^2·K: $T_s = 25 + \dfrac{100}{10} = 35$°C

(b) for $h_c = 100$ W/m^2·K: $T_s = 25 + \dfrac{100}{100} = 26$°C

(c) for $h_c = 1000$ W/m^2·K: $T_s = 25 + \dfrac{100}{1000} = 25.1$°C

Note, as h_c increases, the surface temperature approaches the ambient temperature.

3.6.3 Radiation

Radiation: *The transfer of energy as photons.*

All surfaces emit energy as radiation, and all surfaces absorb some of the radiant energy that strikes the surface emanating from the surroundings. Physically, **radiation** is the transfer of energy as photons. Thus, radiation can be transferred through a transparent substance, such as air, and it differs from the other forms of heat transfer in that it can be transferred through a perfect vacuum.

Black body: *The surface which possesses an upper limit to the amount of radiation that can be emitted.*

The rate of radiation that is emitted per unit area of a surface is called the *emissive power* q''_ε with units of W/m^2. For any given temperature, there is an upper limit to the amount of radiation that can be emitted; the associated surface is called a **black body**. The radiation emitted by such a black body is determined by the *Stefan–Boltzmann law*:

$$q''_{\varepsilon b} = \sigma T^4 \tag{3.26}$$

where $q''_{\varepsilon b}$ represents the black body radiation (W/m^2), T is the absolute temperature (K) of the surface, and σ is the *Stefan–Boltzmann constant* and is equal to 5.67×10^8 W/m^2·K^4. (Absolute temperatures must be used due to the nonlinearity of temperature: 300 $K^4 \neq 27°C^4$) As a result of the fourth power, radiation becomes very important at high temperatures.

Since a black body represents an upper limit, real surfaces emit a fraction of this limit. Thus, the flux emitted per unit area of a real surface can be expressed in terms of the black body emission as

$$q''_\varepsilon = \varepsilon q''_{\varepsilon b} = \varepsilon \sigma T^4 \tag{3.27}$$

where ε is a dimensionless property called the **emissivity**, with values ranging from 0 to 1. The emissivity is an indication of how efficiently the surface emits radiative energy compared to a perfect emitter (a black body).

Emissivity: *An indication of how efficiently the surface emits radiative energy.*

When conducting radiative heat transfer analyses, we must not only consider the radiation emitted by a given surface, we must also consider the amount of radiation absorbed by the surface. The rate of energy absorbed by a surface can be represented as a fraction of the energy incident on a surface. We can express this fraction as

$$q''_\alpha = \alpha_R q''_I \tag{3.28}$$

where q''_α (W/m^2) is the rate of radiation per unit area absorbed by the surface, while the **irradiation** q''_I is the total radiation flux (W/m^2) incident on the surface. The constant α_R is another dimensionless radiative property called *absorbtivity*, with values ranging from 0 to 1. The incident radiation not absorbed is either transmitted through the material (if the material is not opaque) or reflected from the surface. Thus, as shown in Fig. 3.12, the total flux incident on a surface can be expressed as the fraction of radiative flux (W/m^2) absorbed q''_α, reflected q''_ρ, and/or transmitted q''_τ:

Irradiation: *The total radiation flux incident on the surface.*

$$q''_I = q''_\alpha + q''_\rho + q''_\tau = \alpha_R q''_I + \rho_R q''_I + \tau_R q''_I \tag{3.29}$$

The dimensionless radiative properties ρ_R and τ_R are the *reflectivity* and *transmissivity*, respectively, with values again ranging from 0 to 1. After dividing by q''_I, there results

$$1 = \alpha_R + \rho_R + \tau_R \tag{3.30}$$

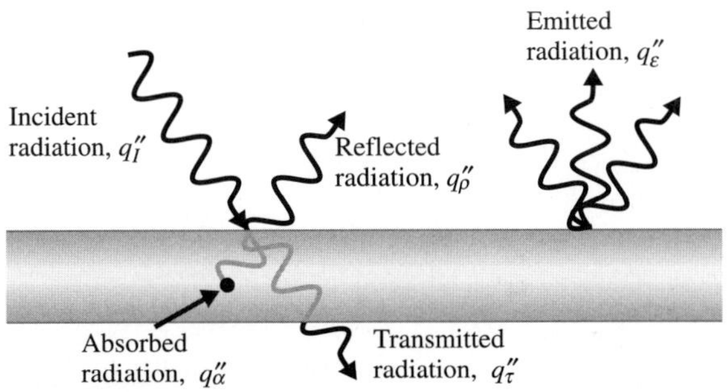

FIGURE 3.12 Illustrations of emitted radiation q''_ε on a surface and incident radiation q''_I on a surface, with the resulting reflected (q''_ρ), transmitted (q''_τ), and absorbed (q''_α) radiation.

Opaque material: *A material that does not transmit radiation through it.*

A material that does not transmit radiation through it is **opaque**, and its transmissivity τ_R is equal to zero. Thus, for an opaque surface, Eq. 3.29 becomes

$$q_I'' = \alpha_R q_I'' + \rho_R q_I'' \tag{3.31}$$

or

$$1 = \alpha_R + \rho_R \qquad (\textit{opaque surface}) \tag{3.32}$$

Furthermore, if a material absorbs all radiation striking it (i.e., all irradiation is absorbed), then both its transmissivity τ_R and its reflectivity ρ_R are equal to zero. An object consisting of this material is a black body, and the surface of this material is a *black surface*. Equation (3.30) becomes

$$\alpha_R = 1 \qquad (\textit{black body surface}) \tag{3.33}$$

Key Concept: *A surface that appears black is not necessarily a black surface.*

A black surface would appear black at room temperatures since no radiation would be reflected from it, and at this low temperature, essentially no radiation in the visible wavelength spectrum would be emitted by it. However, a surface that appears black is not necessarily a black surface, because it may not have black surface absorbtivity for radiation outside of the visible spectrum. Black body radiation closely approximates the actual thermal radiation existing in many real situations.

In many situations, we are interested in radiation heat transfer from the surface of an object to a surrounding uniform temperature where the temperature of the surroundings are much higher than the temperature of the object in question. In this case, the net radiation flux q''_{rad} (the emitted radiation from the surface minus the incident radiation from the surroundings) is

$$q''_{\text{rad}} = \varepsilon\sigma(T^4 - T_{\text{sur}}^4) \tag{3.34}$$

Example 3.12

Radiation is emitted from a surface at 300 K with an emissivity of 0.6. What is the radiation emitted from the surface, and what is the change in radiation emitted if the surface temperature is increased to 1000 K?

Solution

From Eq. 3.27, the radiation emitted at 300 K is

$$q''_\varepsilon = \varepsilon\sigma T^4 = 0.6 \cdot 5.67 \times 10^{-8} \times 300^4 = 276 \text{ W/m}^2$$

If the temperature is increased to 1000 K, the resulting emissive power is

$$q''_\varepsilon = \varepsilon\sigma T^4 = 0.6 \times 5.67 \times 10^{-8} \times 1000^4 = 34\,020 \text{ W/m}^2$$

Thus, due to the nonlinear nature of radiation, a 333 percent increase in temperature led to a 12 200 percent increase in emissive power.

3.7 SUMMARY

In this chapter, we defined the following terms:

Work: Work is done by a system on its surroundings if the effect on the surroundings could be the raising of a weight. Its equivalent is a force times the distance moved in the direction of the force.
Exact differential: The integral of an exact differential depends only on the end points.
Inexact differential: The integral of an inexact differential depends on the path between the end points.
Heat transfer: Energy transferred between a system and its surroundings due to a temperature difference.
Adiabatic process: A process during which there is no heat transfer between the system and the surroundings.
Conduction: Heat transfer due to a temperature difference in a material.
Convection: Energy transferred from a solid surface due to a fluid in motion.
Radiation: The transfer of energy as photons.
Thermal conductivity: The rate at which a material can transport energy, with units of W/m·K.
Black body: Contains the upper limit to the amount of radiation that can be emitted at a given temperature.
Emissivity: A dimensionless property that indicates how efficiently the surface emits radiative energy.
Irradiation: The total radiation flux incident on a surface.
Opaque material: A material that does not transmit radiation.

Several important equations were introduced. The first allows us to calculate work due to pressure acting on the system's boundary:

$$W_{1-2} = \int_{V_1}^{V_2} p\, dV \tag{3.35}$$

The work transferred by a rotating shaft during the time interval Δt is

$$W = T\omega\Delta t \tag{3.36}$$

The work needed to stretch a linear spring, one in which $F = Kx$, is

$$W = \frac{1}{2}K(x_2^2 - x_1^2) \tag{3.37}$$

The work done when a current flows through a resistor during the time interval Δt is

$$W = Vi\Delta t \tag{3.38}$$

where V here represents the voltage.

There were three types of heat transfer introduced. Conduction is described by Fourier's law. For a one-dimensional plane wall, it is

$$q_x'' = k\frac{T_1 - T_2}{L} = \frac{T_1 - T_2}{R_{\text{mat}}} \tag{3.39}$$

where k is the thermal conductivity and R_{mat} is the thermal resistivity. Heat transferred by convection is given by Newton's law of cooling:

$$q'' = h_c(T_s - T_\infty) \tag{3.40}$$

where h_c is called the convective heat transfer coefficient. Black body radiation emitted from a surface is given by

$$q''_{\varepsilon b} = \sigma T^4 \tag{3.41}$$

where σ is the Stefan–Boltzmann constant equal to 5.67×10^8 (W/m^2·K^4). The net radiation flux is given by

$$q''_{\text{rad}} = \varepsilon\sigma(T^4 - T^4_{\text{sur}}) \tag{3.42}$$

REVIEW PROBLEMS FOR THE FE/EIT EXAMINATION

3.1FE A person caries a 200-N weight from one end of a 100-m hall to the other end. How much work the person has done?
A. 0 J **B.** 20 J
C. 20 kJ **D.** 20 kW

3.2FE Which work mode is a nonequilibrium work mode?
A. Compressing a spring
B. Transmitting torque with a rotating shaft
C. Energizing an electrical resistor
D. Compressing gas in a cylinder

3.3FE Ten kilograms of saturated steam at 800 kPa are heated at a constant pressure to 400°C. The work required is:
A. 1150 kJ **B.** 960 kJ
C. 660 kJ **D.** 115 kJ

3.4FE Ten kilograms of air at 800 kPa are heated at a constant pressure from 170°C to 400°C. The work required is nearest:
A. 1150 kJ **B.** 960 kJ
C. 660 kJ **D.** 115 kJ

3.5FE A stop is located 20 mm above the piston shown in Fig. P3.5FE. If the mass of the frictionless piston is 64 kg, what work must the air do on the piston to increase the pressure in the cylinder to 500 kPa?
A. 13 J **B.** 28 J
C. 41 J **D.** 53 J

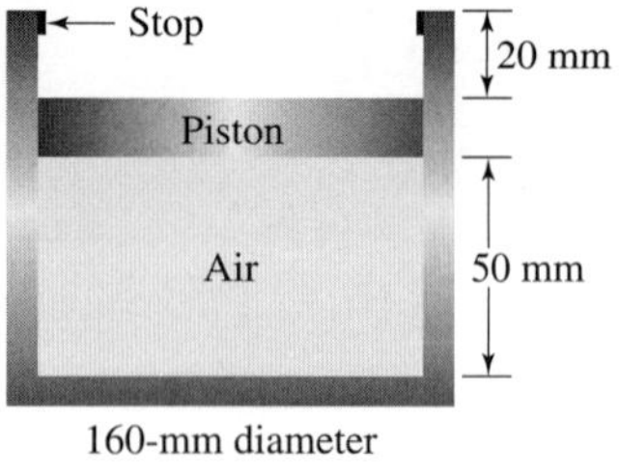

FIGURE P3.5FE

3.6FE Which of the following does not transfer work to or from a system?
A. A moving piston
B. The expanding membrane of a balloon
C. An electrical-resistance heater
D. A membrane that bursts

3.7FE Calculate the work needed to expand a piston in a cylinder at a constant pressure if saturated liquid water is completely vaporized at 200°C.
A. 2790 kJ/kg **B.** 1940 kJ/kg
C. 850 kJ/kg **D.** 196 kJ/kg

3.8FE Which of the following statements about work for a quasi-equilibrium process is incorrect?
A. The differential of work is inexact
B. Work is the area under a p-T diagram
C. Work is a path function
D. Work is always zero for a constant volume process

3.9FE A paddle wheel and an electric heater operate in a system. If the torque is 20 N·m, the rotational speed is 400 rpm, the voltage is 20 V, and the amperage is 10 A, what is the work rate?
A. −9200 J **B.** −9200 kJ
C. −2040 W **D.** −2040 kW

3.10FE A 200-mm-diameter piston is lowered by increasing the pressure from 100 kPa to 800 kPa such that the p-$V\!\!\!\!-$ relationship is $pV\!\!\!\!-^2 =$ const. What is the work done on the system if $V\!\!\!\!-_1 = 0.1$ m^3?
A. −18.3 kJ **B.** −24.2 kJ
C. −31.6 kJ **D.** −42.9 kJ

3.11FE A 120-V electric resistance heater draws 10 A. It operates for 10 minutes in a rigid volume. Calculate the work done on the air in the volume.

A. 720 000 kJ **B.** 720 kJ
C. 12 000 J **D.** 12 kJ

3.12FE Find the work if air expands from 0.2 m^3 to 0.8 m^3 while $p = 0.2 + 0.4V$ kPa.

A. 0.48 kJ **B.** 0.42 kJ
C. 0.36 kJ **D.** 0.24 kJ

3.13FE Ten kilograms of saturated liquid water expands until $T_2 = 200°C$, while the pressure remains constant at 500 kPa. Find W_{1-2}.

A. 230 kJ **B.** 926 kJ
C. 1080 kJ **D.** 2120 kJ

3.14FE Heat can be transferred by conduction through

A. Only solids
B. Only liquids
C. Only gases
D. All of the above

3.15FE Convective heat transfer involves both

A. Conduction and advection
B. Conduction and radiation
C. Radiation and advection
D. Work and conduction

3.16FE Radiation is emitted by

A. Only liquids
B. Only opaque solids
C. Only gases
D. All materials at a finite temperature

3.17FE A house has a large planar outer wall, 20-cm thick. The outside temperature is −5°C, and heat is lost through the wall at a rate of 15 W/m^2. What would the overall effective thermal conductivity of the wall need to be if the inside of the house is to be maintained at 20°C? (Assume no other losses or heat sources.)

A. 0.12 W/m·K **B.** 0.33 W/m·K
C. 18.75 W/m·K **D.** 12 W/m·K

PROBLEMS

In the following problem set, work each problem using the appropriate equations and/or Appendix tables. Use TK Solver for property evaluation when requested.

Quasi-equilibrium Work

3.1 Four kg of saturated liquid water is maintained at a constant pressure of 600 kPa while heat is added until the temperature reaches 600°C. Determine the work done by the water using **(a)** the steam tables and **(b)** TK Solver.

3.2 Two kg of saturated steam at 400 kPa is contained in a piston-cylinder arrangement. The steam is heated at constant pressure to 300°C. Calculate the work done by the steam.

3.3 A mass of 0.025 kg of steam at a quality of 10 percent and a pressure of 200 kPa is heated in a rigid container until the temperature reaches 200°C. Find the final quality and the work done by the steam using **(a)** the steam tables and **(b)** TK Solver.

3.4 The frictionless piston shown in Fig. P3.4 has a mass of 16 kg. Heat is added until the temperature reaches 400°C. If the initial quality is 20 percent, find **(a)** the initial pressure, **(b)** the mass of water, **(c)** the quality when the piston hits the stops, **(d)** the final pressure, and **(e)** the work done on the piston.

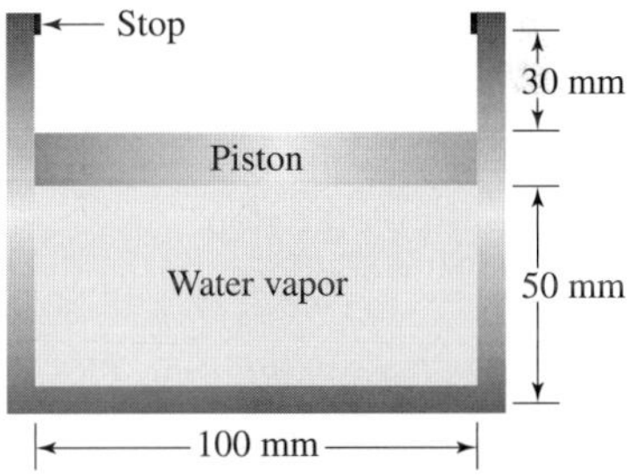

FIGURE P3.4

3.5 The frictionless piston shown in equilibrium has a mass of 64 kg (Fig. P3.5). Energy is added until the temperature reaches 220°C. The atmospheric pressure is 100 kPa. Determine **(a)** the initial pressure, **(b)** the initial quality, **(c)** the quality when the piston just hits the stops, **(d)** the final quality (or pressure if superheat), and **(e)** the work done on the piston. **(f)** Answer parts **(c)**, **(d)**, and **(e)** using TK Solver.

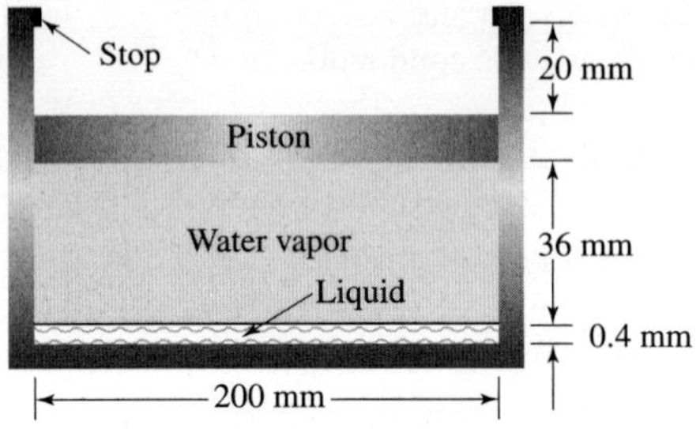

FIGURE P3.5

3.6 Saturated water vapor at 180°C is contained in a piston-cylinder arrangement at an initial volume of 0.1 m^3. Energy is added and the piston withdrawn so that the temperature remains constant until the pressure is 100 kPa.
(a) Find the work done. (Since there is no equation that relates p and $\mathcal{V}$, this must be done graphically.)
(b) Use the ideal-gas law and calculate the work.
(c) What is the percent error in using the ideal-gas law?

3.7 A 75-lb piston and weights rest on a stop (Fig. P3.7). The volume of the cylinder at this point is 40 in^3. Energy is added to the 0.4 lbm of water mixture until the temperature reaches 300°F. Atmospheric pressure is 14 psia.
(a) What is the initial specific volume of the mixture of vapor and liquid?
(b) What is the temperature in the cylinder when the piston just lifts off the stop?
(c) Determine the work done during the entire process.
(d) Answer parts **(b)** and **(c)** using TK Solver.

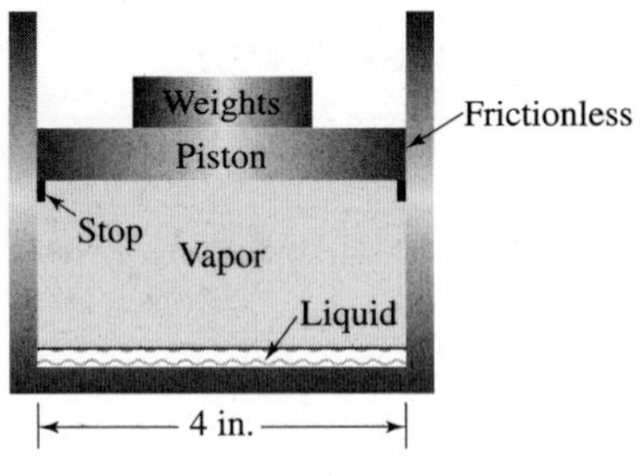

FIGURE P3.7

3.8 Air is compressed in a cylinder such that the volume changes from 100 in^3 to 10 in^3. The initial pressure is 50 psia, and the temperature is held constant at 100°F. Calculate the work.

3.9 Air is compressed in a cylinder such that the volume changes from 0.2 to 0.02 m^3. The pressure at the beginning of the process is 200 kPa. Calculate the work if **(a)** the pressure is constant, and **(b)** the temperature is constant at 50°C. Sketch each process on a p-$\mathcal{V}$ diagram.

3.10 Air contained in a circular cylinder (Fig. P3.10) is heated until a 100-kg weight is raised 0.4 m. Calculate the work done by the expanding air on the weight. Atmospheric pressure is 80 kPa.

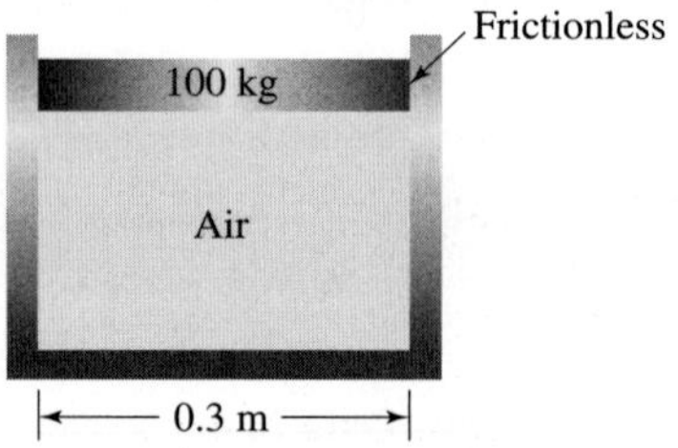

FIGURE P3.10

3.11 Six grams of air is contained in the cylinder shown in Fig. P3.11. The air is heated until the piston raises 50 mm. The spring just touches the piston initially. Calculate **(a)** the temperature when the piston leaves the stops, and **(b)** the work done by the air on the piston.

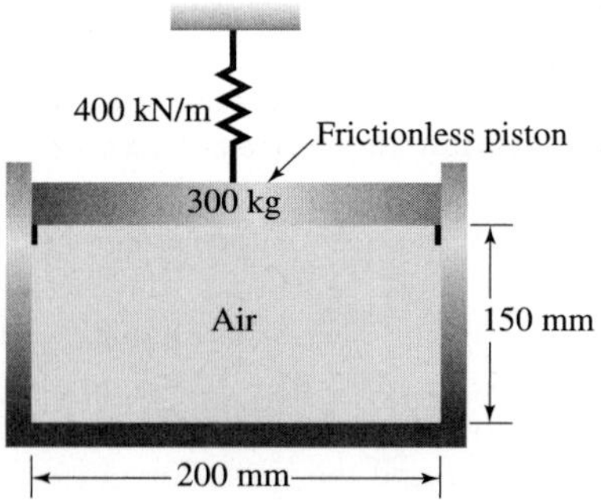

FIGURE P3.11

3.12 A process for an ideal gas is represented by $p\mathcal{V}^n =$ const, where n takes on a particular value for a given process. Show that the expression for the work done for a process between states 1 and 2 is given by

$$W = \frac{p_2 V_2 - p_1 V_1}{1 - n}$$

Is this valid for an isothermal process? If not, determine the correct expression.

3.13 Air is expanded in a piston-cylinder arrangement at a constant pressure of 200 kPa from a volume of 0.1 m^3 to a volume of 0.3 m^3. Then, the temperature is held constant during an expansion of 0.5 m^3. Determine the total work done by the air.

3.14 A 60-ft-diameter balloon is to be filled with helium from a pressurized tank. The balloon is initially empty at an elevation where the atmospheric pressure is 12 psia. Determine the work done by the helium while the balloon is being filled. The pressure varies with radius according to $p = 0.04(r - 30)^2 + 12$, where p is in psi.

3.15 Estimate the work necessary to compress the air in an air-compressor cylinder from a pressure of 100 kPa to 2000 kPa. The initial volume is 1000 cm^3. An isothermal process is to be assumed.

3.16 Estimate the work done by a gas during an unknown equilibrium process. Data obtained that relates pressure and volume are:

p	200	250	300	350	400	450	500	kPa
V	800	650	550	475	415	365	360	cm^3

3.17 The pressure in the gas contained in a piston-cylinder arrangement changes according to $p = a + 30/V$ where p is in psi and V is in ft^3. Initially, the pressure is 7 psia and the volume is 3 ft^3. Determine the work done if the final pressure is 50 psia. Show the area that represents the work on a p-V diagram.

3.18 Two kg of air experiences the three-process cycle shown in Fig. P3.18. Calculate the net work.

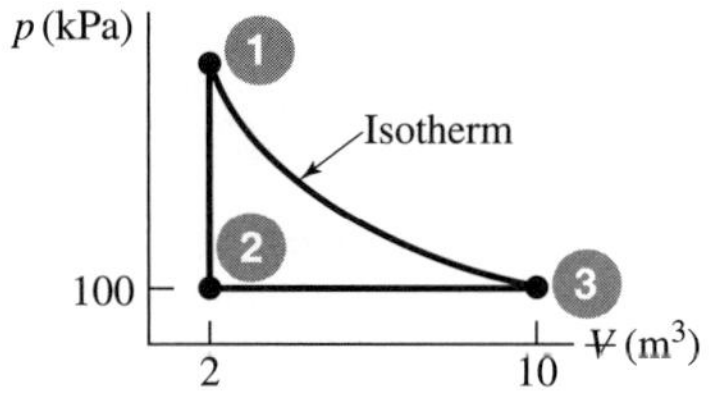

FIGURE P3.18

3.19 Air undergoes a three-process cycle. Find the net work done for 2 kg of air if the processes are:

1 → 2: constant-pressure expansion
2 → 3: constant volume
3 → 1: constant-temperature compression

The necessary information is $T_1 = 100°C$, $T_2 = 600°C$, and $p_1 = 200$ kPa. Sketch the cycle on a p-V diagram.

Other Work Modes

3.20 A paddle wheel (Fig. P3.20) requires a torque of 20 ft-lbf to rotate it at 100 rpm. If it rotates for 20 s, calculate the net work done by the air if the frictionless piston raises 2 ft during this time.

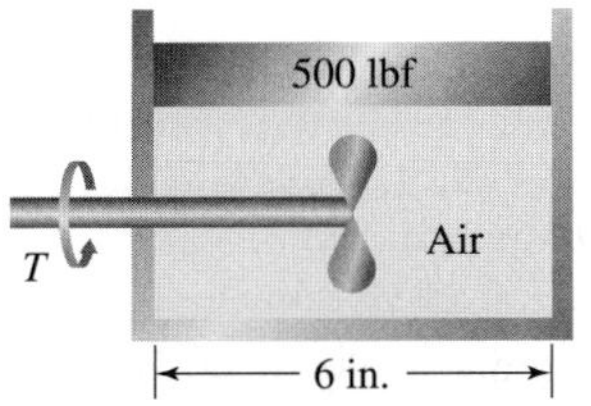

FIGURE P3.20

3.21 An unstretched spring is attached to a horizontal piston (Fig. P3.21). Energy is added to the gas until the pressure in the cylinder is 400 kPa. Find the work done by the gas on the piston. Use $p_{atm} = 75$ kPa.

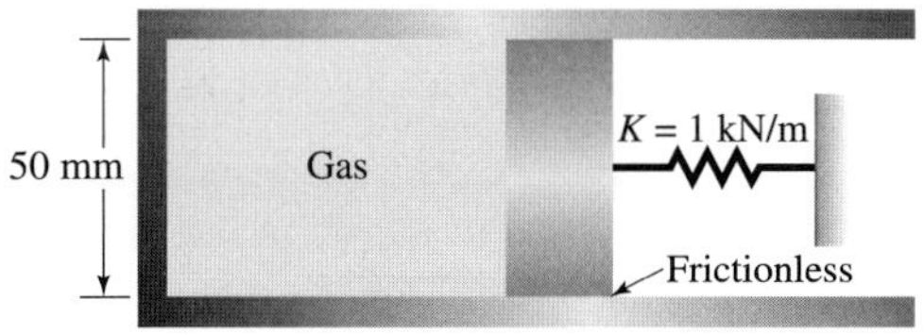

FIGURE P3.21

3.22 The force needed to compress a nonlinear spring is given by the expression $F = 200x + 30x^2$ N, where x is the displacement of the spring from its unstretched length measured in meters. Determine the work needed to compress the spring a distance of 60 cm.

3.23 An electric motor draws 3 A from the 12-V battery (Fig. P3.23). Ninety percent of the energy is used to spin the paddle wheel shown. After 50 s of operation the 30-kg piston is raised a distance of 100 mm. Determine the net work done by the gas on the surroundings. Use $p_{atm} = 95$ kPa.

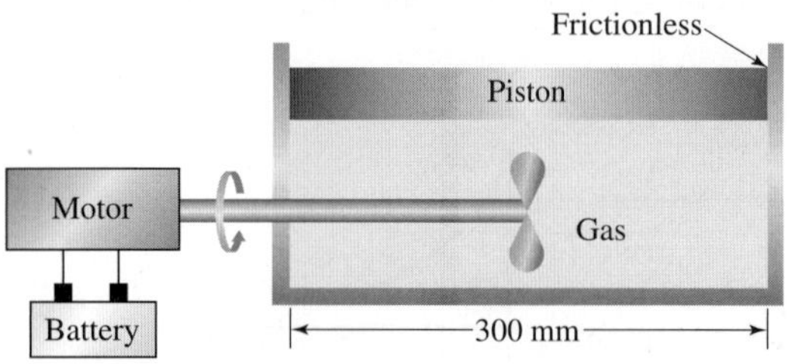

FIGURE P3.23

3.24 A torque of 2 ft-lbf is necessary to rotate a paddle wheel at a rate of 20 rad/s. The paddle wheel is located in a rigid vessel containing gas. What is the net work done on the gas during 10 min of operation?

3.25 Wind is blowing at 80 kph around a 250-mm-diameter tower that is 100-m high, shown in Fig. P3.25. The drag coefficient is 0.4 (see Example 3.4). Calculate the total force acting on the tower and the rate at which the wind does work on the tower.

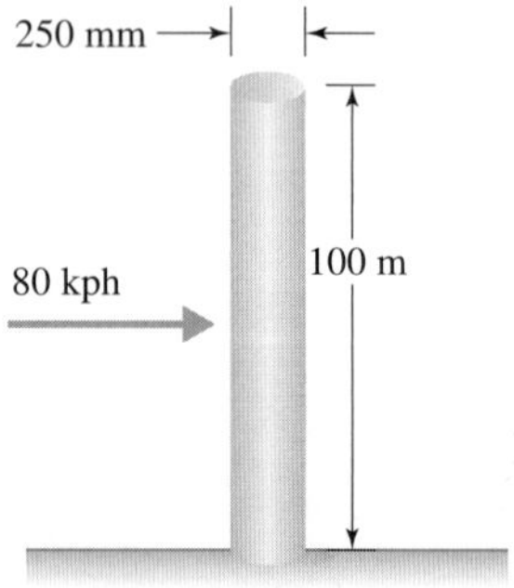

FIGURE P3.25

3.26 Derive an expression for the work required to stretch an unstretched length of wire a relatively small distance l. The force is related to the amount of stretch x by $F = EAx/L$, where L is the original length of the wire, A is the cross-sectional area, and E is a material constant (the modulus of elasticity).

3.27 A linear spring with a free length of 0.8 ft requires a work input of 4 ft-lbf to extend it to its maximum usable length. If the spring constant is 100 lbf/ft, determine the maximum length of the spring.

3.28 A linear spring requires 20 J of work to compress it from an unstretched length of 100 mm to a length of 20 mm. Find the spring constant.

3.29 The force necessary to compress a nonlinear spring is given by $F = 10x^2$ N, where x is the distance the spring is compressed, measured in meters. Calculate the work needed to compress the spring from 0.2 m to 0.8 m.

3.30 An automobile engine develops 100 hp, 96 percent of which is transferred to the drive shaft. Calculate the torque transferred by the drive shaft if it is rotating at 300 rpm.

3.31 A paddle wheel is placed in a small creek in an attempt to generate electricity as in Fig. P3.31. The water causes the tip of the 2-ft-radius paddles to travel at 4 ft/sec, while a force of 100 lbf acts at an average distance of 1.2 ft from the hub. Determine the maximum continuous amperage output that could be used to charge a bank of 12-V batteries.

FIGURE P3.31

3.32 An electrical voltage of 110 V is applied across a resistor with the result that a current of 12 A flows through the resistor. Determine **(a)** the power necessary to accomplish this and **(b)** the work done during a period of 10 min.

3.33 A gasoline engine drives a small generator that is to supply sufficient electrical energy for a motor home. What is the minimum horsepower engine that would be necessary if a maximum of 200 A is anticipated from the 12-V system?

Heat Transfer

3.34 A house has a large 0.25-in. thick single-pain glass window 6.5-ft tall and 6-ft wide. If the inside temperature of the window is 55°F, and the outside window surface temperature is 20°F, calculate the total heat transfer rate through the window in units of Btu/h. (Assume the thermal conductivity of the glass is 0.8 Btu/h·ft·°F.)

3.35 Two different materials are being considered to insulate a coat. Material *A* was shown to maintain a 20°C difference across a 7-mm thickness for a heat flux of 50 W/m^2. **(a)** Calculate the thermal conductivity of material *A*. **(b)** If material *B* has a thermal conductivity of 0.03 W/m·K, determine the thickness of material *B* required to maintain the same temperature difference for the same heat flux.

3.36 An electronic component with a surface area of 50 mm^2 generates heat at the rate of 2 W. It is cooled by convection to air at 30°C. If the surface temperature of the component cannot exceed 110°C, calculate the minimum heat transfer coefficient needed to cool the component.

3.37 A wall has a thermal conductivity of 2 W/m·K and a thickness of 5 cm. The temperature on the inner surface of the wall is 21°C, and the temperature on the outer surface is 10°C. The heat is then transferred by convection from the outer wall to the surroundings at 0°C. Considering the heat flux by conduction to be equal to the heat flux by convection, what is the convective heat transfer coefficient at the outer wall?

3.38 A highly polished copper wire has an emissivity of 0.04. Neglecting convection, determine the net heat flux dissipated to the surroundings at 300 K by radiation if the wire is maintained at 800 K.

3.39 A satellite dissipates 500 W/m^2 into space at 0 K. If the surface temperature of the satellite is not to exceed 125°C, determine the minimum required emissivity at the surface.

The potential energy in water at a relatively high elevation can be used to power equipment. (Image courtesy of Getty Images.)

4

The First Law of Thermodynamics

Outline

Chapter Objectives

The objectives of the chapter are to:

- ▲ Apply the first law to a process involving a system.
- ▲ Present the four common processes that most often occur for a system.
- ▲ Introduce enthalpy and latent heats.
- ▲ Derive the conservation of mass and energy equations for a control volume.
- ▲ Present the most common control volume devices encountered in applications.
- ▲ Introduce two transient flow situations.
- ▲ Present several applications involving heat transfer.
- ▲ Present numerous examples and practice problems that illustrate the concepts discussed.

Nomenclature

The following is the nomenclature introduced in this chapter:

c_p	Constant-pressure specific heat	k	Ratio of specific heats
c_v	Constant-volume specific heat	$\dot{m}$	Mass flow rate
c.s.	Control surface	n	Polytropic exponent
c.v.	Control volume	u	Specific internal energy
E	Total energy	V_n	Normal velocity component
h	Specific enthalpy	w_S	Shaft work per unit mass
H	Enthalpy	W_S	Shaft work
		$\dot{W}_S$	Shaft power

4.1 INTRODUCTION

The first law of thermodynamics is commonly called the law of conservation of energy. In elementary physics courses, the study of conservation of energy emphasizes changes in kinetic and potential energy and their relationship to work. A more general form of conservation of energy includes the effects of heat transfer and internal energy changes. This more general form is usually called the *first law of thermodynamics*. Other forms of energy could also be included, such as electrostatic, magnetic, strain, and surface energy; but in a first course in thermodynamics, problems involving these forms of energy are typically not included. We will present the first law for a system and then for a control volume.

4.2 THE FIRST LAW APPLIED TO A CYCLE

Law: *A statement based on observations of many experiments.*

The first law: *The net heat transfer is equal to the net work done for a system undergoing a cycle.*

Having discussed the concepts of work and heat, we are now ready to present the first law of thermodynamics. Recall that a **law** is not derived or proved from basic principles but is simply a statement that we write based on our observations of many experiments. If an experiment shows a law to be violated, either the law must be revised or additional conditions must be placed on the applicability of the law. Historically, **the first law** *of thermodynamics* was stated for a cycle: the net heat transfer is equal to the net work done for a system undergoing a cycle. This is expressed in equation form by

$$\sum W = \sum Q \quad \text{or} \quad \oint \delta W = \oint \delta Q \tag{4.1}$$

where the symbol $\oint$ implies an integration around a complete cycle. A cycle results when a system undergoes two or more processes and returns to the initial state.

The first law can be illustrated by considering the following experiment. Let a weight be attached to a pulley-paddle-wheel setup, such as that shown in Fig. 4.1a. Let the weight fall a certain distance, thereby doing work on the system contained in the tank shown, equal to the weight multiplied by the distance

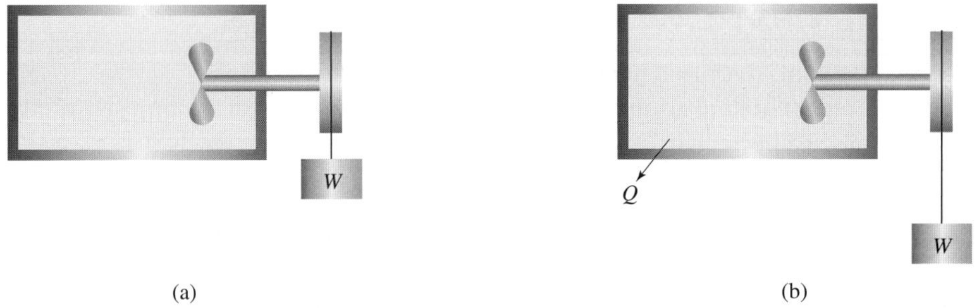

FIGURE 4.1 The first law applied to a cycle.

dropped. The temperature of the system (the fluid in the tank) will immediately rise an amount ΔT. Now, the system is returned to its initial state (the completion of the cycle) by transferring heat to the surroundings, as implied by the Q in Fig. 4.1b. This reduces the temperature of the system to its initial temperature. The first law states that this heat transfer will be exactly equal to the work that was done by the falling weight.

Example 4.1

A linear spring is stretched a distance of 0.8 m and attached to a paddle wheel (Fig. E4.1). The paddle wheel then rotates until the spring is unstretched. Calculate the heat transfer necessary to return the system to its initial state.

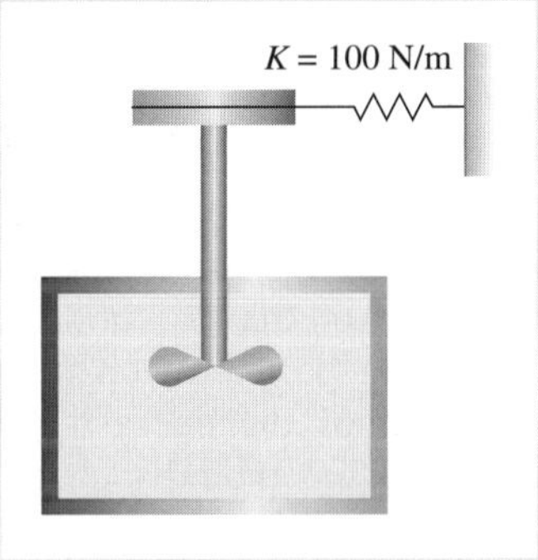

FIGURE E4.1

Solution

The work done by the linear spring on the system is given by

$$W_{1\text{-}2} = \int_0^{0.8} F\,dx = \int_0^{0.8} 100x\,dx = 100 \times \frac{0.8^2}{2} = 32\ \text{N·m}$$

Since the heat transfer returns the system to its initial state, a cycle results. The first law then states that

$$Q_{1\text{-}2} = W_{1\text{-}2} = 32\ \text{J}$$

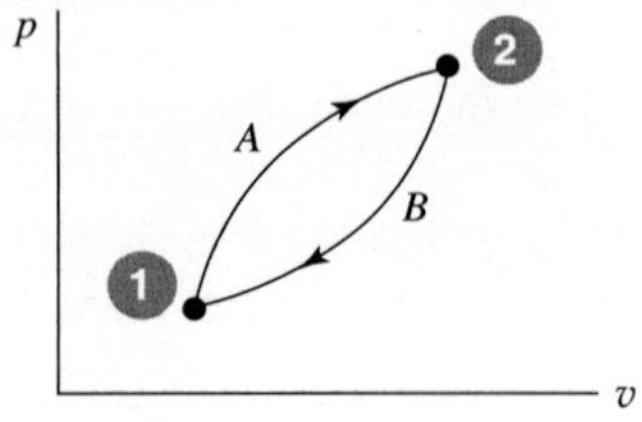

FIGURE 4.2 A cycle composed of two processes.

4.3 THE FIRST LAW APPLIED TO A PROCESS

The first law of thermodynamics is often applied to a process as the system changes from one state to another. We could consider a cycle composed of the two processes represented by A and B in Fig. 4.2. Applying the first law to this cycle, Eq. 4.1b takes the form

$$\int_1^2 \delta Q_A + \int_2^1 \delta Q_B = \int_1^2 \delta W_A + \int_2^1 \delta W_B \tag{4.2}$$

We interchange the limits on the process from 1 to 2 along B and write this as

$$\int_1^2 \delta Q_A - \int_1^2 \delta Q_B = \int_1^2 \delta W_A - \int_1^2 \delta W_B \tag{4.3a}$$

or, equivalently

$$\int_1^2 \delta(Q - W)_A = \int_1^2 \delta(Q - W)_B \tag{4.3b}$$

That is, the change in the quantity $Q - W$ from state 1 to state 2 is the same along path A as along path B; since this change is independent of the path between states 1 and 2 we let

$$\delta Q - \delta W = dE \tag{4.4a}$$

where dE is an exact differential.

The quantity E is an extensive property of the system and can be shown experimentally to represent the energy of the system at a particular state. Equation 4.4a can be integrated to yield

$$Q_{1\text{-}2} - W_{1\text{-}2} = E_2 - E_1 \tag{4.4b}$$

Key Concept: *More often than not, the subscripts will be dropped on Q and W when working problems.*

where Q_{1-2} is the heat transferred to the system during the process from state 1 to state 2, W_{1-2} is the work done by the system on the surroundings during the process, and E_2 and E_1 are the values of the property E. More often than not, the subscripts will be dropped on Q and W when working problems.

The property E represents all of the energy: kinetic energy KE, potential energy PE, and internal energy U that includes chemical energy and the energy associated with the molecules and atoms. Any other form of energy is also included in the total energy E. Its associated intensive property is designated e.

The first law of thermodynamics then takes the form

$$\begin{aligned} Q_{1\text{-}2} - W_{1\text{-}2} &= KE_2 - KE_1 + PE_2 - PE_1 + U_2 - U_1 \\ &= \frac{1}{2}m(V_2^2 - V_1^2) + mg(z_2 - z_1) + U_2 - U_1 \end{aligned} \tag{4.5}$$

If we apply the first law to an isolated system, one which is not in contact with its surroundings so that $Q_{1-2} = W_{1-2} = 0$, the first law becomes the conservation of energy, that is,

$$E_2 = E_1 \tag{4.6}$$

The internal energy U is an extensive property. Its associated intensive property is the specific internal energy u; that is, $u = U/m$. For simple systems in equilibrium, only two properties are necessary to establish the state of a pure substance (i.e., air or steam). Since internal energy is a property, it depends only on pressure and temperature; or, for steam in the quality region, it depends on quality and temperature (or pressure). For steam, its value for a particular quality would be

$$u = u_f + x(u_g - u_f) \tag{4.7}$$

Key Concept: *Two properties are necessary to establish the state of a pure substance.*

We can now apply the first law to systems involving working fluids with tabulated property values. Before we apply the first law to systems involving substances such as ideal gases or solids, we will introduce several additional properties that will simplify that task in the next three articles, but first some examples.

Example 4.2

A 5-hp fan is used in a large room to provide air circulation. Assuming a well-insulated, sealed room, determine the internal energy increase after 1 hour of operation.

Solution

By assumption, $Q = 0$. With $\Delta PE = \Delta KE = 0$ the first law becomes $-W = \Delta U$. The work input is

$$W = (-5 \text{ hp})(746 \text{ W/hp})(1 \text{ h})(3600 \text{ s/h}) = -1.343 \times 10^7 \text{ J}$$

The negative sign results because the work is input to the system. Finally, the internal energy increase is

$$\Delta U = -(-1.343 \times 10^7) = 1.343 \times 10^7 \text{ J} \quad \text{or} \quad 13.43 \text{ MJ}$$

Example 4.3

A rigid volume contains 6 ft^3 of steam originally at a pressure of 400 psia and a temperature of 900°F. Estimate the final temperature if 800 Btu of heat is added.

Solution

The first law of thermodynamics, with $\Delta KE = \Delta PE = 0$, is $Q - W = \Delta U$. For a rigid container, the work is zero. Thus,

$$Q = \Delta U = m(u_2 - u_1)$$

From the steam tables in Appendix C, we find $u_1 = 1324$ Btu/lbm and $v_1 = 1.978$ ft^3/lbm. The mass is then

$$m = \frac{V}{v_1} = \frac{6}{1.978} = 3.033 \text{ lbm}$$

The energy transferred to the volume by heat is given. Thus,

$$800 = 3.033(u_2 - 1324) \qquad \therefore u_2 = 1588 \text{ Btu/lbm}$$

(continued)

From Appendix Table C.3E, we must find the temperature for which $v_2 = 1.978 \text{ ft}^3/\text{lbm}$ and $u_2 = 1588$ Btu/lbm. This is not a simple task since we do not know the pressure. At 500 psia, if $v = 1.978 \text{ ft}^3/\text{lbm}$, then $u = 1459$ Btu/lbm and $T = 1221°\text{F}$. At 600 psia, if $v = 1.978 \text{ ft}^3/\text{lbm}$, then $u = 1603$ Btu/lbm and $T = 1546°\text{F}$. Now, we linearly interpolate to find the temperature at $u_2 = 1588$ Btu/lbm:

$$T = 1546 - \frac{1603 - 1588}{1603 - 1459}(1546 - 1221) = 1512°\text{F}$$

Example 4.4

A frictionless piston is used to provide a constant pressure of 400 kPa in a cylinder containing steam originally at 200°C with a volume of 2 m^3. Calculate the final temperature if 3500 kJ of heat is added. The first law of thermodynamics, using $\Delta PE = \Delta KE = 0$, is $Q - W = \Delta U$. The work done during the motion of the piston is

$$W = \int p\,dV = p(V_2 - V_1) = 400(V_2 - V_1)$$

The mass before and after remains unchanged. Using the steam tables in Appendix C, this is expressed as

$$m = \frac{V_1}{v_1} = \frac{2}{0.5342} = 3.744 \text{ kg}$$

The volume V_2 is written as $V_2 = mv_2 = 3.744v_2$. The first law is then, finding u_1 from the steam tables,

$$3500 - 400(3.744v_2 - 2) = (u_2 - 2647) \times 3.744$$

Solution

This requires a trial-and-error process. One plan for obtaining a solution is to guess a value for v_2 and calculate u_2 from the equation above. If this value checks with the u_2 from the steam tables at the same temperature, then the guess is the correct one. For example, guess $v_2 = 1.0 \text{ m}^3/\text{kg}$. Now, the equation gives $u_2 = 3395$ kJ/kg. From the steam tables, with $p = 0.4$ MPa, the u_2 value allows us to interpolate $T_2 = 654°\text{C}$, and the v_2 gives $T_2 = 600°\text{C}$. Therefore, the guess must be revised. Try $v_2 = 1.06 \text{ m}^3/\text{kg}$. The equation now gives $u_2 = 3372$ kJ/kg. The tables are interpolated to give $T_2 = 640°\text{C}$; for v_2, $T_2 = 647°\text{C}$. The actual v_2 is a little less than 1.06 m^3/kg, with the final temperature being approximately

$$T_2 = 644°\text{C}$$

4.4 ENTHALPY

In the solution of problems involving systems, certain products or sums of properties occur with regularity. One such combination of properties can be demonstrated by considering the addition of heat to the constant-pressure situation shown in Fig. 4.3. Heat is added slowly to the system (the gas in the cylinder),

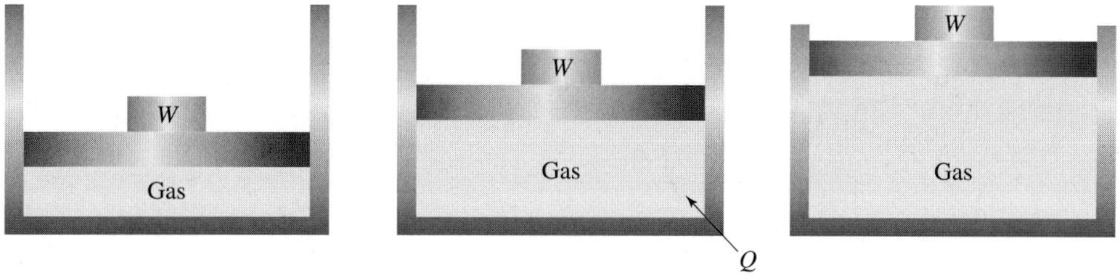

FIGURE 4.3 Constant-pressure heat addition.

which is maintained at constant pressure by assuming a frictionless seal between the piston and the cylinder. If the kinetic energy changes and potential energy changes of the system are negligible and all other work modes are absent, the first law of thermodynamics requires that

$$Q - W = U_2 - U_1 \tag{4.8}$$

The work done raising the weight for the constant-pressure process is given by

$$W = p(V\!\!\!/_2 - V\!\!\!/_1) \tag{4.9}$$

The first law can then be written as

$$Q = (U + pV\!\!\!/)_2 - (U + pV\!\!\!/)_1 \tag{4.10}$$

The quantity in parentheses is a combination of properties and, thus, a property itself. It is called the *enthalpy* H of the system; that is,

Key Concept: *Enthalpy is simply a combination of properties.*

$$H = U + pV\!\!\!/ \tag{4.11}$$

The specific enthalpy h is found by dividing by the mass. It is

$$h = u + pv \tag{4.12}$$

Enthalpy is a property of a system and is also found in the Appendix steam tables. The energy equation can now be written for a constant-pressure equilibrium process as

$$Q_{1-2} = H_2 - H_1 \tag{4.13}$$

The enthalpy was defined using a constant-pressure system with the difference in enthalpies between two states being the heat transfer. For a variable-pressure process, the difference in enthalpy loses its physical significance when considering a system. But enthalpy remains a property, as defined by Eq. 4.11. In a non-equilibrium constant-pressure process, ΔH would not equal the heat transfer.

It is only the changes in enthalpy or internal energy that are important; hence, we can arbitrarily choose the datum from which to measure h and u. We choose saturated liquid at 0°C to be the datum point for water; there $h = 0$ and $u = 0$.

Example 4.5

Using the concept of enthalpy, solve the problem presented in Example 4.4 using (a) the steam tables and (b) TK Solver.

Solution

The energy equation for a constant-pressure process is (with the subscript on the heat transfer omitted)

$$Q = H_2 - H_1 \qquad \text{or} \qquad 3500 = m(h_2 - 2860)$$

(a) Using the steam tables in Appendix C, as in Example 4.4, the mass is

$$m = \frac{V}{v} = \frac{2}{0.5342} = 3.744 \text{ kg}$$

Thus,

$$h_2 = \frac{3500}{3.744} + 2860 = 3795 \text{ kJ/kg}$$

From the steam tables, this interpolates to

$$T_2 = 600 + \frac{92.6}{224} \times 100 = 641°\text{C}$$

Obviously, enthalpy was very useful in solving this constant-pressure problem. Trial and error was unnecessary, and the solution was rather straightforward. We illustrated that the quantity we made up, enthalpy, is not necessary, but it is quite handy. We will use it often in our calculations.

(b) Using TK Solver with $p_1 = 400$ kPa and $T_1 = 200°$C, we find $v_1 = 0.53417$ m^3/kg and $h_1 = 2860.1$ kJ/kg. Then $m = 2/0.53417 = 3.7441$ kg, resulting in

$$h_2 = \frac{Q}{m} + h_1 = \frac{3500}{3.7441} + 2860.1 = 3794.9 \text{ kJ/kg}$$

TK Solver then gives $T_2 = 641.4°$C.

4.5 LATENT HEAT

Latent heat: *The energy that must be transferred, in the form of heat, to a substance held at a constant pressure in order that a phase change occurs.*

Key Concept: *The heat of fusion and the heat of sublimation are insensitive to pressure or temperature changes.*

The amount of energy that must be transferred, in the form of heat, to a substance held at constant pressure in order that a phase change occurs is called the **latent heat**. It is the change in enthalpy of the substance at the saturated conditions of the two phases. The heat that is necessary to melt a unit mass of a substance at constant pressure is the *heat of fusion* and is equal to $h_{if} = h_f - h_i$, where h_i is the enthalpy of saturated solid and h_f is the enthalpy of saturated liquid. The *heat of vaporization* is the heat required to completely vaporize a unit mass of saturated liquid; it is equal to $h_{fg} = h_g - h_f$. When a solid changes phase directly to a gas, sublimation occurs; the *heat of sublimation* is equal to $h_{ig} = h_g - h_i$.

The heat of fusion and the heat of sublimation are relatively insensitive to pressure or temperature changes. For ice, the heat of fusion is approximately 330 kJ/kg (143 Btu/lbm) and the heat of sublimation is about 2040 kJ/kg (880 Btu/lbm). The heat of vaporization of water is included as h_{fg} in Appendix Tables C.1 and C.2 and is very sensitive to pressure and temperature.

4.6 SPECIFIC HEATS

For a simple system, only two independent variables are necessary to establish the state of the system. Consequently, we can consider the specific internal energy to be a function of temperature and specific volume; that is,

$$u = u(T, v) \tag{4.14}$$

Using the chain rule from calculus, we express the differential, in terms of the partial derivatives, as

$$du = \left.\frac{\partial u}{\partial T}\right|_v dT + \left.\frac{\partial u}{\partial v}\right|_T dv \tag{4.15}$$

Since u, v, and T are all properties, the partial derivative is also a property and is called the *constant-volume specific heat* c_v; that is

$$c_v = \left.\frac{\partial u}{\partial T}\right|_v \tag{4.16}$$

One of the classical experiments of thermodynamics, first performed by Joule in 1843, is illustrated in Fig. 4.4. Pressurize volume A with an ideal gas, and evacuate volume B. After equilibrium is attained, open the valve. Even though the pressure and volume of the ideal gas have changed markedly, the temperature does not change. Because there is no change in temperature, there is no net heat transfer to the water. Observing that no work is done, we conclude, from the first law, that the internal energy of an ideal gas does not depend on pressure or volume and is only a function of the temperature.

Key Concept: *The internal energy of an ideal gas is only a function of the temperature.*

For such a gas, which behaves as an ideal gas, we have

$$\left.\frac{\partial u}{\partial v}\right|_T = 0 \tag{4.17}$$

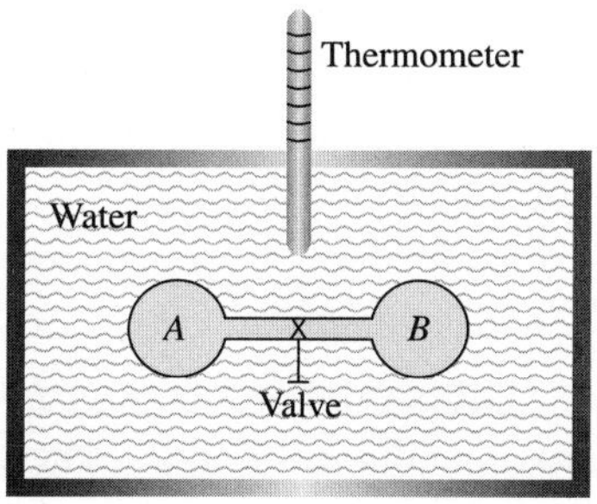

FIGURE 4.4 Joule's experiment.

Combining Eqs. 4.15, 4.16, and 4.17, we have

$$du = c_v\,dT \tag{4.18}$$

This can be integrated to give

$$u_2 - u_1 = \int_{T_1}^{T_2} c_v\,dT \tag{4.19}$$

For a known $c_v(T)$, this can be integrated to find the change in internal energy over any temperature interval for an ideal gas.

Likewise, considering specific enthalpy to be dependent on the two variables T and p, we have

$$dh = \left.\frac{\partial h}{\partial T}\right|_p dT + \left.\frac{\partial h}{\partial p}\right|_T dp \tag{4.20}$$

The *constant-pressure specific heat* c_p is defined as

$$c_p = \left.\frac{\partial h}{\partial T}\right|_p \tag{4.21}$$

For an ideal gas, returning to the definition of enthalpy in Eq. 4.12, we have

$$h = u + pv = u + RT \tag{4.22}$$

Key Concept: *Enthalpy is function of T only for an ideal gas.*

where we have used the ideal-gas equation of state. Since u is only a function of T, we see that h is also a function of T only for an ideal gas. Hence, for an ideal gas

$$\left.\frac{\partial h}{\partial p}\right|_T = 0 \tag{4.23}$$

and we have, from Eq. 4.20,

$$dh = c_p\,dT \tag{4.24}$$

Over the temperature range of T_1 to T_2, this is integrated to give

$$h_2 - h_1 = \int_{T_1}^{T_2} c_p\,dT \tag{4.25}$$

for an ideal gas.

It is often convenient to specify specific heats on a per-mole, rather than a per-unit-mass, basis; these *molar specific heats* are $\bar{c}_v$ and $\bar{c}_p$. Clearly, we have the relations

$$\bar{c}_v = Mc_v \qquad \text{and} \qquad \bar{c}_p = Mc_p \tag{4.26}$$

Key Concept: *The overbar notation for a molar quantity is used throughout thermodynamics.*

where M is the molar mass. Thus, values of $\bar{c}_v$ and $\bar{c}_p$ may be determined from the values of c_v and c_p listed in Appendix Table B.2. (The common overbar notation for a molar quantity is used throughout thermodynamics.)

The equation for enthalpy can be used to relate, for an ideal gas, the specific heats and the gas constant. In differential form, Eq. 4.12 takes the form

$$dh = du + d(pv) \tag{4.27}$$

Introducing the specific-heat relations and the ideal-gas equation, we have

$$c_p\,dT = c_v\,dT + R\,dT \tag{4.28}$$

which, after dividing by dT, gives

$$c_p = c_v + R \tag{4.29}$$

This relationship, or its molar equivalent $\bar{c}_p = \bar{c}_v + \bar{R}$, allows c_v to be determined from tabulated values or expressions for c_p. Note that the difference between c_p and c_v for an ideal gas is always a constant, even though both are functions of temperature.

Key Concept: *The difference between c_p and c_v for an ideal gas is always constant.*

The *specific heat ratio* k is also a property of particular interest; it is defined as

$$k = \frac{c_p}{c_v} \tag{4.30}$$

This can be substituted into Eq. 4.29 to give

$$c_p = R\frac{k}{k-1} \tag{4.31}$$

or

$$c_v = R\frac{1}{k-1} \tag{4.32}$$

Obviously, since R is a constant for an ideal gas, the specific heat ratio k will depend only on the temperature.

For gases, the specific heats slowly increase with increasing temperature. Since they do not vary significantly over fairly large temperature differences, it is often acceptable to treat c_v and c_p as constants. For such situations there results

Key Concept: *It is often acceptable to treat c_p and c_v as constants.*

$$u_2 - u_1 = c_v(T_2 - T_1) \tag{4.33}$$

$$h_2 - h_1 = c_p(T_2 - T_1) \tag{4.34}$$

For air, we will use[1] $c_v = 0.717$ kJ/kg·°C (0.171 Btu/lbm-°R) and $c_p = 1.00$ kJ/kg·°C (0.24 Btu/lbm-°R), unless otherwise stated. For more accurate calculations, with air or other gases, one should consult ideal-gas tables, such as those in Appendix F, which tabulate $h(T)$ and $u(T)$, or integrate using expressions for $c_p(T)$ found in Appendix Table B.5. Once c_p is found, c_v can be calculated using $c_v = R - c_p$.

Key Concept: *Once c_p is found, c_v can be calculated using $c_v = R - c_p$.*

For liquids and solids, the specific heat c_p is tabulated in Appendix Table B.4. Since it is quite difficult to maintain constant volume while the temperature is

[1]The units on specific heat can be written as kJ/kg·K or kJ/kg·°C since the Kelvin degree and the Celsius degree have the same magnitude.

Key Concept: *For liquids and solids, the difference $c_p - c_v$ is quite small and usually ignored.*

changing, c_v values are usually not tabulated for liquids and solids; the difference $c_p - c_v$ is quite small. For most liquids, the specific heat is relatively insensitive to temperature change. For water, we will use the nominal value of 4.18 kJ/kg·°C (1.00 Btu/lbm-°R). For ice, the specific heat in kJ/kg·°C is approximately $c_p = 2.1 + 0.0069T$, where T is measured in °C; and in English units of Btu/lbm-°F, it is $c_p = 0.47 + 0.001T$, where T is measured in °F. The variation of specific heat with pressure is usually quite slight except for special situations.

Example 4.6

The specific heat of superheated steam at approximately 150 kPa can be determined by the equation

$$c_p = 2.07 + \frac{T - 400}{1480} \text{ kJ/kg·°C}$$

(a) What is the enthalpy change between 300°C and 700°C for 3 kg of steam? Compare with the Appendix steam tables and with TK Solver.

(b) What is the average value of c_p between 300°C and 700°C, based on the equation and based on the tabulated data?

Solution

(a) The enthalpy change is found to be

$$\Delta H = m\int_{T_1}^{T_2} c_p\,dT = 3\int_{300}^{700}\left(2.07 + \frac{T - 400}{1480}\right)dT = 2565 \text{ kJ}$$

From the Appendix tables, we find, using $p = 150$ kPa,

$$\Delta H = 3 \times (3928 - 3073) = 2565 \text{ kJ}$$

Using TK Solver, we find $h_1 = 3072.7$ kJ/kg and $h_2 = 3928.6$ kJ/kg. Hence,

$$\Delta H = 3(3928.6 - 3072.7) = 2567.7 \text{ kJ}$$

(b) The average value $c_{p,\text{avg}}$ is found by using the relation

$$mc_{p,\text{avg}}\Delta T = m\int_{T_1}^{T_2} c_p\,dT$$

$$3c_{p,\text{avg}} \times 400 = 3\int_{300}^{700}\left(2.07 + \frac{T - 400}{1480}\right)dT$$

The integral was evaluated in part (a); hence, we have

$$c_{p,\text{avg}} = \frac{2565}{3 \times 400} = 2.14 \text{ kJ/kg·°C}$$

Using the values from the steam table, we have

$$c_{p,\text{avg}} = \frac{\Delta h}{\Delta T} = \frac{3928 - 3073}{400} = 2.14 \text{ kJ/kg·°C}$$

(continued)

Because the steam tables give the same values as the linear equation of this example, we can assume safely that the $c_p(T)$ relationship for steam over this temperature range is approximated closely by a linear relation. This linear relation would change, however, for each pressure chosen; hence, the steam tables are essential.

Example 4.7

Determine the value of c_p for steam at $T = 800°F$ and $p = 800$ psia using (a) the steam tables and (b) TK Solver.

Solution

(a) To determine c_p, we use a finite-difference approximation to Eq. 4.21. We use the entries at $T = 900°F$ and $T = 700°F$, which provide a better approximation to the slope compared to using the values at 800°F and 750°F or at 900°F and 800°F. Appendix Table C.3E provides us with

$$c_p = \frac{\Delta h}{\Delta T} = \frac{1455.6 - 1338.0}{200} = 0.588 \text{ Btu/lbm-°F}$$

Figure E4.7 shows why it is better to use values on either side of the position of interest. If values at 900°F and 800°F are used (a forward difference), c_p is too low. If values at 800°F and 750°C are used, (a backward difference), c_p is too high. Thus, both a forward value and a backward value (a central difference) should be used, resulting in a more accurate estimate of the slope.

(b) Select $T_2 = 850°F$ and $T_1 = 750°F$. Use TK Solver to find $h_2 = 1427.1$ Btu/lbm and $h_1 = 1368.5$ Btu/lbm. Finally,

$$c_p = \frac{\Delta h}{\Delta T} = \frac{1427.1 - 1368.5}{100} = 0.586 \text{ Btu/lbm-°F}$$

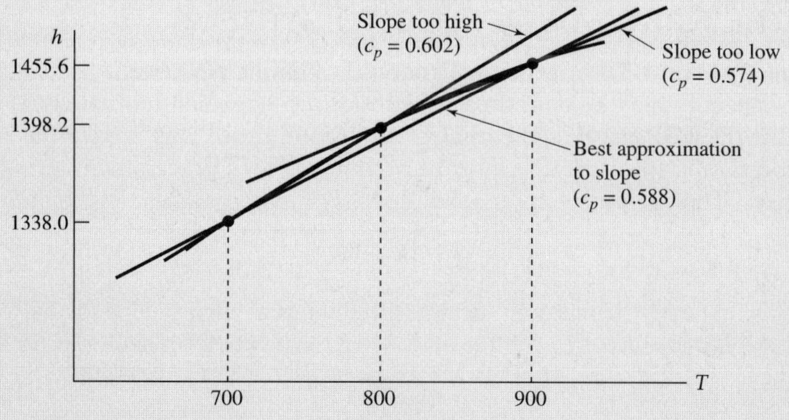

FIGURE E4.7

Example 4.8

Determine the enthalpy change for 1 kg of nitrogen which is heated from 300 to 1200 K by (a) using the gas tables, (b) integrating $c_p(T)$, and (c) assuming constant specific heat. Use $M = 28$ kg/kmol.

Solution

(a) Using the gas table in Appendix F, find the enthalpy change to be

$$\Delta h = 36\,777 - 8723 = 28\,054 \text{ kJ/kmol} \qquad \text{or} \qquad 28\,054/28 = 1002 \text{ kJ/kg}$$

(b) The expression for $c_p(T)$ is found in Appendix Table B.5. The enthalpy change is

$$\Delta h = \int_{300}^{1200} \left[39.06 - 512.79\left(\frac{T}{100}\right)^{-1.5} + 1072.7\left(\frac{T}{100}\right)^{-2} - 820.4\left(\frac{T}{100}\right)^{-3} \right] dT$$

$$= 39.06 \times (1200 - 300) - 512.79 \times \frac{100}{-0.5}(12^{-.5} - 3^{-.5}) + 1072.7 \times \frac{100}{-1}(12^{-1} - 3^{-1})$$

$$- 820.4 \times \frac{100}{-2}(12^{-2} - 3^{-2}) = 28\,093 \text{ kJ/kmol} \quad \text{or} \quad 1003 \text{ kJ/kg}$$

(c) Assuming constant specific heat (found in Appendix Table B.2), the enthalpy change is found to be

$$\Delta h = c_p \Delta T = 1.042 \times (1200 - 300) = 938 \text{ kJ/kg}$$

Note: The value found by integrating is essentially the same as that found from the gas tables. However, the enthalpy change found by assuming constant specific heat is in error by over six percent. If T_2 were closer to 300 K (i.e., 600 K), the error would be much smaller.

4.7 THE FIRST LAW APPLIED TO SYSTEMS

The Constant-Temperature Process. For the isothermal process, tables in the Appendix may be consulted for substances for which tabulated values are available. Internal energy and enthalpy vary slightly with pressure for the isothermal process, and this variation must be accounted for in processes involving many substances. The energy equation 4.5 with $\Delta KE = \Delta PE = 0$ is

$$Q - W = \Delta U \tag{4.35}$$

For a gas that approximates an ideal gas, the internal energy depends only on the temperature, and thus, $\Delta U = 0$ for an isothermal process; for such a process,

$$Q = W \tag{4.36}$$

Using the ideal-gas equation $P\mathcal{V} = mRT$, the work for a quasi-equilibrium isothermal process involving a system can be found to be

$$W = \int_{V_1}^{V_2} p\,dV$$

$$= mRT \int_{V_1}^{V_2} \frac{dV}{V} = mRT \ln \frac{V_2}{V_1} = mRT \ln \frac{p_1}{p_2} \tag{4.37}$$

where we used $p_1 V_1 = p_2 V_2$.

The Constant-Volume Process. The work for a constant-volume quasi-equilibrium process involving a system is zero, since dV is zero. For such a process the first law with $\Delta KE = \Delta PE = 0$ becomes

$$Q = \Delta U \tag{4.38}$$

If tabulated values are available for a substance, we may directly determine ΔU. For a gas, approximated by an ideal gas, we would have

$$Q = m \int_{T_1}^{T_2} c_v\,dT \tag{4.39}$$

or, for a process for which c_v is essentially constant,

$$Q = mc_v(T_2 - T_1) \qquad \text{or} \qquad q = c_v(T_2 - T_1) \tag{4.40}$$

If nonequilibrium work, such as paddle-wheel work, is present, that work must be accounted for in the first law.

Equation 4.40 provides the motivation for the name "specific heat" for c_v. Historically, this equation was used to define c_v; it was defined as the heat necessary to raise the temperature of one unit of substance one degree in a constant-volume process. Today, scientists prefer the definition of c_v to be in terms of properties only, without reference to heat transfer, as in Eq. 4.16.

The Constant-Pressure Process. The first law, for a constant-pressure quasi-equilibrium process, was shown in Sec. 4.4 to be

$$Q = \Delta H \tag{4.41}$$

Hence, the heat transfer for such a process can easily be found using tabulated values (see Appendix), if available.

For a gas that behaves as an ideal gas, we have

$$Q = m \int_{T_1}^{T_2} c_p\,dT \tag{4.42}$$

For a process involving an ideal gas for which c_p is constant, there results

$$Q = mc_p(T_2 - T_1) \tag{4.43}$$

For a nonequilibrium process, the work must be accounted for directly in the first law, and it cannot be expressed as $p(V_2 - V_1)$. For such a process, the above three equations would not be valid.

The Adiabatic Process. There are numerous examples of processes for which there is no, or negligibly small, heat transfer (e.g., the compression and expansion of air in the cylinders of an automobile engine). The study of such processes is, however, often postponed until after the second law of thermodynamics is presented. This postponement is not necessary for an ideal gas, and because the adiabatic quasi-equilibrium process is quite common, it is presented here.

The differential form of the first law for the adiabatic process, with no kinetic or potential energy change, is

$$-\delta W = du \tag{4.44}$$

or, for a quasi-equilibrium process in which $\delta W = p\,dv$,

$$du + p\,dv = 0 \tag{4.45}$$

To utilize this equation for substances such as water or refrigerants, we must use tabulated properties, an impossible task at this point because we do not know a property that remains constant for this process. However, if an ideal gas can be assumed, Eq. 4.45 can be solved as follows.

Let us investigate the adiabatic process for an ideal gas with constant specific heats. For such a process, Eq. 4.45 takes the form

$$c_v\,dT + \frac{RT}{v}\,dv = 0 \tag{4.46}$$

Rearranging, we have

$$\frac{c_v}{R}\frac{dT}{T} = -\frac{dv}{v} \tag{4.47}$$

This is integrated, assuming constant c_v between states 1 and 2, to give

$$\frac{c_v}{R}\ln\frac{T_2}{T_1} = -\ln\frac{v_2}{v_1} \tag{4.48}$$

which can be put in the form

$$\frac{T_2}{T_1} = \left(\frac{v_1}{v_2}\right)^{R/c_v} = \left(\frac{v_1}{v_2}\right)^{k-1} \tag{4.49}$$

referring to Eq. 4.32. Using the ideal-gas law, this can also be written as

$$\frac{T_2}{T_1} = \left(\frac{p_2}{p_1}\right)^{(k-1)/k} \qquad \frac{p_2}{p_1} = \left(\frac{v_1}{v_2}\right)^{k} \tag{4.50}$$

Finally, the above three relations can be put in general forms, without reference to particular points. For the adiabatic quasi-equilibrium process involving an ideal gas with constant c_p and c_v, we have

$$Tv^{k-1} = \text{const} \qquad Tp^{(1-k)/k} = \text{const} \qquad pv^k = \text{const} \tag{4.51}$$

The adiabatic process for a real substance (such as water, steam, or a solid) will be treated in detail in Chapter 5.

The Polytropic Process. A careful inspection of the special quasi-equilibrium processes for an ideal gas presented in this section suggests that each process can be expressed as

$$p\mathcal{V}^n = \text{const} \tag{4.52}$$

The work is calculated

$$W = \int_{\mathcal{V}_1}^{\mathcal{V}_2} p\, d\mathcal{V} = p_1\mathcal{V}_1^n \int_{\mathcal{V}_1}^{\mathcal{V}_2} \mathcal{V}^{-n}\, d\mathcal{V}$$

$$= \frac{p_1\mathcal{V}_1^n}{1-n}\left(\mathcal{V}_2^{1-n} - \mathcal{V}_1^{1-n}\right) = \frac{p_2\mathcal{V}_2 - p_1\mathcal{V}_1}{1-n} \tag{4.53}$$

Except, Eq. 4.37 is used if $n = 1$, since Eq. 4.53 is undefined for $n = 1$. The heat transfer follows from the first law.

Each quasi-equilibrium process of an ideal gas is associated with a particular value for n as follows:

Isothermal:	$n = 1$
Constant-volume:	$n = \infty$
Constant-pressure:	$n = 0$
Adiabatic (c_p = const):	$n = k$

The processes are displayed on a $(\ln p)$ versus $(\ln \mathcal{V})$ plot in Fig. 4.5. The slope of each straight line is the exponent on $\mathcal{V}$ in Eq. 4.52. If the slope is none of the values ∞, k, 1, or 0, then the process can be referred to as a *polytropic process*. For such a process involving an ideal gas, any of the Equations (Eqs. 4.49, 4.50, or 4.51) can be used with k simply replaced by n.

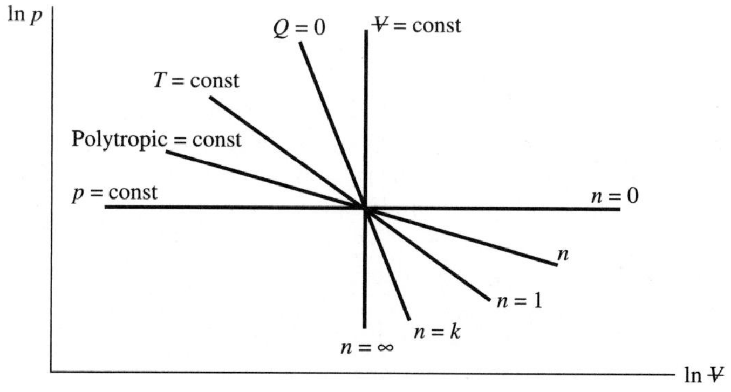

FIGURE 4.5 Polytropic exponents for various processes.

Example 4.9

Determine the heat transfer necessary to increase the pressure of 70 percent quality steam from 200 to 800 kPa, maintaining the volume constant at 2 m^3. Assume a quasi-equilibrium process and use (a) the Appendix steam tables and (b) TK Solver.

Solution

For the constant-volume quasi-equilibrium process, the work is zero. The first law reduces to $Q = m(u_2 - u_1)$.

(a) The mass must be found. It is

$$m = \frac{V\!\!\!/}{v} = \frac{2}{0.0011 + 0.7(0.8857 - 0.0011)} = 3.224 \text{ kg}$$

The internal energy at state 1 is

$$u_1 = 504.5 + 0.7(2529.5 - 504.5) = 1922 \text{ kJ/kg}$$

The constant-volume process demands that $v_2 = v_1 = 0.6203 \text{ m}^3\text{/kg}$. From the steam tables in the Appendix, at 800 kPa we find, by extrapolation, that

$$u_2 = \frac{0.6203 - 0.6181}{0.6181 - 0.5601}(3661 - 3476) = 3668 \text{ kJ/kg}$$

Note: Extrapolation was necessary, since the temperature at state 2 exceeds the highest tabulated temperature of 800°C. The heat transfer is then

$$Q = m(u_2 - u_1) = 3.224 \times (3668 - 1922) = 5629 \text{ kJ}$$

(b) Using TK Solver, we find $v = 0.62039 \text{ m}^3\text{/kg}$ so that the mass is $m = 2/0.62039 = 3.2238$ kg. Calculate the heat transfer as follows:

$$u_1 = h_1 - p_1 v_1 = 2046 - 200 \times 0.62039 = 1921.9 \text{ kJ/kg}$$

$$u_2 = h_2 - p_2 v_2 = 4166.3 - 800 \times 0.62039 = 3670 \text{ kJ/kg}$$

$$\therefore\ Q = m(u_2 - u_1) = 3.2238 \times (3670 - 1921.9) = 5636 \text{ kJ}$$

Example 4.10

A piston-cylinder arrangement contains 0.02 m^3 of air at 50°C and 400 kPa. Heat is added in the amount of 50 kJ and work is done by a paddle wheel until the temperature reaches 700°C. If the pressure is held constant, how much paddle-wheel work must be added to the air? Assume constant specific heats.

Solution

The process cannot be approximated by a quasi-equilibrium process because of the paddle-wheel work. Thus, the heat transfer is not equal to the enthalpy change. The first law may be written as

(continued)

$$Q - W_{\text{paddle}} = m(h_2 - h_1)$$
$$= mc_p(T_2 - T_1)$$

To find m, we use the ideal-gas equation. It gives us

$$m = \frac{p\mathcal{V}}{RT} = \frac{400\,000 \times 0.02}{287 \times (50 + 273)} = 0.0863 \text{ kg}$$

From the first law, the paddle-wheel work is found to be

$$W_{\text{paddle}} = Q - mc_p(T_2 - T_1)$$
$$= 50 - 0.0863 \times 1.0 \times (700 - 50) = -6.096 \text{ kJ}$$

Note: We could have used the first law as $Q - W_{\text{net}} = m(u_2 - u_1)$ and then let $W_{\text{paddle}} = W_{\text{net}} - p(\mathcal{V}_2 - \mathcal{V}_1)$. We would then need to calculate $\mathcal{V}_2$.

Example 4.11

Calculate the work necessary to compress air in an insulated cylinder from a volume of 6 ft^3 to a volume of 1.2 ft^3. The initial temperature and pressure are 50°F and 30 psia, respectively.

Solution

We will assume that the compression process is approximated by a quasi-equilibrium process, which is acceptable for most compression processes, and that the process is adiabatic due to the presence of the insulation. The first law is then written as

$$-W = m(u_2 - u_1) = mc_v(T_2 - T_1)$$

The mass is found from the ideal-gas equation to be

$$m = \frac{p\mathcal{V}}{RT} = \frac{(30 \times 144) \times 6}{53.3 \times (50 + 460)} = 0.9535 \text{ lbm}$$

The final temperature T_2 is found for the adiabatic quasi-equilibrium process from Eq. 4.49; it is

$$T_2 = T_1\left(\frac{\mathcal{V}_1}{\mathcal{V}_2}\right)^{k-1} = 510\left(\frac{6}{1.2}\right)^{1.4-1} = 970.9^\circ\text{R}$$

Finally,

$$W = -0.9535 \text{ lbm}\left(0.171 \frac{\text{Btu}}{\text{lbm-}^\circ\text{R}}\right)(971 - 510)^\circ\text{R} = -75.2 \text{ Btu}$$

4.8 GENERAL FORMULATION FOR CONTROL VOLUMES

In the application of the various laws, we have thus far restricted ourselves to systems; no mass crosses the boundary of a system. This restriction is acceptable for many problems of interest and may, in fact, be imposed on the power plant

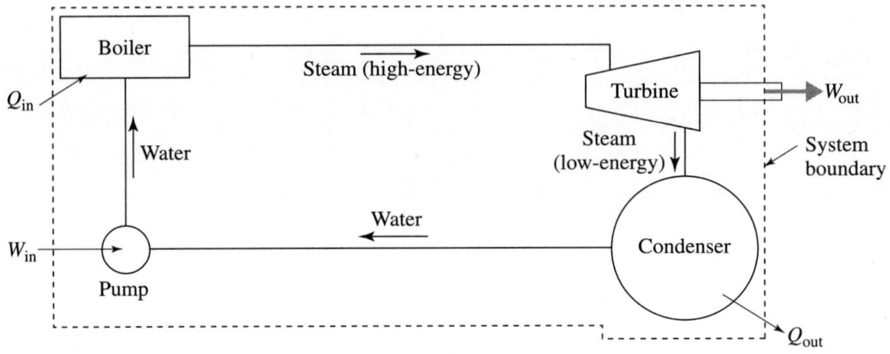

FIGURE 4.6 A schematic for a power plant.

schematic shown in Fig. 4.6. However, if the first law is applied to this system, only an incomplete analysis can be accomplished. For a more complete analysis, we must relate W_{in}, Q_{in}, W_{out}, and Q_{out} to the pressure and temperature changes for the pump, boiler, turbine, and condenser, respectively. To do this, we must consider each device of the power plant as a control volume into which and from which a fluid flows. For example, water flows into the pump at a low pressure and leaves the pump at a high pressure; the work input into the pump is obviously related to this pressure rise. We must formulate equations that allow us to make the necessary calculations. For most applications that we will consider, it will be acceptable to assume both a **steady flow** (the flow variables do not change with time) and a **uniform flow** (the velocity, pressure, and density are constant over the cross-sectional area). We will, however, develop the unsteady, nonuniform flow case that will find some application in our study of thermodynamics. Fluid mechanics treats the more general unsteady, nonuniform situations in much greater detail.

Steady flow: *The flow variables do not change with time.*

Uniform flow: *The velocity, pressure, and density are constant over the cross-sectional area.*

The Conservation of Mass. When applying the energy equation to control volumes, it is often necessary to apply the conservation of mass principle to the mass as it moves through a volume in space. This may allow us to determine needed information, such as the diameter of an outlet pipe leaving a pump, or the volume of flow required for a specified power output of a power plant. So first, we will derive the equation that results from a conservation of mass.

Consider the system composed of the mass contained in volume 1 plus that contained in the fixed control volume shown in Fig. 4.7a. The dashed lines bound the control volume. During the time increment Δt, the mass Δm_1 in the short pipe increment enters the control volume, and the mass Δm_2 in Fig. 4.7b leaves the control volume so that, at time $t + \Delta t$, the system then occupies the control volume plus volume 2. The conservation of mass principle requires that the mass in the system remains constant. The mass in the control volume may or may not remain constant; the control volume may gain or lose mass. This is expressed as

$$\Delta m_1(t) + m_{c.v.}(t) = m_{c.v.}(t + \Delta t) + \Delta m_2(t + \Delta t) \tag{4.54a}$$

This can be rearranged as

$$\Delta m_1(t) = \Delta m_2(t + \Delta t) + m_{c.v.}(t + \Delta t) - m_{c.v.}(t) \tag{4.54b}$$

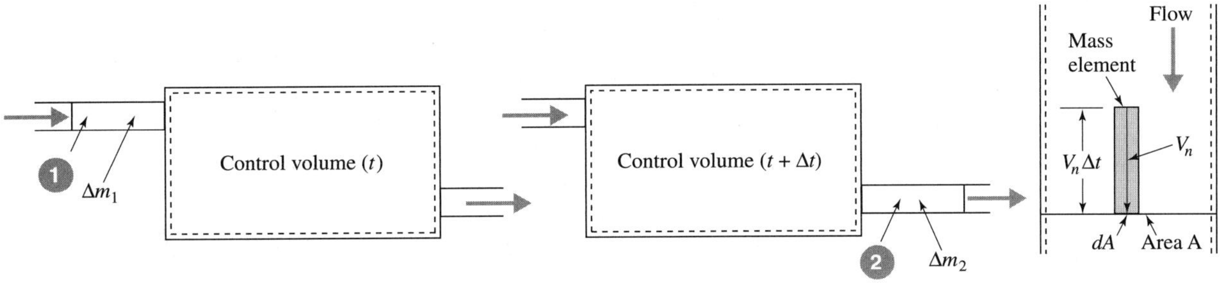

(a) System at time t (b) System at time $t + \Delta t$ (c) Mass element

FIGURE 4.7 System to control volume transformation for mass conservation.

If we divide by Δt and let $\Delta t \to 0$ then we have $\Delta m/\Delta t = dm/dt = \dot{m}$ and $[m_{\text{c.v.}}(t + \Delta t) - m_{\text{c.v.}}(t)]/\Delta t = dm_{\text{c.v.}}/dt$, following the definitions of a derivative from calculus. The conservation of mass equation then becomes

$$\dot{m}_1 = \dot{m}_2 + \frac{dm_{\text{c.v.}}}{dt} \tag{4.55a}$$

or, for multiple inlets and/or outlets, there results

$$\sum_{\text{inlets}} \dot{m}_i = \sum_{\text{outlets}} \dot{m}_i + \frac{dm_{\text{c.v.}}}{dt} \tag{4.55b}$$

which states that the entering mass flow rate equals the exiting mass flow rate plus the rate of mass stored in the control volume.

Now let's find an expression for $\dot{m}$, the **mass flow rate** (also called the mass flux) across an area. Consider the incremental mass element sketched in Fig. 4.7c. The length of the element is $V_n \Delta t$, the volume is $V_n \Delta t \, dA$, and the mass contained in the differential volume is $\rho V_n \Delta t \, dA$, where V_n is the velocity component normal to the area dA. Divide by Δt, and find the mass flow rate through the differential area to be $\rho V_n \, dA$. Integrate over the area to find the mass flow rate crossing an area A to be $\dot{m} = \int \rho V \, dA$, where the velocity is assumed to be the normal component (the tangential component carries no mass across the area). If we assume the density ρ and velocity V to be constant over the area (velocity is most often uniform over the inlet and exit areas especially for turbulent flow in pipes), the usual assumptions in our applications, then the mass flow rate across an area takes the form

Mass flow rate: *Equals ρAV for a uniform flow.*

Key Concept: *Velocity is approximately uniform for turbulent flow in pipes.*

$$\dot{m} = \rho AV \tag{4.56}$$

Let's consider control volumes that have one inlet and one outlet, the usual situation for pumps, turbines, condensers, compressors, etc. Substitute Eq. 4.56 into Eq. 4.55a and obtain

$$\rho_1 A_1 V_1 = \rho_2 A_2 V_2 + \frac{dm_{\text{c.v.}}}{dt} \tag{4.57}$$

If there is more than one inlet and one exit, we simply add the additional inlets and exits on the appropriate side of the above equation.

For the steady-flow situation with one inlet and one outlet in which the mass in the control volume remains constant, the continuity equation reduces to

$$\rho_1 A_1 V_1 = \rho_2 A_2 V_2 \tag{4.58}$$

Volume flow rate: *Equals AV for a uniform flow.*

which is the most common form of the continuity equation for the various devices. We also occasionally introduce the **volume flow rate** which is defined as Q_f and is equal to AV. It will be used quite often in our Fluid Mechanics part, since with a liquid flow $\rho_1 = \rho_2$.

Example 4.12

Water is flowing in a pipe that changes diameter from 20 to 40 mm. If the water in the 20-mm section has a velocity of 40 m/s, determine the velocity in the 40-mm section. Also, calculate the mass flow rate.

Solution

The continuity equation (Eq. 4.58) is used. There results, using $\rho_1 = \rho_2$,

$$A_1 V_1 = A_2 V_2 \qquad \pi \times 0.01^2 \times 40 = \pi \times 0.02^2 \times V_2 \qquad \therefore\ V_2 = 10 \text{ m/s}$$

The mass flow rate is found to be

$$\dot{m} = \rho A_1 V_1 = 1000 \times \pi \times 0.01^2 \times 40 = 12.57 \text{ kg/s}$$

where we have used $\rho = 1000$ kg/m^3, the standard value for water.

The Energy Equation. Consider again a fixed control volume as sketched in Fig. 4.8, where the system occupies volume 1 plus the control volume at time t and then the control volume plus volume 2 at time $t + \Delta t$. (The dashed lines enclose the control volume.) The first law of thermodynamics for the system can be stated as

$$\begin{aligned} Q - W &= E_{\text{sys}}(t + \Delta t) - E_{\text{sys}}(t) \\ &= \Delta E_2(t + \Delta t) + E_{\text{c.v.}}(t + \Delta t) - E_{\text{c.v.}}(t) - \Delta E_1(t) \\ &= \rho_2 A_2 V_2 \Delta t\left(u_2 + \tfrac{1}{2}V_2^2 + gz_2\right) + E_{\text{c.v.}}(t + \Delta t) - E_{\text{c.v.}}(t) \\ &\quad - \rho_1 A_1 V_1 \Delta t\left(u_1 + \tfrac{1}{2}V_1^2 + gz_1\right) \end{aligned} \tag{4.59}$$

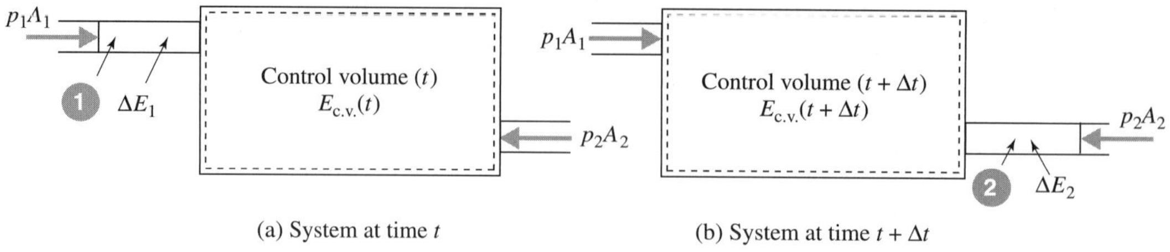

FIGURE 4.8 The system to control volume transformation for the first law.

where $\rho AV\Delta t$ is the mass in an incremental volume and Q and W are transferred to and from the system during the time increment Δt. Note that the energy consists of internal energy, kinetic energy, and potential energy. The work W is composed of two parts: the work due to the pressure needed to move the fluid, sometimes called **flow work**, and the work that results from a shaft that is rotating, called **shaft work** W_S. This is expressed as

Flow work: *Work due to the pressure needed to move the fluid.*

Shaft work: *Work that results from a shaft that is rotating.*

$$W = p_2 A_2 V_2 \Delta t - p_1 A_1 V_1 \Delta t + W_S \tag{4.60}$$

where pA is the pressure force and $V\Delta t$ is the distance it moves during the time increment Δt. The negative sign results because the work done on the system is negative when moving the fluid into the control volume. Substitute the expression for work W into Eq. 4.59, and express the flow work term as $\rho AV(p/\rho)\Delta t$. The first law equation is then arranged as

$$\begin{aligned} Q - W_S - \rho_2 A_2 V_2 \left(\frac{V_2^2}{2} + gz_2 + u_2 + \frac{p_2}{\rho_2} \right) \Delta t \\ - \rho_1 A_1 V_1 \left(\frac{V_1^2}{2} + gz_1 + u_1 + \frac{p_1}{\rho_1} \right) \Delta t + \Delta E_{\text{c.v.}} \end{aligned} \tag{4.61}$$

Divide through by Δt to obtain *the energy equation*:

$$\begin{aligned} \dot{Q} - \dot{W}_S = \dot{m}_2 \left(\frac{V_2^2}{2} + gz_2 + u_2 + \frac{p_2}{\rho_2} \right) \\ - \dot{m}_1 \left(\frac{V_1^2}{2} + gz_1 + u_1 + \frac{p_1}{\rho_1} \right) + \frac{dE_{\text{c.v.}}}{dt} \end{aligned} \tag{4.62}$$

where we have used

$$\dot{Q} = \frac{Q}{\Delta t} \qquad \dot{W}_S = \frac{W_S}{\Delta t} \qquad \dot{m} = \rho AV \tag{4.63}$$

If more than one inlet and one outlet are present, the equation becomes

$$\begin{aligned} \dot{Q} - \dot{W}_S = \frac{dF_{\text{c.v.}}}{dt} + \sum_{\text{outlets}} \dot{m}_i \left(\frac{V_i^2}{2} + gz_i + u_i + \frac{p_i}{\rho_i} \right) \\ - \sum_{\text{inlets}} \dot{m}_i \left(\frac{V_i^2}{2} + gz_i + u_i + \frac{p_i}{\rho_i} \right) \end{aligned} \tag{4.64}$$

If the flow is a steady flow, which it is in most of the control volume problems encountered in thermodynamics, the energy equation takes the form

$$\dot{Q} - \dot{W}_S = \dot{m}\left[h_2 - h_1 + g(z_2 - z_1) + \tfrac{1}{2}(V_2^2 - V_1^2)\right] \tag{4.65}$$

where the enthalpy of Eq. 4.12 has been introduced recognizing that $v = 1/\rho$ and the conservation of mass allows $\dot{m}_2 = \dot{m}_1 = \dot{m}$. This is the form most often used when a gas or a vapor is flowing.

Quite often, the kinetic energy and potential energy changes are negligible. The first law then takes the simplified form

$$\dot{Q} - \dot{W}_S = \dot{m}(h_2 - h_1) \tag{4.66}$$

or

$$q - w_S = h_2 - h_1 \tag{4.67}$$

where $q = \dot{Q}/\dot{m}$ and $w_S = \dot{W}_S/\dot{m}$. These simplified forms of the energy equation have a surprisingly large number of applications.

For a control volume through which a liquid flows, it is most convenient to return to Eq. 4.62. For a steady flow with $\rho_2 = \rho_1 = \rho$, neglecting heat transfer and changes in internal energy, the energy equation takes the form

$$-\dot{W}_S = \dot{m}\left[\frac{p_2 - p_1}{\rho} + \frac{V_2^2 - V_1^2}{2} + g(z_2 - z_1)\right] \tag{4.68}$$

This is the form to use for a pump or a hydroturbine. If $\dot{Q}$ and Δu are not zero, simply include them.

4.9 THE FIRST LAW APPLIED TO CONTROL VOLUMES

Key Concept: *Control surfaces should be chosen so that the flow variables are uniform or known functions over the areas where the fluid enters and exits.*

There are several points that must be considered in the analysis of most problems in which the energy equation is used. As a first step, it is very important to identify the control volume selected in the solution of the problems; dashed lines are used to outline the control surface. If at all possible, the control surface should be chosen so that the flow variables are uniform or known functions over the areas where the fluid enters and exits the control volume. The control surface should be chosen sufficiently far downstream from an abrupt area change (an entrance or a sudden contraction), so that the velocity and pressure can be approximated by uniform distributions.

It is also necessary to specify the process by which the flow variables change. Is it incompressible? isothermal? constant-pressure? adiabatic? A sketch of the process on a suitable diagram is often of use in the calculations. If the working substance behaves as an ideal gas, then the appropriate equations may be used; if not, tabulated values must be used, such as those provided for steam. For real gases that do not behave as ideal gases, specialized equations are available for calculations; they can be found in handbooks.

Key Concept: *For many devices, the losses are included as an efficiency of the device.*

Often heat transfer from a device or the internal energy change across a device, such as a pump, is not desired. For such situations, the heat transfer and internal energy change may be lumped together as *losses*. In a pipeline, losses occur because of friction; in a pump, losses occur because of separated fluid flow around the rotating blades. For many devices, the losses are included as an efficiency of the device. Later examples will illustrate this.

Kinetic energy or potential energy changes can often be neglected in comparison with other terms in the energy equation. Potential energy changes are usually included only in situations where liquid is involved and where a large vertical dis-

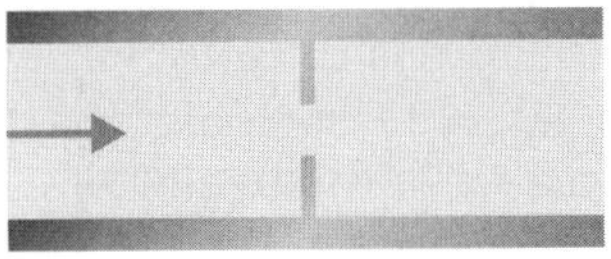

(a) Orifice plate

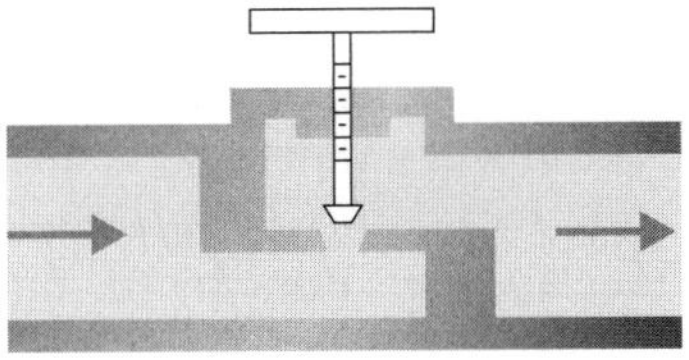

(b) Globe valve

FIGURE 4.9 Throttling devices.

tance separates the inlet and exit areas. The following applications will illustrate many of the above points.

Throttling Devices. A throttling device involves a steady-flow adiabatic process that provides a pressure drop with no significant potential energy or kinetic energy changes. The process occurs relatively rapidly, with the result that negligible heat transfer occurs. Two such devices are sketched in Fig. 4.9. If the energy equation is applied to such a device, obviously there is no work done; neglecting kinetic and potential energy changes, we have, for this adiabatic process (see Eq. 4.67),

$$h_2 = h_1 \tag{4.69}$$

where section 1 is upstream and section 2 is downstream. Most valves are throttling devices, for which the energy equation takes the form of Eq. 4.69. They are also used in many refrigeration units in which the sudden drop in pressure causes a change in phase of the working substance.

Example 4.13

Steam enters a throttling valve at 8000 kPa and 300°C and leaves at a pressure of 1600 kPa. Determine the final temperature and specific volume of the steam using (a) the steam tables and (b) TK Solver.

Solution

(a) The enthalpy of the steam as it enters is found from the superheat steam table to be $h_1 = 2785$ kJ/kg. This must equal the exiting enthalpy as demanded by Eq. 4.69. The exiting steam is in the quality region, since at 1600 kPa $h_g = 2794$ kJ/kg. Thus, the final temperature is $T_2 = 201.4$°C.

To find the specific volume, we must know the quality. It is found from

$$h_2 = h_f + xh_{fg} \qquad 2785 = 859 + 1935x \qquad \therefore\ x_2 = 0.995$$

The specific volume is then

$$v_2 = v_f + x(v_g - v_f)$$
$$= 0.0012 + 0.995 \times (0.1238 - 0.0012) = 0.1232 \text{ m}^3/\text{kg}$$

(b) At $T_1 = 300$°C and $p_1 = 8000$ kPa, TK Solver provides $h_1 = 2784.6$ kJ/kg. Using $h_2 = h_1$ and $p_2 = 1600$ kPa, we find

$$T_2 = 201.4\text{°C} \qquad \text{and} \qquad v_2 = 0.12319 \text{ m}^3/\text{kg}$$

Compressors, Pumps, and Turbines. A pump is a device that transfers energy to a liquid flowing through the pump with the result that the pressure is increased. Compressors and blowers also fall into this category but have the primary purpose of increasing the pressure in a gas. A turbine, on the other hand, is a device in which work is done by the fluid on a set of rotating blades. As a result, there is a pressure drop from the inlet to the outlet of the turbine. In some situations, there may be heat transferred from the device to the surroundings, but often the heat transfer is negligible. In addition, the kinetic and potential energy changes are often negligible. For such devices operating in a steady-state mode, the energy equation takes the form (see Eq. 4.66)

$$-\dot{W}_S = \dot{m}(h_2 - h_1) \qquad \text{or} \qquad -w_S = h_2 - h_1 \tag{4.70}$$

where $\dot{W}_S$ is negative for a compressor and positive for a gas or steam turbine. In the event that heat transfer does occur, from perhaps a high-temperature working fluid, it must, of course, be included in the above equation.

For liquids, such as water, the energy equation (Eq. 4.68), neglecting kinetic and potential energy changes, becomes

$$-\dot{W}_S = \dot{m}\,\frac{p_2 - p_1}{\rho} \tag{4.71}$$

which is used for a pump or a hydroturbine. If kinetic or potential energy changes are significant, they must be included.

Example 4.14

Steam enters a turbine at 4000 kPa and 500°C and leaves, as shown in Fig. E4.14. For an inlet velocity of 200 m/s, calculate the turbine power output. Neglect any heat transfer and kinetic energy change. Show that the kinetic energy change is negligible. Use (a) the steam tables and (b) TK Solver.

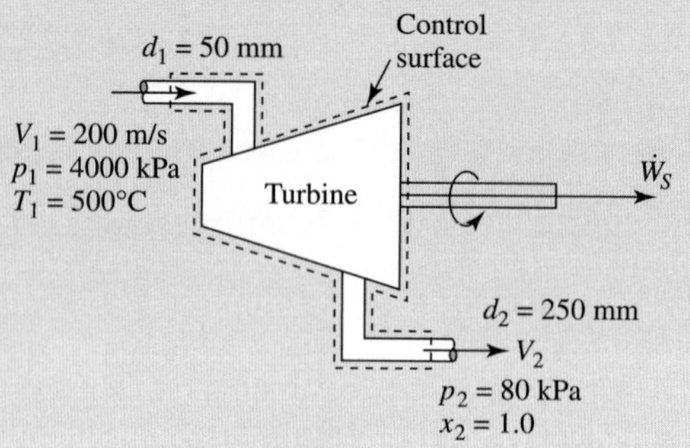

FIGURE E4.14

(continued)

Solution

(a) The energy equation in the form of Eq. 4.70 is $-W_T = \dot{m}(h_2 - h_1)$. We find $\dot{m}$ as follows:

$$\dot{m} = \rho_1 A_1 V_1 = \frac{1}{v_1} A_1 V_1 = \frac{1}{0.08643} \times \pi \times 0.025^2 \times 200 = 4.544 \text{ kg/s}$$

The enthalpies are found from Appendix Table C.3 to be

$$h_1 = 3445.2 \text{ kJ/kg} \qquad h_2 = 2665.7 \text{ kJ/kg}$$

The maximum power output is then

$$\dot{W}_T = -4.544 \times (2665.7 - 3445.2) = 3542 \text{ kJ/s} \quad \text{or} \quad 3.542 \text{ MW}$$

To show that the kinetic energy change is negligible, we must calculate the exiting velocity:

$$V_2 = \frac{A_1 V_1 \rho_1}{A_2 \rho_2} = \frac{\pi \times 0.025^2 \times 200/0.08643}{\pi \times 0.125^2/2.087} = 193 \text{ m/s}$$

The kinetic energy change is then

$$\Delta KE = \dot{m}\left(\frac{V_2^2 - V_1^2}{2}\right) = 4.544 \times \frac{193^2 - 200^2}{2} = -6250 \text{ J/s} \quad \text{or} \quad -6.25 \text{ kJ/s}$$

This is less than 0.2 percent of the power output and is indeed negligible. Kinetic energy changes are usually omitted in the analysis of devices (but not in a nozzle or a diffuser).

(b) Using TK Solver at $T_1 = 500°\text{C}$ and $p_1 = 4000$ kPa, we find $v_1 = 0.08642 \text{ m}^3/\text{kg}$, so that

$$\dot{m} = \rho A V = \frac{1}{0.08642} \times \pi \times 0.025^2 \times 200 = 4.545 \text{ kg/s}$$

The enthalpies are $h_1 = 3445.4$ kJ/kg and $h_2 = 2665.3$ kJ/kg. Then

$$\dot{W}_T = 4.545 \times (3445.4 - 2655.3) = 3546 \text{ kJ/s} \quad \text{or} \quad 3.546 \text{ kW}$$

The exiting velocity is

$$V_2 = \frac{A_1 V_1 / v_1}{A_2 / v_2} = \frac{\pi \times 0.025^2 \times 200/0.08642}{\pi \times 0.125^2/2.0859} = 193.1 \text{ m/s}$$

The kinetic energy change is

$$\Delta KE = 4.545 \times \frac{193.1^2 - 200^2}{2} = -6164 \text{ J/s}$$

which is less than 0.2% of the power output. It is indeed negligible.

Example 4.15

Determine the maximum pressure increase across the 10-hp pump shown in Fig. E4.15. The inlet velocity of the water is 30 ft/sec.

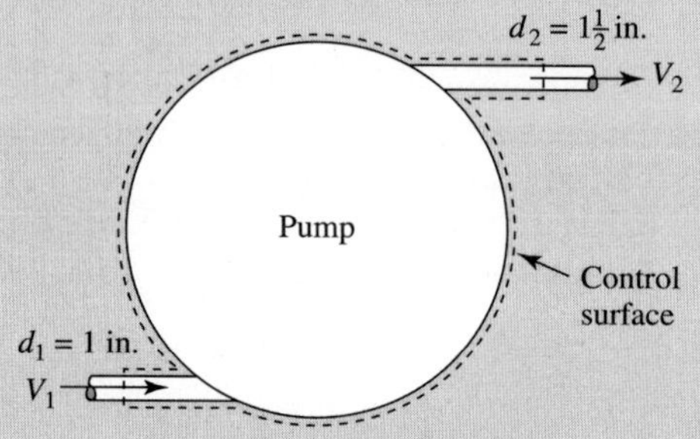

FIGURE E4.15

Solution

The energy equation (Eq. 4.68) is used. By neglecting the heat transfer and assuming no increase in internal energy, we establish the maximum pressure rise. Neglecting the potential-energy change, the energy equation takes the form

$$-\dot{W}_S = \dot{m}\left(\frac{p_2 - p_1}{\rho} + \frac{V_2^2 - V_1^2}{2}\right)$$

The velocity V_1 is given and V_2 is found from the continuity equation as follows:

$$A_1 V_1 = A_2 V_2 \qquad \pi \times 0.5^2 \times 30 = \pi \times 0.75^2 \times V_2 \qquad \therefore\ V_2 = 13.33 \text{ ft/sec}$$

The mass flow rate needed in the energy equation is, using $\rho = 62.4$ lbm/ft^3,

$$\dot{m} = \rho A_1 V_1 = 62.4 \times \pi \times 0.5^2 \times 30 = 10.21 \text{ lbm/sec}$$

Recognizing that the pump work is negative, the energy equation is

$$-(-10) \times 550\ \frac{\text{ft-lbf}}{\text{sec}} = 10.21\ \frac{\text{lbm}}{\text{sec}}\left[\frac{p_2 - p_1}{62.4 \text{ lbm/ft}^3} + \frac{(13.33^2 - 30^2)\ \text{ft}^2/\text{sec}^2}{2 \times 32.2 \text{ lbm-ft/(sec}^2\text{-lbf)}}\right]$$

where the factor 32.2 lbm-ft/(sec^2-lbf) is needed in the kinetic energy term. This predicts a pressure rise of

$$p_2 - p_1 = 62.4 \times \left[\frac{5500}{10.21} - \frac{13.33^2 - 30^2}{2 \times 32.2}\right] = 34\,310 \text{ lbf/ft}^2 \quad \text{or} \quad 238.2 \text{ psi}$$

Note that, in this example, the kinetic energy terms are retained because of the difference in inlet and exit areas; if they were omitted, only a two percent error would result. In most applications, the inlet and exit areas will be equal so that $V_2 = V_1$; but even with different areas, as in this example, kinetic energy changes are usually ignored in a pump or turbine, and Eq. 4.71 is used.

Nozzles and Diffusers. A nozzle is a device that is used to increase the velocity of a flowing fluid. It does this by reducing the pressure. A diffuser is a

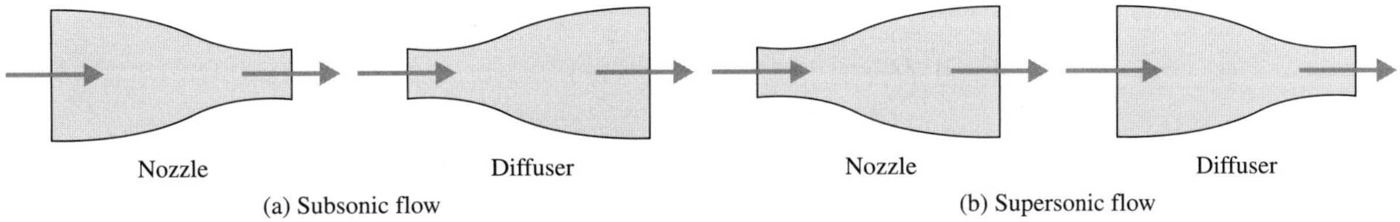

FIGURE 4.10 Nozzles and diffusers.

device that increases the pressure in a flowing fluid by reducing the velocity. There is no work input into the devices and usually negligible heat transfer. With the additional assumption of negligible potential energy changes, the energy equation takes the form

$$0 = \frac{V_2^2 - V_1^2}{2} + h_2 - h_1 \tag{4.72}$$

Based on our intuition, we expect a nozzle to have a decreasing area in the direction of flow and a diffuser to have an increasing area in the direction of flow. This is indeed the case for a subsonic flow of a gas in which $V < \sqrt{kRT}$. For a supersonic gas flow, in which $V > \sqrt{kRT}$, the opposite is true: a nozzle has an increasing area, and a diffuser has a decreasing area. This is shown in Fig. 4.10.

Key Concept: *For a supersonic gas flow, a nozzle has an increasing area, and a diffuser has a decreasing area.*

Three equations may be used for nozzle and diffuser flow: energy, continuity, and a process equation, such as for an adiabatic quasi-equilibrium flow. Thus, we may have three unknowns at the exit, given the entering conditions. There may also be shock waves in supersonic flows or "choked" subsonic flows. These more complicated flows are included in the fluid mechanics part. Only the more simple situations will be included here.

Example 4.16

Air flows through the supersonic nozzle shown in Fig. E4.16. The inlet conditions are 7 kPa and 420°C. The nozzle exit diameter is adjusted such that the exiting velocity is 700 m/s. Calculate (a) the exit temperature, (b) the mass flow rate, and (c) the exit diameter. Assume an adiabatic quasi-equilibrium flow.

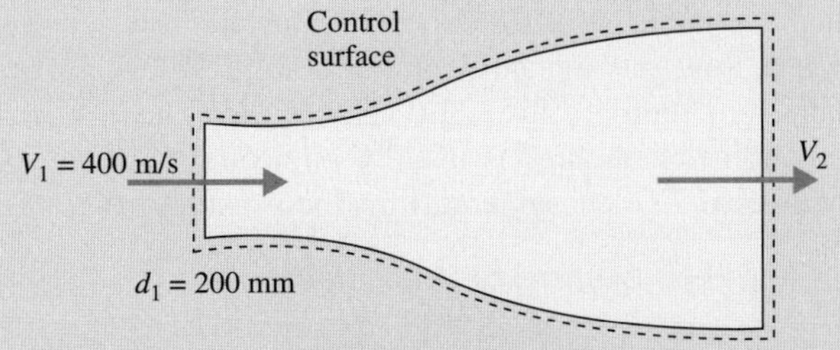

FIGURE E4.16

(continued)

Solution

(a) To find the exit temperature, the energy equation (Eq. 4.72) is used. It is, using $\Delta h = c_p \Delta T$,

$$\frac{V_1^2}{2} + c_p T_1 = \frac{V_2^2}{2} + c_p T_2$$

We then have, using $c_p = 1000$ J/kg·K,

$$T_2 = \frac{V_1^2 - V_2^2}{2c_p} + T_1 = \frac{400^2 - 700^2}{2 \times 1000} + 420 = 255°\text{C}$$

(b) To find the mass flow rate, we must find the density at the entrance. From the inlet conditions, we have

$$\rho_1 = \frac{p_1}{RT_1} = \frac{7000}{287 \times 693} = 0.0352 \text{ kg/m}^3$$

The mass flow rate is then

$$\dot{m} = \rho_1 A_1 V_1 = 0.0352 \times \pi \times 0.1^2 \times 400 = 0.442 \text{ kg/s}$$

(c) To find the exit diameter, we would use the continuity equation $\rho_1 A_1 V_1 = \rho_2 A_2 V_2$. This requires the density at the exit. It is found by assuming adiabatic quasi-equilibrium flow. Referring to Eq. 4.49, with $\rho = 1/v$, we have

$$\rho_2 = \rho_1 \left(\frac{T_2}{T_1}\right)^{1/(k-1)} = 0.0352 \left(\frac{528}{693}\right)^{1/(1.4-1)} = 0.01784 \text{ kg/m}^3$$

Hence,

$$d_2^2 = \frac{\rho_1 d_1^2 V_1}{\rho_2 V_2} = \frac{0.0352 \times 0.2^2 \times 400}{0.01784 \times 700} = 0.0451 \qquad \therefore d_2 = 0.212 \text{ m} \quad \text{or} \quad 212 \text{ mm}$$

Key Concept: *Heat exchangers are used to transfer energy from a hotter body to a colder body.*

Heat Exchangers. An important device that has many applications in engineering is the heat exchanger. Heat exchangers are used to transfer energy from a hotter body to a colder body or to the surroundings by means of heat transfer. Energy is transferred from the hot gases after combustion in a power plant to the water in the pipes of the boiler and from the hot water that leaves an automobile engine to the atmosphere by use of a radiator, and electrical generators are cooled by water flowing through internal flow passages. Many heat exchangers utilize a flow passage into which a fluid enters and from which the fluid exits at a different temperature. The velocity does not normally change, the pressure drop through the passage is usually neglected, and the potential energy change is assumed zero. The energy equation then results in

$$\dot{Q} = \dot{m}(h_2 - h_1) \tag{4.73}$$

since no work occurs in the heat exchanger .

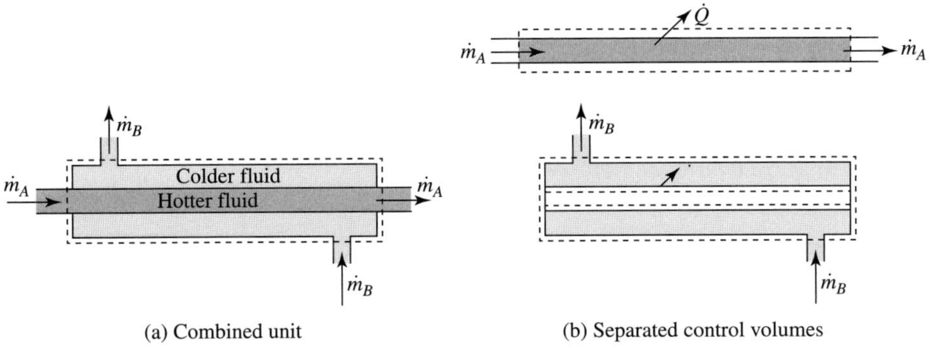

FIGURE 4.11 A heat exchanger.

Energy may be exchanged between two moving fluids, as shown schematically in Fig. 4.11. For a control volume including the combined unit, which is assumed to be insulated, the energy equation, as applied to the control volume of Fig. 4.11a, would be

$$0 = \dot{m}(h_{A2} - h_{A1}) + \dot{m}_B(h_{B2} - h_{B1}) \tag{4.74}$$

The energy that leaves fluid A is transferred to fluid B by means of the heat transfer $\dot{Q}$. For the control volumes shown in Fig. 4.11b, we have

$$\dot{Q} = \dot{m}_B(h_{B2} - h_{B1}) \qquad -\dot{Q} = \dot{m}_A(h_{A2} - h_{A1}) \tag{4.75}$$

Example 4.17

Liquid sodium, flowing at 100 kg/s, enters a heat exchanger at 450°C and exits at 350°C. The specific heat of sodium is 1.25 kJ/kg·°C. Water enters at 5000 kPa and 20°C. Determine the minimum mass flow rate of the water so that the water just vaporizes. Neglect the pressure drop through the exchanger. Also, calculate the rate of heat transfer.

Solution

The energy equation (Eq. 4.75) is used as

$$\dot{m}_s(h_{s1} - h_{s2}) = \dot{m}_w(h_{w2} - h_{w1}) \qquad \text{or} \qquad \dot{m}_s c_p(T_{s1} - T_{s2}) = \dot{m}_w(h_{w2} - h_{w1})$$

Using the given values, we have (use Appendix Table C.4 to find h_{w1})

$$100 \times 1.25 \times (450 - 350) = \dot{m}_w(2792.8 - 88.7) \qquad \therefore \dot{m}_w = 4.623 \text{ kg/s}$$

We have assumed a saturated-vapor state for the exiting steam to obtain the maximum allowable exiting enthalpy. The heat transfer is found using the energy equation (Eq. 4.75) applied to one of the separate control volumes:

$$\dot{Q} = \dot{m}_w(h_{w2} - h_{w1}) = 4.623 \times (2792.8 - 88.7) = 12\,500 \text{ kW} \quad \text{or} \quad 12.5 \text{ MW}$$

4.10 TRANSIENT FLOW

The Filling of a Tank. If the steady-flow assumption of the preceding sections is not valid, then the time dependence of the various properties must be included. The filling of a rigid tank with a gas and the release of gas from a pressurized tank are examples that we will consider.

The energy equation is written as

$$\dot{Q} - \dot{W}_S = \frac{dE_{\text{c.v.}}}{dt} + \dot{m}_2\left(\frac{V_2^2}{2} + gz_2 + h_2\right) - \dot{m}_1\left(\frac{V_1^2}{2} + gz_1 + h_1\right) \quad (4.76)$$

We will consider the kinetic energy and potential energy terms to be negligible, so that $E_{\text{c.v.}}$ will consist of internal energy only. The first problem we wish to study is the filling of a rigid tank, as sketched in Fig. 4.12. In the tank, there is only an entrance. With no shaft work present, the energy equation reduces to

$$\dot{Q} = \frac{d}{dt}(um) - \dot{m}_1 h_1 \quad (4.77)$$

where m is the mass in the control volume. If we multiply this equation by dt and integrate from an initial time t_i to some final time t_f we have

$$Q = u_f m_f - u_i m_i - m_1 h_1 \quad (4.78)$$

where m_1 = mass that enters
m_f = final mass in control volume
m_i = initial mass in control volume

In addition, for the filling process, the enthalpy h_1 is assumed constant (see Eq. 4.69) over the time interval. If the tank is insulated, Q would be zero.

The continuity equation for the unsteady-flow situation may be necessary in the solution process. Since the final mass is equal to the initial mass plus the mass that entered, this is expressed as

$$m_f = m_i + m_1 \quad (4.79)$$

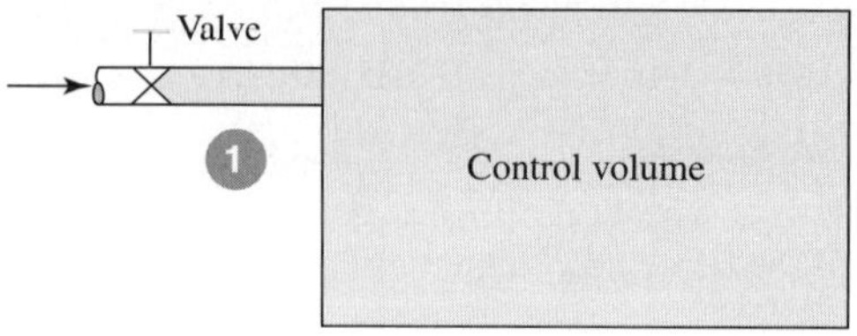

FIGURE 4.12 The filling of a rigid tank.

The Discharge of a Tank. Now consider the discharge of a pressurized tank. This second problem is more complicated than the filling of a tank in that the properties at the exiting area are not constant over the time interval of interest; we must include the variation of the variables with time. We will assume an insulated tank, so that no heat transfer occurs, and again neglect kinetic energy and potential energy. The energy equation becomes, assuming no shaft work,

$$0 = \frac{d}{dt}(um) + \dot{m}_2(p_2 v_2 + u_2) \tag{4.80}$$

where m is the mass in the control volume. The continuity equation takes the form

$$\frac{dm}{dt} = -\dot{m}_2 \tag{4.81}$$

If this is substituted into Eq. 4.80, we have

$$d(um) = (p_2 v_2 + u_2)\, dm \tag{4.82}$$

We will assume that the gas escapes through a small valve opening, as shown in Fig. 4.13. Just upstream of the valve is area A_2 with properties p_2, v_2, and u_2. The velocity at this exiting area is assumed to be quite small, so that p_2, v_2, and u_2 are approximately the same as the respective quantities in the control volume. With this assumption, Eq. 4.82 becomes

$$d(um) = (pv + u)\, dm \tag{4.83}$$

Letting $d(um) = u\, dm + m\, du$, there results

$$m\, du = pv\, dm \tag{4.84}$$

Now, we will restrict ourselves to a gas that behaves as an ideal gas. For such a gas, $du = c_v\, dT$ and $pv = RT$, and we obtain

$$mc_v\, dT = RT\, dm \tag{4.85}$$

This is put in the form

$$\frac{c_v}{R}\frac{dT}{T} = \frac{dm}{m} \tag{4.86}$$

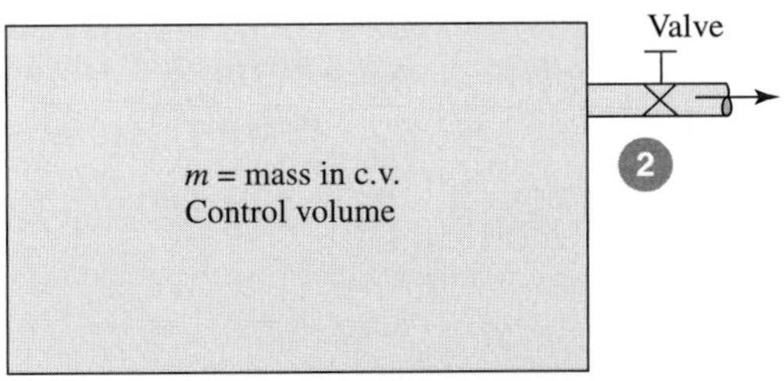

FIGURE 4.13 The discharge of a pressurized tank.

which can be integrated from the initial state, signified by the subscript i, to the final state, signified by the subscript f. There results

$$\frac{c_v}{R}\ln\frac{T_f}{T_i} = \ln\frac{m_f}{m_i} \quad \text{or} \quad \frac{m_f}{m_i} = \left(\frac{T_f}{T_i}\right)^{1/(k-1)} \tag{4.87}$$

where we have used $c_v/R = 1/(k-1)$; see Eq. 4.32. In terms of the pressure ratio, Eq. 4.50 allows us to write

$$\frac{m_f}{m_i} = \left(\frac{p_f}{p_i}\right)^{1/k} \tag{4.88}$$

Remember that these equations are applicable if there is no heat transfer from the volume; the process is quasistatic in that the properties are assumed uniformly distributed throughout the control volume (this requires a relatively slow discharge velocity, such as 100 m/s or less), and the gas must behave as an ideal gas with constant specific heats.

Example 4.18

A completely evacuated, insulated, rigid tank with a volume of 300 ft^3 is filled from a steam line transporting steam at 800°F and 500 psia. Determine (a) the temperature of steam in the tank when its pressure is 500 psia and (b) the mass of steam that flows into the tank.

Solution

(a) The energy equation used is Eq. 4.78. With $Q = 0$ and $m_i = 0$, we have $u_f m_f = m_1 h_1$. The continuity equation (Eq. 4.79) allows us to write $m_f = m_1$, which states that the final mass m_f in the tank is equal to the mass m_1 that entered the tank. Thus, there results $u_f = h_1$. From Appendix Table C.3E, h_1 is found, at 800°F and 500 psia, to be 1412.1 Btu/lbm. Using $p_f = 500$ psia as the final tank pressure, we can interpolate for the temperature, using $u_f = 1412.1$ Btu/lbm, and find

$$T_f = \frac{1412.1 - 1406.0}{1449.2 - 1406.0} \times 100 + 1100 = 1114.1°\text{F}$$

(b) We recognize that $m_1 = m_f = V\!\!\!/_{\text{tank}}/v_f$. The specific volume of the steam in the tank at 500 psia and 1114.1°F is

$$v_f = \frac{1114.1 - 1100}{100} \times (1.9518 - 1.8271) + 1.8271 = 1.845 \text{ ft}^3/\text{lbm}$$

This gives

$$m_f = 300/1.845 = 162.6 \text{ lbm}$$

Example 4.19

An air tank with a volume of 20 m^3 is pressurized to 10 MPa. The tank eventually reaches room temperature of 25°C. If the air is allowed to escape with no heat transfer until $p_f = 200$ kPa, determine the mass of air remaining in the tank and the final temperature of air in the tank.

Solution

The initial mass of air in the tank is found to be

$$m_i = \frac{p_i \forall_i}{RT_i} = \frac{10 \times 10^6 \times 20}{287 \times 298} = 2338 \text{ kg}$$

Equation (4.88) gives, using $k = 1.4$,

$$m_f = m_i \left(\frac{p_f}{p_i}\right)^{1/k} = 2338 \left(\frac{2 \times 10^5}{10 \times 10^6}\right)^{1/1.4} = 143.0 \text{ kg}$$

The final temperature is found using Eq. 4.87:

$$T_f = T_i \left(\frac{m_f}{m_i}\right)^{k-1} = 298 \times \left(\frac{143}{2338}\right)^{0.4} = 97.46 \text{ K} \quad \text{or} \quad -175.5°\text{C}$$

A person who accidentally comes in contact with a flow of gas from a pressurized tank faces immediate freezing.

4.11 THE FIRST LAW WITH HEAT TRANSFER APPLICATIONS

In the previous sections, we determined heat transfer in terms of work and changes in internal energy using the first law. Although this was sufficient for many situations, it did not allow us to calculate heat transfer from temperature measurements. In this section, we will consider the basic relationships between heat transfer and temperature, so that we can determine the rate of heat transfer from a surface and the resulting temperature distribution within a given material. Heat transfer can be a very important factor in a vast range of applications from the design of large- and small-scale systems, such as power plants and circuit boards, to the analysis of different processes in biological systems, such as the use of hyperthermia to destroy cancerous tumors and heating to preserve foods. Regardless of the application, we will continue to focus on macroscopic systems. First, we will use the first law to analyze the effects of conduction, convection, and radiation. In doing so, we will introduce basic methods of engineering heat transfer analyses and observe the effects of different material properties on the resulting temperature of the system.

The Energy Equation Revisited. Previously, we developed a general formulation for the energy equation and illustrated its use for various applications in which the heat transfer was either known or could be determined from other

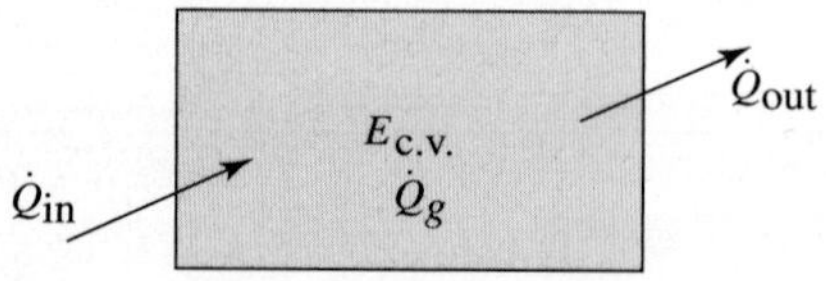

FIGURE 4.14 A control volume with only heat transfer.

known factors. We will again use conservation of energy principles, but, here, our aim will be to determine the rate of heat transfer and the resulting relationship with the temperature distribution within the system.

Consider the control volume of Fig. 4.8 in which the only energy transfer occurs due to heat transfer, as illustrated in Fig. 4.14. The energy equation (Eq. 4.62) would take the form

$$\dot{Q}_{in} - \dot{Q}_{out} + \dot{Q}_g = \frac{dE_{c.v.}}{dt} = \dot{E}_{c.v.} \tag{4.89}$$

where the additional term $\dot{Q}_g$ allows for volumetric heat generation of energy due to conversion from one energy form to another (e.g., chemical or nuclear). The heat transfer at the boundaries is in the form of conduction, convection, and/or radiation. The last term on the right-hand side of Eq. 4.89 is the energy storage term.

We will consider two special cases, which have many practical applications. In the first, we will apply the first law to the system boundary. In this case, we have a control surface rather than a control volume, and hence, the volumetric terms (i.e., the time derivative and the heat generation terms) do not enter the equation. In the second case, the temperature of a material is assumed uniform in space but not in time; this is known as the *lumped capacitance* assumption.

Conservation of Energy at a Surface. In some cases, it is convenient to apply the conservation of energy to the surface of a material. In this special case, the surface has no volume or mass, so that the general first law equation reduces to

$$\dot{Q}_{in} - \dot{Q}_{out} = 0 \tag{4.90}$$

where $\dot{Q}_{in}$ and $\dot{Q}_{out}$ are a result of conduction, convection, and/or radiation. Consider the plane wall shown in Fig. 4.15 with a fixed temperature T_1 at $x = 0$ and T_2 at $x = L$. Cooling occurs at $x = L$ by convection to an air stream at T_∞

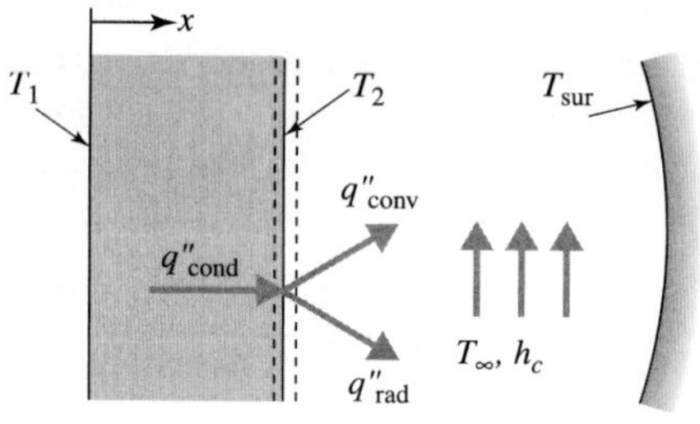

FIGURE 4.15 Conservation of energy at the surface of a wall.

and radiation to the surroundings at T_{sur}. If we apply the energy equation (Eq. 4.90) to the surface at $x = L$, we observe that the energy in is due to conduction, whereas the energy out is due to convection and radiation. Using the nomenclature of Section 3.6, Eq. 4.90 can be written as

$$q''_{cond} = q''_{conv} + q''_{rad} \tag{4.91}$$

where Eqs. 3.22, 3.25, and 3.34 allow us to write

$$q''_{cond} = k\frac{T_1 - T_2}{L} \qquad q''_{conv} = h_c(T_2 - T_\infty) \qquad q''_{rad} = \sigma\varepsilon(T_2^4 - T_{sur}^4) \tag{4.92}$$

Thus, we have the expression

$$k\frac{T_1 - T_2}{L} = h_c(T_2 - T_\infty) + \sigma\varepsilon(T_2^4 - T_{sur}^4) \tag{4.93}$$

where k is the conductivity of the wall, h_c is the heat transfer coefficient for the air stream, and ε is the emissivity of the wall at $x = L$. If T_1 is the unknown, the solution is straightforward. If T_2 is the unknown, an iterative solution can be used as illustrated in Example 4.20.

Example 4.20

Consider a vessel with stainless steel walls ($k = 15.1$ W/m·K) containing saturated steam at 1.0 MPa. The outside of the vessel is cooled by an air stream at 20°C, with a convective heat transfer coefficient of 100 W/m^2·K and radiation to the surrounding wall at 20°C, as shown in Fig. E4.20. If the walls are 1-cm thick, and the emissivity of the outer surface is 0.2, find the temperature of the outer wall.

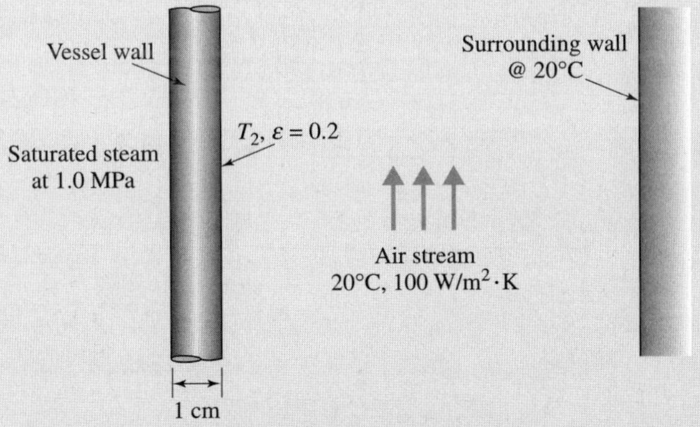

FIGURE E4.20

(continued)

Solution

From Eqs. 4.91 and 4.93:

$$q''_{\text{cond}} = q''_{\text{conv}} + q''_{\text{rad}}$$

and

$$k\frac{T_1 - T_2}{L} = h_c(T_2 - T_\infty) + \sigma\varepsilon(T_2^4 - T_{\text{sur}}^4)$$

where T_1 is equal to the saturation temperature of the steam (179.91°C = 453 K at 1 MPa), $k = 15.1$ W/m·K, $L = 0.01$ m, $h_c = 100$ W/m^2·K, $T_\infty = T_{\text{sur}} = 20$°C = 293 K, $\varepsilon = 0.2$, and $\sigma = 5.67 \times 10^{-8}$ W/m^2·K^4. Rearranging, we can write:

$$T_2^4 + \left(\frac{h_c + \frac{k}{L}}{\sigma\varepsilon}\right)T_2 = \left(\frac{h_c T_\infty + \frac{k}{L}T_1}{\sigma\varepsilon}\right) + T_{\text{sur}}^4$$

Since the right hand side is known, we can assume a value for T_2, and calculate the left hand side to determine whether or not our assumed value is correct. Assuming a value of 440 K (remember all temperatures should be absolute), we have

$$6.25 \times 10^{13} < 6.29 \times 10^{13}$$

Our first assumption was too low; therefore, try $T_2 = 450$ K:

$$6.39 \times 10^{13} > 6.29 \times 10^{13}$$

Since this value is too high, we can interpolate between 440 K and 450 K to find our solution, $T_2 = 443$ K:

$$6.29 \times 10^{13} = 6.29 \times 10^{13}$$

Thus, $T_2 \cong 443$ K. Note that we could have solved this problem using a root-finding software routine.

The Lumped Capacitance Method. Consider steady-state, one-dimensional heat transfer through the plane wall discussed in Section 3.6 and shown in Fig. 3.10. Now, assume that the wall temperature at $x = 0$ is fixed at T_1, and that the wall at $x = L$ is cooled by convection only (no radiation) as shown in

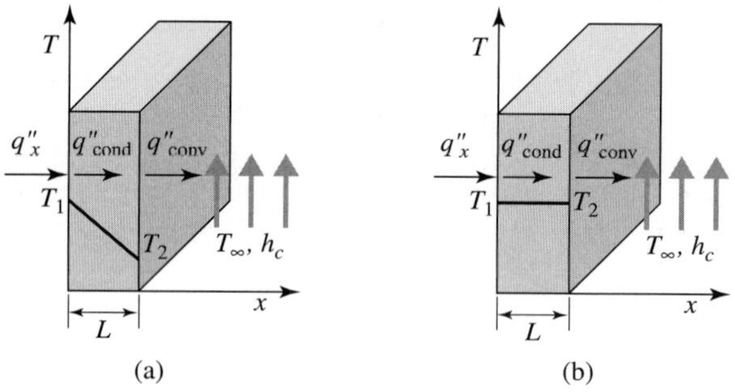

FIGURE 4.16 Steady-state, one-dimensional heat transfer through a plane wall, with T_1 fixed at $x = 0$ and convection at $x = L$.

Fig. 4.16a. Using an energy balance at $x = L$ as was done in Eq. 4.93, we have $q''_{\text{cond}} = q''_{\text{conv}}$. Thus, Eq. 4.93 reduces to

$$k\frac{T_1 - T_2}{L} = h_c(T_2 - T_\infty) \tag{4.94}$$

If the temperature of the material is uniform, T_1 must be approximately equal to T_2, or $T_1 - T_2 \approx 0$, as shown in Fig. 4.16b. There are several ways this could be accomplished:

1. Select a material with a large k.
2. Select a small L.
3. Make h_c small.

Referring to Eq. 4.94, another way of stating this is that

$$\frac{T_1 - T_2}{T_2 - T_\infty} = \frac{h_c L}{k} \ll 1 \tag{4.95}$$

Hence, we define the **Biot number** Bi, a dimensionless parameter, and use it to determine whether or not a uniform temperature or *lumped capacitance* assumption is appropriate. Thus, if

Biot number: *A parameter used to determine whether or not a uniform-temperature assumption is appropriate.*

$$Bi = \frac{h_c L}{k} < 0.1 \tag{4.96}$$

the errors associated with the lumped capacitance assumption are assumed to be small. If this assumption holds true, our analysis is simplified.

Consider an object experiencing no energy in and no energy generation. The first law expressed by Eq. 4.89, is

$$-\dot{Q}_{\text{out}} = \frac{dE_{\text{c.v.}}}{dt} \tag{4.97}$$

where $\dot{Q}_{\text{out}} = h_c A_S(T - T_\infty)$ (from Eq. 3.25 with $T_S = T$), and $E_{\text{c.v.}} = mcT = \rho\forall cT$. Then we have

$$-h_c A_S(T - T_\infty) = \rho\forall c\frac{dT}{dt} \tag{4.98}$$

Thus, we can separate the variables to solve for temperature as a function of time. Rearranging and integrating, assuming constant properties from time $t = 0$ when the temperature $T = T_i$, to the final temperature $T(t)$, there results

$$-\frac{h_c A_S}{\rho\forall c}\int_0^t dt = \int_{T_i}^{T} \frac{dT}{T - T_\infty} \tag{4.99}$$

Noting that $dT = d(T - T_\infty)$, and carrying out the integration:

$$-\frac{h_c A_S}{\rho\forall c}t = \ln\frac{T - T_\infty}{T_i - T_\infty} \tag{4.100}$$

or, equivalently,

$$T(t) = T_\infty + (T_i - T_\infty)e^{-(h_c A_S/(\rho \forall c))t} \tag{4.101}$$

An application of the lumped capacitance method will be illustrated in Example 4.21.

Example 4.21

A hot copper coin at 60°C is cooled by an air stream at 20°C with a convective heat transfer coefficient of 50 W/m^2·K. The coin has a thickness of 0.13 cm and a diameter of 1.95 cm. (a) Can the lumped capacitance assumption be used? (b) Determine how long will it take the coin to cool to 25°C. (For copper $k = 401$ W/m·K, $c = 385$ J/kg·K, and $\rho = 8933$ kg/m^3.)

Solution

(a) For the lumped capacitance assumption to be applied, Eq. 4.96 must hold true:

$$Bi = \frac{h_c L}{k} = \frac{50 \times 0.13/100}{400} = 1.6 \times 10^{-4} \ll 0.1$$

The assumption is valid.

(b) Therefore, we can use Eq. 4.100 and solve for time:

$$t = \frac{\rho \forall c}{h_c A_S} \ln \frac{T_i - T_\infty}{T - T_\infty}$$

where the surface area $A_S = 2 \times \pi d^2/4 + \pi d\,L$ and the volume $\forall = \pi d^2 L/4$. Using these expressions and the data given, we solve for the time:

$$t = \frac{\rho \pi d^2 Lc/4}{h_c \pi d(d/2 + L)} \ln \frac{T_i - T_\infty}{T - T_\infty} = \frac{\rho d L c}{2h_c(d + 2L)} \ln \frac{T_i - T_\infty}{T - T_\infty}$$

$$= \frac{8933 \times 0.0195 \times 0.0013 \times 385}{2 \times 100(0.0195 + 2 \times 0.0013)} \ln \frac{60 - 20}{25 - 20} = 41 \text{ s}$$

Conservation of Energy and Conduction Heat Transfer. Although the application of the energy equation at the surface of a material and the use of the lumped capacitance assumption provide the means for relatively simple analyses in many situations, there are many other cases in which these simplified analyses are not sufficient. A simple example is the case where the lumped capacitance assumption does not hold true, and we want to find the temperature distribution within the material. In this section, we will consider the problem of finding the temperature distribution within a material as a result of one-dimensional, steady-state conduction.

One-dimensional, Steady-state Conduction. Consider once again the planar wall, shown in Fig. 4.16a, subject to one-dimensional, steady-state conduction with no heat generation. If we apply the conservation of energy to a differential control volume as shown in Fig. 4.17, we have

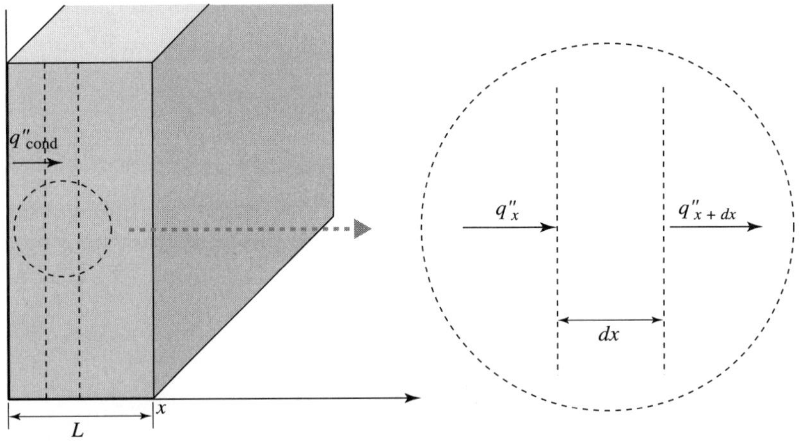

FIGURE 4.17 Differential volume for steady-state, one-dimensional heat transfer through a plane wall.

$$\dot{Q}_{\text{in}} - \dot{Q}_{\text{out}} = 0 \qquad \text{or} \qquad (q''_x - q''_{x+dx})A = 0 \tag{4.102}$$

The Taylor series

$$q''_{x+dx} = q''_x + \frac{dq''_x}{dx}dx + \text{higher order terms} \tag{4.103}$$

leads to, neglecting the higher order terms and using $q''_x = q''_{x+dx}$ from Eq. 4.102,

$$\frac{dq''_x}{dx} = \frac{d}{dx}k\left(\frac{dT}{dx}\right) = 0 \tag{4.104}$$

where q''_x is found from Fourier's law (Eq. 3.21). Thus, q''_x is constant and T is linear with x for constant k, and the general solution for Eq. 4.104 is

$$T(x) = c_1 + c_2 x \tag{4.105}$$

where c_1 and c_2 are constants that can be determined from the boundary conditions. For the example shown in Fig. 4.16a, c_1 and c_2 can be determined from the temperatures at $x = 0$ and $x = L$:

$$T(0) = c_1 = T_1 \qquad \text{and} \qquad T(L) = c_1 + c_2 L = T_2 \tag{4.106}$$

The temperature distribution is then

$$T(x) = T_1 + \frac{T_2 - T_1}{L}x \tag{4.107}$$

Fourier's law (Eq. 3.21) and the definition of $\dot{Q}$ in Eq. 3.18 then provide the rate of heat transfer

$$\dot{Q} = kA\frac{T_1 - T_2}{L} \tag{4.108}$$

Resistance Model. Equation (4.108) suggests an analogy between the heat rate and a current through an electrical circuit according to Ohm's law:

$$I = \frac{E_1 - E_2}{R_e} \tag{4.109}$$

Key Concept: *Heat rate and temperatures are analogous to current and electrical potential.*

where I is the current, E is the electrical potential, and R_e is the electrical resistance. Comparing Equation 4.108 with Eq. 4.109, we can see that the heat rate and temperature in Eq. 4.108 are analogous to current and electrical potential in Eq. 4.109. This suggests that we can define a thermal resistance for conduction analogous to R_e as

$$R_{\text{cond}} = \frac{L}{kA} \tag{4.110}$$

with units of ($m^2 \cdot K/W$). Likewise, we can define a thermal resistance for convection as

$$R_{\text{conv}} = \frac{1}{h_c A} \tag{4.111}$$

The advantage of these analogies is that we can apply the same general rules for series and parallel circuits to our 'thermal circuit.' Thus, for thermal resistances in series, we have

$$R_{\text{total}} = \sum_i R_i \tag{4.112}$$

and for thermal resistances in parallel, we can write

$$R_{\text{total}} = \left(\sum_i R_i^{-1}\right)^{-1} \tag{4.113}$$

Thus, for one-dimensional, steady-state conditions, we can apply the thermal-circuit concept in the analysis of rather complex systems as shown in Example 4.22.

Example 4.22

Consider the composite outer wall for a home, as shown in Fig. E4.22a. The wall is composed of an outer-brick face with a thickness of 10 cm, a 1.5-cm plywood layer, 15-cm by 4-cm hardwood studs with blown insulation, and an 1.5-cm-thick inner-wall board. The hardwood studs are placed every 60 cm along the wall, as shown in Figure E4.22a. The thermal conductivities for the brick face, the plywood layer, the hardwood studs, the insulation, and the inner-wall board are 1.3 W/m·K, 0.12 W/m·K, 0.16 W/m·K, 0.03 W/m·K, and 0.17 W/m·K, respectively. Assuming one-dimensional, steady-state heat transfer, determine the overall effective thermal resistance for one section of the wall.

(continued)

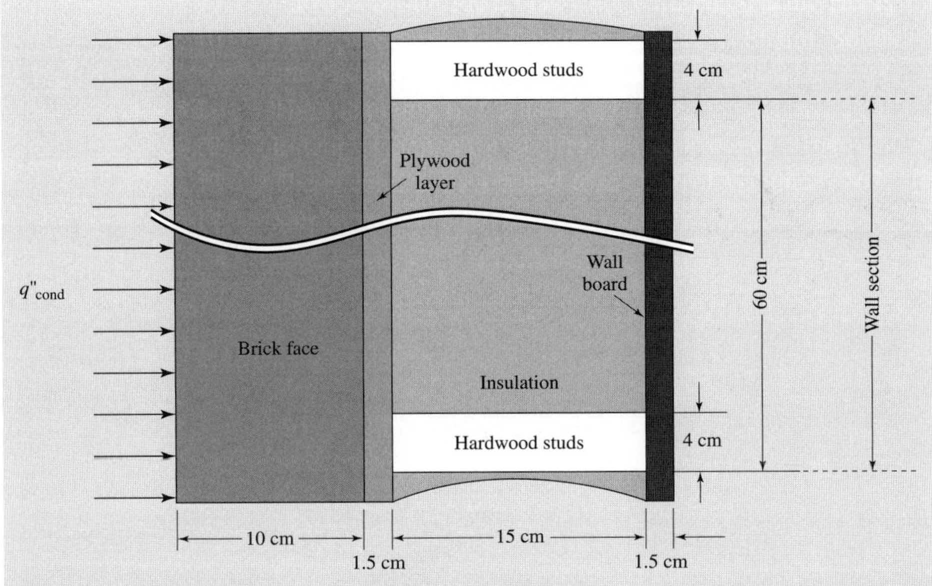

FIGURE E4.22(a)

Solution

To find the overall effective thermal resistance for the composite-wall section, we must first identify the thermal circuit. Note that we have a combination of thermal resistances in series and in parallel, as indicated in Fig. E4.22b.

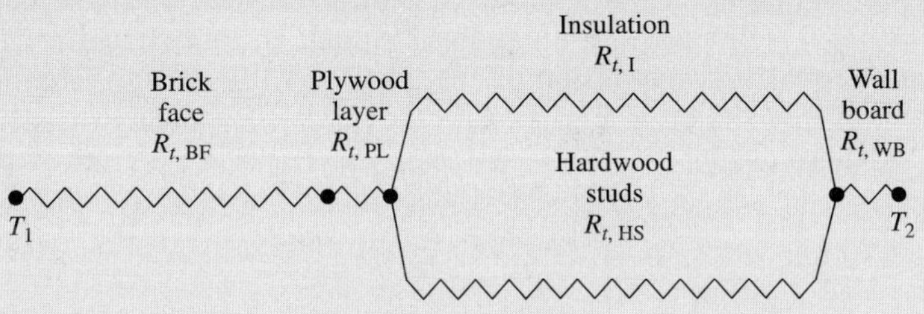

FIGURE E4.22(b)

Therefore, we can write the total resistance from the individual resistances, using subscripts "BF" for brick face, "PL" for plywood layer, "I" for insulation, "HS" for hardwood stud, and "WB" for wall board:

$$R_{\text{total}} = R_{\text{BF}} + R_{\text{PL}} + \frac{1}{R_I^{-1} + R_{\text{HS}}^{-1}} + R_{\text{WB}}$$

(continued)

where

$$R_{\text{BF}} = \frac{L_{\text{BF}}}{k_{\text{BF}} A_{\text{BF}}} \qquad R_{\text{PL}} = \frac{L_{\text{PL}}}{k_{\text{PL}} A_{\text{PL}}} \qquad R_I = \frac{L_I}{k_I A_I} \qquad R_{\text{HS}} = \frac{L_{\text{HS}}}{k_{\text{HS}} A_{\text{HS}}} \qquad R_{\text{WB}} = \frac{L_{\text{WB}}}{k_{\text{WB}} A_{\text{WB}}}$$

Substitution results in

$$R_{\text{total}} = \frac{1}{A_{\text{WB}}} \left[\frac{L_{\text{BF}}}{k_{\text{BF}}} + \frac{L_{\text{PL}}}{k_{\text{PL}}} + \frac{L_{\text{WB}}}{k_{\text{WB}}} \right] + \left[\left(\frac{L_I}{k_I A_I} \right)^{-1} + \left(\frac{L_{\text{HS}}}{k_{\text{HS}} A_{\text{HS}}} \right)^{-1} \right]^{-1}$$

$$= \frac{1}{60 \times 100 \times 1} \left[\frac{10 \times 100}{1.3} + \frac{1.5 \times 100}{0.16} + \frac{1.5 \times 100}{0.17} \right]$$

$$+ \left[\left(\frac{15 \times 100}{0.06 \times 56 \times 100 \times 1} \right)^{-1} + \left(\frac{15 \times 100}{0.16 \times 4 \times 100 \times 1} \right)^{-1} \right]^{-1}$$

$$= 4.18 \text{ m}^2\text{·°C/W} = 4.18 \text{ m}^2\text{·°C/W}$$

Note: If we ignored the studs, we would simply add the resistances, which is often done, and obtain $R_{\text{total}} = 4.60$ m^2·°C/W, an error of 10%.

Cylindrical and Spherical Coordinates. We can apply a similar analysis to that shown for steady-state, one-dimensional heat transfer through a planar wall to a cylinder and a sphere. For both cylindrical coordinates and spherical coordinates, we can write Fourier's law as

$$\dot{Q} = -kA \frac{dT}{dt} \tag{4.114}$$

where A is the surface area. For a circular cylinder, the area is $A_{\text{cyl}} = 2\pi rL$ and for a sphere it is $A_{\text{sph}} = 4\pi r^2$. Using an analysis similar to that shown for the planar case, we can apply the conservation of energy (Eq. 4.102) to a differential control volume, such as that shown in Fig. 4.18 for a cylinder. Following the steps leading to Eq. 4.105, there results

$$T(r) = c_1 \ln r + c_2 \quad (\textit{cylinder}) \tag{4.115}$$

and

$$T(r) = c_1 + \frac{c_2}{r} \quad (\textit{sphere}) \tag{4.116}$$

Given the boundary conditions, we can solve both Eqs. 4.115 and 4.116 for the temperature distribution as was done for the planar case (i.e., Eq. 4.107). We assume a hollow cylinder or sphere and constant temperatures at each boundary, such that

$$T(r_1) = T_1 \quad \text{and} \quad T(r_2) = T_2 \tag{4.117}$$

Substituting these boundary conditions into Eqs. 4.115 and 4.116 results in

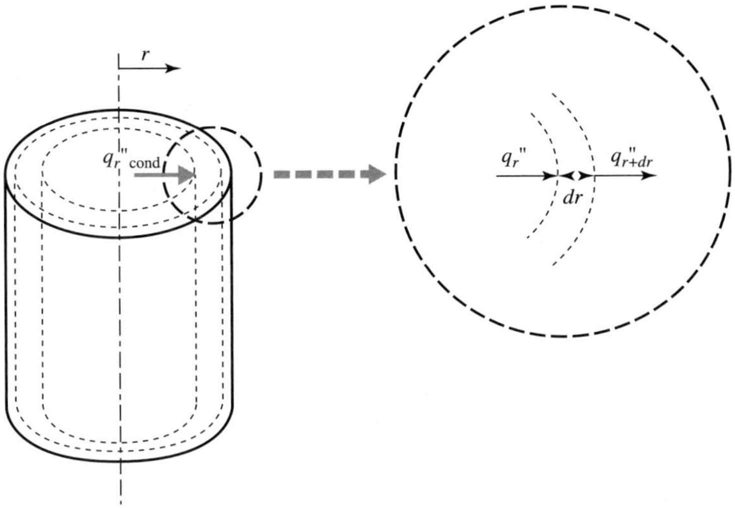

FIGURE 4.18 Differential volume for steady-state, one-dimensional heat transfer using cylindrical coordinates.

$$T(r) = T_2 + (T_2 - T_1)\frac{\ln(r/r_2)}{\ln(r_1/r_2)} \quad (cylinder) \tag{4.118}$$

and

$$T(r) = T_2 + (T_2 - T_1)\frac{1 - r_1/r}{1 - r_1/r_2} \quad (sphere) \tag{4.119}$$

We can also define a thermal resistance for each case by substituting Eqs. 4.118 and 4.119 into Eq. 4.114 and performing the differentiation of T with respect to r. For the cylinder, the thermal resistance for conduction is defined as

$$R_{cyl} = \frac{1}{2\pi Lk} \ln \frac{r_2}{r_1} \quad (cylinder) \tag{4.120}$$

while for the sphere, we find

$$R_{sph} = \frac{1}{4\pi k}\left(\frac{1}{r_1} - \frac{1}{r_2}\right) \quad (sphere) \tag{4.121}$$

Therefore, we can apply the concepts used for the thermal circuit in the planar case to these cases. This is illustrated in Example 4.23 for the cylinder.

Example 4.23

Consider modifying Example 4.20 for a cylindrical pipe rather than a planar wall. The pipe has stainless steel walls ($k = 15.1$ W/m·K) containing saturated steam at 1.0 MPa. The outside of the pipe is cooled by an air stream at 20°C, with a convective heat transfer coefficient of 100 W/m^2·K, and radiation to the surrounding walls at 20°C, as shown in Figure E4.23. The pipe wall is 10-m long with an inner radius of 4 cm and a thickness of 1 cm. The emissivity of the outer-pipe surface is 0.2. Find the temperature T_2 of the outer wall.

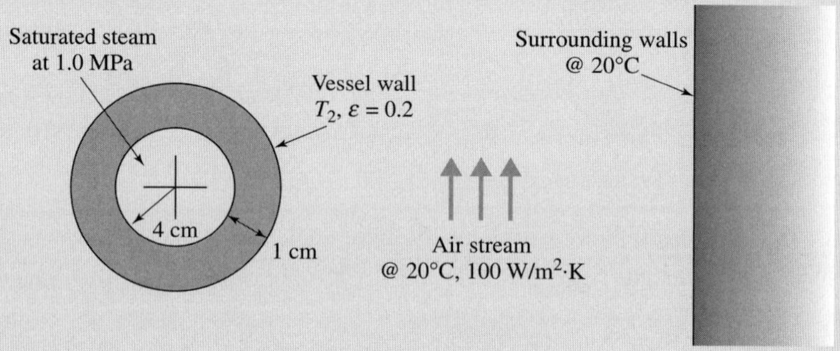

FIGURE E4.23

Solution

We can write conservation of energy (see Eq. 4.91) as:

$$\dot{Q}_{\text{cond}} = \dot{Q}_{\text{conv}} + \dot{Q}_{\text{rad}}$$

where $\dot{Q}$ has units of watts. Rewriting the conduction term in Eq. 4.93 in terms of the cylindrical resistance and heat transfer rate (rather than flux), we have

$$\frac{T_1 - T_2}{R_{\text{cyl}}} = h_c PL(T_2 - T_\infty) + \sigma\varepsilon PL(T_2^4 - T_{\text{sur}}^4)$$

where P and L are the outer perimeter and length of the pipe, and R_{cyl} is the cylindrical resistance, given by (Eq. 4.120):

$$R_{\text{cyl}} = \frac{1}{2\pi Lk} \ln \frac{r_2}{r_1}$$

where $r_1 = 0.04$ m, $r_2 = 0.05$ m, $L = 10$ m, and all other variables remain the same as in Example 4.20. Using the equations above for the cylindrical case and rearranging as before, we can write:

$$T_2^4 + \left(\frac{h_c PL + \dfrac{2\pi Lk}{\ln(r_2/r_1)}}{\sigma\varepsilon PL} \right) T_2 = \left(\frac{h_c PLT_\infty + \dfrac{2\pi Lk}{\ln(r_2/r_1)} T_1}{\sigma\varepsilon PL} \right) + T_{\text{sur}}^4$$

(continued)

Simplifying with $P = 2\pi r_2$, we have

$$T_2^4 + (\sigma\varepsilon)^{-1}\left(h_c + \frac{k}{r_2 \ln(r_2/r_1)}\right)T_2 = (\sigma\varepsilon)^{-1}\left(h_c T_\infty + \frac{k}{r_2 \ln(r_2/r_1)}T_1\right) + T_{\text{sur}}^4$$

We can solve the above equation iteratively, as in Example 4.20. Assuming a value of 440 K (remember all temperatures should be absolute), we have

Left-hand side		*Right-hand side*
5.64×10^{13}	$<$	5.67×10^{13}

As this assumption was too low, try $T_2 = 450$ K:

$$5.77 \times 10^{13} > 5.67 \times 10^{13}$$

Since this value is too high, we can interpolate between 440 K and 450 K to find our solution, $T_2 = 441.8$ K:

$$5.67 \times 10^{13} = 5.67 \times 10^{13}$$

Thus, $T_2 \cong 442$ K, which is slightly lower than our solution for the plane wall in Example 4.20. Note, also, that the solution was independent of the length of the pipe.

4.12 SUMMARY

In this chapter, we introduced and applied the first law of thermodynamics to applications that included both systems and control volumes. The first law was applied to a system undergoing a cycle then to a system undergoing a process. The first law applied to a control volume followed. The quantities introduced in this chapter were:

Latent heat: The heat that must be transferred at constant pressure to change the phase of a unit of mass of a substance.

Heat of fusion: The heat needed to melt a unit of mass of a substance.

Heat of vaporization: The heat needed to vaporize a unit of mass of a substance.

Heat of sublimation: The heat needed to change the phase of a unit of mass of a substance from a solid to a vapor.

Steady flow: Flow variables in a control volume do not vary with time.

Uniform flow: Flow variables do not vary across a section of a conduit.

Biot number: Used to determine if the lumped capacitance assumption can be used.

The equations that are of primary importance in our study will be listed. The first law applied to a system undergoing a cycle is

$$\sum Q = \sum W \tag{4.122}$$

For a system undergoing a process, the first law takes the form

$$Q - W = U_2 - U_1 + \frac{1}{2}m(V_2^2 - V_1^2) \quad \text{or} \quad q - w = u_2 - u_1 + \frac{1}{2}(V_2^2 - V_1^2) \tag{4.123}$$

if we neglect the potential energy change.

Enthalpy is a quantity that is useful in our calculations. It is defined to be

$$H = U + pV\!\!\!/ \quad \text{or} \quad h = u + pv \tag{4.124}$$

For a quasi-equilibrium constant-pressure process, neglecting the kinetic-energy change and using $w = p(v_2 - v_1)$, the first law takes the form

$$Q = H_2 - H_1 \quad \text{or} \quad q = h_2 - h_1 \tag{4.125}$$

Enthalpy change and internal energy change can be found for an ideal gas assuming constant specific heats:

$$\Delta h = c_p(T_2 - T_1) \quad \text{and} \quad \Delta u = c_v(T_2 - T_1) \tag{4.126}$$

The ratio of specific heats is

$$k = \frac{c_p}{c_v} \tag{4.127}$$

For an ideal gas undergoing a quasi-equilibrium isothermal process,

$$q_{1-2} = w_{1-2} = RT \ln \frac{v_2}{v_1} = RT \ln \frac{p_1}{p_2} \tag{4.128}$$

For an ideal gas undergoing a quasi-equilibrium adiabatic process with constant specific heats,

$$\frac{T_2}{T_1} = \left(\frac{p_2}{p_1}\right)^{(k-1)/k} = \left(\frac{v_1}{v_2}\right)^{k-1} \quad \text{and} \quad \frac{p_2}{p_1} = \left(\frac{v_1}{v_2}\right)^{k} \tag{4.129}$$

For a control volume, the conservation of mass for steady flow takes the form

$$\dot{m} = \rho_1 A_1 V_1 = \rho_2 A_2 V_2 \tag{4.130}$$

The first law (the energy equation) for a steady flow with negligible potential energy change is

$$\dot{Q} - \dot{W}_S = \dot{m}\left[h_2 - h_1 + \tfrac{1}{2}(V_2^2 - V_1^2)\right] \quad \text{or} \quad q - w_S = h_2 - h_1 + \tfrac{1}{2}(V_2^2 - V_1^2) \tag{4.131}$$

For a liquid flow, we often neglect kinetic energy and potential energy changes, heat transfer, and internal energy change resulting in

$$\dot{W}_S = \dot{m}\,\frac{p_2 - p_1}{\rho} \tag{4.132}$$

If a fluid is filling an insulated tank, the energy equation takes the form

$$u_f m_f = u_i m_i + m_1 h_1 \tag{4.133}$$

where the subscript "f" refers to the final state in the control volume and "i" refers to the initial state in the control volume. The subscript "1" refers to the inlet condition. If an insulated tank is discharging an ideal gas with constant specific heats, the energy equation becomes

$$\frac{m_f}{m_i} = \left(\frac{p_f}{p_i}\right)^{1/k} \tag{4.134}$$

At a surface, the energy transferred to the surface by conduction is transported away from the surface by convection and radiation:

$$k\frac{T_1 - T_2}{L} = h_c(T_2 - T_\infty) + \sigma\varepsilon(T_2^4 - T_{\text{sur}}^4) \tag{4.135}$$

where k is the conductivity, h_c is the convective-heat transfer coefficient, $\sigma = 5.67 \times 10^8$ W/m^2·K^4, and ε is the emissivity. The temperature of a body that loses energy only by convection is given by

$$T(t) = T_\infty + (T_i - T_\infty)e^{-(h_c A_S/(\rho\forall c))t} \tag{4.136}$$

where A_S is its surface area, $\forall$ is its volume, ρ is its density, c is its specific heat, T_∞ is the temperature of the surrounding fluid, T_i is its initial temperature, and t is the elapsed time.

REVIEW PROBLEMS FOR THE FE/EIT EXAMINATION

4.1FE Select a correct statement of the first law. (*Note:* Neglect kinetic and potential energy changes).
- **A.** The heat transfer equals the work done for a process.
- **B.** The net heat transfer equals the net work for a cycle.
- **C.** The net heat transfer minus net work equals internal-energy change for a cycle.
- **D.** The heat transfer minus work equals internal energy for a process.

4.2FE Select the incorrect statement of the first law. (*Note:* Neglect kinetic and potential energy changes).
- **A.** The heat transfer equals the internal energy change for an adiabatic process.
- **B.** The heat transfer and the work have the same magnitude for a constant-volume quasi-equilibrium process in which the internal energy remains constant.
- **C.** The total energy input must equal the total work output for an engine operating on a cycle.
- **D.** The internal energy change plus the work must equal zero for an adiabatic quasi-equilibrium process.

4.3FE Ten kilograms of hydrogen are contained in a rigid, insulated tank at 20°C. Estimate the final temperature if a 400-W resistance heater operates in the hydrogen for 40 minutes.

A. 116°C **B.** 84°C
C. 29°C **D.** 27°C

4.4FE Saturated-water vapor at 400 kPa is heated in a rigid volume until $T_2 = 400$°C. The heat transfer is nearest:
- **A.** 407 kJ/kg
- **B.** 508 kJ/kg
- **C.** 604 kJ/kg
- **D.** 702 kJ/kg

4.5FE Find the work needed to compress 2 kg of air in an insulated cylinder from 100 kPa to 600 kPa if $T_1 = 20°C$.
A. −469 kJ **B.** −390 kJ
C. −280 kJ **D.** −220 kJ

4.6FE Find the temperature rise after five minutes in the volume of Fig. P4.6FE.
A. 423°C **B.** 378°C
C. 313°C **D.** 287°C

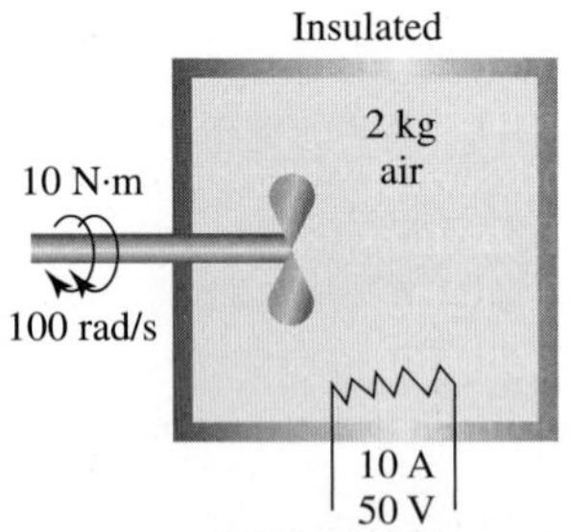

FIGURE P4.6FE

4.7FE One kilogram of air is compressed at constant temperature of 100°C until the volume is halved. How much heat is rejected?
A. 42 kJ **B.** 53 kJ
C. 67 kJ **D.** 74 kJ

4.8FE Energy is added to 5 kg of air with a paddle wheel until $\Delta T = 100°C$. Find the paddle wheel work if the rigid container is insulated.
A. 524 kJ **B.** 482 kJ
C. 412 kJ **D.** 358 kJ

4.9FE Initially $p_1 = 400$ kPa and $T_1 = 400°C$, as shown in Fig. P4.9FE. What is T_2 when the frictionless piston hits the stops?
A. 315°C **B.** 316°C
C. 317°C **D.** 318°C

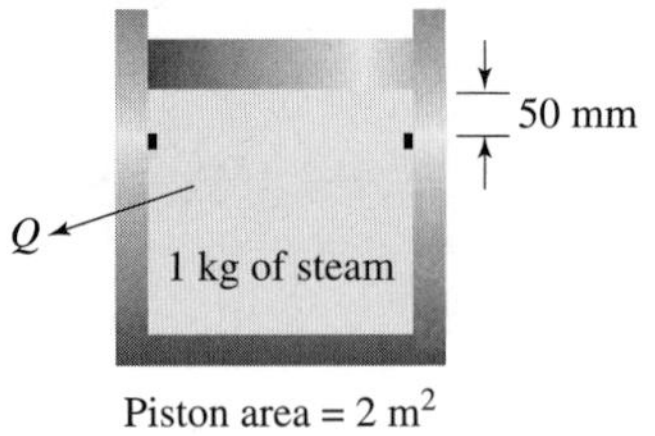

FIGURE P4.9FE

4.10FE Calculate the heat released during the process of Problem 4.9FE.
A. 190 kJ **B.** 185 kJ
C. 180 kJ **D.** 175 kJ

4.11FE After the piston hits the stops of Fig. P4.9FE, how much additional heat is released until $p_3 = 100$ kPa?
A. 1580 kJ **B.** 1260 kJ
C. 930 kJ **D.** 730 kJ

4.12FE The pressure of 10 kg of air is increased isothermally at 60°C from 100 kPa to 800 kPa. How much heat is rejected?
A. 1290 kJ **B.** 1610 kJ
C. 1810 kJ **D.** 1990 kJ

4.13FE Saturated water is heated at constant pressure of 400 kPa until $T_2 = 400°C$. How much heat must be removed?
A. 2070 kJ/kg **B.** 2370 kJ/kg
C. 2670 kJ/kg **D.** 2870 kJ/kg

4.14FE One kilogram of steam in a cylinder requires 170 kJ of heat transfer while the pressure remains constant at 1000 kPa. Estimate the temperature T_2 if $T_1 = 320°C$.
A. 420°C **B.** 410°C
C. 400°C **D.** 390°C

4.15FE Estimate the work required for the process of Problem 4.14FE.
A. 89 kJ **B.** 85 kJ
C. 45 kJ **D.** 39 kJ

4.16FE The pressure of steam at 400°C and $u =$ 2949 kJ/kg is most nearly:
A. 2000 kPa **B.** 1900 kPa
C. 1800 kPa **D.** 1700 kPa

4.17FE The enthalpy of steam at $p = 500$ kPa and $v =$ 0.7 m^3/kg is nearest:
A. 3480 kJ/kg **B.** 3470 kJ/kg
C. 3460 kJ/kg **D.** 3450 kJ/kg

4.18FE Estimate c_p for steam at 4 MPa and 350°C.
A. 2.48 kJ/kg·°C **B.** 2.71 kJ/kg·°C
C. 2.53 kJ/kg·°C **D.** 2.31 kJ/kg·°C

4.19FE Methane is heated at constant pressure from 0°C to 300°C. How much heat is needed if $p_1 =$ 200 kPa?
A. 731 kJ/kg **B.** 692 kJ/kg
C. 676 kJ/kg **D.** 623 kJ/kg

4.20FE Estimate the equilibrium temperature if 20 kg of copper at 0°C and 10 L of water at 30°C are placed in an insulated container.
A. 27.2°C **B.** 25.4°C
C. 22.4°C **D.** 20.3°C

4.21FE Estimate the equilibrium temperature if 10 kg of ice at 0°C is mixed with 60 kg of water at 20°C in an insulated container.
A. 12°C **B.** 5.8°C
C. 2.1°C **D.** 1.1°C

4.22FE The table shows a three-process cycle; determine c.
A. 140 **B.** 100
C. 80 **D.** 40

Process	Q	W	ΔU
1 → 2	100	a	0
2 → 3	b	60	40
3 → 1	40	c	d

4.23FE Find w_{1-2} for the process in Fig. P4.23FE.
A. 219 kJ/kg **B.** 166 kJ/kg
C. 113 kJ/kg **D.** 53 kJ/kg

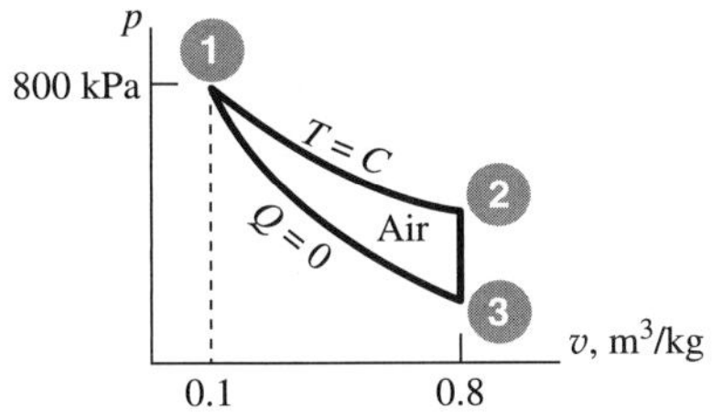

FIGURE P4.23FE

4.24FE Find w_{3-1} for the process in Fig. P4.23FE.
A. −219 kJ/kg **B.** −166 kJ/kg
C. −113 kJ/kg **D.** −53 kJ/kg

4.25FE Find q_{cycle} for the cycle in Fig. P4.23FE.
A. 219 kJ/kg **B.** 166 kJ/kg
C. 113 kJ/kg **D.** 53 kJ/kg

4.26FE Clothes are hung on a clothesline to dry on a cold winter day. The clothes dry due to:
A. Sublimation **B.** Evaporation
C. Vaporization **D.** Melting

4.27FE Air is compressed adiabatically from 100 kPa and 20°C to 800 kPa. Estimate T_2.
A. 440°C **B.** 360°C
C. 290°C **D.** 260°C

4.28FE The work required to compress 2 kg of air in an insulated cylinder from 100°C and 100 kPa to 600 kPa is nearest:
A. 460 kJ **B.** 360 kJ
C. 280 kJ **D.** 220 kJ

4.29FE One hundred people are at a meeting in a 10 m × 20 m × 3 m room when the air conditioning fails. Estimate the temperature increase if the system is off for 15 minutes. Each person emits 400 kJ/hr of heat and the lights add 300 W of energy. Neglect all other forms of energy input.
A. 15°C **B.** 18°C
C. 21°C **D.** 25°C

4.30FE Air undergoes a three-process cycle with a constant pressure process, a constant temperature process, and a constant volume process. Select the correct statement for a piston-cylinder arrangement.
A. The work associated with the constant-pressure process is zero.
B. The heat transfer associated with the constant-volume process is zero.
C. The heat transfer associated with the constant-temperature process is zero.
D. The work associated with the constant-volume process is zero.

4.31FE The term $\dot{m}\Delta h$ in the control volume equation $\dot{Q} - \dot{W}_S = \dot{m}\Delta h$
A. Accounts for the rate of change in energy in the control volume.
B. Represents the rate of change of energy between the inlet and outlet.
C. Is often neglected in control-volume applications.
D. Includes the work rate due to the pressure forces.

4.32FE Select an assumption that is made when deriving the continuity equation $\dot{m}_1 = \dot{m}_2$.
A. Constant density **B.** Steady flow
C. Uniform flow **D.** Constant velocity

4.33FE A nozzle accelerates air from 20 m/s to 200 m/s. What temperature change is expected?
A. 40°C **B.** 30°C
C. 20°C **D.** 10°C

4.34FE Steam enters a valve at 10 MPA and 550°C and exits at 0.8 MPa. What exiting temperature is expected?
A. 590°C **B.** 535°C
C. 520°C **D.** 510°C

4.35FE Air enters an insulated compressor at 100 kPa and 20°C and exits at 800 kPa. Estimate the exiting temperature.
A. 530°C **B.** 462°C
C. 323°C **D.** 258°C

4.36FE If $\dot{m}_1 = 2$ kg/s for the compressor of Problem 4.35FE and $d_1 = 20$ cm, calculate V_1.
A. 62 m/s **B.** 53 m/s
C. 41 m/s **D.** 33 m/s

4.37FE Ten kg/s of saturated steam at 10 kPa is to be completely condensed using 400 kg/s of cooling water. Estimate the temperature change of the cooling water.
A. 32°C **B.** 24°C
C. 18°C **D.** 14°C

4.38FE One hundred kg/min of air enters a relatively short, constant-diameter tube at 25°C and leaves at 20°C. Determine the heat loss.
A. 750 kJ/min **B.** 670 kJ/min
C. 500 kJ/min **D.** 360 kJ/min

4.39FE The minimum power needed by a water pump that increases the pressure of 4 kg/s from 100 kPa to 6 MPa is:
A. 250 kW **B.** 95 kW
C. 24 kW **D.** 6 kW

4.40FE A key concept in analyzing the filling of an evacuated tank is:
A. The mass flow rate into the tank remains constant.
B. The enthalpy across a valve remains constant.
C. The internal energy in the tank remains constant.
D. The temperature in the tank remains constant.

4.41FE A given volume of material, initially at 100°C, cools to 60°C in 40 seconds. Assuming no phase change and only convective cooling to air at 20°C, how long would it take for the same material to cool to 60°C if the heat transfer coefficient were doubled?
A. 4.4 seconds **B.** 3.0 seconds
C. 20 seconds **D.** 80 seconds

4.42FE Consider a constant heat flux q'' through a planar material A, of thickness L_A and thermal conductivity k_A, at steady-state. The temperatures at $x = 0$ and $x = L_A$ are equal to T_{A1} and T_{A2}, where x is in the direction parallel to the heat flux through the thickness of the material. Material A is replaced by Material B, which has a thermal conductivity of $k_B = \frac{1}{4}k_A$, and respective thickness and temperatures of L_B, T_{B1}, and T_{B2}. Which of the scenarios below will result in the same heat flux q'' through material B?
A. L_B is equal to L_A; $T_{B1} = \frac{1}{4}T_{A1}$; and $T_{B2} = \frac{1}{4}T_{A2}$
B. L_B is equal to L_A; $T_{B1} = T_{A1}$; and $T_{B2} = \frac{1}{4}T_{A2}$
C. L_B is equal to $\frac{1}{2}L_A$; $T_{B1} = 2T_{A1}$; and $T_{B2} = 2T_{A2}$
D. L_B is equal to $2L_A$; $T_{B1} = \frac{1}{2}T_{A1}$; and $T_{B2} = \frac{1}{2}T_{A2}$

PROBLEMS

In the following problem set, work each problem using the appropriate equations and/or Appendix tables. Use TK Solver for property evaluation when requested.

4.1 An unknown mass is attached by a pulley to a paddle wheel that is inserted in the volume of water of Fig. P4.1. The mass is then dropped a distance of 3 m. If 100 J of heat must be transferred from the water in order to return the water to its initial state, determine the mass in kilograms.

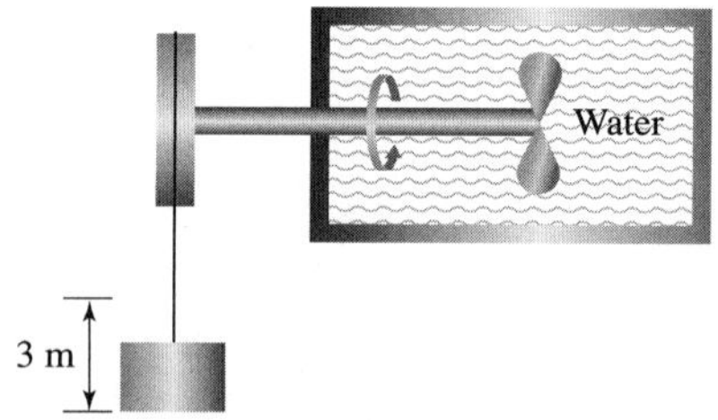

FIGURE P4.1

4.2 A 1500-kg automobile traveling at 30 m/s is brought to rest by impacting a shock absorber composed of a piston with small holes that moves in a cylinder containing water. How much heat must be removed from the water to return it to its original temperature?

4.3 While 300 J of heat is added to the air in the cylinder of Fig. P4.3, the piston raises a distance of 0.2 m. Determine the change in internal energy.

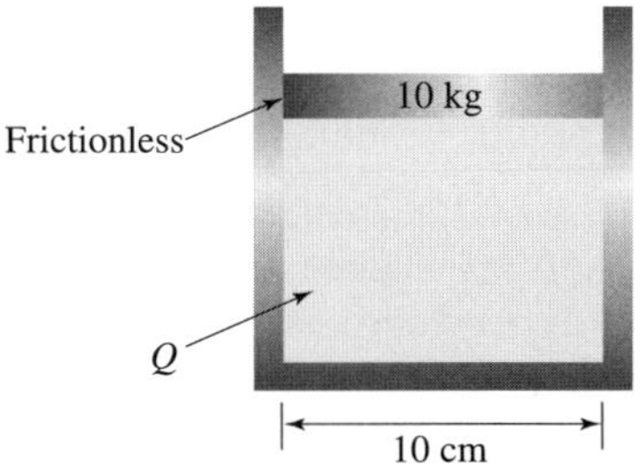

FIGURE P4.3

4.4 A constant force of 600 lbf is required to move the piston shown in Fig. P4.4. If 2 Btu of heat is transferred from the cylinder when the piston moves the entire length, what is the change in internal energy?

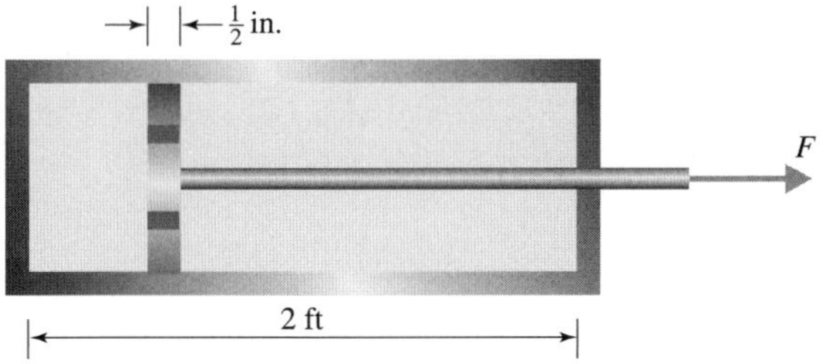

FIGURE P4.4

4.5 Each of the letters **(a)** through **(e)** in the accompanying table represents a process. Supply the missing values, in kJ.

	Q	W	ΔE	E_2	E_1
(a)	20	5			7
(b)		−3	6		8
(c)	40			30	15
(d)	−10		20	10	
(e)		10		−8	6

4.6 A system undergoes a cycle consisting of four processes. Some of the values of the energy transfers and energy changes are given in the table. Fill in all the missing values. All units are kJ.

Process	Q	W	ΔU
1 ⟶ 2	−200	a	0
2 ⟶ 3	800	b	c
3 ⟶ 4	d	600	400
4 ⟶ 1	0	e	−1200

4.7 A system undergoes a cycle consisting of the three processes listed in the table. Compute the missing values. All quantities are in kJ.

Process	Q	W	ΔE
1 ⟶ 2	a	100	100
2 ⟶ 3	b	−50	c
3 ⟶ 1	100	d	−200

4.8 A frictionless piston moves upward a distance of 5 cm while 200 J of heat is added (Fig. P4.8). Calculate the change in internal energy of the vapor if the spring is originally unstretched.

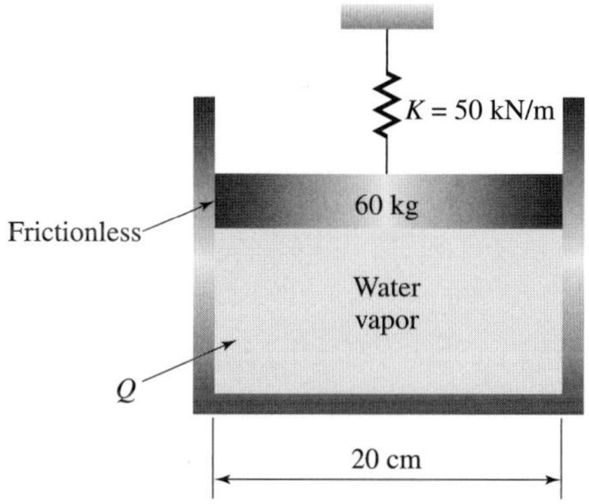

FIGURE P4.8

4.9 A 6-V insulated battery delivers a 5-A current over a period of 20 min. Calculate the heat transfer that must occur to return the battery to its initial temperature.

4.10 A 12-V battery is charged by supplying 3 A over a period of 6 h. If a heat loss of 400 kJ occurs from the battery during the charging period, what is the change in energy stored within the battery?

4.11 A 12-V battery delivers a current of 10 A over a 30-min time period. The stored energy decreases by 300 kJ. Determine the heat lost during the time period.

4.12 A 110-V heater draws 15 A while heating a particular air space. During a 2-h period, the internal energy in the space increases by 8000 Btu. Calculate the amount of heat lost in Btu.

4.13 A refrigerator is situated in an insulated room; it has a 2-hp motor that drives a compressor. Over a 30-minute period of time it provides 5300 kJ of cooling to the refrigerated space and 8000 kJ of heating from the coils on the back of the refrigerator. Calculate the increase in internal energy in the room.

4.14 How much heat must be added to a 0.3-m^3 rigid volume containing water at 200°C in order that the final temperature is raised to 800°C? The initial pressure is 1 MPa.

4.15 A 0.2-m^3 rigid volume shown in Fig. P4.15 contains steam at 600 kPa and a quality of 0.8. If 1000 kJ of heat is added, determine the final temperature using **(a)** the steam tables and **(b)** TK Solver.

FIGURE P4.15

4.16 A 2-ft^3 rigid volume contains water at 120°F with a quality of 0.5. Calculate the final temperature if 8 Btu of heat is added using **(a)** the steam tables and **(b)** TK Solver.

4.17 A piston-cylinder arrangement provides a constant pressure of 120 psi on steam which has an initial quality of 0.95 and an initial volume of 100 in^3. Determine the heat transfer necessary to raise the temperature to 1000°F. Work this problem without using enthalpy.

4.18 Steam is contained in a 4-liter volume at a pressure of 1.5 MPa and a temperature of 200°C. If the pressure is held constant by expanding the volume while 40 kJ of heat is added, find the final temperature. Work this problem without using enthalpy.

Enthalpy

4.19 Work Problem 4.17 using enthalpy. Use **(a)** the steam tables and **(b)** TK Solver.

4.20 Work Problem 4.18 using enthalpy. Use **(a)** the steam tables and **(b)** TK Solver.

4.21 A frictionless piston provides a constant pressure of 400 kPa in a cylinder containing R134a with an initial quality of 80 percent. Calculate the final temperature if 80 kJ/kg of heat is transferred to the cylinder.

4.22 Calculate the heat transfer necessary to raise the temperature of the 2 kg of steam, shown in Fig. P4.22, at a constant pressure of 100 kPa **(a)** from 50°C to 400°C and **(b)** from 400°C to 750°C, using the steam tables. **(c)** Answer parts **(a)** and **(b)** using TK Solver.

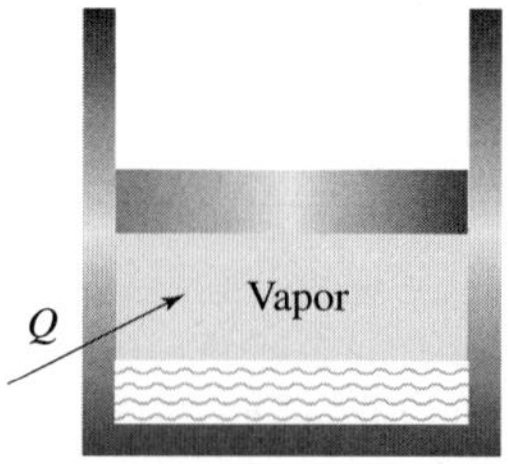

FIGURE P4.22

4.23 Steam is contained in a 1.2-m^3 volume at a pressure of 3 MPa and a quality of 0.8. The pressure is held constant. What is the final temperature if **(a)** 3 MJ and **(b)** 30 MJ of heat is added? Sketch the process on a T-v diagram.

4.24 A piston-cylinder arrangement contains 2 kg of steam originally at 200°C and 90 percent quality. The volume triples while the temperature is held constant. Calculate the heat that must be transferred and the final pressure using **(a)** the steam tables and **(b)** TK Solver.

4.25 Estimate the constant-pressure specific heat for steam at 400°C if the pressure is **(a)** 10 kPa, **(b)** 100 kPa, and **(c)** 30 000 kPa, using the steam tables. **(d)** Answer parts **(a)**, **(b)**, and **(c)** using TK Solver.

4.26 Determine approximate values for the constant-volume specific heat for steam at 800°F if the pressure is **(a)** 1 psia, **(b)** 14.7 psia, and **(c)** 3000 psia.

4.27 Estimate the constant-pressure specific heat and the constant-volume specific heat for R134a at 30 psia and 100°F using **(a)** the steam tables and **(b)** TK Solver.

4.28 Calculate the change in enthalpy of air that is heated from 300 K to 700 K if
(a) $c_p = 1.006$ kJ/kg·°C
(b) $c_p = 0.946 + 0.213 \times 10^{-3}T - 0.031 \times 10^{-6}T^2$ kJ/kg·°C
(c) The gas tables are used.
(d) Compare the calculations of **(a)** and **(b)** with that of **(c)**.

4.29 Calculate the change in enthalpy of 2 kg of air which is heated from 400 K to 600 K if **(a)** $c_p = 1.006$ kJ/kg·K, **(b)** $c_p = 0.946 + 0.213 \times 10^{-3}T - 0.031 \times 10^{-6}T^2$ kJ/kg·°C, and **(c)** the gas tables are used.

4.30 Compare the enthalpy change of 2 kg of water for a temperature change from 10°C to 60°C with that of 2 kg of ice for a temperature change from −60°C to −10°C.

Latent Heat and Specific Heats

4.31 Sixteen ice cubes, each with a temperature of −10°C and a volume of 8 milliliters, are added to 1 liter of water at 20°C in the insulated container of Fig. P4.31. What is the equilibrium temperature? Use $(c_p)_{ice} = 2.1$ kJ/kg·°C.

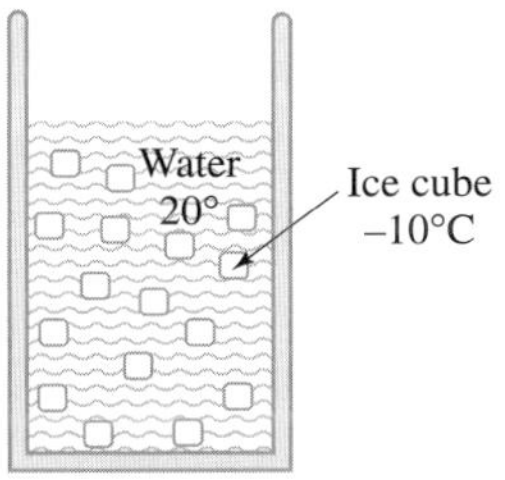

FIGURE P4.31

4.32 Two MJ of heat is added to 2.3 kg of ice held at a constant pressure of 200 kPa at **(a)** −60°C and **(b)** 0°C. What is the final temperature? Sketch the process on a T-v diagram.

4.33 What is the heat transfer required to raise the temperature of 10 lbm of water from 0°F (ice) to 600°F (vapor) at a constant pressure of 30 psia? Sketch the process on a T-v diagram.

4.34 Five ice cubes (4 cm × 2 cm × 2 cm) at −20°C are added to an insulated glass of cola at 20°C. Estimate the final temperature (if above 0°C) or the percentage of ice melted (if at 0°C) if the cola volume is **(a)** 2 liters and **(b)** 0.25 liters. Use $\rho_{ice} = 917$ kg/m^3.

4.35 A 5-kg block of copper at 300°C is submerged in 20 liters of water at 0°C contained in an insulated tank. Estimate the final equilibrium temperature.

4.36 A 40-lbm block of copper at 200°F is dropped in the insulated tank of Fig. P4.36 containing 3 ft^3 of water at 60°F. Calculate the final equilibrium temperature.

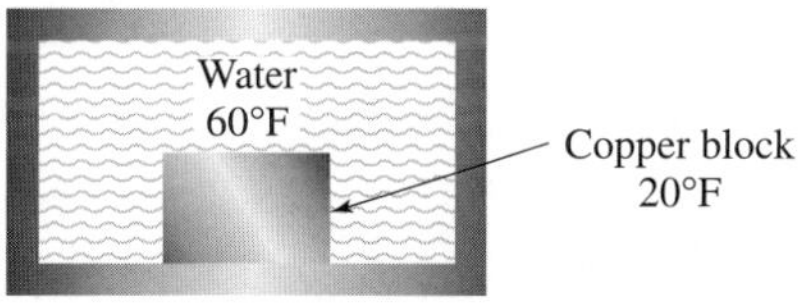

FIGURE P4.36

4.37 A 50-kg block of copper at 0°C and a 100-kg block of iron at 200°C are brought into contact in an insulated space. Predict the final equilibrium temperature.

4.38 Determine the enthalpy change and the internal energy change for 4 kg of air if the temperature changes from 100°C to 400°C. Assume constant specific heats.

The First Law Applied to Various Processes

4.39 For each of the following quasi-equilibrium processes, supply the missing information. The working fluid is 0.4 kg of air in a cylinder.

Process	Q (kJ)	W (kJ)	ΔU (kJ)	ΔH (kJ)	T_2 (°C)	T_1 (°C)	p_2 (kPa)	p_1 (kPa)	V_2 (m^3)	V_1 (m^3)
a) $T = C$	60				100		50			
b) $V = C$				60	300		200			
c) $p = C$	100					200		500		
d) $Q = 0$						50			0.1	0.48

4.40 For the first three quasi-equilibrium processes presented in the table in Problem 4.39, supply the missing information if the working fluid is 0.4 kg of steam. (*Note:* For process **(a)**, it is necessary to integrate graphically.)

4.41 Two lb of air is compressed from 20 psia to 200 psia while maintaining the temperature constant at 100°F. Calculate the heat transfer needed to accomplish this process.

4.42 Helium is contained in a 2-m^3 rigid volume at 50°C and 200 kPa. Calculate the heat transfer needed to increase the pressure to 800 kPa.

4.43 One thousand Btu of heat is added to 2 lbm of steam maintained at 60 psia. Calculate the final temperature if the initial temperature of the steam is **(a)** 600°F and **(b)** 815°F, using the steam tables. **(c)** Answer parts **(a)** and **(b)** using TK Solver.

4.44 Fifty kJ of heat is transferred and the air maintained at 400 kPa in the cylinder of Fig. P4.44 with an initial volume of 0.2 m^3. Determine the final temperature if the initial temperature is **(a)** 0°C and **(b)** 200°C.

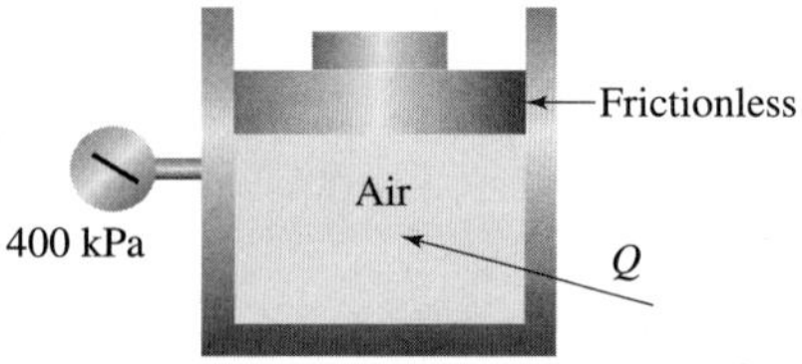

FIGURE P4.44

4.45 The initial temperature and pressure of 8000 cm^3 of air are 100°C and 800 kPa, respectively. Determine the necessary heat transfer if the volume does not change and the final pressure is **(a)** 200 kPa and **(b)** 3000 kPa.

4.46 The air in the cylinder of an air compressor is compressed from 100 kPa to 10 MPa. Estimate the final temperature and the work required if the air is initially at 100°C.

4.47 Nitrogen at 100°C and 600 kPa expands in such a way that it can be approximated by a polytropic process with $n = 1.2$ (see Eq. 4.52). Calculate the work and the heat transfer if the final pressure is 100 kPa.

4.48 Calculate the heat transfer necessary to raise the temperature of air, initially at 10°C and 100 kPa, to a temperature of 27°C if the air is contained in an initial volume with dimensions 3 m × 5 m × 2.4 m. The pressure is held constant.

4.49 How much work must be input by the paddle wheel in Fig. P4.49 to raise the piston 5 in? The initial temperature is 100°F.

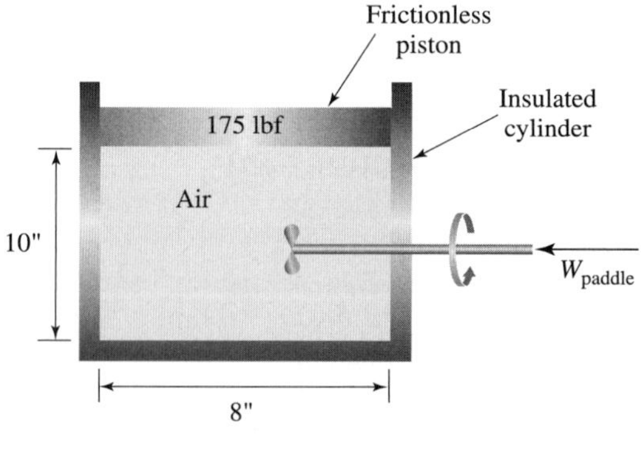

FIGURE P4.49

4.50 Heat is added to a fixed 0.15-m^3 volume of steam initially at a pressure of 400 kPa and a quality of 0.5. Determine the final pressure and temperature if **(a)** 800 kJ and **(b)** 200 kJ of heat is added. Sketch the process on a *p-v* diagram. **(c)** Answer part **(a)** using TK Solver.

4.51 Two hundred Btu of heat is added to a rigid air tank that has a volume of 3 ft^3. Find the final temperature if initially **(a)** $p = 60$ psia and $T = 30°F$ and **(b)** $p = 600$ psia and $T = 820°F$. Use the air tables.

4.52 A system consisting of 5 kg of air is initially at 300 kPa and 20°C. Determine the heat transfer necessary to **(a)** increase the volume by a factor of two at constant pressure, **(b)** increase the pressure by a factor of two at constant volume, **(c)** increase the pressure by a factor of two at constant temperature, and **(d)** increase the absolute temperature by a factor of two at constant pressure.

4.53 Heat is added to a container holding 0.5 m^3 of steam initially at a pressure of 400 kPa and a quality of 80 percent (Fig. P4.53). If the pressure is held constant, find the heat transfer necessary if the final temperature is **(a)** 500°C and **(b)** 675°C. Also determine the work done. Sketch the process on a *T-v* diagram. **(c)** Answer part **(b)** using TK Solver.

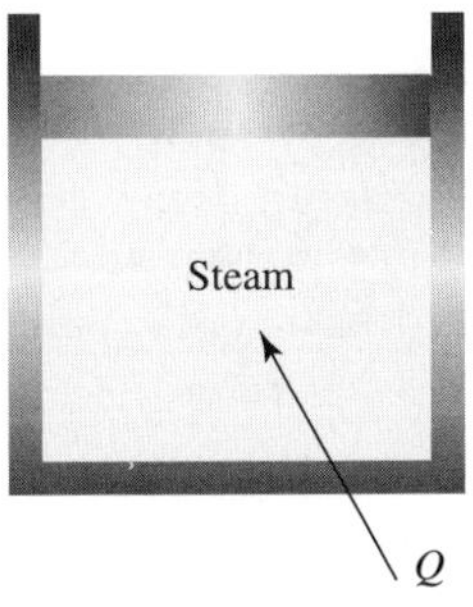

FIGURE P4.53

4.54 A rigid 1.5-m^3 tank at a pressure of 200 kPa contains 5 liters of liquid water and the remainder steam. Calculate the heat transfer necessary to **(a)** completely vaporize the water, **(b)** raise the temperature to 400°C, and **(c)** raise the pressure to 800 kPa.

4.55 Ten Btu of heat is added to a rigid container holding 4 lbm of air in a volume of 100 ft^3. Determine ΔH.

4.56 Eight thousand cm^3 of air in a piston-cylinder arrangement is compressed isothermally at 30°C from a pressure of 200 kPa to a pressure of 800 kPa. Find the heat transfer.

4.57 Two kg of air is compressed in an insulated cylinder from 400 kPa to 15 000 kPa. Determine the final temperature and the work necessary if the initial temperature is **(a)** 200°C and **(b)** 350°C.

4.58 Air is compressed in an insulated cylinder from the position shown in Fig. P4.58 so that the pressure increases to 5000 kPa from atmospheric pressure of 100 kPa. What is the required work if the mass of the air is 0.2 kg?

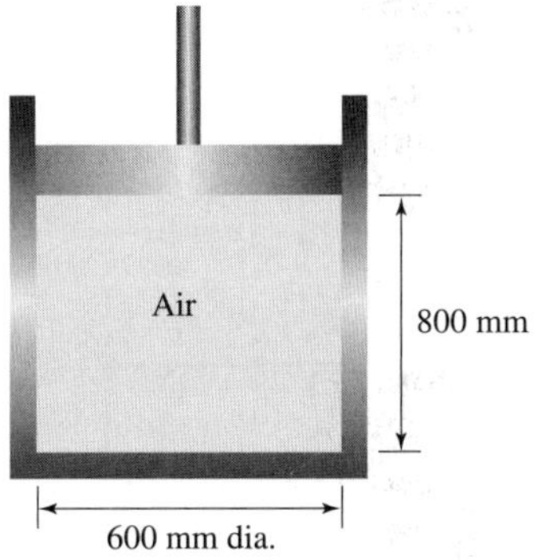

FIGURE P4.58

4.59 The average person emits approximately 400 Btu of heat per hour. There are 1000 people in an unventilated room 10 ft × 75 ft × 150 ft. Approximate the increase in temperature after 15 min, assuming **(a)** constant pressure and **(b)** constant volume. **(c)** Which assumption is the more realistic?

4.60 Two hundred kJ of work is transferred to the air by means of a paddle wheel inserted into an insulated volume (Fig. P4.60). If the initial pressure and temperature are 200 kPa and 100°C, respectively, determine the final temperature and pressure.

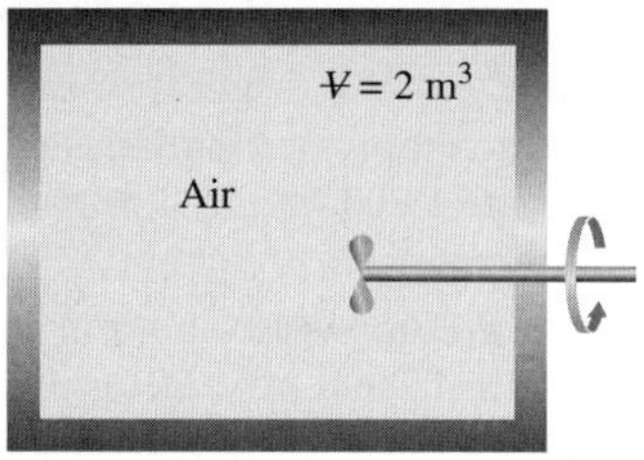

FIGURE P4.60

4.61 A 2-kg rock falls from 10 m and lands in a 10-liter container of water. Neglecting friction during the fall, calculate the maximum temperature increase in the water.

4.62 A torque of 10 N·m is required to turn a paddle wheel at the rate of 100 rad/s. During a 45-s time period, a volume of air in which the paddle wheel rotates is increased from 0.1 to 0.4 m^3. The pressure is maintained constant at 400 kPa. Determine the heat transfer necessary if the initial temperature is **(a)** 0°C and **(b)** 300°C.

The First Law Applied to Cycles

4.63 For the cycle shown in Fig. P4.63, find the work output and the net heat transfer if 0.8 lbm of air is contained in a cylinder with $T_1 = 800°\text{F}$ assuming the process from 3 to 1 is **(a)** an isothermal process and **(b)** an adiabatic process.

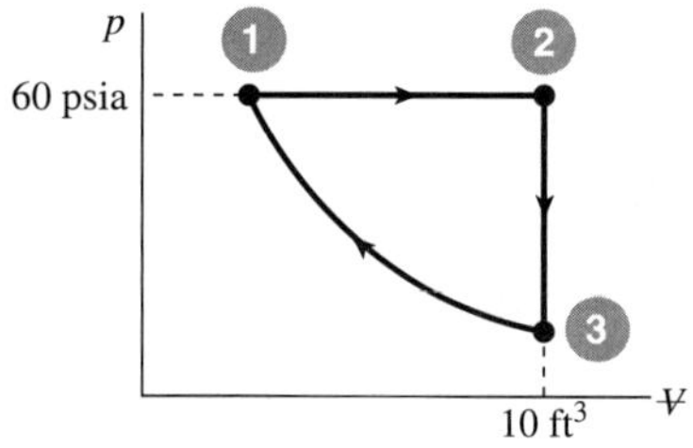

FIGURE P4.63

4.64 For the cycle in Fig. P4.64, find the work output and the net heat transfer if the 0.1 kg of air is contained in a piston-cylinder arrangement.

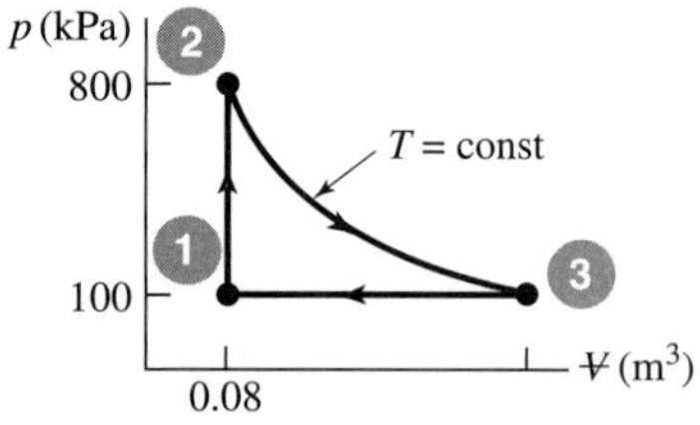

FIGURE P4.64

4.65 For the cycle shown in Fig. P4.65, find the net heat transfer and work output if steam is contained in a cylinder using **(a)** the steam tables and **(b)** TK Solver.

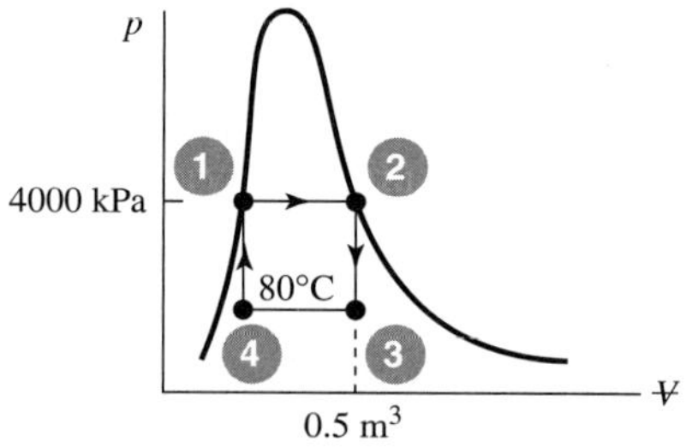

FIGURE P4.65

4.66 If 0.03 kg of air undergoes the cycle shown in Fig. P4.66 (a piston-cylinder arrangement), calculate the work output.

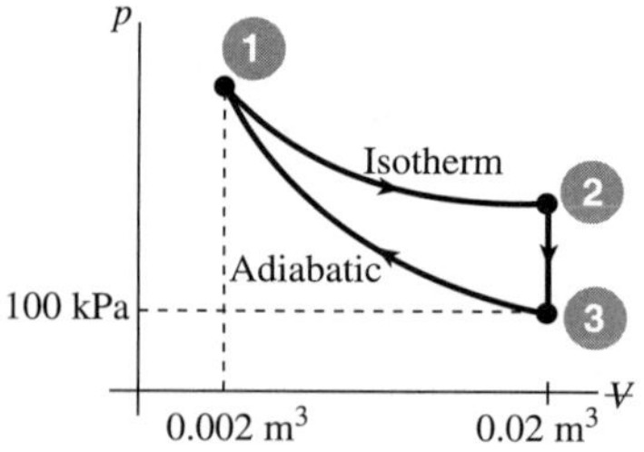

FIGURE P4.66

The Conservation of Mass Applied to Control Volumes

4.67 Air is flowing at an average speed of 100 m/s through a 10-cm-diameter pipe. If the pipe undergoes an enlargement to 20 cm in diameter, determine the average speed in the enlarged pipe.

4.68 Water enters a radiator through a 4-cm-diameter hose at 0.02 kg/s. It travels down through all the rectangular passageways on its way to the water pump. The passageways are each 10 mm × 1 mm, and there are 800 of them in a cross section. How long does it take water to traverse from the top to the bottom of the 60-cm-high radiator?

4.69 Air enters a vacuum cleaner through a 2-in-diameter pipe at a speed of 150 ft/sec. It passes through a rotating impeller (Fig. P4.69) of thickness 0.5″ through which the air exits. Determine the average velocity exiting normal to the impeller.

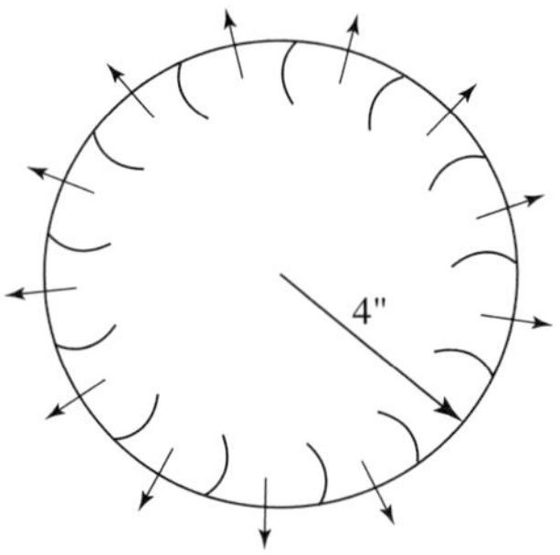

FIGURE P4.69

4.70 Air enters a device at 4 MPa and 300°C with a velocity of 150 m/s. The inlet area is 10 cm^2 and the outlet area is 50 cm^2. Determine the mass flow rate and the outlet velocity if the air exits at 0.4 MPa and 100°C.

4.71 A 10-m^3 tank is being filled with steam at 800 kPa and 400°C. It enters the tank through a 10-cm-diameter pipe. Determine the rate at which the density in the tank is varying when the velocity of the steam in the pipe is 20 m/s using **(a)** the steam tables and **(b)** TK Solver.

4.72 Air enters the device shown in Fig. P4.72 at 2 MPa and 350°C with a velocity of 125 m/s. At one outlet area, the conditions are 150 kPa and 150°C with a velocity of 40 m/s. Determine the mass flow rate and the velocity at the second outlet for conditions of 0.45 MPa and 200°C.

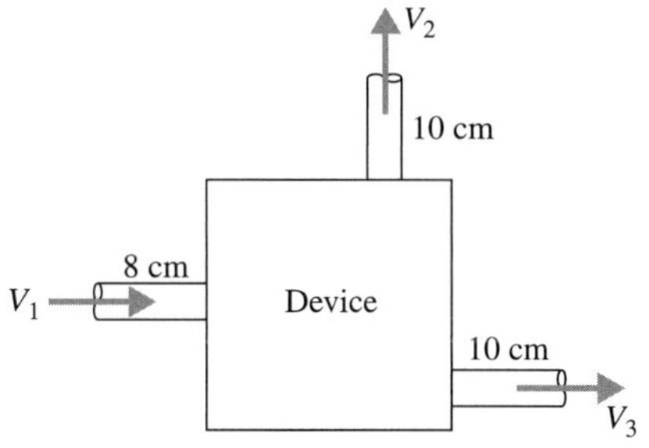

FIGURE P4.72

4.73 Water enters a 4-ft-wide, $\frac{1}{2}$-in.-high channel with a mass flow rate of 15 lbm/sec. It leaves with a parabolic distribution $V(y) = V_{max}(1 - y^2/h^2)$, where h is half the channel height. Calculate V_{max} and V_{avg} (the average velocity over any cross-section of the channel). Assume that the water completely fills the channel.

4.74 Steam at 400 kPa and 250°C is being transferred through a 50-cm-diameter pipe at a speed of 30 m/s. It splits into two pipes with equal diameters of 25 cm. Calculate the mass flow rate and the velocity in each of the smaller pipes if the pressure and temperature in each are 200 kPa and 200°C, respectively. Use **(a)** the steam tables and **(b)** TK Solver.

4.75 Steam enters a device through a 2-in^2 area at 500 psia and 600°F. It exits through a 10-in^2 area at 20 psia and 400°F with a velocity of 800 ft/sec. What are the mass flow rate and the entering velocity?

4.76 Steam enters a 10-m^3 tank at 2 MPa and 600°C through an 8-cm-diameter pipe with a velocity of 20 m/s. It leaves at 1 MPa and 400°C through a 12-cm-diameter pipe with a velocity of 10 m/s. Calculate the rate at which the density in the tank is changing using **(a)** the steam tables and **(b)** TK Solver.

4.77 Water flows into a 1.2-cm-diameter pipe with a uniform velocity of 0.8 m/s. At some distance down the pipe, a parabolic velocity profile is established. Determine the maximum velocity in the pipe and the mass flow rate. The parabolic profile can be expressed as $V(r) = V_{max}(1 - r^2)$ m/s, where r is the radius of the pipe.

4.78 Water enters the contraction shown in Fig. P4.78 with a parabolic profile $V(r) = 2(1 - r^2)$ m/s, where r is measured in centimeters. The exiting profile after the contraction is essentially uniform. Determine the mass flow rate and the exit velocity.

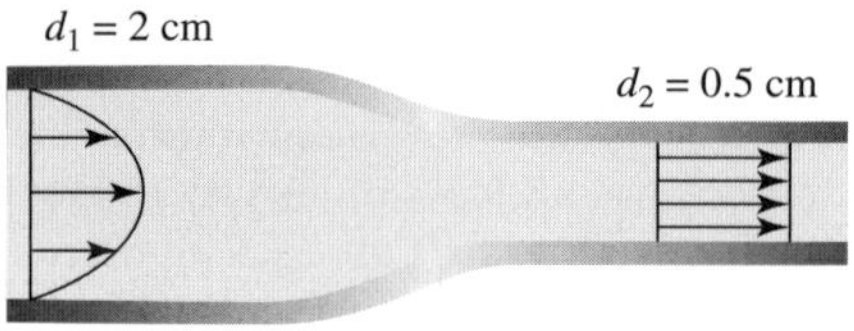

FIGURE P4.78

The First Law Applied to Control Volumes

4.79 Air enters a 4-in. constant-diameter pipe at 100 ft/sec with a pressure of 60 psia and a temperature of 100°F. Heat is added to the air, causing it to pass a downstream area at 70 psia, 300°F. Calculate the downstream velocity and the heat transfer rate. (*Note:* Neglect kinetic energy changes.)

4.80 R134a enters a valve at 800 kPa and 30°C. The pressure downstream of the valve is measured to be 60 kPa. Estimate the internal energy downstream using **(a)** the tables and **(b)** TK Solver.

4.81 Water at 9000 kPa and 300°C flows through a partially open valve. The pressure immediately after the valve is measured to be 600 kPa. Estimate the specific internal energy of the water leaving the valve. (*Note:* The enthalpy of slightly compressed liquid is essentially equal to the enthalpy of saturated liquid at the same temperature.)

4.82 Steam at 9000 kPa and 600°C passes through a throttling process so that the pressure is suddenly reduced to 400 kPa. **(a)** What is the expected temperature after the throttle? **(b)** What area ratio is necessary for the kinetic energy change to be zero? **(c)** If air at the same conditions passed through the throttle, what would be the expected temperature after the throttle?

4.83 Water at 70°F flows through the partially open valve shown in Fig. P4.83. The area before and after the valve is the same. Determine the specific internal energy downstream of the valve.

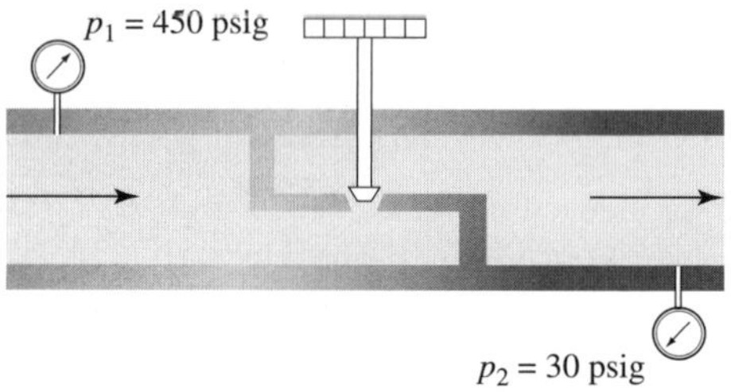

FIGURE P4.83

4.84 The pressure of 200 kg/s of water is to be increased by 4 MPa. The water enters through a 20-cm-diameter pipe and exits through a 12-cm-diameter pipe. Calculate the minimum horsepower required to operate the pump.

4.85 The inlet conditions on an air compressor are 50 kPa and 20°C. To compress the air to 400 kPa, 5 kW of energy is needed. Neglecting heat transfer and kinetic and potential energy changes, estimate the mass flow rate.

4.86 The air compressor shown in Fig. P4.86 draws air from the atmosphere and discharges it at 500 kPa. Determine the minimum power required to drive the insulated compressor. Assume atmospheric conditions of 25°C and 80 kPa.

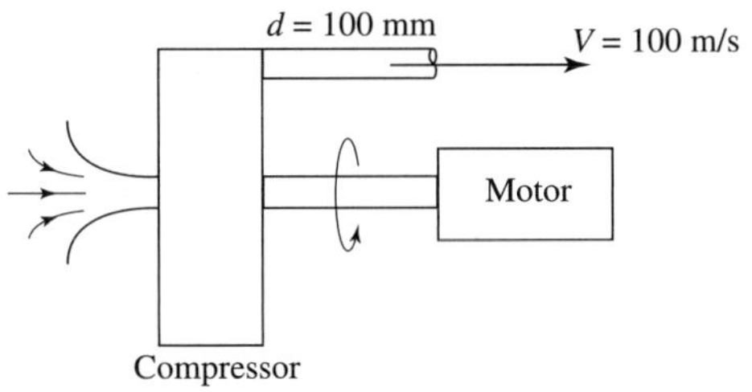

FIGURE P4.86

4.87 The power required to compress 0.01 kg/s of steam from a saturated vapor state at 50°C to a pressure of 800 kPa at 200°C is 6 kW. Find the rate of heat transfer from the compressor using **(a)** the steam tables and **(b)** TK Solver.

4.88 Two thousand lbm/hr of saturated liquid water at 2 psia is compressed by a pump to a pressure of 2000 psia. Neglecting heat transfer and kinetic energy change, estimate the power required by the pump.

4.89 The pump in Fig. P4.89 increases the pressure in the water from 200 to 4000 kPa. What is the minimum horsepower motor required to drive the pump for a volume flow rate of 0.1 m^3/s?

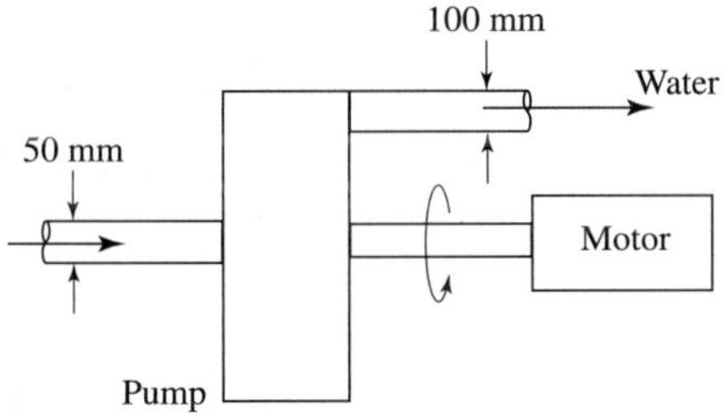

FIGURE P4.89

4.90 A turbine at a hydroelectric plant accepts 20 m^3/s of water at a gage pressure of 300 kPa and discharges it to the atmosphere. Determine the maximum power output.

4.91 A hydroturbine operates on a stream in which 100 kg/s of water flows. Estimate the maximum power output if the turbine is in a dam with a distance of 40 m from the surface of the reservoir to the surface of the backwater.

4.92 Water flows in a creek at 1.5 m/s. It has cross-sectional dimensions of 0.6 m × 1.2 m upstream of a proposed dam that would be capable of developing a head of 2 m above the outlet of a turbine. Determine the maximum power output of the turbine.

4.93 Superheated steam at 800 psia and 1000°F enters a turbine at a power plant at the rate of 30 lb/sec. Saturated steam exits at 5 psia. If the power output is 10 MW, determine the heat transfer rate using **(a)** the steam tables and **(b)** TK Solver.

4.94 An insulated turbine accepts superheated steam at 800 psia and 1200°F and rejects it as saturated vapor at 2 psia (Fig. P4.94). Predict the horsepower output if the mass flow rate is 1000 lbm/min. Also, calculate the velocity at the exit.

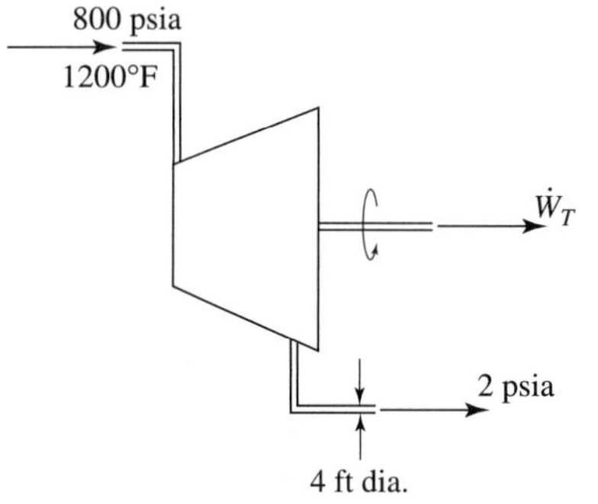

FIGURE P4.94

4.95 Superheated steam enters an insulated turbine (Fig. P4.95) at 4000 kPa and 500°C and leaves at 20 kPa with $x_2 = 0.9$. If the mass flow rate is 6 kg/s, determine the power output and the exiting velocity.

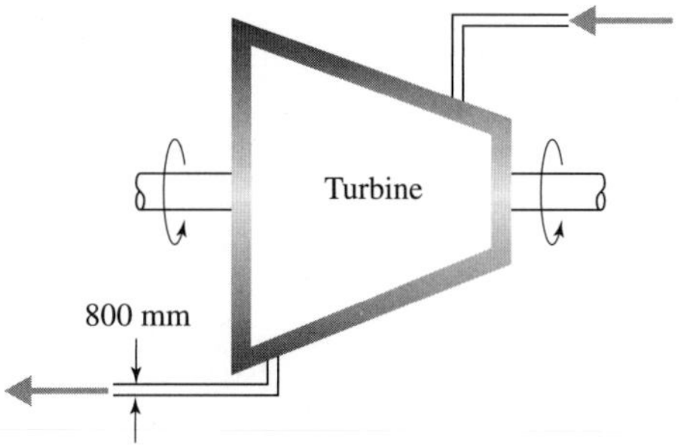

FIGURE P4.95

4.96 Air enters a turbine at 600 kPa and 100°C through a 100-mm-diameter pipe at a speed of 100 m/s. The air exits at 140 kPa and −20°C through a 400-mm-diameter pipe. Calculate the power output, neglecting heat transfer.

4.97 Air enters a compressor at atmospheric conditions of 20°C and 80 kPa and exits at 800 kPa and 200°C. Calculate the rate of heat transfer if the power input is 400 kW. The air exits at 20 m/s through an exit diameter of 10 cm.

4.98 A turbine delivers 500 kW of power by extracting energy from air at 450 kPa and 100°C flowing in a 120-mm-diameter pipe at 150 m/s. For an exit pressure of 120 kPa and a temperature of 20°C determine the heat transfer rate.

4.99 Water flows through a nozzle that converges from 4 in. to 0.8 in. in diameter. For a mass flow rate of 30 lbm/sec, calculate the upstream pressure if the downstream pressure is 14.7 psia.

4.100 Air travels through the 4 m × 2 m test section of a wind tunnel at 20 m/s. The gage pressure in the test section is measured to be −20 kPa and the temperature 20°C. After the test section, a diffuser leads to a 6-m-diameter exit pipe. Estimate the velocity and temperature in the exit pipe.

4.101 Air enters a nozzle like that shown in Fig. P4.101 at a temperature of 195°C and a velocity of 100 m/s. If the air exits to the atmosphere where the pressure is 85 kPa, find **(a)** the exit temperature, **(b)** the exit velocity, and **(c)** the exit diameter. Assume an adiabatic quasi-equilibrium process.

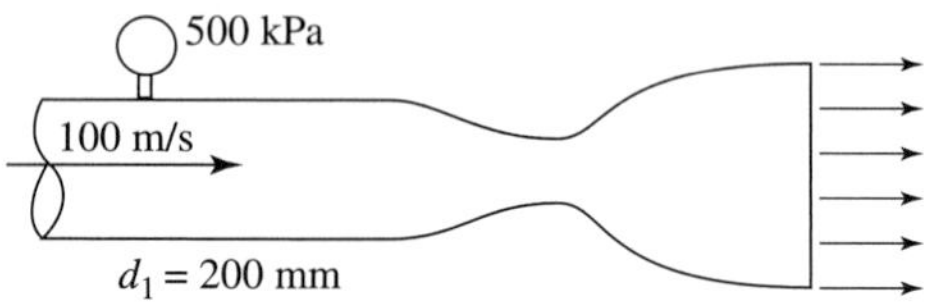

FIGURE P4.101

4.102 Nitrogen enters a diffuser at 200 m/s with a pressure of 80 kPa and a temperature of −20°C. It leaves with a velocity of 15 m/s at an atmospheric pressure of 95 kPa. If the inlet diameter is 100 mm, calculate **(a)** the mass flow rate and **(b)** the exit temperature.

4.103 Steam enters a diffuser as a saturated vapor at 220°F with a velocity of 600 ft/sec. It leaves with a velocity of 50 ft/sec at 20 psia. Estimate the exit temperature using **(a)** the steam tables and **(b)** TK Solver.

4.104 Steam with a mass flow rate of 600 lbm/min exits a turbine as saturated steam at 2 psia and passes through a condenser (a heat exchanger). What mass flow rate of cooling water is needed if the steam is to exit the condenser as saturated liquid and the cooling water is allowed a 15°F temperature rise?

4.105 Water is used in a heat exchanger (Fig. P4.105) to cool 5 kg/s of air from 400°C to 200°C. Calculate **(a)** the minimum mass flow rate of the water and **(b)** the quantity of heat transferred to the water each second.

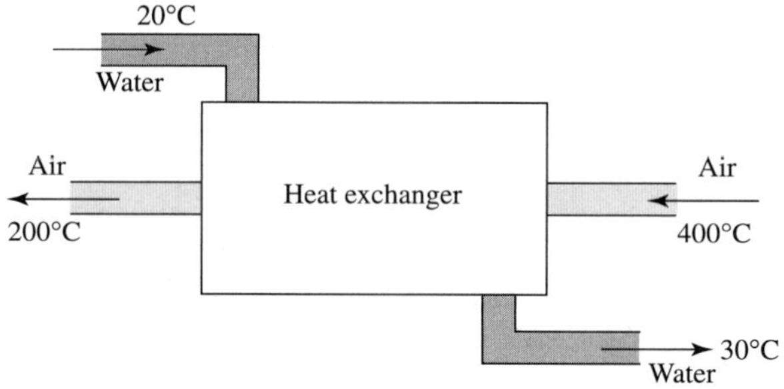

FIGURE P4.105

Transient Flow

4.106 An insulated, 4-m^3-evacuated tank is connected to a 4 MPa, 600°C steam line. A valve is opened and the steam fills the tank. Estimate the final temperature of the steam in the tank and the final mass of the steam in the tank.

4.107 A steam line containing superheated steam at 1000 psia and 1200°F is connected to a 50-ft^3 evacuated, insulated tank by a small line with a valve. The valve is closed when the pressure in the tank just reaches 800 psia. Calculate **(a)** the final temperature in the tank and **(b)** the mass of steam that entered the tank.

4.108 Air is contained in a 3-m^3 tank at 250 kPa and 25°C. Heat is added to the tank as the air escapes, thereby maintaining the temperature constant at 25°C. How much heat is required if the air escapes until the final pressure is atmospheric. Assume $p_{atm} = 80$ kPa.

4.109 An air line carries air at 800 kPa (Fig. P4.109). An insulated tank initially contains 20°C air at an atmospheric pressure of 90 kPa. The valve is opened, and air flows into the tank. Determine the final temperature of the air in the tank and the mass of air that enters the tank if the valve is left open.

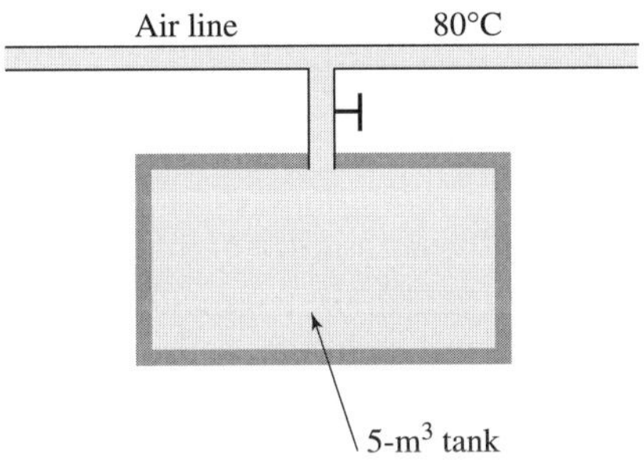

FIGURE P4.109

4.110 An insulated tank is evacuated. Air from the atmosphere at 12 psia and 70°F is allowed to flow into the 100-ft^3 tank. Calculate **(a)** the final temperature and **(b)** the final mass of air in the tank just after the flow ceases.

4.111 **(a)** An insulated tank contains pressurized air at 2000 kPa and 30°C. The air is allowed to escape to the atmosphere ($p_{atm} = 95$ kPa, $T_{atm} = 30$°C) until the flow ceases. Determine the final temperature in the tank. **(b)** Eventually, the air in the tank will reach atmospheric temperature. If a valve was closed after the initial flow ceased, calculate the pressure that is eventually reached in the tank.

4.112 An insulated tank with a volume of 4 m^3 is pressurized to 800 kPa at a temperature of 30°C. An automatic valve allows the air to leave at a constant rate of 0.02 kg/s. **(a)** What is the temperature after 5 min? **(b)** What is the pressure after 5 min? **(c)** How long will it take for the temperature to drop to −20°C?

4.113 A tank with a volume of 2 m^3 contains 90 percent liquid water and 10 percent water vapor by volume at 100 kPa. Heat is transferred to the tank at 10 kJ/min. A relief valve attached to the top of the tank allows vapor to discharge when the gage pressure reaches 600 kPa. The pressure is maintained at that value as more heat is transferred. **(a)** What is the temperature in the tank at the instant the relief valve opens? **(b)** How much mass is discharged when the tank contains 50 percent vapor by volume? **(c)** How long does it take for the tank to contain 75 percent vapor by volume?

Heat Transfer

4.114 A power module includes a thin transistor that generates 1.0 W on the top surface, which has an area of 10^{-4} m^2, as shown in Fig. P4.114. The power module is placed on a heat sink with the same area, and air flows both over and under the module and heat sink at 325 K with a convective heat transfer coefficient of 50 W/m^2·K. The emissivity of the transistor is 0.5, and the surroundings are at 300 K. The overall resistance due to conduction from the transistor to the bottom surface of the heat sink is 50 K/W. (*Note:* Neglect radiation from the bottom surface.)

(a) Determine whether or not the transistor is less than its maximum allowable temperature of 125°C.

(b) Determine the fraction of heat loss due to radiation.

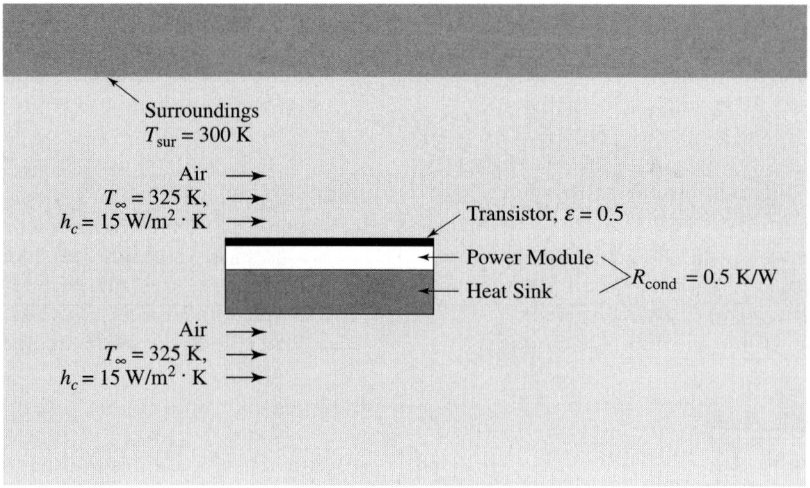

FIGURE P4.114

4.115 A thin pancake at 80°C is cooled by air at 20°C with a convective heat transfer coefficient of 8 W/m^2·K. The pancake has a thickness of 0.3 cm and a diameter of 7.5 cm. **(a)** Can the lumped capacitance assumption be used? **(b)** Determine how long will it take the pancake to cool to 25°C. Assume the pancake has the following properties: $k = 0.25$ W/m·K, $c = 3000$ J/kg·K, and $\rho =$ 500 kg/m^3.

4.116 Consider the composite outer wall of a home, as shown in Fig. P4.116. The wall is composed of 2.5-cm siding, 20 cm × 4 cm hardwood studs with blown insulation between the studs, a 1.5-cm-thick inner-plaster board, and 0.5-cm-thick paneling. The hardwood studs are placed every 60 cm along the wall. The thermal conductivities for the siding, the hardwood studs, the insulation, the plaster board, and the paneling are 0.094 W/m·K, 0.16 W/m·K, 0.03 W/m·K, 0.17 W/m·K, and 0.12 W/m·K, respectively. The home is maintained at 20°C, and the outside temperature is 5°C. The convective heat transfer coefficients on the inside and outside of the wall are 8 W/m^2·K and 208 W/m^2·K, respectively. Assuming one-dimensional, steady-state heat transfer, what is the heat transfer through one section of the wall?

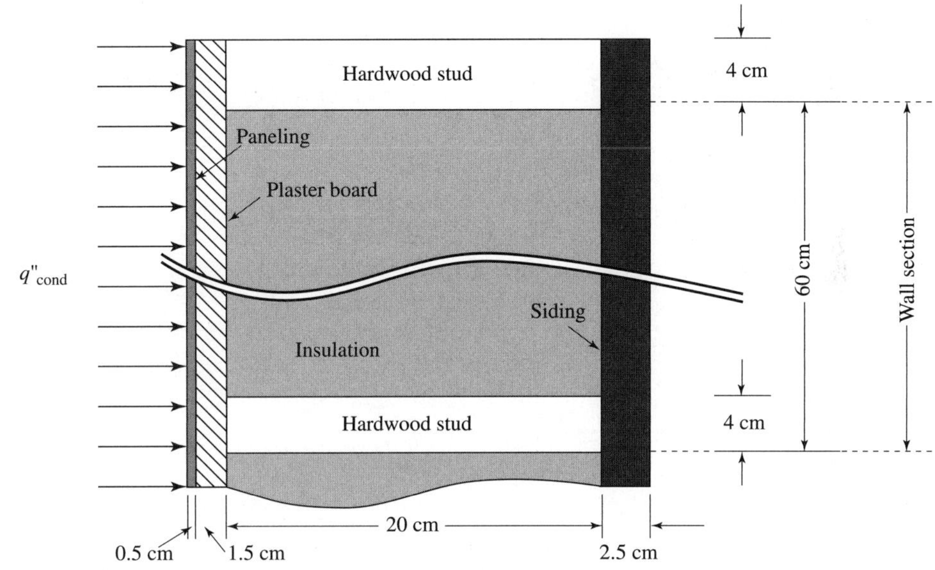

FIGURE P4.116

4.117 A 7-m long stainless-steel (AISI 316) pipe contains saturated steam at 200°C under steady-state conditions. The inner diameter of the pipe is 7.5 cm and the outer diameter is 8 cm. The inside wall of the pipe is maintained at the saturation temperature of the steam. The outer surface of the insulation is exposed to air at 10°C with a convective heat transfer coefficient of 30 W/m^2·K.

(a) Assuming the thermal conductivity of the stainless steel is 16.5 W/m·K and neglecting radiation, determine the total heat loss from the pipe.

(b) If the pipe is covered with 5-cm-thick insulation, what is the percent reduction in the total heat loss from the pipe? (Assume the thermal conductivity of the insulation is 0.055 W/m·K.)

4.118 The inner wall of a hollow sphere is maintained at 400°C at steady-state. The outer wall is exposed to air at 20°C with a convective heat transfer coefficient of 100 W/m^2·K. The sphere is 1-cm thick, with an inner diameter of 5 cm. If the total heat loss through the sphere to the air is 500 W, what is the thermal conductivity of the sphere?

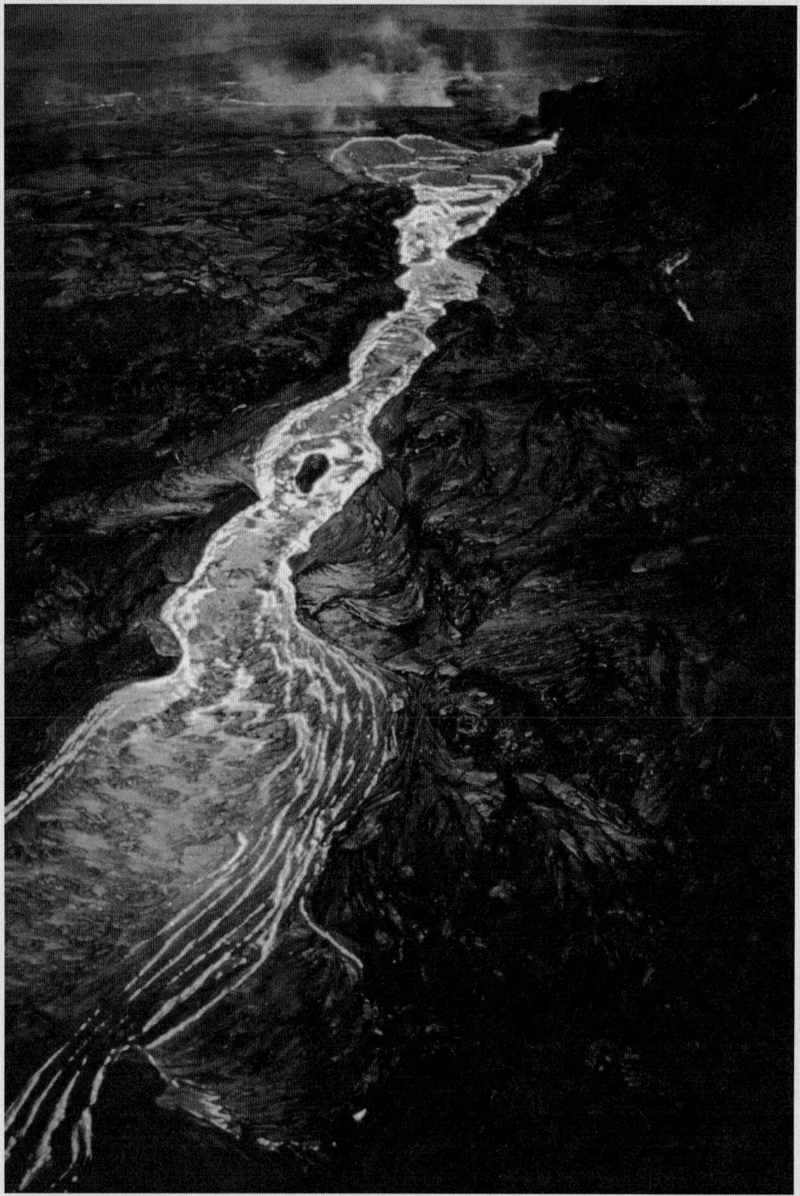

The huge amount of energy in the very hot lava leaving a volcano results in a large increase in entropy in the atmosphere. (Image courtesy of Getty Images.)

5

The Second Law of Thermodynamics

Outline

Chapter Objectives

The objectives of this chapter are to:

- ▲ Introduce the basic concepts and definitions required by the second law.
- ▲ Define and discuss the concept of reversibility.
- ▲ Present the Carnot cycle and engine.
- ▲ Introduce the concept of entropy.
- ▲ Develop equations needed to calculate entropy changes for various substances.
- ▲ Derive the inequality of Clausius.
- ▲ Apply the second law to both systems and control volumes.
- ▲ Present numerous examples and practice problems that illustrate the concepts discussed.

Nomenclature

The following is a list of the nomenclature introduced in this chapter:

c	Specific heat	T_L	Temperature of low-temperature reservoir
COP	Coefficient of performance	v_r	Relative specific volume
p_r	Relative pressure	w_a	Actual shaft work output
Q_H	Heat transfer from high-temperature reservoir	W_{irr}	Irreversible work
Q_L	Heat transfer from low-temperature reservoir	W_{net}	The net work from a cycle
rev	Refers to a reversible process	w_s	Isentropic shaft work output
s	Specific entropy	η	Efficiency of an engine or device
S	Entropy	ϕ	A tabulated function related to entropy
surr	Refers to the surroundings	σ	Entropy production
T_H	Temperature of high-temperature reservoir		

5.1 INTRODUCTION

Key Concept: *The first law relates the several variables involved in a process but does not give the direction of the process.*

Water flows down a hill; heat flows from a hot body to a cold one; rubber bands unwind; fluid flows from a high-pressure region to a low-pressure region, and we all get old! Our experiences in life suggest that processes have a definite direction. The first law of thermodynamics relates the several variables involved in a physical process but does not give any information as to the direction of the process. It is the second law of thermodynamics that helps us establish that direction.

Consider, for example, the situation illustrated in Fig. 5.1. Here, the first law states that the work done by the falling weight increases the internal energy of the air contained in the fixed volume, provided the volume is insulated so that $Q = 0$. It would not be a violation of the first law if we postulated that an internal energy decrease of the air is used to turn the paddle and raise the weight. This, however, would be a violation of the second law of thermodynamics and would thus be an impossibility.

In the beginning of this chapter, we will state the second law as it applies to a cycle. Several devices will be studied. A statement of the second law will follow

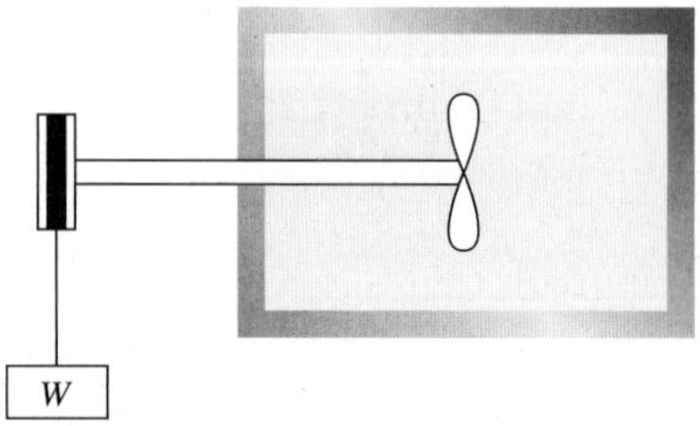

FIGURE 5.1 Paddle-wheel work.

this, as it applies to a process and a control volume, treating the second law in the same way we treated the first law.

5.2 HEAT ENGINES, HEAT PUMPS, AND REFRIGERATORS

We refer to a device operating on a cycle as a heat engine, a heat pump, or a refrigerator, depending on the objective of the particular device. If the objective of the device is to perform work, it is a **heat engine**; if its objective is to transfer heat to a body, it is a **heat pump**; if its objective is to transfer heat from a body, it is a **refrigerator**. Generically, a heat pump and a refrigerator are collectively referred to as a refrigerator. A schematic diagram of a simple heat engine is shown in Fig. 5.2.

Heat engine: *When the objective of the device is to perform work.*

Heat pump: *When the objective of the device is to transfer heat to a body.*

Refrigerator: *When the objective of the device is to transfer heat from a body.*

An engine or a refrigerator operates between two *thermal energy* **reservoirs**: entities that are capable of providing or accepting heat without changing temperatures. The atmosphere and lakes serve as *heat sinks*; furnaces, solar collectors, and burners serve as *heat sources*. Temperatures T_H and T_L identify the respective temperatures of a source and a sink.

Reservoirs: *Entities that are capable of providing or accepting heat without changing temperatures.*

The net work produced by the engine in one cycle would be equal to the net heat transfer and is a consequence of the first law (see Eq. 4.1):

$$W = Q_H - Q_L \tag{5.1}$$

If the cycle of Fig. 5.2 were reversed, a net work input would be required, as shown in Fig. 5.3. A heat pump would provide energy as heat Q_H to the warmer body (e.g., a house), and a refrigerator would extract energy as heat Q_L from the cooler body (e.g., a freezer). This work would also be given by Eq. 5.1. Here, we use magnitudes only.

The *thermal efficiency* of the heat engine and the *coefficients of performance* (abbreviated COP) of the refrigerator and the heat pump are defined as follows:

$$\eta = \frac{W}{Q_H} \qquad \text{COP}_{\text{refrig}} = \frac{Q_L}{W} \qquad \text{COP}_{\text{h.p.}} = \frac{Q_H}{W} \tag{5.2}$$

where W is the net work output of the engine or the work input to the refrigerator, Q_H is the heat transfer to or from the high-temperature reservoir, and Q_L is the heat transfer to or from the low-temperature reservoir. Each of the performance parameters represents the output divided by the input.

Key Concept: *Performance parameters represent the output divided by the input.*

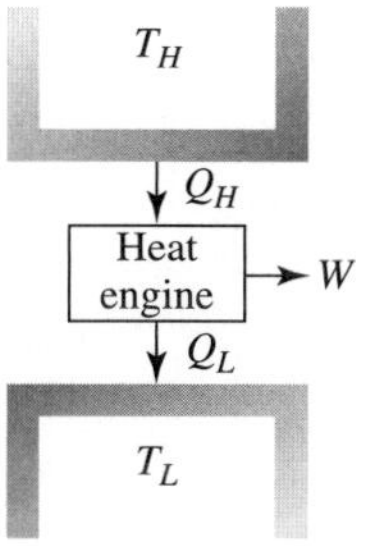

FIGURE 5.2 A heat engine.

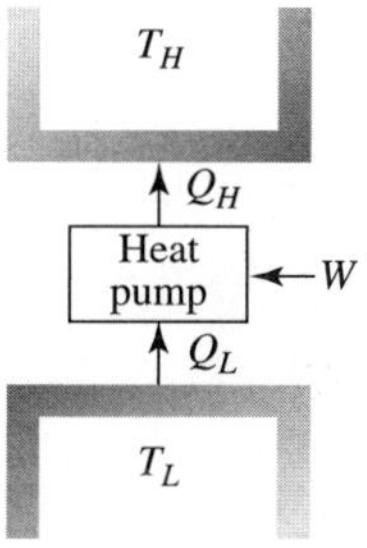

FIGURE 5.3 A refrigerator.

The second law of thermodynamics will place limits on the above measures of performance. The first law would allow a maximum of unity for the thermal efficiency and an infinite coefficient of performance. The second law, however, establishes limits that are surprisingly low, limits that cannot be exceeded regardless of the cleverness of proposed designs.

One additional note concerning heat engines is appropriate. There are devices that we will refer to as heat engines that do not strictly meet our definition; they do not operate on a thermodynamic cycle but instead exhaust the working fluid and then intake new fluid. The internal combustion engine is an example of this process. Thermal efficiency, as defined above, remains a quantity of interest for such devices.

5.3 STATEMENTS OF THE SECOND LAW OF THERMODYNAMICS

As with the other basic laws presented, we do not derive a basic law but merely observe that such a law is never violated. The second law of thermodynamics can be stated in a variety of ways. Here we present two: the *Clausius statement* and the *Kelvin–Planck statement*. Neither is presented in mathematical terms. We will, however, provide a property (entropy) of the system, which can be used to determine whether the second law is being violated for any particular situation. The first general statement of the second law is:

Clausius Statement *It is impossible to construct a device that operates in a cycle and whose sole effect is the transfer of heat from a cooler body to a hotter body.*

This statement relates to a refrigerator (or a heat pump). It states that it is impossible to construct a refrigerator that transfers energy from a cooler body to a hotter body without the input of work; this violation is shown in Fig. 5.4a. The second statement of the second law takes the following form:

Kelvin–Planck Statement *It is impossible to construct a device that operates in a cycle and produces no other effect than the production of work and the transfer of heat from a single body.*

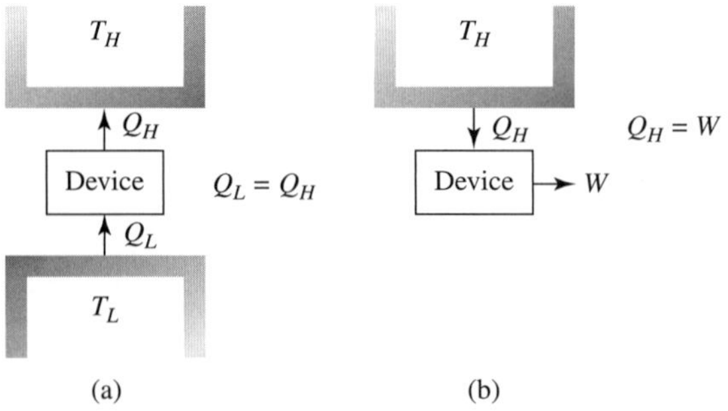

FIGURE 5.4 Violations of the second law.

In other words, it is impossible to construct a heat engine that extracts energy from a reservoir, does work, and does not transfer heat to a low-temperature reservoir. This rules out any heat engine that is 100 percent efficient, like the one shown in Fig. 5.4b.

Note that the two statements of the second law are negative statements. They are expressions of experimental observations. No experimental evidence has ever been obtained that violates either statement of the second law. It should also be noted that the two statements are equivalent. This will be demonstrated with the following example.

Example 5.1

Show that the Clausius and Kelvin–Planck statements of the second law are equivalent.

Solution

We will show that a violation of the Clausius statement implies a violation of the Kelvin–Planck statement, and vice versa, demonstrating that the two statements are equivalent. Consider the system shown in Fig. E5.1a. The device on the left transfers heat and violates the Clausius statement, since it has no work input. Let the heat engine transfer the same amount of heat Q_L. Then Q'_H is greater than Q_L by the amount W. If we simply transfer the heat Q_L directly from the engine to the device, as shown in Fig. E5.1b, there is no need for the low-temperature reservoir and the net result is a conversion of energy $(Q'_H - Q_H)$ from the high-temperature reservoir into an equivalent amount of work, a violation of the Kelvin–Planck statement of the second law. Conversely (see Problem 5.6), a violation of the Kelvin–Planck statement is equivalent to a violation of the Clausius statement.

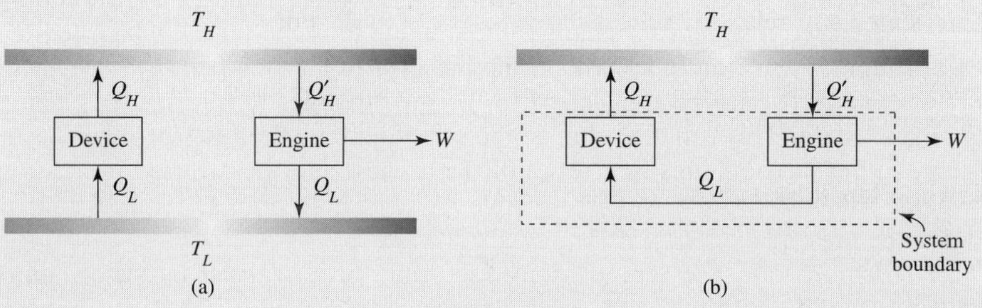

FIGURE E5.1

5.4 REVERSIBILITY

In our study of the first law, we made use of the concept of equilibrium, and we defined equilibrium and quasi-equilibrium with reference to the system only. We must now introduce the concept of *reversibility* so that we can discuss the most efficient engine that can possibly be constructed; it is an engine that operates with reversible processes only: a *reversible engine*.

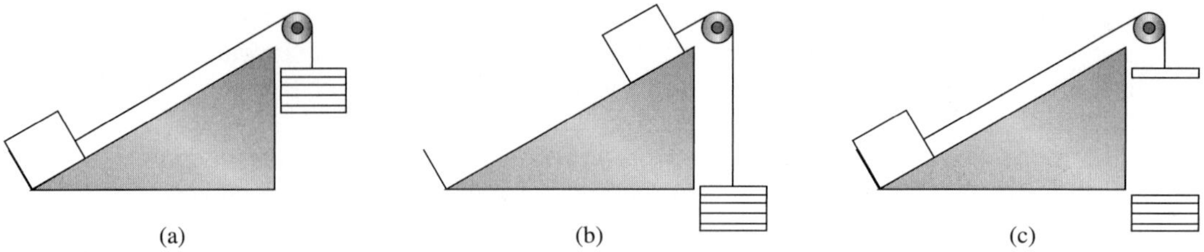

FIGURE 5.5 Irreversibility due to friction.

Reversible process: *A process that can be reversed, and in so doing, leaves no change in either the system or the surroundings.*

A **reversible process** is defined as a process that, having taken place, can be reversed, and in so doing, leaves no change in either the system or the surroundings. Observe that our definition of a reversible process refers to both the system and the surroundings. The process obviously has to be a quasi-equilibrium process; additional requirements are:

1. No friction is involved in the process.
2. Heat transfer occurs due to an infinitesimal temperature difference only.
3. Unrestrained expansion does not occur.

The mixing of different substances and combustion also lead to irreversibilities, but the above three are the ones of concern in the study of the devices of interest.

To illustrate the obvious fact that friction makes a process irreversible, consider the system of a block plus an inclined plane shown in Fig. 5.5. Weights are added until the block is raised to the position shown in part (b). Now, to return the system to its original state, some weight must be removed so that the block will slide back down the plane, as shown in part (c). Note that the surroundings have experienced a significant change; the weights must be raised, which requires a work input. Also, the block and plane are at a higher temperature due to the friction, and heat must be transferred to the surroundings to return the system to its original state. This will also change the surroundings. Because there has been a change in the surroundings as a result of the process and the reversed process, we conclude that the process was irreversible. A reversible process requires that no friction be present.

Key Concept: *Heat transfer across a finite temperature difference makes a process irreversible.*

To demonstrate the less obvious fact that heat transfer across a finite-temperature difference makes a process irreversible, consider a system composed of two blocks, one at a higher temperature than the other. Bringing the blocks together results in a heat transfer process; the surroundings are not involved in this process. To return the system to its original state, we must refrigerate the block that had its temperature raised. This will require a work input, demanded by the second law, resulting in a change in the surroundings since the surroundings must supply the work. Hence, the heat transfer across a finite temperature difference is an irreversible process. A reversible process requires that all heat transfer occurs across an infinitesimal temperature difference.

For an example of unrestrained expansion, consider the high-pressure gas contained in the cylinder of Fig. 5.6a. Pull the pin, and let the piston suddenly move to the stops shown. Note that the only work done by the gas on the sur-

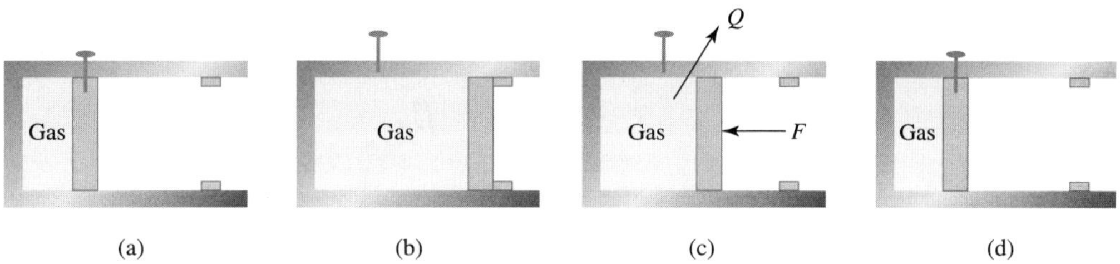

FIGURE 5.6 Unrestrained expansion.

roundings is to move the piston against atmospheric pressure. Now, to reverse this process, it is necessary to exert a force on the piston. If the force is sufficiently large, we can move the piston to its original position, shown in part (d). This will demand a considerable amount of work, to be supplied by the surroundings. In addition, the temperature will increase substantially, and this heat must be transferred to the surroundings to return the temperature to its original value. The net result is a significant change in the surroundings, a consequence of irreversibility. Unrestrained expansion cannot occur in a reversible process.

Key Concept: *Unrestrained expansion cannot occur in a reversible process.*

5.5 THE CARNOT ENGINE

The heat engine that operates the most efficiently between a high-temperature reservoir and a low-temperature reservoir is the **Carnot engine**. This is an ideal engine that uses reversible processes to form its cycle of operation; thus it is also called a *reversible engine*. We will determine the efficiency of the Carnot engine and also evaluate its reverse operation. The Carnot engine is very useful, since its efficiency establishes the maximum possible efficiency of any real engine. If the efficiency of a real engine is significantly lower than the efficiency of a Carnot engine operating between the same limits, then additional improvements may be possible.

Carnot engine: *An ideal engine that uses reversible processes to form its cycle.*

Key Concept: *The Carnot engine establishes the maximum possible efficiency of any real engine.*

The cycle associated with the Carnot engine is shown in Fig. 5.7, using an ideal gas as the working substance. It is composed of the following four reversible processes:

1 ⟶ 2: *Isothermal Expansion.* Heat is transferred reversibly from the high-temperature reservoir at the constant temperature T_H. The piston in the cylinder is withdrawn, and the volume increases.

2 ⟶ 3: *Adiabatic Reversible Expansion.* The cylinder is completely insulated, so that no heat transfer occurs during this reversible process. The piston continues to be withdrawn, with the volume increasing.

3 ⟶ 4: *Isothermal Compression.* Heat is transferred reversibly to the low-temperature reservoir at the constant temperature T_L. The piston compresses the working substance, with the volume decreasing.

4 ⟶ 1: *Adiabatic Reversible Compression.* The completely insulated cylinder allows no heat transfer during this reversible process. The pis-

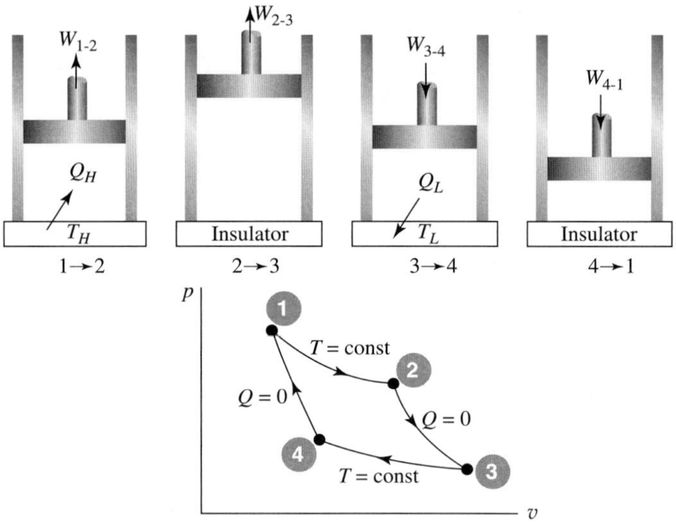

FIGURE 5.7 The Carnot cycle.

ton continues to compress the working substance until the original volume, temperature, and pressure are reached, thereby completing the cycle.

Applying the first law to the cycle, we note that

$$Q_H - Q_L = W_{\text{net}} \tag{5.3}$$

where Q_L is assumed to be a positive value for the heat transfer to the low-temperature reservoir. This allows us to write the thermal efficiency (see Eq. 5.2) for the Carnot cycle as

$$\eta = \frac{Q_H - Q_L}{Q_H} = 1 - \frac{Q_L}{Q_H} \tag{5.4}$$

The following examples and a home problem will be used to prove the following three postulates:

Postulate 1 *It is impossible to construct an engine, operating between two given temperature reservoirs, that is more efficient than the Carnot engine.*

Postulate 2 *The efficiency of a Carnot engine is not dependent on the working substance used or any particular design feature of the engine.*

Postulate 3 *All reversible engines, operating between two given temperature reservoirs, have the same efficiency as a Carnot engine operating between the same two temperature reservoirs.*

Example 5.2

Show that the efficiency of a Carnot engine is the maximum possible efficiency.

Solution

Assume that an engine exists, operating between two reservoirs, that has an efficiency greater than that of a Carnot engine; also, assume that a Carnot engine operates as a refrigerator between the same two reservoirs, as sketched in Fig. E5.2a. Let the heat transferred from the high-temperature reservoir to the engine be equal to the heat rejected by the refrigerator. Then the work produced by the engine will be greater than the work required by the refrigerator (that is, $Q_L' < Q_L$), since the efficiency of the engine is greater than that of a Carnot engine. Now, our system can be organized as shown in Fig. E5.2b. The engine drives the refrigerator using the rejected heat from the refrigerator. But there is some net work ($W' - W$) that leaves the system. The net result is the conversion of energy from a single reservoir into work: a violation of the second law. Thus, the Carnot engine is the most efficient engine operating between two particular reservoirs.

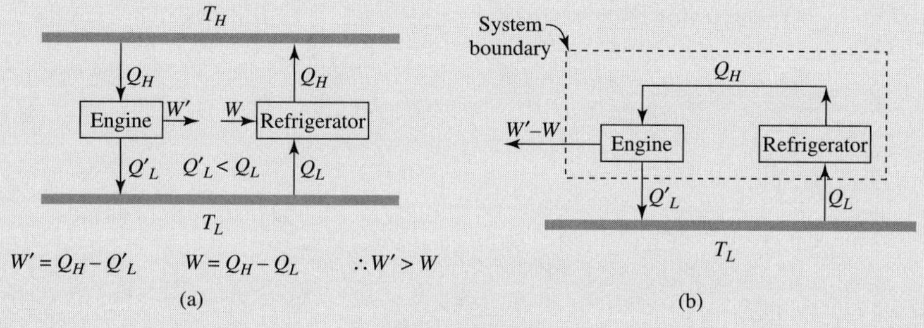

FIGURE E5.2

Example 5.3

Show that the efficiency of a Carnot engine operating between two reservoirs is independent of the working substance used by the engine.

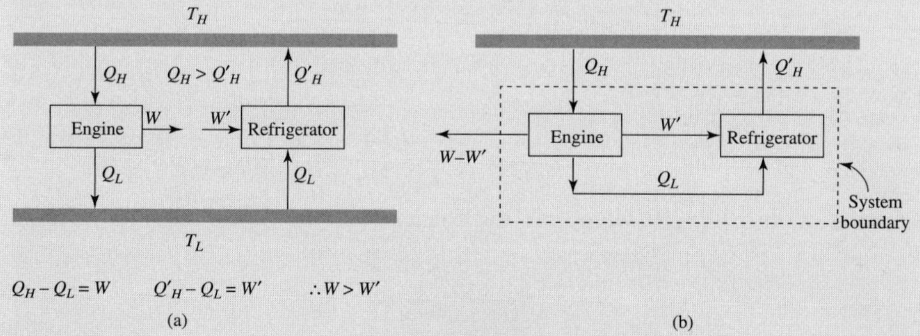

FIGURE E5.3

(continued)

Solution

Suppose that a Carnot engine drives a Carnot refrigerator as shown in Fig. E5.3a. Let the heat rejected by the engine equal the heat required by the refrigerator. Suppose the working fluid in the engine results in Q_H being greater than Q'_H; then W would be greater than W' (a consequence of the first law), and we would have the equivalent system shown in Fig. E5.3b. The net result is a transfer of heat $(Q_H - Q'_H)$ from a single reservoir and the production of work: a clear violation of the second law. Thus, the efficiency of a Carnot engine is not dependent on the working substance.

5.6 CARNOT EFFICIENCY

Since the efficiency of a Carnot engine is dependent only on the two reservoir temperatures, the objective of this article will be to determine that relationship. We will assume the working substance to be an ideal gas (refer to Postulate 2) and simply perform the required calculations for the four processes of Fig. 5.7.

The heat transfer for each of the four processes is as follows:

$$\begin{aligned}
&\mathbf{1 \longrightarrow 2:}\ Q_H = W_{1\text{-}2} = \int_{\mathcal{V}_1}^{\mathcal{V}_2} p\,d\mathcal{V} = mRT_H \ln \frac{\mathcal{V}_2}{\mathcal{V}_1} \\
&\mathbf{2 \longrightarrow 3:}\ Q_{2\text{-}3} = 0 \\
&\mathbf{3 \longrightarrow 4:}\ Q_L = -W_{3\text{-}4} = -\int_{\mathcal{V}_3}^{\mathcal{V}_4} p\,d\mathcal{V} = -mRT_L \ln \frac{\mathcal{V}_4}{\mathcal{V}_3} \\
&\mathbf{4 \longrightarrow 1:}\ Q_{4\text{-}1} = 0
\end{aligned} \tag{5.5}$$

Note that we want Q_L to be a positive quantity, as in the thermal efficiency relationship; hence, the negative sign. The thermal efficiency, using Eq. 5.4, is then

$$\eta = 1 - \frac{Q_L}{Q_H} = 1 + \frac{T_L}{T_H} \frac{\ln \mathcal{V}_4/\mathcal{V}_3}{\ln \mathcal{V}_2/\mathcal{V}_1} \tag{5.6}$$

During the reversible adiabatic processes $2 \rightarrow 3$ and $4 \rightarrow 1$, Eq. 4.49 provides

$$\frac{T_L}{T_H} = \left(\frac{\mathcal{V}_2}{\mathcal{V}_3}\right)^{k-1} \qquad \frac{T_L}{T_H} = \left(\frac{\mathcal{V}_1}{\mathcal{V}_4}\right)^{k-1} \tag{5.7}$$

Thus we see that

$$\frac{\mathcal{V}_2}{\mathcal{V}_3} = \frac{\mathcal{V}_1}{\mathcal{V}_4} \quad \text{or} \quad \frac{\mathcal{V}_4}{\mathcal{V}_3} = \frac{\mathcal{V}_1}{\mathcal{V}_2} \tag{5.8}$$

Substituting into Eq. 5.5, we obtain the efficiency of the Carnot engine (recognizing that $\ln \mathcal{V}_2/\mathcal{V}_1 = -\ln \mathcal{V}_1/\mathcal{V}_2$):

Key Concept: *The thermal efficiency of a Carnot engine is dependent only on the absolute temperatures of the reservoirs.*

$$\eta = 1 - \frac{T_L}{T_H} \tag{5.9}$$

We have simply replaced Q_L/Q_H with T_L/T_H. We can make this replacement for all reversible engines or refrigerators. We see that the thermal efficiency of a

Carnot engine is dependent only on the high and low absolute temperature of the reservoirs. The fact that we used an ideal gas to perform the calculations is not important, since we have shown that Carnot efficiency is independent of the working substance. Consequently, the relationship (Eq. 5.9) is applicable for all working substances or for all reversible engines, regardless of the particular design characteristics.

The Carnot engine, when operated in reverse, becomes a heat pump or a refrigerator, depending on the desired heat transfer. The coefficient of performance for a Carnot heat pump becomes

$$\mathrm{COP} = \frac{Q_H}{W_{\mathrm{net}}} = \frac{Q_H}{Q_H - Q_L} = \frac{1}{1 - T_L/T_H} \tag{5.10}$$

The coefficient of performance for a Carnot refrigerator takes the form

$$\mathrm{COP} = \frac{Q_L}{W_{\mathrm{net}}} = \frac{Q_L}{Q_H - Q_L} = \frac{1}{T_H/T_L - 1} \tag{5.11}$$

The above measures of performance set limits that real devices can only approach. The reversible cycles assumed are obviously unrealistic, but the fact that we have limits that we know cannot be exceeded is often very helpful in evaluating proposed designs and determining the direction for further effort.

Key Concept: *The fact that we have limits that we know cannot be exceeded is very helpful in evaluating proposed designs.*

Example 5.4

A Carnot engine operates between two temperature reservoirs maintained at 200°C and 20°C, respectively. If the desired output of the engine is 15 kW, as shown in Fig. E5.4, determine the heat transfer from the high-temperature reservoir and the heat transfer to the low-temperature reservoir.

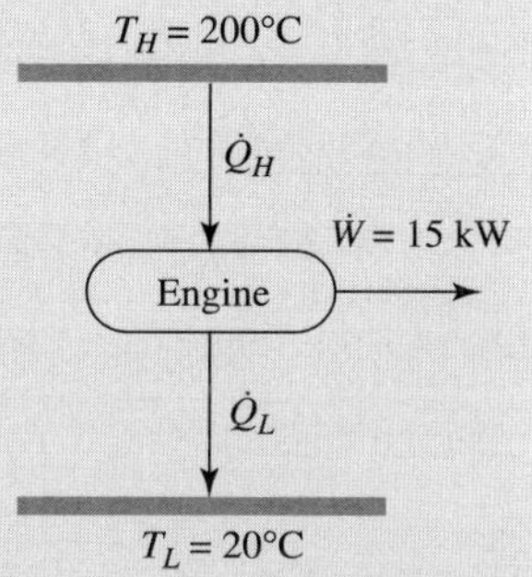

FIGURE E5.4

Solution

The efficiency of a Carnot engine is given by

$$\eta = \frac{\dot{W}}{\dot{Q}_L} = 1 - \frac{T_L}{T_H}$$

This gives, converting the temperatures to absolute temperatures,

$$\dot{Q}_H = \frac{\dot{W}}{1 - T_L/T_H} = \frac{15}{1 - 293/473} = 39.42 \text{ kW}$$

Using the first law, we have

$$\dot{Q}_L = \dot{Q}_H - \dot{W} = 39.42 - 15 = 24.42 \text{ kW}$$

Example 5.5

A refrigeration unit is cooling a space to −5°C by rejecting energy to the atmosphere at 20°C. It is desired to reduce the temperature in the refrigerated space to −25°C. Calculate the minimum percentage increase in work required, by assuming a Carnot refrigerator, for the same amount of energy removed.

Solution

For a Carnot refrigerator, we know that

$$\text{COP} = \frac{Q_L}{W} = \frac{1}{T_H/T_L - 1}$$

For the first situation, we have

$$W_1 = Q_L\left(\frac{T_L}{T_H} - 1\right) = Q_L\left(\frac{293}{268} - 1\right) = 0.0933 Q_L$$

For the second situation,

$$W_2 = Q_L\left(\frac{293}{248} - 1\right) = 0.181 Q_L$$

The percentage increase in work is then

$$\frac{W_2 - W_1}{W_1} \times 100 = \frac{0.181 Q_L - 0.0933 Q_L}{0.0933 Q_L} \times 100 = 94.0\%$$

Note the large increase in energy required to reduce the temperature in a refrigerated space. This is a minimum percentage increase, since we have assumed an ideal refrigerator.

Example 5.6

A Carnot engine operates with air, using the cycle shown in Fig. E5.6. Determine the thermal efficiency and the work output for each cycle of operation.

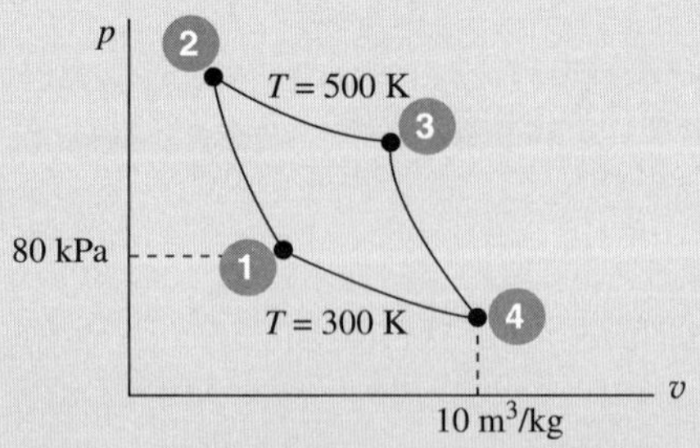

FIGURE E5.6

(continued)

Solution

The thermal efficiency is found to be

$$\eta = 1 - \frac{T_L}{T_H} = 1 - \frac{300}{500} = 0.4 \quad \text{or} \quad 40\%$$

To find the work output, we can determine the heat added during the constant temperature expansion and determine w from $\eta = W/Q_H = w/q_H$. We find q_H from the first law using $\Delta u = 0$:

$$q_H = w_{2\text{-}3} = \int p\,dv = RT_H \int_{v_2}^{v_3} \frac{dv}{v} = RT_H \ln \frac{v_3}{v_2}$$

To find v_2, first we must find v_1:

$$v_1 = \frac{RT_1}{p_1} = \frac{287 \times 300}{80\,000} = 1.076 \text{ m}^3/\text{kg}$$

Using Eq. 4.49, we have

$$v_2 = v_1 \left(\frac{T_1}{T_2}\right)^{1/(k-1)} = 1.076 \times \left(\frac{300}{500}\right)^{1/0.4} = 0.300 \text{ m}^3/\text{kg}$$

Likewise,

$$v_3 = v_4 \left(\frac{T_4}{T_3}\right)^{1/(k-1)} = 10 \times \left(\frac{300}{500}\right)^{1/0.4} = 2.789 \text{ m}^3/\text{kg}$$

Finally, the work for each cycle is

$$w = \eta q_H = 0.4 \times 0.287 \times 500 \ln 2.789/0.300 = 128 \text{ kJ/kg}$$

5.7 ENTROPY

To allow us to apply the second law of thermodynamics to a process, we will identify a property called entropy. This will parallel our discussion of the first law; first we stated the first law for a cycle and then derived a relationship for a process.

Consider the reversible Carnot engine operating on a cycle consisting of the processes described in Sec. 5.5. The quantity $\oint \delta Q/T$ is the cyclic integral of the heat transfer divided by the absolute temperature at which the heat transfer occurs. Since the temperature T_H is constant during the heat transfer Q_H, and T_L is constant during heat transfer Q_L, the integral is given by

$$\oint \frac{\delta Q}{T} = \frac{Q_H}{T_H} - \frac{Q_L}{T_L} \tag{5.12}$$

where the heat Q_L leaving the Carnot engine is considered to be positive. Using Eqs. 5.4 and 5.9 we see that, for the Carnot cycle,

$$\frac{Q_L}{Q_H} = \frac{T_L}{T_H} \quad \text{or} \quad \frac{Q_H}{T_H} = \frac{Q_L}{T_L} \tag{5.13}$$

Substituting this into Eq. 5.12, we find the interesting result

$$\oint \frac{\delta Q}{T} = 0 \tag{5.14}$$

Thus, the quantity $\delta Q/T$ is a perfect differential, since its cyclic integral is zero. We let this perfect differential be denoted by dS, where S represents a scalar function that depends only on the state of the system. This, in fact, was our definition of a property of a system. We shall call this extensive property *entropy*; its differential is given by

$$dS = \left.\frac{\delta Q}{T}\right|_{\text{rev}} \tag{5.15}$$

where the subscript "rev" emphasizes the reversibility of the process. This can be integrated for a process to give

$$\Delta S = \int_1^2 \left.\frac{\delta Q}{T}\right|_{\text{rev}} \tag{5.16}$$

Key Concept: *If a process is irreversible but adiabatic, it is not generally true that $\Delta S = 0$.*

From the above equation, we see that the entropy change for a reversible process can be either positive or negative, depending on whether energy is added to or extracted from the system during the heat transfer process. For a reversible adiabatic process ($Q = 0$), the entropy change is zero. If the process is irreversible but adiabatic, it is not generally true that $\Delta S = 0$.

We often sketch a temperature-entropy diagram for cycles or processes of interest. The Carnot cycle provides a simple display when plotting temperature versus entropy; it is shown in Fig. 5.8. The change in entropy for the first isothermal process from state 1 to state 2 is

$$S_2 - S_1 = \int_1^2 \frac{\delta Q}{T} = \frac{Q_H}{T_H} \tag{5.17}$$

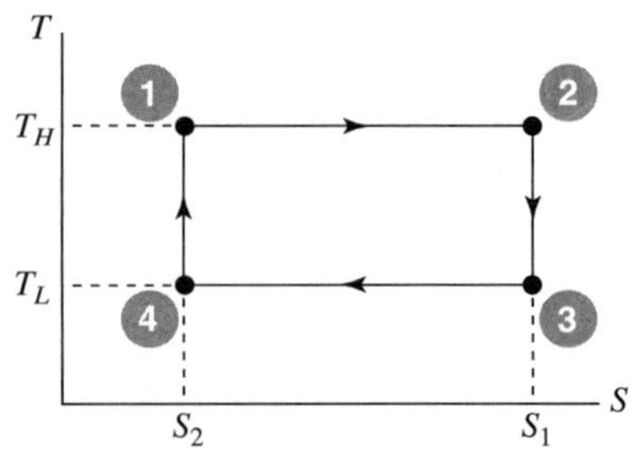

FIGURE 5.8 The Carnot cycle.

The entropy change for the reversible adiabatic process from state 2 to state 3 is zero. For the isothermal process from state 3 to state 4, the entropy change is negative that of the first process from state 1 to state 2; the process from state 4 to state 1 is also a reversible adiabatic process and is accompanied with a zero entropy change.

The heat transfer during a reversible process can be expressed in differential form (see Eq. 5.15) as

$$\delta Q = T\,dS \qquad \text{or} \qquad Q = \int_1^2 T\,dS \tag{5.18}$$

Hence, the area under the curve in the T-S diagram represents the heat transfer during any reversible process. The rectangular area in Fig. 5.8 thus represents the net heat transfer during the Carnot cycle. Since the heat transfer is equal to the work done for a cycle, the area also represents the net work accomplished by the system during the cycle. For this Carnot cycle, $Q_{\text{net}} = W_{\text{net}} = \Delta T \Delta S$.

Key Concept: *The area under the T-S diagram represents the heat transfer during any reversible process.*

The first law of thermodynamics for a reversible infinitesimal change (refer to Eq. 4.4a) becomes, using Eq. 5.18,

$$T\,dS - p\,d\mathcal{V} = dU \tag{5.19}$$

This is an important relationship in our study of simple systems. We arrived at it assuming a reversible process. However, since it involves only properties of the system, it holds for any process including any irreversible process. If we have an irreversible process, in general, $\delta W \neq p\,d\mathcal{V}$ and $\delta Q \neq T\,dS$, but Eq. 5.19 still holds as a relationship between the properties since changes in properties do not depend on the process. Dividing by the mass, we have

$$T\,ds - p\,dv = du \tag{5.20}$$

where the specific entropy is defined to be

$$s = \frac{S}{m} \tag{5.21}$$

To relate the entropy change to the enthalpy change, we differentiate Eq. 4.12 and obtain

$$dh = du + p\,dv + v\,dp \tag{5.22}$$

Substituting into Eq. 5.20 for du, we have

$$T\,ds = dh - v\,dp \tag{5.23}$$

Equations 5.20 and 5.23 will be used in subsequent sections of our study of thermodynamics for various reversible and irreversible processes.

5.8 ENTROPY CHANGE FOR AN IDEAL GAS WITH CONSTANT SPECIFIC HEATS

Assuming an ideal gas, Eq. 5.20 becomes

$$ds = \frac{du}{T} + \frac{p\,dv}{T} = c_v \frac{dT}{T} + R\frac{dv}{v} \tag{5.24}$$

where we have used

$$du = c_v\,dT \qquad pv = RT \tag{5.25}$$

Equation 5.24 is integrated, assuming constant specific heat, to yield

$$s_2 - s_1 = c_v \ln \frac{T_2}{T_1} + R \ln \frac{v_2}{v_1} \tag{5.26}$$

Similarly, Eq. 5.23 is rearranged and integrated to give

$$s_2 - s_1 = c_p \ln \frac{T_2}{T_1} - R \ln \frac{p_2}{p_1} \tag{5.27}$$

Note, again, that the above equations were developed assuming a reversible process; however, they relate the change in entropy to other thermodynamic properties at the two end states. Since the change of a property is independent of the process used in going from one state to another, the above relationships hold for any process, reversible or irreversible, providing the working substance can be approximated by an ideal gas with constant specific heats.

If the entropy change is zero, as in a reversible adiabatic process, Eqs. 5.26 and 5.27 can be used to obtain

$$\frac{T_2}{T_1} = \left(\frac{v_1}{v_2}\right)^{k-1} \qquad \frac{T_2}{T_1} = \left(\frac{p_2}{p_1}\right)^{(k-1)/k} \tag{5.28}$$

These two equations are combined to give

$$\frac{p_2}{p_1} = \left(\frac{v_1}{v_2}\right)^{k} \tag{5.29}$$

Isentropic process: *A quasi-equilibrium adiabatic process.*

These are, of course, identical to the equations obtained in Chapter 4 for an ideal gas undergoing a quasi-equilibrium adiabatic process. We now refer to such a process as an **isentropic process**.

Example 5.7

Air is contained in an insulated, rigid volume at 20°C and 200 kPa. A paddle wheel, inserted in the volume, does 720 kJ of work on the air. If the volume is 2 m^3, calculate the entropy increase assuming constant specific heats.

Solution

To determine the final state of the process, we use the energy equation, assuming zero heat transfer. We have $-W = \Delta U = mc_v\Delta T$. The mass m is found from the ideal-gas equation to be

$$m = \frac{pV}{RT} = \frac{200 \times 2}{0.287 \times 293} = 4.76 \text{ kg}$$

The first law, taking the paddle-wheel work as negative, is then

$$-W = mc_v\Delta T$$

$$-(-720) = 4.76 \times 0.717 \times (T_2 - 293) \qquad \therefore\ T_2 = 504.0 \text{ K}$$

Using Eq. 5.26 for this constant-volume process there results

$$\Delta S = mc_v \ln\frac{T_2}{T_1} = 4.76 \times 0.717 \times \ln\frac{504}{293} = 1.851 \text{ kJ/K}$$

Example 5.8

After a combustion process in a cylinder, the pressure is 1200 kPa and the temperature is 350°C. The gases are expanded to 140 kPa with a reversible adiabatic process. Calculate the work done by the gases, assuming they can be approximated by air with constant specific heats.

Solution

The first law can be used, with zero heat transfer, to give $-w = \Delta u = c_v(T_2 - T_1)$. The temperature T_2 is found from Eq. 5.28 to be

$$T_2 = T_1\left(\frac{p_2}{p_1}\right)^{(k-1)/k} = 623\left(\frac{140}{1200}\right)^{0.4/1.4} = 337 \text{ K}$$

This allows the specific work to be calculated:

$$w = c_v(T_2 - T_1) = 0.717 \times (623 - 337) = 205 \text{ kJ/kg}$$

Since neither the mass nor the volume was given, only the specific work could be determined.

5.9 ENTROPY CHANGE FOR AN IDEAL GAS WITH VARIABLE SPECIFIC HEATS

If the specific heats for an ideal gas cannot be assumed constant over a particular temperature range, we return to Eq. 5.23 and write

$$ds = \frac{dh}{T} - \frac{v\,dp}{T} = \frac{c_p}{T}dT - \frac{R}{p}dp \tag{5.30}$$

The gas constant R can be removed from the integral, but $c_p = c_p(T)$ cannot be removed. Hence, we integrate Eq. 5.30 and obtain

$$s_2 - s_1 = \int_{T_1}^{T_2} \frac{c_p}{T}dT - R\ln\frac{p_2}{p_1} \tag{5.31}$$

The integral in the above equation depends only on temperature, and we can evaluate its magnitude from the gas tables in the Appendix. It is found, using the tabulated function ϕ from Appendix F, to be

$$\phi_2 - \phi_1 = \int_{T_1}^{T_2} \frac{c_p}{T}dT \tag{5.32}$$

Thus, the entropy change is (in some textbooks s^o is used rather than ϕ)

$$s_2 - s_1 = \phi_2 - \phi_1 - R\ln\frac{p_2}{p_1} \tag{5.33}$$

This more exact expression for the entropy change is used only when improved accuracy is desired.

For an isentropic process, we cannot use Eqs. 5.28 and 5.29 if the specific heats are not constant. However, we can use Eq. 5.33 and obtain, for an isentropic process,

$$\frac{p_2}{p_1} = \exp\left(\frac{\phi_2 - \phi_1}{R}\right) = \frac{\exp(\phi_2/R)}{\exp(\phi_1/R)} = \frac{f(T_2)}{f(T_1)} \tag{5.34}$$

Thus, we define a *relative pressure* p_r, which depends only on the temperature, as

$$p_r = e^{\phi/R} \tag{5.35}$$

It[1] is included as an entry in the gas tables in the Appendix. The pressure ratio for an isentropic process is then

$$\frac{p_2}{p_1} = \frac{p_{r2}}{p_{r1}} \tag{5.36}$$

The volume ratio can be found using the ideal-gas equation of state. It is

$$\frac{v_2}{v_1} = \frac{p_1}{p_2}\frac{T_2}{T_1} \tag{5.37}$$

[1] The values in the ideal-gas tables are actually a constant multiple of this value in order to create a more convenient range of numbers. The actual value is not of interest, only the ratio.

where we would assume an isentropic process when using the relative pressure ratio. Consequently, we define a *relative specific volume* v_r, dependent solely on the temperature, as

$$v_r = \frac{T}{p_r} \tag{5.38}$$

Using its value from the Appendix gas tables, we find the specific-volume ratio for an isentropic process; it is

$$\frac{v_2}{v_1} = \frac{v_{r2}}{v_{r1}} \tag{5.39}$$

With the entries from the gas tables, we can perform the calculations required when working problems involving an ideal gas with variable specific heats.

Example 5.9

Repeat Example 5.7 assuming variable specific heats.

Solution

Using the gas tables, we write the first law as $-W = \Delta U = m(u_2 - u_1)$. The mass is found from the ideal-gas equation to be

$$m = \frac{pV}{RT} = \frac{200 \times 2}{0.287 \times 293} = 4.76 \text{ kg}$$

The first law is then written as

$$u_2 = -\frac{W}{m} + u_1 = -\frac{-720}{4.76} + 209.1 = 360.4 \text{ kJ/kg}$$

where u_1 is found at 293 K in the gas tables by interpolation. Now, using this value for u_2, we can interpolate to find

$$T_2 = 501.2 \text{ K} \qquad \phi_2 = 2.222$$

The value for ϕ_1 is interpolated to be $\phi_1 = 1.678$. The pressure at state 2 is found using the ideal-gas equation for our constant-volume process:

$$\frac{p_2}{T_2} = \frac{p_1}{T_1} \qquad p_2 = p_1 \frac{T_2}{T_1} = 200 \times \frac{501.2}{293} = 342.1 \text{ kPa}$$

Finally, the entropy change is

$$\Delta S = m\left(\phi_2 - \phi_1 - R \ln \frac{p_2}{p_1}\right)$$

$$= 4.76 \times \left(2.222 - 1.678 - 0.287 \ln \frac{342.1}{200}\right) = 1.856 \text{ kJ/K}$$

The approximate result of Example 5.7 is seen to be less than 0.3% in error. If T_2 were substantially greater than 500 K, the error could become much more significant.

Example 5.10

After a combustion process in a cylinder, the pressure is 1200 kPa and the temperature is 350°C. The gases are expanded to 140 kPa in a reversible, adiabatic process. Calculate the work done by the gases, assuming they can be approximated by air with variable specific heats.

Solution

At 623 K, the relative pressure p_{r1} is interpolated to be

$$p_{r1} = \frac{3}{20}(20.64 - 18.36) + 18.36 = 18.70$$

For an isentropic process,

$$p_{r2} = p_{r1}\frac{p_2}{p_1} = 18.70 \times \frac{140}{1200} = 2.182$$

With this value for the relative pressure at state 2,

$$T_2 = \frac{2.182 - 2.149}{2.626 - 2.149} \times 20 + 340 = 341 \text{ K}$$

The work is found from the first law to be

$$\begin{aligned} w &= u_1 - u_2 \\ &= \left[\frac{3}{20}(465.5 - 450.1) + 450.1\right] - \left[\frac{2.182 - 2.149}{2.626 - 2.149}(257.2 - 242.8) + 242.8\right] \\ &= 208.6 \text{ kJ/kg} \end{aligned}$$

The approximate result of Example 5.8 is seen to have an error less than 1.5%. The temperature difference $(T_2 - T_1)$ is not very large, hence the small error.

5.10 ENTROPY CHANGE FOR SUBSTANCES SUCH AS STEAM, SOLIDS, AND LIQUIDS

The entropy change has been found for an ideal gas with constant specific heats and for an ideal gas with variable specific heats. For pure substances, such as steam, entropy is included as an entry in the Appendix tables. In the quality region, it is found using the relation

$$s = s_f + x s_{fg} \tag{5.40}$$

Key Concept: *Only the change in entropy is of interest; the arbitrary datum for entropy is of no consequence.*

Note that the entropy of saturated-liquid water at 0°C is arbitrarily set equal to zero. It is only the change in entropy that is of interest; hence, this arbitrary datum for entropy is of no consequence. In the superheated region, it is tabulated as a function of temperature and pressure along with the other properties.

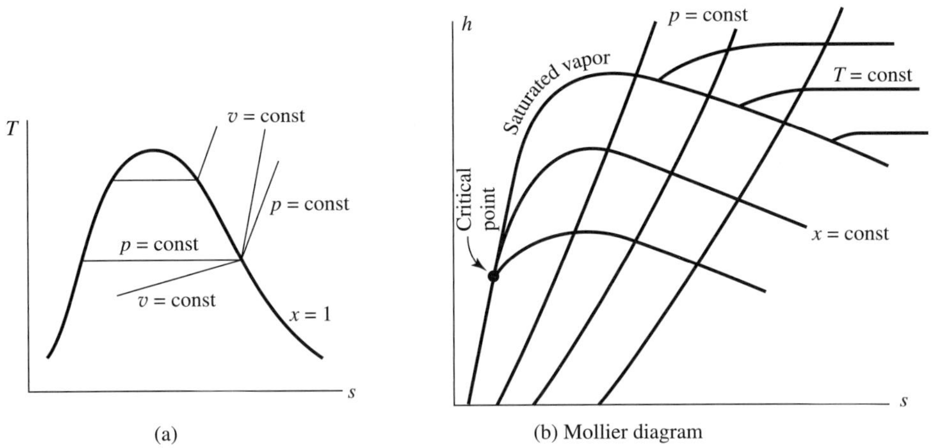

FIGURE 5.9 The *T*-*s* and *h*-*s* diagrams for steam.

For a compressed liquid, it is included as an entry in Appendix Table C.4, the compressed liquid table, or it can be approximated by the saturated liquid values s_f at the given temperature (ignoring the pressure). From the compressed liquid table at 10 MPa and 100°C, $s = 1.30$ kJ/kg·K, and from the saturated steam table at 100°C, $s_f = 1.31$ kJ/kg·K; this is an insignificant difference.

The temperature-entropy diagram is of particular interest and is often sketched during the problem solution. A *T*-*s* diagram is sketched in Fig. 5.9a; for steam, it is essentially symmetric about the critical point. Note that the high-pressure lines in the compressed liquid region are indistinguishable from the saturated liquid line. It is often helpful to visualize a process on a *T*-*s* diagram, since such a diagram illustrates assumptions regarding irreversibilities.

In addition to a *T*-*s* diagram, an *h*-*s* diagram, which is called a *Mollier diagram*, is sometimes used when solving particular types of problems. The general shape of an *h*-*s* diagram is sketched in Fig. 5.9b.

For a solid or a liquid, the entropy change can be found quite easily if we can assume the specific heat to be constant. Returning to Eq. 5.20, we can write, assuming the solid or liquid to be incompressible so that $dv = 0$,

$$T\,ds = du = c\,dT \tag{5.41}$$

where we have dropped the subscript on the specific heat since for solids and liquids $c_p \cong c_v$. Tables usually list values for c_p; these are assumed to be equal to c. Assuming a constant specific heat, we find that

$$\Delta s = \int_{T_1}^{T_2} c\frac{dT}{T} = c\ln\frac{T_2}{T_1} \tag{5.42}$$

If the specific heat is a known function of temperature, the integration can be performed. Specific heats for solids and liquids are listed in Appendix Table B.4.

Example 5.11

Steam is contained in a rigid container at an initial pressure of 100 psia and 600°F. The pressure is reduced to 10 psia by removing energy via heat transfer. Calculate the entropy change and the heat transfer and sketch a *T-s* diagram. Use (a) the steam tables and (b) TK Solver.

Solution

(a) From the steam tables, $v_1 = v_2 = 6.216$ ft^3/lbm. State 2 is in the quality region. Using the above value for v_2, the quality is found as follows:

$$v_2 = [v_f + x(v_g - v_f)]_2$$

$$6.216 = 0.0166 = x_2(38.42 - 0.0166) \qquad \therefore\ x_2 = 0.1614$$

The entropy at state 2 is $s_2 = [s_f + x(s_g - s_f)]_2 = 0.2836 + 0.1614 \times 1.5041 = 0.5264$ Btu/lbm-°R. The entropy change is then

$$\Delta s = s_2 - s_1 = 0.5264 - 1.7582 = -1.232 \text{ Btu/lbm-°R}$$

The heat transfer is found from the first law using $w = 0$ and $u_1 = 1214.2$ Btu/lbm:

$$q = u_2 - u_1 = (161.2 + 0.1614 \times 911.01) - 1214.2 = -906 \text{ Btu/lbm}$$

The process is displayed in the *T-s* diagram shown in Fig. E5.11.

(b) Using TK Solver at 100 psia and 600°F, we find $v_1 = 6.2104$ ft^3/lbm, $s_1 =$ 1.759 Btu/lbm-°R, and $h_1 = 1329.1$ Btu/lbm. At $p_2 = 10$ psia and $v_2 = 6.2104$ ft^3/lbm, we find $s_2 = 0.52615$ Btu/lbm-°R and $h_2 = 319.6$ Btu/lbm. Finally,

$$\Delta s = s_2 - s_1 = 0.52615 - 1.759 = -1.233 \text{ Btu/lbm-°R}$$

$$q = u_2 - u_1 = h_2 - h_1 - (p_2 - p_1)v_1$$

$$= 319.6 - 1329.1 - (10 - 100) \times 144 \times 6.2104/778 = -906 \text{ Btu/lbm}$$

The factor 778 converts ft-lbf to Btu.

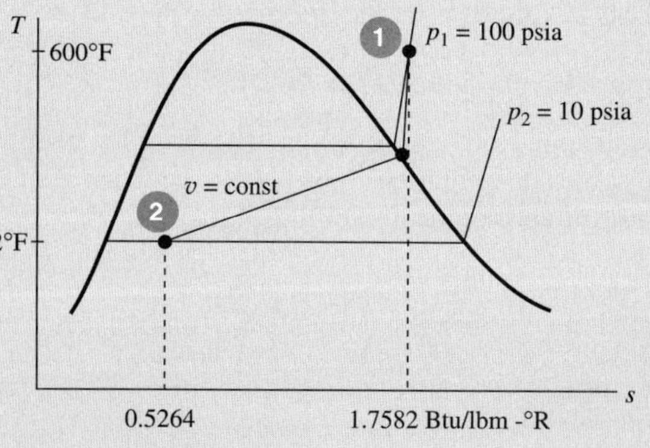

FIGURE E5.11

5.11 THE INEQUALITY OF CLAUSIUS

The Carnot cycle is a reversible cycle and produces work which we will refer to as W_{rev}. Consider an irreversible cycle operating between the same two reservoirs, shown in Fig. 5.10. Obviously, since the Carnot cycle possesses the maximum possible efficiency, the efficiency of the irreversible cycle must be less than that of the Carnot cycle. In other words, for the same amount of heat addition Q_H, we must have

$$W_{irr} < W_{rev} \tag{5.43}$$

From the first law applied to a cycle ($W = Q_H - Q_L$), we see that, assuming that $(Q_H)_{irr}$ and $(Q_H)_{rev}$ are the same,

$$(Q_L)_{rev} < (Q_L)_{irr} \tag{5.44}$$

This requires, referring to Eqs. 5.12 and 5.14,

$$\oint \left(\frac{\delta Q}{T}\right)_{irr} < 0 \tag{5.45}$$

since the above integral for a reversible cycle is zero.

If we were considering an irreversible refrigerator rather than an engine, we would require more work for the same amount of refrigeration Q_L. By applying the first law to refrigerators, we would arrive at the same inequality as that of Eq. 5.45. Hence, for all cycles, reversible or irreversible, we can write

$$\oint \left(\frac{\delta Q}{T}\right) \leq 0 \tag{5.46}$$

This is known as *the inequality of Clausius*. It is a consequence of the second law of thermodynamics. It will be used to establish entropy relationships for real processes.

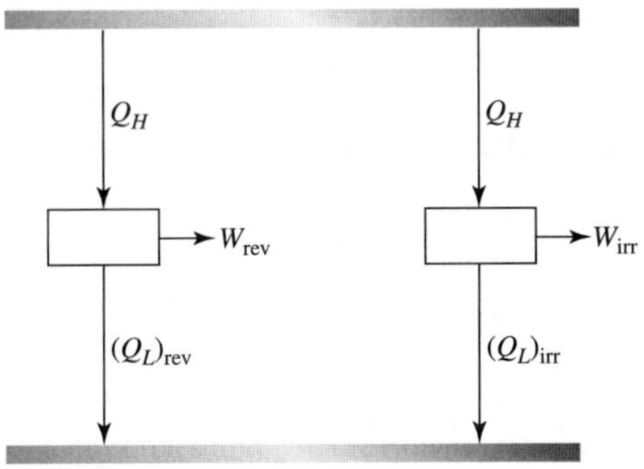

FIGURE 5.10 A reversible and an irreversible engine operating between two reservoirs.

Example 5.12

It is proposed to operate a simple steam power plant, as shown in Fig. E5.12. The water is completely vaporized in the boiler, so that the heat transfer Q_B takes place at constant temperature T_B. Does this proposal comply with the inequality of Clausius? Assume no heat transfer occurs from the pump or the turbine.

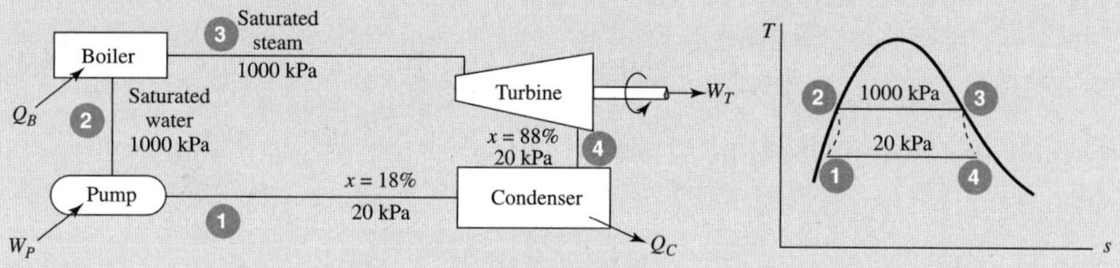

FIGURE E5.12

Solution

The quantity that we seek is $\oint \delta Q/T$. Since the proposed heat transfer occurs at constant temperature, this takes the form

$$\oint \frac{\delta Q}{T} = \frac{Q_B}{T_B} - \frac{Q_C}{T_C}$$

From the steam tables, we can find the following for each kilogram of water ($m = 1$ kg):

$$T_B = 179.9°\text{C} \qquad T_C = 60.1°\text{C}$$

$$Q_B = m(h_3 - h_2) = 2778 - 763 = 2015 \text{ kJ}$$

$$Q_C = m(h_4 - h_1) = (251 + 0.88 \times 2358) - (251 + 0.18 \times 2358) = 1651 \text{ kJ}$$

Thus, we have

$$\oint \frac{\delta Q}{T} = \frac{2015}{452.9} - \frac{1651}{333.1} = -0.507 \text{ kJ/K}$$

This is negative, so the proposed power plant satisfies the inequality of Clausius.

5.12 ENTROPY CHANGE FOR AN IRREVERSIBLE PROCESS

Consider a cycle to be composed of two reversible processes, as shown in Fig. 5.11. Suppose that we can also return from state 2 to state 1 along the irreversible process marked by path C. For the system experiencing the reversible cycle we have

$$\underbrace{\int_1^2 \frac{\delta Q}{T}}_{\text{along } A} + \underbrace{\int_2^1 \frac{\delta Q}{T}}_{\text{along } B} = 0 \tag{5.47a}$$

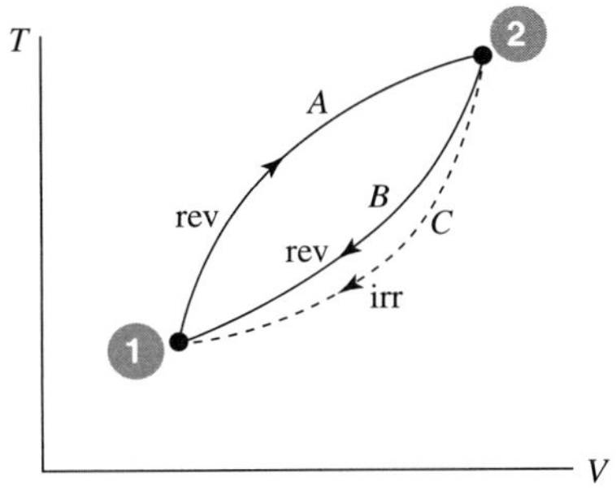

FIGURE 5.11 A cycle that includes the possibility of an irreversible process.

For the system experiencing the cycle involving the irreversible process, the Clausius inequality demands that

$$\underset{\text{along } A}{\int_1^2 \frac{\delta Q}{T}} + \underset{\text{along } C}{\int_2^1 \frac{\delta Q}{T}} < 0 \tag{5.47b}$$

Subtracting Eq. 5.47a from Eq. 5.47b, there results

$$\underset{\text{along } B}{\int_2^1 \frac{\delta Q}{T}} > \underset{\text{along } C}{\int_2^1 \frac{\delta Q}{T}} \tag{5.48}$$

But, along the reversible path B, $\delta Q/T = dS$. Thus, for any path representing any process,

$$\Delta S \geq \int \frac{\delta Q}{T} \quad \text{or} \quad dS \geq \frac{\delta Q}{T} \tag{5.49}$$

The equality holds for a system experiencing a reversible process and the inequality for an irreversible process.

We can also write Eq. 5.49 introducing the *entropy production* σ:

$$\Delta S - \int \frac{\delta Q}{T} = \sigma \quad \text{or} \quad dS - \frac{\delta Q}{T} = \delta\sigma \tag{5.50}$$

where we have used $\delta\sigma$, since it is a path dependent function. It is obvious that entropy production is zero for a reversible process and positive for a real process, that is, any process with irreversibilities.

Key Concept: *Entropy production is positive for a real process.*

Relationship (Eq. 5.49) leads to an important conclusion in thermodynamics. Consider a infinitesimal process of a system at absolute temperature T. If the process is reversible, the differential change in entropy is $\delta Q/T$; if the process is irreversible, the change in entropy is greater than $\delta Q/T$. We thus conclude that the effect of irreversibility (e.g., friction) is to increase the entropy of a system. Observing that the entropy production σ, in Eq. 5.50, must be positive, states this alternatively.

Now, in our application of the second law to a process, Eq. 5.50 can summarize our results. If we wish to investigate whether a proposed process satisfies the second law, we can check for a positive σ using Eq. 5.50. We see that entropy and the second law are related in the similar way that energy and the first law are related.

Finally, consider an *isolated* system: a system that exchanges no work or heat with its surroundings. For such a system, the first law demands that $E_2 = E_1$ for any process. Equation 5.49 takes the form

$$\Delta S \geq 0 \tag{5.51}$$

Key Concept: *The entropy of an isolated system undergoing a real process must increase.*

demanding that the entropy of an isolated system either remains constant or increases, depending on whether the process is reversible or irreversible. Hence, for any real process, the entropy of an isolated system must increase. Alternatively, we can state that the entropy production associated with a real process of an isolated system must be positive, since, for an isolated system, $\Delta S = \sigma$.

We can restate Eq. 5.51 by considering an isolated system to be composed of any system under consideration plus the surroundings to that system. The heat and work transferred to the system under consideration are negative the heat and work transferred from the surroundings. This combination of the system and surroundings is often referred to as the *universe*. For the universe we can write

$$\Delta S_{\text{sys}} + \Delta S_{\text{surr}} = \sigma \geq 0 \tag{5.52}$$

where $\sigma = 0$ applies to a reversible (ideal) process and the inequality to an irreversible (real) process. Relation (Eq. 5.52), the *principle of entropy increase*, is often used as the mathematical statement of the second law. In Eq. 5.52, σ has also been referred to as ΔS_{gen}, the entropy generated in the universe. It should be noted that σ is not a property of the system; thus, we chose not to use as its symbol the symbol S for entropy.

Example 5.13

Air is contained in one half of the insulated tank shown in Fig. E5.13. The other side is completely evacuated. The membrane is punctured, and the air quickly fills the entire volume. Calculate the specific entropy change of this isolated system.

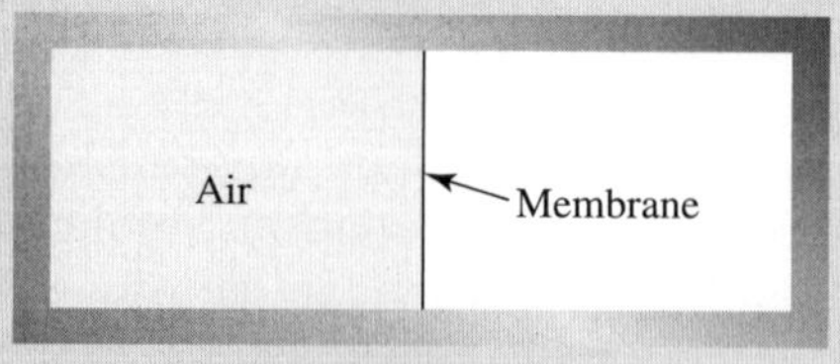

FIGURE E5.13

(continued)

Solution
The entire tank is chosen as the system boundary. No heat transfer occurs across the boundary, and the air does no work. The first law then takes the form $\Delta U = mc_v(T_2 - T_1) = 0$. Hence, the final temperature is equal to the initial temperature. Using Eq. 5.26 for the entropy change, we have, with $T_1 = T_2$,

$$\Delta s = R \ln \frac{v_2}{v_1} = \frac{53.3}{778} \ln 2 = 0.04749 \text{ Btu/lbm-°R}$$

Note that this satisfies Eq. 5.49, since for this example $Q = 0$, so that $\int \delta Q/T = 0 < m\Delta s$.

Example 5.14

Two kg of superheated steam at 400°C and 600 kPa is cooled at constant pressure by transferring heat from a cylinder until the steam is completely condensed. The surroundings are at 25°C. Determine the net entropy production due to this process.

Solution
The entropy of the steam that defines our system decreases, since heat is transferred from the system to the surroundings. From the steam tables, this change is found to be

$$\Delta S_{sys} = m(s_2 - s_1) = 2(1.9316 - 7.7086) = -11.55 \text{ kJ/K}$$

The heat transfer to the surroundings occurs at constant temperature. Hence, the entropy change of the surroundings (see Eq. 5.16) is

$$\Delta S_{surr} = \frac{Q}{T}$$

The heat transfer for the constant-pressure process needed to condense the steam is

$$Q = m\Delta h = 2(3270.2 - 670.6) = 5199 \text{ kJ}$$

giving $\Delta S_{surr} = 5199/298 = 17.45$ kJ/K and

$$\sigma = \Delta S_{surr} + \Delta S_{sys} = 17.45 - 11.55 = 5.90 \text{ kJ/K}$$

5.13 THE SECOND LAW APPLIED TO A CONTROL VOLUME

The second law has been applied thus far in this chapter to a system: a particular collection of mass particles. We now wish to apply the second law to a control volume, following the same strategy used in our study of the first law (refer to Section 4.8). In Fig. 5.12, the system at time t includes the control volume

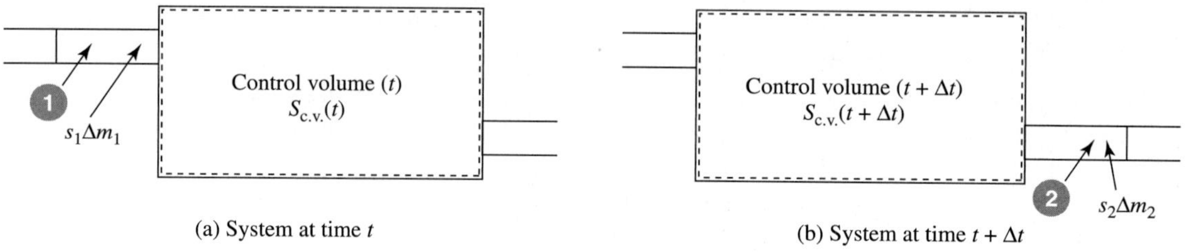

FIGURE 5.12 The system to control-volume transformation for the second law analysis.

(enclosed by the dashed control surface) and volume 1; after a small time increment Δt, the system moves so that it includes the control volume plus volume 2. The second law for the system, Eq. 5.49, takes the form

$$S_{\text{sys}}(t+\Delta t) - S_{\text{sys}}(t) \geq \int \frac{\delta Q}{T}$$
$$S_{\text{c.v.}}(t+\Delta t) + s_2\Delta m_2 - s_1\Delta m_1 - S_{\text{c.v.}}(t) \geq \int \frac{\delta Q}{T} \tag{5.53}$$

This is put in the form

$$\Delta S_{\text{c.v.}} + s_2\Delta m_2 - s_1\Delta m_1 - \int_{\text{c.v.}} \frac{\delta Q_{\text{c.s.}}}{T_{\text{c.s.}}} \geq 0 \tag{5.54}$$

where $Q_{\text{c.s.}}$ is the heat transfer into the control volume at a boundary location, where the temperature is $T_{\text{c.s.}}$. Introducing the entropy production of Eq. 5.50,

$$\Delta S_{\text{c.v.}} + s_2\Delta m_2 - s_1\Delta m_1 - \int_{\text{c.v.}} \frac{\delta Q_{\text{c.s.}}}{T_{\text{c.s.}}} = \sigma_{\text{c.v.}} \tag{5.55}$$

Let's assume that the entire control surface is at a constant temperature $T_{\text{c.s.}}$, a rather common assumption. If we divide Eq. 5.55 by Δt and use dots to denote rates, we arrive at the rate equation

$$\dot{S}_{\text{c.v.}} + \dot{m}_2 s_2 - \dot{m}_1 s_1 - \frac{\dot{Q}}{T_{\text{c.s.}}} = \dot{\sigma}_{\text{c.v.}} \tag{5.56}$$

Zero rate of entropy production is associated with a reversible process, whereas positive entropy production is associated with the many irreversibilities that may exist within the control volume. These include viscous effects that are always present in the flow of the working fluid, separations of the flow from turbine and pump blades or from boundaries where abrupt changes in geometry occur, large temperature differences existing within the control volume, and shock waves in high-speed compressible flow, to name a few.

For a steady-flow process, the entropy of the control volume remains constant with time. We can then write, recognizing that $\dot{m}_2 = \dot{m}_1 = \dot{m}$,

$$\dot{m}(s_2 - s_1) - \frac{\dot{Q}}{T_{\text{c.s.}}} = \dot{\sigma}_{\text{c.v.}} \tag{5.57}$$

By transferring energy to the control volume via heat transfer, we can obviously increase the entropy of the fluid flowing from the control volume so that $s_2 > s_1$. If irreversibilities are present within the control volume, $\dot{\sigma}_{\text{c.v.}}$ will be positive and s_2 will be greater than the exiting entropy for the same rate of heat transfer without irreversibilities.

We also note that for an adiabatic steady-flow process the entropy also increases from inlet to exit due to irreversibilities, since, for that case, Eq. 5.57 shows that

$$s_2 \geq s_1 \tag{5.58}$$

since $\dot{\sigma}_{\text{c.v.}}$ must be positive or zero. For the reversible adiabatic process, the inlet entropy and exit entropy are equal since $\dot{\sigma}_{\text{c.v.}} = 0$; this is the isentropic process. We use this fact when studying a reversible adiabatic process involving, for example, steam or a refrigerant flowing through an ideal turbine or a compressor.

We have calculated the efficiency of a cycle earlier in this chapter. Now we desire a quantity that can be used as a measure of the irreversibilities that exist in a particular device. The *efficiency* of a device is one such measure; it is defined as the ratio of the actual performance of a device to the ideal performance. The ideal performance is often that associated with an isentropic process. For example, the efficiency of a turbine would be

$$\eta = \frac{w_a}{w_s} \tag{5.59}$$

where w_a is the actual (specific) shaft work output and w_s is the (specific) shaft work output associated with an isentropic process. In general, the efficiency is defined using the optimum output as the measure; for a nozzle, we would use the maximum kinetic energy increase. For a compressor, the actual work required is greater than the ideal work requirement of an isentropic process. For a compressor or pump, the efficiency is defined to be

$$\eta = \frac{w_s}{w_a} \tag{5.60}$$

Each efficiency above is also called the *adiabatic efficiency*, since each efficiency is based on an adiabatic process.

Example 5.15

A preheater is used to preheat water in a power plant cycle, as shown in Fig. E5.15. The superheated steam is at a temperature of 250°C, and the entering water is subcooled at 45°C. All pressures are 600 kPa. Calculate the rate of entropy production.

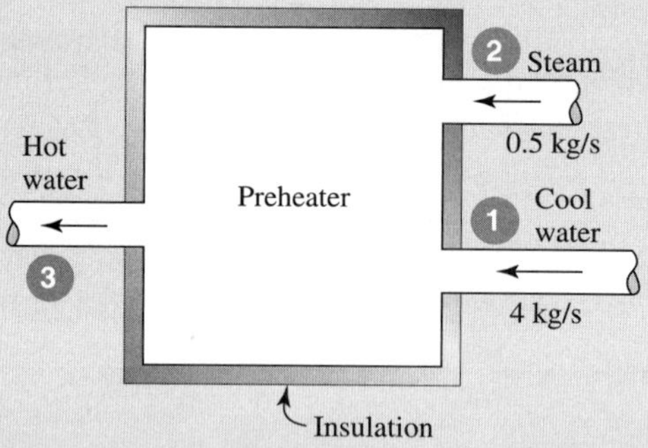

FIGURE E5.15

Solution

From the conservation of mass, $\dot{m}_3 = \dot{m}_2 + \dot{m}_1 = 0.5 + 4 = 4.5$ kg/s. The first law allows us to calculate the temperature of the exiting water. Neglecting kinetic energy and potential energy changes and assuming zero heat transfer, the first law takes the form $\dot{m}_3 h_3 = \dot{m}_2 h_2 + \dot{m}_1 h_1$. Using the steam tables (h_1 is the enthalpy of saturated liquid at 45°C),

$$\dot{m}_3 h_3 = \dot{m}_2 h_2 + \dot{m}_1 h_1$$

$$4.5 h_3 = 0.5 \times 2957.2 + 4 \times 188.4 \qquad \therefore\ h_3 = 496 \text{ kJ/kg}$$

This enthalpy is less than that of saturated liquid at 600 kPa. Thus, the exiting water is also subcooled. Its temperature is interpolated (from the saturated steam tables find T that gives $h_f = 496$ kJ/kg), to be

$$T_3 = \frac{496 = 461.3}{503.7 - 461.3} \times 10 + 110 = 118°\text{C}$$

The entropy at this temperature is then interpolated (using s_f) to be $s_3 = 1.508$ kJ/kg·K. The entropy of the entering superheated steam is found to be $s_2 = 7.182$ kJ/kg·K. The entering entropy of the subcooled water is s_f at $T_1 = 45°$C, or $s_1 = 0.639$ kJ/kg·K. Finally, modifying Eq. 5.57 to account for two inlets, we have, with no heat transfer,

$$\dot{\sigma} = \dot{m}_3 s_3 - \dot{m}_2 s_2 - \dot{m}_1 s_1 = 4.5 \times 1.508 - 0.5 \times 7.182 - 4 \times 0.639 = 0.639 \text{ kW/K}$$

This is positive, indicating that entropy is produced: a consequence of the second law. The mixing process between the superheated steam and the subcooled water is indeed an irreversible process.

Example 5.16

Superheated steam enters a turbine, as shown in Fig. E5.16a, and exits at 2 psia. If the mass flow rate is 4 lbm/sec, determine the maximum power output using (a) the steam tables and (b) TK Solver. Sketch the process on a T-s diagram.

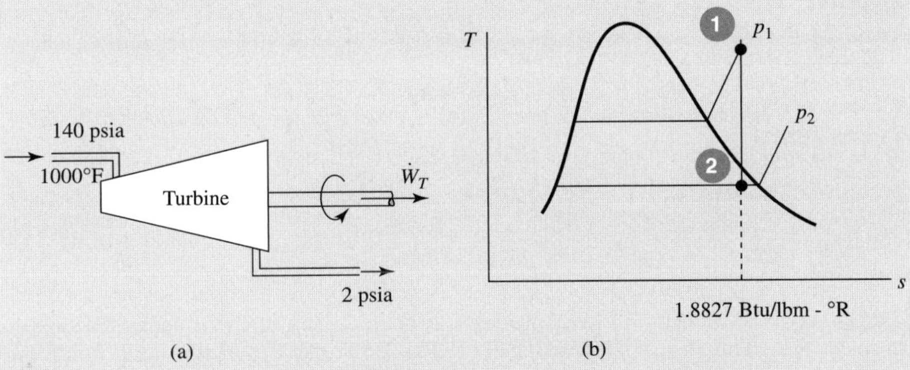

FIGURE E5.16

Solution

If we neglect kinetic energy and potential energy changes, the first law, for an adiabatic process, is $-W_T = m(h_2 - h_1)$. Since we desire the maximum power output, the process is assumed to be reversible. The entropy exiting is the same as the entropy entering, as shown in Fig. E5.16b (such a sketch is quite useful in visualizing the process).

(a) From the steam tables, at 1000°F and 140 psia,

$$h_1 = 1531 \text{ Btu/lbm} \qquad s_1 = s_2 = 1.8827 \text{ Btu/lbm-°R}$$

With the above value for s_2, we see that state 2 is in the quality region. The quality is determined as follows:

$$s_2 = s_f + x_2 s_{fg}$$

$$1.8827 = 0.1750 + 1.7448 x_2 \qquad \therefore\ x_2 = 0.9787$$

Then

$$h_2 = h_f + x_2 h_{fg} = 94.02 + 0.9787 \times 1022.1 = 1094 \text{ Btu/lbm}$$

so that

$$\dot{W}_T = \dot{m}(h_2 - h_1)$$

$$= 4 \times (1531 - 1094) = 1748 \text{ Btu/sec} \quad \text{or} \quad 2473 \text{ hp}$$

(b) Using TK Solver with $T_1 = 1000$°F and $p_1 = 140$ psia, we find $h_1 = 1531.1$ Btu/lbm and $s_1 = 1.883$ Btu/lbm-°R. Then, with $p_2 = 2$ psia and $s_2 = s_1$, we find $h_2 =$ 1094.5 Btu/lbm. Finally,

$$\dot{W}_T = \dot{m}(h_2 - h_1)$$

$$= 4 \times (1531.1 - 1094.5) = 1746.4 \text{ Btu/sec} \quad \text{or} \quad 2470 \text{ hp}$$

Example 5.17

The turbine of Example 5.16 is assumed to be 80 percent efficient. Determine the entropy and temperature of the final state. Sketch the real process on a *T-s* diagram.

Solution

Using the definition of efficiency, the actual power output is found to be

$$\dot{W}_a = \eta_T \dot{W}_s = 0.8 \times 1748 = 1398 \text{ Btu/sec}$$

From the first law, $-\dot{W}_a = \dot{m}(h_{2'} - h_1)$, we have

$$h_{2'} = h_1 - \frac{\dot{W}_a}{\dot{m}} = 1521 - \frac{1398}{4} = 1182 \text{ Btu/lbm}$$

Using this value and $p_{2'} = 2$ psia, we see that state 2′ lies in the superheated region, since $h_{2'} > h_g$. This is shown schematically in Fig. E5.17. At $p_2 = 2$ psia and $h_{2'} = 1182$, we interpolate to find the value of $T_{2'}$:

$$T_{2'} = \frac{1186 - 1182}{1186 - 1168}(280 - 240) + 280 = 271°\text{F}$$

The entropy is interpolated to be $s_{2'} = 2.053$ Btu/lbm-°R.

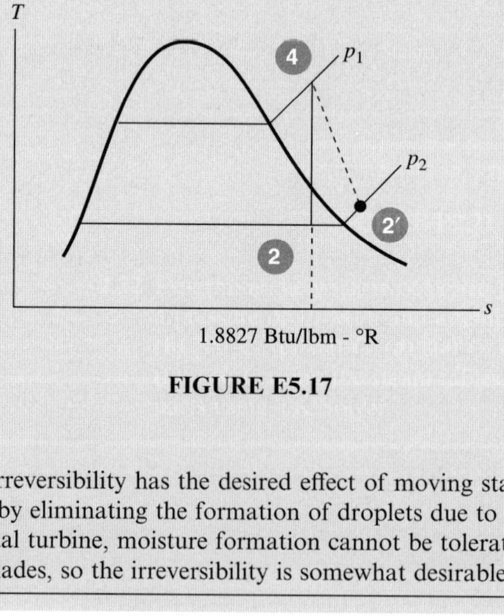

FIGURE E5.17

Note that the irreversibility has the desired effect of moving state 2 into the superheated region, thereby eliminating the formation of droplets due to the condensation of moisture. In an actual turbine, moisture formation cannot be tolerated because of damage to the turbine blades, so the irreversibility is somewhat desirable.

5.14 SUMMARY

In this chapter, we defined the following terms:

Heat engine: A device that performs work.
Refrigerator: A device that transfers heat from a body.
Heat pump: A device that transfers heat to a body.
Thermal-energy reservoir: An entity that transfers heat with no change in temperature.
Reversible process: A process which, having taken place, can be reversed and in so doing leaves no changes in the system or surroundings.
Reversible engine: An engine that operates with reversible processes. It has no losses.
Carnot engine: A reversible engine.
Entropy: An abstract quantity used in the application of the second law of thermodynamics.
Entropy production: The net entropy produced due to the occurrence of a process.

A number of equations associated with the second law were introduced. These include the net work W produced by an engine during one cycle:

$$W = Q_H - Q_L \tag{5.61}$$

where Q_H is the heat transferred from the high-temperature reservoir and Q_L is the heat transferred to the low-temperature reservoir. The efficiency of an engine and the coefficients of performance of a refrigerator and heat pump are, respectively,

$$\eta = \frac{W}{Q_H}, \qquad \text{COP}_{\text{refrig}} = \frac{Q_L}{W}, \qquad \text{COP}_{\text{h.p.}} = \frac{Q_H}{W} \tag{5.62}$$

The maximum efficiency of an engine and the maximum coefficients of performance are those of the Carnot cycles:

$$\eta_{\max} = 1 - \frac{T_L}{T_H}, \qquad \text{COP}_{\text{refrig, max}} = \frac{T_L}{T_H - T_L}, \qquad \text{COP}_{\text{h.p., max}} = \frac{T_H}{T_H - T_L} \tag{5.63}$$

Heat transfer is related to entropy by

$$Q = \int T\,dS \tag{5.64}$$

For an ideal gas with constant specific heats, the entropy change is

$$\Delta s = c_v \ln \frac{T_2}{T_1} + R \ln \frac{v_2}{v_1} = c_p \ln \frac{T_2}{T_1} - R \ln \frac{p_2}{p_1} \tag{5.65}$$

For an ideal gas with variable specific heats, we use

$$\Delta s = \phi_2 - \phi_1 + R \ln \frac{p_2}{p_1} \tag{5.66}$$

where ϕ is found in Appendix F. The entropy change for a liquid or a solid is given by

$$\Delta s = c_p \ln \frac{T_2}{T_1} \tag{5.67}$$

For a reservoir, the entropy change is

$$\Delta S = \frac{Q}{T_{\text{reservoir}}} \tag{5.68}$$

The principle of entropy increase can be expressed by referring to the entropy production σ:

$$\sigma = S_{\text{sys}} + S_{\text{surr}} \geq 0 \tag{5.69}$$

where $\sigma = 0$ is for a reversible process. For a control volume, the rate of entropy production is

$$\dot{\sigma}_{\text{c.v.}} = \dot{m}(s_2 - s_1) + \frac{\dot{Q}}{T_{\text{c.s.}}} \tag{5.70}$$

For an adiabatic reversible process, entropy production is zero, and $s_2 = s_1$.

The efficiency of a device is defined where the subscript "s" refers to an isentropic process. For a turbine and a compressor, the efficiencies are, respectively,

$$\eta_{\text{turbine}} = \frac{w_a}{w_s} \qquad \eta_{\text{comp}} = \frac{w_s}{w_a} \tag{5.71}$$

where w_a refers to the actual work.

REVIEW PROBLEMS FOR THE FE/EIT EXAMINATION

5.1FE Select an acceptable paraphrase of the Kelvin–Planck statement of the second law.

A. No process can produce more work than the heat that it accepts.
B. No engine can produce more work than the heat that it intakes.
C. An engine cannot produce work without accepting heat.
D. An engine has to reject heat.

5.2FE Which of the following can be assumed to be reversible?

A. A paddle wheel
B. A burst membrane
C. A resistance heater
D. A piston compressing gas in a race engine

5.3FE An inventor claims a thermal engine operates between ocean layers at 27°C and 10°C. It produces 10 kW and discharges 9900 kJ/min. Such an engine is:

A. Impossible **B.** Reversible
C. Possible **D.** Probable

5.4FE A Carnot engine operates between reservoirs at 20°C and 200°C. If 10 kW of power is produced, find the rejected heat rate.

A. 26.3 kJ/s **B.** 20.2 kJ/s
C. 16.3 kJ/s **D.** 12.0 kJ/s

5.5FE A Carnot cycle is a cycle of special interest because:
A. It establishes a lower limit on cycle efficiency.
B. It operates between two constant temperature reservoirs.
C. It provides the maximum efficiency for any cycle.
D. When it is carefully constructed in a laboratory, it provides an upper limit on efficiency.

5.6FE Select an incorrect statement relating to a Carnot cycle.
A. There are two adiabatic processes.
B. There are two constant-pressure processes.
C. Work occurs for all four processes.
D. Each process is a reversible process.

5.7FE A Carnot refrigerator requires 10 kW to remove 20 kJ/s from a 20°C reservoir. The temperature of the high-temperature reservoir is:
A. 440 K **B.** 400 K
C. 360 K **D.** 340 K

5.8FE A heat pump is to provide 2000 kJ/hr to a house maintained at 20°C. If it is −20°C outside, what is the minimum power requirement?
A. 385 kJ/hr **B.** 316 kJ/hr
C. 273 kJ/hr **D.** 184 kJ/hr

5.9FE An engine operates on 100°C geothermal water. It exhausts to a 20°C stream. Its maximum efficiency is nearest:
A. 21% **B.** 32%
C. 58% **D.** 80%

5.10FE Which of the following entropy relationships for a process is wrong?
A. Air, $V\!\!\!/ =$ const: $\Delta s = c_v \ln T_2/T_1$
B. Water: $\Delta s = c_p \ln T_2/T_1$
C. Reservoir: $\Delta s = c_p \ln T_2/T_1$
D. Copper: $\Delta s = c_p \ln T_2/T_1$

5.11FE One kilogram of air is heated in a rigid container from 20°C to 300°C. The entropy change is nearest:
A. 0.64 kJ/K **B.** 0.54 kJ/K
C. 0.48 kJ/K **D.** 0.34 kJ/K

5.12FE Ten kilograms of air is expanded isentropically from 500°C and 6 MPa to 400 kPa. The work accomplished is nearest:
A. 7400 kJ **B.** 6200 kJ
C. 4300 kJ **D.** 2990 kJ

5.13FE Calculate the total entropy change if 10 kg of ice at 0°C are mixed in an insulated container with 20 kg of water at 20°C. The heat of melting for ice is 340 kJ/kg.
A. 6.1 kJ/K **B.** 3.9 kJ/K
C. 1.2 kJ/K **D.** 0.21 kJ/K

5.14FE Ten kilograms of iron at 300°C are chilled in a large volume of ice and water. Find the total entropy change.
A. 0.88 kJ/K **B.** 1.01 kJ/K
C. 1.26 kJ/K **D.** 1.61 kJ/K

5.15FE Which of the following second law statements is incorrect?
A. The entropy of an isolated system must remain constant or increase.
B. The entropy of a hot copper block decreases as it cools.
C. If ice is melted in water in an insulated container, the net entropy decreases.
D. Work must be input if energy is transferred from a cold body to a hot body.

5.16FE Find the work needed to isentropically compress 2 kg of steam in a cylinder at 400 kPa and 400°C to 2 MPa.
A. 1020 kJ **B.** 940 kJ
C. 780 kJ **D.** 560 kJ

5.17FE Find w_T of the insulated turbine shown in Fig. P5.17FE.
A. 1410 kJ/kg **B.** 1360 kJ/kg
C. 1200 kJ/kg **D.** 1020 kJ/kg

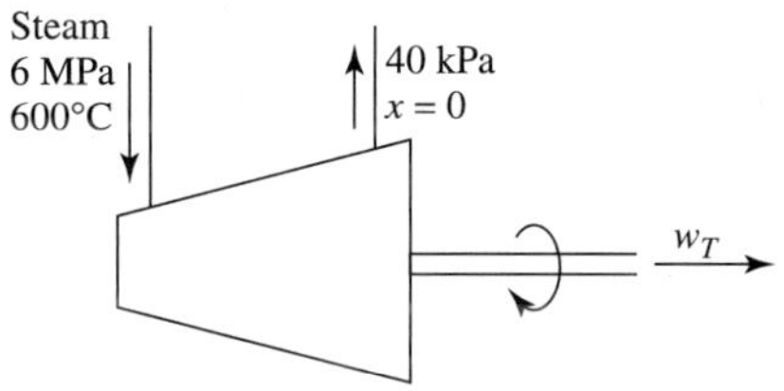

FIGURE P5.17FE

5.18FE For the turbine of Problem 5.17FE, find $(w_T)_{max}$.
A. 1410 kJ/kg **B.** 1360 kJ/kg
C. 1200 kJ/kg **D.** 1020 kJ/kg

5.19FE For the turbine of Fig. P5.17FE, find T_2.
A. 64°C **B.** 76°C
C. 88°C **D.** 104°C

5.20FE Calculate the efficiency of the turbine of Fig. P5.17FE.
A. 85% **B.** 89%
C. 91% **D.** 93%

PROBLEMS

In the following problem set, work each problem using the appropriate equations and/or Appendix tables. Use TK Solver for property evaluation when requested.

Heat Engines, Heat Pumps, and Refrigerators

5.1 A refrigerator is rated at a COP of four. The refrigerated space that it cools requires a peak cooling rate of 30 000 kJ/h. What size electrical motor (rated in horsepower) is required for the refrigerator?

5.2 A heat pump provides 75 MJ/h to a house. If the compressors require an electrical energy input of 4 kW, calculate the COP.

5.3 A Carnot heat engine produces 10 hp by transferring energy between two reservoirs at 40°F and 212°F. Calculate the rate of heat transfer from the high-temperature reservoir.

5.4 A power plant burns 1000 kg of coal each hour and produces 500 kW of power. Calculate the overall thermal efficiency if each kg of coal produces 6 MJ of energy.

5.5 An automobile that has a gas mileage of 13 km/L is traveling at 100 km/h. At this speed, essentially all of the power produced by the engine is used to overcome air drag. If the air drag force is given by $\frac{1}{2}\rho V^2 A C_D$, determine the thermal efficiency of the engine at this speed using the projected area $A =$ 2 m^2, drag coefficient $C_D = 0.28$, and heating value of gasoline 9000 kJ/kg. Gasoline has a density of 740 kg/m^3.

5.6 Show that a violation of the Kelvin–Planck statement of the second law implies a violation of the Clausius statement.

5.7 A battery does work by producing an electric current while transferring heat with a constant-temperature atmosphere. Is this a violation of the second law? Explain.

The Carnot Cycle and Reversibility

5.8 Show that all reversible engines, operating between two given temperature reservoirs, have the same efficiency as a Carnot engine operating between the same two temperature reservoirs.

5.9 An inventor proposes an engine that operates between the 27°C warm surface layer of the ocean and a 10°C layer several meters down. The inventor claims that the engine produces 100 kW by pumping 20 kg/s of seawater. Is this possible?

5.10 A power utility company desires to use the hot groundwater from a hot spring to power a heat engine. If the groundwater is at 95°C, estimate the maximum power output if a mass flow rate of 0.2 kg/s is possible. The atmosphere is at 20°C.

5.11 A Carnot cycle operates between 200°C and 1200°C. Calculate **(a)** its thermal efficiency if it operates as a power cycle, **(b)** its COP if it operates as a refrigerator, and **(c)** its COP if it operates as a heat pump.

5.12 A Carnot engine rejects 80 MJ of energy every hour by transferring heat to a reservoir at 10°C. Determine the temperature of the high-temperature reservoir and the power produced if the rate of energy addition is 40 kW.

5.13 Two Carnot engines operate in series between two reservoirs maintained at 600°F and 100°F, respectively. The energy rejected by the first engine is input into the second engine. If the first engine's efficiency is 20 percent greater than the second engine's efficiency, calculate the intermediate temperature.

5.14 A Carnot engine operating on air accepts 50 kJ/kg of heat and rejects 20 kJ/kg. Calculate the high- and low-reservoir temperatures if the maximum specific volume is 10 m^3/kg and the pressure after the isothermal expansion is 200 kPa.

5.15 A proposed power cycle is designed to operate between temperature reservoirs, as shown in Fig. P5.15. It is supposed to produce 43 hp from the 2500 kJ of energy extracted each minute. Is the proposal feasible?

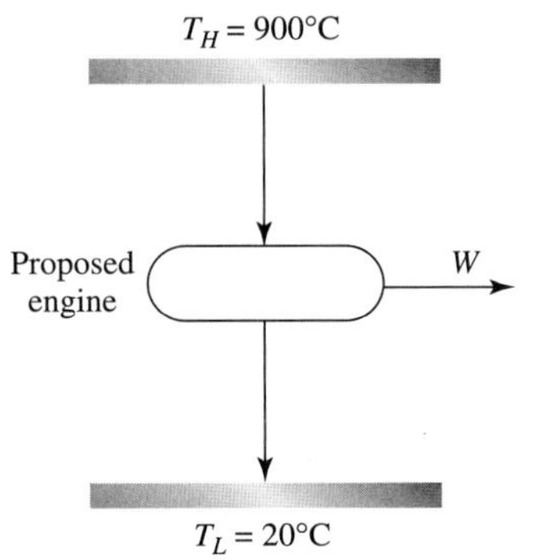

FIGURE P5.15

5.16 **(a)** What is the maximum efficiency that can result from an engine that operates on the thermal gradients in the ocean? The surface waters at the proposed location are at 85°F and those at a reasonable depth are at 50°F. **(b)** What would be the maximum COP of a heat pump, operating between the two layers, used to heat an offshore oil rig?

5.17 A Carnot engine operates between reservoirs at temperatures T_1 and T_2, and a second Carnot engine operates between reservoirs maintained at T_2 and T_3. Express the efficiency η_3 of a third engine operating between T_1 and T_3 in terms of the efficiencies η_1 and η_2 of the other two engines.

5.18 A heat engine operates on a Carnot cycle with an efficiency of 75 percent. What COP would a refrigerator operating on the same cycle have? The low temperature is 0°C.

5.19 Two Carnot engines operate in series between two reservoirs maintained at 500°C and 40°C, respectively. The energy rejected by the first engine is utilized as energy input to the second engine. Determine the temperature of this intermediate reservoir between the two engines if the efficiencies of both engines are the same.

5.20 Two Carnot refrigerators operate in series between two reservoirs maintained at 20°C and 200°C, respectively. The energy output by the first refrigerator is used as the heat-energy input to the second refrigerator. If the COPs of the two refrigerators are the same, what should the intermediate temperature be?

5.21 A Carnot engine operates on air with the cycle shown in Fig. P5.21. If there are 30 kJ/kg of heat added from the high-temperature reservoir maintained at 200°C, determine the work produced.

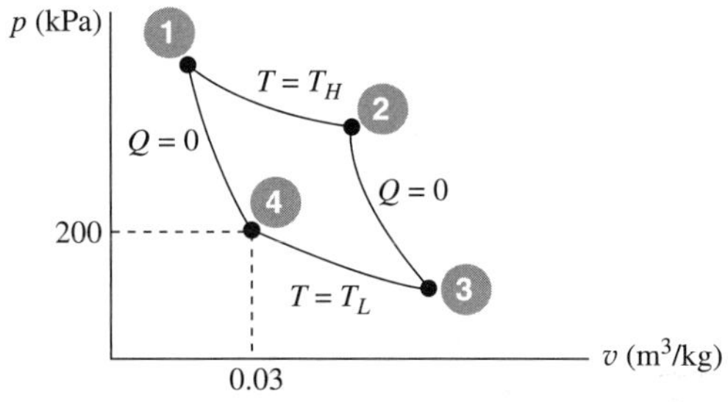

FIGURE P5.21

5.22 A Carnot engine operates between a low pressure of 15 psia and a high pressure of 300 psia. The corresponding volumes are 250 in^3 and 25 in^3. If there is 0.01 lbm of air, calculate the work output.

5.23 A Carnot engine using hydrogen gas operates with the high-temperature reservoir maintained at 600 K. The pressure ratio for the adiabatic compression is 15 to 1 and the volume during the heat-addition process is tripled. If the minimum pressure is 100 kPa, determine the thermal efficiency and work produced.

5.24 A heat pump is proposed in which 50°F groundwater is used to heat a house to 70°F. The groundwater is to experience a temperature drop of 12°F, and the house requires 75,000 Btu/hr. Calculate the minimum mass flow rate of the groundwater and the minimum horsepower required.

5.25 A heat pump is to maintain a house at 20°C when the outside air is at −25°C. It is determined that 1800 kJ is required each minute to accomplish this. Calculate the minimum horsepower required.

5.26 If the heat pump of Problem 5.25 is to be used as an air conditioner, calculate the maximum outside temperature for which the inside temperature can be maintained at 23°C. Assume a linear relationship between temperature difference and heat flux, using the information from Problem 5.25.

5.27 A heat pump uses a 5-hp compressor while extracting 500 Btu of energy from groundwater each minute. What is the COP **(a)** if the purpose is to cool the groundwater and **(b)** if the purpose is to heat a building?

5.28 A Carnot refrigeration cycle is used to estimate the energy requirement in an attempt to reduce the temperature of a specimen to absolute zero. Suppose that we wish to remove 0.01 J of energy from the specimen when it is at 2×10^{-6} K. How much work is necessary if the high-temperature reservoir is at 20°C?

5.29 A refrigerator is proposed that will require 10 hp to extract 3 MJ of energy each minute from a space which is maintained at −18°C. The outside air is at 20°C. Is this possible?

5.30 A refrigeration unit is used to cool a space to 5°C by transferring heat to the surroundings, which are at 25°C. The same unit is then used to cool the space to −20°C. Estimate the maximum cooling rate for the second condition if the cooling rate for the first is 5 tons.

Entropy

5.31 A Carnot engine delivers 100 kW of power by operating between temperature reservoirs at 100°C and 1000°C. Calculate the entropy change of each reservoir and the net entropy change of the two reservoirs after 20 min of operation.

5.32 A Carnot engine extracts 100 kJ of heat from an 800°C reservoir and rejects heat to the surroundings at 20°C. Calculate the entropy change **(a)** of the reservoir and **(b)** of the surroundings.

5.33 A Carnot refrigerator removes 200 kJ of heat from a refrigerated space maintained at −10°C. Its COP is 10. Calculate the entropy change **(a)** of the refrigerated space and **(b)** of the high-temperature reservoir.

5.34 A reversible heat pump requires 4 hp while providing 50,000 Btu/hr to heat a space maintained at 70°F. Calculate the entropy change of the space and the low-temperature reservoir after 10 min of operation.

5.35 Compare the entropy increase of the high-temperature reservoir and the entropy decrease of the specimen of Problem 5.28.

Entropy Change of an Ideal Gas

5.36 Verify that Eqs. 5.28 result from Eqs. 5.26 and 5.27.

5.37 Two kg of air are heated at constant pressure of 200 kPa to 500°C. Calculate the entropy change if the initial volume is 0.8 m^3.

5.38 A gas of mass 0.2 kg is compressed slowly from 150 kPa and 40°C to 600 kPa in an adiabatic process. Determine the final volume if the gas is **(a)** air, **(b)** carbon dioxide, **(c)** nitrogen, and **(d)** hydrogen.

5.39 Air is compressed in an automobile engine cylinder from 14.7 to 2000 psia. If the initial temperature is 60°F, estimate the final temperature.

5.40 Two kg of gas is compressed from 120 kPa and 27°C to 600 kPa in a rigid container. Calculate the entropy change if the gas is **(a)** air, **(b)** carbon dioxide, **(c)** nitrogen, and **(d)** hydrogen.

5.41 A piston allows air to expand from 6 MPa to 200 kPa. The initial volume and temperature are 500 cm^3 and 800°C. If the temperature is held constant, calculate the heat transfer and the entropy change.

5.42 Determine the entropy change of a gas in a rigid container that is heated from the conditions shown in Fig. P5.42 to 100 psia, if the gas is **(a)** air, **(b)** carbon dioxide, **(c)** nitrogen, and **(d)** hydrogen. Atmospheric pressure is 13 psia.

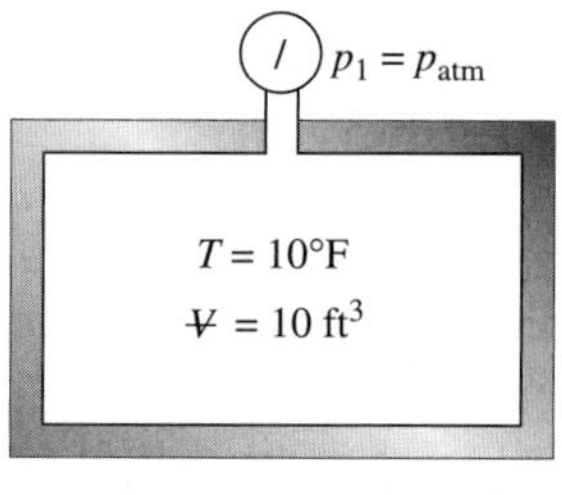

FIGURE P5.42

5.43 A paddle wheel provides 200 kJ of work to the air contained in a 0.2-m^3 rigid volume, initially at 400 kPa and 40°C. Determine the entropy change if the volume is insulated.

5.44 The entropy change in a certain expansion process is 5.2 kJ/K. The gas, initially at 80 kPa, 27°C, and 4 m^3 achieves a final temperature of 127°C. Calculate the final volume if the gas is **(a)** air, **(b)** carbon dioxide, **(c)** nitrogen, and **(d)** hydrogen.

5.45 Nine kJ of heat is added to the 20-cm-diameter cylinder shown in Fig. P5.45. Initially, the frictionless piston is 20 cm above the cylinder bottom. If the initial conditions are 200 kPa and 47°C, compute the work done and the entropy change for **(a)** air, **(b)** carbon dioxide, **(c)** nitrogen, and **(d)** hydrogen.

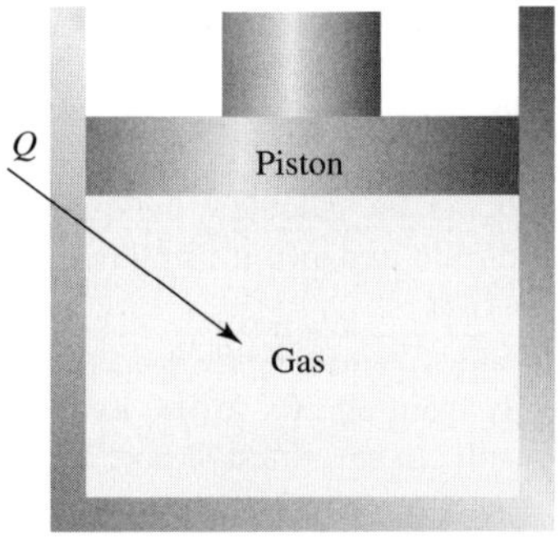

FIGURE P5.45

5.46 A piston is inserted into a cylinder causing the pressure to change from 50 to 4000 kPa while the temperature remains constant at 27°C. To accomplish this, heat transfer must occur. Determine the heat transfer and the entropy change if the working substance is **(a)** air, **(b)** carbon dioxide, **(c)** nitrogen, and **(d)** hydrogen.

5.47 The temperature of a gas changes from 60°F to 900°F while the pressure remains constant at 16 psia. Compute the heat transfer and the entropy change if the gas is **(a)** air, **(b)** carbon dioxide, **(c)** nitrogen, and **(d)** hydrogen.

5.48 A rigid, insulated 4-m^3 volume is divided in half by a membrane. One chamber is pressurized with air to 100 kPa and the other is completely evacuated. The membrane is ruptured, and after a period of time, equilibrium is restored. What is the entropy change?

5.49 Four hundred kJ of paddle-wheel work are transferred to air in a rigid, insulated 2-m^3 volume, initially at 100 kPa and 57°C. Calculate the entropy change if the working substance is **(a)** air, **(b)** carbon dioxide, **(c)** nitrogen, and **(d)** hydrogen.

5.50 Air is compressed in an automobile cylinder from 14.7 to 2000 psia. Predict the final temperature if the initial temperature is 60°F. Do not assume constant specific heat.

5.51 A torque of 40 N·m is needed to rotate a shaft at 40 rad/s. It is attached to a paddle wheel located in a rigid 2-m^3 volume. Initially, the temperature is 47°C, and the pressure is 200 kPa. If the paddle wheel rotates for 10 min and 500 kJ of heat is transferred to the air in the volume, determine the entropy increase **(a)** assuming constant specific heats and **(b)** using the gas table.

5.52 Two pounds of air are contained in an insulated piston-cylinder arrangement. The air is compressed from 16 psia and 60°F by applying 2×10^5 ft-lbf of work. Compute the final pressure and temperature **(a)** assuming constant specific heats and **(b)** using the gas table.

5.53 Air expands from 200 cm^3 to 1000 cm^3 in a cylinder while the pressure is held constant at 600 kPa. If the initial temperature is 20°C, calculate the heat transfer assuming **(a)** constant specific heat and **(b)** variable specific heat.

5.54 A piston-cylinder arrangement is used to compress 0.2 kg of air isentropically from initial conditions of 120 kPa and 27°C to 2000 kPa. Calculate the work necessary **(a)** assuming constant specific heats and **(b)** using the gas table.

5.55 Four kg of air expands in an insulated cylinder from 500 kPa and 227°C to 20 kPa. What is the work output **(a)** assuming constant specific heats and **(b)** using the gas table?

Entropy Change for Water and Steam

5.56 Steam, at a quality of 85 percent, is expanded in a cylinder at a constant pressure of 800 kPa by adding 2000 kJ/kg of heat. Compute the entropy increase and the final temperature using **(a)** the steam tables and **(b)** TK Solver.

5.57 Two lbm of steam, initially at a quality of 40 percent and a pressure of 600 psia, is expanded in a cylinder at constant temperature until the pressure is halved. Determine the entropy change and the heat transfer.

5.58 Water is maintained at a constant pressure of 400 kPa while the temperature changes from 20°C to 400°C. Calculate the heat transfer and the entropy change using **(a)** the steam tables and **(b)** TK Solver.

5.59 Two kg of water are contained in a 6-liter, rigid tank at 60°C. If one MJ of heat is added, calculate the final entropy.

5.60 A mass of 0.1-kg water is expanded in a cylinder at a constant pressure of 4 MPa from saturated liquid until the temperature is 600°C. Calculate the work necessary and the entropy change using **(a)** the steam tables and **(b)** TK Solver.

5.61 Two kg of saturated steam at 100°C is contained in a cylinder. If the steam undergoes an isentropic expansion to 20 kPa, determine the work output.

5.62 Five kg of steam contained in a 2-m^3 cylinder at 40 kPa is compressed isentropically to 5000 kPa. Calculate the work required using **(a)** the steam tables and **(b)** TK Solver.

5.63 Ten lb of water at 14.7 psia is heated at constant pressure from 40°F to saturated vapor. Compute the heat transfer necessary and the entropy change.

5.64 Five kg of ice at −20°C is mixed with water initially at 20°C. If there is no significant heat transfer from the container, determine the final temperature and the net entropy change if the initial mass of water is **(a)** 10 kg and **(b)** 40 kg.

5.65 Five ice cubes (each 1.2 in^3) at 0°F are placed in a full 16-oz glass of water at 60°F. Calculate the final equilibrium temperature and the net entropy change, assuming an insulated glass.

5.66 A Carnot engine operates with steam on the cycle shown in Fig. P5.66. What is the thermal efficiency? If the work output is 300 kJ/kg, what is the quality of state 1?

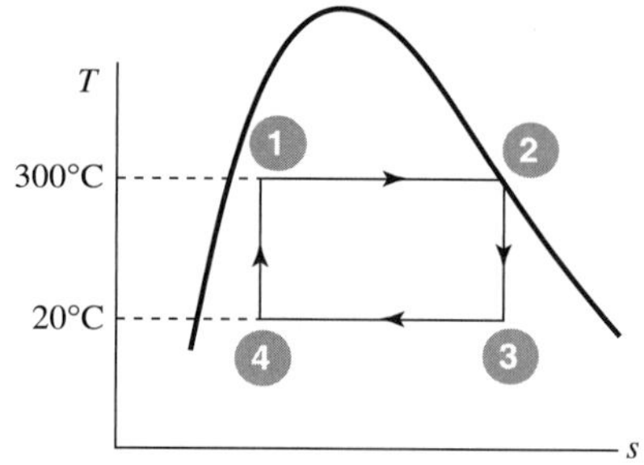

FIGURE P5.66

5.67 The steam in a Carnot engine is compressed adiabatically from 10 kPa to 6 MPa with saturated liquid occurring at the end of the process. If the work output is 500 kJ/kg, calculate the quality at the end of the isothermal expansion using **(a)** the steam tables and **(b)** TK Solver.

5.68 The steam in a Carnot engine is compressed adiabatically from 20 kPa to 800 kPa. The heat addition results in saturated vapor. If the quality of state 1 in Fig. P5.66 (ignore the temperatures in the figure) is 15 percent, calculate the net work per cycle and the thermal efficiency.

5.69 A Carnot engine that operates with steam has a quality of 20 percent at the beginning of the adiabatic compression process, as shown in Fig. P5.69. If the adiabatic expansion process begins with a saturated

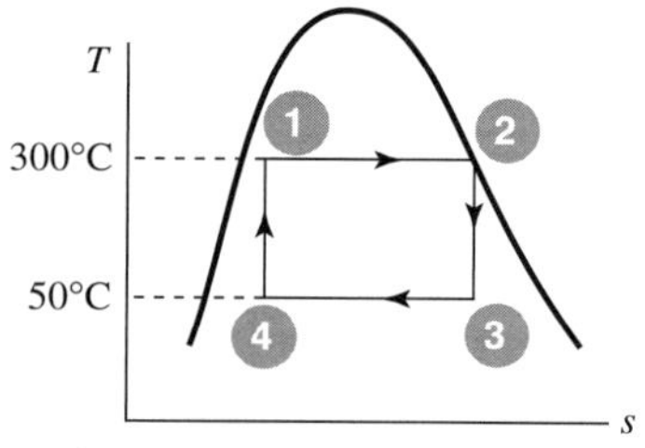

FIGURE 5.69

vapor, determine the heat added using **(a)** the steam tables and **(b)** TK Solver.

5.70 A Carnot engine operates with 0.02 kg/s of steam, as shown in Fig. P5.69. If the quality of state 4 is 15 percent: **(a)** What is the power output? **(b)** What is the quality of state 3?

5.71 The R134a in the Carnot refrigerator of Fig. P5.71 operates between saturated liquid and vapor during the heat-rejection process. If the cycle has a high temperature of 50°C and a low temperature of −20°C, calculate the heat transfer from the refrigerated space and the quality at the beginning of the heat addition process using **(a)** the R134a tables and **(b)** TK Solver.

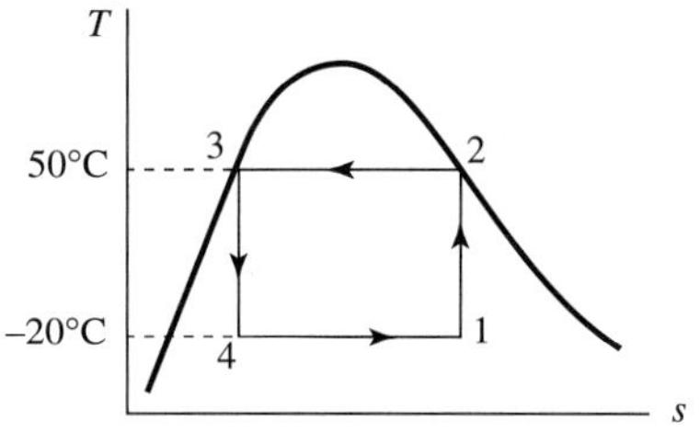

FIGURE P5.71

The Inequality of Clausius

5.72 For a Carnot engine operating under the conditions of Problem 5.12, show that the inequality of Clausius is satisfied.

5.73 Show that a Carnot engine operating with steam between pressures of 40 kPa and 4 MPa satisfies the inequality of Clausius. The work output is 350 kJ/kg, and saturated vapor enters the adiabatic expansion process. Refer to Fig. P5.69, but ignore the temperatures.

5.74 Using the information given in Problem 5.21, verify that the inequality of Clausius is satisfied.

5.75 For the steam cycle of Problem 5.66, show that the inequality of Clausius is satisfied. Use **(a)** the steam tables and **(b)** TK Solver.

Entropy Change for Irreversible Processes

5.76 A 5-lb block of copper at 200°F is submerged in 10 lbm of water at 50°F, and after a period of time, equilibrium is established. If the container is insulated, calculate the entropy change of the universe.

5.77 Two kg of saturated steam are contained in 0.2-m^3 rigid volume. Heat is transferred to the surroundings at 30°C until the quality reaches 20 percent. Calculate the entropy change of the universe using **(a)** the steam tables and **(b)** TK Solver.

5.78 One lb of air is contained in a 6-ft^3 volume at a pressure of 30 psia. Heat is transferred to the air from a high-temperature reservoir until the temperature is tripled in value while the pressure is held constant. Determine the entropy change of **(a)** the air, **(b)** the high-temperature reservoir which is at 1000°F, and **(c)** the universe.

5.79 Two kg of air is stored in a rigid volume of 2 m^3 with the temperature initially at 300°C. Heat is transferred from the air until the pressure reaches 120 kPa. Calculate the entropy change of **(a)** the air and **(b)** the universe if the surroundings are at 27°C.

5.80 Three kg of saturated steam at 200°C is cooled at constant pressure until the steam is completely condensed. Calculate the net entropy change of the universe if the surroundings are at 20°C using **(a)** the steam tables and **(b)** TK Solver.

5.81 Steam at a quality of 80 percent is contained in a rigid vessel of a volume of 400 cm^3. The initial pressure is 200 kPa. Energy is added to the steam by heat transfer from a source maintained at 700°C until the pressure is 600 kPa. What is the entropy change of the universe?

The Second Law Applied to Control Volumes

5.82 The feedwater heater shown in Fig. P5.82 is used to preheat water in a power plant cycle. Saturated water leaves the preheater. Calculate the entropy production if all pressures are 60 psia.

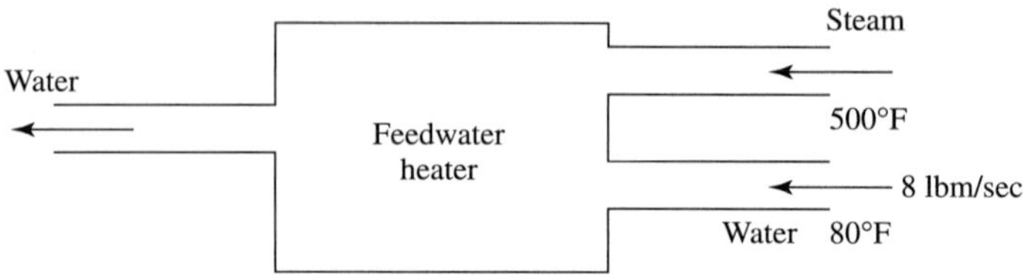

FIGURE P5.82

5.83 A steam turbine accepts 2 kg/s of steam at 6 MPa and 600°C and exhausts saturated steam at 20 kPa while producing 2000 kW of power. If the surroundings are at 30°C and the flow is steady, calculate the rate of entropy production using **(a)** the steam tables and **(b)** TK Solver.

5.84 A rigid tank is sealed when the temperature is 0°C. On a hot day, the temperature in the tank reaches 50°C. If a small hole is drilled in the tank, estimate the velocity of the escaping air.

5.85 Air flows from a tank maintained at 140 kPa and 27°C from a 25-mm-diameter hole. Estimate the mass flow rate from the hole assuming an isentropic process.

5.86 Air flows from a nozzle in which the diameter is reduced from 100 mm to 40 mm. The inlet conditions are 130 kPa and 150°C with a velocity of 40 m/s. Assuming an isentropic process, calculate the exit velocity if the exit pressure is 85 kPa.

5.87 The gases flowing through a turbine have essentially the same properties as air. The inlet gases are at 800 kPa and 900°C and the exit pressure is atmospheric at 90 kPa. Compute the work output assuming an isentropic process if **(a)** the specific heats are constant and **(b)** the gas tables are used.

5.88 Saturated steam at 300°F is compressed to a pressure of 800 psia. The device used for the compression process is well insulated. Assuming the process to be reversible, calculate the power needed if 6 lbm/sec of steam is flowing using **(a)** the steam tables and **(b)** TK Solver.

5.89 Steam expands isentropically through a turbine from 6 MPa and 600°C to 10 kPa. Calculate the power output if the mass flow rate is 2 kg/s.

5.90 A steam turbine produces 3000 hp from a mass flow rate of 20,000 lbm/hr. The steam enters at 1000°F and 800 psia and exits at 2 psia. Calculate the efficiency of the turbine using **(a)** the steam tables and **(b)** TK Solver.

5.91 Every second 3.5 kg of superheated steam flows through the turbine shown in Fig. P5.91. Assuming an isentropic process, calculate the maximum power rating of this turbine.

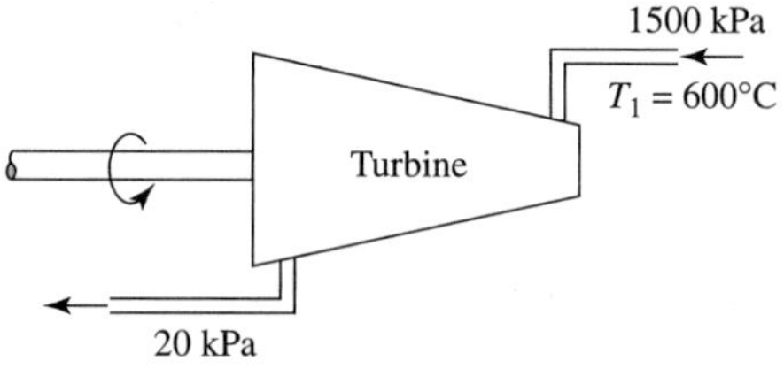

FIGURE P5.91

5.92 Two hundred kW is to be produced by a steam turbine. The outlet steam is to be saturated at 80 kPa and the steam entering will be at 600°C. For an isentropic process, determine the mass flow rate of steam using **(a)** the steam tables and **(b)** TK Solver.

5.93 A turbine produces 3 MW by extracting energy from 4 kg of steam which flows through the turbine every second. The steam enters at 250°C and 1500 kPa and exits as saturated steam at 2 kPa. Calculate the turbine efficiency.

5.94 A steam turbine is 85% efficient. Steam enters at 900°F and 300 psia and leaves at 4 psia. **(a)** How much energy can be produced? **(b)** If 3000 hp must be produced, what must the mass flow rate be?

5.95 Calculate the efficiency of the nozzle of Fig. 5.95 that accelerates air at 20°C and 115 kPa from 10 m/s to 150 m/s at atmospheric pressure of 100 kPa.

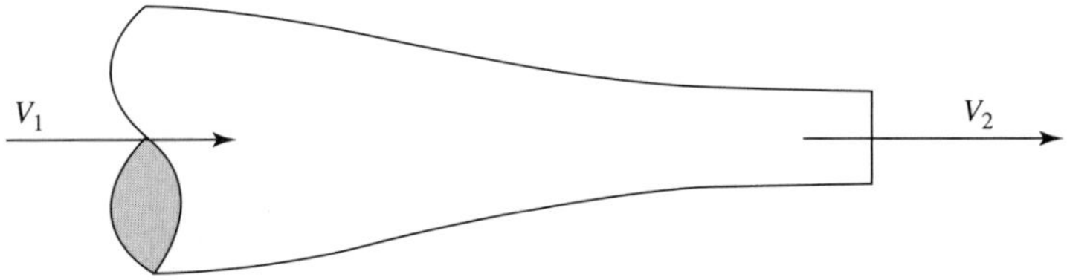

FIGURE P5.95

5.96 A nozzle accelerates steam at 120 kPa and 200°C from 20 m/s to the atmosphere. If it is 85% efficient, estimate the exiting velocity using **(a)** the steam tables and **(b)** TK Solver.

The steam, exhausted by geysers from the Earth's core, can be used to provide the energy input to a power plant. (Image courtesy of Getty Images.)

6

Power and Refrigeration Vapor Cycles

Outline

Chapter Objectives

The objectives of this chapter are to:

- ▲ Present the Rankine power cycle.
- ▲ Introduce improvements to the Rankine cycle.
- ▲ Discuss losses in the Rankine cycle.
- ▲ Present the vapor-refrigeration cycle.
- ▲ Demonstrate how the refrigeration cycle provides heat.
- ▲ Provide numerous examples and practice problems that illustrate the concepts introduced.

Nomenclature

The following is the nomenclature introduced in this chapter:

Q_C	Heat transfer from the condenser	W_T	Work from the turbine
		W_P	Work required by the pump
Q_E	Heat transfer from the evaporator	η_P	Pump efficiency
Q_B	Heat transfer to the boiler	η_T	Turbine efficiency

6.1 INTRODUCTION

The ideal Carnot cycle is used as a model against which all real and all other ideal cycles are compared. The efficiency of a Carnot power cycle is the maximum possible for any power cycle; it was shown in Chapter 5 to be given by

$$\eta = 1 - \frac{T_L}{T_H} \tag{6.1}$$

Note that the efficiency is increased by raising the temperature T_H at which heat is added or by lowering the temperature T_L at which heat is rejected. We will observe that this carries over to real cycles: using the highest maximum temperature and the lowest minimum temperature maximizes the cycle efficiency.

Key Concept: *Using the highest maximum temperature and the lowest minimum temperature maximizes the cycle efficiency.*

We will discuss vapor cycles that are used to generate power, then vapor cycles that are used to refrigerate or heat a space. In the next chapter, we will examine both power and refrigeration cycles that use air as the working fluid.

6.2 THE RANKINE CYCLE

The first class of power cycles that we consider are those utilized by the electric power generating industry, namely, power cycles that operate in such a way that the working fluid changes phase from a liquid to a vapor. The simplest vapor-power cycle is called the *Rankine cycle*, shown schematically in Fig. 6.1a. A major feature of such a cycle is that the pump requires very little work to deliver high-pressure water to the boiler. A possible disadvantage is that the expansion process in the turbine usually enters the quality region, resulting in the formation of liquid droplets that may damage the turbine blades.

The Rankine cycle is an idealized cycle in which friction losses in each of the four components are neglected. The losses usually are quite small and will be neglected completely in our initial analysis. The Rankine cycle is composed of the four ideal processes shown on the *T*-*s* diagram in Fig. 6.1b:

1 ⟶ 2: Isentropic compression in a pump
2 ⟶ 3: Constant-pressure heat addition in a boiler

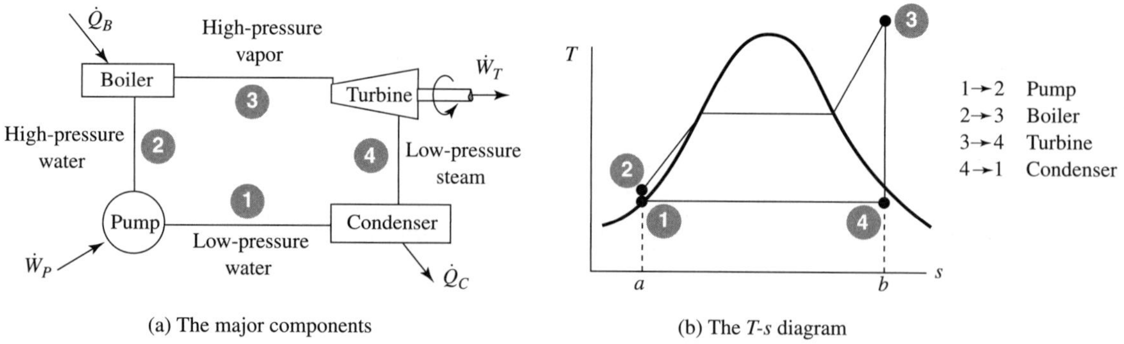

FIGURE 6.1 The Rankine cycle.

3 ⟶ 4: Isentropic expansion in a turbine
4 ⟶ 1: Constant-pressure heat rejection in a condenser

The pump is used to increase the pressure of the saturated liquid. Actually, states 1 and 2 are essentially the same, since the high-pressure lines are extremely close to the saturation curve; they are shown separated for illustration only. The *boiler* (also called a *steam generator*) and the *condenser* are heat exchangers that neither require nor produce any work. An energy balance was applied to similar devices of the Rankine cycle in Section 4.9.

Key Concept: *The* boiler *and the* condenser *are heat exchangers.*

If we neglect kinetic energy and potential energy changes, the net work output is the area under the *T-s* diagram, represented by area 1-2-3-4-1 of Fig. 6.1b; this is true since the first law requires that $W_{net} = Q_{net}$. The heat transfer to the working substance is represented by area *a*-2-3-*b*-*a*. Thus, the thermal efficiency η of the Rankine cycle is

$$\eta = \frac{\text{area 1-2-3-4-1}}{\text{area } a\text{-2-3-}b\text{-}a} \tag{6.2}$$

that is, the desired output divided by the energy input (the purchased energy). Obviously, the thermal efficiency can be improved by increasing the numerator or by decreasing the denominator. This can be done by increasing the pump outlet pressure p_2, increasing the boiler outlet temperature T_3, or decreasing the turbine outlet pressure p_4. Examples will illustrate.

Note that the efficiency of the Rankine cycle is significantly less than that of a Carnot cycle operating between the high temperature T_3 and the low temperature T_1, since most of the heat transfer from a high-temperature reservoir occurs across large temperature differences, a situation that occurs in most heat exchangers. Recall that heat transfer across a finite temperature difference is irreversible, i.e., a source of loss.

Let us now find expressions for the work and the heat transfer of the four processes of Fig. 6.1. Using Section 4.9 and referring to Fig. 6.1b, the pump, the boiler, the turbine, and the condenser have the following respective expressions:

$$w_P = v_1(p_2 - p_1) \qquad q_B = h_3 - h_2 \qquad w_T = h_3 - h_4 \qquad q_C = h_4 - h_1 \tag{6.3}$$

where w_P and q_C are expressed as positive quantities. In terms of the above, the thermal efficiency is

$$\eta = \frac{w_T - w_P}{q_B} \tag{6.4}$$

The pump work is usually quite small, however, compared to the turbine work and can most often be neglected. With this approximation, there results

Key Concept: *The pump work is small and can most often be neglected.*

$$\eta = \frac{w_T}{q_B} \tag{6.5}$$

This is the relation most often used for the thermal efficiency of the Rankine cycle.

Example 6.1

A steam power plant is proposed to operate between the pressures of 10 kPa and 2 MPa with a maximum temperature of 400°C, as shown in Fig. E6.1. Determine the maximum efficiency possible from the power cycle using (a) the steam tables and (b) TK Solver.

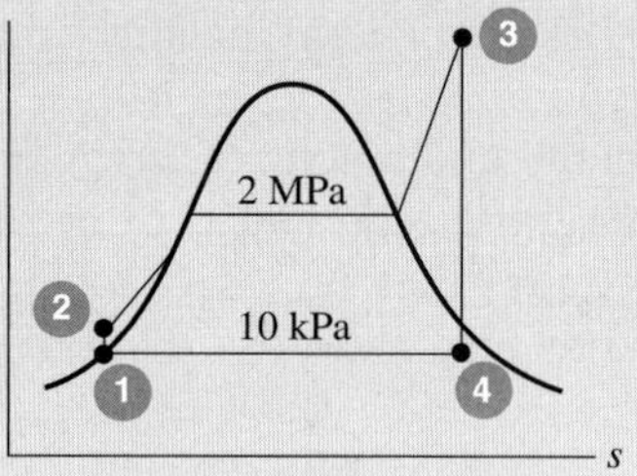

FIGURE E6.1

Solution

Let us include the pump work in the calculation and show that it is negligible. Also, we will assume a unit mass of working fluid, since we are only interested in the efficiency. The pump work is (see Eq. 4.71)

$$w_P = \frac{p_2 - p_1}{\rho} = \frac{2000 - 10}{1000} = 1.99 \text{ kJ/kg}$$

(a) From the steam tables in the Appendix, we find the values listed above. Using Eq. 4.67, we find that

$$h_2 = h_1 + w_P = 191.8 + 1.99 = 193.8 \text{ kJ/kg}$$

The heat input is found to be

$$q_B = h_3 - h_2 = 3248 - 193.8 = 3054 \text{ kJ/kg}$$

To locate state 4, we recognize that $s_4 = s_3 = 7.1279$. Hence,

$$s_4 = s_f + x_4 s_{fg} \qquad \therefore\ 7.1279 = 0.6491 + 7.5019 x_4$$

giving the quality of state 4 as $x_4 = 0.8636$. This allows us to find h_4 to be

$$h_4 = h_f + x_4 h_{fg} = 192 + 0.8636 \times 2393 = 2259 \text{ kJ/kg}$$

The work output from the turbine is

$$w_T = h_3 - h_4 = 3248 - 2259 = 989 \text{ kJ/kg}$$

Consequently, the efficiency is

$$\eta = \frac{w_T - w_P}{q_B} = \frac{989 - 1.99}{3054} = 0.3232 \quad \text{or} \quad 32.32\%$$

(continued)

Obviously, the work required in the pumping process is negligible, being only 0.2 percent of the turbine work. In engineering applications, we often neglect quantities that have an influence of less than 3 percent, since invariably there is some quantity in the calculations that is known to only ±3 percent (i.e., the mass flow rate, the dimensions of a pipe, or the density of the fluid).

(b) Using TK Solver, we find at states 1 and 3: $v_1 = 0.00101\ \text{m}^3/\text{kg}$, $h_1 = 191.9$ kJ/kg, $h_3 = 3247.5$ kJ/kg, and $s_3 = 7.1269$ kJ/kg·K. Then, with $s_4 = s_3$ and $p_4 = 10$ kPa, $h_4 = 2258.1$ kJ/kg. Using these values,

$$w_P = 0.00101 \times (2000 - 10) = 2.01\ \text{kJ/kg}$$

$$q_B = 3247.5 - (191.9 + 2.01) = 3053.6\ \text{kJ/kg}$$

$$w_T = 3247.5 - 2258.1 = 989.4\ \text{kJ/kg}$$

$$\therefore\ \eta = \frac{989.4 - 2.01}{3053.6} \times 100 = 32.34\%$$

6.3 A POSSIBLE STEAM CARNOT CYCLE

It is possible for the efficiency of a modified Rankine cycle to be equal to that of a Carnot cycle if the cycle is designed to operate as shown in Fig. 6.2a. However, the pump would be required to pump a mixture of liquid and vapor, a rather difficult and work-consuming task compared to pumping all liquid. In addition, the condensation of liquid droplets in the turbine would result in severe damage. To avoid the damage from droplets, one could propose superheating the steam at constant temperature, as shown in Fig. 6.2b. This, however, requires that the pressure for the constant-temperature superheated portion of the process decrease from the saturated-vapor point to state 3. To achieve such a decrease, the flow in the boiler pipes would have to be accelerated, a task that would require pipes of decreasing diameter. This would be prohibitively expensive, should it be attempted. Thus, it is proposed that we pump only liquid and that p_2 and T_3 be quite large (T_3 being limited by the temperature-resistance characteristics of the pipe metal and the first rows of blades in the turbine, typically about 600°C although modern research on heat-resistant materials is pushing that limit up considerably) as shown in Fig. 6.2c. It is also proposed that the condenser outlet

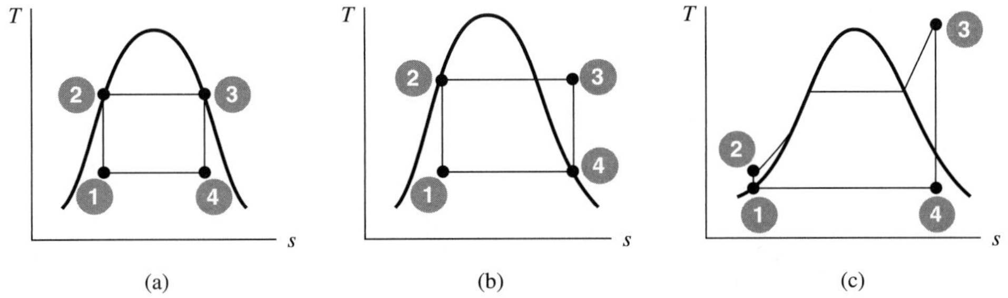

FIGURE 6.2 Modified Rankine cycles in (a) and (b). Conventional Rankine cycle in (c).

pressure be very low (it can be quite close to absolute zero). This would, however, result in state 4 being in the quality region (a quality of 90 percent is too low), causing water droplets to form. Because of these problems, the Carnot steam cycle is not practical.

6.4 RANKINE CYCLE EFFICIENCY

The efficiency of the Rankine cycle can be improved by increasing the boiler pressure while maintaining the maximum temperature and the minimum pressure. The net increase in work output is the crosshatched area minus the shaded area of Fig. 6.3a, a relatively small change; the added heat, however, decreases by the shaded area minus the crosshatched area of Fig. 6.3b. This is a significant decrease, and it leads to a significant increase in efficiency. Example 6.2 illustrates this effect. The disadvantage of raising the boiler pressure is that the quality of the steam exiting the turbine may become too low (less than 95 percent), resulting in severe water-droplet damage to the turbine blades and impaired turbine efficiency.

Increasing the maximum temperature also results in an improvement in thermal efficiency of the Rankine cycle. In Fig. 6.4a, the net work is increased by the

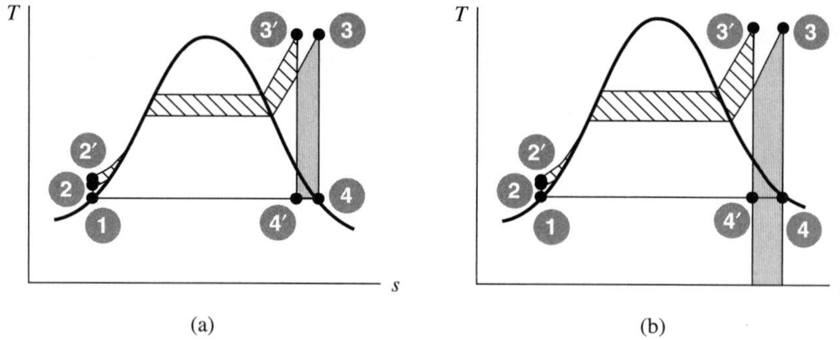

FIGURE 6.3 Effect of increased pressure on the Rankine cycle.

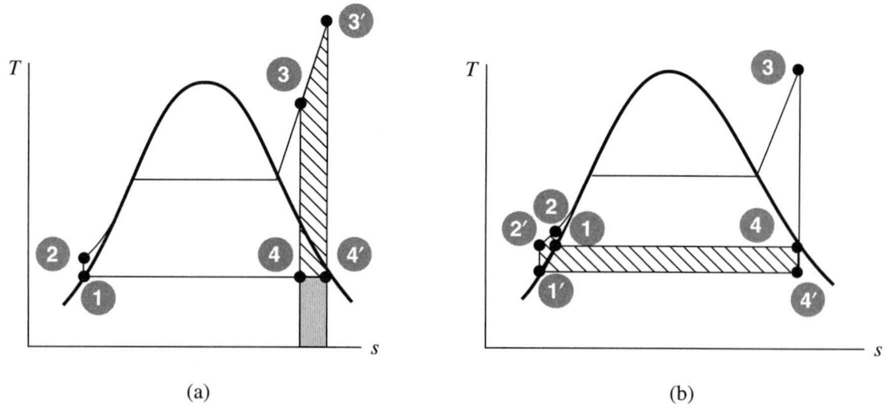

FIGURE 6.4 Effect of increased temperature and decreased condenser pressure.

crosshatched area, and the heat input is increased by the sum of the crosshatched area and the shaded area (a smaller percentage increase than the work increase). Since the numerator of Eq. 6.5 realizes a larger percentage increase than the denominator, there will be a resulting increase in efficiency. This will be illustrated in Example 6.3. Of course, metallurgical considerations limit the maximum temperature that can be attained in the boiler. Temperatures are limited to about 600°C by the turbine blades near the turbine inlet; ceramic blades with imbedded metal may allow much higher temperatures in the future. Another advantage of raising the boiler temperature is that the quality of state 4 is obviously increased; this reduces water-droplet formation in the turbine.

A decrease in condenser pressure, illustrated in Fig. 6.4b, will also result in increased Rankine cycle efficiency. The net work will increase a significant amount, represented by the crosshatched area, and the heat input will increase a slight amount because state 1′ will move to a slightly lower enthalpy than that of state 1; the net result will be an increase in the Rankine cycle efficiency. The heat transfer process that occurs in the condenser limits the low pressure. The heat is rejected by transferring heat to cooling water or to air which enters the condenser at about 20°C; the heat transfer process requires a temperature differential between the cooling water and the steam of at least 10°C. Hence, a temperature of at least 30°C is required in the condenser; this corresponds to a minimum condenser pressure (see the saturated steam tables in the Appendix) of approximately 4 kPa. This is, of course, dependent on the temperature of the cooling water and the temperature differential required in the heat exchanger.

Example 6.2

Increase the boiler pressure of Example 6.1 to 4 MPa while maintaining the maximum temperature and the minimum pressure. Neglect the work of the pump and calculate the percentage increase in the thermal efficiency using (a) the steam tables and (b) TK Solver.

Solution

(a) The enthalpy h_2 is essentially equal to h_1: $h_2 = 192$ kJ/kg. At 400°C and 4 MPa the enthalpy and entropy are $h_3 = 3214$ kJ/kg and $s_3 = 6.7698$ kJ/kg·K. State 4 is in the quality region. Using $s_4 = s_3$, the quality is found to be

$$x_4 = \frac{s_4 - s_f}{s_{fg}} = \frac{6.7698 - 0.6491}{7.5019} = 0.8159$$

Observe that the moisture content has increased to 18.4 percent, an undesirable result. The enthalpy of state 4 is then

$$h_4 = h_f + x_4 h_{fg} = 192 + 0.8159 \times 2393 = 2144 \text{ kJ/kg}$$

The heat addition is

$$q_B = h_3 - h_2 = 3214 - 192 = 3022 \text{ kJ/kg}$$

(continued)

and the turbine work output is

$$w_T = h_3 - h_4 = 3214 - 2144 = 1070 \text{ kJ/kg}$$

Finally, the thermal efficiency is

$$\eta = \frac{w_T}{q_B} = \frac{1070}{3022} = 0.3541$$

The percentage increase in efficiency from that of Example 6.1 is

$$\% \text{ increase} = \frac{0.3541 - 0.3232}{0.3232} \times 100 = 9.55\%$$

(b) Using TK Solver, we find the following: $h_1 = 191.8$ kJ/kg, $h_3 = 3213.4$ kJ/kg, $s_3 = 6.7688$ kJ/kg·K. Then, $h_4 = 2143.9$ kJ/kg and we have

$$q_B = 3213.4 - 191.8 = 3021.6 \text{ kJ/kg} \qquad w_T = 3213.4 - 2143.9 = 1069.5 \text{ kJ/kg}$$

$$\therefore \eta = \frac{1069.5}{3021.6} \times 100 = 35.40\%$$

The percent increase is

$$\% \text{ increase} = \frac{0.354 - 0.3234}{0.3234} \times 100 = 9.45\%$$

Example 6.3

Increase the maximum temperature in the cycle of Example 6.1 to 600°C, while maintaining the boiler pressure and condenser pressure, and determine the percentage increase in thermal efficiency using (a) the steam tables and (b) TK Solver.

Solution

(a) At 600°C and 2 MPa, the enthalpy and entropy are $h_3 = 3690$ kJ/kg and $s_3 = 7.7032$ kJ/kg·K. State 4 remains in the quality region, and using $s_4 = s_3$, we have

$$x_4 = \frac{s_4 - s_f}{s_{fg}} = \frac{7.7032 - 0.6491}{7.5019} = 0.9403$$

Note here that the moisture content has been decreased to 6.0 percent, a desirable result. The enthalpy of state 4 is now found to be $h_4 = 192 + (0.9403)(2393) = 2442$ kJ/kg. This allows us to calculate the thermal efficiency as

$$\eta = \frac{w_T}{q_B} = \frac{h_3 - h_4}{h_3 - h_2} = \frac{3690 - 2442}{3690 - 192} = 0.3568$$

where h_2 is taken from Example 6.1. The percentage increase is

$$\% \text{ increase} = \frac{0.3568 - 0.3232}{0.3232} \times 100 = 10.4\%$$

(continued)

In addition to a significant increase in efficiency, the quality of the steam exiting the turbine is 94%, a much-improved value.

(b) Using TK Solver, we find $h_1 = 191.8$ kJ/kg, $h_3 = 3690.2$ kJ/kg, $s_3 = 7.7024$ kJ/kg·K, and using $s_4 = s_3$, $h_4 = 2441.7$ kJ/kg. We then calculate

$$q_B = 3690.2 - 191.8 = 3498.4 \text{ kJ/kg} \qquad w_T = 3690.2 - 191.8 = 1248.5 \text{ kJ/kg}$$

$$\therefore \eta = \frac{1248.5}{3498.4} = 0.3569 \quad \text{so that} \quad \% \text{ increase} = \frac{0.3569 - 0.3234}{0.3234} \times 100 = 10.36\%$$

Example 6.4

Decrease the condenser pressure of Example 6.1 to 4 kPa while maintaining the boiler pressure and maximum temperature, and determine the percentage increase in thermal efficiency using (a) the steam tables and (b) TK Solver.

Solution

(a) The enthalpies $h_2 = 192$ kJ/kg and $h_3 = 3248$ kJ/kg remain as stated in Example 6.1. Using $s_3 = s_4 = 7.1279$, with $p_4 = 4$ kPa, we find the quality to be

$$x_4 = \frac{s_4 - s_f}{s_{fg}} = \frac{7.1279 - 0.4225}{8.0529} = 0.8327$$

Note that the moisture content of 16.7 percent has increased from that of Example 6.1, as expected. The enthalpy of state 4 is $h_4 = 121 + (0.8327)(2433) = 2147$ kJ/kg. The thermal efficiency is then

$$\eta = \frac{h_3 - h_4}{h_3 - h_2} = \frac{3248 - 2147}{3248 - 192} = 0.3603$$

The percentage increase is found to be

$$\% \text{ increase} = \frac{0.3603 - 0.3232}{0.3232} \times 100 = 11.5\%$$

This is the largest percentage increase, which shows why the lowest possible condenser pressure is used.

(b) Using TK Solver, we find $h_1 = 191.8$ kJ/kg, $h_3 = 3247.5$ kJ/kg, $s_3 = 7.1269$ kJ/kg·K, and $h_4 = 2146.9$ kJ/kg. Then

$$q_B = 3247.5 - 191.8 = 3055.7 \text{ kJ/kg} \qquad w_T = 3247.5 - 191.8 = 1100.6 \text{ kJ/kg}$$

$$\therefore \eta = \frac{1100.6}{3055.7} = 0.3602 \quad \text{so that} \quad \% \text{ increase} = \frac{0.3602 - 0.3234}{0.3234} \times 100 = 11.38\%$$

6.5 THE REHEAT CYCLE

It is apparent from the previous section that when operating a Rankine cycle with a high boiler pressure or a low condenser pressure it is difficult to prevent liquid droplets from forming in the low-pressure portion of the turbine. Since most

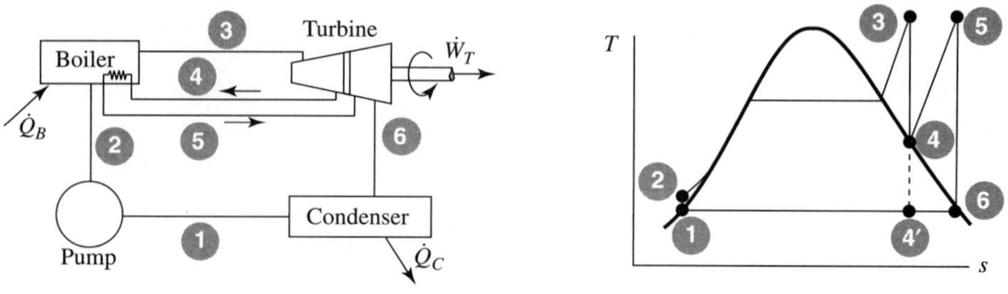

FIGURE 6.5 The reheat cycle.

Key Concept: *The* **reheat** *cycle is used to prevent liquid-droplet formation.*

metals cannot withstand temperatures above about 600°C, the *reheat cycle* is often used to prevent liquid-droplet formation: the steam passing through the turbine is reheated at some intermediate pressure, thereby raising the temperature to state 5 in the *T-s* diagram of Fig. 6.5. The steam then passes through the low-pressure section of the turbine and enters the condenser at state 6. This controls or completely eliminates the moisture problem in the turbine. Often, the turbine is separated into a high-pressure turbine and a low-pressure turbine. The reheat cycle does not significantly influence the thermal efficiency of the cycle, but it does result in a significant additional work output, represented in the figure by area 4-5-6-4′-4 of Fig. 6.5. The reheat cycle demands a significant investment in additional equipment, and the use of such equipment must be economically justified by the increased work output. If reheat is not used to avoid droplet formation, the condenser pressure must be quite high, resulting in a relatively low cycle efficiency. In that sense, reheat significantly increases cycle efficiency when compared to the cycle with no reheat but with the higher condenser pressure.

Example 6.5

High-pressure steam enters a turbine at 600 psia and 1000°F. It is reheated at a pressure of 40 psia to 600°F and then expanded to 2 psia. Determine the cycle efficiency using (a) the steam tables and (b) TK Solver. Refer to Fig. 6.5.

Solution

(a) At 2 psia, saturated water has an enthalpy of (refer to Appendix Table C.2E) $h_1 = h_2 = 94$ Btu/lbm. From Appendix Table C.3E, we find $h_3 = 1518$ Btu/lbm and $s_3 = 1.716$ Btu/lbm-°R. Setting $s_4 = s_3$, we interpolate, obtaining

$$h_4 = \frac{1.716 - 1.712}{1.737 - 1.712}(1217 - 1197) + 1197 = 1200 \text{ Btu/lbm}$$

At 40 psia and 600°F, we have

$$h_5 = 1333 \text{ Btu/lbm} \quad \text{and} \quad s_5 = 1.862 \text{ Btu/lbm-°R}$$

In the quality region, use $s_6 = s_5$ and find

$$x_6 = \frac{s_6 - s_f}{s_{fg}} = \frac{1.862 - 0.175}{1.745} = 0.9668$$

(continued)

Thus, $h_6 = 94 + (0.9668)(1022) = 1082$ Btu/lbm. The energy input and output are

$$q_B = (h_5 - h_4) + (h_3 - h_2) = 1333 - 1200 + 1518 - 94 = 1557 \text{ Btu/lbm}$$

$$w_T = (h_5 - h_6) + (h_3 - h_4) = 1333 - 1082 + 1518 - 1200 = 569 \text{ Btu/lbm}$$

The thermal efficiency is then calculated to be

$$\eta = \frac{w_T}{q_B} = \frac{569}{1557} = 0.365 \quad \text{or} \quad 36.5\%$$

Note: The fact that state 6 is in the mixture region is not of concern, since x_6 is quite close to the saturated-vapor state. As we will see in Section 6.7, turbine losses will move state 6 into the superheat region.

(b) Using TK Solver, we find $h_1 = h_2 = 94.05$ Btu/lbm, $h_3 = 1517.8$ Btu/lbm, $s_4 = s_3 = 1.7157$ Btu/lbm-°R, $h_4 = 1198.4$ Btu/lbm, $h_5 = 1333.3$ Btu/lbm, $s_6 = s_5 = 1.8634$ Btu/lbm-°R, and $h_6 = 1083$ Btu/lbm. Consequently,

$$q_B = 1333.3 - 1198.4 + 1517.8 - 94.05 = 1558.6 \text{ kJ/kg}$$

$$w_T = 1333.3 - 1083 + 1517.8 - 1198.4 = 569.7 \text{ kJ/kg}$$

The cycle efficiency is then

$$\eta = \frac{569.7}{1558.6} \times 100 = 36.49\%$$

6.6 THE REGENERATIVE CYCLE

In the conventional Rankine cycle, as well as in the reheat cycle, a considerable percentage of the total energy input is used to heat the high-pressure water from T_2 to its saturation temperature. The crosshatched area in Fig. 6.6a represents this necessary energy. To reduce this energy, the water could be preheated before it enters the boiler by intercepting some of the steam as it expands in the turbine (for example, at state 5 of Fig. 6.6b) and mixing it with the water as it exits the first of the pumps, thereby preheating the water from T_2 to T_6. This would avoid the necessity of condensing all of the steam, thereby reducing the amount of energy

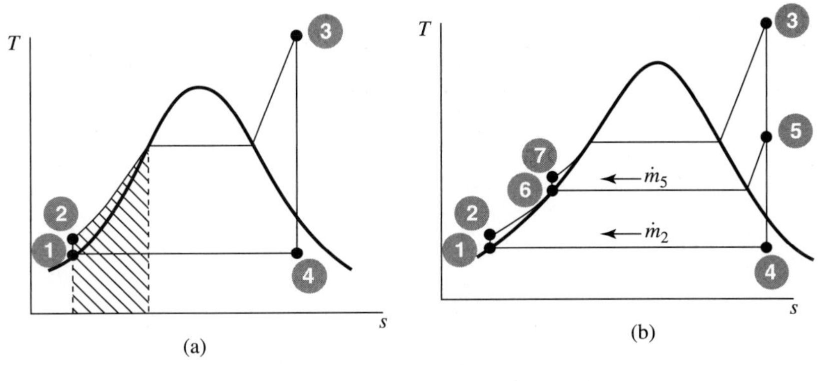

FIGURE 6.6 The regenerative cycle.

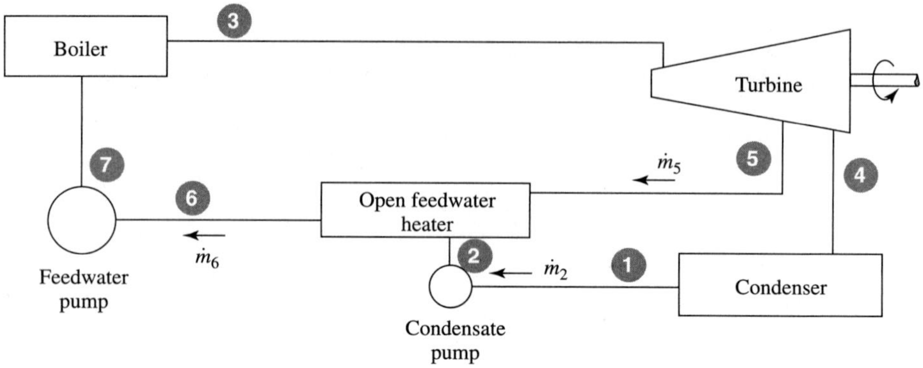

FIGURE 6.7 The major elements of the regenerative cycle.

Regenerative cycle: *A Rankine cycle that utilizes preheated water.*

Feedwater: *The water entering the boiler.*

lost from the condenser. (Note that the use of cooling towers would allow smaller towers for a given energy output.) A cycle that utilizes this type of heating is a **regenerative cycle**, and the process is referred to as *regeneration*. A schematic representation of the major elements of such a cycle is shown in Fig. 6.7. The water entering the boiler is often referred to as **feedwater**, and the device used to mix the extracted steam and the condenser water is called a *feedwater heater*. When the condensate is mixed directly with the steam, it is done so in an *open* feedwater heater, as sketched in Fig. 6.8.

When analyzing a regenerative cycle, we must consider a control volume surrounding the feedwater heater; see Fig. 6.8. A mass balance would result in

$$\dot{m}_6 = \dot{m}_5 + \dot{m}_2 \tag{6.6}$$

An energy balance, assuming an insulated heater, neglecting kinetic and potential energy changes, gives

$$\dot{m}_6 h_6 = \dot{m}_5 h_5 + \dot{m}_2 h_2 \tag{6.7}$$

Combining the above two equations gives

$$\dot{m}_5 = \frac{h_6 - h_2}{h_5 - h_2} \times \dot{m}_6 \tag{6.8}$$

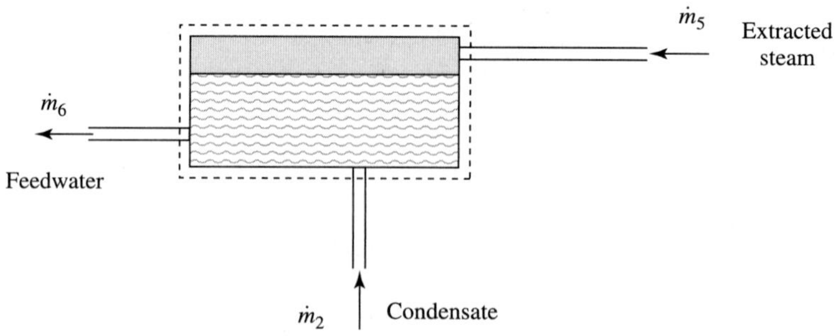

FIGURE 6.8 An open feedwater heater.

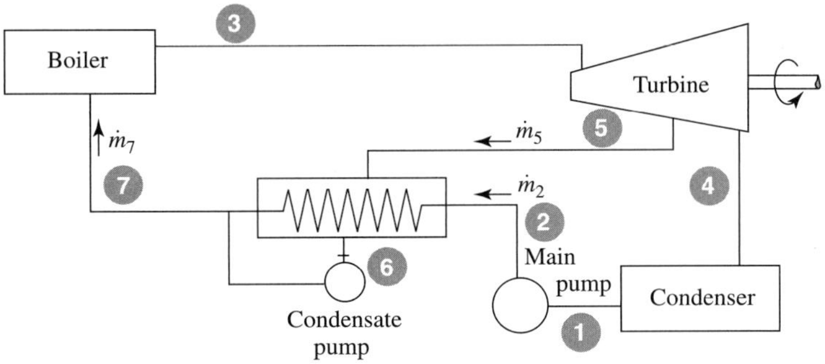

FIGURE 6.9 A closed feedwater heater.

A *closed* feedwater heater, which can be designed into a system using only one main pump, is also a possibility. Figure 6.9 is a schematic of a system using a closed feedwater heater. The disadvantages of such a system are that it is more expensive and its heat transfer characteristics are not as desirable as heat transfer in which the steam and water are simply mixed, as in the open heater.

The closed feedwater heater is a heat exchanger in which the water passes through in tubes and the steam surrounds the tubes, condensing on the outer surfaces. The condensate thus formed, at temperature T_6, is pumped with a small condensate pump into the main feedwater line, as shown in Fig. 6.9, or it passes through a trap (a device that permits only liquid to pass through) and is fed back to the condenser or back to a lower-pressure feedwater heater. Mass and energy balances are also required when analyzing a closed feedwater heater; if the pump energy requirement is neglected in the analysis, the same relationship (see Eq. 6.8) results.

The pressure at which the steam should be extracted from the turbine is approximated as follows. For one heater, the steam should be extracted at the point that allows the exiting feedwater temperature T_6 to be midway between the saturated steam temperature in the boiler and the condenser temperature. For several heaters, the temperature difference should be divided as equally as possible.

Obviously, if one feedwater heater improves thermal efficiency, two should improve it more. This is, in fact, true, but two heaters cost more initially and are more expensive to maintain. With a large number of heaters, it is possible to approach the Carnot efficiency but at an unjustifiably high cost. Small power plants may have two heaters; large power plants have as many as six.

The regenerative cycle is afflicted by the moisture problem in the low-pressure portions of the turbine; hence, it is not uncommon to combine a reheat cycle and a regenerative cycle, thereby avoiding the moisture problem and increasing the thermal efficiency. A possible combination cycle is shown in Fig. 6.10. Ideal efficiencies significantly higher than for nonregenerative cycles can be realized with this combination cycle.

Key Concept: *Combine a reheat cycle and a regenerative cycle to avoid the moisture problem and increase the thermal efficiency.*

There is an additional technique to effectively increase the efficiency of a power plant. The efficiency of a cycle is calculated using the turbine work output as the desired output and considering the rejected heat from the condenser as lost

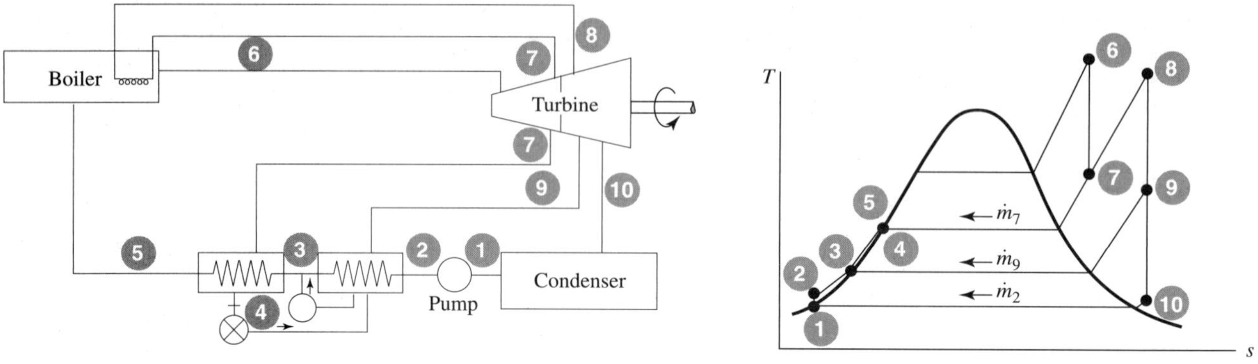

FIGURE 6.10 A combined reheat-regenerative cycle.

Key Concept: *Often, one-half of the rejected heat can be effectively used, almost doubling the "efficiency" of a power plant.*

energy. There are special situations where a power plant can be located strategically so that the rejected steam can be utilized to heat or cool buildings or the steam can be used in various industrial processes. This is often referred to as *cogeneration*. Often, one-half of the rejected heat can be effectively used, almost doubling the "efficiency" of a power plant. Steam or hot water cannot be transported very far; thus, the power plant must be located close to an industrial area or a densely populated area. A college campus is an obvious candidate for cogeneration, as are most large industrial plants. Several major cities are using rejected heat in downtown areas for cooling (absorption refrigeration) as well as heating.

Example 6.6

The high-temperature situation of Example 6.3 is to be modified by inserting an open feedwater heater with an extraction pressure of 200 kPa. Determine the percentage increase in thermal efficiency using (a) the steam tables and (b) TK Solver. Refer to Fig. 6.6b and Fig. 6.7.

Solution

(a) We have, from Example 6.3 and the steam tables,

$$h_1 \cong h_2 = 192 \text{ kJ/kg} \qquad h_6 \cong h_7 = 505 \text{ kJ/kg} \qquad h_3 = 3690 \text{ kJ/kg} \qquad h_4 = 2442 \text{ kJ/kg}$$

Now, locate state 5. Using $s_5 = s_3 = 7.7032$ kJ/kg·K, we interpolate and find, at 200 kPa,

$$h_5 = \frac{7.7032 - 7.5074}{7.7094 - 7.5074}(2971 - 2870) + 2870 = 2968 \text{ kJ/kg}$$

We now apply the conservation of mass and the first law to a control volume surrounding the feedwater heater. For every kilogram of steam passing through the boiler (i.e., $m_6 = 1$ kg), we use Eq. 6.8 to find m_5:

$$m_5 = \frac{h_6 - h_2}{h_5 - h_2} \times m_6 = \frac{505 - 192}{2968 - 192} \times 1 = 0.1128 \text{ kg} \qquad \text{and} \qquad m_2 = 0.8872 \text{ kg}$$

(continued)

The work output from the turbine is

$$w_T = h_3 - h_5 + (h_5 - h_4)m_2 = 3690 - 2968 + (2968 - 2442) \times 0.8872 = 1189 \text{ kJ}$$

The energy input to the boiler is $q_B = h_3 - h_7 = 3690 - 505 = 3185$ kJ/kg. The thermal efficiency is calculated to be

$$\eta = \frac{w_T}{q_B} = \frac{1189}{3185} = 0.3733$$

The increase in efficiency is

$$\% \text{ increase} = \frac{0.3733 - 0.3568}{0.3568} \times 100 = 4.62\%$$

(b) Using TK Solver, we find $h_1 = h_2 = 191.8$ kJ/kg, $h_3 = 3690.2$ kJ/kg, $s_3 = s_4 = s_5 = 7.7024$ kJ/kg·K. Then, $h_5 = 2967.8$ kJ/kg and $h_6 = h_7 = 504.8$ kJ/kg. Assuming $m_6 = 1$ kg, use Eq. 6.8 to find

$$m_5 = \frac{504.8 - 191.8}{2967.8 - 191.8} = 0.1128 \text{ kg} \qquad \text{and} \qquad m_2 = 0.8872 \text{ kg}$$

There results

$$q_B = 3690.2 - 504.8 = 3185.4 \text{ kJ/kg}$$

$$w_T = 3690.2 - 2967.8 + (2967.8 - 2441.7) \times 0.8872 = 1189.2 \text{ kJ/kg}$$

The increase in efficiency is found as follows:

$$\eta = \frac{1189.2}{3185.4} = 0.3733 \quad \text{so that} \quad \% \text{ increase} = \frac{0.3733 - 0.3568}{0.3568} \times 100 = 4.62\%$$

Example 6.7

An open feedwater heater is added to the reheat cycle of Example 6.5. Steam is extracted where the reheater interrupts the turbine flow. Determine the efficiency of this reheat-regeneration cycle.

Solution

A *T-s* diagram (Fig. E6.7a) is sketched below to aid in the calculations. From the steam tables or from Example 6.5,

$$h_1 \cong h_2 = 94 \text{ Btu/lbm} \qquad h_7 \cong h_8 = 236 \text{ Btu/lbm} \qquad h_3 = 1518 \text{ Btu/lbm}$$

$$h_5 = 1333 \text{ Btu/lbm} \qquad h_6 = 1082 \text{ Btu/lbm} \qquad h_4 = 1200 \text{ Btu/lbm}$$

Continuity using $m_8 = 1$ lbm and the first law applied to the heater give (see Eq. 6.8 and Fig. E6.7b)

$$m_4 = \frac{h_8 - h_2}{h_4 - h_2} \times m_8 = \frac{236 - 94}{1200 - 94} \times 1 = 0.128 \text{ lbm} \qquad \text{and} \qquad m_2 = 0.872 \text{ lbm}$$

(continued)

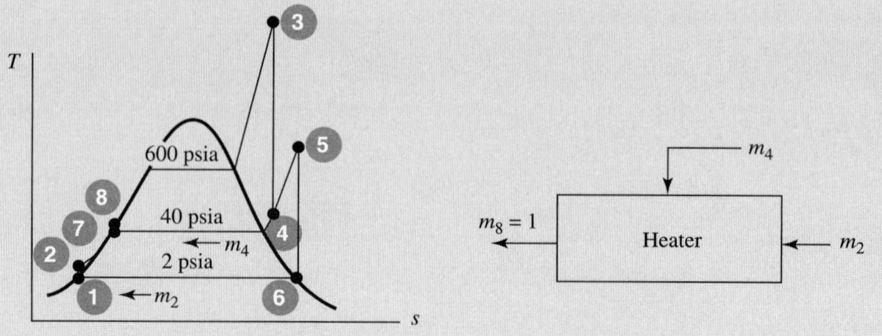

(a) (b)

FIGURE E6.7

The turbine work output is then

$$W_T = h_3 - h_4 + (h_5 - h_6) \times m_2 = 1518 - 1200 + (1333 - 1082) \times 0.872 = 537 \text{ Btu/lbm}$$

The energy input is $q_B = h_3 - h_5 = 1518 - 236 + (1333 - 1200) \times 0.872 = 1398$ Btu/lbm. The efficiency is calculated to be

$$\eta = \frac{w_T}{q_B} = \frac{537}{1398} = 0.384 \quad \text{or} \quad 38.4\%$$

Note the significant improvement of 5.2 percent in cycle efficiency.

6.7 EFFECT OF LOSSES ON POWER CYCLE EFFICIENCY

Key Concept: *The losses in the combustion process and in the fluid in the pipes of the boiler are not included here.*

Key Concept: *Only one loss is accounted for when we calculate cycle efficiency: the loss in the turbine.*

The preceding sections dealt with ideal cycles, assuming no pressure drop through the pipes in the boiler, no losses as the superheated steam passes over the blades in the turbine, no subcooling of the water leaving the condenser, and no pump losses during the compression process. The losses in the combustion process and the inefficiencies in the subsequent heat transfer to the fluid in the pipes of the boiler are not included here; those losses, which are in the neighborhood of 15 percent of the energy contained in the coal or oil, would be included in the overall *plant efficiency*.

There is actually only one substantial loss that must be accounted for when we calculate the actual cycle efficiency: the loss that occurs when the steam is expanded through the rows of turbine blades in the turbine. As the steam passes over a turbine blade, there is friction on the blade, and the steam may separate from the rear portion of the blade. In addition, heat transfer from the turbine may occur, although this is usually quite small. These losses result in a turbine efficiency of 80 to 89 percent. Turbine efficiency is defined as

$$\eta_T = \frac{w_a}{w_s} \tag{6.9}$$

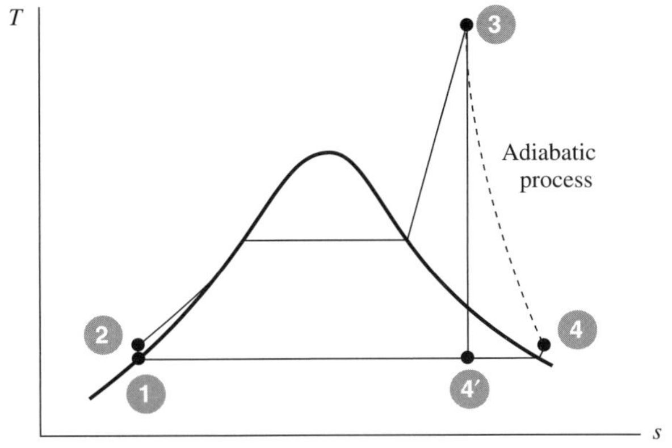

FIGURE 6.11 The Rankine cycle with turbine losses.

where w_a is the actual work and w_s is the isentropic work. The definition of pump efficiency is

$$\eta_P = \frac{w_s}{w_a} \tag{6.10}$$

where the isentropic work input is obviously less than the actual input.

There is a substantial loss in pressure, probably 10 to 20 percent, as the fluid flows through the many meters of pipe from the pump exit through the boiler to the turbine inlet. The loss can be overcome by simply increasing the exit pressure from the pump. This does require more pump work, but the pump work is still less than 1 percent of the turbine output and is thus negligible. Consequently, we ignore the boiler pipe losses. The condenser can be designed to operate such that the exiting water is very close to the saturated liquid condition. This will minimize the condenser losses so that they can also be neglected. The resulting actual Rankine cycle is shown on the *T-s* diagram in Fig. 6.11; the only significant loss is the turbine loss. Note the increase in entropy of state 4 as compared to state 3. Also, note the desirable effect of the decreased moisture content of state 4; in fact, state 4 may even move into the superheated region, as shown.

Example 6.8

A Rankine cycle operates between pressures of 2 MPa and 10 kPa with a maximum temperature of 600°C. If the insulated turbine has an efficiency of 80 percent, calculate the cycle efficiency and the temperature of steam at the turbine outlet using (a) the steam tables and (b) TK Solver.

Solution

(a) From the steam tables, we find $h_1 \cong h_2 = 192$ kJ/kg, $h_3 = 3690$ kJ/kg, and $s_3 = 7.7032$ kJ/kg·K. Setting $s_{4'} = s_3$, we find the quality and enthalpy of state 4′ (see Fig. 6.11) to be

(continued)

$$x_{4'} = \frac{s_{4'} - s_f}{s_{fg}} = \frac{7.7032 - 0.6491}{7.5019} = 0.9403$$

$$\therefore\ h_{4'} = h_f + x_{4'} h_{fg} = 192 + 0.9403 \times 2393 = 2442\ \text{kJ/kg}$$

From the definition of turbine efficiency,

$$\eta_T = \frac{w_a}{w_s} \qquad \text{or} \qquad 0.8 = \frac{w_a}{3690 - 2442} \qquad \therefore\ w_a = 998\ \text{kJ/kg}$$

The cycle efficiency is then

$$\eta = \frac{w_T}{q_B} = \frac{998}{3690 - 192} = 0.285 \quad \text{or} \quad 28.5\%$$

Note the substantial reduction from the ideal cycle efficiency of 35.7 percent as calculated in Example 6.3.

If we neglect kinetic and potential energy changes, the adiabatic process from state 3 to state 4 allows us to write

$$w_a = h_3 - h_4 \qquad \text{or} \qquad 998 = 3690 - h_4 \qquad \therefore\ h_4 = 2692\ \text{kJ/kg}$$

At 10 kPa, we find state 4 in the superheated region. The temperature is interpolated to be

$$T_4 = \frac{2692 - 2688}{2783 - 2688}(150 - 100) + 100 = 102^\circ\text{C}$$

Obviously, the moisture problem has been eliminated by the losses in the turbine; the losses tend to act as a small reheater.

(b) Using TK Solver, we find $h_1 = h_2 = 191.8$ kJ/kg, $h_3 = 3690.2$ kJ/kg, $s_4 = s_3 = 7.7024$ kJ/kg·K, and thus, $h_{4'} = 2441.7$ kJ/kg. Then

$$q_B = 3690.2 - 191.8 = 3498.4\ \text{kJ/kg}$$

$$w_a = \eta_T w_s = 0.8 \times (3690.2 - 2441.7) = 998.8\ \text{kJ/kg}$$

$$\therefore\ \eta = \frac{998.5}{3498.4} \times 100 = 28.55\%$$

The enthalpy of the actual state 4 is

$$h_4 = h_3 - w_a = 3690.2 - 998.8 = 2691.4\ \text{kJ/kg}$$

Using $p_4 = 10$ kPa, TK Solver is used to find state 4 to be in the superheat region with $T_4 = 102.3^\circ$C.

6.8 THE VAPOR-REFRIGERATION CYCLE

It is possible to extract heat from a space by operating a vapor cycle, similar to the Rankine cycle, in reverse. Work input is, of course, required in the operation of such a cycle, as shown in Fig. 6.12a. The work is input by a compressor that increases the pressure, and thereby the temperature, through an isentropic com-

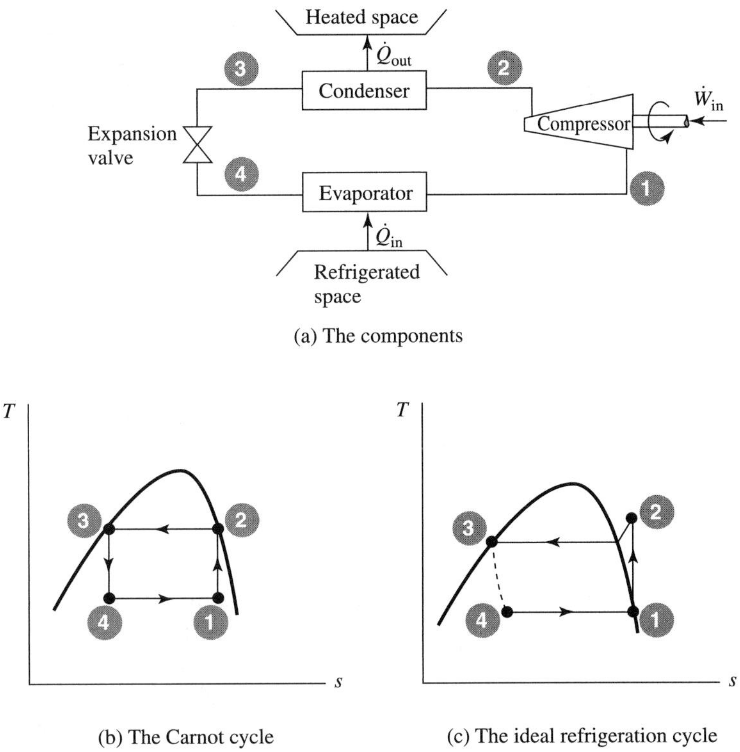

FIGURE 6.12 The vapor refrigeration cycle.

pression process in the ideal cycle. The working fluid (often R134a) then enters a condenser in which heat is extracted, resulting in saturated liquid. The pressure is then reduced in an expansion process so that the fluid can be evaporated with the addition of heat from the refrigerated space.

The most efficient cycle, a Carnot cycle, is shown in Fig. 6.12b. There are, however, two major drawbacks when an attempt is made to put such a cycle into actual operation. First, it is not advisable to compress the mixture of liquid and vapor as represented by state 1 in Fig. 6.12b, since the liquid droplets would cause excessive wear; in addition, equilibrium between the liquid phase and the vapor phase is difficult to maintain in such a process. Hence, in the ideal refrigeration cycle, a saturated vapor state is assumed at the end of the evaporation process; this allows superheated vapor to exist in the compressor, as shown by process 1–2 in Fig. 6.12c. Second, it would be quite expensive to construct a device to be used in the expansion process that would be nearly isentropic (no losses allowed). It is much simpler to reduce the pressure irreversibly by using an expansion valve that employs a throttling process in which enthalpy remains constant, as shown by the dotted line in Fig. 6.12c. Even though this expansion process is characterized by losses, it is considered to be part of the ideal vapor refrigeration cycle. Because the expansion process is a nonequilibrium process, the area under the *T-s* diagram does not represent the net work input.

Key Concept: *In the ideal refrigeration cycle, a saturated vapor state is assumed at the end of the evaporation process.*

Key Concept: *Because the expansion process is a nonequilibrium process, the area under the T-s diagram does not represent the net work input.*

The performance of the refrigeration cycle, when used as a refrigerator, is measured by

$$\mathrm{COP}_{\mathrm{ref}} = \frac{\dot{Q}_{\mathrm{in}}}{\dot{W}_{\mathrm{in}}} \tag{6.11}$$

When the cycle is used as a heat pump, the performance is measured by

$$\mathrm{COP}_{\mathrm{h.p.}} = \frac{\dot{Q}_{\mathrm{out}}}{\dot{W}_{\mathrm{in}}} \tag{6.12}$$

The coefficient of performance can attain values of perhaps 5 for properly designed heat pumps and 4 for refrigerators.

The condensation and evaporation temperatures, and hence, the pressures, are established by the particular situation that motivates the design of the refrigeration unit. For example, in a home refrigerator that is designed to cool the freezer space to −18°C (0°F), it is necessary to design the evaporator to operate at approximately −25°C to allow for effective heat transfer between the space and the cooling coils. The refrigerant condenses by transferring heat to air maintained at about 20°C; consequently, to allow for effective heat transfer from the coils that transport the refrigerant, the refrigerant must be maintained at a temperature of at least 27°C. This is shown in Fig. 6.13.

To accomplish refrigeration for most spaces, it is necessary that the evaporation temperature be quite low (in the neighborhood of −25°C, perhaps). This, of course, rules out water as a possible refrigerant. Two common refrigerants in use today are ammonia (NH_3) and R134a. The thermodynamic properties of R134a are presented in Appendix D and of ammonia in Appendix E. The selection of a refrigerant depends on the high and low design temperatures. For example, temperatures well below −100°C are required to liquefy many gases. Obviously, neither ammonia nor R134a may be used at such low temperatures since they do

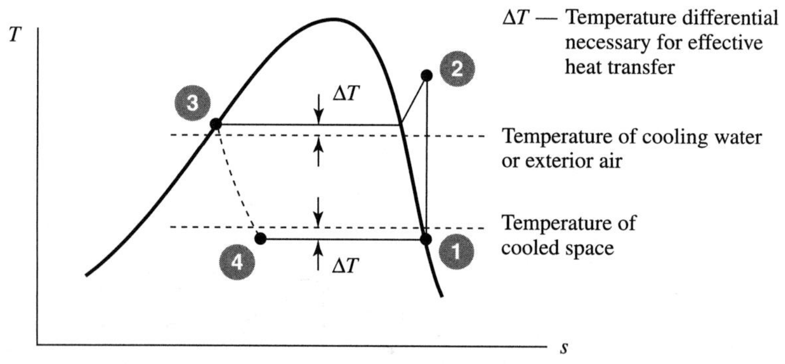

FIGURE 6.13 The refrigeration cycle showing design temperatures.

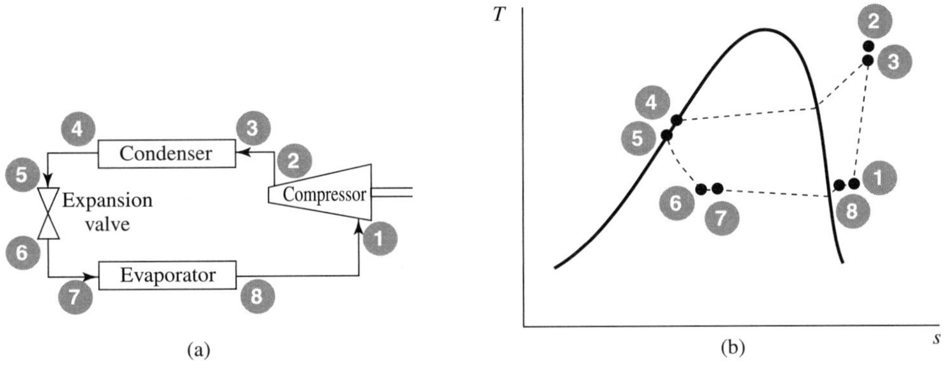

FIGURE 6.14 The actual refrigeration cycle showing losses.

not exist in a liquid form below −100°C. Also, it is desirable to operate a refrigeration cycle such that the low pressure is above atmospheric pressure, thereby avoiding air contamination should a leak occur. In addition, for most applications, the refrigerant must be nontoxic, stable, environmentally benign, and relatively inexpensive.

Deviations from the ideal vapor refrigeration cycle are shown on the *T-s* diagram of Fig. 6.14b. These include:

- Pressure drops due to friction in connecting pipes.
- Heat transfer occurs from or to the refrigerant through the pipes connecting the components.
- Pressure drops occur through the condenser and evaporator tubes.
- Heat transfer occurs from the compressor.
- Frictional effects and flow separation occur on the compressor blades.
- The vapor entering the compressor may be slightly superheated.
- The temperature of the liquid exiting the condenser may be below the saturation temperature.

Some of these effects are small and can be neglected, depending on the location of the components and whether the components and pipes are insulated. Also, not all of the effects are undesirable; the subcooling of the condensate in the condenser allows state 4 in Fig. 6.12c to move to the left, thereby increasing the refrigerant effect. Example 6.10 illustrates the difference between an ideal refrigeration cycle and an actual refrigeration cycle.

Quite often, the size of an air conditioner is rated in tons (a rather interesting unit of rate of heat transfer). A "ton" of refrigeration is supposedly the heat rate necessary to melt a ton of ice in 24 hours. By definition, 1 ton of refrigeration equals 3.52 kW (12,000 Btu/hr). Also, the performance of an air conditioner is often expressed as the **EER** (*the* **Energy Efficiency Rating**): the output in Btu's divided by the energy purchased in kWh. We multiply the COP by 3.412 to obtain the EER.

EER (Energy Efficient Rating): *The output in Btu's divided by the energy purchased in kWh.*

Example 6.9

R134a is used in an ideal vapor refrigeration cycle, operating between saturation temperatures of −26°C in the evaporator and 39.39°C in the condenser. Calculate the rate of refrigeration, the coefficient of performance, and the rating in horsepower per ton if the refrigerant flows at 0.6 kg/s. Also, calculate the coefficient of performance if the cycle is operated as a heat pump. Use (a) the tables and (b) TK Solver.

Solution

(a) The *T-s* diagram in Fig. E6.9 is sketched as an aid in the solution. The enthalpy of each state is needed. From Appendix D, we find that $h_1 = 231.6$ kJ/kg, $h_3 = h_4 = 105.3$ kJ/kg, and $s_1 = 0.939$ kJ/kg·K. Using $s_1 = s_2$, we interpolate at a pressure of 1.0 MPa, which is the pressure associated with the saturation temperature of 39.39°C, and find that

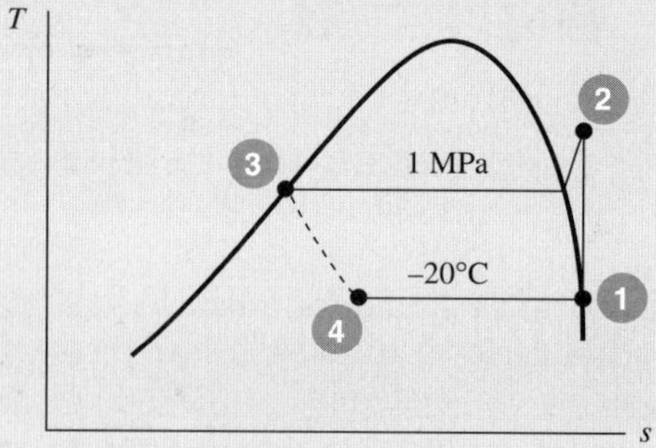

FIGURE E6.9

$$h_2 = \frac{0.939 - 0.9066}{0.9428 - 0.9066}(280.19 - 268.68) + 268.68 = 279.0 \text{ kJ/kg}$$

The rate of refrigeration is measured by the heat transfer rate needed in the evaporation process, namely,

$$\dot{Q}_E = \dot{m}(h_1 - h_4) = 0.6(231.6 - 105.3) = 75.8 \text{ kJ/s}$$

The power needed to operate the compressor is

$$\dot{W}_C = \dot{m}(h_2 - h_1) = 0.6(279.0 - 231.6) = 28.4 \text{ kW}$$

The coefficient of performance is then calculated to be

$$\text{COP} = \frac{\dot{Q}_{\text{in}}}{\dot{W}_C} = \frac{75.8}{28.4} = 2.67$$

The horsepower per ton of refrigeration is determined, with the appropriate conversion of units, as follows:

$$\text{hp/ton} = \frac{28.4/0.746}{75.8/3.52} = 1.77$$

If the above cycle were operated as a heat pump, the coefficient of performance would be

$$\text{COP}_{\text{h.p.}} = \frac{\dot{Q}_{\text{out}}}{\dot{W}_{\text{in}}} = \frac{h_2 - h_3}{h_2 - h_1} = \frac{279.0 - 105.3}{279.0 - 231.6} = 3.67$$

Obviously, the COP for a heat pump is greater than the COP for a refrigerator, since $\dot{Q}_{\text{out}}$ must always be greater than $\dot{Q}_{\text{in}}$. Note, however, that the heat pump in this problem heats the space with 4 times the energy input to the device.

(continued)

(b) Using TK Solver and selecting R134a, we find, at $T_1 = -26°C$ and $x_1 = 1.0$, that $h_1 = 234.8$ kJ/kg, $s_1 = 0.9520$ kJ/kg·K, and $p_1 = 101.6$ kPa. Then, at $T_3 = 39.39°C$ and $x_3 = 0.0$, we find $h_3 = 107.3$ kJ/kg and $p_3 = 1001$ kPa. At $p_2 = 1001$ kPa and $s_2 = 0.9520$ kJ/kg·K, we find $h_2 = 283.9$ kJ/kg. Using $h_3 = h_4$, there results

$$\dot{Q}_E = \dot{m}(h_1 - h_4) = 0.6 \times (234.8 - 107.3) = 76.5 \text{ kW}$$

$$\dot{W}_C = \dot{m}(h_2 - h_1) = 0.6 \times (283.9 - 234.8) = 29.5 \text{ kW}$$

$$\therefore \text{COP}_{\text{ref}} = \frac{76.5}{29.5} = 2.60$$

and

$$\text{COP}_{\text{h.p.}} = \text{COP}_{\text{ref}} + 1 = 3.60$$

Also,

$$\text{hp/ton} = \frac{29.5/0.746}{76.5/3.52} = 1.82$$

Example 6.10

The ideal refrigeration cycle of Example 6.9 is used in the operation of an actual refrigerator. It experiences the following real effects:

- The refrigerant leaving the evaporator is superheated to $-20°C$.
- The refrigerant leaving the condenser is subcooled to $40°C$.
- The compressor is 80 percent efficient.

Calculate the actual rate of refrigeration and the coefficient of performance.

Solution

From Appendix D, we find, using $T_3 = 40°C$, that $h_3 = h_4 = 106.2$ kJ/kg. Also, from Table D.1, we observe that $p_1 = 0.102$ MPa. From Table D.3, at $p_1 = 0.10$ MPa and $T_1 = -20°C$,

$$h_1 = 236.5 \text{ kJ/kg} \qquad s_1 = 0.960 \text{ kJ/kg·K}$$

If the compressor were isentropic, then, with $s_{2'} = s_1$ and $p_2 = 1.0$ MPa,

$$h_{2'} = \frac{0.960 - 0.9428}{0.9768 - 0.9428}(291.36 - 280.19) + 180.19 = 285.8 \text{ kJ/kg}$$

From the definition of efficiency, $\eta = w_s/w_a$, we have

$$0.8 = \frac{h_{2'} - h_1}{h_2 - h_1} = \frac{285.8 - 236.5}{h_2 - 236.5} \qquad \therefore h_2 = 298.1 \text{ kJ/kg}$$

The rate of refrigeration is

$$\dot{Q}_E = 0.6(185 - 74.5) = 66.3 \text{ kW}$$

(continued)

The coefficient of performance becomes

$$\mathrm{COP} = \frac{q_{\mathrm{in}}}{w_C} = \frac{78.2}{298.1 - 236.5} = 2.12$$

The 42 percent decrease in the COP occurs because the power input to the compressor has increased substantially.

6.9 THE HEAT PUMP

The heat pump utilizes the vapor refrigeration cycle discussed in Sec. 6.8. It can be used to heat a house in cool weather or to cool a house in warm weather, as shown schematically in Fig. 6.15. Note that, in the heating mode, the house gains heat from the condenser, whereas, in the cooling mode, the house loses heat to the evaporator. This is possible since the evaporator and the condenser are similar heat exchangers. In an actual situation, valving is used to perform the desired switching of the heat exchangers.

A heat pump system is sized to meet the heating load or the cooling load, whichever is greater. In southern areas where the cooling loads are extremely large, the system may be oversized for the small heating demand of a chilly night; an air conditioner with an auxiliary heating system may be advisable in those cases. In a northern area where the large heating load demands a relatively large heat pump, the cooling load on a warm day may be too low for effective use of the heat pump; the large cooling capacity would quickly reduce the temperature of the house without a simultaneous reduction in the humidity (a necessary feature of any cooling system). In that case, a furnace that provides the heating with an auxiliary cooling system is usually advisable. Or, the heat pump could be designed based on the cooling load with an auxiliary heater for times of heavy heating demands.

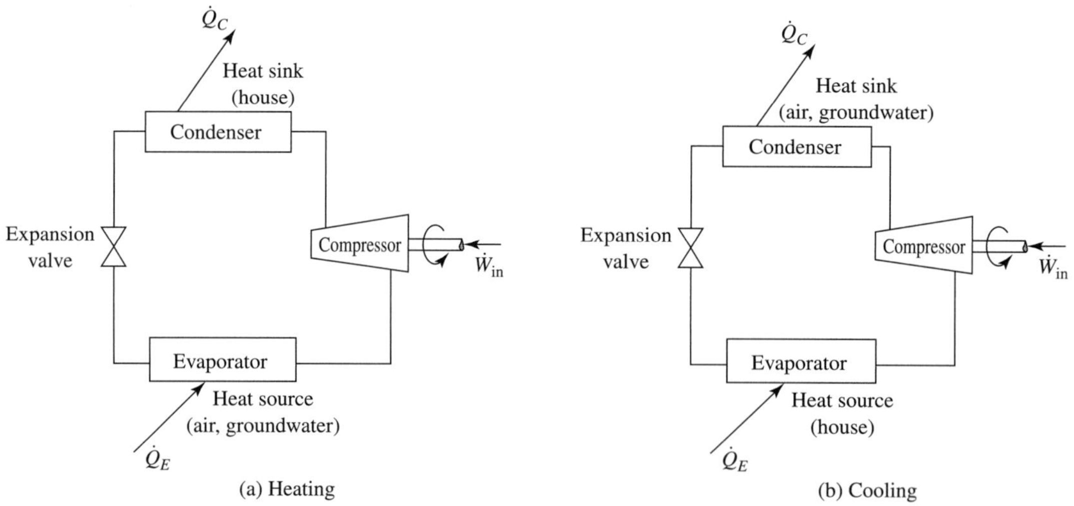

FIGURE 6.15 The heat pump.

Example 6.11

A heat pump using R134a is proposed for heating a home that requires a maximum heating load of 300 kW. The evaporator operates at −12°C and the condenser at 900 kPa. Assume an ideal cycle.

(a) Determine the COP.
(b) Determine the cost of electricity at \$0.07/kWh at peak load.
(c) Compare the R134a system with the cost of operating a furnace at a peak load using natural gas at \$0.50/therm if there are 100 000 kJ/therm of natural gas.

Solution

(a) The *T-s* diagram (Fig. E6.11) is sketched for reference. From Appendix D, we find $h_1 = 240.2$ kJ/kg, $s_1 = s_2 = 0.9267$ kJ/kg·K, and $h_3 = h_4 = 99.56$ kJ/kg. Interpolating at state 2, there results

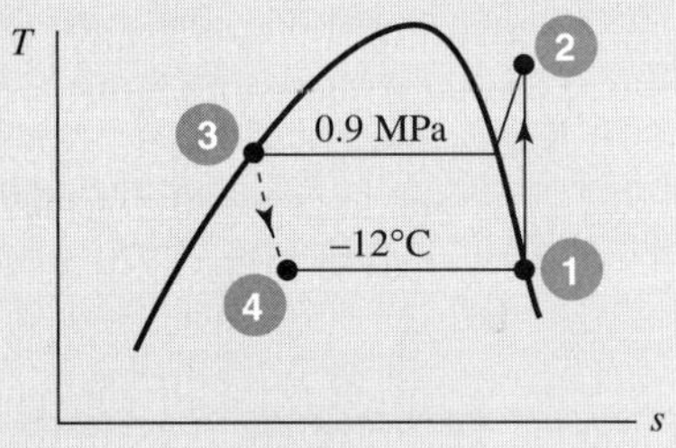

FIGURE E6.11

$$h_2 = \frac{0.9267 - 0.9217}{0.9566 - 0.9217}(282.34 - 271.25) + 271.25 = 272.8 \text{ kJ/kg}$$

The heat rejected by the condenser is

$$\dot{Q}_C = \dot{m}(h_2 - h_3) \qquad 300 = \dot{m}(272.8 - 99.56)$$

This gives the refrigerant mass-flow rate as $\dot{m} = 1.731$ kg/s. The required power by the compressor is then

$$\dot{W}_{\text{in}} = \dot{m}(h_2 - h_1) = 1.731 \times (272.8 - 240.2) = 56.4 \text{ kW}$$

This results in a coefficient of performance of

$$\text{COP} = \frac{\dot{Q}_C}{\dot{W}_{\text{in}}} = \frac{300}{56.4} = 5.32$$

(b) Cost of electricity = 56.4 kW × \$0.07/kWh = \$3.56/h

(c) Assuming the gas furnace to be ideal (i.e., it converts all of the energy of the gas into usable heat), we have

$$\text{Cost of gas} = \frac{300 \times 3600}{100\,000} \times 0.50 = \$5.40/\text{h}$$

The above uses a heat pump that is assumed to operate ideally. A heat pump operates significantly below the ideal COP, whereas a gas furnace can operate very near its ideal. The two systems are often quite competitive, depending on the cost of electricity compared with the price of gas.

6.10 SUMMARY

In this chapter, we defined the following terms:

Rankine cycle: The simplest vapor-power cycle.
Reheat cycle: The steam passes through much of the turbine and is reheated by passing it again through the boiler, and returning it to the turbine.
Regenerative cycle: Some of the steam is intercepted in the turbine and mixed with the water leaving the condenser thereby preheating the water before it enters the boiler.
Feedwater: The water entering the boiler from the pump.
Open feedwater heater: The steam and water are mixed together.
Closed feedwater heater: The steam heats the water in a heat exchanger. The steam and water do not mix.
Cogeneration: When the rejected heat from the condenser is used in some beneficial way such as in heating and cooling buildings, heating green houses, and supplying heat to industrial processes.
Refrigeration cycle: The cycle used to cool or heat a space.

The equations introduced in this chapter are mostly those developed in earlier chapters. They are simply applied to the vapor cycles of this chapter. Some are presented here. The efficiency of the Rankine cycle is

$$\eta = \frac{w_T - w_P}{q_B} \tag{6.13}$$

where the work of the pump can most often be neglected. The mass flow rate required from the turbine by the regenerative cycle (see Fig. 6.8) is

$$\dot{m}_5 = \dot{m}_6 \times \frac{h_6 - h_2}{h_5 - h_2} \tag{6.14}$$

REVIEW PROBLEMS FOR THE FE/EIT EXAMINATION

6.1FE The Rankine power cycle is idealized. Which of the following is not one of the idealizations?
A. Friction is absent.
B. Heat transfer does not occur across a finite temperature difference.
C. Pressure drops in pipes are neglected.
D. Pipes connecting components are insulated.

6.2FE The component of the Rankine cycle that leads to a relatively low cycle efficiency is:
A. The pump **B.** The boiler
C. The turbine **D.** The condenser

6.3FE Water passes through four primary components as it traverses a cycle in a power plant: a turbine, a condenser, a pump, and a boiler. Estimate the heat rejected in the condenser if $\dot{Q}_{boiler} = 60$ MJ/s, $\dot{W}_{turbine} = 20 \times 10^6$ N·m/s, and $\dot{W}_{pump} = 100$ hp.
A. 174 MJ/s **B.** 115 MJ/s
C. 52 MJ/s **D.** 40 MJ/s

6.4FE Find q_B in the Rankine cycle of Fig. P6.4FE.
A. 3410 kJ/kg **B.** 3070 kJ/kg
C. 1050 kJ/kg **D.** 860 kJ/kg

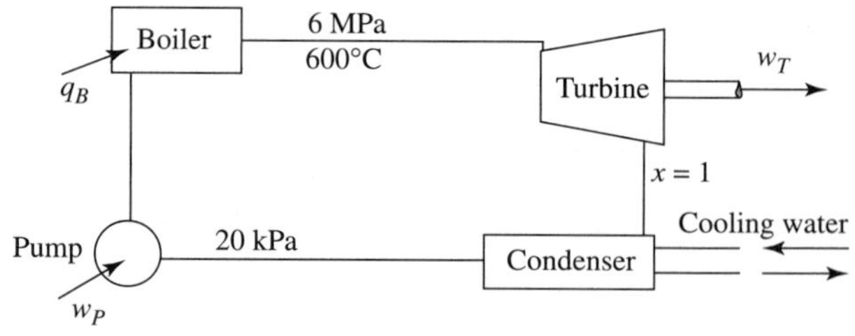

FIGURE P6.4FE

6.5FE Find $w_{turbine}$ in Fig. P6.4FE.
A. 3410 kJ/kg **B.** 3070 kJ/kg
C. 1050 kJ/kg **D.** 860 kJ/kg

6.6FE Find w_{pump} in Fig. P6.4FE.
A. 6 kW **B.** 6 kJ/kg
C. 6 kJ **D.** 6 kW/kg

6.7FE The turbine efficiency of the turbine in Fig. P6.4FE is nearest:
A. 64% **B.** 72%
C. 76% **D.** 81%

6.8FE The cooling water in Fig. P6.4FE is allowed a 10°C increase. If $\dot{m}_{steam} = 2$ kg/s, determine $\dot{m}_{cooling\ water}$ if the quality of the steam leaving the condenser is just 0.0.
A. 80 kg/s **B.** 91 kg/s
C. 102 kg/s **D.** 113 kg/s

6.9FE An open feedwater heater accepts superheated steam at 400 kPa and 200°C and mixes the steam with condensate at 20 kPa from an ideal condenser. If 20 kg/s flows out of the condenser, what is the mass flow rate of the superheated steam?
A. 2.26 kg/s **B.** 2.71 kg/s
C. 3.25 kg/s **D.** 3.86 kg/s

6.10FE How much heat is transferred from the superheated steam to the condensate of Problem 6.9FE?
A. 3860 kJ/s **B.** 4390 kJ/s
C. 5090 kJ/s **D.** 7070 kJ/s

6.11FE Refrigerant flows through the coil on the back of most refrigerators. This coil is the:
A. Evaporator **B.** Intercooler
C. Reheater **D.** Condenser

6.12FE An ideal refrigeration cycle operates between a low temperature of −12°C and a high pressure of 800 kPa. If the mass flow rate is 2 kg/s of R134a, the compressor power requirement is nearest:
A. 38 kW **B.** 34 kW
C. 50 kW **D.** 60 kW

6.13FE The quality of the refrigerant of the cycle of Problem 6.12FE entering the evaporator is nearest:
A. 0.14 **B.** 0.26
C. 0.29 **D.** 0.34

6.14FE How much heat could the cycle of Problem 6.12FE supply?
A. 1270 MJ/hr **B.** 1060 MJ/hr
C. 932 MJ/hr **D.** 865 MJ/hr

6.15FE The COP of the cycle of Problem 6.12FE, when used as a refrigerator, is nearest:
A. 2.4 **B.** 3.4
C. 4.0 **D.** 4.8

6.16FE Which of the following refrigeration cycle deviations from the ideal is considered desirable?
A. Pressure drops through the connecting tubes.
B. Friction on the compressor blades.
C. Subcooling of the refrigerant in the condenser.
D. Superheating of the vapor entering the compressor.

PROBLEMS

In the following problem set, work each problem using the appropriate equations and/or Appendix tables. Use TK Solver for property evaluation when requested.

Rankine Cycles

6.1 A power plant operating on an ideal Rankine cycle has steam exiting the turbine at 500°C and 2 MPa. If the steam enters the pump at 10 kPa, calculate the thermal efficiency with pump work included, the thermal efficiency neglecting pump work, and the percentage error in efficiency neglecting pump work using **(a)** the steam tables and **(b)** TK Solver.

6.2 A steam power plant is designed to operate on a Rankine cycle with a condenser outlet temperature of 80°C and boiler outlet temperature of 500°C. If the pump outlet pressure is 2 MPa, calculate the maximum possible thermal efficiency of the cycle. Compare with the efficiency of a Carnot engine operating between the same temperature limits.

6.3 For the ideal Rankine cycle, shown in Fig. P6.3, determine the mass flow rate of steam and the cycle efficiency using **(a)** the steam tables and **(b)** TK Solver.

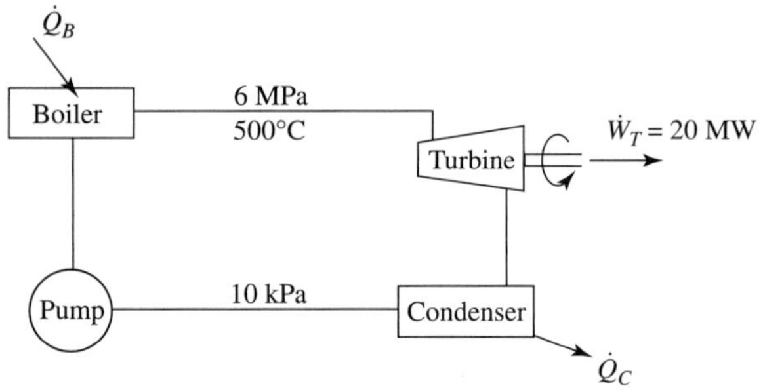

FIGURE P6.3

6.4 A solar bank of collectors with an area of 8000 ft^2 supplies energy to the boiler of an ideal Rankine-cycle power plant. At peak load, the collectors provide 200 Btu/ft^2/hr to the working fluid. The R134a working fluid leaves the boiler at 300 psia and 240°F and enters the pump at 100°F. Determine the pump work, the cycle efficiency, the mass-flow rate of the R134a, and the maximum power output.

6.5 An ideal Rankine cycle operates between temperatures of 500°C and 60°C. Determine the cycle efficiency and the quality of the turbine outlet steam if the pump outlet pressure is **(a)** 2 MPa, **(b)** 6 MPa, and **(c)** 10 MPa. **(d)** Answer parts **(a)**, **(b)**, and **(c)** using TK Solver.

6.6 The influence of maximum temperature on the efficiency of an ideal Rankine cycle is desired. Holding the maximum and minimum pressures constant at 1000 psia and 2 psia, respectively, what is the thermal efficiency if the boiler-outlet steam temperature is **(a)** 800°F, **(b)** 1000°F, and **(c)** 1200°F?

6.7 A power plant is to be operated on an ideal Rankine cycle with the superheated steam exiting the boiler at 4 MPa and 500°C. Calculate the thermal efficiency and the quality at the turbine outlet if the condenser pressure is **(a)** 20 kPa, **(b)** 10 kPa, and **(c)** 4 kPa. **(d)** Answer parts **(a)**, **(b)**, and **(c)** using TK Solver.

6.8 A power plant operates on a Rankine cycle between temperatures of 600°C and 40°C. The maximum pressure is 8 MPa and the turbine output is 20 MW. Determine the minimum mass flow rate of cooling water through the condenser if a maximum temperature differential of 10°C is allowed.

6.9 Oil, with a heating value of 30 MJ/kg, is used in the boiler shown schematically in Fig. P6.9. If 85 percent of the energy is transferred to the working fluid in the boiler, calculate the minimum amount of oil that is needed per hour.

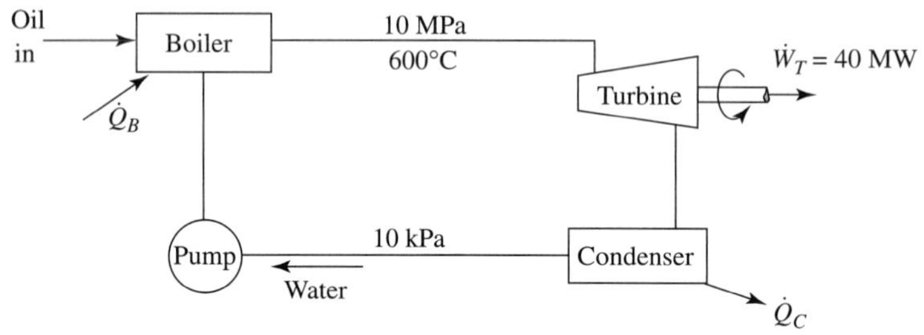

FIGURE P6.9

6.10 Hot geyser water at 95°C is available to supply energy to the boiler of a Rankine-cycle power plant. R134a is the working fluid. The maximum possible mass flow rate of hot water is 2.0 kg/s. The R134a exits the boiler as saturated vapor at 80°C, and the condenser temperature is 40°C. Calculate the pump work rate, the maximum possible power output, and its thermal efficiency using **(a)** the steam tables and **(b)** TK Solver. Assume that the hot water can equal the R134a temperature as it leaves the boiler.

6.11 Coal, with a heating value of 2500 Btu/lbm, is used to provide energy to the working fluid in a boiler that is 85 percent efficient. Determine the minimum mass flow rate of coal, in lbm/hr, that would be necessary for the turbine output to be 100 MW. The pump receives water at 2 psia in the ideal Rankine cycle, and delivers it to the boiler at 2000 psia. Superheated steam is to leave the boiler at 1000°F.

Reheat Cycles

6.12 The steam of a Rankine cycle, operating between 4 MPa and 10 kPa, is reheated at 400 kPa to 400°C. Determine the cycle efficiency if the maximum temperature is 600°C using **(a)** the steam tables and **(b)** TK Solver.

6.13 For the ideal reheat cycle shown in Fig. P6.13, calculate the thermal efficiency and the pump mass flow rate.

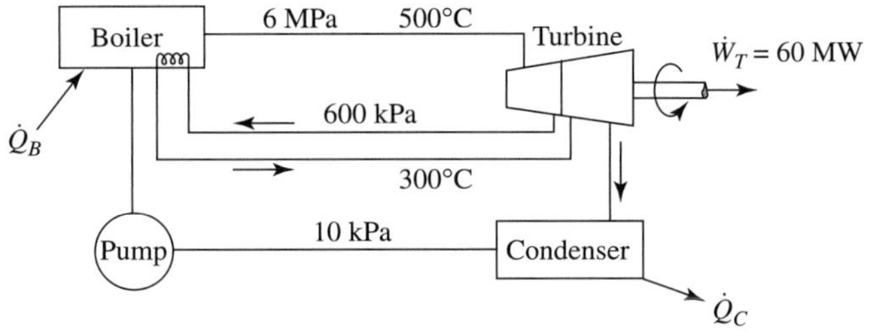

FIGURE P6.13

6.14 The steam passing through the turbine of the power cycle of Problem 6.1 is reheated at 100 kPa to 400°C. Find the thermal efficiency using **(a)** the steam tables and **(b)** TK Solver.

6.15 The steam passing through the turbine of Problem 6.5b is reheated to 300°C at an extraction pressure of **(a)** 100 kPa, **(b)** 400 kPa, and **(c)** 600 kPa. Calculate the thermal efficiency.

6.16 The power cycle of Problem 6.6b is proposed for reheat. Calculate the thermal efficiency if the steam is reheated to 1000°F after being extracted at a pressure of **(a)** 400 psia, **(b)** 200 psia, and **(c)** 100 psia. **(d)** Answer part **(b)** using TK Solver.

6.17 The steam passing through the ideal turbine of Problem 6.9 is reheated at 600 kPa to 400°C and at 50 kPa to 400°C. What is the resulting thermal efficiency? Also, calculate the oil needed per hour for the same power output of the turbine of Problem 6.9.

6.18 For the ideal reheat cycle shown in Fig. P6.18, find the thermal efficiency and the mass flow rate of steam.

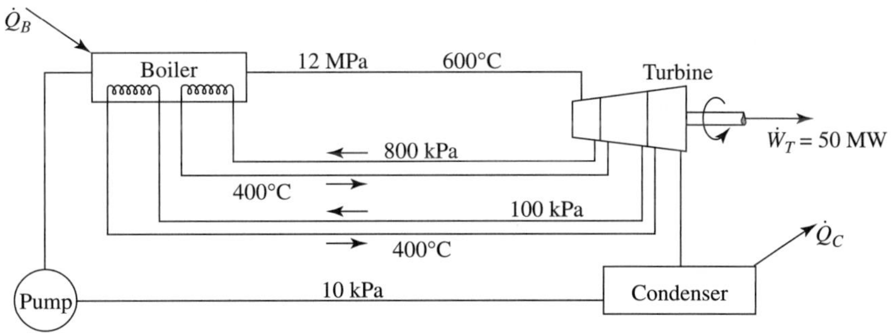

FIGURE P6.18

6.19 An ideal reheat Rankine cycle operates between 8 MPa and 4 kPa with a maximum temperature of 600°C (Fig. P6.19). Two reheat stages, each with a maximum temperature of 600°C, are to be added at 1 MPa and 100 kPa. Calculate the resulting cycle efficiency using **(a)** the steam tables and **(b)** TK Solver.

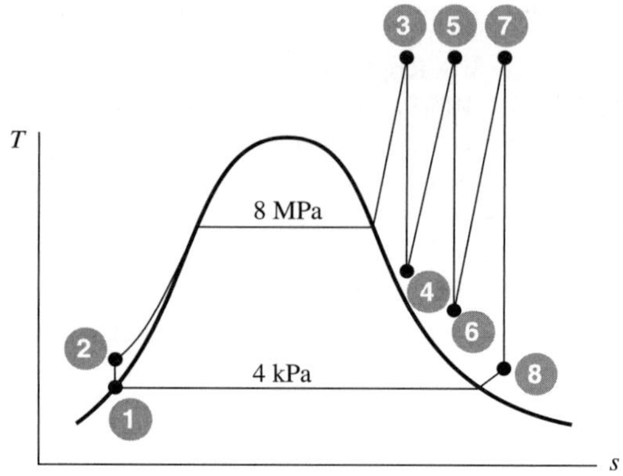

FIGURE P6.19

Regenerative Cycles

6.20 An open feedwater heater is to be designed for the power cycle of Problem 6.1 by extracting steam from the turbine at 400 kPa. Determine the thermal efficiency of the ideal regenerative cycle.

6.21 A portion of the steam passing through the turbine of Problem 6.5b is extracted and fed into an open feedwater heater. Calculate the thermal efficiency if it is extracted at a pressure of **(a)** 600 kPa, **(b)** 800 kPa, and **(c)** 1000 kPa. **(d)** Answer part **(b)** using TK Solver.

6.22 The condenser pressure of a regenerative cycle is 3 kPa, and the feedwater pump provides a pressure of 6 MPa to the boiler. Calculate the cycle efficiency if one open feedwater heater is to be used. The maximum temperature is 600°C.

6.23 An open feedwater heater extracts steam from the turbine of Problem 6.6b at 100 psia. Determine the thermal efficiency if the superheated steam enters the turbine at **(a)** 700°F, **(b)** 800°F, and **(c)** 1000°F. **(d)** Answer part **(b)** using TK Solver.

6.24 A closed feedwater heater extracts steam from the turbine of Problem 6.5b at 800 kPa. What is the thermal efficiency of the resulting ideal regenerative cycle?

6.25 For the ideal regenerative cycle shown in Fig. P6.25, determine the thermal efficiency, the mass flow rate of steam, and the ratio of rejected heat to added heat using **(a)** the steam tables and **(b)** TK Solver. (*Note:* Neglect pump work)

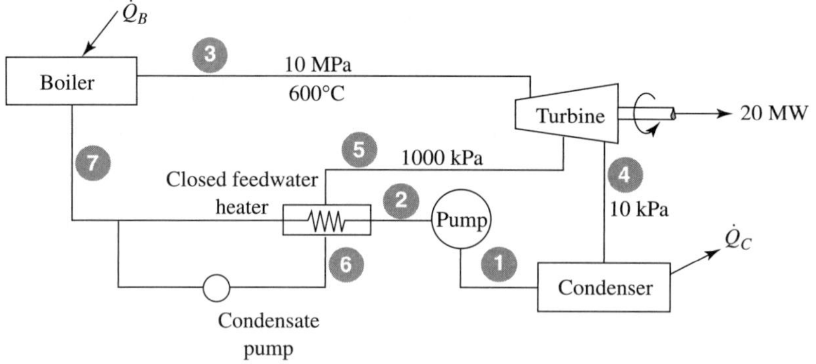

FIGURE P6.25

6.26 Part of the steam passing through the ideal turbine of Problem 6.9 is extracted at 1000 kPa and fed into a closed feedwater heater. Calculate **(a)** the thermal efficiency and **(b)** the mass flow rate of oil for the same power output.

Reheat-regenerative Cycles

6.27 To avoid a moisture problem in the turbine of Problem 6.8 the steam is extracted at 600 kPa and reheated to 400°C, and an open feedwater heater, using extracted steam at the same pressure, is inserted into the cycle. What are the resulting thermal efficiency and the mass flow rate of water flowing through the feedwater pump?

6.28 A power plant operates on an ideal reheat-regenerative cycle in which steam at 1000°F and 2000 psia enters the turbine. It is reheated at a pressure of 400 psia to 800°F and has two open feedwater heaters, one using extracted steam at 400 psia and the other using extracted steam at 80 psia. Determine the thermal efficiency if the condenser operates at 2 psia using **(a)** the steam tables and **(b)** TK Solver.

6.29 For the ideal reheat-regenerative cycle shown in Fig. P6.29 calculate **(a)** the thermal efficiency, **(b)** the mass flow rate of water fed to the boiler, and **(c)** the mass flow rate of the condenser cooling water.

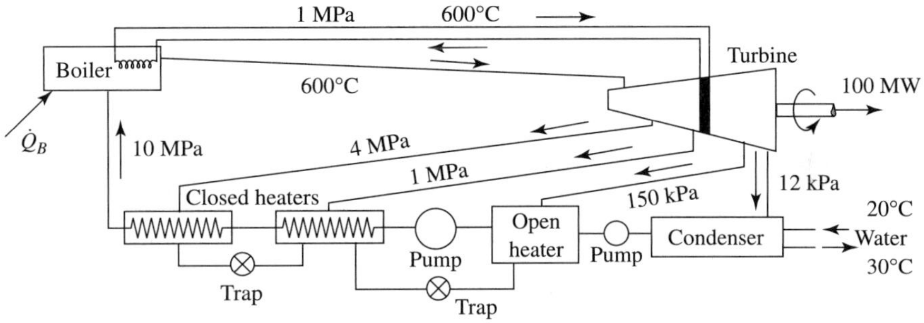

FIGURE P6.29

6.30 For the ideal steam power cycle, operating as shown in the *T-s* diagram of Fig. P6.30, two open feedwater heaters are employed. Calculate the thermal efficiency.

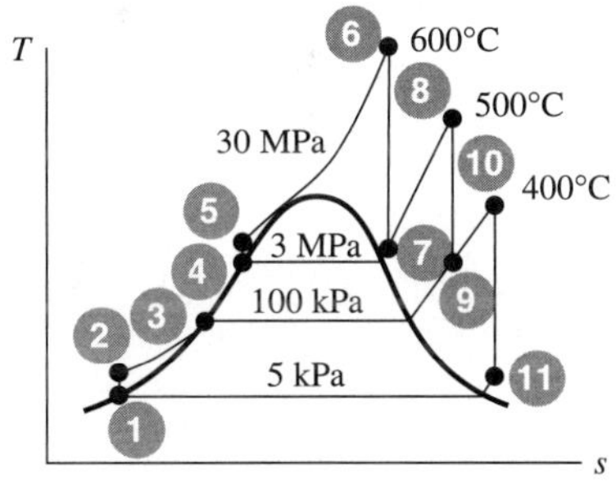

FIGURE P6.30

Rankine Cycle Efficiency

6.31 If the turbine is 85 percent efficient, determine the cycle thermal efficiency in **(a)** Problem 6.1, **(b)** Problem 6.5a, **(c)** Problem 6.6b, and **(d)** Problem 6.8. **(e)** Answer parts **(a)** and **(c)** using TK Solver.

6.32 The turbine of Problem 6.3 is 87 percent efficient. Determine the mass flow rate and the cycle efficiency with $W_T = 20$ MW.

6.33 The turbine of a Rankine cycle operating between 4 MPa and 10 kPa is 84 percent efficient. If the steam is reheated at 400 kPa to 400°C, determine the cycle efficiency. The maximum temperature is 600°C. Also, calculate the mass flow rate of condenser cooling water if it increases 10°C as it passes through the condenser when the cycle mass flow rate of steam is 10 kg/s using **(a)** the steam tables and **(b)** TK Solver.

6.34 If the turbine of Problem 6.9 is 80 percent efficient, determine the mass flow rate of oil needed to maintain the same power output.

6.35 Assume a turbine efficiency of 85 percent for Problem 6.10, and calculate the thermal efficiency and the expected power output.

6.36 For the simple Rankine cycle shown in Fig. P6.36, the turbine efficiency is 85 percent. Determine the thermal efficiency, the mass flow rate of steam, the diameter of the inlet pipe to the turbine if a maximum velocity of 100 m/s is allowed, and the mass flow rate of condenser cooling water using **(a)** the steam tables and **(b)** TK Solver.

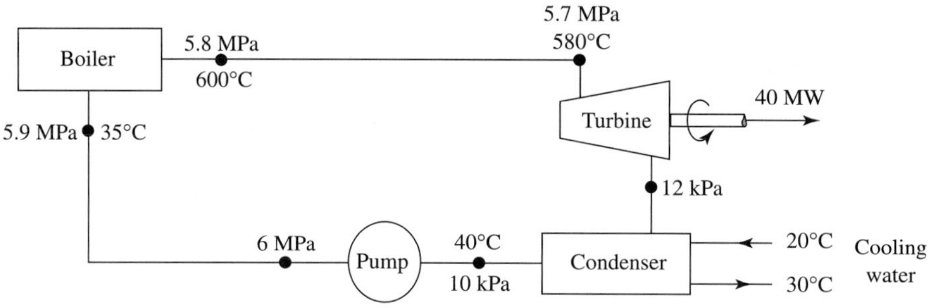

FIGURE P6.36

6.37 The actual turbine of Problem 6.13 has an efficiency of 85 percent in the high-pressure side of the turbine and 80 percent in the low-pressure side. Calculate the cycle thermal efficiency and the pump mass flow rate for the same power output.

6.38 Calculate the cycle thermal efficiency if the turbine is 85 percent efficient for the cycle of Problem 6.18.

6.39 Calculate the cycle thermal efficiency if the turbine is 87 percent efficient for the cycle of **(a)** Problem 6.20, **(b)** Problem 6.23b, and **(c)** Problem 6.24.

6.40 Determine the thermal efficiency for the cycle shown in Fig. P6.40 if the turbine is 85 percent efficient using **(a)** the steam tables and **(b)** TK Solver.

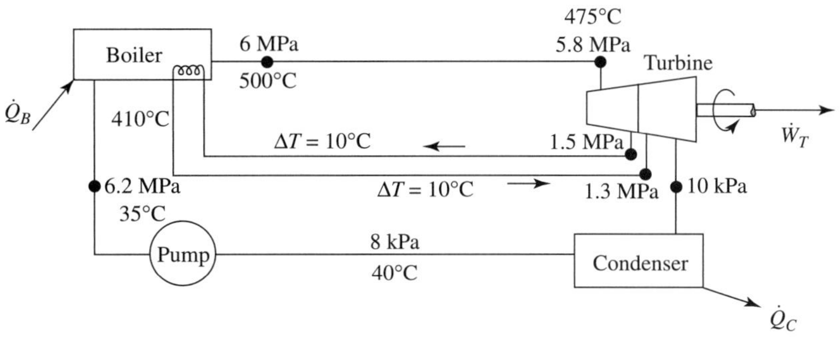

FIGURE P6.40

Refrigeration Cycles

6.41 An ideal vapor refrigeration cycle utilizes R134a as the working fluid between saturation temperatures of −28°C and 40°C. For a flow of 0.6 kg/s, determine the rate of refrigeration, the coefficient of performance, and the coefficient of performance if used as a heat pump using **(a)** the steam tables and **(b)** TK Solver.

6.42 An ideal refrigeration cycle uses R134a as the working fluid between saturation temperatures of −40°F and 50°F. If the refrigerant mass flow rate is 2.0 lbm/sec, determine the rate of refrigeration and the coefficient of performance.

6.43 R134a is used in an ideal refrigeration cycle between pressures of 120 kPa and 1000 kPa. If the compressor requires 10 hp, calculate the rate of refrigeration, the coefficient of performance, and the coefficient of performance if used as a heat pump using **(a)** the steam tables and **(b)** TK Solver.

6.44 An ideal refrigeration cycle using R134a produces 10 tons of refrigeration. If it operates between saturation temperatures of −10°F and 120°F, determine the COP, the power input needed for the compressor, and the volume rate of flow into the compressor.

6.45 An ideal refrigeration cycle operates using R134a between saturation temperatures of −20°C and 30°C. If the compressor requires 20 horsepower, determine the COP and the tons of refrigeration provided.

6.46 For 20 tons of refrigeration, calculate the minimum power input to the compressor for the cycle shown in Fig. P6.46 if the working fluid is **(a)** R134a and **(b)** water. **(c)** Answer part **(a)** using TK Solver.

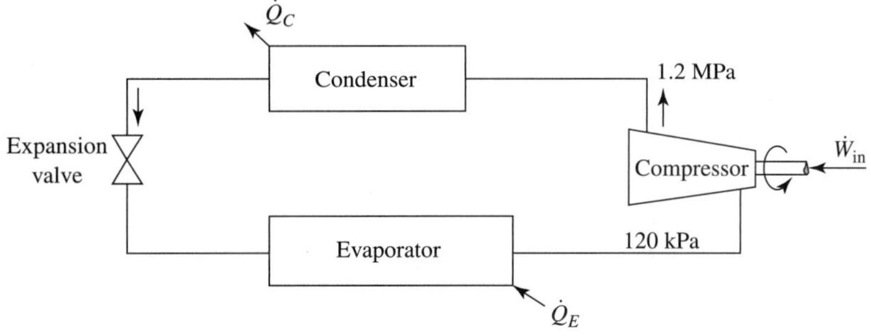

FIGURE P6.46

6.47 The turbine of the ideal Rankine cycle, shown in Fig. P6.47, produces just enough power to operate the compressor of the ideal refrigeration cycle. The R134a is mixed in the condenser and is then separated into mass flow rates $\dot{m}_p$ and $\dot{m}_r$. Determine $\dot{m}_p/\dot{m}_r$ and $\dot{Q}_B/\dot{Q}_E$.

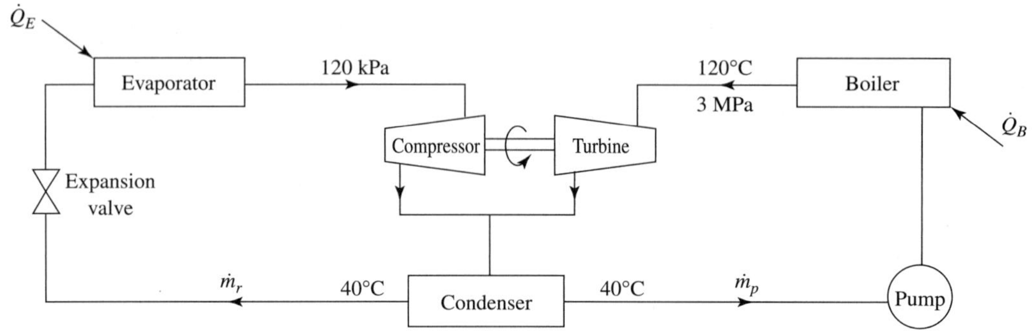

FIGURE P6.47

6.48 R134a is compressed from 200 kPa to 1.0 MPa in an 80 percent efficient compressor (Fig. P6.48). The condenser exiting temperature is 40°C. Calculate the COP and the refrigerant mass flow rate for 100 tons (352 kW) of refrigeration.

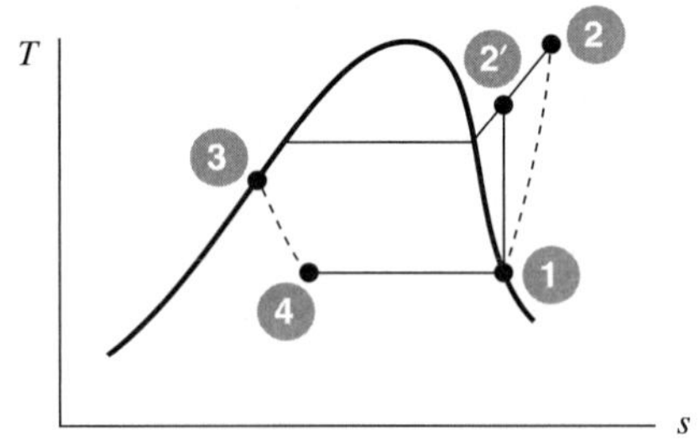

FIGURE P6.48

6.49 Assume that the refrigerant leaving the condenser of Problem 6.41 is subcooled to 36°C, while the compressor efficiency remains at 100%. Calculate the coefficient of performance using **(a)** the steam tables and **(b)** TK Solver.

6.50 The compressor of a refrigeration cycle accepts R134a as saturated vapor at 200 kPa and compresses it to 1200 kPa; it is 80 percent efficient. The R134a leaves the condenser at 40°C. Determine the COP and the mass flow rate of R134a for 10 tons of refrigeration.

6.51 R134a enters a compressor at 15 psia and 0°F and leaves at 180 psia and 200°F. If it exits the condenser as saturated liquid and the system produces 12 tons of refrigeration, calculate the COP, the mass flow rate of refrigerant, the power input to the compressor, the compressor efficiency, and the volume rate of flow entering the compressor using **(a)** the steam tables and **(b)** TK Solver.

6.52 A refrigeration cycle circulates 0.2 kg/s of R134a. Saturated vapor enters the compressor at 140 kPa and leaves at 1200 kPa and 80°C. The temperature at the condenser exit is 45°C. Determine the COP, the tons of refrigeration, the required power input, the efficiency of the compressor, and the mass flow rate of condenser cooling water if a temperature rise of 10°C is allowed.

6.53 A refrigeration cycle utilizes a compressor that is 80 percent efficient; it accepts R134a as saturated vapor at −24°C. The liquid leaving the condenser is at 800 kPa and 30°C. For a mass flow rate of 0.1 kg/s, calculate the COP, the tons of refrigeration, and the mass flow rate of condenser cooling water for a temperature rise of 10°.

Heat Pumps

6.54 A heat pump using R134a as the refrigerant provides 80 MJ/h to a building. The cycle operates between pressures of 1000 and 200 kPa. Assuming an ideal cycle, determine the COP, the compressor horsepower, and the volume flow rate into the compressor using **(a)** the steam tables and **(b)** TK Solver.

6.55 A heat pump uses groundwater at 12°C as an energy source. If the energy delivered by the heat pump is to be 60 MJ/h, estimate the minimum mass flow rate of groundwater if the compressor operates with R134a between pressures of 100 kPa and 1.0 MPa. Also, calculate the minimum compressor horsepower.

6.56 A home-heating system uses a heat pump with R134a as the refrigerant. The maximum heating load results when the temperature of 1000 ft^3/min of circulation air is raised 45°F. If the compressor increases the pressure from 30 psia to 160 psia, calculate the COP, the mass flow rate of R134a, and the compressor power needs. Assume an ideal cycle.

The jet engines on a large commercial aircraft operate on a gas cycle. (Image courtesy of Getty Images.)

7

Power and Refrigeration Gas Cycles

Outline

Chapter Objectives

The objectives of the chapter are to:

- ▲ Introduce the nomenclature for piston-cylinder cycles.
- ▲ Present the four most common gas power cycles.
- ▲ Analyze the gas refrigeration cycle.
- ▲ Present numerous examples and practice problems that illustrate the concepts discussed.

Nomenclature

The following is a list of the nomenclature introduced in this chapter:

BDC	Bottom dead center
BWR	Back work ratio
c	Percent clearance
MEP	Mean effective pressure
r	Compression ratio
r_c	Cutoff ratio
r_p	Pressure ratio
TDC	Top dead center
W_{GT}	Work output of the gas turbine
W_{ST}	Work output of the steam turbine
η_{reg}	Effectiveness of a regenerator

7.1 INTRODUCTION

Several cycles utilize a gas as the working substance, the most common being the Otto cycle and the diesel cycle used in internal combustion engines. The word "cycle" used in reference to an internal combustion engine is technically incorrect, since the working fluid does not undergo a thermodynamic cycle; air enters the engine, mixes with a fuel, undergoes combustion, and exits the engine as exhaust gases. This is often referred to as an *open cycle*, but we should keep in mind that a thermodynamic cycle does not really occur; the engine itself operates in what we could call a *mechanical cycle*. We do, however, analyze an internal combustion engine as though the working fluid operated on a cycle; it is an approximation that allows us to predict influences of engine design on such quantities as efficiency and fuel consumption. It also allows us to compare the features of the various cycles.

7.2 THE AIR-STANDARD CYCLE

In this section, we introduce engines that utilize a gas as the working fluid. Spark-ignition engines that burn gasoline and compression-ignition (diesel) engines that burn fuel oil are the two most common engines of this type.

The operation of a gas engine can be analyzed by assuming that the working fluid does indeed go through a complete thermodynamic cycle. The cycle is often called an *air-standard cycle*. All the air-standard cycles we will consider have certain features in common:

- Air is the working fluid throughout the entire cycle. The mass of the small quantity of injected fuel is negligible.
- There is no inlet process or exhaust process.
- A heat transfer process replaces the combustion process with energy transferred from an external source.
- The exhaust process, used to restore the air to its original state, is replaced with a constant-volume process, transferring heat to the surroundings so that no work is accomplished.
- All processes are assumed to be in quasi-equilibrium.
- The air is assumed to be an ideal gas with constant specific heats.

Stroke: *The distance the piston travels in one direction.*

Clearance volume: *The volume occupied by the air when the piston is at top dead center* (TDC).

Displacement volume: *The difference between the maximum volume and the clearance volume.*

Compression ratio: *The ratio of the volume with the piston at* BDC *to the clearance volume.*

A number of the engines we will consider make use of a closed system with a piston-cylinder arrangement, as shown in Fig. 7.1. The cycle shown on the p-v and T-s diagrams in the figure is representative. The diameter of the piston is called the *bore*, and the distance the piston travels in one direction is the **stroke**. When the piston is at top dead center (TDC), the volume occupied by the air in the cylinder is at a minimum; this volume is the **clearance volume**. When the piston moves to bottom dead center (BDC), the air occupies the maximum volume. The difference between the maximum volume and the clearance volume is the **displacement volume**. The clearance volume is often implicitly presented as the *percent clearance* c (the ratio of the clearance volume to the displacement volume). The **compression ratio** r is defined to be the ratio of the volume in the cylinder with the piston at

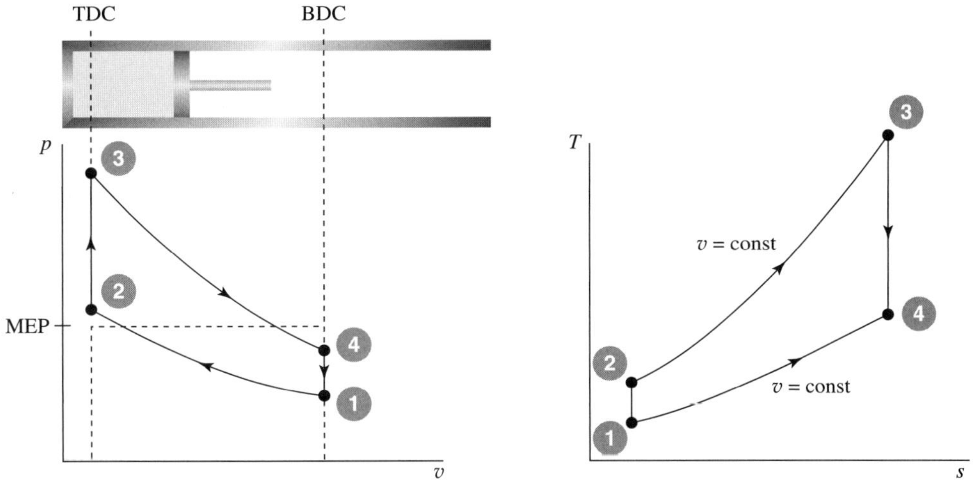

FIGURE 7.1 The cycle of a piston-cylinder gasoline engine.

BDC to the clearance volume, that is, referring to Fig. 7.1,

$$r = \frac{V_1}{V_2} \tag{7.1}$$

The **mean effective pressure** (MEP) is another quantity that is often used when rating piston-cylinder engines; it is the pressure that, if acting on the piston during the power stroke, would produce an amount of work equal to that actually done during the entire cycle. Thus,

$$W_{\text{cycle}} = \text{MEP} \times (V_{\text{BDC}} - V_{\text{TDC}}) \tag{7.2}$$

In Fig. 7.1, this means that the enclosed area of the actual cycle is equal to the area under the MEP dotted line.

Example 7.1

An engine operates with air on the cycle shown in Fig. 7.1 with isentropic processes $1 \rightarrow 2$ and $3 \rightarrow 4$. If the compression ratio is 12, the minimum pressure is 200 kPa, and the maximum pressure is 10 MPa, determine the percent clearance and the MEP.

Solution

The percent clearance is given by

$$c = \frac{V_2}{V_1 - V_2} \times 100$$

(continued)

But the compression ratio is $r = V_1/V_2 = 12$. Thus,

$$c = \frac{V_2}{12V_2 - V_2} \times 100 = 9.09\%$$

To determine the MEP, we must calculate the area under the p-V diagram; this is equivalent to calculating the work. The work from $3 \to 4$ is, using $pV^k = C$,

$$W_{3\text{-}4} = \int p\,dV = C\int_{V_3}^{V_4} \frac{dV}{V^k} = \frac{C}{1-k}(V_4^{1-k} - V_3^{1-k}) = \frac{p_4V_4 - p_3V_3}{1-k}$$

where $C = p_4V_4^k = p_3V_3^k$. But we know that $V_4/V_3 = 12$, so

$$W_{3\text{-}4} = \frac{V_3}{1-k}(12p_4 - p_3)$$

Likewise, the work from $1 \to 2$ is

$$W_{1\text{-}2} = \frac{V_2}{1-k}(p_2 - 12p_1)$$

Since no work occurs in the two constant-volume processes, we find, using $V_2 = V_3$,

$$W_{\text{cycle}} = \frac{V_2}{1-k}(12p_4 - p_3 + p_2 - 12p_1)$$

The pressures p_2 and p_4 are found as follows:

$$p_2 = p_1\left(\frac{V_1}{V_2}\right)^k = 200 \times 12^{1.4} = 1665 \text{ kPa}$$

$$p_4 = p_3\left(\frac{V_3}{V_4}\right)^k = 10\,000 \times \left(\frac{1}{12}\right)^{1.4} = 308 \text{ kPa}$$

so that

$$W_{\text{cycle}} = \frac{V_2}{-0.4}(12 \times 308 - 10\,000 + 1665 - 12 \times 200) = 20\,070V_2$$

But

$$W_{\text{cycle}} = \text{MEP} \times (V_1 - V_2) = \text{MEP} \times (12V_2 - V_2)$$

Equating the two expressions yields

$$\text{MEP} = \frac{20\,070}{11} = 1824 \text{ kPa}$$

7.3 THE CARNOT CYCLE

This ideal cycle was treated in detail in Chapter 5. Recall that the thermal efficiency of a Carnot engine,

$$\eta_{\text{Carnot}} = 1 - \frac{T_L}{T_H} \tag{7.3}$$

exceeds that of any real engine operating between the given temperatures. We will use this efficiency as an upper limit for all engines operating between T_L and T_H.

7.4 THE OTTO CYCLE

The four processes that form the *Otto cycle* are displayed in the *T*-*s* and *p*-*v* diagrams of Fig. 7.2. The piston starts at state 1 at BDC and compresses the air until it reaches TDC at state 2. Instantaneous combustion then occurs, due to spark ignition, resulting in a sudden jump in pressure to state 3, while the volume remains constant (this combustion process is simulated with a quasi-equilibrium heat-addition process). The process that follows is the power stroke as the air (simulating the combustion products) expands isentropically to state 4. In the final process, heat transfer to the surroundings occurs and the cycle is completed.

The thermal efficiency of the Otto cycle is found from

$$\eta = \frac{\dot{W}_{\text{net}}}{\dot{Q}_{\text{in}}} = \frac{\dot{Q}_{\text{in}} - \dot{Q}_{\text{out}}}{\dot{Q}_{\text{in}}} = 1 - \frac{\dot{Q}_{\text{out}}}{\dot{Q}_{\text{in}}} \tag{7.4}$$

Noting that the two heat transfer processes occur during constant-volume processes, for which the work is zero, there results

$$\dot{Q}_{\text{in}} = \dot{m}(u_3 - u_2) = \dot{m}c_v(T_3 - T_2) \qquad \dot{Q}_{\text{out}} = \dot{m}(u_3 - u_1) = \dot{m}c_v(T_4 - T_1) \tag{7.5}$$

where we have assumed each quantity to be positive. Then

$$\eta = 1 - \frac{T_4 - T_1}{T_3 - T_2} \tag{7.6}$$

This can be written as

$$\eta = 1 - \frac{T_1}{T_2}\left(\frac{T_4/T_1 - 1}{T_3/T_2 - 1}\right) \tag{7.7}$$

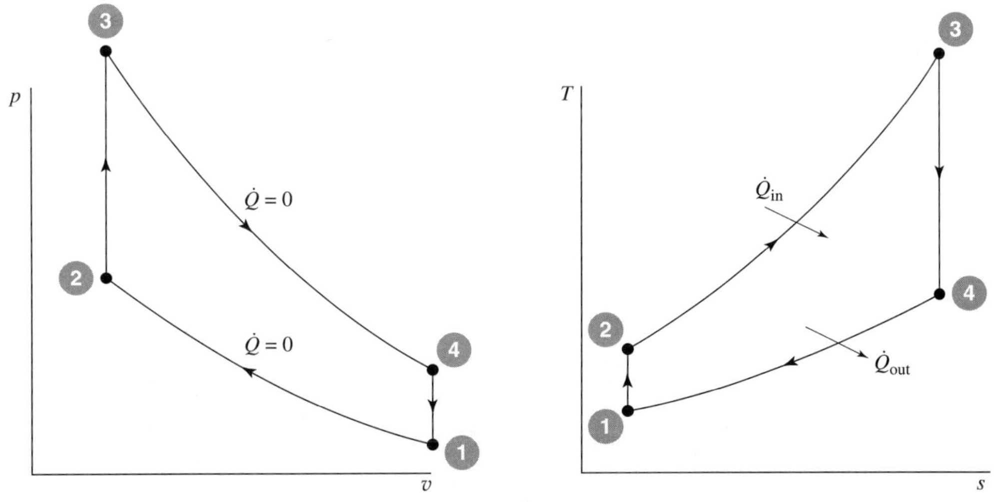

FIGURE 7.2 The Otto cycle.

For the isentropic processes, we have

$$\frac{T_2}{T_1} = \left(\frac{V_1}{V_2}\right)^{k-1} \quad \text{and} \quad \frac{T_3}{T_4} = \left(\frac{V_4}{V_3}\right)^{k-1} \tag{7.8}$$

But, using $V_1 = V_4$ and $V_3 = V_2$, we see that

$$\frac{T_2}{T_1} = \frac{T_3}{T_4} \tag{7.9}$$

Thus, Eq. 7.7 gives the thermal efficiency as

$$\eta = 1 - \frac{T_1}{T_2} = 1 - \left(\frac{V_2}{V_1}\right)^{k-1} = 1 - \frac{1}{r^{k-1}} \tag{7.10}$$

Key Concept: *The thermal efficiency of the Otto cycle is dependent only on the compression ratio r.*

We see, then, that the thermal efficiency in this idealized cycle is dependent only on the compression ratio r: the higher the compression ratio, the higher the thermal efficiency.

Example 7.2

A spark-ignition engine is proposed to have a compression ratio of 10, while operating with a low air temperature of 227°C and a low pressure of 200 kPa. If the work output is to be 1000 kJ/kg, calculate the maximum possible thermal efficiency and compare with that of a Carnot cycle. Also calculate the MEP.

Solution

The Otto cycle provides the model for this engine. The maximum possible thermal efficiency for the engine would be

$$\eta = 1 - \frac{1}{r^{k-1}} = 1 - \frac{1}{10^{0.4}} = 0.602 \quad \text{or} \quad 60.2\%$$

Since process $1 \rightarrow 2$ is isentropic, we find that

$$T_2 = T_1\left(\frac{v_1}{v_2}\right)^{k-1} = 500 \times 10^{0.4} = 1256 \text{ K}$$

The net work for the cycle is given by

$$\begin{aligned} w_{\text{net}} &= w_{1\text{-}2} + \not{w}_{2\text{-}3} + w_{3\text{-}4} + \not{w}_{4\text{-}1} \\ &= c_v(T_1 - T_2) + c_v(T_3 - T_4) \\ 1000 &= 0.717 \times (500 - 1256 + T_3 - T_4) \end{aligned}$$

But, for the isentropic process $3 \rightarrow 4$,

$$T_3 = T_4\left(\frac{v_4}{v_3}\right)^{k-1} = T_4 \times 10^{0.4} = 2.512T_4$$

(continued)

Solving the last two equations simultaneously, we find $T_3 = 3573$ K and $T_4 = 1422$ K, so that

$$\eta_{\text{Carnot}} = 1 - \frac{T_L}{T_H} = 1 - \frac{500}{3573} = 0.860 \quad \text{or} \quad 86.0\%$$

The Otto cycle efficiency is less than that of a Carnot cycle operating between the limiting temperatures because the heat-transfer processes in the Otto cycle are highly irreversible.

The MEP is found by using the equation

$$w_{\text{net}} = \text{MEP} \times (v_1 - v_2)$$

The ideal gas law and the compression ratio provide

$$v_1 = \frac{RT_1}{p_1} = \frac{0.287 \times 500}{200} = 0.7175 \text{ m}^3/\text{kg} \qquad \text{and} \qquad v_2 = \frac{v_1}{10}$$

Thus

$$\text{MEP} = \frac{w_{\text{net}}}{v_1 - v_2} = \frac{1000}{0.7175 - 0.07175} = 1549 \text{ kPa}$$

7.5 THE DIESEL CYCLE

If the compression ratio is large enough, the temperature of the air in the cylinder when the piston approaches TDC will exceed the ignition temperature of diesel fuel. This will occur if the compression ratio is about 14 or greater. No external spark is needed; the diesel fuel is simply injected into the cylinder and combustion occurs because of the high temperature of the compressed air. This type of engine is referred to as a **compression-ignition** *engine*. The ideal cycle used to model the compression-ignition engine is the *diesel cycle*, as shown in Fig. 7.3. The difference between this cycle and the Otto cycle is that, in the diesel cycle, the heat is added during a constant-pressure process.

Compression ignition: *When combustion occurs because of the high temperature of the compressed air.*

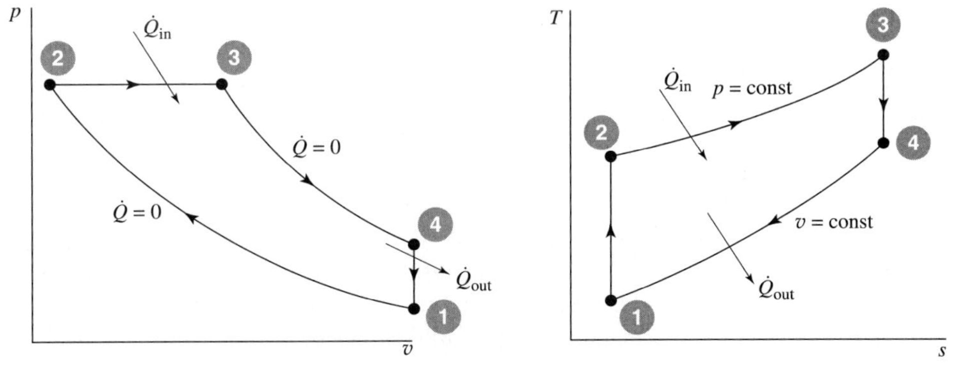

FIGURE 7.3 The diesel cycle.

The cycle begins with the piston at BDC in state 1; compression of the air occurs isentropically to state 2 at TDC; heat addition takes place (this models the injection and combustion of fuel) at constant pressure until state 3 is reached; expansion occurs isentropically to state 4 at BDC; constant-volume heat rejection completes the cycle and returns the air to the original state. Note that the power stroke includes the heat-addition process and the expansion process.

The thermal efficiency of the diesel cycle is expressed as

$$\eta = \frac{\dot{W}_{\text{net}}}{\dot{Q}_{\text{in}}} = 1 - \frac{\dot{Q}_{\text{out}}}{\dot{Q}_{\text{in}}} \tag{7.11}$$

For the constant-volume process and the constant-pressure process

$$\dot{Q}_{\text{out}} = \dot{m}c_v(T_4 - T_1) \qquad \dot{Q}_{\text{in}} = \dot{m}c_p(T_3 - T_2) \tag{7.12}$$

The efficiency is then

$$\eta = 1 - \frac{c_v(T_4 - T_1)}{c_p(T_3 - T_2)} = 1 - \frac{T_4 - T_1}{k(T_3 - T_2)} \tag{7.13}$$

This can be put in the form

$$\eta = 1 - \frac{T_1}{kT_2}\left(\frac{T_4/T_1 - 1}{T_3/T_2 - 1}\right) \tag{7.14}$$

Cutoff ratio: *The ratio of volumes from TDC to the end of combustion.*

This expression for the thermal efficiency is often written in terms of the compression ratio r and the **cutoff ratio** r_c, which is the ratio of volumes from TDC to the end of combustion (i.e., V_3/V_2); there results

Key Concept: *For a given compression ratio r, the efficiency of the diesel cycle is less than that of an Otto cycle.*

$$\eta = 1 - \frac{1}{r^{k-1}}\left(\frac{r_c^k - 1}{k(r_c - 1)}\right) \tag{7.15}$$

From this expression, we see that, for a given compression ratio r, the efficiency of the diesel cycle is less than that of an Otto cycle. For example, if $r = 10$ and $r_c = 2$, the Otto cycle efficiency is 60.2 percent and the diesel cycle efficiency is 53.4 percent. As r_c increases, the diesel cycle efficiency decreases. In practice, however, a compression ratio of 20 or so can be achieved in a diesel engine; using $r = 20$ and $r_c = 2$, we would find $\eta = 64.7$ percent. Thus, because of the higher compression ratios, a diesel engine typically operates at a significantly higher efficiency than a gasoline engine.

The decrease in diesel cycle efficiency with an increase in r_c can also be observed by considering the T-s diagram shown in Fig. 7.4. If we increase r_c, the end of the heat-input process moves to state 3′. The increased work output is then represented by area 3-3′-4′-4-3. The heat input increases considerably, as represented by area 3-3′-*a*-*b*-3. The net effect is a decrease in cycle efficiency, shown by the convergence of the constant-pressure and constant-volume lines on the T-s diagram. The two constant-volume lines diverge for the Otto cycle, resulting in an increase in cycle efficiency with increasing T_3.

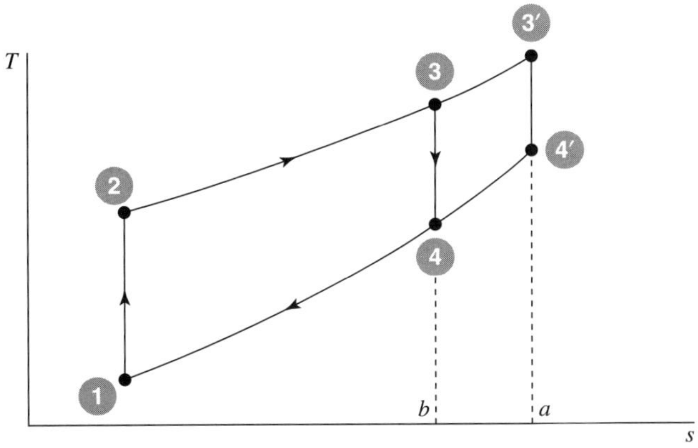

FIGURE 7.4 Diagram showing an increased cutoff ratio.

Example 7.3

A diesel cycle, with a compression ratio of 18, operates on air with a low pressure of 200 kPa and a low temperature of 200°C. If the high temperature is limited to 2200 K, determine the thermal efficiency and the MEP. Also, calculate the efficiency of an Otto cycle operating with the same compression ratio.

Solution

The cutoff ratio r_c is found first. We have

$$v_1 = \frac{RT_1}{p_1} = \frac{0.287 \times 473}{200} = 0.6788 \text{ m}^3/\text{kg} \quad \text{and} \quad v_2 = v_1/18 = 0.03771 \text{ m}^3/\text{kg}$$

Since process $1 \to 2$ is isentropic, we find

$$T_2 = T_1\left(\frac{v_1}{v_2}\right)^{k-1} = 473 \times 18^{0.4} = 1503 \text{ K} \quad \text{and}$$

$$p_3 = p_2 = p_1\left(\frac{v_1}{v_2}\right)^{k} = 200 \times 18^{1.4} = 11\,440 \text{ kPa}$$

The cutoff ratio results:

$$v_3 = \frac{RT_3}{p_3} = \frac{0.287 \times 2200}{11\,440} = 0.05519 \text{ m}^3/\text{kg} \quad \text{so that} \quad r_c = \frac{v_3}{v_2} = \frac{0.05519}{0.03771} = 1.464$$

The thermal efficiency is now calculated as

$$\eta = 1 - \frac{1}{r^{k-1}}\frac{r_c^k - 1}{k(r_c - 1)} = 1 - \frac{1}{18^{0.4}}\frac{1.464^{1.4} - 1}{1.4(1.464 - 1)} = 0.658 \quad \text{or} \quad 65.8\%$$

The temperature of state 4 is found to be

$$T_4 = T_3\left(\frac{v_3}{v_4}\right)^{k-1} = 2200 \times \left(\frac{0.05519}{0.6788}\right)^{0.4} = 806.2 \text{ K}$$

(continued)

The first law then allows us to calculate the MEP:

$$q_{\text{in}} = c_p \times (T_3 - T_2) = 1.0 \times (2200 - 1503) = 697 \text{ kJ/kg}$$

$$q_{\text{out}} = c_v \times (T_4 - T_1) = 0.717 \times (806.2 - 473) = 238.9 \text{ kJ/kg}$$

$$\therefore \text{ MEP} = \frac{w_{\text{net}}}{v_1 - v_2} = \frac{697 - 238.9}{0.6788 - 0.03771} = 714.6 \text{ kPa}$$

For the same compression ratio, the Otto cycle efficiency is calculated as follows:

$$\eta_{\text{Otto}} = 1 - \frac{1}{r^{k-1}} = 1 - \frac{1}{18^{0.4}} = 0.685 \quad \text{or} \quad 68.5\%$$

Example 7.4

Rework Example 7.3, but do not assume constant specific heats (i.e., use the air table, in Appendix Table F.1). Calculate the thermal efficiency, the MEP, and the errors in both quantities as calculated in Example 7.3.

Solution

Equation 7.10, that gives the thermal efficiency of the Otto cycle, was obtained assuming constant specific heats; so it cannot be used. We must calculate the work output and the heat added. Refer to Table F.1 in the Appendix:

$$T_1 = 473 \text{ K} \quad \therefore v_1 = \frac{0.287 \times 473}{200} = 0.6788 \text{ m}^3\text{/kg}, \ u_1 = 339 \text{ kJ/kg, and } v_{r1} = 198.3$$

The process from state 1 to state 2 is isentropic:

$$v_{r2} = v_{r1}\frac{v_2}{v_1} = 198.3 \times \frac{1}{18} = 11.02 \quad \therefore T_2 = 1310 \text{ K}, \ p_{r2} = 342, \text{ and } h_2 = 1407.8 \text{ kJ/kg}$$

The specific volume and pressure at state 2 are

$$v_2 = v_1/18 = 0.6788/18 = 0.03771 \text{ m}^3\text{/kg} \quad \text{and} \quad p_3 = p_2 = \frac{0.287 \times 1310}{0.03771} = 9970 \text{ kPa}$$

At state 3, where $T_3 = 2200$ K, $h_3 = 2503$ kJ/kg and

$$v_3 = \frac{RT_3}{p_3} = \frac{0.287 \times 2200}{9970} = 0.06333 \text{ m}^3\text{/kg} \qquad \text{and} \qquad v_{r3} = 2.012$$

$$\therefore v_{r4} = v_{r3} \times \frac{v_4}{v_3} = 2.012 \times \frac{0.6788}{0.06333} = 21.57$$

Hence, at state 4, we find from Table F.1 that $T_4 = 1053$ K and $u_4 = 804.8$ kJ/kg. The first law results in

$$q_{\text{in}} = h_3 - h_2 = 2503 - 1408 = 1095 \text{ kJ/kg}$$

$$q_{\text{out}} = u_4 - u_1 = 804.8 - 339 = 465.8 \text{ kJ/kg}$$

(continued)

Finally,

$$\eta = \frac{w_{net}}{q_{in}} = \frac{1095 - 465.8}{1095} = 0.575 \quad \text{or} \quad 57.5\%$$

$$\text{MEP} = \frac{w_{net}}{v_1 - v_2} = \frac{1095 - 465.8}{0.6788 - 0.03771} = 981 \text{ kPa}$$

The percent errors in the answers to Example 7.3 are

$$\% \text{ error in } \eta = \frac{65.8 - 57.5}{57.5} \times 100 = 14.4\%$$

and

$$\% \text{ error in MEP} = \frac{714.6 - 981}{981} \times 100 = -27.2\%$$

The errors are substantial due primarily to large temperature differences experienced in the diesel cycle. However, the constant specific heat analyses are of interest particularly in parametric studies.

7.6 THE BRAYTON CYCLE

The gas turbine is another mechanical system that produces power. It may operate on an open cycle when used as a tank or truck engine or on a closed cycle when used in a nuclear power plant. In open cycle operation, air enters the compressor, passes through a constant-pressure combustion chamber, passes through a turbine, and then exits as products of combustion to the atmosphere, as shown in Fig. 7.5a. In a closed-cycle operation, the combustion chamber is replaced with a heat exchanger in which energy enters the cycle from some exterior source; an additional heat exchanger transfers heat from the cycle so that the air is returned to its initial state, as shown in Fig. 7.5b.

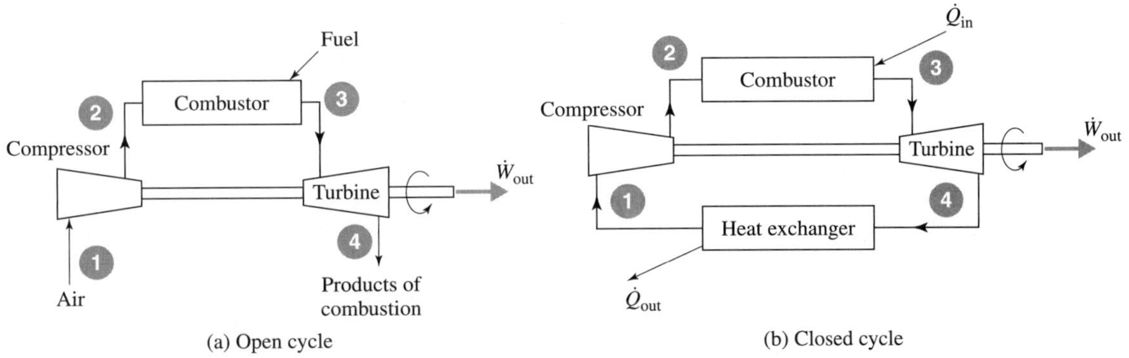

FIGURE 7.5 The Brayton cycle components.

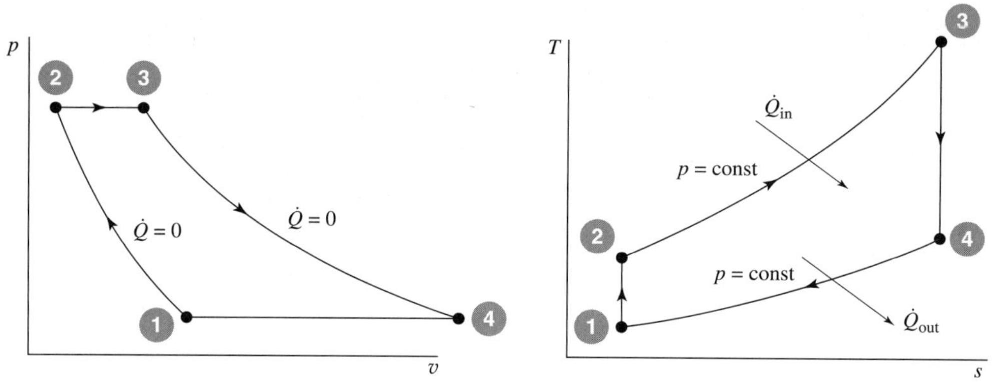

FIGURE 7.6 The Brayton cycle.

Brayton cycle: *The cycle used to model the gas turbine.*

The ideal cycle used to model the gas turbine is the **Brayton cycle**. It utilizes isentropic compression and expansion, as indicated in Fig. 7.6. The efficiency of such a cycle is given by

$$\eta = 1 - \frac{\dot{Q}_{out}}{\dot{Q}_{in}} = 1 - \frac{c_p(T_4 - T_1)}{c_p(T_3 - T_2)} = 1 - \frac{T_1}{T_2}\left(\frac{T_4/T_1 - 1}{T_3/T_2 - 1}\right) \quad (7.16)$$

Using the isentropic relations,

$$\frac{p_2}{p_1} = \left(\frac{T_2}{T_1}\right)^{k/(k-1)} \qquad \frac{p_3}{p_4} = \left(\frac{T_3}{T_4}\right)^{k/(k-1)} \quad (7.17)$$

and observing that $p_2 = p_3$ and $p_1 = p_4$, we see that

$$\frac{T_2}{T_1} = \frac{T_3}{T_4} \quad \text{or} \quad \frac{T_4}{T_1} = \frac{T_3}{T_2} \quad (7.18)$$

Hence, the thermal efficiency can be written as

$$\eta = 1 - \frac{T_1}{T_2} = 1 - \left(\frac{p_1}{p_2}\right)^{(k-1)/k} \quad (7.19)$$

In terms of the pressure ratio $r_p = p_2/p_1$, the thermal efficiency is

$$\eta = 1 - \frac{1}{r_p^{(k-1)/k}} \quad (7.20)$$

Key Concept: *Losses significantly reduce the efficiency of the gas-turbine engine.*

Of course, this expression for thermal efficiency was obtained using constant specific heats. For more accurate calculations, the gas tables in the Appendix should be used.

In an actual gas turbine, the compressor and the turbine are not isentropic; some losses do occur. These losses, usually in the neighborhood of 15 percent, significantly reduce the efficiency of the gas-turbine engine.

Another important feature of the gas turbine that seriously limits thermal efficiency is the high work requirement of the compressor, measured by the *back work ratio*, BWR $= \dot{W}_{comp}/\dot{W}_{turb}$. The compressor may require up to 80 percent of the turbine's output (a back work ratio of 0.8), leaving only 20 percent for net work output. This relatively high limit is experienced when the efficiencies of the compressor and turbine are too low. The following examples illustrate this point.

Example 7.5

Air enters the compressor of a gas turbine at 100 kPa and 25°C. For a pressure ratio of 5 and a maximum temperature of 850°C, determine the back work ratio (BWR) and the thermal efficiency for this Brayton cycle using (a) the ideal-gas equations of this section and (b) TK Solver.

Solution

(a) To find the back work ratio we observe that

$$\text{BWR} = \frac{w_{comp}}{w_{turb}} = \frac{c_p(T_2 - T_1)}{c_p(T_3 - T_4)} = \frac{T_2 - T_1}{T_3 - T_4}$$

The temperatures are $T_1 = 298$ K, $T_3 = 1123$ K, and

$$T_2 = T_1\left(\frac{p_2}{p_1}\right)^{(k-1)/k} = 298 \times 5^{0.2857} = 472.0$$

$$T_4 = T_3\left(\frac{p_4}{p_3}\right)^{(k-1)/k} = 1123 \times 5^{-0.2857} = 709.1 \text{ K}$$

The back work ratio is then

$$\text{BWR} = \frac{w_{comp}}{w_{turb}} = \frac{472.0 - 298}{1123 - 709.1} = 0.420 \quad \text{or} \quad 42.0\%$$

The thermal efficiency is

$$\eta = 1 - r^{(1-k)/k} = 1 - 5^{-0.2857} = 0.369 \quad \text{or} \quad 36.9\%$$

(b) To use TK Solver, we select "Applications" from the top bar, then "Thermodynamics on TK," then "Data," then "Air" with SI units. Select "OK" on the bottom bar, then "OK" on the next window. On the "Variables" window, input T1 = 25°C. Click the light bulb and find h1 = 300.4 kJ/kg and pr1 = 1.3806. At state 2, we know that $p_{r2} = p_{r1} \times 5 = 6.903$. Under "Status" type "Guess" for T2. Click the light bulb and find T2 = 197.7°C and h2 = 475.1 kJ/kg. Now, input T3 = 850°C and find h3 = 1189.8 kJ/kg and pr3 = 184.8. For the isentropic process, $p_{r4} = p_{r3}/5 = 184.8/5 = 36.96$. Input pr4 = 36.96 and, under "Status," type "Guess" for T4. Click the light bulb and find T4 = 471.4°C and h4 = 763.4 kJ/kg. Finally, we calculate

$$\text{BWR} = \frac{w_{comp}}{w_{turb}} = \frac{h_2 - h_1}{h_3 - h_4} = \frac{475.1 - 300.4}{1189.8 - 763.4} = 0.410 \quad \text{or} \quad 41.0\%$$

$$\eta = \frac{w_{turb} - w_{comp}}{q_{in}} = \frac{1189.8 - 763.4 - (475.1 - 300.40)}{1189.8 - 475.1} = 0.352 \quad \text{OR} \quad 35.2\%$$

Example 7.6

Assume the compressor and the gas turbine in Example 7.5 both have an efficiency of 75 percent. Determine the back work ratio (BWR) and the thermal efficiency for the Brayton cycle.

Solution

We can calculate the desired quantities if we determine w_{comp}, w_{turb}, and q_{in}. The compressor work is

$$w_{\text{comp}} = \frac{w_{s,\text{comp}}}{\eta_{\text{comp}}} = \frac{c_p(T_{2'} - T_1)}{\eta_{\text{comp}}}$$

where $w_{s,\text{comp}}$ is the isentropic work and $T_{2'}$ is the temperature of state $2'$, assuming an isentropic process where state 2 is the actual state. We then have, using $T_{2'} = T_2$ from Example 7.5,

$$w_{\text{comp}} = \frac{1.00}{0.75}(472 - 298) = 232 \text{ kJ/kg}$$

Likewise, there results

$$w_{\text{turb}} = \eta_{\text{turb}} w_{s,\text{turb}} = \eta_{\text{turb}} c_p(T_3 - T_{4'})$$
$$= 0.75 \times 1.0 \times (1123 - 709.1) = 310.4 \text{ kJ/kg}$$

where $T_{4'}$ is T_4 as calculated in Example 7.5. State 4 is the actual state, and state $4'$ is the isentropic state. The back work ratio is then

$$\text{BWR} = \frac{w_{\text{comp}}}{w_{\text{turb}}} = \frac{232}{310.4} = 0.747 \quad \text{or} \quad 74.7\%$$

The heat-transfer input necessary in this cycle is $q_{\text{in}} = h_3 - h_2 = c_p(T_3 - T_2)$, where T_2 is the actual temperature of the air leaving the compressor. It is found by returning to the compressor:

$$w_{\text{comp}} = c_p(T_2 - T_1) \qquad 232 = 1.00 \times (T_2 - 298) \qquad \therefore\ T_2 = 530 \text{ K}$$

Thus,

$$q_{\text{in}} = 1.00 \times (1123 - 530) = 593 \text{ kJ/kg}$$

The thermal efficiency of the cycle can then be written as

$$\eta = \frac{w_{\text{net}}}{q_{\text{in}}} = \frac{w_{\text{turb}} - w_{\text{comp}}}{q_{\text{in}}} = \frac{310.4 - 232}{593} = 0.132 \quad \text{or} \quad 13.2\%$$

It is obvious why the Brayton cycle cannot operate efficiently with relatively low efficiencies of the compressor and turbine. Efficiencies had to be raised into the 90's before the Brayton cycle became a cycle that could effectively be used to produce power.

7.7 THE REGENERATIVE BRAYTON CYCLE

The heat transfer from the simple gas-turbine cycle of the previous section is simply lost to the surroundings—either directly, with the products of combustion, or from a heat exchanger. Some of this exit energy can be utilized, since the temperature of the flow exiting the turbine is greater than the temperature of the flow entering the compressor. A counterflow heat exchanger, called a **regenerator**, is used to transfer some of this energy to the air leaving the compressor, as shown in Fig. 7.7. For an ideal regenerator, the exit temperature T_3 would equal the entering temperature T_5, and, similarly, T_2 would equal T_6. Since less energy is rejected from the cycle, the thermal efficiency is expected to increase. It is given by

Regenerator: *A heat exchanger used to transfer energy to the air leaving the compressor.*

$$\eta = \frac{w_{\text{turb}} - w_{\text{comp}}}{q_{\text{in}}} \tag{7.21}$$

Using the first law, expressions for q_{in} and w_{turb} are found to be

$$q_{\text{in}} = c_p(T_4 - T_3) \qquad w_{\text{turb}} = c_p(T_4 - T_5) \tag{7.22}$$

Hence, for the ideal regenerator in which $T_3 = T_5$ and $q_{\text{in}} = w_{\text{turb}}$, the thermal efficiency can be written as

$$\eta = 1 - \frac{w_{\text{comp}}}{w_{\text{turb}}} = 1 - \frac{T_2 - T_1}{T_4 - T_5} = 1 - \frac{T_1}{T_4}\left(\frac{T_2/T_1 - 1}{1 - T_5/T_4}\right) \tag{7.23}$$

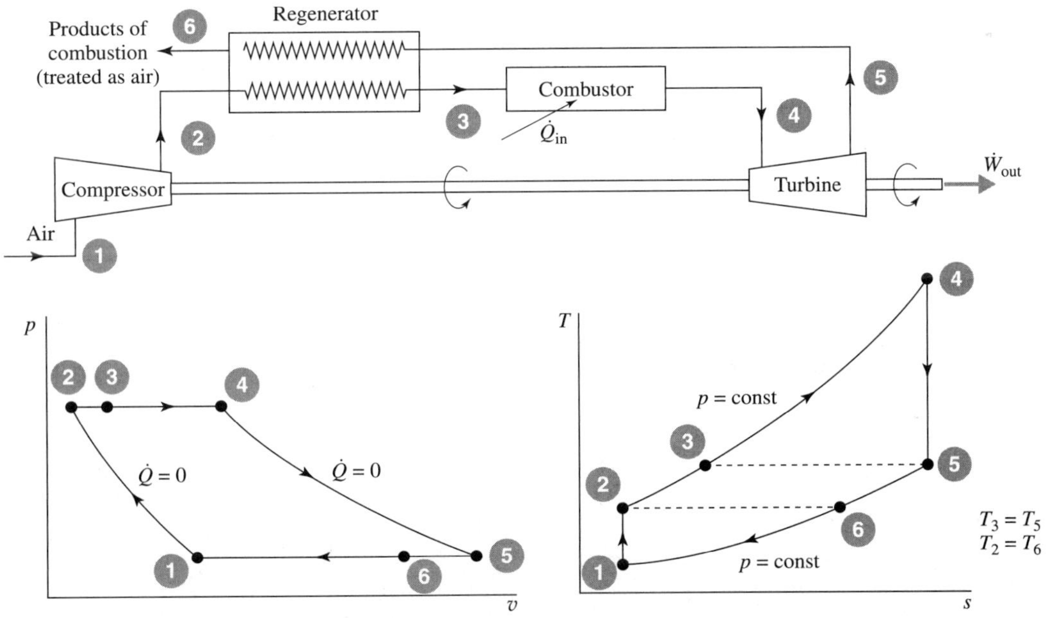

FIGURE 7.7 The regenerative Brayton cycle.

Using the appropriate isentropic relation (after some algebra), this can be written in the form

$$\eta = 1 - \frac{T_1}{T_4} r_p^{(k-1)/k} \tag{7.24}$$

Note that this expression for thermal efficiency is quite different from that for the Brayton cycle. For a given pressure ratio, the efficiency increases as the ratio of minimum to maximum temperature decreases. But, perhaps more surprisingly, as the pressure ratio increases the efficiency decreases, an effect opposite to that of the Brayton cycle. Hence, it is not surprising that for a given regenerative-cycle temperature ratio, there is a particular pressure ratio for which the efficiency of the Brayton cycle will equal the efficiency of the regenerative cycle. This is shown for a temperature ratio of 0.25 in Fig. 7.8.

Key Concept: *There is a pressure ratio for which the efficiency of the Brayton cycle will equal the efficiency of the regenerative cycle.*

In practice, the temperature of the air leaving the regenerator at state 3 must be less than the temperature of the air entering at state 5. Also, $T_6 > T_2$. The effectiveness or efficiency of a regenerator is measured by

$$\eta_{\text{reg}} = \frac{h_3 - h_2}{h_5 - h_2} \tag{7.25}$$

This is equivalent to

$$\eta_{\text{reg}} = \frac{T_3 - T_2}{T_5 - T_2} \tag{7.26}$$

if we assume an ideal gas with constant specific heats. For the ideal regenerator, $T_3 = T_5$ and $\eta_{\text{reg}} = 1$. Regenerator efficiencies exceeding 80 percent are common.

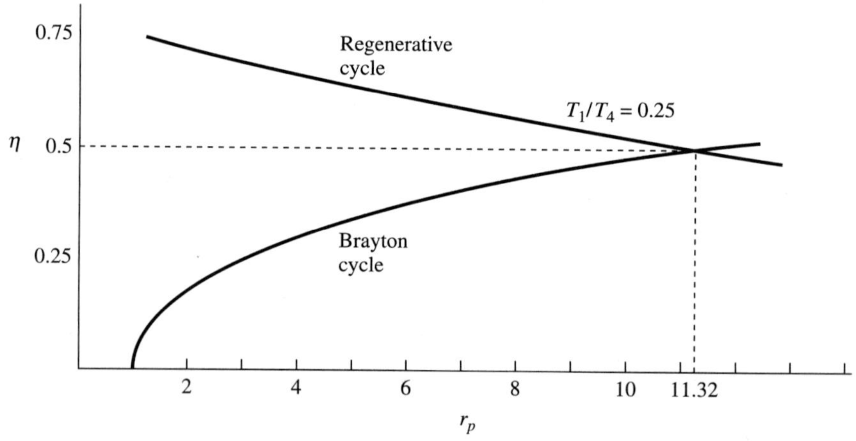

FIGURE 7.8 Efficiencies of the Brayton and regenerative cycles.

Example 7.7

Add an ideal regenerator to the gas-turbine cycle of Example 7.5 and calculate the thermal efficiency and the back work ratio.

Solution

The thermal efficiency is found using Eq. 7.24:

$$\eta = 1 - \frac{T_1}{T_4}\left(\frac{p_2}{p_1}\right)^{(k-1)/k} = 1 - \frac{298}{1123} \times 5^{0.2857} = 0.58 \quad \text{or} \quad 58\%$$

This represents a 57 percent increase in efficiency (a rather large effect). Note that, for the information given, the back work ratio does not change; hence, $\text{BWR} = w_{\text{comp}}/w_{\text{turb}} = 0.420$.

7.8 THE COMBINED BRAYTON–RANKINE CYCLE

The Brayton cycle efficiency is quite low primarily because a substantial amount of the energy input is exhausted to the surroundings. This exhausted energy is usually at a relatively high temperature, and thus, it can be used effectively to produce power. One possible application is the **combined** *Brayton–Rankine* **cycle**, in which the high-temperature exhaust gases exiting the gas turbine are used to supply energy to the boiler of the Rankine cycle, as illustrated in Fig. 7.9. Note

Combined cycle: *The exhaust gases exiting the gas turbine are used to supply energy to the boiler of the Rankine cycle.*

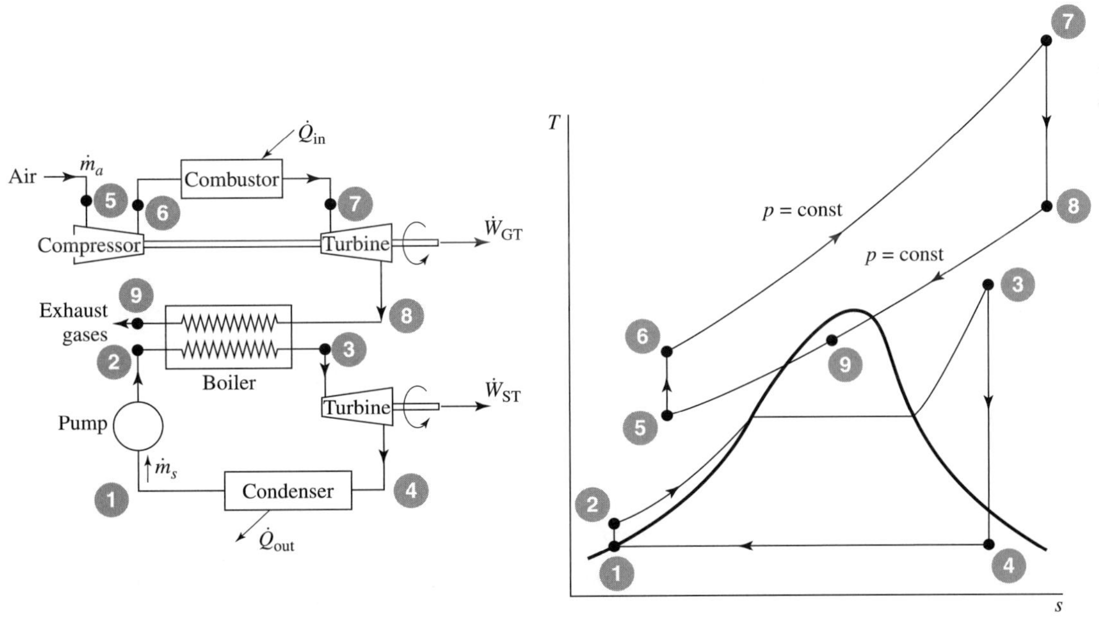

FIGURE 7.9 The combined Brayton–Rankine cycle.

that the temperature T_9 of the Brayton cycle gases exiting the boiler is less than the temperature T_3 of the Rankine cycle steam exiting the boiler; this is possible in the counterflow heat exchanger, the boiler.

To relate the air mass flow rate $\dot{m}_a$ of the Brayton cycle to the steam mass flow rate $\dot{m}_s$ of the Rankine cycle, we use an energy balance in the boiler; it gives (see Fig. 7.9),

$$\dot{m}_a(h_8 - h_9) = \dot{m}_s(h_3 - h_2) \tag{7.27}$$

assuming no additional energy addition in the boiler, which would be possible with an oil burner, for example. The cycle efficiency would be found by considering the purchased energy as $\dot{Q}_{\text{in}}$, the energy input in the combustor. The output is the sum of the net output $\dot{W}_{\text{GT}}$ from the gas turbine and the output $\dot{W}_{\text{ST}}$ from the steam turbine. The combined cycle efficiency is thus given by

$$\eta = \frac{\dot{W}_{\text{GT}} + \dot{W}_{\text{ST}}}{\dot{Q}_{\text{in}}} \tag{7.28}$$

The following example illustrates the increase in efficiency of such a combined cycle.

Example 7.8

A simple steam power plant operates between pressures of 10 kPa and 4 MPa with a maximum temperature of 400°C. The power output from the steam turbine is 100 MW. A gas turbine provides the energy to the boiler; it accepts air at 100 kPa and 25°C, has a pressure ratio of 5, and has a maximum temperature of 850°C. The exhaust gases exit the boiler at 350 K. Determine the thermal efficiency of the combined Brayton–Rankine cycle. Refer to Fig. 7.9 in the solution.

Solution

If we neglect the work of the pump, the enthalpy remains unchanged across the pump. Hence, $h_2 = h_1 = 192$ kJ/kg. At 400°C and 4 MPa, we have $h_3 = 3214$ kJ/kg and $s_3 = 6.7698$ kJ/kg·K. State 4 is located by noting that $s_4 = s_3$, so that the quality is

$$x_4 = \frac{s_4 - s_f}{s_{fg}} = \frac{6.798 - 0.6491}{7.5019} = 0.8159$$

Thus,

$$h_4 = h_f + x_4 h_{fg} = 192 + 0.8159 \times 2393 = 2144 \text{ kJ/kg}$$

The steam mass flow rate is found using the turbine output as follows:

$$\dot{w}_{\text{ST}} = \dot{m}_s(h_3 - h_4) \qquad 100\,000 = \dot{m}_s(3214 - 2144) \qquad \therefore\ \dot{m}_s = 93.46 \text{ kg/s}$$

Considering the gas-turbine cycle,

(continued)

$$T_6 = T_5\left(\frac{p_6}{p_5}\right)^{(k-1)/k} = 298 \times 5^{0.2857} = 472.0 \text{ K}$$

$$T_8 = T_7\left(\frac{p_8}{p_7}\right)^{(k-1)/k} = 1123 \times 5^{-0.2857} = 709.1 \text{ K}$$

Thus, we have, for the boiler,

$$\dot{m}_s(h_3 - h_2) = \dot{m}_a c_p(T_8 - T_9) \qquad 93.46(3214 - 192) = \dot{m}_a \times 1.00(709.1 - 350)$$

$$\therefore \dot{m}_a = 786.5 \text{ kg/s}$$

The power produced by the gas turbine is (*Note:* This is not $\dot{W}_{GT}$)

$$\dot{W}_{\text{turb}} = \dot{m}_a c_p(T_7 - T_8) = 786.5 \times 1.00(1123 - 709.1) = 325\,500 \text{ kW}$$

The energy needed by the compressor is

$$\dot{W}_{\text{comp}} = \dot{m}_a c_p(T_6 - T_5) = 786.5 \times 1.00(472 - 298) = 136\,900 \text{ kW}$$

Hence, the net gas turbine output is $\dot{W}_{GT} = \dot{W}_{\text{turb}} - \dot{W}_{\text{comp}} = 325.5 - 136.9 = 188.6$ MW. The energy input by the combustor is

$$\dot{Q}_{\text{in}} = \dot{m}_a c_p(T_7 - T_6) = 786.5 \times 1.00(1123 - 472) = 512\,000 \text{ kW}$$

The above calculations allow us to determine the combined cycle efficiency as

$$\eta = \frac{\dot{W}_{ST} + \dot{W}_{GT}}{\dot{Q}_{\text{in}}} = \frac{100 + 188.6}{512} = 0.564 \quad \text{or} \quad 56.4\%$$

Note that this efficiency is 59.3 percent higher than the Rankine cycle (see Example 6.2) and 52.8 percent higher than the Brayton cycle (see Example 7.5). Using steam reheaters, steam regenerators, gas intercoolers, and gas reheaters could increase cycle efficiency even more.

7.9 THE GAS-REFRIGERATION CYCLE

If the flow of the gas is reversed in the Brayton cycle of Sec. 7.6, the gas undergoes an isentropic-expansion process as it flows through the turbine, resulting in a substantial reduction in temperature, as shown in Fig. 7.10. The gas with the low turbine exit temperature can be used to refrigerate a space to a temperature T_2 by extracting heat at a rate $\dot{Q}_{\text{in}}$ from the refrigerated space.

Figure 7.10 illustrates a *closed*-refrigeration cycle. (An *open*-cycle system is used in aircraft; air is extracted from the atmosphere at state 2 and inserted into the passenger compartment at state 1. This provides both fresh air and cooling.) An additional heat exchanger may be used, like the regenerator of the Brayton power cycle, to provide an even lower turbine exit temperature, as illustrated in Fig. 7.11. The gas does not enter the expansion process (the turbine) at state 5; rather, it passes through an *internal heat exchanger* (it does not exchange heat with the surroundings). This allows the temperature of the gas entering the turbine to be much lower than that of Fig. 7.10. The temperature T_1 in Fig. 7.11, after the expansion, is so low that gas liquefication is possible. It should be noted, however,

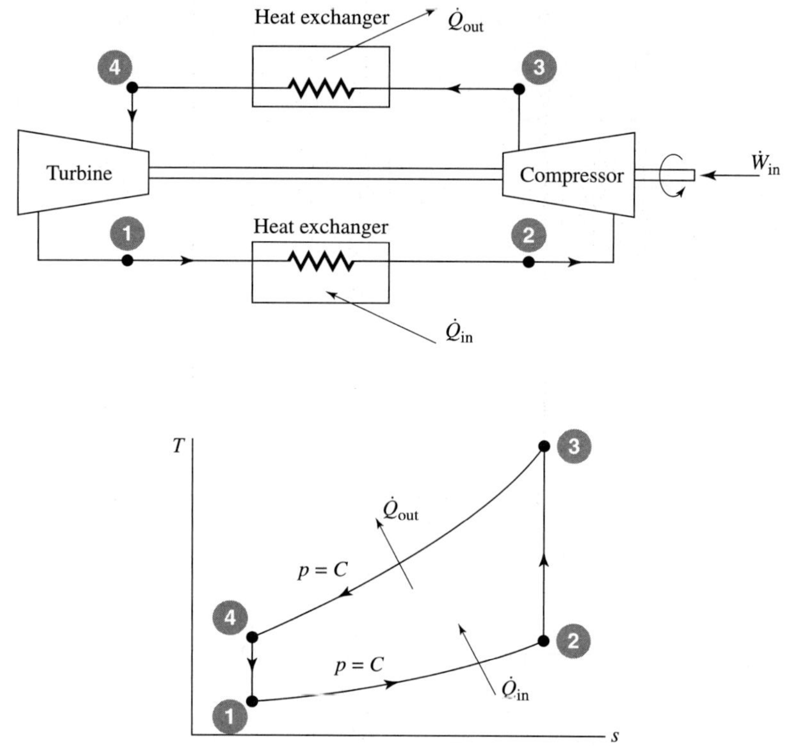

FIGURE 7.10 The gas-refrigeration cycle.

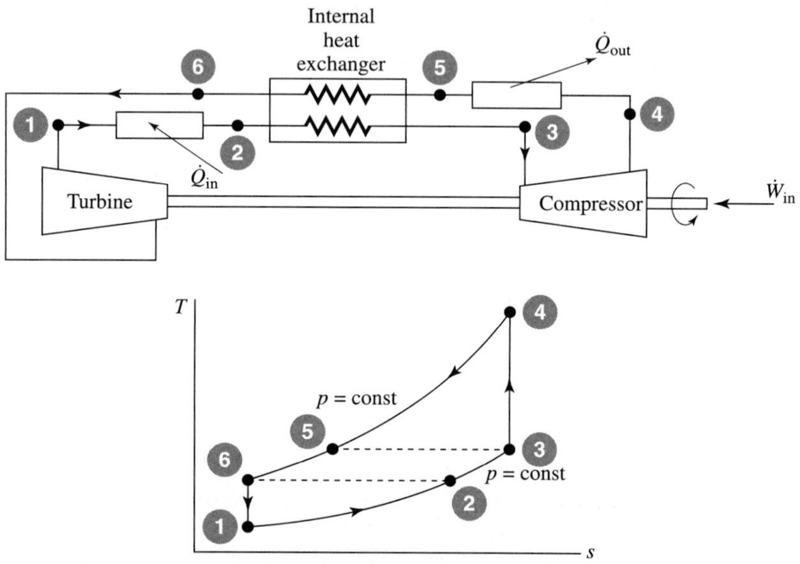

FIGURE 7.11 The gas-refrigeration cycle with an internal heat exchanger.

that the coefficient of performance is actually reduced by the inclusion of an internal heat exchanger.

A reminder: when the purpose of a thermodynamic cycle is to cool or heat a space, we do not define a cycle's efficiency; rather, we define its *coefficient of performance*:

Key Concept: *The coefficient of performance is reduced by the inclusion of an internal heat exchanger.*

$$\text{COP} = \frac{\text{desired effect}}{\text{energy purchased}} = \frac{\dot{Q}_{\text{in}}}{\dot{W}_{\text{in}}} \tag{7.47}$$

where $\dot{W}_{\text{in}} = \dot{m}(h_{\text{comp}} - h_{\text{turb}})$.

Example 7.9

Air enters the compressor of a simple gas-refrigeration cycle at −10°C and 100 kPa. For a compression ratio of 10 and a turbine inlet temperature of 30°C, calculate the minimum cycle temperature and the coefficient of performance using (a) the ideal-gas equations of this section and (b) TK Solver.

Solution

(a) Assuming isentropic compression and expansion processes, we find

$$T_3 = T_2\left(\frac{p_3}{p_2}\right)^{(k-1)/k} = 263 \times 10^{0.2857} = 508 \text{ K}$$

$$T_1 = T_4\left(\frac{p_1}{p_4}\right)^{(k-1)/k} = 303 \times 10^{-0.2857} = 157 \text{ K} = -116°\text{C}$$

The COP is now calculated as follows:

$$q_{\text{in}} = c_p(T_2 - T_1) = 1.00(263 - 157) = 106 \text{ kJ/kg}$$

$$w_{\text{comp}} = c_p(T_3 - T_2) = 1.00(508 - 263) = 245 \text{ kJ/kg}$$

$$w_{\text{turb}} = c_p(T_4 - T_1) = 1.00(303 - 157) = 146 \text{ kJ/kg}$$

$$\therefore \text{COP} = \frac{q_{\text{in}}}{w_{\text{comp}} - w_{\text{turb}}} = \frac{106}{245 - 146} = 1.07$$

This coefficient of performance is quite low when compared with that of a vapor-refrigeration cycle. Thus, gas-refrigeration cycles are used only for special applications.

(b) Use TK Solver by following the steps outlines in Example 7.5b. At $T_2 = -10°\text{C}$, we find $h_2 = 265.3$ kJ/kg and $p_{r2} = 0.89224$. At $p_{r3} = 0.89224 \times 10 = 8.9224$, we input "Guess" under "Status" and find $T_3 = 232.7°\text{C}$ and $h_3 = 511.1$ kJ/kg. At $T_4 = 30°\text{C}$, we find $h_4 = 305.45$ kJ/kg and $p_{r4} = 1.4633$. Then, with $p_{r1} = p_{r4}/10 = 0.14633$, we find $T_1 = -115.9°\text{C}$ and $h_1 = 158.66$ kJ/kg. The COP is found as follows:

$$q_{\text{in}} = h_2 - h_1 = 265.3 - 158.66 = 106.6 \text{ kJ/kg}$$

$$w_{\text{comp}} = h_3 - h_2 = 511.1 - 265.3 = 245.8 \text{ kJ/kg}$$

$$w_{\text{turb}} = h_4 - h_1 = 305.45 - 158.66 = 146.8 \text{ kJ/kg}$$

$$\therefore \text{COP} = \frac{106.6}{245.8 - 146.8} = 1.077$$

Example 7.10

Use the given information for the compressor of the refrigeration cycle of Example 7.9, but add an ideal internal heat exchanger (a regenerator, as illustrated in Fig. 7.11), so that the air temperature entering the turbine is −40°C. Calculate the minimum cycle temperature and the coefficient of performance.

Solution

Assuming isentropic compression, we again have

$$T_4 = T_3\left(\frac{p_4}{p_3}\right)^{(k-1)/k} = 263 \times 10^{0.2857} = 508 \text{ K}$$

For an ideal internal heat exchanger, we would have $T_5 = T_3 = 263$ K and $T_6 = T_2 = 233$ K. The minimum cycle temperature is

$$T_1 = T_6\left(\frac{p_1}{p_6}\right)^{(k-1)/k} = 233 \times 10^{-0.2857} = 121 \text{ K} = -152°\text{C}$$

The COP follows:

$$q_{\text{in}} = c_p(T_2 - T_1) = 1.00(233 - 121) = 112 \text{ kJ/kg}$$

$$w_{\text{comp}} = c_p(T_4 - T_3) = 1.00(508 - 263) = 245 \text{ kJ/kg}$$

$$w_{\text{turb}} = c_p(T_6 - T_1) = 1.00(233 - 121) = 112 \text{ kJ/kg}$$

$$\therefore \text{COP} = \frac{q_{\text{in}}}{w_{\text{comp}} - w_{\text{turb}}} = \frac{112}{245 - 112} = 0.842$$

Obviously, the COP is lower than that of the cycle with no internal heat exchanger. The objective is not to increase the COP but to provide extremely low refrigeration temperatures.

7.10 SUMMARY

In this chapter, we defined the following terms:

Air-standard cycle: A power cycle in which air is the working fluid, there is no inlet or exhaust, a heat transfer process replaces combustion, a constant-volume process replaces exhaust, and a constant specific heats are assumed for all processes.

Bore: The diameter of the piston.

Stroke: The distance traveled in one direction by the piston.

Clearance volume: The volume occupied by the air in the cylinder when the piston is at top dead center.

Displacement volume: The difference between the maximum volume and the clearance volume.

Compression ratio: The ratio of the displacement volume to the clearance volume.

Mean effective pressure: The pressure that, if acting on the piston during the power stroke, would produce the net work done during the entire cycle.

Otto cycle: The cycle that models that of the spark-ignition engine.

Compression ignition: When ignition is due to the high temperature that results from the high pressure in the cylinder.

Diesel cycle: The cycle that models that of the compression-ignition engine.

Cutoff ratio: The ratio of the volume in the cylinder at the end of combustion to the clearance volume in the diesel cycle.

Brayton cycle: The cycle that models that of the gas-turbine engine.

Back work ratio: The ratio of the work of the compressor to the work of the turbine in the Brayton cycle.

Regenerator: A heat exchanger that preheats the air before it enters the burner by using the hotter air that leaves the turbine.

Gas-refrigeration cycle: A refrigeration cycle that operates on a gas, not a phase-change refrigerant.

Most of the equations used in this chapter are presented in earlier chapters. There are several additional equations, however, that were developed in this chapter. They include the compression ratio:

$$r = \frac{\forall_{\text{displacement}}}{\forall_{\text{clearance}}} \tag{7.29}$$

The mean effective pressure is

$$\text{MEP} = \frac{W_{\text{cycle}}}{\forall_{\text{displacement}} - \forall_{\text{clearance}}} \tag{7.30}$$

The efficiencies of the Otto, diesel, and Brayton cycles are

$$\eta_{\text{Otto}} = 1 - \frac{1}{r^{k-1}}, \quad \eta_{\text{diesel}} = 1 - \frac{1}{r^{k-1}}\left(\frac{r_c^k - 1}{k(r_c - 1)}\right), \quad \eta_{\text{Brayton}} = 1 - \frac{1}{r_p^{(k-1)/k}} \tag{7.31}$$

The back work ratio of the Brayton cycle is

$$\text{BWR} = \frac{w_{\text{comp}}}{w_{\text{turb}}} \tag{7.32}$$

The efficiency of the regenerative Brayton cycle is

$$\eta_{\text{regen Brayton}} = 1 - \frac{T_1}{T_4} r_p^{(k-1)/k} \tag{7.33}$$

where T_1 and T_4 are the low and high temperatures in the cycle.

REVIEW PROBLEMS FOR THE FE/EIT EXAMINATION

7.1FE Select the primary reason why an Otto cycle is less efficient than a Carnot cycle.
A. The temperature after compression is too high.
B. Heat transfer occurs across a large temperature difference.
C. Friction exists between the piston and the cylinder.
D. The sudden expansion process results in very large losses.

7.2FE Which of the following statements is not true of the diesel cycle?
A. The expansion process is an isentropic process.
B. The combustion process is a constant-volume process.
C. The exhaust process is a constant-volume process.
D. The compression process is an adiabatic process.

7.3FE The engines on a commercial jet aircraft operate on which of the basic cycles?
A. Otto **B.** Diesel
C. Carnot **D.** Brayton

7.4FE The exhaust process in the Otto and diesel cycles is replaced with a constant-volume process for what primary reason?
A. To simulate the zero work of the actual exhaust process.
B. To simulate the zero heat transfer of the actual process.
C. To restore the air to its original state.
D. To insure that the first law is satisfied.

7.5FE The heat rejected in the Otto cycle shown in Fig. P7.5FE is:
A. $c_p(T_4 - T_1)$ **B.** $c_p(T_3 - T_4)$
C. $c_v(T_4 - T_1)$ **D.** $c_v(T_3 - T_4)$

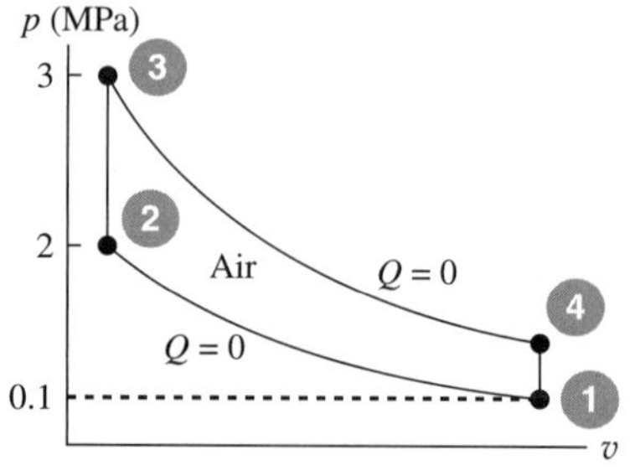

FIGURE P7.5FE

7.6FE The cycle efficiency of Fig. P7.5FE is:
A. $1 - \dfrac{T_1}{T_3}$ **B.** $\dfrac{T_3 - T_2}{T_4 - T_1}$
C. $1 - \dfrac{T_4 - T_1}{T_3 - T_2}$ **D.** $\dfrac{T_3 - T_2}{T_4 - T_1} - 1$

7.7FE If $T_1 = 27°C$ in Fig. P7.5FE, what mass of air exists in the 1000 cm^3 cylinder?
A. 0.00187 kg **B.** 0.00116 kg
C. 0.00086 kg **D.** 0.00062 kg

7.8FE If $T_1 = 27°C$ in Fig. P7.5FE, calculate T_2 assuming constant specific heats.
A. 700°C **B.** 510°C
C. 480°C **D.** 430°C

7.9FE Find T_2 in the Brayton cycle of Fig. P7.9FE.
A. 531°C **B.** 446°C
C. 327°C **D.** 258°C

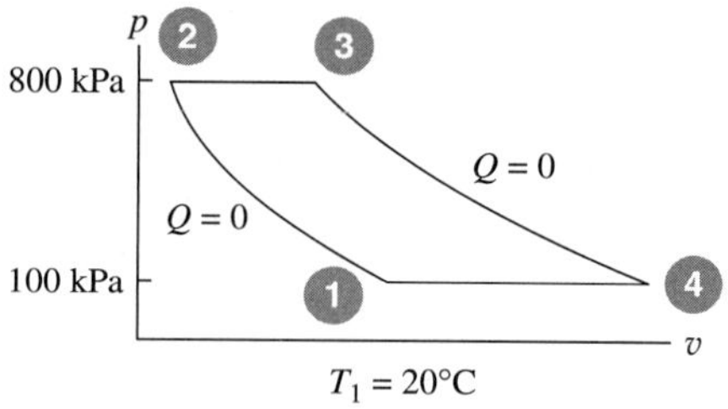

FIGURE P7.9FE

7.10FE If $T_{high} = 1200°C$, find w_T (not w_{net}) of the turbine of Fig. P7.9FE.
A. 720 kJ/kg **B.** 660 kJ/kg
C. 590 kJ/kg **D.** 540 kJ/kg

7.11FE Find the back work ratio for the cycle of Fig. P7.9FE.
A. 0.40 **B.** 0.38
C. 0.36 **D.** 0.34

7.12FE If an ideal regenerator is added to the cycle of Fig. P7.9FE, what is the temperature entering the burner?
A. T_1 **B.** T_2
C. T_3 **D.** T_4

7.13FE If $\dot{m} = 0.2$ kg/s in Fig. P7.9FE, what should be the inlet diameter of the compressor if a velocity of 80 m/s is permitted?
A. 5.2 cm **B.** 6.4 cm
C. 8.6 cm **D.** 12.4 cm

7.14FE Air enters the compressor of an ideal gas-refrigeration cycle at 10°C and 80 kPa. If the maximum and minimum temperatures are 250°C and −50°C, the compressor work is nearest:
A. 170 kJ/kg
B. 190 kJ/kg
C. 220 kJ/kg
D. 240 kJ/kg

7.15FE The pressure ratio across the compressor in the cycle of Problem 7.14FE is nearest:
A. 8.6 **B.** 8.2
C. 7.8 **D.** 7.4

7.16FE The COP for the refrigeration cycle of Problem 7.14FE is nearest:
A. 1.04 **B.** 1.18
C. 1.22 **D.** 1.49

PROBLEMS

In the following problem set, work each problem using the appropriate equations and/or Appendix tables. Use TK Solver for property evaluation when requested.

The Air-Standard Cycle

7.1 An engine with a bore and a stroke of 0.2 m × 0.2 m and a clearance of 5 percent experiences a minimum pressure of 120 kPa and a maximum pressure of 12 MPa. If it operates with air on the cycle of Fig. 7.1, determine the displacement volume, the compression ratio, and the MEP.

7.2 An air-standard cycle operates in a piston-cylinder arrangement with the following four processes: 1 → 2: isentropic compression from 100 kPa and 15°C to 2 MPa; 2 → 3: constant-pressure heat addition to 1200°C; 3 → 4: isentropic expansion; and 4 → 1: constant-volume heat rejection. Show the cycle on *p-v* and *T-s* diagrams, calculate the heat addition, and calculate the cycle efficiency.

7.3 An air-standard cycle operates in a piston-cylinder arrangement with the following four processes: 1 → 2: constant-temperature compression from 12 psia and 70°F to 400 psia; 2 → 3: constant-pressure expansion to 1400°F; 3 → 4: isentropic expansion; and 4 → 1: a constant-volume process. Show the cycle on *p-v* and *T-s* diagrams, calculate the work output, and calculate the cycle efficiency.

The Carnot Cycle

7.4 A Carnot piston engine operates with air between 20°C and 600°C with a low pressure of 100 kPa. If it is to deliver 800 kJ/kg of work, calculate the thermal efficiency, the compression ratio, and the MEP. (See Fig. 5.8.)

7.5 A Carnot piston engine operates on air between high and low pressures of 3 MPa and 100 kPa and with a low temperature of 20°C. For a compression ratio of 15, calculate the thermal efficiency, the MEP, and the work output. (See Fig. 5.8.)

7.6 A Carnot engine operates on air as shown in Fig. P7.6. Find the thermal efficiency, the power output, and the MEP. (See Fig. 5.8.)

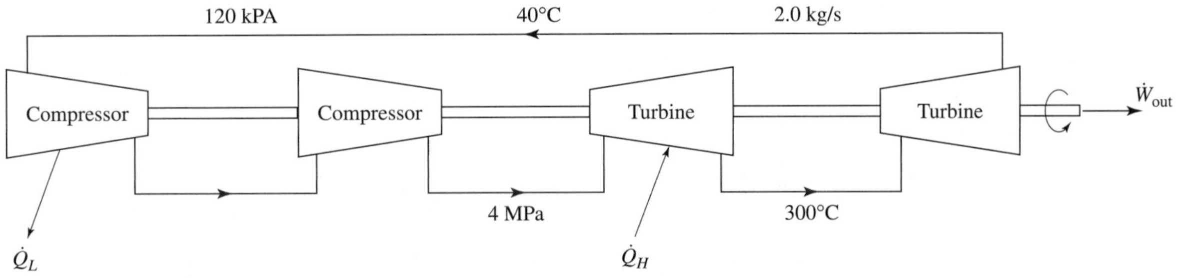

FIGURE P7.6

7.7 A Carnot engine similar to the one of Fig. P7.6 has heat addition during the combustion process of 4000 Btu/sec. If the temperature limits are 1200°F and 30°F, with high and low pressures of 1500 psia and 10 psia, determine the mass flow rate of air and the MEP. (See Fig. 5.8.)

7.8 A Carnot engine, similar to the one of Fig. P7.6, operates between the temperatures of 100°C and 600°C with pressure limits of 150 kPa and 10 MPa. Calculate the mass flow rate of air if the rejected heat flux is **(a)** 100 kW, **(b)** 400 kW, and **(c)** 2 MW. (See Fig. 5.8.)

7.9 An inventor proposes a reciprocating engine with a compression ratio of 10, operating on 1.6 kg/s of atmospheric air at 20°C that produces 50 hp. After combustion, the temperature is 400°C. Is the proposed engine feasible?

7.10 A piston engine with a cylinder that has a 0.2 m × 0.2 m bore and stroke is modeled as a Carnot engine. It operates on 0.5 kg/s of air between temperatures of 20°C and 500°C with a low pressure of 85 kPa and a clearance of 2 percent. Find the power delivered, the compression ratio, the MEP, and the volume at top dead center. (See Fig. 5.7.)

The Otto Cycle

7.11 A spark-ignition engine operates on an Otto cycle with a compression ratio of 9 and temperature limits of 30°C and 1000°C. If the power output is 500 kW, calculate the thermal efficiency and the mass flow rate of air.

7.12 A six-cylinder engine with a compression ratio of 8 and a total volume at TDC of 600 mL intakes atmospheric air at 20°C. The maximum temperature during a cycle is 1500°C. Assuming an Otto cycle, calculate the heat supplied per cycle, the thermal efficiency, and the power output for 4000 rpm.

7.13 An Otto cycle operates with air entering the compression process at 15 psia and 90°F. If 600 Btu/lbm of energy is added during combustion and the compression ratio is 10, determine the work output and the MEP.

7.14 The maximum allowable pressure in an Otto cycle is 8 MPa. Conditions at the beginning of the air compression are 85 kPa and 22°C. Calculate the required heat addition and the MEP if the compression ratio is 8.

7.15 A maximum temperature of 1600°C is possible in an Otto cycle in which air enters the compression process at 85 kPa and 30°C. Find the heat addition and the MEP if the compression ratio is 6.

7.16 If the Otto cycle shown in Fig. P7.16 operates on air, calculate the thermal efficiency and the MEP.

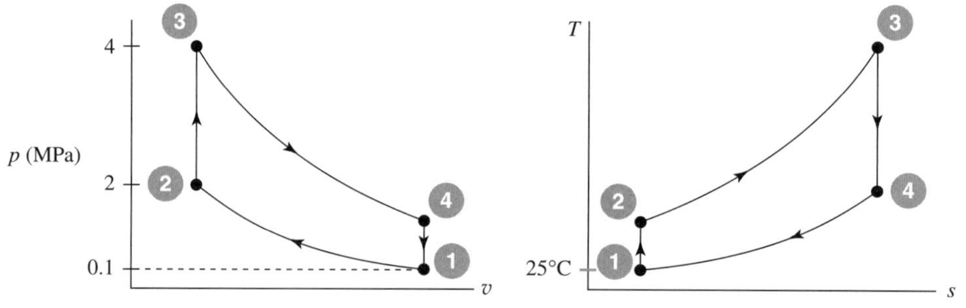

FIGURE P7.16

7.17 A spark-ignition engine with a compression ratio of 8 operates on an Otto cycle using air with a low temperature of 60°F and a low pressure of 14.7 psia. If the energy addition during combustion is 800 Btu/lbm, determine the work output and the maximum pressure.

7.18 Use the air tables (Appendix F) to solve **(a)** Problem 7.13 and **(b)** Problem 7.16. Do not assume constant specific heats.

The Diesel Cycle

7.19 A diesel engine is designed to operate with a compression ratio of 16 and air entering the compression stroke at 110 kPa and 20°C. If the energy added during combustion is 1800 kJ/kg, calculate the cutoff ratio, the thermal efficiency, and the MEP.

7.20 A diesel engine intakes atmospheric air at 60°F and adds 800 Btu/lbm of energy. If the maximum pressure is 1200 psia, calculate the cutoff ratio, the thermal efficiency, and the power output for an airflow of 0.2 lbm/sec.

7.21 A diesel cycle operates on air which enters the compression process at 85 kPa and 30°C. If the compression ratio is 16, the power output is 500 hp, and the maximum temperature is 2000°C, calculate the cutoff ratio, the thermal efficiency, and the mass flow rate of air.

7.22 Air enters the compression process of a diesel cycle at 120 kPa and 15°C. The pressure after compression is 8 MPa and 1500 kJ/kg is added during combustion. What are the cutoff ratio, the thermal efficiency, and the MEP?

7.23 For the cycle shown in Fig. P7.23, find the thermal efficiency and the work output.

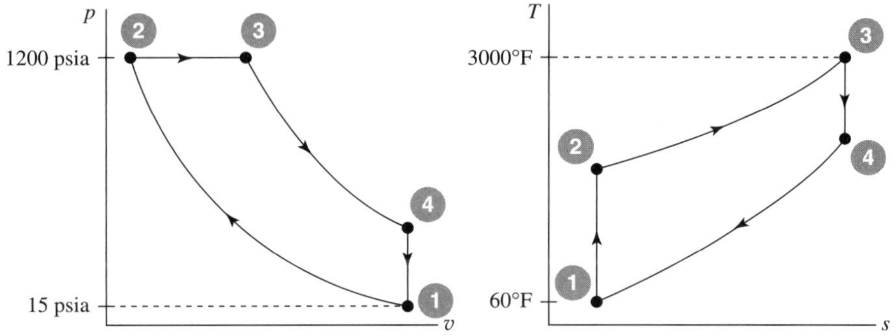

FIGURE P7.23

7.24 A diesel engine has a 0.6 m × 1.2 m bore and stroke and operates with 5 percent clearance. For a power output of 5000 hp, calculate the compression ratio and the rate of heat input if the cutoff ratio is 2.5.

7.25 Use the air tables (Appendix F) to solve **(a)** Problem 7.19 and **(b)** Problem 7.23. Do not assume constant specific heats.

The Stirling and Ericsson Cycles

7.26 Calculate the work output and thermal efficiency for the cycles shown in Fig. P7.26. Air is the operating fluid. Assume that the heat needed to go from state 2 to state 3 is transferred internally by passing from state 4 to state 1. Compare each cycle efficiency with that of a Carnot cycle operating between the same temperature limits.

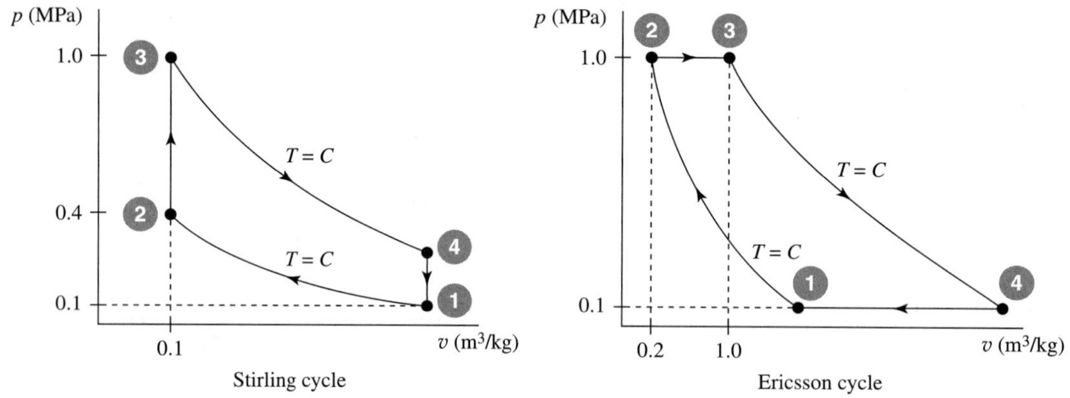

FIGURE P7.26

The Brayton Cycle

7.27 Air enters the compressor of a gas turbine at 85 kPa and 0°C. If the pressure ratio is 6 and the maximum temperature is 1000°C, find the thermal efficiency and the back work ratio for the associated Brayton cycle using **(a)** the ideal-gas equations of this chapter and **(b)** TK Solver.

7.28 A gas-turbine power plant is to produce 800 kW of power by compressing atmospheric air at 20°C to 800 kPa. If the maximum temperature is 800°C, calculate the minimum mass flow rate of the air.

7.29 Three kg of air enters the compressor of a gas turbine each second at 100 kPa and 10°C. If the pressure ratio is 5 and the maximum temperature is 800°C, determine the thermal efficiency, the horsepower output, and the back work ratio for the associated Brayton cycle.

7.30 Ten pounds of air enters the compressor of a gas turbine each second at 14.7 psia and 90°F. If the pressure ratio is 6 and the maximum temperature is 1500°F, determine the thermal efficiency, the horsepower output, and the back work ratio for the associated Brayton cycle using **(a)** the ideal-gas equations of this chapter and **(b)** TK Solver.

7.31 Determine the compressor outlet pressure that will result in maximum work output for a Brayton cycle in which the compressor inlet air conditions are 14.7 psia and 65°F and the maximum temperature is 1500°F.

7.32 Air enters the compressor of a Brayton cycle at 80 kPa and 30°C and compresses it to 500 kPa. If 1800 kJ/kg of energy is added in the combustor, calculate the compressor work requirement, the net turbine output, and the back work ratio.

7.33 Find the cycle efficiency, the horsepower output, and the back work ratio of the cycle shown in Fig. P7.33 using **(a)** the ideal-gas equations of this chapter and **(b)** TK Solver.

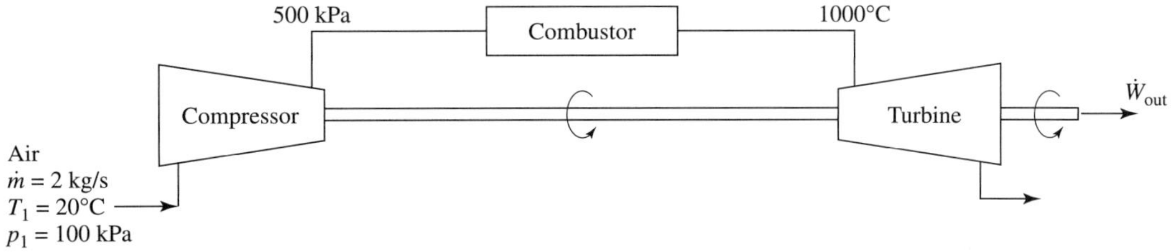

FIGURE P7.33

7.34 Calculate the thermal efficiency and the back work ratio of the gas turbine of Problem 7.27 if the respective compressor and turbine efficiencies are **(a)** 80%, 80% and **(b)** 83%, 86%, respectively.

7.35 If the efficiency of the turbine of Problem 7.28 is 85 percent and that of the compressor is 80 percent, calculate the mass flow rate of air needed, keeping the other quantities unchanged. Also calculate the cycle efficiency.

7.36 Determine the efficiency of the compressor and turbine (the efficiencies are equal) that would result in a zero thermal efficiency for the gas turbine of Problem 7.27.

7.37 Calculate the thermal efficiency and the back work ratio of the Brayton cycle of Problem 7.32 if the respective compressor and turbine efficiencies are **(a)** 83%, 83% and **(b)** 81%, 88%, respectively. **(c)** Answer part **(b)** using TK Solver.

7.38 Determine the efficiency of the compressor and turbine (the efficiencies are equal) of the Brayton cycle of Problem 7.32 that would result in no net work output.

7.39 The efficiency of the turbine of Problem 7.32 is 83 percent. What compressor efficiency would reduce the Brayton-cycle thermal efficiency to zero?

7.40 Use the air tables to find the thermal efficiency and the back work ratio for **(a)** Problem 7.27, **(b)** Problem 7.32, and **(c)** Problem 7.33. Do not assume constant specific heats.

The Regenerative Brayton Cycle

7.41 A regenerator is installed in the gas turbine of Problem 7.32. Determine the cycle efficiency if its effectiveness is **(a)** 100 percent and **(b)** 80 percent.

7.42 Assuming the ideal-gas turbine and regenerator shown in Fig. P7.42, find $\dot{Q}_{in}$ (the cycle efficiency) and the back work ratio using **(a)** the ideal-gas equations of this chapter and **(b)** TK Solver.

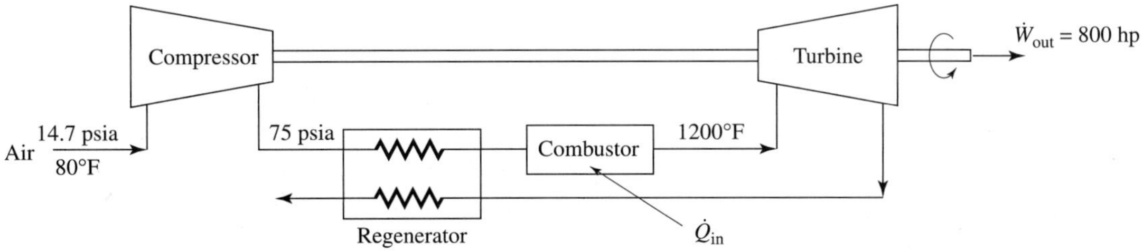

FIGURE P7.42

7.43 For the ideal-gas turbine with regenerator shown in Fig. P7.43, find $\dot{W}_{out}$, the cycle efficiency, and the back work ratio.

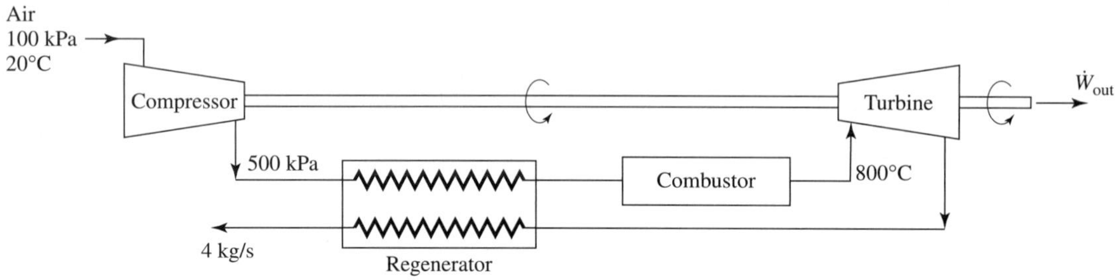

FIGURE P7.43

7.44 Temperatures for the ideal regenerative gas-turbine cycle of Fig. 7.7 are $T_1 = 60°F$, $T_2 = 500°F$, $T_3 = 700°F$, and $T_4 = 1600°F$. Calculate the thermal efficiency and the back work ratio if air is the working fluid using **(a)** the ideal-gas equations of this chapter and **(b)** TK Solver.

7.45 Assume that the efficiencies of the compressor and turbine of Problem 7.43 are 83 percent and 86 percent, respectively, and that the effectiveness of the regenerator is 90 percent. Determine the power output and the back work ratio.

The Combined Brayton–Rankine Cycle

7.46 Calculate the maximum thermal efficiency of the combined cycle shown in Fig. P7.46.

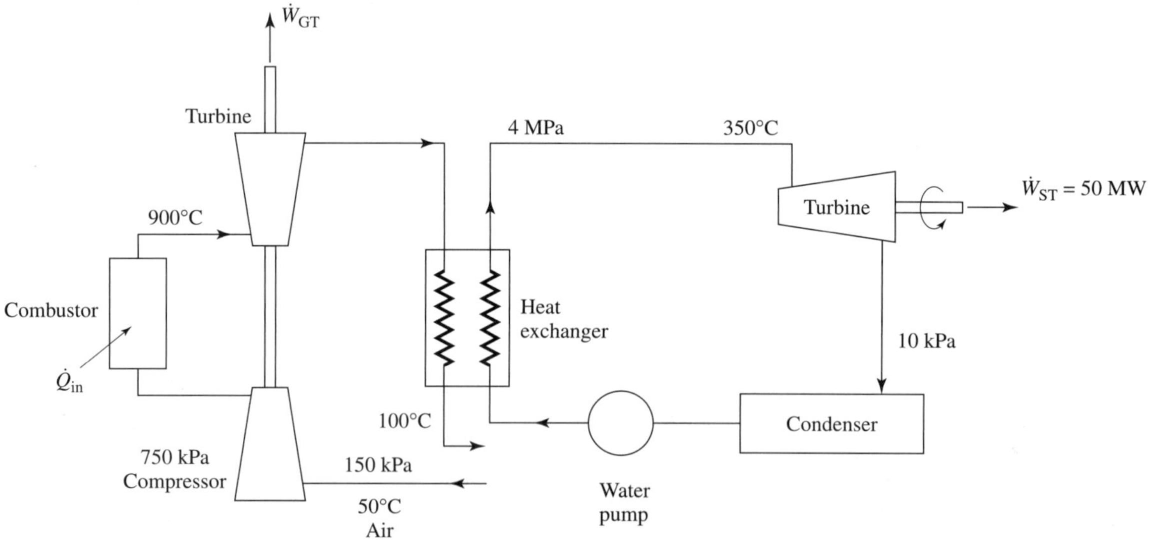

FIGURE P7.46

7.47 A gas-turbine cycle inlets 20 kg/s of atmospheric air at 15°C, compresses it to 1200 kPa, and heats it to 1200°C in a combustor. The gases leaving the turbine heat the steam of a Rankine cycle to 350°C and exit the heat exchanger (boiler) at 100°C. The pump of the Rankine cycle operates between 10 kPa and 6 MPa. Calculate the maximum power output of the combined cycle and the combined cycle efficiency.

7.48 An ideal gas-turbine cycle intakes 50 kg/s of air at 100 kPa and 20°C. It compresses it by a factor of 6 and the combustor heats it to 900°C. It then enters the boiler of a simple Rankine cycle power plant that operates on steam between 8 kPa and 4 MPa. The heat exchanger–boiler outlets steam at 400°C and exhaust gases at 300°C. Determine the total power output and the overall cycle efficiency.

7.49 The compressor and turbine of the gas cycle of Problem 7.48 are 85% efficient, and the steam turbine is 87% efficient. Calculate the combined cycle power output and efficiency.

The Gas-Refrigeration Cycle

7.50 Air flows at the rate of 2.0 kg/s through the compressor of an ideal gas-refrigeration cycle where the pressure increases to 500 kPa from 100 kPa. The maximum and minimum cycle temperatures are 300°C and −20°C, respectively. Calculate the COP and the power needed to drive the compressor fluid using **(a)** the ideal-gas equations of this chapter and **(b)** TK Solver.

7.51 Rework Problem 7.50, assuming the efficiencies of the compressor and turbine are 84% and 88%, respectively.

7.52 A simple gas cycle produces 10 tons of refrigeration by compressing air from 200 kPa to 2 MPa. If the maximum and minimum temperatures are 300°C and −90°C, respectively, find the compressor power and the cycle COP. The compressor is 82 percent efficient and the turbine is 87 percent efficient.

7.53 An ideal internal heat exchanger is added to the cycle of Problem 7.50 (see Fig. 7.11) so that the low temperature is reduced to −60°C while the maximum temperature remains at 300°C. Determine the COP and the compressor power requirement fluid using **(a)** the ideal-gas equations of this chapter and **(b)** TK Solver.

7.54 Air enters the compressor of a gas-refrigeration cycle at −10°C and is compressed from 200 kPa to 800 kPa. The high-pressure air is then cooled to 0°C by transferring energy to the surroundings and then to −30°C with an internal heat exchanger before it enters the turbine. Calculate the minimum possible temperature of the air leaving the turbine, the coefficient of performance, and the mass flow rate for 8 tons of refrigeration. Assume ideal components.

7.55 What is the COP for the ideal air cycle shown in Fig. P7.55 if it is **(a)** used to refrigerate a space and **(b)** used to heat a space? **(c)** Answer part **(a)** using TK Solver.

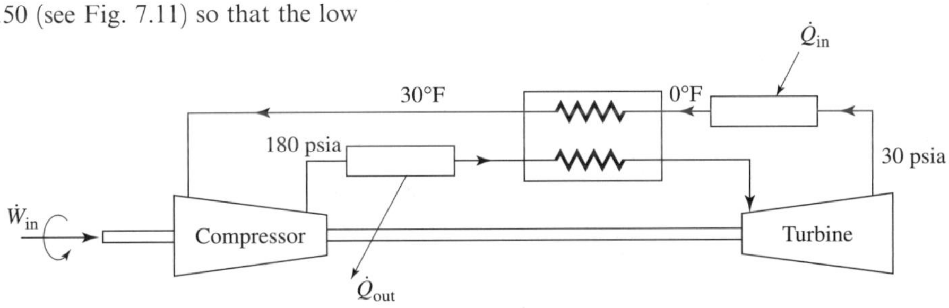

FIGURE P7.55

7.56 Rework Problem 7.55 assuming the compressor and turbine are each 87 percent efficient and the effectiveness of the internal heat exchanger is 90 percent.

Fans provide cooling to animals as well as humans. (Image courtesy of Getty Images.)

8

Psychrometrics

Outline

Chapter Objectives

The objectives of the chapter are to:

- ▲ Present the equations needed to include water vapor in our calculations.
- ▲ Define dry-bulb, wet-bulb, and dew-point temperatures.
- ▲ Introduce the psychrometric chart.
- ▲ Outline the most common air-conditioning processes.
- ▲ Present numerous examples and practice problems that illustrate the concepts discussed.

Nomenclature

The following list of nomenclature is introduced in this chapter:

m_a	mass of dry air	T_{db}	dry-bulb temperature
m_v	mass of water vapor	T_{dp}	dew-point temperature
N	number of moles	T_{wb}	wet-bulb temperature
p_a	partial pressure of dry air		
p_v	partial pressure of water vapor	ϕ	relative humidity
		ω	humidity ratio
Q_f	volume flow rate		

8.1 INTRODUCTION

Thus far in our study of thermodynamics, we have considered only single-component systems. In this chapter, we will study the two-component system of air and water vapor, generally called **psychrometrics**. The more general presentation

Psychrometrics: *The two-component system of air and water vapor.*

in which several components make up the system is not considered in this text. The two-component mixture of air and water vapor must be considered when designing air-conditioning systems, whether they are for heating, for cooling, or simply for dehumidification. A cooling tower used to remove heat from a power plant is an example of an air-conditioning system as is the usual heating and cooling system in an office building.

8.2 GAS-VAPOR MIXTURES

Dry air: *When water vapor is not included.*

Dalton model: *Where each component occupies the same volume and has the same temperature as the mixture.*

Partial pressures: *When the total pressure is the sum of the component pressures.*

Air is a mixture of nitrogen, oxygen, and argon (plus traces of some other gases). When water vapor is not included, we refer to it as **dry air**. If water vapor is included, as in *atmospheric air*, we must be careful to properly account for it. At the relatively low atmospheric temperature, we can treat dry air as an ideal gas with constant specific heats. It is also possible to treat the water vapor in the air as an ideal gas, even though the water vapor may be at the saturation state. Consequently, we can consider atmospheric air to be a mixture of two ideal gases.

Two models are used to study a mixture of ideal gases. The *Amagat model* treats each component as though it exists separately at the same pressure and temperature of the mixture; the total volume is the sum of the volumes of the components. In this chapter, we use the **Dalton model** in which each component occupies the same volume and has the same temperature as the mixture; the total pressure is the sum of the component pressures (termed the **partial pressures**). For the Dalton model,

$$p = p_a + p_v \tag{8.1}$$

where the total pressure p is the sum of the partial pressure p_a of the dry air and the partial pressure p_v of the water vapor (called the *vapor pressure*). We will assume the total pressure to be 100 kPa (14.7 psia) unless stated otherwise.

To show that the water vapor in the atmosphere can be treated as an ideal gas, consider an atmosphere at 20°C. From Appendix Table C.1 at 20°C, we see that $p_v = 2.338$ kPa (the pressure needed to vaporize water at 20°C). The ideal-gas law gives

$$v = \frac{RT}{p_v} = \frac{0.4165 \times 293}{2.338} = 57.8 \text{ m}^3/\text{kg} \tag{8.2}$$

Observe that this is the same value found in Table C.1 for v_g at 20°C. Hence, we conclude that water vapor found in the atmosphere can be treated as an ideal gas.

Since we assume that the water vapor is an ideal gas, its enthalpy is dependent on temperature only. Hence, we use the enthalpy of the water vapor to be the enthalpy of saturated water vapor at the temperature of the air, expressed as

$$h_v(T) = h_g(T) \tag{8.3}$$

In Fig. 8.1, this means that $h_1 = h_2$, where $h_2 = h_g$ from the steam tables at $T = T_1$. This is acceptable for situations in which the total pressure is rela-

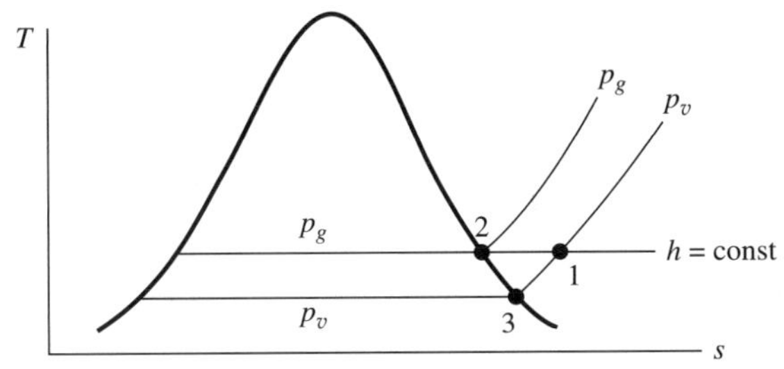

FIGURE 8.1 *T-s* diagram for the water-vapor component.

tively low (near atmospheric pressure) and the temperature is below about 60°C (140°F).

The amount of water vapor in the air of state 1 in Fig. 8.1 is related to the relative humidity and the humidity ratio. The **relative humidity** ϕ (often simply *humidity*) is defined as the ratio of the mass of the water vapor m_v to the maximum amount of water vapor m_g the air can hold at the same temperature:

Relative humidity: *The ratio of the mass of the water vapor to the maximum amount of water vapor the air can hold.*

$$\phi = \frac{m_v}{m_g} \tag{8.4}$$

Using the ideal-gas law, we find

$$\phi = \frac{p_v V/RT}{p_g V/RT} = \frac{p_v}{p_g} \tag{8.5}$$

where the constant-pressure lines for p_v and p_g are shown in Fig. 8.1.

The **humidity ratio** ω (also referred to as *specific humidity*) is the ratio of the mass of water vapor to the mass of dry air:

Humidity ratio: *The ratio of the mass of water vapor to the mass of dry air.*

$$\omega = \frac{m_v}{m_a} \tag{8.6}$$

Using the ideal-gas law for air and water vapor, this becomes

$$\begin{aligned}\omega &= \frac{p_v V/R_v T}{p_a V/R_a T} = \frac{p_v/R_v}{p_g/R_a} \\ &= \frac{p_v/0.4615}{p_a/0.287} = 0.622\frac{p_v}{p_a}\end{aligned} \tag{8.7}$$

Combining Eqs. 8.5 and 8.7, we relate the above two quantities as

$$\omega = 0.622\frac{\phi p_g}{p_a} \qquad \phi = 1.608\frac{\omega p_a}{p_g} \tag{8.8}$$

Note that at state 3 in Fig. 8.1, the relative humidity is 1.0 (100%). Also note that for a given mass of water vapor in the air, ω remains constant but ϕ varies, depending on the temperature.

When making an energy balance in an air-conditioning system, we must use enthalpy as with all control volumes involving gases. The total enthalpy is the sum of the enthalpy of the air and the enthalpy of the water vapor:

$$\begin{aligned} H &= H_a + H_v \\ &= m_a h_a + m_v h_v \end{aligned} \tag{8.9}$$

Divide by m_a and obtain

$$h = h_a + \omega h_v \tag{8.10}$$

where h is the specific enthalpy of the mixture per unit mass of dry air.

Dry-bulb temperature: *The temperature of the air as measured by a conventional thermometer.*

Dew-point temperature: *The temperature at which condensation begins.*

The temperature of the air, as measured by a conventional thermometer, is referred to as the **dry-bulb temperature** T (T_1 in Fig. 8.1). The temperature at which condensation begins, if air is cooled at constant pressure, is the **dew-point temperature** T_{dp} (T_3 in Fig. 8.1). If the temperature falls below the dew-point temperature, condensation occurs and the amount of water vapor in the air decreases. This may occur on a cool evening; it also may occur on the cool coils of an air conditioner, the inside of a pane of glass when it's cold outside, or the outside of your glass filled with water and ice sitting on a table.

Example 8.1

The air at 25°C and 100 kPa in a 150-m^3 room has a relative humidity of 60%. Calculate the humidity ratio, the dew point, the mass of water vapor in the air, and the volume percentage of the room that is water vapor, using (a) the equations in this section and (b) TK Solver.

Solution

(a) By Eq. 8.5, the partial pressure of the water vapor is

$$p_v = p_g \phi = 3.169 \times 0.6 = 1.90 \text{ kPa}$$

where p_g is the saturation pressure at 25°C found in Table C.1. The partial pressure of the air is then

$$p_a = p - p_v = 100 - 1.9 = 98.1$$

where we have used the total pressure of the air in the room to be at 100 kPa. The humidity ratio is then

$$\omega = 0.622 \frac{p_v}{p_a} = 0.622 \times \frac{1.9}{98.1} = 0.01205 \text{ kg water/kg dry air}$$

The dew point is the temperature T_3 of Fig. 8.1 associated with the partial pressure p_v. It is found by interpolation in Table C.1 or Table C.2, whichever appears to be easier: interpolation gives $T_{\text{dp}} = 16.6$°C.

(continued)

From the definition of the humidity ratio, the mass of water vapor is found to be

$$m_v = \omega m_a = \omega \frac{p_a V}{R_a T} = 0.01205\left(\frac{98.1 \times 150}{0.287 \times 298}\right) = 2.07 \text{ kg}$$

We have assumed that both the dry air and the water vapor fill the entire 150 m^3.

To find the volume percentage of the water vapor, we must compute the moles of each component:

$$N_v = \frac{m_v}{M_v} = \frac{2.07}{18} = 0.1152 \text{ mol} \qquad N_a = \frac{m_a}{M_a} = \frac{98.1 \times 150/0.287 \times 298}{28.47} = 5.94 \text{ mol}$$

The fraction of the volume occupied by the water vapor is

$$\frac{N_v}{N_a + N_v} = \frac{0.1152}{5.94 + 0.1152} = 0.0194 \quad \text{or} \quad 1.94\%$$

This demonstrates that air with 60% humidity is about 2% water vapor by volume. We usually ignore this when analyzing air and consider air to be dry air as we have done in the problems up to this point. Ignoring the water vapor does not lead to significant errors in most engineering applications. It must be included, however, when considering problems involving, for example, combustion and air conditioning.

(b) To use TK Solver, select "Applications" on the top bar, the "Thermodynamics on TK," then "Basic equations," and then "Psychrometric data." Input T1 = 25°C, pm1 = 100 kPa, and phi1 = 0.6. Click the light bulb, and find $\omega_1 = 0.01204$, $T_{dp} =$ dp1 = 16.7°C, $p_v = 1.8993$ kPa, and $p_a = 98.101$ kPa.

The mass of water vapor and the volume percentage of water vapor are found as in part (a).

Example 8.2

The air in Example 8.1 is cooled below the dew point to 10°C. Estimate the percentage of water vapor that will condense. Reheat the air to 25°C and calculate the relative humidity.

Solution

At 10°C, the air is saturated, with $\phi = 100\%$. In Fig. 8.1, we are at a state on the saturation line that lies below state 3. At 10°C, we find from Table C.1 that $p_v = 1.228$ kPa, so that

$$p_a = p - p_v = 100 - 1.228 = 98.77 \text{ kPa}$$

The humidity ratio is then

$$\omega = 0.622\frac{p_v}{p_a} = 0.622\frac{1.288}{98.77} = 0.00773 \text{ kg water/kg dry air}$$

The difference in the humidity ratio just calculated and the humidity ratio of Example 8.1 is $\Delta\omega = 0.01205 - 0.00773 = 0.00432$ kg water/kg dry air. The mass of water vapor removed (condensed) is found to be

(continued)

$$\Delta m_v = \Delta\omega \times m_a = 0.00432\left(\frac{98.1 \times 150}{0.287 \times 298}\right) = 0.743 \text{ kg water}$$

where we have used the initial mass of dry air. The percentage that condenses is $(0.743/2.07) \times 100 = 35.9\%$.

As we reheat the air back to 25°C, the ω remains constant at 0.00773. Using Eq. 8.8, the relative humidity is then reduced to

$$\phi = 1.608\frac{\omega p_a}{p_g} = 1.608\frac{0.00773 \times 98.77}{3.169} = 0.387 \quad \text{or} \quad 38.7\%$$

where p_g is used as the saturation pressure at 25°C from Table C.1.

8.3 ADIABATIC SATURATION AND WET-BULB TEMPERATURES

It is quite difficult to measure the relative humidity ϕ and the humidity ratio ω directly, at least with any degree of accuracy. This section presents two indirect methods for accurately determining these quantities.

Consider a relatively long insulated channel, shown in Fig. 8.2. Air with an unknown relative humidity enters, moisture is added to the air by the sufficiently long pool of water, and saturated air exits. This process involves no heat transfer because the channel is insulated, and hence, the exiting temperature is called the *adiabatic saturation temperature*. Let us find an expression for the desired humidity ratio ω_1. Consider that the liquid water added is at temperature T_2. An energy balance on this control volume, neglecting kinetic and potential energy changes, is done considering the dry-air and the water-vapor components. With $\dot{Q} = \dot{W} = 0$ we have

$$\dot{m}_{v1}h_{v1} + \dot{m}_{a1}h_{a1} + \dot{m}_f h_{f2} = \dot{m}_{a2}h_{a2} + \dot{m}_{v2}h_{v2} \tag{8.11}$$

But, from conservation of mass for both the dry air and the water vapor,

$$\dot{m}_{a1} = \dot{m}_{a2} = \dot{m}_a \qquad \dot{m}_{v1} + \dot{m}_f = \dot{m}_{v2} \tag{8.12}$$

Using the definition of ω in Eq. 8.6, the above equations allow us to write

$$\dot{m}_a\omega_1 + \dot{m}_f = \omega_2\dot{m}_a \tag{8.13}$$

Substituting this into Eq. 8.11 for $\dot{m}_f$, there results, using $h_v = h_g$,

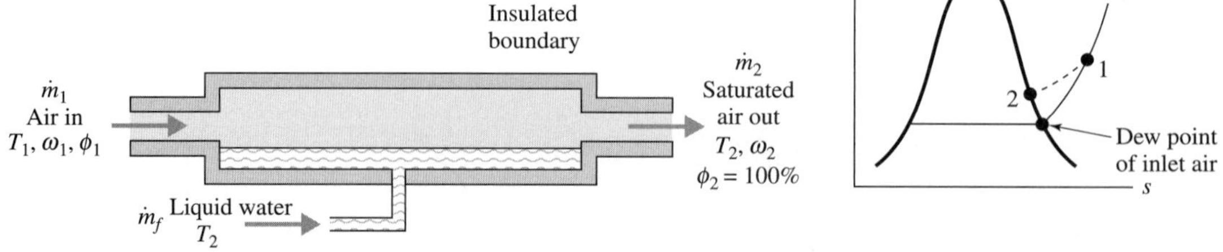

FIGURE 8.2 Setup used to find ω of air.

$$\dot{m}_a\omega_1 h_{g1} + \dot{m}_a h_{a1} + (\omega_2 - \omega_1)\dot{m}_a h_{f2} = \dot{m}_a h_{a2} + \omega_2\dot{m}_a h_{g2} \tag{8.14}$$

At state 2, we know that $\phi_2 = 1.0$, and using Eq. 8.8,

$$\omega_2 = 0.622\frac{p_{g2}}{p - p_{g2}} \tag{8.15}$$

Thus, Eq. 8.14 becomes

$$\omega_1 = \frac{\omega_2 h_{fg2} + c_p(T_2 - T_1)}{h_{g1} - h_{f2}} \tag{8.16}$$

where $h_{a2} - h_{a1} = c_p(T_2 - T_1)$ for the dry air, and $h_{fg2} = h_{g2} - h_{f2}$. Consequently, if we measure the temperatures T_2 and T_1 and the total pressure p, we can find ω_2 from Eq. 8.15 and then ω_1 from Eq. 8.16, since the remaining quantities are given in Appendix C.

Because T_2 may be significantly less than T_1, the apparatus sketched in Fig. 8.2 can be used to cool an air stream. This is done in relatively dry climates, so that T_2 is reduced but usually not to the saturation temperature. Such a device is often referred to as a "swamp cooler." A fan blowing air through a series of wicks that stand in water is quite effective at cooling low-humidity air.

Using the device of Fig. 8.2 to obtain the adiabatic saturation temperature is a rather involved process. A much simpler approach is to wrap the bulb of a thermometer with a cotton wick saturated with water and then either to blow air over the wick or to swing the thermometer overhead through the air until the temperature reaches a steady-state value. This **wet-bulb temperature** T_{wb} and the adiabatic saturation temperature are essentially the same for water if the pressure is approximately atmospheric.

Wet-bulb temperature: *The temperature reached when air is blown over the wet bulb of a thermometer.*

Example 8.3

The dry-bulb and wet-bulb temperatures of a 14.7-psia air stream are measured to be 100°F and 80°F, respectively. Determine the humidity ratio, the relative humidity, and the specific enthalpy of the air.

Solution

We use Eq. 8.16 to find ω_1. But first, ω_2 is found using Eq. 8.15:

$$\omega_2 = 0.622\frac{p_{g2}}{p - p_{g2}} = 0.622\frac{0.5073}{14.7 - 0.5073} = 0.0222$$

where p_{g2} is the saturation pressure at 80°F. Second, Eq. 8.16 gives

$$\omega_1 = \frac{\omega_2 h_{fg2} + c_p(T_2 - T_1)}{h_{g1} - h_{f2}} = \frac{0.0222 \times 1048 + 0.24(80 - 100)}{1105 - 48.09}$$

$$= 0.01747 \text{ kg water/kg dry air}$$

(continued)

The partial pressure of the water vapor is found using Eq. 8.7:

$$\omega_1 = 0.622\frac{p_{v1}}{p_{a1}} \qquad 0.01747 = 0.622\frac{p_{v1}}{14.7 - p_{v1}} \qquad \therefore\ p_{v1} = 0.402 \text{ psia}$$

The relative humidity is obtained from Eq. 8.5:

$$\phi = \frac{p_{v1}}{p_{g1}} = \frac{0.402}{0.9503} = 0.423 \quad \text{or} \quad 42.3\%$$

The specific enthalpy is found by assuming a zero value for air at $T = 0°\text{F}$ (note that Tables C.1 and C.2 select $h = 0$ at $T = 0°\text{F}$, so that establishes the datum). Using Eq. 8.10, the enthalpy for the mixture is

$$h = h_a + \omega h_v = c_p T + \omega h_g = 0.24 \times 100 + 0.01747 \times 1105 = 43.3 \text{ Btu/lb dry air}$$

where we have used $h_v = h_g$ at 100°F (see Fig. 8.1). The enthalpy is always expressed per mass unit of dry air.

8.4 THE PSYCHROMETRIC CHART

Psychrometric chart: *A way of relating the various properties associated with a water vapor-air mixture.*

A convenient way of relating the various properties associated with a water vapor-air mixture is to plot these quantities on a **psychrometric chart**, such as Fig. 8.3, or (for standard atmospheric pressure) Fig. G.1 in the Appendix. Any two of the properties establishes a state from which the other properties are determined. As an example, consider a state A that is located by specifying the dry-bulb temperature and the relative humidity. The wet-bulb temperature would be read at 1, the dew-point temperature at 2, the enthalpy at 3, and the humidity ratio at 4. Referring to Fig. G.l, a dry-bulb temperature of 30°C and a relative humidity of 80%

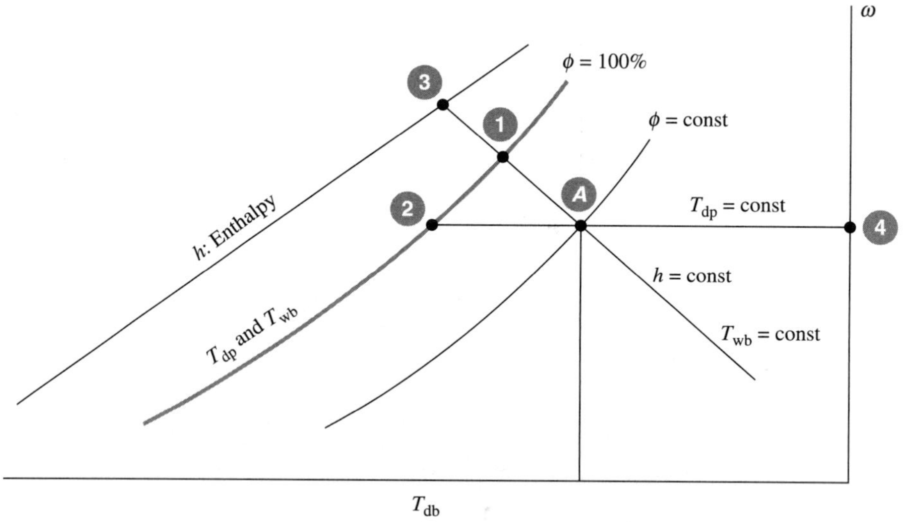

FIGURE 8.3 The psychrometric chart.

would provide the following: $T_{dp} = 26°C$, $T_{wb} = 27°C$, $h = 85$ kJ/kg dry air, and $\omega = 0.0215$ kg water/kg dry air. The chart provides us with a quick, relatively accurate method (large charts are used in the industry) for finding the quantities of interest. If the pressure is significantly different from standard atmospheric pressure, the equations presented in the preceding sections must be used.

Example 8.4

Using Appendix Fig. G.1E, rework Example 8.3 ($T_{db} = 100°F$ and $T_{wb} = 80°F$) to find ω, ϕ, and h.

Solution

Using the chart, the intersection of $T_{db} = 100°F$ and $T_{wb} = 80°F$ gives

$$\omega = 0.0175 \text{ lbm water/lbm dry air} \qquad \phi = 42\% \qquad h = 44 \text{ Btu/lbm dry air}$$

These values are less accurate than those calculated in Example 8.3 but certainly are acceptable. The simplicity of using the psychrometric chart is obvious.

8.5 AIR-CONDITIONING PROCESSES

Generally, people feel most comfortable when the air is in the "comfort zone": the temperature is between 22°C (72°F) and 27°C (80°F) and the relative humidity is between 40% and 60%. In Fig. 8.4, the area enclosed by the dotted lines represents the comfort zone. There are several situations in which air must be conditioned to put it in the comfort zone:

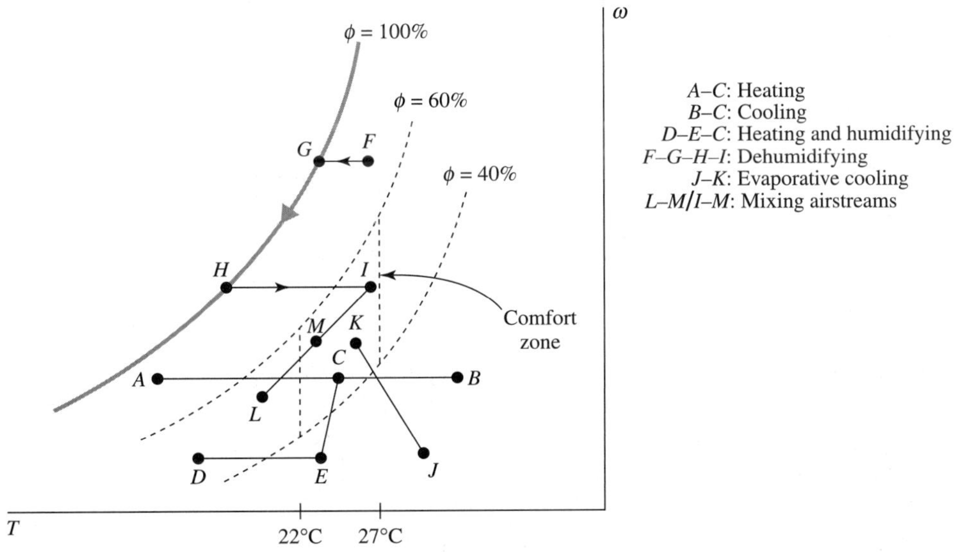

FIGURE 8.4 The conditioning of air.

- The air is too cold or too hot. Heat is simply added or extracted. This is represented by *A*–C and *B*–*C* in Fig. 8.4.
- The air is too cold, and the humidity is too low. The air can first be heated, and then moisture added, as in *D*–*E*–*C*.
- The temperature is acceptable, but the humidity is too high. The air is first cooled from *F*–G. Moisture is removed from G–*H*. Heat is added from *H*–*I*.
- The air is too hot, and the humidity is low. Moisture is added, and the process represented by *J*–*K* results.
- The air is too hot. An air stream from the outside is mixed with an air stream from the inside to provide natural cooling or fresh air. Process *I*–*M* represents the warmer inside air mixed with the outside air represented by *L*–*M*. State *M* represents the mixed air. State *M* lies along a line connecting *L* and *I*.

Each of these situations will be considered in the following examples. The first law will be used to predict the heating or cooling needed or to establish the final state.

Often the *volume flow rate* is used in air-conditioning problems; it is the average velocity in a conduit times the area of the conduit. It is customary to use the symbol Q_f to represent the volume flow rate. Referring to Eq. 4.58, we observe that

$$Q_f = AV \qquad \text{and} \qquad \dot{m} = \rho Q_f \tag{8.17}$$

The units on Q_f are m^3/s or ft^3/sec. The English units are often stated as cfs for ft^3/sec or cfm for ft^3/min.

Example 8.5

Outside air at 5°C and 70% relative humidity is heated to 25°C. Calculate the rate of heat transfer needed if the incoming volume flow rate is 50 m^3/min. Also, find the final relative humidity. Assume $p = 100$ kPa.

(a) Use the psychrometric chart and (b) TK Solver.

Solution

(a) First, let's find the mass flow rate of air. The density of dry air is found using the partial pressure p_{a1} in the ideal-gas law:

$$p_{a1} = p - p_{v1} = p - \phi p_{g1} = 100 - 0.7 \times 0.872 = 99.4 \text{ kPa}$$

$$\therefore \rho_{a1} = \frac{p_{a1}}{R_a T_1} = \frac{99.4}{0.287 \times 278} = 1.246 \text{ kg/m}^3$$

(continued)

The mass flow rate of dry air is then

$$\dot{m}_a = Q_f \rho_a = \frac{50 \text{ m}^3/\text{min}}{60 \text{ s/min}} \times 1.246 \frac{\text{kg}}{\text{m}^3} = 1.038 \text{ kg/s}$$

Using the psychrometric chart at state 1 ($T_1 = 5°\text{C}$, $\phi_1 = 70\%$), we find $h_1 = 14$ kJ/kg air. Since ω remains constant (no moisture is added or removed), we follow curve A–C in Fig. 8.4; at state 2 we find that $h_2 = 35$ kJ/kg air. Hence,

$$\dot{Q} = \dot{m}_a(h_2 - h_1) = 1.038(35 - 14) = 22 \text{ kJ/s}$$

At state 2, we also note from the chart that $\phi_2 = 19\%$ (a relatively low value). Only two significant digits are used since the chart can be read to only two significant digits.

(b) Using TK Solver (see Example 8.1), input T1 = 5°C, pm1 = 100 kPa, and phi1 = 0.7. We find $\omega_1 = 0.0038089$, $h_1 = 14.584$ kJ/kg, and $v_{a1} = 0.80318$ m^3/kg. The mass flux of dry air is

$$\dot{m} = Q_f / v_{a1} = \frac{50/60}{0.80318} = 1.0375 \text{ kg/s}$$

At state 2, input omega2 = 0.0038089, T2 = 25°C, and pm2 = 100 kPa. We find $h_2 =$ 34.823 kJ/kg. Then,

$$\dot{Q} = \dot{m}_a(h_2 - h_1) = 1.0375 \times (34.823 - 14.584) = 21.0 \text{ kJ/s}$$

To find ϕ_2, we input "Guess" for "Status": try 0.2 (if it doesn't work try a different guess) and find $\phi_2 = 0.1923$ or 19.23%.

Example 8.6

Outside air at 5°C and 40% relative humidity is heated to 25°C, and the final relative humidity is raised to 40%, while the temperature remains constant by introducing steam at 400 kPa into the air stream. Find the needed rate of heat transfer if the incoming volume flow rate of air is 60 m^3/min. Calculate the rate of steam supplied and the state of the steam introduced.

Solution

The process we must follow is first simple heating and then humidification, sketched as D–E–C in Fig. 8.4. The partial pressure of dry air is

$$p_{a1} = p - p_{v1} = p - \phi p_{g1} = 100 - 0.4 \times 0.872 = 99.7 \text{ kPa}$$

where we have assumed standard atmospheric pressure. The dry air density is

$$\rho_{a1} = \frac{p_{a1}}{R_a T_1} = \frac{99.7}{0.287 \times 278} = 1.25 \text{ kg/m}^3$$

so that the mass flow rate of dry air is

$$\dot{m}_a = Q_f \rho_a = \frac{60 \text{ m}^3/\text{min}}{60 \text{ s/min}} \times 1.25 \frac{\text{kg}}{\text{m}^3} = 1.25 \text{ kg/s}$$

(continued)

The rate of heat addition is found using h_1 and h_2 from the psychrometric chart:

$$\dot{Q} = \dot{m}_a(h_2 - h_1) = 1.25 \times (31 - 10) = 26 \text{ kJ/s}$$

We assume that all the heating is done in the D–E process and that humidification takes place in a process in which the steam is mixed with the air flow. Assuming a constant temperature in the mixing process, conservation of mass demands that

$$\dot{m}_s = (\omega_3 - \omega_2)\dot{m}_a = (0.008 - 0.0021) \times 1.25 = 0.0074 \text{ kg/s}$$

where the air enters the humidifier at state 2 and leaves at state 3.

An energy balance around the humidifier provides us with $h_s\dot{m}_s = (h_3 - h_2)\dot{m}_a$. Hence,

$$h_s = \frac{\dot{m}_a}{\dot{m}_s}(h_3 - h_2) = \frac{1.25}{0.0074} \times (45 - 33) = 2400 \text{ kJ/kg}$$

This is less than h_g at 400 kPa. Consequently, the temperature of the steam is 144°C, and the quality is

$$x_s = \frac{2365 - 604.7}{2133.8} = 0.82$$

Only two significant figures are used because of inaccuracy using the relatively small psychrometric chart provided.

Example 8.7

Outside air at 80°F and 90% relative humidity is conditioned so that it enters a building at 75°F and 40% relative humidity. Estimate the amount of moisture removed, the heat removed, and the necessary added heat.

Solution

The overall process is sketched as F–G–H–I in Fig. 8.4. Heat is removed during the F–H process, moisture is removed during the G–H process, and heat is added during the H–I process. Using the psychrometric chart, we find the moisture removed to be

$$\Delta\omega = \omega_3 - \omega_2 = 0.0075 - 0.0177 = -0.010 \text{ lbm water/lbm dry air}$$

where states 2 and 3 are at G and H, respectively. (The negative sign implies moisture removal.)

The heat that must be removed to cause the air to follow the F–G–H process is

$$q = h_3 - h_1 = 20 - 39.5 = -18.5 \text{ Btu/lbm dry air}$$

We neglected the small amount of heat due to the liquid state.

The heat that must be added to change the state of the air from the saturated state at H to the desired state at I is

$$q = h_4 - h_3 = 26.5 - 20 = 6.5 \text{ Btu/lbm dry air}$$

Example 8.8

Hot, dry air at 40°C and 10% relative humidity passes through an evaporative cooler. Water is added as the air passes through a series of wicks, and the mixture exits at 27°C. Find the outlet relative humidity, the amount of water added, and the lowest temperature that could be realized using (a) the psychrometric chart and (b) TK Solver.

Solution

(a) The heat transfer is negligible in an evaporative cooler, so that $h_2 = h_1$. A constant enthalpy line is shown in Fig. 8.4 and is represented by *J–K*. From the psychrometric chart, we find, at 27°C, $\phi_2 = 45\%$.

The added water is found to be

$$\omega_2 - \omega_1 = 0.010 - 0.0046 = 0.0054 \text{ kg water/kg dry air}$$

The lowest possible temperature occurs when $\phi = 100\%$: $T_{\text{min}} = 18.5°\text{C}$.

(b) Using TK Solver, we input T1 = 40°C, phi1 = 0.1, and pm1 = 100 kPa. We find $\omega_1 = 0.0046245$ and $h_1 = 52.101$ kJ/kg. For state 2, we input T2 = 27°C, h2 = 52.101 kJ/kg, and pm2 = 100 kPa. We input "Guess" for "Status" in order to find phi2 and find $\omega_2 = 0.0097907$ and $\phi_2 = 0.43487$ or 43.487%.

The added water is

$$\omega_2 - \omega_1 = 0.0097907 - 0.0046245 = 0.005176$$

The lowest possible temperature is found by inputting "Guess" for T2 (try 20°C), with hm2 = 52.1 kJ/kg (the enthalpy remains constant). We find $T_2 = 18.32°\text{C}$.

Example 8.9

Outside, cool air at 15°C and 40% relative humidity (air stream 1) is mixed with inside air taken near the ceiling at 32°C and 70% relative humidity (air stream 2). Determine the relative humidity and temperature of the resultant air stream 3 if the outside flow rate is 40 m^3/min and the inside flow rate is 20 m^3/min.

Solution

An energy and mass balance of the mixing of air stream 1 with air stream 2 to produce air stream 3 would reveal the following facts relative to the psychrometric chart (see line *L–M–I* in Fig. 8.4):

- State 3 lies on a straight line connecting state 1 and state 2.
- The ratio of the distance 2–3 to the distance 3–1 is equal to $\dot{m}_{a1}/\dot{m}_{a2}$.
- State 1 and state 2 can easily be located on the psychrometric chart. We must determine $\dot{m}_{a1}$ and $\dot{m}_{a2}$:

$$h_1 = 37 \text{ kJ/kg dry air} \qquad \omega_1 = 0.0073 \text{ kg water/kg dry air}$$

$$h_2 = 110 \text{ kJ/kg dry air} \qquad \omega_2 = 0.0302 \text{ kg water/kg dry air}$$

$$p_{a1} = p - p_{v1} = 100 - 1.7 = 98.3 \text{ kPa} \qquad p_{a2} = p - p_{v2} = 100 - 4.8 = 95.2 \text{ kPa}$$

(continued)

$$\therefore\ \rho_{a1} = \frac{98.3}{0.287 \times 288} = 1.19\ \text{kg/m}^3 \qquad \rho_{a2} = \frac{95.2}{0.287 \times 305} = 1.09\ \text{kg/m}^3$$

$$\therefore\ \dot{m}_{a1} = Q_{f1}\rho_{a1} = 40 \times 1.19 = 47.6\ \text{kg/min} \qquad \dot{m}_{a2} = Q_{f2}\rho_{a2} = 20 \times 1.09 = 21.8\ \text{kg/min}$$

State 3 is located by the ratio

$$\frac{d_{2-3}}{d_{3-1}} = \frac{\dot{m}_{a1}}{\dot{m}_{a2}} = \frac{47.6}{21.8} = 2.18$$

where d_{2-3} is the distance from state 2 to state 3. State 3 is positioned on the psychrometric chart, and we find

$$\phi_3 = 63\% \qquad \text{and} \qquad T_3 = 20.2°\text{C}$$

Note: State 3 is located nearer state 1, since the greater outside flow dominates.

Example 8.10

Water is used to remove the heat from the condenser of a power plant; 10 000 kg/min of 40°C water enters a cooling tower, as shown in Fig. E8.10. Water leaves at 25°C. Air enters at 20°C and leaves at 32°C. Estimate the volume flow rate of air into the cooling tower and the mass flow rate of water that leaves the cooling tower from the bottom.

Solution

An energy balance for the cooling tower provides

$$\dot{m}_{a1}h_1 + \dot{m}_{w3}h_3 = \dot{m}_{a2}h_2 + \dot{m}_{w4}h_4$$

where $\dot{m}_{a1} = \dot{m}_{a2} = \dot{m}_a$ is the mass flow rate of dry air. From the psychrometric chart, we find

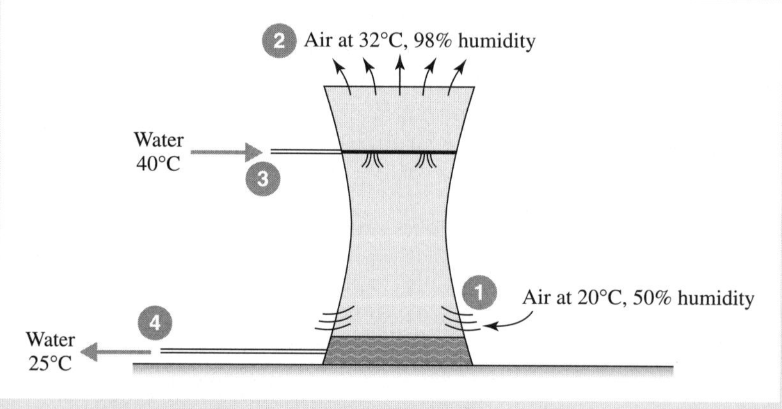

FIGURE E8.10

(continued)

$$h_1 = 37 \text{ kJ/kg dry air} \qquad \omega_1 = 0.0073 \text{ kg water/kg dry air}$$

$$h_2 = 110 \text{ kJ/kg dry air} \qquad \omega_2 = 0.0302 \text{ kg water/kg dry air}$$

From the Appendix steam tables, we use h_f at 40°C and find $h_3 = 167.5$ kJ/kg and $h_4 = 104.9$ kJ/kg. A mass balance on the water results in $\dot{m}_{w4} = \dot{m}_{w3} - (\omega_2 - \omega_1)\dot{m}_a$. Substituting this into the energy balance, with $\dot{m}_{a1} = \dot{m}_{a2} = \dot{m}_a$, results in

$$\dot{m}_a = \frac{\dot{m}_{w3}(h_4 - h_3)}{h_1 - h_2 + (\omega_2 - \omega_1)h_4} = \frac{10\,000 \times (104.9 - 167.5)}{37 - 110 + (0.0302 - 0.0073)104.9} = 8870 \text{ kg/min}$$

From the psychrometric chart, we find that $v_1 = 0.84$ m^3/kg dry air. This allows us to find the volume flow rate:

$$Q_f = \dot{m}_a v_1 = 8870 \times 0.84 = 7450 \text{ m}^3\text{/min}$$

This air flow rate requires fans, although there is some "chimney effect", since the hotter air tends to rise.

Returning to the mass balance provides

$$\dot{m}_4 = \dot{m}_{w3} - (\omega_2 - \omega_1)\dot{m}_a = 10\,000 - (0.0302 - 0.0073) \times 8870 = 9800 \text{ kg/min}$$

If the exiting water is returned to the condenser, it must be augmented by 200 kg/min, so that 10 000 kg/min is furnished. The added water is called *makeup water*.

8.6 SUMMARY

In this chapter, we defined the following terms:

Atmospheric air: Air with water vapor included.
Dry air: Air with the water vapor not included.
Dalton model: All components of a gas have the same volume and temperature but different partial pressures.
Amagat model: All components of a gas have the same pressure and temperature but different partial volumes.
Partial pressure: The sum of the partial pressures equals the pressure of the whole.
Vapor pressure: The partial pressure of the water vapor component.
Relative humidity: The ratio of the water vapor mass to the maximum water vapor mass the air can hold at the given temperature. (It is also called simply "humidity".)
Humidity ratio: The ratio of water vapor mass to the dry air mass. (It is also call "specific humidity".)
Dry-bulb temperature: The temperature measured by a conventional thermometer.
Dew-point temperature: The temperature at which condensation begins if air is cooled at constant pressure.
Adiabatic saturation temperature: The temperature reached when air is blown over a long reach of water in an insulated container.

Wet-bulb temperature: The temperature measured when air is blown over the bulb of a thermometer that is covered with a wick saturated with water. (It is approximately equal to the adiabatic saturation temperature.)
Psychrometric chart: A chart showing the various properties plotted as a function of dry-bulb temperature and humidity ratio.
Volume flow rate: The area multiplied by the average velocity normal to the area.
Evaporative cooling: When air is cooled by blowing the air over wet wicks.
Cooling tower: A tower used to remove heat from hot water, most often from the condenser of a power plant.

Several equations are unique to this chapter. They include the equation relating the partial pressures of air and water vapor:

$$p = p_a + p_v \tag{8.18}$$

The relative humidity is given by

$$\phi = \frac{p_v}{p_g} \tag{8.19}$$

where p_g is the saturation pressure of the water vapor at the dry-bulb temperature (it is found in Appendix Table C.1 next to the temperature). The humidity ratio is

$$\omega = 0.622 \times \frac{p_v}{p_a} \tag{8.20}$$

The enthalpy per unit mass is

$$h = h_a + \omega h_v \tag{8.21}$$

The humidity ratio can be calculated from measured and tabulated quantities with

$$\omega_1 = \frac{\omega_2 h_{fg2} + c_p(T_2 - T_1)}{h_{g1} - h_{f2}} \tag{8.22}$$

where ω_2 is calculated assuming 100% humidity at T_2.

The volume flow rate is related to the average velocity V over a cross-sectional area A by the relationship

$$Q_f = AV \tag{8.23}$$

If we desire the mass flow rate, it is related to the volume flow rate by

$$\dot{m} = \rho Q_f \tag{8.24}$$

REVIEW PROBLEMS FOR THE FE/EIT EXAMINATION

8.1FE The air in a conference room is to be conditioned from its present state of 18°C and 40% humidity. Select the appropriate conditioning strategy.
A. Heat and dehumidify.
B. Cool, dehumidify, and then heat.
C. Heat, dehumidify, and then cool.
D. Heat and humidify.

8.2FE For which of the following situations would you expect condensation to not occur?
A. On a glass of ice water sitting on the kitchen table.
B. On the grass on a cool evening in Tucson, Arizona.
C. On your glasses when you come into a warm room on a cold winter day.
D. On the inside of a windowpane in an apartment that has high air infiltration on a cold winter day.

8.3FE Which temperature is most different from the outside temperature as measured with a conventional thermometer?
A. Wet-bulb temperature
B. Dry-bulb temperature
C. Dew-point temperature
D. Ambient temperature

8.4FE An evaporative cooler inlets air at 40°C and 20% humidity. If the exiting air is at 80% humidity, the exiting temperature is nearest:
A. 29°C
B. 27°C
C. 25°C
D. 23°C

8.5FE The dry-bulb temperature is 35°C, and the wet-bulb temperature is 27°C. What is the relative humidity?
A. 65%
B. 60%
C. 55%
D. 50%

8.6FE Estimate the liters of water in the air in a 3 m × 10 m × 20 m room if the temperature is 20°C and the humidity is 70%.
A. 7.3 L
B. 6.2 L
C. 5.1 L
D. 4.0 L

8.7FE Air at 5°C and 80% humidity is heated to 25°C in an enclosed room. The final humidity is nearest:
A. 36%
B. 32%
C. 26%
D. 22%

8.8FE Air at 30°C and 80% humidity is to be conditioned to 20°C and 40% humidity. How much heating is required?
A. 30 kJ/kg
B. 25 kJ/kg
C. 20 kJ/kg
D. 15 kJ/kg

8.9FE Air at 35°C and 70% humidity in a 3 m × 10 m × 20 m classroom is cooled to 25°C and 40% humidity. How much water is removed?
A. 4 kg
B. 6 kg
C. 11 kg
D. 16 kg

8.10FE Outside air at 15°C and 40% humidity is mixed with inside air at 32°C and 70% humidity taken near the ceiling. Determine the relative humidity of the mixed stream if the outside flow rate is 40 m^3/min and the inside flow rate is 20 m^3/min.
A. 63% **B.** 58%
C. 53% **D.** 49%

8.11FE The temperature of the mixed stream of Problem 8.10FE is nearest:
A. 28°C **B.** 25°C
C. 23°C **D.** 20°C

PROBLEMS

In the following problem set, work each problem using the appropriate equations and/or Appendix tables. Use TK Solver for property evaluation when requested.

Gas-Vapor Mixtures

8.1 Atmospheric air at 30°C and 100 kPa has a relative humidity of 40%. Determine the humidity ratio, the dew-point temperature, and the specific volume of the dry air using **(a)** the equations and **(b)** TK Solver.

8.2 Atmospheric air at 90°F and 14.2 psia has a humidity ratio of 0.02. Calculate the relative humidity, the dew-point temperature, the specific volume of the dry air, and the enthalpy ($h = 0$ at 0°F) per unit mass of dry air.

8.3 Outside air at 30°C and 100 kPa is observed to have a dew point of 20°C. Find the relative humidity, the partial pressure of the dry air, and the humidity ratio using **(a)** the equations and **(b)** TK Solver.

8.4 The air in a 12 m × 15 m × 3 m room is at 25°C and 100 kPa, with a 50% relative humidity. Estimate the humidity ratio, the mass of dry air, the mass of water vapor in the room, and the enthalpy in the room ($h = 0$ at 0°C).

8.5 A tank contains 0.4 kg of dry air and 0.1 kg of saturated water vapor at 30°C. Calculate the volume of the tank and the pressure in the tank.

8.6 The partial pressure of water vapor is 1 psia in atmospheric air at 14.5 psia and 110°F. Find the relative humidity, the humidity ratio, the dew-point temperature, the specific volume of the dry air, and the enthalpy per unit mass of dry air.

8.7 A person wearing glasses comes from outside, where the temperature is 10°C, into a room with 40% relative humidity. Determine the room temperature at which the glasses just begin to fog up using **(a)** the equations and **(b)** TK Solver.

8.8 The outer surface temperature of a glass of cola, in a room at 28°C, is 5°C. At what relative humidity will water begin to collect on the outside of the glass?

8.9 A cold-water pipe at 50°F runs through a basement where the temperature is 70°F. At what relative humidity will water begin to condense on the pipe?

8.10 On a cold winter day, the temperature on the inside of a thermopane window is 10°C. If the inside temperature is 21°C, what relative humidity is needed to just cause condensation on the inside of window?

8.11 Atmospheric air has a dry-bulb temperature of 30°C and a wet-bulb temperature of 20°C. Calculate the humidity ratio, the relative humidity, and the enthalpy per kg of dry air ($h = 0$ at 0°C) using **(a)** the equations and **(b)** TK Solver.

8.12 Outside air at 25°C has a relative humidity of 60%. What would be the expected wet-bulb temperature?

Psychrometric Chart

8.13 Use the psychrometric chart (Appendix G) to provide the missing values in the following table.

8.14 Rework Problem 8.12 using **(a)** the psychrometric chart and **(b)** TK Solver.

	Dry-bulb temperature	Wet-bulb temperature	Relative humidity	Humidity ratio	Dew-point temperature	Specific enthalpy
(a)	20°C			0.012		
(b)	20°C	10°C				
(c)	70°F		60%			
(d)		60°F	70%			
(e)		25°C			15°C	
(f)		70°F		0.015		

Air-Conditioning Processes

8.15 Atmospheric air at 10°C and 60% relative humidity is heated to 21°C. Use the psychrometric chart to estimate the final humidity and the rate of heat transfer needed if the mass flow rate of dry air is 50 kg/min.

8.16 Heat is removed from a room without condensing out any of the water vapor. Calculate the final relative humidity if the air is initially at 35°C and 50% relative humidity and the temperature is reduced to 25°C using **(a)** the psychrometric chart and **(b)** TK Solver.

8.17 Outside air at 0°C and 40% relative humidity enters through the cracks in a house and is heated to 20°C. Estimate the final relative humidity of the air if no other sources of water vapor are available.

8.18 Air at 90°F and 20% relative humidity is cooled to 75°F. Assuming standard atmospheric pressure, calculate the required rate of energy transfer if the inlet flow rate is 1500 cfm, and find the final humidity using **(a)** the psychrometric chart and **(b)** the equations.

8.19 Air at 90°F and 90% relative humidity is cooled to 75°F. Calculate the required rate of energy transfer if this inlet flow rate is 1500 ft^3/min. Also, find the final humidity. Compare with the results of Problem 8.18. Use the psychrometric chart.

8.20 A rigid 2-m^3 tank contains air at 160°C and 400 kPa and a relative humidity of 20%. Heat is removed until the final temperature is 20°C. Determine the temperature at which condensation begins, the mass of water condensed during the process, and the heat transfer using **(a)** the psychrometric chart and **(b)** TK Solver.

8.21 Atmospheric air at 10°C and 40% relative humidity is heated to 25°C in the heating section of an air-conditioning device. Then steam is introduced to increase the relative humidity to 50%, while the temperature increases to 26°C, as displayed in Fig. P8.21. Calculate the mass flow rate of water vapor added and the rate of heat transfer needed in the heating section if the volume flow rate of inlet air is 50 m^3/min.

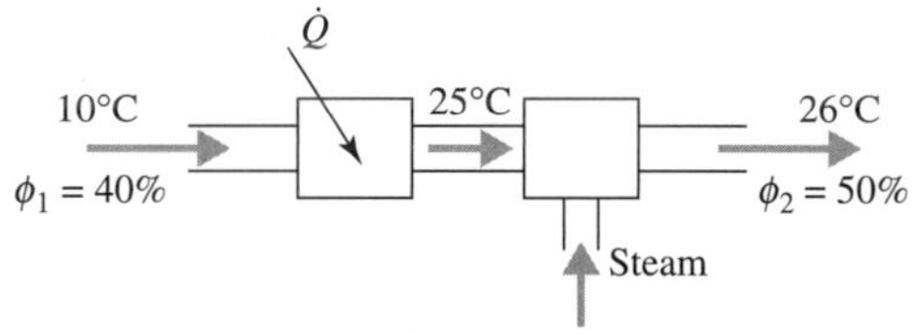

FIGURE P8.21

8.22 Atmospheric air at 40°F and 50% relative humidity enters the heating section of an air-conditioning device at a volume flow rate of 100 cfm. Water vapor is added to the heated air to increase the relative humidity to 55%. Estimate the rate of heat transfer needed in the heating section and the mass flow rate of water vapor added if the temperature at the heating section is 72°F and the temperature at the exit is 74°F.

8.23 Hot, dry air at 40°C and 20% humidity passes through an evaporative cooler until the humidity is 40%; it is then cooled to 25°C. For an inlet airflow of 50 m^3/min, determine the amount of water added per hour and the rate of cooling using **(a)** the psychrometric chart and **(b)** TK Solver.

8.24 Outside air in a dry climate enters an air conditioner at 40°C and 10% relative humidity and is cooled to 22°C. Calculate the heat removed and the total energy required to condition outside (humid) air at 30°C and 90% relative humidity to 22°C and 10% relative humidity. (*Hint:* Sum the energy removed and the energy added.)

8.25 One hundred m^3/min of outside air at 36°C and 80% relative humidity (Fig. P8.25) is conditioned for an office building by cooling and heating. Estimate both the rate of cooling and the rate of heating required if the final state of the air is 25°C and 40% relative humidity.

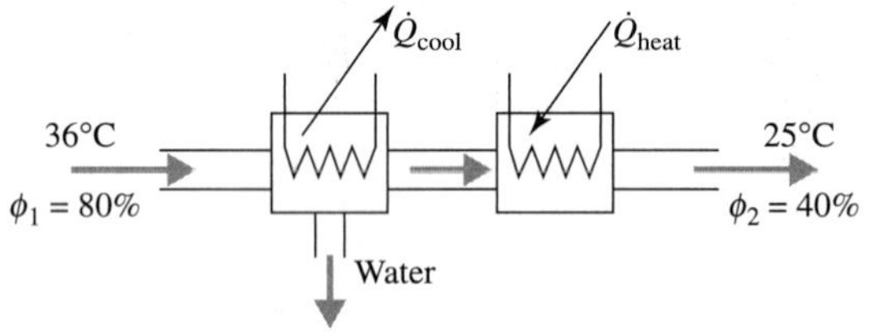

FIGURE P8.25

8.26 Room air at 29°C and 70% relative humidity is cooled by passing it over coils through which chilled water (at 5°C) flows. The mass flow rate of the chilled water is 0.5 kg/s, and it experiences a 10°C temperature rise. If the room air exits the conditioner at 18°C and 100% relative humidity, estimate the mass flow rate of the room air and the heat transfer rate using **(a)** the psychrometric chart and **(b)** TK Solver.

8.27 Atmospheric air at 100°F and 15% relative humidity enters an evaporative cooler at 900 cfm and leaves with a relative humidity of 60% as shown in Fig. P8.27. Estimate the exit temperature and the mass flow rate at which water must be supplied to the cooler.

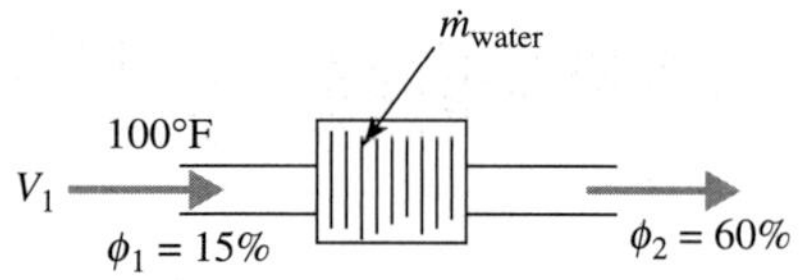

FIGURE P8.27

8.28 Outside air at 40°C and 20% relative humidity is to be cooled by using an evaporative cooler. If the flow rate of the air is 40 m^3/min, estimate the minimum possible temperature of the exit stream and the maximum mass flow rate needed for the water supply using **(a)** the psychrometric chart and **(b)** TK Solver.

8.29 Thirty m^3/min of outside air at 0°C and 40% relative humidity is heated and then passed through an evaporative cooler so that the final state is 25°C and 50% relative humidity, as shown in Fig. P8.29. Determine the temperature T_2 of the air when it enters the cooler, the heat transfer rate needed during the heating process, and the mass flow rate of water required by the cooler.

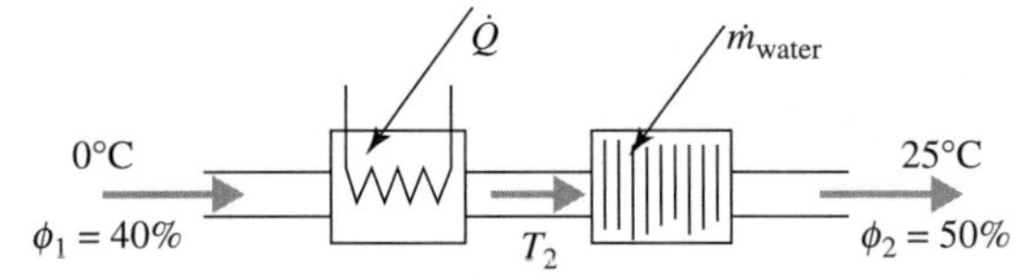

FIGURE P8.29

8.30 Outside air at 10°C and 60% relative humidity mixes with 50 m^3/min of inside air at 28°C and 40% relative humidity. If the outside flow rate is 30 m^3/min, estimate the relative humidity, the temperature, and the mass flow rate of the exiting stream using **(a)** the psychrometric chart and **(b)** TK Solver.

8.31 Outside air at 10°C and 30% relative humidity is available to mix with inside air at 30°C and 60% humidity. The inside-flow rate is 50 m^3/min. Use the equations to determine what the outside-flow rate should be to provide a mixed stream at 22°C.

8.32 Inside air at 80°F and 80% relative humidity is mixed with 900 cfm of outside air at 40°F and 20% relative humidity. If the relative humidity of the exiting stream is 60%, estimate the flow rate of the inside air, the temperature of the exiting stream, and the heat transfer rate from the outside air to the inside air.

8.33 Cooling water leaves the condenser of a power plant at 38°C with a mass flow rate of 40 kg/s. It is cooled to 24°C in a cooling tower that receives atmospheric air at 25°C and 60% relative humidity. Saturated air exits the tower at 32°C. Estimate the required volume flow rate of entering air and the mass flow rate of the makeup water.

8.34 A cooling tower cools 40 lbm/sec of water from 80°F to 60°F by moving 800 cfs of atmospheric air with dry-bulb and wet-bulb temperatures of 75°F and 55°F, respectively, through the tower. Saturated air exits the tower. Find the temperature of the exiting air stream and the mass flow rate of the makeup water using **(a)** the psychrometric chart and **(b)** TK Solver.

8.35 A cooling tower cools water from 35°C to 27°C, as shown in Fig. P8.35. The tower receives 200 m^3/s of atmospheric air at 30°C and 40% relative humidity. The air exits the tower at 33°C and 95% relative humidity. Estimate the mass flow rate of water that is cooled and the mass flow rate of makeup water.

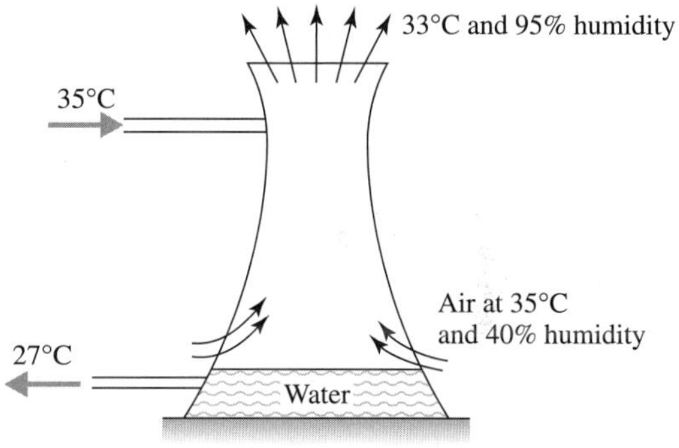

FIGURE P8.35

Combustion occurs in many ways in our modern society: engines, furnaces, forests, power plants, camp fires, weapons, gas refrigerators, dryers, piles of leaves, to mention a few. (Image courtesy of Getty Images.)

9 Combustion

Outline

Chapter Objectives

The objectives of the chapter are to:

- ▲ Review the method to balance chemical-reaction equations.
- ▲ Introduce the terminology associated with complete and incomplete combustion.
- ▲ Apply the first law to combustion problems.
- ▲ Calculate the temperature of the flame after combustion occurs.
- ▲ Present numerous examples and practice problems that illustrate the concepts discussed.

Nomenclature

The following is a list of nomenclature introduced in this chapter:

AF	Air–fuel ratio
AF_{th}	Theoretical air–fuel ratio
FA	Fuel–air ratio
g	A symbol that refers to gas
h°	Enthalpy at a reference state
h_f°	Enthalpy of formation
l	A symbol that refers to liquid
U	Overall heat-transfer coefficient

9.1 COMBUSTION EQUATIONS

Combustion involves the burning of a fuel with oxygen or a substance containing oxygen such as air. A chemical-reaction equation relates the components before and after the chemical process takes place. Let us begin our study of this particular variety of chemical-reaction equations by considering the combustion of propane in a pure oxygen environment. The chemical reaction is represented by

$$C_3H_8 + 5O_2 \rightarrow 3CO_2 + 4H_2O \tag{9.1}$$

Note that the number of moles of the elements on the left-hand side may not equal the number of moles on the right-hand side. However, the number of atoms of an element must remain the same before, after, and during a chemical reaction; this demands that the mass of each element be conserved during combustion.

In writing the equation, we have demonstrated some knowledge of the products of the reaction. Unless otherwise stated, we will assume *complete combustion*: the products of the combustion of a hydrocarbon fuel will include H_2O and CO_2 but no CO, OH, C, or H_2. *Incomplete combustion* results in products that contain H_2, CO, C, and/or OH.

For a simple chemical reaction, such as that of Eq. 9.1, we can immediately write down a balanced chemical equation. For more complex reactions the following systematic steps prove useful:

1. Set the number of moles of fuel equal to 1.
2. Balance CO_2 with number of C from the fuel.
3. Balance H_2O with H from the fuel.
4. Balance O_2 from CO_2 and H_2O.

For the combustion of propane, we assumed that the process occurred in a pure oxygen environment. Actually, such a combustion process would normally occur in air. For our purposes, we assume that air consists of 21% O_2 and 79% N_2 by volume (we ignore the 1% argon), so that for each mole of O_2 in a reaction we will have

$$\frac{V\!\!\!/_{N_2}}{V\!\!\!/_{O_2}} = \frac{79}{21} = 3.76 \ \frac{\text{mol } N_2}{\text{mol } O_2} \tag{9.2}$$

Thus, on the (simplistic) assumption that N_2 will not undergo any chemical reaction, Eq. 9.1 is replaced by

$$C_3H_8 + 5(O_2 + 3.76N_2) \rightarrow 3CO_2 + 4H_2O + 18.8N_2 \tag{9.3}$$

Theoretical air: *The minimum amount of air that supplies sufficient O_2 for the complete combustion of the fuel.*

The minimum amount of air that supplies sufficient O_2 for the complete combustion of the fuel is called **theoretical air** or *stoichiometric air*. When complete combustion is achieved with theoretical air, the products contain no O_2, as in the reaction of Eq. 9.3. In practice, it is found that if complete combustion is to occur, air must be supplied in an amount greater than theoretical air. This is due to the chemical kinetics and molecular activity of the reactants and products. Thus, we often speak in terms of *percent theoretical air* or *percent excess air*, where

$$\%\ \text{theoretical air} = 100\% + \%\ \text{excess air} \tag{9.4}$$

Key Concept: *Slightly insufficient air results in CO being formed.*

Slightly insufficient air results in CO being formed; some hydrocarbons may result from larger deficiencies.

The parameter that relates the amount of air used in a combustion process is the *air–fuel ratio* (AF), which is the ratio of the mass of air to the mass of fuel.

The reciprocal is the *fuel–air ratio* (FA). Thus

$$AF = \frac{m_{\text{air}}}{m_{\text{fuel}}} \qquad FA = \frac{m_{\text{fuel}}}{m_{\text{air}}} \tag{9.5}$$

Again, considering propane combustion with theoretical air as in Eq. 9.3, the air–fuel ratio is

$$AF = \frac{m_{\text{air}}}{m_{\text{fuel}}} = \frac{5 \times 4.76 \times 29}{1 \times 44} = 15.69 \ \frac{\text{kg air}}{\text{kg fuel}} \tag{9.6}$$

where we have used the molecular weight of air as 29 kg/kmol and that of propane as 44 kg/kmol. If, for the combustion of propane, $AF > 15.69$, a **lean mixture** occurs; if $AF < 15.69$, a *rich mixture* results.

Lean mixture: *Excess air is present.*

The combustion of hydrocarbon fuels involves H_2O in the products of combustion. The calculation of the dew point of the products is often of interest; it is the saturation temperature at the partial pressure of the water vapor. If the temperature drops below the dew point, the water vapor begins to condense. The condensate usually contains corrosive elements, and thus, it is often important to ensure that the temperature of the products does not fall below the dew point. This is especially necessary in the chimney of a house; the corrosive elements can eat through the building materials and cause substantial damage. For that reason, mid-efficiency furnaces (80% or so) require chimney inserts that are able to resist the corrosive elements.

When analyzing a mixture of gases, either a gravimetric analysis based on mass (or weight) or a volumetric analysis based on moles (or volume) may be used. If the analysis is presented on a dry basis, the water vapor is not included. An Orsat analyzer and a gas chromatograph are two of the devices used to measure the products of combustion.

Example 9.1

Butane is burned with dry air at an air–fuel ratio of 20. Calculate the percent excess air, the volume percentage of CO_2 in the products, and the dew-point temperature of the products.

Solution

The reaction equation for theoretical air is (see Appendix Table B.3 for the composition of butane)

$$C_4H_{10} + 6.5(O_2 + 3.76N_2) \rightarrow 4CO_2 + 5H_2O + 24.44N_2$$

The air–fuel ratio for theoretical air is

$$AF_{\text{th}} = \frac{m_{\text{air}}}{m_{\text{fuel}}} = \frac{6.5 \times 4.76 \times 29}{1 \times 58} = 15.47 \ \frac{\text{kg air}}{\text{kg fuel}}$$

(continued)

This represents 100% theoretical air. The actual air–fuel ratio is 20. The excess air is then

$$\% \text{ excess air} = \frac{AF_{\text{actual}} - AF_{\text{th}}}{AF_{\text{th}}} \times 100 = \frac{20 - 15.47}{15.47} \times 100 = 29.28\%$$

The reaction equation with 129.28% theoretical air is

$$C_4H_{10} + 6.5 \times 1.2928 \times (O_2 + 3.76N_2) \rightarrow 4CO_2 + 5H_2O + 1.903O_2 + 31.6N_2$$

The volume percentage is obtained using the total moles in the products of combustion. For CO_2, we have

$$\% \text{ CO}_2 = \frac{4}{42.5} \times 100 = 9.41\%$$

To find the dew-point temperature of the products, we need the partial pressure p_v of the water vapor. We know that

$$p_v = \frac{N_v RT}{V\!\!\!\!-} \qquad \text{and} \qquad p = \frac{NRT}{V\!\!\!\!-}$$

where N is the number of moles. Then the ratio of the above two equations provides

$$p_v = p\frac{N_v}{N} = 100 \times \frac{5}{42.5} = 11.76 \text{ kPa} \quad \text{or} \quad 0.01176 \text{ MPa}$$

where we have assumed an atmospheric pressure of 100 kPa. Using Appendix Table C.2, we find the dew-point temperature to be $T_{\text{dp}} = 49°\text{C}$.

Example 9.2

Butane is burned with 90% theoretical air. Calculate the volume percentage of CO in the products and the air–fuel ratio. Assume no hydrocarbons in the products.

Solution

For incomplete combustion, we add CO to the products of combustion. Using the reaction equation from Example 9.1,

$$C_4H_{10} + 0.9 \times 6.5 \times (O_2 + 3.76N_2) \rightarrow aCO_2 + 5H_2O + 22N_2 + bCO$$

With atomic balances on the carbon and oxygen, we find:

$$\left.\begin{array}{l} \text{C: } 4 = a + b \\ \text{O: } 11.7 = 2a + 5 + b \end{array}\right\} \quad \therefore a = 2.7, b = 1.3$$

The volume percentage of CO is then

$$\% \text{ CO} = \frac{1.3}{31} \times 100 = 4.19\%$$

The air–fuel ratio is

$$AF = \frac{m_{\text{air}}}{m_{\text{fuel}}} = \frac{0.9 \times 6.5 \times 4.76 \times 29}{1 \times 58} = 13.92 \ \frac{\text{kg air}}{\text{kg fuel}}$$

Example 9.3

Butane is burned with dry air, and volumetric analysis of the products on a dry basis (the water vapor is not measured) gives 11.0% CO_2, 1.0% CO, 3.5% O_2, and 84.5% N_2. Determine the percent theoretical air.

Solution

The problem is solved assuming that there are 100 moles of dry products. The chemical equation is

$$aC_4H_{10} + b(O_2 + 3.76N_2) \rightarrow 11CO_2 + 1CO + 3.5O_2 + 84.5N_2 + cH_2O$$

We perform the following balances:

$$\left.\begin{aligned} &C: 4a = 11 + 1 \\ &H: 10a = 2c \\ &O: 2b = 22 + 1 + 7 + c \end{aligned}\right\} \quad \therefore a = 3, c = 15, b = 22.5$$

A balance on the nitrogen allows a check: $3.76b = 84.5$, or $b = 22.47$. This is quite close to 22.5, so the above values are acceptable. Dividing through the chemical equation by the value of a so that we have 1 mol fuel, provides

$$C_4H_{10} + 7.5(O_2 + 3.76N_2) \rightarrow 3.67CO_2 + 0.33CO + 1.17O_2 + 28.17N_2 + 5H_2O$$

Comparing this with the combustion equation of Example 9.1 using theoretical air, we find

$$\%\text{ theoretical air} = \frac{7.5}{6.5} \times 100 = 115.4\%$$

Example 9.4

Volumetric analysis of the products of combustion of an unknown hydrocarbon, measured on a dry basis, gives 10.4% CO_2, 1.2% CO, 2.8% O_2, and 85.6% N_2. Determine the composition of the hydrocarbon and the percent excess air.

Solution

The chemical equation for 100 mol of dry products is

$$C_aH_b + c(O_2 + 3.76N_2) \rightarrow 10.4CO_2 + 1.2CO + 2.8O_2 + 85.6N_2 + dH_2O$$

Balancing each element,

$$\left.\begin{aligned} &C: a = 10.4 + 1.2 \\ &N: 3.76c = 85.6 \\ &O: 2c = 20.8 + 1.2 + 5.6 + d \\ &H: b = 2d \end{aligned}\right\} \quad \therefore a = 11.6, c = 22.8, d = 17.9, b = 35.9$$

(continued)

The chemical formula for the fuel is $C_{11.6}H_{35.9}$. This could represent a mixture of hydrocarbons, but it is not any species listed in Appendix B, since the ratio of hydrogen atoms to carbon atoms is $35.9/11.6 = 3.09$.

To find the percent theoretical air, we must have the chemical equation using 100% theoretical air:

$$C_{11.6}H_{35.9} + 20.57(O_2 + 3.76N_2) \rightarrow 11.6CO_2 + 17.95H_2O + 77.34N_2$$

Using the number of moles of air from the actual chemical equation, we find

$$\%\ \text{excess air} = \%\ \text{theoretical air} - 100 = \frac{22.8}{20.57} \times 100 - 100 = 10.8\%$$

9.2 ENTHALPY OF FORMATION, ENTHALPY OF COMBUSTION, AND THE FIRST LAW

When a chemical reaction occurs, there may be a considerable change in the chemical composition of a system. The problem this creates is that, for a control volume, the mixture that exits is different from the mixture that enters. Since various tables use different zeros for the enthalpy, it is necessary to establish a standard reference state, chosen as 25°C (77°F) and 1 atm and denoted by the superscript "°", for example, $h°$.

Consider the combustion of H_2 with O_2, resulting in H_2O:

$$H_2 + \frac{1}{2}O_2 \rightarrow H_2O(l) \tag{9.7}$$

where the symbol (l) after a chemical compound implies the liquid phase, and the symbol (g) would imply the gaseous phase. If no symbol is given, a gas is implied. If H_2 and O_2 enter a combustion chamber at 25°C (77°F) and 1 atm and $H_2O(l)$ leaves the chamber at 25°C (77°F) and 1 atm, then the measured heat transfer will be −285 830 kJ for each kmol of $H_2O(l)$ formed. The negative sign on the heat transfer means that the energy has left the control volume, as shown schematically in Fig. 9.1.

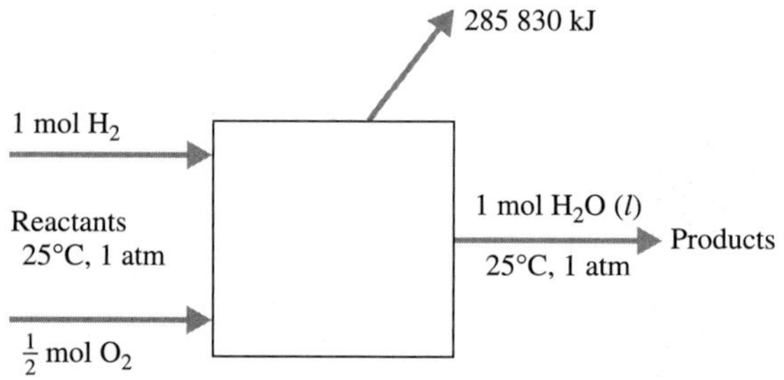

FIGURE 9.1 The control volume used during combustion of hydrogen and oxygen.

The first law applied to a combustion process in a control volume with no shaft work or kinetic and potential energy changes is

$$Q = H_P - H_R \tag{9.8}$$

where H_P is the enthalpy of the *products of combustion* that leave the combustion chamber, and H_R is the enthalpy of the *reactants* that enter. If the reactants are stable elements, as in our example in Fig. 9.1, and the process is at constant temperature and constant pressure, then the enthalpy change is called the *enthalpy of formation*, denoted by h_f°. The enthalpies of formation of numerous compounds are listed in Appendix Table B.6. Note that some compounds have a positive h_f°, indicating that they require energy to form (an **endothermic** *reaction*); others have a negative h_f°, indicating that they give off energy when they are formed (an **exothermic** *reaction*).

Endothermic: *Compounds that require energy to form.*

Exothermic: *Compounds that give off energy when they are formed.*

The enthalpy of formation is the enthalpy change when a compound is formed. The enthalpy change when a compound undergoes complete combustion at constant temperature and pressure is called the *enthalpy of combustion.* For example, the enthalpy of formation of H_2 is zero, yet when 1 mol H_2 undergoes complete combustion to $H_2O(l)$, it gives off 285 830 kJ heat; the enthalpy of combustion of H_2 is 285 830 kJ/kmol. Values are listed for several compounds in Appendix Table B.7. If the products contain liquid water, the enthalpy of combustion is the **higher heating value** (HHV); if the products contain water vapor, the enthalpy of combustion is the *lower heating value.* The difference between the higher heating value and the lower heating value is the heat of vaporization h_{fg} at standard conditions.

Higher heating value (HHV): *The enthalpy of combustion when products contain liquid water.*

If the reactants and products consist of several components, the first law, allowing for shaft work and neglecting kinetic and potential energy changes, is

$$Q - W_S = \sum_{\text{prod}} N_i(\bar{h}_f^\circ + \bar{h} - \bar{h}^\circ)_i - \sum_{\text{react}} N_i(\bar{h}_f^\circ + \bar{h} - \bar{h}^\circ)_i \tag{9.9}$$

where N_i represents the number of moles of substance i. The work is often zero, but not for example, in a combustion turbine.

If combustion occurs in a rigid chamber, like a bomb calorimeter, so that no work results, the first law is (see Eq. 4.38)

$$Q = U_P - U_R = \sum_{\text{prod}} N_i(\bar{h}_f^\circ + \bar{h} - \bar{h}^\circ - pv)_i - \sum_{\text{react}} N_i(\bar{h}_f^\circ + \bar{h} - \bar{h}^\circ - pv)_i \tag{9.10}$$

where we have used enthalpy (see Eq. 4.12) since the h_i values are tabulated. Since the volume of any liquid or solid is negligible compared to the volume of the gases, we write Eq. 9.10 as

$$Q = \sum_{\text{prod}} N_i(\bar{h}_f^\circ + \bar{h} - \bar{h}^\circ - \bar{R}T)_i - \sum_{\text{react}} N_i(\bar{h}_f^\circ + \bar{h} - \bar{h}^\circ - \bar{R}T)_i \tag{9.11}$$

If $N_{prod} = N_{react}$, then Q for the rigid volume is equal to Q for the control volume for the isothermal process.

In the above relations, we employ one of the following methods to find $(\bar{h} - \bar{h}^\circ)$:

- **For a solid or liquid**

  Use $\bar{h} - \bar{h}^\circ = \bar{c}_p \Delta T$.

- **For gases**

  Method 1: Assume an ideal gas with constant specific heat so that $\bar{h} - \bar{h}^\circ = \bar{c}_p \Delta T$.

  Method 2: Assume an ideal gas, and use tabulated values for $\bar{h}$.

Which method to use (especially for gases) is left to the judgment of the engineer. In our examples, we'll usually use method 2 for gases with Appendix Tables F.1–F.6, since temperature changes for combustion processes are often quite large and method 1 introduces substantial error.

Example 9.5

Calculate the enthalpy of combustion of gaseous propane and of liquid propane, assuming the reactants and products to be at 25°C and 1 atm. Assume liquid water in the products exiting the steady-flow combustion chamber.

Solution

Assuming theoretical air (the use of excess air would not influence the result since the process is isothermal), the chemical equation is

$$C_3H_8 + 5(O_2 + 3.76N_2) \rightarrow 3CO_2 + 4H_2O(l) + 18.8N_2$$

where, for the HHV, a liquid is assumed for H_2O. The first law becomes, for the isothermal process $h = h^\circ$,

$$Q = H_P - H_R = \sum_{\text{prod}} N_i(\bar{h}_f^\circ)_i - \sum_{\text{react}} N_i(\bar{h}_f^\circ)_i$$

$$= 3(-393\,520) + 4(-285\,830) - (-103\,850) = -2\,220\,000 \text{ kJ/kmol fuel}$$

This is the enthalpy of combustion; it is stated with the negative sign. The sign is dropped for the HHV; for gaseous propane it is 2220 MJ for each kmol of fuel.

For liquid propane, we find

$$Q = 3(-393\,520) + 4(-285\,830) - (-103\,850 - 15\,060) = -2\,205\,000 \text{ kJ/kmol fuel}$$

This is slightly less than the HHV for gaseous propane, because some energy is needed to vaporize the liquid fuel.

Example 9.6

Calculate the heat transfer required if propane and air enter a steady-flow combustion chamber at 25°C and 1 atm and the products leave at 600 K and 1 atm. Use theoretical air.

Solution

The combustion equation is written using H_2O in the vapor form due to the high exit temperature:

$$C_3H_8 + 5(O_2 + 3.76N_2) \rightarrow 3C_2 + 4H_2O(g) + 18.8N_2$$

The first law takes the form (see Eq. 9.9)

$$Q = \sum_{\text{prod}} N_i(\bar{h}_f^\circ + \bar{h} - \bar{h}^\circ)_i - \sum_{\text{react}} N_i(\bar{h}_f^\circ + \bar{h} - \bar{h}^\circ)_i$$

$$= 3(-393\,520) + 22\,280 - 9360 + 4(-241\,810 + 20\,400 - 9900)$$

$$+ 18.8(17\,560 - 8670) - (-103\,850) = -1\,796\,000 \text{ kJ/kmol fuel}$$

where we have used method 2 listed for gases. This heat transfer is less than the enthalpy of combustion of propane, as it should be, since some energy is needed to heat the products to 600 K.

Example 9.7

Liquid octane, at 25°C, fuels a jet engine. Air, at 600 K, enters the insulated combustion chamber and the products leave at 1000 K. The pressure is assumed constant at 1 atm. Estimate the exit velocity using theoretical air.

Solution

The equation is

$$C_8H_{18}(l) + 12.5(O_2 + 3.76N_2) \rightarrow 8CO_2 + 9H_2O + 47N_2$$

The first law, with $Q = W_S = 0$ and including the kinetic energy change (neglect V_{inlet}) is

$$0 = H_P - H_R + \frac{V^2}{2}M_P \qquad \text{or} \qquad V^2 = \frac{2}{M_P}(H_R - H_P)$$

where M_P is the mass of the products per kmol fuel. For the products,

$$H_P = 8(-393\,520 + 42\,770 - 9360) + 9(-241\,810 + 35\,880 - 9900)$$

$$+ 47(30\,130 - 8670)$$

$$= -3\,814\,700 \text{ kJ/kmol fuel}$$

For the reactants,

$$H_R = (-249\,910) + 12.5(17\,930 - 8680) + 47(17\,560 - 8670) = 283\,540 \text{ kJ/kmol}$$

The mass of the products is $M_P = 8 \times 44 + 9 \times 18) + 47 \times 28 = 1830$ kg/kmol fuel, and so

$$V^2 = \frac{2}{1830}(0.28354 + 3.8147) \times 10^9 \qquad \therefore\ V = 2120 \text{ m/s}$$

Example 9.8

Liquid octane is burned with 300% excess air. The octane and air enter the steady-flow combustion chamber at 25°C and 1 atm and the products exit at 1000 K and 1 atm. Estimate the heat transfer. Neglect the kinetic energy change.

Solution

The reaction with theoretical air is

$$C_8H_{18} + 12.5(O_2 + 3.76N_2) \rightarrow 8CO_2 + 9H_2O + 47N_2$$

For 300% excess air (400% theoretical air), the reaction is

$$C_8H_{18}(l) + 50(O_2 + 3.76N_2) \rightarrow 8CO_2 + 9H_2O + 37.5O_2 + 188N_2$$

The first law applied to the combustion chamber is

$$Q = H_P - H_R = 8(-393\,520 + 42\,770 - 9360) + (9)(-241\,810 + 35\,880 - 9900)$$

$$+ 37.5(31\,390 - 8680) + 188(30\,130 - 8670) - (-249\,910) = 312\,500 \text{ kJ/kmol fuel}$$

In this situation, heat must be added to obtain the desired exit temperature.

Example 9.9

A constant-volume bomb calorimeter is surrounded by water at 77°F. Liquid propane is burned with pure oxygen in the calorimeter, and the heat transfer is determined to be −874 000 Btu/lbmol. Calculate the enthalpy of formation and compare with that given in Appendix Table B.6E.

Solution

The complete combustion of propane follows:

$$C_3H_8 + 5O_2 \rightarrow 3CO_2 + 4H_2O(g)$$

The surrounding water sustains a constant-temperature process, so that Eq. 9.11 becomes

$$Q = \sum_{\text{prod}} N_i(\bar{h}_f^\circ)_i - \sum_{\text{react}} N_i(\bar{h}_f^\circ)_i + (N_R - N_P)\bar{R}T$$

$$-874\,000 = 3(-169\,000) + 4(-104\,000) - (\bar{h}_f^\circ)_{C_3H_8} + (6 - 7) \times 1.987 \times 537$$

$$\therefore (\bar{h}_f^\circ)_{C_3H_8} = -51\,130 \text{ Btu/lbmol}$$

This compares with $\bar{h}_f^\circ$ from the Table B.6E of $(-44\,680 - 6480) = -51\,160$ Btu/lbmol.

9.3 ADIABATIC FLAME TEMPERATURE

Adiabatic flame temperature: *The temperature of the products if a combustion takes places adiabatically.*

If we consider a combustion process that takes place adiabatically, with no work or changes in kinetic and potential energy, then the temperature of the products is referred to as the **adiabatic flame temperature**. We find that the maximum adiabatic flame temperature that can be achieved occurs at theoretical air. This fact allows us to control the adiabatic flame temperature by the amount of excess air involved in the process: The greater the amount of excess air, the lower the adiabatic flame temperature. If the blades in a turbine can withstand a certain maximum temperature, we can determine the excess air needed so that the maximum allowable blade temperature is not exceeded. We will find that an iterative (trial-and-error) procedure is needed to find the adiabatic flame temperature. A quick approximation to the adiabatic flame temperature is found by assuming the products to be completely N_2. Examples will illustrate this.

The adiabatic flame temperature is calculated assuming complete combustion, no heat transfer from the combustion chamber, and no dissociation of the products into other chemical species. Each of these effects tends to reduce the adiabatic flame temperature. Consequently, the adiabatic flame temperature that we will calculate represents the maximum possible flame temperature for the specified percentage of theoretical air.

If a significant amount of heat transfer does occur from the combustion chamber, we can account for it by including the following term in the energy equation:

$$\dot{Q} = UA(T_P - T_E) \tag{9.12}$$

where U = overall heat-transfer coefficient (specified)
T_E = temperature of environment
T_P = temperature of products
A = surface area of combustion chamber

[Note that the units on U are kW/m^2·K or Btu/sec-ft^2-°R.]

Example 9.10

Propane is burned with 250% theoretical air; both are at 25°C and 1 atm. Predict the adiabatic flame temperature in the steady-flow combustion chamber.

Solution

The combustion with theoretical air is

$$C_3H_8 + 5(O_2 + 3.76N_2) \rightarrow 3CO_2 + 4H_2O + 18.8N_2$$

For 250% theoretical air, we have

$$C_3H_8 + 12.5(O_2 + 3.76N_2) \rightarrow 3CO_2 + 4H_2O + 7.5O_2 + 47N_2$$

Since $Q = 0$ for an adiabatic process, we demand that $H_R = H_P$. The enthalpy of the reactants, at 25°C, is $H_R = -103\,850$ kJ/kmol fuel.

The temperature of the products is the unknown, and we cannot obtain the enthalpies of the components of the products from the tables without knowing the temperatures. This requires a trial-and-error solution. To obtain an initial guess, we assume the products to be composed entirely of nitrogen:

(continued)

$$H_R = H_P = -103\,850 = 3(-393\,520) + 4(-241\,820) + 61.5(h_P - 8670)$$

where we have noted that the products contain 61.5 mol of gas. This gives $h_P =$ 43 400 kJ/kmol, which suggests a temperature of about 1380 K (take T_P a little less than that predicted by the all-nitrogen assumption). Using this temperature, we check using the actual products:

$$-103\,850 \stackrel{?}{=} 3(-393\,520 + 64\,120 - 9360) + 4(-241\,820 + 52\,430 - 9900)$$
$$+ 7.5(44\,920 - 8680) + 47(42\,920 - 8670) = 68\,110$$

The temperature is obviously too high. We select a lower value; $T_P = 1300$ K. There results:

$$-103\,850 \stackrel{?}{=} 3(393\,520 + 59\,520 - 9360) + 4(-241\,820 + 48\,810 - 9900)$$
$$+ 7.5(44\,030 - 8680) + 47(40\,170 - 8670) = -96\,100$$

We use the above two results for 1380 K and 1300 K and, assuming a linear relationship, predict that T_P is

$$T_P = 1300 - \left[\frac{103\,850 - 96\,100}{68\,110 - (-96\,100)}\right](1380 - 1300) = 1296 \text{ K}$$

Example 9.11

Propane is burned with theoretical air; both are at 25°C and 1 atm in a steady-flow combustion chamber. Predict the adiabatic flame temperature.

Solution

The combustion with theoretical air is

$$C_3H_8 + 5(O_2 + 3.76N_2) \rightarrow 3CO_2 + 4H_2O + 18.8N_2$$

For the adiabatic process, the first law takes the form $H_R = H_P$. Hence, assuming the products to be composed entirely of nitrogen,

$$-103\,850 = 3(-393\,520) + 4(-241\,820) + 25.8(h_P - 8670)$$

where the products contain 25.8 mol gas. This gives $h_P = 87\,900$ kJ/kmol, which suggests a temperature of about 2600 K.

With this temperature, we find, using the actual products,

$$-103\,850 \stackrel{?}{=} 3(-393\,520 + 137\,400 - 9360) + 4(-241\,820 + 114\,300 - 9900)$$
$$+ 18.8(86\,600 - 8670) = 119\,000$$

At 2400 K there results:

$$-103\,850 \stackrel{?}{=} 3(-393\,520 + 125\,200 - 9360) + 4(-241\,820 + 103\,500 - 9900)$$
$$+ 18.8(79\,320 - 8670) = -97\,700$$

A straight line extrapolation gives $T_P = 2394$ K.

Example 9.12

The overall heat-transfer coefficient of a steady-flow combustion chamber with a 2-m^2 surface area is determined to be 0.5 kW/m^2·K. Propane is burned with theoretical air, both at 25°C and 1 atm. Predict the temperature of the products of combustion if the propane mass flow rate is 0.2 kg/s.

Solution

The molar influx is $\dot{m}_{\text{fuel}} = 0.2/44 = 0.004545$ kmol/s, where the molecular weight of propane, 44 kg/kmol, is used. Referring to the chemical reaction given in Example 9.11, the mole fluxes of the products are given by:

$$\dot{m}_{CO_2} = 3 \times 0.004545 = 0.01364 \text{ kmol/s}$$

$$\dot{m}_{H_2O} = 4 \times 0.004545 = 0.02273 \text{ kmol/s}$$

$$\dot{m}_{N_2} = 18.8 \times 0.004545 = 0.1068 \text{ kmol/s}$$

We can write the energy equation (the first law) as

$$\dot{Q} + \dot{H}_R = \dot{H}_P$$

Using Eq. 9.12, the energy equation becomes

$$\begin{aligned} -0.5 \times 2(T_P - 298) &+ 0.004545(-103\,850) \\ &= 0.01364(-393\,520 + \bar{h}_{CO_2} - 9360) + 0.02273(-241\,820 + \bar{h}_{H_2O} - 9900) \\ &\quad + 0.1068(\bar{h}_{N_2} - 8670) \end{aligned}$$

For a first guess at T_P, let us assume a somewhat lower temperature than that of Example 9.11, since energy is leaving the combustion chamber. The guesses follow:

$$T_P = 1600 \text{ K}:$$

$$-1774 \stackrel{?}{=} -4446 - 4295 + 4475 = -4266$$

$$T_P = 2000 \text{ K}:$$

$$-2174 \stackrel{?}{=} -4120 - 3844 + 5996 = -1968$$

$$T_P = 1900 \text{ K}:$$

$$-2074 \stackrel{?}{=} -4202 - 3960 + 5612 = -2550$$

Interpolation between the last two entries gives $T_P = 1970$ K. Checking,

$$T_P = 1970 \text{ K}:$$

$$-2144 \stackrel{?}{=} -4145 - 3879 + 5881 = -2143$$

Hence, $T_P = 1970$ K. If we desire the temperature of the products to be less than this, we can increase the overall heat-transfer coefficient or add excess air.

9.4 SUMMARY

In this chapter, we defined the following terms:

Chemical-reaction equation: The chemical balance equation that relates the components before and after a chemical process.
Complete combustion: The products of combustion include H_2O and CO_2 but no CO, C, H_2, or OH.
Incomplete combustion: The products of combustion include H_2, C, CO, and/or OH.
Theoretical (stoichiometric) air: The minimum amount of air that allows for complete combustion.
Percent theoretical air: The percent of excess air added to 100%.
Air–fuel ratio: The ratio of the mass of air to the mass of fuel.
Fuel–air ratio: The reciprocal of the air–fuel ratio.
Lean mixture: If the air–fuel ratio exceeds that associated with theoretical air.
Rich mixture: If the air–fuel ratio is less than that associated with theoretical air.
Volumetric analysis: The analysis of a mixture of gases based on the moles (or volume) of each component.
Gravimetric analysis: The analysis of a mixture of gases based on the mass (or weight) of each component.
Reactants: The components that enter a combustion chamber.
Products of combustion: The components that leave a combustion chamber.
Enthalpy of formation: The energy absorbed or released when a compound is formed from its components while maintaining the reference temperature and pressure.
Endothermic reaction: A reaction that requires energy to form a compound.
Exothermic reaction: A reaction that releases energy when a compound is formed.
Enthalpy of combustion: The energy released when a compound undergoes complete combustion at constant temperature and pressure.
Higher heating value: When the products of combustion contain liquid water.
Lower heating value: When the products of combustion contain water vapor.
Adiabatic flame temperature: The temperature of the products that results when the combustion process takes place adiabatically with no work or kinetic and potential energy changes. The maximum temperature occurs at theoretical air.

The equations introduced in this chapter are few in number. They include the first law applied to a control-volume combustion chamber with negligible kinetic and potential energy changes and no work:

$$Q = H_P - H_R \tag{9.13}$$

where H_P and H_R are the enthalpy of the products and the reactants, respectively. If we allow shaft work to take place, this takes the form

$$Q - W_S = \sum_{\text{prod}} N_i(\bar{h}_f^\circ + \bar{h} - \bar{h}^\circ)_i - \sum_{\text{react}} N_i(\bar{h}_f^\circ + \bar{h} - \bar{h}^\circ)_i \tag{9.14}$$

If combustion takes place in a rigid bomb calorimeter, there results

$$Q = \sum_{\text{prod}} N_i(\bar{h}_f^\circ + \bar{h} - \bar{h}^\circ - pv)_i - \sum_{\text{react}} N_i(\bar{h}_f^\circ + \bar{h} - \bar{h}^\circ - pv)_i \tag{9.15}$$

Or, assuming the products and reactants are gases, we have

$$Q = \sum_{\text{prod}} N_i(\bar{h}_f^\circ + \bar{h} - \bar{h}^\circ - \bar{R}T)_i - \sum_{\text{react}} N_i(\bar{h}_f^\circ + \bar{h} - \bar{h}^\circ - \bar{R}T)_i \tag{9.16}$$

REVIEW PROBLEMS FOR THE FE/EIT EXAMINATION

(There are no multiple-choice problems for this chapter since the probability is very low that a combustion problem will be included on the FE/EIT exam or the GRE/Engineering exam.)

PROBLEMS

9.1 The following fuels combine with stoichiometric air: **(a)** C_2H_4, **(b)** C_3H_6, **(c)** C_4H_{10}, **(d)** C_5H_{12}, **(e)** C_8H_{18} and **(f)** CH_3OH. Provide the correct values for x, y, z in the reaction equation

$$C_aH_b + w(O_2 + 3.76N_2) \rightarrow xCO_2 + yH_2O + zN_2$$

9.2 Methane (CH_4) is burned with stoichiometric air, and the products are cooled to 20°C, assuming complete combustion at 100 kPa. Calculate the air–fuel ratio, the percentage of CO_2 by weight of the products, the dew-point temperature of the products, and the percentage of water vapor condensed.

9.3 Repeat Problem 9.2 for ethane (C_2H_6).

9.4 Repeat Problem 9.2 for propane (C_3H_8).

9.5 Repeat Problem 9.2 for butane (C_4H_{10}).

9.6 Repeat Problem 9.2 for pentane (C_5H_{12}).

9.7 Ethane (C_2H_6) is burned with dry air that contains 5 mol O_2 for each mole of fuel. Calculate the percent of excess air, the air–fuel ratio, and the dew-point temperature.

9.8 Ethane (C_2H_6) undergoes complete combustion at 95 kPa with 180% theoretical air. Find the air–fuel ratio, the percentage of CO_2 by volume in the products, and dew-point temperature.

9.9 Repeat Problem 9.8 for propane (C_3H_8).

9.10 Repeat Problem 9.8 for butane (C_4H_{10}).

9.11 Repeat Problem 9.8 for pentane (C_5H_{12}).

9.12 A fuel mixture of 60% methane, 30% ethane, and 10% propane by volume is burned with stoichiometric air. Calculate the volume flow rate of air required if the fuel mass is 12 lbm/h, assuming the air to be at 70°F and 14.7 psia.

9.13 Calculate the mass flow rate of fuel required if the inlet air flow rate is 20 m^3/min at 20°C and 100 kPa, using stoichiometric air with **(a)** methane (CH_4), **(b)** ethane (C_2H_6), **(c)** propane (C_3H_8), and **(d)** butane (C_4H_{10}).

9.14 Propane (C_3H_8) undergoes complete combustion at 90 kPa and 20°C with 130% theoretical air. Calculate the air–fuel ratio and the dew-point temperature if the relative humidity of the combustion air is **(a)** 90%, **(b)** 80%, **(c)** 60%, and **(d)** 40%.

9.15 Butane (C_4H_{10}) is burned with 20°C air at 70% relative humidity. The air–fuel ratio is 20. Calculate the dew-point temperature of the products. Compare with Example 9.1.

9.16 An air–fuel ratio of 25 is used in an engine that burns octane (C_8H_{18}). Find the percentage of excess air required and the percentage of CO_2 by volume in the products.

9.17 Butane (C_4H_{10}) is burned with 50% excess air. If 5% of the carbon in the fuel is converted to CO, calculate the air–fuel ratio and the dew point of the products. Combustion takes place at 100 kPa.

9.18 A fuel that is 60% ethane and 40% octane by volume undergoes complete combustion with 200% theoretical air. Find the air–fuel ratio, the percent by volume of N_2 in the products, and the dew-point temperature of the products if the pressure is 98 kPa.

9.19 One lbm of butane, 2 lbm of methane, and 2 lbm of octane undergo complete combustion with 20 lbm of air. Calculate the air–fuel ratio, the percent theoretical air, and the dew-point temperature of the products if the combustion process occurs at 14.7 psia.

9.20 Each minute, 1 kg of methane, 2 kg of butane, and 2 kg of octane undergo complete combustion with stoichiometric 20°C air. Calculate the flow rate of air required if the process takes place at 100 kPa.

9.21 Methane is burned with dry air, and volumetric analysis of the products on a dry basis gives 10% CO_2, 1% CO, 1.8% O_2, and 87.6% N_2. Calculate the air–fuel ratio, the percent excess air, and the percentage of water vapor that condenses if the products are cooled to 30°C.

9.22 An unknown hydrocarbon fuel combusts with dry air; the resulting products have the following dry volumetric analysis: 12% CO_2, 1.5% CO, 3% O_2, and 83.5% N_2. Calculate the percent excess air.

9.23 A volumetric analysis of the products of butane (C_4H_{10}) on a dry basis yields 7.6% CO_2, 8.2% O_2, 83.2% N_2, and 1.4% CO. What percent excess air was used?

9.24 A volumetric analysis of the products of combustion of octane (C_8H_{18}) on a dry basis yields 9.1% CO_2, 7.0% O_2, 83.4% N_2, and 0.9% CO. Calculate the air–fuel ratio.

9.25 Three moles of a mixture of hydrocarbon fuels, denoted by C_xH_y, is burned and a volumetric analysis on a dry basis of the products yields 10% CO_2, 8% O_2, 1.2% CO, and 80.8% N_2. Estimate the values for x and y and the percent theoretical air utilized.

9.26 Producer gas, created from coal, has a volumetric analysis of 3% CH_4, 14% H_2, 50.9% N_2, 0.6% O_2, 27% CO, and 4.5% CO_2. Complete combustion occurs with 150% theoretical air at 100 kPa. What percentage of the water vapor will condense out if the temperature of the products is 20°C?

Enthalpy of Formation and Combustion

9.27 Using the enthalpy of formation data from Appendix Table B.6, calculate the enthalpy of combustion for a steady-flow process, assuming liquid water in the products. Inlet and outlet temperatures are 25°C and the pressure is 100 kPa. (Compare with the value listed in Table B.7.) The fuel is **(a)** methane, **(b)** acetylene, **(c)** propane gas, and **(d)** liquid pentane.

9.28 Carbon reacts with oxygen to form carbon dioxide in a steady-flow chamber. Calculate the energy involved and state the type of reaction. Assume the reactants and products are at 25°C and 1 atm.

9.29 Methane enters a steady-flow combustion chamber at 77°F and 1 atm with 80% excess air which is at 800°R and 1 atm. Calculate the heat transfer if the products leave at 1600°R and 1 atm.

9.30 Propane gas (C_3H_8) undergoes complete combustion with stoichiometric air; both are at 77°F and 1 atm. Calculate the heat transfer if the products from a steady-flow combustor are at **(a)** 77°F, **(b)** 1540°F, and **(c)** 2540°F.

9.31 Liquid propane (C_3H_8) undergoes complete combustion with air; both are at 25°C and 1 atm. Calculate the heat transfer if the products from a steady-flow combustor are at 1000 K and the percent theoretical air is **(a)** 100%, **(b)** 150%, and **(c)** 200%.

9.32 Ethane gas (C_2H_6) at 25°C is burned with 150% theoretical air at 500 K and 1 atm. Find the heat transfer from a steady-flow combustor if the products are at 1000 K and **(a)** complete combustion occurs or **(b)** 95% of the carbon is converted to CO_2 and 5% to CO.

9.33 Ethane at 25°C is burned in a steady-flow combustion chamber with 20% excess air at 127°C, but only 95% of the carbon is converted to CO_2. If the products leave at 1200 K, calculate the heat transfer. The pressure remains constant at 1 atm.

9.34 Complete combustion occurs between butane gas (C_4H_{10}) and air; both are at 25°C and 1 atm. If the steady-flow combustion chamber is insulated, what percent theoretical air is needed to maintain the products at **(a)** 1000 K and **(b)** 1500 K?

9.35 Complete combustion occurs between ethylene gas (C_2H_4) and air; both are at 77°F and 1 atm. If 150 000 Btu of heat is removed per lbmol of fuel from the steady-flow combustor, estimate the percent theoretical air required to maintain the products at 1500°R.

9.36 Butane gas (C_4H_{10}) at 25°C is burned in a steady-flow combustion chamber with 150% theoretical air at 500 K and 1 atm. If 90% of the carbon is converted to CO_2 and 10% to CO, estimate the heat transfer if the products are at 1200 K.

9.37 Butane gas (C_4H_{10}) undergoes complete combustion with 40% excess air; both are at 25°C and 100 kPa. Calculate the heat transfer from the steady-flow combustor if the products are at 1000 K and the humidity of the combustion air is **(a)** 90%, **(b)** 70%, and **(c)** 50%.

9.38 A rigid volume contains 0.2 lbm of propane gas and 0.8 lbm of oxygen at 77°F and 30 psia. The propane burns completely, and the final temperature, after a period of time, is observed to be 1600°R. Calculate the final pressure and the heat transfer.

9.39 A rigid tank contains a mixture of 0.2 kg of ethane gas (C_2H_6) and 1.2 kg of O_2 at 25°C and 100 kPa. The mixture is ignited and complete combustion occurs. If the final temperature is 1000 K, find the heat transfer and the final pressure.

9.40 A mixture of 1 lbmol methane gas (CH_4) and stoichiometric air at 77°F and 20 psia is contained in a rigid tank. If complete combustion occurs, calculate the heat transfer and the final pressure if the final temperature is 1540°F.

9.41 A mixture of octane gas (C_8H_{18}) and 20% excess air at 25°C and 200 kPa is contained in a 50-liter cylinder. Ignition occurs, and the pressure remains constant until the temperature reaches 800 K. Assuming complete combustion, estimate the heat transfer during the expansion process.

9.42 A mixture of butane gas (C_4H_{10}) and stoichiometric air is contained in a rigid tank at 25°C and 100 kPa. If 95% of the carbon is burned to CO_2 and the remainder to CO, calculate the heat transfer from the tank and the volume percent of the water that condenses out if the final temperature is 25°C.

Adiabatic Flame Temperature

9.43 Propane is burned in a steady-flow combustion chamber with 80% theoretical air, both at 25°C and 1 atm. Estimate the adiabatic flame temperature and compare with that of Examples 9.10 and 9.11.

9.44 Butane gas (C_4H_{10}) mixes with air, both at 25°C and 1 atm, and undergoes complete combustion in a steady-flow insulated combustion chamber. Calculate the adiabatic flame temperature for **(a)** 100% theoretical air, **(b)** 150% theoretical air, and **(c)** 100% excess air.

9.45 Ethane (C_2H_6) at 25°C undergoes complete combustion with air at 400 K and 1 atm in a steady-flow insulated combustor. Determine the exit temperature for 50% excess air.

9.46 Hydrogen gas and air, both of 400 K and 1 atm, undergo complete combustion in a steady-flow insulated combustor. Estimate the exit temperature for 200% theoretical air.

9.47 Liquid methyl alcohol (CH_3OH) at 25°C reacts with 150% theoretical air. Find the exit temperature, assuming complete combustion, from a steady-flow insulated combustor if the air enters at **(a)** 25°C, **(b)** 400 K, and **(c)** 600 K. Assume atmospheric pressure.

9.48 Ethylene (C_2H_4) at 77°F undergoes complete combustion with stoichiometric air at 77°F and 70% humidity in an insulated steady-flow combustion chamber. Estimate the exit temperature assuming a pressure of 14.5 psia.

9.49 Ethane (C_2H_6) at 25°C combusts with 90% theoretical air at 400 K and 1 atm in an insulated steady-flow combustor. Determine the exit temperature.

9.50 An insulated, rigid 0.7-m^3 tank contains 0.05 kg of ethane and 100% theoretical air at 25°C. The fuel is ignited and complete combustion occurs. Estimate **(a)** the final temperature and **(b)** the final pressure.

9.51 A mixture of liquid propane (C_3H_8) and stoichiometric air at 25°C and 100 kPa undergoes complete combustion in a rigid container. Determine the maximum temperature and pressure (the *explosion pressure*) immediately after combustion.

Part II
Fluid Mechanics

Left: Contemporary windmills are used to generate electricity at many locations in the United States. They are located in areas where consistent prevailing winds exist. (California Department of Water Resources) *Top right*: Hurricane Bonnie, Atlantic Ocean about 800 km from Bermuda. At this stage in its development, the storm has a well-developed center, or "eye," where air currents are relatively calm. Vortex-like motion occurs away from the eye. (U.S. National Aeronautics and Space Administration) *Bottom right*: The Space Shuttle Discovery leaves the Kennedy Space Center on October 29, 1998. In 6 seconds the vehicle cleared the launch tower with a speed of 160 km/h, and in about two minutes it was 250 km down range from the Space Center, 47 km above the ocean with a speed of 6150 km/h. The wings and rudder on the tail are necessary for successful re-entry as it enters the earth's atmosphere upon completion of its mission. (U.S. National Aeronautics and Space Administration)

10

Basic Considerations

Outline

Chapter Objectives

The objectives of this chapter are to:

- ▲ Introduce many of the quantities encountered in fluid mechanics including their dimensions and units.
- ▲ Identify the liquids to be considered in this text.
- ▲ Introduce the fluid properties of interest.
- ▲ Present the thermodynamic laws and associated quantities.

Nomenclature

The following variables are introduced in this chapter:

a	acceleration	B	constant, bulk modulus
$\mathbf{a}$	acceleration vector	c	speed of sound, specific heat
A	area, constant		

c_p	constant pressure specific heat	t	tangential direction, time
c_v	constant volume specific heat	T	temperature
D	diameter	u	x-component velocity
d	diameter, differential	$\tilde{u}$	specific internal energy
E	total energy	$\forall$	volume
F	force	W	weight, work
$\mathbf{F}$	force vector	x	coordinate direction
g	gravity	y	coordinate direction
H	enthalpy	z	coordinate direction
h	height, length, specific enthalpy	β	angle
k	ratio of specific heats	γ	specific weight
l	length	ε	small volume
m	mass	Δ	incremental amount
n	normal direction	λ	mean free path
p	pressure	μ	viscosity
Q	heat transfer	ν	kinematic viscosity
r	radial direction	ρ	density
R	radius, gas constant	σ	surface tension
s	entropy	$\sum$	summation symbol
S	specific gravity	τ	shear stress
		Ω	angular velocity

10.1 INTRODUCTION

A proper understanding of the mechanics of fluids is extremely important in many areas of engineering. In biomechanics, the flow of blood and cerebral fluid are of particular interest; in meteorology and ocean engineering, an understanding of the motions of air movements and ocean currents requires a knowledge of the mechanics of fluids; chemical engineers must understand fluid mechanics to design the many different kinds of chemical-processing equipment; aeronautical engineers use their knowledge of fluids to maximize lift and minimize drag on aircraft and to design fan-jet engines; mechanical engineers design pumps, turbines, internal combustion engines, air compressors, air-conditioning equipment, pollution-control equipment, and power plants using a proper understanding of fluid mechanics; and civil engineers must also utilize the results obtained from a study of the mechanics of fluids to design piping systems, sewage treatment plants, irrigation channels, flood control systems, dams, and domed athletic stadiums.

Key Concept: *We will present the fundamentals of fluids so that engineers are able to understand the role that fluid plays in particular applications.*

It is not possible to present fluid mechanics in such a way that all of the foregoing subjects can be treated specifically; it is possible, however, to present the fundamentals of fluid mechanics so that engineers are able to understand the role that the fluid plays in a particular application. This role may involve the proper sizing of a pump (the horsepower and flow rate) or the calculation of a force acting on a structure.

We will present the general equations that result from the conservation of mass principle, Newton's second law, and the first law of thermodynamics. From these, a number of particular situations will be considered that are of special interest. After studying this book, the engineer should be able to apply the basic principles of the mechanics of fluids to new and different situations.

In this chapter, topics are presented that are directly or indirectly relevant to all subsequent chapters. We include a macroscopic description of fluids, fluid properties, physical laws dominating fluid mechanics, and a summary of units and dimensions of important physical quantities. Before we can discuss quantities of interest, we must present the units and dimensions that will be used in our study of fluid mechanics.

Some of the material presented in this Fluid Mechanics Part was presented in the Thermodynamics Part. There may be students who did not study from the Thermodynamics Part who are now studying this Part. Consequently, we have designed this Fluid Mechanics Part to not depend on the first part of this text, but only on any course in thermodynamics. It also may help to repeat some of the basic ideas that are needed in our study of fluid mechanics.

10.2 DIMENSIONS, UNITS, AND PHYSICAL QUANTITIES

Before we begin the more detailed studies of fluid mechanics, let us discuss the dimensions and units that will be used throughout this part. Physical quantities require quantitative descriptions when solving an engineering problem. Density is one such physical quantity. It is a measure of the mass contained in a unit volume. Density does not, however, represent a fundamental dimension. There are nine quantities that are considered to be fundamental dimensions: length, mass, time, temperature, amount of a substance, electric current, luminous intensity, plane angle, and solid angle. The dimensions of all other quantities can be expressed in terms of the fundamental dimensions. For example, the quantity "force" can be related to the fundamental dimensions of mass, length, and time. To do this, we use Newton's second law expressed in simplified form in one direction as

$$F = ma \tag{10.1}$$

Using brackets to denote "the dimension of," this is written dimensionally as

$$\begin{aligned} [F] &= [m][a] \\ F &= M\frac{L}{T^2} \end{aligned} \tag{10.2}$$

where F, M, L, and T are the dimensions of force mass, length, and time, respectively. If force had been selected as a fundamental dimension rather than mass, a common alternative, mass would have dimensions of

$$\begin{aligned} [m] &= \frac{[F]}{[a]} \\ M &= F\frac{T^2}{L} \end{aligned} \tag{10.3}$$

where F is the dimension[1] of force.

[1] Unfortunately, the quantity force F and the dimension of force $[F]$ use the same symbol.

TABLE 10.1 Fundamental Dimensions and Their Units

Quantity	Dimensions	SI units		English units	
Length l	L	meter	m	foot	ft
Mass m	M	kilogram	kg	slug	slug
Time t	T	second	s	second	sec
Electric current i		ampere	A	ampere	A
Temperature T	Θ	kelvin	K	Rankine	°R
Amount of substance	M	kg-mole	kg-mol	lb-mole	lb-mol
Luminous intensity		candela	cd	candela	cd
Plane angle		radian	rad	radian	rad
Solid angle		steradian	sr	steradian	sr

There are also systems of dimensions in which both mass and force are selected as fundamental dimensions. In such systems, conversion factors, such as a gravitational constant, are required; we do not consider these types of systems in our study of fluid mechanics, so they will not be discussed here.

The fundamental dimensions and their units are presented in Table 10.1 as was done in the Thermodynamics Part; some derived units appropriate to fluid mechanics are given in Table 10.2. Other units that are acceptable are the hectare (ha), which is 10 000 m^2, used for large areas; the metric ton (t), which is 1000 kg,

TABLE 10.2 Derived Units

Quantity	Dimensions	SI units	English units
Area A	L^2	m^2	ft^2
Volume $\forall$	L^3	m^3 L (liter)	ft^3
Velocity V	L/T	m/s	ft/sec
Acceleration a	L/T^2	m/s^2	ft/sec^2
Angular velocity Ω	T^{-1}	s^{-1}	sec^{-1}
Force F	ML/T^2	kg·m/s^2 N (newton)	slug-ft/sec^2 lb (pound)
Density ρ	M/L^3	kg/m^3	slug/ft^3
Specific weight γ	M/L^2T^2	N/m^3	lb/ft^3
Frequency f	T^{-1}	s^{-1}	sec^{-1}
Pressure p	M/LT^2	N/m^2 Pa (pascal)	lb/ft^2
Stress τ	M/LT^2	N/m^2 Pa (pascal)	lb/ft^2
Surface tension σ	M/T^2	N/m	lb/ft
Work W	ML^2/T^2	N·m J (joule)	ft-lb
Energy E	ML^2/T^2	N·m J (joule)	ft-lb
Heat rate $\dot{Q}$	ML^2/T^3	J/s	Btu/sec
Torque T	ML^2/T^2	N·m	ft-lb
Power P	ML^2/T^3	J/s W (watt)	ft-lb/sec
Viscosity μ	M/LT	N·s/m^2	lb-sec/ft^2
Mass flux $\dot{m}$	M/T	kg/s	slug/sec
Flow rate Q_f	L^3/T	m^3/s	ft^3/sec
Specific heat c	$L^2/T^2\Theta$	J/kg·K	Btu/slug-°R

TABLE 10.3 SI Prefixes

Multiplication factor	Prefix	Symbol
10^{12}	tera	T
10^{9}	giga	G
10^{6}	mega	M
10^{3}	kilo	k
10^{-2}	centi[a]	c
10^{-3}	milli	m
10^{-6}	micro	μ
10^{-9}	nano	n
10^{-12}	pico	p

[a]Permissible if used alone as cm, cm^2, or cm^3.

used for large masses; and the liter (L), which is 0.001 m^3. Also, density is occasionally expressed as grams per liter (g/L).

When expressing a quantity with a numerical value and a unit, prefixes have been defined so that the numerical value may be between 0.1 and 1000. These prefixes are presented in Table 10.3. Using scientific notation, however, we use powers of 10 rather than prefixes (e.g., 2×10^6 N rather than 2 MN). If larger numbers are written, the comma is not used; twenty thousand would be written as 20 000 with a space and no comma.

Key Concept: *When using SI units, if larger numbers are written (5 digits or more), the comma is not used. The comma is replaced by a space (i.e., 20 000).*

Newton's second law relates a net force acting on a rigid body to its mass and acceleration. This is expressed as

$$\sum \mathbf{F} = ma \tag{10.4}$$

Consequently, the force needed to accelerate a mass of 1 kilogram at 1 meter per second squared in the direction of the net force is 1 newton; using English units, the force needed to accelerate a mass of 1 slug at 1 foot per second squared in the direction of the net force is 1 pound. This allows us to relate the units by

$$\text{N} = \text{kg}\cdot\text{m/s}^2 \qquad \text{lb} = \text{slug-ft/sec}^2 \tag{10.5}$$

Key Concept: *The relationship* $\text{N} = \text{kg}\cdot\text{m/s}^2$ *is often used in the conversion of units.*

which are included in Table 10.2. Note that it is customary to not use the unit lbf in our study of fluids. The unit lb is always a force unit. In the English system used here, mass is always measured in slugs, so there is no need for the unit lbm that was used in thermodynamics. The above relationships between units are often used in the conversion of units. In the SI system, weight is always expressed in newtons, never in kilograms. In the English system, mass is always expressed in slugs, never in pounds. To relate weight to mass, we use

$$W = mg \tag{10.6}$$

where g is the local gravity. The standard value for gravity is 9.80665 m/s^2 (32.174 ft/sec^2), and it varies from a minimum of 9.77 m/s^2 at the top of Mt. Everest to a maximum of 9.83 m/s^2 in the deepest ocean trench. A nominal value of 9.81 m/s^2 (32.2 ft/sec^2) will be used unless otherwise stated.

Finally, a note on significant figures. In engineering calculations, we often do not have confidence in a calculation beyond three significant digits since the information given in the problem statement is often not known to more than three significant digits; in fact, viscosity and other fluid properties may not be known to even three significant digits. The diameter of a pipe may be stated as 2 cm; this would, in general, not be as precise as 2.000 cm would imply. If information used in the solution of a problem is known to only two significant digits, it is incorrect to express a result to more than two significant digits. In the examples and problems, we will assume that all information given is known to three or four significant digits, and the results will be expressed accordingly. If the numeral 1 begins a number, it is not counted in the number of significant digits (i.e., the number 1.210 has three significant digits).

Key Concept: *We will assume that all information given is known to three or four significant digits.*

Example 10.1

A mass of 100 kg is acted on by a 400-N force acting vertically upward and a 600-N force acting upward at a 45° angle. Calculate the vertical component of the acceleration. The local acceleration of gravity is 9.81 m/s^2.

Solution

The first step in solving a problem involving forces is to draw a free-body diagram with all forces acting on it, as shown in Fig. E10.1.

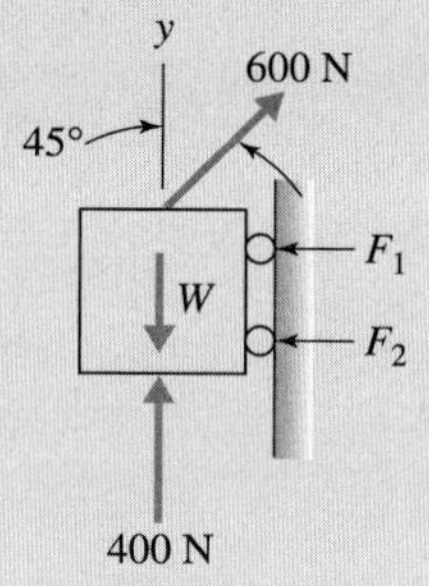

FIGURE E10.1

Second, apply Newton's second law (Eq. 10.4). It relates the net force acting on a mass to the acceleration and is expressed as

$$\sum F_y = ma_y$$

Using the appropriate components in the y-direction, we have

$$400 + 600 \sin 45° - 100 \times 9.81 = 100a_y$$

$$a_y = -1.567 \text{ m/s}^2$$

Note: We have used only three significant digits in the answer since the information given in the problem is assumed known to three significant digits. (The number 1.567 has three significant digits. A leading "1" is not counted as a significant digit.)

10.3 CONTINUUM VIEW OF GASES AND LIQUIDS

Substances referred to as fluids may be **liquids** or **gases**. In our study of the mechanics of fluids, we restrict the liquids that are studied. Before we state the restriction, we must define a shearing stress. A force ΔF that acts on an area ΔA can be decomposed into a normal component ΔF_n and a tangential component ΔF_t, as shown in Fig. 10.1. The force divided by the area upon which it acts is called a *stress*. The force vector divided by the area is a *stress vector*, the normal component of force divided by the area is a *normal stress*, and the tangential force divided by the area is a **shear stress**. In this discussion, we are interested in the shear stress τ. Mathematically, it is defined as

$$\tau = \lim_{\Delta A \to 0} \frac{\Delta F_t}{\Delta A} \tag{10.7}$$

Our restricted family of fluids may now be identified; the fluids considered in this book are *those liquids and gases that move under the action of a shear stress, no matter how small that shear stress may be*. This means that even a very small shear stress results in motion in the fluid. Gases obviously fall within this category of fluids, as do water and tar. Some substances, such as plastics and catsup, may resist small shear stresses without moving; a study of these substances is included in the subject of *rheology* and is not included in this book.

Liquid: *A state of matter in which the molecules are relatively free to change their positions with respect to each other but restricted by cohesive forces so as to maintain a relatively fixed volume.*[2]

Gas: *A state of matter in which the molecules are practically unrestricted by cohesive forces. A gas has neither definite shape nor volume.*

Shear stress: *The tangential force divided by the area.*

Key Concept: *Fluids considered in this text are those that move under the action of a shear stress.*

It is worthwhile to consider the microscopic behavior of fluids in more detail. Consider the molecules of a gas in a container. These molecules are not stationary but move about in space with very high velocities. They collide with each other and strike the walls of the container in which they are confined, giving rise to the pressure exerted by the gas. If the volume of the container is increased while the temperature is maintained constant, the number of molecules impacting on a given area is decreased and as a result the pressure decreases. If the temperature of a gas in a given volume increases (i.e., the velocities of the molecules increase), the pressure increases due to increased molecular activity.

Molecular forces in liquids are relatively high, as can be inferred from the following example. The pressure necessary to compress 20 grams of water vapor at 20°C into 20 cm^3, assuming that no molecular forces exist, can be shown by the ideal gas law to be approximately 1340 times the atmospheric pressure. Of course, this pressure is not required because 20 g of water occupies 20 cm^3. It follows that the cohesive forces in the liquid phase must be very large.

Despite the high molecular attractive forces in a liquid, some of the molecules at the surface escape into the space above. If the liquid is contained, an equilib-

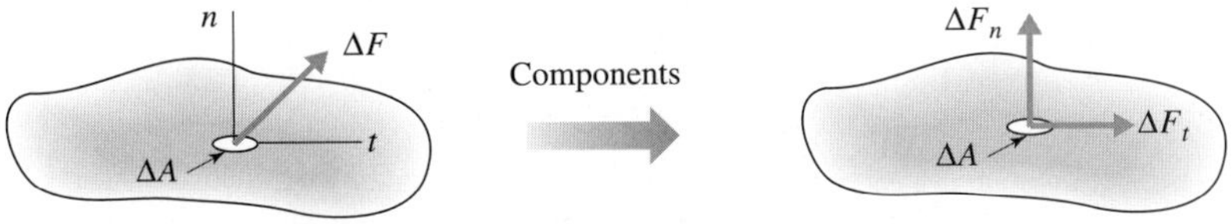

FIGURE 10.1 Normal and tangential components of a force.

[2] *Handbook of Chemistry and Physics*, 40th ed. CRC Press, Boca Raton, Fla.

rium is established between outgoing and incoming molecules. The presence of molecules above the liquid surface leads to a vapor pressure. This pressure increases with temperature. For water at 20°C, this pressure is 0.02 times the atmospheric pressure.

Continuum: *Continuous distribution of a liquid or gas throughout a region of interest.*

In our study of the mechanics of fluids, it is convenient to assume that both gases and liquids are continuously distributed throughout a region of interest (that is, the fluid is treated as a **continuum**). The primary property used to determine if the continuum assumption is appropriate is the *density* ρ, defined by

$$\rho = \lim_{\Delta v \to 0} \frac{\Delta m}{\Delta V} \tag{10.8}$$

Standard atmospheric conditions: *A pressure of* 101.3 kPa *and temperature of* 15°C.

where Δm is the incremental mass contained in the incremental volume ΔV. The density for air at **standard atmospheric conditions**, that is, at a pressure of 101.3 kPa (14.7 psi) and a temperature of 15°C (59°F), is 1.23 kg/m^3 (0.00238 slug/ft^3). For water, the nominal value of density is 1000 kg/m^3 (1.94 slug/ft^3).

Key Concept: *To determine if the continuum model is acceptable, compare a length l with the mean free path.*

Physically, we cannot let $\Delta V \to 0$, since as ΔV gets extremely small, the mass contained in ΔV would vary discontinuously depending on the number of molecules in ΔV; this is shown graphically in Fig. 10.2. Actually, the zero in the definition of density should be replaced by some small volume ε, below which the continuum assumption fails. For most engineering applications, the small volume ε shown in Fig. 10.2 is extremely small. For example, there are 2.7×10^{16} molecules contained in a cubic millimeter of air at standard conditions; hence, ε is much smaller than a cubic millimeter. An appropriate way to determine if the continuum model is acceptable is to compare a characteristic length l (e.g., the diameter of a rocket) of the device or object of interest with the **mean free path** λ, the average distance a molecule travels before it collides with another molecule; if $l \gg \lambda$, the continuum model is acceptable. The mean free path is derived in molecular theory. It is

Mean free path: *The average distance a molecule travels before it collides with another molecule.*

$$\lambda = 0.225 \frac{m}{\rho d^2} \tag{10.9}$$

where m is the mass (kg) of a molecule, ρ the fluid density (kg/m^3) and d the diameter (m) of a molecule. For air, $m = 4.8 \times 10^{-26}$ kg and $d = 3.7 \times 10^{-10}$ m. At standard atmospheric conditions, the mean free path is approximately 6×10^{-6} cm;

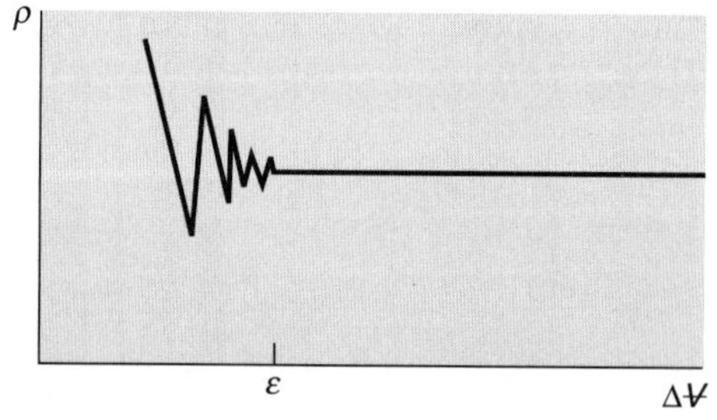

FIGURE 10.2 Density at a point in a continuum.

at an elevation of 100 km it is 10 cm, and at 160 km it is 5000 cm. Obviously, at higher elevations the continuum assumption is not acceptable, and the theory of rarefied gas dynamics (or free molecular flow) must be utilized. Satellites are able to orbit the earth if the primary dimension of the satellite is of the same order of magnitude as the mean free path.

With the continuum assumption, fluid properties can be assumed to apply uniformly at all points in the region at any particular instant in time. For example, the density ρ can be defined at all points in the fluid; it may vary from point to point and from instant to instant; that is, in Cartesian coordinates, ρ is a continuous function of x, y, z, and t written as $\rho(x, y, z, t)$.

10.4 PRESSURE AND TEMPERATURE SCALES

In fluid mechanics, pressure results from a normal compressive force acting on an area. The *pressure* p is defined as (see Fig. 10.3)

$$p = \lim_{\Delta A \to 0} \frac{\Delta F_n}{\Delta A} \tag{10.10}$$

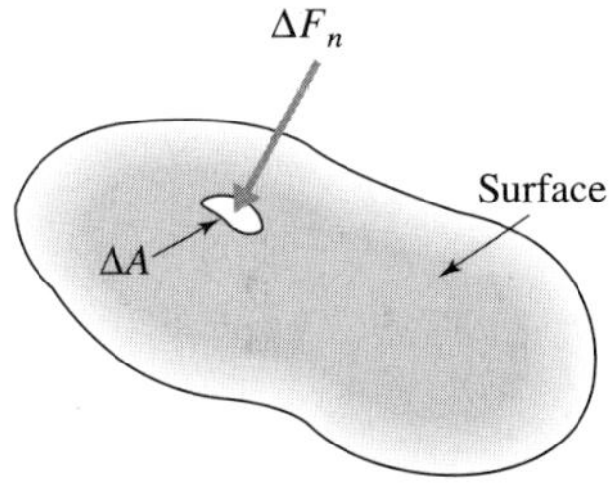

FIGURE 10.3 Definition of pressure.

where ΔF_n is the incremental normal compressive force acting on the incremental area ΔA. The metric units to be used on pressure are newtons per square meter (N/m^2) or pascal (Pa). Since the pascal is a very small unit of pressure, it is more conventional to express pressure in units of kilopascal (kPa). For example, standard atmospheric pressure at sea level is 101.3 kPa. The English units for pressure are pounds per square inch (psi) or pounds per square foot (lb/ft^2).

Both pressure and temperature are physical quantities that can be measured using different scales. There exist absolute scales for pressure and temperature and there are scales that measure these quantities relative to selected reference points. In many thermodynamic relationships (see Section 10.7), absolute scales must be used for pressure and temperature. Figures 10.4 and 10.5 summarize the commonly used scales. Note that atmospheric pressure is often expressed as inches of mercury or feet of water, as shown in Fig. 10.4; such a column of fluid creates the pressure at the bottom of the column.

Key Concept: *In many relationships, absolute scales must be used for pressure and temperature.*

The **absolute pressure** reaches zero when an ideal vacuum is achieved, that is, when no molecules are left in a space; consequently, a negative absolute pressure is an impossibility. A second scale is defined by measuring pressures relative to the local atmospheric pressure. This pressure is called **gage pressure**. A conversion from gage pressure to absolute pressure can be carried out using

Absolute pressure: *A scale measuring pressure, where zero is reached when an ideal vacuum is achieved.*

Gage pressure: *A scale measuring pressure relative to the local atmospheric pressure.*

$$p_{\text{absolute}} = p_{\text{atmospheric}} + p_{\text{gage}} \tag{10.11}$$

Note that the atmospheric pressure in Eq. 10.11 is the local atmospheric pressure, which may change with time, particularly when a weather "front" moves through. However, if the local atmospheric pressure is not given, we use the value given for a particular elevation, as given in Table B.1 of Appendix B, and assume zero ele-

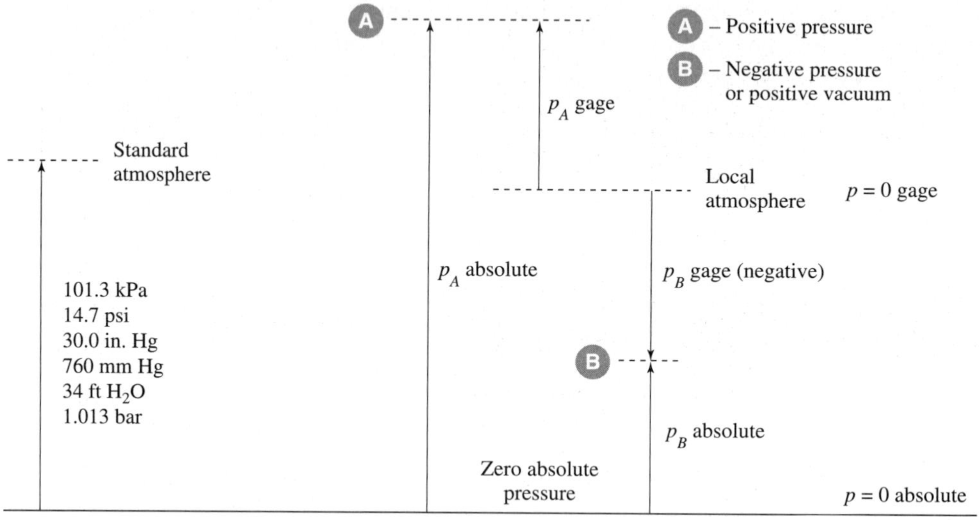

FIGURE 10.4 Gage pressure and absolute pressure.

Vacuum: *Whenever the absolute pressure is less than the atmospheric pressure.*

vation if the elevation is also unknown. The gage pressure is negative whenever the absolute pressure is less than atmospheric pressure; it may then be called a **vacuum**. In this book, the word "absolute" will generally follow the pressure value if the pressure is given as an absolute pressure (e.g., $p = 50$ kPa absolute). If it were stated as $p = 50$ kPa, the pressure would be taken as a gage pressure, except that atmospheric pressure is always an absolute pressure. Most often in fluid mechanics, gage pressure is used.

Two temperature scales are commonly used, the Celsius (C) and Fahrenheit (F) scales. Both scales are based on the ice point and steam point of water at an atmospheric pressure of 101.3 kPa (14.7 psi). Figure 10.5 shows that the ice and steam point are 0 and 100°C on the Celsius scale and 32 and 212°F on the Fahrenheit scale. There are two corresponding absolute temperature scales. The absolute scale corresponding to the Celsius scale is the kelvin (K) scale. The relation between these scales is

$$K = °C + 273.15 \tag{10.12}$$

	°C	K	°F	°R
Steam point	100°	373	212°	672°
Ice point	0°	273	32°	492°
	−18°	255	0°	460°
Zero absolute temperature				

FIGURE 10.5 Temperature scales.

The absolute scale corresponding to the Fahrenheit scale is the Rankine scale (°R). The relation between these scales is

$$°R = °F + 459.67 \tag{10.13}$$

Note that in the SI system we do not write 100°K but simply 100 K, which is read "100 kelvins," similar to other units.

Key Concept: *In the SI system, we write* 100 K, *which is read "100 kelvins."*

Reference will often be made to "standard atmospheric conditions" or "standard temperature and pressure." This refers to sea-level conditions at 40° latitude, which are taken to be 101.3 kPa (14.7 psi) for pressure and 15°C (59°F) for temperature.

Example 10.2

A pressure gage attached to a rigid tank measures a vacuum of 42 kPa inside the tank shown in Fig. E10.2, which is situated at a site in Colorado where the elevation is 2000 m. Determine the absolute pressure inside the tank.

FIGURE E10.2

Solution

To determine the absolute pressure, the atmospheric pressure must be known. If the elevation were not given, we would assume the standard atmospheric pressure of 101 kPa. However, with the elevation given, the atmospheric pressure is found from Table B.3 in Appendix B to be 79.5 kPa. Thus

$$p = -42 + 79.5 = 37.5 \text{ kPa absolute}$$

Note: A vacuum is always a negative gage pressure.

10.5 FLUID PROPERTIES

In this section, we present several of the more common fluid properties. If density variation or heat transfer is significant, several additional properties, not presented here, become important. Some will be presented in subsequent chapters.

TABLE 10.4 Density, Specific Weight, and Specific Gravity of Air and Water at Standard Conditions

	Density ρ		Specific weight γ		Specific gravity S
	kg/m^3	slug/ft^3	N/m^3	lb/ft^3	
Air	1.23	0.0024	12.1	0.077	0.00123
Water	1000	1.94	9810	62.4	1

10.5.1 Density and Specific Weight

Specific weight: *Weight per unit volume ($\gamma = \rho g$).*

Fluid density was defined in Eq. 10.8 as mass per unit volume. A fluid property directly related to density is the **specific weight** γ or weight per unit volume. It is defined by

$$\gamma = \rho g \tag{10.14}$$

where g is the local gravity. The units of specific weight are N/m^3 (lb/ft^3). For water, we use the nominal value of 9810 N/m^3 (62.4 lb/ft^3).

Specific gravity: *The ratio of the density of a substance to that of water.*

The **specific gravity** S is often used to determine the specific weight or density of a fluid (usually a liquid). It is defined as the ratio of the density of a substance to that of water at a reference temperature of 4°C:

$$S = \frac{\rho}{\rho_{\text{water}}} = \frac{\gamma}{\gamma_{\text{water}}} \tag{10.15}$$

Key Concept: *Specific gravity is often used to determine the density of a fluid.*

For example, the specific gravity of mercury is 13.6, a dimensionless number; that is, the mass of mercury is 13.6 times that of water for the same volume. The density, specific weight, and specific gravity of air and water at standard conditions are given in Table 10.4.

10.5.2 Viscosity

Viscosity: *The internal stickiness of a fluid.*

Viscosity can be thought of as the internal stickiness of a fluid. It is one of the properties that influences the power needed to move an airfoil through the atmosphere. It accounts for the energy losses associated with the transport of fluids in ducts, channels, and pipes. Further, viscosity plays a primary role in the generation of turbulence. Needless to say, viscosity is an extremely important fluid property in our study of fluid flows.

Key Concept: *Viscosity accounts for energy losses in a pipe.*

The rate of deformation of a fluid is directly linked to the viscosity of the fluid. For a given stress, a highly viscous fluid deforms at a slower rate than a fluid with a low viscosity. Consider the flow of Fig. 10.6 in which the fluid particles move in the x-direction at different speeds, so that particle velocities u vary with the y-coordinate. Two particle positions are shown at different times; observe how the particles move relative to one another. For such a simple flow field, in which $u = u(y)$, we can define the **viscosity** μ of the fluid by the relationship

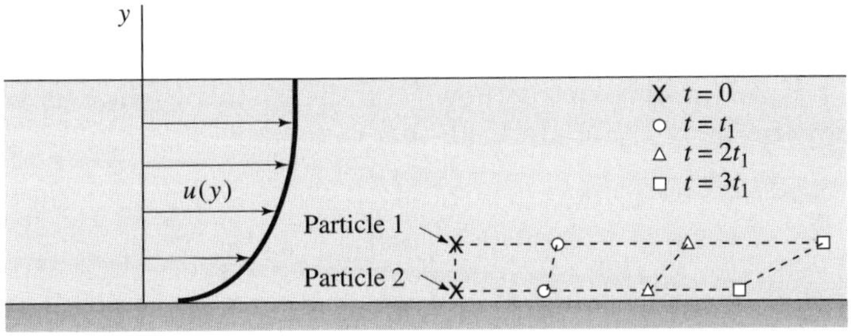

FIGURE 10.6 Relative movement of two fluid particles in the presence of shear stresses.

$$\tau = \mu \frac{du}{dy} \tag{10.16}$$

where τ is the shear stress of Eq. 10.7 and u is the velocity in the x-direction. The units of τ are N/m^2 or Pa (lb/ft^2), and of μ are N·s/m^2 (lb-sec/ft^2). The quantity du/dy is a velocity gradient and can be interpreted as a **strain rate**. Stress velocity-gradient relationships for more complicated flow situations are presented in Appendix K.2.

Strain rate: *The rate at which a fluid element deforms.*

The concept of viscosity and velocity gradients can also be illustrated by considering a fluid within the small gap between two concentric cylinders, as shown in Fig. 10.7. A torque is necessary to rotate the inner cylinder at constant

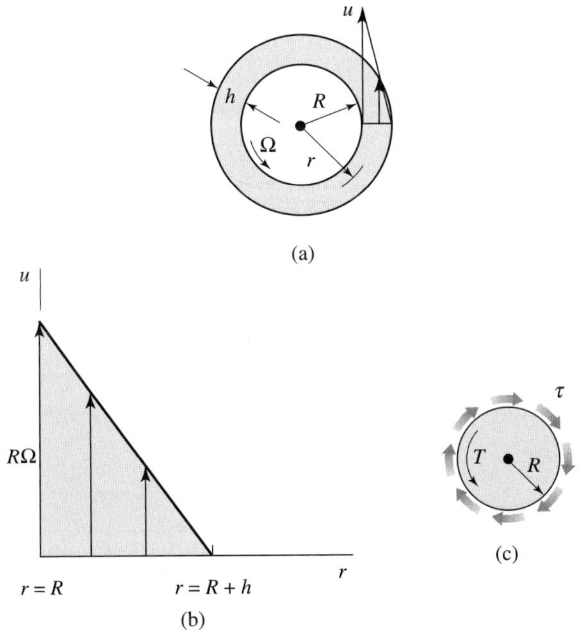

FIGURE 10.7 Fluid being sheared between cylinders with a small gap: (a) rotating inner cylinder; (b) velocity distribution; (c) the inner cylinder. The outer cylinder is fixed and the inner cylinder is rotating.

rotational speed Ω, while the outer cylinder remains stationary. This resistance to the rotation of the cylinder is due to viscosity. The only stress that exists to resist the applied torque for this simple flow is a shear stress, which is observed to depend directly on the velocity gradient; that is,

$$\tau = \mu \left| \frac{du}{dr} \right| \tag{10.17}$$

where du/dr is the velocity gradient and u is the tangential velocity component, which depends only on r. For a small gap ($h \ll R$), this gradient can be approximated by assuming a linear velocity distribution[3] in the gap. Thus

$$\left| \frac{du}{dr} \right| = \frac{\Omega R}{h} \tag{10.18}$$

where h is the gap width. We can thus relate the applied torque T to the viscosity and other parameters by the equation

$$\begin{aligned} T &= \text{stress} \times \text{area} \times \text{moment arm} \\ &= \tau \times 2\pi RL \times R \\ &= \mu \frac{\Omega R}{h} \times 2\pi RL \times R = \frac{2\pi R^3 \Omega L \mu}{h} \end{aligned} \tag{10.19}$$

where we have neglected the shearing stress acting on the ends of the cylinder; L represents the length of the rotating cylinder. Note that the torque depends directly on the viscosity, thus the cylinders could be used as a *viscometer*, a device that measures the viscosity of a fluid.

Newtonian fluid: *A fluid in which the shear stress is directly proportional to the velocity gradient.*

If the shear stress of a fluid is directly proportional to the velocity gradient, as was assumed in Eqs. 10.16 and 10.17, the fluid is said to be a **Newtonian fluid**. Fortunately, many common fluids, such as air, water, and oil, are Newtonian. *Non-Newtonian fluids*, with shear stress versus strain rate relationships as shown in Fig. 10.8, often have a complex molecular composition.

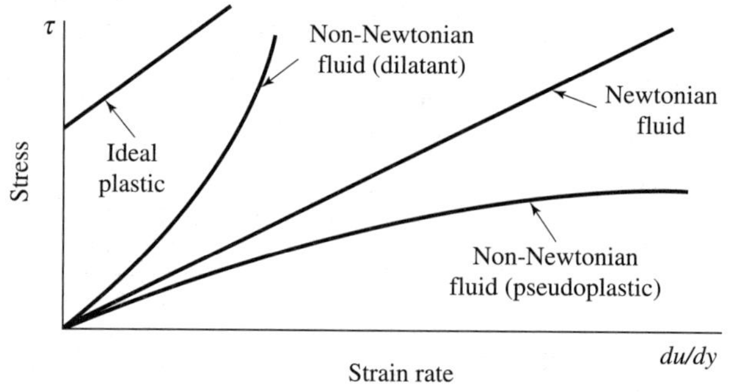

FIGURE 10.8 Newtonian and non-Newtonian fluids.

[3] If the gap is not small relative to R, the velocity distribution will not be linear (see Section 15.5). The distribution will also not be linear for relatively small values of Ω.

Dilatants (quicksand, slurries) become more resistant to motion as the strain rate increases, and *pseudoplastics* become less resistant to motion with increased strain rate. *Ideal plastics* (or *Bingham fluids*) require a minimum shear stress to cause motion. Clay suspensions and toothpaste are examples that also require a minimum shear to cause motion, but they do not have a linear stress-strain rate relationship.

An important effect of viscosity is to cause the fluid to adhere to the surface; this is known as the **no-slip condition**. This was assumed in the example of Fig. 10.7. The velocity of the fluid at the rotating cylinder was taken to be ΩR and the velocity of the fluid at the stationary cylinder was set equal to zero, as shown in Fig. 10.7b.

No-slip condition: *Condition where viscosity causes fluid to adhere to the surface.*

The viscosity is very dependent on temperature in liquids in which cohesive forces play a dominant role; note that the viscosity of liquids decreases with increased temperature, as shown in Fig. B.1 in Appendix B. The curves are often approximated by the equation

$$\mu = Ae^{Bt} \tag{10.20}$$

known as *Andrade's equation*; the constants A and B would be determined from measured data. For a gas, it is molecular collisions that provide the internal stresses, so that as the temperature increases, resulting in increased molecular activity, the viscosity increases. This can be observed in the bottom curve for a gas of Fig. B.1 in Appendix B. Note, however, that the percentage change of viscosity in a liquid is much greater than in a gas for the same temperature difference. Also, one can show that cohesive forces and molecular activity are quite insensitive to pressure, so that $\mu = \mu(T)$ only for both liquids and gases.

Since the viscosity is often divided by the density in the derivation of equations, it has become useful and customary to define *kinematic viscosity* to be

$$\nu = \frac{\mu}{\rho} \tag{10.21}$$

where the units of ν are m^2/s (ft^2/sec). Note that for a gas, the kinematic viscosity will also depend on the pressure, since the density is pressure sensitive.

Example 10.3

A viscometer is constructed with two 30-cm-long concentric cylinders, one 20.0 cm in diameter and the other 20.2 cm in diameter. A torque of 0.13 N·m is required to rotate the inner cylinder at 400 rpm (revolutions per minute). Calculate the viscosity.

Solution

The applied torque is just balanced by a resisting torque due to the shear stresses (see Fig. 10.7c). This is expressed by the small gap equation, Eq. 10.19. The radius is $R = d/2 = 10$ cm; the gap is $h = (d_2 - d_1)/2 = 0.1$ cm; the rotational speed, expressed as rad/s, is $\Omega = 400 \times 2\pi/60 = 41.89$ rad/s.

(continued)

Equation 10.19 provides:

$$\mu = \frac{Th}{2\pi R^3 \Omega L}$$

$$= \frac{0.13(0.001)}{2\pi(0.1)^3(41.89)(0.3)} = 0.001646 \text{ N·s/m}^2$$

Note: All lengths are in meters so that the desired units on μ are obtained. The units can be checked by substitution (recall that radians are dimensionless):

$$[\mu] = \frac{\text{N·m·m}}{\text{m}^3(\text{rad/s})\text{m}} = \frac{\text{N·s}}{\text{m}^2}$$

10.5.3 Compressibility

Bulk modulus of elasticity: *The ratio of change in pressure to relative change in density.*

In the preceding section, we discussed the deformation of fluids that results from shear stresses. In this section, we discuss the deformation that results from pressure changes. All fluids compress if the pressure increases, resulting in an increase in density. A common way to describe the compressibility of a fluid is by the following definition of the **bulk modulus of elasticity** B:

$$B = \lim_{\Delta \mathcal{V} \to 0} \left[-\frac{\Delta p}{\Delta \mathcal{V}/\mathcal{V}} \right]_T = \lim_{\Delta \rho \to 0} \frac{\Delta p}{\Delta \rho/\rho}\bigg|_T$$

$$= -\mathcal{V} \frac{\partial p}{\partial \mathcal{V}}\bigg|_T = \rho \frac{\partial p}{\partial \rho}\bigg|_T \quad (10.22)$$

In words, the bulk modulus, also called the *coefficient of compressibility*, is defined as the ratio of the change in pressure (Δp) to relative change in density ($\Delta \rho/\rho$) while the temperature remains constant. The bulk modulus obviously has the same units as pressure.

The bulk modulus for water at standard conditions is approximately 2100 MPa (310,000 psi), or 21 000 times the atmospheric pressure. For air at standard conditions, B is equal to 1 atm. In general, B for a gas is equal to the pressure of the gas. To cause a 1% change in the density of water, a pressure of 21 MPa (210 atm) is required. This is an extremely large pressure needed to cause such a small change; thus liquids are often assumed to be incompressible. For gases, if significant changes in density occur, say 4%, they should be considered as compressible; for small density changes under 3%, they may also be treated as incompressible. This occurs for atmospheric airspeeds under about 100 m/s, which includes many airflows of engineering interest: air flow around automobiles, landing and take-off of aircraft, and air flow in and around buildings.

Key Concept: *Gases with small density changes under 3% may be treated as incompressible.*

The bulk modulus can also be used to calculate the speed of sound in a liquid; it is given by

$$c = \sqrt{\left.\frac{\Delta p}{\Delta \rho}\right|_T} = \sqrt{\frac{B}{\rho}} \tag{10.23}$$

This yields approximately 1450 m/s (4800 ft/sec) for the speed of sound in water at standard conditions.

10.5.4 Surface Tension

Surface tension: *A property resulting from the attractive forces between molecules.*

Surface tension is a property that results from the attractive forces between molecules. As such, it manifests itself only in liquids at an interface, usually a liquid–gas interface. The forces between molecules in the bulk of a liquid are equal in all directions, and as a result, no net force is exerted on the molecules. However, at an interface, the molecules exert a force that has a resultant in the interface layer. This force holds a drop of water suspended on a rod and limits the size of the drop that may be held. It also causes the small drops from a sprayer or atomizer to assume spherical shapes. It may also play a significant role when two immiscible liquids (e.g., oil and water) are in contact with each other.

Key Concept: *Force due to surface tension results from a length multiplied by the surface tension.*

Surface tension has units of force per unit length, N/m (lb/ft). The force due to surface tension results from a length multiplied by the surface tension; the length to use is the length of fluid in contact with a solid or the circumference in the case of a bubble. A surface-tension effect can be illustrated by considering the free-body diagrams of half a droplet and half a bubble as shown in Fig. 10.9. The droplet has one surface and the bubble is composed of a thin film of liquid with an inside surface and an outside surface. The pressure inside the droplet and bubble can now be calculated.

The pressure force $p\pi R^2$ in the droplet balances the surface-tension force around the circumference. Hence

$$p\pi R^2 = 2\pi R\sigma$$

$$\therefore p = \frac{2\sigma}{R} \tag{10.24}$$

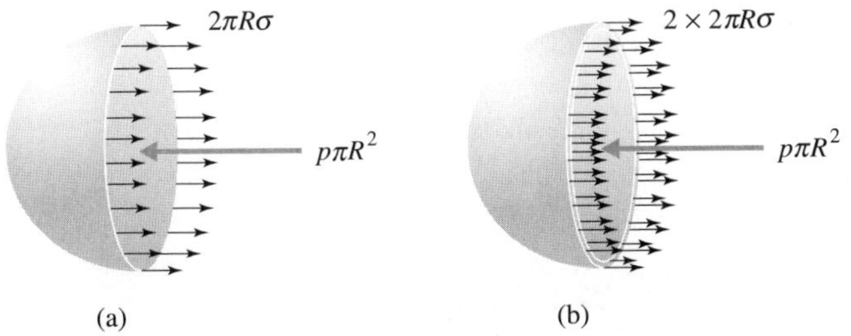

FIGURE 10.9 Internal forces in (a) a droplet and (b) a bubble.

Similarly, the pressure force in the bubble is balanced by the surface-tension forces on the two circumferences. Therefore,

$$p\pi R^2 = 2(2\pi R\sigma)$$

$$\therefore\ p = \frac{4\sigma}{R} \tag{10.25}$$

From Eqs. 10.24 and 10.25, we can conclude that the internal pressure in a bubble is twice as large as that in a droplet of the same size, providing σ is the same for both.

Figure 10.10 shows the rise of a liquid in a clean glass capillary tube due to surface tension. The liquid makes a contact angle β with the glass tube. Experiments have shown that this angle for water and most liquids in a clean glass tube is zero. There are also cases for which this angle is greater than 90° (e.g., mercury in a clean glass tube); such liquids have a capillary drop. If h is the capillary rise, D the diameter, ρ the density, and σ the surface-tension, h can be determined from equating the vertical component of the surface-tension force to the weight of the liquid column:

$$\sigma\pi D\cos\beta = \gamma\frac{\pi D^2}{4}h \tag{10.26}$$

or

$$h = \frac{4\sigma\cos\beta}{\gamma D} \tag{10.27}$$

Surface-tension may influence engineering problems when, for example, laboratory modeling of waves is conducted at a scale that surface-tension forces are of the same order of magnitude as gravitational forces.

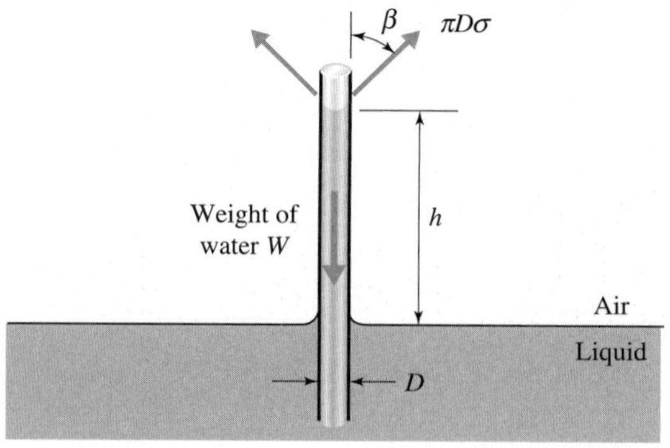

FIGURE 10.10 Rise in a capillary tube.

Example 10.4

A 2-mm-diameter clean glass tube is inserted in water at 15°C (Fig. E10.4). Determine the height that the water will climb up the tube. The water makes a contact angle of 0° with the clean glass.

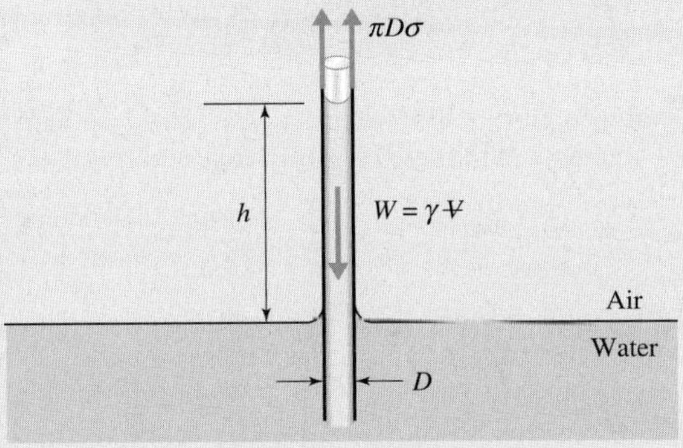

FIGURE E10.4

Solution

A free-body diagram of the water shows that the upward surface-tension force is equal and opposite to the weight. Writing the surface-tension force as surface tension times distance, we have

$$\sigma \pi D = \gamma \frac{\pi D^2}{4} h$$

or

$$h = \frac{4\sigma}{\gamma D} = \frac{4(0.0741)}{9800(0.002)} = 0.01512 \text{ m} \quad \text{or} \quad 15.12 \text{ mm}$$

The numerical values for σ and ρ were obtained from Table B.9 in the Appendix B. Note that the nominal value used for the specific weight of water is $\gamma = \rho g = 9800 \text{ N/m}^3$. Here it would more accurately be $\gamma = 999.1 \times 9.81 = 9801 \text{ N/m}^3$, but 9800 is sufficiently accurate for engineering calculations.

10.5.5 Vapor Pressure

When a small quantity of liquid is placed in a closed container, a certain fraction of the liquid will vaporize. Vaporization will terminate when equilibrium is reached between the liquid and gaseous states of the substance in the container—in other words, when the number of molecules escaping from the water surface is equal to the number of incoming molecules. The pressure resulting from molecules in the gaseous state is the **vapor pressure**.

Vapor pressure: *The pressure resulting from molecules in a gaseous state.*

The vapor pressure is different from one liquid to another. For example, the vapor pressure of water at standard conditions (15°C, 101.3 kPa) is 1.70 kPa absolute and for ammonia it is 33.8 kPa absolute.

FIGURE 10.11 Cooking food in boiling water takes a longer amount of time at a high altitude. It would take longer to boil these eggs in Denver than in New York City. (Courtesy Food Pix, Inc.)

The vapor pressure is highly dependent on pressure and temperature; it increases significantly when the temperature increases. For example, the vapor pressure of water increases to 101.3 kPa (14.7 psi) if the temperature reaches 100°C (212°F). Water vapor pressures for other temperatures at atmospheric pressure are given in Appendix B.9.

Boiling: *The point where vapor pressure is equal to the atmospheric pressure.*

Key Concept: *Cavitation can be very damaging.*

Cavitation: *Bubbles form in a liquid when the local pressure falls below the vapor pressure of the liquid.*

It is, of course, no coincidence that the water vapor pressure at 100°C is equal to the standard atmospheric pressure. At that temperature the water is **boiling**, that is, the liquid state of the water can no longer be sustained because the attractive forces are not sufficient to contain the molecules in a liquid phase. In general, a transition from the liquid state to the gaseous state occurs if the local absolute pressure is less than the vapor pressure of the liquid. At high elevations where the atmospheric pressure is relatively low, boiling occurs at temperatures less than 100°C (see Fig. 10.11).

In liquid flows, conditions can be created that lead to a pressure below the vapor pressure of the liquid. When this happens, bubbles are formed locally. This phenomenon, called **cavitation**, can be very damaging when these bubbles are transported by the flow to higher pressure regions. What happens is that the bubbles collapse upon entering the higher pressure region, and this collapse produces local pressure spikes which have the potential of damaging a pipe wall or a ship's propeller.

Example 10.5

Calculate the vacuum necessary to cause cavitation in a water flow at a temperature of 80°C in Colorado where the elevation is 2500 m.

Solution

The vapor pressure of water at 80°C is given in Appendix Table B.9. It is 47.3 kPa absolute. The atmospheric pressure is found by interpolation using Table B.1 to be 79.48 – (79.48 – 61.64)500\2000 = 75.02. The required pressure is then

$$p = 47.3 - 75.0 = -27.7 \text{ kPa} \quad \text{or} \quad 27.7 \text{ kPa vacuum}$$

10.6 CONSERVATION LAWS

From experience, it has been found that fundamental laws exist that appear exact; that is, if experiments are conducted with the utmost precision and care, deviations from these laws are very small, and in fact, the deviations would be even smaller if improved experimental techniques were employed. Three such laws form the basis for our study of fluid mechanics. The first is the **conservation of mass**, which states that matter is indestructible. Even though Einstein's theory of relativity postulates that, under certain conditions, matter is convertible into energy and leads to the statement that the extraordinary quantities of radiation from the sun are associated with a conversion of 3.3×10^{14} kg of matter per day into energy, the destructibility of matter under typical engineering conditions is not measurable and does not violate the conservation of mass principle.

Conservation of mass: *Matter is indestructible.*

For the second and third laws, it is necessary to introduce the concept of a system. A **system** is defined as a fixed quantity of matter upon which attention is focused. Everything external to the system is separated by the system boundaries. These boundaries may be fixed or movable, real or imaginary. With this definition, we can now present our second fundamental law, the **conservation of momentum**: The momentum of a system remains constant if no external forces are acting on the system. A more specific law based on this principle is *Newton's second law*: The sum of all external forces acting on a system is equal to the time rate of change of linear momentum of the system. A parallel law exists for the moment of momentum: The rate of change of angular momentum is equal to the sum of all torques acting on the system.

System: *A fixed quantity of matter.*

Conservation of momentum: *The momentum of a system remains constant if no external forces act on the system.*

The third fundamental law is the **conservation of energy**, which is also known as the *first law of thermodynamics*: The total energy of an isolated system remains constant. If a system is in contact with the surroundings, its energy increases only if the energy of the surroundings experiences a corresponding decrease. It is noted that the total energy consists of potential, kinetic, and internal energy, the latter being the energy content due to the temperature of the system. Other forms of energy are not considered in fluid mechanics. The first law of thermodynamics and other thermodynamic relationships are presented in the following section.

Conservation of energy: *The total energy of an isolated system remains constant. Also known as the first law of thermodynamics.*

10.7 THERMODYNAMIC PROPERTIES AND RELATIONSHIPS

For incompressible fluids, the three laws mentioned in the preceding section suffice. This is usually true for liquids but also for gases if relatively small pressure, density, and temperature changes occur. However, for a compressible fluid, it may be necessary to introduce other relationships, so that density, temperature, and pressure changes are properly taken into account. An example is the prediction of changes in density, pressure, and temperature when compressed gas is released from a container. Such problems were considered in the Thermodynamics Part of this book. Several equations that were presented there (that will be used in our study of fluids), will be reviewed here.

Thermodynamic properties (quantities that define the state of a system), either depend on the system's mass or are independent of the mass. The former is called an **extensive property** and the latter is called an **intensive property**. An intensive

Extensive property: *A property that depends on the system's mass.*

Intensive property: *A property that is independent of the system's mass.*

property can be obtained by dividing the extensive property by the mass of the system. Temperature and pressure are intensive properties; momentum and energy are extensive properties.

10.7.1 Properties of an Ideal Gas

The behavior of gases in most engineering applications can be described by the ideal-gas law, also called the perfect-gas law. When the temperature is relatively low and/or the pressure relatively high, caution should be exercised and real-gas laws should be applied. For air with temperatures higher than −50°C (−58°F), the ideal-gas law approximates the behavior of air to an acceptable degree, provided that the pressure is not extremely high.

The *ideal-gas law* is given by

$$p = \rho RT \tag{10.28}$$

where p is the absolute pressure, ρ the density, T the absolute temperature, and R the gas constant. Another form that the ideal-gas law takes is

$$p\mathcal{V} = mRT \tag{10.29}$$

Example 10.6

A tank with a volume of 0.2 m^3 contains 0.5 kg of nitrogen. The temperature is 20°C. What is the pressure?

Solution

Assume this is an ideal gas. Apply Eq. 10.28 (R can be found in Appendix B.2.) Solving the equation, $p = \rho RT$, we obtain

$$p = \frac{0.5\text{ kg}}{0.2\text{ m}^3} \times 0.2968 \frac{\text{kJ}}{\text{kg}\cdot\text{K}} (273 + 20)\text{ K} = 218\text{ kPa absolute}$$

Note: The resulting units are kJ/m^3 = kN·m/m^3 = kN/m^2 = kPa. The ideal-gas law requires that pressure and temperature be in absolute units.

10.7.2 First Law of Thermodynamics

In the study of incompressible fluids, the first law of thermodynamics is particularly important. The first law of thermodynamics states that when a system, which is a fixed quantity of fluid, changes from state 1 to state 2, its energy content changes from E_1 to E_2 by energy exchange with its surroundings. The energy exchange is in the form of heat transfer or work. If we define heat transfer to the system as positive and work done by the system as positive,[4] the first law of thermodynamics can be expressed as

[4]In some presentations, the work done on the system is positive, so that Eq. 10.30 would appear as $Q + W = \Delta E$. Either choice is acceptable.

$$Q_{1\text{-}2} - W_{1\text{-}2} = E_2 - E_1 \tag{10.30}$$

where $Q_{1\text{-}2}$ is the amount of heat transfer to the system and $W_{1\text{-}2}$ is the amount of work done by the system. The energy E represents the total energy, which consists of kinetic energy $(mV^2/2)$, potential energy (mgz), and internal energy $(m\tilde{u})$, where $\tilde{u}$ is the internal energy[5] per unit mass; hence

Key Concept: *Energy exchange with surroundings is heat transfer or work.*

$$E = m\left(\frac{V^2}{2} + gz + \tilde{u}\right) \tag{10.31}$$

Note that $V^2/2$, gz, and $\tilde{u}$ are all intensive properties and E is an extensive property.

For an isolated system, one that is thermodynamically disconnected from the surroundings (i.e., $Q_{1\text{-}2} = W_{1\text{-}2} = 0$), Eq. 10.30 becomes

$$E_1 = E_2 \tag{10.32}$$

This equation represents the conservation of energy.

The work term in Eq. 10.30 results from a force F moving through a distance as it acts on the system's boundary; if the force is due to pressure, it is given by

Key Concept: *Work results from a force moving through a distance.*

$$\begin{aligned} W_{1\text{-}2} &= \int_{l_1}^{l_2} F\,dl \\ &= \int_{l_1}^{l_2} pA\,dl = \int_{\text{\unicode{x2C65}}_1}^{\text{\unicode{x2C65}}_2} p\,d\text{\unicode{x2C65}} \end{aligned} \tag{10.33}$$

where $A\,dl = d\text{\unicode{x2C65}}$. An example that demonstrates an application of the first law of thermodynamics follows.

Example 10.7

A cart with a mass of 2 slug is pushed up a ramp with an initial force of 100 lb (Fig. E10.7). The force decreases according to

$$F = 5(20 - l)\text{ lb}$$

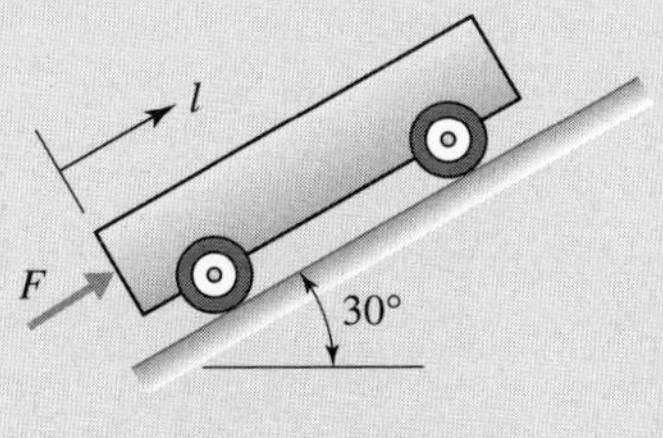

FIGURE E10.7

(continued)

[5]In our study of fluids, we use u as a velocity component, so $\tilde{u}$ is used to denote specific internal energy.

If the cart starts from rest at $l = 0$, determine its velocity after it has traveled 20 ft along the ramp. Neglect friction.

Solution

The energy equation (Eq. 10.30) allows us to relate the quantities of interest. Neglecting heat transfer we have

$$-W_{1\text{-}2} = E_2 - E_1$$

Recognizing that the force is doing work on the system, the work is negative. Hence, the energy equation becomes

$$-\left[-\int_0^{20} 5(20 - l)\, dl\right] = m\left(\frac{V_2^2}{2} + gz_2\right) - m\left(\overset{0}{\cancel{\frac{V_1^2}{2}}} + \overset{0}{\cancel{gz_1}}\right)$$

Taking the datum as $z_1 = 0$, we have $z_2 = 20 \sin 30° = 10$ ft. Thus

$$100 \times 20 - 5 \times \frac{20^2}{2} = 2\left(\frac{V_2^2}{2} + 32.2 \times 10\right)$$

$$\therefore\ V_2 = 18.9 \text{ ft/sec}$$

Note: We have assumed no internal energy change and no heat transfer.

10.7.3 Other Thermodynamic Quantities

Enthalpy: *A property created to aid in thermodynamic calculations.*

In compressible fluids, it is sometimes useful to define thermodynamic quantities that are combinations of other thermodynamic quantities. One such combination is the sum $(m\tilde{u} + p\forall)$, which can be considered a system property; it is encountered in numerous thermodynamic processes. This property is defined as **enthalpy** H:

$$H = m\tilde{u} + p\forall \tag{10.34}$$

The corresponding intensive property (H/m) is

$$h = \tilde{u} + \frac{p}{\rho} \tag{10.35}$$

Key Concept: *Constant-volume specific heat and constant-pressure specific heat are used to calculate the enthalpy and internal energy changes.*

Other useful thermodynamic quantities are the constant-pressure specific heat c_p and the constant-volume specific heat c_v; they are used to calculate the enthalpy and the internal energy changes in an ideal gas as follows:

$$\Delta h = \int c_p\, dT \tag{10.36}$$

and

$$\Delta \tilde{u} = \int c_v \, dT \tag{10.37}$$

For many situations, we can assume constant specific heats in the foregoing relationships. Specific heats for common gases are listed in Appendix Table B.2. For an ideal gas, c_p is related to c_v by using Eq. 10.35 in differential form:

$$c_p = c_v + R \tag{10.38}$$

The **ratio of specific heats** k is often of use for an ideal gas; it is expressed as

Ratio of specific heats: *The ratio of c_p to c_v.*

$$k = \frac{c_p}{c_v} \tag{10.39}$$

For liquids and solids, we use $\Delta u = c_p \, \Delta T$, where c_p is the specific heat of the substance. For water, $c_p \cong 4.18$ kJ/kg·°C (1 Btu/lb-°F). For other liquids consult Table B.4.

A process in which pressure, temperature, and other properties are essentially constant at any instant throughout the system is called a **quasi-equilibrium** or *quasi-static* process. An example of such a process is the compression and expansion in an internal combustion engine. If, in addition, no heat is transferred ($Q_{1\text{-}2} = 0$), the process is called an *adiabatic*, quasi-equilibrium or *isentropic* process. For such an isentropic[6] process involving an ideal gas with constant specific heats, we may use

Quasi-equilibrium process: *A process in which properties are essentially constant at any instant throughout a system.*

$$\frac{p_1}{p_2} = \left(\frac{\rho_1}{\rho_2}\right)^k \qquad \frac{T_1}{T_2} = \left(\frac{p_1}{p_2}\right)^{(k-1)/k} \qquad \frac{T_1}{T_2} = \left(\frac{\rho_1}{\rho_2}\right)^{k-1} \tag{10.40}$$

For a small pressure wave traveling in a gas at relatively low frequency, the wave speed is given by an isentropic process so that

$$c = \sqrt{\left.\frac{dp}{d\rho}\right|_s} = \sqrt{kRT} \tag{10.41}$$

If the frequency is relatively high, entropy is not constant and we use

$$c = \sqrt{\left.\frac{dp}{d\rho}\right|_T} = \sqrt{RT} \tag{10.42}$$

[6]An isentropic process occurs when the entropy is constant. We will not define or calculate entropy here; it is discussed in Section 17.1.

Example 10.8

A cylinder fitted with a piston has an initial volume of 0.5 m^3. It contains 2.0 kg of air at 400 kPa absolute. Heat is transferred to the air while the pressure remains constant until the temperature is 300°C. Calculate the heat transfer and the work done. Assume constant specific heats.

Solution

Using the first law (Eq. 10.30), Eq. 10.33 with $p = \text{const}$, and the definition of enthalpy (Eq. 10.34), we see that

$$\begin{aligned} Q_{1\text{-}2} &= p_2 V_2 - p_1 V_1 + m\tilde{u}_2 - m\tilde{u}_1 \\ &= m\tilde{u}_2 + p_2\mathcal{V}_2 - (m\tilde{u}_1 + p_1\mathcal{V}_1) \\ &= H_2 - H_1 = m(h_2 - h_1) = mc_p(T_2 - T_1) \end{aligned}$$

where Eq. 10.36 is used assuming c_p to be constant. The initial temperature is

$$T_1 = \frac{p_1\mathcal{V}_1}{mR} = \frac{400 \text{ kN/m}^2 \times 0.5 \text{ m}^3}{2.0 \text{ kg} \times 0.287 \text{ kJ/kg·K}} = 348.4 \text{ K}$$

(Use kJ = kN·m to check the units.) Thus, the heat transfer is (c_p is found in Table B.2)

$$Q_{1\text{-}2} = 2.0 \times 1.00[(300 + 273) - 348.4] = 449 \text{ kJ}$$

The final volume is found using the ideal-gas law:

$$\mathcal{V}_2 = \frac{mRT_2}{p_2} = \frac{2 \times 0.287 \times 573}{400} = 0.822 \text{ m}^3$$

The work done for the constant-pressure process is, using Eq. 10.33 with $p = \text{const}$,

$$\begin{aligned} W_{1\text{-}2} &= p(\mathcal{V}_2 - \mathcal{V}_1) \\ &= 400(0.822 - 0.5) = 129 \text{ kJ} \end{aligned}$$

Example 10.9

The temperature on a cold winter day in the mountains of Wyoming is −22°F at an elevation of 10,000 ft. Calculate the density of the air assuming the same pressure as in the normal atmosphere; also find the speed of sound.

Solution

From Appendix Table B.1, we find the atmospheric pressure at an elevation of 10,000 ft to be 10.1 psi. The absolute temperature is found to be

$$T = -22 + 460 = 438°\text{R}$$

(continued)

Using the ideal-gas law, the density is calculated as

$$\rho = \frac{p}{RT}$$

$$= \frac{10.1 \text{ lb/in}^2 \times 144 \text{ in}^2/\text{ft}^2}{1716 \text{ ft-lb/slug-°R} \times 438\text{°R}} = 0.00194 \text{ slug/ft}^3$$

The speed of sound, using Eq. 10.41, is determined to be

$$c = \sqrt{kRT}$$

$$= \sqrt{1.4 \times 1716 \times 438} = 1026 \text{ ft/sec}$$

Note: The gas constant in the foregoing equations has units of ft-lb/slug-°R, so that the appropriate units result. Express slug = lb-sec^2/ft (from $m = F/a$) to observe that this is true.

10.8 SUMMARY

To relate units, we often use Newton's second law which allows us to write

$$\text{N} = \text{kg·m/s}^2 \tag{10.43}$$

When making engineering calculations, an answer should have the same number of significant digits as the least accurate number used in the calculations. Most fluid properties are known to at most four significant digits. Hence, answers should be expressed to at most four significant digits, and often to only three significant digits.

In fluid mechanics, pressure is expressed as gage pressure unless stated otherwise. This is unlike thermodynamics, in which pressure is assumed to be absolute. If absolute pressure is needed, add 101 kPa if the atmospheric pressure is not given in the problem statement.

The density, or specific weight, of a fluid is known if the specific gravity is known:

$$\rho_x = S_x \rho_{\text{water}} \qquad \gamma_x = S_x \gamma_{\text{water}} \tag{10.44}$$

The shear stress due to viscous effects in a simple flow where $u = u(y)$ is given by

$$\tau = \mu \frac{du}{dy} \tag{10.45}$$

This stress can be used to calculate the torque needed to rotate a shaft in a bearing.

Many air flows, and other gases too, are assumed to be incompressible at low speeds, speeds under about 100 m/s for atmospheric air.

The three fundamental laws used in our study of fluid mechanics are the conservation of mass, Newton's second law, and the first law of thermodynamics. These will take on various forms depending on the problem being studied. Much of our study of fluid mechanics will be expressing these laws in mathematical forms so that quantities of interest can be calculated.

REVIEW PROBLEMS FOR THE FE/EIT EXAMINATION

10.1FE If force, length, and time are selected as the three fundamental dimensions, the units of mass in the SI system could be written as:
A. FT^2/L **B.** FL/T^2
C. N·s^2/m **D.** N·m/s^2

10.2FE Select the dimensions of viscosity using the *F-L-T* system:
A. FT^2/L **B.** FT/L^2
C. N·s/m^2 **D.** N·s^2/m

10.3FE The quantity 2.36×10^{-8} Pa can be written as:
A. 23.6 nPa **B.** 236 μPa
C. 236×10^{-3} mPa **D.** 236 nPa

10.4FE A body that weighs 250 N on earth would weigh how much on the moon where $g \cong 1.6$ m/s^2?
A. 5030 N **B.** 250 N
C. 40.77 N **D.** 6.2 N

10.5FE A 4200-N force acts on a 250-cm area at an angle of 30° to the normal. The shear stress acting on the area is:
A. 84 Pa **B.** 84 mPa
C. 84 kPa **D.** 84 MPa

10.6FE The temperature at 11 000 m in the standard atmosphere, using a straight line interpolation of the entries in Appendix Table B.1, is nearest:
A. −52.2°C **B.** −53.0°C
C. −52.7°C **D.** −53.6°C

10.7FE Using an equation, estimate the density of water at 80°C:
A. 980 kg/m^3 **B.** 972 kg/m^3
C. 972 kg/m^3 **D.** 968 kg/m^3

10.8FE The velocity distribution in a 4-cm-diameter pipe transporting 20°C water is given by $u(r) = 10(1 - 2500r^2)$ m/s. The shearing stress at the wall is nearest:
A. 1.0 Pa **B.** 0.1 Pa
C. 0.01 Pa **D.** 0.001 Pa

10.9FE The distance 20°C water would climb in a long 10-μm-diameter, clean glass tube is nearest:
A. 50 cm **B.** 100 cm
C. 200 cm **D.** 300 cm

10.10FE Which of the following is an intensive property?
A. Kinetic energy
B. Enthalpy
C. Density
D. Momentum

10.11FE The mass of propane contained in a 4-m^3 tank maintained at 800 kPa and 10°C is nearest:
A. 100 kg **B.** 80 kg
C. 60 kg **D.** 20 kg

10.12FE Five 40-cm^3 ice cubes completely melt in 2 liters of warm water. (It takes 320 kJ to melt a kilogram of ice.) The temperature drop in the water is nearest:
A. 10°C **B.** 8°C
C. 6°C **D.** 4°C

10.13FE The speed of sound of a dog whistle in the atmosphere at a location where the temperature is 50°C is nearest:
A. 396 m/s **B.** 360 m/s
C. 332 m/s **D.** 304 m/s

PROBLEMS

10.1 State the three basic laws that are used in the study of the mechanics of fluids. State at least one global (integral) quantity that occurs in each. State at least one quantity that may be defined at a point that occurs in each.

Dimensions, Units, and Physical Quantities

10.2 Verify the dimensions given in Table 10.2 for the following quantities:
(a) Density **(b)** Pressure
(c) Power **(d)** Energy
(e) Mass flux **(f)** Flow rate

10.3 Express the dimensions of the following quantities using the *F-L-T* system:
(a) Density **(b)** Pressure
(c) Power **(d)** Energy
(e) Mass flux **(f)** Flow rate

10.4 Recognizing that all terms in an equation must have the same dimensions, determine the dimensions on the constants in the following equations:
(a) $d = 4.9t^2$ where d is distance and t is time.
(b) $F = 9.8m$ where F is a force and m is mass.
(c) $Q_f = 80AR^{2/3}S_0^{1/2}$ where A is area, R is a radius, S_0 is a slope, and Q_f is a flow rate with dimensions of L^3/T.

10.5 Determine the units on each of the constants in the following equations, recognizing that all terms in an equation have the same dimensions:
(a) $d = 4.9t^2$ where d is in meters and t is in seconds.
(b) $F = 9.8m$ where F is in newtons and m is in kilograms.
(c) $Q_f = 80AR^{2/3}S_0^{1/2}$ where A is in meters squared, R is in meters, S_0 is the slope, and Q_f has units of meters cubed per second.

10.6 State the SI units of Table 10.1 on each of the following:
(a) Pressure **(b)** Energy
(c) Power **(d)** Viscosity
(e) Heat flux **(f)** Specific heat

10.7 Write the following with the use of prefixes:
(a) 2.5×10^5 N **(b)** 5.72×10^{11} Pa
(c) 4.2×10^{-8} Pa **(d)** 1.76×10^{-5} m^3
(e) 1.2×10^{-4} m^2 **(f)** 7.6×10^{-8} m^3

10.8 Write the following with the use of powers; do not use a prefix:
(a) 125 MN **(b)** 32.1 μs
(c) 0.67 GPa **(d)** 0.0056 mm^3
(e) 520 cm^2 **(f)** 7.8 km^3

10.9 Rewrite Eq. 10.9 using the English units of Table 10.1.

10.10 Express each of the following in the SI units of Table 10.2:
(a) 20 cm/hr **(b)** 2000 rpm
(c) 500 hp **(d)** 100 ft^3/min
(e) 2000 kN/cm^2 **(f)** 4 slug/min
(g) 500 g/L **(h)** 500 kWh

10.11 What net force is needed to accelerate a 10-kg mass at the rate of 40 m/s^2:
(a) Horizontally?
(b) Vertically upward?
(c) On an upward slope of 30°?

10.12 A particular body weighs 60 lb on earth. Calculate its weight on the moon, where $g \cong 5.4$ ft/sec^2.

10.13 Calculate the mean free path in the atmosphere using Eq. 10.9 and Table B.1 in the Appendix at an elevation of:
(a) 30 000 m **(b)** 50 000 m **(c)** 80 000 m

Pressure and Temperature

10.14 A gage pressure of 52.3 kPa is read on a gage. Find the absolute pressure if the elevation is:
(a) At sea level **(b)** 1000 m
(c) 5000 m **(d)** 10 000 m
(e) 30 000 m

10.15 A vacuum of 31 kPa is measured in an airflow at sea level. Find the absolute pressure in:
(a) kPa **(b)** mm Hg
(c) psi **(d)** ft H_2O
(e) in. Hg

10.16 For a constant-temperature atmosphere, the pressure as a function of elevation is given by $p(z) = p_0 e^{-gz/RT}$, where g is gravity, $R = 287$ J/kg·K, and T is the absolute temperature. Use this equation and estimate the pressure at 4000 m assuming that $p_0 = 101$ kPa and $T = 15$°C. What is the error (refer to Appendix Table B.1)?

10.17 Estimate the pressure and temperature at an elevation of 22,560 ft using Table B.1E. Employ:
(a) A linear interpolation: $f \cong f_0 + n(f_1 - f_0)$.
(b) A parabolic interpolation: $f \cong f_0 + n(f_1 - f_0) + (n/2)(n-1)(f_2 - 2f_1 + f_0)$.

10.18 Estimate the temperature in °C and °F at 33,000 ft (an elevation at which many commercial airplanes fly). Use Appendix Table B.1E.

10.19 An applied force of 26.5 MN is distributed uniformly over a 152-cm^2 area; however, it acts at an angle of 42° with respect to a normal vector (see Fig. P10.19). If it produces a compressive stress, calculate the resulting pressure.

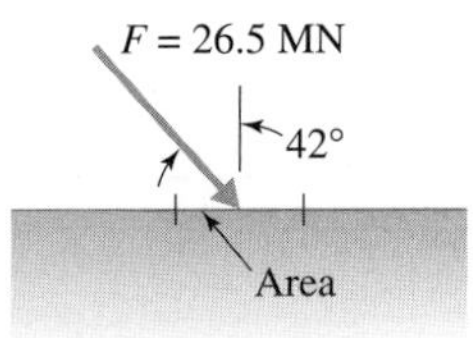

FIGURE P10.19

10.20 The force on an area of 0.2 cm^2 is due to a pressure of 120 kPa and a shear stress of 20 Pa, as shown in Fig. P10.20. Calculate the magnitude of the force acting on the area and the angle of the force with respect to a normal coordinate.

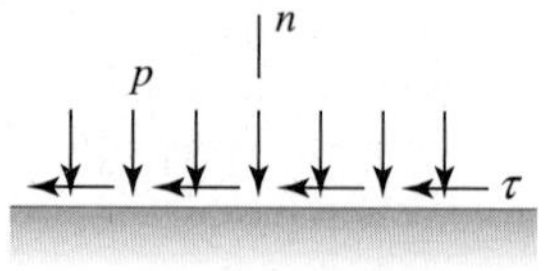

FIGURE P10.20

Density and Specific Weight

10.21 Calculate the density and specific weight of water if 0.2 slug occupies 180 in^3.

10.22 The specific weight of an unknown liquid is 12 400 N/m^3. What mass of the liquid is contained in a volume of 500 cm^3? Use:
- **(a)** The standard value of gravity.
- **(b)** The minimum value of gravity on the earth.
- **(c)** The maximum value of gravity on the earth.

10.23 A liquid with a specific gravity of 1.2 fills a volume. If the mass in the volume is 10 slug, what is the magnitude of the volume?

10.24 A velocity distribution in a 2-in.-diameter pipe is measured to be $u(r) = 30(1 - r^2/r_0^2)$ ft/sec, where r_0 is the radius of the pipe. Calculate the shear stress at the wall if water at 75°F is flowing.

10.25 For two 0.2-m-long rotating concentric cylinders, the velocity distribution is given by $u(r) = 0.4/r - 1000r$ m/s. If the diameters of the cylinders are 2 cm and 4 cm, respectively, calculate the fluid viscosity if the torque on the inner cylinder is measured to be 0.0026 N·m.

10.26 A 4-ft-long 1-in.-diameter shaft rotates inside an equally long cylinder that is 1.02 in. in diameter. Calculate the torque required to rotate the inner shaft at 2000 rpm if SAE-30 oil at 70°F fills the gap. Also, calculate the horsepower required. Assume symmetric motion.

10.27 A 60-cm-wide belt moves as shown in Fig. P10.27. Calculate the horsepower requirement assuming a linear velocity profile in the 10°C water.

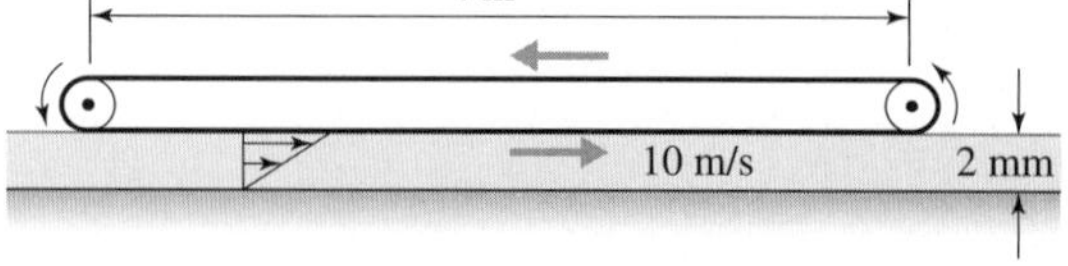

FIGURE P10.27

10.28 A 6-in.-diameter horizontal disk rotates a distance of 0.08 in. above a solid surface. Water at 60°F fills the gap. Estimate the torque required to rotate the disk at 400 rpm.

10.29 The velocity distribution in a 1.0-cm-diameter pipe is given by $u(r) = 16(1 - r^2/r_0^2)$ m/s, where r_0 is the pipe radius. Calculate the shearing stress at the centerline, at $r = 0.25$ cm and at the wall if water at 20°C is flowing.

10.30 Calculate the torque needed to rotate the cone shown in Fig. P10.30 at 2000 rpm if SAE-30 oil at 40°C fills the gap. Assume a linear velocity profile.

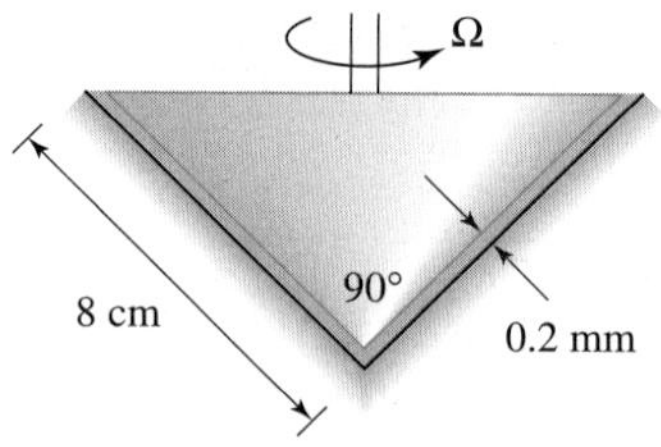

FIGURE P10.30

10.31 The viscosity of water at 20°C is 0.001 N·s/m^2, and at 80°C, it is 3.57×10^{-4} N·s/m^2. Using Andrade's equation $\mu = Ae^{B/T}$, estimate the viscosity of water at 40°C. Determine the percent error.

Compressibility

10.32 Show that $d\rho/\rho = -d\mathcal{V}/\mathcal{V}$, as was assumed in Eq. 10.22.

10.33 What is the volume change of 2 m^3 of water at 20°C due to an applied pressure of 10 MPa?

10.34 Two engineers wish to estimate the distance across a lake. One pounds two rocks together under water on one side of the lake and the other submerges his head and hears a small sound 0.62 s later, as indicated by a very accurate stopwatch. What is the distance between the two engineers?

10.35 A pressure is applied to 20 L of water. The volume is observed to decrease to 18.7 L. Calculate the applied pressure.

10.36 Calculate the speed of propagation of a small-amplitude wave through water at:
(a) 40°F
(b) 100°F
(c) 200°F

10.37 The change in volume of a liquid with temperature is given by $\Delta\mathcal{V} = \alpha_T \mathcal{V} \Delta T$, where α_T is the **coefficient of thermal expansion**. For water at 40°C, $\alpha_T = 3.8 \times 10^{-4}\ K^{-1}$. What is the volume change of 1 m^3 of 40°C water if $\Delta T = -20$°C? What pressure change would be needed to cause that same volume change?

Surface Tension

10.38 Calculate the pressure in the small 10-μm-diameter droplets that are formed by spray machines. Assume the properties to be the same as water at 15°C. Calculate the pressure for bubbles of the same size.

10.39 A small 1/16-in.-diameter bubble is formed by a stream of 60°F water. Estimate the pressure inside the bubble.

10.40 Determine the height that 20°C water would climb in a vertical 0.02-cm-diameter tube if it attaches to the wall with an angle β of 30° to the vertical.

10.41 Mercury makes an angle of 130° (β in Fig. 10.10) when in contact with clean glass. What distance will mercury depress in a vertical 0.8-in.-diameter glass tube? Use $\sigma = 0.032$ lb/ft.

10.42 Find an expression for the rise of liquid between two parallel plates a distance t apart. Use a contact angle β and surface tension σ.

10.43 Write an expression for the maximum diameter d of a needle of length L that can float in a liquid with surface tension σ. The density of the needle is ρ.

10.44 Would a 7-cm-long 4-mm-diameter steel needle be able to float in 15°C water? Use $\rho_{steel} = 7850$ kg/m^3.

10.45 Find an expression for the maximum vertical force F needed to lift a thin wire ring of diameter D slowly from a liquid with surface tension σ.

Vapor Pressure

10.46 Water is transported through the pipe of Fig. P10.46 such that a vacuum of 80 kPa exists at a particular location. What is the maximum possible temperature of the water? Use $p_{atm} = 92$ kPa.

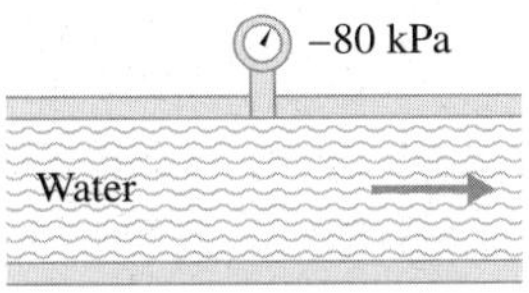

FIGURE P10.46

10.47 A group of explorers desired their elevation. An engineer boiled water and measured the temperature to be 82°C. They found a fluid mechanics book in a backpack and the engineer told the group their elevation! What elevation should the engineer have quoted?

10.48 Water is forced through a contraction causing low pressure. The water is observed to boil at a pressure of −11.5 psi. If atmospheric pressure is 14.5 psi, what is the temperature of the water?

10.49 Oil is transported through a pipeline by a series of pumps that can produce a pressure of 10 MPa in the oil leaving each pump. The losses in the pipeline cause a pressure drop of 600 kPa each kilometer. What is the maximum possible spacing of the pumps?

Ideal Gas

10.50 Determine the density and the specific gravity of air at standard conditions (i.e., 15°C and 101.3 kPa absolute).

10.51 Calculate the density of air inside a house and outside a house using 20°C inside and −25°C outside. Use an atmospheric pressure of 85 kPa. Do you think there would be a movement of air from the inside to the outside (infiltration), even without a wind? Explain.

10.52 A 15-ft^3 air tank is pressurized to 750 psia. When the temperature reaches 10°F, calculate the density and the air mass.

10.53 Estimate the weight of air contained in a classroom that measures 10 m × 20 m × 4 m. Assume reasonable values for the variables.

10.54 A car tire is pressurized to 35 psi in Michigan when the temperature is −10°F. The car is driven to Arizona, where the temperature on the highway, and in the tire, reaches 150°F. Estimate the maximum pressure in the tire.

10.55 The mass of all the air in the atmosphere contained above a 1-m^2 area is to be contained in a spherical volume. Estimate the diameter of the sphere if the air is at standard conditions.

First Law

10.56 Determine the final velocity of the 15-kg mass of Fig. P10.56 moving horizontally if it starts at 10 m/s and moves a distance of 10 m while the following net force acts in the direction of motion (where s is the distance in the direction of motion):
(a) 200 N **(b)** $20s$ N **(c)** $200 \cos s\pi/20$ N

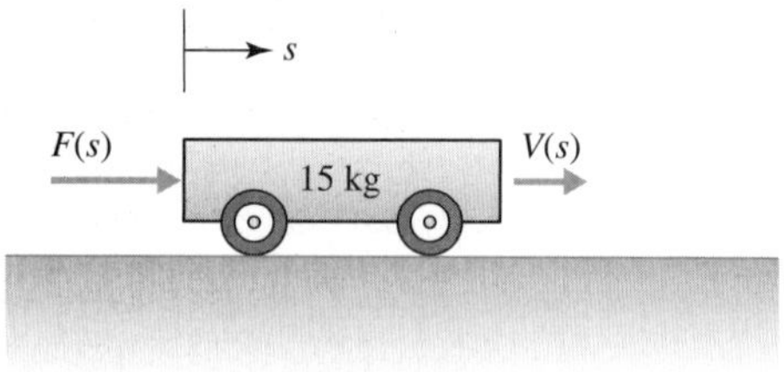

FIGURE P10.56

10.57 The 10-kg mass shown in Fig. P10.57 is traveling at 40 m/s and strikes a plunger that is attached to a piston. The piston compresses 0.2 kg of air contained in a cylinder. If the mass is brought to rest, calculate the maximum rise of temperature in the air. What effects could lead to a lower temperature rise?

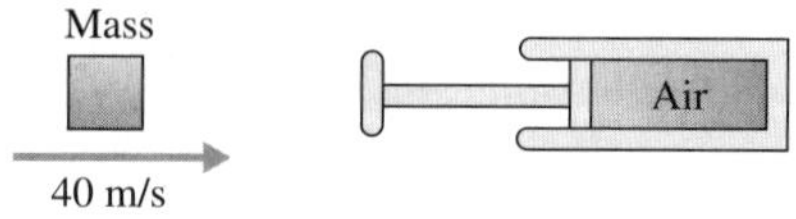

FIGURE P10.57

10.58 A 1500-kg automobile traveling at 100 km/h is suddenly grabbed by a hook and all of its kinetic energy is dissipated in a hydraulic absorber containing 2000 cm^3 of water. Calculate the maximum temperature rise in the water.

10.59 A fuel mass of 0.2 kg contains 40 MJ/kg of energy. Calculate the temperature rise of 100 kg of water if complete combustion occurs and the water, which surrounds the fuel, is completely insulated from the surroundings.

10.60 Four pounds of air is compressed in a cylinder-piston arrangement while the temperature remains constant at 70°F. If the initial pressure is 30 psi absolute, calculate the work needed to compress the air such that the absolute pressure doubles. Also, calculate the heat transfer.

10.61 Determine the heat transfer needed to double the absolute pressure in a fixed 2-m^3 volume containing air at 200 kPa absolute if the initial temperature is:
(a) 20°C **(b)** 100°C **(c)** 200°C

10.62 Heat is transferred to 2 kg of air in a cylinder so that the temperature doubles while the pressure remains constant. What work is required if the initial temperature is:
(a) 60°C? **(b)** 150°C? **(c)** 200°C?

Isentropic Flow

10.63 Air flows from a tank maintained at 5 MPa absolute and 20°C. It exits a hole and reaches a pressure of 500 kPa absolute. Assuming an adiabatic, quasi-equilibrium process, calculate the exiting temperature.

Speed of Sound

10.64 Calculate the speed of sound at 20°C in:
(a) Air **(b)** Carbon dioxide
(c) Nitrogen **(d)** Hydrogen
(e) Steam

10.65 Compare the speed of sound in the atmosphere at an elevation of 10 000 m with that at sea level by calculating a percentage decrease.

10.66 A lumberman, off in the distance, is chopping with an axe. An observer, using a digital stopwatch, measures a time of 8.32 s from the instant the axe strikes the tree until the sound is heard. How far is the observer from the lumberman if:
(a) $T = -20°C$? **(b)** $T = 20°C$?
(c) $T = 45°C$?

The Morrow Point Dam is an example of an arch-type dam. The curved wall enables large hydrostatic loads to exist on the upstream face while minimizing the required thickness of the structure. (U.S. Bureau of Reclamation)

11

Fluid Statics

Outline

Chapter Objectives

The objectives of this chapter are to:

- ▲ Establish the variation of pressure in a fluid at rest.
- ▲ Learn how to use manometers to measure pressure.
- ▲ Calculate forces on plane and curved surfaces including buoyant forces.
- ▲ Calculate pressures and forces in accelerating and rotating containers.
- ▲ Present numerous examples and problems that demonstrate how pressures and forces are calculated in fluids at rest.

Nomenclature

The following variables are introduced in this chapter:

C	centroid	F_W	weight
c.p.	center of pressure	G	center of gravity
F_B	buoyant force	H	a length
F_H	horizontal force component	I	second moment about centroid axis
F_V	vertical force component		

I_{xy}	product of inertia	x_p	distance to location of force
s	a length, subscript denoting stratosphere	$\bar{y}$	distance to centroid
		y_p	distance to location of force
S_x	unknown specific gravity		
T	tension force	α	lapse rate, angle
$\mathcal{V}_W$	volume	Ω	angular velocity
$\bar{x}$	distance to centroid		

11.1 INTRODUCTION

Fluid statics: *The study of fluids in which there is no relative motion between fluid particles.*

Key Concept: *The only stress that exists where there is no motion is pressure.*

Fluid statics is the study of fluids in which there is no relative motion between fluid particles. If there is no relative motion, no shearing stresses exist, since velocity gradients, such as du/dy, are required for shearing stresses to be present. The only stress that exists is a normal stress, the pressure, so it is the pressure that is of primary interest in fluid statics.

Three situations, depicted in Fig. 11.1, involving fluid statics will be investigated. These include fluids at rest, such as water pushing against a dam, fluids contained in devices that undergo linear acceleration, and fluids contained in rotating cylinders. In each of these three situations the fluid is in static equilibrium with respect to a reference frame attached to the boundary surrounding the fluid. In addition to the examples shown for fluids at rest, we consider instruments called manometers, and investigate the forces of buoyancy.

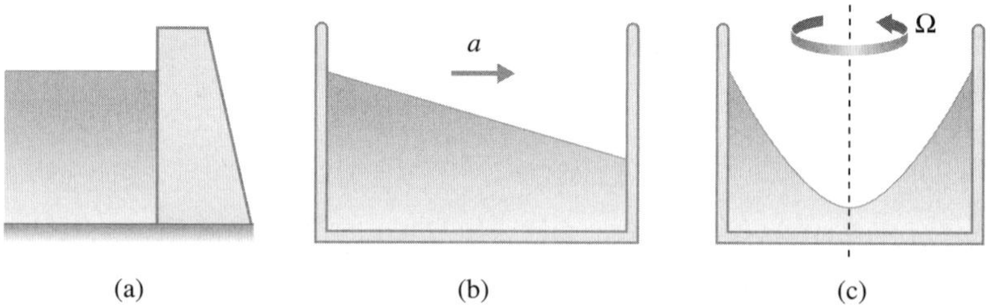

FIGURE 11.1 Examples included in fluid statics: (a) liquids at rest; (b) linear acceleration; (c) angular rotation.

11.2 PRESSURE AT A POINT

We have defined pressure as being the infinitesimal normal compressive force divided by the infinitesimal area over which it acts. This defines the pressure at a point. One might question whether the pressure at a given point varies as the normal to the area changes direction. To show that this is not the case, even for fluids in motion with no shear, consider the wedge-shaped element of unit depth (in the z-direction) shown in Fig. 11.2. Assume that a pressure p acts on the hypotenuse and that a different pressure acts on each of the other areas, as shown. Since the forces on the two end faces are in the z-direction, we have not included them on the element. Now, let us apply Newton's second law (Eq. 10.4) to the element, for both the x- and y-directions:

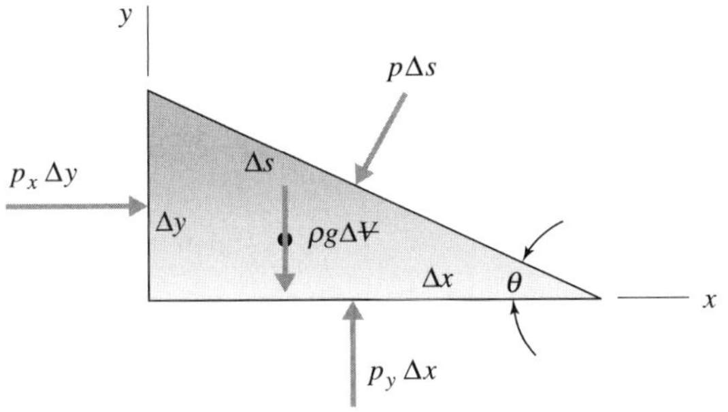

FIGURE 11.2 Pressure at a point in a fluid.

$$\sum F_x = ma_x: \qquad p_x\Delta y - p\Delta s \sin\theta = \rho\frac{\Delta x\Delta y}{2}a_x$$
$$\sum F_y = ma_y: \quad p_y\Delta x - \rho g\frac{\Delta x\Delta y}{2} - p\Delta s\cos\theta = \rho\frac{\Delta x\Delta y}{2}a_y \tag{11.1}$$

where we have used $\Delta V = \Delta x\Delta y/2$ (we could include Δz in each term to account for the depth). The pressures shown are due to the surrounding fluid and are the average pressure on the areas. Substituting

$$\Delta s\sin\theta = \Delta y \qquad \Delta s\cos\theta = \Delta x \tag{11.2}$$

we see that Eqs. 11.1 take the forms

$$p_x - p = \frac{\rho a_x\Delta x}{2}$$
$$p_y - p = \frac{\rho(a_y + g)\Delta y}{2} \tag{11.3}$$

Note that in the limit as the element shrinks to a point, $\Delta x \to 0$ and $\Delta y \to 0$. Hence, the right-hand sides in the equations above go to zero, even for fluids in motion, providing us with the result that, at a point,

$$p_x = p_y = p \tag{11.4}$$

Since θ is arbitrary, this relationship holds for all angles at a point. We could have analyzed an element in the xz-plane and concluded that $p_x = p_z = p$. Thus, we conclude that the pressure in a fluid is constant at a point; that is, pressure is a scalar function. It acts equally in all directions at a given point for both a static fluid and a fluid that is in motion in the absence of shear stress.

Key Concept: *Pressure in a fluid acts equally in all directions at a given point in the absence of shear stress.*

11.3 PRESSURE VARIATION

A general equation is derived to predict the pressure variation of fluids at rest or fluids undergoing an acceleration, while the relative position of fluid elements to one another remains the same (this eliminates shear stress). To determine the pressure variation in such fluids, consider the infinitesimal element displayed in

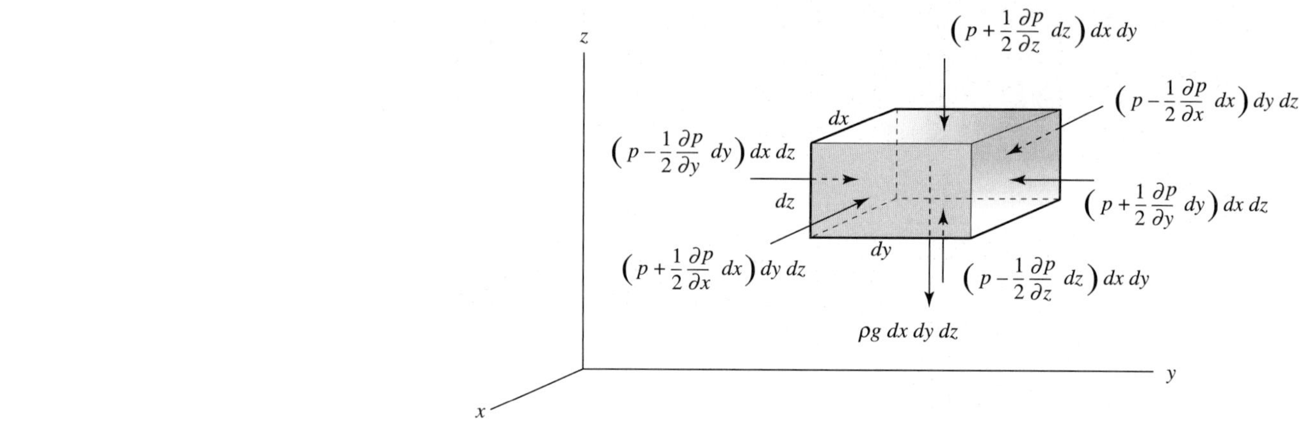

FIGURE 11.3 Forces acting on an infinitesimal element that is at rest in the *xyz*-reference frame. The reference frame may be accelerating or rotating.

Fig. 11.3, where the z-axis is in the vertical direction. The pressure variation from one point to another will be determined by applying Newton's second law; that is, the sum of the forces acting on the fluid element is equal to the mass times the acceleration of the element.

If we assume that a pressure p exists at the center of this element, the pressures at each of the sides can be expressed by using the chain rule from calculus with $p(x, y, z)$:

$$dp = \frac{\partial p}{\partial x}dx + \frac{\partial p}{\partial y}dy + \frac{\partial p}{\partial z}dz \tag{11.5}$$

If we move from the center to a face a distance $(dx/2)$ away, we see that the pressure is

$$p\left(x+\frac{dx}{2}, y, z\right) = p(x, y, z) + \frac{\partial p}{\partial x}\frac{dx}{2} \tag{11.6}$$

The pressures at all faces are expressed in this manner, as shown in Fig. 11.3. Newton's second law can be applied to the element of Fig. 11.3, resulting in the three component equations, assuming z to be vertical and using the mass as $\rho\,dx\,dy\,dz$,

$$\begin{aligned}
\textstyle\sum F_x = ma_x: &\quad -\frac{\partial p}{\partial x}dx\,dy\,dz = \rho a_x\,dx\,dy\,dz \\
\textstyle\sum F_y = ma_y: &\quad -\frac{\partial p}{\partial y}dx\,dy\,dz = \rho a_y\,dx\,dy\,dz \\
\textstyle\sum F_z = ma_z: &\quad -\frac{\partial p}{\partial z}dx\,dy\,dz = \rho(a_z+g)\,dx\,dy\,dz
\end{aligned} \tag{11.7}$$

Division by the element's volume $dx\,dy\,dz$ yields

$$\frac{\partial p}{\partial x} = -\rho a_x$$
$$\frac{\partial p}{\partial y} = -\rho a_y \qquad (11.8)$$
$$\frac{\partial p}{\partial z} = -\rho(a_z + g)$$

The pressure differential in any direction can now be determined from Eq. 11.5 as

$$dp = -\rho a_x\,dx - \rho a_y\,dy - \rho(a_z + g)\,dz \qquad (11.9)$$

where z is always vertical. Pressure differences between specified points can be found by integrating Eq. 11.9. This equation is useful in a variety of problems, as will be demonstrated in the remaining sections of this chapter.

11.4 FLUIDS AT REST

A fluid at rest does not undergo any acceleration. Therefore, set $a_x = a_y = a_z = 0$ and Eq. 11.9 reduces to

$$dp = -\rho g\,dz \qquad (11.10)$$

or

$$\frac{dp}{dz} = -\gamma \qquad (11.11)$$

This equation implies that there is no pressure variation in the x- and y-directions, that is, in the horizontal plane. The pressure varies in the z-direction only. Also note that dp is negative if dz is positive; that is, the pressure decreases as we move up and increases as we move down (a rather obvious result).

11.4.1 Pressures in Liquids at Rest

If the density can be assumed constant, Eq. 11.11 is integrated to yield

$$\Delta p = -\gamma\Delta z \quad \text{or} \quad p + \gamma z = \text{constant} \quad \text{or} \quad \frac{p}{\gamma} + z = \text{constant} \qquad (11.12)$$

so that pressure increases with depth. Note that z is positive in the upward direction. If the point of interest were a distance h below a **free surface** (a surface separating a gas from a liquid), as shown in Fig. 11.4, Eq. 11.12 would result in

$$p = \gamma h \qquad (11.13)$$

Free surface: *A surface separating a gas from a liquid.*

Key Concept: *The equation $p = \gamma h$ is used to convert pressure to a height of liquid.*

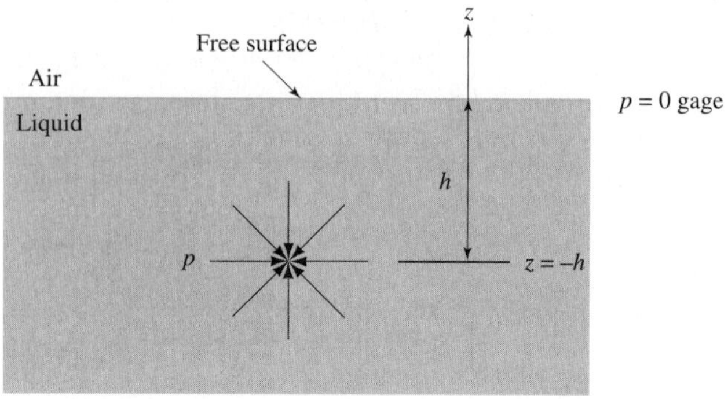

FIGURE 11.4 Pressure below a free surface.

where $p = 0$ at $h = 0$. This equation will be quite useful in converting pressure to an equivalent height of liquid. For example, atmospheric pressure is often expressed as millimeters of mercury; that is, the atmospheric pressure is equal to the pressure at a certain depth in a mercury column. By knowing the specific weight of mercury, we can then determine that depth using Eq. 11.13.

11.4.2 Pressures in the Atmosphere

Standard atmosphere: *Position at 40° latitude where calculations are standardized.*

For the atmosphere where the density depends on height [i.e., $\rho = \rho(z)$], we must integrate Eq. 11.10 along a vertical path. The atmosphere is divided into four layers: the *troposphere* (nearest Earth), the *stratosphere*, the *ionosphere*, and the *exosphere*. Because conditions change with time and latitude in the atmosphere with the layers being thicker at the equator and thinner at the poles, we base calculations on the **standard atmosphere**, which is at 40° latitude. In the standard atmosphere, the temperature in the troposphere varies linearly with elevation, $T(z) = T_0 - \alpha z$, where the *lapse rate* $\alpha = 0.0065$ K/m (0.00357°R/ft), and T_0 is 288 K (518°R). In the part of the stratosphere between 11 and 20 km, the temperature is constant at −56.5°C. (The lower part of this constant-temperature region is where commercial aircraft usually fly.) The temperature then increases again and reaches a maximum near 50 km; it then decreases to the edge of the ionosphere. The standard atmosphere is sketched in Fig. 11.5. Because the density of the air in the ionosphere and the exosphere is so low, it is possible for satellites to orbit the earth in either of these layers.

To determine the pressure variation of the troposphere, we can use the ideal-gas law $p = \rho RT$ and Eq. 11.10; there results

$$dp = -\frac{pg}{RT}dz \tag{11.14}$$

or

$$\frac{dp}{p} = -\frac{g}{RT}dz \tag{11.15}$$

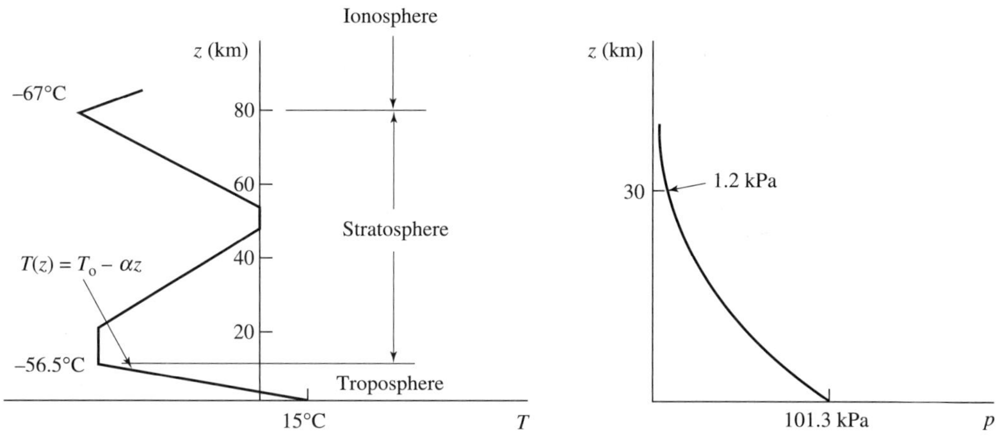

FIGURE 11.5 Standard atmosphere.

This can be integrated, between sea level and an elevation z:

$$\int_{p_{atm}}^{p} \frac{dp}{p} = -\frac{g}{R}\int_{0}^{z} \frac{dz}{T_0 - \alpha z} \tag{11.16}$$

Upon integration this gives

$$\ln \frac{p}{p_{atm}} = \frac{g}{\alpha R} \ln \frac{T_0 - \alpha z}{T_0} \tag{11.17}$$

which can be put in the form

$$p = p_{atm}\left(\frac{T_0 - \alpha z}{T_0}\right)^{g/\alpha R} \tag{11.18}$$

If we use standard atmospheric conditions in Eq. 11.18, we find that $p/p_{atm} = 0.999$ at $z = 10$ m. Consequently, we ignore changes in pressure in a gas such as air unless z is relatively large.

In that part of the stratosphere, where the temperature is constant, Eq. 11.15 is integrated again as follows:

$$\int_{p_s}^{p} \frac{dp}{p} = -\frac{g}{RT_s}\int_{z_s}^{z} dz \tag{11.19}$$

$$\ln \frac{p}{p_s} = -\frac{g}{RT_s}(z - z_s) \tag{11.20}$$

or

$$p = p_s \exp\left[\frac{g}{RT_s}(z_s - z)\right] \tag{11.21}$$

The subscript s denotes conditions at the troposphere–stratosphere interface. Properties of the standard atmosphere up to 80 km are listed in Appendix B.1.

Example 11.1

The atmospheric pressure is given as 680-mm Hg at a mountain location. Convert this to kilopascals and meters of water. Also, calculate the pressure decrease due to a 500-m elevation increase, starting at 2000-m elevation, assuming constant density.

Solution

Use Eq. 11.13 and find, using $S_{Hg} = 13.6$ with Eq. 10.15,

$$\begin{aligned} p &= \gamma_{Hg} h \\ &= (9810 \times 13.6) \times 0.680 = 90\,700 \text{ Pa} \quad \text{or} \quad 90.7 \text{ kPa} \end{aligned}$$

To convert this to meters of water, we have

$$\begin{aligned} h &= \frac{p}{\gamma_{H_2O}} \\ &= \frac{90\,700}{9810} = 9.25 \text{ m of water} \end{aligned}$$

To find the pressure decrease, we use Eq. 11.12 and find the density in Appendix Table B.1:

$$\begin{aligned} \Delta p &= -\gamma \Delta z = -\rho g \Delta z \\ &= -1.007 \times 9.81 \times 500 = -4940 \text{ Pa} \end{aligned}$$

Note: Since gravity is known to three significant digits, we express the answer to three significant digits.

Example 11.2

Assume an isothermal atmosphere and approximate the pressure at 10 000 m. Calculate the percent error when compared with the values using Eq. 11.18 and from Appendix B.1. Use a temperature of 256 K (the temperature at 5000 m).

Solution

Integrate Eq. 11.15, assuming that T is constant, as follows:

$$\int_{101}^{p} \frac{dp}{p} = -\frac{g}{RT} \int_{0}^{z} dz$$

$$\ln \frac{p}{101} = -\frac{gz}{RT} \quad \text{or} \quad p = 101 e^{-gz/RT}$$

Substituting $z = 10\,000$ m and $T = 256$ K, there results

$$\begin{aligned} p &= 101 e^{-9.81 \times 10\,000/(287 \times 256)} \\ &= 26.57 \text{ kPa} \end{aligned}$$

(continued)

Using Eq. 11.18, we have

$$p = p_{atm}\left(\frac{T_0 - \alpha z}{T_0}\right)^{g/\alpha R}$$

$$= 101\left(\frac{288 - 0.0065 \times 10\,000}{288}\right)^{9.81/0.0065\times 287} = 26.3 \text{ kPa}$$

The actual pressure at 10 000 m is found from Table B.1 to be 26.50 kPa. Hence, the percent errors are

$$\% \text{ error} = \left(\frac{26.57 - 26.3}{26.3}\right) \times 100 = 1.03\%$$

$$\% \text{ error} = \left(\frac{26.57 - 26.50}{26.50}\right) \times 100 = 0.26\%$$

Because the error is so small, we often assume the atmosphere to be isothermal. *Note:* When evaluating gz/RT, we use $R = 287$ J/kg·K not 0.287 kJ/kg·K. To observe that gz/R is dimensionless, use N = kg·m/s^2 so that

$$\left[\frac{gz}{RT}\right] = \frac{(\text{m/s}^2)\text{m}}{(\text{J/kg·K})\text{K}} = \frac{\text{m}^2/\text{s}^2}{\text{N·m/kg}} = \frac{\text{m}^2/\text{s}^2}{(\text{kg·m}^2/\text{s}^2)/\text{kg}} = \frac{\text{m}^2/\text{s}}{\text{m}^2/\text{s}}$$

11.4.3 Manometers

Manometers are instruments that use columns of liquids to measure pressures. Two such instruments, shown in Fig. 11.6, are discussed to illustrate their use. Fig. 11.6a displays a U-tube manometer used to measure relatively small pressures. In this case, the pressure in the pipe can be determined by defining a point 1 at the center of the pipe and a point 2 at the surface of the right column. Then, using Eq. 11.12,

$$p_1 + \gamma z_1 = p_2 + \gamma z_2 \tag{11.22}$$

where the datum from which z_1 and z_2 are measured is located at any desired position, such as through point 1. Since $p_2 = 0$ (gage pressure is selected; if absolute pressure is desired, we would select $p_2 = p_{atm}$) and $z_2 - z_1 = h$,

$$p_1 = \gamma h \tag{11.23}$$

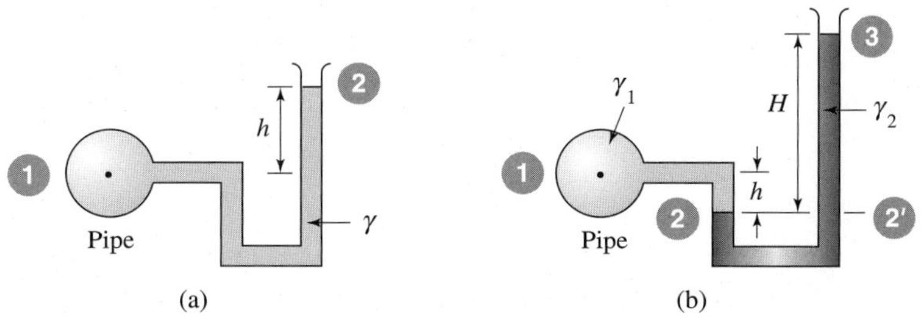

FIGURE 11.6 Manometers: (a) U-tube manometer (small pressures); (b) U-tube manometer (large pressures).

Figure 11.6b shows a U-tube manometer used to measure relatively large pressures, since we can select γ_2 to be quite large. For example, we could select γ_2 to be that of mercury so that $\gamma_2 = 13.6\ \gamma_{\text{water}}$. The pressure can be determined by introducing the points indicated. This is necessary because Eq. 11.12 applies throughout one fluid; γ must be constant. The value of γ changes abruptly at point 2. The pressure at point 2 and at point 2′ is the same, since the points are at the same elevation in the same fluid. Hence,

$$p_2 = p_2'$$
$$p_1 + \gamma_1 h = p_3 + \gamma_2 H \tag{11.24}$$

Setting $p_3 = 0$ (gage pressure is used) results in

$$p_1 = -\gamma_1 h + \gamma_2 H \tag{11.25}$$

Note that in the equations above, for the two manometers, we have identified all interfaces with a point. This is always necessary when analyzing a manometer.

Example 11.3

Water and oil flow in horizontal pipelines. A double U-tube manometer is connected between the pipelines, as shown in Fig. E11.3. Calculate the pressure difference between the water pipe and the oil pipe.

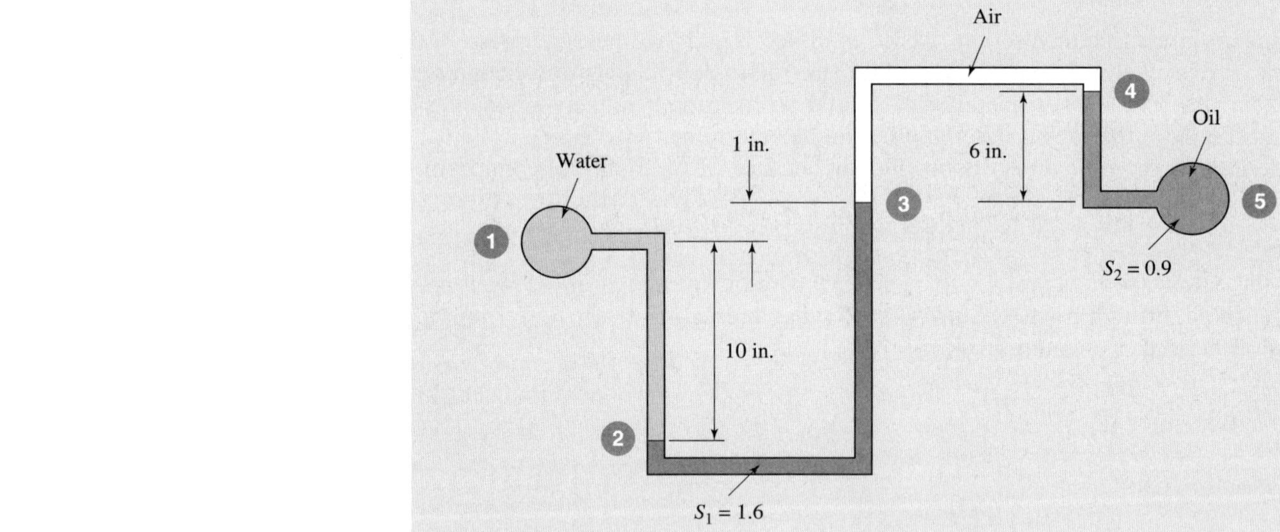

FIGURE E11.3

Solution

We first identify the relevant points as shown in the figure. Begin at point 1 and add pressure when the elevation decreases and subtract pressure when the elevation increases, until point 5 is reached:

(continued)

$$p_1 + \gamma(z_1 - z_2) - \gamma S_1(z_3 - z_2) - \gamma S_{\text{air}}(z_4 - z_3) + \gamma S_2(z_4 - z_5) = p_5$$

where $\gamma = 62.4 \text{ lb/ft}^3$, $S_1 = 1.6$, $S_2 = 0.9$, and $S_{\text{air}} \approx 0$. Thus

$$p_1 - p_5 = 62.4\left(-\frac{10}{12} + 1.6 \times \frac{11}{12} + 0 \times \frac{6}{12} - 0.9 \times \frac{6}{12}\right)$$

$$= 11.44 \text{ lb/ft}^2 \quad \text{or} \quad 0.0794 \text{ psi}$$

Note that by neglecting the weight of the air, the pressure at point 3 is equal to the pressure at point 4.

11.4.4 Forces on Plane Areas

In the design of devices and objects that are submerged (e.g., dams, flow obstructions, surfaces on ships, and holding tanks), it is necessary to calculate the magnitudes and locations of forces that act on both plane and curved surfaces. In this section, we consider only plane surfaces, such as the plane surface of general shape shown in Fig. 11.7. Note that a side view is given as well as a view showing the shape of the plane. The total force of the liquid on the plane surface is found by integrating the pressure over the area; that is,

$$F = \int_A p\, dA \tag{11.26}$$

where we usually use gage pressure. Atmospheric pressure cancels out since it acts on both sides of the area. The x and y coordinates are in the plane of the plane

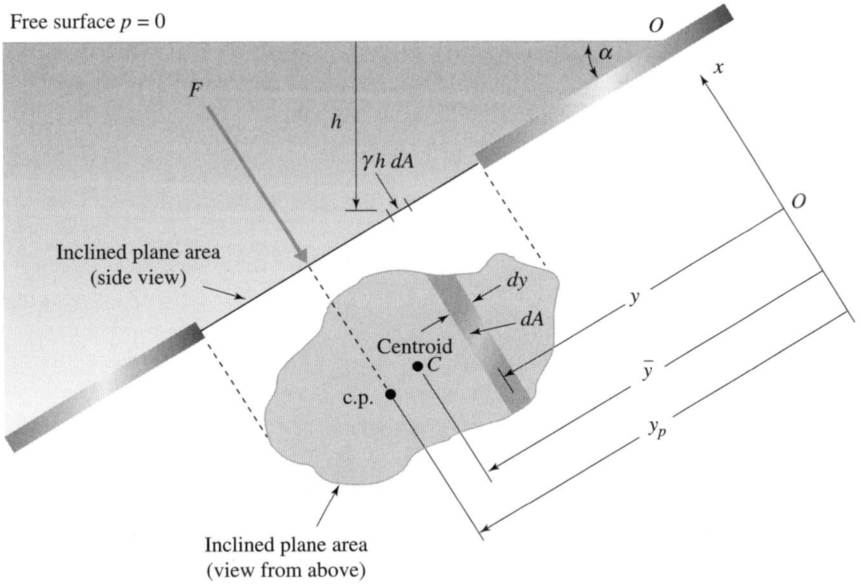

FIGURE 11.7 Force on an inclined plane area.

surface, as shown. Assuming that $p = 0$ at $h = 0$, we know that

$$\begin{aligned} p &= \gamma h \\ &= \gamma y \sin \alpha \end{aligned} \qquad (11.27)$$

where h is measured vertically down from the free surface to the elemental area dA and y is measured from point O on the free surface. The force may then be expressed as

$$\begin{aligned} F &= \int_A \gamma h \, dA \\ &= \gamma \sin \alpha \int_A y \, dA \end{aligned} \qquad (11.28)$$

The distance to a centroid is defined as

$$\bar{y} = \frac{1}{A} \int_A y \, dA \qquad (11.29)$$

The expression for the force then becomes

$$\begin{aligned} F &= \gamma \bar{y} A \sin \alpha \\ &= \gamma \bar{h} A = p_C A \end{aligned} \qquad (11.30)$$

Key Concept: *The force on a plane surface is the pressure at the centroid multiplied by the area.*

where $\bar{h}$ is the vertical distance from the free surface to the centroid of the area and p_C is the pressure at the centroid. Thus, we see that the magnitude of the force on a plane surface is the pressure at the centroid multiplied by the area. The force does not, in general, act at the centroid.

Center of pressure: *The point where the resultant force acts.*

To find the location of the resultant force F, we note that the sum of the moments of all the infinitesimal pressure forces acting on the area A must equal the moment of the resultant force. Let the force F act at the point (x_p, y_p), which is the **center of pressure** (c.p.). The value of y_p can be obtained by equating moments about the x-axis:

$$\begin{aligned} y_p F &= \int_A y p \, dA \\ &= \gamma \sin \alpha \int_A y^2 \, dA = \gamma I_x \sin \alpha \end{aligned} \qquad (11.31)$$

where the second moment of the area about the x-axis is

$$I_x = \int_A y^2 \, dA \qquad (11.32)$$

The second moment of an area is related to the second moment of an area $\bar{I}$ about the centroidal axis by the parallel-axis-transfer theorem,

$$I_x = \bar{I} + A\bar{y}^2 \qquad (11.33)$$

Substitute Eqs. 11.30 and 11.33 into Eq. 11.31, and obtain

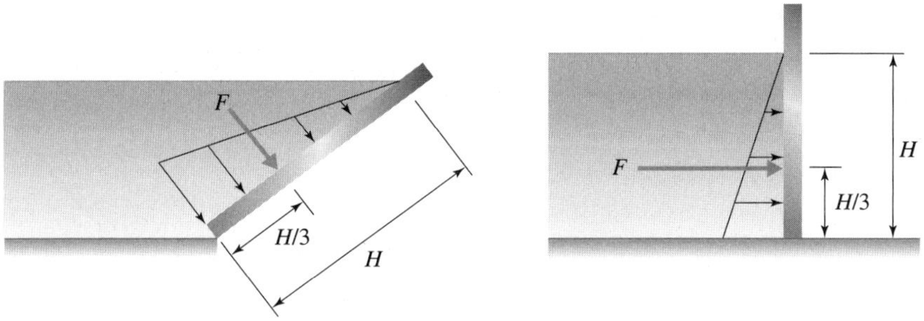

FIGURE 11.8 Force on a plane area with the top edge in a free surface.

$$y_p = \frac{\gamma(\bar{I} + A\bar{y}^2)\sin\alpha}{\gamma\bar{y}A\sin\alpha} = \bar{y} + \frac{\bar{I}}{A\bar{y}} \tag{11.34}$$

where $\bar{y}$ is measured parallel to the plane area along the y-axis.

Centroids and moments for several areas are presented in Appendix J. Using the expression above, we can show that the force on a rectangular gate, with the top edge even with the liquid surface as shown in Fig. 11.8, acts two-thirds of the way down. This is also obvious considering the triangular pressure distribution acting on the gate. Note that Eq. 11.34 shows that y_p is always greater than $\bar{y}$; that is, the resultant force of the liquid on a plane surface always acts below the centroid of the area (except on a horizontal area for which $\bar{y} = \infty$; then the center of pressure and the centroid coincide).

Key Concept: *The force on a rectangular gate, with the top edge even with the liquid surface, acts two-thirds of the way down.*

Similarly, to locate the x-coordinate x_p of the c.p., we write

$$x_pF = \int_A xp\,dA = \gamma\sin\alpha\int_A xy\,dA = \gamma I_{xy}\sin\alpha \tag{11.35}$$

where the product of inertia of the area A is

$$I_{xy} = \int_A xy\,dA \tag{11.36}$$

Using the transfer theorem for the product of inertia,

$$I_{xy} = \bar{I}_{xy} + A\bar{x}\bar{y} \tag{11.37}$$

Equation 11.35 becomes

$$x_p = \bar{x} + \frac{\bar{I}_{xy}}{A\bar{y}} \tag{11.38}$$

We now have expressions for the coordinates locating the center of pressure.

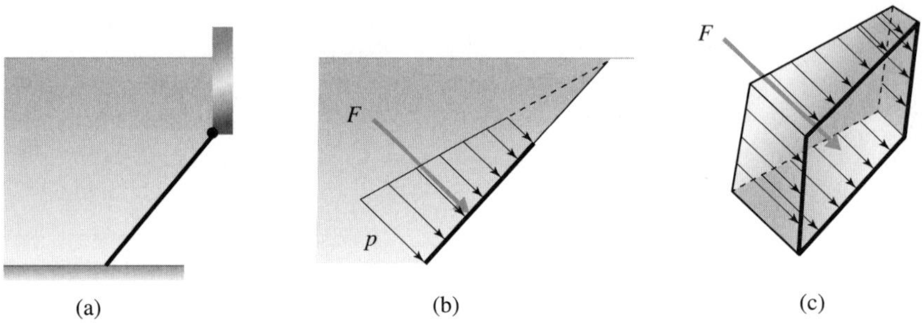

FIGURE 11.9 Pressure prism: (a) rectangular area; (b) pressure distribution on the area; (c) pressure prism.

Finally, we should note that the force F in Fig. 11.7 is the result of a *pressure prism* acting on the area. For the rectangular area shown in Fig. 11.9, the pressure increases, as shown by the pressure distribution in Fig. 11.9b. If we form the integral $\int p\,dA$, we obtain the volume of the pressure prism, which equals the force F acting on the area, shown in Fig. 11.9c. The force acts through the centroid of the volume. For the rectangular area shown in Fig. 11.9a, the volume could be divided into two volumes: a rectangular volume with centroid at its center, and a triangular volume with centroid one third the distance from the appropriate base. The location of the force is then found by locating the centroid of the composite volume.

Example 11.4

A plane area of 80 cm × 80 cm acts as a window on a submersible in the Great Lakes. If it is on a 45° angle with the horizontal, what force applied normally to the window at the bottom edge is needed to just open the window, if it is hinged at the top edge when the top edge is 10 m below the surface? The pressure inside the submersible is assumed to be atmospheric.

Solution

First, a sketch of the window would be very helpful, as in Fig. E11.4. The force of the water acting on the window is

$$F = \gamma \bar{h} A$$
$$= 9810(10 + 0.4 \times \sin 45°)(0.8 \times 0.8) = 64\,560 \text{ N}$$

The distance $\bar{y}$ is

$$\bar{y} = \frac{\bar{h}}{\sin 45°} = \frac{10 + 0.4 \times \sin 45°}{\sin 45°} = 14.542 \text{ m}$$

so that

(continued)

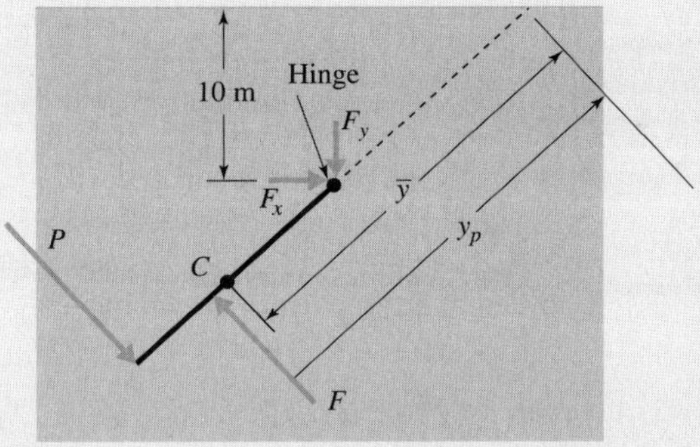

FIGURE E11.4

$$y_p = \bar{y} + \frac{\bar{I}}{A\bar{y}}$$

$$= 14.542 + \frac{0.8 \times 0.8^3/12}{(0.8 \times 0.8) \times 14.542} = 14.546 \text{ m}$$

Taking moments about the hinge provides the needed force P to open the window:

$$0.8P = (y_p - \bar{y} + 0.4)F$$

$$\therefore P = \frac{14.546 - 14.542 + 0.4}{0.8} \times 64\,560 = 32\,610 \text{ N}$$

Alternatively, we could have sketched the pressure prism, composed of a rectangular volume and a triangular volume. Moments about the top hinge would provide the desired force.

Example 11.5

Find the location of the resultant force F of the water on the triangular gate and the force P necessary to hold the gate in the position shown in Fig. E11.5a.

Solution

First we draw a free-body diagram of the gate, including all of the forces acting on the gate (Fig. E11.5c). We have neglected the weight of the gate. The centroid of the gate is shown in Fig. E11.5b. The y-coordinate of the location of the resultant F can be found using Eq. 11.34 as follows:

$$\bar{y} = 2 + 5 = 7$$

$$y_p = \bar{y} + \frac{\bar{I}}{A\bar{y}}$$

$$= 7 + \frac{2 \times 3^3/36}{3 \times 7} = 7.071 \text{ m}$$

(continued)

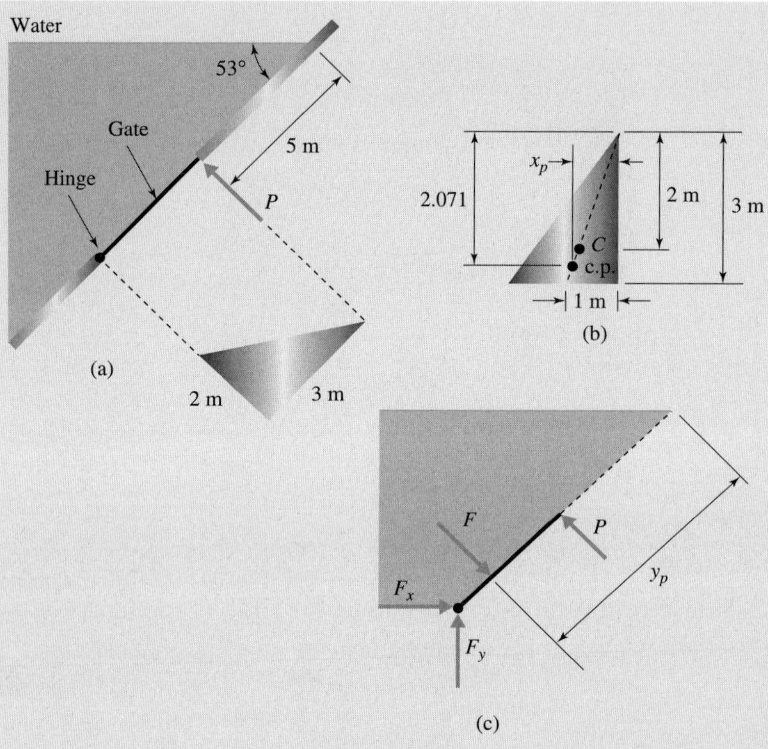

FIGURE E11.5

To find x_p, we could use Eq. 11.38. Rather than that, we recognize that the resultant force must act on a line connecting the vertex and the midpoint of the opposite side, since each infinitesimal force acts on this line (the moment of the resultant must equal the moment of its components). Thus, using similar triangles we have

$$\frac{x_p}{1} = \frac{2.071}{3}$$

$$\therefore\ x_p = 0.690 \text{ m}$$

The coordinates x_p and y_p locate where the force due to the water acts on the gate.

If we take moments about the hinge, assumed to be frictionless, we can determine the force P necessary to hold the gate in the position shown:

$$\sum M_{\text{hinge}} = 0$$

$$\therefore\ 3 \times P = (3 - 2.071)F$$

$$= 0.929 \times \gamma \bar{h} A$$

$$= 0.929 \times 9810 \times (7 \sin 53°) \times 3$$

where $\bar{h}$ is the vertical distance from the centroid to the free surface. Hence

$$P = 50\,900 \text{ N} \quad \text{or} \quad 50.9 \text{ kN}$$

11.4.5 Forces on Curved Surfaces

We do not use a direct method of integration to find the force due to the hydrostatic pressure on a curved surface. Rather, a free-body diagram that contains the curved surface and the liquids directly above or below the curved surface is identified. Such a free-body diagram contains only plane surfaces upon which unknown fluid forces act; these unknown forces can be found using methods as in the preceding section.

As an example, let us determine the force of the curved gate on the stop, as shown in Fig. 11.10a. The free-body diagram, which includes the gate and some of the water contained directly above the gate, is shown in Fig. 11.10b. The forces F_x and F_y are the horizontal and vertical components, respectively, of the force acting on the hinge. F_1 and F_2 are due to the surrounding water and are the resultant forces of the pressure distributions shown. The body force F_W is due to the weight of the water shown. By summing moments about an axis passing through the hinge, we can determine the force P acting on the stop.

If the curved surface is a quarter circle, the problem can be greatly simplified. This is observed by considering a free-body diagram of the gate only (see Fig. 11.10c). The horizontal force F_H acting on the gate is equal to F_1 of Fig. 11.10b, and the component F_V is equal to the combined force $F_2 + F_W$ of Figure 11.10b. Now, F_H and F_V are due to the differential pressure forces acting on the circular arc; each differential pressure force acts through the center of the circular arc. Hence, the resultant force $\mathbf{F}_H + \mathbf{F}_V$ (this is a vector addition) must act through the center. Consequently, we can locate the components F_H and F_V at the center of the quarter circle, resulting in a much simpler problem. Example 11.6 will illustrate.

Key Concept: *The resultant force $\mathbf{F}_H + \mathbf{F}_V$ must act through the center of the circular arc.*

If the pressure on the free surface is p_0, we can simply add a depth of liquid necessary to provide p_0 at the location of the free surface, and then work the resulting problem, with a fictitious free surface located the appropriate distance above the original free surface. Or, the pressure force p_0A is added to the force F_2 of Fig. 11.10b.

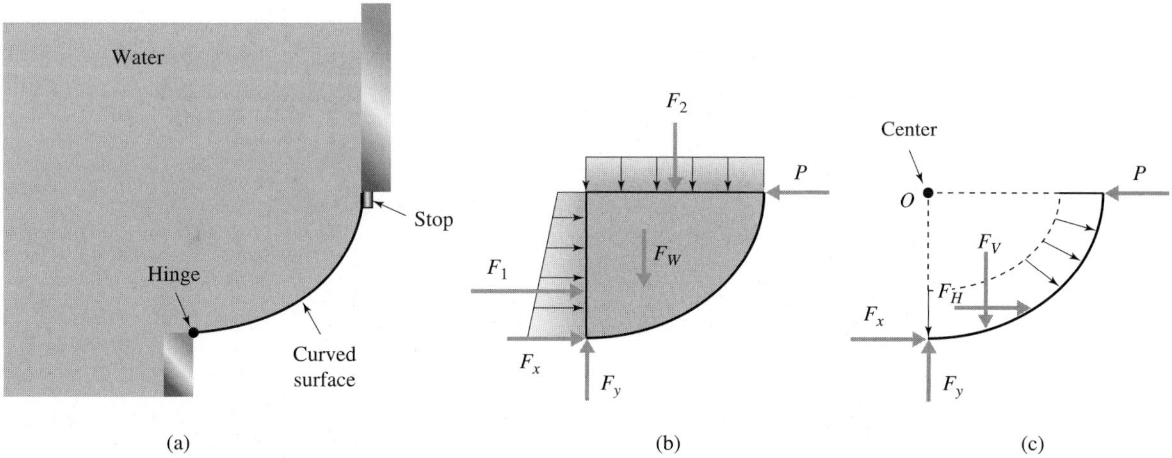

FIGURE 11.10 Forces acting on a curved surface: (a) curved surface; (b) free-body diagram of water and gate; (c) free-body diagram of gate only.

Example 11.6

Calculate the force P necessary to hold the 4-m-wide gate in the position shown in Fig. E11.6a. Neglect the weight of the gate.

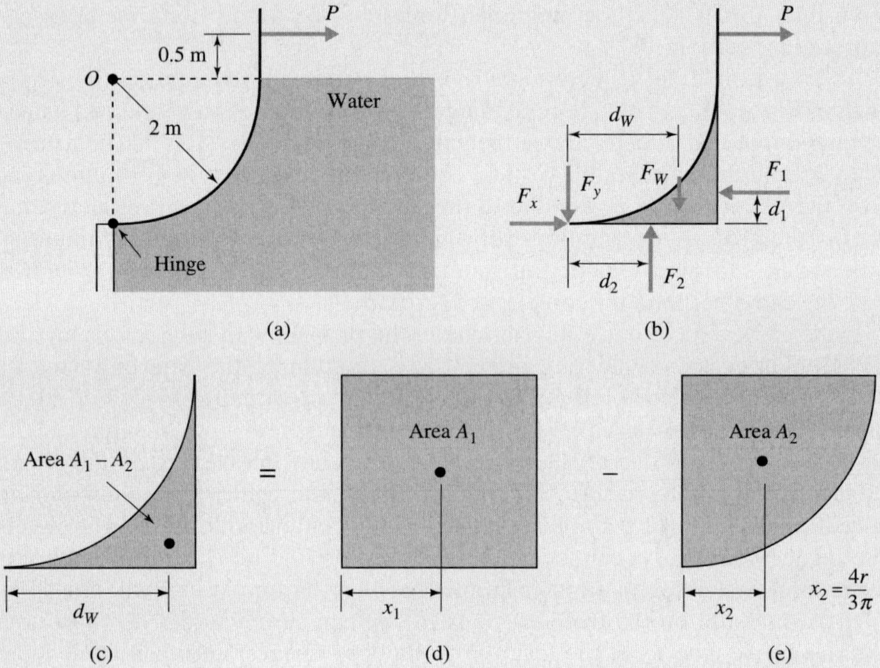

FIGURE E11.6

Solution

The first step is to draw a free-body diagram of the gate. One choice is to select the gate and the water directly below the gate as shown in Fig. E11.6b. To calculate P, we must determine F_1, F_2, F_W, d_1, d_2, and d_W; then moments about the hinge will allow us to find P. The force components are given by

$$\begin{aligned} F_1 &= \gamma \bar{h}_1 A_1 \\ &= 9810 \times 1 \times 8 = 78\,480 \text{ N} \\ F_2 &= \gamma \bar{h}_2 A_2 \\ &= 9810 \times 2 \times 8 = 156\,960 \text{ N} \\ F_W &= \gamma \mathcal{V}_{\text{water}} \\ &= 9810 \times 4\left(4 - \frac{\pi \times 2^2}{4}\right) = 33\,700 \text{ N} \end{aligned}$$

(continued)

The distance d_W is the distance to the centroid of the volume. It can be determined by considering the area as the difference of a square and a quarter circle as shown in Fig. E11.6c–e. Moments of areas yield

$$d_W(A_1 - A_2) = x_1A_1 - x_2A_2$$

$$d_W = \frac{x_1A_1 - x_2A_2}{A_1 - A_2}$$

$$= \frac{1 \times 4 - (4 \times 2/3\pi) \times \pi}{4 - \pi} = 1.553 \text{ m}$$

The distance $d_2 = 1$ m. Because F_1 is due to a triangular pressure distribution (see Fig. 11.8), d_1 is given by

$$d_1 = \frac{1}{3}(2) = 0.667 \text{ m}$$

Summing moments about the frictionless hinge gives

$$2.5P = d_1F_1 + d_2F_2 - d_WF_W$$

$$P = \frac{0.667 \times 78.5 + 1 \times 157.0 - 1.553 \times 33.7}{2.5} = 62.8 \text{ kN}$$

Rather than the somewhat tedious procedure above, we could observe that all the infinitesimal forces that make up the resultant force $(\mathbf{F}_H + \mathbf{F}_V)$ acting on the circular arc pass through the center O, as noted in Fig. 11.10c. Since each infinitesimal force passes through the center, the resultant force must also pass through the center. Hence we could have located the resultant force $(\mathbf{F}_H + \mathbf{F}_V)$ at point O. If F_V and F_H were located at O, F_V would pass through the hinge, producing no moment about the hinge. Then, realizing that $F_H = F_1$ and summing moments about the hinge gives

$$2.5P = 2F_H$$

Therefore,

$$P = 2 \times \frac{78.48}{2.5} = 62.8 \text{ kN}$$

This was obviously much simpler. All we needed to do was calculate F_H and then sum moments!

Example 11.7

Find the force P needed to hold the gate in the position shown in Fig. E11.7a if P acts 3 m from the y-axis. The parabolic gate is 150-cm wide.

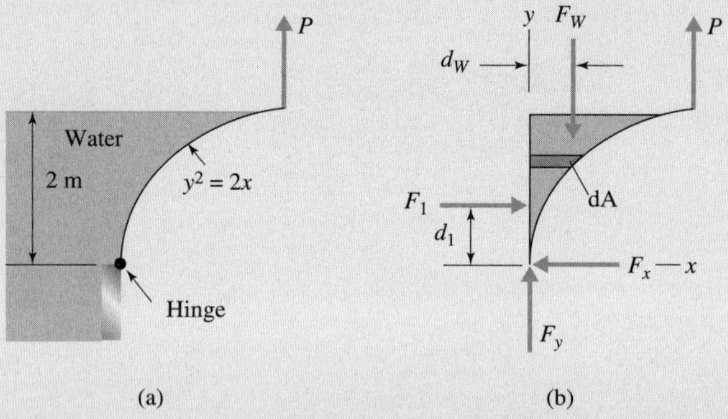

FIGURE E11.7

Solution

A free-body diagram of the gate and the water directly above the gate is shown in Fig. E11.7b. The forces are found to be

$$F_1 = \gamma \bar{h} A$$
$$= 9810 \times 1 \times (2 \times 1.5) = 29\,430 \text{ N}$$
$$F_W = \gamma \forall$$
$$= 9810 \int_0^2 1.5x\,dy = 14\,715 \int_0^2 \frac{y^2}{2}\,dy = 14\,715 \frac{2^3}{6} = 19\,620 \text{ N}$$

The distance d_1 is $\frac{1}{3}(2) = 0.667$ m since the top edge is in the free surface. The distance d_W through the centroid is found using a horizontal strip:

$$d_W = \frac{\int_0^2 x(x/2)\,dy}{\int_0^2 x\,dy} = \frac{\frac{1}{8}\int_0^2 y^4\,dy}{\frac{1}{2}\int_0^2 y^2\,dy} = \frac{1}{4}\frac{2^5/5}{2^3/3} = 0.6 \text{ m}$$

Sum moments about the hinge and find P as follows:

$$3P = d_1 F_1 + d_W F_W$$
$$= 0.667 \times 29\,430 + 0.6 \times 19\,620 \qquad \therefore P = 10\,470 \text{ N}$$

11.4.6 Buoyancy

The law of buoyancy, known as Archimedes' principle, dates back some 2200 years to the Greek philosopher Archimedes. Legend has it that Hiero, king of Syracuse, suspected that his new gold crown may have been constructed of mate-

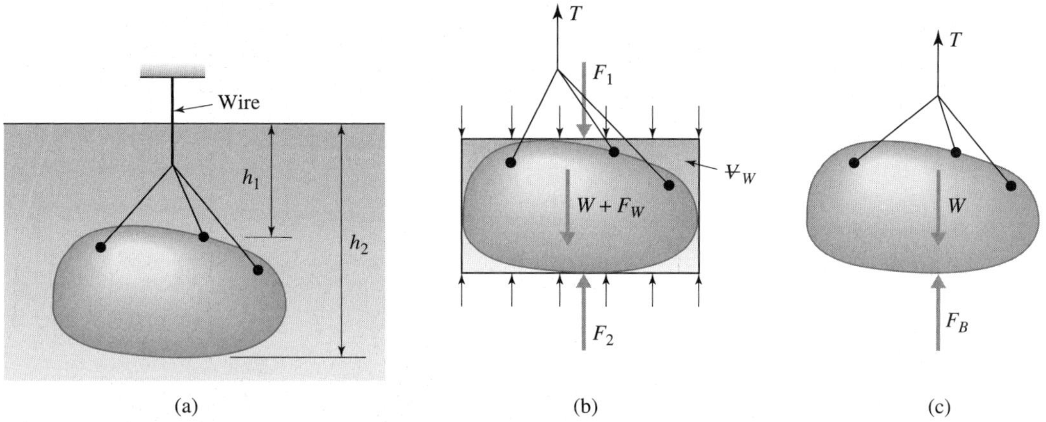

FIGURE 11.11 Forces on a submerged body: (a) submerged body; (b) free-body diagram; (c) free body showing F_B.

rials other than pure gold, so he asked Archimedes to test it. Archimedes probably made a lump of pure gold that weighed the same as the crown. The lump was discovered to weigh more in water than the crown weighed in water, thereby proving to Archimedes that the crown was not pure gold. The fake material possessed a larger volume to have the same weight as gold, hence it displaced more water. *Archimedes' principle* is: There is a buoyancy force on an object equal to the weight of displaced liquid.

Key Concept:
Archimedes' principle: the buoyancy force on an object equals the weight of displaced liquid.

To prove the law of buoyancy, consider the submerged body shown in Fig. 11.11a. In part (b) a cylindrical free-body diagram is shown that includes the submerged body with weight W and liquid having a weight F_W; the cross-sectional area A is the maximum cross-sectional area of the body. From the diagram we see that the resultant vertical force acting on the free-body diagram due to the water only (do not include W) is equal to

$$\sum F = F_2 - F_1 - F_W \tag{11.39}$$

This resultant force is by definition the *buoyant force* F_B. It can be expressed as

$$F_B = \gamma(h_2 A - h_1 A - \mathcal{V}_W) \tag{11.40}$$

where $\mathcal{V}_W$ is the liquid volume included in the free-body diagram. Recognizing that the volume of the submerged body is

$$\mathcal{V}_B = (h_2 - h_1)A - \mathcal{V}_W \tag{11.41}$$

we see from Eq. 11.40 that

$$F_B = \gamma \mathcal{V}_{\text{displaced liquid}} \tag{11.42}$$

thereby proving the law of buoyancy.

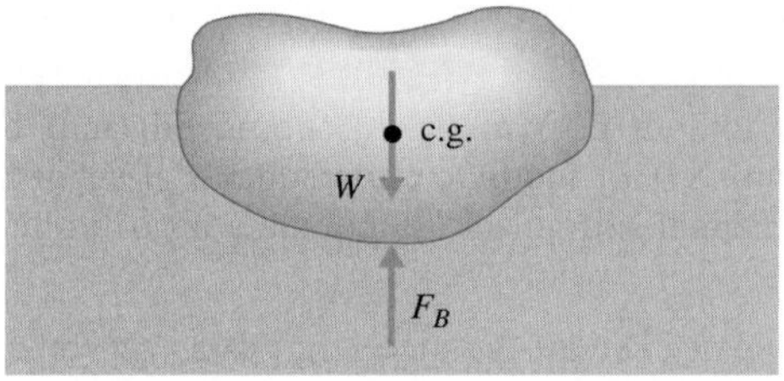

FIGURE 11.12 Forces on a floating object.

The force necessary to hold the submerged body in place (see Fig. 11.11c) is equal to

$$T = W - F_B \tag{11.43}$$

where W is the weight of the submerged body.

For a floating object, as in Fig. 11.12, the buoyant force is

$$F_B = \gamma \forall_{\text{displaced liquid}} \tag{11.44}$$

Obviously, $T = 0$, so that Eq. 11.43 gives

$$F_B = W \tag{11.45}$$

where W is the weight of the floating object.

Key Concept: *The buoyant force acts through the centroid of the displaced liquid volume.*

From the foregoing analysis, it is apparent that the buoyant force F_B acts through the centroid of the displaced liquid volume. For the floating object, the weight of the object acts through its center of gravity, so the center of gravity of the object must lie on the same vertical line as the centroid of the liquid volume.

Hydrometer: *An instrument used to measure the specific gravity of liquids.*

A **hydrometer**, an instrument used to measure the specific gravity of liquids, operates on the principle of buoyancy. A sketch is shown in Fig. 11.13. The upper part, the stem, has a constant diameter. When placed in pure water the specific gravity is marked to read 1.0. The force balance is

$$W = \gamma_{\text{water}} \forall \tag{11.46}$$

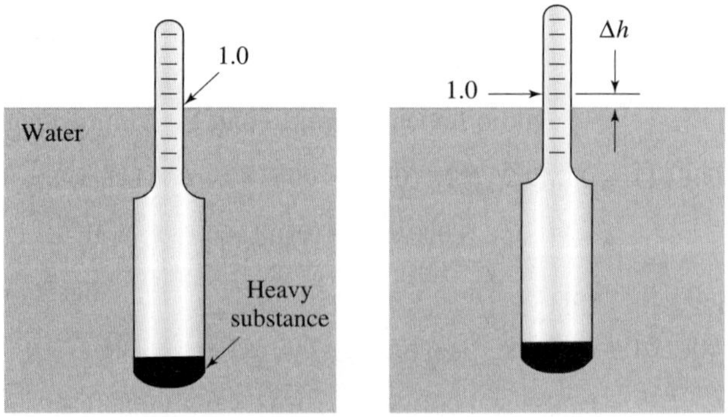

(a) (b)

FIGURE 11.13 Hydrometer: (a) in water; (b) in an unknown liquid.

where W is the weight of the hydrometer and $\mathcal{V}$ is the submerged volume below the $S = 1.0$ line. In an unknown liquid of specific weight γ_x, a force balance would be

$$W = \gamma_x(\mathcal{V} - A\Delta h) \tag{11.47}$$

where A is the cross-sectional area of the stem. Equating these two expressions gives

$$\Delta h = \frac{\mathcal{V}}{A}\left(1 - \frac{1}{S_x}\right) \tag{11.48}$$

where $S_x = \gamma_x/\gamma_{\text{water}}$. For a given hydrometer, $\mathcal{V}$ and A are fixed so that the quantity Δh is dependent only on the specific gravity S_x. Thus the stem can be calibrated to read S_x directly. Hydrometers are used to measure the amount of antifreeze in the radiator of an automobile, or the charge in a battery since the density of the fluid changes as H_2SO_4 is consumed or produced.

Example 11.8

The specific weight and the specific gravity of a body of unknown composition are desired. Its weight in air is found to be 200 lb, and in water, it weighs 150 lb.

Solution

The volume is found from a force balance when submerged as follows (see Fig. 11.11c):

$$T = W - F_B$$

$$150 = 200 - 62.4\mathcal{V} \qquad \therefore \mathcal{V} = 0.801 \text{ ft}^3$$

The specific weight is then

$$\gamma = \frac{W}{\mathcal{V}} = \frac{200}{0.801} = 250 \text{ lb/ft}^3$$

The specific gravity is found to be

$$S = \frac{\gamma}{\gamma_{\text{water}}} = \frac{250}{62.4} = 4.00$$

11.5 LINEARLY ACCELERATING CONTAINERS

In this section, the fluid will be at rest relative to a reference frame that is linearly accelerating with a horizontal component a_x and a vertical component a_z. Then Eq. 11.9 simplifies to

$$dp = -\rho a_x\, dx - \rho(g + a_z)\, dz \tag{11.49}$$

Integrating between two arbitrary points 1 and 2 results in

$$p_2 - p_1 = -\rho a_x(x_2 - x_1) - \rho(g + a_z)(z_2 - z_1) \tag{11.50}$$

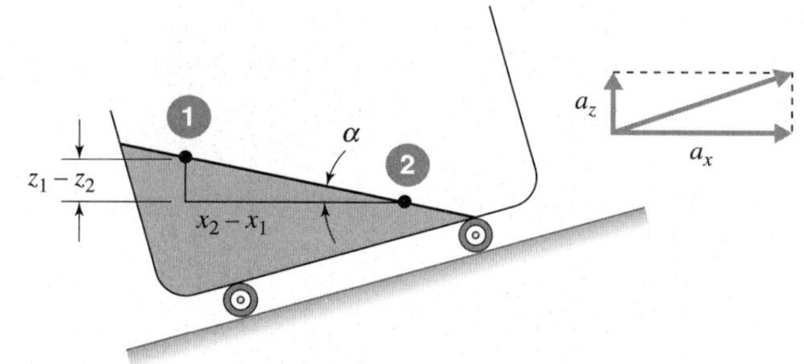

FIGURE 11.14 Linearly accelerating tank.

If points 1 and 2 lie on a constant-pressure line, such as the free surface in Fig. 11.14, $p_2 - p_1 = 0$ and we have

$$\frac{z_1 - z_2}{x_2 - x_1} = \tan \alpha = \frac{a_x}{g + a_z} \tag{11.51}$$

where α is the angle that the constant-pressure line makes with the horizontal.

Key Concept: *We often utilize the conservation of mass and equate the volumes before and after the acceleration is applied.*

In the solution of problems involving liquids, we must often utilize the conservation of mass and equate the volumes before and after the acceleration is applied. After the acceleration is initially applied, sloshing may occur. Our analysis will assume that sloshing is not present; either sufficient time passes to dampen out time-dependent motions, or the acceleration is applied in such a way that such motions are minimal.

Example 11.9

The tank shown in Fig. E11.9a is accelerated to the right. Calculate the acceleration a_x needed to cause the free surface, shown in Fig. E11.9b, to touch point A. Also, find p_B and the total force acting on the bottom of the tank if the width is 1 m.

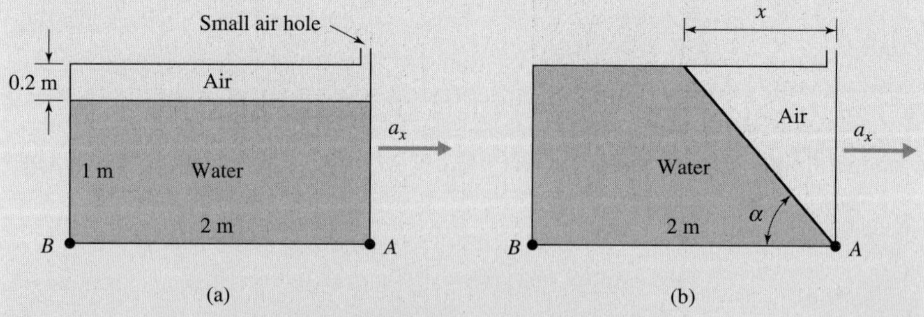

FIGURE E11.9

(continued)

Solution

The angle the free surface takes is found by equating the air volume (actually, areas since the width is constant) before and after since no water spills out,

$$0.2 \times 2 = \tfrac{1}{2}(1.2x)$$

$$x = 0.667 \text{ m}$$

The quantity tan α is now known. It is

$$\tan\alpha = \frac{1.2}{0.667} = 1.8$$

Using Eq. 11.51, we find a_x to be, letting $a_z = 0$,

$$\begin{aligned} a_x &= g\tan\alpha \\ &= 9.81 \times 1.8 = 17.66 \text{ m/s}^2 \end{aligned}$$

We can find the pressure at B by noting the pressure dependence on x. At A, the pressure is zero. Hence, Eq. 11.50 yields

$$\begin{aligned} p_B - \overset{0}{\cancel{p_A}} &= -\rho a_x(x_B - x_A) \\ p_B &= -1000 \times 17.66(-2) \\ &= 35\,300 \text{ Pa} \quad \text{or} \quad 35.3 \text{ kPa} \end{aligned}$$

To find the total force acting on the bottom of the tank, we realize that the pressure distribution is decreasing linearly from $p = 35.3$ kPa at B to $p = 0$ kPa at A. Hence, we can use the average pressure over the bottom of the tank:

$$\begin{aligned} F &= \frac{p_B + p_A}{2} \times \text{area} \\ &= \frac{35\,300 + 0}{2} \times 2 \times 1 = 35\,300 \text{ N} \end{aligned}$$

11.6 ROTATING CONTAINERS

In this section we consider the situation of a liquid contained in a rotating container, such as that shown in Fig. 11.15. After a relatively short time the liquid reaches static equilibrium with respect to the container and the rotating rz-reference frame. The horizontal rotation will not alter the pressure distribution in the vertical direction. There will be no variation of pressure with respect to the θ-coordinate. Applying Newton's second law ($\sum F_r = ma_r$) in the r-direction to the element shown, using $\sin d\theta/2 \cong d\theta/2$, yields

Key Concept: *Horizontal rotation will not alter the pressure distribution in the vertical direction.*

$$\begin{aligned} &-\frac{\partial p}{\partial r} dr\, r d\theta\, dz - pr\, d\theta\, dz - p\, dr\, d\theta\, dz - \frac{\partial p}{\partial r}(dr)^2\, d\theta\, dz \\ &+ 2\left(p\, dr\, dz \frac{d\theta}{2}\right) + pr d\theta\, dz = -\rho\, r d\theta\, dr\, dz\, r\Omega^2 \end{aligned} \quad (11.52)$$

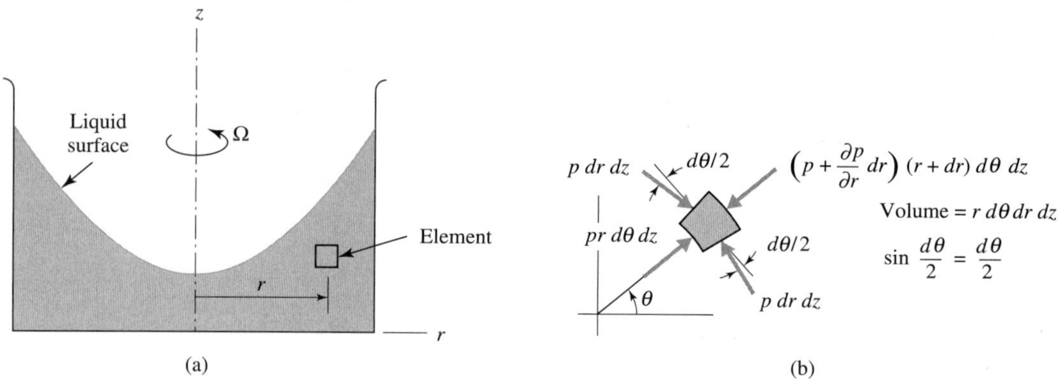

FIGURE 11.15 Rotating container: (a) liquid cross section; (b) top view of element.

where the acceleration is $r\Omega^2$ toward the center of rotation (in the negative r-direction). Simplify and divide by the volume $rd\theta\, dr\, dz$; then

$$\frac{\partial p}{\partial r} = \rho r \Omega^2 \tag{11.53}$$

where we have neglected the higher-order term that contains the differential dr. The pressure differential then becomes

$$\begin{aligned} dp &= \frac{\partial p}{\partial r} dr + \frac{\partial p}{\partial z} dz \\ &= \rho r \Omega^2\, dr - \rho g\, dz \end{aligned} \tag{11.54}$$

where we have used the static pressure variation given by the third part of Eq. 11.8 with $a_z = 0$. We can now integrate between any two points (r_1, z_1) and (r_2, z_2) to obtain

$$p_2 - p_1 = \frac{\rho \Omega^2}{2}(r_2^2 - r_1^2) - \rho g(z_2 - z_1) \tag{11.55}$$

If the two points are on a constant-pressure surface, such as the free surface, locating point 1 on the z-axis so that $r_1 = 0$, there results

$$\frac{\Omega^2 r_2^2}{2} = g(z_2 - z_1) \tag{11.56}$$

Key Concept: *The free surface is a paraboloid of revolution.*

which is the equation of a parabola. Hence the free surface is a paraboloid of revolution. The equations above can now, with the conservation of mass, be used to solve problems of interest.

Example 11.10

The cylinder shown in Fig. E11.10 is rotated about its centerline. Calculate the rotational speed that is necessary for the water to just touch the origin O and also the pressures at A and B.

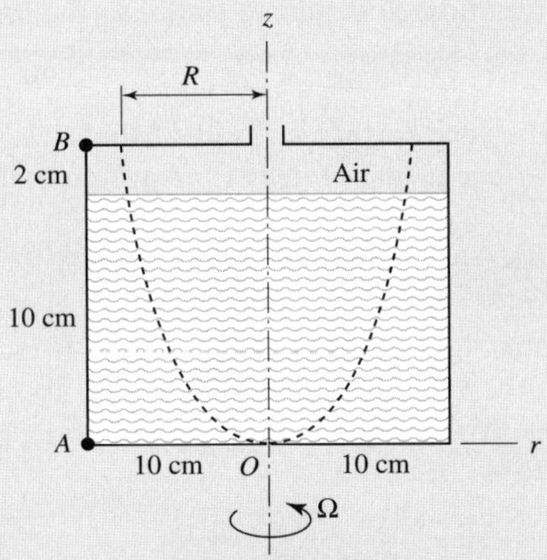

FIGURE E11.10

Solution

Since no water spills from the container, the air volume remains constant, that is,

$$\pi \times 10^2 \times 2 = \tfrac{1}{2}\pi R^2 \times 12$$

where we have used the fact that the volume of a paraboloid of revolution is one-half that of a circular cylinder with the same height and radius. This gives the value

$$R = 5.77 \text{ cm}$$

Using Eq. 11.56 with $r_2 = R$, we have

$$\frac{\Omega^2 \times 0.0577^2}{2} = 9.81 \times 0.12$$

$$\Omega = 26.6 \text{ rad/s}$$

To find the pressure at point A, we simply calculate the pressure difference between A and O. Using Eq. 11.55 with $r_2 = r_A = 0.1$ m, $r_1 = r_O = 0$, and $p_1 = p_O = 0$, there results

$$p_A = \frac{\rho\Omega^2}{2}(r_A^2 - \overset{0}{\cancel{r_O^2}}) = \frac{1000 \times 26.6^2}{2} \times 0.1^2 = 3540 \text{ Pa} \quad \text{or} \quad 3.54 \text{ kPa}$$

The pressure at B can be found by applying Eq. 11.55 to points A and B. This equation simplifies to

$$p_B - p_A = -\rho g(z_B - z_A)$$

Hence

$$p_B = 3540 - 1000 \times 9.81 \times 0.12 = 2360 \text{ Pa} \quad \text{or} \quad 2.36 \text{ kPa}$$

11.7 SUMMARY

The pressure variation in the vertical z-direction in a constant density fluid is found using

$$\Delta p = -\gamma \Delta z \tag{11.57}$$

This is used to interpret manometers and to establish the force on a plane as

$$F = \gamma \bar{h} A \tag{11.58}$$

where $\bar{h}$ is the vertical distance to the centroid of the area. The force is located a distance from the free surface to the center of pressure parallel to the area given by

$$y_p = \bar{y} + \frac{\bar{I}}{A\bar{y}} \tag{11.59}$$

where $\bar{I}$ is about the centroidal axis. Forces on curved surfaces are found using the above relationships and the weight of liquid contained above the surface.

Pressures and forces in linearly accelerating containers are determined using the angle α of a constant-pressure line:

$$\tan\alpha = \frac{a_x}{g + a_z} \tag{11.60}$$

Quite often the acceleration a_z in the vertical direction is zero.

In a container rotating with angular velocity Ω, a constant-pressure surface is described by

$$\tfrac{1}{2}\Omega^2 r_2^2 = g(z_2 - z_1) \tag{11.61}$$

where point 1 is on the axis of rotation and point 2 is anywhere on the constant-pressure surface.

REVIEW PROBLEMS FOR THE FE/EIT EXAMINATION

11.1FE A meteorologist states that the barometric pressure is 28.5 inches of mercury. Convert this pressure to kilopascals.

A. 98.6 kPa **B.** 97.2 kPa
C. 96.5 kPa **D.** 95.6 kPa

11.2FE The pressure in the foothills of the Rockies near Boulder, Colorado is 84 kPa. The pressure, assuming a constant density of 1.00 kg/m^3, at the top of a nearby 4000-m-high mountain is closest to:

A. 60 kPa **B.** 55 kPa
C. 50 kPa **D.** 45 kPa

11.3FE Estimate the pressure in the water pipe shown in Fig. P11.3FE. The manometor is open to the atmosphere.

A. 10 kPa **B.** 9 kPa
C. 8 kPa **D.** 7 kPa

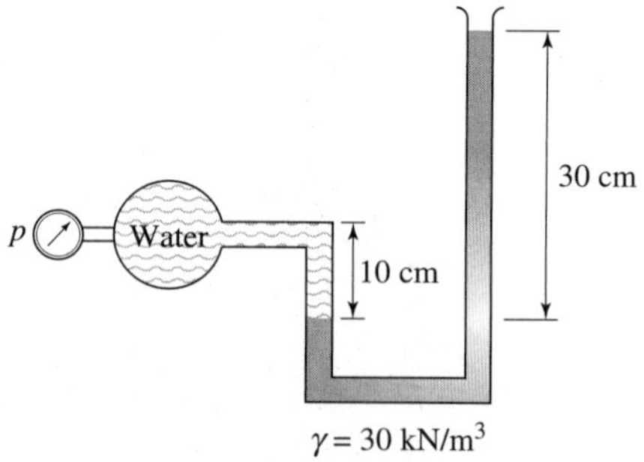

FIGURE P11.3FE

11.4FE If the pressure in the air of Fig. P11.4FE is increased by 10 kPa, the magnitude of H will be nearest:

A. 8.5 cm
B. 10.5 cm
C. 16 cm
D. 24.5 cm

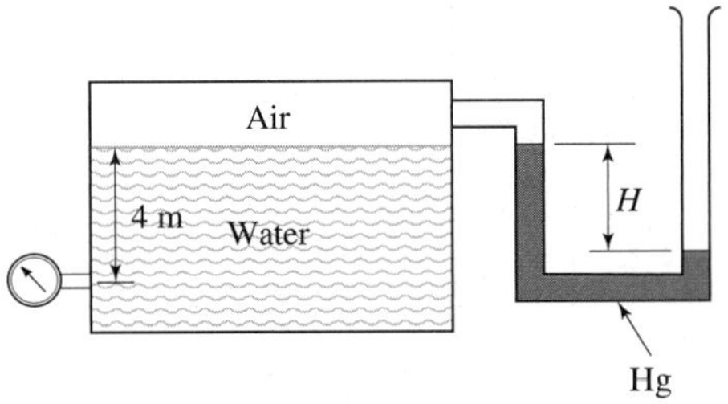

FIGURE P11.4FE

11.5FE The rectangular gate shown in Fig. P11.5FE is 3 m wide. The force P needed to hold the gate in the position shown is most nearly:

A. 24.5 kN
B. 32.7 kN
C. 98 kN
D. 147 kN

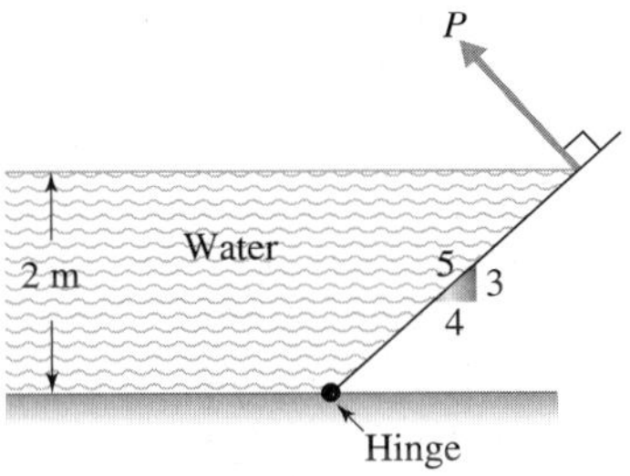

FIGURE P11.5FE

11.6FE The rigid gate hinged at a central point as shown in Fig. P11.6FE opens when $H = 5$ m. How far is the hinge above the water bottom?

A. 1.08 m **B.** 1.10 m
C. 1.12 m **D.** 1.14 m

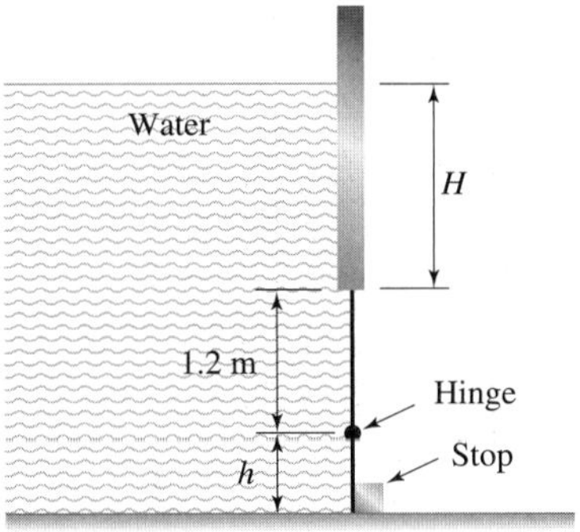

FIGURE P11.6FE

11.7FE A force $P = 300$ kN is needed to just open the gate of Fig. P11.7FE with $R = 1.2$ m and $H = 4$ m. How wide is the gate?

A. 2.98 m **B.** 3.67 m
C. 4.32 m **D.** 5.16 m

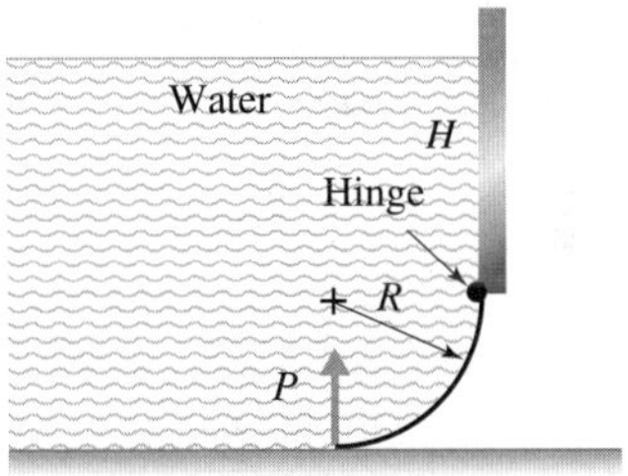

FIGURE P11.7FE

11.8FE The rectangular barge of Fig. P11.8FE is known to be 15 m long. A load having a mass of 900 kg is added to the barge causing it to sink 10 mm. How wide is the barge?

A. 6 m **B.** 9.2 m
C. 7.5 m **D.** 0.62 m

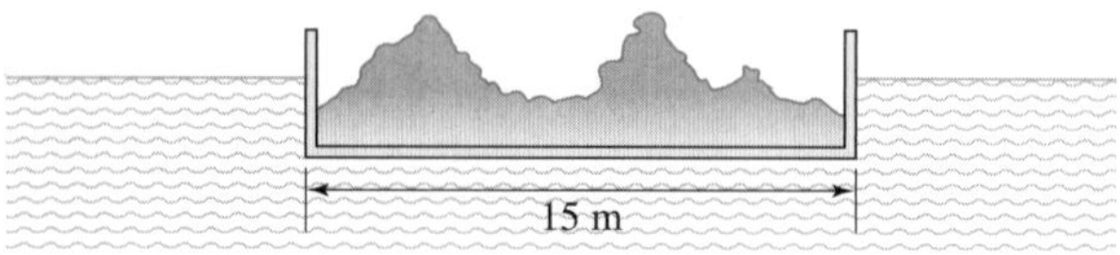

FIGURE P11.8FE

11.9FE The tank, with an initial pressure of $p = 20$ kPa, is accelerated, as shown in Fig. P11.9FE, at the rate of 5 m/s^2. The force on the 4-cm-diameter plug is nearest:

A. 30 N **B.** 50 N
C. 130 N **D.** 420 N

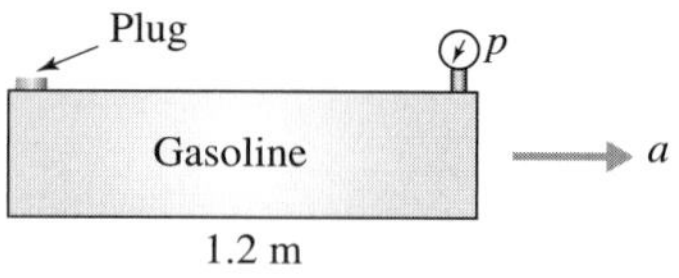

FIGURE P11.9FE

PROBLEMS

Pressure

11.1 Assume the element of Fig. 11.2 to be in the yz-plane with unit depth in the x-direction. Find a result similar to that of Eq. 11.4. Assume gravity to be in the z-direction.

11.2 Calculate the pressure at a depth of 10 m in a liquid with specific gravity of:
(a) 1.0 **(b)** 0.8
(c) 13.6 **(d)** 1.59
(e) 0.68

11.3 What depth is necessary in a liquid to produce a pressure of 250 kPa if the specific gravity is:
(a) 1.0? **(b)** 0.8?
(c) 13.6? **(d)** 1.59?
(e) 0.68?

11.4 A pressure of 20 psi is measured at a depth of 20 ft. Calculate the specific gravity and the density of the liquid if $p = 0$ on the surface.

11.5 How many meters of water are equivalent to:
(a) 760 mm Hg? **(b)** 75 cm Hg?
(c) 10 mm Hg?

11.6 Determine the pressure at the bottom of an open tank if it contains layers of:
(a) 20 cm of water and 2 cm of mercury
(b) 52 mm of water and 26 mm of carbon tetrachloride
(c) 3 m of oil, 2 m of water, and 10 cm of mercury

11.7 Assuming the density of air to be constant at 0.0024 slug/ft^3, calculate the pressure change from the top of a mountain to its base if the elevation change is 10,000 ft.

11.8 Assume that the air pressure is 100 kPa absolute at the top of a 3-m-high wall. Assuming constant density, estimate the difference in pressure at the base of the wall if outside the wall the temperature is −20°C and inside the wall it is 20°C. (This pressure difference induces an infiltration even though no wind is present.)

11.9 The specific gravity of a liquid varies linearly from 1.0 at the surface to 1.1 at a depth of 10 m. Calculate the pressure at $h = 10$ m.

11.10 Use Eq. 11.18 to determine the pressure at the top of a building 300 m tall. Then assume that the density is constant at the $z = 0$ value and estimate p at 300 m; also calculate the percent error in this second calculation. Use standard conditions at $z = 0$. Comment as to the advisability of an engineer assuming the atmosphere to have constant density up to heights of 300 m.

11.11 Assume the bulk modulus to be constant and find an expression for the pressure as a function of depth h in the ocean. Use this expression and estimate the pressure assuming $\rho_0 = 2.00$ slug/ft^3. Now, assume a constant density of 2.00 slug/ft and calculate the pressure and the percent error, assuming the estimate in the first calculation to be correct. Use depths of **(a)** 1500 ft, **(b)** 5000 ft, and **(c)** 15,000 ft.

11.12 Estimate the pressure at 10 000 m assuming an isothermal atmosphere with temperature:
(a) 0°C **(b)** 15°C
(c) −15°C

11.13 The temperature in the atmosphere is approximated by $T(z) = 15 - 0.0065z$°C for elevations less than 11 000 m. Calculate the pressure at elevations of:
(a) 3000 m **(b)** 6000 m
(c) 9000 m **(d)** 11 000 m

11.14 Estimate the elevation where $p = 0.001$ psia, assuming an isothermal atmosphere with $T = -5$°F.

Manometers

11.15 Calculate the pressure in a pipe transporting air if a U-tube manometer measures 25-cm Hg. Note that the weight of air in the manometer is negligible.

11.16 If the pressure of air in a pipe is 450 kPa, what will a U-tube manometer with mercury read? Use $h = 1.5$ cm in Fig. 11.6b.
(a) Neglect the weight of the air column.
(b) Include the weight of the air column, assuming that $T_{air} = 20$°C, and calculate the percent error of part (a).

11.17 A U-tube manometer is attached to a pipe transporting a liquid. It is known that the pressure in the pipe at the location of the manometer is 2.4 kPa. Select the liquid from Appendix Table B.10 that is most likely being transported if the manometer indicates the following height of liquid above the pipe:
(a) 36.0 cm **(b)** 27.2 cm
(c) 24.5 cm **(d)** 15.4 cm

11.18 It is known that the pressure at the nose of an airplane traveling at relatively low speed is related to its speed by $p = \frac{1}{2}\rho V^2$, where ρ is the density of air. Determine the speed of an airplane traveling near the earth's surface if a U-tube manometer, that measures the nose pressure, reads:
(a) 6 cm of water **(b)** 3 in. of water
(c) 10 cm of water **(d)** 5 in. of water

11.19 Oil with $S = 0.86$ is being transported in a pipe. Calculate the pressure if a U-tube manometer reads 9.5-in. Hg. The oil in the manometer is depressed 5 in. below the pipe centerline.

11.20 Several liquids are layered inside a tank with pressurized air at the top. If the air pressure is 3.2 kPa, calculate the pressure at the bottom of the tank if the layers include 20 cm of SAE-10 oil, 10 cm of water, 15 cm of glycerin, and 18 cm of carbon tetrachloride.

11.21 For the setup shown in Fig. P11.21, calculate the manometer reading H.

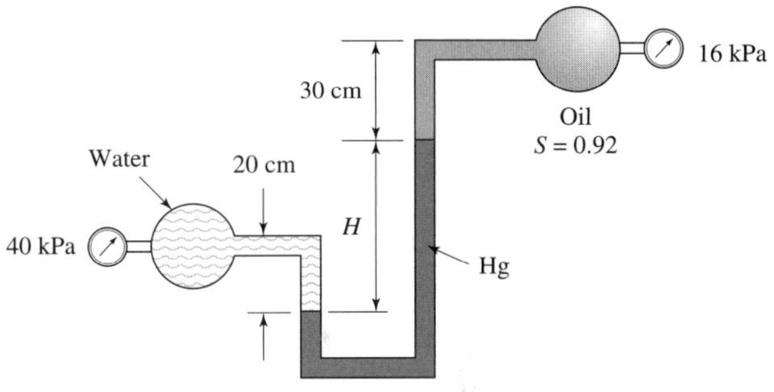

FIGURE P11.21

11.22 What is the pressure in the water pipe shown in Fig. P11.22?

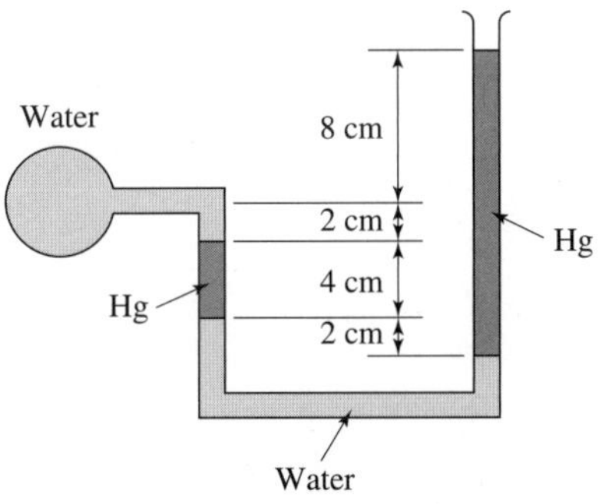

FIGURE P11.22

11.23 Determine the pressure difference between the water pipe and the oil pipe shown in Fig. P11.23.

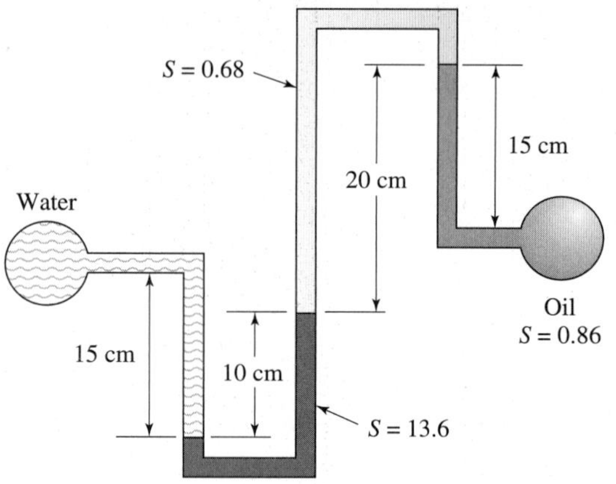

FIGURE P11.23

11.24 What is the pressure in the oil pipe shown in Fig. P11.24 if the pressure in the water pipe is 15 kPa?

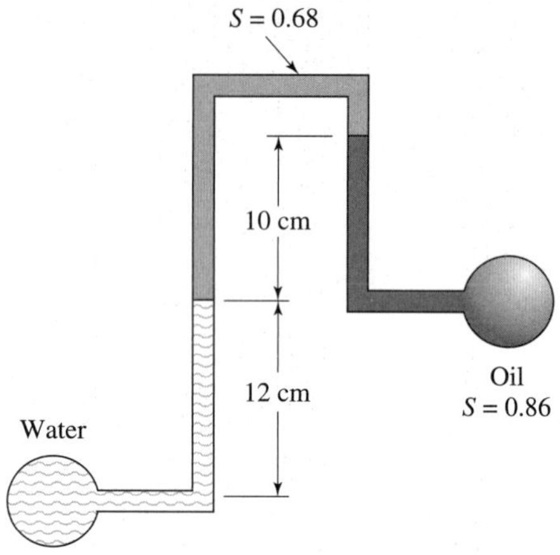

FIGURE P11.24

11.25 For the tank shown in Fig. P11.25, determine the reading of the pressure gage if:

(a) $H = 2$ m, $h = 10$ cm
(b) $H = 0.8$ m, $h = 20$ cm
(c) $H = 6$ ft, $h = 4$ in.
(d) $H = 2$ ft, $h = 8$ in.

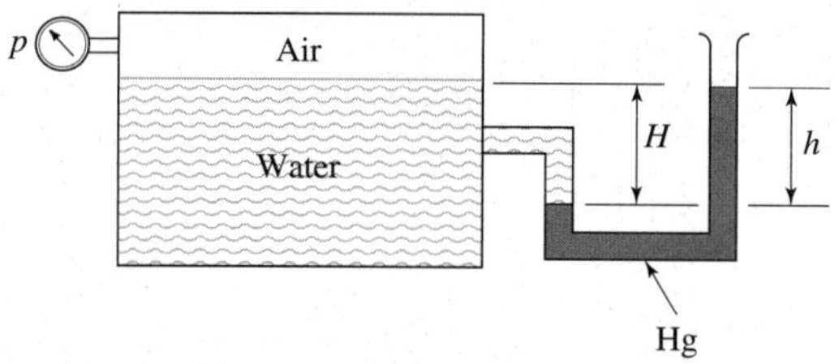

FIGURE P11.25

11.26 The pressure in the water pipe of Fig. 11.6b is 8.2 kPa, with $h = 25$ cm and $S_2 = 1.59$. Find the pressure in the water pipe if the H reading increases by 27.3 cm.

11.27 Find the pressure in the water pipe shown in Fig. P11.27.

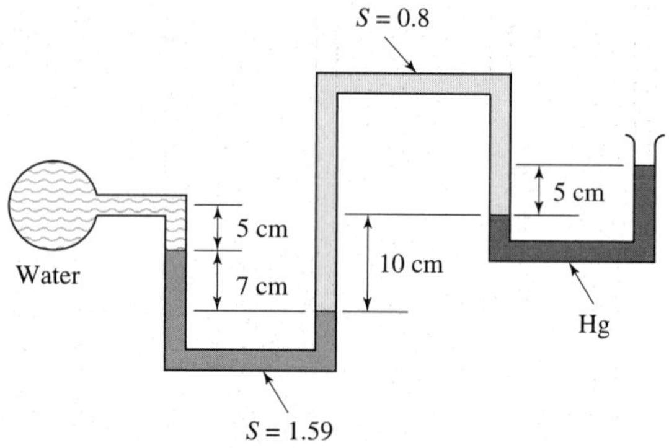

FIGURE P11.27

11.28 In Fig. P11.28, with the manometer top open, the mercury level is 8 in. below the air pipe; there is no pressure in the air pipe. The manometer top is then sealed. Calculate the manometer reading H for a pressure of 30 psi in the air pipe. Assume an isothermal process for the air in the sealed tube.

11.29 Determine the new manometer reading h if the air pressure is increased by 10% in:
(a) Problem 11.25a **(b)** Problem 11.25b
(c) Problem 11.25c **(d)** Problem 11.25d

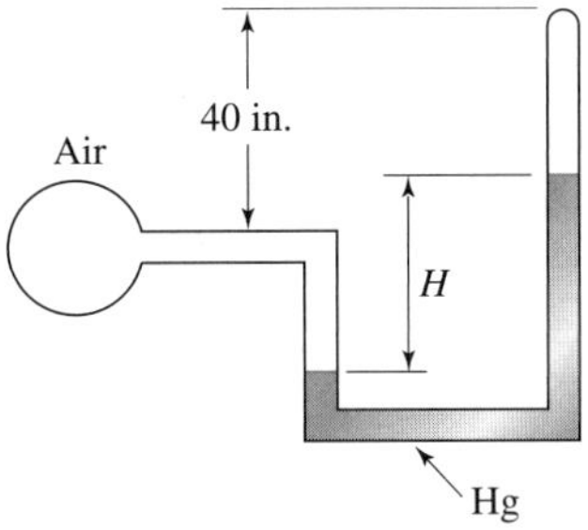

FIGURE P11.28

Forces on Plane Areas

11.30 Calculate the force acting on a 30-cm-diameter porthole of a ship if the center of the porthole is 10 m below water level.

11.31 A swimming pool is filled with 2 m of water. Its bottom is square and measures 4 m on a side. Two opposite sides are vertical; one end is at 45° and the other makes an angle of 60° with the horizontal. Calculate the force of the water on:
(a) The bottom **(b)** A vertical side
(c) The 45° end **(d)** The 60° end

11.32 An enclosed rectangular concrete vault with outside dimensions 2 m × 1 m × 1.5 m and wall thickness 10 cm is buried with the top surface even with the ground surface. Will the empty vault tend to rise out of the ground if the ground becomes completely saturated with water? Use $S_{concrete} = 2.4$.

11.33 Gasoline fills a tank 4 m in diameter and 6-m long. Calculate the force that the gasoline exerts on one end of the tank. Assume that the tank is not pressurized and the ends are vertical.

11.34 The sides of a triangular area measure 2 m, 3 m, and 3 m, respectively. Calculate the force of water on one side of the area if the 2-m side is horizontal and 10 m below the surface and the triangle is:
(a) Vertical
(b) Horizontal
(c) On a 60° angle sloped upward.

11.35 The triangular gate shown in Fig. P11.35 has its 6-ft side parallel and 30 ft below the water surface. Calculate the magnitude and location of the force acting on the gate if it is:
(a) Vertical
(b) Horizontal
(c) On a 45° angle sloped upward

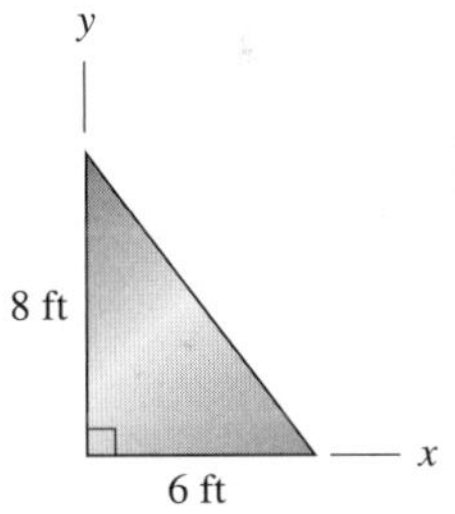

FIGURE P11.35

11.36 The top of each gate shown in Fig. P11.36 lies 4 m below the water surface. Find the location and magnitude of the force acting on one side assuming a vertical orientation.

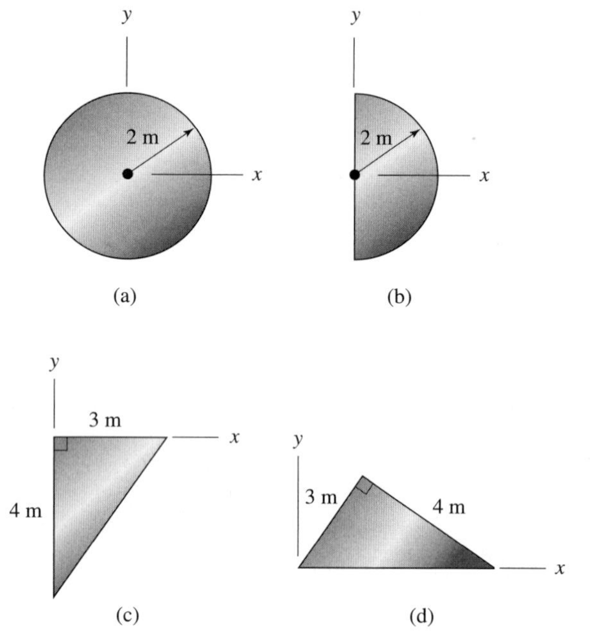

FIGURE P11.36

11.37 A 6-ft-wide 10-ft-high vertical rectangular gate has its top edge 6 ft below the water level. It is hinged along its lower edge. What force, acting on the top edge, is necessary to hold the gate shut?

11.38 Determine the force P needed to hold the 4-m-wide gate in the position shown in Fig. P11.38.

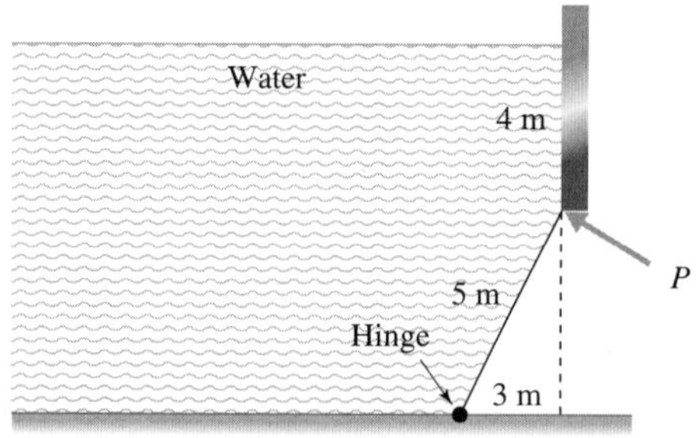

FIGURE P11.38

11.39 Use Eq. 11.34 and show that the force F in Fig. 11.8 acts one-third the way up on a vertical rectangular area and also on a sloped rectangular area. Assume the sloped gate to be on an angle α with the horizontal.

11.40 Find the force P needed to hold the 3-m-wide rectangular gate as shown in Fig. P11.40 if:
(a) $l = 2$ m **(b)** $l = 4$ m **(c)** $l = 5$ m

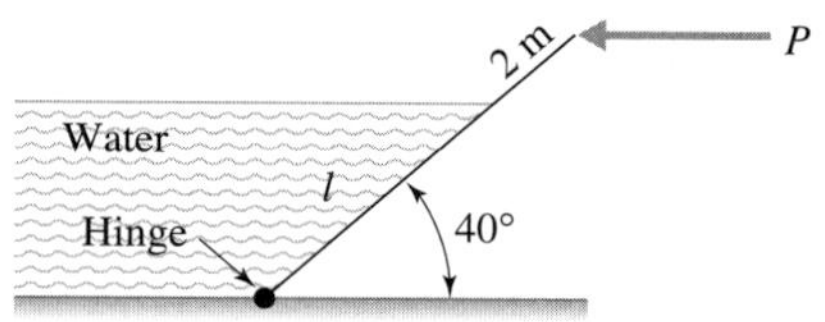

FIGURE P11.40

11.41 A vertical gate at the end of a channel (Fig. P11.41) opens when the water above the hinge produces a moment greater than the moment of the water below the hinge. What height h of water is needed to open the gate if:
(a) $H = 0.9$ m? **(b)** $H = 1.2$ m? **(c)** $H = 1.5$ m?

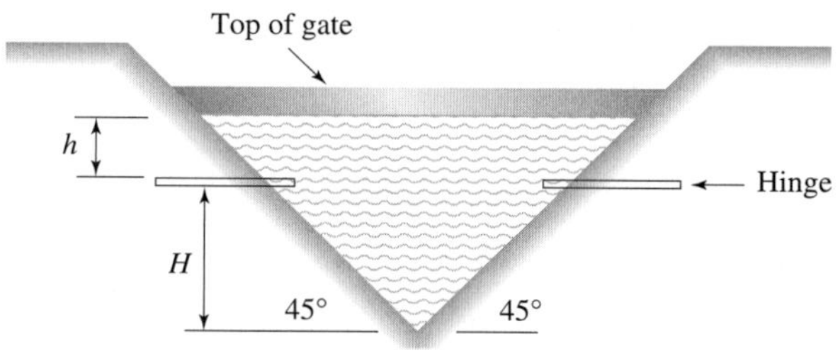

FIGURE P11.41

11.42 At what height H will the rigid gate, hinged at a central point as shown in Fig. P11.42, open if h is:
(a) 0.6 m? **(b)** 0.8 m? **(c)** 1.0 m?

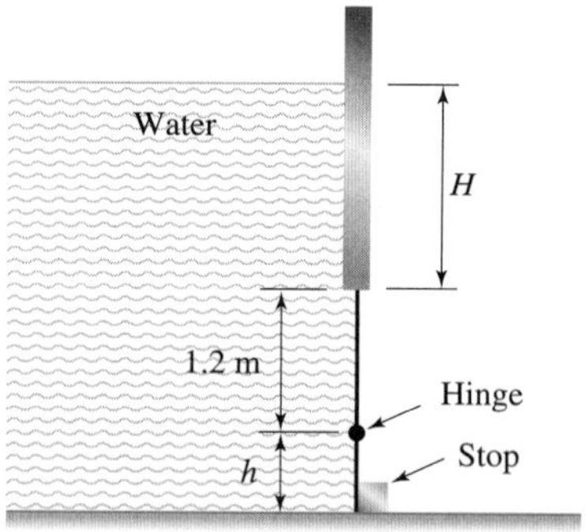

FIGURE P11.42

11.43 For the gate shown in Fig. P11.43, calculate the height H that will result in the gate opening automatically if (neglect the weight of the gate):
(a) $l = 2$ m **(b)** $l = 1$ m
(c) $l = 6$ ft **(d)** $l = 3$ ft

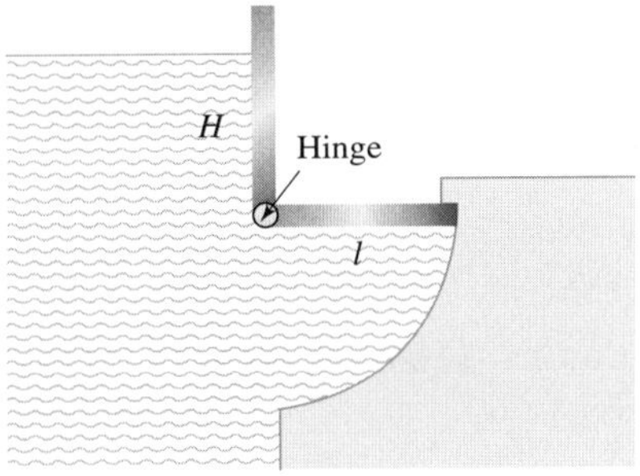

FIGURE P11.43

11.44 Assume a linear pressure distribution over the base of the concrete ($S = 2.4$) dam shown in Fig. P11.44. Will the dam topple over (sum moments about the bottom, right-hand corner)? Use:
(a) $H = 40$ ft **(b)** $H = 60$ ft **(c)** $H = 80$ ft

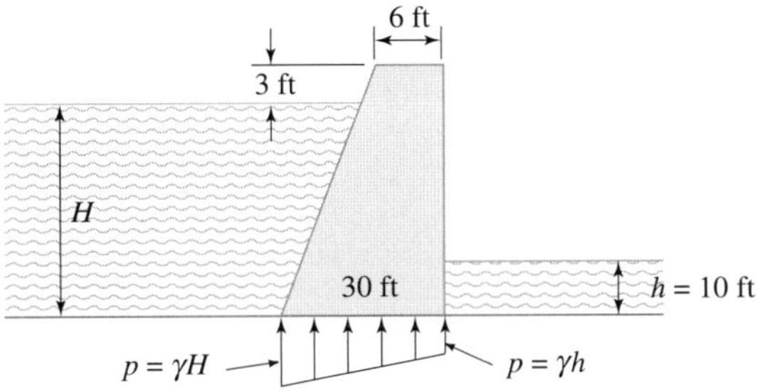

FIGURE P11.44

Forces on Curved Surfaces

11.45 In Example 11.6 assume that the water is above the gate rather than below the gate. Water above the gate will produce the same pressure distribution on the gate and hence the same forces (except the forces will be in opposite directions). Consequently, the force P will be numerically the same (it will act to the left). With water above the gate, draw a free-body diagram and calculate P. Compare with the details of the first method of Example 11.6.

11.46 Find the force P needed to hold the 10-m-long cylindrical object in the position shown in Fig. P11.46.

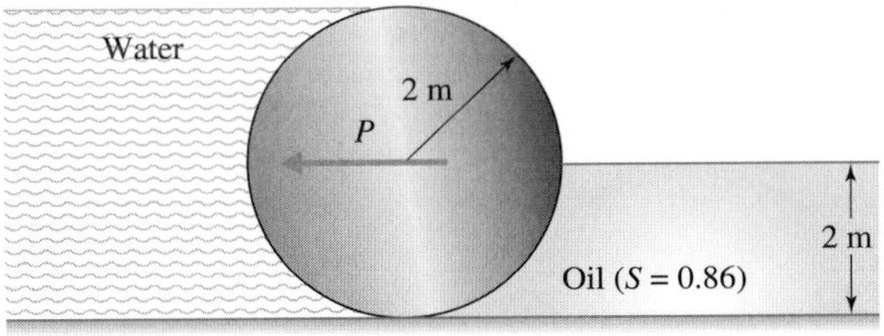

FIGURE P11.46

11.47 Find the force P needed to just open the gate shown in Fig. P11.47 if:
(a) $H = 6$ m, $R = 2$ m, and the gate is 4 m wide.
(b) $H = 20$ ft, $R = 6$ ft, and the gate is 12 ft wide.

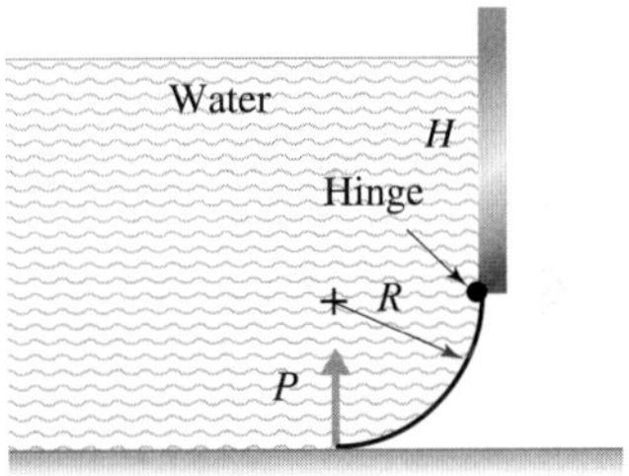

FIGURE P11.47

11.48 What P is needed to hold the 4-m-wide gate shown in Fig. P11.48 closed?

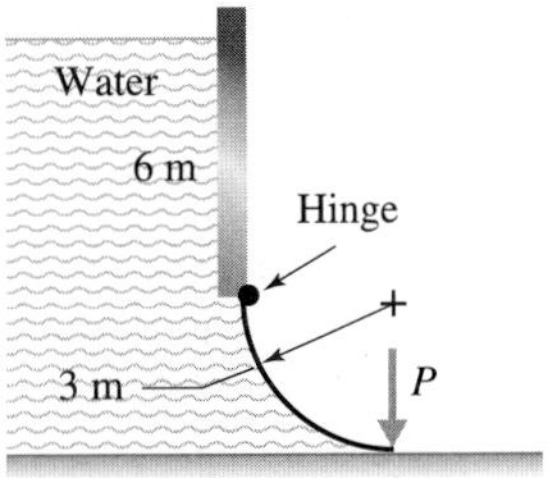

FIGURE P11.48

11.49 Find the force P required to hold the gate in the position shown in Fig. P11.49. The gate is 5 m wide.

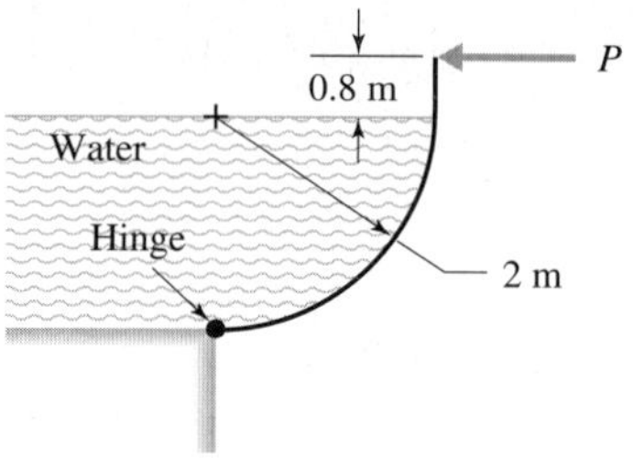

FIGURE P11.49

11.50 The 3-m-wide circular gate shown in Fig. P11.50 weighs 400 N with center of gravity 0.9 m to the left of the hinge. Estimate the force P needed to open the gate.

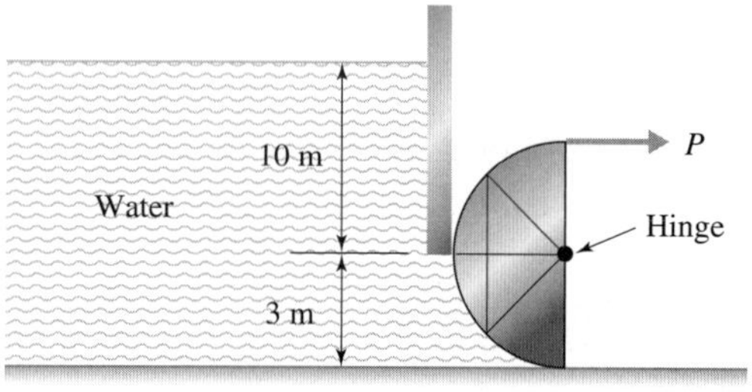

FIGURE P11.50

11.51 The quarter-circle cylindrical gate (Fig. P11.51; $S = 0.2$) is in equilibrium, as shown. Calculate the value of γ_x using:
(a) SI units **(b)** English units

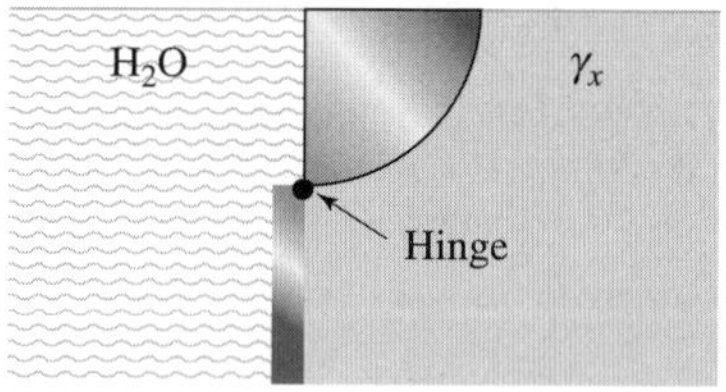

FIGURE P11.51

11.52 A log is in equilibrium, as shown in Fig. P11.52. Calculate the force pushing it against the dam and the specific gravity of the log if:
(a) Its length is 6 m and $R = 0.6$ m.
(b) Its length is 20 ft and $R = 2$ ft.

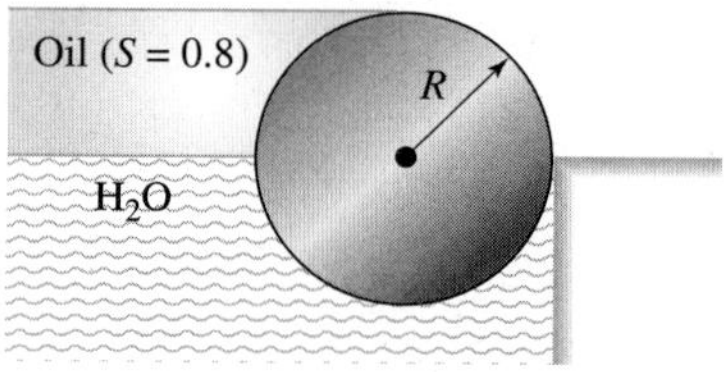

FIGURE P11.52

11.53 Find the force P if the parabolic gate shown in Fig. P11.53 is:
(a) 2 m wide and $H = 2$ m
(b) 4 ft wide and $H = 8$ ft

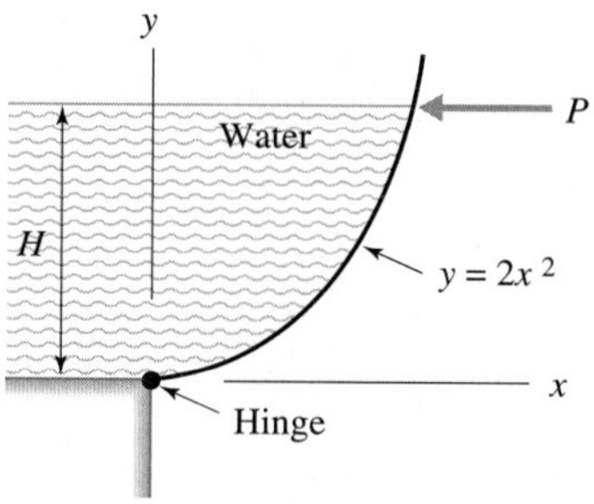

FIGURE P11.53

Buoyancy

11.54 The 3-m-wide barge shown in Fig. P11.54 weighs 20 kN empty. It is proposed that it carry a 250-kN load. Predict the draft in:
(a) Fresh water **(b)** Salt water ($S = 1.03$)

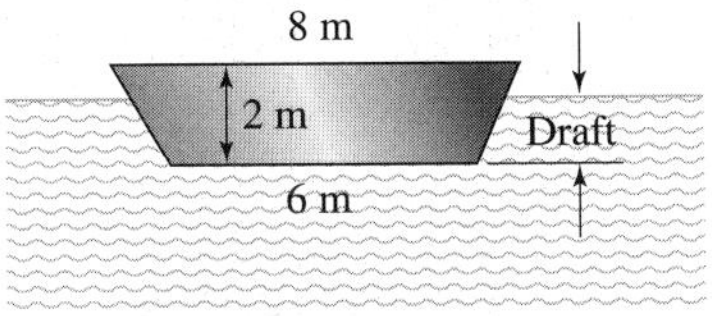

FIGURE P11.54

11.55 An object weighs 100 N in the air and 25 N when submerged in water. Calculate its volume and specific weight.

11.56 A car ferry is essentially rectangular with dimensions 25 ft wide and 300 ft long. If 60 cars, with an average weight per car of 3000 lb, are loaded on the ferry, how much farther will it sink into the water?

11.57 A 30-m-long vessel, with cross section shown in Fig. P11.57, is to carry a load of 6000 kN. How far will the water level be from the top of the vessel if its mass is 100 000 kg?

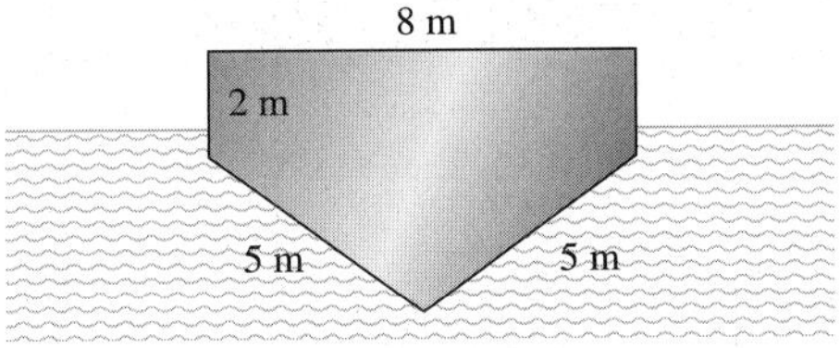

FIGURE P11.57

11.58 A body, with a volume of 2 m^3, weighs 40 kN. Determine its weight when submerged in a liquid with $S = 1.59$.

11.59 A hot-air balloon carries a load of 1000 N, including its own weight. If its spherical shape is about 10 m in diameter, estimate the average temperature of the air inside if the air outside is at 20°C.

11.60 A large blimp is proposed to travel near the Earth's surface. If the blimp resembles a large cylinder 1500 m long with a diameter of 300 m, estimate the payload if its own weight is 10% of the payload. How many 800-N people could it carry? The blimp is filled with helium and standard conditions prevail. (This vehicle will not cause seasickness, and sunsets are spectacular!)

11.61 An object is constructed of a material lighter than water. It weighs 50 N in air and a force of 10 N is required to hold it under water. What is its density, specific weight, and specific gravity?

11.62 The plug and empty cylinder shown in Fig. P11.62 weigh 1500 lb. Calculate the height h needed to lift the plug if the radius R of the 10-ft-long cylinder is:
(a) 12 in. **(b)** 16 in. **(c)** 20 in.

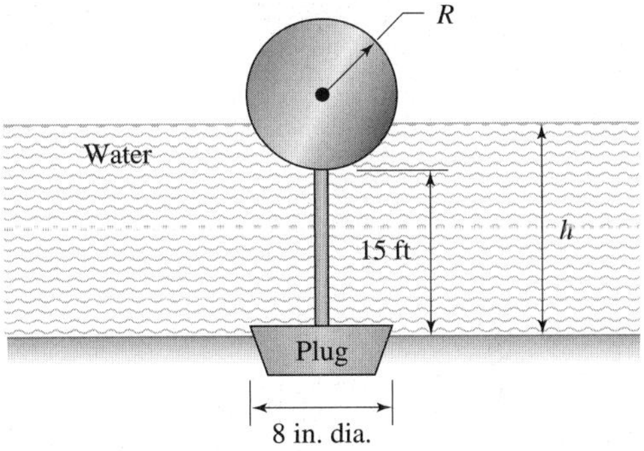

FIGURE P11.62

11.63 The hydrometer shown in Fig. P11.63, with no mercury, has a mass of 0.01 kg. It is designed to float at the midpoint of the 12-cm-long stem in pure water.
(a) Calculate the mass of mercury needed.
(b) What is the specific gravity of the liquid if the hydrometer is just submerged?
(c) What is the specific gravity of the liquid if the stem of the hydrometer is completely exposed?

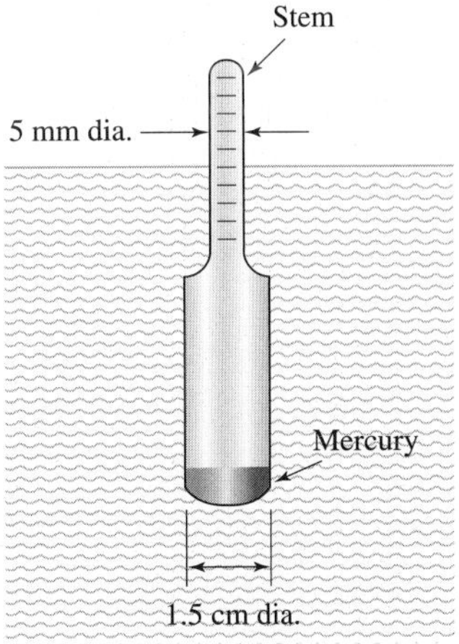

FIGURE P11.63

Linearly Accelerating Containers

11.64 The tank shown in Fig. P11.64 is completely filled with water and accelerated. Calculate the maximum pressure in the tank if:
(a) $a_x = 20\ \text{m/s}^2$, $a_z = 0$, $L = 2$ m
(b) $a_x = 0$, $a_z = 20\ \text{m/s}^2$, $L = 2$ m
(c) $a_x = 60\ \text{ft/sec}^2$, $a_z = 60\ \text{ft/sec}^2$, $L = 6$ ft
(d) $a_x = 0$, $a_z = 60\ \text{ft/sec}^2$, $L = 6$ ft

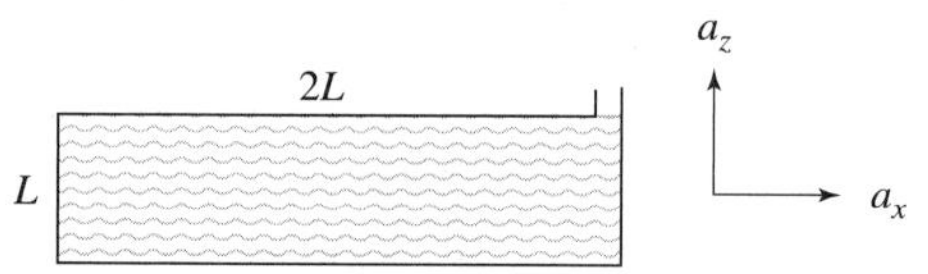

FIGURE P11.64

11.65 The tank shown in Fig. P11.65 is accelerated to the right at $10\ \text{m/s}^2$. Find:
(a) p_A
(b) p_B
(c) p_C

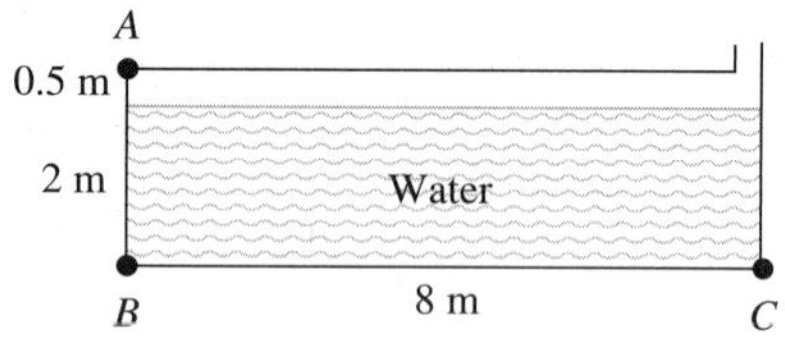

FIGURE P11.65

11.66 The tank of Problem 11.65 is accelerated so that $p_B = 60$ kPa. Find a_x assuming that:
(a) $a_z = 0$
(b) $a_z = 10\ \text{m/s}^2$
(c) $a_z = 5\ \text{m/s}^2$

11.67 The tank shown in Fig. P11.67 is filled with water and accelerated. Find the pressure at A if:
(a) $a = 20\ \text{m/s}^2$, $L = 1$ m
(b) $a = 10\ \text{m/s}^2$, $L = 1.5$ m
(c) $a = 60\ \text{ft/sec}^2$, $L = 3$ ft
(d) $a = 30\ \text{ft/sec}^2$, $L = 4$ ft

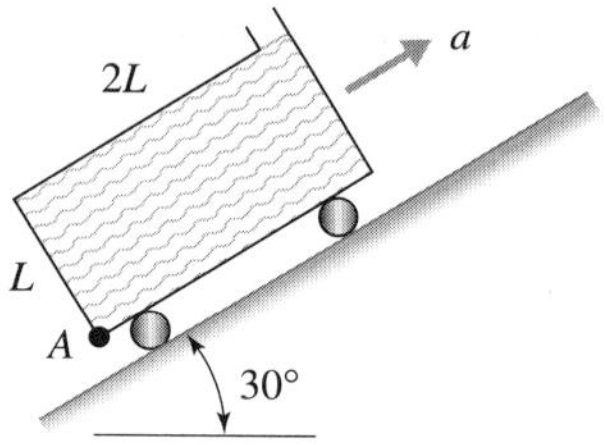

FIGURE P11.67

11.68 The tank of Problem 11.65 is 4 m wide. Find the force acting on:
(a) The end AB **(b)** The bottom **(c)** The top

11.69 For the U-tube shown in Fig. P11.69, determine the pressure at points A, B, and C if:
(a) $a_x = 0$, $a_z = 10\ \text{m/s}^2$, $L = 60$ cm
(b) $a_x = 20\ \text{m/s}^2$, $a_z = 0$, $L = 60$ cm
(c) $a_x = 20\ \text{m/s}^2$, $a_z = 10\ \text{m/s}^2$, $L = 60$ cm
(d) $a_x = 0$, $a_z = -60\ \text{ft/sec}^2$, $L = 25$ in
(e) $a_x = 60\ \text{ft/sec}^2$, $a_z = 0$, $L = 25$ in
(f) $a_x = -30\ \text{ft/sec}^2$, $a_z = 30\ \text{ft/sec}^2$, $L = 25$ in

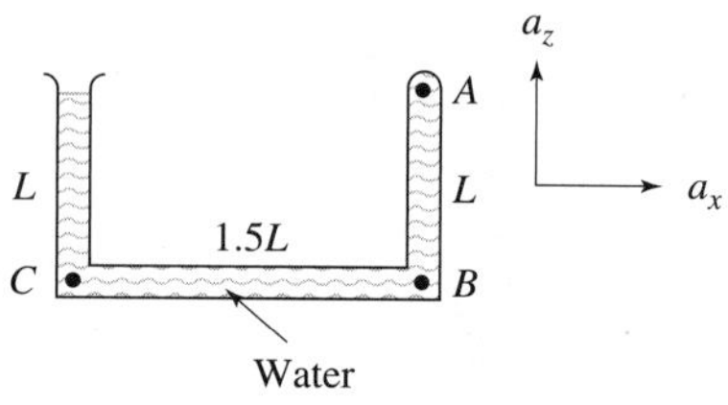

FIGURE P11.69

Rotating Containers

11.70 The U-tube of Problem 11.69 is rotated about the left leg at 50 rpm. Find p_A, p_B, and p_C if:
(a) $L = 60$ cm
(b) $L = 40$ cm
(c) $L = 25$ in
(d) $L = 15$ in

11.71 The U-tube of Problem 11.69 is rotated about the right leg at 10 rad/s. Find the pressures at points A, B, and C if:
(a) $L = 60$ cm
(b) $L = 40$ cm
(c) $L = 25$ in
(d) $L = 15$ in

11.72 The U-tube of Problem 11.69 is rotated about the center of the horizontal leg so that the pressure at the center is zero. Calculate Ω if:
(a) $L = 60$ cm
(b) $L = 40$ cm
(c) $L = 25$ in
(d) $L = 15$ in

11.73 For the cylinder shown in Fig. P11.73, determine the pressure at point A for a rotational speed of:
(a) 5 rad/s **(b)** 7 rad/s
(c) 10 rad/s **(d)** 20 rad/s

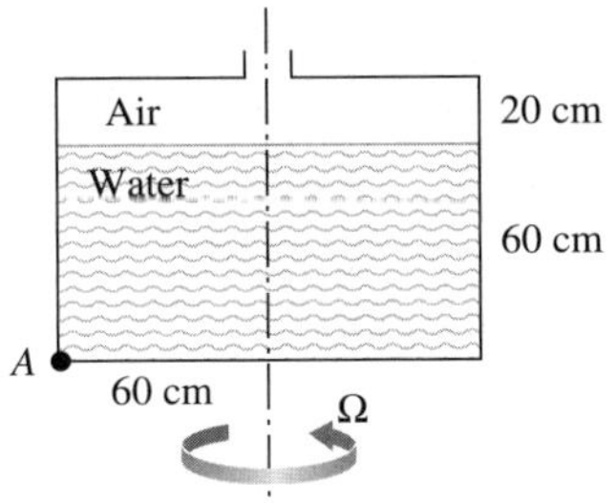

FIGURE P11.73

11.74 The hole in the cylinder of Problem 11.73 is closed and the air pressurized to 25 kPa. Find the pressure at point A if the rotational speed is:
(a) 5 rad/s **(b)** 7 rad/s
(c) 10 rad/s **(d)** 20 rad/s

11.75 Find the force on the bottom of the cylinder of:
(a) Problem 11.73a **(b)** Problem 11.73b
(c) Problem 11.73c **(d)** Problem 11.73d

Whitewater rafting on the Ottawa River, Ontario. Rafting is a popular sport in North America. It embodies the thrill of experiencing rapidly moving water in a craft that demands quick thinking and manipulative skills. (Courtesy of Ottawa River Rafting Co.)

12

Introduction to Fluids in Motion

Outline

Chapter Objectives

The objectives of this chapter are to:

- ▲ Mathematically describe the motion of a fluid.
- ▲ Express the acceleration of a fluid particle given the velocity components.
- ▲ Describe the deformation of a fluid particle.
- ▲ Classify various fluid flows. Is a flow viscous, is it turbulent, is it incompressible, is it a uniform flow?
- ▲ Derive the Bernoulli equation and identify its restrictions.
- ▲ Present several examples and numerous problems that demonstrate how fluid flows are described, how flows are classified, and how Bernoulli's equation is used to estimate flow variables.

Nomenclature

The following variables are introduced in this chapter:

$\mathbf{A}$	absolute acceleration	v_r, v_θ, v_z	cylindrical velocity components
a_r, a_θ, a_z	cylindrical acceleration components	v_r, v_θ, v_ϕ	spherical velocity components
a_r, a_θ, a_ϕ	spherical acceleration components		
a_x, a_y, a_z	rectangular acceleration components	ϵ_{ij}	rate-of-stain components
$\hat{\mathbf{i}}, \hat{\mathbf{j}}, \hat{\mathbf{k}}$	unit vectors	$\epsilon_{xx}, \epsilon_{xy}, \epsilon_{xz}$	rate-of-strain components
M	Mach number		
$\mathbf{r}$	position vector	θ	coordinate angle
Re	Reynolds number	ϕ	coordinate angle
s	tangential direction	$\omega_x, \omega_y, \omega_z$	vorticity components
S	position of reference frame	$\boldsymbol{\omega}$	vorticity vector
u, v, w	rectangular velocity components	$\boldsymbol{\Omega}$	angular velocity vector
$\mathbf{V}$	velocity vector	$\Omega_x, \Omega_y, \Omega_z$	angular velocity components

12.1 INTRODUCTION

This chapter serves as an introduction to all the following chapters that deal with fluid motions. Fluid motions manifest themselves in many different ways. Some can be described very easily, while others require a thorough understanding of physical laws. In engineering applications, it is important to describe the fluid motions as simply as can be justified. This usually depends on the required accuracy. Often, accuracies of $\pm 10\%$ are acceptable, although in some applications higher accuracies have to be achieved. The general equations of motion are very difficult to solve; consequently, it is the engineer's responsibility to know which simplifying assumptions can be made. This, of course, requires experience and, more important, an understanding of the physics involved.

Key Concept: *Under certain conditions, viscous effects can be neglected.*

Some common assumptions used to simplify a flow situation are related to fluid properties. For example, under certain conditions, the viscosity can affect the flow significantly; in others, viscous effects can be neglected greatly simplifying the equations without significantly altering the predictions. It is well known that the compressibility of a gas in motion should be taken into account if the velocities are very high. But compressibility effects do not have to be taken into account to predict wind forces on buildings or to predict any other physical quantity that is a direct effect of wind. Wind speeds are simply not high enough. Numerous examples could be cited. After our description of fluid motions, the appropriate assumptions used should become more obvious.

This chapter has three additional sections. In the first section we introduce the reader to some important general approaches used to analyze fluid mechanics problems. In the second section we give a brief overview of different types of flow, such as compressible and incompressible flows, and viscous and inviscid flows. Detailed discussions of each of these flow types follow in later chapters. The third section introduces the reader to the commonly used Bernoulli equation, an equation that establishes how pressures and velocities vary in a flow field. The use of this equation, however, requires many simplifying assumptions, and its application is, therefore, limited.

12.2 DESCRIPTION OF FLUID MOTION

The analysis of complex fluid flow problems is often aided by the visualization of flow patterns, which permit the development of a better intuitive understanding and help in formulating the mathematical problem. The flow in a washing machine is a good example. An easier, yet difficult problem is the flow in the vicinity of where a wing attaches to a fuselage, or where a bridge support interacts with the water at the bottom of a river. In Section 12.2.1 we discuss the description of physical quantities as a function of space and time coordinates. The second topic in this section introduces the different flow lines that are useful in our objective of describing a fluid flow. Finally, the mathematical description of motion is presented.

12.2.1 Lagrangian and Eulerian Descriptions of Motion

In the description of a flow field, it is convenient to think of individual particles each of which is considered to be a small mass of fluid, consisting of a large number of molecules, that occupies a small volume $\Delta V\!\!\!/$ that moves with the flow. If the fluid is incompressible, the volume does not change in magnitude but may deform. If the fluid is compressible, as the volume deforms, it also changes its magnitude. In both cases the particles are considered to move as an entity through a flow field.

In the study of particle mechanics, where attention is focused on individual particles, motion is observed as a function of time. The position, velocity, and acceleration of each particle are listed as $\mathbf{r}(x_0, y_0, z_0, t)$, $\mathbf{V}(x_0, y_0, z_0, t)$, and $\mathbf{a}(x_0, y_0, z_0, t)$ and quantities of interest can be calculated. The point (x_0, y_0, z_0) locates the starting point—the name—of each particle. This is the **Lagrangian** description of motion that is used in a course on dynamics. In the Lagrangian description many particles can be followed and their influence on one another noted. This becomes, however, a difficult task as the number of particles becomes extremely large, as in a fluid flow.

Lagrangian: *Description of motion where individual particles are observed as a function of time.*

An alternative to following each fluid particle separately is to identify points in space and then observe the velocity of particles passing each point; we can observe the rate of change of velocity as the particles pass each point, that is, $\partial\mathbf{V}/\partial x$, $\partial\mathbf{V}/\partial y$, and $\partial\mathbf{V}/\partial z$, and we can observe if the velocity is changing with time at each particular point, that is, $\partial\mathbf{V}/\partial t$. In this **Eulerian** description of motion, the flow properties, such as velocity, are functions of both space and time. In rectangular, Cartesian coordinates the velocity is expressed as $\mathbf{V} = \mathbf{V}(x, y, z, t)$. The region of flow that is being considered is called a **flow field**.

Eulerian: *Description of motion where the flow properties are functions of both space and time.*

Flow field: *The region of flow of interest.*

An example may clarify these two ways of describing motion. An engineering firm is hired to make recommendations that would improve the traffic flow in a large city. The engineering firm has two alternatives: Hire college students to travel in automobiles throughout the city recording the appropriate observations (the Lagrangian approach), or hire college students to stand at the intersections and record the required information (the Eulerian approach). A correct interpretation of each set of data would lead to the same set of recommendations, that is, the same solution. In this example it may not be obvious which approach would be preferred; in an introductory course in fluids, however, the Eulerian description is used exclusively since the physical laws using the Eulerian description are easier to apply to actual situations. Yet, there are examples where a Lagrangian description is needed, such as drifting buoys used to study ocean currents.

Steady flow: *Where quantities do not depend on time.*

If the quantities of interest do not depend on time, that is, $\mathbf{V} = \mathbf{V}(x, y, z)$, the flow is said to be a **steady flow**. Most of the flows of interest in this introductory textbook are steady flows. For a steady flow all flow quantities at a particular point are independent of time, that is,

$$\frac{\partial \mathbf{V}}{\partial t} = 0 \qquad \frac{\partial p}{\partial t} = 0 \qquad \frac{\partial \rho}{\partial t} = 0 \tag{12.1}$$

to list a few. It is implied that x, y, and z are held fixed in the above. Note that the properties of a fluid particle do, in general, vary with time; the velocity and pressure vary with time as a fluid particle progresses along its path in a flow, even in a steady flow. In a steady flow, however, properties do not vary with time at a fixed point.

12.2.2 Pathlines, Streaklines, and Streamlines

Pathline: *A history of the particle's locations.*

Three different lines help us in describing a flow field. A **pathline** is the locus of points traversed by a given particle as it travels in a field of flow; the pathline provides us with a "history" of the particle's locations. A photograph of a pathline would require a time exposure of an illuminated particle. A photograph showing pathlines of particles below a water surface with waves is given in Fig. 12.1.

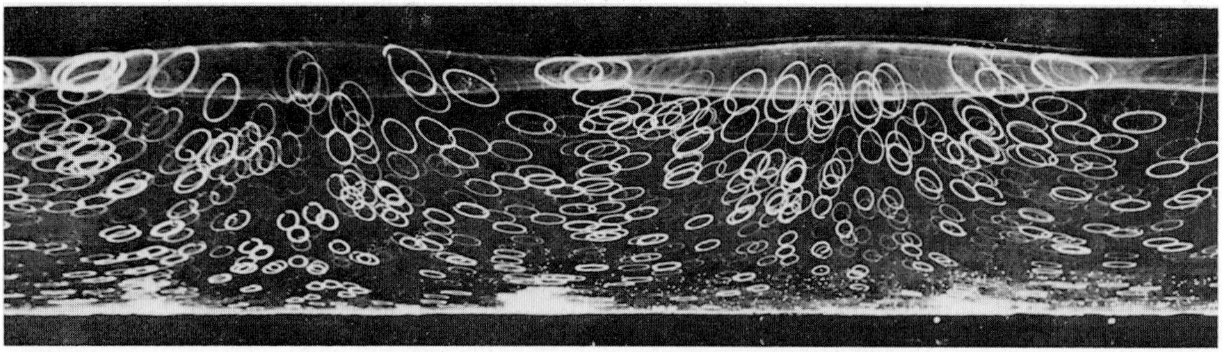

FIGURE 12.1 Pathlines underneath a wave in a tank of water. (Photograph by Wallet and Ruellan. Courtesy of M. C. Vasseur.)

FIGURE 12.2 Streaklines in the unsteady flow around a cylinder. (Photograph by Sadatashi Taneda.)

A **streakline** is defined as an instantaneous line whose points are occupied by all particles originating from some specified point in the flow field. Streaklines tell us where the particles are "right now." A photograph of a streakline would be a snapshot of the set of illuminated particles that passed a certain point. Figure 12.2 shows streaklines produced by the continuous release of a small-diameter stream of smoke as it moves around a cylinder.

Streakline: *An instantaneous line of particles passing a specified point.*

A **streamline** is a line in the flow possessing the following property: the velocity vector of each particle occupying a point on the streamline is tangent to the streamline. This is shown graphically in Fig. 12.3. An equation which expresses that the velocity vector is tangent to a streamline is

Streamline: *The velocity vector is tangent to the streamline.*

$$\mathbf{V} \times d\mathbf{r} = \mathbf{0} \tag{12.2}$$

since $\mathbf{V}$ and $d\mathbf{r}$ are in the same direction, as shown in the figure; recall that the cross product of two vectors in the same direction is zero. This equation will be used in future chapters as the mathematical expression of a streamline. A photograph of a streamline cannot be made directly. For a general unsteady flow, the streamlines can be inferred from photographs of short pathlines of a large number of particles.

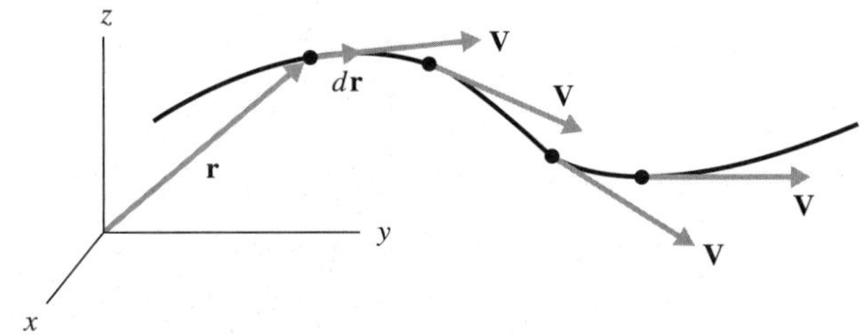

FIGURE 12.3 Streamline in a flow field.

Streamtube: *A tube whose walls are streamlines.*

A **streamtube** is a tube whose walls are streamlines. Since the velocity is tangent to a streamline, no fluid can cross the walls of a streamtube. The streamtube is of particular interest in fluid mechanics. A pipe is a streamtube since its walls are streamlines; an open channel is a streamtube since no fluid crosses the walls of the channel. We often sketch a streamtube with a small cross section in the interior of a flow for demonstration purposes.

Key Concept: *In a steady flow, pathlines, streaklines, and streamlines are all coincident.*

In a steady flow, pathlines, streaklines, and streamlines are all coincident. All particles passing a given point will continue to trace out the same path since nothing changes with time; hence the pathlines and streaklines coincide. In addition, the velocity vector of a particle at a given point will be tangent to the line that the particle is moving along; thus the line is also a streamline. Since the flows that we observe in laboratories are invariably steady flows, we call the lines that we observe streamlines even though they may actually be streaklines, or for the case of time exposures, pathlines.

12.2.3 Acceleration

The acceleration of a fluid particle is found by considering a particular particle shown in Fig. 12.4. Its velocity changes from $\mathbf{V}(t)$ at time t to $\mathbf{V}(t+dt)$ at time $t+dt$. The acceleration is, by definition,

$$\mathbf{a} = \frac{d\mathbf{V}}{dt} \tag{12.3}$$

where $d\mathbf{V}$ is shown in Fig. 12.4. The velocity vector $\mathbf{V}$ is given in component form as

$$\mathbf{V} = u\hat{\mathbf{i}} + v\hat{\mathbf{j}} + w\hat{\mathbf{k}} \tag{12.4}$$

where (u, v, w) are the velocity components in the x-, y-, and z-directions, respectively, and $\hat{\mathbf{i}}$, $\hat{\mathbf{j}}$ and $\hat{\mathbf{k}}$ are the unit vectors. The quantity $d\mathbf{V}$ is, using the chain rule from calculus with $\mathbf{V} = \mathbf{V}(x, y, z, t)$,

$$d\mathbf{V} = \frac{\partial \mathbf{V}}{\partial x}dx + \frac{\partial \mathbf{V}}{\partial y}dy + \frac{\partial \mathbf{V}}{\partial z}dz + \frac{\partial \mathbf{V}}{\partial t}dt \tag{12.5}$$

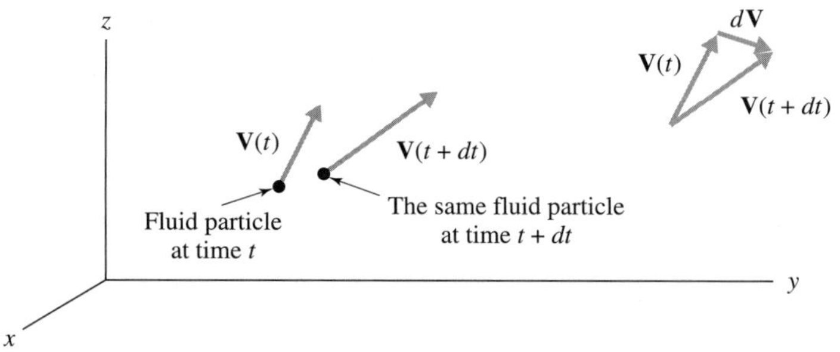

FIGURE 12.4 Velocity of a fluid particle.

This gives the acceleration as

$$\mathbf{a} = \frac{\partial \mathbf{V}}{\partial x}\frac{dx}{dt} + \frac{\partial \mathbf{V}}{\partial y}\frac{dy}{dt} + \frac{\partial \mathbf{V}}{\partial z}\frac{dz}{dt} + \frac{\partial \mathbf{V}}{\partial t} \tag{12.6}$$

Since we have followed a particular particle as in Fig. 12.4, we recognize that

$$\frac{dx}{dt} = u \qquad \frac{dy}{dt} = v \qquad \frac{dz}{dt} = w \tag{12.7}$$

The acceleration is then expressed as

$$\mathbf{a} = u\frac{\partial \mathbf{V}}{\partial x} + v\frac{\partial \mathbf{V}}{\partial y} + w\frac{\partial \mathbf{V}}{\partial z} + \frac{\partial \mathbf{V}}{\partial t} \tag{12.8}$$

The scalar component equations of the above vector equation for rectangular coordinates are written as

$$\begin{aligned} a_x &= \frac{\partial u}{\partial t} + u\frac{\partial u}{\partial x} + v\frac{\partial u}{\partial y} + w\frac{\partial u}{\partial z} \\ a_y &= \frac{\partial v}{\partial t} + u\frac{\partial v}{\partial x} + v\frac{\partial v}{\partial y} + w\frac{\partial v}{\partial z} \\ a_z &= \frac{\partial w}{\partial t} + u\frac{\partial w}{\partial x} + v\frac{\partial w}{\partial y} + w\frac{\partial w}{\partial z} \end{aligned} \tag{12.9}$$

We often return to Eq. 12.3 and write Eq. 12.8 in a simplified form as

$$\mathbf{a} = \frac{D\mathbf{V}}{Dt} \tag{12.10}$$

where, in rectangular coordinates,

$$\frac{D}{Dt} = u\frac{\partial}{\partial x} + v\frac{\partial}{\partial y} + w\frac{\partial}{\partial z} + \frac{\partial}{\partial t} \tag{12.11}$$

This derivative is called the **substantial derivative**, or **material derivative**. It is given a special name and special symbol (D/Dt instead of d/dt) because we followed a particular fluid particle, that is, we followed the substance (or material). It represents the relationship between a Lagrangian derivation in which a quantity depends on time t and a Eulerian derivation in which a quantity depends on position (x, y, z) and time t. The substantial derivative can be used with other dependent variables; for example, DT/Dt would represent the rate of change of the temperature of a fluid particle as we followed the particle along.

Substantial or material derivative: *The derivative D/Dt.*

The substantial derivative and acceleration components in cylindrical and spherical coordinates are presented in Table 12.1.

The time-derivative term on the right side of Eqs. 12.8 and 12.9 for the acceleration is called the **local acceleration** and the remaining terms on the right

Local acceleration: *The time-derivative term $\partial V/\partial t$ for acceleration.*

TABLE 12.1 The Substantial Derivative, Acceleration, and Vorticity in Cartesian, Cylindrical and Spherical Coordinates

Substantive derivative

Cartesian

$$\frac{D}{Dt} = u\frac{\partial}{\partial x} + v\frac{\partial}{\partial y} + w\frac{\partial}{\partial z} + \frac{\partial}{\partial t}$$

Cylindrical

$$\frac{D}{Dt} = v_r\frac{\partial}{\partial r} + \frac{v_\theta}{r}\frac{\partial}{\partial \theta} + v_z\frac{\partial}{\partial z} + \frac{\partial}{\partial t}$$

Spherical

$$\frac{D}{Dt} = v_r\frac{\partial}{\partial r} + \frac{v_\theta}{r}\frac{\partial}{\partial \theta} + \frac{v_\phi}{r\sin\theta}\frac{\partial}{\partial \phi} + \frac{\partial}{\partial t}$$

Acceleration

Cartesian

$$a_x = \frac{\partial u}{\partial t} + u\frac{\partial u}{\partial x} + v\frac{\partial u}{\partial y} + w\frac{\partial u}{\partial z}$$

$$a_y = \frac{\partial v}{\partial t} + u\frac{\partial v}{\partial x} + v\frac{\partial v}{\partial y} + w\frac{\partial v}{\partial z}$$

$$a_z = \frac{\partial w}{\partial t} + u\frac{\partial w}{\partial x} + v\frac{\partial w}{\partial y} + w\frac{\partial w}{\partial z}$$

Cylindrical

$$a_r = \frac{\partial v_r}{\partial t} + v_r\frac{\partial v_r}{\partial r} + \frac{v_\theta}{r}\frac{\partial v_r}{\partial \theta} + v_z\frac{\partial v_r}{\partial z} - \frac{v_\theta^2}{r}$$

$$a_\theta = \frac{\partial v_\theta}{\partial t} + v_r\frac{\partial v_\theta}{\partial r} + \frac{v_\theta}{r}\frac{\partial v_\theta}{\partial \theta} + v_z\frac{\partial v_\theta}{\partial z} + \frac{v_r v_\theta}{r}$$

$$a_z = \frac{\partial v_z}{\partial t} + v_r\frac{\partial v_z}{\partial r} + \frac{v_\theta}{r}\frac{\partial v_z}{\partial \theta} + v_z\frac{\partial v_z}{\partial z}$$

Spherical

$$a_r = \frac{\partial v_r}{\partial t} + v_r\frac{\partial v_r}{\partial r} + \frac{v_\theta}{r}\frac{\partial v_r}{\partial \theta} + \frac{v_\phi}{r\sin\theta}\frac{\partial v_r}{\partial \phi} - \frac{v_\phi^2 + v_\theta^2}{r}$$

$$a_\theta = \frac{\partial v_\theta}{\partial t} + v_r\frac{\partial v_\theta}{\partial t} + \frac{v_\theta}{r}\frac{\partial v_\theta}{\partial \theta} + \frac{v_\phi}{r\sin\phi}\frac{\partial v_\theta}{\partial \phi} + \frac{v_r v_\theta - v_\phi^2\cot\theta}{r}$$

$$a_\phi = \frac{\partial v_\phi}{\partial t} + v_r\frac{\partial v_\phi}{\partial r} + \frac{v_\theta}{r}\frac{\partial v_\phi}{\partial \theta} + \frac{v_\phi}{r\sin\theta}\frac{\partial v_\phi}{\partial \phi} + \frac{v_r v_\phi + v_\theta v_\phi\cot\theta}{r}$$

Vorticity

Cartesian

$$\omega_x = \frac{\partial w}{\partial y} - \frac{\partial v}{\partial z} \qquad \omega_y = \frac{\partial u}{\partial z} - \frac{\partial w}{\partial x} \qquad \omega_z = \frac{\partial v}{\partial x} - \frac{\partial u}{\partial y}$$

Cylindrical

$$\omega_r = \frac{1}{r}\left(\frac{\partial v_z}{\partial \theta}\right) - \frac{\partial v_\theta}{\partial z} \qquad \omega_\theta = \frac{\partial v_r}{\partial z} - \frac{\partial v_z}{\partial r} \qquad \omega_z = \frac{1}{r}\left(\frac{\partial (r v_\theta)}{\partial r} - \frac{\partial v_r}{\partial \theta}\right)$$

Spherical

$$\omega_r = \frac{1}{r\sin\theta}\left[\frac{\partial}{\partial \theta}(v_\phi \sin\theta) - \frac{\partial v_\theta}{\partial \phi}\right] \qquad \omega_\phi = \frac{1}{r}\left[\frac{\partial}{\partial r}(r v_\theta) - \frac{\partial v_r}{\partial \theta}\right]$$

$$\omega_\theta = \frac{1}{r}\left[\frac{1}{\sin\theta}\frac{\partial v_r}{\partial \phi} - \frac{\partial}{\partial r}(r v_\phi)\right]$$

Convective acceleration: *All other terms other than the local acceleration term.*

Key Concept:
Convective acceleration occurs in the vicinity of a change in the geometry.

side in each equation form the **convective acceleration**. Hence the acceleration of a fluid particle is the sum of the local acceleration and convective acceleration. In a pipe, local acceleration results if, for example, a valve is being opened or closed; and convective acceleration occurs in the vicinity of a change in the pipe geometry, such as a pipe contraction or an elbow. In both cases fluid particles change speed, but for very different reasons.

We must note that the foregoing expressions for acceleration give the acceleration relative to an observer in the observer's reference frame only. In certain situations the observer's reference frame may be accelerating; then the accelera-

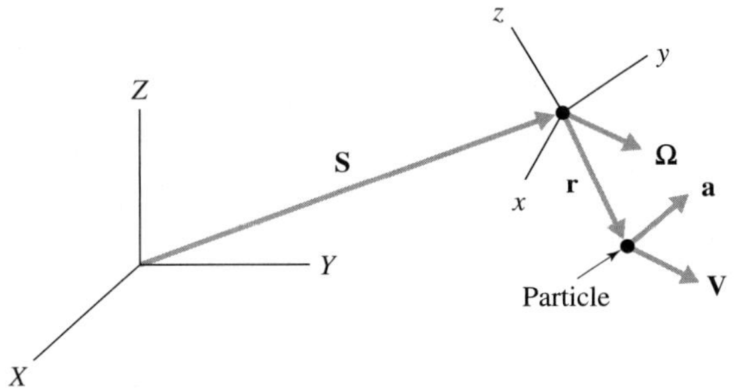

FIGURE 12.5 Motion relative to a noninertial reference frame.

tion of a particle relative to a fixed reference frame may be needed. It is given by

$$\mathbf{A} = \mathbf{a} + \underbrace{\frac{d^2\mathbf{S}}{dt^2}}_{\substack{\text{acceleration of}\\ \text{reference frame}}} + \underbrace{2\boldsymbol{\Omega}\times\mathbf{V}}_{\substack{\text{Coriolis}\\ \text{acceleration}}} + \underbrace{\boldsymbol{\Omega}\times(\boldsymbol{\Omega}\times\mathbf{r})}_{\substack{\text{normal}\\ \text{acceleration}}} + \underbrace{\frac{d\boldsymbol{\Omega}}{dt}\times\mathbf{r}}_{\substack{\text{angular}\\ \text{acceleration}}} \tag{12.12}$$

where **a** is given by Eq. 12.8, $d^2\mathbf{S}/dt^2$ is the acceleration of the observer's reference frame, **V** and **r** are the velocity and position vectors of the particle, respectively, in the observer's reference frame, and $\boldsymbol{\Omega}$ is the angular velocity of the observer's reference frame (see Fig. 12.5). For most engineering applications, reference frames attached to the earth yield $\mathbf{A} = \mathbf{a}$, since the other terms in Eq. 12.12 are often negligible with respect to **a**. We may decide, however, to attach the *xyz*-reference frame to an accelerating device (a rocket), or to a rotating device (a sprinkler arm); then certain terms of Eq. 12.12 must be included along with **a** of Eq. 12.8.

If the acceleration of all fluid particles is given by $\mathbf{A} = \mathbf{a}$ in a selected reference frame, that is an *inertial* reference frame. If $\mathbf{A} \neq \mathbf{a}$, it is a *noninertial* reference frame. A reference frame that moves with constant velocity without rotating is an inertial reference frame. When analyzing flow about, for example, an airfoil, we attach the reference frame to the airfoil so that a steady flow is observed in that reference frame.

12.2.4 Angular Velocity and Vorticity

A fluid flow may be thought of as the motion of a collection of fluid particles. As a particle travels along it may rotate or deform. The rotation and deformation of the fluid particles are of particular interest in our study of fluid mechanics. There are certain flows, or regions of a flow, in which the fluid particles do not rotate; such flows are of special importance, particularly in flows around objects, and are referred to as **irrotational flows**. Flow outside a thin boundary layer on airfoils, outside the separated flow region around autos and other moving vehicles, in the flow around submerged objects, and many other flows are examples of irrotational flows. Irrotational flows are extremely important.

Let us consider a small fluid particle that occupies an infinitesimal volume that has the *xy*-face as shown in Fig. 12.6. The **angular velocity** about the *z*-axis,

Irrotational flows: *Flows where the fluid particles do not rotate.*

Angular velocity: *The average velocity of two perpendicular line segments of a fluid particle.*

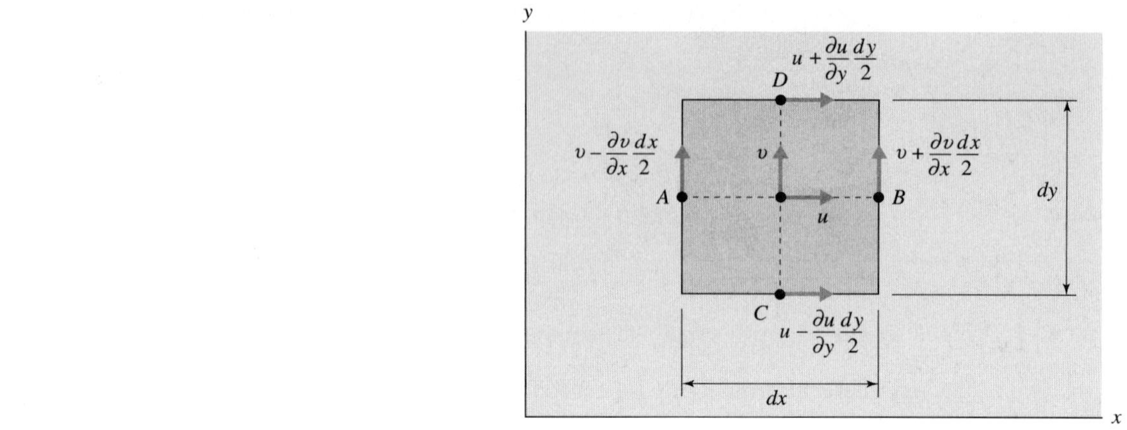

FIGURE 12.6 Fluid particle occupying an infinitesimal parallelepiped at a particular instant.

Ω_z, is the average of the angular velocity of line segment AB and line segment CD. The two angular velocities, counterclockwise being positive, are

$$\Omega_{AB} = \frac{v_B - v_A}{dx}$$

$$= \left[v + \frac{\partial v}{\partial x}\frac{dx}{2} - \left(v - \frac{\partial v}{\partial x}\frac{dx}{2} \right) \right] \Big/ dx = \frac{\partial v}{\partial x} \tag{12.13}$$

$$\Omega_{CD} = -\frac{u_D - u_C}{dy}$$

$$= -\left[u + \frac{\partial u}{\partial y}\frac{dy}{2} - \left(u - \frac{\partial u}{\partial y}\frac{dy}{2} \right) \right] \Big/ dy = \frac{\partial u}{\partial y} \tag{12.14}$$

Consequently, the angular velocity Ω_z of the fluid particle is

$$\Omega_z = \tfrac{1}{2}(\Omega_{AB} + \Omega_{CD})$$

$$= \frac{1}{2}\left(\frac{\partial v}{\partial x} - \frac{\partial u}{\partial y} \right) \tag{12.15}$$

If we had considered the xz-face, we would have found the angular velocity about the y-axis to be

$$\Omega_y = \frac{1}{2}\left(\frac{\partial u}{\partial z} - \frac{\partial w}{\partial x} \right) \tag{12.16}$$

and the yz-face would provide us with the angular velocity about the x-axis:

$$\Omega_x = \frac{1}{2}\left(\frac{\partial w}{\partial y} - \frac{\partial v}{\partial z} \right) \tag{12.17}$$

These are the three components of the angular velocity vector. A cork placed in a water flow in a wide channel (the xy-plane) would rotate with the angular velocity given by Eq. 12.15.

It is common to define the **vorticity** $\boldsymbol{\omega}$ to be twice the angular velocity; its three components are then

Vorticity: *Twice the angular velocity.*

$$\omega_x = \frac{\partial w}{\partial y} - \frac{\partial v}{\partial z} \qquad \omega_y = \frac{\partial u}{\partial z} - \frac{\partial w}{\partial x} \qquad \omega_z = \frac{\partial v}{\partial x} - \frac{\partial u}{\partial y} \tag{12.18}$$

The vorticity components in cylindrical and spherical coordinates are included in Table 12.1.

An irrotational flow possesses no vorticity; the cork mentioned above would not rotate in an irrotational flow. We consider this special flow in Section 16.5.

The deformation of the particle of Fig. 12.6 is the rate of change of the angle that line segment AB makes with line segment CD. If AB is rotating with an angular velocity different from that of CD, the particle is deforming. The deformation is represented by the **rate-of-strain tensor**; its component ϵ_{xy} in the xy-plane is given by

Rate-of-strain tensor: *The rate at which deformation occurs.*

$$\begin{aligned}\epsilon_{xy} &= \tfrac{1}{2}(\Omega_{AB} - \Omega_{CD}) \\ &= \frac{1}{2}\left(\frac{\partial v}{\partial x} + \frac{\partial u}{\partial y}\right)\end{aligned} \tag{12.19}$$

For the xz-plane and the yz-plane we have

$$\epsilon_{xz} = \frac{1}{2}\left(\frac{\partial w}{\partial x} + \frac{\partial u}{\partial z}\right) \qquad \epsilon_{yz} = \frac{1}{2}\left(\frac{\partial w}{\partial y} + \frac{\partial v}{\partial z}\right) \tag{12.20}$$

Observe that $\epsilon_{xy} = \epsilon_{yx}$, $\epsilon_{xz} = \epsilon_{zx}$, and $\epsilon_{yz} = \epsilon_{zy}$. The rate-of-strain tensor is symmetric.

The fluid particle could also deform by being stretched or compressed in a particular direction. For example, if point B of Fig. 12.6 is moving faster than point A, the particle would be stretching in the x-direction. This normal rate of strain is measured by

$$\begin{aligned}\epsilon_{xx} &= \frac{u_B - u_A}{dx} \\ &= \left[u + \frac{\partial u}{\partial x}\frac{dx}{2} - \left(u - \frac{\partial u}{\partial x}\frac{dx}{2}\right)\right] \Big/ dx = \frac{\partial u}{\partial x}\end{aligned} \tag{12.21}$$

Similarly, in the y- and z-directions we would find that

$$\epsilon_{yy} = \frac{\partial v}{\partial y} \qquad \epsilon_{zz} = \frac{\partial w}{\partial z} \tag{12.22}$$

The rate-of-strain tensor can be displayed as

$$\epsilon_{ij} = \begin{pmatrix} \epsilon_{xx} & \epsilon_{xy} & \epsilon_{xz} \\ \epsilon_{yx} & \epsilon_{yy} & \epsilon_{yz} \\ \epsilon_{zx} & \epsilon_{zy} & \epsilon_{zz} \end{pmatrix} \tag{12.23}$$

where the subscripts i and j take on numerical values 1, 2, or 3. Then ϵ_{12} represents ϵ_{xy} in row 1 column 2.

The normal and shear stress components in a flow are related to the foregoing rate-of-strain components; they are displayed in Appendix K.2. In fact, in the one-dimensional flow of Fig. 10.6, the shear stress was related to $\partial u/\partial y$ with Eq. 10.16; note that $\partial u/\partial y$ is twice the rate-of-strain component given by Eq. 12.19 with $v = 0$.

Example 12.1

The velocity field is given by $\mathbf{V} = 2x\hat{\mathbf{i}} - yt\hat{\mathbf{j}}$ m/s, where x and y are in meters and t is in seconds. Find the equation of the streamline passing through $(2, -1)$ and a unit vector normal to the streamline at $(2, -1)$ at $t = 4$ s.

Solution

The velocity vector is tangent to a streamline so that $\mathbf{V} \times d\mathbf{r} = 0$ (the cross product of two parallel vectors is zero). For the given velocity vector we have, at $t = 4$ s,

$$(2x\hat{\mathbf{i}} - 4y\hat{\mathbf{j}}) \times (dx\hat{\mathbf{i}} + dy\hat{\mathbf{j}}) = (2x\,dy + 4y\,dx)\hat{\mathbf{k}} = 0$$

where we have used $\hat{\mathbf{i}} \times \hat{\mathbf{j}} = \hat{\mathbf{k}}$, $\hat{\mathbf{j}} \times \hat{\mathbf{i}} = -\hat{\mathbf{k}}$, and $\hat{\mathbf{i}} \times \hat{\mathbf{i}} = 0$. Consequently,

$$2x\,dy = -4y\,dx \qquad \text{or} \qquad \frac{dy}{y} = -2\frac{dx}{x}$$

Integrate both sides:

$$\ln y = -2 \ln x + \ln C$$

where we used $\ln C$ for convenience. This is written as

$$\ln y = \ln x^{-2} + \ln C = \ln(Cx^{-2})$$

Hence

$$x^2 y = C$$

At $(2, -1)$ $C = -4$, so that the streamline passing through $(2, -1)$ has the equation

$$x^2 y = -4$$

A normal vector is perpendicular to the streamline, hence the velocity vector, so that using $\hat{\mathbf{n}} = n_x\hat{\mathbf{i}} + n_y\hat{\mathbf{j}}$ we have at $(2, -1)$ and $t = 4$ s

$$\mathbf{V} \cdot \hat{\mathbf{n}} = (4\hat{\mathbf{i}} + 4\hat{\mathbf{j}}) \cdot (n_x\hat{\mathbf{i}} + n_y\hat{\mathbf{j}}) = 0$$

Using $\hat{\mathbf{i}} \cdot \hat{\mathbf{i}} = 1$ and $\hat{\mathbf{i}} \cdot \hat{\mathbf{j}} = 0$, this becomes

$$4n_x + 4n_y = 0 \qquad \therefore\ n_x = -n_y$$

Then, because $\hat{\mathbf{n}}$ is a unit vector, $n_x^2 + n_y^2 = 1$ and we find that

$$n_x^2 = 1 - n_x^2 \qquad \therefore\ n_x = \frac{\sqrt{2}}{2}$$

A unit vector normal to the streamline is written as

$$\hat{\mathbf{n}} = \frac{\sqrt{2}}{2}(\hat{\mathbf{i}} - \hat{\mathbf{j}})$$

Example 12.2

A velocity field in a particular flow is given by $\mathbf{V} = 20y^2\hat{\mathbf{i}} - 20xy\hat{\mathbf{j}}$ m/s. Calculate the acceleration, the angular velocity, the vorticity vector, and any nonzero rate-of-strain components at the point $(1, -1, 2)$.

Solution

We could use Eq. 12.9 and find each component of the acceleration, or we could use Eq. 12.8 and find a vector expression. Using Eq. 12.8, we have

$$\begin{aligned}\mathbf{a} &= u\frac{\partial \mathbf{V}}{\partial x} + v\frac{\partial \mathbf{V}}{\partial y} + \overset{0}{\cancel{w\frac{\partial \mathbf{V}}{\partial z}}} + \overset{0}{\cancel{\frac{\partial \mathbf{V}}{\partial t}}}\\ &= 20y^2(-20y\hat{\mathbf{j}}) - 20xy(40y\hat{\mathbf{i}} - 20x\hat{\mathbf{j}})\\ &= -800xy^2\hat{\mathbf{i}} - 400(y^3 - x^2y)\hat{\mathbf{j}}\end{aligned}$$

where we have used $u = 20y^2$ and $v = -20xy$, as given by the velocity vector. All particles at the point $(1, -1, 2)$ have the acceleration

$$\mathbf{a} = -800\hat{\mathbf{i}} \text{ m/s}^2$$

The angular velocity has two zero components:

$$\Omega_x = \frac{1}{2}\left(\overset{0}{\cancel{\frac{\partial w}{\partial y}}} - \overset{0}{\cancel{\frac{\partial v}{\partial z}}}\right) = 0, \qquad \Omega_y = \frac{1}{2}\left(\overset{0}{\cancel{\frac{\partial u}{\partial z}}} - \overset{0}{\cancel{\frac{\partial w}{\partial x}}}\right) = 0$$

The non-zero z-component is, at the point $(1, -1, 2)$,

$$\begin{aligned}\Omega_z &= \left(\frac{1}{2}\frac{\partial v}{\partial x} - \frac{\partial u}{\partial y}\right)\\ &= \tfrac{1}{2}(-20y - 40y) = 30 \text{ rad/s}\end{aligned}$$

The vorticity vector is twice the angular velocity vector:

$$\boldsymbol{\omega} = 2\Omega_z\hat{\mathbf{k}} = 60\hat{\mathbf{k}} \text{ rad/s}$$

The nonzero rate-of-strain components are at $(1, -1, 2)$

$$\begin{aligned}\epsilon_{xy} &= \frac{1}{2}\left(\frac{\partial v}{\partial x} + \frac{\partial u}{\partial y}\right)\\ &= \tfrac{1}{2}(-20y + 40y) = -10 \text{ rad/s}\\ \epsilon_{yy} &= \frac{\partial v}{\partial y}\\ &= -20x = -20 \text{ rad/s}\end{aligned}$$

All other rate-of-strain components are zero.

12.3 CLASSIFICATION OF FLUID FLOWS

In this section we provide an overview of some of the aspects of fluid mechanics that are considered in more depth in subsequent chapters and sections. Although most of the notions presented here are redefined and discussed in detail later, it will be helpful at this point to introduce the general classification of fluid flows.

12.3.1 One-, Two-, and Three-Dimensional Flows

Three-dimensional flow: *The velocity vector depends on three space variables.*

In the Eulerian description of motion the velocity vector, in general, depends on three space variables and time, that is, $\mathbf{V} = \mathbf{V}(x, y, z, t)$. Such a flow is a **three-dimensional flow**, because the velocity vector depends on three space coordinates. The solutions to problems in such a flow are very difficult and are beyond the scope of an introductory course. Even if the flow could be assumed to be steady [i.e., $\mathbf{V} = \mathbf{V}(x, y, z)$], it would remain a three-dimensional flow. A particular flow is shown in Fig. 12.7. This flow is normal to a plane surface; the fluid decelerates and comes to rest at the **stagnation point**. The velocity components, u, v, and w depend on x, y and z; that is, $u = u(x, y, z)$, $v = v(x, y, z)$ and $w = w(x, y, z)$.

Stagnation point: *The point where the fluid comes to rest.*

Often a three-dimensional flow can be approximated as a two-dimensional flow. For example, the flow over a wide dam is three-dimensional because of the end conditions, but the flow in the central portion away from the ends can be treated as two-dimensional. In general, a **two-dimensional flow** is a flow in which the velocity vector depends on only two space variables. An example is a **plane flow**, in which the velocity vector depends on two spatial coordinates, x and y, but not z [i.e., $\mathbf{V} = \mathbf{V}(x, y)$].

Two-dimensional flow: *The velocity vector is dependent on only two space variables.*

Plane flow: *The velocity vector depends on the two coordinates x and y.*

A **one-dimensional flow** is a flow in which the velocity vector depends on only one space variable. Such flows occur in long, straight pipes or between parallel plates, as shown in Fig. 12.8. The velocity in the pipe varies only with r, i.e., $u = u(r)$. The velocity between parallel plates varies only with the coordinate y, i.e., $u = u(y)$. Even if the flow is unsteady so that $u = u(y, t)$, as would be the situation during startup, the flow is one-dimensional.

One-dimensional flow: *The velocity vector depends on only one space variable.*

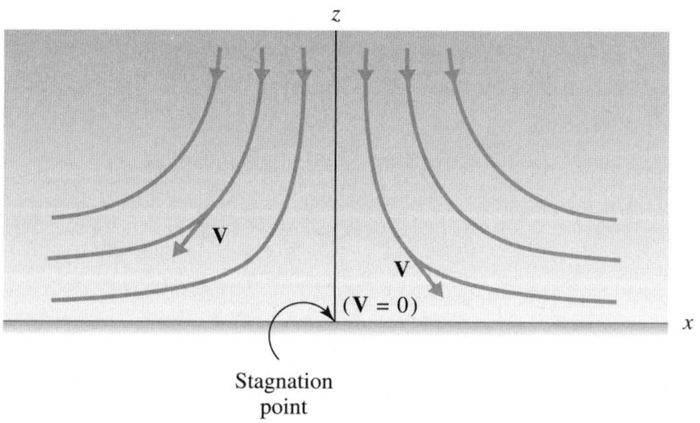

FIGURE 12.7 A stagnation point flow.

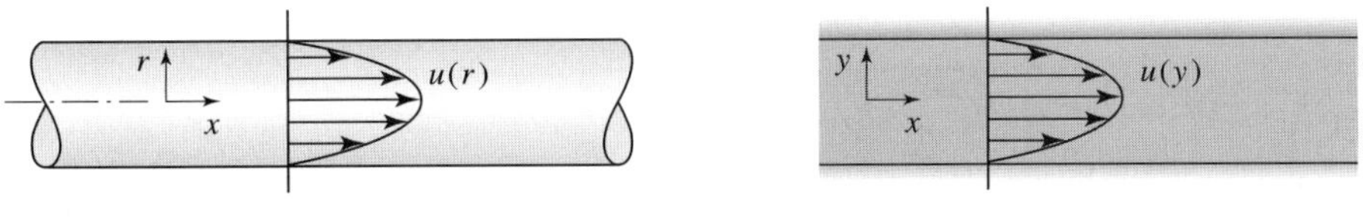

FIGURE 12.8 One-dimensional flow: (a) flow in a pipe; (b) flow between parallel plates.

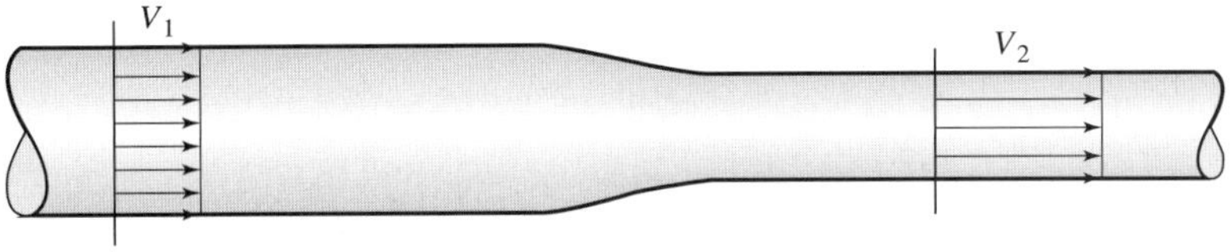

FIGURE 12.9 Uniform velocity profiles.

The flows shown in Fig. 12.8 may also be referred to as **developed flows**; that is, the velocity profiles do not vary with respect to the space coordinate in the direction of flow. This demands that the region of interest be a substantial distance from an entrance or a sudden change in geometry.

Developed flows: *The velocity profiles do not vary with respect to the space coordinate in the direction of flow.*

There are many engineering problems in fluid mechanics in which a flow field is simplified to a **uniform flow**: the velocity, and other fluid properties, are constant over the area as in Fig. 12.9. This simplification is made when the velocity is essentially constant over the area, a rather common occurrence. Examples of such flows are relatively high speed flow in a pipe section, and flow in a stream. The average velocity may change from one section to another; the flow conditions depend only on the space variable in the flow direction. For large conduits, however, it may be necessary to consider hydrostatic variation in the pressure normal to the streamlines.

Uniform flow: *The fluid properties are constant over the area.*

12.3.2 Viscous and Inviscid Flows

A fluid flow may be broadly classified as either a viscous flow or an inviscid flow. An **inviscid flow** is one in which viscous effects do not significantly influence the flow and are thus neglected. In a **viscous flow** the effects of viscosity are important and cannot be ignored.

Inviscid flow: *Viscous effects do not significantly influence the flow.*

Viscous flow: *The effects of viscosity are significant.*

To model an inviscid flow analytically, we can simply let the viscosity be zero; this will obviously make all viscous effects zero. It is more difficult to create an inviscid flow experimentally, because all fluids of interest (such as water and air) possess viscosity. The question then becomes: Are there flows of interest in which the viscous effects are negligibly small? The answer is "yes, if the shear stresses in the flow are small and act over such small areas that they do not significantly affect the flow field." This statement is very general, of course, and it will take considerable analysis to justify the inviscid flow assumption.

Based on experience, it has been found that the primary class of flows, which can be modeled as inviscid flows, is **external flows**, that is, flows which exist exte-

External flows: *Flows which exist exterior to a body.*

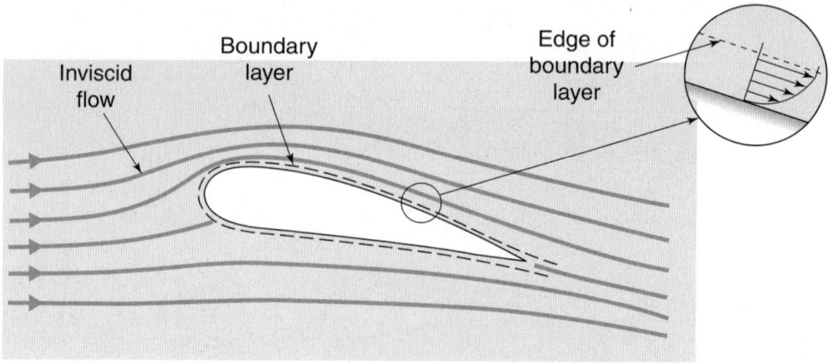

FIGURE 12.10 Flow around an airfoil.

rior to a body. Inviscid flows are of primary importance in flows around streamlined bodies, such as flow around an airfoil or a hydrofoil. Any viscous effects that may exist are confined to a thin layer, called a **boundary layer**, that is attached to the boundary, such as that shown in Fig. 12.10; the velocity in a boundary layer is always zero at a fixed wall, a result of viscosity. For many flow situations, boundary layers are so thin that they can simply be ignored when studying the gross features of a flow around a streamlined body. For example, the inviscid flow solution provides an excellent prediction to the flow around the airfoil, except inside the boundary layer and possibly near the trailing edge. Inviscid flow is also encountered in contractions inside piping systems and in short regions of internal flows where viscous effects are negligible.

Boundary layer: *A thin layer attached to the boundary in which viscous effects are concentrated.*

Key Concept: *Inviscid flow provides an excellent prediction to the flow around an airfoil.*

Viscous flows include the broad class of internal flows, such as flows in pipes and conduits and in open channels. In such flows viscous effects cause substantial "losses" and account for the huge amounts of energy that must be used to transport oil and gas in pipelines. The no-slip condition resulting in zero velocity at the wall, and the resulting shear stresses, lead directly to these losses.

12.3.3 Laminar and Turbulent Flows

A viscous flow can be classified as either a laminar flow or a turbulent flow. In a **laminar flow** the fluid flows with no significant mixing of neighboring fluid particles. If dye were injected into the flow, it would not mix with the neighboring fluid except by molecular activity; it would retain its identity for a relatively long period of time. Viscous shear stresses always influence a laminar flow. The flow may be highly time dependent, due to the erratic motion of a piston as shown by the output of a velocity probe in Fig. 12.11a, or it may be steady, as shown in Fig. 12.11b.

Laminar flow: *A flow with no significant mixing of particles but with significant viscous shear stresses.*

In a **turbulent flow** fluid motions vary irregularly so that quantities such as velocity and pressure show a random variation with time and space coordinates. The physical quantities are often described by statistical averages. In this sense we can define a "steady" turbulent flow: a flow in which the time-average physical quantities do not change in time. Figure 12.12 shows instantaneous velocity

Turbulent flow: *Flow varies irregularly so that flow quantities show random variation.*

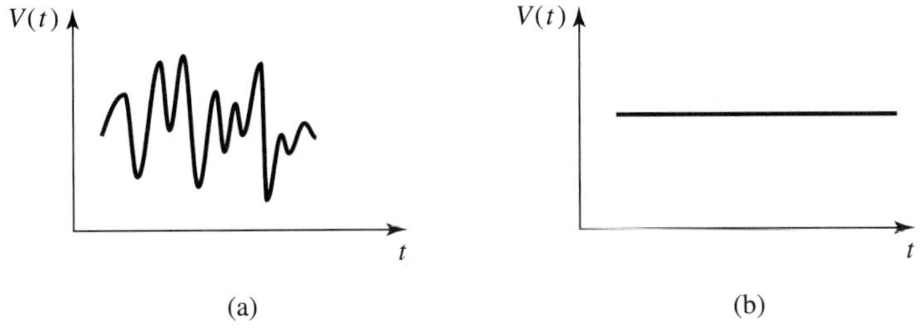

FIGURE 12.11 Velocity as a function of time in a laminar flow: (a) unsteady flow; (b) steady flow.

Key Concept: *A dye injected into a turbulent flow would mix immediately.*

measurements in an unsteady and a steady turbulent flow. A dye injected into a turbulent flow would mix immediately by the action of the randomly moving fluid particles; the dye would quickly lose its identity in this diffusion process.

A laminar flow and a turbulent flow can be observed by performing a simple experiment with a water faucet. Turn the faucet on so the water flows out very slowly as a silent stream. This is laminar flow. Open the faucet slowly and observe the flow becoming turbulent. Note that a turbulent flow develops with a relatively low flow rate.

The reason why a flow can be laminar or turbulent has to do with what happens to a small flow disturbance, a perturbation to the velocity components. A flow disturbance can either increase or decrease in size. If a flow disturbance in a laminar flow increases (i.e., the flow is unstable), the flow may become turbulent; if the disturbance decreases, the flow remains laminar. In certain situations the flow may develop into a different laminar flow, as is the case between concentrically rotating cylinders shown in Fig. 12.13. At low rotational speed the flow would be in simple circles. But at a sufficiently high speed the flow becomes unstable and the vortices suddenly appear; it is a much more complex laminar flow.

The flow regime depends on three physical parameters describing the flow conditions. The first parameter is a length scale of the flow field, such as the thickness of a boundary layer or the diameter of a pipe. If this length scale is

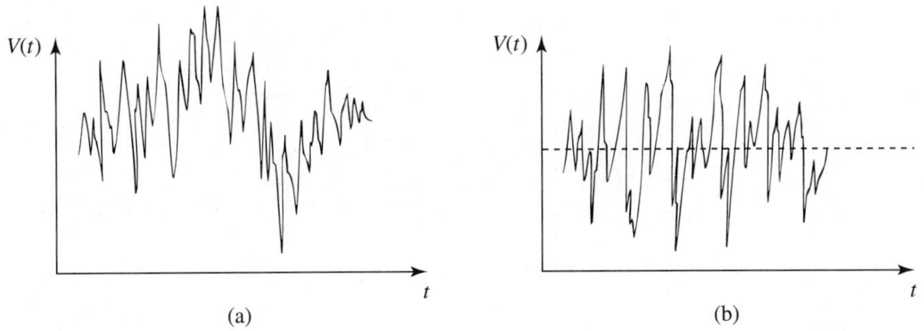

FIGURE 12.12 Velocity as a function of time in a turbulent flow: (a) unsteady flow; (b) "steady" flow.

sufficiently large, a flow disturbance may increase and the flow may be turbulent. The second parameter is a velocity scale such as a spatial average of the velocity; for a large enough velocity the flow may be turbulent. The third parameter is the kinematic viscosity; for a small enough viscosity the flow may be turbulent.

Reynolds number: *Parameter combining a length of scale, a velocity scale, and the kinematic viscosity into a dimensionless parameter.*

The three parameters can be combined into a single parameter that can serve as a tool to predict the flow regime. This quantity is the **Reynolds number**, a dimensionless parameter, defined as

$$\mathrm{Re} = \frac{VL}{\nu} \tag{12.24}$$

Critical Reynolds number: *The number above which a primary laminar flow ceases to exist.*

where L and V are a characteristic length and velocity, respectively, and ν is the kinematic viscosity; for example, in a pipe flow L could be the pipe diameter and V could be the average velocity. If the Reynolds number is relatively small, the flow is laminar as shown in Figs. 12.13 and 12.14; if it is large, the flow is turbulent. This is more precisely stated by defining a **critical Reynolds number**, $\mathrm{Re}_{\mathrm{crit}}$,

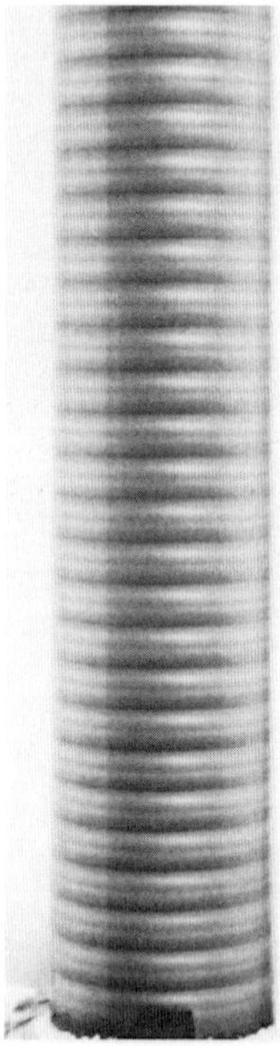

FIGURE 12.13 Laminar flow between rotating cylinders. A secondary flow occurs as regularly spaced toroidal vortices. (Photograph by Burkhalter and Koschmieder.)

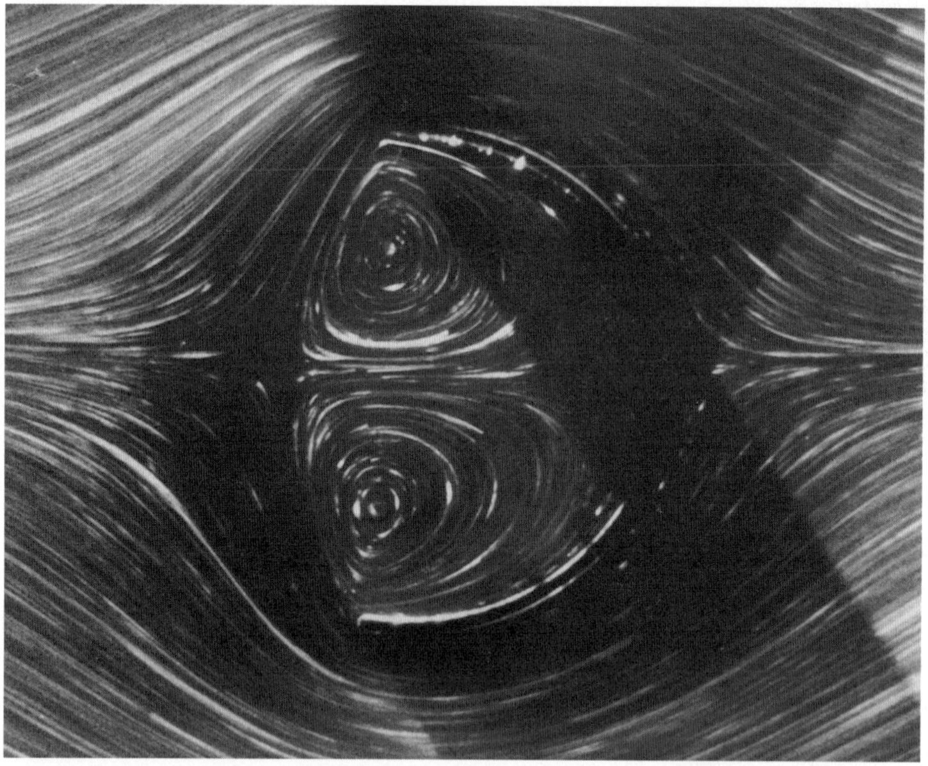

FIGURE 12.14 Streamlines around a semicircular arc. At this Reynolds number of 0.031, the centers of the pair of eddies in the cavity are separated by 0.52 diameter, in good agreement with a numerical solution. Aluminum powder dispersed in glycerine is illuminated by a slit of light. (Courtesy of The Parabolic Press, Stanford, California. Reprinted with permission.)

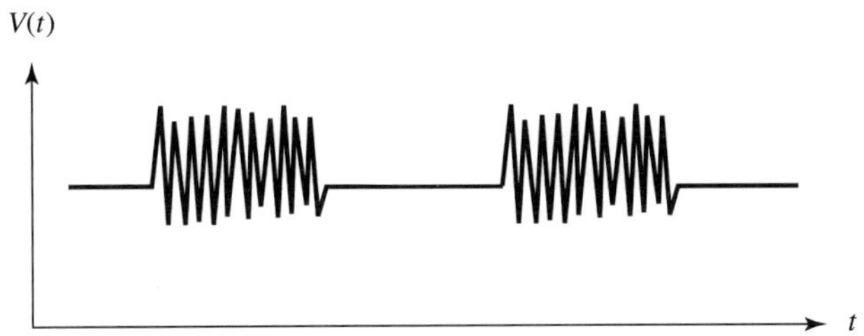

FIGURE 12.15 Velocity versus time signal from a velocity probe in an intermittent flow.

so that the flow is laminar if $\text{Re} < \text{Re}_{\text{crit}}$. For example, in a flow inside a rough-walled pipe it is found that $\text{Re}_{\text{crit}} \approx 2000$. This is the minimum critical Reynolds number and is used for most engineering applications. If the pipe wall is extremely smooth and free of vibration, the critical Reynolds number can be increased as the fluctuation level in the flow is decreased; values in excess of 40 000 have been measured. The critical Reynolds number is different for every geometry, e.g, it is 1500 for flow between parallel plates using the average velocity and the distance between the plates.

The flow can also be intermittently turbulent and laminar; this is called an *intermittent flow*. This phenomenon can occur when the Reynolds number is close to Re_{crit}. Figure 12.15 shows the output of a velocity probe for such a flow.

In a boundary layer that exists on a flat plate, due to a constant-velocity fluid stream, as shown in Fig. 12.16, the length scale changes with distance from the upstream edge. A Reynolds number is calculated using the length x as the characteristic length. For a certain x_T, Re becomes Re_{crit} and the flow undergoes transition from laminar to turbulent. For a smooth rigid plate in a uniform flow with a low free-stream fluctuation level, values as high as $\text{Re}_{\text{crit}} = 10^6$ have been observed. In most engineering applications we assume a rough wall, or high free-stream fluctuation level, with an associated critical Reynolds number of approximately 3×10^5.

Key Concept: *In most applications, we assume a critical Reynolds number of 3×10^5 in flow on a flat plate.*

It is not appropriate to refer to an inviscid flow as laminar or as turbulent. The inviscid flow of Fig. 12.10 is often called the **free stream**. The free stream can be irrotational or it can possess vorticity; most often it is irrotational.

Free stream: *The inviscid flow outside the boundary layer in an external flow.*

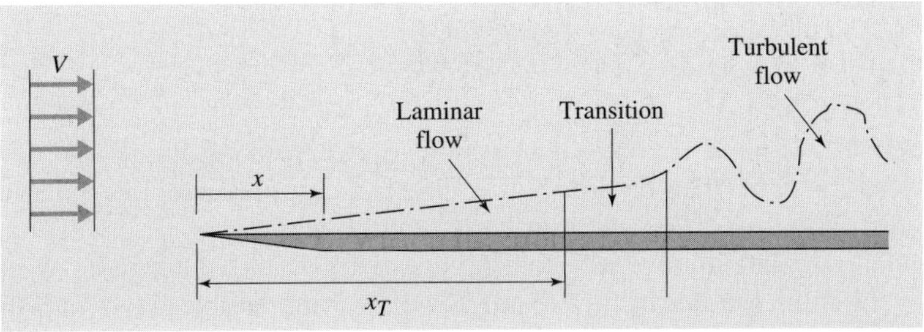

FIGURE 12.16 Boundary layer flow on a flat plate.

Example 12.3

The 2-cm-diameter pipe of Fig. E12.3 is used to transport water at 20°C. What is the maximum average velocity that may exist in the pipe for which laminar flow is guaranteed?

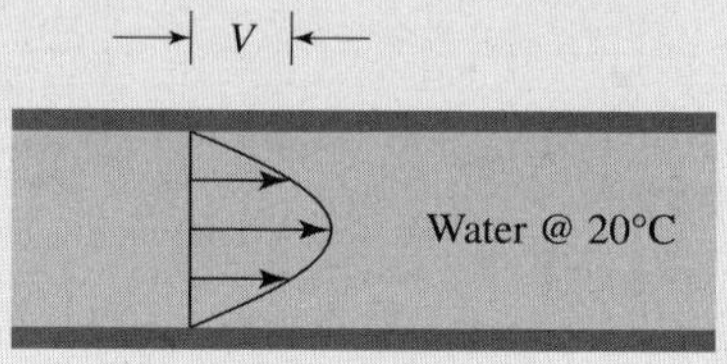

FIGURE E12.3

Solution

The kinematic viscosity is found in Appendix B.9 to be $\nu = 10^{-6}$ m^2/s. Using a Reynolds number of 2000 so that a laminar flow is guaranteed, we find that

$$V = \frac{2000\nu}{D}$$

$$= \frac{2000 \times 10^{-6}}{0.02} = 0.1 \text{ m/s}$$

This average velocity is quite small. Velocities this small are not usually encountered in actual situations; hence laminar flow is seldom of engineering interest except for specialized topics such as lubrication. Most internal flows are turbulent flows and thus the study of turbulence gains much attention.

12.3.4 Incompressible and Compressible Flows

Incompressible flow: *The density of each fluid particle remains constant.*

The last major classification of fluid flows to be considered in this chapter separates flows into incompressible and compressible flows. An **incompressible flow** exists if the density of each fluid particle remains relatively constant as it moves through the flow field, that is,

$$\frac{D\rho}{Dt} = 0 \tag{12.25}$$

Key Concept: *Constant density is more restrictive than incompressibility.*

This does not demand that the density is everywhere constant. If the density is everywhere constant, then obviously, the flow is incompressible, but that would be a more restrictive condition. Atmospheric flow, in which $\rho = \rho(z)$, where z is vertical, and flows that involve adjacent layers of fresh and salt water, as happens when rivers enter the ocean, are examples of incompressible flows in which the density varies.

In addition to liquid flows, low-speed gas flows, such as the atmospheric flow referred to above, are also considered to be incompressible flows. The *Mach number* is defined as

$$M = \frac{V}{c} \tag{12.26}$$

where V is the gas speed and the wave speed $c = \sqrt{kRT}$. Equation 12.26 is useful in deciding whether a particular gas flow can be studied as an incompressible flow. If $M < 0.3$, density variations are at most 3% and the flow is assumed to be incompressible; for standard air this corresponds to a velocity below about 100 m/s or 300 ft/sec. If $M > 0.3$, the density variations influence the flow and compressibility effects should be accounted for; such flows are **compressible flows** and are considered in Chapter 17.

Compressible flow: *Density variations influence the flow.*

Incompressible gas flows include atmospheric flows, the aerodynamics of landing and takeoff of commercial aircraft, heating and air-conditioning airflows, flow around automobiles and through radiators, and the flow of air around buildings, to name a few. Compressible flows include the aerodynamics of high-speed aircraft, airflow through jet engines, steam flow through the turbine in a power plant, airflow in a compressor, and the flow of the air-gas mixture in an internal combustion engine.

12.4 THE BERNOULLI EQUATION

In this section we present an equation that is probably used more often in fluid flow applications than any other equation. It is also often misused; it is thus important to understand its limitations. Its limitations are a result of several assumptions made in the derivation. One of the assumptions is that viscous effects are neglected. In other words, in view of Eq. 10.16, shear stresses introduced by velocity gradients are not taken into consideration. These stresses are often very small compared with pressure differences in the flow field. Locally, these stresses have little effect on the flow field and the assumption is justified. However, over long distances or in regions of high-velocity gradients, these stresses may affect the flow conditions so that viscous effects must be included.

Key Concept: *Shear stresses are often very small compared with pressure differences.*

The derivation of this important equation, the Bernoulli equation, starts with the application of Newton's second law to a fluid particle. Let us use an infinitesimal cylindrical particle positioned as shown in Fig. 12.17, with length ds and cross-sectional area dA. The forces acting on the particle are the pressure forces and the weight, as shown. Summing forces in the direction of motion, the s-direction, there results

$$p\,dA - \left(p + \frac{\partial p}{\partial s} ds\right) dA - \rho g\, ds\, dA \cos\theta = \rho\, ds\, dA\, a_s \tag{12.27}$$

where a_s is the acceleration of the particle in the s-direction. It is given by[1]

[1]This can be verified by considering Eq. 12.9a, assuming that $v = w = 0$. Think of the x-direction being tangent to the streamline at the instant shown, so that $u = V$.

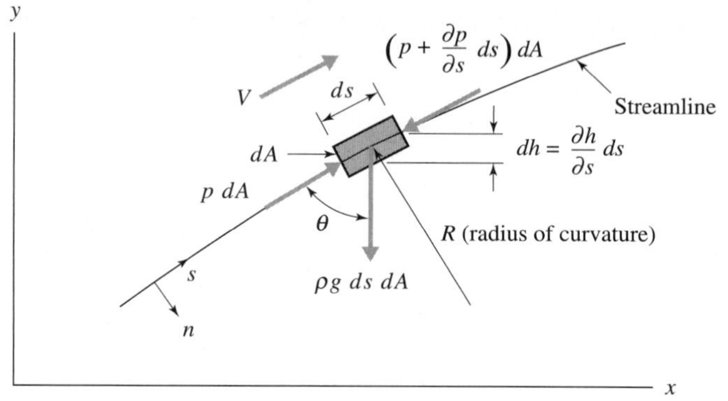

FIGURE 12.17 Particle moving along a streamline.

$$a_s = V\frac{\partial V}{\partial s} + \frac{\partial V}{\partial t} \tag{12.28}$$

where $\partial V/\partial t = 0$ since we will assume steady flow. Also, we see that

$$dh = ds \cos\theta = \frac{\partial h}{\partial s} ds \tag{12.29}$$

so that

$$\cos\theta = \frac{\partial h}{\partial s} \tag{12.30}$$

Then, after dividing by $ds\,dA$, and using the above equations for a_s and $\cos\theta$, Eq. 12.27 takes the form

$$-\frac{\partial p}{\partial s} - \rho g\frac{\partial h}{\partial s} = \rho V\frac{\partial V}{\partial s} \tag{12.31}$$

Now, we assume constant density and note that $V\,\partial V/\partial s = \partial(V^2/2)/\partial s$; then we can write Eq. 12.31 as

$$\frac{\partial}{\partial s}\left(\frac{V^2}{2} + \frac{p}{\rho} + gh\right) = 0 \tag{12.32}$$

This is satisfied if, along the streamline,

$$\frac{V^2}{2} + \frac{p}{\rho} + gh = \text{const} \tag{12.33}$$

where the constant may have a different value on a different streamline. Between two points on the same streamline,

$$\frac{V_1^2}{2} + \frac{p_1}{\rho} + gh_1 = \frac{V_2^2}{2} + \frac{p_2}{\rho} + gh_2 \tag{12.34}$$

This is the well-known *Bernoulli equation*. Note the assumptions:

— Inviscid flow (no shear stresses)
— Steady flow ($\partial V/\partial t = 0$)
— Along a streamline ($a_s = V\,\partial V/\partial s$)
— Constant density ($\partial \rho/\partial s = 0$)
— Inertial reference frame ($\mathbf{A} = \mathbf{a}$ in Eq. 12.12)

If Eq. 12.34 is divided by g, this equation becomes

$$\frac{V_1^2}{2g} + \frac{p_1}{\gamma} + h_1 = \frac{V_2^2}{2g} + \frac{p_2}{\gamma} + h_2 \tag{12.35}$$

Key Concept: *The total pressure is* $p + \rho V^2/2$.

The sum of the three terms is called the *total head*. The pressure p is often referred to as **static pressure** and the sum of the two terms

$$p + \rho\frac{V^2}{2} = p_T \tag{12.36}$$

is called the *total pressure* p_T or **stagnation pressure**, the pressure at a stagnation point (see Fig. 12.7) in the flow.

Static pressure: *The pressure p, usually expressed as gage pressure.*

Stagnation pressure: *The pressure that exists at a stagnation point.*

The static pressure in a pipe can be measured simply by installing a so-called **piezometer**, shown[2] in Fig. 12.18a. A device, known as a **pitot probe**, sketched in Fig. 12.18b, is used to measure the total pressure in a fluid flow. Point 2 just inside the pitot tube is a stagnation point; the velocity there is zero. The difference between the readouts can be used to determine the velocity at point 1. A **pitot-static probe** is also used to measure the difference between total and static pressure with one probe (Fig. 12.18c). The velocity at point 1 (using the readings of the piezometer and pitot probes, or the reading from the pitot-static probe) can be determined by applying the Bernoulli equation between points 1 and 2:

Piezometer: *Gauge designed to measure static pressure.*

Pitot probe: *Gauge designed to measure total pressure.*

Pitot-static probe: *Gauge designed to measure the difference between total and static pressure.*

$$\frac{V_1^2}{2g} + \frac{p_1}{\gamma} = \frac{p_2}{\gamma} \tag{12.37}$$

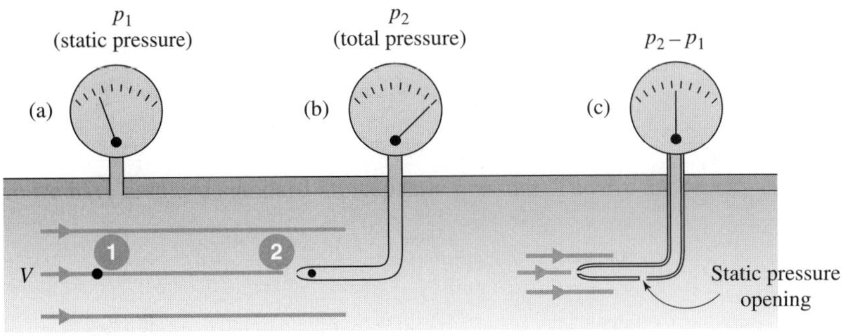

FIGURE 12.18 Pressure probes: (a) piezometer; (b) pitot probe; (c) pitot-static probe.

[2]When drilling the hole in the wall necessary for the piezometer, burrs are often formed on the inner surface. It is important that such burrs be removed since they may cause errors as high as 30% in the pressure reading.

where we have assumed point 2 to be a stagnation point so that $V_2 = 0$. This gives

$$V_1 = \sqrt{\frac{2}{\rho}(p_2 - p_1)} \tag{12.38}$$

Key Concept: *Never confuse Bernoulli's equations with the energy equation.*

We will find many uses for Bernoulli's equation in our study of fluids. We must be careful, however, never to use it in an unsteady flow or if viscous effects are significant (the primary reasons for making Bernoulli's equation inapplicable). We must also never confuse Bernoulli's equation with the energy equation; they are independent equations as Example 12.6 illustrates.

The Bernoulli equation can be used to determine how high the water from a fireman's hose will reach, to find the pressure on the surface of a low-speed airfoil,[3] and to find the wind force on the window in a house. These examples are all external flows, flows around objects submerged in the fluid.

Key Concept: *Viscous effects usually require substantial areas in order to be significant.*

Another class of problems where inviscid flow can be assumed and where the Bernoulli equation finds frequent application involves internal flows over relatively short distances, for example, flow through a contraction, as shown in Fig. 12.19a, or flow from a plenum, as shown in Fig. 12.19b. For a given velocity profile entering the short contraction, the pressure drop $(p_1 - p_2)$ and the velocity profile at section 2 can be approximated assuming an inviscid flow. Viscous effects are typically very small and require substantial distances and areas over which to operate in order to become significant; so in situations such as those shown in Fig. 12.19, viscous effects can often be neglected.

Inviscid flow does not always give a good approximation to the actual flow that exists around a body. Consider the inviscid flow around the sphere, shown in Fig. 12.20. A stagnation point where $V = 0$ exists at both the front and the back of the sphere. Bernoulli's equation predicts a maximum pressure at the stagnation points A and C because the velocity is zero at such points. A maximum velocity, and thus a minimum pressure, would exist at point B. In the inviscid flow of part (a) the fluid flowing from B to C must flow from the low-pressure region near B

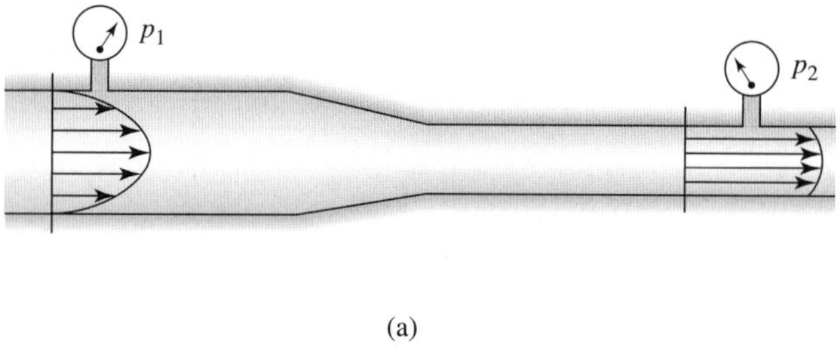

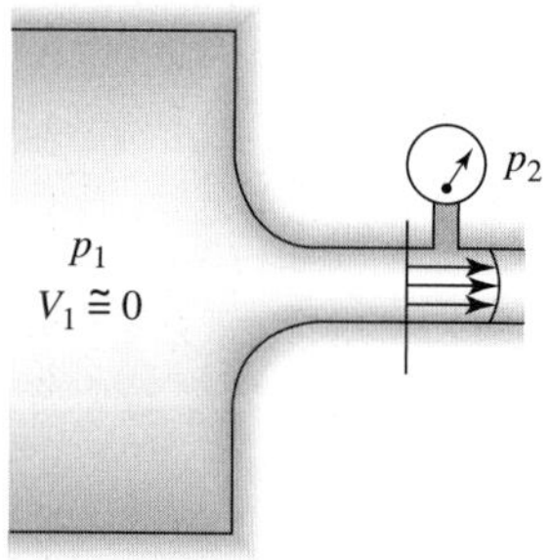

FIGURE 12.19 Internal inviscid flows: (a) flow through a contraction; (b) flow from a plenum.

[3]To consider the flow around an aircraft as a steady flow, we simply make the aircraft stationary and move the air, as is done in model studies using a wind tunnel. The pressures and forces remain unchanged.

to the high-pressure region near *C*. In the actual flow there exists a thin boundary layer in which the velocity goes to zero at the surface of the sphere. This slow-moving fluid near the boundary does not have sufficient momentum to move into the higher-pressure region near *C*; the result is that the fluid *separates* from the boundary—the boundary streamline leaves the boundary—creating a **separated region**, a region of recirculating flow, as shown in the actual flow sketched in part (b). The pressure does not increase but remains relatively low over the rear part of the sphere. The high pressure that exists near the front stagnation point is never recovered on the rear of the sphere, resulting in a relatively large drag force in the direction of flow. A similar situation occurs in the flow around an automobile.

Separated region: *A region of recirculating flow due to the fluid separating from the boundary.*

Key Concept: *Pressure remains relatively low over the rear part of a sphere.*

The flow on the front of the sphere is well approximated by an inviscid flow; but it is obvious that the flow over the rear of the sphere deviates radically from an inviscid flow. Viscous effects in the boundary layer have led to a separated flow, a phenomenon that is often undesirable. For example, separated flow on an airfoil is called **stall** and must never occur, except on the wings of special stunt planes. On the blades of a turbine separated flows lead to substantially reduced efficiency. The air deflector on the roof of the cab of a semitruck reduces the separated region, thereby reducing drag and fuel consumption.

Key Concept: *The flow on the front of a sphere is approximated by an inviscid flow.*

Stall: *Separated flow on an airfoil.*

If viscous effects are negligible in a steady liquid flow, we can use Bernoulli's equation to locate points of possible *cavitation*. This condition occurs when the local pressure becomes equal to the vapor pressure of the liquid. It is to be avoided, if at all possible, because of damage to solid surfaces or because the vaporized liquid may cause devices to not operate effectively. Figure 12.21 shows cavitating flow just downstream of a contraction in a pipe. At the point where cavitation occurs, small vapor bubbles are generated and these bubbles collapse when they enter a higher-pressure region. The collapse is accompanied by very large local pressures that last for only a small fraction of a second. These pressure spikes may reach a wall, where they can, after repeated applications, result in significant damage.

Key Concept: *Cavitation occurs when the local pressure equals the vapor pressure.*

An important observation regarding pressure changes in a fluid needs to be made regarding entrances and exits to a pipe or conduit. Consider a flow from a reservoir through a pipe, as sketched in Fig. 12.22. At the entrance the streamlines are curved and the pressure is not constant across section 1, so the pressure at

Key Concept: *At a pipe entrance, the streamlines are curved and the pressure is not constant across the inlet area.*

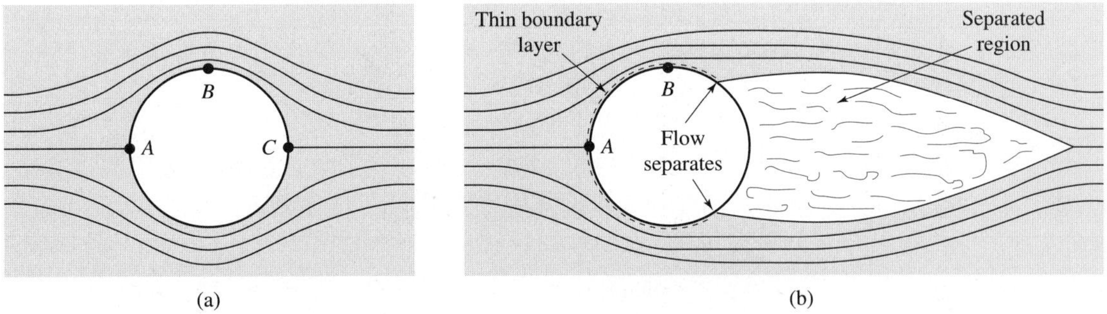

FIGURE 12.20 Flow around a sphere: (a) inviscid flow; (b) actual flow.

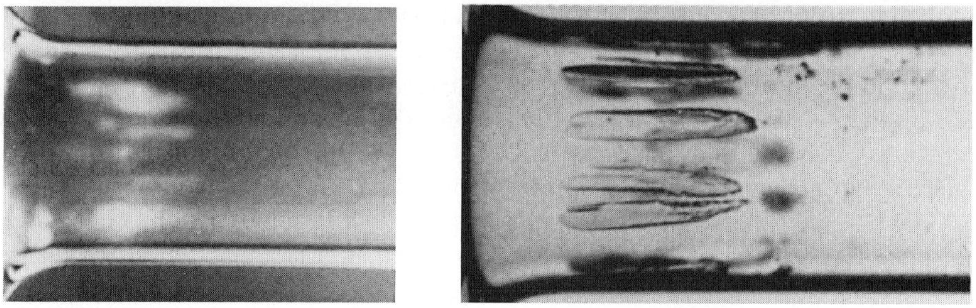

(a) (b)

FIGURE 12.21 Cavitation in a nozzle, with water flowing at a velocity of 15 m/s: (a) incandescent lamp, exposure time $\frac{1}{30}$ s; (b) strobe light exposure time 5 μs. (Photograph courtesy of the Japan Society of Mechanical Engineers and Pergamon Press.)

section 1 cannot be assumed to be uniform. At the exit, however, the streamlines are straight, so that no acceleration exists normal to the streamlines; hence the pressure forces acting on the ends of the small cylindrical control volume must be equal. We write this as $p_2 = p_{atm}$ or $p_2 = 0$ gage pressure.

In Fig. 12.17 we summed forces on the fluid element along the streamline and derived Bernoulli's equation. We can gain additional insight into the pressure field if we sum forces normal to the streamline. Let us consider the fluid particle to be a parallelpiped with thickness dn in the n-direction and area dA_s on the side with length ds. Applying Newton's second law in the n-direction provides

$$p\,dA_s - \left(p + \frac{\partial p}{\partial n}dn\right) dA_s = \rho\,dA_s\,dn\frac{V^2}{R} \tag{12.39}$$

where we have neglected the weight since we are not intending to integrate over significant distances. We have assumed the acceleration in the normal direction to be V^2/R, where R is the radius of curvature in this plane flow (in a three-dimensional flow there would be a principal radius of curvature and a binormal radius of curvature). Equation (12.39) reduces to

$$-\frac{\partial p}{\partial n} = \rho\frac{V^2}{R} \tag{12.40}$$

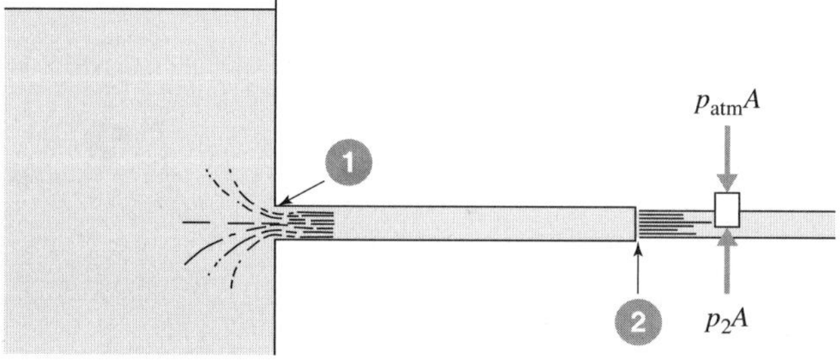

FIGURE 12.22 Exit flow into the atmosphere.

From this equation we can qualitatively describe how the pressure changes normal to a streamline (Bernoulli's equation predicts the pressure changes along a streamline). If we replace $\partial p/\partial n$ with $\Delta p/\Delta n$, the incremental pressure change Δp over the short distance Δn normal to the streamline is given by

$$\Delta p = -\rho \frac{V^2}{R} \Delta n \tag{12.41}$$

This says that the pressure decreases in the n-direction; this decrease is directly proportional to ρ and V^2 and inversely proportional to R. Consequently, a tornado, with $p = 0$ outside the tornado, will have a very low pressure at its center where R is relatively small and V is quite large.

Key Concept: *The pressure decreases in the n-direction.*

In Fig. 12.22 the pressure would be relatively low at the corner of section 1 and relatively high at the center of section 1. Such qualitative descriptions can be quite helpful in understanding fluid flow behavior.

Example 12.4

The wind reaches a speed of 65 mph in a storm. Calculate the force acting on the 3 ft × 6 ft window of Fig. E12.4 facing the storm. The window is in a high-rise building, so the wind speed is not reduced due to ground effects. Use $\rho = 0.0024$ slug/ft^3.

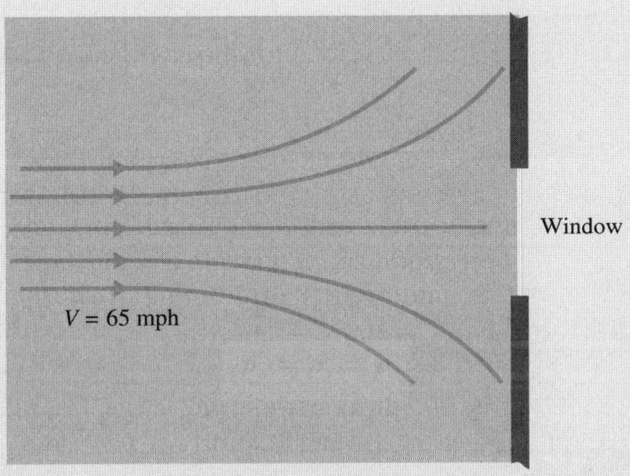

FIGURE E12.4

Solution

The window facing the storm will be in a stagnation region where the wind speed is brought to zero. Working with gage pressures, the pressure p upstream in the wind is zero. The velocity V must have units of ft/sec. It is

$$V = 65\ \frac{\text{mi}}{\text{hr}} \times \frac{1\ \text{hr}}{3600\ \text{sec}} \times \frac{5280\ \text{ft}}{1\ \text{mi}} = 95.3\ \text{ft/sec}$$

(continued)

Bernoulli's equation can be used in this situation since we can neglect viscous effects, and steady flow occurs along a streamline at constant density (air is incompressible at speeds below about 250 mph). We calculate the pressure on the window selecting state 1 in the free stream and state 2 on the window, as follows:

$$\frac{V_1^2}{2g} + \frac{\cancelto{0}{p_1}}{\gamma} + h_1 = \frac{\cancelto{0}{V_2^2}}{2g} + \frac{p_2}{\gamma} + h_2$$

$$\therefore\ p_2 = \frac{\rho V_1^2}{2}$$

$$= \frac{0.0024 \times 95.3^2}{2} = 10.9\ \text{lb/ft}^2$$

where we have used $\gamma = \rho g$, $h_2 = h_1$, $p_1 = 0$, and $V_2 = 0$. Multiply by the area and find the force to be

$$F = pA$$

$$= 10.9 \times 3 \times 6 = 196\ \text{lb}$$

We recommend that you verify the units of lb/ft^2 on the pressure calculation above. To do this, use $F = ma$ which provides slug $= \text{lb}\cdot\text{sec}^2/\text{ft}$. When using English units, always use mass in slugs, length in feet, force in pounds, and time in seconds.

Example 12.5

The static pressure head in an air pipe (Fig. E12.5) is measured with a piezometer as 16 mm of water. A pitot probe at the same location indicates 24 mm of water. Calculate the velocity of the 20°C air. Also, calculate the Mach number and comment as to the compressibility of the flow.

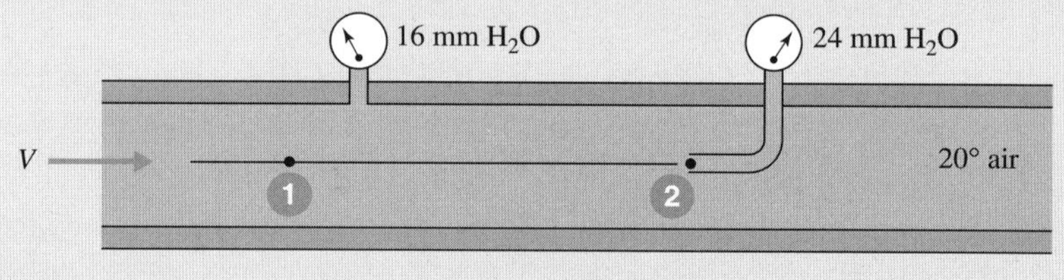

FIGURE E12.5

Solution

Bernoulli's equation is applied between two points on the streamline that terminates at the stagnation point of the pitot probe. Point 1 is upstream and p_2 is the total pressure at point 2; then, with no elevation change,

$$\frac{V_1^2}{2g} + \frac{p_1}{\gamma} = \frac{p_T}{\gamma}$$

(continued)

The pressure measured with the piezometer is $p_1 = \gamma h = 9810 \times 0.016 = 157$ Pa. We use the ideal gas law to calculate the density:

$$\rho = \frac{p}{RT}$$

$$= \frac{157 + 101\,000}{287 \times (273 + 20)} = 1.203 \text{ kg/m}^3$$

where standard atmospheric pressure, which is 101 000 Pa (if no elevation is given, assume standard conditions), is added since absolute pressure is needed in the preceding equation. The units are checked by using Pa = N/m^2 and J = N·m. The velocity is then

$$V_1 = \sqrt{\frac{2}{\rho}(p_T - p_1)}$$

$$= \sqrt{\frac{2(0.024 - 0.016) \times 9810}{1.203}} = 11.42 \text{ m/s}$$

where the units can be verified by using kg = N·s^2/m. To find the Mach number, we must calculate the speed of sound. It is

$$c = \sqrt{kRT}$$

$$= \sqrt{1.4 \times 287 \times 293} = 343 \text{ m/s}$$

The Mach number is then

$$\text{M} = \frac{V}{c} = \frac{11.44}{343} = 0.0334$$

Obviously, the flow can be assumed to be incompressible since M < 0.3. The velocity would have to be much higher before compressibility would be significant.

Example 12.6

Bernoulli's equation, in the form of Eq. 12.34, looks very much like the energy equation developed in thermodynamics for a control volume. Discuss the differences between the two equations.

Solution

From thermodynamics we recall that the steady-flow energy equation for a control volume with one inlet and one outlet takes the form

$$\dot{Q} - \dot{W}_s = \dot{m}\left(\frac{V_2^2}{2} + \frac{p_2}{\rho_2} + \tilde{u}_2 + gz_2\right) - \dot{m}\left(\frac{V_1^2}{2} + \frac{p_1}{\rho_1} + \tilde{u}_1 + gz_1\right)$$

(continued)

This becomes, after dividing through by g,

$$\frac{V_2^2}{2g} + \frac{p_2}{\gamma} + z_2 = \frac{V_1^2}{2g} + \frac{p_1}{\gamma} + z_1$$

where we have made the following assumptions:

No heat transfer ($\dot{Q} = 0$)
No shaft work ($\dot{W}_s = 0$)
No temperature change ($\tilde{u}_2 = \tilde{u}_1$, i.e., no losses due to shear stresses)
Uniform velocity profiles at the two sections
Steady flow
Constant density ($\gamma_2 = \gamma_1$)

Even though several of these assumptions are the same as those made in the derivation of the Bernoulli equation (steady flow, constant density, and no shear stress), we must not confuse the two equations; the Bernoulli equation is derived from Newton's second law and is valid along a streamline, whereas the energy equation is derived from the first law of thermodynamics and is valid between two sections in a fluid flow. The energy equation can be used across a pump to determine the horsepower required to provide a particular pressure rise; the Bernoulli equation can be used along a stagnation streamline to determine the pressure at a stagnation point, a point where the velocity is zero. The equations are quite different, and just because the energy equation degenerates to the Bernoulli equation for particular situations, the two should not be used out of context.

Example 12.7

Explain why a burr on the upstream side of the piezometer opening of Fig. 12.18a will result in a low reading of the pressure.

Solution

A burr on the upstream side of the piezometer opening would result in a flow in the vicinity of the burr somewhat like that shown in Fig. E12.7. A streamline pattern would develop so that a relatively high pressure would occur on the upstream side of the burr and a relatively low pressure on the downstream side at the opening of the piezometer tube. Consequently, since the center of the curvature of the streamline is in the vicinity of the opening, a low reading of the pressure will be recorded.

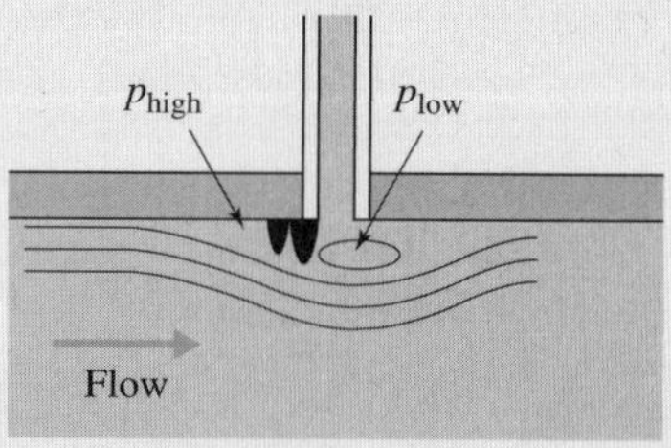

FIGURE E12.7

12.5 SUMMARY

The Eulerian description of motion was used to express the acceleration as

$$\mathbf{a} = u\frac{\partial \mathbf{V}}{\partial x} + v\frac{\partial \mathbf{V}}{\partial y} + w\frac{\partial \mathbf{V}}{\partial z} + \frac{\partial \mathbf{V}}{\partial t} \tag{12.42}$$

Fluid motion may result in fluid particles being rotated and/or deformed. For a flow in the xy-plane a particle would rotate with angular velocity

$$\Omega_z = \frac{1}{2}\left(\frac{\partial v}{\partial x} - \frac{\partial u}{\partial y}\right) \tag{12.43}$$

and deform with

$$\epsilon_{xx} = \frac{\partial u}{\partial x}, \qquad \epsilon_{yy} = \frac{\partial v}{\partial y}, \qquad \epsilon_{xy} = \frac{1}{2}\left(\frac{\partial v}{\partial x} + \frac{\partial u}{\partial y}\right) \tag{12.44}$$

Fluid flows are classified as steady or unsteady; viscous or inviscid; laminar, turbulent, or free stream; and incompressible or compressible. Any of these can be uniform, one-, two-, or three-dimensional flows. Experience and practice are needed to appropriately classify a particular flow of interest. It is only for the simpler flows (e.g., a steady, laminar, incompressible one-dimensional) that we hope to obtain a relatively simple solution.

Finally, the Bernoulli equation

$$\frac{V_1^2}{2g} + \frac{p_1}{\gamma} + gh_1 = \frac{V_2^2}{2g} + \frac{p_2}{\gamma} + gh_2 \tag{12.45}$$

was presented for a steady, inviscid, constant-density flow along a streamline in an inertial reference frame. The estimate of the pressure change normal to a streamline,

$$\Delta p = -\rho\frac{V^2}{R}\Delta n \tag{12.46}$$

was also obtained.

REVIEW PROBLEMS FOR THE FE/EIT EXAMINATION

12.1FE Determine the unit vector normal to the streamline at a point where $\mathbf{V} = 3\hat{\imath} - 4\hat{\jmath}$ in a plane flow.

A. $0.6\hat{\imath} + 0.8\hat{\jmath}$
B. $-0.6\hat{\imath} + 0.8\hat{\jmath}$
C. $0.8\hat{\imath} - 0.6\hat{\jmath}$
D. $0.8\hat{\imath} + 0.6\hat{\jmath}$

12.2FE A velocity field is given by $\mathbf{V} = 2xy\hat{\imath} - y^2\hat{\jmath}$ m/s. The magnitude of the acceleration at (−1 m, 2 m) is nearest:

A. 11.21 m/s^2 **B.** 14.69 m/s^2
C. 17.89 m/s^2 **D.** 1.2 m/s^2

12.3FE The velocity shown in Fig. P12.3FE is given by $V(x) = 10/(4 - x)^2$ m/s. The acceleration at $x = 2$ m is nearest:

A. 52.5 m/s^2 **B.** 42.5 m/s^2
C. 25 m/s^2 **D.** 6.25 m/s^2

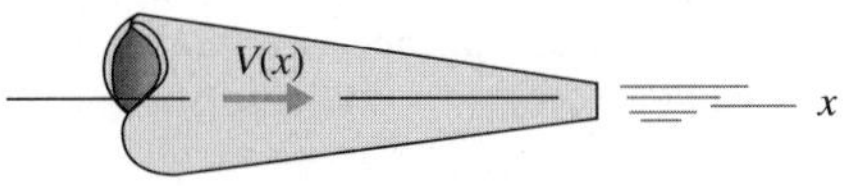

FIGURE P12.3FE

12.4FE The flow in the section of the conduit shown in Fig. P12.3FE is a:
A. Developed flow
B. Uniform flow
C. One-dimensional flow
D. Two-dimensional flow

12.5FE The velocity of an airplane is measured with a pitot tube. If the pitot tube measures 800 mm of water, estimate the speed of the airplane. Use $\rho_{air} = 1.23\ kg/m^3$.
A. 125 m/s
B. 113 m/s
C. 80 m/s
D. 36 m/s

12.6FE A pitot tube measures 600 mm of water in a pipe transporting water. A static pressure probe at the same location measures 200 mm of water. The velocity of water in the pipe is nearest:
A. 1.10 m/s **B.** 1.98 m/s
C. 2.8 m/s **D.** 3.43 m/s

12.7FE A manometer, utilizing a pitot probe, measures 10 mm of mercury. If the velocity is desired in a pipe transporting water to which the manometer is attached, what additional information in the following list is needed?

I. The temperature of the water	**A.** I and II
II. The pressure in the pipe	**B.** II and III
III. The density of the mercury	**C.** III and IV
IV. The diameter of the pipe	**D.** III and IV

12.8FE A water hose is pressurized to 800 kPa with a nozzle in the off position. If the nozzle is opened a small amount, as shown in Fig. P12.8FE, estimate the exiting velocity of the water. Assume the velocity inside the hose to be negligible.
A. 40 m/s **B.** 30 m/s
C. 20 m/s **D.** 10 m/s

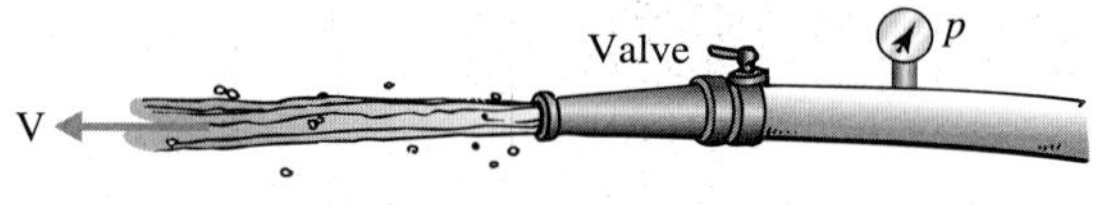

FIGURE P12.8FE

12.9FE Benzene flows out through the disks shown in Fig. P12.9FE. If $V_2 = 30$ m/s, the pressure p_1 is nearest:
A. 150 kPa **B.** 200 kPa
C. 250 kPa **D.** 300 kPa

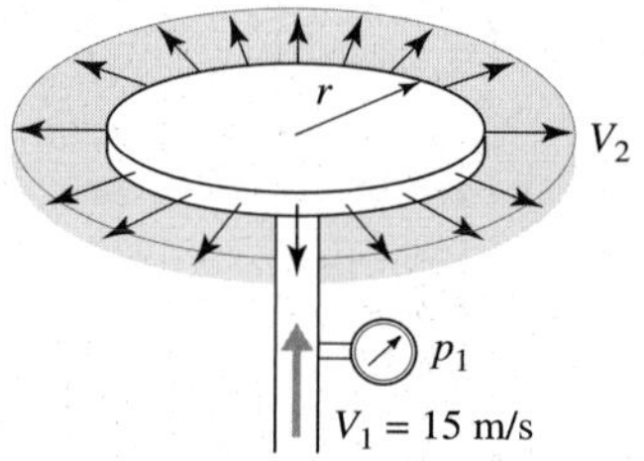

FIGURE P12.9FE

PROBLEMS

Flow Fields

12.1 A fire is started and the smoke from the chimney goes straight up; no wind is present. After a few minutes a wind arises but the smoke continues to rise slowly. Sketch the streakline of the smoke, the pathline of the first particles leaving the chimney, and a few streamlines, assuming that the wind blows parallel to the ground in a constant direction.

12.2 A researcher has a large number of small flotation devices each equipped with a battery and light bulb. Explain how she would determine the pathlines and streaklines near the surface of a stream with some unknown currents that vary with time.

12.3 A little boy chases his dad around the yard with the water hose of Fig. P12.3. Sketch a pathline and a streakline if the boy is running perpendicular to the water jet.

FIGURE P12.3

12.4 The hot-air balloon of Fig. P12.4 travels with the wind. The wind velocity vector is $\mathbf{V} = 6\hat{\mathbf{i}} + 10\hat{\mathbf{j}}$ m/s for the first hour and then it is $10\hat{\mathbf{i}} + 5\hat{\mathbf{j}}$ m/s for 2 hr. On xy-coordinates, sketch the pathline of the balloon and streamlines at $t = 2$ hrs. If a number of hot-air balloons started from the same location, sketch the streakline formed by the balloons at $t = 3$ hr. The balloons start at the origin.

FIGURE P12.4

12.5 A velocity field is given by $\mathbf{V} = (2t + 2)\hat{\mathbf{i}} + 2t\hat{\mathbf{j}}$ m/s. Sketch the pathlines of two particles up to $t = 5$ s, one that originates from the origin at $t = 0$, and another that originates from the origin at $t = 2$ s. Also, sketch the streamlines at $t = 5$ s.

12.6 Using rectangular coordinates, express the z-component of Eq. 12.2.

12.7 The traffic situation on Mackinac Island, Michigan, where no automobiles are allowed (bikes are permitted), is to be studied. Comment on how such a study could be performed using a Lagrangian approach and an Eulerian approach.

12.8 Determine the speed of a fluid particle at the origin and at the point $(1, -2, 0)$ for each of the following velocity fields when $t = 2$ s. All distances are in meters and t is in seconds.
(a) $\mathbf{V} = (x + 2)\hat{\mathbf{i}} + xt\hat{\mathbf{j}} - z\hat{\mathbf{k}}$ m/s
(b) $\mathbf{V} = xy\hat{\mathbf{i}} - 2y^2\hat{\mathbf{j}} + tyz\hat{\mathbf{k}}$ m/s
(c) $\mathbf{V} = x^2t\hat{\mathbf{i}} - (xz + 2t)\hat{\mathbf{j}} + xyt\hat{\mathbf{k}}$ m/s

12.9 Calculate the angle that the velocity vector makes with the x-axis; and a unit vector normal to the streamline at $(1, -2)$ for the following velocity fields when $t = 2$ s. All distances are in meters and t is in seconds.
(a) $\mathbf{V} = (x + 2)\hat{\mathbf{i}} + xt\hat{\mathbf{j}}$ m/s
(b) $\mathbf{V} = xy\hat{\mathbf{i}} - 2y^2\hat{\mathbf{j}}$ m/s
(c) $\mathbf{V} = (x^2 + 4)\hat{\mathbf{i}} - y^2t\hat{\mathbf{j}}$ m/s

12.10 Find the equation of the streamline that passes through $(1, -2)$ at $t = 2$ s for the flow of:
(a) Problem 12.9a **(b)** Problem 12.9b
(c) Problem 12.9c

12.11 Find the acceleration vector field for a fluid flow that possesses the following velocity field where x, y, z are in meters. Evaluate the acceleration at $(2, -1, 3)$ at $t = 2$ s.
(a) $\mathbf{V} = 20(1 - y^2)\hat{\mathbf{i}}$ m/s
(b) $\mathbf{V} = 2x\hat{\mathbf{i}} + 2y\hat{\mathbf{j}}$ m/s
(c) $\mathbf{V} = x^2t\hat{\mathbf{i}} + 2xyt\hat{\mathbf{j}} + 2yzt\hat{\mathbf{k}}$ m/s
(d) $\mathbf{V} = x\hat{\mathbf{i}} - 2xyz\hat{\mathbf{j}} + tz\hat{\mathbf{k}}$ m/s

12.12 Find the angular velocity vector for the following flow fields. Evaluate the angular velocity at $(2, -1, 3)$ at $t = 2$ s.
(a) Problem 12.11a **(b)** Problem 12.11b
(c) Problem 12.11c **(d)** Problem 12.11d

12.13 Find the vorticity vector for the following flow fields. Evaluate the vorticity at $(2, -1, 3)$ at $t = 2$ s.
(a) Problem 12.11a **(b)** Problem 12.11b
(c) Problem 12.11c **(d)** Problem 12.11d

12.14 Determine the components of the rate-of-strain tensor for the velocity field of the following at $(2, -1, 3)$ at $t = 2$ s.
(a) Problem 12.11a **(b)** Problem 12.11b
(c) Problem 12.11c **(d)** Problem 12.11d

12.15 The velocity components in cylindrical coordinates are given by

$$v_r = \left(10 - \frac{40}{r^2}\right) \cos\theta \text{ m/s},$$

$$v_\theta = -\left(10 + \frac{40}{r^2}\right) \sin\theta \text{ m/s}$$

(a) Calculate the acceleration of a fluid particle occupying the point (4 m, 180°).
(b) Calculate the vorticity component at (4 m, 180°).

12.16 The velocity components in spherical coordinates are given in m/s by

$$v_r = \left(10 - \frac{80}{r^3}\right) \cos\theta, \qquad v_\theta = -\left(10 + \frac{80}{r^3}\right) \sin\theta$$

(a) Calculate the acceleration of a fluid particle occupying the point (4 m, 180°).
(b) Calculate the vorticity component at (4 m, 180°).

12.17 An unsteady flow occurs between parallel plates such that $u = u(y, t)$, $v = 0$, and $w = 0$. Write an expression for the acceleration. What is the acceleration if the flow is steady, that is, $u = u(y)$, $v = 0$, and $w = 0$?

12.18 Consider a symmetrical steady flow in a pipe with the axial and radial velocity components designated $u(r, x)$ and $v(r, x)$, respectively. Write the equations for the two acceleration components a_r and a_x. Use equations from Table 12.1. See Fig. P12.18 for the coordinates.

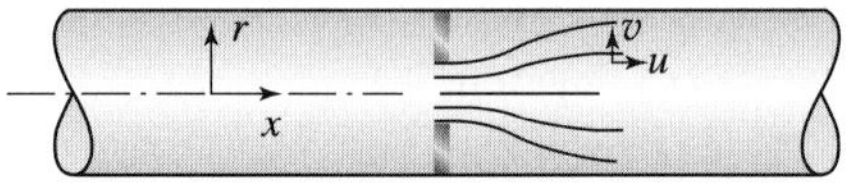

FIGURE P12.18

12.19 The velocity in the 2-cm-diameter pipe of Fig. P12.19 has only one nonzero velocity component given by $u(r, t) = 2(1 - r^2/r_0^2)(1 - e^{-t/10})$ m/s, where r_0 is the radius of the pipe and t is in seconds. Calculate the maximum velocity and the maximum acceleration:

(a) Along the centerline of the pipe.

(b) Along a streamline at $r = 0.5$ cm.

(c) Along a streamline just next to the pipe wall.

Hint: Let $v_z = u(r, t)$, $v_r = 0$, and $v_\theta = 0$ in the appropriate equations in Table 12.1.

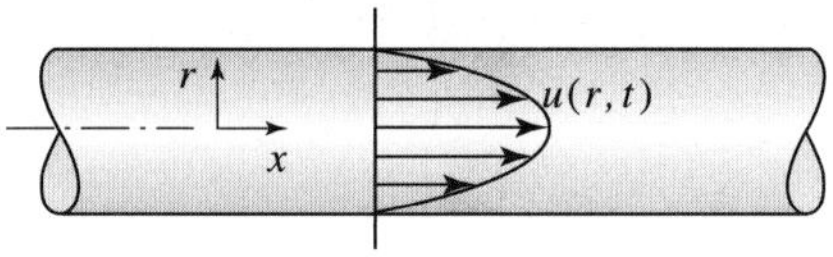

FIGURE P12.19

12.20 The temperature changes periodically in a flow according to $T(y, t) = 20(1 - y^2)\cos \pi t/100$°C. If the velocity is given by $u = 2(1 - y^2)$ m/s, determine the rate of change of the temperature of a fluid particle located at $y = 0$ if $t = 20$ s.

12.21 The density of air in the atmosphere varies according to $\rho(z) = 1.23e^{-10^{-4}z}$ kg/m^3. Air flowing over the mountain of Fig. P12.21 has the velocity vector $\mathbf{V} = 20\hat{\mathbf{i}} + 10\hat{\mathbf{k}}$ m/s at a location of interest where $z = 3000$ m. Find the rate at which a particle's density is changing at that location.

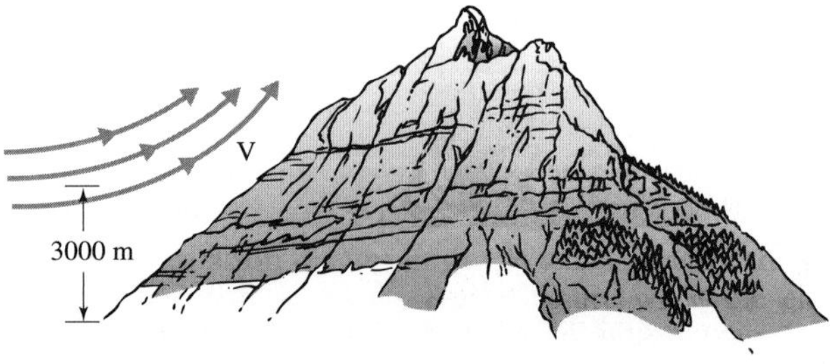

FIGURE P12.21

12.22 The density variation with elevation is given by $\rho(z) = 1000(1 - z/4)$ kg/m^3. At a location where $\mathbf{V} = 10\hat{\mathbf{i}} + 10\hat{\mathbf{k}}$ m/s, find $D\rho/Dt$.

12.23 Salt is slowly added to water in a pipe so that $\partial\rho/\partial x = 0.01$ kg/m^4. Determine $D\rho/Dt$ if the velocity is uniform at 4 m/s.

12.24 Express the substantial derivative in terms of the gradient ∇ and the velocity vector $\mathbf{V}$. Recall from calculus that in rectangular coordinates

$$\nabla = \frac{\partial}{\partial x}\hat{\mathbf{i}} + \frac{\partial}{\partial y}\hat{\mathbf{j}} + \frac{\partial}{\partial z}\hat{\mathbf{k}}.$$

12.25 We can write Eqs. 12.9 in a simplified vector form. The gradient is a vector operator expressed in rectangular coordinates as $\nabla = \frac{\partial}{\partial x}\hat{\mathbf{i}} + \frac{\partial}{\partial y}\hat{\mathbf{j}} + \frac{\partial}{\partial z}\hat{\mathbf{k}}$. Write the dot product of the velocity vector $\mathbf{V}$ and the gradient ∇ and then write Eqs. 12.9 as one vector equation expressing the acceleration $\mathbf{a}$ as the sum of the local acceleration and the convective acceleration.

12.26 For the flow shown in Fig. P12.26, relative to a fixed reference frame, find the acceleration of a fluid particle at:

(a) Point A

(b) Point B

The water at B makes an angle of 30° with respect to the ground and the sprinkler arm is horizontal.

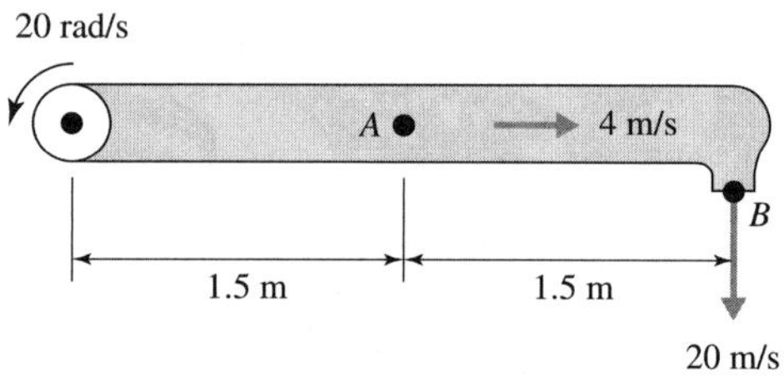

FIGURE P12.26

12.27 A river is flowing due south at 5 m/s at a latitude of 45°. Calculate the acceleration of a particle floating with the river relative to a fixed reference frame. The radius of the Earth is 6000 km.

Classification of Fluid Flows

12.28 Consider each of the following flows and state whether it could be approximated as a one-, two-, or three-dimensional flow or as a uniform flow:
- **(a)** Flow from a vertical pipe striking a horizontal wall
- **(b)** Flow in the waves of the ocean near a beach
- **(c)** Flow near the entrance of a pipe
- **(d)** Flow around a rocket with a blunt nose
- **(e)** Flow around an automobile
- **(f)** Flow in an irrigation channel
- **(g)** Flow through an artery
- **(h)** Flow through a vein

12.29 Which of the flows in Problem 12.28 could be assumed to be a steady flow? Which must be modeled as an unsteady flow?

12.30 Which flow in Problem 12.28 could best be modeled as a plane flow?

12.31 Select the flows in Problem 12.28 that would possess a stagnation point. Sketch each of the selected flows indicating the location of the stagnation point.

12.32 Which of the flows in Problem 12.28 could be modeled as a developed flow?

12.33 State whether each of the flows in Problem 12.28 could be considered as primarily an inviscid flow or a viscous flow.

12.34 Select the flows in Problem 12.28 that are external flows. Select those external flows that possess a stagnation point.

12.35 Sketch the flow around a razor blade positioned parallel to the flow showing the boundary layers.

12.36 The 32°C water exiting the 1.5-cm-diameter faucet of Fig. P12.36 has an average velocity of 2 m/s. Would you expect the flow to be laminar or turbulent?

FIGURE P12.36

12.37 The Red Cedar River flows placidly through Michigan State University's campus. In a certain section the depth is 0.8 m and the average velocity is 0.2 m/s. Is the flow laminar or turbulent?

12.38 Air at 40°C flows in a rectangular 30 cm × 6 cm heating duct at an average velocity of 4 m/s. Is the flow laminar or turbulent?

12.39 The sphere of diameter D of Fig. P12.39 is traveling with a velocity V of 1.2 m/s in 20°C atmospheric air. If $\text{Re} = VD/\nu$ is less than 4×10^4, the boundary layer around the front of the sphere is completely laminar. Determine if the boundary layer is completely laminar on a sphere of diameter:
(a) 1 cm **(b)** 1 m

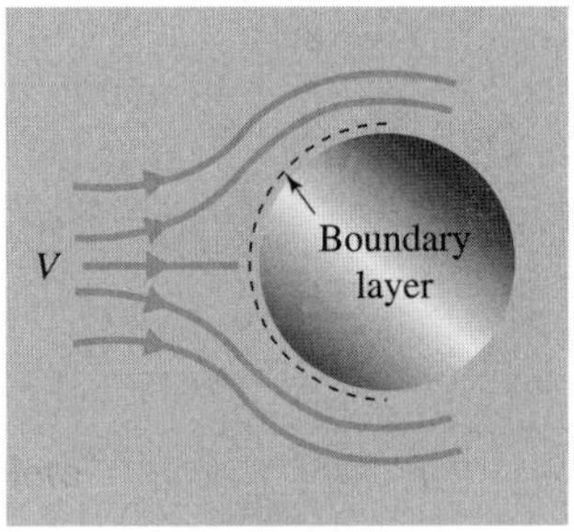

FIGURE P12.39

12.40 The airfoil on a commercial airliner is approximated as a flat plate sketched in Fig. P12.40. How long would you expect the laminar portion of the boundary layer to be if it is flying:
- **(a)** At an altitude of 10 000 m and a speed of 900 km/h?
- **(b)** At an altitude of 30,000 ft and a speed of 600 mph?

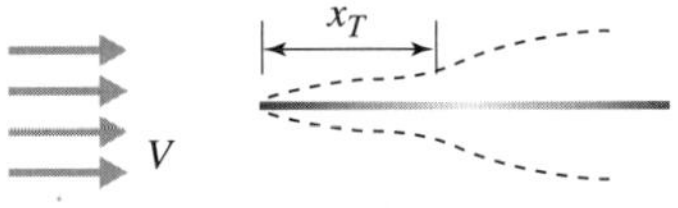

FIGURE P12.40

12.41 A leaf keeps cool by transpiration, a process in which water flows from the leaf to the atmosphere. An experimenter wonders if the boundary layer on a leaf influences the transpiration, so an "experimental" leaf is set up in the laboratory and air is blown over it at 6 m/s. Comment as to whether the boundary layer is expected to be laminar or turbulent.

12.42 For the following situations, state whether a compressible flow is required or if the flow can be approximated with an incompressible flow:
(a) An aircraft flying at 100 m/s at an elevation of 8000 m
(b) A golf ball traveling at 80 m/s
(c) Flow around an object being studied in a high-temperature wind tunnel if the temperature is 100°C and the air velocity is 100 m/s

12.43 Write out Eq. 12.25 using Eq. 12.11. For a steady, plane flow what relationship must exist for an incompressible flow in which the density is allowed to vary?

12.44 If $\rho = \rho_0(1 + cz)$ models the density variation in a channel (there is heavy salt water at the bottom and fresh water at the top) in which $u(y, z)$ is the only velocity component, is the flow incompressible?

Bernoulli's Equation

12.45 A pitot tube is used to measure the velocity of a small aircraft flying at 3000 ft. Calculate its velocity if the pitot tube measures:
(a) 0.3 psi **(b)** 0.9 psi
(c) 0.09 psi

12.46 Approximate the force acting on the 15-cm-diameter headlight shown in Fig. P12.46 of an automobile traveling at 120 kph.

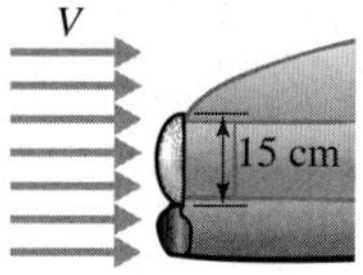

FIGURE P12.46

12.47 A vacuum cleaner is capable of creating a vacuum of 2 kPa just inside the hose of Fig. P12.47. What maximum average velocity would be expected in the hose?

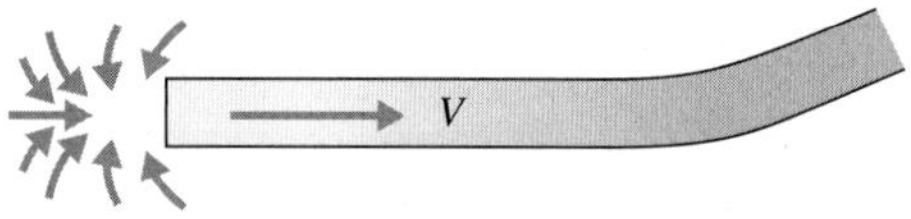

FIGURE P12.47

12.48 The inviscid, incompressible flow in the vicinity of a stagnation point (Fig. P12.48) is approximated by $u = -10x$, $v = 10y$. If the pressure at the origin is p_0, find an expression for the pressure neglecting gravity effects:
(a) Along the negative x-axis
(b) Along the positive y-axis

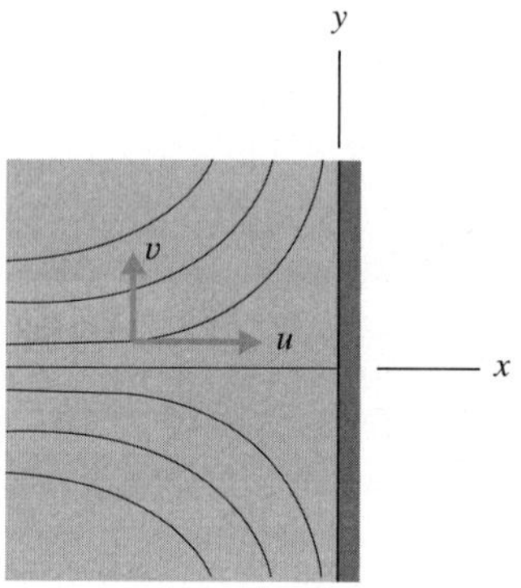

FIGURE P12.48

12.49 The inviscid, incompressible flow field exterior to the cylinder shown in Fig. P12.49 is given by

$$v_r = U_\infty \left(1 - \frac{r_c^2}{r^2}\right) \cos\theta \text{ and } v_\theta = U_\infty \left(1 + \frac{r_c^2}{r^2}\right) \sin\theta$$

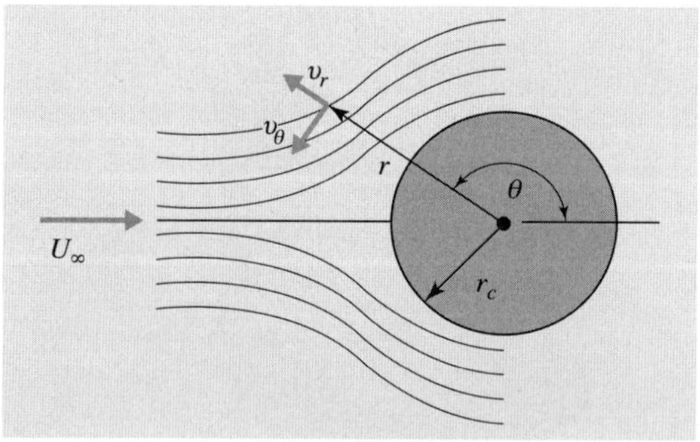

FIGURE P12.49

If the pressure at $r = \infty$ is zero (i.e., $p_\infty = 0$), find an expression for the pressure neglecting gravity effects:
(a) Along the negative x-axis
(b) At the stagnation point
(c) On the surface of the cylinder
(d) On the surface of the cylinder at $\theta = 90°$

12.50 The inviscid, incompressible flow field exterior to a sphere (see Fig. P12.49) is given by

$$v_r = U_\infty \left(1 - \frac{r_c^3}{r^3}\right) \cos\theta \text{ and } v_\theta = U_\infty \left(1 + \frac{r_c^3}{r^3}\right) \sin\theta$$

If the pressure at $r = \infty$ is zero, find an expression for the pressure neglecting gravity effects:
(a) Along the negative x-axis
(b) At the stagnation point
(c) On the surface of the sphere
(d) On the surface of the sphere at $\theta = 90°$

12.51 The velocity along the negative x-axis in the inviscid, incompressible flow field exterior to the body shown in Fig. P12.51 is given by $u(x) = U_\infty + q/2\pi x$. If the pressure at $x = -\infty$ is zero, find an expression for the pressure neglecting gravity effects:
(a) Along the negative x-axis if $U_\infty = 10$ m/s and $q = 20\pi$ m^2/s
(b) At the stagnation point if $U_\infty = 10$ m/s and $q = 20\pi$ m^2/s
(c) Along the negative x-axis if $U_\infty = 30$ ft/sec and $q = 60\pi$ ft^2/sec
(d) At the stagnation point if $U_\infty = 30$ ft/sec and $q = 60\pi$ ft^2/sec

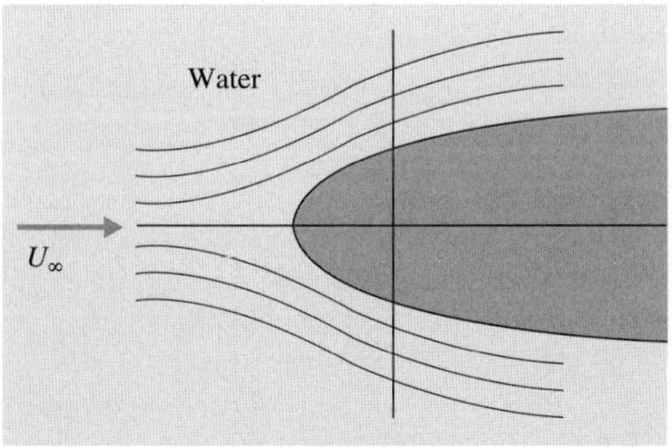

FIGURE P12.51

12.52 The incompressible flow of water through the short contraction of Fig. 12.19a is assumed to be inviscid. If a pressure drop of 20 kPa is measured, estimate the velocity at the wall at section 2 just downstream of the contraction. (Actually, a boundary layer would develop, and the velocity calculated at the wall would be the velocity at the edge of the boundary layer; see the inset of Fig. 12.10.)

12.53 Air flows from a relatively large plenum in a furnace out a relatively small rectangular duct. If the pressure in the plenum measures 60 Pa and in the duct 10.2 Pa, estimate the velocity of the 40°C air in the duct.

12.54 What is the velocity of the water in the pipe if the manometer shown in Fig. P12.54 reads:
(a) 4 cm?
(b) 10 cm?
(c) 2 in.?
(d) 4 in.?

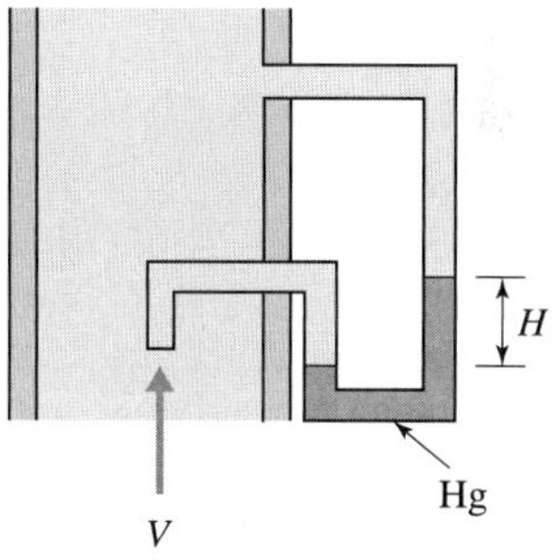

FIGURE P12.54

12.55 A manometer, positioned inside a cylinder as shown in Fig. P12.55, reads 4 cm of water. Estimate U_∞ assuming an inviscid flow. Refer to the velocity field in Problem 12.49.

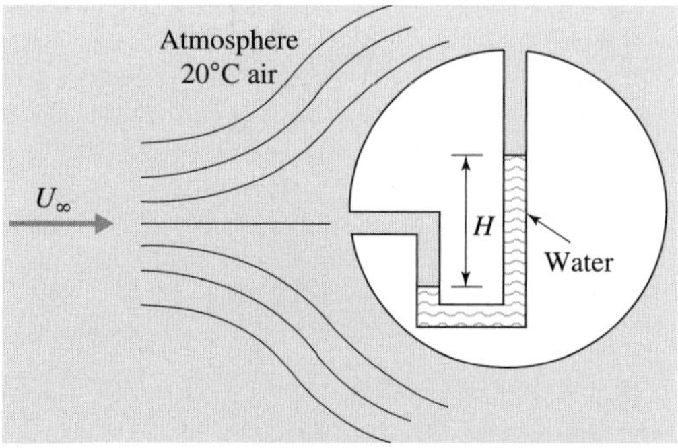

FIGURE P12.55

12.56 A manometer, positioned as shown in Figure P12.55 inside a sphere, reads 4 cm of water. Estimate U_∞ assuming an inviscid flow. Refer to the velocity field in Problem 12.50.

12.57 Air at 20°C is drawn into the hose of a vacuum cleaner through a head that is relatively free of obstructions (the flow can be assumed to be inviscid). Estimate the velocity in the hose if the vacuum in the hose measures:
(a) 2 cm of water
(b) 8 cm of water
(c) 1 in. of water
(d) 4 in. of water

12.58 A wind tunnel is designed to draw in air from the atmosphere and produce a velocity of 100 m/s in the test section. The fan is located downstream of the test section. What pressure is to be expected in the test section if the atmospheric temperature and pressure are:
(a) −20°C, 90 kPa?
(b) 0°C, 95 kPa?
(c) 20°C, 92 kPa?
(d) 40°C, 100 kPa?

12.59 The pump shown in Fig. P12.59 creates a flow such that $V = 14$ m/s. Predict the pressure at the gage shown assuming an inviscid flow in the entrance and a uniform flow at the gage. Use a streamline starting at:
(a) Point A.
(b) Point B.

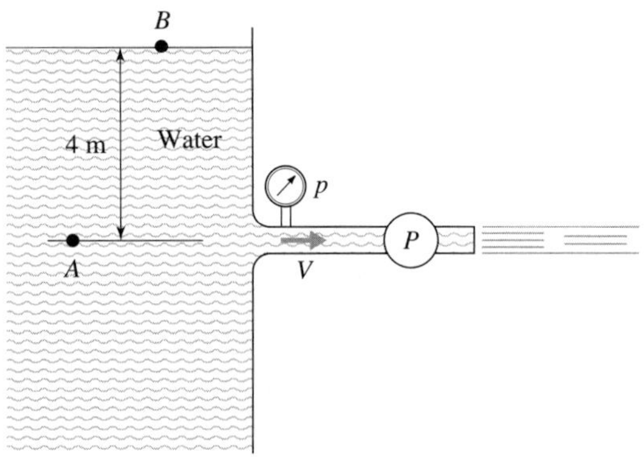

FIGURE P12.59

12.60 For the flow shown in Fig. P12.60, estimate the pressure p_1 and velocity V_1 if $V_2 = 20$ m/s and:
(a) $H = 1$ cm **(b)** $H = 5$ cm
(c) $H = 10$ cm

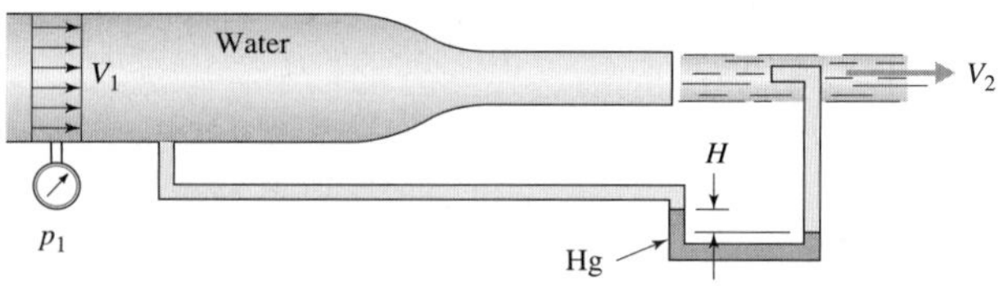

FIGURE P12.60

12.61 A fireman reduces the exit area on a nozzle so that the velocity inside the hose is quite small relative to the exiting velocity. What is the maximum exiting velocity and what is the maximum height the water can reach if the pressure inside the hose is:
(a) 700 kPa? **(b)** 1400 kPa?
(c) 100 psi? **(d)** 200 psi?

12.62 The velocity downstream of a sluice gate is assumed to be uniform (Fig. P12.62). Express V in terms of H and h for this inviscid flow. Use a streamline:
(a) Along the top **(b)** Along the bottom

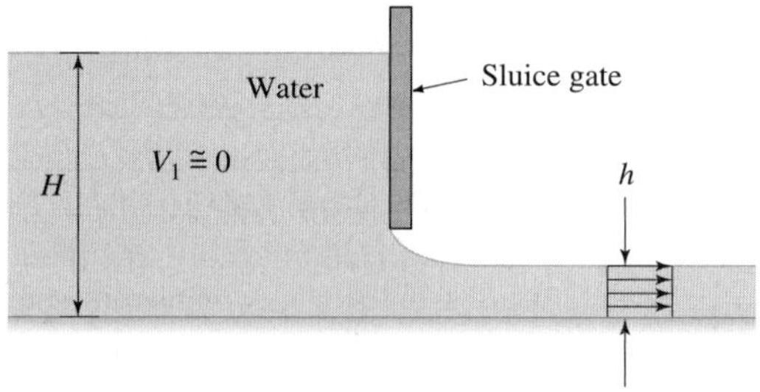

FIGURE P12.62

12.63 To what maximum velocity can water be accelerated before it reaches the turbine blades of a hydroturbine if it enters with relatively low velocity at:
(a) 600 kPa? **(b)** 300 kPa?
(c) 80 psi? **(d)** 40 psi?

12.64 Water exists in a city's water system at a pressure of 500 kPa at a particular location. The water pipe must traverse over a hill. How high could the hill be, above that location, for the system to possibly supply water to the other side of the hill?

12.65 Fluid flows between the radial disks shown in Fig. P12.65. Estimate the pressure in the 2-cm-diameter pipe if the fluid exits to the atmosphere. Neglect viscous effects. The fluid is:

(a) water **(b)** benzene
(c) gasoline **(d)** air

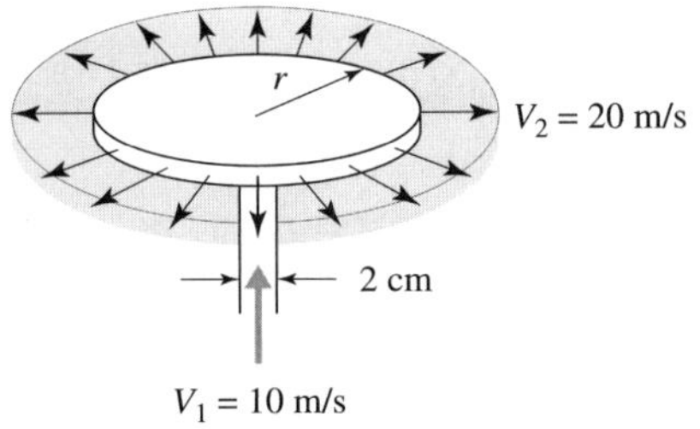

FIGURE P12.65

12.66 Estimate the pressure at $r = 10$ cm if the velocity there is 8 m/s in Problem 12.65d.

12.67 It is proposed that air blown through a tube attached to a metal disk be used to pick up envelopes, as shown in Fig. P12.67. Would such a setup actually pick up an envelope? Explain. Assume an inviscid flow with the air slowing down as it moves radially outward.

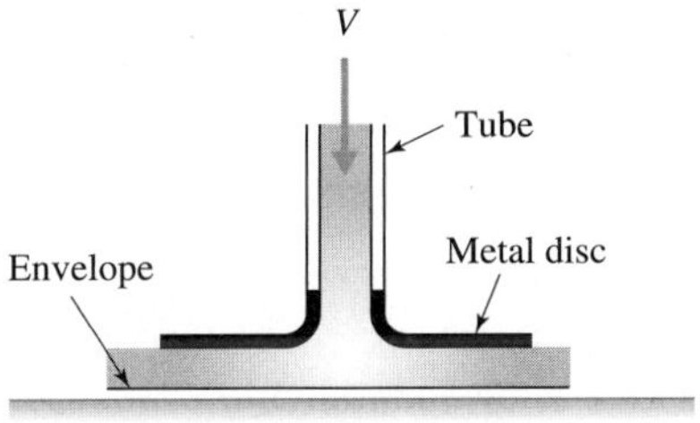

FIGURE P12.67

12.68 From which of the following objects would you expect the flow to separate and form a substantial separated region?

(a) A golf ball
(b) A telephone wire
(c) A wind machine blade
(d) A 2-mm-diameter wire in a low-speed wind tunnel
(e) An automobile
(f) An aircraft

Note: Separation occurs whenever the Reynolds number exceeds a value around 20 over a blunt object.

12.69 Explain, with the use of a sketch, why a burr on the downstream side of a piezometer opening in a pipe (see Fig. 12.18a) will result in a reading of the pressure that is too high.

12.70 An incompressible, inviscid flow of water enters a bend with a uniform velocity of $V_1 = 10$ m/s (Fig. P12.70). Estimate the pressure difference between points A and B if the average radius of curvature at the bend is 5 cm. Sketch an anticipated velocity profile along AB. Assume that $p_A < p_1$ and $p_B > p_1$.

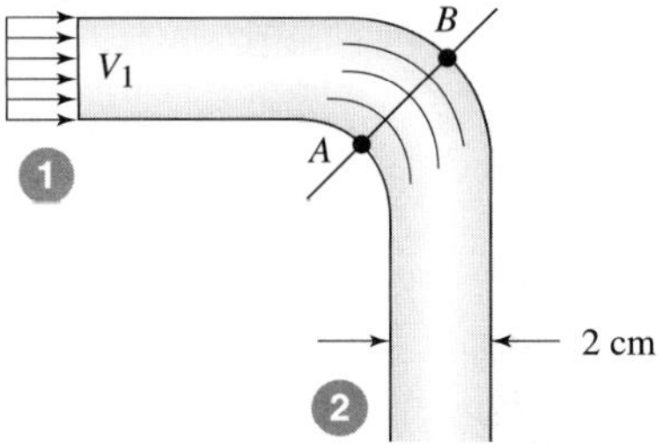

FIGURE P12.70

12.71 Viscosity causes a fluid to stick to a surface. If the fluid in Problem 12.70 sticks to the surface, explain why, in a viscous flow, a secondary flow is caused by the bend. Sketch such a flow on a circular cross section at section 2.

12.72 In Fig. P12.72, assuming an inviscid flow, insert one of these signs between the pressures: >, <, ≅.

$p_A \quad p_B$
$p_C \quad p_D$
$p_B \quad p_D$

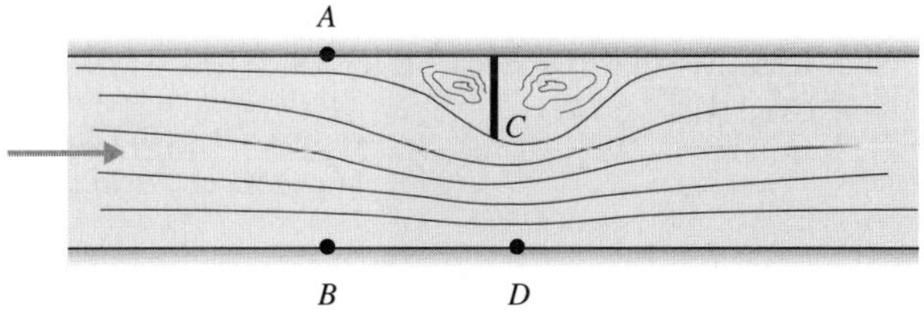

FIGURE P12.72

Collegiate Rowing Regatta on the Schuylkill River, Philadelphia, Pennsylvania. With each stroke, the work performed by the crew is transferred to the hull to overcome drag forces. (Courtesy of Corbis)

13

The Integral Forms of the Fundamental Laws

Outline

Chapter Objectives

The objectives of this chapter are to:

- ▲ Derive an equation that will allow us to convert the three basic laws that are stated for a system to a form that is applicable to a control volume.
- ▲ Apply the conservation of mass to control volumes of interest.
- ▲ Analyze the work-rate term of the energy equation.
- ▲ Apply the energy equation to numerous engineering situations.
- ▲ Apply Newton's second law to control volumes of interest.
- ▲ Express the basic laws in their most general control-volume form so that complex problems encountered in engineering applications can be properly analyzed and hopefully solved.

▲ Present numerous examples of the basic laws applied to control volumes so that students will be able to correctly solve fluid flow problem involving many of the control volumes of interest to engineers.

Nomenclature

The following variables are introduced in this chapter:

c.s.	control surface	V_B	blade velocity
c.v.	control volume	V_I	velocity measured in a fixed ref. frame
e	specific energy		
h_L	head loss	V_r	relative speed
H_P, H_T	pump head, turbine head	w	width
I	subscript referring to inertial	$\dot{W}$	work rate
K	loss coefficient	$\dot{W}_P$, $\dot{W}_T$	pump power, turbine power
$\dot{m}$	mass flux (mass rate of flow)		
$\dot{m}_r$	mass flux due to relative velocity	$\dot{W}_S$	shaft power
N	an extensive property, number of jets	α	kinetic-energy-correction factor
$\hat{\mathbf{n}}$	normal unit vector	β	momentum-correction factor
Q_f	flow rate		
R, R_x, R_y	reaction force and components	η	an intensive property
		η_P, η_T	pump and turbine efficiencies
$\dot{S}$	control volume velocity		
sys	system	τ	stress vector
$\bar{V}$	average velocity over area		

13.1 INTRODUCTION

Key Concept: *To determine an integral quantity the integrand must be known, or information must be available so that a good approximation to the integrand can be made.*

Quantities of interest to engineers can often be expressed in terms of integrals. For example, volume flow rate is the integral of the velocity over an area; heat transfer is the integral of the heat flux over an area; force is the integral of a stress over an area; mass is the integral of the density over a volume; and kinetic energy is the integral of $V^2/2$ over each mass element in a volume. There are, of course, many other integral quantities. To determine an integral quantity, the integrand must be known, or information must be available so that a good approximation to the integrand can be made. If the velocity is not known or cannot be approximated with any degree of certainty, appropriate partial differential equations called the Navier-Stokes equations, must be solved yielding the velocity; the integration may then be performed giving the engineer the desired integral quantity. The Navier-Stokes equations are presented in Appendix K.3.

In this chapter we present the integral quantities of interest, develop equations that relate the integral quantities, and work a number of problems for which the integrands are given or can be approximated. This includes a surprisingly large variety of problems. There are, however, many integral quantities that cannot be determined since the integrands are unknown. These would include the lift and drag on an airfoil, the torque on the blades of a wind machine, and the kinetic energy in the wake of a submarine. To determine such integrands, it would be necessary to solve the Navier-Stokes equations, a task that is often quite difficult; some relatively simple situations that do not require the Navier-Stokes equations are considered in subsequent chapters.

Also, there are many quantities of interest that are not integral in nature. Included would be the point of separation of the flow around a body, the concentration of a pollutant in a stream at a certain location, the pressure distribution on the side of a building, and the wave-shore interaction along a lake. To study subjects such as these, it is necessary to consider the differential equations that describe the flow situation. Most of the topics mentioned are relegated to specialized graduate courses.

13.2 THE THREE BASIC LAWS

The integral quantities of primary interest in fluid mechanics are contained in three basic laws: conservation of mass, first law of thermodynamics, and Newton's second law. These basic laws are expressed using a Lagrangian description in terms of a *system*, a fixed collection of material particles. For example, if we consider flow through a pipe, we could identify a fixed quantity of fluid at time t as the system (Fig. 13.1); this system would then move due to velocity to a downstream location at time $t + \Delta t$. Any of the three basic laws could be applied to this system. This is, however, not an easy task. First let us state the basic laws in their general form.

Conservation of Mass: The law stating that mass must be conserved is:

The mass of a system remains constant.

The mass of a fluid particle is $\rho\, d\maltese$, where $d\maltese$ is the volume occupied by the particle and ρ is its density. Knowing that the density can change from point to point in the system, the conservation of mass can be expressed in integral form as

$$\frac{D}{Dt}\int_{\text{sys}} \rho\, d\maltese = 0 \tag{13.1}$$

where D/Dt is used since we are following a specified group of material particles, a system.

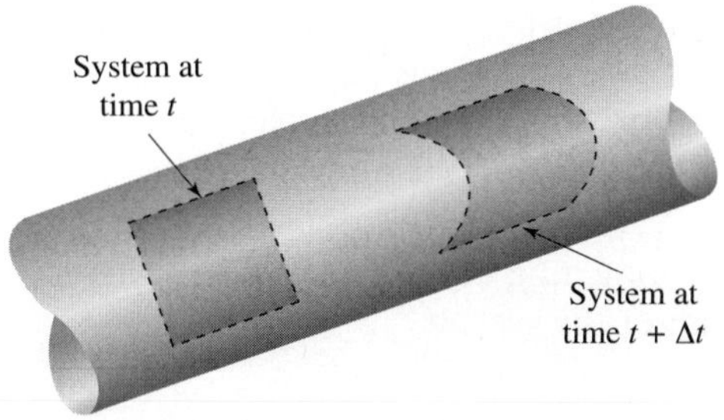

FIGURE 13.1 Example of a system in fluid mechanics.

First Law of Thermodynamics: The law that relates heat transfer, work, and energy change is the first law of thermodynamics; it states:

> The rate of heat transfer to a system minus the rate at which the system does work equals the rate at which the energy of the system is changing.

Recognizing that both density and specific energy may change from point to point in the system, it may be expressed as

$$\dot{Q} - \dot{W} = \frac{D}{Dt}\int_{\text{sys}} e\rho \, dV \tag{13.2}$$

Specific energy: *Accounts for kinetic energy, potential energy, and internal energy per unit mass.*

where the **specific energy** e accounts for kinetic energy, potential energy, and internal energy per unit mass. Equation 13.2 is often referred to as the energy equation.

Other forms of energy—chemical, electrical, nuclear—are not included in an elementary course in fluid mechanics. In its basic form stated here, the first law of thermodynamics applies only to a system, a collection of fluid particles; therefore, D/Dt is used. We study $\dot{Q}$ and $\dot{W}$ in Section 13.5, where we consider the energy equation in detail.

Newton's Second Law: Newton's second law, also called the momentum equation, states:

> The resultant force acting on a system equals the rate at which the momentum of the system is changing.

The momentum of a fluid particle of mass is a vector quantity given by $\mathbf{V}\rho \, dV$; consequently, Newton's second law may be expressed in an inertial reference frame as

$$\sum \mathbf{F} = \frac{D}{Dt}\int_{\text{sys}} \mathbf{V}\rho \, dV \tag{13.3}$$

recognizing that both density and velocity may change from point to point in the system. This equation reduces to $\sum \mathbf{F} = m\mathbf{a}$ if $\mathbf{V}$ and ρ are uniform throughout the entire system; ρ is often a constant, but in fluid mechanics the velocity vector invariably changes from point to point. Again, D/Dt is used to provide the rate-of-change since Newton's second law is applied to a system.

Moment-of-Momentum Equation: The moment-of-momentum equation results from Newton's second law; it states:

> The resultant moment acting on a system equals the rate of change of the angular momentum of the system.

In equation form this becomes, relative to an inertial reference frame,

$$\sum \mathbf{M} = \frac{D}{Dt}\int_{\text{sys}} \mathbf{r} \times \mathbf{V}\rho \, d\forall \tag{13.4}$$

where $\mathbf{r} \times \mathbf{V}\rho \, d\forall$ represents the angular momentum of a fluid particle with mass $\rho \, d\forall$. The vector $\mathbf{r}$ locates the volume element $d\forall$ and is measured from the origin of the coordinate axes, the point relative to which the resultant moment is measured.

Key Concept: *In each of the basic laws, the integral quantity is an extensive property of the system.*

Note that in each of the basic laws the integral quantity is an extensive property of the system (see Section 10.7). We will use the symbol N_{sys} to denote this extensive property; for example, N_{sys} could be the mass, the momentum, or the energy of the system. The left-hand side of Eq. 13.1 and the right-hand sides of Eqs. 13.2, 13.3, and 13.4 may all be expressed as

$$\frac{DN_{\text{sys}}}{Dt} \tag{13.5}$$

where N_{sys} represents an integral quantity, either a scalar quantity or a vector quantity.

It is also useful to introduce the variable η for the intensive property, the property of the system per unit mass. The relation between N_{sys} and η is given by

$$N_{\text{sys}} = \int_{\text{sys}} \eta\rho \, d\forall \tag{13.6}$$

As an example, the extensive property of Newton's second law is the momentum

$$\textbf{momentum}_{\text{system}} = \int_{\text{sys}} \mathbf{V}\rho \, d\forall \tag{13.7}$$

which is a vector quantity. The corresponding intensive property would be the velocity vector $\mathbf{V}$. Note that the density and velocity, which may vary from point to point within the system, may be functions of time, as in unsteady flow.

Control volume: *A region of space, into which fluid enters and/or from which fluid leaves.*

Our interest is most often focused on a device, or a region of space, into which fluid enters and/or from which fluid leaves; we identify this region as a **control volume**. An example of a fixed control volume is shown in Fig. 13.2a. A control

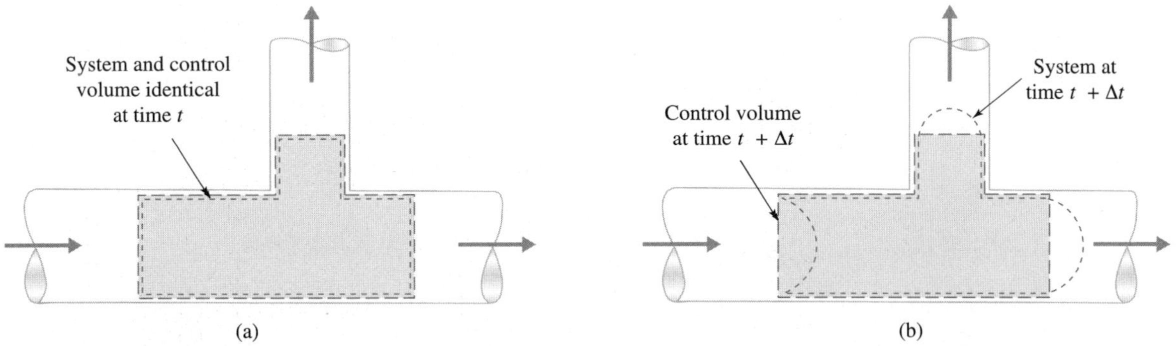

FIGURE 13.2 Example of a fixed control volume and a system: (a) time t; (b) time $t + \Delta t$.

volume need not be fixed; it could deform as in a piston-cylinder during exhaust or in a balloon as it deflates. We will, however, consider only fixed control volumes in this book. This will not limit us in most problem situations.

The difference between a control volume and a system is illustrated in Fig. 13.2b. The figure indicates that the system occupies the control volume at time t and has partially moved out of it at time $t + \Delta t$. Since it is often more convenient to focus on a control volume (e.g., a pump) rather than on a system, the first order of business is to find a transformation that will allow us to express the substantial derivative of a system (a Lagrangian description) in terms of quantities associated with a control volume (an Eulerian description) so that the basic laws can be applied directly to a control volume. This will be done in general and then applied to the specific laws.

13.3 SYSTEM-TO-CONTROL-VOLUME TRANSFORMATION

We are interested in the time rate of change of the extensive property N_{sys} as we follow the system along, that is, DN_{sys}/Dt, and we would like to express this in terms of quantities that pertain to the control volume. In this section we present the derivation of the transformation.

The derivation involves fluxes of the extensive property in and out of the control volume. A flux is a measure of the rate at which an extensive property crosses an area; for example, a mass flux is the rate at which mass crosses an area. It is useful to introduce vector notation to describe these fluxes. Consider an area element dA of the **control surface**, the surface area that completely encloses the control volume. The property flux across an elemental area dA (see Fig. 13.3) may be expressed by

Control surface: *The surface area that completely encloses the control volume.*

$$\text{flux across } dA = \eta\rho\hat{\mathbf{n}} \cdot \mathbf{V}\, dA \tag{13.8}$$

where $\hat{\mathbf{n}}$, a unit vector normal to the area element dA, always points out of the control volume, and η represents the intensive property associated with N_{sys}. Note

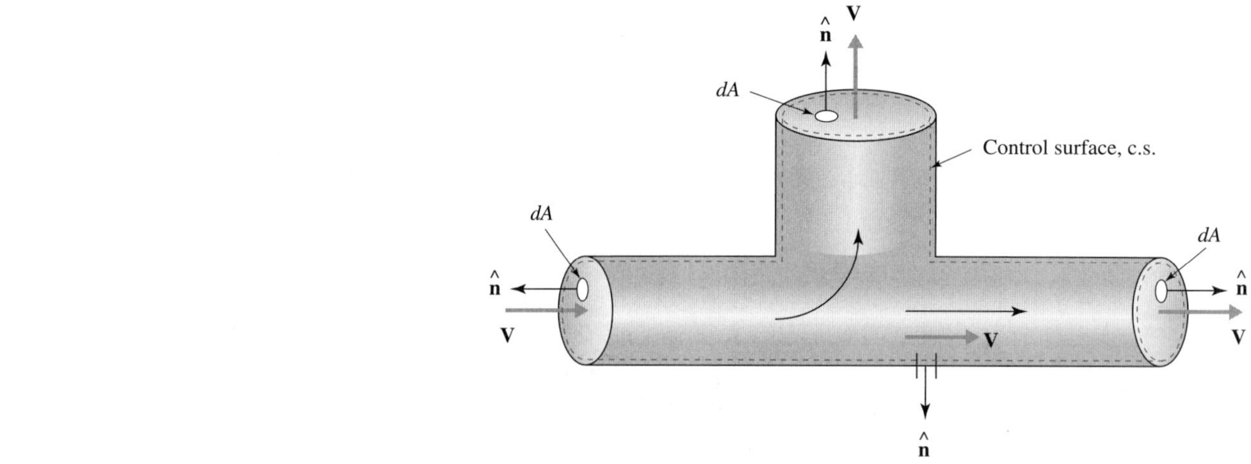

FIGURE 13.3 Illustration showing the flux of an extensive property.

that this expression yields a negative value if it concerns a property influx. Only the normal component $\hat{\mathbf{n}} \cdot \mathbf{V}$ of the velocity vector contributes to this flux term. If there is no normal component of velocity on a particular area, such as the wall of a pipe, no flux occurs across that area. A positive $\hat{\mathbf{n}} \cdot \mathbf{V}$ indicates a flux out of the volume; a negative $\hat{\mathbf{n}} \cdot \mathbf{V}$, that is, $\mathbf{V}$ has a component in the opposite direction of $\hat{\mathbf{n}}$, indicates a flux into the volume. We must always use $\hat{\mathbf{n}}$ pointing out of the volume. The velocity vector $\mathbf{V}$ may be at some angle to the unit vector $\hat{\mathbf{n}}$; the dot product $\hat{\mathbf{n}} \cdot \mathbf{V}$ accounts for the appropriate component of $\mathbf{V}$ that produces a flux through the area.

The net property flux out of the control surface is then obtained by integrating over the entire control surface:

$$\text{net flux of property} = \int_{\text{c.s.}} \eta\rho\hat{\mathbf{n}} \cdot \mathbf{V}\, dA \tag{13.9}$$

If the net flux is positive, the flux out is larger than the flux in.

Let us return now to the derivative DN_{sys}/Dt. The definition of a derivative allows us to write

$$\frac{DN_{\text{sys}}}{Dt} = \lim_{\Delta t \to 0} \frac{N_{\text{sys}}(t + \Delta t) - N_{\text{sys}}(t)}{\Delta t} \tag{13.10}$$

The system is shown in Fig. 13.4 at times t and $t + \Delta t$. Assume that the system occupies the full control volume at time t; if we were considering a device, such as a pump, the particles of the system would just fill the device at time t. Since the device, the control volume shown in Fig. 13.4, is assumed to be fixed in space, the system will move through the device. Equation 13.10 can then be written

Key Concept: *The system occupies the full control volume at time t.*

$$\begin{aligned}\frac{DN_{\text{sys}}}{Dt} &= \lim_{\Delta t \to 0} \frac{N_3(t + \Delta t) + N_2(t + \Delta t) - N_2(t) - N_1(t)}{\Delta t} \\ &= \lim_{\Delta t \to 0} \frac{N_2(t + \Delta t) + N_1(t + \Delta t) - N_2(t) - N_1(t)}{\Delta t} \\ &\quad + \lim_{\Delta t \to 0} \frac{N_3(t + \Delta t) - N_1(t + \Delta t)}{\Delta t}\end{aligned} \tag{13.11}$$

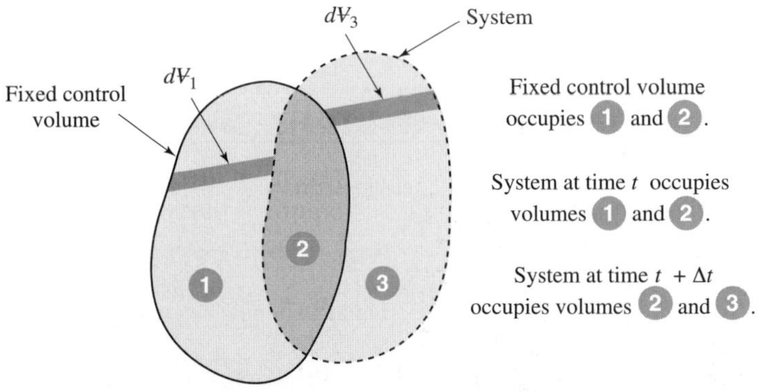

FIGURE 13.4 System and fixed control volume.

where, in this second expression, we have simply added and subtracted $N_1(t+\Delta t)$ in the numerator. In the equations above, the numerical subscript denotes the region; for example, $N_2(t)$ signifies the extensive property in region 2 at time t. Now, we observe that the first limit on the right-hand side refers to the control volume, so we can write

$$\frac{DN_{\text{sys}}}{Dt} = \lim_{\Delta t \to 0} \frac{N_{\text{c.v.}}(t+\Delta t) - N_{\text{c.v.}}(t)}{\Delta t} + \lim_{\Delta t \to 0} \frac{N_3(t+\Delta t) - N_1(t+\Delta t)}{\Delta t} \tag{13.12}$$

The first ratio on the right-hand side is $dN_{\text{c.v.}}/dt$, where we use an ordinary derivative since we are not following specific fluid particles. Thus there results

$$\frac{DN_{\text{sys}}}{Dt} = \frac{dN_{\text{c.v.}}}{dt} + \lim_{\Delta t \to 0} \frac{N_3(t+\Delta t) - N_1(t+\Delta t)}{\Delta t} \tag{13.13}$$

Key Concept: *The unit vector* $\hat{\mathbf{n}}$ *always points out of the volume.*

Now, we must find expressions for the extensive quantities $N_3(t+\Delta t)$ and $N_1(t+\Delta t)$. They, of course, depend on the mass contained in the volume elements shown in Fig. 13.4 and enlarged in Fig. 13.5. Note that the unit vector $\hat{\mathbf{n}}$ always points out of the volume, and hence to obtain a positive differential volume a negative sign is required for region 1. Also, note that the cosine of the angle between the velocity vector and the normal vector is required,[1] thus the presence of the dot product. Referring to Fig. 13.5, we have

$$\begin{aligned} N_3(t+\Delta t) &= \int_{A_3} \eta\rho\hat{\mathbf{n}}\cdot\mathbf{V}\,\Delta t\,dA_3 \\ N_1(t+\Delta t) &= -\int_{A_1} \eta\rho\hat{\mathbf{n}}\cdot\mathbf{V}\,\Delta t\,dA_1 \end{aligned} \tag{13.14}$$

Recognizing that A_3 plus A_1 completely surrounds the control volume, we combine the two integrals into one integral. That is,

$$N_3(t+\Delta t) - N_1(t+\Delta t) = \int_{\text{c.s.}} \eta\rho\hat{\mathbf{n}}\cdot\mathbf{V}\,\Delta t\,dA \tag{13.15}$$

where the control surface, denoted by c.s., is an area that completely surrounds the control volume. Substituting Eq. 13.15 back into Eq. 13.13 yields the desired

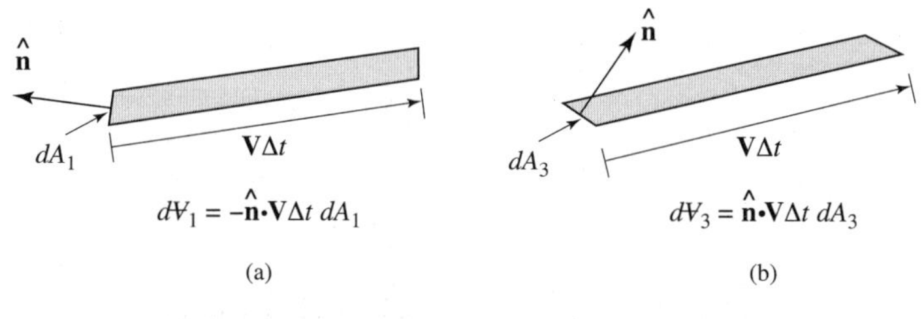

FIGURE 13.5 Differential volume elements.

[1] To obtain the volume of a box, we multiply the height by the area of the base, provided that the box is upright. If it is completely collapsed, its volume is zero. Hence for some intermediate position the volume is the height times the area of the base times the cosine of the appropriate angle.

result, the system-to-control-volume transformation, or equivalently, the **Reynolds transport theorem**:

Reynolds transport theorem: *The system-to-control-volume transformation.*

$$\frac{DN_{\text{sys}}}{Dt} = \frac{d}{dt}\int_{\text{c.v.}} \eta\rho \, d\forall + \int_{\text{c.s.}} \eta\rho\hat{\mathbf{n}} \cdot \mathbf{V} \, dA \tag{13.16}$$

Equation 13.16 is a Lagrangian-to-Eulerian transformation of the rate of change of an extensive integral quantity.

The first integral represents the rate of change of the extensive property in the control volume. The second integral represents the flux of the extensive property across the control surface; it is nonzero only where fluid crosses the control surface. We study this flux term in considerable detail in the following sections. Thus we can now express the basic laws in terms of a fixed volume in space. We will do this in subsequent sections for each of the basic laws.

We can move the time derivative of the control volume term inside the integral since, for a fixed control volume, the limits on the volume integral are independent of time; we then write

Key Concept: *The time derivative of the control volume term can be moved inside the integral for a fixed control volume.*

$$\frac{DN_{\text{sys}}}{Dt} = \int_{\text{c.v.}} \frac{\partial}{\partial t}(\rho\eta) \, d\forall + \int_{\text{c.s.}} \eta\rho\hat{\mathbf{n}} \cdot \mathbf{V} \, dA \tag{13.17}$$

In this form we have used $\partial/\partial t$ since ρ and η are, in general, dependent on the position variables.

13.3.1 Simplifications of the System-to-Control-Volume Transformation

Many flows of interest are steady flows, so that $\partial(\eta\rho)/\partial t = 0$. Our system-to-control-volume transformation then takes the form

$$\frac{DN_{\text{sys}}}{Dt} = \int_{\text{c.s.}} \eta\rho\hat{\mathbf{n}} \cdot \mathbf{V} \, dA \tag{13.18}$$

Furthermore, there is often only one area A_1 across which fluid enters the control volume and one area A_2 across which fluid leaves the control volume; assuming that the velocity vector is normal to the area (see Fig. 13.6), we can write $\hat{\mathbf{n}} \cdot \mathbf{V}_1 = -V_1$ over area A_1 and $\hat{\mathbf{n}} \cdot \mathbf{V}_2 = V_2$ over area A_2. Then Eq. 13.18 becomes

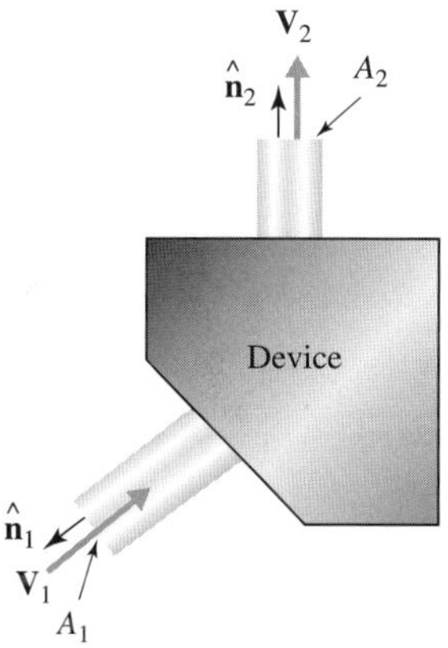

FIGURE 13.6 Flow into and from a device.

$$\frac{DN_{\text{sys}}}{Dt} = \int_{A_2} \eta_2\rho_2 V_2 \, dA - \int_{A_1} \eta_1\rho_1 V_1 \, dA \tag{13.19}$$

Finally, there are many situations that are modeled acceptably by assuming uniform properties over each plane area (see Fig. 12.9); then the equation simplifies to

Key Concept: *Many situations are modeled by assuming uniform properties over the inlet and exit areas.*

$$\frac{DN_{\text{sys}}}{Dt} = \eta_2\rho_2 V_2 A_2 - \eta_1\rho_1 V_1 A_1 \tag{13.20}$$

We will find that the system-to-control-volume transformation in this simplified form is most often used in the application of the basic laws to problems of interest in an introductory course in fluid mechanics. Some applications will, however, be included that will illustrate nonuniform distributions and unsteady flows.

If we generalize Eq. 13.20 to include several areas across which the fluid flows, we could write

$$\frac{DN_{\text{sys}}}{Dt} = \sum_{i=1}^{N} \eta_i \rho_i \mathbf{V}_i \cdot \hat{\mathbf{n}}_i A_i \tag{13.21}$$

Key Concept: *For an inlet area, $\hat{\mathbf{n}} \cdot \mathbf{V}$ introduces a negative sign.*

where N is the number of areas. The dot product $\hat{\mathbf{n}} \cdot \mathbf{V}$ would provide us with the appropriate sign at each area; for an inlet area, $\hat{\mathbf{n}} \cdot \mathbf{V}$ introduces a negative sign, and for an exit area, $\hat{\mathbf{n}} \cdot \mathbf{V}$ introduces a positive sign.

For an unsteady flow in which flow properties are assumed to be uniform throughout the control volume, the system-to-control-volume equation takes the form

$$\frac{DN_{\text{sys}}}{Dt} = \maltese_{\text{c.v.}} \frac{d(\eta\rho)}{dt} + \eta_2\rho_2 V_2 A_2 - \eta_1 \rho_1 V_1 A_1 \tag{13.22}$$

for one inlet and one outlet with uniform properties.

13.4 CONSERVATION OF MASS

A system is a given collection of fluid particles; hence its mass remains fixed:

$$\frac{Dm_{\text{sys}}}{Dt} = \frac{D}{Dt}\int_{\text{sys}} \rho \, d\maltese = 0 \tag{13.23}$$

In Eq. 13.6, N_{sys} represents the mass of the system, so we simply let $\eta = 1$. Thus the conservation of mass, referring to Eq. 13.16, becomes

$$0 = \frac{d}{dt}\int_{\text{c.v.}} \rho \, d\maltese + \int_{\text{c.s.}} \rho \hat{\mathbf{n}} \cdot \mathbf{V} \, dA \tag{13.24}$$

or, since the c.v. is fixed, it may take the alternate form

$$0 = \int_{\text{c.v.}} \frac{\partial \rho}{\partial t} d\maltese + \int_{\text{c.s.}} \rho \hat{\mathbf{n}} \cdot \mathbf{V} \, dA \tag{13.25}$$

If the flow is steady, there results

$$\int_{\text{c.s.}} \rho \hat{\mathbf{n}} \cdot \mathbf{V} \, dA = 0 \tag{13.26}$$

which, for a uniform flow with one entrance and one exit, takes the form

$$\rho_2 A_2 V_2 = \rho_1 A_1 V_1 \tag{13.27}$$

Key Concept: *$\hat{\mathbf{n}}$ always points out of the control volume.*

where for an inlet we have used $\hat{\mathbf{n}}_1 \cdot \mathbf{V}_1 = -V_1$ and for an exit $\hat{\mathbf{n}} \cdot \mathbf{V}_2 = V_2$. Recall $\hat{\mathbf{n}}$ always points out of the control volume.

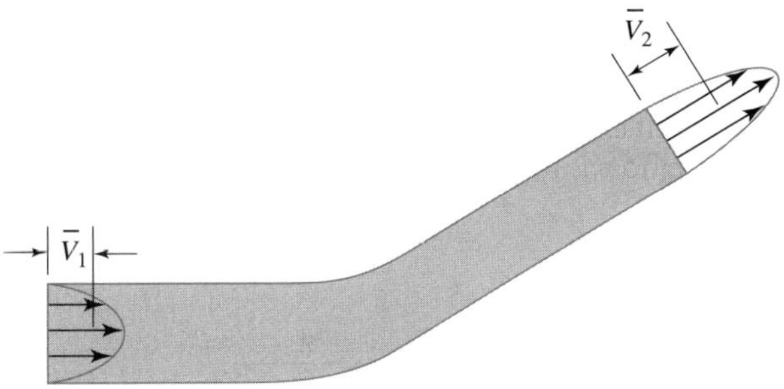

FIGURE 13.7 Nonuniform velocity profiles.

If the density is constant in the control volume, the derivative $\partial\rho/\partial t = 0$ even if the flow is unsteady. The continuity equation (13.27) then reduces to

$$A_1 V_1 = A_2 V_2 \tag{13.28}$$

This form of the continuity equation is used quite often, particularly with liquids and low-speed gas flows.

At this point we wish to discuss again the use of uniform velocity profiles (see also Section 12.3.1). Suppose that the velocity profiles at the entrance and the exit are not uniform, such as sketched in Fig. 13.7. Furthermore, suppose that the density is uniform over each area. Then the continuity equation takes the form

$$\rho_1 \int_{A_1} V_1 \, dA = \rho_2 \int_{A_2} V_2 \, dA \tag{13.29}$$

or, letting an overbar denote an average, we can write

$$\rho_1 \overline{V}_1 A_1 = \rho_2 \overline{V}_2 A_2 \tag{13.30}$$

where $\overline{V}_1$ and $\overline{V}_2$ are the *average velocities* over the areas at sections 1 and 2, respectively. In examples and problems the overbar is often omitted. It should be kept in mind, however, that actual velocity profiles are usually not uniform; Eqs. 13.27 and 13.28 are used with the velocities representing average velocities.

Any one of the above equations (13.24) through (13.30) is referred to as the *continuity equation*.

Before presenting some examples applying the continuity equation, two fluxes are defined that will be useful in specifying the quantity of flow. The *mass flux* $\dot{m}$, or the mass rate of flow, is

$$\dot{m} = \int_A \rho V_n \, dA \tag{13.31}$$

and has units of kg/s (slug/sec); V_n is the normal component of velocity. The *flow rate* Q_f, or the volume rate of flow, is

$$Q_f = \int_A V_n \, dA \tag{13.32}$$

and has units of m^3/s (ft^3/sec) or sometimes L/s. The mass flux is usually used in specifying the quantity of flow for a compressible flow and the flow rate for an incompressible flow. We often refer to the flow rate as **discharge**.

Discharge: *Another term for flow rate.*

In terms of average velocity, we have

$$Q_f = A\bar{V} \tag{13.33}$$

$$\dot{m} = \rho A\bar{V} \tag{13.34}$$

where for the mass flux we assume a uniform density profile; we also assume that the velocity is normal to the area.

The following examples are solved by first selecting a control volume. If you study the examples carefully, you will notice that often there is only one proper choice for the control volume. We must position the inlet and exit areas at locations where the integrands are either known or where they can be approximated; also, the quantity being sought is often included at an inlet or exit area. In a few cases there may be more freedom in the selection of the control volume, as in Example 13.5.

Key Concept: *Position the inlet and exit areas at locations where the integrands are known or where the quantity being sought is located.*

This first example represents the primary use of the continuity equation. It allows us to calculate the velocity at one section if it is known at another section.

Example 13.1

Water flows at a uniform velocity of 3 m/s into a nozzle that reduces the diameter from 10 cm to 2 cm (Fig. E13.1). Calculate the water's velocity leaving the nozzle and the flow rate.

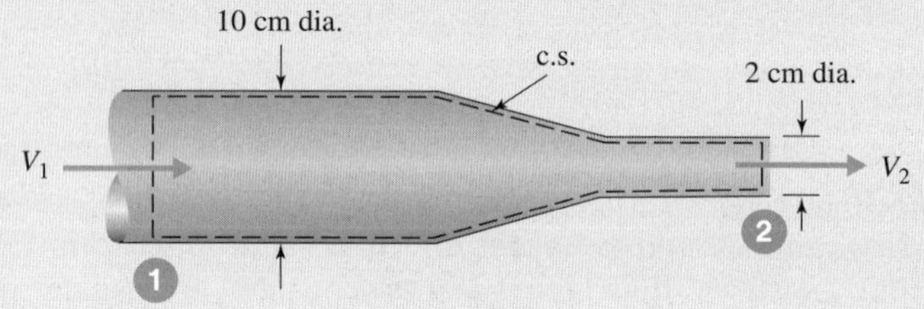

FIGURE E13.1

Solution

The control volume is selected to be the inside of the nozzle as shown. Flow enters the control volume at section 1 and leaves at section 2. The simplified continuity equation (13.28) is used since the density is assumed constant and the velocity profiles are uniform:

$$A_1V_1 = A_2V_2$$

$$\therefore\ V_2 = V_1\frac{A_1}{A_2} = 3\frac{\pi \times 0.1^2/4}{\pi \times 0.02^2/4} = 75 \text{ m/s}$$

The flow rate, or discharge, is found to be

$$Q_f = V_1A_1$$
$$= 3 \times \pi \times 0.1^2/4 = 0.0236 \text{ m}^3/\text{s}$$

Example 13.2

Water flows in and out of a device as shown in Fig. E13.2a. Calculate the rate of change of the mass of water (dm/dt) in the device.

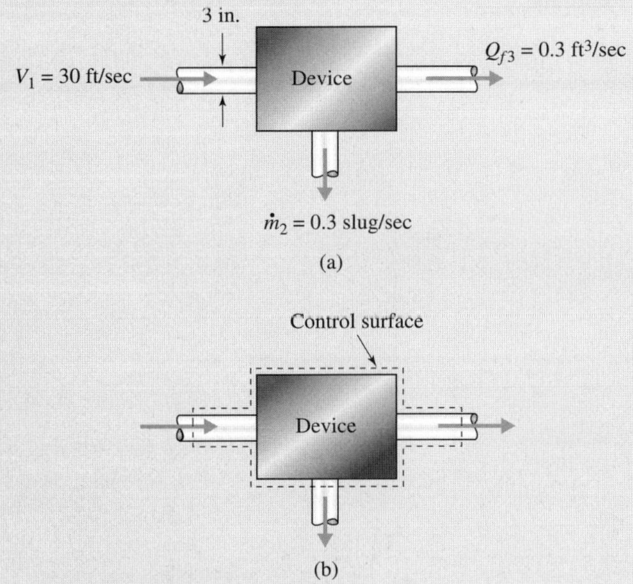

FIGURE E13.2

Solution

The control volume selected is shown in Fig. E13.2b. For the control surface surrounding the device, the continuity equation (13.24), with three surfaces across which water flows, takes the following form:

$$0 = \frac{d}{dt}\int_{\text{c.v.}} \rho\, dV\!\!\!/ + \int_{\text{c.s.}} \rho \hat{\mathbf{n}} \cdot \mathbf{V}\, dA$$

$$= \frac{dm}{dt} - \rho_1 A_1 V_1 + \rho_2 A_2 V_2 + \rho_3 A_3 V_3$$

where we have assumed the density to be constant over the volume and we have used $\mathbf{V}_1 \cdot \mathbf{n} = -V_1$ since $\hat{\mathbf{n}}_1$ points out of the volume, opposite to the direction of $\mathbf{V}_1$. The last three terms come from the area integral. In terms of the quantities given, the above can be expressed as

$$0 = \frac{dm}{dt} - \rho_1 A_1 V_1 + \dot{m}_2 + \rho_3 Q_{f3}$$

$$= \frac{dm}{dt} - 1.94 \times \pi \times \frac{1.5^2}{144} \times 30 + 0.3 + 1.94 \times 0.3$$

This is solved to yield

$$\frac{dm}{dt} = 1.975 \text{ slug/sec}$$

Hence the mass is increasing at the rate of 1.975 slug/sec. To accomplish this, the device could have a spongelike material that absorbs water.

Example 13.3

Uniform flow of air approaches a cylinder as shown in Fig. E13.3a. The symmetrical velocity distribution at the location shown downstream in the wake of the cylinder is approximated by

$$u(y) = 1.25 + \frac{y^2}{4} \qquad -1 < y < 1$$

where $u(y)$ is in m/s and y is in meters. Determine the mass flux across the surface AB per meter of depth. Use $\rho = 1.23$ kg/m^3.

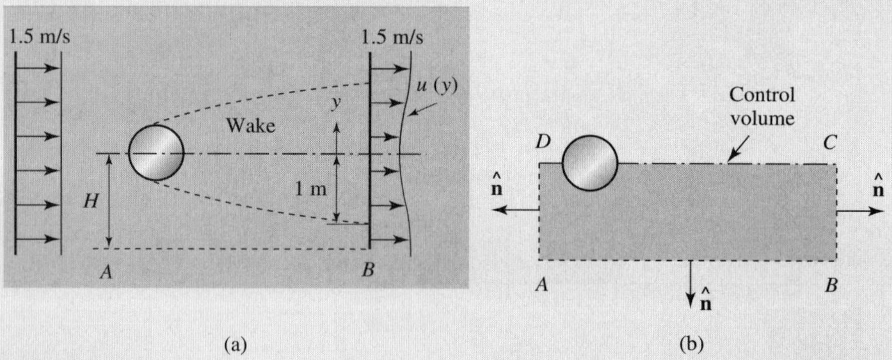

FIGURE E13.3

Solution

Select $ABCD$ as the control volume (Fig. E13.3b). Outside the wake (a region of retarded flow) the velocity is constant at 1.5 m/s. Hence the velocity normal to plane AD is 1.5 m/s. Obviously, no mass flux crosses the surface CD because of symmetry. Assuming a steady flow, the continuity equation (13.25) becomes

$$0 = \int_{\text{c.s.}} \rho \mathbf{V} \cdot \hat{\mathbf{n}}\, dA$$

Mass flux occurs across three surfaces: AB, BC, and AD. Thus the equation above takes the form

$$0 = \int_{A_{AB}} \rho \mathbf{V} \cdot \hat{\mathbf{n}}\, dA + \int_{A_{BC}} \rho \mathbf{V} \cdot \hat{\mathbf{n}}\, dA + \int_{A_{AD}} \rho \mathbf{V} \cdot \hat{\mathbf{n}}\, dA$$

$$= \dot{m}_{AB} + \int_0^H \rho u(y) 1 \times dy - \rho H \times 1 \times 1.5$$

where the negative sign for surface AD results from the fact that the unit vector points out of the volume to the left while the velocity vector points to the right. Recall that a negative sign in the steady-flow continuity equation is always associated with an influx and a positive sign with an outflux. Now, we integrate out to 1 m instead of H, since the mass that enters on the left beyond 1 m simply leaves on the right with no net gain or loss. So, letting $H = 1$ m, we have

$$0 = \dot{m}_{AB} + \int_0^1 1.23\left(1.25 + \frac{y^2}{4}\right) dy - 1.23 \times 1 \times 1.5$$

Perform the integration and there results

$$\dot{m}_{AB} = 0.205 \text{ kg/s per meter}$$

Example 13.4

A balloon is being inflated with a water supply of 0.6 m^3/s (Fig. E13.4). Find the rate of growth of the radius at the instant when $R = 0.5$ m.

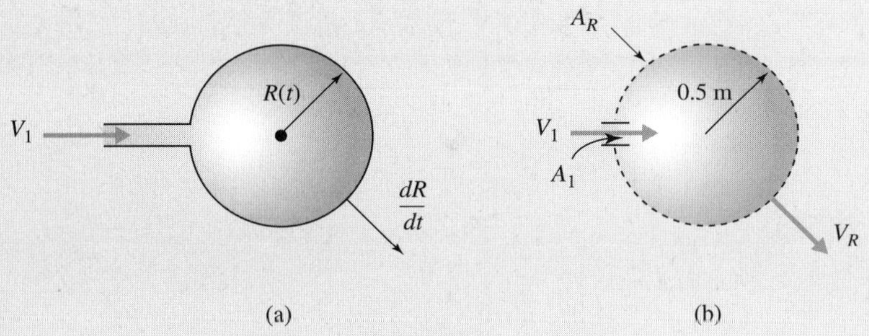

FIGURE E13.4

Solution

The objective is to find dR/dt when the radius $R = 0.5$ m. This growth rate dR/dt is the same as the water velocity normal to the wall of the balloon. Therefore, we select as our fixed control volume a sphere with a constant radius of 0.5 m so that we can calculate the velocity of the water at the surface at the instant shown moving radially out at $R = 0.5$ m. The continuity equation is written as

$$0 = \int_{\text{c.v.}} \underbrace{\frac{\partial \rho}{\partial t}}_{0} d\forall + \int_{\text{c.s.}} \rho \mathbf{V} \cdot \hat{\mathbf{n}}\, dA$$

The first term is zero because the density of water inside the control volume does not change with time. Further, the water crosses two areas: the inlet area A_1 with a velocity V_1 and the remainder of the sphere surface A_R with a velocity V_R. We will assume that $A_1 \ll A_R$. The continuity equation then takes the form

$$0 = -\rho A_1 V_1 + \rho A_R V_R$$

Since the flow rate into the volume is $A_1 V_1 = 0.6\ m^3/s$ and $A_R \simeq 4\pi R^2$ assuming that A_1 is quite small, we can solve for V_R. At $R = 0.5$ m

$$V_R = \frac{A_1 V_1}{4\pi R^2} = \frac{0.6}{4\pi \times 0.5^2} = 0.191 \text{ m/s}$$

$$\therefore \quad \frac{dR}{dt} = 0.191 \text{ m/s}$$

We have used a fixed control volume and allowed the moving surface of the balloon to pass through it at the instant considered. With this approach it is possible to model situations in which surfaces, such as a piston, are allowed to move.

Example 13.5

This example shows that there may be more than one good choice for a control volume. We want to determine the rate at which the water level rises in an open container if the water coming in through a 0.10-m^2 pipe has a velocity of 0.5 m/s and the flow rate going out is 0.2 m^3/s (Fig. E13.5a). The container has a circular cross section with a diameter of 0.5 m.

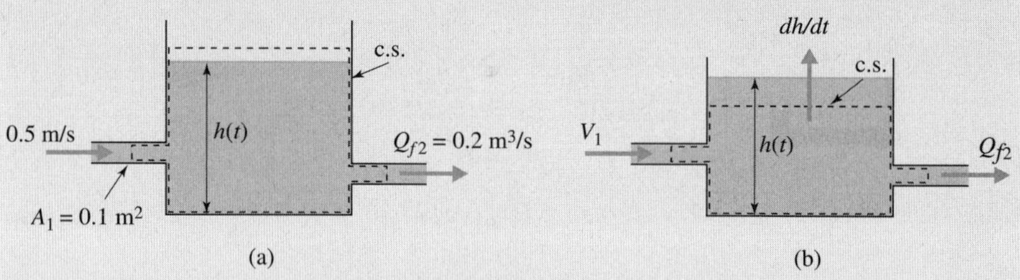

FIGURE E13.5

Solution

First we select a control volume that extends above the water surface as shown in Fig. E13.5a. Apply the continuity equation (Eq. 13.24):

$$\frac{d}{dt}\int_{\text{c.v.}} \rho\, d\mathcal{V} + \rho(-V_1)A_1 + \rho V_2 A_2 = 0$$

in which the first term describes the rate of change of mass in the control volume. Hence, neglecting the airmass above the water, we have

$$\frac{d(\rho h \pi D^2/4)}{dt} - \rho V_1 A_1 + \rho Q_{f2} = 0$$

Divide by the constant ρ,

$$\frac{\pi D^2}{4}\frac{dh}{dt} - V_1 A_1 + Q_{f2} = 0$$

The rate at which the water level is rising is then

$$\frac{dh}{dt} = \frac{V_1 A_1 - Q_{f2}}{\pi D^2/4}$$

Thus

$$\frac{dh}{dt} = \frac{0.5 \times 0.1 - 0.2}{\pi \times 0.5^2/4} = -0.764 \text{ m/s}$$

The negative sign indicates that the water level is actually decreasing.

Let's solve this problem again but with another choice for the control volume, one with its top surface below the water level (Fig. E13.5b). The velocity at the top surface is then equal to the rate at which the surface rises. The flow condition inside the control

(continued)

volume is steady. Hence we can apply Eq. 13.26. There are three areas across which fluid flows. On the third area, the velocity is dh/dt; hence the continuity equation takes the form

$$\rho(-V_1)A_1 + \rho Q_{f2} + \rho\frac{dh}{dt}\frac{\pi}{4}D^2 = 0$$

so that

$$\frac{dh}{dt} = \frac{V_1A_1 - Q_{f2}}{\pi D^2/4}$$

This is the same result as given above.

13.5 ENERGY EQUATION

Many problems involving fluid motion demand that the first law of thermodynamics, often referred to as the *energy equation*, be used to relate quantities of interest. If the heat transferred to a device (a boiler or compressor), or the work done by a device (a pump or turbine), is desired, the energy equation is obviously needed. It is also used to relate pressures and velocities when Bernoulli's equation is not applicable; this is the case whenever viscous effects cannot be neglected, such as flow through a piping system or in an open channel. Let us express the energy equation in control volume form. For a system it is

$$\dot{Q} - \dot{W} = \frac{D}{Dt}\int_{\text{sys}} e\rho\, dV\!\!\!/ \tag{13.35}$$

where the specific energy e includes specific kinetic energy $V^2/2$, specific potential energy gz, and specific internal energy $\tilde{u}$; that is,

$$e = \frac{V^2}{2} + gz + \tilde{u} \tag{13.36}$$

We will not include other forms of energy, such as energy due to magnetic or electric field-flow field interactions or those due to chemical reactions. In terms of a control volume, Eq. 13.35 becomes

$$\dot{Q} - \dot{W} = \frac{d}{dt}\int_{\text{c.v.}} e\rho\, dV\!\!\!/ + \int_{\text{c.s.}} \rho e\mathbf{V}\cdot\hat{\mathbf{n}}\, dA \tag{13.37}$$

This can be put in simplified forms for certain restricted flows, but first let us discuss the rate-of-heat transfer term $\dot{Q}$ and the work-rate term $\dot{W}$.

The term $\dot{Q}$ represents the rate-of-energy transfer across the control surface due to a temperature difference. (Do not confuse this term with the flow rate Q_f.) The rate-of-heat transfer term is usually given or results from using Eq. 13.37. However, we will calculate $\dot{Q}$ for certain simplified flow situations in Section 16.6.6. The work-rate term is discussed in detail in the following section.

Key Concept: $\dot{Q}$ *represents the rate-of-energy transfer across the control surface due to a temperature difference.*

13.5.1 Work-Rate Term

Key Concept: *The rate of doing work is given by the dot product of a force* **F** *with its velocity.*

The work-rate term results from work being done by the system. Or, since we consider the instant that the system occupies the control volume, we can also state that the work-rate term results from work being done by the control volume. Work is due to a force moving through a distance while it acts on the control volume. The rate of doing work $\dot{W}$, or power, is given by the dot product of a force **F** with its velocity:

$$\dot{W} = -\mathbf{F} \cdot \mathbf{V}_I \tag{13.38}$$

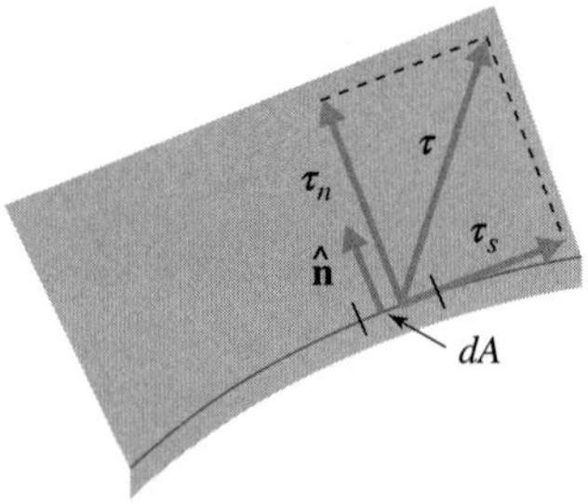

FIGURE 13.8 Stress vector acting on the control surface.

where $\mathbf{V}_I$ is the velocity measured with respect to a fixed reference frame. The negative sign results because we have selected the convention that work done on the control volume is negative.

If the force results from a variable stress acting over the control surface, we must integrate,

$$\dot{W} = -\int_{\text{c.s.}} \boldsymbol{\tau} \cdot \mathbf{V}_I \, dA \tag{13.39}$$

where $\boldsymbol{\tau}$ is the stress vector acting on the elemental area dA, the differential force being represented by $d\mathbf{F} = \boldsymbol{\tau}\, dA$, as shown in Fig. 13.8.

For a moving control volume, such as a car, we have to evaluate the velocity with respect to a fixed reference frame. For example, let us consider an automobile traveling at constant speed. If we want to apply the energy equation, we could make the car the control volume. In that case, the velocity in Eq. 13.38 would be measured relative to a fixed reference and not relative to the car. If the velocity relative to the car were used, the drag force would have a zero velocity which would result in no work done; but we know that at high speed, energy from the gasoline goes primarily to overcome the drag. Thus the stationary reference frame is needed.

In general, for moving control volumes the velocity vector $\mathbf{V}_I$ is related to a relative velocity $\mathbf{V}$, observed in a reference frame attached to the control volume by

$$\mathbf{V}_I = \mathbf{V} + \dot{\mathbf{S}} + \boldsymbol{\Omega} \times \mathbf{r} \tag{13.40}$$

where $\dot{\mathbf{S}}$ is the velocity of the control volume (see Fig. 12.5). We can now write the work rate as

$$\dot{W} = -\int \boldsymbol{\tau} \cdot \mathbf{V}\, dA + \dot{W}_I \tag{13.41}$$

where the "inertial work-rate" term is given by

$$\dot{W}_I = -\int_{\text{c.s.}} \boldsymbol{\tau} \cdot (\dot{\mathbf{S}} + \boldsymbol{\Omega} \times \mathbf{r})\, dA \tag{13.42}$$

Next, express the stress vector as the sum of a normal component and a shear component, that is,

$$\boldsymbol{\tau} = -p\hat{\mathbf{n}} + \boldsymbol{\tau}_s \tag{13.43}$$

where the pressure p is assumed to be positive in a compressive state. Then

$$\dot{W} = \int_{\text{c.s.}} p\hat{\mathbf{n}} \cdot \mathbf{V}\, dA - \int_{\text{c.s.}} \boldsymbol{\tau}_s \cdot \mathbf{V}\, dA + \dot{W}_I \qquad (13.44)$$

We will allow the shear stress term to consist of two parts. One part accounts for work called *shaft work* $\dot{W}_S$ transmitted by a rotating shaft that is cut by the control surface; this term is important when we deal with flows in pumps and turbines. The other part will be denoted *shear work* $\dot{W}_{\text{shear}}$ and results from moving boundaries; this term is required if the control surface itself moves relative to the control volume as occurs with a moving belt.

Key Concept: *Shaft work $\dot{W}_S$ is transmitted by a rotating shaft that is cut by the control surface.*

Hence the work-rate term becomes

$$\dot{W} = \int_{\text{c.s.}} p\hat{\mathbf{n}} \cdot \mathbf{V}\, dA + \dot{W}_S + \dot{W}_{\text{shear}} + \dot{W}_I \qquad (13.45)$$

The terms are summarized as follows:

$\int p\hat{\mathbf{n}} \cdot \mathbf{V}\, dA$	Work rate resulting from the force due to pressure moving at the control surface. It is often referred to as **flow work**.
$\dot{W}_S$	Work rate resulting from rotating shafts such as that of a pump or turbine, or the equivalent electric power.
$\dot{W}_{\text{shear}}$	Work rate due to the shear acting on a moving boundary such as a moving belt.
$\dot{W}_I$	Work rate that occurs when the control volume moves relative to a fixed reference frame.

Flow work: *Work rate resulting from the force due to pressure moving at the control surface.*

We should note that the work-rate terms $\dot{W}_{\text{shear}}$ and $\dot{W}_I$ are seldom encountered in problems in an introductory course and are often omitted from textbooks. They are included here for completeness.

13.5.2 General Energy Equation

When the work-rate term of Eq. 13.45 is substituted into Eq. 13.37, we obtain the energy equation in the form

$$\dot{Q} - \dot{W}_S - \dot{W}_{\text{shear}} - \dot{W}_I = \frac{d}{dt}\int_{\text{c.v.}} e\rho\, d\mathcal{V} + \int_{\text{c.s.}} \left(e + \frac{p}{\rho}\right)\rho\hat{\mathbf{n}} \cdot \mathbf{V}\, dA \qquad (13.46)$$

Note that the work-rate term needed to move the pressure force has been moved to the right-hand side and is treated like an energy flux term.

Substitution of Eq. 13.36 results in

$$\dot{Q} - \dot{W}_S - \dot{W}_{\text{shear}} - \dot{W}_I = \frac{d}{dt}\int_{\text{c.v.}} \left(\frac{V_I^2}{2} + gz + \tilde{u}\right)\rho\, d\mathcal{V} + \int_{\text{c.s.}} \left(\frac{V_I^2}{2} + gz + \tilde{u} + \frac{p}{\rho}\right)\rho\mathbf{V} \cdot \hat{\mathbf{n}}\, dA \qquad (13.47)$$

This general form of the energy equation is useful in analyzing fluid flow problems that may include time-dependent effects and nonuniform profiles. We will now drop the $\dot{W}_{\text{shear}}$ and $\dot{W}_I$ terms and relegate them to special topics in more advanced courses. We will consequently assume an inertial reference frame so that $V_I = V$. Before we simplify the equation for steady flow and uniform profiles, let us introduce the notion of "losses."

In many fluid flows, useful forms of energy (kinetic energy and potential energy) and flow work are converted into unusable energy forms (internal energy or heat transfer). If we assume that the temperature of the control volume remains unchanged, the internal energy does not change and the losses are balanced by heat transfer across the control surface. This heat transfer can be the result of convection, radiation, or conduction at the control surfaces. The theory of heat transfer is aimed at the detailed description of these effects. However, the sum of these effects is lumped together and denoted $\dot{Q}$. Thus we define **losses** as the sum of all the terms representing unusable forms of energy:

Losses: *The sum of all the terms representing unusable forms of energy.*

$$\text{losses} = -\dot{Q} + \frac{d}{dt}\int_{\text{c.v.}} \tilde{u}\rho\, dV + \int_{\text{c.s.}} \tilde{u}\rho \mathbf{V}\cdot\hat{\mathbf{n}}\, dA \tag{13.48}$$

We can now rewrite the energy equation as

$$-\dot{W}_S = \frac{d}{dt}\int_{\text{c.v.}} \left(\frac{V^2}{2} + gz\right)\rho\, d\forall + \int_{\text{c.s.}} \left(\frac{V^2}{2} + gz + \frac{p}{\rho}\right)\rho\mathbf{V}\cdot\hat{\mathbf{n}}\, dA + \text{losses} \tag{13.49}$$

Losses are due to two primary effects:

Key Concept: *Losses are due primarily to internal friction and separated flows.*

1. Viscosity causes internal friction that results in increased internal energy (temperature increase) or heat transfer.
2. Changes in geometry result in separated flows that require useful energy to maintain the resulting secondary motions in which viscous dissipation occurs.

In a conduit, the losses due to viscous effects are distributed over the entire length, whereas the loss due to a geometry change (a valve, an elbow, an enlargement) is concentrated in the vicinity of the geometry change.

It turns out that the analytical calculation of losses is rather difficult, particularly when the flow is turbulent. In general, the prediction of losses is based on empirical formulas. Such formulas will be given in subsequent chapters. In this chapter we discuss losses qualitatively and, in examples and problems, losses will be given. For a pump or a turbine the losses are expressed in terms of the efficiency. For example, if the efficiency of a pump is 80%, the losses would be 20% of the energy input to the pump.

Key Concept: *For a pump or a turbine, losses are expressed in terms of the efficiency.*

It may be that the objective in a particular fluid flow is to change the internal energy of the fluid, such as in the steam generator (boiler) of a power plant, by the transfer of heat; then the definition of losses above must be altered so that the loss term includes only the dissipative effects of the viscosity of the fluid. Generally, for problems of interest in fluid mechanics, Eq. 13.49 is acceptable.

13.5.3 Steady Uniform Flow

Consider a steady-flow situation in which there is one entrance and one exit across which uniform profiles can be assumed. For such a flow the term $(V^2/2 + gz + p/\rho)$ in Eq. 13.49 is constant across the cross section because V is constant (we assume a uniform velocity profile) and the sum of $p/\rho + gz$ is constant if the streamlines at each section are parallel. The energy equation (Eq. 13.49) then simplifies to

$$-\dot{W}_S = \rho_2 V_2 A_2\left(\frac{V_2^2}{2} + \frac{p_2}{\rho_2} + gz_2\right) - \rho_1 V_1 A_1\left(\frac{V_1^2}{2} + \frac{p_1}{\rho_1} + gz_1\right) + \text{losses} \qquad (13.50)$$

where the subscripts 1 and 2 refer to the entrance and exit, respectively. The mass flux is given by $\dot{m} = \rho_1 A_1 V_1 = \rho_2 A_2 V_2$. After dividing by $\dot{m}g$ we have

$$-\frac{\dot{W}_S}{\dot{m}g} = \frac{V_2^2 - V_1^2}{2g} + \frac{p_2}{\gamma_2} - \frac{p_1}{\gamma_1} + z_2 - z_1 + h_L \qquad (13.51)$$

where we have introduced the *head loss* h_L, defined to be

$$h_L = \frac{\tilde{u}_2 - \tilde{u}_1}{g} - \frac{\dot{Q}}{\dot{m}g} \qquad (13.52)$$

It is often written in terms of a *loss coefficient* K as

$$h_L = K\frac{V^2}{2g} \qquad (13.53)$$

where V may be either V_1 or V_2; if it is not obvious, it will be specified. Loss coefficients are discussed in some detail in Chapter 15 and are tabulated in Table 15.2.

The head loss is referred to as a "head" since it has dimensions of length. We may also refer to $V^2/2g$ as the *velocity head* and p/γ as the *pressure head* since those terms also have dimensions of length. Further, the sum $V^2/2g + p/\gamma + z$ is called the *total head*.

The energy equation, in the form of Eq. 13.51, is useful in many applications and is, perhaps, the most often used form of the energy equation. If the losses are negligible and if there is no shaft work, we note that the energy equation takes the form

$$\frac{V_2^2}{2g} + \frac{p_2}{\gamma_2} + z_2 = \frac{V_1^2}{2g} + \frac{p_1}{\gamma_1} + z_1 \qquad (13.54)$$

Key Concept: *Bernoulli's equation is applicable along a streamline and the energy equation is applied between two sections.*

Observe that the energy equation has been reduced to a form identical with Bernoulli's equation when $\gamma_2 = \gamma_1$ (a constant density flow). We must remember, however, that Bernoulli's equation is a momentum equation applicable along a streamline and the equation above is an energy equation applied between two

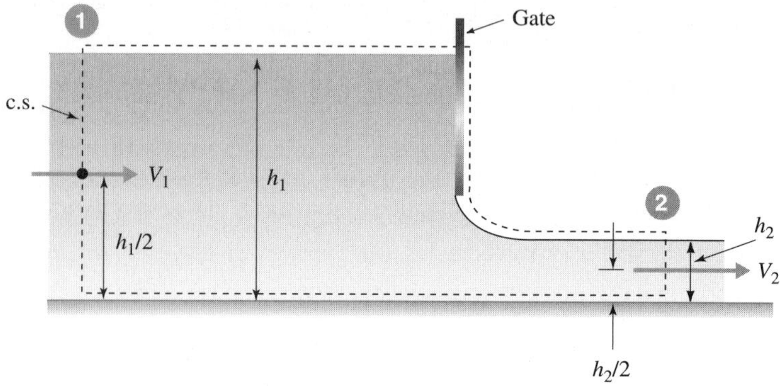

FIGURE 13.9 Application of the energy equation to a gate in an open channel.

sections of a flow. It is not surprising that both should predict identical results from the conditions stated because the velocity head is constant over a cross section and the sum of pressure head and elevation remains constant over a cross section.

The energy equation (13.51) may be applied to any steady, uniform flow with one entrance and one exit. The control volume is usually selected such that the entrance and exit sections have a uniform total head. For example, it may be applied to water flow through a long pipeline; the total head at the entrance and exit may then be evaluated conveniently at the center of the pipe entrance and exit. The energy equation may be applied to the flow passing a gate (Fig. 13.9); an appropriate control volume is shown. The total head at the entrance and exit can be evaluated at any point at the entrance and exit, respectively. However, a convenient choice would be the points at the water surface. Thus the energy equation becomes

$$\frac{V_1^2}{2g} + \overset{0}{\cancel{\frac{p_1}{\gamma}}} + h_1 = \frac{V_2^2}{2g} + \overset{0}{\cancel{\frac{p_2}{\gamma}}} + h_2 + h_L \tag{13.55}$$

where the shaft work has been set to zero. If we had picked the centroids of the entrance and exit, as shown in Fig. 13.9, we would have obtained

$$\frac{V_1^2}{2g} + \frac{p_1}{\gamma} + \frac{h_1}{2} = \frac{V_2^2}{2g} + \frac{p_2}{\gamma} + \frac{h_2}{2} + h_L \tag{13.56}$$

We see that this result is the same as in Eq. 13.55 if we substitute $p_1 = \gamma h_1/2$ and $p_2 = \gamma h_2/2$. For completeness, it is noted that the losses between 1 and 2 in Fig. 13.9 could be neglected because internal viscous effects occur only over a relatively short distance and no significant secondary flows are generated.

The energy equation (13.49) can be applied to any control volume. For example, consider steady, uniform incompressible flow through a T-section in a pipe (Fig. 13.10) in which there is one entrance and two exits. The energy equation can be applied to each of two control volumes, one for the mass flux that exits section 2 and the other for the mass flux that exits section 3:

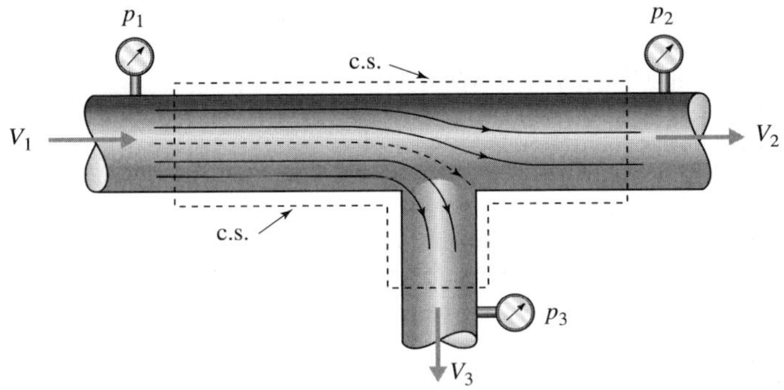

FIGURE 13.10 Application of the energy equation to a T-section.

$$\frac{V_1^2}{2g} + \frac{p_1}{\gamma} + z_1 = \frac{V_2^2}{2g} + \frac{p_2}{\gamma} + z_2 + h_{L_{1\text{-}2}}$$
$$\frac{V_1^2}{2g} + \frac{p_1}{\gamma} + z_1 = \frac{V_3^2}{2g} + \frac{p_3}{\gamma} + z_3 + h_{L_{1\text{-}3}} \qquad (13.57)$$

The loss terms in Eqs. 13.57 include the losses between the inlet and the respective exits. If the losses are negligible, the energy equation reduces to a form similar to the Bernoulli equation being applied along a streamline going from 1 to 2 or a streamline going from 1 to 3.

A final note to this section regards nomenclature for pumps and turbines in a flow system. It is often conventional to call the energy term $(\dot{W}_S/\dot{m}g)$ associated with a pump the *pump head* H_P, and the term $(\dot{W}_S/\dot{m}g)$ associated with a turbine the *turbine head* H_T. Then the energy equation, for an incompressible flow, takes the form

$$H_P + \frac{V_1^2}{2g} + \frac{p_1}{\gamma} + z_1 = H_T + \frac{V_2^2}{2g} + \frac{p_2}{\gamma} + z_2 + h_L \qquad (13.58)$$

In this form we have equated the energy at the inlet plus added energy to the energy at the exit plus extracted energy (energy per unit weight, of course). If any of the quantities is zero (e.g., there is no pump), the appropriate term is simply omitted. The terms H_P and H_T above represent the energy that is transferred to and from the fluid, respectively. If the energy delivered by the turbine or required by the pump is desired, the efficiency of each device must be used.

Key Concept: *H_P and H_T represent the energy that is transferred to and from the fluid, respectively.*

The power generated by the turbine with an efficiency of η_T is simply

$$\dot{W}_T = \dot{m}gH_T\eta_T = \gamma Q_f H_T \eta_T \qquad (13.59)$$

The power requirement by a pump with an efficiency of η_P would be

$$\dot{W}_P = \frac{\dot{m}gH_P}{\eta_P} = \frac{\gamma Q_f H_P}{\eta_P} \qquad (13.60)$$

We will calculate power in watts, ft-lb/sec, or horsepower. Recall that one horsepower is equivalent to 746 W or 550 ft-lb/sec.

13.5.4 Steady Nonuniform Flow

If the assumption of uniform velocity profiles is not acceptable for a problem of interest, as is sometimes the situation, we have to consider the control-surface integral in Eq. 13.49 with the proper expression for the velocity distribution. In practice, a velocity distribution can be accounted for by introducing the *kinetic-energy correction factor* α, defined by

$$\alpha = \frac{\int V^3\, dA}{\overline{V}^3 A} \tag{13.61}$$

where $\overline{V}$ is the average velocity over the area A, given by Eq. 13.33. Then the term that accounts for the flux of kinetic energy in Eq. 13.49 is

$$\frac{1}{2}\rho \int_A V^3\, dA = \frac{1}{2}\alpha\rho\overline{V}^3 A \tag{13.62}$$

where we have used $\mathbf{V}\cdot\hat{\mathbf{n}} = V$. Using this factor, we can account for nonuniform velocity distributions by modifying Eq. 13.58 to read

$$H_P + \alpha_1\frac{\overline{V}_1^2}{2g} + \frac{p_1}{\gamma} + z_1 = H_T + \alpha_2\frac{\overline{V}_2^2}{2g} + \frac{p_2}{\gamma} + z_2 + h_L \tag{13.63}$$

Key Concept: *For most internal turbulent flows, we let $\alpha = 1$.*

where $\overline{V}_1$ and $\overline{V}_2$ are the average velocities at sections 1 and 2, respectively. For a flow with a parabolic profile in a pipe we can calculate $\alpha = 2.0$ (see Example 13.9). For most internal turbulent flows, however, the profile is nearly uniform with $\alpha \simeq 1.05$. Hence we simply let $\alpha = 1$ since it is so close to unity; this will always be done unless otherwise stated, since most of the internal flows that we encounter are, in fact, turbulent flows.

Example 13.6

The pump of Fig. E13.6 is to increase the pressure of 0.2 m^3/s of water from 200 kPa to 600 kPa. If the pump is 85% efficient, how much electrical power will the pump require? The exit area is 20 cm above the inlet area. Assume inlet and exit areas are equal.

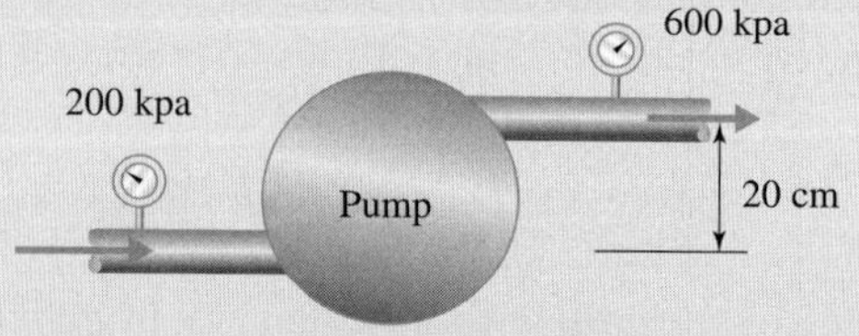

FIGURE E13.6

(continued)

Solution

Equation (13.58) across the pump provides

$$H_P = \frac{p_2 - p_1}{\gamma} + z_2 - z_1$$

$$= \frac{600\,000 - 200\,000}{9810} + 0.2 = 41.0 \text{ m}$$

where $V_2 = V_1$ since the inlet and exit areas are equal, and any losses are accounted for with the efficiency of Eq. 13.60. That equation provides the power:

$$\dot{W}_P = \frac{\gamma Q_f H_P}{\eta_P}$$

$$= \frac{9810 \times 0.2 \times 41.0}{0.85} = 94\,600 \text{ W} \quad \text{or} \quad 94.6 \text{ kW}$$

Note: The units work out if length is in meters ($H_P = 41$ m), weight is in newtons ($\gamma = 9810$ N/m^3), and time is in seconds ($Q_f = 0.2$ m^3/s). Check the units to make sure they provide the answer in watts, i.e., J/s.

Example 13.7

Water flows from a reservoir through the 2.5-ft-diameter pipeline of Fig. E13.7 to a turbine-generator unit and exits to a river that is 100 ft below the reservoir surface. If the flow rate is 90 ft^3/sec, and the turbine-generator efficiency is 88%, calculate the power output. Assume the loss coefficient in the pipeline (including the exit) to be $K = 2$.

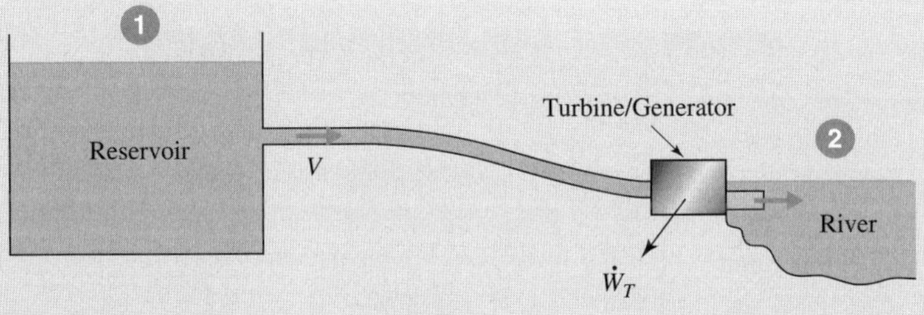

FIGURE E13.7

Solution

Referring to Fig. E13.7, we select the control volume to extend from section 1 to section 2 on the reservoir and river surfaces, where we know the velocities, pressures, and elevations; we consider the water surface of the left reservoir to be the entrance and the water surface of the river to be the exit. The velocity in the pipe is

$$V = \frac{Q_f}{A} = \frac{90}{\pi \times 2.5^2/4} = 18.3 \text{ ft/sec}$$

(continued)

Now, consider the energy equation. We will use gage pressures so that $p_1 = p_2 = 0$; the datum is placed through the lower section 2 so that $z_2 = 0$; the velocities V_1 and V_2 on the reservoir surfaces are negligibly small; K is assumed to be based on the 2.5-ft-diameter pipe velocity. The energy equation (13.58) then becomes

$$\overset{0}{\cancel{H_P}} + \overset{0}{\cancel{\frac{V_1^2}{2g}}} + \overset{0}{\cancel{\frac{p_1}{\gamma}}} + z_1 = H_T + \overset{0}{\cancel{\frac{V_2^2}{2g}}} + \overset{0}{\cancel{\frac{p_2}{\gamma}}} + \overset{0}{\cancel{z_2}} + K\frac{V^2}{2g}$$

$$100 = H_T + 2\frac{18.3^2}{2 \times 32.2}$$

$$\therefore\ H_T = 89.6 \text{ ft}$$

From this the power output is found using Eq. 13.59 to be

$$\dot{W}_T = Q_f \gamma H_T \eta_T$$

$$= 90 \times 62.4 \times 89.6 \times 0.88 = 443{,}000 \text{ ft-lb/sec} \quad \text{or} \quad 805 \text{ hp}$$

In this example we have used gage pressure; the potential-energy datum was assumed to be placed through section 2, V_1 and V_2 were assumed to be insignificantly small, and K was assumed to be based on the 2.5-ft-diameter pipe velocity.

Example 13.8

The venturi meter shown reduces the pipe diameter from 10 cm to a minimum of 5 cm (Fig. E13.8). Calculate the flow rate and the mass flux assuming ideal conditions.

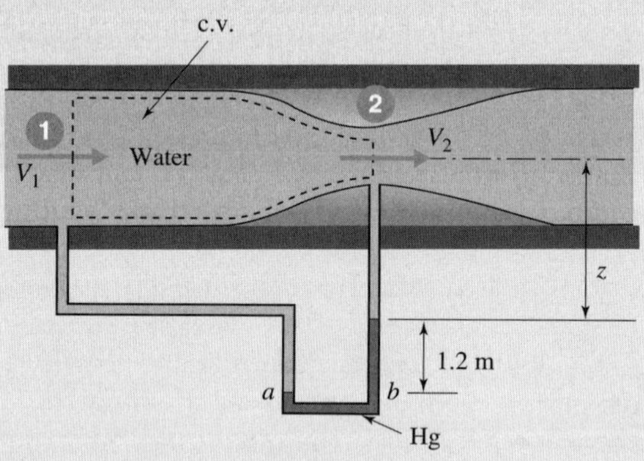

FIGURE E13.8

Solution

The control volume is selected as shown such that the entrance and exit correspond to the sections where the pressure information of the manometer can be applied. The manometer's reading is interpreted as follows:

(continued)

$$p_a = p_b$$

$$p_1 + \gamma(z + 1.2) = p_2 + \gamma z + 13.6\gamma \times 1.2$$

where z is the distance from the pipe centerline to the top of the mercury column. The manometer then gives

$$\frac{p_1 - p_2}{\gamma} = 15.12 \text{ m}$$

Continuity allows us to relate V_2 to V_1 by

$$V_1 A_1 = V_2 A_2$$

$$\therefore\ V_2 = \frac{A_1}{A_2} V_1 = 4V_1$$

The energy equation (13.51) assuming ideal conditions (no losses and uniform flow) with $h_L = \dot{W}_S = 0$ takes the form

$$0 = \frac{V_2^2 - V_1^2}{2g} + \frac{p_2 - p_1}{\gamma} + (\cancelto{0}{z_2 - z_1})$$

$$= \frac{16V_1^2 - V_1^2}{2g} - 15.12$$

$$\therefore\ V_1 = 4.45 \text{ m/s}$$

The flow rate is

$$Q_f = A_1 V_1 = \pi \times 0.05^2 \times 4.45 = 0.0350 \text{ m}^3/\text{s}$$

The mass flux is

$$\dot{m} = \rho Q_f = 1000 \times 0.035 = 35.0 \text{ kg/s}$$

Example 13.9

The velocity distribution for a certain flow in a pipe is $V(r) = V_{\max}(1 - r^2/r_0^2)$, where r_0 is the pipe radius (Fig. E13.9). Determine the kinetic-energy correction factor.

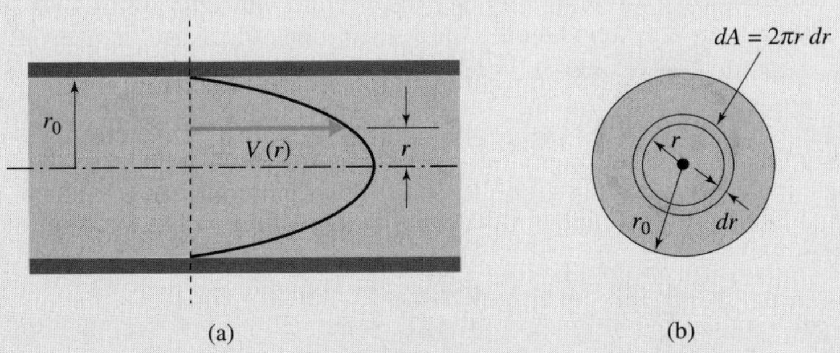

FIGURE E13.9

(continued)

Solution

To find the kinetic-energy correction factor α, we must know the average velocity. It is (combine Eqs. 13.32 and 13.33)

$$\bar{V} = \frac{\int V\,dA}{A}$$

$$= \frac{1}{\pi r_0^2}\int_0^{r_0} V_{\max}\left(1 - \frac{r^2}{r_0^2}\right)2\pi r\,dr = \frac{2\pi V_{\max}}{\pi r_0^2}\int_0^{r_0}\left(r - \frac{r^3}{r_0^2}\right)dr$$

$$= \frac{2V_{\max}}{r_0^2}\left(\frac{r_0^2}{2} - \frac{r_0^4}{4r_0^2}\right) = \frac{1}{2}V_{\max}$$

Using Eq. 13.61, there results

$$\alpha = \frac{\int V^3\,dA}{\bar{V}^3 A}$$

$$= \frac{\int_0^{r_0} V_{\max}^3 (1 - r^2/r_0^2)^3 2\pi r\,dr}{\left(\frac{1}{2}V_{\max}\right)^3 \pi r_0^2} = \frac{16}{r_0^2}\int_0^{r_0}\left(1 - \frac{3r^2}{r_0^2} + \frac{3r^4}{r_0^4} - \frac{r^6}{r_0^6}\right) r\,dr$$

$$= \frac{16}{r_0^2}\left(\frac{r_0^2}{2} - \frac{3r_0^2}{4} + \frac{3r_0^2}{6} - \frac{r_0^2}{8}\right) = 2$$

Consequently, the kinetic energy flux associated with a parabolic velocity distribution across a circular area is given by

$$\int \rho \mathbf{V}\cdot\hat{\mathbf{n}}\frac{V^2}{2}\,dA = 2 \times \frac{\dot{m}\bar{V}^2}{2}$$

Parabolic velocity distributions are encountered in laminar flows in pipes and between parallel plates (for parallel plates $\alpha = 1.54$). The Reynolds number must be quite small, usually less than about 2000.

13.6 MOMENTUM EQUATION

13.6.1 General Momentum Equation

Newton's second law, often called the *momentum equation*, states that the resultant force acting on a system equals the rate of change of momentum of the system when measured in an inertial reference frame; that is,

$$\sum \mathbf{F} = \frac{D}{Dt}\int_{\text{sys}} \rho \mathbf{V}\,d\forall \tag{13.64}$$

Using Eq. 13.16, with η replaced by $\mathbf{V}$, this is written for a control volume as

$$\sum \mathbf{F} = \frac{d}{dt}\int_{\text{c.v.}} \rho \mathbf{V}\,d\forall + \int_{\text{c.s.}} \rho \mathbf{V}(\mathbf{V}\cdot\hat{\mathbf{n}})\,dA \tag{13.65}$$

where $\mathbf{V}\cdot\hat{\mathbf{n}}$ is simply a scalar for each differential area dA. The control surface integral on the right represents the net momentum flux across the control surface of the fluid entering and/or leaving the control volume.

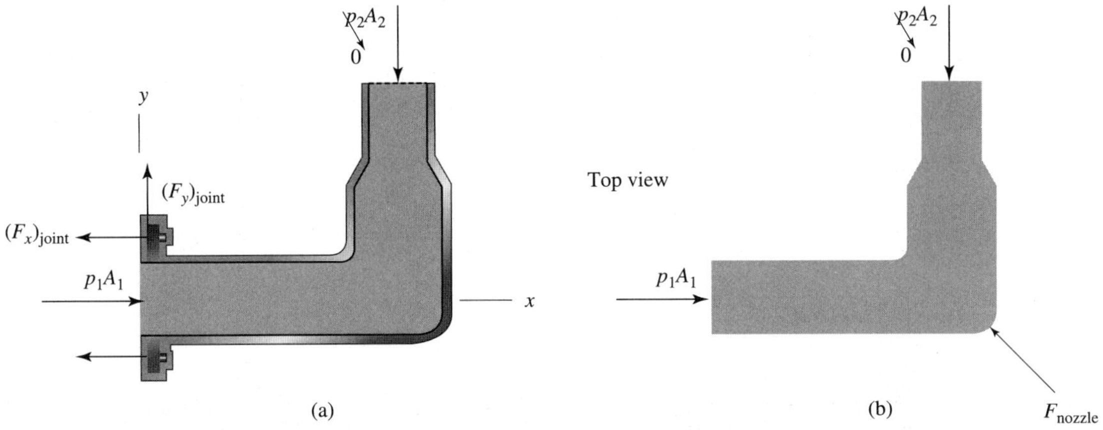

FIGURE 13.11 Forces acting on the control volume of a horizontal nozzle: (a) control volume includes nozzle and fluid in nozzle; (b) control volume includes fluid in nozzle only. We have neglected body forces.

When applying Newton's second law the quantity $\sum \mathbf{F}$ represents all forces acting on the control volume. The forces include the surface forces resulting from the surroundings acting on the control surface and body forces that result from gravity and magnetic fields. The momentum equation is often used to determine the forces induced by the flow. For example, the equation allows us to calculate the force on the support of an elbow in a pipeline or the force on a submerged body in a free-surface flow.

Key Concept: *The momentum equation is used primarily to determine the forces induced by the flow.*

When we apply the momentum equation the surrounding fluid and sometimes the entire conduit or container is separated from the control volume. For example, in the horizontal nozzle of Fig. 13.11a, the nozzle and the fluid in the nozzle are isolated. Thus care must be taken to include the pressure forces shown and the force $\mathbf{F}_{\text{joint}}$. It is convenient to use gage pressures so that the pressure acting on the exterior of the pipe is then zero. Alternatively, we could have selected a control volume that includes only the fluid in the nozzle (Fig. 13.11b). In that case we have to consider the pressure forces at the entrance and exit and the resultant pressure force $\mathbf{F}_{\text{nozzle}}$ of the interior wall of the nozzle on the fluid. Of course, the force $\mathbf{F}_{\text{joint}}$ and $\mathbf{F}_{\text{nozzle}}$ are equal in magnitude, as is obvious from a free body of the nozzle excluding the fluid. If the problem is to determine the force exerted by the flow on the nozzle (Fig. 13.11b), we have to reverse the direction of the calculated force $\mathbf{F}_{\text{nozzle}}$. Examples at the end of this section illustrate this.

13.6.2 Steady Uniform Flow

Equation 13.65 can be simplified considerably if a device has entrances and exits across which the flow may be assumed to be uniform and if the flow is steady. Then there results

$$\sum \mathbf{F} = \sum_{i=1}^{N} \rho_i A_i \mathbf{V}_i (\mathbf{V}_i \cdot \hat{\mathbf{n}}) \qquad (13.66)$$

where N is the number of flow exit/entrance areas.

At an entrance $\mathbf{V} \cdot \hat{\mathbf{n}} = -V$ since the unit vector points out of the volume and at the exit $\mathbf{V} \cdot \hat{\mathbf{n}} = V$. If there is only one entrance and one exit, as in Fig. 13.11, the momentum equation becomes

$$\sum \mathbf{F} = \rho_2 A_2 V_2 \mathbf{V}_2 - \rho_1 A_1 V_1 \mathbf{V}_1 \tag{13.67}$$

Using continuity,

$$\dot{m} = \rho_1 A_1 V_1 = \rho_2 A_2 V_2 \tag{13.68}$$

the momentum equation takes the simplified form

$$\sum \mathbf{F} = \dot{m}(\mathbf{V}_2 - \mathbf{V}_1) \tag{13.69}$$

Key Concept: *The momentum equation is a vector equation which represents three scalar equations.*

Note that the momentum equation is a vector equation which represents the following three scalar equations:

$$\begin{aligned} \sum F_x &= \dot{m}(V_{2x} - V_{1x}) \\ \sum F_y &= \dot{m}(V_{2y} - V_{1y}) \\ \sum F_z &= \dot{m}(V_{2z} - V_{1z}) \end{aligned} \tag{13.70}$$

If we consider the nozzle of Fig. 13.11a and we want to determine the x-component of the force of the joint on the nozzle, $(\mathbf{V}_1)_x = V_1$ and $(\mathbf{V}_2)_x = 0$ so that the momentum equation for the x-direction becomes

$$\sum F_x = -(F_x)_{\text{joint}} + p_1 A_1 = -\dot{m} V_1 \tag{13.71}$$

Similarly, we could write the y-component equation and find an expression for $(F_y)_{\text{joint}}$.

An example of a free-surface flow in a rectangular channel is shown in Fig. 13.12. If we want to determine the force of the gate on the flow, the following expression can be derived from the momentum equation:

$$\sum F_x = -F_{\text{gate}} + F_1 - F_2 = \dot{m}(V_2 - V_1) \tag{13.72}$$

where F_1 and F_2 are pressure forces (see Fig. 13.12).

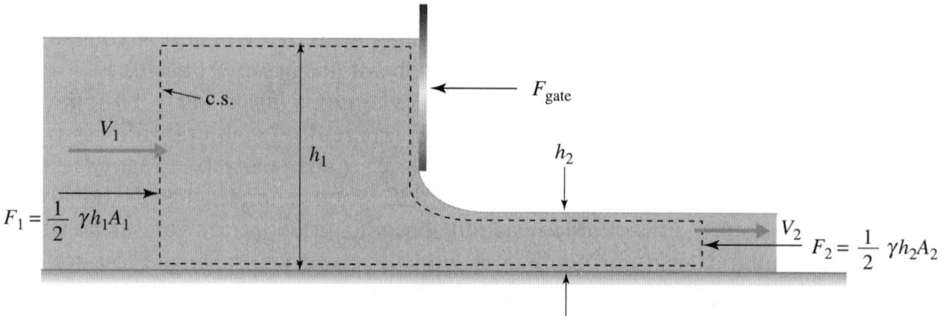

FIGURE 13.12 Force of the flow on a gate in a free-surface flow.

Example 13.10

Water flows through a horizontal pipe bend and exits into the atmosphere (Fig. E13.10a). The flow rate is 0.3 ft^3/sec. Calculate the force in each of the rods holding the pipe bend in position. Neglect body forces and viscous effects and shear force in the rods.

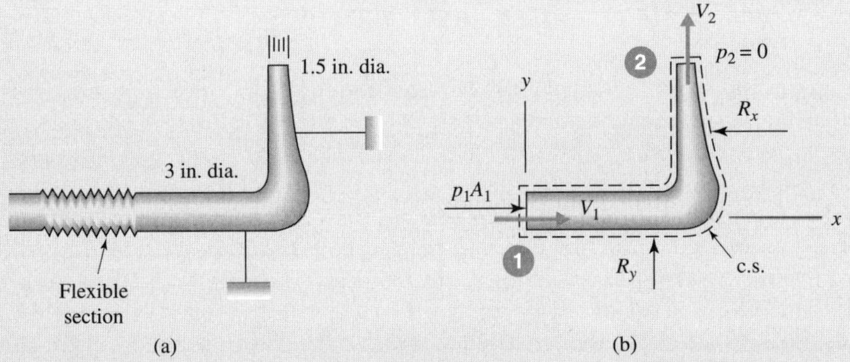

FIGURE E13.10

Solution

We have selected a control volume that surrounds the bend, as shown in Fig. E13.10b. Since the rods have been cut, the forces that the rods exert on the control volume are included. The pressure force at the entrance of the control volume is also shown. The flexible section is capable of resisting the interior pressure but it transmits no axial force or moment. The body force (weight of the control volume) does not act in the *x*- or *y*-direction but normal to it. Therefore, no other forces are shown. The average velocities are found to be

$$V_1 = \frac{Q_f}{A_1} = \frac{0.3}{\pi \times (3/12)^2/4} = 6.11 \text{ ft/sec}; \qquad V_2 = \frac{Q_f}{A_2} = \frac{0.3}{\pi \times (1.5/12)^2/4} = 24.4 \text{ ft/sec}$$

Before we can calculate the forces R_x and R_y we need to find the pressures p_1 and p_2. The pressure p_2 is zero because the flow exits into the atmosphere. The pressure at section 1 can be determined using the energy equation or the Bernoulli equation. Neglecting losses between sections 1 and 2, the energy equation gives

$$\frac{V_1^2}{2g} + \frac{p_1}{\gamma} = \frac{V_2^2}{2g} + \overset{0}{\cancel{\frac{p_2}{\gamma}}}$$

$$\therefore p_1 = \frac{\gamma}{2g}(V_2^2 - V_1^2) = \frac{62.4}{2 \times 32.2}(24.4^2 - 6.11^2) = 541 \text{ psf}$$

(continued)

Now we can apply the momentum equation (13.69) in the x-direction to find R_x and in the y-direction to find R_y:

$$x\text{-direction:} \qquad p_1A_1 - R_x = \dot{m}(\overset{0}{\cancel{V_{2x}}} - V_{1x})$$

$$541 \times \frac{\pi}{4}\left(\frac{3}{12}\right)^2 - R_x = 1.94 \times 0.3 \times (-6.11)$$

$$\therefore\ R_x = 30.1 \text{ lb}$$

$$y\text{-direction:} \qquad R_y = \dot{m}(V_{2y} - \overset{0}{\cancel{V_{1y}}})$$

$$= 1.94 \times 0.3 \times 24.4 = 14.2 \text{ lb}$$

Note that we have assumed uniform profiles and steady flow and used $\dot{m} = \rho Q_f$. These are the usual assumptions if information is not given otherwise.

Example 13.11

When the velocity of a flow in an open rectangular channel of width w is relatively large, it is possible for the flow to "jump" from a depth y_1 to a depth y_2 over a relatively short distance, as shown in Fig. E13.11; this is referred to as a *hydraulic jump*. Express y_2 in terms of y_1 and V_1; assume a horizontal uniform flow.

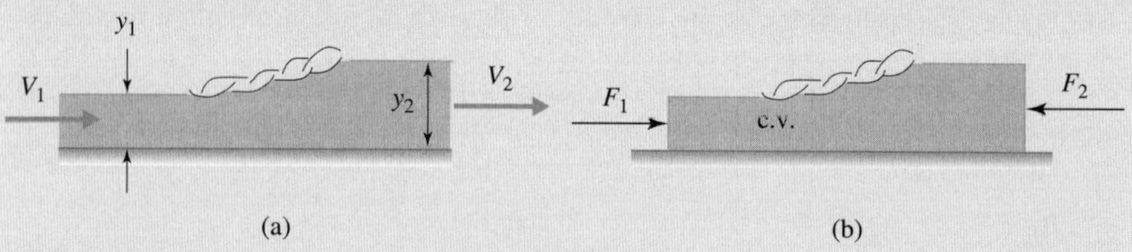

FIGURE E13.11

Solution

A control volume is selected as shown with inlet and exit areas upstream and downstream of the "jump" sufficiently far that the streamlines are parallel to the wall with hydrostatic pressure distributions. Neglecting the drag that is present on the walls (if the distance between sections is relatively small, the drag force should be negligible), the momentum equation can be manipulated as follows:

$$\sum F_x = \dot{m}(V_{2x} - V_{1x})$$

$$F_1 - F_2 = \rho A_1 V_1 (V_2 - V_1)$$

$$\gamma\frac{y_1}{2}(y_1 w) - \gamma\frac{y_2}{2}(y_2 w) = \rho y_1 w V_1\left(V_1\frac{y_1}{y_2} - V_1\right)$$

(continued)

where we have expressed F_1 and F_2 using Eq. 11.30, and continuity in the form of Eq. 13.28, so that

$$V_2 = \frac{y_1}{y_2} V_1$$

The above momentum equation can be simplified to

$$\frac{\gamma}{2}(y_1^2 - y_2^2) = \rho y_1 V_1^2 \frac{y_1 - y_2}{y_2}$$

or

$$\frac{g}{2}(y_1 - y_2)(y_1 + y_2) = \frac{y_1}{y_2} V_1^2 (y_1 - y_2)$$

The factor $(y_1 - y_2)$ is divided out and y_2 is found assuming y_1 and V_1 are known as follows:

$$\frac{g}{2}(y_1 + y_2) = \frac{y_1}{y_2} V_1^2$$

$$y_2^2 + y_1 y_2 - \frac{2}{g} y_1 V_1^2 = 0$$

$$\therefore\ y_2 = \frac{1}{2}\left(-y_1 + \sqrt{y_1^2 + \frac{8}{g} y_1 V_1^2}\right)$$

where the quadratic formula has been used. The energy equation could now be used to provide an expression for the losses in the hydraulic jump.

Example 13.12

Consider the symmetrical flow of air around the cylinder. The control volume, excluding the cylinder, is shown in Fig. E13.12. The velocity distribution downstream of the

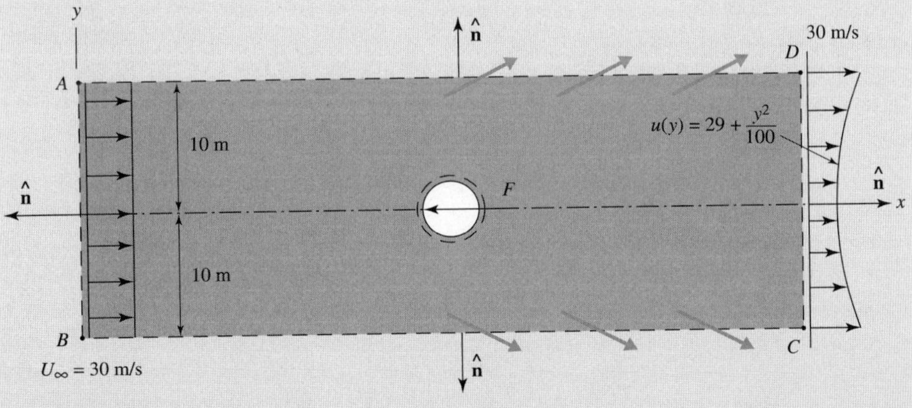

FIGURE E13.12

(continued)

cylinder is approximated with the parabola, as shown. Determine the drag force per meter of length acting on the cylinder. Use $\rho = 1.23\ \text{kg/m}^3$.

Solution

First, we must recognize that not all of the mass flux entering through AB exits through CD; consequently, some mass flux must exit AD and BC, as shown. The momentum equation (13.65) for the steady flow, applied to the control volume $ABCD$, takes the form

$$
\begin{aligned}
-F = \int_{\text{c.s.}} \rho V_x \mathbf{V}\cdot\hat{\mathbf{n}}\,dA &= \int_{A_{CD}} \rho u \mathbf{V}\cdot\hat{\mathbf{n}}\,dA + \int_{A_{AD}} \rho u \mathbf{V}\cdot\hat{\mathbf{n}}\,dA + \int_{A_{BC}} \rho u \mathbf{V}\cdot\hat{\mathbf{n}}\,dA \\
&\quad + \int_{A_{AB}} \rho u \mathbf{V}\cdot\hat{\mathbf{n}}\,dA \\
&= \int_{A_{CD}} \rho u^2\,dA + U_\infty \dot{m}_{AD} + U_\infty \dot{m}_{BC} - \int_{A_{AB}} \rho u^2\,dA \\
&= 2\int_0^{10} 1.23\left(29 + \frac{y^2}{100}\right)^2 dy + 2\times 30\dot{m}_{AD} - 1.23\times 30^2\times 20
\end{aligned}
$$

where $\dot{m}_{BC} = \dot{m}_{AD}$ is the mass flux crossing BC and AD with the x-component velocity equal to 30 m/s. We have used Eq. 13.31 for $\dot{m}_{AD}$ and $\dot{m}_{BC}$ recognizing that $\mathbf{V}\cdot\hat{\mathbf{n}} = V_n$ which would be the small y-component velocity. We now use continuity to find $\dot{m}_{AD}$:

$$
\begin{aligned}
0 = \int_{\text{c.s.}} \rho\hat{\mathbf{n}}\cdot\mathbf{V}\,dA &= \int_{A_{AD}} \rho\hat{\mathbf{n}}\cdot\mathbf{V}\,dA + \int_{A_{BC}} \rho\hat{\mathbf{n}}\cdot\mathbf{V}\,dA + \int_{A_{CD}} \rho\hat{\mathbf{n}}\cdot\mathbf{V}\,dA + \int_{A_{AB}} \rho\hat{\mathbf{n}}\cdot\mathbf{V}\,dA \\
&= \dot{m}_{AD} + \dot{m}_{BC} + 2\int_0^{10} \rho u(y)\,dy - \rho\times 20\times 30 \\
&= 2\dot{m}_{AD} + 2\int_0^{10} 1.23\times\left(29 + \frac{y^2}{100}\right) dy - 1.23\times 20\times 30
\end{aligned}
$$

$$\therefore\ \dot{m}_{AD} = 8.2\ \text{kg/s per meter of length}$$

Evaluating the terms in the momentum equation above gives us

$$
\begin{aligned}
F &= -21\,170 - 60\times 8.2 + 22\,140 \\
&= 478\ \text{N/m}
\end{aligned}
$$

Example 13.13

Find an expression for the head loss in a sudden expansion in a pipe in terms of V_1 and the area ratio (Fig. E13.13a). Assume uniform velocity profiles and assume that the pressure at the sudden enlargement is p_1.

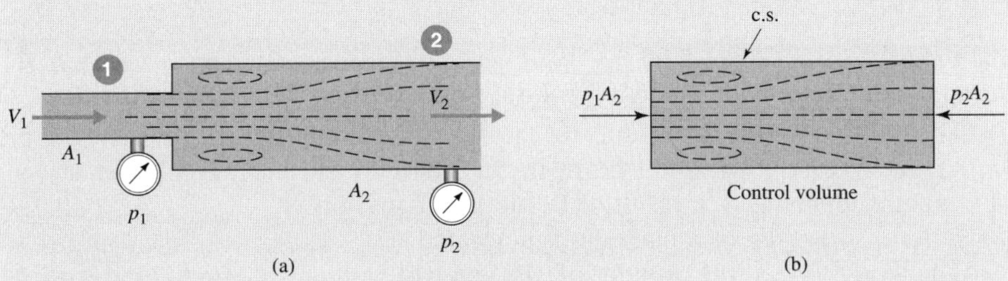

FIGURE E13.13

Solution

Figure E13.13a shows a sudden expansion with the diameter changing from d_1 to d_2. The pressure at the sudden enlargement is closest to p_1 since the streamlines are approximately parallel as shown (there is no pressure variation normal to parallel streamlines); they take some distance to again fill the pipe. Hence the force acting on the left end of the control volume shown in Fig. E13.13b is p_1A_2. Newton's second law applied to the control volume yields, assuming uniform profiles,

$$\sum F_x = \dot{m}(V_2 - V_1)$$

$$(p_1 - p_2)A_2 = \rho A_2 V_2 (V_2 - V_1)$$

$$\therefore \quad \frac{p_1 - p_2}{\rho} = V_2(V_2 - V_1)$$

The energy equation (13.51) provides

$$0 = \frac{V_2^2 - V_1^2}{2g} + \frac{p_2 - p_1}{\gamma} + \cancelto{0}{z_2 - z_1} + h_L$$

$$\therefore h_L = \frac{p_1 - p_2}{\gamma} - \frac{V_2^2 - V_1^2}{2g}$$

$$= \frac{V_2(V_2 - V_1)}{g} - \frac{(V_2 + V_1)(V_2 - V_1)}{2g} = \frac{(V_1 - V_2)^2}{2g}$$

To express this in terms of only V_1, we can use continuity and relate

$$V_2 = \frac{A_1}{A_2} V_1$$

Then the expression above for the head loss becomes

$$h_L = \left(1 - \frac{A_1}{A_2}\right)^2 \frac{V_1^2}{2g}$$

13.6.3 Momentum Equation Applied to Deflectors

The application of the momentum equation to deflectors forms an integral part of the analysis of many turbomachines, such as turbines, pumps, and compressors. In this section we illustrate the steps in such an analysis. It will be separated into two parts: fluid jets deflected by stationary deflectors and fluid jets deflected by moving deflectors. For both problems we will assume the following:

Key Concept: *The pressure in the fluid as it moves over a deflector remains constant.*

Key Concept: *The relative speed between the deflector surface and the jet stream remains unchanged.*

- The pressure external to the fluid jets is everywhere constant so that the pressure in the fluid as it moves over a deflector remains constant.
- The frictional resistance due to the fluid-deflector interaction is negligible so that the relative speed between the deflector surface and the jet stream remains unchanged, a result of Bernoulli's equation.
- Lateral spreading of a plane jet is neglected.
- The body force, the weight of the control volume, is small and will be neglected.

Stationary Deflector. Let us first consider the stationary deflector, illustrated in Fig. 13.13. Bernoulli's equation allows us to conclude that the magnitudes of the velocity vectors are equal (i.e., $V_2 = V_1$), since the pressure is assumed to be constant external to the fluid jet and elevation changes are negligible (see Eq. 12.35). Assuming steady, uniform flow the momentum equation takes the form of Eq. 13.69, which for the x- and y-directions becomes

$$\begin{aligned} -R_x &= \dot{m}(V_2 \cos\alpha - V_1) = \dot{m}V_1(\cos\alpha - 1) \\ R_y &= \dot{m}V_2 \sin\alpha = \dot{m}V_1 \sin\alpha \end{aligned} \qquad (13.73)$$

For given jet conditions the reactive force components can be calculated.

Moving Deflectors. The situation involving a moving deflector depends on whether a single deflector is moving (a water scoop used to slow a high-speed train) or whether a series of deflectors is moving (the vanes on a turbine). Let us first consider a single deflector shown in Fig. 13.14 to be moving in the positive x-direction with the speed V_B. In a reference frame attached to the stationary nozzle, from which the fluid jet issues, the flow is unsteady; that is, at a particular

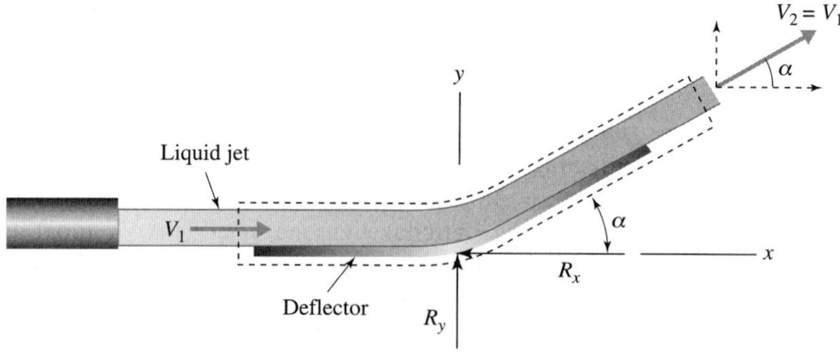

FIGURE 13.13 Stationary deflector.

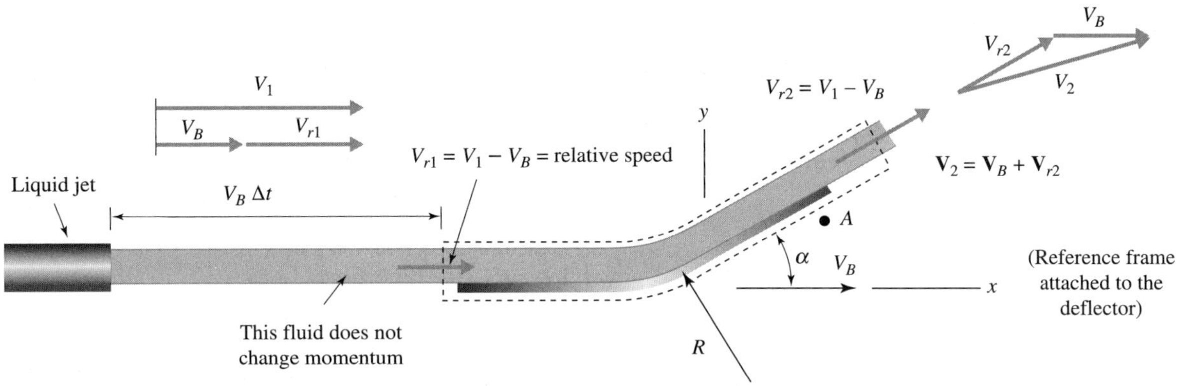

FIGURE 13.14 Moving deflector.

point in space, the flow situation varies with time.[2] A steady flow is observed, however, from a reference frame attached to the deflector. From this inertial reference frame, moving with the constant velocity V_B, we observe the relative speed V_{r1} entering the control volume to be $V_1 - V_B$, as shown. It is this relative speed that remains constant as the fluid flows relative to the deflector; it does not change since the pressure does not change. Hence, from this moving frame, the momentum equation takes the forms

$$\begin{aligned} -R_x &= \dot{m}_r(V_1 - V_B)(\cos\alpha - 1) \\ R_y &= \dot{m}_r(V_1 - V_B)\sin\alpha \end{aligned} \tag{13.74}$$

where $\dot{m}_r$ represents only that part of the mass flux exiting the fixed jet that has its momentum changed. Since the deflector moves away from the fixed jet some of the fluid that exits the fixed jet never experiences a momentum change; this fluid is represented by the distance $V_B\Delta t$, shown in Fig. 13.14. Hence

$$\dot{m}_r = \rho A(V_1 - V_B) \tag{13.75}$$

where the relative speed $(V_1 - V_B)$ is used in the calculation; the mass flux $\rho A V_B$ is subtracted from the exiting mass flux $\rho A V_1$ to provide the mass flux $\dot{m}_r$ that experiences a momentum change.

For a series of vanes (a cascade) the jets may be oriented to the side, as shown in Fig. 13.15. The actual force on a particular vane would be zero until the jet strikes the vane; then the force would increase to a maximum and decrease to zero as the vane leaves the jet. We will idealize the situation as follows: Assume that, on the average, the jet is deflected by the vanes as shown in Figs. 13.15 and 13.16a as viewed from a stationary reference frame; the fluid jet enters the vanes with an angle β_1 and exits with an angle β_2. What is desired, however, is that the relative velocity enter the vanes tangent to the leading edge of the vanes, that is, V_{r1} in

[2]This is a rather subtle point. To determine whether a flow is steady, we observe the flow at a given point in space. If a flow property changes with time at that point, the flow is unsteady. In this situation, if we focus our attention on a particular point just before the blade, such as point A in Fig. 13.14, first there is no flow, then the blade and the jet pass through the point; then there is again no flow. This is an unsteady flow.

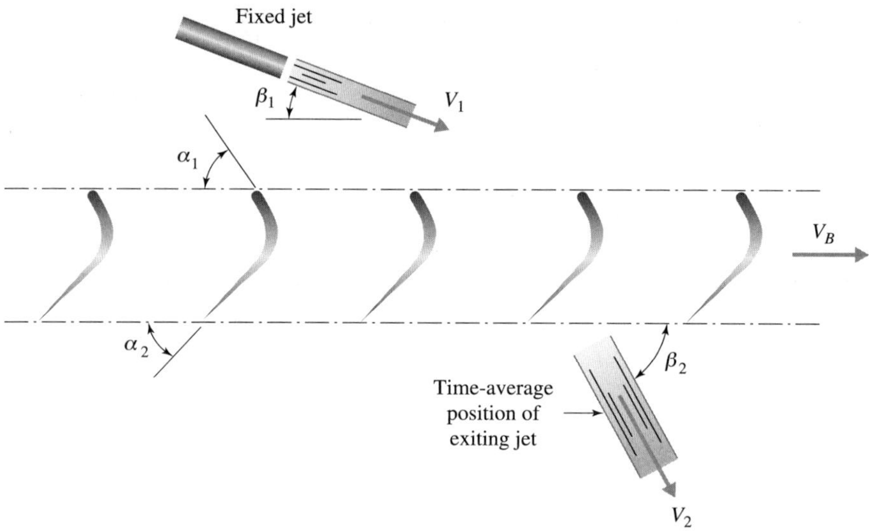

FIGURE 13.15 Fluid striking a series of vanes.

Key Concept: *The relative speed remains constant as the fluid travels over a moving vane.*

Fig. 13.16b is at the angle α_1. The relative speed then remains constant as the fluid travels over the vane with the exiting relative velocity V_{r2} leaving with the vane angle α_2. The relative and absolute velocities are related with the velocity equations which are displayed by the velocity polygons of Fig. 13.16b and c. Assuming that all of the mass exiting the fixed jet has its momentum changed, we can write the momentum equation as

Key Concept: *Only the x-component of force is related to the power output.*

$$-R_x = \dot{m}(V_{2x} - V_{1x}) \tag{13.76}$$

Example 13.16 will illustrate the details.

Interest is usually focused on the x-component of force since it is this component that is related to the power output (or requirement). The power would be

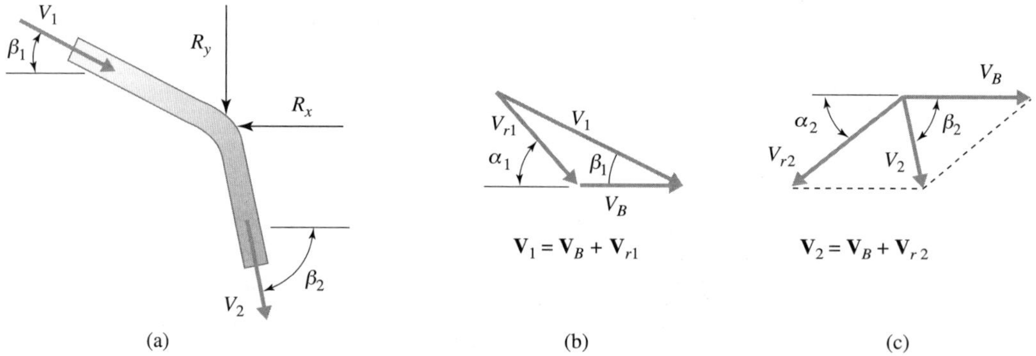

FIGURE 13.16 Detail of the flow situation involving a series of vanes: (a) average position of jet; (b) entrance velocity polygon; (c) exit velocity polygon.

found by multiplying the x-component force by the blade speed for each jet; this takes the form

$$\dot{W} = NR_x V_B \tag{13.77}$$

where N represents the number of jets. The y-component force does not move in the y-direction so it does no work.

Example 13.14

A deflector turns a sheet of water through an angle of 30° as shown in Fig. E13.14. What force is necessary to hold the deflector in place if $\dot{m} = 32$ kg/s?

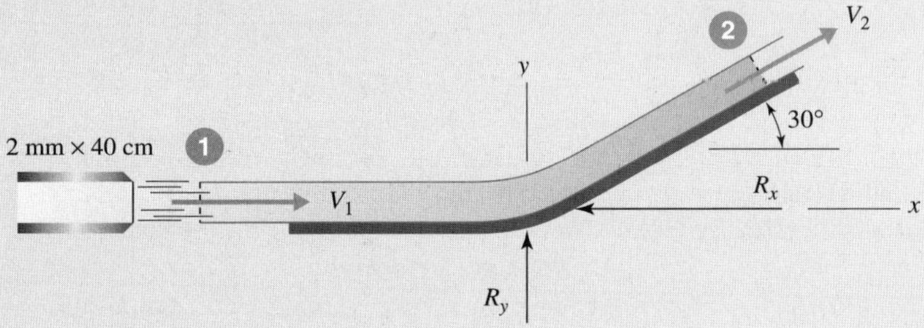

FIGURE E13.14

Solution

The control volume we have selected includes the deflector and the water adjacent to it. The only force that is acting on the control volume is due to a support needed to hold the deflector. This force has been decomposed into R_x and R_y.

The velocity V_1 is found to be

$$V_1 = \frac{\dot{m}}{\rho A_1}$$

$$= \frac{32}{1000 \times 0.002 \times 0.4} = 40 \text{ m/s}$$

Bernoulli's equation (12.34) shows that if the pressure does not change, then the magnitude of the velocity does not change, provided that there is no significant change in elevation and that viscous effects are negligible; thus we can conclude that $V_2 = V_1$ since $p_2 = p_1$. Next, the momentum equation is applied in the x-direction to find R_x and then in the y-direction for R_y:

x-direction: $$-R_x = \dot{m}(V_{2x} - V_{1x})$$

$$= 32(40 \cos 30° - 40) \qquad \therefore R_x = 172 \text{ N}$$

y-direction: $$R_y = \dot{m}(V_{2y} - \overset{0}{\cancel{V_{1y}}})$$

$$= 32(40 \sin 30°) = 640 \text{ N}$$

$$\therefore R = \sqrt{R_x^2 + R_y^2} = 663 \text{ N}$$

Example 13.15

The deflector shown in Fig. E13.15 moves to the right at 30 m/s while the nozzle remains stationary. Determine (a) the force components needed to support the deflector, (b) $\mathbf{V}_2$ as observed from a fixed observer, and (c) the power generated by the vane. The jet velocity is 80 m/s.

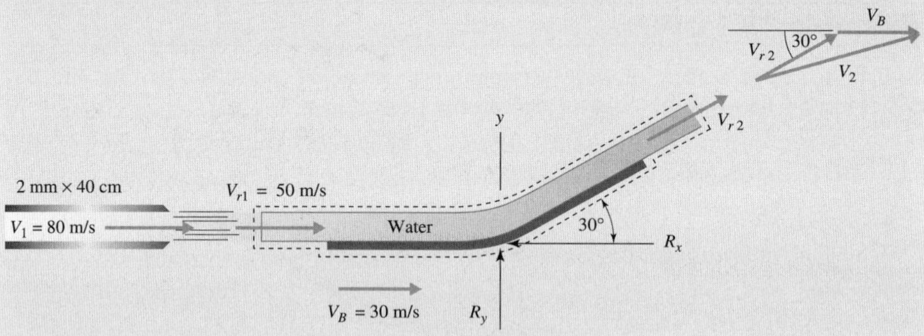

FIGURE E13.15

Solution

(a) To solve the problem of a moving deflector, we observe the flow from a reference frame attached to the deflector. In this moving reference frame the flow is steady and Bernoulli's equation with $p_1 = p_2$ can then be used to show that $V_{r1} = V_{r2} = 50$ m/s, the velocity of the sheet of water as observed from the deflector. Note that we cannot apply Bernoulli's equation in a fixed reference frame since the flow would not be steady. Applying the momentum equation to the moving control volume, which is indicated again by the dashed line, we obtain the following:

$$x\text{-direction:} \quad -R_x = \dot{m}_r[(V_{r2})_x - (V_{r1})_x]$$

$$= 1000 \times 0.002 \times 0.4 \times 50(50 \cos 30° - 50)$$

$$\therefore R_x = 268 \text{ N}$$

$$y\text{-direction:} \quad R_y = \dot{m}_r[(V_{r2})_y - \overset{0}{\cancel{(V_{r1})_y}}]$$

$$= 1000 \times 0.002 \times 0.4 \times 50(50 \sin 30°) = 1000 \text{ N}$$

When calculating $\dot{m}_r$ we must use only that water which has its momentum changed; hence the velocity used is 50 m/s.

(continued)

(b) Observed from a fixed observer the velocity $\mathbf{V}_2$ of the fluid after the deflection is $\mathbf{V}_2 = \mathbf{V}_{r2} + \mathbf{V}_B$, where $\mathbf{V}_{r2}$ is directed tangential to the deflector at the exit and has a magnitude equal to V_{r1} (see the velocity diagram above). Thus

$$(V_2)_x = V_{r2} \cos 30° + V_B$$

$$= 50 \times 0.866 + 30 = 73.3 \text{ m/s}$$

$$(V_2)_y = V_{r2} \sin 30°$$

$$= 50 \times 0.5 = 25 \text{ m/s}$$

Finally,

$$\mathbf{V}_2 = 73.3\hat{\mathbf{i}} + 25\hat{\mathbf{j}} \text{ m/s}$$

(c) The power generated by the moving vane is equal to the velocity of the vane times the force the vane exerts in the direction of the motion. Therefore,

$$\dot{W} = V_B \times R_x = 30 \times 268 = 8040 \text{ W}$$

Example 13.16

High-speed air jets strike the blades of a turbine rotor tangentially while the 1.5-m-diameter rotor rotates at 140 rad/s (Fig. E13.16a). There are 10 such 4-cm-diameter jets. Calculate the maximum power output. The air density is 2.4 kg/m^3.

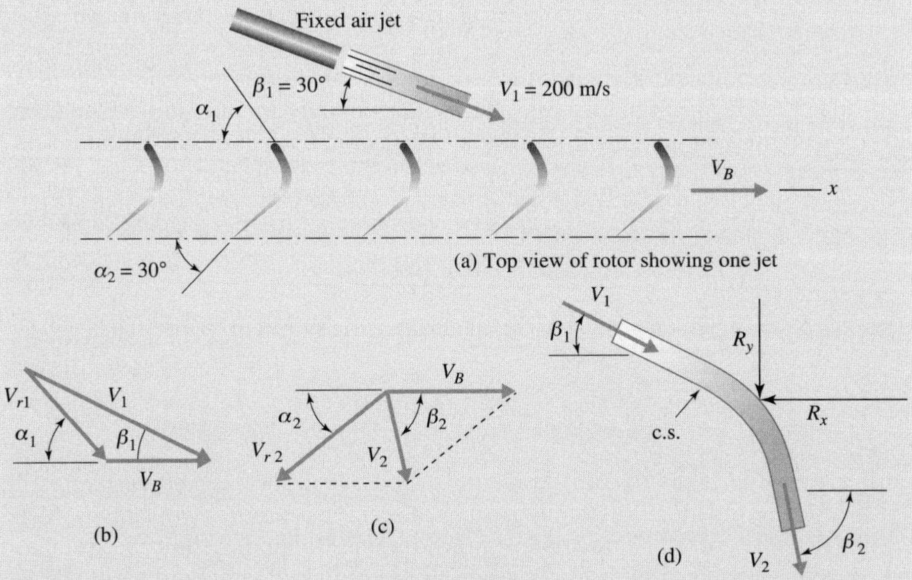

FIGURE E13.16

(continued)

Solution

The blade angle α_1 is set by demanding that the air jet enter the blades tangentially, as observed from the moving blade; that is, the relative velocity vector $\mathbf{V}_r$ must make the angle α_1 with respect to the blade velocity $\mathbf{V}_B$. This is shown in Fig. E13.16b. The relative entrance velocity is $\mathbf{V}_{r1}$ (Fig. E13.16b) and the relative exit velocity is $\mathbf{V}_{r2}$ (Fig. E13.16c). Both velocity polygons are presented by the vector equation

$$\mathbf{V} = \mathbf{V}_r + \mathbf{V}_B$$

which states that the absolute velocity equals the relative velocity plus the blade velocity. From the polygon at the entrance we have

$$V_1 \sin \beta_1 = V_{r1} \sin \alpha_1$$

$$V_1 \cos \beta_1 = V_{r1} \cos \alpha_1 + V_B$$

$$\therefore\ 200 \sin 30° = V_{r1} \sin \alpha_1$$

$$200 \cos 30° = V_{r1} \cos \alpha_1 + 0.75 \times 140$$

where V_B is the radius multiplied by the angular velocity. A simultaneous solution yields

$$V_{r1} = 121 \text{ m/s} \qquad \alpha_1 = 55.7°$$

The friction between the air and the blade is quite small and can be neglected when calculating the maximum output. This allows us to assume $V_{r2} = V_{r1}$. From the exiting velocity polygon we can write

$$V_B - V_{r2} \cos \alpha_2 = V_2 \cos \beta_2$$

$$V_{r2} \sin \alpha_2 = V_2 \sin \beta_2$$

$$\therefore\ 0.75 \times 140 - 121 \cos 30° = V_2 \cos \beta_2$$

$$121 \sin 30° = V_2 \sin \beta_2$$

A simultaneous solution results in

$$V_2 = 60.5 \text{ m/s} \qquad \beta_2 = 89.8°$$

The momentum equation applied to the control volume, shown in Fig. E13.16d, gives

$$-R_x = \dot{m}(V_{2x} - V_{1x})$$

$$= 2.4 \times \pi \times 0.02^2 \times 200(60.5 \cos 89.8° - 200 \cos 30°)$$

$$\therefore\ R_x = 104.3 \text{ N}$$

There are 10 jets, each producing the force above. The maximum power output is then

$$\dot{W} = 10 \times R_x \times V_B$$

$$= 10 \times 104.3 \times (0.75 \times 140) = 109\,600 \text{ W} \quad \text{or} \quad 109.6 \text{ kW}$$

13.6.4 Steady Nonuniform Flow

If we cannot assume uniform velocity profiles, we can let

$$\int_A V^2\, dA = \beta \overline{V}^2 A \tag{13.78}$$

where we have introduced the *momentum-correction factor* β, expressed explicitly as

$$\beta = \frac{\int V^2\, dA}{\overline{V}^2 A} \tag{13.79}$$

The momentum equation (13.69), for a steady flow with one inlet and one exit, can then be written as

$$\sum \mathbf{F} = \dot{m}(\beta_2 \mathbf{V}_2 - \beta_1 \mathbf{V}_1) \tag{13.80}$$

For a laminar flow with a parabolic profile in a circular pipe, $\beta = 4/3$. If a profile is given, however, the integral is usually simply integrated and Eq. 13.65 is used.

Example 13.17

Calculate the momentum correction factor for a parabolic profile (a) between parallel plates and (b) in a circular pipe. The parabolic profiles are shown in Fig. E13.17.

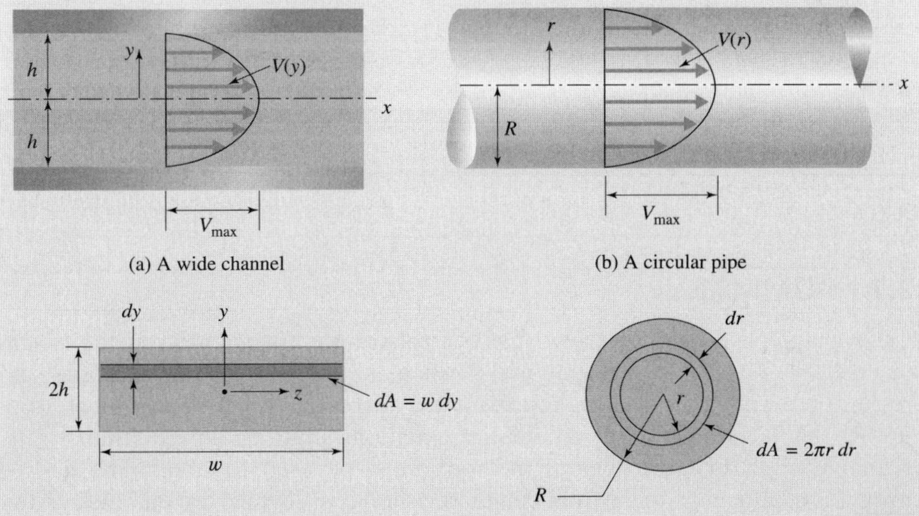

FIGURE E13.17

(continued)

Solution

(a) A parabolic profile between parallel plates can be expressed as

$$V(y) = V_{\max}\left(1 - \frac{y^2}{h^2}\right)$$

where y is measured from the centerline, the velocity is zero at the walls where $y = \pm h$, and $V_{\max}$ is the centerline velocity at $y = 0$. First, let us find the average velocity. It is

$$\overline{V} = \frac{1}{A}\int V\, dA$$

$$= \frac{1}{hw}\int_0^h V_{\max}\left(1 - \frac{y^2}{h^2}\right) w\, dy = \frac{V_{\max}}{h}\left(h - \frac{1}{3}h\right) = \frac{2}{3}V_{\max}$$

where we have integrated over the top half of the cross-section. Then

$$\beta = \frac{\int V^2\, dA}{\overline{V}^2 A} = \frac{2}{\frac{4}{9}V_{\max}^2 \times 2hw}\int_0^h V_{\max}^2\left(1 - \frac{y^2}{h^2}\right)^2 w\, dy = \frac{6}{5}$$

(b) For a circular pipe a parabolic profile can be written as

$$V(r) = V_{\max}\left(1 - \frac{r^2}{R^2}\right)$$

where R is the pipe radius and $V = 0$ at $r = R$. The average velocity is found to be

$$\overline{V} = \frac{1}{A}\int V\, dA = \frac{1}{\pi R^2}\int_0^R V_{\max}\left(1 - \frac{r^2}{R^2}\right) 2\pi r\, dr = \frac{1}{2}V_{\max}$$

The momentum correction factor is then

$$\beta = \frac{\int V^2\, dA}{\overline{V}^2 A} = \frac{1}{\frac{1}{4}V_{\max}^2 \pi R^2}\int_0^R V_{\max}^2\left(1 - \frac{r^2}{R^2}\right)^2 2\pi r\, dr = \frac{4}{3}$$

The correction factors above can be used to express the momentum flux across a cross-sectional area as $\beta\rho A\overline{V}^2$.

13.7 SUMMARY

In this chapter we have presented the control-volume formulation of the fundamental laws. This formulation is useful when the integrands (the velocities and pressure) are known or can be approximated with an acceptable degree of accuracy. If this is not the case, differential equations must be solved (numerically or analytically), or experimental methods must be used to obtain the desired information; much of the remainder of this book is devoted to this task. After unknown velocities and pressures are determined, we often return to the control-volume formulation and calculate integral quantities of interest. Examples would include the lift and drag on an airfoil, the torque on a row of turbine blades, and the oscillating force on a suspension cable of a bridge.

As we have observed in the examples and problems of this chapter, the task of applying the control-volume equations is highly dependent on the appropriate selection of the boundaries of the control volume. Such boundaries are selected at locations where either information is known or where the unknowns appear. Experience is often required in the selection of a control volume, as it is in the selection of a free-body diagram in dynamics and solid mechanics. The student has undoubtedly gained some of this experience while working through the sections of this chapter. Table 13.1 presents the various forms of the fundamental laws to aid the user in selecting an appropriate form for a particular problem.

TABLE 13.1 Integral Forms of the Fundamental Laws

Continuity	Energy	Momentum
	General Form	
$0 = \frac{d}{dt}\int_{\text{c.v.}} \rho\, d\mathcal{V} + \int_{\text{c.s.}} \rho \mathbf{V}\cdot\hat{\mathbf{n}}\, dA$	$-\sum \dot{W} = \frac{d}{dt}\int_{\text{c.v.}} \left(\frac{V^2}{2} + gz\right)\rho\, d\mathcal{V} + \int_{\text{c.s.}} \left(\frac{V^2}{2} + \frac{p}{\rho} + gz\right)\rho\mathbf{V}\cdot\hat{\mathbf{n}}\, dA + \text{losses}$	$\sum \mathbf{F} = \frac{d}{dt}\int_{\text{c.v.}} \rho\mathbf{V}\, d\mathcal{V} + \int_{\text{c.s.}} \rho\mathbf{V}(\mathbf{V}\cdot\hat{\mathbf{n}})\, dA$
	Steady Flow	
$0 = \int_{\text{c.s.}} \rho\mathbf{V}\cdot\hat{\mathbf{n}}\, dA$	$-\sum \dot{W} = \int_{\text{c.s.}} \left(\frac{V^2}{2} + \frac{p}{\rho} + gz\right)\rho\mathbf{V}\cdot\hat{\mathbf{n}}\, dA + \text{losses}$	$\sum \mathbf{F} = \int_{\text{c.s.}} \rho\mathbf{V}(\mathbf{V}\cdot\hat{\mathbf{n}})\, dA$
	Steady Nonuniform Flow[a]	
$\dot{m} = \rho_1 A_1 \bar{V}_1 = \rho_2 A_2 \bar{V}_2$	$\frac{-\sum \dot{W}}{\dot{m}g} = \alpha_2 \frac{\bar{V}_2^2}{2g} + \frac{p_2}{\gamma_2} + z_2 - \alpha_1 \frac{\bar{V}_1^2}{2g} - \frac{p_1}{\gamma_1} - z_1 + h_L$	$\sum F_x = \dot{m}(\beta_2 \bar{V}_{2x} - \beta_1 \bar{V}_{1x})$ $\sum F_y = \dot{m}(\beta_2 \bar{V}_{2y} - \beta_1 \bar{V}_{1y})$
	Steady Uniform Form[a]	
$\dot{m} = \rho_1 A_1 V_1 = \rho_2 A_2 V_2$	$-\frac{\sum \dot{W}}{\dot{m}g} = \frac{V_2^2}{2g} + \frac{p_2}{\gamma_2} + z_2 - \frac{V_1^2}{2g} - \frac{p_1}{\gamma_1} - z_1 + h_L$	$\sum \mathbf{F} = \dot{m}(\mathbf{V}_2 - \mathbf{V}_1)$
	Steady Uniform Incompressible Flow[a]	
$Q_f = A_1 V_1 = A_2 V_2$	$-\frac{\sum \dot{W}}{\dot{m}g} = \frac{V_2^2}{2g} + \frac{p_2}{\gamma} + z_2 - \frac{V_1^2}{2g} - \frac{p_1}{\gamma} - z_1 + h_L$ or $H_P + \frac{V_1^2}{2g} + \frac{p_1}{\gamma} + z_1 = H_T + \frac{V_2^2}{2g} + \frac{p_2}{\gamma} + z_2 + h_L$	$\sum \mathbf{F} = \dot{m}(\mathbf{V}_2 - \mathbf{V}_1)$
$\dot{m}$ = mass flux	α = kinetic energy correction factor	h_L = head loss
Q_f = flow rate	$= \frac{\int V^3\, dA}{\bar{V}^3 A}$	$\sum \dot{W} = \dot{W}_S + \dot{W}_{\text{shear}} + \dot{W}_I$
$\bar{V}$ = average velocity	β = momentum correction factor	H_P = pump head $= -\eta_P \dot{W}_P/\dot{m}g$
$= \frac{\int V\, dA}{A}$	$= \frac{\int V^2\, dA}{\bar{V}^2 A}$	H_T = turbine head $= \dot{W}_T/\eta_T \dot{m}g$

[a]The control volume has one entrance (section 1) and one exit (section 2).

REVIEW PROBLEMS FOR THE FE/EIT EXAMINATION

13.1FE Select the extensive property from the following:
A. Temperature **B.** Volume
C. Pressure **D.** Density

13.2FE Air flows through an 8-cm-diameter pipe with an average velocity of 70 m/s with a temperature of 20°C and a pressure of 200 kPa. The mass flux is nearest:
A. 3.7 kg/s **B.** 2.37 kg/s
C. 1.26 kg/s **D.** 0.84 kg/s

13.3FE Water flows with a velocity of 3 m/s in an 80-cm-diameter sewer pipe at a depth of 30 cm. The flow rate is nearest:
A. 516 L/s **B.** 721 L/s
C. 938 L/s **D.** 1262 L/s

13.4FE What is the energy requirement of an 85% efficient pump that transports 40 L/s of water if the pressure increases from 200 kPa to 1200 kPa?
A. 4.8 kW **B.** 14.2 kW
C. 34.0 kW **D.** 47.1 kW

13.5FE A high-speed water jet is used to cut a material. If the velocity issuing from the 2-mm-diameter jet is 120 m/s, the maximum pressure on the material at the point of impact is nearest:
A. 7200 kPa **B.** 3600 kPa
C. 735 kPa **D.** 452 kPa

13.6FE Estimate V_1 of Fig. P13.6FE. Assume the air to be incompressible with $\rho = 1.2\ \text{kg/m}^3$.
A. 62 m/s **B.** 40 m/s
C. 18 m/s **D.** 10 m/s

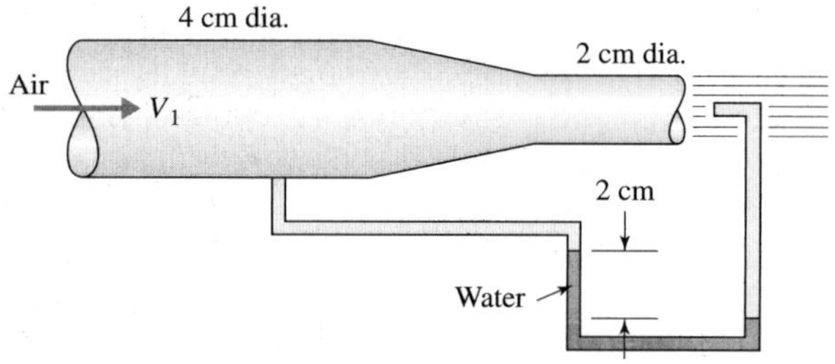

FIGURE P13.6FE

13.7FE The pressure drop across a valve, through which 40 L/s of water flows is measured to be 100 kPa. Estimate the loss coefficient if the nominal diameter of the valve is 8 cm.
A. 0.79 **B.** 3.2
C. 8.7 **D.** 31

13.8FE An 89%-efficient pump is inserted in a 4-cm-diameter line transporting 40 L/s of water. A pressure rise of 400 kPa is desired. The power required by the pump is nearest:
A. 12 kW **B.** 16 kW
C. 18 kW **D.** 22 kW

13.9FE A hydroturbine generates power by transporting 0.2 m³/s of water from a dam. The water surface is 10 m above the turbine outlet. the overall loss coefficient for the 24-cm-diameter connecting pipe is 3.2. The maximum turbine output is nearest:
A. 42 kW **B.** 21 kW
C. 18 kW **D.** 13 kW

13.10FE A 75%-efficient pump delivers 0.1 m³/s of water from a reservoir to a device at an elevation of 50 m above the reservoir. The pressure at the 8-cm-diameter entrance to the device is 180 kPa. If the piping loss coefficient is 5.6, the necessary power input to the pump is nearest:
A. 263 kW **B.** 203 kW
C. 121 kW **D.** 91.3 kW

13.11FE A strong wind blows directly against a window on a building. The force on the window can be approximated using:
A. Bernoulli's equation
B. The continuity equation
C. The momentum equation
D. All of the above

13.12FE A nozzle with an exit diameter of 4 cm is attached to a 10-cm-diameter pipe transporting 0.1 m³/s of water. The force needed to hold the nozzle on the pipe is nearest:
A. 6.7 kN **B.** 12.2 kN
C. 17.5 kN **D.** 24.2 kN

13.13FE A 1 cm × 20 cm sheet of water is deflected as shown in Fig. P13.13FE. The magnitude of the

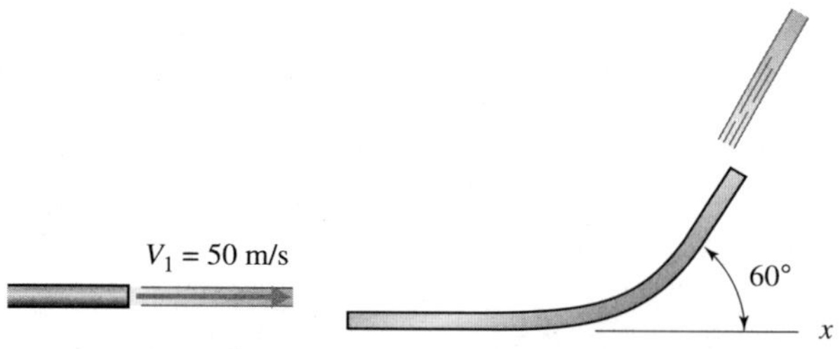

FIGURE P13.13FE

total force acting on the stationary deflector is nearest:
A. 6830 N
B. 5000 N
C. 4330 N
D. 2500 N

13.14FE The water impacts one of the turbine blades as shown in Fig. P13.14FE. For a blade speed of 20 m/s, the maximum power output for a single jet is nearest:
A. 18 kW **B.** 154 kW
C. 206 kW **D.** 309 kW

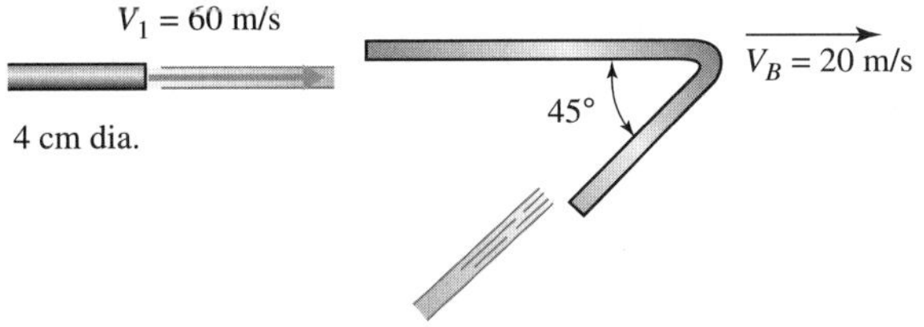

FIGURE P13.14FE

13.15FE The large vehicle of Fig. P13.15FE is slowed by lowering a 2-m-wide scoop into a reservoir of water. Estimate the force exerted on the scoop when the vehicle is travelling at 60 m/s and it scoops off 5 cm of water? The scoop diverts the water through 180°.
A. 720 kN **B.** 360 kN
C. 12 kN **D.** 7.2 kN

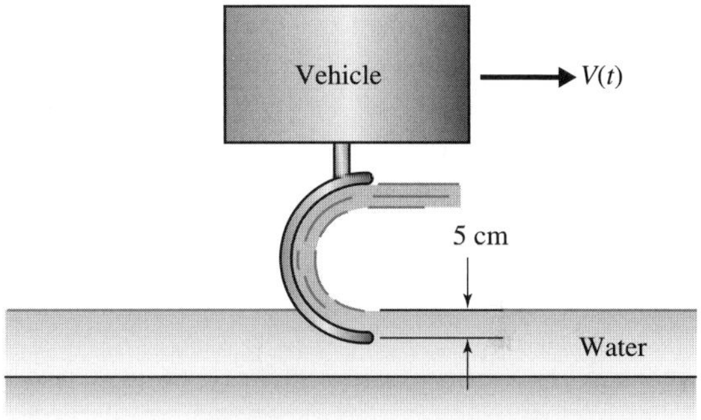

FIGURE P13.15FE

PROBLEMS

Basic Laws

13.1 **(a)** State the necessary conditions for momentum of a system to remain constant.
(b) State the necessary conditions for energy of a system to remain constant.
(c) Show the detailed steps and state the assumptions that allow Eq. 13.3 to be reduced to $\sum \mathbf{F} = m\mathbf{a}$.

13.2 Make a list of five extensive properties that are of interest in fluid mechanics. Also, list their associated intensive properties. In addition, list five other intensive properties.

13.3 A control volume is identified as the volume interior to a balloon. At an instant the system is also identified as the air inside the balloon. Air escapes during a short time increment Δt. Sketch the system and the control volume at t and at $t + \Delta t$.

13.4 At an instant in time the control volume and the system occupy the volume inside the pump shown in Fig. P13.4 and a few diameters of the pipe at the inlet side. Sketch the system and the control volume at times t and $t + \Delta t$.

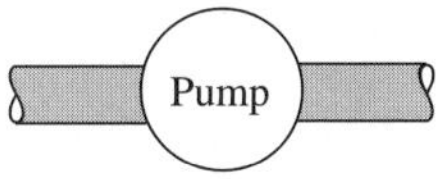

FIGURE P13.4

13.5 Sketch the unit vector $\hat{\mathbf{n}}$ and the velocity vector $\mathbf{V}$ on each of the areas listed:
(a) The exit area of a fireman's nozzle
(b) The inlet area of a pump
(c) The wall area of a pipe
(d) The porous bottom area of a river into which a small amount of water flows
(e) The cylindrical exit area of a rotating impeller

13.6 Sketch the unit vector $\hat{\mathbf{n}}$ and the velocity vector $\mathbf{V}$ at several positions on a rectangular box surrounding the airfoil shown in Fig. P13.6.

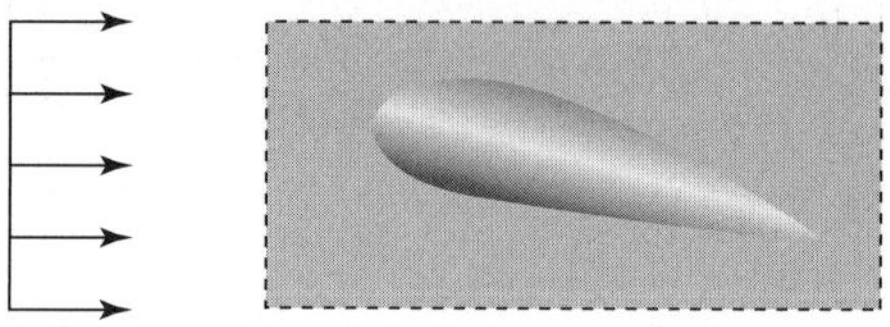

FIGURE P13.6

System-to-Control-Volume Transformation

13.7 Assume that $V_1 = V_2 = V_3 = 10$ m/s for the control volume shown in Fig. P13.7. Write $\hat{\mathbf{n}}_1$, $\hat{\mathbf{n}}_2$, and $\hat{\mathbf{n}}_3$ in terms of $\hat{\mathbf{i}}$, $\hat{\mathbf{j}}$, and $\hat{\mathbf{k}}$, and calculate the normal component of the velocity vector on each of the three plane areas. The volume is of uniform depth in the z-direction.

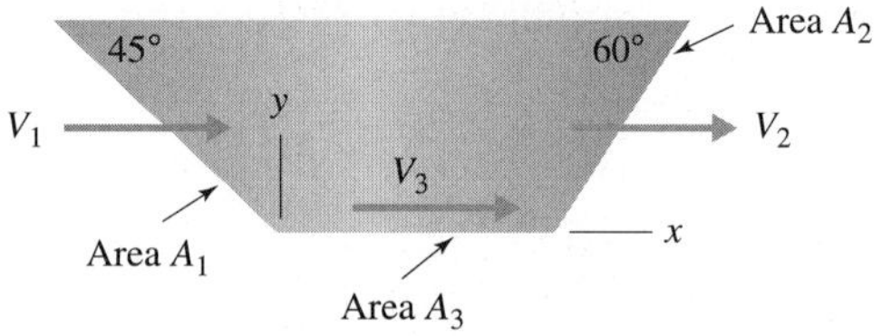

FIGURE P13.7

13.8 We recognize that

$$\frac{d}{dt}\int_{\text{c.v.}} \rho\eta \, d\mathcal{V} = \int_{\text{c.v.}} \frac{\partial}{\partial t}\rho\eta \, d\mathcal{V}$$

What condition allows this equivalence? Why is an ordinary derivative used on the left and a partial derivative on the right?

13.9 The air inside the lungs at the end of an inhaled breath is identified as the system at $t = 0$. Select a control volume and sketch both the system and control volume at $t = \Delta t$ if air is exhaled through the nose only.

13.10 The control volume selected to analyze flow around an airfoil is the rectangular box as sketched in the figure of Problem 13.6. The system occupies the box at t. Sketch the system at $t + \Delta t$.

Conservation of Mass

13.11 Show that Eq. 13.27 results from Eq. 13.26 by assuming one inlet and one exit and uniform flow (constant properties).

13.12 An incompressible fluid enters a volume filled with an absorbing material with a mass flux $\dot{m}$ and leaves the volume with a flow rate Q_f. Determine an expression for the rate of change of mass in the volume.

13.13 A liquid of density ρ flows into a volume filled with a sponge at a flow rate Q_{f1}. It exits one area with a mass flux $\dot{m}_2$ and a second area A_3 with an average velocity V_3 as shown in Fig. P13.13. Write an expression for dm_{sponge}/dt, the rate of change of mass in the sponge.

FIGURE P13.13

13.14 Water is flowing in a 2.5-in.-diameter pipe at 60 ft/sec. If the pipe enlarges to a diameter of 5 in., calculate the reduced velocity. Also, calculate the mass flux and the flow rate. In addition, express the answers using SI units.

13.15 Water flows in the 5-cm-diameter pipe shown in Fig. P13.15 with an average velocity of 10 m/s. It turns a 90° angle and flows radially between two parallel plates. What is the velocity at a radius of 60 cm? What are the mass flux and the discharge?

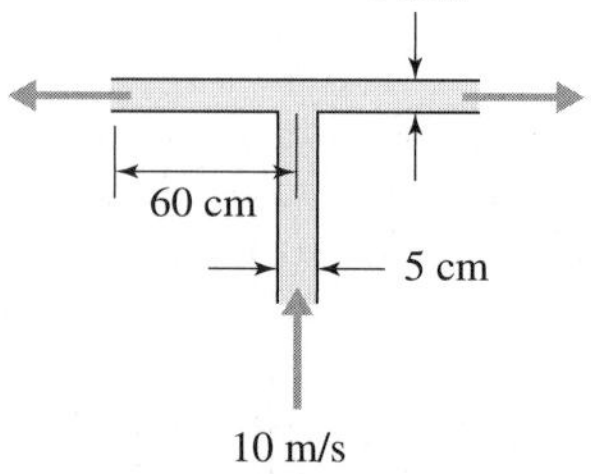

FIGURE P13.15

13.16 A pipe transports 200 kg/s of water. The pipe tees into a 5-cm-diameter pipe and a 7-cm-diameter pipe (Fig. P13.16). If the average velocity in the smaller-diameter pipe is 25 m/s, calculate the flow rate in the larger pipe.

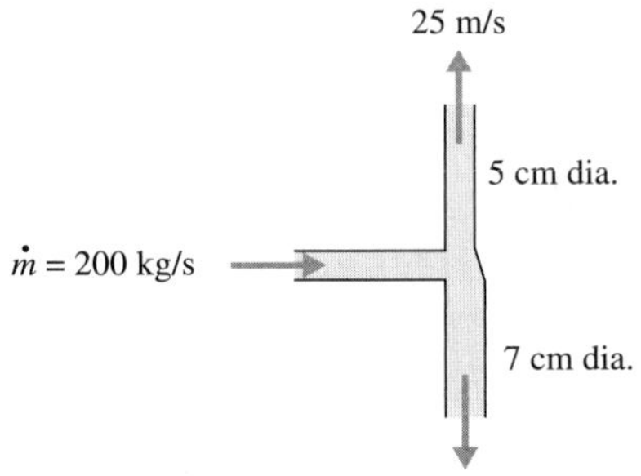

FIGURE P13.16

13.17 Air at 60°F and 40 psia flows in a 4-in.-diameter pipe with a mass flux of 0.2 slug/sec. The pipe undergoes a conversion to a 2 in. by 3 in. rectangular duct in which $T = 150°\text{F}$ and $p = 7$ psia. Calculate the velocity in each section.

13.18 Air at 120°C and 500 kPa absolute is flowing in a pipe at 600 m/s and suddenly undergoes an abrupt change to 249°C and 1246 kPa absolute at a location where the diameter is 10 cm. Calculate the velocity after the sudden change (a shock wave) illustrated in Fig. P13.18. Also, calculate the mass flux and the flow rates before and after the abrupt change.

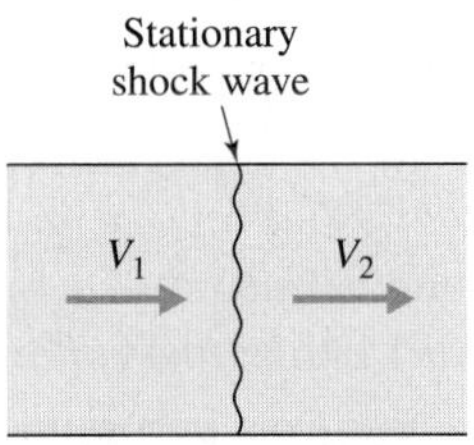

FIGURE P13.18

13.19 Water flows with a velocity of 3 m/s at a depth of 1.5 m in a trapezoidal channel with a 2-m base and sides sloping at 45°. It empties into a circular pipe and flows at 2 m/s. What is the diameter if:

(a) The pipe flows full?
(b) The pipe flows half full?
(c) The water in the pipe flows at a depth of one-half the radius?

13.20 Water flows in a 8-cm-diameter pipe with the profiles shown in Fig. P13.20. Find the average velocity, the mass flux, and the flow rate.

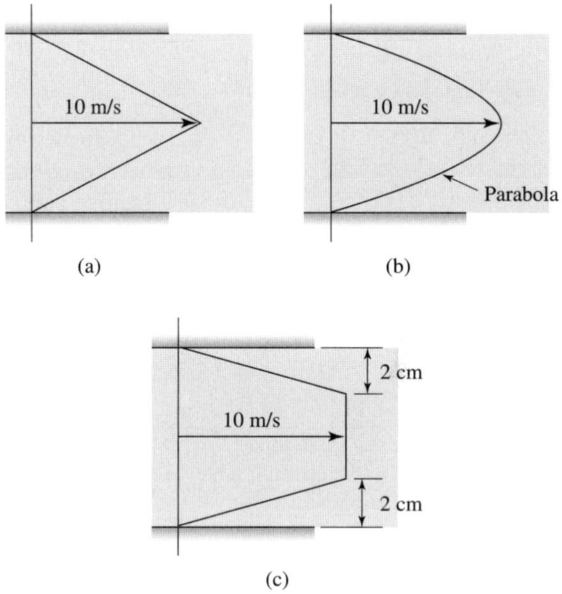

FIGURE P13.20

13.21 The profiles above are assumed to exist in a wide rectangular channel, 8 cm high and 80 cm wide. Find the average velocity, the mass flux, and the flow rate.

13.22 A constant-density fluid flows as shown in Fig. P13.22. Find the equation of the parabola if the conduit is:

(a) A pipe with $d = 1$ in. and $V = 6$ fps
(b) A wide rectangular channel with $d = 1$ in. and $V = 6$ fps
(c) A pipe with $d = 2$ cm and $V = 2$ m/s
(d) A wide rectangular channel with $d = 2$ cm and $V = 2$ m/s

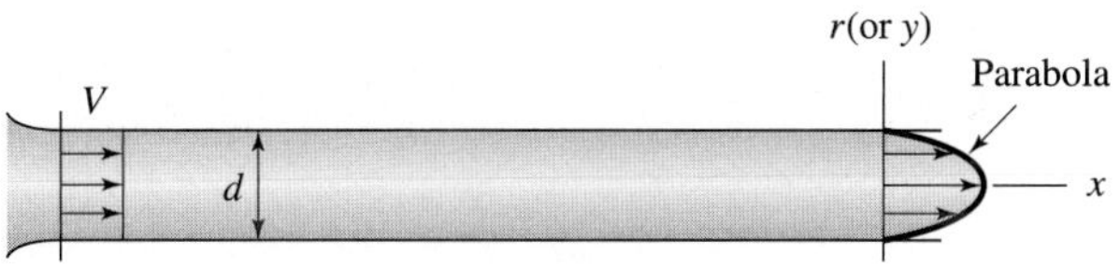

FIGURE P13.22

13.23 A parabolic profile exists in a 10-mm-diameter pipe. The pipe contracts to a 5-mm-diameter pipe in which the velocity profile is essentially uniform at 2 m/s. Write the equation for the parabola. Assume an incompressible flow.

13.24 As air flows as shown in Fig. P13.24 over a flat plate, the velocity is reduced to zero at the wall. If $u(y) = 10(20y - 100y^2)$ m/s, find the mass flux $\dot{m}$ through a surface parallel to the plate and 0.2 m above the plate. The plate is 2 m wide and $\rho = 1.23$ kg/m^3.

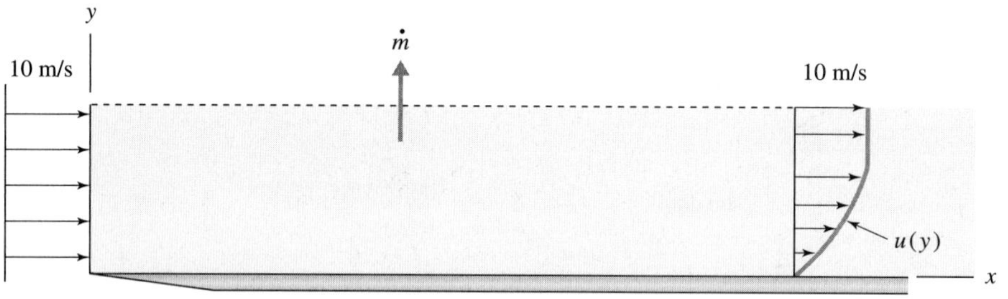

FIGURE P13.24

13.25 A streamline is 5 cm above the plate shown in Fig. P13.24 at the leading edge. How far from the plate is that same streamline at the location of the profile $u(y) = 10(20y - 100y^2)$?

13.26 Stratified salt water flows at a depth of 4 in. in a channel with a velocity distribution $2(6y - 9y^2)$ ft/sec, where y is measured in feet. If the density varies linearly from 2.2 slug/ft^3 at the bottom to clear water at the top, find $\dot{m}$. Also, show that $\dot{m} \neq \bar{\rho}\overline{VA}$. The channel is 5 ft wide.

13.27 Water flows as shown in Fig. P13.27. Calculate V_2.

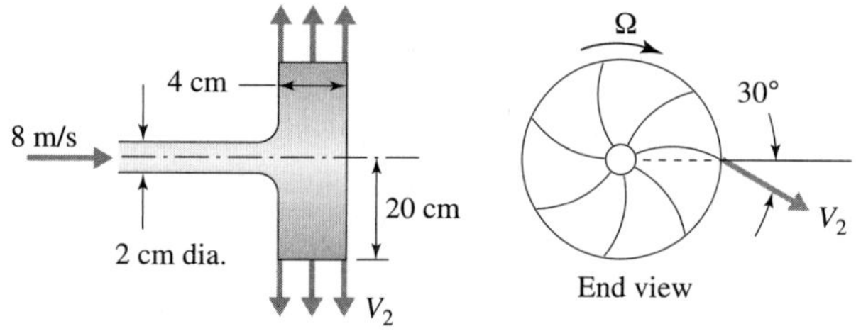

FIGURE P13.27

13.28 Air at 37 psi gage and 60°F is being forced into a tire, which has a volume of 17 ft^3, at a velocity of 180 ft/sec through a 1/4-in.-diameter opening. Approximate the rate of change of density in the tire.

13.29 In Fig. P13.29, if the mass of the control volume is not changing, find $\bar{V}_3$.

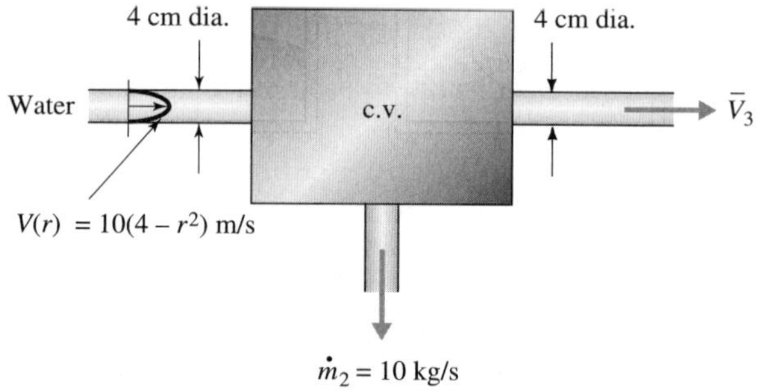

FIGURE P13.29

13.30 The average velocity $\bar{V}_3 = 10$ m/s in Problem 13.29. Find the rate at which the mass of the control volume is changing.

13.31 Find the velocity of the gas-fuel interface shown in Fig. P13.31. Use $R_{gas} = 0.28$ kJ/kg·K and $d_e = 30$ cm.

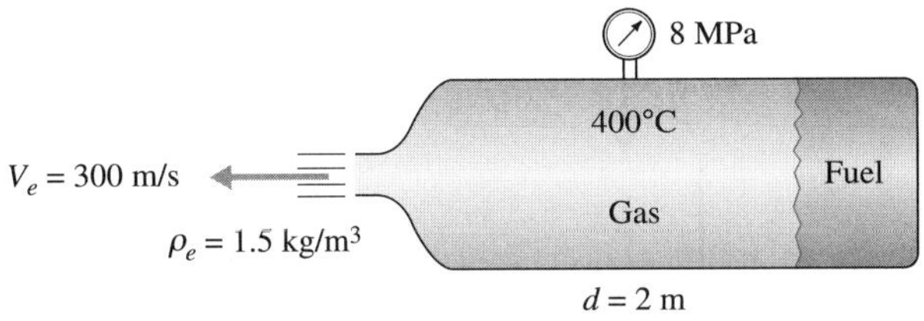

FIGURE P13.31

13.32 Conditioned air is delivered to a large conference room through four inlets, each transferring 1500 cfm. If the air is returned to the conditioner through a single 2 ft × 4 ft rectangular duct, estimate the average velocity in the duct. Make any needed assumptions.

13.33 The jet pump operates by inducing a flow due to the high velocity in the 5-cm-diameter pipe as shown in Fig. P13.33. The velocity in the small pipe is $200[1 - (r/R)^2]$. Estimate the average velocity at the exit.

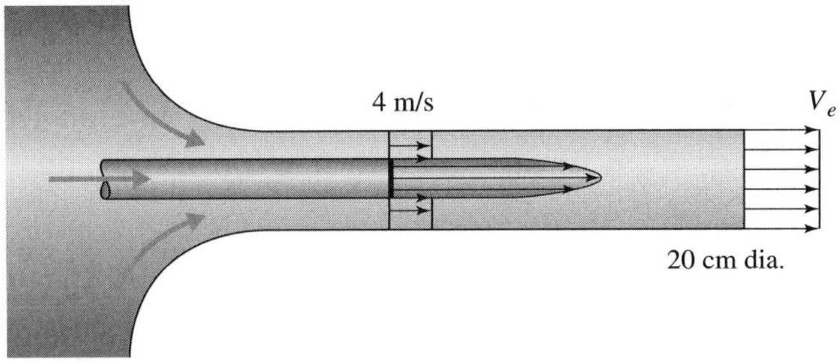

FIGURE P13.33

13.34 The water is 4 m deep behind the sluice gate of Fig. P13.34 in a rectangular channel that is suddenly opened. Find the initial dh/dt if $V_2 =$ 8 m/s and $V_1 = 0.2$ m/s. The upstream channel length is 100 m.

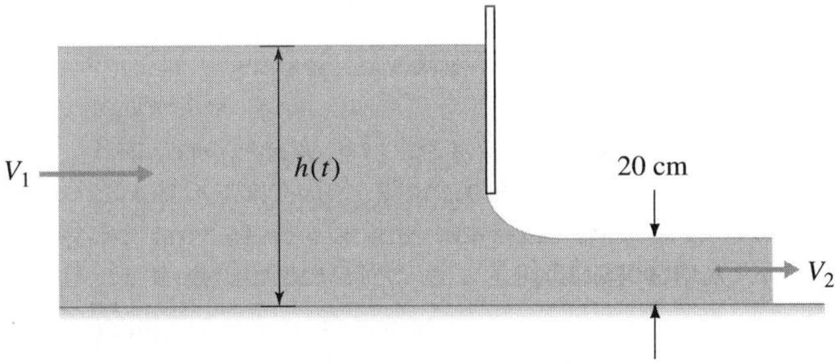

FIGURE P13.34

13.35 Water enters a kidney through a tube at 10 mL/min. It exits through a 6-mm-diameter tube at 20 mm/s. What is the rate of change of mass of water in the kidney?

13.36 In Fig. P13.36, find the rate of change of $h(t)$ if water is the fluid at all locations:

(a) $V_1 = 10$ m/s, $\dot{m}_2 = 10$ kg/s, $Q_{f3} = 600$ L/min
(b) $V_1 = 0$, $\dot{m}_2 = 20$ kg/s, $Q_{f3} = 10$ L/s
(c) $V_1 = 5$ m/s, $\dot{m}_2 = 10$ kg/s, $Q_{f3} = 1000$ L/min

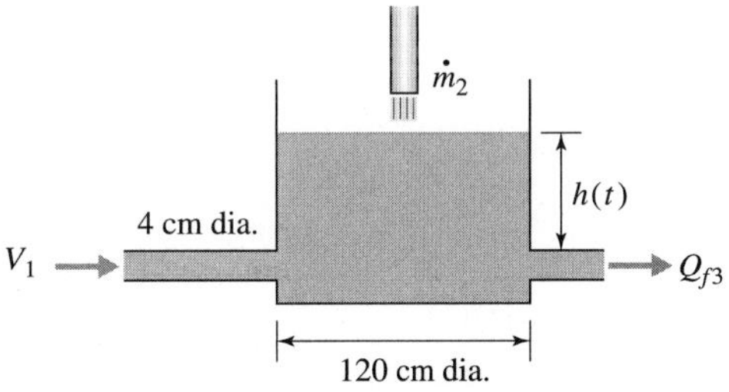

FIGURE P13.36

Energy Equation

13.37 In Fig. P13.37, determine the work rate by the air at the instant shown if $V_{piston} = 10$ m/s, the torque $T = 20$ N·m, the velocity gradient at the belt surface is 100 s^{-1}, and the pressure acting on the piston is 400 Pa. The belt is 80 cm × 50 cm and the 40-cm-high piston is 50 cm deep (into the paper).

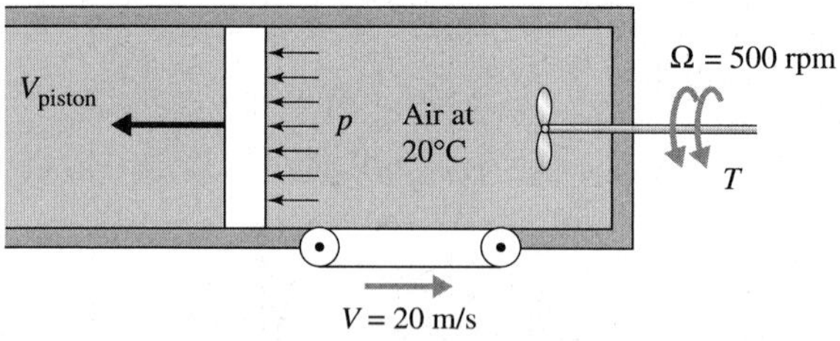

FIGURE P13.37

13.38 An insulated water pump requires 500 W while pumping 0.02 m^3/s with an efficiency of 80%. What is the temperature rise of the water from inlet to exit of the pump assuming equal inlet and exit areas? The specific heat of water is 4.18 kJ/kg·°C.

13.39 A water pump requires 5 hp to create a pump head of 20 m. If its efficiency is 87%, what is the flow rate of water?

13.40 An 89% efficient hydro turbine operates with a turbine head of 40 m. What is the turbine output if the mass flux is:
(a) 200 kg/s? **(b)** 90 000 kg/min?
(c) 8×10^6 kg/h?

13.41 The desired output from a set of 89% efficient turbines on a river is 10 MW. If the maximum turbine head attainable is 50 m, determine the average velocity at a location where the river is 60 m wide and 3 m deep.

13.42 Water flows in an open rectangular channel at a depth of 3 ft with a velocity of 12 ft/sec. The bottom of the channel drops over a short length a distance of 3 ft. Calculate the two possible depths of flow after the drop. Neglect all losses.

13.43 Find the velocity V_1 of the water in the vertical pipe shown in Fig. P13.43. Assume no losses.

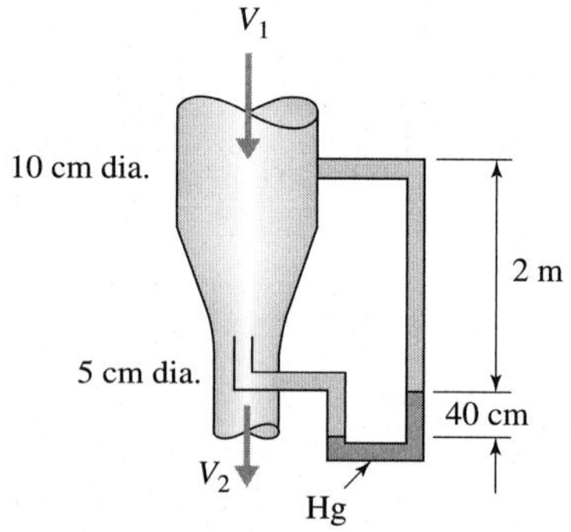

FIGURE P13.43

13.44 If the head-loss coefficient (based on V_2) between sections 1 and 2 of Problem 13.43 is 0.05, determine V_1 of the water.

13.45 The flow rate of water in a 2-in.-diameter horizontal pipeline at a pressure of 60 psi is 120 gal/min. If the pipeline increases to 3 in. diameter, calculate the increased pressure if the loss coefficient (based on V_1) is 0.37.

13.46 Water flows at the rate of 600 L/min in a 4-cm-diameter horizontal pipeline with a pressure of 690 kPa. If the pressure after an enlargement to 6 cm diameter is measured to be 700 kPa, calculate the head loss across the enlargement.

13.47 Calculate the pressure p_1 shown in Fig. P13.47 needed to maintain a flow rate of 0.08 m^3/s of water in a 6-cm-diameter horizontal pipe leading to a nozzle if a loss coefficient based on V_1 is 0.2 between the pressure gage and the exit.

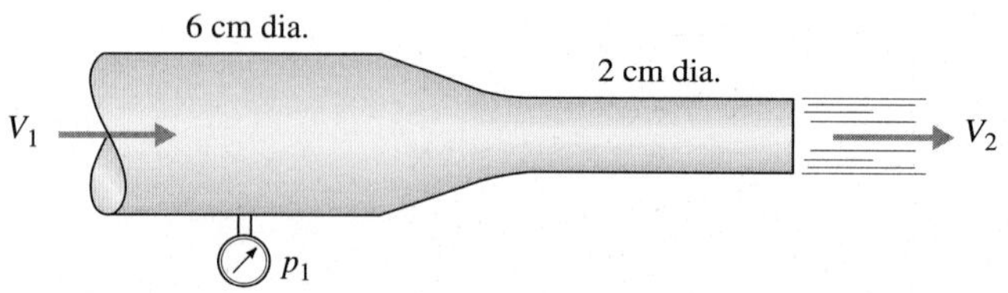

FIGURE P13.47

13.48 In Fig. P13.48, neglect all losses and predict the value of H and p if:
(a) $h = 15$ cm
(b) $h = 20$ cm

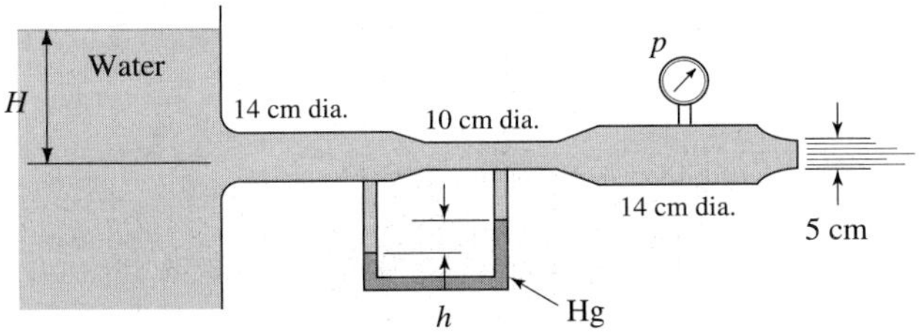

FIGURE P13.48

13.49 The overall loss coefficient for the pipe shown in Fig. P13.49 is 5; up to A it is 0.8, from A to B it is 1.2, from B to C it is 0.8, from C to D it is 2.2. Estimate the flow rate and the pressures at A, B, C, and D. The elevations are shown.

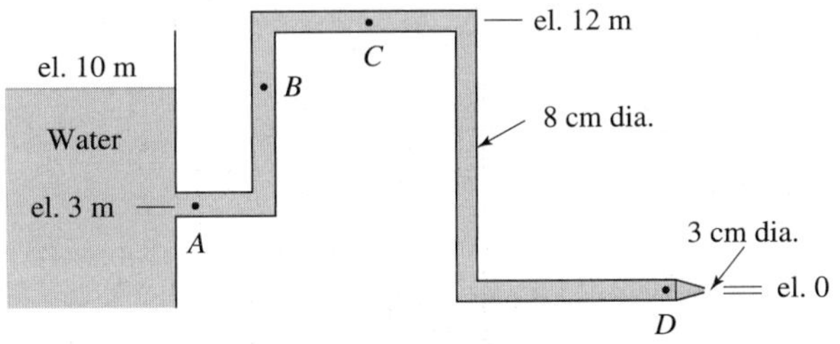

FIGURE P13.49

13.50 Water exits from a pressurized reservoir as shown in Fig. P13.50. Calculate the flow rate if on section A we:
(a) Attach a nozzle with exit diameter 5 cm
(b) Attach a diffuser with exit diameter 18 cm
(c) Leave as an open pipe as shown
Neglect losses for all cases.

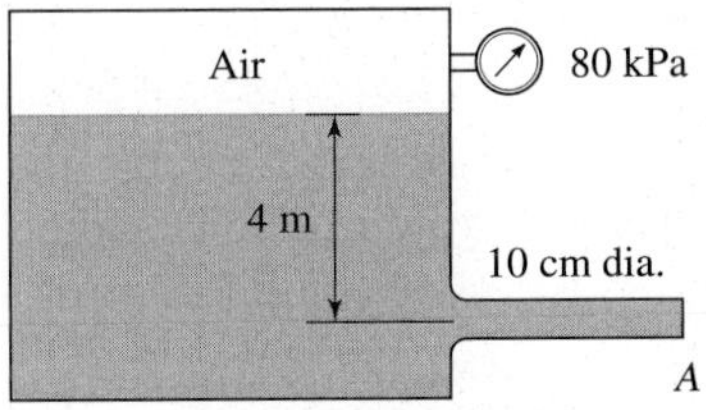

FIGURE P13.50

13.51 Relate the flow rate of the water through the venturi meter shown in Fig. P13.51 to the diameters and the manometer reading. Assume no losses.

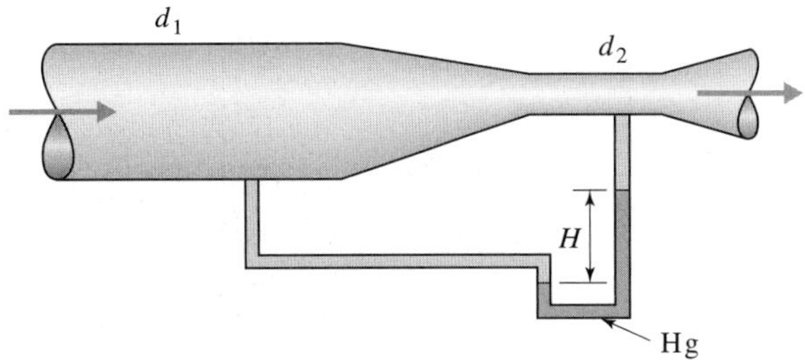

FIGURE P13.51

13.52 In the venturi meter of Problem 13.51, calculate the flow rate if:
(a) $H = 20$ cm, $d_1 = 2d_2 = 16$ cm
(b) $H = 40$ cm, $d_1 = 3d_2 = 24$ cm
(c) $H = 10$ in., $d_1 = 2d_2 = 6$ in
(d) $H = 15$ in., $d_1 = 3d_2 = 12$ in

13.53 In Fig. P13.53, determine the maximum possible height H if cavitation is to be avoided. Let:
(a) $d = 10$ cm and $T = 20°$C
(b) $d = 4$ in. and $T = 70°$F
Neglect all losses and assume $P_{atm} = 100$ kPa (14.7 psi).

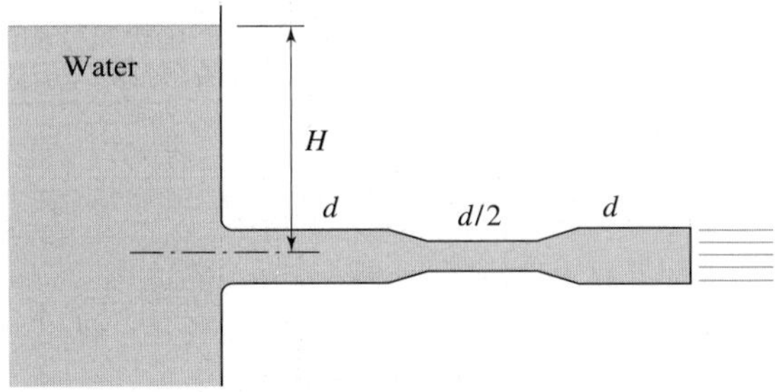

FIGURE P13.53

13.54 Cavitation is observed in the small section of pipe of Problem 13.53 when $H = 65$ cm. Estimate the temperature of the water. Neglect all losses and assume $p_{atm} = 100$ kPa. Use:
(a) $d = 10$ cm
(b) $d = 12$ cm

13.55 In Fig. P13.55, what is the minimum possible depth H if cavitation is to be avoided? Assume a vapor pressure of 6 kPa absolute and an overall loss coefficient of 8 based on V_2 and including the exit loss. Neglect losses up to the enlargement.

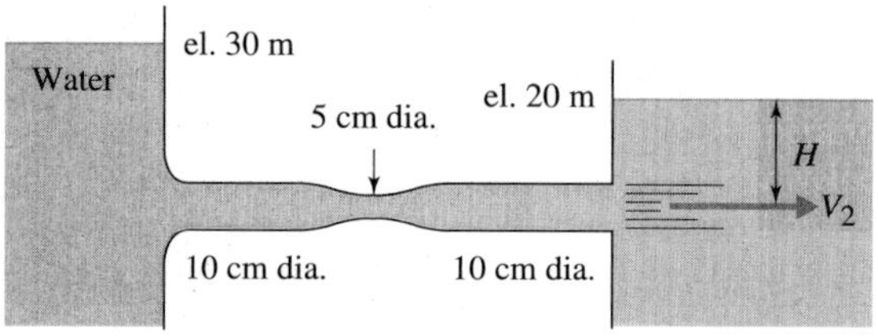

FIGURE P13.55

13.56 In Fig. P13.56, calculate the maximum diameter D such that cavitation will be avoided if:
(a) $d = 20$ cm, $H = 5$ m, and $T_{water} = 20°C$
(b) $d = 8$ in., $H = 15$ ft, and $T_{water} = 70°F$
Neglect all losses and use $p_{atm} = 100$ kPa (14.7 psi).

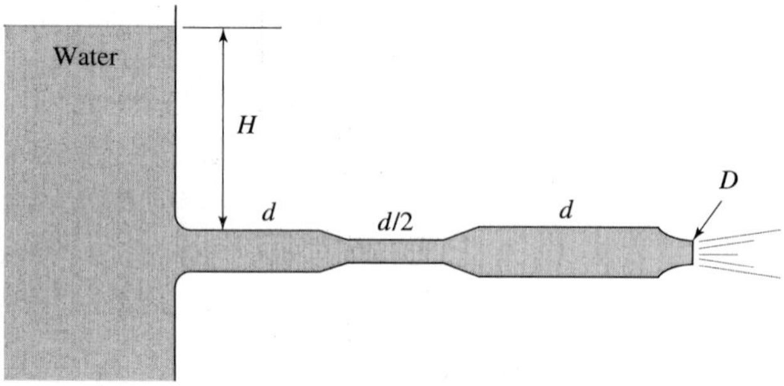

FIGURE P13.56

13.57 The overall loss coefficient in the siphon shown in Fig. P13.57 is 4; up to section A it is 1.5. At what height H will the siphon cease to function?

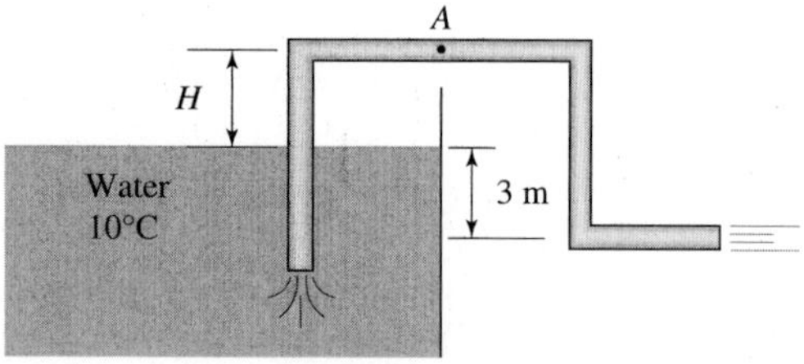

FIGURE P13.57

13.58 The pump shown in Fig. P13.58 is 85% efficient. If the pressure rise is 120 psi, calculate the required energy input in horsepower.

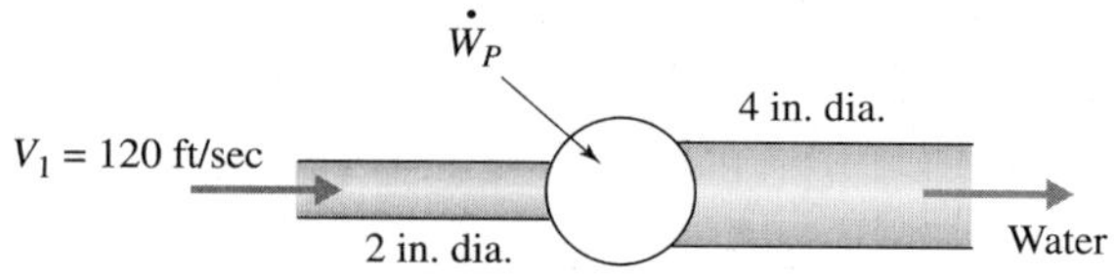

FIGURE P13.58

13.59 The pump shown in Fig. P13.58 is powered by a 20-kW motor. If the pump is 82% efficient, determine the pressure rise.

13.60 A turbine, which is 87% efficient, accepts 2 m^3/s of water from a 50-cm-diameter pipe. The pressure drop is 600 kPa and the exit velocity is small. What is the turbine output?

13.61 A turbine receives 450 ft^3/sec of water from a 6-ft-diameter pipe at a pressure of 120 psi and delivers 10 000 kW. The pressure in the $7\frac{1}{2}$-ft-diameter exit pipe is 18 psi. Calculate the turbine efficiency.

13.62 Air enters a compressor with negligible velocity at 85 kPa absolute and 20°C. It leaves with a velocity of 200 m/s at 600 kPa absolute. For a mass flux of 5 kg/s, calculate the exit temperature if the power required is 1500 kW and:
(a) No heat transfer occurs
(b) The heat transfer rate is 60 kW

13.63 Air enters a compressor at standard conditions with negligible velocity. At the 1-in.-diameter exit the pressure, temperature, and velocity are 60 psia, 300°F, and 600 ft/sec, respectively. If the heat transfer is 10 Btu/lb of air, find the power required by the compressor.

13.64 A small river with a flow rate of 15 m^3/s feeds the reservoir shown in Fig. P13.64. Calculate the energy that is available continuously if the turbine is 80% efficient. The loss coefficient for the overall piping system is $K = 4.5$.

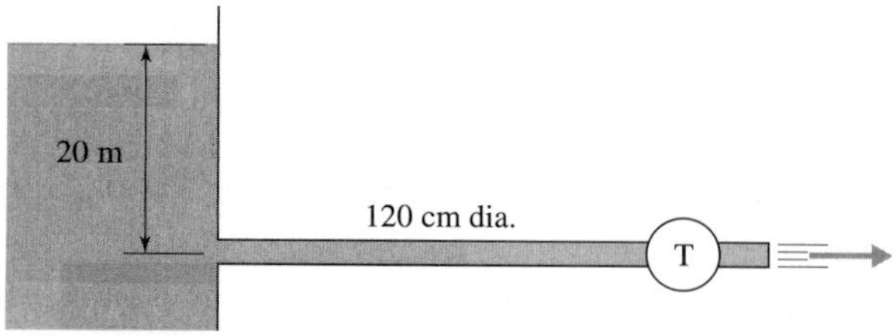

FIGURE P13.64

13.65 For the system shown in Fig. P13.65 the average velocity in the pipe is 10 m/s. Up to point A, $K = 1.5$, from B to C, $K = 6.2$, and the pump is 80% efficient. If $p_C = 200$ kPa find p_A and p_B and the power required by the pump.

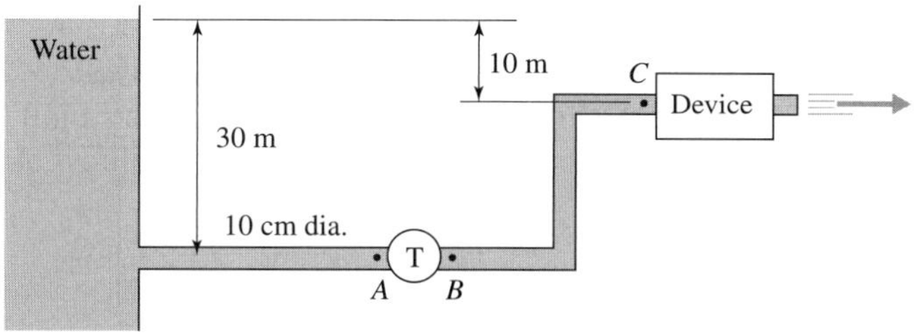

FIGURE P13.65

13.66 Determine the power output of the turbine shown in Fig. P13.66 for a water flow rate of 18 ft³/sec. The turbine is 90% efficient.

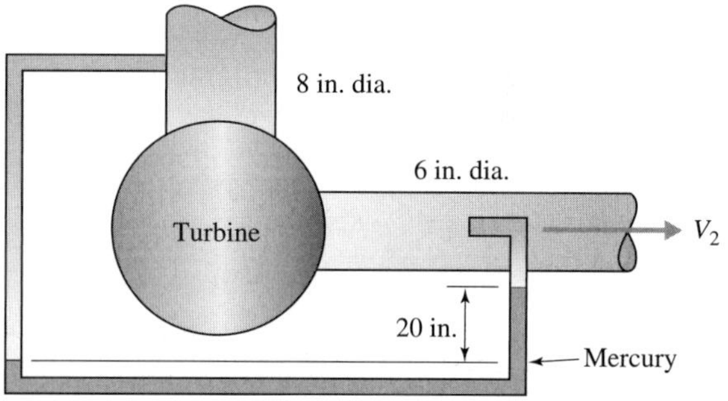

FIGURE P13.66

13.67 Find the power requirement of the 85%-efficient pump shown in Fig. P13.67 if the loss coefficient up to A is 3.2, and from B to C, $K = 1.5$. Neglect the losses through the exit nozzle. Also, calculate p_A and p_B.

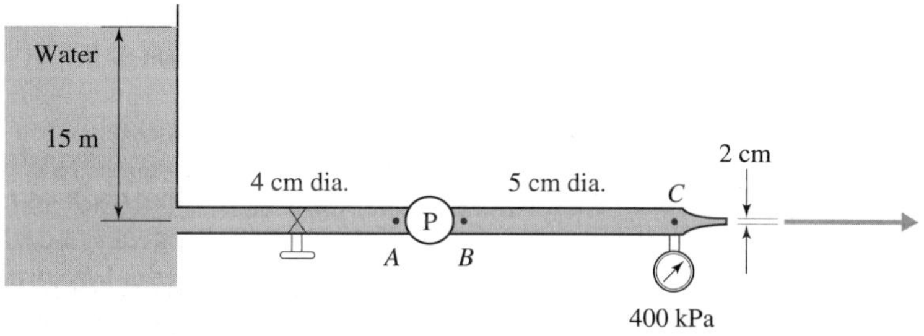

FIGURE P13.67

13.68 Calculate the head loss between the two sections shown in Fig. P13.68. Assume a pipe with $d = 1.2$ cm.

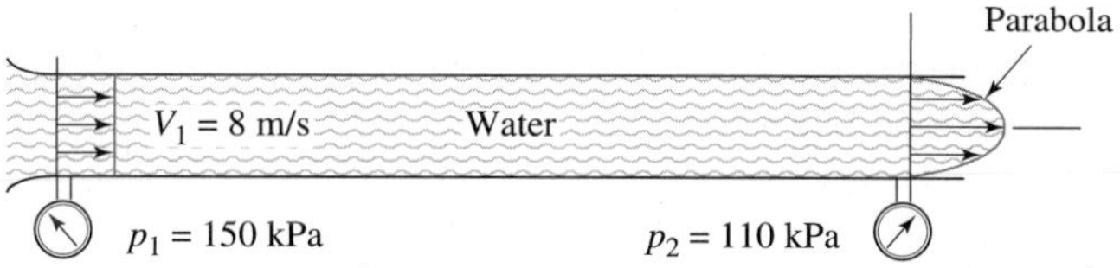

FIGURE P13.68

13.69 Determine the kinetic-energy correction factor if:
(a) $u(r) = 10(1 - r^2/R^2)$ in a 2-cm-diameter pipe.
(b) $u(y) = 10(1 - y^2/h^2)$ in a 2-cm-high channel.

13.70 A turbulent velocity profile in a pipe is often written as $u(r) = u_{\max}(1 - r/R)^{1/n}$, where n ranges between 5 and 9, 7 being the most common value. Calculate an expression for the kinetic energy passing a section of pipe and the kinetic-energy correction factor if:
(a) $n = 5$
(b) $n = 7$
(c) $n = 9$

13.71 An automobile moves at 100 km/hr with a drag force of 1340 N. The gas consumption is noted to be 5 km/L. If the engine efficiency is 15%, determine the energy released per kilogram of fuel. The fuel density is 680 kg/m^3.

13.72 The manufacturer's pump curve for the pump in the flow system shown in Fig. P13.72a is provided in Fig. P13.72b. Estimate the flow rate. The overall loss coefficient is:
(a) $K = 5$ **(b)** $K = 20$
The solution involves a trial-and-error procedure, or the energy equation can be written as $H_p = H_p(Q_f)$ and plotted on the pump curve.

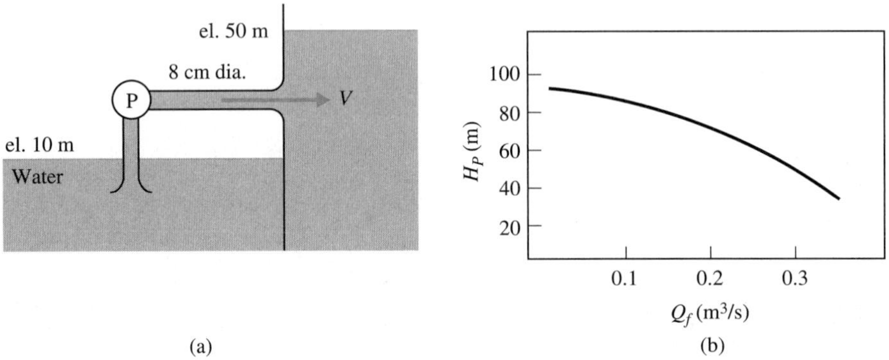

FIGURE P13.72

Momentum Equation

13.73 Water flows in a pipe of diameter d at a pressure p. It flows out a nozzle of diameter $d/2$ to the atmosphere. Calculate the force of the water on the nozzle if:
(a) $d = 6$ cm, $p = 200$ kPa
(b) $d = 12$ cm, $p = 200$ kPa
(c) $d = 3$ in., $p = 30$ psi
(d) $d = 6$ in., $p = 30$ psi

13.74 A nozzle and hose are attached to the ladder of a fire truck. What force is needed to hold a nozzle supplied by a 9-cm-diameter hose with a pressure of 2000 kPa? The nozzle outlet diameter is 3 cm.

13.75 Water flows in a 10-cm-diameter pipe at a pressure of 400 kPa out a straight nozzle. Calculate the force of the water on the nozzle if the outlet diameter is:
(a) 8 cm **(b)** 6 cm **(c)** 4 cm **(d)** 2 cm

13.76 Find the horizontal force of the water on the horizontal bend shown in Fig. P13.76.

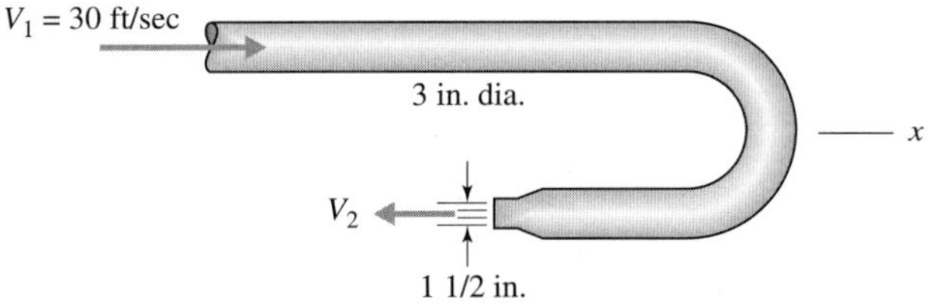

FIGURE P13.76

13.77 Find the horizontal force components of the water on the horizontal bend shown in Fig. P13.77 if p_1 is:
(a) 200 kPa **(b)** 400 kPa **(c)** 800 kPa

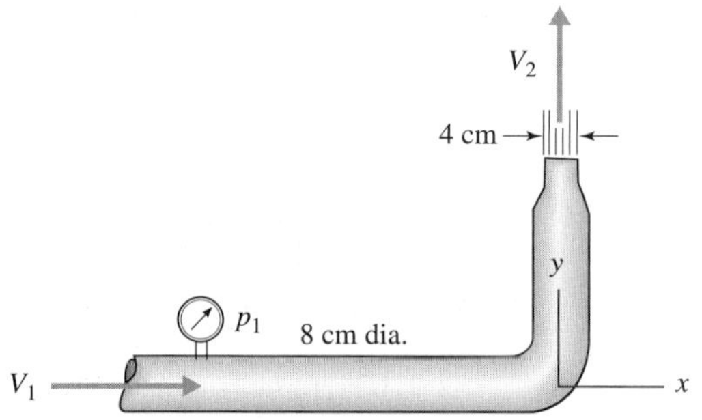

FIGURE P13.77

13.78 What is the net force needed to hold the orifice plate shown in Fig. P13.78 onto the pipe?

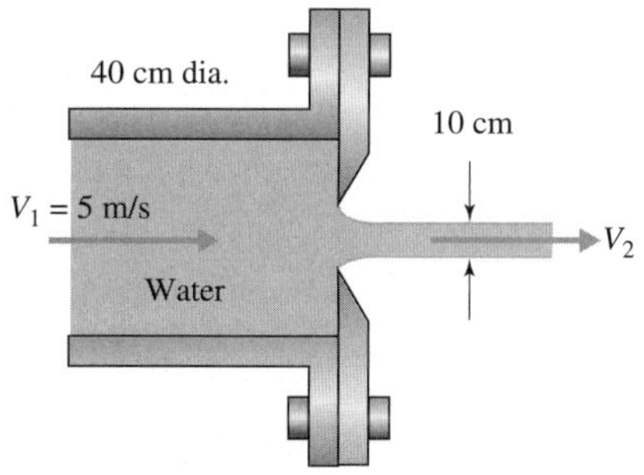

FIGURE P13.78

13.79 Assuming uniform velocity profiles, find F needed to hold the plug in the pipe shown in Fig. P13.79. Neglect viscous effects.

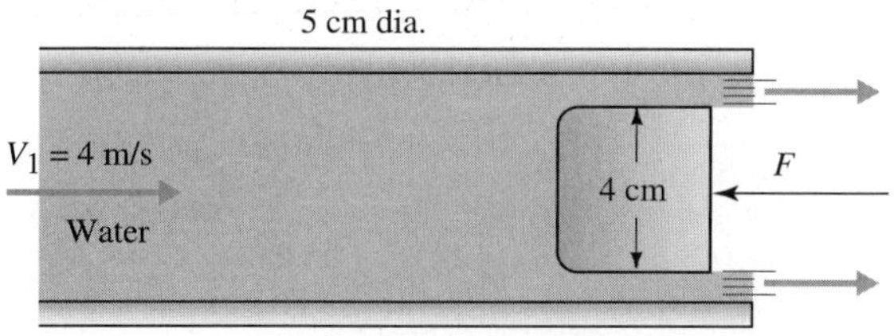

FIGURE P13.79

13.80 Assuming hydrostatic pressure distributions, uniform velocity profiles, and negligible viscous effects, find the horizontal force needed to hold the sluice gate in the position shown in Fig. P13.80.

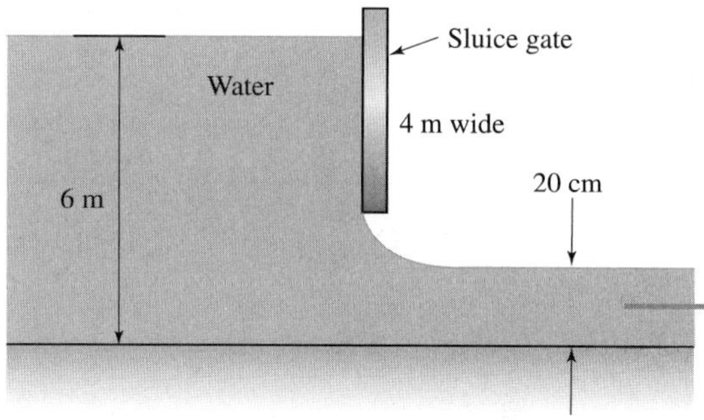

FIGURE P13.80

13.81 A sudden jump (a hydraulic jump) occurs in a rectangular channel as shown in Fig. P13.81. Find y_2 and V_2 if:
(a) $V_1 = 8$ m/s, $y_1 = 60$ cm
(b) $V_1 = 12$ m/s, $y_1 = 40$ cm
(c) $V_1 = 20$ ft/sec, $y_1 = 2$ ft
(d) $V_1 = 30$ ft/sec, $y_1 = 3$ ft

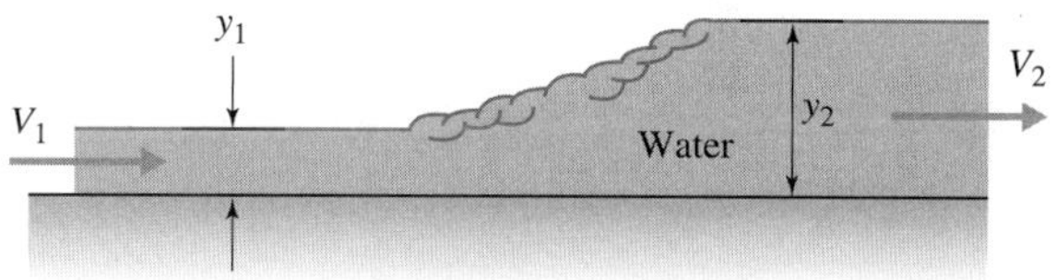

FIGURE P13.81

13.82 A hydraulic jump, as shown in Fig. P13.81, occurs so that $V_2 = \frac{1}{4} V_1$. Find V_1 and y_2 if:
(a) $y_1 = 80$ cm
(b) $y_1 = 2$ ft

13.83 The velocity is measured to be 10 ft/sec downstream of a hydraulic jump where the depth is 6 ft. Calculate the velocity and depth prior to the jump.

13.84 For the system shown in Fig. P13.84, estimate the down stream pressure p_2 if $p_1 = 60$ kPa and $V_1 = 20$ m/s. Neglect losses. (*Note:* The pressure immediately after the pipe expansion is p_1.)

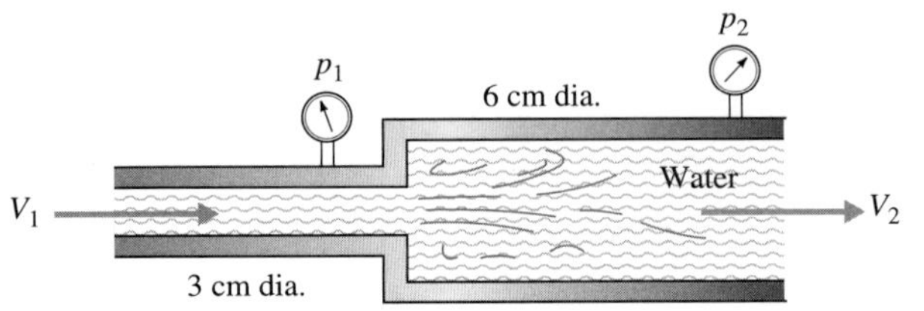

FIGURE P13.84

13.85 Water flows at 15 m/s in a 10-cm-diameter stem of a horizontal T-section that branches into 5-cm-diameter pipes. Find the force of the water on the T-section if the branches exit to the atmosphere. Neglect viscous effects.

13.86 Find the x- and y-force components on the horizontal T-section shown in Fig. P13.86. Neglect viscous effects.

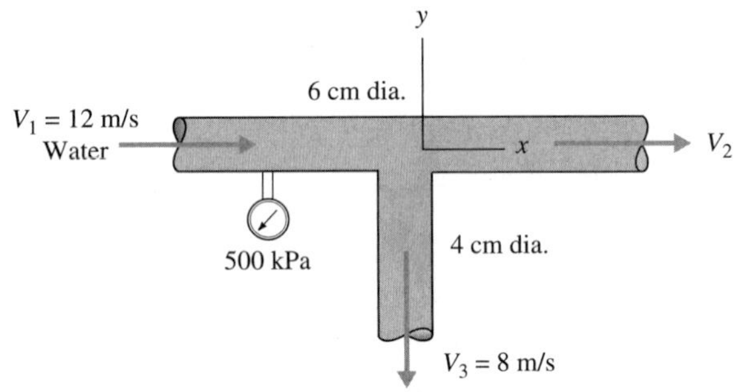

FIGURE P13.86

13.87 A horizontal 10-cm-diameter jet of water with $\dot{m} = 300$ kg/s strikes a vertical plate. Calculate:
(a) The force needed to hold the plate stationary
(b) The force needed to move the plate away from the jet at 10 m/s
(c) The force needed to move the plate into the jet at 10 m/s

13.88 A horizontal $2\frac{1}{2}$-in.-diameter jet of water strikes a vertical plate. Determine the velocity issuing from the jet if a force of 200 lb is needed to:
(a) Hold the plate stationary
(b) Move the plate away from the jet at 30 ft/sec
(c) Move the plate into the jet at 30 ft/sec

13.89 Determine the mass flux issuing from the jet shown in Fig. P13.89 if a force of 700 N is needed to:
(a) Hold the cone stationary
(b) Move the cone away from the jet at 8 m/s
(c) Move the cone into the jet at 8 m/s

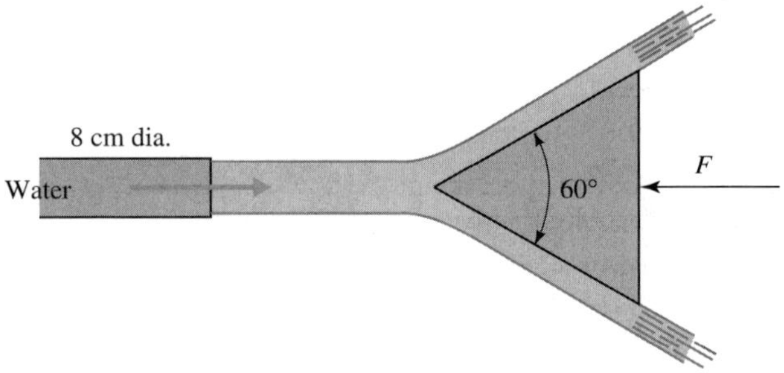

FIGURE P13.89

13.90 Calculate the force components of the water acting on the deflecting blade shown in Fig. P13.90 if:
(a) The blade is stationary
(b) The blade moves to the right at 60 ft/sec
(c) The blade moves to the left at 60 ft/sec

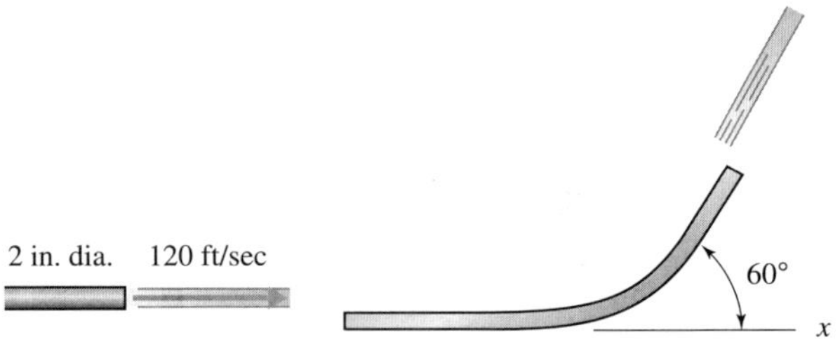

FIGURE P13.90

13.91 The blade of Problem 13.90 is one of a series of blades that are attached to a 50-cm-radius rotor that has a rotational speed of 30 rad/s. If there are 10 such water jets, find the power output.

13.92 Determine the force components of superheated steam acting on the blade shown in Fig. P13.92 if:
(a) The blade is stationary
(b) The blade moves to the right at 100 m/s
(c) The blade moves to the left at 100 m/s

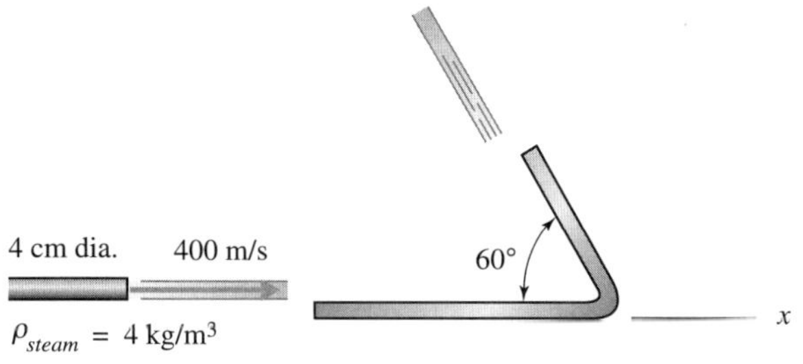

FIGURE P13.92

13.93 The blade of Problem 13.92 is one of a series of blades that are attached to a 1.2-m-radius rotor that rotates at 150 rad/s. Calculate the power output if there are 15 such steam jets.

13.94 Twelve high-speed water jets strike the blades as shown in Fig. P13.94. Find the power output and the blade angles if V_B is:
(a) 20 m/s
(b) 40 m/s
(c) 50 m/s

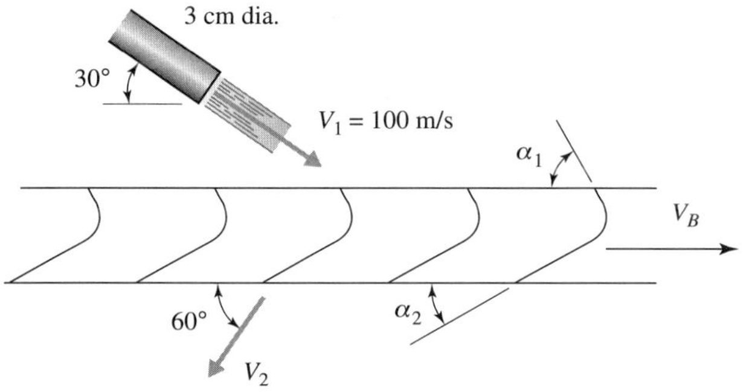

FIGURE P13.94

13.95 Fifteen water jets strike the blades of a turbine as shown in Fig. P13.95. Calculate the power output and the blade angles if β_2 is:
(a) 60°
(b) 70°
(c) 80°

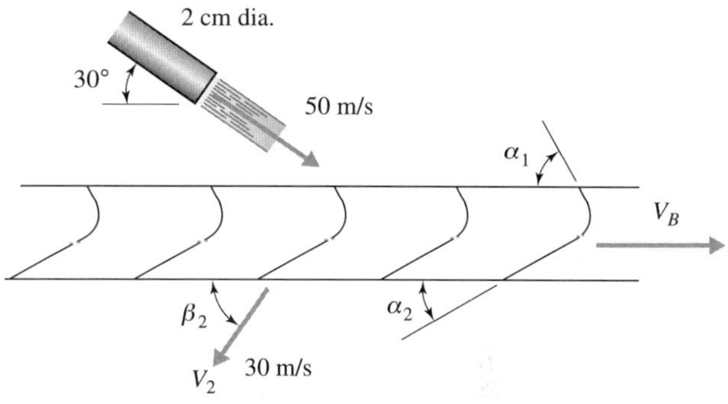

FIGURE P13.95

13.96 Water flows from the rectangular jet as shown in Fig. P13.96 against the fixed plate. Find the force F and the mass fluxes $\dot{m}_2$ and $\dot{m}_3$ if
(a) $b = 2$ cm, $h = 40$ cm, $V_1 = 40$ m/s
(b) $b = 1$ in., $h = 20$ in., $V_1 = 120$ fps

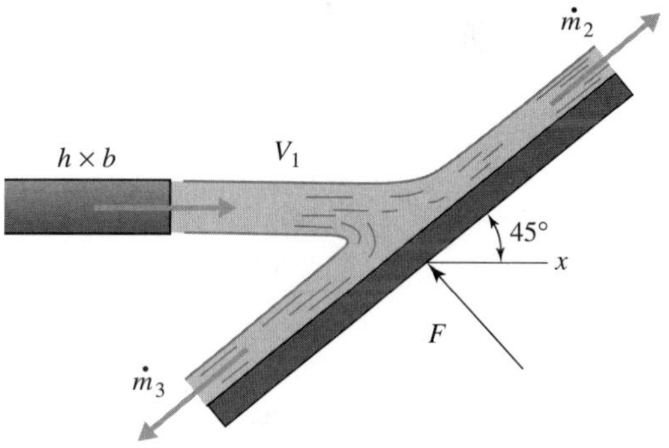

FIGURE P13.96

13.97 The plate of Problem 13.96a moves to the left at 20 m/s. Find the power required.

13.98 Calculate the velocity that the plate of Problem 13.96a must move with (in the x-direction) to produce the maximum power output.

13.99 A 2.5-m-wide snowplow travels at 50 km/h plowing snow at a depth of 0.8 m. The snow leaves the blade normal to the direction of motion of the snowplow. What power does the plowing operation require if the density of the snow is 90 kg/m^3?

13.100 A vehicle with a mass of 5000 kg is traveling at 900 km/h. It is decelerated by lowering a 20-cm-wide scoop into water a depth of 6 cm (Fig. P13.100). If the water is deflected through 180°, calculate the distance the vehicle must travel for the speed to be reduced to 100 km/h.

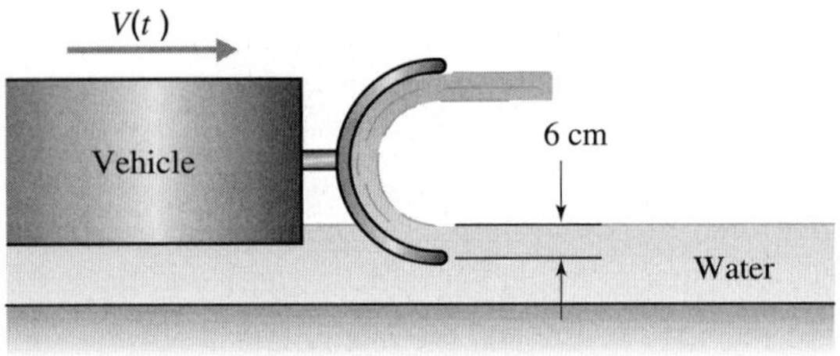

FIGURE P13.100

13.101 A 20-slug vehicle is required to have an initial acceleration of 6 ft/sec^2. It is proposed that a $2\frac{1}{2}$-in.-diameter water jet strike a vane built into the rear of the vehicle that will divert the water through an angle of 180°. What jet velocity is necessary? What speed will be achieved after 2 sec?

13.102 Calculate the change in the momentum flux of the water flowing through the plane contraction shown in Fig. P13.102 if the flow rate is 0.2 m^3/s. The slope of the two profiles is the same. The upstream profile is created by a plate containing slots of various widths.

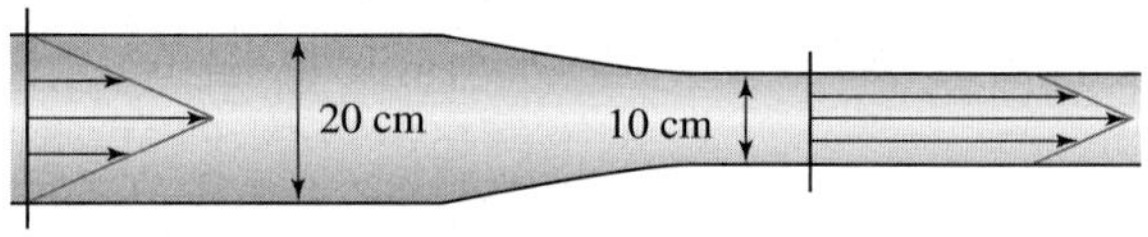

FIGURE P13.102

13.103 Determine the momentum correction factor for the following profile shown in Problem 13.102:

(a) The entering profile

(b) The exiting profile

13.104 Water at 60°F flows in a $1\frac{1}{2}$-in.-diameter horizontal pipe and experiences a pressure drop of 0.03 psi over a 30-ft length of pipe. Calculate the velocity gradient at the wall. Recall that $\tau = \mu|du/dr|$.

13.105 Find the drag force on the walls between the two sections of the horizontal pipe shown in Fig. P13.105.

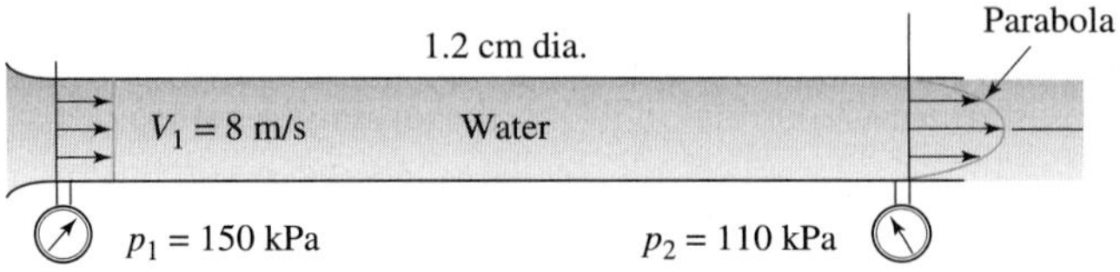

FIGURE P13.105

13.106 The velocity distribution downstream of a 10-m-long circular cylinder is as shown in Fig. P13.106. Determine the force of the air on the cylinder. Use $\rho = 1.23$ kg/m^3.

13.107 Calculate the drag force acting on the flat 2-m-wide plate shown in Fig. P13.107. Outside the viscous region the velocity is uniform. Select (use $\rho = 1.23$ kg/m^3):

(a) A rectangular control volume extending outside the viscous region (mass flux crosses the top).

(b) A control volume with the upper boundary a streamline (no mass flux crosses a streamline).

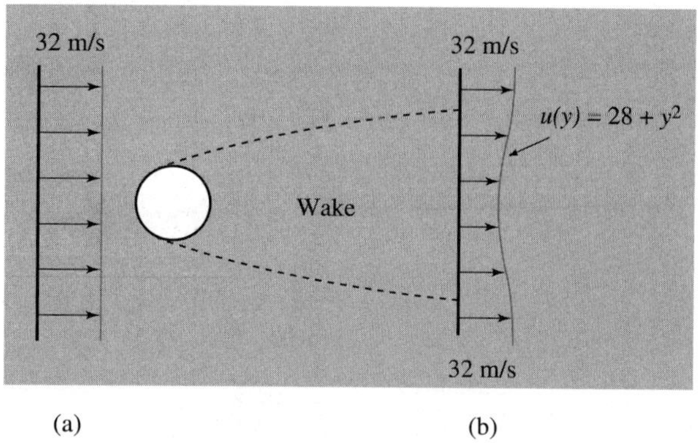

FIGURE P13.106

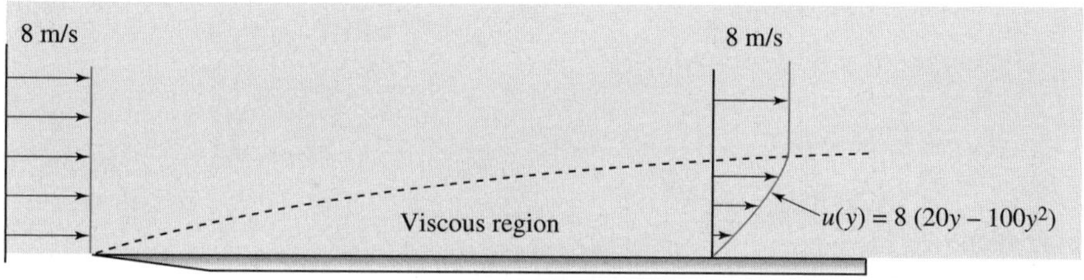

FIGURE P13.107

Momentum and Energy

13.108 Determine the power lost in the hydraulic jump of:

(a) Problem 13.81a.

(b) Problem 13.83.

13.109 Calculate the loss coefficient for the expansion of Problem 13.84. Base the coefficient on the velocity V_1.

U.S. Olympian ski jumpers have used wind tunnels to improve their technique. The jumper is suspended by cables attached to his chest and thighs, enabling his lift and drag to be measured. The sport has recently been revolutionized by opening skis like a *V* into the wind. (Calspan Corporation, Buffalo, New York)

14

Dimensional Analysis and Similitude

Outline

Chapter Objectives

The objectives of this chapter are to:

- ▲ Establish the parameters necessary to guide experimental studies.
- ▲ Present the technique used to apply the results of model studies to prototypes for a variety of flow situations.
- ▲ Provide examples and problems that illustrate how dimensionless flow parameters are used, and how model studies allow us to predict quantities of interest on a prototype.

Nomenclature

The following variables are introduced in this chapter:

C_D	drag coefficient	p	subscript that refers to "prototype"
Eu	Euler number	St	Strouhal number
f	frequency	We	Weber number
Fr	Froude number	π	a dimensionless term
m	subscript that refers to "model"		

14.1 INTRODUCTION

Dimensional homogeneity: *Condition where all terms in an equation have the same dimensions.*

There are many problems of interest in the field of fluid mechanics in the real world of design that cannot be solved using equations only. It is often necessary to resort to experimental methods to establish relationships between the variables of interest. Since experimental studies are usually quite expensive, it is necessary to keep the required experimentation to a minimum. This is done using a technique called *dimensional analysis*, which is based on the notion of **dimensional homogeneity**—that all terms in an equation must have the same dimensions. For example, if we write Bernoulli's equation in the form

$$\frac{V_1^2}{2g} + \frac{p_1}{\gamma} + z_1 = \frac{V_2^2}{2g} + \frac{p_2}{\gamma} + z_2 \tag{14.1}$$

we note that the dimension of each term is length. Furthermore, if we factored out z_1 from the left-hand side and z_2 from the right-hand side, we would have

$$\frac{V_1^2}{2gz_1} + \frac{p_1}{\gamma z_1} + 1 = \left(\frac{V_2^2}{2gz_2} + \frac{p_2}{\gamma z_2} + 1\right)\frac{z_2}{z_1} \tag{14.2}$$

In this form of Bernoulli's equation the terms are all dimensionless and we have written the equation as a combination of dimensionless parameters, the basic idea in dimensional analysis that will be presented in the next section.

Buckingham π-theorem: *A theory which organizes steps to ensure dimensional homogeneity.*

We use the **Buckingham π-theorem**, which organizes the steps of ensuring dimensional homogeneity; it requires some knowledge of the phenomenon being studied in order that the appropriate quantities of interest are included.

Often in experimental work we are required to perform experiments on objects that are quite large, too large to experiment with for a reasonable cost. This would include flows over weirs and dams; wave interactions with piers and breakwaters; flows around submarines and ships; subsonic and supersonic flows around aircraft; flows around stadiums and buildings, as shown in Fig. 14.1; flows through large pumps and turbines; and flows around automobiles and trucks. Such flows are usually studied in laboratories using models that are smaller than the prototype, the actual device. This substantially reduces the costs when compared with full-scale studies and allows for the study of various configurations or flow conditions.

Key Concept: *Fluid flows are often studied using models.*

FIGURE 14.1 Scale model of the large buildings in a city. Air flow around the buildings is studied. The roughening elements on the floor generate the desired wall turbulence. (Courtesy of Fluid Mechanics and Diffusion Laboratory, Colorado State University.)

There are also flows of interest that involve rather small dimensions, such as flow around a turbine blade, flow into a capillary tube, flow around a microorganism, flow through a small control valve, and flow around and inside a falling droplet. These flows would require that the model be larger than the prototype so that observations could be made with an acceptable degree of accuracy.

Similitude is the study of predicting prototype conditions from model observations. This will be presented following dimensional analysis. Similitude involves the use of the dimensionless parameters obtained in dimensional analysis.

Similitude: *The study of predicting prototype conditions from model observations.*

14.2 DIMENSIONAL ANALYSIS

14.2.1 Motivation

In the study of phenomena involving fluid flows, either analytically or experimentally, there are invariably many flow and geometric parameters involved. In the interest of saving time and money the fewest possible combinations of parameters should be utilized. For example, consider the pressure drop across the slider valve of Fig. 14.2. We may suspect that the pressure drop depends on such parameters as pipe mean velocity V, the density ρ of the fluid, the fluid viscosity μ, the pipe diameter d, and the gap height h. This could be expressed as

Key Concept: *The fewest possible combinations of parameters should be utilized.*

$$\Delta p = f(V, \rho, \mu, d, h) \tag{14.3}$$

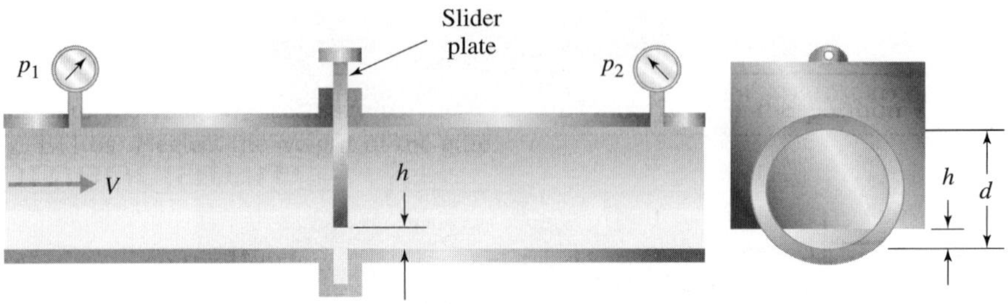

FIGURE 14.2 Flow around a slider valve.

Now, if we attempt an experimental study of this problem, consider the strategy for finding the dependence of the pressure drop on the parameters involved. We could fix all parameters except the velocity and investigate the dependence of the pressure drop on the average velocity. Then the diameter could be changed and the experiment repeated. This would lead to the set of results shown in Fig. 14.3a. Following that set of experiments the gap height h could be changed, leading to the curves of Fig. 14.3b. Again, different fluids could be studied, leading to curves with ρ and μ changing values.

Key Concept: *Any equation can be written in terms of dimensionless parameters.*

Consider next the notion that any equation that relates a certain set of variables, such as Eq. 14.3, can be written in terms of dimensionless parameters, as was done with the Bernoulli equation (14.2). We can organize the variables of Eq. 14.3 into dimensionless parameters (the steps needed to do this will be presented in a subsequent section) as follows:

$$\frac{\Delta p}{\rho V^2} = f\left(\frac{V\rho d}{\mu}, \frac{h}{d}\right) \tag{14.4}$$

Obviously, this is a much simpler relationship. We could perform an experiment with a fixed h/d (say, $h/d = 0.1$) by varying $V\rho d/\mu$ (this is done by simply varying V), resulting in a curve as shown in Fig. 14.4. The quantity h is changed so that $h/d = 0.5$ and the testing is repeated. Finally, the entire experiment is presented

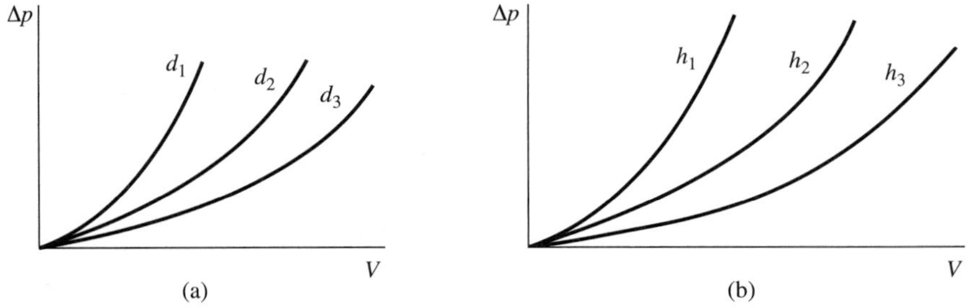

FIGURE 14.3 Pressure drop versus velocity curves: (a) ρ, μ, h fixed; (b) ρ, μ, d fixed.

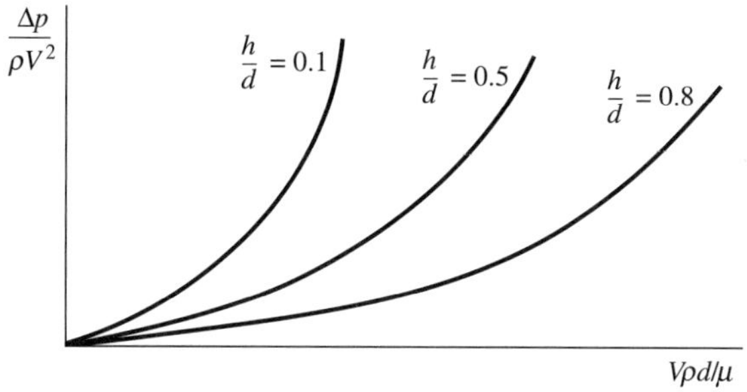

FIGURE 14.4 Dimensionless pressure drop versus dimensionless velocity.

in one figure, as in Fig. 14.4. This has greatly reduced the effort and the cost in determining the actual form of $f(V\rho d/\mu, h/d)$; we would use only one pipe and one valve, and we would use only one fluid.

It is not always clear, however, which parameters should be included in an equation such as Eq. 14.3. The selection of these parameters requires a detailed understanding of the physics involved. In the selection of the parameters that affect the pressure drop across the slider valve, it was assumed that density and viscosity are important parameters, while parameters such as pipe pressure and fluid compressibility are not. It should be kept in mind that the selection of the proper parameters is a first crucial step in the application of dimensional analysis.

14.2.2 Review of Dimensions

Before we present the dimensional analysis technique let's review the dimensions of the quantities of interest in an introductory course in fluid mechanics. All of the quantities have some combination of dimensions of length, time, mass, and force which are related by Newton's second law,

$$\sum \mathbf{F} = m\mathbf{a} \tag{14.5}$$

In terms of dimensions, it is written as

$$F = \frac{ML}{T^2} \tag{14.6}$$

where F, M, L, and T are the dimensions of force, mass, length, and time, respectively. Thus we see that it is sufficient to use only three basic dimensions. We will choose the M-L-T system because we can eliminate the force dimension with Eq. 14.6.

Key Concept: *Throughout this chapter, we will use the M-L-T system.*

If we were considering more complicated flow situations such as those involving electromagnetic field interactions, or those involving temperature gradients, we would need to include the appropriate additional dimensions. However, in this book such phenomena will not be introduced, except for the compressible

TABLE 14.1 Symbols and Dimensions of Quantities Used in Fluid Mechanics

Quantity	Symbol	Dimensions
Length	l	L
Time	t	T
Mass	m	M
Force	F	ML/T^2
Velocity	V	L/T
Acceleration	a	L/T^2
Angular velocity	Ω	T^{-1}
Gravity	g	L/T^2
Area	A	L^2
Flow rate	Q_f	L^3/T
Mass flux	$\dot{m}$	M/T
Pressure	p	M/LT^2
Stress	τ	M/LT^2
Density	ρ	M/L^3
Specific weight	γ	M/L^2T^2
Viscosity	μ	M/LT
Kinematic viscosity	ν	L^2/T
Work	W	ML^2/T^2
Power, heat flux	$\dot{W}, \dot{Q}$	ML^2/T^3
Surface tension	σ	M/T^2
Bulk modulus	B	M/LT^2

flow of an ideal gas; for that case an equation of state relates the thermal effects to the dimensions above. That is,

$$p = \rho RT \tag{14.7}$$

where T represents temperature. This allows us to write

$$[RT] = [p/\rho] = \frac{F}{L^2} \cdot \frac{L^3}{M} = \frac{ML/T^2}{L^2} \cdot \frac{L^3}{M} = \frac{L^2}{T^2} \tag{14.8}$$

where the brackets represent "the dimensions of." Notice that the equation of state does not introduce additional dimensions.

The quantities of interest in fluid mechanics are listed with their respective dimensions in Table 14.1. Reference to this table will simplify writing the dimensions of the quantities introduced in the problems.

14.2.3 Buckingham π-Theorem

In a given physical problem the dependent variable x_1 can be expressed in terms of the independent variables as

$$x_1 = f(x_2, x_3, x_4, \ldots, x_n) \tag{14.9}$$

where n represents the total number of variables. Referring to Eq. 14.3, Δp is the dependent variable and V, ρ, μ, d, and h are the independent variables. The *Buckingham π-theorem*, states that $(n - m)$ dimensionless groups of variables, called π-

terms, where m is the number[1] of basic dimensions included in the variables, can be related by

$$\pi_1 = f_1(\pi_2, \pi_3, \ldots, \pi_{n-m}) \qquad (14.10)$$

where π_1 includes the dependent variable and the remaining π-terms include only independent variables, as in Eq. 14.4.

Further, it is noted that a requirement for a successful application of dimensional analysis is that a dimension must occur at least twice or not at all. For example, the equation $\Delta p = f(V, l, d)$ is ill-stated since pressure involves the dimensions of force and $V, l,$ and d do not contain such a dimension.

Key Concept: *A dimension must occur at least twice or not at all.*

The procedure used in applying the π-theorem is summarized as follows:

1. Write the functional form of the dependent variable depending on the $(n-1)$ independent variables. This step requires knowledge of the phenomenon being studied. All variables that effect the dependent variable must be included. These include geometric variables, fluid properties, and external effects that influence the variable being studied. Quantities that have no influence on the dependent variable should not be included. Also, do not include variables that depend on each other; e.g., both radius and area would not be included. The variables on the right-hand side of Eq. 14.9 should be independent.
2. Identify m **repeating variables**, variables that will be combined with each remaining variable to form the π-terms. The repeating variables selected from the independent variables must include all of the basic dimensions, but they must not form a π-term by themselves. An angle cannot be a repeating variable since it is dimensionless and forms a π-term itself.
3. Form the π-terms by combining the repeating variables with each of the remaining variables.
4. Write the functional form of the $(n-m)$ dimensionless π-terms.

Repeating variables: *Variables that will be combined with each remaining variable to form the π-terms.*

Step 3 can be accomplished by a relatively simple algebraic procedure; we will also illustrate a procedure in the examples that utilizes simple observation.

The algebraic procedure will now be illustrated with an example. Suppose that we desire to combine the variables surface tension σ, velocity V, density ρ, and length l into a π-term; this can be written as

$$\pi = \sigma^a V^b \rho^c l^d \qquad (14.11)$$

The objective is to determine $a, b, c,$ and d so that the grouping is dimensionless. In terms of dimensions, Eq. 14.11 is

$$M^0 L^0 T^0 = \left(\frac{M}{T^2}\right)^a \left(\frac{L}{T}\right)^b \left(\frac{M}{L^3}\right)^c L^d \qquad (14.12)$$

[1]There are situations where m is less than the number of basic dimensions. Example 14.2 will illustrate.

Equating exponents on each of the basic dimensions:

$$\begin{aligned} M:& \quad 0 = a + c \\ L:& \quad 0 = b - 3c + d \\ T:& \quad 0 = -2a - b \end{aligned} \tag{14.13}$$

The three algebraic equations are solved simultaneously to yield

$$a = -c \qquad b = 2c \qquad d = c \tag{14.14}$$

so that the π-term becomes

$$\pi = \left(\frac{\rho l V^2}{\sigma}\right)^c \tag{14.15}$$

Key Concept: *A dimensionless parameter raised to any power remains dimensionless.*

A dimensionless parameter raised to any power remains dimensionless; consequently, we can select c to be any number other than zero. It is usually selected as $c = 1$, depending on the ratio desired. Selecting $c = 1$, the π-term is

$$\pi = \frac{\rho l V^2}{\sigma} \tag{14.16}$$

Actually, we could have selected $c = 1$ in Eq. 14.11 and proceeded with only three unknowns. Or if it were desired to have σ in the numerator to the first power, we could have set $a = 1$ and let b, c, and d be unknowns.

A final note: If only one π-term results, the functional form would state that the π-term must be a constant since the right-hand side of Eq. 14.10 would contain no additional π-terms. This would result in an expression that includes an arbitrary constant that could be determined through analysis or experimentation.

Example 14.1

The drag force F_D on a cylinder of diameter d and length l is to be studied. What functional form relates the dimensionless variables if a fluid with velocity V flows normal to the cylinder?

Solution

First, we must determine the variables that have some influence on the drag force. If we include variables that do not influence the drag force, we would have additional π-terms that experimentation would show to be unimportant; if we do not include a variable that does influence the drag force, experimentation would also reveal that problem. Experience is essential in choosing the correct variables; in this example we will include as influential variables the free stream velocity V, the viscosity μ, the density ρ of the fluid, in addition to the diameter d and the length l of the cylinder, resulting in $n = 6$ variables. This is written as

$$F_D = f(d, l, V, \mu, \rho)$$

(continued)

The variables are observed to include $m = 3$ dimensions (consult Table 14.1 for the dimensions):

$$[F_D] = \frac{ML}{T^2} \qquad [V] = \frac{L}{T} \qquad [\mu] = \frac{M}{LT} \qquad [d] = L \qquad [l] = L \qquad [\rho] = \frac{M}{L^3}$$

Consequently, we can expect $n - m = 6 - 3 = 3$ π-terms.

We choose repeating variables with the simplest combinations of dimensions such that they do not form a π-term by themselves (we could not include d and l as repeating variables); the repeating variables are chosen to be d, V, and ρ. These three variables are combined with each of the remaining variables to form the π-terms. Rather than writing equations similar to Eq. 14.11 for the π-terms, let us form the π-terms by inspection. When the repeating variables are combined with F_D we observe that only F_D and ρ have the mass dimension; hence F_D must be divided by ρ. Only F_D and V have the time dimension; hence, F_D must be divided by V^2. Thus F_D divided by ρ has L^4 in the numerator; when divided by V^2 this results in L^2 remaining in the numerator. Hence we must have d^2 in the denominator resulting in

$$\pi_1 = \frac{F_D}{\rho V^2 d^2}$$

When d, V, and ρ are combined with l there obviously results

$$\pi_2 = \frac{l}{d}$$

The last π-term results from combining μ with d, V, and ρ. The mass dimension disappears if we divide μ by ρ. The time dimension disappears if we divide μ by V. This leaves one length dimension in the numerator; hence d is needed in the denominator resulting in

$$\pi_3 = \frac{\mu}{\rho V d}$$

The dimensionless, functional relationship relating the π-terms is

$$\pi_1 = f_1(\pi_2, \pi_3) \qquad \text{or} \qquad \frac{F_D}{\rho V^2 d^2} = f_1\left(\frac{l}{d}, \frac{\mu}{\rho V d}\right)$$

Rather than the original relationship of six variables we have reduced the problem to one involving three π-terms, a much simpler problem. To determine the particular form of the functional relationship above, we would actually have to solve the problem; experimentation would be needed if analytical or numerical methods were not available. This is often the case in fluid mechanics.

Note that we could have included several additional variables in our original list, such as gravity g, the angle θ that the velocity makes with the cylinder, and the roughness e of the cylinder surface. To not include variables that are significant, or to include variables that are not significant is a matter of experience. The novice must learn how to identify significant variables; however, even the experienced researcher is often at a loss to correlate certain phenomena; much experimentation is often needed to discover the appropriate parameters.

Example 14.2

The rise of liquid in a capillary tube is to be studied. It is anticipated that the rise h will depend on surface tension σ, tube diameter d, liquid specific weight γ, and angle β of attachment between the liquid and tube. Write the functional form of the dimensionless variables.

Solution

The expression relating the variables is

$$h = f(\sigma, d, \gamma, \beta)$$

The dimensions of the variables are

$$[h] = L \qquad [\gamma] = \frac{M}{L^2T^2} \qquad [\beta] = 1 \text{ (dimensionless)} \qquad [\sigma] = \frac{M}{T^2} \qquad [d] = L$$

By observation we see that M/T^2 occurs as that combination in both σ and γ; hence M and T are not independent dimensions in this problem. There are only two independent groupings of basic dimensions, L and M/T^2. Thus $m = 2$ and we choose σ and d as the repeating variables. When combined with h, the first π-term is

$$\pi_1 = \frac{h}{d}$$

When σ and d are combined with γ, the second π-term is

$$\pi_2 = \frac{\gamma d^2}{\sigma}$$

Finally, since the angle β is dimensionless, it forms a π-term by itself; that is,

$$\pi_3 = \beta$$

The final functional form relating the π-terms is

$$\pi_1 = f_1(\pi_2, \pi_3) \qquad \text{or} \qquad \frac{h}{d} = f_1\left(\frac{\gamma d^2}{\sigma}, \beta\right)$$

Note: In this example we could not have chosen the angle β as a repeating variable since it already is a dimensionless π-term. Also, we could not have chosen three repeating variables since M and T were not independent.

Also, note that we may have thought that gravity should have been included in the problem. If it had been included above, it would not have appeared in any of the π-terms, indicating that it should not have been included. If density and gravity, rather than specific weight, had been included the relationship above would have resulted since $\gamma = \rho g$; this, by the way, would have avoided the necessity of observing that M/T^2 was a dimensional grouping.

A final note regarding the functional form of the π-terms: The relationship above could equally have been written as

$$\frac{h}{d} = f_1\left(\frac{\sigma}{\gamma d^2}, \beta\right)$$

Also, occasionally a different set of repeating variables could be selected. This simply expresses the final functional equation in a different but equivalent form. Actually, a second form can be shown to be a combination of the π-terms from an initial form.

14.2.4 Common Dimensionless Parameters

Consider a relatively general relationship between the pressure drop Δp, a characteristic length l, a characteristic velocity V, the density ρ, the viscosity μ, the gravity g, the surface tension σ, the speed of sound c, and an angular velocity Ω, written as

$$\Delta p = f(l, V, \rho, \mu, g, c, \Omega, \sigma) \tag{14.17}$$

The π-theorem applied to this problem, with l, V, and ρ as repeating variables, results in

$$\frac{\Delta p}{\rho V^2} = f_1\left(\frac{V\rho l}{\mu}, \frac{V^2}{lg}, \frac{V}{c}, \frac{l\Omega}{V}, \frac{V^2\rho l}{\sigma}\right) \tag{14.18}$$

Each of the π-terms in this expression is a common dimensionless parameter that appears in numerous fluid flow situations. They are identified as follows:

$$\begin{aligned}
\text{Euler number, Eu} &= \frac{\Delta p}{\rho V^2} \\
\text{Reynolds number, Re} &= \frac{V\rho l}{\mu} \\
\text{Froude number, Fr} &= \frac{V}{\sqrt{lg}} \\
\text{Mach number, M} &= \frac{V}{c} \\
\text{Strouhal number, St} &= \frac{l\Omega}{V} \\
\text{Weber number, We} &= \frac{V^2 l\rho}{\sigma}
\end{aligned} \tag{14.19}$$

The physical significance of each parameter can be determined by observing that each dimensionless number can be written as the ratio of two forces. The forces are observed to be

$$\begin{aligned}
F_P &= \text{pressure force} = \Delta p A \sim \Delta p l^2 \\
F_I &= \text{inertial force} = mV\frac{dV}{ds} \sim \rho l^3 V\frac{V}{l} = \rho l^2 V^2 \\
F_\mu &= \text{viscous force} = \tau A = \mu\frac{du}{dy}A \sim \mu\frac{V}{l}l^2 = \mu l V \\
F_g &= \text{gravity force} = mg \sim \rho l^3 g \\
F_B &= \text{compressibility force} = BA \sim \rho\frac{dp}{d\rho}l^2 = \rho c^2 l^2 \\
F_\omega &= \text{centrifugal force} = mr\Omega^2 \sim \rho l^3 l\Omega^2 = \rho l^4\Omega^2 \\
F_\sigma &= \text{surface tension force} = \sigma l
\end{aligned} \tag{14.20}$$

where the symbol $\sim$ means "proportional to." The area A is proportional to a length squared. Thus we see that

$$\begin{aligned}
\text{Eu} &\propto \frac{\text{pressure force}}{\text{inertial force}} \\
\text{Re} &\propto \frac{\text{inertial force}}{\text{viscous force}} \\
\text{Fr} &\propto \frac{\text{inertial force}}{\text{gravity force}} \\
\text{M} &\propto \frac{\text{inertial force}}{\text{compressibility force}} \\
\text{St} &\propto \frac{\text{centrifugal force}}{\text{inertial force}} \\
\text{We} &\propto \frac{\text{inertial force}}{\text{surface tension force}}
\end{aligned} \qquad (14.21)$$

Key Concept: *The ratios of forces allows us to anticipate the significant parameters in a flow.*

Thinking of the dimensionless parameters in terms of the ratios of forces allows us to anticipate the significant parameters in a particular flow of interest. If viscous forces are important, such as in the pipe flow of Fig. 12.8 or the boundary layer flow of Fig. 12.10, we know that the Reynolds number is a significant dimensionless parameter. If surface tension forces are instrumental in affecting the flow, as in droplet formation or flow over a weir with a small head, we expect the Weber number to be important. Similar analysis can be applied to other fluid flow situations.

Obviously, all of the effects included in the general relationship (14.18) would not be of interest in any one situation. It would be very unlikely that both compressibility effects and surface tension effects would influence a flow simultaneously. In addition, there is often more than one length of importance, thereby introducing additional geometric, dimensionless ratios. We have, however, introduced the more common dimensionless flow parameters of interest in fluid mechanics. Table 14.2 summarizes this section.

TABLE 14.2 Common Dimensionless Parameters in Fluid Mechanics

Parameter	Expression	Flow situations where parameter is important
Euler number	$\frac{\Delta p}{\rho V^2}$	Flows in which pressure drop is significant: most flow situations
Reynolds number	$\frac{\rho l V}{\mu}$	Flows that are influenced by viscous effects: internal flows, boundary layer flows
Froude number	$\frac{V}{\sqrt{lg}}$	Flows that are influenced by gravity: primarily free surface flows
Mach number	$\frac{V}{c}$	Compressibility is important in these flows, usually if $V > 0.3c$
Strouhal number	$\frac{l\Omega}{V}$	Flow with an unsteady component that repeats itself periodically
Weber number	$\rho\frac{V^2 l}{\sigma}$	Surface tension influences the flow; flow with an interface may be such a flow

14.3 SIMILITUDE

14.3.1 General Information

As stated in the introduction, *similitude* is the study of predicting prototype conditions from model observations. When an analytical or numerical solution is not practical, or when calculations are based on a simplified model so that uncertainty is introduced, it is usually advisable to perform tests on a model if testing is not practical on a full-scale prototype, be it too large or too small.

If it is decided that a model study is to be performed, it is necessary to develop the means whereby a quantity measured on the model can be used to predict the associated quantity on the prototype. We can develop such a means if we have **dynamic similarity** between model and prototype, that is, if the forces which act on corresponding masses in the model flow and the prototype flow are in the same ratio throughout the entire flow fields. Suppose that inertial forces, pressure forces, viscous forces, and gravity forces are present; then dynamic similarity requires that, at corresponding points in the flow fields,

Dynamic similarity: *Forces which act on corresponding masses in the model flow and prototype flow are in the same ratio throughout the entire flows.*

$$\frac{(F_I)_m}{(F_I)_p} = \frac{(F_P)_m}{(F_P)_p} = \frac{(F_\mu)_m}{(F_\mu)_p} = \frac{(F_g)_m}{(F_g)_p} = \text{const.} \tag{14.22}$$

These can be rearranged to read

$$\left(\frac{F_I}{F_P}\right)_m = \left(\frac{F_I}{F_P}\right)_p \qquad \left(\frac{F_I}{F_\mu}\right)_m = \left(\frac{F_I}{F_\mu}\right)_p \qquad \left(\frac{F_I}{F_g}\right)_m = \left(\frac{F_I}{F_g}\right)_p \tag{14.23}$$

which, in the preceding section, have been shown to be

$$\text{Eu}_m = \text{Eu}_p \qquad \text{Re}_m = \text{Re}_p \qquad \text{Fr}_m = \text{Fr}_p \tag{14.24}$$

If the forces above were the only ones present, we could write

$$F_I = f(F_P, F_\mu, F_g) \tag{14.25}$$

Recognizing that there is only one basic dimension, namely force, dimensional analysis would allow us to write (see Eq. 14.10) the equation above in terms of force ratios or

$$\text{Eu} = f(\text{Re}, \text{Fr}) \tag{14.26}$$

Hence we could conclude that if the Reynolds number and the Froude number are the same on the model and prototype, the Euler number must also be the same. Thus dynamic similarity between model and prototype is guaranteed by equating the Reynolds number and the Froude number of the model to those on the prototype, respectively. If compressibility forces were included here, the analysis above would result in the Mach number being included in Eq. 14.26.

We can write the inertial force ratio as

$$\frac{(F_I)_m}{(F_I)_p} = \frac{a_m m_m}{a_p m_p} = \text{const.} \tag{14.27}$$

showing that the acceleration ratio between corresponding points on the model and prototype is a constant provided that the mass ratio of corresponding fluid elements is a constant. We can write the acceleration ratio as

$$\frac{a_m}{a_p} = \frac{V_m^2/l_m}{V_p^2/l_p} = \text{const.} \tag{14.28}$$

Kinematic similarity: *Condition where the velocity ratio is a constant between all corresponding points in the flows.*

Key Concept: *The streamline pattern around the model is the same as that around the prototype.*

Geometric similarity: *A condition where the model has the same shape as the prototype.*

showing that the velocity ratio between corresponding points is a constant providing the length ratio is a constant. The velocity ratio being a constant between all corresponding points in the flow fields is the statement of **kinematic similarity**. This would result in the streamline pattern around the model being the same as that around the prototype except for a scale factor. The length ratio being constant between all corresponding points in the flow fields is the demand of **geometric similarity** which results in the model having the same shape as the prototype. Hence, to ensure complete similarity between model and prototype, we demand that:

- Geometric similarity be satisfied
- The mass ratio of corresponding fluid elements be a constant
- The appropriate dimensionless parameters of Eq. 14.18 be equal

Assuming that complete similarity between model and prototype exists, we can now predict quantities of interest on a prototype from measurements on a model. If we measure a drag force F_D on a model and wish to predict the corresponding drag on the prototype, we could equate the ratio of the drag forces to the ratio of the inertial forces (see the inertial force of Eq. 14.20) as

$$\frac{(F_D)_m}{(F_D)_p} = \frac{(F_I)_m}{(F_I)_p} = \frac{\rho_m V_m^2 l_m^2}{\rho_p V_p^2 l_p^2} \tag{14.29}$$

If we measure the power input to a model and wish to predict the power requirement of the prototype, we would recognize that power is force times velocity and write

$$\frac{\dot{W}_m}{\dot{W}_p} = \frac{(F_I)_m V_m}{(F_I)_p V_p} = \frac{\rho_m V_m^2 l_m^2 V_m}{\rho_p V_p^2 l_p^2 V_p} \tag{14.30}$$

Hence we can predict a prototype quantity if we select the model fluid (this provides ρ_m/ρ_p), the scale ratio (this provides l_m/l_p), and the appropriate dimensionless number from Table 14.2 (this provides V_m/V_p). Examples will illustrate.

14.3.2 Confined Flows

A confined flow is a flow that has no free surfaces (a liquid-gas surface) or interfaces (two different liquids forming an interface). It is confined to move within a specified region; such flows include external flows around objects, such as aircraft, buildings, and submarines, as well as internal flows in pipes and conduits.

Gravity does not influence the flow pattern in confined flows; that is, if gravity could be changed in magnitude, the flow pattern and associated flow quantities would not change. The dominant effect is that of viscosity in incompressible confined flows (all liquid flows and gas flows in which $M < 0.3$). Surface tension is obviously not a factor, as it would be in bubble formation, and for steady flows there would be no unsteady effects due to oscillations in the flow. The three relevant forces are pressure forces, inertial forces, and viscous forces. Therefore, in confined flows dynamic similarity is achieved if the ratios of viscous forces, inertial forces, and pressure forces between model and prototype are the same. This leads to the conclusion (see Eq. 14.26) that $\text{Eu} = f(\text{Re})$, so that it is only necessary to consider the Reynolds number as the dominant dimensionless parameter in a confined incompressible flow. If compressibility effects are significant, the Mach number would also be important.

Key Concept: *Gravity does not influence the flow pattern in confined flows.*

Key Concept: *The Reynolds number is the dominant dimensionless parameter in a confined incompressible flow.*

14.3.3 Free-Surface Flows

A free-surface flow is one in which part of the boundary involves a pressure boundary condition. This includes flows over weirs and dams, as shown in Fig. 14.5, flows in channels and spillways, flows involving two fluids separated by an interface, and flows around floating objects with waves and around submerged objects with cavitation present. In all of these flows the location of the free surface is unknown and the velocity at the free surface is unknown; it is the pressure that must be the same[2] on either side of the interface. In free-surface flows, gravity controls both the location and the motion of the free surface. This introduces the Froude number because of the influence of the gravity forces. If we consider flows that do not exhibit periodic motions, which have negligible surface tension and compressibility effects, we can ignore the influence of St, M, and We. That leaves only the viscous effects to be considered. There are many free-surface flows in which viscous effects are significant. Consider, however, that in most model studies water is the only economical fluid to use; if the prototype fluid is also water, as it often is, we would find from the Froude numbers

Key Concept: *We use the Froude number when modeling a free-surface flow.*

$$\frac{V_m^2}{l_m g_m} = \frac{V_p^2}{l_p g_p} \qquad \therefore \quad \frac{V_m}{V_p} = \left(\frac{l_m}{l_p}\right)^{1/2} \tag{14.31}$$

[2]The surface tension, if significant, results in a pressure difference across the interface.

FIGURE 14.5 Model of the Bonneville lock and dam on the Columbia River. (Courtesy of the U.S. Army Corps of Engineers Waterways Experiment Station.)

assuming that $g_m = g_p$. Matching the Reynolds numbers (using $\nu_m = \nu_p$):

$$\frac{V_m l_m}{\nu_m} = \frac{V_p l_p}{\nu_p} \qquad \therefore \frac{V_m}{V_p} = \frac{l_p}{l_m} \tag{14.32}$$

Thus we have a conflict. If we use the same fluid in the model study as in the prototype flow, we cannot satisfy both the Froude number criterion and the Reynolds number criterion. If we demand that both criteria be satisfied by using different fluids for model and prototype ($\nu_m \neq \nu_p$), we must select a model fluid with a viscosity $\nu_m = \nu_p(l_m/l_p)^{3/2}$ (this results from a matching of the Froude and Reynolds numbers). A fluid with this viscosity is probably either an impossibility or an impracticality. Hence, when modeling free-surface flows in which viscous effects are important, we equate the Froude numbers and include viscous effects by some other technique. For example, if we measured the total drag on the model of a ship, we would approximate the viscous drag (using some technique not included here) and subtract it from the total drag, leaving the drag due to wave resistance. The wave drag on the prototype would then be predicted using similitude, and the approximated viscous drag would be added to the wave drag, yielding the expected drag on the ship. For the better designs the viscous drag on ship hulls can be of the same order of magnitude as the wave drag.

Example 14.3

A test is to be performed on a proposed design for a large pump that is to deliver 1.5 m^3/s of water from a 40-cm-diameter impeller with a pressure rise of 400 kPa. A model with an 8-cm-diameter impeller is to be used. What flow rate should be used and what pressure rise is to be expected? The model fluid is water at the same temperature as the water in the prototype.

Solution

For similarity to exist in this confined incompressible flow problem, the Reynolds numbers must be equal, that is,

$$\text{Re}_m = \text{Re}_p$$

$$\frac{V_m d_m}{\nu_m} = \frac{V_p d_p}{\nu_p}$$

Recognizing that $\nu_m = \nu_p$ if the temperatures are equal, we see that

$$\frac{V_m}{V_p} = \frac{d_p}{d_m}$$

$$= \frac{0.4}{0.08} = 5$$

The ratio of flow rates is found recognizing that $Q_f = VA$:

$$\frac{(Q_f)_m}{(Q_f)_p} = \frac{V_m d_m^2}{V_p d_p^2}$$

$$= 5 \times \left(\frac{1}{5}\right)^2 = \frac{1}{5}$$

Thus we find that

$$(Q_f)_m = \frac{(Q_f)_p}{5} = \frac{1.5}{5} = 0.3 \text{ m}^3/\text{s}$$

The dimensionless pressure rise is found using the Euler number:

$$\left(\frac{\Delta p}{\rho V^2}\right)_m = \left(\frac{\Delta p}{\rho V^2}\right)_p$$

Hence the pressure rise for the model is

$$\Delta p_m = \Delta p_p \frac{\rho_m}{\rho_p} \frac{V_m^2}{V_p^2}$$

$$= 400 \times 1 \times 5^2 = 10\,000 \text{ kPa}$$

Note that in this example we see that the velocity in the model is equal to the velocity in the prototype multiplied by the length ratio, and the pressure rise in the model is equal to the pressure rise in the prototype multiplied by the length ratio squared. If the length ratio were very large, it is obvious that to maintain Reynolds number equivalence would be quite difficult, indeed. This observation is discussed in more detail in Section 14.3.4.

Example 14.4

A 1:20 scale model of a surface vessel is used to test the influence of a proposed design on the wave drag. A wave drag of 6.2 lb is measured at a model speed of 8.0 ft/sec. What speed does this correspond to on the prototype, and what wave drag is predicted for the prototype? Neglect viscous effects, and assume the same fluid for model and prototype.

Solution

The Froude number must be equated for both model and prototype. Thus

$$\text{Fr}_m = \text{Fr}_p \qquad \frac{V_m}{\sqrt{l_m g}} = \frac{V_p}{\sqrt{l_p g}}$$

This yields, recognizing that g does not vary significantly on the surface of the earth,

$$V_p = V_m \left(\frac{l_p}{l_m}\right)^{1/2} = 8.0\sqrt{20} = 35.8 \text{ ft/sec}$$

To find the wave drag on the prototype, we equate the drag ratio to the inertia force ratio:

$$\frac{(F_D)_m}{(F_D)_p} = \frac{\rho_m V_m^2 l_m^2}{\rho_p V_p^2 l_p^2}$$

This allows us to calculate the wave drag on the prototype as, using $\rho_p = \rho_m$,

$$(F_D)_p = (F_D)_m \frac{\rho_p V_p^2 l_p^2}{\rho_m V_m^2 l_m^2}$$

$$= 6.2 \times \frac{35.8^2}{8^2} \times 20^2 = 49,700 \text{ lb}$$

Note: We could have used the gravity force ratio rather than the inertial force ratio, but we could not have used the viscous force ratio since viscous forces were assumed negligible.

14.3.4 High-Reynolds-Number Flows

Key Concept: *When the Reynolds numbers are matched the model velocity is often prohibitively large.*

In a confined flow in which the Reynolds number is the dimensionless parameter that guarantees dynamic similarity, we note that if the same fluid is used in model and prototype, the velocity in the model study is $V_m = V_p l_p / l_m$; the velocity in the model is the velocity in the prototype multiplied by the scale factor. This often results in velocities that are prohibitively large in the model study. Also, the pressures encountered in the model study are large, as shown in Example 14.3, and the energy consumption is also very large. Because of these problems the Reynolds numbers may not be matched in studies involving large Reynolds numbers.

There is, however, some justification for not matching the Reynolds number in model studies. Consider a typical drag coefficient C_D versus Reynolds number curve, as shown in Fig. 14.6 (the complete curve is presented in Fig. 16.9). The drag coefficient is a dimensionless drag, defined as $C_D = \text{drag}/\frac{1}{2}\rho V^2 A$. At a sufficiently high Reynolds number, between 10^3 and 10^5, the flow is insensitive to changes in the Reynolds number; note that the drag coefficient is essentially con-

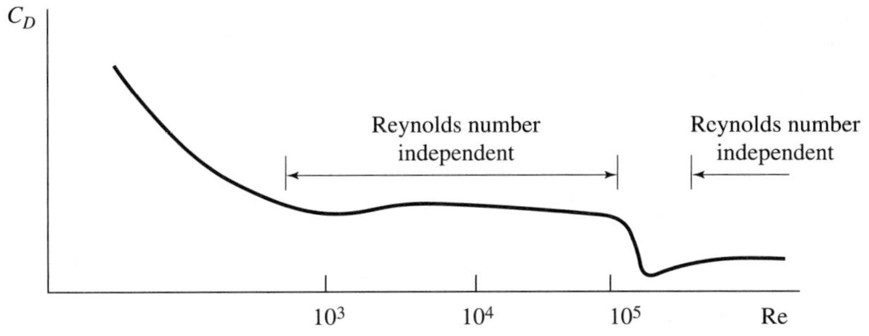

FIGURE 14.6 Drag coefficient versus Reynolds number.

stant and independent of Re. That implies that the flow field is similar at Re $= 10^3$ to that at Re $= 10^5$. Thus if $\text{Re}_p = 10^5$, it is only necessary that $10^3 < \text{Re}_m < 10^5$ for the viscous effects to have the same effect on model and prototype. This often allows another parameter of interest to be matched, such as the Froude number or the Mach number. There are, however, high-Reynolds-number flows in which compressibility effects and free-surface effects are negligible, so that neither the Froude number nor the Mach number are applicable. Examples include flow around automobiles, large smoke stacks, and dirigibles. For such flows we must only ensure that the Reynolds number be within the range where the drag coefficient is constant. We should note that for Reynolds numbers quite large (Re $>$ 5×10^5 in Fig. 14.6) the flow may also become Reynolds number independent. If that is the case, it is only necessary that Re_m be sufficiently large.

Example 14.5

A 1:10 scale model of an automobile is used to measure the drag on a proposed design. It is to simulate a prototype speed of 90 km/h. What speed should be used in the wind tunnel if Reynolds numbers are equated? For this condition, what is the ratio of drag forces?

Solution

The same fluid exists on model and prototype, thus equating the Reynolds numbers results in

$$\frac{V_m l_m}{\nu_m} = \frac{V_p l_p}{\nu_p} \qquad \therefore\ V_m = V_p \frac{l_p}{l_m}$$

$$= 90 \times 10 = 900 \text{ km/h}$$

This speed would, of course, introduce compressibility effects, effects that do not exist in the prototype. Hence the proposed model study would be inappropriate.

If we did use this velocity in the model, the drag force ratio would be

$$\frac{(F_D)_p}{(F_D)_m} = \frac{\rho_p V_p^2 l_p^2}{\rho_m V_m^2 l_m^2} \qquad \therefore\ \frac{(F_D)_p}{(F_D)_m} = 1$$

Thus we see that the drag force on the model is the same as the drag force on the prototype if the same fluids are used when we equate Reynolds numbers.

Example 14.6

In Example 14.5, if the Reynolds numbers were equated, the velocity in the model study was observed to be in the compressible flow regime (i.e., M > 0.3 or $V_m > 360$ km/h). To conduct an acceptable model study, could we use a velocity of 90 km/h on a model with a characteristic length of 10 cm? Assume that the drag coefficient ($C_D = F_D/\frac{1}{2}\rho V^2 A$, where A is the projected area), is independent of Re for Re > 10^5. If so, what drag force on the prototype would correspond to a drag force of 1.2 N measured on the model?

Solution

The proposed model study in a wind tunnel is to be conducted with $V_m = 90$ km/h and $l_m = 0.1$ m. Using $\nu = 1.6 \times 10^{-5}$ m²/s, the Reynolds number is

$$\text{Re}_m = \frac{V_m l_m}{\nu_m} = \frac{\dfrac{90 \times 1000}{3600} \times 0.1}{1.6 \times 10^{-5}} = 1.56 \times 10^5$$

This Reynolds number is greater than 10^5, so we will assume that similarity exists between model and prototype. The velocity of 90 km/h is sufficiently high.

The drag force on the prototype traveling at 90 km/h corresponding to 1.2 N on the model is found from

$$\frac{(F_D)_p}{(F_D)_m} = \frac{\rho_p V_p^2 l_p^2}{\rho_m V_m^2 l_m^2} \qquad \therefore\ (F_D)_p = (F_D)_m \overset{1}{\frac{\rho_p}{\rho_m}} \overset{1}{\frac{V_p^2}{V_m^2}} \frac{l_p^2}{l_m^2} = 1.2 \times 10^2 = 120 \text{ N}$$

Note that in this example we have assumed that the drag coefficient is independent of Re for Re > 10^5. If the drag coefficient continued to vary above Re = 10^5 (this would be evident from experimental data), the foregoing analysis would have to be modified accordingly.

14.3.5 Compressible Flows

For most compressible flow situations the Reynolds number is so large (refer to Fig. 14.6) that it is not a parameter of significance; the compressibility effects lead to the Mach number as the primary dimensionless parameter for model studies. Thus for a particular model study we require

$$\text{M}_m = \text{M}_p \qquad \text{or} \qquad \frac{V_m}{c_m} = \frac{V_p}{c_p} \tag{14.33}$$

If the model study is carried out in a wind tunnel and the prototype fluid is air, we can assume that $c_m \simeq c_p$ if the temperature is the same in the two flows. For such a case, the velocity in the model study is equal to the velocity associated with the prototype. Of course, if the speeds of sound are different, the velocity ratio will be different from unity, accordingly.

Example 14.7

The pressure rise from free stream to the nose of a fusilage section of an aircraft is measured in a wind tunnel at 20°C to be 34 kPa with a wind-tunnel airspeed of 900 km/h. If the test is to simulate flight at an elevation of 12 km, what is the prototype velocity and the expected nose pressure rise?

Solution

To find the prototype velocity corresponding to a wind-tunnel airspeed of 900 km/h, we equate the Mach numbers

$$\mathrm{M}_m = \mathrm{M}_p \qquad \text{or} \qquad \frac{V_m}{\sqrt{kRT_m}} = \frac{V_p}{\sqrt{kRT_p}}$$

Thus

$$V_p = V_m \left(\frac{kRT_p}{kRT_m} \right)^{1/2} = 900 \left(\frac{216.7}{293} \right)^{1/2} = 774 \text{ km/h}$$

The pressure at the nose of the prototype fusilage is found using the Euler number as follows:

$$\frac{\Delta p_m}{\rho_m V_m^2} = \frac{\Delta p_p}{\rho_p V_p^2}$$

$$\therefore \Delta p_p = \Delta p_m \frac{\rho_p}{\rho_m} \frac{V_p^2}{V_m^2}$$

$$= 34 \times \frac{0.3119}{1.225} \times \frac{774^2}{900^2} = 6.4 \text{ kPa}$$

The densities and temperature T_p were found in Appendix B.1.

14.3.6 Periodic Flows

In many flow situations there are regions of the flows in which periodic motions occur. Such flows include the periodic fluid motion (in Section 16.3.2 this is called vortex shedding) that takes place when a fluid flows past a cylindrical object such as a bridge, a TV tower, a cable, or a tall building; the flow past a wind machine; and the flow through turbomachinery. In flows such as these it is necessary to equate the Strouhal numbers, which can be written as

$$\frac{V_m}{\Omega_m l_m} = \frac{V_p}{\Omega_p l_p} \tag{14.34}$$

in order that the periodic motion be properly modeled.

In addition to the Strouhal number, there may be additional dimensionless parameters that must be equated: in viscous flows, the Reynolds number; in free-surface flows, the Froude number; and in compressible flows, the Mach number.

Example 14.8

A large wind turbine, designed to operate at 50 km/h, is to be tested in a laboratory by constructing a 1:15 scale model. What airspeed should be used in the wind tunnel, what angular velocity should be used to simulate a prototype speed of 5 rpm, and what power output is expected from the model if the prototype output is designed to be 500 kW?

Solution

The speed in the wind tunnel can be any speed above that needed to provide a sufficiently large Reynolds number. Let us select the same speed with which the prototype is to operate, namely, 50 km/h, and calculate the minimum characteristic length that a Reynolds number of 10^5 would demand; this gives

$$\text{Re} = \frac{Vl}{\nu} \qquad 10^5 = \frac{(50 \times 1000/3600) \times l}{1.6 \times 10^{-5}} \qquad \therefore l = 0.12 \text{ m}$$

Obviously, in a reasonably large wind tunnel we can maintain a characteristic length (e.g., the blade length) that large.

The angular velocity is found by equating the Strouhal numbers. There results

$$\frac{V_m}{\Omega_m l_m} = \frac{V_p}{\Omega_p l_p} \qquad \therefore \Omega_m = \Omega_p \overset{1}{\cancel{\frac{V_m}{V_p}}} \frac{l_p}{l_m} = 5 \times 1 \times 15 = 75 \text{ rpm}$$

assuming that the wind velocities are equal.

The power is found by observing that power is force times velocity.

$$\frac{\dot{W}_m}{\dot{W}_p} = \frac{\rho_m V_m^3 l_m^2}{\rho_p V_p^3 l_p^2}$$

or

$$\dot{W}_m = \dot{W}_p \frac{\rho_m}{\rho_p} \overset{1}{\cancel{\frac{V_m^3}{V_p^3}}} \frac{l_m^2}{l_p^2} = 500 \times \left(\frac{1}{15}\right)^2 = 2.22 \text{ kW}$$

14.4 SUMMARY

Experimental studies are greatly simplified by reducing the number of variables that influence the phenomenon being studied. We demonstrated how the relationship

$$\Delta p = f(V, \rho, \mu, d, h) \tag{14.35}$$

could be written in the simplified form,

$$\frac{\Delta p}{\rho V^2} = f\left(\frac{V\rho d}{\mu}, \frac{h}{d}\right) \tag{14.36}$$

The most common flow parameters are, using l as the characteristic length,

$$\mathrm{Re} = \frac{V\rho l}{\mu}, \quad \mathrm{Fr} = \frac{V}{\sqrt{lg}}, \quad \mathrm{M} = \frac{V}{c}, \quad \mathrm{Eu} = \frac{\Delta p}{\rho V^2}, \quad \mathrm{St} = \frac{l\Omega}{V}, \quad \mathrm{We} = \frac{V^2 l\rho}{\sigma} \qquad (14.37)$$

The parameters, used to guide model studies, that are of primary significence in a particular flow are identified as follows:

Confined flows: $$\mathrm{Re} = \frac{V\rho l}{\mu} \qquad (14.38)$$

Free-surface flows: $$\mathrm{Fr} = \frac{V}{\sqrt{lg}} \qquad (14.39)$$

High-Reynolds number flows: $$\mathrm{Re} > (\mathrm{Re})_{\text{minimum}} \qquad (14.40)$$

Compressible flows: $$\mathrm{M} = \frac{V}{c} \qquad (14.41)$$

Periodic flows: $$\mathrm{St} = \frac{V}{l\Omega} \qquad (14.42)$$

REVIEW PROBLEMS FOR THE FE/EIT EXAMINATION

14.1FE Combine power $\dot{W}$, diameter d, pressure drop Δp, and average velocity into a dimensionless group.

A. $\dfrac{\dot{W}}{d^2 V\Delta p}$ **B.** $\dfrac{\dot{W}}{V\Delta p}$

C. $\dfrac{\dot{W}d}{V\Delta p}$ **D.** $\dfrac{\dot{W}}{Vd\Delta p}$

14.2FE It is proposed that the velocity of a flow depends on a diameter d, a length l, gravity g, rotational speed Ω, and viscosity μ. Select the variable(s) that would not influence the velocity.

A. μ **B.** Ω

C. g **D.** both d and l

14.3FE What velocity should be selected in a wind tunnel where a 9:1 scale model of an automobile is to simulate a speed of 12 m/s? Neglect compressibility effects.

A. 108 m/s **B.** 12 m/s

C. 4 m/s **D.** 1.33 m/s

14.4FE Flow around an underwater structural component is to be studied in a 20°C-wind tunnel with a 10:1 scale model. What speed should be selected in the wind tunnel to simulate a 10°C-water speed of 4 m/s?

A. 4.61 m/s **B.** 40 m/s

C. 31.6 m/s **D.** 461 m/s

14.5FE What upstream velocity should be selected in a 16:1 scale model of a levee which has an upstream velocity of 2 m/s?

A. 2 m/s **B.** 1 m/s

C. 0.5 m/s **D.** 0.25 m/s

14.6FE A force of 10 N is measured on a 25:1 scale model of a ship tested in a water channel. What force should be expected on the prototype ship? Neglect viscous effects.

A. 156 kN **B.** 62.5 kN

C. 6250 N **D.** 250 N

PROBLEMS

14.1 Write Bernoulli's equation in dimensionless form by dividing Eq. 14.1 by V_1^2 and multiplying by g. Express the equation in a form similar to that of Eq. 14.2.

14.2 If the F-L-T system of units were used, what would the dimensions be on each of the following?
(a) Mass flux
(b) Pressure
(c) Density
(d) Viscosity
(e) Work
(f) Power
(g) Surface tension

Dimensional Analysis

14.3 Verify that the dimensions of power are ML^2/T^3 as listed in Table 14.1.

14.4 If the velocity V in a fluid flow depends on a dimension l, the fluid density ρ, and the viscosity μ, show that this implies that the Reynolds number $Vl\rho/\mu$ is constant.

14.5 If the velocity V in a fluid depends on the surface tension σ, the density ρ, and a diameter d, show that this implies that the Weber number $V^2 d\rho/\sigma$ is constant.

14.6 Assume that the velocity V of fall of an object depends on the height H through which it falls, the gravity g, and the mass m of the object. Find an expression for V.

14.7 Include the density ρ and viscosity μ of the surrounding fluid and repeat Problem 14.6. This would account for the resistance (drag) of the fluid.

14.8 Select l, V, and ρ as the repeating variables in Example 14.1 and find an expression for F_D. Show that this is an equivalent form to that of Example 14.1.

14.9 Select d, μ, and V as the repeating variables in Example 14.1 and find an expression for F_D. Show that this is equivalent to the expression of Example 14.1.

14.10 Include gravity g in the list of variables of Example 14.2 and determine the final expression that results for h.

14.11 Find an expression for the centrifugal force F_c if it depends on the mass m, the angular velocity Ω, and the radius R of an impeller.

14.12 The normal stress σ in a beam depends on the bending moment M, the distance y from the neutral axis, and the moment of inertia I. Relate σ to the other variables if we know that σ varies linearly with y.

14.13 Find an expression for the average velocity in a smooth pipe if it depends on the viscosity, the diameter, and the pressure gradient $\partial p/\partial x$.

14.14 It is suggested that the velocity of the water flowing from a hole in the side of an open tank depends on the height H of the hole from the surface, the gravity, and the density of the water. What expression relates the variables?

14.15 Derive an expression for the velocity of liquid issuing from a hole in the side of an open tank if the velocity depends on the height H of the hole from the free surface, the fluid viscosity and density, gravity, and the hole diameter.

14.16 The pressure drop Δp in the pipe of Fig. P14.16 depends on the average velocity, the pipe diameter, the kinematic viscosity, the length L of pipe, the wall roughness height e, and the fluid density. Find an expression for Δp.

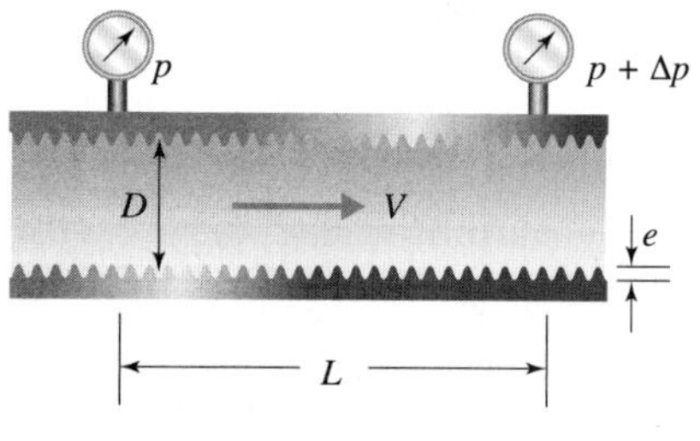

FIGURE P14.16

14.17 Select an appropriate set of variables that influence the drag force F_D on an airfoil (Fig. P14.17) and write the final form in terms of dimensionless parameters.

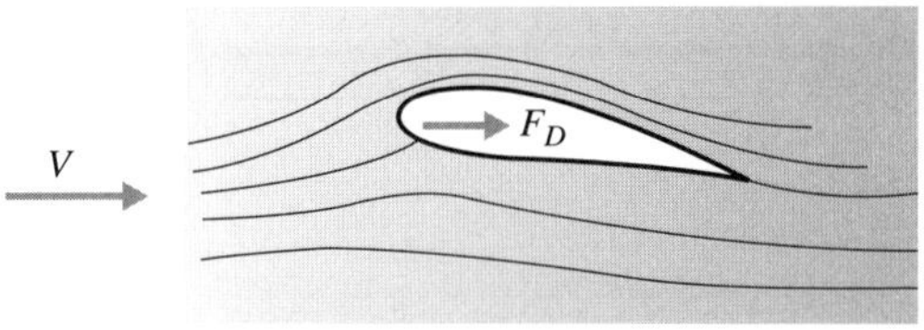

FIGURE P14.17

14.18 The flow rate Q_f in an open channel depends on the hydraulic radius R, the cross-sectional area A, the wall roughness height e, gravity g, and the slope S. Relate Q_f to the other variables using **(a)** the M-L-T system, and **(b)** the F-L-T system.

14.19 The velocity of propagation of ripples in a shallow liquid depends on the liquid depth h, gravity g, surface tension σ, and the liquid density ρ. Find an expression for the propagation velocity V. See Fig. P14.19.

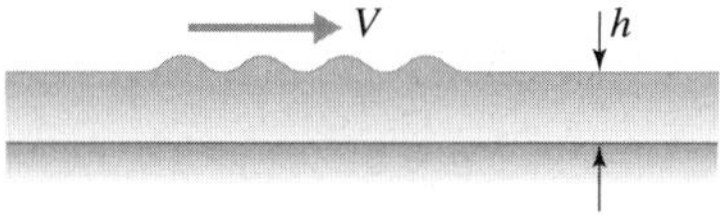

FIGURE P14.19

14.20 The drag force F_D on a sphere depends on the velocity V, the viscosity μ, the density ρ, the surface roughness height e, the freestream fluctuation intensity I (a dimensionless quantity), and the diameter D. Find an expression for F_D.

14.21 The drag force F_D on the smooth sphere of Fig. P14.21 falling in a liquid depends on the constant sphere speed V, the solid density ρ_s, the liquid density ρ and viscosity μ, the sphere diameter D, and gravity g. Find an expression for F_D using **(a)** the M-L-T system, and **(b)** the F-L-T system.

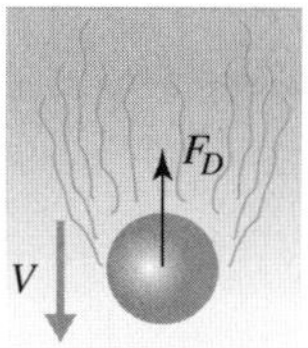

FIGURE P14.21

14.22 The drag force F_D on a golf ball depends on the velocity, the viscosity, the density, the diameter, the dimple depth, the dimple radius, and the concentration C of dimples measured by the number of dimples per unit area. What expression relates F_D to the other variables?

14.23 The frequency f with which vortices are shed from a cylinder depends on the viscosity, density, velocity, and diameter. Derive an expression for f.

14.24 The lift F_L on an airfoil depends on the aircraft speed V, the speed of sound c, the fluid density, the airfoil chord length l_c, the airfoil thickness t, and the angle of attack α. Find an expression for F_L.

14.25 Derive an expression for the torque T needed to rotate a disk of diameter d at the angular velocity Ω in a fluid with density ρ and viscosity μ if the disk is a distance t from a wall. Also, find an expression for the power requirement.

14.26 The cables holding a suspension bridge (Fig. P14.26) are observed to undergo large vibrations under certain wind conditions. Select a set of variables that influence the periodic force acting on a cable and write a simplified relationship of dimensionless parameters.

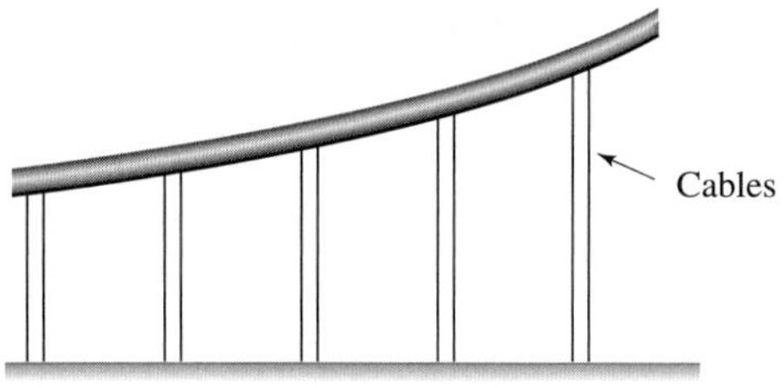

FIGURE P14.26

14.27 A vacuum cleaner creates a pressure drop Δp across its fan. Relate this pressure drop to the impeller diameter D and width h, its rotational speed Ω, the air density ρ, and the inlet and outlet diameters d_i and d_o. Also, find an expression for the power requirement of the fan.

14.28 Derive an expression for the maximum torque needed to rotate an agitator if it depends on the frequency of oscillation, the angular velocity with which the agitator rotates during an oscillation, the diameter of the agitator, the nominal height of the paddles, the length of the paddles, the number of paddles, the depth of liquid, and the liquid density.

14.29 The flow rate of water over a weir depends on the head of water, the width of the weir, gravity, viscosity, density, and surface tension. Relate the flow rate to the other variables.

14.30 The size of droplets in the spray of a fruit sprayer depends on the air velocity, the spray jet velocity, the spray jet exit diameter, the liquid spray surface tension, density, and viscosity, and the air density. Relate the droplet diameter to the other variables.

14.31 Relate the torque T to the other variables shown in Fig. P14.31.

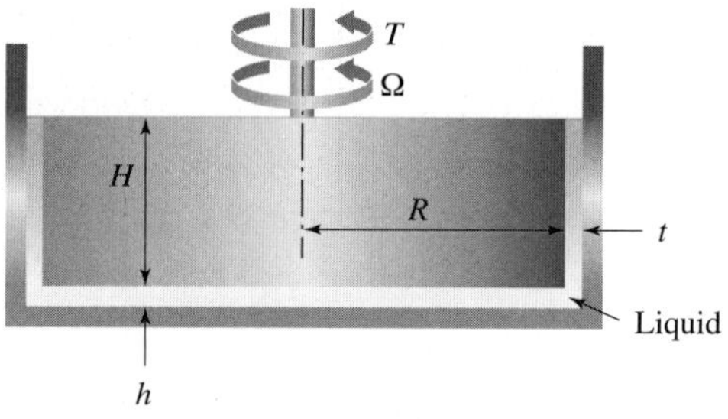

FIGURE P14.31

14.32 A viscometer is composed of an open tank in which the liquid level is held constant. A small-diameter tube empties into a calibrated volume. Find an expression for the viscosity using the relevant parameters.

14.33 Derive an expression for the torque T necessary to rotate the shaft of Fig. P14.33.

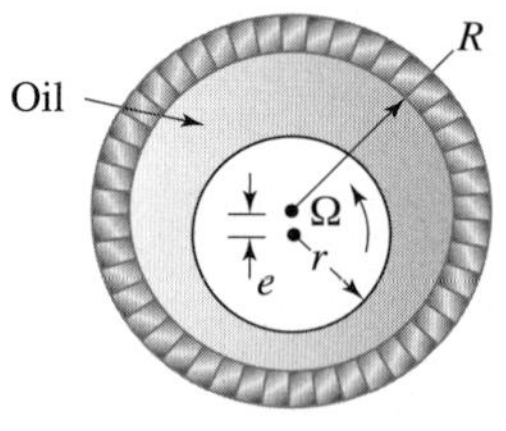

FIGURE P14.33

14.34 Derive an expression for the depth y_2 in the hydraulic jump of Fig. P14.34.

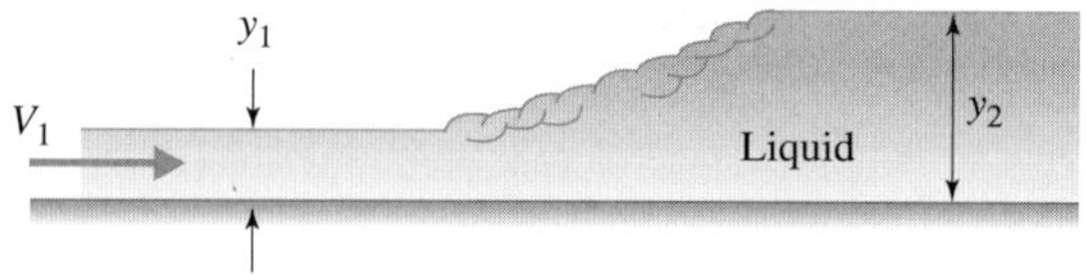

FIGURE P14.34

14.35 Derive an expression for the frequency with which a cylinder, suspended by a string in a liquid flow, oscillates.

Similitude

14.36 After a model study has been performed, it is necessary to predict quantities of interest that are to be expected on the prototype. Write expressions, in terms of density, velocity, and length, for the ratio of model to prototype of each of the following: flow rate Q_f, pressure drop Δp, pressure force F_p, shear stress τ_0, torque T, and heat transfer rate $\dot{Q}$.

14.37 A 1:7 scale model simulates the operation of a large turbine that is to generate 200 kW with a flow rate of 1.5 m^3/s. What flow rate should be used in the model, and what power output is expected?

(a) Water at the same temperature is used in both model and prototype.

(b) The model water is at 25°C and the prototype water is at 10°C.

14.38 A 1:5 scale model of a large pump is used to test a proposed change. The prototype pump produces a pressure rise of 600 kPa at a mass flux of 800 kg/s. Determine the mass flux to be used in the model and the expected pressure rise.

(a) Water at the same temperature is used in both model and prototype.

(b) The water in the model study is at 30°C and the water in the prototype is at 15°C.

14.39 A force on a component of a 1:10 scale model of a large pump is measured to be 10 lb. What force is expected on the prototype component if water is used for both model and prototype:

(a) With the water at the same temperature?

(b) With the prototype water at 50°F and the model water at 70°F?

14.40 A 1:10 scale model study of an automobile is proposed. A prototype speed of 100 km/h is desired. What wind-tunnel speed should be selected for the model study? Comment on the advisability of such a speed selection.

14.41 A 1:10 scale model study of a torpedo is proposed. Prototype speeds of 90 km/h are to be studied. Should a wind tunnel or a water facility be used?

14.42 It is proposed that a model study be carried out on a proposed low-speed airfoil that is to fly at low altitudes at a speed of 50 m/s. If a 1:10 scale model is to be constructed, what velocity should be used in a wind tunnel? Comment as to the advisability of such a test. Would it be more advisable to perform the test in a 20°C water channel? If a water channel study were undertaken, calculate the drag ratio between the model and prototype.

14.43 A model study of oil (SAE-10W) at 30°F flowing in a 2.5-ft-diameter pipe is performed by using water at 70°F. Pipe of what diameter should be used if the average velocities are to be the same? What pressure drop ratio is expected? $S_{\text{oil}} = 0.9$.

14.44 A 0.025-mm-long microorganism moves through water at the rate of 0.1 body length per second. Could a model study be performed for such a prototype in a water channel or a wind tunnel?

14.45 A 1:30 scale model of the ship of Fig. P14.45 is to be tested with complete similarity. What should be the viscosity of the model fluid? Is such a liquid possible? What conclusion can be stated?

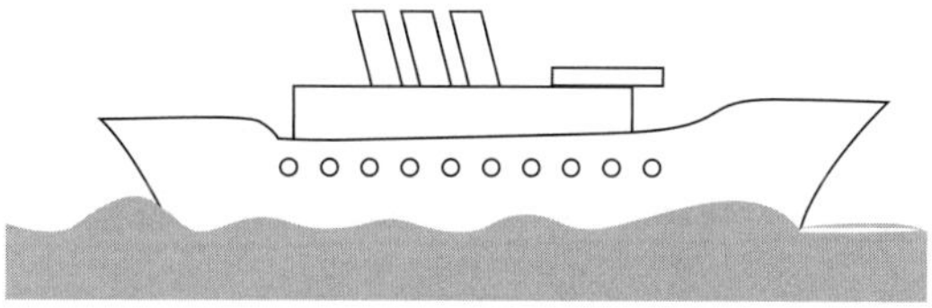

FIGURE P14.45

14.46 A 1:60 scale model of a ship is used in a water tank to simulate a ship speed of 10 m/s. What should be the model speed? If a towing force of 10 N is measured on the model, what force is expected on the prototype? Neglect viscous effects.

14.47 The flow rate over a weir is 2 m^3/s of water. A 1:10 scale model of the weir is tested in a water channel.
(a) What flow rate should be used?
(b) If a force of 12 N is measured on the model, what force would be expected on the prototype?

14.48 A 1:10 scale model of a hydrofoil is tested in a water channel with a force of 0.8 lb measured at a speed of 20 ft/sec. Determine the speed and force predicted for the prototype. Neglect viscous effects.

14.49 The propeller of a ship is to be studied with a 1:10 scale model.
(a) Assuming that the propeller operates close to the surface, select the propeller speed of the model if the prototype speed is 600 rpm.
(b) What torque would be expected if 1.2 N·m is measured on the model?

14.50 The pontoon of a seaplane is to be studied in a water channel that has a channel speed capability of 6 m/s. If the seaplane is to lift off at 100 m/s, what model scale should be selected?

14.51 A 1:30 scale model study of a submarine is proposed in an attempt to study the influence of a suggested shape modification. The prototype is 2 m in diameter and it is designed to travel at 15 m/s. The model is towed in a water tank at 2 m/s and a drag force of 2.15 N is measured. Does similitude exist for this test? If so, predict the power needed for the prototype.

14.52 The smoke emanating from the stacks on a ship has a tendency to make its way to the deck down the outside of the stacks causing discomfort to the passengers. This problem is studied with a 1:20 scale model of a 4-m-diameter stack. The ship is traveling at 10 m/s. What range of wind-tunnel airspeeds could be used in this study?

14.53 A model study of the dirigible of Fig. P14.53 is to be performed. The 10-m-diameter dirigible is to travel at 20 m/s. If a 40-cm-diameter model is proposed for use in a wind tunnel, or a 10-cm-diameter model for use in a 20°C water channel, which should be selected? Suppose that the wind-tunnel model is used with a speed of 15 m/s and a drag force of 3.2 N is measured. What force would be expected on the water-channel model with a speed of 2.4 m/s in the water channel? What horsepower would be predicted to overcome drag on the prototype? Assume that the flow is Reynolds number independent for Re > 10^5.

FIGURE P14.53

14.54 A model study is to be performed of a 1000-ft-tall, 45-ft-diameter smokestack of a power plant. It is known that the smokestack is submerged in a ground boundary layer that is 1200 ft thick. Could the study be performed in a wind tunnel that produces a 4-ft-thick boundary layer?

14.55 A 1:20 scale model of an aircraft is tested in a 23°C wind tunnel. A speed of 200 m/s is used in the model study. A drag force of 10 N is measured. What prototype speed and drag force does this simulate if the elevation is:
(a) Sea level?
(b) 5000 m?
(c) 10 000 m?

14.56 A 1:10 scale model of an airfoil is tested in a wind tunnel that uses outside air at 0°C. The test is to simulate an aircraft speed of 250 m/s at an elevation of 10 000 m. What wind-tunnel speed should be used? If a velocity of 290 m/s and a pressure of 80 kPa absolute are measured on the model at a particular location at a 5° angle of attack, what speed and pressure are expected on the prototype at the corresponding location, and what is the angle of attack?

14.57 A 1:10 scale model of a propeller on a ship is to be tested in a water channel. What should be the rotational speed of the model if the rotational speed of the propeller is 2000 rpm and:

(a) The Froude number governs the study?

(b) The Reynolds number governs the study?

14.58 A torque of 12 N·m is measured on a 1:10 scale model of a large wind machine with a wind tunnel speed of 60 m/s. This is to simulate a wind velocity of 15 m/s since viscous effects are considered negligible. What torque is expected on the prototype? If the model rotates at 500 rpm, what prototype angular velocity is simulated?

14.59 An underwater study of a porpoise is to be performed by using a 1:10 scale model. A porpoise swimming at 10 m/s and making one swimming motion each second is to be simulated. What speed could be used in the water channel, and for that speed, how many swimming motions per second should be used?

The Trans Alaska Pipeline conveys crude oil over long distances and through a variety of terrain. Pumping stations are spaced along its length to overcome pressures lost due to viscous forces and elevation changes. (Alaska Division of Tourism)

15

Internal Flows

Outline

Chapter Objectives

The objectives of this chapter are to:

- Establish the length of the entrance region for both laminar and turbulent flows.
- Determine the laminar flow solution for pipes, parallel plates, and rotating cylinders.
- Present the quantities of interest for turbulent pipe flow. Losses are of particular interest.
- Calculate the flow rate with a centrifugal pump in a single piping system.
- Determine flow rates in open channels.
- Provide numerous examples and problems that demonstrate entrance and developed flow solutions for both laminar and turbulent flows, including losses due to wall friction and various devices. Open channel flows are also analyzed.

Nomenclature

The following variables are introduced in this chapter:

A	constant of integration	l_m	mixing length
B	constant of integration	n	an exponent, Manning n
C	a constant	P	wetted perimeter
EGL	energy grade line	R	hydraulic radius
f	friction factor	S	slope
HGL	hydraulic grade line	u_τ	shear velocity
K_{uv}	correlation coefficient	$\overline{u}, \overline{v}, \overline{w}$	time-average velocities
L_d	profile development length	u', v', w'	velocity fluctuations
L_e	equivalent length		
L_E	entrance length	η	eddy viscosity
L_i	inviscid core length	λ	a constant

15.1 INTRODUCTION

In this chapter the effects of viscosity on an incompressible, internal flow will be studied. Such flows are of particular importance to engineers. Flow in a circular pipe is undoubtedly the most common internal fluid flow. It is encountered in the veins and arteries in a body, in a city's water system, in a farmer's irrigation system, in the piping systems transporting fluids in a factory, in the hydraulic lines of an aircraft, and in the ink jet of a computer's printer. Flows in noncircular ducts and open channels must also be included in our study. In Chapter 14 we discovered that the viscous effects in a flow resulted in the introduction of the Reynolds number,

$$\mathrm{Re} = \frac{V\rho l}{\mu} \tag{15.1}$$

Key Concept: *When the surface areas are large, viscous effects often become important.*

The Reynolds number was observed to be a ratio of the inertial force to the viscous force. Hence, when this ratio becomes large, it is expected that the inertial forces may dominate the viscous forces. This is usually true when short, sudden geometric changes occur; for long reaches of pipe or open channels, this is not the situation. When the surface areas, such as the wall area of a pipe, are relatively large, viscous effects become quite important and must be included in our study.

Internal flow between parallel plates, in a pipe, between rotating cylinders, and in an open channel will be considered in detail. For a sufficiently low Reynolds number ($\mathrm{Re} < 2000$ in a pipe and $\mathrm{Re} < 1500$ in a wide channel) a laminar flow results, and at sufficiently high Reynolds number a turbulent flow occurs. See Section 12.3.3 for a more detailed discussion. We consider laminar flow first and then turbulent flow.

15.2 ENTRANCE FLOW AND DEVELOPED FLOW

Developed laminar flow: *A flow where the velocity profile ceases to change in the flow direction.*

When considering internal flows we are interested primarily in developed flows in conduits. Let us first focus our attention on a laminar flow. A **developed laminar flow** results when the velocity profile ceases to change in the flow direction. In the

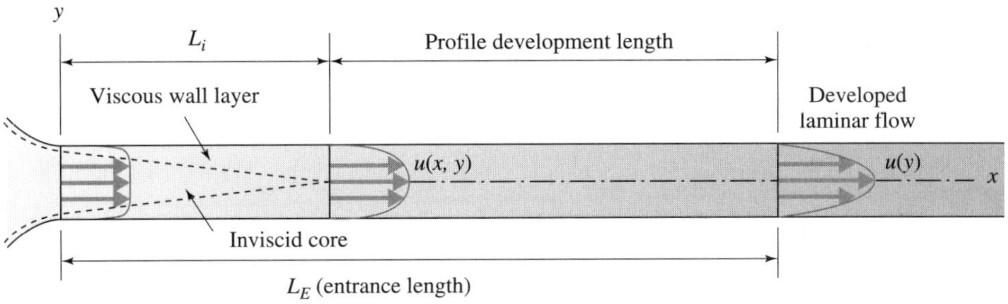

FIGURE 15.1 Entrance region of a laminar flow in a pipe or a wide rectangular channel.

entrance region of a laminar flow the velocity profile changes in the flow direction, as shown in Fig. 15.1. The idealized flow from a reservoir begins at the inlet as a uniform flow (in reality, there is a thin viscous layer on the wall, as shown); the viscous wall layer then grows over the *inviscid core length* L_i until the viscous stresses dominate the entire cross section; the profile then continues to change in the *profile development region* due to viscous effects until a developed flow is achieved. The inviscid core length is one-fourth to one-third of the *entrance length* L_E, depending on the conduit geometry, shape of the inlet velocity profile, and the Reynolds number.

Key Concept: *The inviscid core length in a laminar flow is one-fourth to one-third of the entrance length.*

For a laminar flow in a circular pipe with a uniform profile at the inlet, the entrance length is given by

$$\frac{L_E}{D} = 0.065\ \text{Re} \qquad \text{Re} = \frac{VD}{\nu} \tag{15.2}$$

where the Reynolds number is based on the average velocity and the diameter. Laminar flow in a pipe has been observed at Reynolds numbers in excess of 40 000 for carefully controlled conditions in a smooth pipe. However, for engineering applications a value of about 2000 is the highest Reynolds number for which laminar flow is assured; this is due to vibrations of the pipe, fluctuations in the flow, or roughness elements on the pipe wall.

For a laminar flow in a high-aspect-ratio channel (the aspect ratio is the width divided by the distance between the top and bottom plates) with a uniform profile at the inlet the entrance length is

$$\frac{L_E}{h} = 0.04\ \text{Re} \qquad \text{Re} = \frac{Vh}{\nu} \tag{15.3}$$

where the Reynolds number is based on the average velocity and the distance h between the plates. The inviscid core length is approximately one-third of the entrance length. A laminar flow cannot exist over Re = 7700; for engineering situations a value of 1500 is often used as the upper limit for laminar flow.

Key Concept: *A developed tubulent flow results when all characteristics of the flow cease to change in the flow direction.*

For a turbulent flow the situation is slightly different, as shown in Fig. 15.2 for flow in a pipe. A developed flow results when all characteristics of the flow cease to change in the flow direction; this includes details of the turbulence that

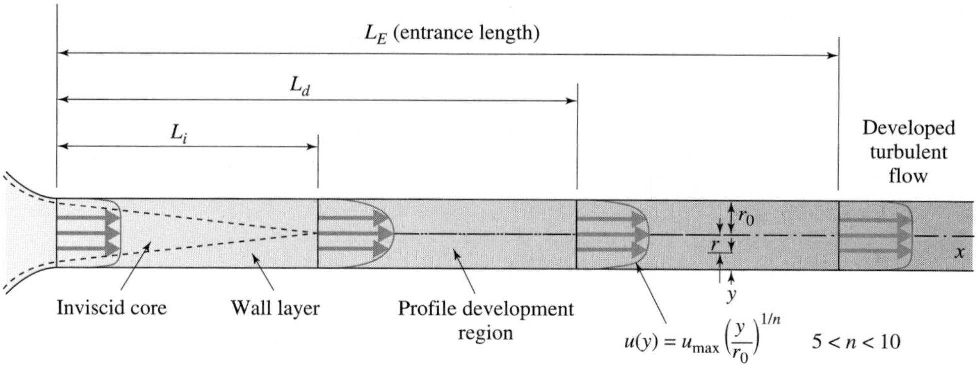

FIGURE 15.2 Velocity profile development in a turbulent pipe flow.

will be introduced later in this chapter. The inviscid core exists followed by the velocity profile development region, which terminates at $x = L_d$. An additional length is needed, however, for the detailed structure of the turbulent flow to develop. The detailed structure is important in certain calculations such as accurate estimates of wall heat transfer. For large Reynolds number flow (Re $> 10^5$) in a pipe, where the turbulence fluctuations initiate near $x = 0$, tests have yielded

Key Concept: *If transition to turbulent flow occurs near the origin, the linear pressure variation begins near L_i.*

$$\frac{L_i}{D} \simeq 10 \qquad \frac{L_d}{D} \simeq 40 \qquad \frac{L_E}{D} \simeq 120 \tag{15.4}$$

Key Concept: *In a developed flow, the pressure decreases linearly with x.*

For turbulent flow with Re $= 4000$ the foregoing developmental lengths would be substantially higher, perhaps five times the values listed. This is true since, for low-Re turbulent flows, transition to a turbulent flow occurs in the profile development region; hence much of the entry region is laminar with relatively low wall shear. Experimental data are not available for low-Reynolds-number turbulent flow.

In Fig. 15.3 the pressure variation is sketched. In the flow beyond a sufficiently large x it is noted that the pressure decreases linearly with x. If transition

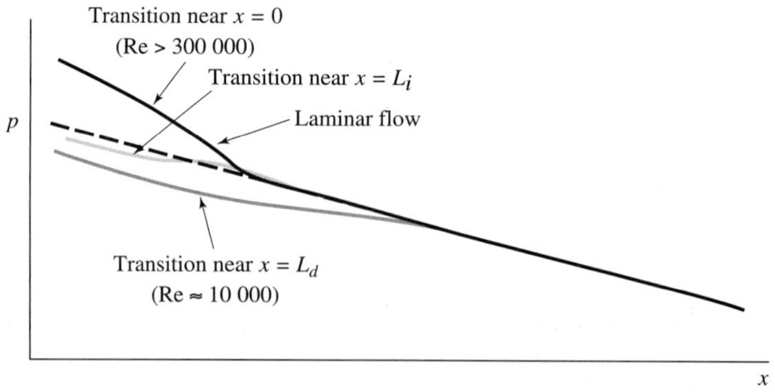

FIGURE 15.3 Pressure variation in a horizontal pipe flow for both laminar and turbulent flows. (From the PhD Thesis of Dr. Jack Backus, Michigan State University.)

to turbulent flow occurs near the origin, the linear pressure variation begins near L_i and the pressure gradient [the slope of the $p(x)$ curve] in the inlet region is higher than in the developed flow region; if transition occurs near L_d, as it does for low Re, the linear variation begins at the end of the transition process and the pressure gradient in the inlet region is less than that of developed flow.

For a laminar flow, the pressure variation qualitatively resembles that associated with a large Reynolds number. The pressure gradient is higher than in the developed flow region due to the higher shear stress at the wall and the increasing momentum flux.

15.3 LAMINAR FLOW IN A PIPE

In this section we investigate incompressible, steady, developed laminar flow in a pipe, as sketched in Fig. 15.4. An elemental volume of the fluid is shown. This volume can be considered an infinitesimal control volume into which and from which fluid flows, or it can be considered an infinitesimal fluid mass upon which forces are acting. If it is considered a control volume, we would apply the momentum equation (13.69); if it is a fluid mass, we would apply Newton's second law. Since the velocity profile does not change in the x-direction, the momentum flux in equals the momentum flux out and the resultant force is zero; since there is no acceleration of the mass element, the resultant force must also be zero. Consequently, a force balance in the x-direction yields

$$p\pi r^2 - (p + dp)\pi r^2 - \tau 2\pi r\, dx + \gamma \pi r^2\, dx \sin\theta = 0 \tag{15.5}$$

which can be simplified to

$$\tau = -\frac{r}{2}\frac{d}{dx}(p + \gamma h) \tag{15.6}$$

where we have used $\sin\theta = -dh/dx$, the vertical direction being denoted by h. Note that Eq. 15.6 can be applied to either a laminar or turbulent flow. The shear

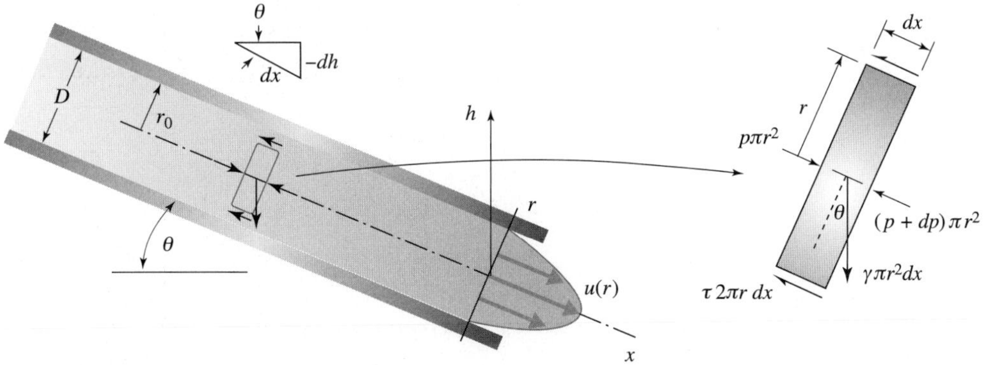

FIGURE 15.4 Developed flow in a circular pipe.

stress[1] in this laminar flow is related to the velocity gradient and the viscosity (see Eq. 10.16), giving

$$-\mu\frac{du}{dr} = -\frac{r}{2}\frac{d}{dx}(p + \gamma h) \tag{15.7}$$

Since $u = u(r)$ for this developed flow, we can conclude from the above that $d/dx(p + \gamma h)$ must be at most a constant; it cannot depend on x, and from fluid statics $(p + \gamma h)$ does not depend on x since there is no acceleration in the radial direction. Equation (15.7) can then be integrated to give the velocity distribution,

$$u(r) = \frac{r^2}{4\mu}\frac{d}{dx}(p + \gamma h) + A \tag{15.8}$$

where A is a constant of integration. Using $u = 0$ at $r = r_0$, we can evaluate A and find the velocity distribution to be

$$u(r) = \frac{1}{4\mu}\frac{d(p + \gamma h)}{dx}(r^2 - r_0^2) \tag{15.9}$$

Poiseuille flow: *A flow with a parabolic profile in a pipe or between parallel plates.*

a parabolic profile. This flow is often referred to as **Poiseuille flow**.

The average velocity V is found by integrating Eq. 15.9:

$$V = \frac{Q_f}{A} = \frac{\int_0^{r_0} u(r)2\pi r\, dr}{\pi r_0^2}$$

$$= \frac{2}{r_0^2}\int_0^{r_0}\frac{1}{4\mu}\frac{d(p + \gamma h)}{dx}(r^2 - r_0^2)r\, dr = -\frac{r_0^2}{8\mu}\frac{d(p + \gamma h)}{dx} \tag{15.10}$$

Or, expressing the pressure drop Δp in terms of the average velocity, we have, for a horizontal[2] pipe,

$$\Delta p = \frac{8\mu V L}{r_0^2} \tag{15.11}$$

Key Concept: *The pressure drop is a positive quantity, whereas the pressure gradient is negative.*

where we have used $\Delta p/L = -dp/dx$ since dp/dx is a constant for developed flow. Note that the pressure drop is a positive quantity, whereas the pressure gradient is negative.

The maximum velocity at $r = 0$ from Eq. 15.9 is

$$u_{\max} = -\frac{r_0^2}{4\mu}\frac{d(p + \gamma h)}{dx} \tag{15.12}$$

[1]The shear stress is a positive quantity, as displayed in Fig. 15.4. The minus sign in $\tau = -\mu\, du/dr$ is necessary since du/dr is negative.

[2]For a pipe on an incline, simply replace p with $(p + \gamma h)$.

so that (see Eq. 15.10)

$$V = \frac{1}{2}u_{\max} \tag{15.13}$$

The shearing stress is determined to be

$$\begin{aligned}\tau &= -\mu\frac{du}{dr}\\ &= -\frac{r}{2}\frac{d(p+\gamma h)}{dx}\end{aligned} \tag{15.14}$$

Letting $\tau = \tau_0$ at $r = r_0$, we see that the pressure drop Δp over a length L of a horizontal section of pipe is

$$\Delta p = \frac{2\tau_0 L}{r_0} \tag{15.15}$$

where we have again used $dp/dx = -\Delta p/L$.

If we introduce the **friction factor** f, a quantity of substantial interest in pipe flow, a dimensionless wall shear, defined by

Friction factor: *A dimensionless wall shear.*

$$f = \frac{\tau_0}{\frac{1}{8}\rho V^2} \tag{15.16}$$

we see that

$$\frac{\Delta p}{\gamma} = h_L = f\frac{L}{D}\frac{V^2}{2g} \tag{15.17}$$

where h_L is the head loss (see Eq. 13.51) with dimension of length. This equation is often referred to as the *Darcy–Weisbach equation.*[3] Combining Eqs. 15.11 and 15.17, we find that

$$f = \frac{64}{\text{Re}} \tag{15.18}$$

for laminar flow in a pipe. Substituting this back into Eq. 15.17, we see that

$$h_L = \frac{32\mu L V}{\gamma D^2} \tag{15.19}$$

The head loss is directly proportional to the average velocity (and hence the flow rate also) to the first power, a result that generally is also applied to developed, laminar flows in conduits of shape other than circular.

Key Concept: *The head loss is directly proportional to the average velocity.*

[3]This equation will be shown to be valid for both laminar and turbulent flows.

Example 15.1

A small-diameter horizontal tube is connected to a supply reservoir as shown in Fig. E15.1. If 6600 mm^3 is captured at the outlet in 10 s, estimate the viscosity of the water.

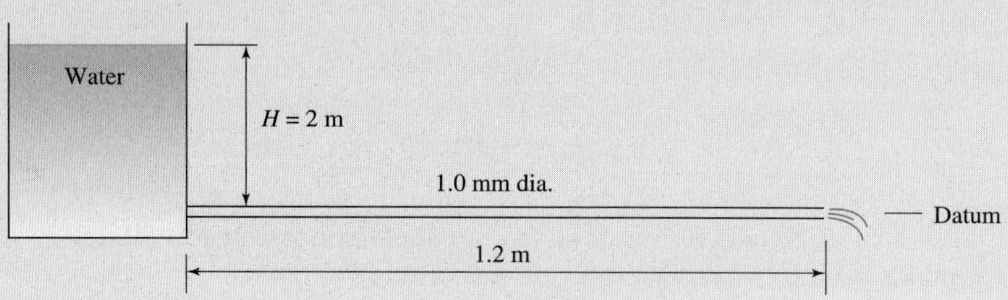

FIGURE E15.1

Solution

The tube is very small, so we expect viscous effects to limit the velocity to a small value. Using Bernoulli's equation from the surface to the entrance to the tube, and neglecting the velocity head, we have, letting 0 be a point on the surface,

$$\overset{0}{\cancel{\frac{p_0}{\gamma}}} + H = \overset{0}{\cancel{\frac{V_1^2}{2g}}} + \frac{p}{\gamma}$$

where we have used gage pressure with $p_0 = 0$. This becomes, assuming $V^2/2g \cong 0$,

$$p = \gamma H = 9800 \times 2 = 19\,600 \text{ Pa}$$

At the exit of the tube the pressure is zero; hence

$$\frac{\Delta p}{L} = \frac{19\,600}{1.2} = 16\,300 \text{ Pa/m}$$

The average velocity is found to be

$$V = \frac{Q_f}{A} = \frac{6600 \times 10^{-9}/10}{\pi \times 0.001^2/4} = 0.840 \text{ m/s}$$

Check to make sure the velocity head is negligible: $V^2/2g = 0.036$ m compared with $p/\gamma = 2$ m, so the assumption of negligible velocity head is valid. Our pressure calculation is acceptable. Using Eq. 15.11, we can find the viscosity of this assumed laminar flow to be

$$\mu = \frac{r_0^2}{8V}\frac{\Delta p}{L} = \frac{0.0005^2}{8 \times 0.84}(16\,300) = 6.06 \times 10^{-4} \text{ N·s/m}^2$$

We should check the Reynolds number to determine if our assumption of a laminar flow was acceptable. It is

$$\text{Re} = \frac{\rho V D}{\mu} = \frac{1000 \times 0.84 \times 0.001}{6.06 \times 10^{-4}} = 1390$$

(continued)

This is obviously a laminar flow since Re < 2000, so the calculations are valid providing the entrance length is not too long. It is

$$L_E = 0.065 \text{ Re} \times D = 0.065 \times 1390 \times 0.001 = 0.09 \text{ m}$$

This is approximately 8% of the total length, a sufficiently small quantity; hence the calculations are assumed reliable.

Example 15.2

Derive an expression for the velocity distribution and the flow rate between horizontal, concentric pipes for a steady, incompressible developed flow (Fig. E15.2).

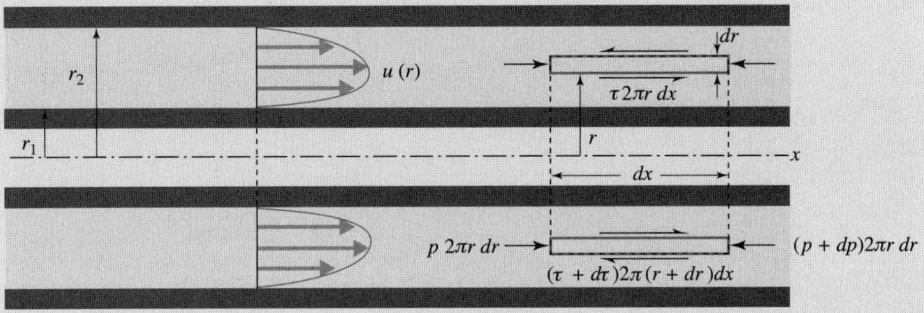

FIGURE E15.2

Solution

The element to be used is a hollow cylindrical shell as sketched in the figure. If we sum forces, we obtain

$$p2\pi r\, dr - (p + dp)2\pi r\, dr + \tau 2\pi r\, dx - (\tau + d\tau)2\pi(r + dr)\, dx = 0$$

Simplifying, there results, neglecting the term of differential magnitude,

$$\frac{dp}{dx} = -\frac{\tau}{r} - \frac{d\tau}{dr} - \underbrace{\frac{d\tau}{r}}_{\text{neglect}}$$

Substituting $\tau = -\mu\, du/dr$ we have

$$\frac{dp}{dx} = \mu\left(\frac{1}{r}\frac{du}{dr} + \frac{d^2u}{dr^2}\right)$$

$$= \frac{\mu}{r}\frac{d}{dr}\left(r\frac{du}{dr}\right)$$

Multiply both sides by $r\, dr$ and divide by μ, then integrate:

$$r\frac{du}{dr} = \frac{1}{2\mu}\frac{dp}{dx}r^2 + A$$

(continued)

Multiply both sides by dr/r and integrate again:

$$u(r) = \frac{1}{4\mu}\frac{dp}{dx}r^2 + A \ln r + B$$

where A and B are arbitrary constants. They are found by setting $u = 0$ at $r = r_1$ and at $r = r_2$; that is,

$$0 = \frac{1}{4\mu}\frac{dp}{dx}r_1^2 + A \ln r_1 + B$$

$$0 = \frac{1}{4\mu}\frac{dp}{dx}r_2^2 + A \ln r_2 + B$$

Solve for A and B:

$$A = \frac{1}{4\mu}\frac{dp}{dx}\frac{r_1^2 - r_2^2}{\ln(r_2/r_1)}$$

$$B = -A \ln r_2 - \frac{r_2^2}{4\mu}\frac{dp}{dx}$$

Thus

$$u(r) = \frac{1}{4\mu}\frac{dp}{dx}\left[r^2 - r_2^2 + \frac{r_2^2 - r_1^2}{\ln(r_1/r_2)}\ln(r/r_2)\right]$$

This is integrated to give the flow rate:

$$Q_f = \int_{r_1}^{r_2} u(r)2\pi r\, dr$$

$$= -\frac{\pi}{8\mu}\frac{dp}{dx}\left[r_2^4 - r_1^4 - \frac{(r_2^2 - r_1^2)^2}{\ln(r_2/r_1)}\right]$$

As $r_1 \to 0$ the velocity distribution approaches the parabolic distribution of pipe flow. As $r_1 \to r_2$ this distribution approaches that of parallel-plate flow presented in the next section. These two conclusions are not obvious and are presented as a problem at the end of this chapter.

15.4 LAMINAR FLOW BETWEEN PARALLEL PLATES

Consider the incompressible, steady, developed flow of a fluid between parallel plates, with the upper plate moving with velocity U as shown in Fig. 15.5. Consider an elemental volume of unit depth in the z-direction, as sketched in Fig. 15.5. If we sum forces in the x-direction, we can write

$$p\,dy - (p + dp)\,dy - \tau\,dx + (\tau + d\tau)\,dx + \gamma\,dx\,dy \sin\theta = 0 \qquad (15.20)$$

since there is no acceleration. The pressure is assumed to depend only on x; variation with y is assumed to be negligible since the dimension a is quite small for most applications. After dividing by $dx\,dy$, the above is simplified to

$$\frac{d\tau}{dy} = \frac{dp}{dx} - \gamma \sin\theta \qquad (15.21)$$

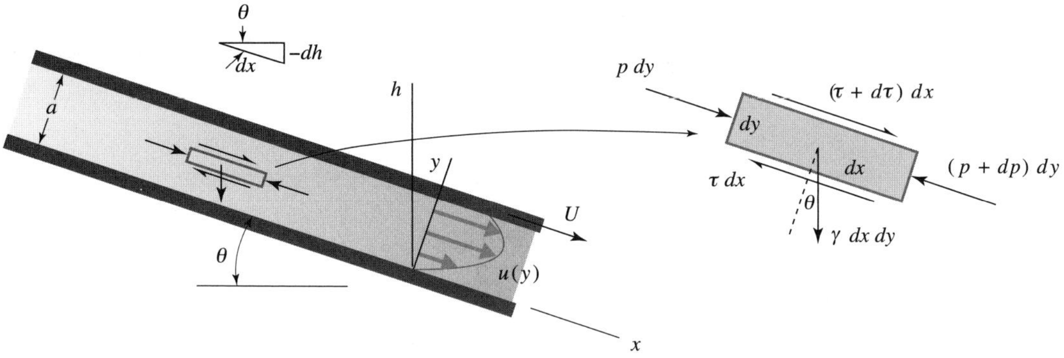

FIGURE 15.5 Developed flow between parallel plates.

Since this is one-dimensional flow, the shear stress is

$$\tau = \mu \frac{du}{dy} \tag{15.22}$$

Using this and $\sin\theta = -dh/dx$, there results

$$\mu \frac{d^2u}{dy} = \frac{dp}{dx} + \gamma \frac{dh}{dx} = \frac{d}{dx}(p + \gamma h) \tag{15.23}$$

Since $u = u(y)$ for this developed flow, the left-hand side is only a function of y; because the right-hand side is a function of x we conclude that it must be a constant. Hence, it can be integrated twice to yield (first, divide by μ)

$$u(y) = \frac{y^2}{2\mu} \frac{d}{dx}(p + \gamma h) + Ay + B \tag{15.24}$$

where A and B are integration constants. If we require $u = 0$ at $y = 0$ and $u = U$ at $y = a$, we have

$$A = \frac{U}{a} - \frac{a}{2\mu} \frac{d}{dx}(p + \gamma h) \qquad \text{and} \qquad B = 0 \tag{15.25}$$

Thus the velocity distribution is the parabola

$$u(y) = \frac{1}{2\mu} \frac{d}{dx}(p + \gamma h)(y^2 - ay) + \frac{U}{a} y \tag{15.26}$$

If the motion is due to the motion of the plate only (a linear profile), the flow is called a **Couette flow**; if the motion is due to the pressure gradient only, i.e., $U = 0$, it is Poiseuille flow.

Couette flow: *A flow with a linear profile that results from the motion of the plate only.*

Using the results derived above, we can write the expression for the velocity distribution between fixed plates (let $U = 0$) as

$$u(y) = \frac{1}{2\mu} \frac{d(p + \gamma h)}{dx} (y^2 - ay) \tag{15.27}$$

Using this distribution, we can find the flow rate per unit width to be

$$Q_f = \int u\, dA$$

$$= \int_0^a \frac{1}{2\mu}\frac{d(p+\gamma h)}{dx}(y^2 - ay)\, dy = -\frac{a^3}{12\mu}\frac{d(p+\gamma h)}{dx} \tag{15.28}$$

The average velocity V is found to be

$$V = \frac{Q_f}{a \times 1}$$

$$= -\frac{a^2}{12\mu}\frac{d(p+\gamma h)}{dx} \tag{15.29}$$

This can be expressed as the pressure drop in terms of the average velocity; for a horizontal[4] channel we have

$$\Delta p = \frac{12\mu VL}{a^2} \tag{15.30}$$

where we have used $\Delta p/L = -dp/dx$ since dp/dx is constant for developed flow.

Observe that the maximum velocity occurs at $y = a/2$ and from Eq. 15.27 is

$$u_{\max} = -\frac{a^2}{8\mu}\frac{dp}{dx} \tag{15.31}$$

Thus the average velocity is related to the maximum velocity by

$$V = \frac{2}{3}u_{\max} \tag{15.32}$$

The shearing stress is found to be

$$\tau = \mu\frac{du}{dy} = \frac{1}{2}\frac{dp}{dx}(2y - a) \tag{15.33}$$

At the wall where $y = 0$ there results

$$\tau_0 = -\frac{a}{2}\frac{dp}{dx} \tag{15.34}$$

The pressure drop Δp over a length L of horizontal channel is found to be

$$\Delta p = \frac{2\tau_0}{a}L \tag{15.35}$$

[4]For plates on an incline, simply replace p with $(p + \gamma h)$.

recognizing that $dp/dx = -\Delta p/L$. This can be expressed in a more convenient form as

$$\frac{\Delta p}{\gamma} = f\frac{L}{2a}\frac{V^2}{2g} \tag{15.36}$$

if we introduce the friction factor f, defined by

$$f = \frac{\tau_0}{\frac{1}{8}\rho V^2} \tag{15.37}$$

In terms of the head loss (see Eq. 13.51), Eq. 15.36 becomes

$$h_L = f\frac{L}{2a}\frac{V^2}{2g} \tag{15.38}$$

If we combine Eqs. 15.30, 15.34, and 15.37, we find that

$$f = \frac{8}{\rho V^2}\left(-\frac{a}{2}\frac{dp}{dx}\right) = \frac{8}{\rho V^2}\left(-\frac{a}{2}\right)\left(-\frac{12\mu V}{a^2}\right) = \frac{48\mu}{\rho a V} = \frac{48}{\text{Re}} \tag{15.39}$$

where the Reynolds number $\text{Re} = \rho a V/\mu$ has been introduced. If this is substituted into Eq. 15.38, we see that

$$h_L = \frac{12\mu L V}{\gamma a^2} \tag{15.40}$$

The head loss is directly proportional to the average velocity, a conclusion that is true for laminar flows, in general.

We have calculated most of the quantities of interest for laminar flow of an incompressible, steady flow between parallel plates. This would, of course, be an acceptable approximation for a wide channel, one for which the width exceeds the height by at least a factor of 8. For channels with smaller aspect ratios, the edge effects become significant and must be accounted for by adding an additional shear stress to the sides of the element in Fig. 15.5. The solution to a set of partial differential equations is required and is beyond the scope of this book.

Key Concept: *A wide channel is one for which the width exceeds the height by at least a factor of 8.*

Example 15.3

Water at 60°F flows with a Reynolds number of 1500 between the 20-inch-wide, horizontal plates shown in Fig. E15.3. Calculate (a) the flow rate, (b) the wall shear stress, (c) the pressure drop over 10 ft, and (d) the velocity at $y = 0.2$ in.

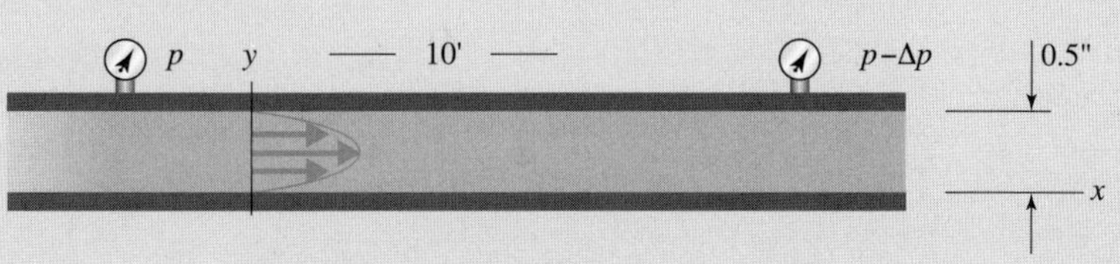

FIGURE E15.3

(continued)

Solution

Since the Reynolds number is 1500, the laminar flow equations are assumed applicable.

(a) Using the definition of the Reynolds number the average velocity is found as follows:

$$1500 = \frac{Va}{\nu}$$

$$\therefore\ V = \frac{1500\nu}{a} = \frac{1500 \times 1.22 \times 10^{-5}}{0.5/12} = 0.439\ \text{ft/sec}$$

Thus

$$Q_f = VA = 0.439 \times \frac{0.5 \times 20}{144} = 0.0305\ \text{ft}^3/\text{sec}$$

(b) Using Eq. 15.30, the pressure gradient is

$$\frac{\Delta p}{L} = \frac{12\mu V}{a^2} = \frac{12 \times 2.36 \times 10^{-5} \times 0.439}{(0.5/12)^2} = 0.0716\ \text{psf/ft}$$

The shearing stress at the wall is found, using Eq. 15.35, to be

$$\tau_0 = \frac{a}{2}\frac{\Delta p}{L} = \frac{0.5/12}{2} \times 0.0716 = 0.00149\ \text{psf}$$

(c) The pressure drop over 10 ft is found to be

$$\Delta p = 0.0716L = 0.0716 \times 10 = 0.716\ \text{psf}$$

(d) The velocity distribution of Eq. 15.27 is

$$\begin{aligned} u(y) &= \frac{1}{2\mu}\frac{dp}{dx}(y^2 - ay) \\ &= \frac{1}{2 \times 2.36 \times 10^{-5}}(-0.0716)\left(y^2 - \frac{0.5}{12y}\right) = -1517(y^2 - 0.0417y) \end{aligned}$$

where we have used $dp/dx = -\Delta p/L$. At $y = 0.2$ in., the velocity is

$$u = -1517\left[\left(\frac{0.2}{12}\right)^2 - 0.0417 \times \frac{0.2}{12}\right] = 0.633\ \text{ft/sec}$$

We have used three significant digits since the fluid properties are known to three significant digits.

Example 15.4

Find an expression for the pressure gradient that results in a zero shear stress at the lower wall, where $y = 0$, of two parallel plates; also, sketch the velocity profiles for a top plate speed of U with various pressure gradients. Assume horizontal plates.

Solution

The velocity distribution for plates with the top plate moving with velocity U is given by Eq. 15.26. Letting $dh/dx = 0$, we have

$$u(y) = \frac{1}{2\mu}\frac{dp}{dx}(y^2 - ay) + \frac{U}{a}y$$

The shear stress is

$$\begin{aligned}\tau &= \mu\frac{du}{dy} \\ &= \frac{1}{2}\frac{dp}{dx}(2y - a) + \mu\frac{U}{a}\end{aligned}$$

If $\tau = 0$ at $y = 0$, then $du/dy = 0$ at $y = 0$ and the pressure gradient is

$$\frac{dp}{dx} = \frac{2\mu U}{a^2}$$

If dp/dx is greater than this value, the slope du/dy at $y = 0$ is negative and thus the velocity u will be negative near $y = 0$. If $dp/dx = 0$, we observe that a linear velocity distribution results, namely,

$$u(y) = \frac{U}{a}y$$

If dp/dx is negative, $u(y)$ is greater at each y-location than the linear distribution since $(y^2 - ay)$ is a negative quantity for all y's of interest.

All of the results above can be qualitatively displayed on a sketch of $u(y)$ for several dp/dx as shown in Fig. E15.4.

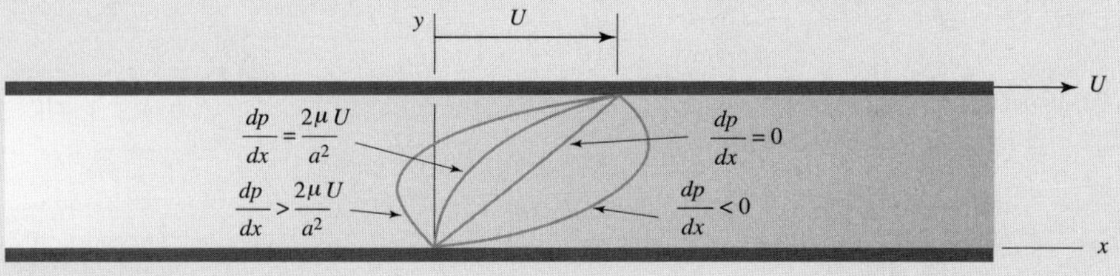

FIGURE E15.4

15.5 LAMINAR FLOW BETWEEN ROTATING CYLINDERS

Key Concept: *The laminar flow solution will be valid up to a Reynolds number of 1700.*

Fully developed, steady flow between concentric, rotating cylinders, as shown in Fig. 15.6a, is another flow that yields a rather simple solution. It has particular application in the field of lubrication, where the fluid may be oil and the inner cylinder a rotating shaft. The laminar flow solution we will find will be valid up to a Reynolds number[5] of 1700, providing the angular velocity of the outer cylinder $\Omega_2 = 0$, as is often the case. Above Re = 1700 a secondary laminar flow (a flow with a different velocity distribution) may develop and eventually a turbulent flow develops. In fact, numerous laminar flows (all different) have been observed for Re > 1700.

Body forces will be neglected in this derivation, or the cylinders will be assumed to be vertical. Since pressure does not vary with θ, an element in the form of a thin cylindrical shell will be used as shown in Fig. 15.6b. The resultant torque acting on this element is zero because it has no angular acceleration; this is expressed as

$$\tau 2\pi r L \times r - (\tau + d\tau)2\pi(r + dr)L \times (r + dr) = 0 \tag{15.41}$$

where L, the length of the cylinders, should be large with respect to the gap width $(r_2 - r_1)$ to avoid three-dimensional end effects. Equation 15.41 reduces to, neglecting the three higher-order terms which approach zero as $dr \to 0$,

$$2\tau + r\frac{d\tau}{dr} = 0 \tag{15.42}$$

The one-dimensional constitutive equation written in cylindrical coordinates (see Appendix K.2) provides the shear stress:

$$\tau = -\mu r \frac{d}{dr}\left(\frac{v_\theta}{r}\right) \tag{15.43}$$

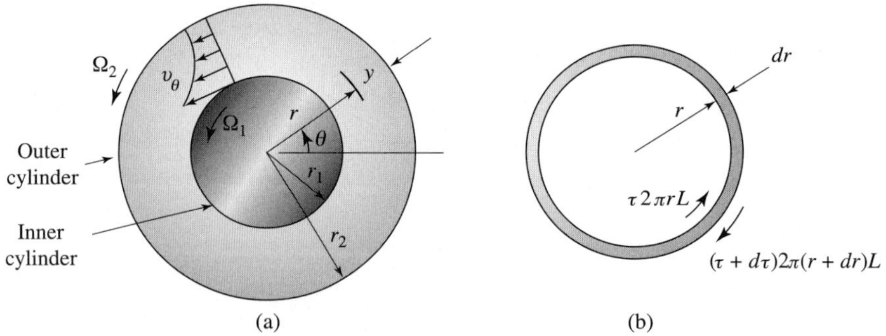

FIGURE 15.6 Flow between concentric cylinders: (a) basic flow variables; (b) element from between the cylinders.

[5]The Reynolds number is defined as $\text{Re} = \Omega_1 r_1 \delta/\nu$, where $\delta = r_2 - r_1$.

where v_θ is the velocity in the θ-direction. The component v_r in the radial direction is assumed to be zero. There results

$$-2\mu r \frac{d}{dr}\left(\frac{v_\theta}{r}\right) - r\mu \frac{d}{dr}\left[r \frac{d}{dr}\left(\frac{v_\theta}{r}\right)\right] = 0 \tag{15.44}$$

Divide by μr, multiply by dr, and integrate to find

$$2\frac{v_\theta}{r} + r\frac{d}{dr}\left(\frac{v_\theta}{r}\right) = A \tag{15.45}$$

This can be rearranged as $\left[\text{perform the differentiation: } \dfrac{d}{dr}\left(\dfrac{v_\theta}{r}\right) = \dfrac{1}{r}\dfrac{dv_\theta}{dr} - \dfrac{v_\theta}{r^2}\right]$

$$\frac{dv_\theta}{dr} + \frac{v_\theta}{r} = A \tag{15.46}$$

or

$$\frac{1}{r}\frac{d}{dr}(rv_\theta) = A \tag{15.47}$$

Multiply by $r\,dr$ and integrate again to find that

$$v_\theta(r) = \frac{A}{2}r + \frac{B}{r} \tag{15.48}$$

The boundary conditions are $v_\theta = r_1\Omega_1$ at $r = r_1$, and $v_\theta = r_2\Omega_2$ at $r = r_2$. These conditions allow the constants to be evaluated as

$$A = 2\frac{\Omega_2 r_2^2 - \Omega_1 r_1^2}{r_2^2 - r_1^2} \qquad B = \frac{r_1^2 r_2^2(\Omega_1 - \Omega_2)}{r_2^2 - r_1^2} \tag{15.49}$$

For a number of situations such as a shaft rotating in a bearing the outer cylinder is fixed. Setting $\Omega_2 = 0$, the velocity distribution is

$$v_\theta = \frac{r_1^2\Omega_1}{r_2^2 - r_1^2}\left(\frac{r_2^2}{r} - r\right) \tag{15.50}$$

The shearing stress τ_1 on the inner cylinder (see Table K.2) is found to be

$$\begin{aligned}\tau_1 &= -\left[\mu r \frac{d}{dr}\left(\frac{v_\theta}{r}\right)\right]_{r=r_1} \\ &= \frac{2\mu r_2^2\Omega_1}{r_2^2 - r_1^2}\end{aligned} \tag{15.51}$$

The torque T necessary to rotate the inner cylinder of length L is

$$\begin{aligned} T &= \tau_1 A_1 r_1 \\ &= \frac{4\pi\mu r_1^2 r_2^2 L\Omega_1}{r_2^2 - r_1^2} \end{aligned} \tag{15.52}$$

Key Concept: *The power is found by multiplying the torque by the rotational speed.*

The power $\dot{W}$ necessary to rotate the shaft is found by multiplying the torque by the rotational speed Ω_1; it is

$$\begin{aligned} \dot{W} &= T\Omega_1 \\ &= \frac{4\pi\mu r_1^2 r_2^2 L\Omega_1^2}{r_2^2 - r_1^2} \end{aligned} \tag{15.53}$$

This power is necessary to overcome the resistance of viscosity and results in an increase in the internal energy and thus the temperature of the fluid. The removal of this energy from the fluid often requires special heat exchangers.

Example 15.5

Estimate the viscosity of an oil contained in the annulus between two 25-cm-long cylinders, as shown in Fig. 15.6. The outer stationary cylinder is 8 cm in diameter. The 7.8-cm-diameter inner cylinder rotates at 3800 rpm when a torque of 0.12 N·m is applied. The specific gravity of the oil is 0.85. Neglect any torque due to the cylinder ends.

Solution

Assuming that the Reynolds number is less than 1700, Eq. 15.52 provides

$$\begin{aligned} \mu &= \frac{T(r_2^2 - r_1^2)}{4\pi r_1^2 r_2^2 L\Omega_1} \\ &= \frac{0.12(0.04^2 - 0.039^2)}{4\pi \times 0.04^2 \times 0.039^2 \times 0.25 \times (3800 \times 2\pi/60)} = 0.00312 \text{ N·s/m}^2 \end{aligned}$$

Check the Reynolds number using $\nu = \mu/\rho$:

$$\text{Re} = \frac{\Omega_1 r_1 \delta}{\nu} = \frac{(3800 \times 2\pi/60) \times 0.039 \times 0.002/2}{0.00312/(1000 \times 0.85)} = 845$$

This is less than 1700 so the calculation is assumed to be acceptable.

Example 15.6

Show that as the inner cylinder radius of Fig. E15.6 approaches the outer cylinder radius the velocity distribution approaches the linear distribution between parallel plates with one plate moving and a zero pressure gradient. This is Couette flow.

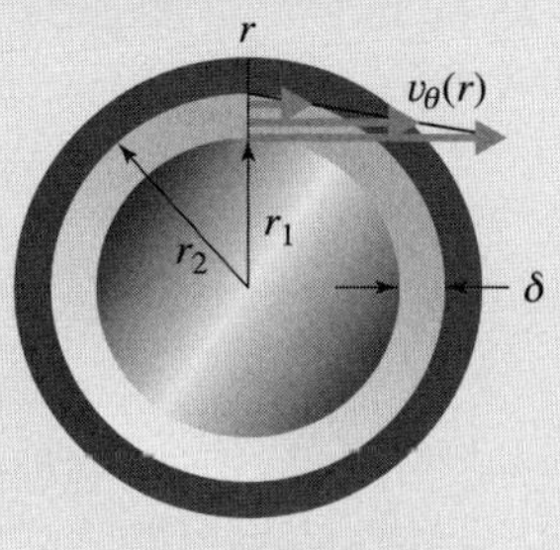

FIGURE E15.6

Solution

For this problem we will let $\Omega_2 = 0$; the velocity distribution (15.50) is

$$v_\theta(r) = \frac{r_1^2\Omega_1}{r_2^2 - r_1^2}\left(\frac{r_2^2}{r} - r\right)$$

$$= \frac{r_1^2\Omega_1}{r_2^2 - r_1^2}\frac{r_2^2 - r^2}{r} = \frac{r_1^2\Omega_1}{r_2^2 - r_1^2}(r_2 - r)\frac{r_2 + r}{r}$$

Introduce the independent variable y, measured from the outer cylinder defined by $r + y = r_2$ (see Fig. 15.6); let $\delta = r_2 - r_1$. Then the above can be written as

$$v_\theta(r) = \frac{r_1^2\Omega_1(r_2 - r)}{(r_2 - r_1)(r_2 + r_1)}\frac{r_2 + r}{r}$$

$$= \frac{r_1^2\Omega_1 y}{\delta(r_2 + r_1)}\frac{2r_2 - y}{r_2 - y}$$

As the inner radius approaches the outer radius we can write $r_1 \simeq r_2$. Letting $r_2 \simeq r_1 \simeq R$ we have $r_2 + r_1 \simeq 2R$ and

$$\frac{2R - y}{R - y} \simeq 2$$

since $y \ll R$. The velocity distribution then simplifies to

$$v_\theta(y) = \frac{R^2\Omega_1 y}{\delta 2R} \times 2 = \frac{R\Omega_1}{\delta}y$$

This is a linear distribution and is a good approximation to the flow whenever $\delta \ll R$.

15.6 TURBULENT FLOW IN A PIPE

The study of developed turbulent flow in a circular pipe is of substantial interest in actual flows since most flows encountered in practical applications are turbulent flows in pipes. Even though for carefully controlled laboratory conditions, laminar flows have been observed up to Reynolds numbers of 40 000 in a pipe flow, turbulent flow is assumed to occur in a pipe under standard operating conditions whenever the Reynolds number

$$\text{Re} = \frac{VD}{\nu} \tag{15.54}$$

exceeds 4000; between 2000 and 4000 the flow is assumed to randomly oscillate between being laminar and turbulent. Consider, for example, water at 20°C flowing in a rather small 5-mm-diameter pipe; the average velocity need only be 0.8 m/s (2 ft/sec in a $\frac{1}{4}$-in. pipe) for turbulent flow to occur. This is the situation whenever we drink water from a drinking fountain. For larger-diameter pipes the average velocity is sufficiently large so that a turbulent flow is produced in most engineering situations.

Key Concept: *Turbulent flow occurs whenever we drink water from a drinking fountain.*

In a turbulent flow all three velocity components are nonzero. If we measure the components as a function of time, graphs similar to those shown in Fig. 15.7 would result for a flow in a pipe where u, v, w are in the x-, r-, and θ-directions, respectively. There is seldom any interest (to the engineer) in the details of the randomly fluctuating velocity components; hence we introduce the notion of a time-average quantity. The velocity components (u, v, w) are written as

$$u = \bar{u} + u' \qquad v = \bar{v} + v' \qquad w = \bar{w} + w' \tag{15.55}$$

Key Concept: *A bar over a quantity denotes a time average.*

where a bar over a quantity denotes a time average and a prime denotes the fluctuating part. Using the component u as an example, the *time average* is defined as,

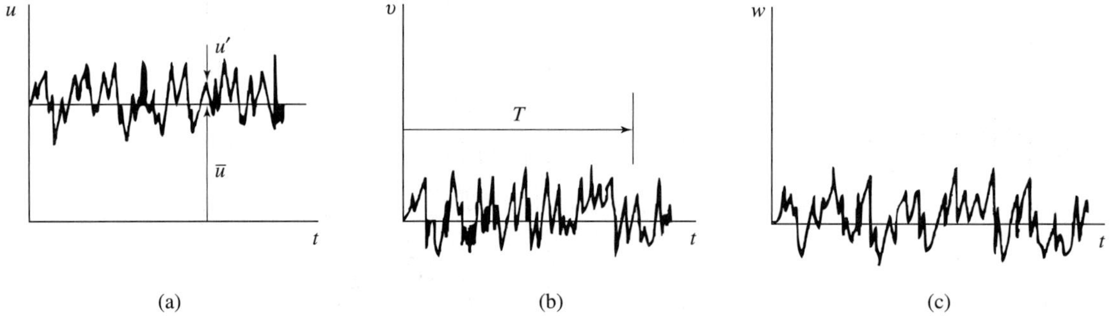

FIGURE 15.7 Velocity components in a turbulent pipe flow: (a) x-component velocity; (b) r-component velocity; (c) θ-component velocity.

$$\bar{u} = \frac{1}{T}\int_0^T u(t)\,dt \tag{15.56}$$

where T is a time increment large enough[6] to eliminate all time dependence from $\bar{u}$. In a developed pipe flow $\bar{u}$ would be nonzero and $\bar{v} = \bar{w} = 0$, as is obvious from Fig. 15.7. If $\bar{u}$ is not a function of time it is a steady turbulent flow.

Example 15.7

Show that $\overline{u'} = 0$ and $\dfrac{\overline{\partial u}}{\partial y} = \dfrac{\partial \bar{u}}{\partial y}$ for a steady turbulent flow.

Solution

To show that $\overline{u'} = 0$ we simply substitute the expression (15.55) for $u(t)$ into Eq. 15.56 and obtain

$$\begin{aligned}\bar{u} &= \frac{1}{T}\int_0^T (\bar{u} + u')\,dt \\ &= \frac{1}{T}\int_0^T \bar{u}\,dt + \frac{1}{T}\int_0^T u'\,dt\end{aligned}$$

Since $\bar{u}$ does not depend on t the aove equation takes the form

$$\begin{aligned}&= \bar{u}\frac{1}{T}\int_0^T dt + \overline{u'} \\ &= \bar{u} + \overline{u'}\end{aligned}$$

Subtracting $\bar{u}$ from both sides results in

$$\overline{u'} = 0$$

Now, let us time average the derivative $\partial u/\partial y$. We have

$$\begin{aligned}\frac{\overline{\partial u}}{\partial y} &= \frac{1}{T}\int_0^T \frac{\partial u}{\partial y}\,dt \\ &= \frac{\partial}{\partial y}\left(\frac{1}{T}\int_0^T u\,dt\right) = \frac{\partial}{\partial y}\bar{u}\end{aligned}$$

since we can move the $\partial/\partial y$ outside the time integral and T is a constant. Thus

$$\frac{\overline{\partial u}}{\partial y} = \frac{\partial \bar{u}}{\partial y}$$

[6] At a given location in a pipe this could be checked experimentally as follows: Start with a relatively large value of T and then decrease T for subsequent runs. If $\bar{u}$ remains unchanged as T decreases, T is sufficiently large. If T is too small, $\bar{u}$ will be different for each run.

15.6.1 Differential Equation

The differential equation that must be solved to provide us with the time-average velocity distribution is derived by using a particle of fluid in a pipe. Consider the situation of a turbulent flow in a horizontal pipe, as shown in Fig. 15.8. We use velocity components u and v in the x- and r-directions, respectively. Fluid particles[7] move randomly throughout the flow. At an instant in time a fluid particle moves through the incremental area dA due to the velocity fluctuation v'; it enters a neighboring layer of fluid which is moving with a higher x-component velocity, thereby providing a retarding effect on the neighboring layer. A fluid particle that moves to a neighboring layer that is traveling with a lower x-component velocity would tend to accelerate the slower moving fluid. The x-component force that results due to the random motion of a fluid particle passing through the incremental area dA would be (see Eq. 13.69)

$$dF = -\rho v'\, dA\, u' \tag{15.57}$$

where u' is the negative change in x-component velocity due to the momentum exchange and $\rho v'\, dA$ is the mass flux through the area; the negative sign provides a positive dF. If we divide both sides by the area dA we obtain a "stress" that we call the *turbulent shear stress*. It is

$$\tau_{\text{turb}} = \frac{dF}{dA} = -\rho u'v' \tag{15.58}$$

Key Concept: *A positive v' produces a negative u'.*

where we know that $(u'v')$ is, on the average, a negative quantity since a positive v' produces a negative u'. This "shear stress" is actually a momentum exchange but since it has the same effect as a stress, we call it a shear stress.

The time-average turbulent shear stress, often called the *apparent shear stress*, which is of primary interest, is

$$\overline{\tau}_{\text{turb}} = -\rho\overline{u'v'} \tag{15.59}$$

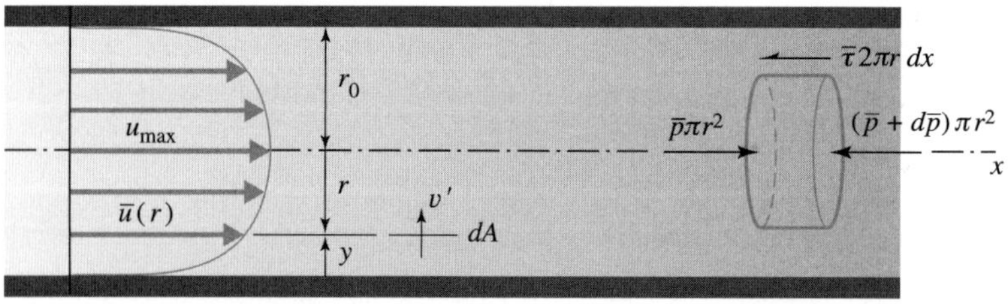

FIGURE 15.8 Turbulent flow in a horizontal pipe.

[7]A particle is a relatively small mass of the fluid, consisting of a large number of molecules, that tends to move as a unit.

where Eq. 15.56 would be used for the time averages. Note that $\overline{u'w'}$ would be zero since a w'-component (in the θ-direction) would not move a fluid particle into a layer of higher or lower x-component velocity. Also, $\overline{v'w'} = 0$ using the same logic.

The total shear stress at a particular location would be due to both the viscosity and the momentum exchange described above; that is,

$$\begin{aligned}\overline{\tau} &= \overline{\tau}_{\text{lam}} + \overline{\tau}_{\text{turb}} \\ &= \mu \frac{\partial \overline{u}}{\partial y} - \rho \overline{u'v'}\end{aligned} \tag{15.60}$$

The shear stress can be related to the pressure gradient by summing forces on the horizontal cylindrical element shown in Fig. 15.8. There results

$$\overline{\tau} = -\frac{r}{2}\frac{d\overline{p}}{dx} = -\frac{r\Delta\overline{p}}{2L} \tag{15.61}$$

which shows that the shear stress distribution is linear for a turbulent flow as well as a laminar flow (see Eq. 15.14). The turbulent shear obviously goes to zero at the wall since the velocity perturbations are zero at the wall, and the total shear is zero at the centerline where $r = 0$ or $y = r_0$, as shown in Fig. 15.9. The viscous shear is nonzero only in a very thin viscous wall layer, of thickness δ_v, near the wall, as shown in part (b). Note that the turbulent shear reaches a maximum near the wall in the viscous wall layer.

Key Concept: *The viscous shear is nonzero only in a very thin viscous wall layer.*

The differential equation that must be solved if the time-average velocity distribution is to be determined is found by combining the two preceding equations; it is

$$\frac{r}{2}\frac{d\overline{p}}{dx} = \mu \frac{d\overline{u}}{dr} + \rho \overline{u'v'} \tag{15.62}$$

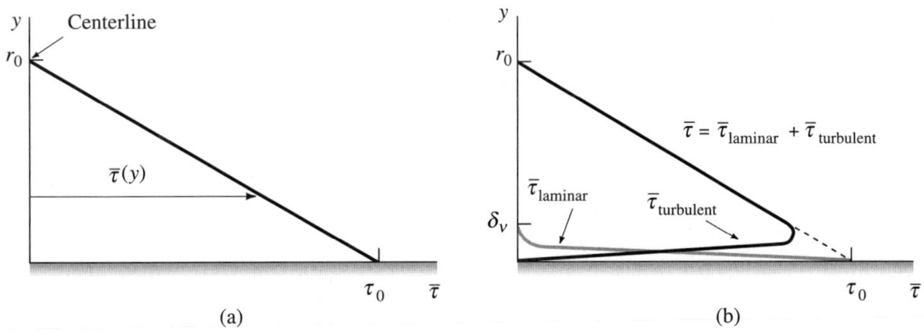

FIGURE 15.9 Shear stress distributions in a developed pipe flow.

Key Concept: *The quantity $\overline{u'v'}$ cannot be determined analytically.*

where we have used $r + y = r_0$ so that $dy = -dr$. For developed flow we know that $d\bar{p}/dx = \text{const.}$; hence if we know how $\overline{u'v'}$ varied with r, the differential equation could be solved. The quantity $\overline{u'v'}$ cannot be determined analytically, however, so the solution to Eq. 15.62 cannot be attempted until an empirical expression is found for $\overline{u'v'}$. Rather than finding an empirical expression for $\overline{u'v'}$ and then solving the differential equation for $\bar{u}$, we will simply present the empirical results obtained for the velocity profile $\bar{u}(y)$. However, before we do this in the next section we should introduce the eddy viscosity, the mixing length, and the correlation coefficient.

Instead of using the quantity $\overline{u'v'}$ as the unknown in Eq. 15.62, we often introduce the *eddy viscosity* η, defined by the relationship

$$\overline{u'v'} = \eta \frac{d\bar{u}}{dy} \tag{15.63}$$

Note that η has the same dimensions as the kinematic viscosity. In terms of the eddy viscosity the differential equation becomes

$$\frac{r}{2}\frac{d\bar{p}}{dx} = \rho(\nu + \eta)\frac{d\bar{u}}{dr} \tag{15.64}$$

If we view the turbulent process as the random and chaotic mixing of particles of fluid, we may choose to introduce the *mixing length* l_m, the distance a particle travels before interacting with another particle. Based on reasoning related to momentum interchange we relate the eddy viscosity to the mixing length with

$$\eta = l_m^2 \left|\frac{d\bar{u}}{dy}\right| \tag{15.65}$$

The *correlation coefficient* K_{uv}, a normalized turbulent shear stress, often used when describing turbulent motions, has limits of ± 1 and is defined by

$$K_{uv} = \frac{\overline{u'v'}}{\sqrt{\overline{u'^2}}\sqrt{\overline{v'^2}}} \tag{15.66}$$

where the time-average quantities would be defined by Eq. 15.56.

Key Concept: *The quantities η, l_m and K_{uv} simply replace the variable $\overline{u'v'}$.*

The quantities η, l_m, and K_{uv} are functions of r (or y) that simply replace the variable $\overline{u'v'}$. They do not simplify the differential equation 15.62; they allow the equation to take slightly different forms. Since we cannot derive an expression for η, l_m, or K_{uv} we cannot find $\bar{u}(r)$ using analytical methods. We must rely on experimental data. It follows in the next section.

Example 15.8

Note that in Fig. 15.9b there is a region near the wall where the turbulent shear is near its maximum and is relatively constant, as shown in Fig. E15.8, and the viscous shear is quite small. Assume that the mixing length is directly proportional to the distance from the wall. With this assumption show that $\overline{u}(y)$ is logarithmic in this region near the wall.

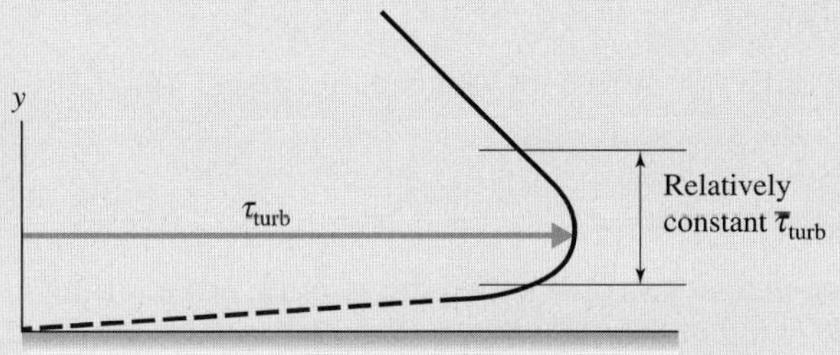

FIGURE E15.8

Solution

If the viscous shear is negligible, we have, combining Eqs. 15.59 and 15.63, and 15.65,

$$\overline{\tau}_{turb} = \rho\eta \frac{d\overline{u}}{dy} = \rho l_m^2 \left(\frac{d\overline{u}}{dy}\right)^2$$

Now, if $\overline{\tau}_{turb} = \text{const.} = c_1$ and we assume that

$$l_m = c_2 y$$

there results

$$c_1 = \rho c_2^2 y^2 \left(\frac{d\overline{u}}{dy}\right)^2$$

or

$$y \frac{d\overline{u}}{dy} = c_3$$

where $c_3 = \sqrt{c_1/\rho c_2^2}$. This is integrated to yield

$$\overline{u}(y) = c_3 \ln y + c_4$$

Hence, with the foregoing assumptions we see that a logarithmic profile is predicted for the region of constant turbulent shear near the wall. This is, in fact, observed from experimental data; so we conclude that the above assumptions are reasonable for a turbulent flow in a pipe.

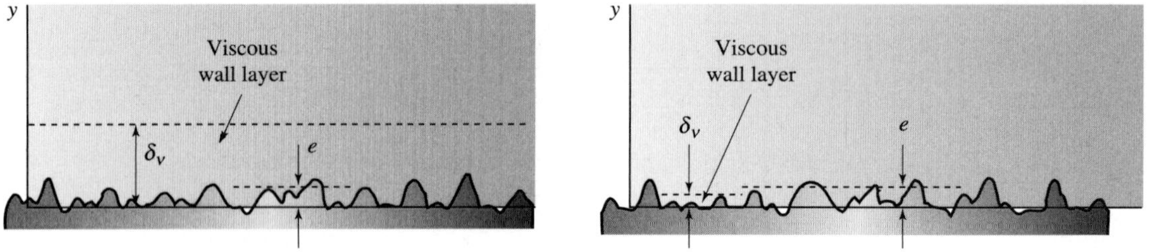

(a) (b)

FIGURE 15.10 (a) A smooth wall and (b) a rough wall.

15.6.2 Velocity Profile

The time-average velocity profile in a pipe is quite sensitive to the magnitude of the average wall roughness height e, as sketched in Fig. 15.10. All materials are "rough" when viewed with sufficient magnification, although glass and plastic are assumed to be smooth with $e = 0$. (Values for e are listed in Fig. 15.13.) As noted in the preceding section, the laminar shear is significant only near the wall in the viscous wall layer with thickness δ_ν. If the thickness δ_ν is sufficiently large, it submerges the wall roughness elements so that they have negligible effect on the flow; it is as if the wall were smooth. Such a condition is often referred to as being *hydraulically smooth*. If the viscous wall layer is relatively thin, the roughness elements protrude out of this layer and the wall is rough. The *relative roughness* e/D and the Reynolds number can be used to determine if a pipe is smooth or rough. This will be observed from the friction factor data presented in the next section.

Key Concept: *If the thickness δ_ν is sufficiently large, it is as if the wall were smooth. The pipe is hydraulically smooth.*

We do not solve for the velocity distribution in a developed turbulent flow since $\overline{u'v'}$ cannot be determined analytically. Thus we will present the empirical data for $\bar{u}(y)$ directly and not solve Eq. 15.62. The two most common expressions for turbulent flow are now presented.

The first method of empirically expressing the velocity distribution involves flows with smooth walls and flows with rough walls. If the flow has a smooth wall, as in Fig. 15.10a, we identify two regions of the flow, the *wall region* and the *outer region*. In the wall region the characteristic velocity and length are the **shear velocity** u_τ defined by[8] $u_\tau = \sqrt{\tau_0/\rho}$, and the **viscous length** ν/u_τ. The dimensionless velocity distribution in the wall region, for a smooth pipe, is

Shear velocity: *The quantity $\sqrt{\tau_0/\rho}$ is the shear velocity u_τ.*

Viscous length: *The length ν/u_τ is a characteristic length in a turbulent flow.*

$$\frac{\bar{u}}{u_\tau} = \frac{u_\tau y}{\nu} \quad \text{(viscous wall layer)} \qquad 0 \le \frac{u_\tau y}{\nu} \lesssim 5 \qquad (15.67)$$

$$\frac{\bar{u}}{u_\tau} = 2.44 \ln \frac{u_\tau y}{\nu} + 4.9 \quad \text{(turbulent region)} \qquad 30 < \frac{u_\tau y}{\nu}, \frac{y}{r_0} < 0.15 \qquad (15.68)$$

[8]The shear velocity u_τ is a fictitious velocity and has no relationship to an actual velocity in the flow. It is a quantity with dimensions of velocity that allows the experimental data to be presented in dimensionless form as *universal profiles* (profiles that are valid for all turbulent pipe flows). The viscous length ν/u_τ is also a fictitious length.

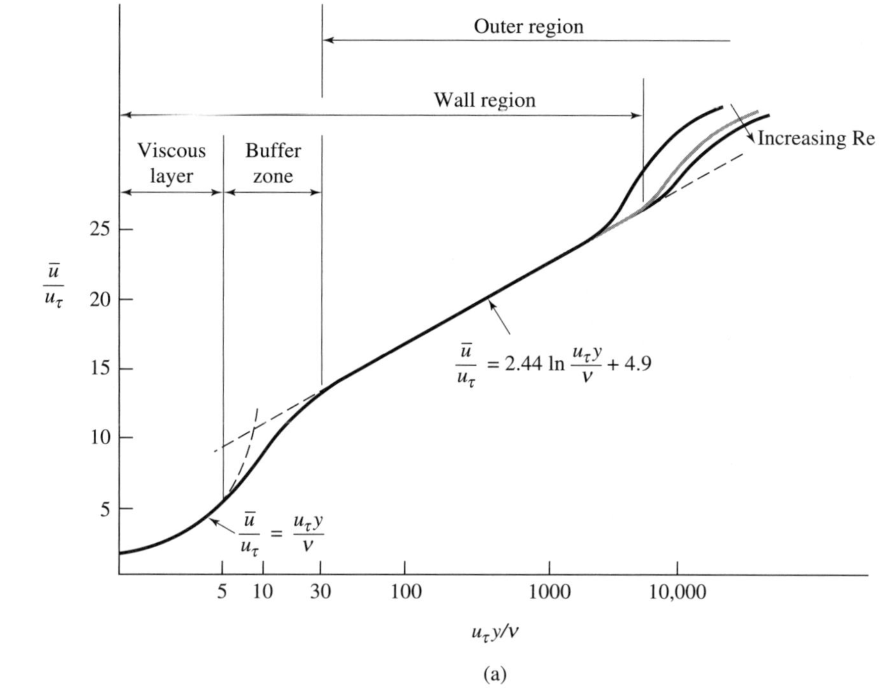

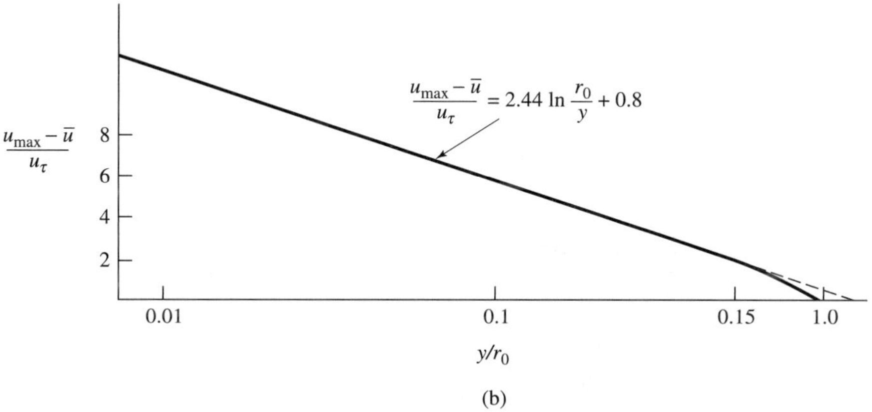

FIGURE 15.11 Empirical relations for turbulent flow in a smooth pipe: (a) wall region; (b) outer region. (Based on data from J. Laufer, The Structure of Turbulence in Fully Developed Pipe Flow, *NACA Report 1174*, 1954.)

In the interval $5 < u_\tau y/\nu < 30$, the *buffer zone*, the experimental data do not fit either of the curves above but merge the two curves as shown in Fig. 15.11a. The viscous wall layer has thickness δ_ν and it is in the viscous layer that turbulence is thought to be initiated; this layer possesses a time-average, linear velocity distribution, but instantaneously the layer is very time dependent. The outer edge of the

Key Concept: *It is in the viscous wall layer that turbulence is thought to be initiated.*

wall region is quite dependent on the Reynolds number, as shown; for low Re it may be located near $u_\tau y/\nu = 3000$.

For rough pipes the viscous wall layer does not play an important role since turbulence initiates from the protruding wall elements, so only a logarithmic profile is necessary in the wall region. The characteristic length is the average roughness height e; the dimensionless velocity profile for the rough pipe is

$$\frac{\bar{u}}{u_\tau} = 2.44 \ln \frac{y}{e} + 8.5 \qquad \frac{y}{r_0} < 0.15 \qquad \text{(wall region)} \tag{15.69}$$

where the constants 8.5 and 2.44 allow a good fit with experimental data.

In the outer region, sketched in Fig. 15.11b, the characteristic length is r_0; the velocity defect $(u_{max} - \bar{u})$ is normalized with u_τ, and the empirical relation, for both smooth and rough pipes, is

$$\frac{u_{max} - \bar{u}}{u_\tau} = 2.44 \ln \frac{r_0}{y} + 0.8 \quad \text{(outer region)} \qquad \frac{y}{r_0} \leq 0.15 \tag{15.70}$$

An additional empirical equation is needed to complete the profile for $0.15 < y/r_0 \leq 1$.

The wall region and the outer region overlap as shown in Fig. 15.11a. In this overlap region we can combine the equations above to obtain an expression for the maximum velocity; for a smooth pipe it is

$$\frac{u_{max}}{u_\tau} = 2.44 \ln \frac{u_\tau r_0}{\nu} + 5.7 \qquad \text{(smooth pipes)} \tag{15.71}$$

and for a rough pipe we find that

$$\frac{u_{max}}{u_\tau} = 2.44 \ln \frac{r_0}{e} + 9.3 \qquad \text{(rough pipes)} \tag{15.72}$$

Although we do not often desire the actual time-average velocity at a specified radial position in a pipe, the distributions above are of occasional use and are presented for completeness. Note, however, that before u_{max} can be found we must know u_τ; before u_τ can be found we must know τ_0. To find τ_0 we use the pressure gradient,

$$\tau_0 = -\frac{r_0}{2}\frac{dp}{dx} \tag{15.73}$$

or the friction factor with Eq. 15.16. If neither dp/dx nor f is known, we may use the power-law form of the profile, described in the following paragraph, to approximate f.

An alternative, and simpler form that adequately describes the turbulent flow velocity distribution in a pipe is the *power-law profile*, namely,

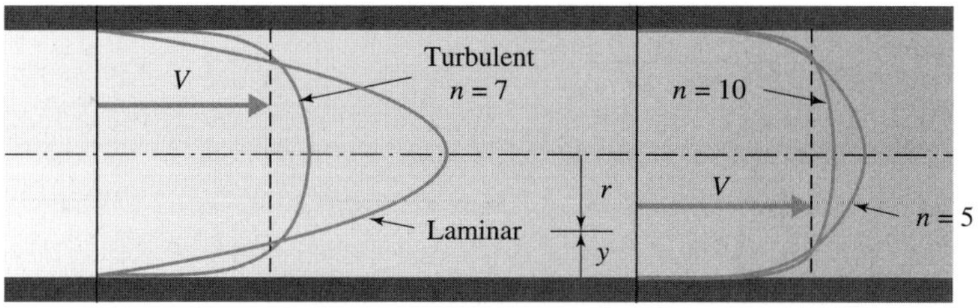

FIGURE 15.12 Turbulent velocity profile.

$$\frac{\overline{u}}{u_{\max}} = \left(\frac{y}{r_0}\right)^{1/n} \tag{15.74}$$

where y is measured from the pipe wall and n is an integer between 5 and 10. Using this distribution the average velocity is found to be

$$V = \frac{\int_0^{r_0} \overline{u}(r) 2\pi r\, dr}{\pi r_0^2} = \frac{2n^2}{(n+1)(2n+1)} u_{\max} \tag{15.75}$$

This distribution is compared with a laminar profile in Fig. 15.12.

The value of n in the exponent is related to the friction factor f by the empirical expression

$$n = \frac{1}{\sqrt{f}} \tag{15.76}$$

The constant n varies from 5 to 10 depending on the Reynolds number and the pipe wall roughness e/D. For smooth pipes the exponent n is related to the Reynolds number as shown in Table 15.1.

The power-law profile cannot be used to obtain the slope at the wall since it will always yield $(d\overline{u}/dy)_{\text{wall}} = \infty$ for all n. Thus it cannot be used to predict the wall shear stress. The wall shear is found by combining Eqs. 15.76 and 15.16. In addition, it gives a positive slope $d\overline{u}/dy$ at the centerline of the pipe, where the slope should be zero, so it is not valid near the centerline.

Key Concept: *The power-law profile cannot be used to obtain the slope at the wall or the slope at the centerline.*

It should be noted that the kinetic-energy-correction factor α (see Eq. 13.63) in pipes is 1.11, 1.06, and 1.03 for $n = 5, 7$, and 10, respectively. Because it is close to unity for $n > 7$, it is often set equal to unity in the energy equation when working problems involving turbulent flow.

TABLE 15.1 Exponent n for Smooth Pipes

$\text{Re} = VD/\nu$	4×10^3	10^5	10^6	$>2 \times 10^6$
n	6	7	9	10

Example 15.9

Water at 20°C flows in a 10-cm-diameter pipe at an average velocity of 1.6 m/s. If the roughness elements are 0.046 mm high, would the wall be rough or smooth? Refer to Fig. 15.10.

Solution

To determine if the wall is rough or smooth, we must compare the viscous wall layer thickness with the height of the roughness elements. So, let's find the viscous wall layer thickness. From Fig. 15.11 the viscous layer thickness is determined by letting $u_\tau y/\nu = 5$, where $y = \delta_\nu$. First, we must find u_τ. The Reynolds number is

$$\text{Re} = \frac{VD}{\nu}$$

$$= \frac{1.6 \times 0.1}{10^{-6}} = 1.6 \times 10^5$$

From Table 15.1 $n \simeq 7.5$, so that, from Eq. 15.76,

$$f = \frac{1}{n^2}$$

$$= \frac{1}{7.5^2} = 0.018$$

The wall shear is calculated from Eq. 15.16:

$$\tau_0 = \tfrac{1}{8}\rho V^2 f$$

$$= \tfrac{1}{8} \times 1000 \times 1.6^2 \times 0.018 = 5.8 \text{ Pa}$$

The friction velocity is

$$u_\tau = \sqrt{\frac{\tau_0}{\rho}}$$

$$= \sqrt{\frac{5.8}{1000}} = 0.076 \text{ m/s}$$

This allows us to calculate the viscous wall layer thickness:

$$\delta_\nu = \frac{5\nu}{u_\tau}$$

$$= \frac{5 \times 10^{-6}}{0.076} = 6.6 \times 10^{-5} \text{ m} \quad \text{or} \quad 0.066 \text{ mm}$$

Since the roughness elements are only 0.046 mm high, they are submerged in the viscous wall layer. Consequently, the wall is smooth (see Fig. 15.10a). If the pipe were made of cast iron with $e = 0.26$ mm, the wall would be rough.

Note that the viscous wall layer, even at this relatively low velocity, is about 0.1% of the radius. The viscous wall layer is usually extremely thin.

Example 15.10

The 4-cm-diameter smooth, horizontal pipe of Fig. E15.10 transports 0.004 m^3/s of water at 20°C . Using the power-law profile, approximate (a) the friction factor, (b) the maximum velocity, (c) the radial position where $u = V$, (d) the wall shear, (e) the pressure drop over a 10-m length, and (f) the maximum velocity using Eq. 15.71.

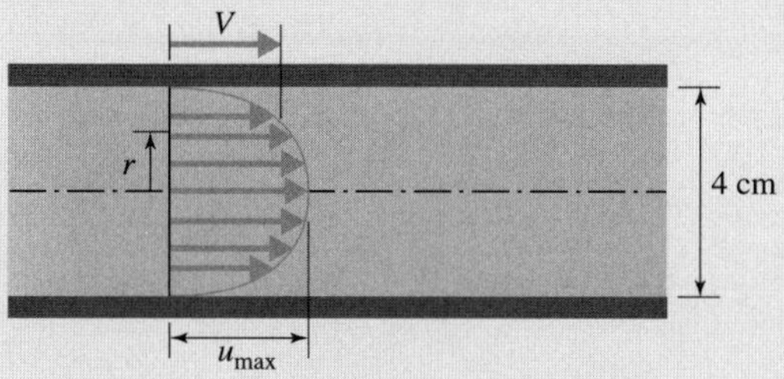

FIGURE E15.10

Solution

(a) The average velocity is calculated to be

$$V = \frac{Q_f}{A} = \frac{0.004}{\pi \times 0.02^2} = 3.18 \text{ m/s}$$

The Reynolds number is

$$\text{Re} = \frac{VD}{\nu} = \frac{3.18 \times 0.04}{10^{-6}} = 1.27 \times 10^5$$

From Table 15.1 we see that $n \cong 7.5$ and from Eq. 15.76,

$$f = \frac{1}{n^2}$$
$$= \frac{1}{7.5^2} = 0.018$$

(b) The maximum velocity is found using Eq. 15.75 to be

$$u_{max} = \frac{(n+1)(2n+1)}{2n^2} V$$
$$= \frac{8.5 \times 16}{2 \times 7.5^2} \times 3.18 = 3.84 \text{ m/s}$$

(c) The distance from the wall where $u = V = 3.18$ m/s is found using Eq. 15.74 as follows:

$$\frac{\bar{u}}{u_{max}} = \left(\frac{y}{r_0}\right)^{1/7.5}$$
$$\therefore y = r_0 \left(\frac{u}{u_{max}}\right)^{7.5}$$
$$= 2\left(\frac{3.18}{3.84}\right)^{7.5} = 0.49 \text{ cm}$$

(continued)

The radial position is thus

$$r = r_0 - y$$
$$= 2 - 0.49 = 1.51 \text{ cm}$$

(d) The wall shear is found using Eq. 15.16 and is

$$\tau_0 = \tfrac{1}{8}\rho V^2 f$$
$$= \tfrac{1}{8} \times 1000 \times 3.18^2 \times 0.018 = 23 \text{ Pa}$$

(e) The pressure drop is calculated using Eq. 15.73 with $\Delta p/L = -dp/dx$ to be

$$\Delta p = \frac{2\tau_0 L}{r_0}$$
$$= \frac{2 \times 23 \times 10}{0.02} = 23\,000 \text{ Pa} \quad \text{or} \quad 23 \text{ kPa}$$

where $\Delta p/L = -\partial p/\partial x$.

(f) To use Eq. 15.71 we must know the shear velocity. It is

$$u_\tau = \sqrt{\frac{\tau_0}{\rho}}$$
$$= \sqrt{\frac{23}{1000}} = 0.152 \text{ m/s}$$

Then we find u_{max} to be

$$u_{max} = 0.152\left(2.44 \ln \frac{0.152 \times 0.02}{10^{-6}} + 5.7\right) = 3.84 \text{ m/s}$$

the same as that given by the power-law formula in part (b). This answer is considered to be more accurate if it differs from that of Eq. 15.75. Note that the experimental data do not allow for accuracy in excess of three significant digits, and often to only two significant digits.

15.6.3 Losses in Developed Pipe Flow

Perhaps the most calculated quantity in pipe flow is the head loss. If the head loss is known in a developed flow, the pressure change can be calculated; for developed flow in a pipe the energy equation (13.51) provides us with

$$h_L = \frac{\Delta(p + \gamma h)}{\gamma} \tag{15.77}$$

The head loss that results from the wall shear in a developed flow is related to the friction factor (see Eq. 15.17) by the Darcy–Weisbach equation, namely,

$$h_L = f\frac{L}{D}\frac{V^2}{2g} \tag{15.78}$$

Key Concept: *If the friction factor is known, we can find the head loss and the pressure drop.*

Consequently, if the friction factor is known, we can find the head loss and then the pressure drop.

The friction factor f depends on the various quantities that affect the flow, written as

$$f = f(\rho, \mu, V, D, e) \tag{15.79}$$

where the average wall roughness height e accounts for the influence of the wall roughness elements. A dimensional analysis following the steps of Section 14.2 provides us with

$$f = f\left(\frac{\rho V D}{\mu}, \frac{e}{D}\right) \tag{15.80}$$

where e/D is the relative roughness.

Experimental data that relate the friction factor to the Reynolds number have been obtained for fully developed pipe flow over a wide range of wall roughnesses. The results of these data are presented in Fig. 15.13, which is commonly referred to as the *Moody diagram*. There are several features of the Moody diagram that should be noted.

Key Concept: *For a sufficiently large value of Re, the friction factor is constant.*

- For a given wall roughness, measured by the relative roughness e/D, there is a sufficiently large value of Re above which the friction factor is constant, thereby defining the *completely turbulent regime*. The average roughness element size e is substantially greater than the viscous wall layer thickness δ_v, so that viscous effects are not significant; the resistance to the flow is produced primarily by the drag of the roughness elements that protrude into the flow.
- For the smaller relative roughness e/D values it is observed that, as Re decreases, the friction factor increases in the *transition zone* and eventually becomes the same as that of a smooth pipe. The roughness elements become submerged in the viscous wall layer so that they produce little effect on the main flow.
- For Reynolds numbers less than 2000, the friction factor of laminar flow is shown. The *critical zone* couples the turbulent flow to the laminar flow and may represent an oscillatory flow that alternately exists between turbulent and laminar flow.
- The e values in this diagram are for new pipes. With age a pipe will corrode and become fouled, changing both the roughness and the pipe diameter, with a resulting increase in the friction factor. Such factors should be included in design considerations; they will not be reviewed here.

The following empirical equations represent the Moody diagram for $\text{Re} > 4000$:

Smooth pipe flow:
$$\frac{1}{\sqrt{f}} = 0.86 \ln \text{Re}\sqrt{f} - 0.8 \tag{15.81}$$

Completely turbulent zone:
$$\frac{1}{\sqrt{f}} = -0.86 \ln \frac{e}{3.7D} \tag{15.82}$$

Transition zone:
$$\frac{1}{\sqrt{f}} = -0.86 \ln\left(\frac{e}{3.7D} + \frac{2.51}{\text{Re}\sqrt{f}}\right) \tag{15.83}$$

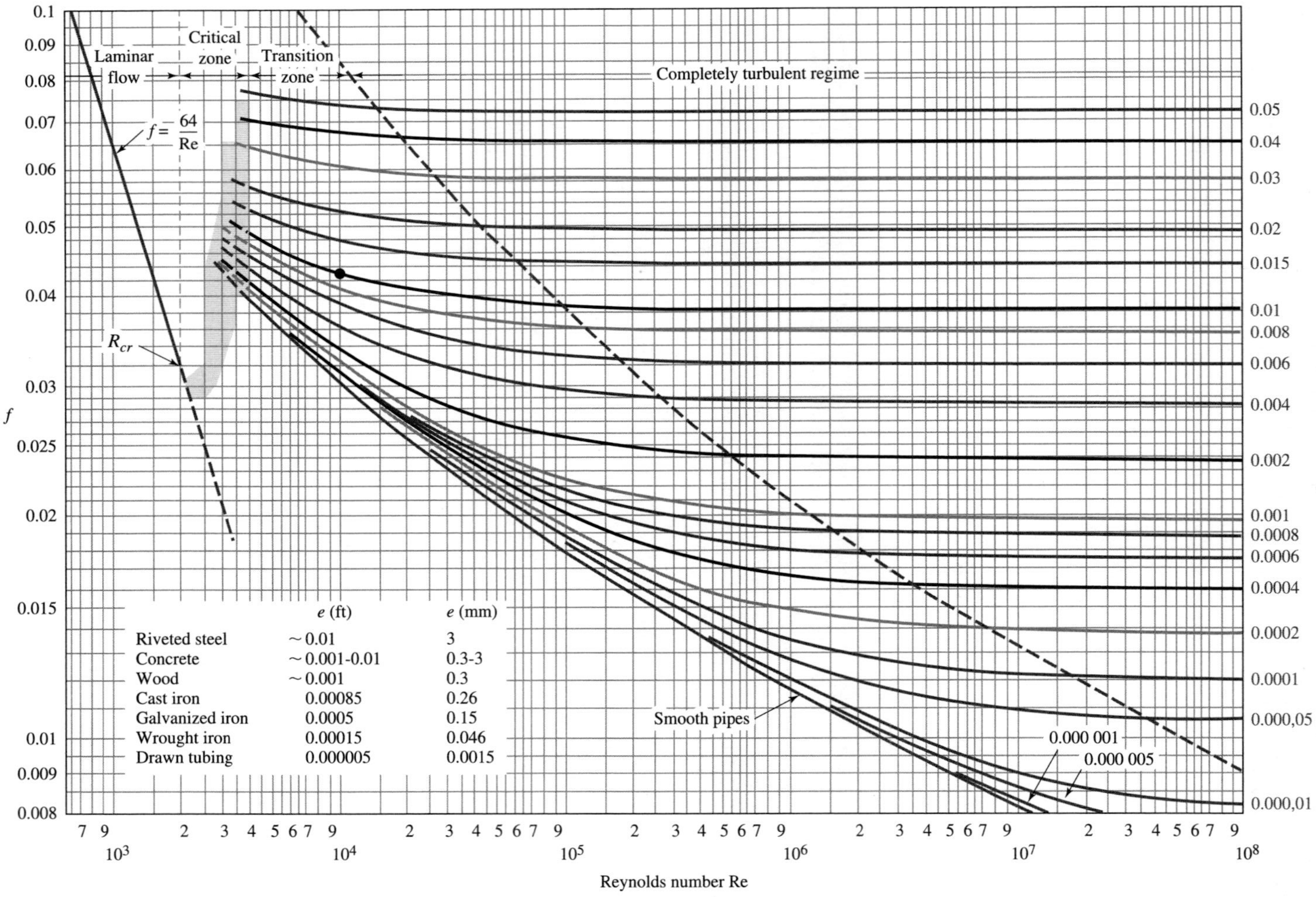

FIGURE 15.13 Moody diagram. (From L. F. Moody, *Trans. ASME*, Vol. 66, 1944.) (Note: If $e/D = 0.006$ and $Re = 10^4$, the dot locates $f = 0.043$.)

The transition zone equation (15.83) that couples the smooth pipe equation to the completely turbulent regime equation is known as the **Colebrook equation**. Note that Eq. 15.81 is the Colebrook equation with $e = 0$, and Eq. 15.82 is the Colebrook equation with $\text{Re} = \infty$.

Colebrook equation: *The equation that couples the smooth pipe equation to the completely turbulent regime equation.*

Three categories of problems can be identified for developed turbulent flow in a straight pipe of length L with no geometric changes:

Category	Known	Unknown
1	Q_f, D, e, ν	h_L
2	D, e, ν, h_L	Q_f
3	Q_f, e, ν, h_L	D

A category 1 problem is straightforward and requires no iteration procedure when using the Moody diagram. Category 2 and 3 problems are more like problems encountered in engineering design situations and require an iterative trial-and-error process when using the Moody diagram. Each of these types will be illustrated with an example.

Key Concept: *Category 2 and 3 problems require an iterative trial-and-error process.*

An alternative to using the Moody diagram that avoids any trial-and-error process is made possible by empirically derived formulas. Perhaps the best of such formulas were presented by Swamee and Jain (1976) for pipe flow; an explicit expression that provides an approximate value for the unknown in each category above is as follows:

$$h_L = 1.07\frac{Q_f^2 L}{gD^5}\left\{\ln\left[\frac{e}{3.7D} + 4.62\left(\frac{\nu D}{Q_f}\right)^{0.9}\right]\right\}^{-2} \qquad \begin{array}{l} 10^{-6} < e/D < 10^{-2} \\ 3000 < \text{Re} < 3 \times 10^8 \end{array} \tag{15.84}$$

$$Q_f = -0.965\left(\frac{gD^5 h_L}{L}\right)^{0.5} \ln\left[\frac{e}{3.7D} + \left(\frac{3.17\nu^2 L}{gD^3 h_L}\right)^{0.5}\right] \qquad \text{Re} > 2000 \tag{15.85}$$

$$D = 0.66\left[e^{1.25}\left(\frac{LQ_f^2}{gh_L}\right)^{4.75} + \nu Q_f^{9.4}\left(\frac{L}{gh_L}\right)^{5.2}\right]^{0.04} \qquad \begin{array}{l} 10^{-6} < e/D < 10^{-2} \\ 5000 < \text{Re} < 3 \times 10^8 \end{array} \tag{15.86}$$

One may use either English or SI units in the equations above.

Equation 15.85 is as accurate as the Moody diagram, and Eqs. 15.84 and 15.86 are accurate to within approximately 2% of the Moody diagram. These tolerances are acceptable for engineering calculations. It is important to realize that the Moody diagram is based on experimental data that likely is accurate to within no more than 5%. Hence the foregoing three formulas of Swamee and Jain, which can easily be input on a programmable hand-held calculator, are often used by design engineers. The following examples also illustrate the use of these approximate formulas.

Key Concept: *The Moody diagram is accurate to within no more than 5%.*

Example 15.11

Water at 74°F is transported for 1500 ft in a $1\frac{1}{2}$-in.-diameter wrought iron horizontal pipe with a flow rate of 0.1 ft^3/sec. Calculate the pressure drop over the 1500-ft length of pipe, using (a) the Moody diagram and (b) the alternate method.

Solution

(a) The average velocity is

$$V = \frac{Q_f}{A} = \frac{0.1}{\pi \times 0.75^2/144} = 8.15 \text{ ft/sec}$$

The Reynolds number is

$$\text{Re} = \frac{VD}{\nu} = \frac{8.15 \times 1.5/12}{10^{-5}} = 1.02 \times 10^5$$

Obtaining e from Fig. 15.13, we have, using $D = 1.5/12$ ft,

$$\frac{e}{D} = \frac{0.00015}{0.125} = 0.0012$$

The friction factor is read from the Moody diagram to be

$$f = 0.023$$

The head loss is calculated as

$$h_L = f\frac{L}{D}\frac{V^2}{2g}$$

$$= 0.023\frac{1500}{1.5/12}\frac{8.15^2}{2 \times 32.2} = 280 \text{ ft}$$

This answer is given to two significant numbers since the friction factor is known to at most two significant numbers. The pressure drop is found by Eq. 15.77 to be

$$\Delta p = \gamma h_L$$

$$= 62.4 \times 280 = 17\,500 \text{ psf} \quad \text{or} \quad 120 \text{ psi}$$

(b) The alternate method for this Category 1 problem uses Eq. 15.84, with $D = 1.5/12 = 0.125$ ft:

$$h_L = 1.07\frac{0.1^2 \times 1500}{32.2 \times 0.125^5}\left\{\ln\left[\frac{0.0012}{3.7} + 4.62\left(\frac{10^{-5} \times 0.125}{0.1}\right)^{0.9}\right]\right\}^{-2}$$

$$= 1.07 \times 15\,265 \times 0.01734 = 280 \text{ ft}$$

This much simpler method provides the same value as that found using the Moody diagram.

Example 15.12

A pressure drop of 700 kPa is measured over a 300-m length of horizontal, 10-cm-diameter wrought iron pipe that transports oil ($S = 0.9$, $\nu = 10^{-5}$ m^2/s). Calculate the flow rate using (a) the Moody diagram, and (b) the alternate method.

Solution

(a) The relative roughness is

$$\frac{e}{D} = \frac{0.046}{100} = 0.00046$$

Assuming that the flow is completely turbulent (Re is not needed), the Moody diagram gives

$$f = 0.0165$$

The head loss is found to be

$$h_L = \frac{\Delta p}{\gamma_{\text{oil}}} = \frac{700\,000}{9800 \times 0.9} = 79.4 \text{ m}$$

The velocity is calculated from Eq. 15.78 to be

$$V = \left(\frac{2gDh_L}{fL}\right)^{1/2} = \left(\frac{2 \times 9.8 \times 0.1 \times 79.4}{0.0165 \times 300}\right)^{1/2} = 5.61 \text{ m/s}$$

This provides us with a Reynolds number of

$$\text{Re} = \frac{VD}{\nu} = \frac{5.61 \times 0.1}{10^{-5}} = 5.61 \times 10^4$$

Using this Reynolds number and $e/D = 0.00046$, the Moody diagram gives the friction factor as

$$f = 0.023$$

This corrects the original value for f. The velocity is recalculated to be

$$V = \left(\frac{2 \times 9.8 \times 0.1 \times 79.4}{0.023 \times 300}\right)^{1/2} = 4.75 \text{ m/s}$$

The Reynolds number is then

$$\text{Re} = \frac{4.75 \times 0.1}{10^{-5}} = 4.75 \times 10^4$$

From the Moody diagram $f = 0.023$ appears to be satisfactory. Thus the flow rate is

$$Q_f = VA = 4.75 \times \pi \times 0.05^2 = 0.037 \text{ m}^3/\text{s}$$

Only two significant numbers are given since f is known to at most two significant numbers.

(b) The alternative method for this Category 2 problem uses the explicit relationship (15.85). We can directly calculate Q_f to be

$$Q_f = -0.965\left(\frac{9.8 \times 0.1^5 \times 79.4}{300}\right)^{0.5} \ln\left[\frac{0.00046}{3.7} + \left(\frac{3.17 \times 10^{-10} \times 300}{9.8 \times 0.1^3 \times 79.4}\right)^{0.5}\right]$$
$$= 0.038 \text{ m}^3/\text{s}$$

This much simpler method produces in value essentially the same as that obtained using the Moody diagram.

Example 15.13

Drawn tubing of what diameter should be selected to transport 0.002 m^3/s of 20°C water over a 400-m length so that the head loss does not exceed 30 m? (a) Use the Moody diagram and (b) the alternative method.

Solution

(a) In this problem we do not know D. Thus, a trial-and-error solution is anticipated. The average velocity is related to D by

$$V = \frac{Q_f}{A} = \frac{0.002}{\pi D^2/4} = \frac{0.00255}{D^2}$$

The friction factor and D are related as follows:

$$h_L = f\frac{L}{D}\frac{V^2}{2g}$$

$$30 = f\frac{400}{D}\frac{(0.00255/D^2)^2}{2 \times 9.8}$$

$$\therefore\ D^5 = 4.42 \times 10^{-6} f$$

The Reynolds number is

$$\text{Re} = \frac{VD}{\nu} = \frac{0.00255D}{D^2 \times 10^{-6}} = \frac{2550}{D}$$

Now, let us simply guess a value for f and check with the relations above and the Moody diagram. The first guess is $f = 0.03$ and the correction is listed in the following table. *Note*: The second guess is the value for f found from the calculations of the first guess.

f	D(m)	Re	e/D	f (Fig. 15.13)
0.03	0.0421	6.06×10^4	0.000036	0.02
0.02	0.0388	6.57×10^4	0.000039	0.02

The value of $f = 0.02$ is acceptable, yielding a diameter of 3.88 cm. Since this diameter would undoubtedly not be standard, a diameter of

$$D = 4 \text{ cm}$$

would be the tube size selected. This tube would have a head loss less than the limit of $h_L = 30$ m imposed in the problem statement. Any larger-diameter tube would also satisfy this criterion but would be more costly, so should not be selected.

(b) The alternative method for this Category 3 problem uses the explicit relationship (15.86). We can directly calculate D to be

$$D = 0.66\left[(1.5 \times 10^{-6})^{1.25}\left(\frac{400 \times 0.002^2}{9.81 \times 30}\right)^{4.75} + 10^{-6} \times 0.002^{9.4}\left(\frac{400}{9.81 \times 30}\right)^{5.2}\right]^{0.04}$$

$$= 0.66[5.163 \times 10^{-33} + 2.102 \times 10^{-31}]^{0.04} = 0.039 \text{ m}$$

Hence $D = 4$ cm would be the tube size selected. This is the same tube size as that selected using the Moody diagram.

15.6.4 Losses in Noncircular Conduits

Good approximations can be made for the head loss in conduits with noncircular cross sections by using the *hydraulic radius R*, defined by

$$R = \frac{A}{P} \tag{15.87}$$

where A is the cross-sectional area and P is the **wetted perimeter**, that perimeter where the fluid is in contact with the solid boundary. For a circular pipe flowing full the hydraulic radius is $R = r_0/2$. Hence we simply replace the radius r_0 with $2R$ and use the Moody diagram with

Wetted perimeter: *That perimeter where the fluid is in contact with the solid boundary.*

$$\text{Re} = \frac{4VR}{\nu} \qquad \begin{matrix}\text{relative}\\ \text{roughness}\end{matrix} = \frac{e}{4R} \tag{15.88}$$

The head loss becomes

$$h_L = f\frac{L}{4R}\frac{V^2}{2g} \tag{15.89}$$

To use this hydraulic radius technique the cross section should be fairly "open," such as a rectangle with aspect ratio less than 4:1, an equilateral triangle, or an oval. For other shapes, such as an annulus, the error would be significant.

Example 15.14

Air at standard conditions is to be transported through 500 m of a smooth, horizontal, 30 cm × 20 cm rectangular duct at a flow rate of 0.24 m^3/s. Calculate the pressure drop.

Solution

The hydraulic radius is

$$R = \frac{A}{P} = \frac{0.3 \times 0.2}{(0.3 + 0.2) \times 2} = 0.06 \text{ m}$$

The average velocity is

$$V = \frac{Q_f}{A} = \frac{0.24}{0.3 \times 0.2} = 4.0 \text{ m/s}$$

This gives a Reynolds number of

$$\text{Re} = \frac{4VR}{\nu} = \frac{4 \times 4 \times 0.06}{1.5 \times 10^{-5}} = 6.4 \times 10^4$$

Using the smooth pipe curve of the Moody diagram, there results

$$f = 0.0196$$

Hence,

$$h_L = f\frac{L}{4R}\frac{V^2}{2g} = 0.0196\frac{500}{4 \times 0.06}\frac{4^2}{2 \times 9.8} = 33.3 \text{ m}$$

The pressure drop is

$$\Delta p = \rho g h_L = 1.23 \times 9.8 \times 33.3 = 402 \text{ Pa}$$

15.6.5 Minor Losses in Pipe Flow

Key Concept: *Minor losses can exceed the frictional losses.*

We now know how to calculate the losses due to a developed flow in a pipe. Pipe systems do, however, include valves, elbows, enlargements, contractions, inlets, outlets, bends, and other fittings that cause additional losses, referred to as *minor losses*, even though such losses can exceed the frictional losses of Eq. 15.78. Each of these devices causes a change in the magnitude and/or the direction of the velocity vectors and hence results in a loss. In general, if the flow is gradually accelerated by a device, the losses are very small; relatively large losses are associated with sudden enlargements or contractions because of the separated regions that result (a separated flow occurs when the primary flow separates from the wall).

A minor loss is expressed in terms of a loss coefficient K, defined by

$$h_L = K\frac{V^2}{2g} \tag{15.90}$$

Values of K have been determined experimentally for the various fittings and geometry changes of interest in piping systems. One exception is the sudden expansion from area A_1 to area A_2, for which the loss can be calculated; this was done in Example 13.13, where we found that

$$h_L = \left(1 - \frac{A_1}{A_2}\right)^2 \frac{V_1^2}{2g} \tag{15.91}$$

Thus, for the sudden expansion

$$K = \left(1 - \frac{A_1}{A_2}\right)^2 \tag{15.92}$$

If A_2 is extremely large (e.g., a pipe exiting into a reservoir), $K = 1.0$, an obvious result since the entire kinetic energy is lost.

Key Concept: *The loss coefficient of an elbow results primarily from the secondary flow.*

A pipe fitting that has a relatively large loss coefficient with no change in cross-sectional area is the pipe bend, or the elbow. This results primarily from the secondary flow caused by the fluid flowing from the high-pressure region to the low-pressure region (see Eq. 12.41), as shown in Fig. 15.14; this secondary flow is

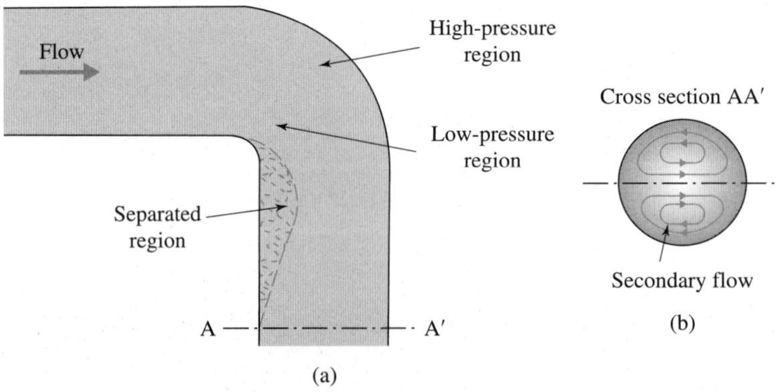

FIGURE 15.14 Flow in an elbow.

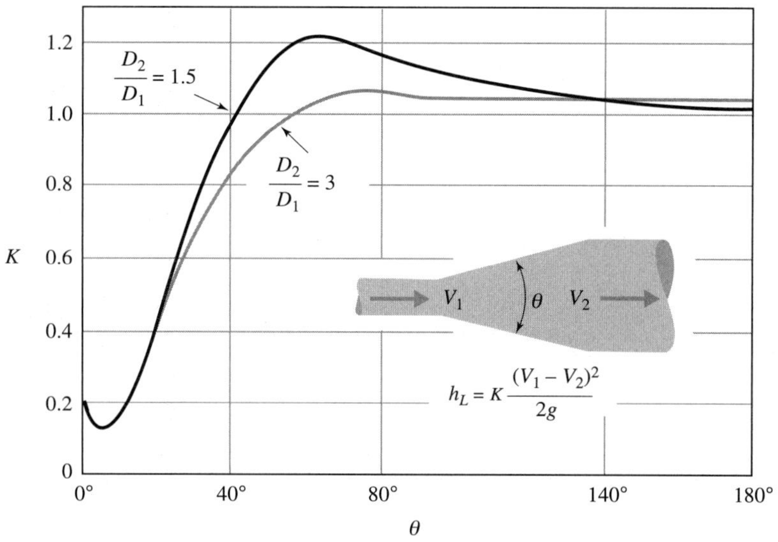

FIGURE 15.15 Loss coefficients in a conical expansion. (From A. H. Gibson, *Engineering*, Vol. 93, 1912.)

eventually dissipated after the fluid leaves the long sweep bend or elbow. In addition, a separated region occurs at the sharp corner in a standard elbow. Energy is needed to maintain a secondary flow and the flow in the separated region. This wasted energy is measured in terms of the loss coefficient.

The loss coefficients for various geometries are presented in Table 15.2 and Fig. 15.15. A globe valve may be used to control the flow rate by introducing large losses by partially closing the valve. The other types of valves should not be used to control the flow since damage could result.

Loss coefficients for sudden contractions and orifice plates can be approximated by neglecting the losses in the converging flow up to the vena contracta and calculating the losses in the diverging flow using the loss coefficient for a sudden expansion. Figure 15.16 provides the information necessary to establish the area of the **vena contracta**, the minimum area; this minimum area results when the converging streamlines begin to expand to fill the downstream area.

Vena contracta: *The minimum area in a sudden contraction.*

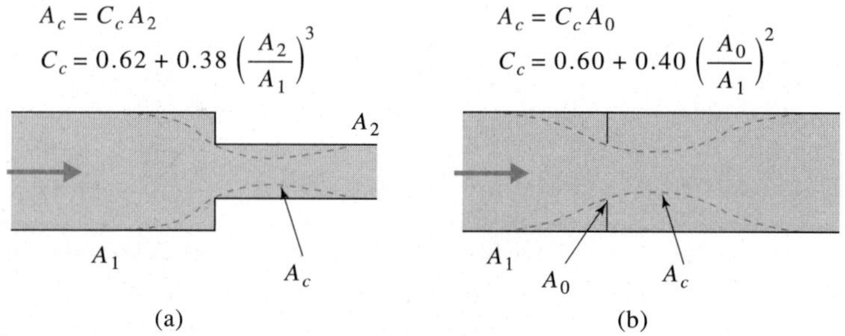

FIGURE 15.16 Vena contractas in contractions and orifices: (a) sudden contraction; (b) concentric orifice.

TABLE 15.2 Nominal Loss Coefficients K (Turbulent Flow)[a]

Type of fitting	Screwed			Flanged		
Diameter	2.5 cm	5 in.	10 cm	5 cm	10 cm	20 cm
Globe valve (fully open)	8.2	6.9	5.7	8.5	6.0	5.8
(half open)	20	17	14	21	15	14
(one-quarter open)	57	48	40	60	42	41
Angle valve (fully open)	4.7	2.0	1.0	2.4	2.0	2.0
Swing check valve (fully open)	2.9	2.1	2.0	2.0	2.0	2.0
Gate valve (fully open)	0.24	0.16	0.11	0.35	0.16	0.07
Return bend	1.5	.95	.64	0.35	0.30	0.25
Tee (branch)	1.8	1.4	1.1	0.80	0.64	0.58
Tee (line)	0.9	0.9	0.9	0.19	0.14	0.10
Standard elbow	1.5	0.95	0.64	0.39	0.30	0.26
Long sweep elbow	0.72	0.41	0.23	0.30	0.19	0.15
45° elbow	0.32	0.30	0.29			
Square-edged entrance 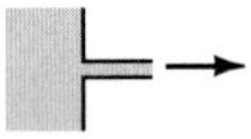			0.5			
Reentrant entrance 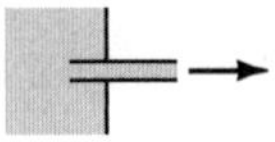			0.8			
Well-rounded entrance 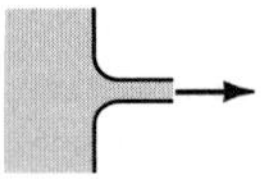			0.03			
Pipe exit			1.0			
	Area ratio					
Sudden contraction[b] 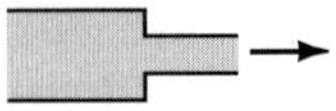	2:1		0.25			
	5:1		0.41			
	10:1		0.46			
	Area ratio A/A_0					
Orifice plate 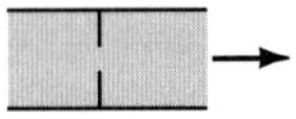	1.5:1		0.85			
	2:1		3.4			
	4:1		29			
	≥6:1		$2.78\left(\frac{A}{A_0}-0.6\right)^2$			
Sudden enlargement[c] 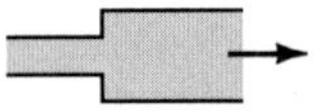			$\left(1-\frac{A_1}{A_2}\right)^2$			
90° miter bend (without vanes) 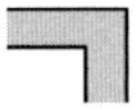			1.1			
(with vanes) 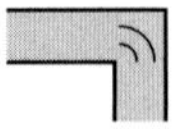			0.2			
General contraction 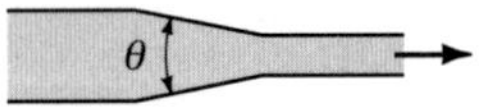	(30° included angle)		0.02			
	(70° included angle)		0.07			

[a]Values for other geometries can be found in *Technical Paper 410*, The Crane Company, 1957.
[b]Based on exit velocity V_2.
[c]Based on entrance velocity V_1.

It is often the practice to express a loss coefficient as an *equivalent length* L_e of pipe. This is done by equating Eq. 15.90 to Eq. 15.78:

$$K\frac{V^2}{2g} = f\frac{L_e}{D}\frac{V^2}{2g} \qquad (15.93)$$

giving the relationship

$$L_e = K\frac{D}{f} \qquad (15.94)$$

Hence the square-edged entrance of a 20-cm-diameter pipe with a friction factor of $f = 0.02$ could be replaced by an equivalent pipe length of $L_e = 5$ m.

Finally, a comment should be made concerning the magnitude of the minor losses. In piping systems involving intermediate lengths (i.e., 100 diameters) of pipe between minor losses, the minor losses may be of the same order of magnitude as the frictional losses; for relatively short lengths the minor losses may be substantially greater than the frictional losses; and for long lengths (e.g., 1000 diameters) of pipe, the minor losses are usually neglected.

Key Concept: *For long lengths of pipe, the minor losses are usually neglected.*

Example 15.15

If the flow rate through a 10-cm-diameter wrought iron pipe (Fig. E15.15) is 0.04 m^3/s, find the difference in elevation H of the two reservoirs.

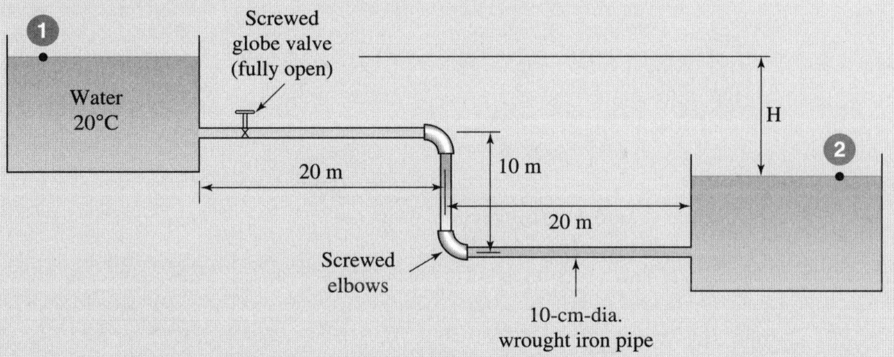

FIGURE E15.15

Solution

The energy equation written for a control volume that contains the two reservoir surfaces (see Eq. 13.51), where $V_1 = V_2 = 0$ and $p_1 = p_2 = 0$, is

$$0 = z_2 - z_1 + h_L$$

Thus, letting $z_1 - z_2 = H$, we have

$$H = (K_{\text{entrance}} + K_{\text{valve}} + 2K_{\text{elbow}} + K_{\text{exit}})\frac{V^2}{2g} + f\frac{L}{D}\frac{V^2}{2g}$$

(continued)

The average velocity, Reynolds number, and relative roughness are

$$V = \frac{Q_f}{A} = \frac{0.04}{\pi \times 0.05^2} = 5.09 \text{ m/s}$$

$$\text{Re} = \frac{VD}{\nu} = \frac{5.09 \times 0.1}{10^{-6}} = 5.09 \times 10^5$$

$$\frac{e}{D} = \frac{0.046}{100} = 0.00046$$

From the Moody diagram we find that

$$f = 0.0173$$

Using the loss coefficients from Table 15.2 for an entrance, a globe valve, screwed 10-cm-diameter standard elbows, and an exit there results

$$H = (0.5 + 5.7 + 2 \times 0.64 + 1.0)\frac{5.09^2}{2 \times 9.8} + 0.0173\frac{50}{0.1}\frac{5.09^2}{2 \times 9.8}$$

$$= 11.2 + 11.4 = 22.6 \text{ m}$$

Note: The minor losses are about equal to the frictional losses as expected, since there are five minor loss elements in 500 diameters of pipe length.

Example 15.16

Approximate the loss coefficient for the sudden contraction $A_1/A_2 = 2$ by neglecting the losses in the contracting portion up to the vena contracta and assuming that all the losses occur in the expansion from the vena contracta to A_2 (see Fig. 15.16). Compare with that given in Table 15.2.

Solution

The head loss from the vena contracta to area A_2 is (see Table 15.2, sudden enlargement)

$$h_L = \left(1 - \frac{A_c}{A_2}\right)^2 \frac{V_c^2}{2g}$$

Continuity allows us to write

$$V_c = \frac{A_2}{A_c} V_2$$

Thus, the head loss based on V_2 is

$$h_L = \left(1 - \frac{A_c}{A_2}\right)^2 \left(\frac{A_2}{A_c}\right)^2 \frac{V_2^2}{2g}$$

so the loss coefficient of Eq. 15.90 is

$$K = \left(1 - \frac{A_c}{A_2}\right)^2 \left(\frac{A_2}{A_c}\right)^2$$

(continued)

Using the expression of C_c given in Fig. 15.16, we have

$$\frac{A_c}{A_2} = C_c = 0.62 + 0.38\left(\frac{1}{2}\right)^3 = 0.67$$

Finally,

$$K = (1 - 0.67)^2 \frac{1}{0.67^2} = 0.24$$

This compares favorably with the value of 0.25 given in Table 15.2.

15.6.6 Hydraulic and Energy Grade Lines

When the energy equation is written in the form of Eq. 13.51, that is,

$$-\frac{\dot{W}_S}{\dot{m}g} = \frac{V_2^2 - V_1^2}{2g} + \frac{p_2 - p_1}{\gamma} + z_2 - z_1 + h_L \qquad (15.95)$$

the terms have dimensions of length. This has led to the conventional use of the hydraulic grade line and the energy grade line. The **hydraulic grade line** (HGL), the dashed line in Fig. 15.17, in a piping system is formed by the locus of points located a distance p/γ above the center of the pipe, or $p/\gamma + z$ above a preselected datum; the liquid in a piezometer tube would rise to the HGL. The **energy grade line** (EGL), the solid line in Fig. 15.17, is formed by the locus of points a distance $V^2/2g$ above the HGL, or the distance $V^2/2g + p/\gamma + z$ above the datum; the

Hydraulic grade line (HGL): *In a piping system, the HGL is located a distance p/γ above the center of the pipe.*

Energy grade line (EGL): *In a piping system, the EGL is located a distance $V^2/2g$ above the HGL.*

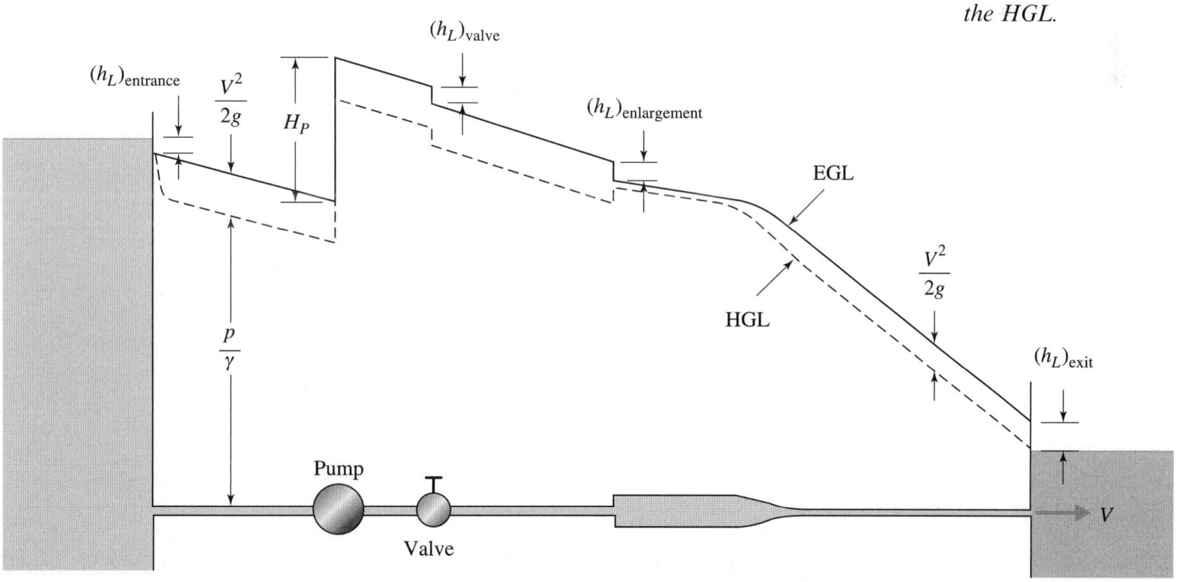

FIGURE 15.17 Hydraulic grade line (HGL) and energy grade line (EGL) for a piping system.

liquid in a pitot tube would rise to the EGL. The following points are noted relating to the HGL and the EGL:

- As the velocity goes to zero, the HGL and the EGL approach each other. Thus, in a reservoir, they are identical and lie on the surface (see Fig. 15.17).
- The EGL and, consequently, the HGL slope downward in the direction of the flow due to the head loss in the pipe. The greater the loss per unit length, the greater the slope. As the average velocity in the pipe increases, the loss per unit length increases.
- A sudden change occurs in the HGL and the EGL whenever a loss occurs due to a sudden geometry change as represented by the valve or the sudden enlargement of Fig. 15.17.
- A jump occurs in the HGL and the EGL whenever useful energy is added to the fluid as occurs with a pump, and a drop occurs if useful energy is extracted from the flow as occurs with a turbine.
- At points where the HGL passes through the centerline of the pipe, the pressure is zero. If the pipe lies above the HGL, there is a vacuum in the pipe, a condition that is often avoided, if possible, in the design of piping systems; an exception would be in the design of a siphon.

Key Concept: *If the pipe lies above the HGL, there is a vacuum in the pipe.*

The concepts of energy grade line and hydraulic grade line may also be applied to open-channel flows. The HGL coincides with the free surface and the EGL is a distance $V^2/2g$ above the free surface. Uniform flows in open channels will be discussed in the following section.

Example 15.17

Water at 20°C flows between two reservoirs at the rate of 0.06 m^3/s as shown in Fig. E15.17. Sketch the HGL and the EGL. What is the minimum diameter D_B allowed to avoid the occurrence of cavitation?

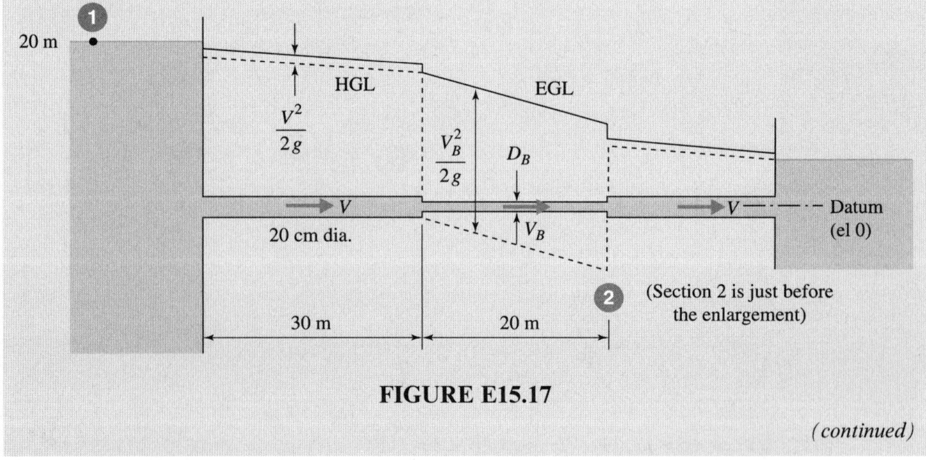

FIGURE E15.17

(continued)

Solution

The EGL and the HGL are sketched on the figure, including sudden changes at the entrance, contraction, enlargement, and the exit. Note the large velocity head (the difference between the EGL and the HGL) in the smaller pipe because of the high velocity. The velocity, Reynolds number, and relative roughness in the 20-cm-diameter pipe are calculated to be

$$V = \frac{Q_f}{A} = \frac{0.06}{\pi \times 0.20^2/4} = 1.91 \text{ m/s}$$

$$\text{Re} = \frac{VD}{\nu} = \frac{1.91 \times 0.2}{10^{-6}} = 3.8 \times 10^5$$

$$\frac{e}{D} = \frac{0.26}{200} = 0.0013$$

Thus $f = 0.022$ from Fig. 15.13. The velocity, Reynolds number, and relative roughness in the smaller pipe are

$$V_B = \frac{0.06}{\pi D_B^2/4} = \frac{0.0764}{D_B^2}$$

$$\text{Re} = \frac{0.0764 \times D_B}{D_B^2 \times 10^{-6}} = \frac{76\,400}{D_B}$$

$$\frac{e}{D_B} = \frac{0.00026}{D_B}$$

The minimum possible diameter is established by recognizing that the water vapor pressure (2450 Pa absolute) at 20°C is the minimum allowable pressure. Since the distance between the pipe and the HGL is an indication of the pressure in the pipe, we can conclude that the minimum pressure will occur at section 2. Hence the energy equation applied between section 1, the reservoir surface, and section 2 gives

$$\overset{0}{\cancel{\frac{V_1^2}{2g}}} + \overset{0}{\cancel{\frac{p_1}{\gamma}}} + z_1 = \frac{V_B^2}{2g} + \frac{p_2}{\gamma} + \overset{0}{\cancel{z_2}} + K_{\text{ent}}\frac{V_A^2}{2g} + K_{\text{cont}}\frac{V_B^2}{2g} + f_A\frac{L_A}{D_A}\frac{V^2}{2g} + f_B\frac{L_B}{D_B}\frac{V_B^2}{2g}$$

where the subscript A refers to the 20-cm-diameter pipe. This simplifies to

$$\frac{101\,000}{9810} + 20 = \frac{(0.0764/D_B^2)^2}{2 \times 9.81}\left(1 + 0.25 + f_B\frac{20}{D_B}\right) + \frac{2450}{9810}$$
$$+ \left(0.5 + 0.022\frac{30}{0.2}\right)\frac{1.91^2}{2 \times 9.81}$$
$$98\,600 = \frac{1.25}{D_B^4} + f_B\frac{20}{D_B^5}$$

where we have used $K_{\text{ent}} = 0.5$ and assumed that $K_{\text{cont}} = 0.25$. This requires a trial-and-error solution. The following illustrates the procedure.

Let $D_B = 0.1$ m. Then $e/D = 0.0026$ and $\text{Re} = 7.6 \times 10^5$. Therefore, $f = 0.026$:

$$98\,600 \overset{?}{=} 12\,500 + 52\,000$$

(continued)

Let $D_B = 0.09$ m. Then $e/D = 0.0029$ and $\text{Re} = 8.4 \times 10^5$. Therefore, $f = 0.027$:

$$98\,600 \stackrel{?}{=} 19\,000 + 91\,000$$

We see that 0.1 m is too large and 0.09 m is too small. In fact, the value of 0.09 m is only slightly too small. Consequently, to be safe we must select the next larger pipe size, of 0.1 m diameter. If there were a pipe size of 9.5 cm diameter, that could be selected. Assuming that that size is not available, we select

$$D_B = 10 \text{ cm}$$

Note that the assumption of a 2:1 area ratio for the contraction is too small. It is actually 4:1. This would give $K_{\text{cont}} \simeq 0.4$. After a quick check we conclude that this value does not significantly influence the result.

15.6.7 Simple Pipe System with a Pump

Key Concept: *If the flow rate is specified, the solution is straightforward.*

The problems we have considered thus far in this section have not involved a pump. If a centrifugal pump is included in the piping system and the flow rate is specified, the solution is straightforward using the techniques we have already developed. If, on the other hand, the flow rate is not specified, as is often the case, a trial-and-error solution involving the centrifugal pump results since the head produced by a centrifugal pump and its efficiency η_P (see Eq. 13.60) depend on the discharge, as shown by the pump characteristic curves, the solid curves in Fig. 15.18. Companies furnish such characteristic curves for each centrifugal pump manufactured. Such a curve provides one equation relating the flow rate Q_f and pump head H_P. The other equation is provided by the energy equation, which can typically be written as (see the energy equation in the following example)

$$H_P = c_1 + c_2 Q_f^2 \tag{15.96}$$

System demand curve: *The energy equation relating the pump head to the unknown flow rate.*

This is the **system demand curve**. This along with the characteristic curve must be solved simultaneously to yield the desired flow rate. To determine the power requirement of the pump, the efficiency η_P must be used.

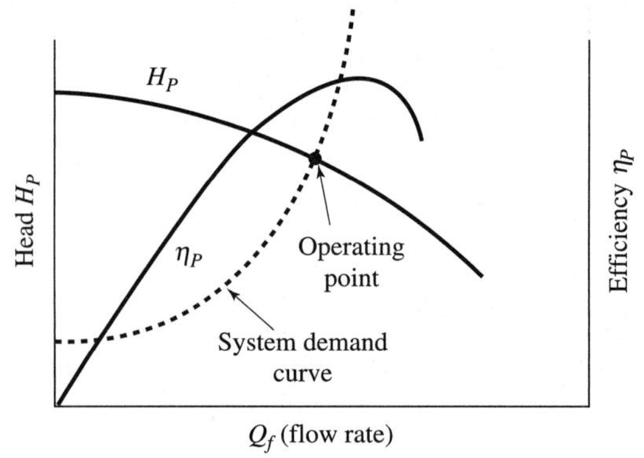

FIGURE 15.18 Pump characteristic curves and the system demand curve.

Note that for the piping system, the required pump energy head H_P demanded by the energy equation increases with Q_f and from the pump characteristic curve we see that H_P decreases with Q_f; hence the two curves will intersect at a point, called the *operating point* of the system. An example will illustrate the solution technique.

Example 15.18

Estimate the flow rate in the simple piping system of Fig. E15.18a if the pump characteristic curves are as shown in Fig. E15.18b. Also, find the pump power requirement.

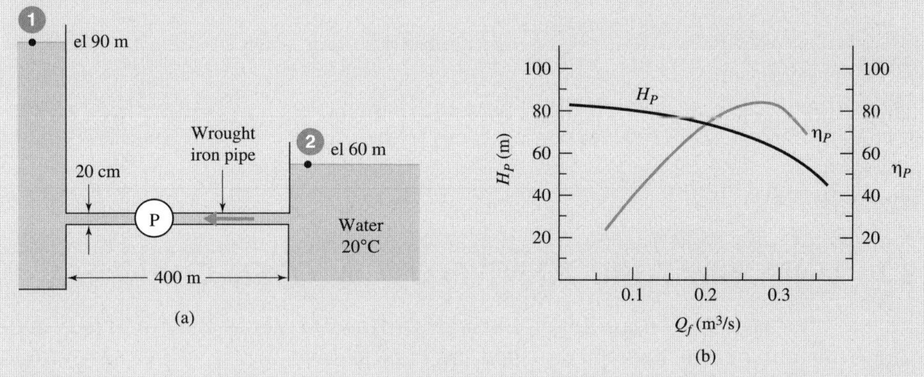

FIGURE E15.18

Solution

We will assume that the Reynolds number is sufficiently large that the flow is completely turbulent. So, using $e/D = 0.046/200 = 0.00023$, the friction factor from the Moody diagram is

$$f = 0.014$$

The energy equation (see Eq. 15.95), with $H_P = -\dot{W}_S/\dot{m}g$, applied between the two surfaces, yields

$$H_P = \frac{V_2^2 - V_1^2}{2g}^{\nearrow 0} + z_2 - z_1 + \frac{p_2 - p_1}{\gamma}^{\nearrow 0} + h_L$$

or

$$H_P = 90 - 60 + \left(K_{\text{entrance}} + K_{\text{exit}} + f\frac{L}{D}\right)\frac{V^2}{2g}$$

$$= 30 + \left(0.5 + 1.0 + 0.014\frac{400}{0.2}\right)\frac{Q_f^2}{2 \times 9.8 \times [\pi \times (0.1)^2]^2}$$

$$= 30 + 1520Q_f^2$$

This equation, the system demand curve, and the characteristic curve $H_P(Q_f)$ of the pump are now solved simultaneously by trial and error. Actually, the curve could be plotted on the same graph as the characteristic curve and the point of intersection, the

(continued)

operating point, would provide Q_f. Try $Q_f = 0.2$ m^3/s: $(H_P)_{\text{energy}} = 91$ m, $(H_P)_{\text{char}} \cong 75$ m. Try $Q_f = 0.15$ m^3/s: $(H_P)_{\text{energy}} = 64$ m, $(H_P)_{\text{char}} \cong 75$ m. Try $Q_f = 0.17$ m^3/s: $(H_P)_{\text{energy}} = 74$ m, $(H_P)_{\text{char}} \cong 76$ m. This is our solution. We have

$$Q_f = 0.17 \text{ m}^3/\text{s}$$

Check the Reynolds number: $\text{Re} = DQ_f/A\nu = 0.2 \times 0.17/\pi \times 0.1^2 \times 10^{-6} = 1.08 \times 10^6$. This is sufficiently large, but marginally so.

The power requirement of the pump is given by:

$$\dot{W}_P = \frac{Q_f \gamma H_P}{\eta_P}$$

$$= \frac{0.17 \times 9800 \times 75}{0.65} = 198\,000 \text{ W} \quad \text{or} \quad 198 \text{ kW}$$

where the efficiency $\eta_P = 0.65$ is found from the characteristic curve at $Q_f = 0.17$ m^3/s.

Note: Since $L/D > 1000$, minor losses due to the entrance and exit could have been neglected.

15.7 UNIFORM TURBULENT FLOW IN OPEN CHANNELS

The last internal flow situation we consider in this chapter is that of steady uniform (constant depth) flow in an open channel, shown in Fig. 15.19. We could already treat this flow by using the Darcy–Weisbach relation presented in Section 15.6.3. In fact, that technique predicts better results than does the more common method we present here. Both methods will be compared in two examples. However, unless otherwise stated, uniform flow in open, rough channels is commonly analyzed using the following less-complicated method.

If the energy equation is applied between two sections of the channel sketched in Fig. 15.19, we obtain

$$0 = \frac{V_2^2 - V_1^2}{2g}\overset{0}{\nearrow} + \frac{p_2 - p_1}{\gamma}\overset{0}{\nearrow} + z_2 - z_1 + h_L \tag{15.97}$$

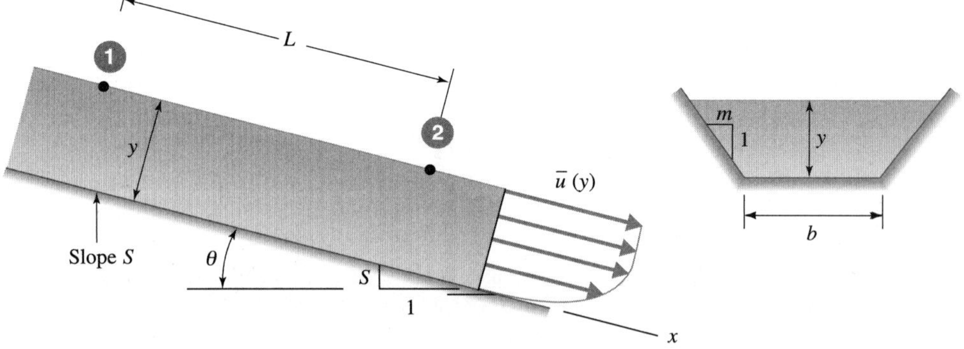

FIGURE 15.19 Uniform flow in an open channel.

which shows that the head loss is

$$h_L = z_1 - z_2$$
$$= L \sin\theta = LS \tag{15.98}$$

where L is the length of the channel between the two sections and S is the slope of the channel, which is assumed small, so that $\sin\theta \simeq S$. (Do not confuse S with specific gravity.)

The Darcy–Weisbach equation (15.78), with $h_L = LS$ from Eq. 15.98, takes the form

$$LS = f\frac{L}{4R}\frac{V^2}{2g} \qquad \therefore\ RS = \frac{f}{8g}V^2 \tag{15.99}$$

where R is the hydraulic radius. Since open channels are usually quite large with large Reynolds numbers, the friction factor is invariably constant (the flow is completely turbulent). Thus the equation above is written as

Key Concept: *Since open channels are quite large, with large Reynolds numbers, the friction factor is constant.*

$$V = C\sqrt{RS} \tag{15.100}$$

where the *Chezy coefficient* C is a dimensional constant; the equation above is referred to as the *Chezy equation*. The Chezy coefficient is related to the channel roughness and the hydraulic radius (much like f is in a pipe) by

$$C = \frac{c_1}{n}R^{1/6} \tag{15.101}$$

where the dimensional constant c_1 has a value of 1.0 if SI units are used and 1.49 if English units are used. The dimensionless constant n is directly related to the wall roughness; it is called the *Manning n*. Values for various wall materials are listed in Table 15.3.

TABLE 15.3 Average Values[a] of the Manning n

Wall material	Manning n
Planed wood	0.012
Unplaned wood	0.013
Finished concrete	0.012
Unfinished concrete	0.014
Sewer pipe	0.013
Brick	0.016
Cast iron, wrought iron	0.015
Concrete pipe	0.015
Riveted steel	0.017
Earth, straight	0.022
Corrugated metal flumes	0.025
Rubble	0.03
Earth with stones and weeds	0.035
Mountain streams	0.05

[a]The values in this table result in flow rates too large for hydraulic radii greater than about 3 m (10 ft). The Manning n should be increased by 10 to 15% for such large conduits.

The flow rate, which is of primary interest in open-channel flow problems, is found to be

$$Q_f = \frac{c_1}{n} A R^{2/3} S^{1/2} \qquad c_1 = \begin{cases} 1.0 \text{ for SI units} \\ 1.49 \text{ for English units} \end{cases} \tag{15.102}$$

This is the *Chezy–Manning equation.*

For smooth-surfaced channels, the use of the Chezy–Manning equation is discouraged since it implicitly assumes a rough wall. Calculations for smooth-surfaced channels such as glass or plastic should be based on the Darcy–Weisbach relation with variable f; see Section 15.6.3.

Example 15.19

The depth of water at 60°F flowing in a 12-ft-wide rectangular, finished concrete channel is measured to be 4 ft. The slope is measured to be 0.0016. Estimate the flow rate using (a) the Chezy–Manning equation and (b) the Darcy–Weisbach equation.

Solution

The hydraulic radius is calculated to be

$$R = \frac{A}{P} = \frac{yb}{2y + b} = \frac{4 \times 12}{2 \times 4 + 12} = 2.4 \text{ ft}$$

(a) Using the Chezy–Manning equation, with $n = 0.012$ from Table 15.3 and $c = 1.49$, we have

$$\begin{aligned} Q_f &= \frac{1.49}{n} A R^{2/3} S^{1/2} \\ &= \frac{1.49}{0.012} \times (4 \times 12) \times 2.4^{2/3} \times 0.0016^{1/2} = 427 \text{ ft}^3/\text{sec} \end{aligned}$$

(b) The relative roughness is, using a low value $e = 0.0015$ ft (it is finished concrete) shown on the Moody diagram:

$$\frac{e}{4R} = \frac{0.0015}{4 \times 2.4} = 0.00016$$

Assuming a completely turbulent flow, the Moody diagram gives the friction factor as

$$f = 0.013$$

The Darcy–Weisbach equation (15.99) then yields the velocity as follows:

$$\begin{aligned} \therefore V &= \left(\frac{8RgS}{f}\right)^{1/2} \\ &= \left(\frac{8 \times 2.4 \times 32.2 \times 0.0016}{0.013}\right)^{1/2} = 8.72 \text{ ft/sec} \end{aligned}$$

The flow rate is calculated as

$$Q_f = VA = 8.72 \times 4 \times 12 = 419 \text{ ft}^3/\text{sec}$$

These two values are within 2%, an acceptable engineering tolerance for this type of problem. That found using the Moody diagram is considered to be more accurate, however.

Example 15.20

The 1.0-m-diameter concrete pipe of Fig. E15.20 transports 20°C water at a depth of 0.4 m. If the slope is 0.001, find the flow rate using (a) Chezy–Manning equation and (b) the Darcy–Weisbach equation.

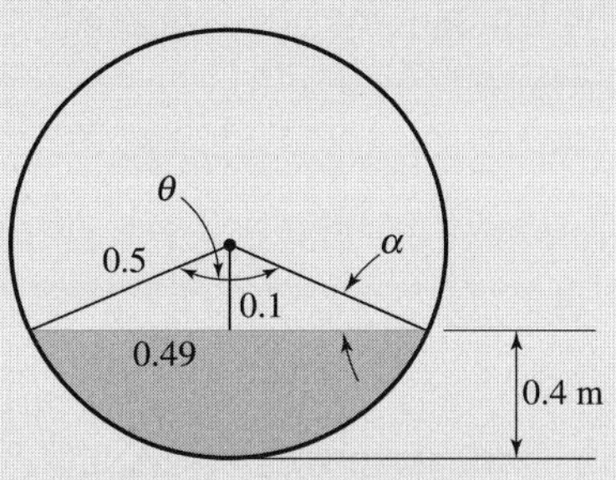

FIGURE E15.20

Solution

From the sketch of the pipe in Fig. E15.20 the following are calculated:

$$\alpha = \sin^{-1}\frac{0.1}{0.5} = 11.54^\circ$$

$$\therefore\ \theta = 156.9^\circ$$

$$\therefore\ A = \pi \times 0.5^2 \times \frac{156.9}{360} - 0.49 \times 0.1 = 0.2933\ \text{m}^2$$

$$P = 2\pi \times 0.5 \times \frac{156.9}{360} = 1.369\ \text{m}$$

The hydraulic radius is found, using the above calculations, to be

$$R = \frac{A}{P} = \frac{0.2933}{1.369} = 0.2142\ \text{m}$$

(a) The Chezy–Manning equation yields, with n from Table 15.3 and $c_1 = 1.0$,

$$Q_f = \frac{1.0}{n}AR^{2/3}S^{1/2} = \frac{1.0}{0.013} \times 0.2933 \times 0.2142^{2/3} \times 0.001^{1/2} = 0.25\ \text{m}^3/\text{s}$$

We used an intermediate value for n since sufficient information is not given.

(b) The relative roughness is, using a relatively rough value for concrete pipe (see Fig. 15.13) of $e = 2.0$ mm,

$$\frac{e}{4R} = \frac{2}{4 \times 214.2} = 0.0023$$

Assuming completely turbulent flow, the Moody diagram yields

$$f = 0.025$$

The Darcy–Weisbach equation (15.99) then gives the following:

$$\therefore\ V = \left(\frac{8RgS}{f}\right)^{1/2} = \left(\frac{8 \times 0.2142 \times 9.81 \times 0.001}{0.025}\right)^{1/2} = 0.820\ \text{m/s}$$

The flow rate is

$$Q_f = VA = 0.820 \times 0.2933 = 0.24\ \text{m}^3/\text{s}$$

This is within 4% of the result above, an acceptable tolerance for this type of problem. The second method, which is more difficult to apply, is considered to be more accurate, however.

15.8 SUMMARY

The laminar entrance lengths for a pipe and wide channel are, respectively,

$$\frac{L_E}{D} = 0.065\ \text{Re} \qquad \frac{L_E}{h} = 0.04\ \text{Re} \tag{15.103}$$

For a high-Reynolds-number turbulent pipe flow the entrance length is

$$\frac{L_E}{D} = 120 \tag{15.104}$$

For laminar flow in a pipe and a wide channel the pressure and friction factor are, respectively,

$$\Delta p = \frac{8\mu VL}{r_0^2} \qquad f = \frac{64}{\text{Re}} \qquad \text{(pipe)} \tag{15.105}$$

$$\Delta p = \frac{12\mu VL}{a^2} \qquad f = \frac{48}{\text{Re}} \qquad \text{(channel)} \tag{15.106}$$

where a is the channel height.

The torque required to rotate an inner cylinder with the outer cylinder fixed is

$$T = \frac{4\pi\mu r_1^2 r_2^2 L\Omega_1}{r_2^2 - r_1^2} \tag{15.107}$$

In a turbulent flow the head loss is calculated using

$$h_L = f\frac{L}{D}\frac{V^2}{2g} \tag{15.108}$$

where f is found using the Moody Diagram of Fig. 15.13. Minor losses are included using

$$h_L = K\frac{V^2}{2g} \tag{15.109}$$

where many loss coefficients K are listed in Table 15.2.

To include a pump in a piping system when the flow rate is not known, it is necessary to have pump characteristic curves, such as those of Fig. E15.18. Example 15.18 illustrated the procedure.

The flow rate in an open channel is most often estimated using the equation

$$Q = \frac{c_1}{n}AR^{2/3}S^{1/2} \qquad c_1 = \begin{cases} 1.0 \text{ for SI units} \\ 1.49 \text{ for English units} \end{cases} \tag{15.110}$$

where n is found in Table 15.3.

REVIEW PROBLEMS FOR THE FE/EIT EXAMINATION

15.1FE A parabolic profile is measured in a pipe flow. The flow is:
I. Laminar
II. Developed
III. Steady
IV. Symmetric
A. I and III
B. I, II, and III
C. I, II, and IV
D. I, II, IV, and IV

15.2FE The velocity profile between parallel plates is calculated to be Vy/a, where y is measured from the lower plate and a is the distance between the plates. We know that
I. The flow is laminar
II. The lower plate is moving with velocity V and the other is stationary.
III. Lower plate is stationary and the other is moving with velocity V
IV. The flow is steady
A. I, III, and IV **B.** II and III
C. I and III **D.** I and IV

15.3FE A liquid flows in a pipe at a Reynolds number of 6000.
A. The flow is laminar
B. The flow is turbulent
C. The flow is transitory, osscilating between laminar and turbulent
D. The flow could be any of the above

15.4FE In a turbulent flow in a pipe, the head loss
A. varies with the velocity squared
B. is directly proportional to the flow rate
C. decreases with increase with Reynolds number
D. is directly proportional to the length of the pipe

15.5FE The curves for the friction factor f on the Moody diagram become horizontal for sufficiently large Reynolds numbers because
A. wall roughness elements project through the wall viscous layer
B. the wall viscous layer completely covers the wall roughness elements
C. the viscous effects become dominant in the flow
D. inertial effects cease to become significant in the flow

15.6FE The pressure drop over 15 m of 2-cm-diameter galvanized iron pipe is measured to be 60 Pa. If the pipe is horizontal, estimate the flow rate of water. Use $\nu = 10^{-6}\ m^2/s$.
A. 6.82 L/s **B.** 2.18 L/s
C. 0.682 L/s **D.** 0.218 L/s

15.7FE Water flows in an 8-cm-diameter pipe down a 30° incline and the pressure remains constant. Estimate the average velocity in the cast iron pipe. Use $\nu = 10^{-6}\ m^2/s$.
A. 0.055 m/s **B.** 0.174 m/s
C. 1.75 m/s **D.** 5.5 m/s

15.8FE A liquid with a density of 900 kg/m^3 flows straight down in a 6-cm-diameter cast iron pipe. Estimate the pressure rise over 20 m of pipe if the average velocity is 4 m/s. Assume $\nu = 8 \times 10^{-6}\ m^2/s$.
A. 250 kPa **B.** 100 kPa
C. 77 kPa **D.** 10.2 kPa

15.9FE The flow of water occurs in a square cast iron conduit, 4 cm on a side. If the horizontal conduit transports 0.02 m^3/s, estimate the pressure drop over 40 m. Use $\nu = 10^{-6}\ m^2/s$.
A. 162 Pa **B.** 703 Pa
C. 1390 Pa **D.** 1590 Pa

15.10FE A new design of a valve is to be tested. Which of the following parameters is the most important if liquid benzene flows through the valve?
A. Froude number **B.** Reynolds number
C. Mach number **D.** Euler number

15.11FE A water system is to be installed in a community that is quite hilly. The design engineer must be certain that:
A. the hydraulic grade line must always be above the pipe line.
B. the energy grade line must always be above the pipe line.
C. the stagmation pressure must remain positive in the pipe line
D. the elevation of the pipe must not be allowed to be negative.

15.12FE Water flows in a 2.4-m-wide, rectanglular, finished concrete channel at a depth of 80 cm. If the slope is 0.002, the flow rate is nearest
A. 2.2 m^3/s **B.** 3.4 m^3/s
C. 4.6 m^3/s **D.** 6.2 m^3/s

PROBLEMS

Laminar or Turbulent Flow

15.1 Calculate the maximum average velocity V with which 20°C water can flow in a pipe in the laminar state if the critical Reynolds number ($\text{Re} = VD/\nu$) at which transition occurs is 2000; the pipe diameter is:
(a) 2 m **(b)** 2 cm **(c)** 2 mm

15.2 Water at 20°C flows in a wide river. Using a critical Reynolds number ($\text{Re} = Vh/\nu$) at which transition occurs of 1500, calculate the average velocity V that will result in a laminar flow if the depth h of the river is:
(a) 4 m **(b)** 1 m **(c)** 0.3 m

15.3 Water at 50°F is flowing in a thin sheet in a parking lot at a depth of 0.2 in. with an average velocity of 1.5 ft/sec. Is the flow laminar or turbulent?

15.4 Water is flowing, apparently quite placidly, in a river that is 20 m wide and 1.4 m deep. A leaf floating in the river is observed to move 1 m in 2 s. Is the flow laminar or turbulent? See Problem 15.2 for the definition of the Reynolds number.

15.5 Flow occurs in a 2-cm-diameter pipe. What is the maximum velocity that can occur for water at 20°C for a laminar flow if:
(a) Re = 2000? **(b)** Re = 40 000?

Entrance and Developed Flow

15.6 Calculate the laminar entrance length in a 4-cm-diameter pipe if 2×10^{-4} m^3/s of water is flowing at:
(a) 10°C **(b)** 20°C
(c) 40°C **(d)** 80°C

15.7 A laminar flow is to occur in an experimental facility with 20°C air flowing through a 4-cm-diameter pipe. Calculate the average velocity, the inviscid core length, and the entrance length if the Reynolds number is:
(a) 1000 **(b)** 80 000

15.8 A 6-cm-diameter pipe originates in a tank and delivers 0.025 m^3/s of water at 20°C to a receiver 50 m away. Is the assumption of developed flow acceptable?

15.9 A laboratory experiment is designed to create a laminar flow in a 2-mm-diameter tube shown in Fig P15.9. Water flows from a reservoir through the tube. If 18 L is collected in 2 hr, can the entrance length be neglected?

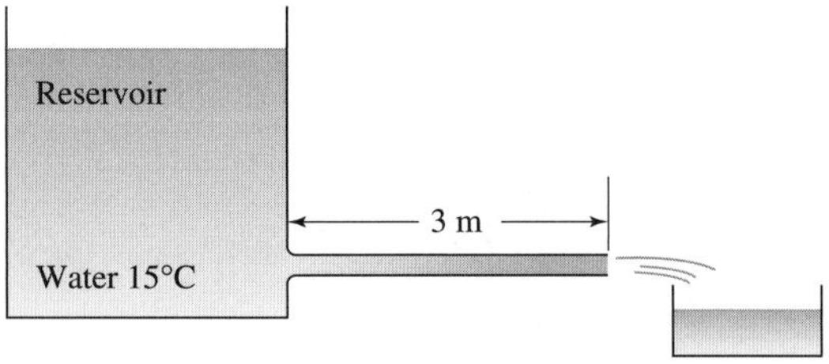

FIGURE P15.9

15.10 Air at 23°C is used as the working fluid in a parallel-plate research project. If the plates are 1.2 cm apart, how long is the longest possible entrance length for laminar flow? What is the shortest entrance length?

15.11 Air at 25°C can exist in either the laminar state or the turbulent state (a trip wire near the entrance is used to make it turbulent) for flow in a 6-cm-diameter pipe in a research lab. If the average velocity is 5 m/s, compare the length of the entrance region for the laminar flow with that of the turbulent flow.

15.12 Water at 20°C flows with an average velocity of 0.2 m/s from a reservoir through a 4-cm-diameter pipe. Estimate the inviscid core length and the entrance length if the flow is:
(a) Laminar
(b) Turbulent

15.13 Draw an incremental control volume with length Δx and radius r_0 and show that for a laminar flow $(\Delta p/\Delta x)_{\text{entrance}} > (\Delta p/\Delta x)_{\text{developed}}$.

15.14 Explain the pressure variations observed for a turbulent flow in Fig. 15.3 for:
(a) A high-Re flow (Re > 300 000)
(b) A low-Re flow (Re ≃ 10 000)
(c) An intermediate-Re flow

Laminar Flow in a Pipe

15.15 Verify that Eq. 15.10 is, in fact, correct.

15.16 A pressure drop of 0.07 psi occurs over a section of 0.8-in.-diameter pipe transporting water at 70°F. Determine the length of the horizontal section if the Reynolds number is 1600. Also, find the shear stress at the wall and the friction factor.

15.17 Find the angle θ of the 10-mm-diameter pipe of Fig. P15.17 in which water at 40°C is flowing with Re = 1500 such that no pressure drop occurs. Also, find the flow rate.

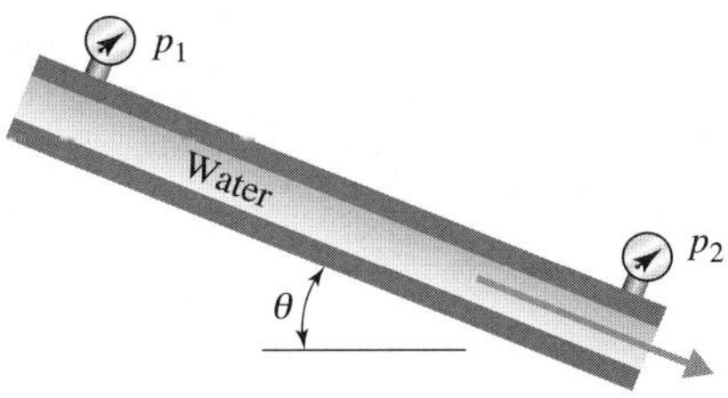

FIGURE P15.17

15.18 A liquid is pumped through a 2-cm-diameter pipe at a flow rate of 12 L/min. Calculate the pressure drop in a 10-m horizontal section if the liquid is:
(a) SAE-10W oil at 20°C
(b) Water at 20°C
(c) Glycerine at 40°C
Is the assumption of laminar flow acceptable?

15.19 A liquid flows with no pressure drop in a vertical 2-cm-diameter pipe. Find the flow rate if, assuming a laminar flow, the liquid is:
(a) Water at 5°C **(b)** SAE-30W oil at 25°C
(c) Glycerine at 20°C
Is the assumption of laminar flow acceptable?

15.20 A laminar flow is to exist in a pipe transporting 0.12 ft^3/sec of SAE-10W oil at 70°F. What is the maximum allowable diameter? What is the pressure drop over 30 ft of horizontal pipe for this diameter?

15.21 Estimate the flow rate through the smooth pipe shown in Fig. P15.21. How long is the entrance region? Assume a laminar flow.

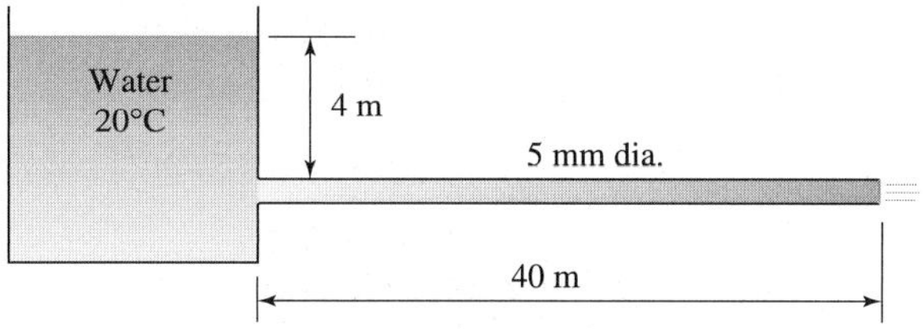

FIGURE P15.21

15.22 A manufacturer of small-diameter tubes wishes to know if the diameters are, in fact, accurate. An experimental setup, like that in Fig. P15.21, is used with a 4-m-long horizontal tube transporting 20°C water with a head of 4 m. If 3.4 L of water is collected in 60 min, what is the inside diameter of the tube, neglecting the effect of the entrance region? Is the effect of entrance region actually negligible?

15.23 Air at 70°F flows in a horizontal 0.8-in.-diameter pipe. Calculate the maximum pressure drop in a 30-ft section for a laminar flow. Assume that $\rho = 0.0024\ slug/ft^3$.

15.24 Water at 20°C flows in the 4-mm-diameter pipe of Fig. P15.24. The pressure rise over the 10-m section is 6 kPa. Find the Reynolds number of the flow and the wall shear stress. Assume laminar flow.

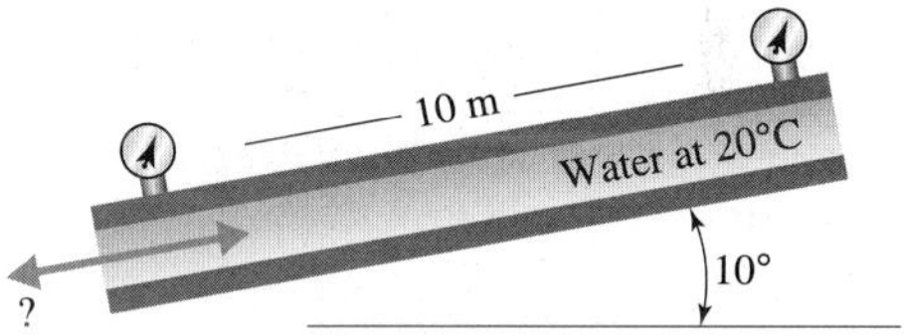

FIGURE P15.24

15.25 A research experiment requires a laminar flow of air at 20°C through a 10-cm-diameter pipe at a Reynolds number of 40 000. What maximum velocity is to be expected? What would be the pressure drop over a 10-m horizontal length of developed flow? How long would the entrance region be? Use $\rho = 1.2\ kg/m^3$.

15.26 Calculate the radius where a pitot probe must be placed in the laminar liquid flow of Fig. P15.26 so that the flow rate is given by $\pi R^2\sqrt{2gH}$.

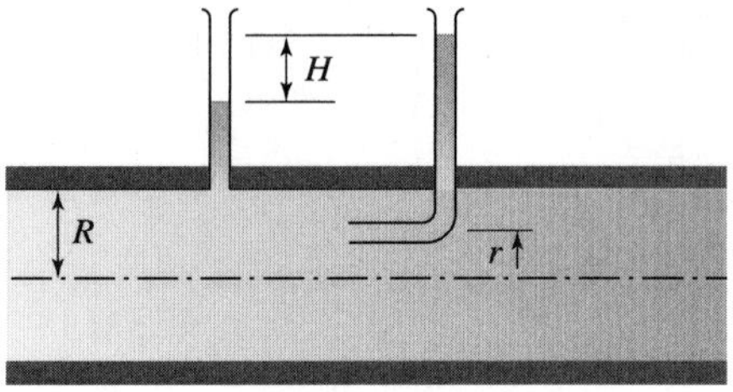

FIGURE P15.26

15.27 A laminar flow of water at 20°C occurs in a vertical 2-mm-diameter pipe. Calculate the flow rate if the pressure is constant. Is it reasonable to assume a laminar flow?

15.28 Find the radius in a developed laminar flow in a pipe where:
(a) The velocity is equal to the average velocity.
(b) The shear stress is equal to one-half the wall shear stress.

15.29 Find the ratio of the total flow rate through a pipe of radius r_0 to the flow rate through an annulus with inner and outer radii of $r_0/2$ and r_0. Assume developed laminar flow with the same pressure gradient.

15.30 A laminar flow of water at 60°F is obtained in a research lab at Re = 20 000 in a horizontal 2-in.-diameter pipe. Calculate the head loss in a 30-ft section of developed flow, the wall shear stress, and the length of the entrance region.

15.31 Water at 20°C flows between the two concentric horizontal pipes of Fig. P15.31 with diameters of 2 cm and 3 cm. A pressure drop of 100 Pa is measured over a 10-m section of developed laminar flow. Find the flow rate and the shear stress on the inner pipe.

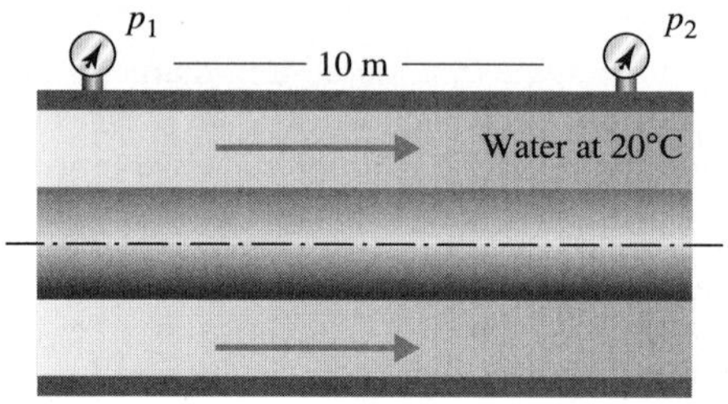

FIGURE P15.31

15.32 Air at 20°C is to flow in the annulus between two concentric, horizontal pipes, with respective diameters of 2 cm and 3 cm, such that a pressure drop of 10 Pa occurs over a 10-m length. Find the average velocity and the shear stress at the inner pipe. Assume a developed laminar flow.

15.33 Fluid flows in the annulus between two concentric horizontal pipes. The inner pipe is maintained at a higher temperature than the outer pipe so that the viscosity in the annulus cannot be assumed to be constant but $\mu = \mu(T)$. What differential equation would be solved to yield $u(r)$ assuming a developed laminar flow?

15.34 Show that the velocity distribution of Example 15.2 approaches that of pipe flow as $r_1 \to 0$ and approaches that of parallel-plate flow as $r_1 \to r_2$.

Laminar Flow between Parallel Plates

15.35 A flow occurs in a horizontal channel $\frac{1}{2}$ in. × 20 in. with Re = 2000. Calculate the flow rate if the fluid is:
(a) Water at 60°F
(b) Atmospheric air at 60°F

15.36 A board 1 m × 1 m that weighs 40 N moves down the incline shown in Fig. P15.36 with a velocity of $V = 0.2$ m/s. Estimate the viscosity of the fluid if θ is:
(a) 20° **(b)** 30°

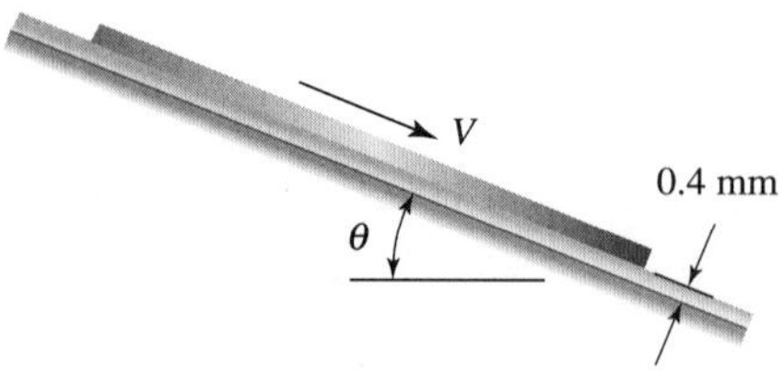

FIGURE P15.36

15.37 Water at 20°C exists between the plate and the surface of Problem 15.36. Calculate the velocity of the plate for an angle θ of:
(a) 20° **(b)** 30°

15.38 Water at 20°C flows down a 20° incline at a depth of 6 mm and a width of 50 m. Calculate the flow rate and the Reynolds number assuming laminar flow. Also, find the maximum velocity and the wall shear.

15.39 Water at 20°C flows down a 100-m-wide parking lot on a slope of 0.00015 at a depth of 10 mm. Determine the flow rate and the Reynolds number assuming laminar flow. Also, calculate the friction factor and the wall shear.

15.40 A pressure drop of 50 Pa is measured over a 60-m length of a 90 cm × 2 cm rectangular horizontal channel transporting 20°C air. Calculate the maximum flow rate and the associated Reynolds number. Use $\rho = 1.2$ kg/m^3.

15.41 In Fig. P15.41, a pressure difference $p_A - p_B$ is measured to be 96 kPa. Find the friction factor for the wide channel assuming laminar flow. The flow direction is unknown.

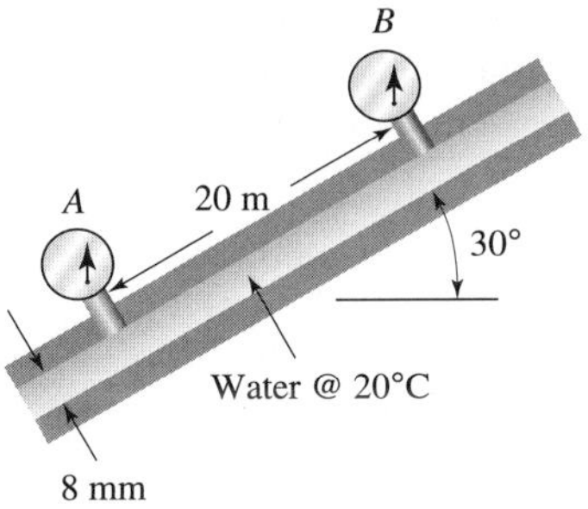

FIGURE P15.41

15.42 A slit with dimensions 0.02 in. × 4 in. exists in the 2-in.-thick side of a pressure vessel containing SAE-10W oil at 80°F and 600 psi. What is the maximum flow rate that can exist from the slit? Assume developed laminar flow.

15.43 Air flows between the parallel plates as shown in Fig. P15.43. Find the pressure gradient such that:
(a) The shear stress at the upper surface is zero.
(b) The shear stress at the lower surface is zero.
(c) The flow rate is zero.
(d) The velocity at $y = 2$ mm is 4 m/s.

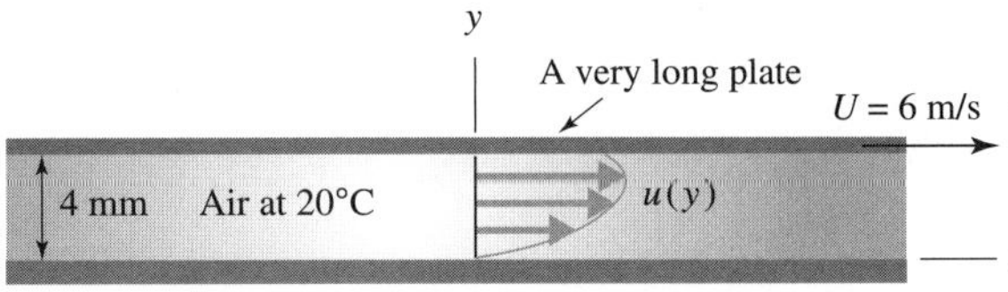

FIGURE P15.43

15.44 A pressure gradient of −20 Pa/m exists in 50°C air flowing between horizontal parallel plates spaced 6 mm apart. Find the velocity of the upper plate so that:
(a) The shear stress at the upper plate is zero.
(b) The shear stress at the lower plate is zero.
(c) The flow rate is zero.
(d) The velocity at $y = 2$ mm is 2 m/s.

15.45 Oil with $\mu = 10^{-4}$ lb-sec/ft^2 fills the concentric space between the rod and the surface shown in Fig. P15.45. Find the force F if $V = 45$ ft/sec. Assume that $dp/dx = 0$.

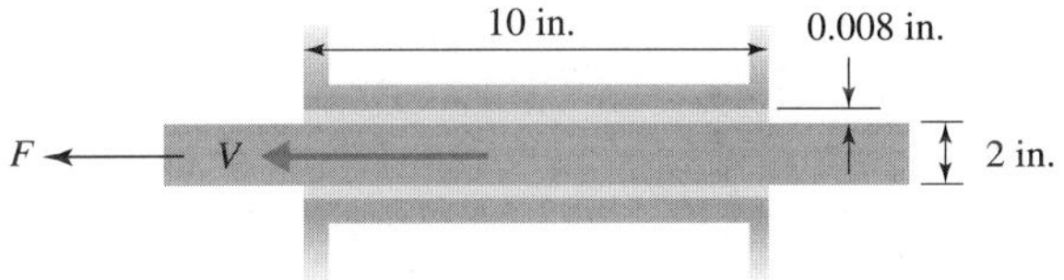

FIGURE P15.45

15.46 Calculate the torque T necessary to rotate the rod shown in Fig. P15.46 at 30 rad/s if the fluid filling the gap is SAE-10W oil at 20°C assuming a linear velocity profile.

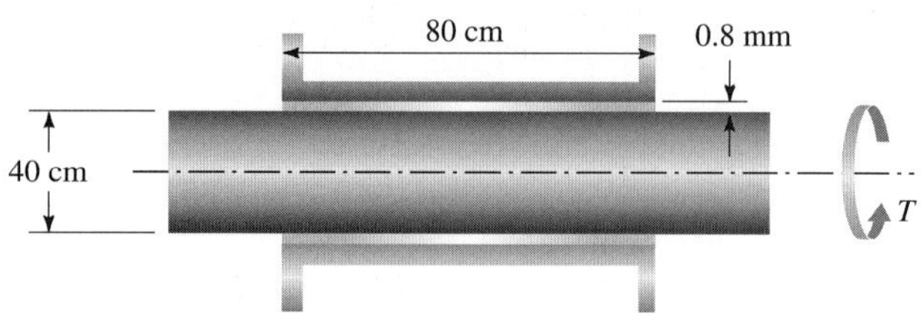

FIGURE P15.46

15.47 Oil with $\mu = 0.01$ N·s/m^2 fills the gap shown in Fig. P15.47. Estimate the torque necessary to rotate the disk shown assuming a linear velocity profile. Is the assumption of laminar flow valid? Use $S = 0.86$.

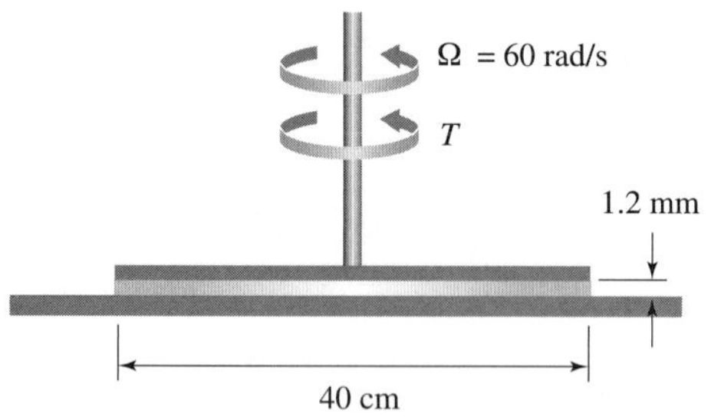

FIGURE P15.47

15.48 Approximate the torque necessary to rotate the inner 20-cm-diameter cylinder shown in Fig. P15.48. SAE-30W oil at 20°C fills the gap. Assume a linear velocity profile.

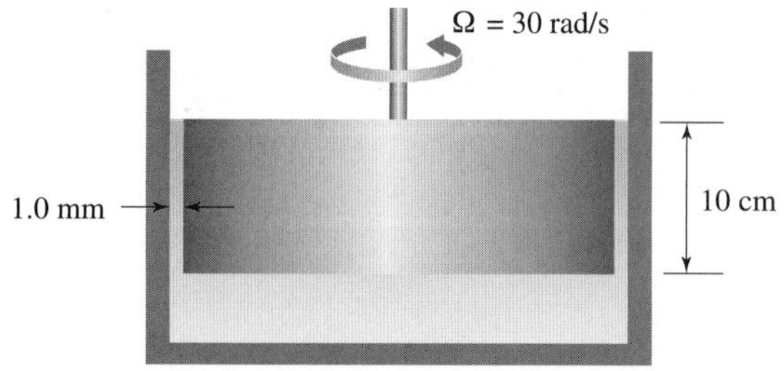

FIGURE P15.48

15.49 Find the torque needed to rotate the cone shown in Fig. P15.49 if oil with $\mu = 0.01$ N·s/m^2 fills the gap as shown. Assume a linear velocity profile.

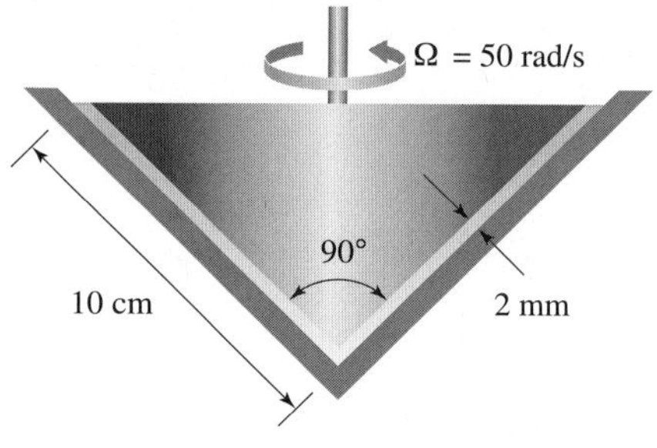

FIGURE P15.49

15.50 To create a high-Reynolds-number flow, the channel setup shown in Fig. P15.50 was proposed by Prof. John Foss of Michigan State University. It is a pressurized channel, thereby avoiding fatal leaks that always are present in a suction channel. (An upstream fan would produce blade vortices that make a high Re impossible to attain.) Estimate the power requirement of the 70%-efficient fan if the channel is 1.2 m wide and Re = 7000.

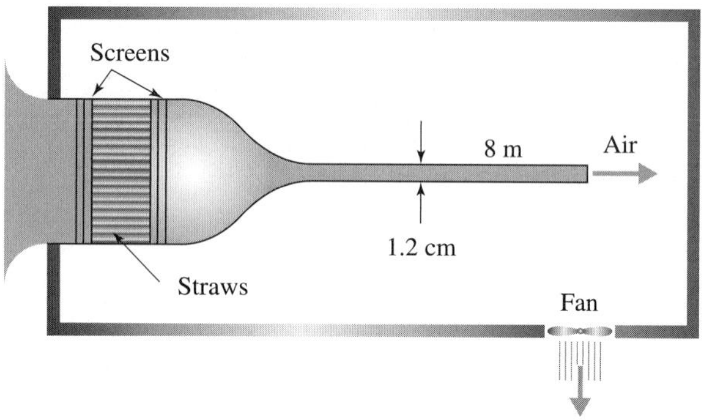

FIGURE P15.50

Laminar Flow between Rotating Cylinders

15.51 A long cylinder of radius R rotates in a large container of liquid. What is the velocity distribution in the liquid? Calculate the torque needed to rotate the 40-in.-long 2-in.-diameter cylinder at 1000 rpm if the liquid is water at 60°F. Assume a laminar flow.

15.52 SAE-10W oil at 40°C fills the gap between two concentric, 40-cm-long cylinders with respective radii of 2 cm and 3 cm. What torque is necessary to rotate the inner cylinder at 3000 rpm if the outer cylinder is fixed? What power is required? Verify that Eq. 15.50 is, in fact, the velocity profile.

15.53 A torque of 0.015 N·m is required to rotate a 4-cm-radius cylinder inside a fixed 5-cm-radius cylinder at 40 rad/s. The concentric cylinders are 50 cm long. Calculate the viscosity of the fluid. Use $S = 0.9$. Verify that Eq. 15.50 is, in fact, the velocity profile.

15.54 Find an expression for the torque necessary to rotate the outer cylinder if the inner cylinder of Fig. 15.6 is fixed.

15.55 Rework Problem 15.53 using the velocity distribution of Eq. 15.50 and calculate the percent error assuming the linear velocity profile.

Turbulent Flow

15.56 The velocity components at a point in a turbulent flow are given in the following table. Find $\bar{u}$, $\bar{v}$, $\overline{u'^2}$, $\overline{v'^2}$, and $\overline{u'v'}$ at that point.

t (s)	u (m/s)	v (m/s)	t (s)	u (m/s)	v (m/s)
0.00	16.1	1.6	0.06	17.1	−1.4
0.01	25.7	−5.4	0.07	28.6	6.7
0.02	10.6	−8.6	0.08	6.7	−5.2
0.03	17.3	3.5	0.09	19.2	−8.2
0.04	5.2	4.1	0.10	21.6	1.5
0.05	10.2	−6.0			

15.57 Over a small radial distance in a developed turbulent flow, the time-average velocity is as given in the following table. The pressure drop in a 30-ft horizontal section is measured to be 8 psf. Find $\overline{u'v'}$ at $r = 0.69$ ft. Air with $\rho = 0.0035$ slug/ft^3 and $\nu = 1.6 \times 10^{-4}$ ft^2/sec is flowing.

r (ft)	0.60	0.63	0.69	0.72
$\bar{u}$ (ft/sec)	81.4	76.9	70.2	60.7

15.58 If at $r = 0.69$ ft for the data of Problem 15.57, we measure $\overline{u'^2} = 316$ ft^2/sec^2 and $\overline{v'^2} = 156$ ft^2/sec^2, what are the magnitudes of the eddy viscosity, the correlation coefficient, and the mixing length?

15.59 The velocity components are measured at a point in a laminar flow to be as shown in Fig. P15.59. Find $\overline{u'v'}$, η, l_m, and K_{uv} if $d\bar{u}/dy = -10$ s^{-1} at the point.

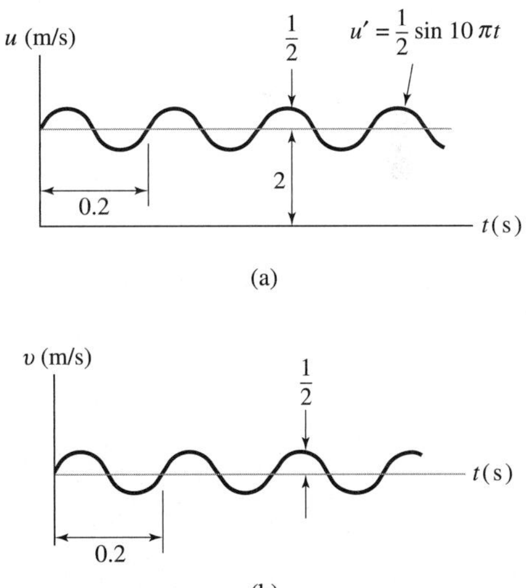

FIGURE P15.59

15.60 A 20-cm-diameter pipe with $e = 0.26$ mm transports water at 20°C . Determine if the pipe is smooth or rough if the average velocity is:
(a) 0.02 m/s **(b)** 0.2 m/s **(c)** 2 m/s

15.61 SAE-30 oil at 40°C is transported in a 10-cm-diameter pipe at an average velocity of 6 m/s. What is the largest size allowed for the roughness element if the pipe is hydraulically smooth?

15.62 Calculate the maximum velocity in the pipe of:
(a) Problem 15.60a **(b)** Problem 15.60c

15.63 The velocity profile for water at 20°C in a turbulent flow in a 10-cm-diameter smooth pipe is approximated by $\bar{u} = 9.2y^{1/7}$ m/s. Find:
(a) The wall shear
(b) The velocity gradient $d\bar{u}/dy$ at the wall
(c) The pressure gradient
(d) The value of η at $r = 2.5$ cm

15.64 Water at 70°F flows in a 5-in.-diameter horizontal pipe at the rate of 2.5 ft^3/sec. Find the constant n in the exponent of Eq. 15.74. What is the maximum velocity?

15.65 Show that the kinetic-energy correction factor is 1.10 for $n = 5$ and 1.03 for $n = 10$ using $\bar{u} = u_{\max}(y/r_0)^{1/n}$ in a circular pipe.

15.66 Water at 20°C flows in a 10-cm-diameter pipe with an average velocity of 10 m/s. Using $\bar{u} = u_{\max}(y/r_0)^{1/n}$ with $n = 7$, plot the viscous shear and the turbulent shear as a function of r. Also, find $d\bar{p}/dx$.

15.67 SAE-10W oil at 10°C is transported in a smooth 80-cm-diameter pipe at the rate of 1.2 m^3/s.
(a) Find the Reynolds number.
(b) Find the friction factor.
(c) Find the maximum velocity using Eq. 15.75.
(d) Find the viscous wall layer thickness.
(e) Compare part (c) with the solution using the logarithmic velocity profile.

15.68 Let the pipe in Problem 15.67 be a cast iron pipe (Fig. 15.13 provides a value for e). Estimate the maximum velocity using the logarithmic velocity profile.

15.69 A pressure drop of 1.5 psi is measured with gages placed 15 ft apart on a smooth horizontal 4-in.-diameter pipe transporting water at 100°F. Estimate:
(a) The wall shear
(b) The maximum velocity
(c) The average velocity
(d) The Reynolds number
(e) The flow rate

15.70 A 12-cm-diameter horizontal pipe transports SAE-10 oil at 10°C . Calculate the wall shear, the average velocity, and the flow rate if the pressure drop over a 10-m section of pipe is measured to be:
(a) 5 kPa **(b)** 20 kPa **(c)** 200 kPa

15.71 Make a linear plot (not a semilog plot) of the velocity profile of the flow in Problem 15.69 using:
(a) The log profile **(b)** The power-law profile

Turbulent Flow in Pipes and Conduits

15.72 Water at 20°C flows in an 8-cm-diameter plastic pipe with a flow rate of 20 L/s. Determine the friction factor using **(a)** the Moody diagram, and **(b)** Eq. 15.81.

15.73 A flow rate of 0.03 m^3/s of 15°C water occurs in a 10-cm-diameter cast iron pipe. Determine the friction factor using **(a)** the Moody diagram, and **(b)** one of the equations (15.81)–(15.83).

15.74 Water at 20°C flows in a 4-cm-diameter cast iron pipe. Determine the friction factor, using the Moody diagram, if the average velocity is:
(a) 0.025 m/s **(b)** 0.25 m/s
(c) 2.5 m/s **(d)** 25 m/s

15.75 A flow rate of 0.02 m^3/s occurs in a 10-cm-diameter wrought iron pipe. Using the Moody diagram, calculate the pressure drop over a 100-m horizontal section if the pipe transports:
(a) Water at 20°C **(b)** Glycerine at 60°C
(c) SAE-30W at 30°C **(d)** Kerosene at 10°C
Compare each answer with that using Eq. 15.84.

15.76 Water at 60°F flows in a 1.5-in.-diameter pipe with a flow rate of 0.06 ft^3/sec. Using the Moody diagram, determine the head loss in a 600-ft section if the pipe is:
(a) Cast iron **(b)** Galvanized iron
(c) Wrought iron **(d)** Plastic

15.77 A mass flux of 1.2 kg/s occurs in a 10-cm-diameter plastic pipe at 20°C and 500 kPa absolute. Assume an incompressible flow and, using the Moody diagram, find the pressure drop in a 100-m section of pipe if the fluid flowing is:
(a) Air **(b)** Carbon dioxide **(c)** Hydrogen
Compare each answer with that using Eq. 15.84.

15.78 SAE-30W oil flows at the rate of 0.08 m^3/s in a 15-cm-diameter horizontal galvanized iron pipe. Find the pressure drop in 100 m if the temperature of the oil is:
(a) 0°C **(b)** 30°C
(c) 60°C **(d)** 90°C
Compare each answer with that using Eq. 15.84.

15.79 Select the material, listed in Fig. 15.13, from which each of the following pipes is probably constructed. Each 5-cm-diameter pipe is tested with water at 20°C using a flow rate of 400 L/min. The following pressure drops are measured over a 10-m length of horizontal pipe:
(a) Pipe 1: 36 kPa **(b)** Pipe 2: 24 kPa
(c) Pipe 3: 19 kPa

15.80 Water at 50°F flows up a 30° incline in a $2\frac{1}{2}$-in.-diameter plastic pipe with a flow rate of 0.3 ft^3/sec. Find the pressure change over a 300-ft length of pipe.

15.81 Water at 40°C flows in a horizontal section of 5-cm-diameter wrought iron pipe with a flow rate of 0.02 m^3/s. Does the pipe behave as a smooth pipe, or is the roughness significant?

15.82 An 80-cm-diameter concrete pipe is to transport storm water at 20°C at a rate of 5 m^3/s. What pressure drop is to be expected over a 100-m section of horizontal pipe?

15.83 A pressure drop of 500 kPa is not to be exceeded over a 200-m length of horizontal 10-cm-diameter cast iron pipe. Calculate the maximum flow rate if the fluid is:
(a) Water at 20°C
(b) Glycerine at 20°C
(c) SAE-10W oil at 20°C
(d) Kerosene at 20°C

15.84 A pressure drop of 200 kPa is not to be exceeded over a 100-m length of horizontal 4-cm-diameter pipe. Estimate the maximum flow rate if water at 20°C is transported and the pipe is:
(a) Cast iron **(b)** Wrought iron **(c)** Plastic

15.85 Neglecting all losses except that due to wall friction, estimate the flow rate through the pipe shown in Fig. P15.85 if the diameter is:
(a) 4 cm **(b)** 8 cm
(c) 12 cm **(d)** 16 cm

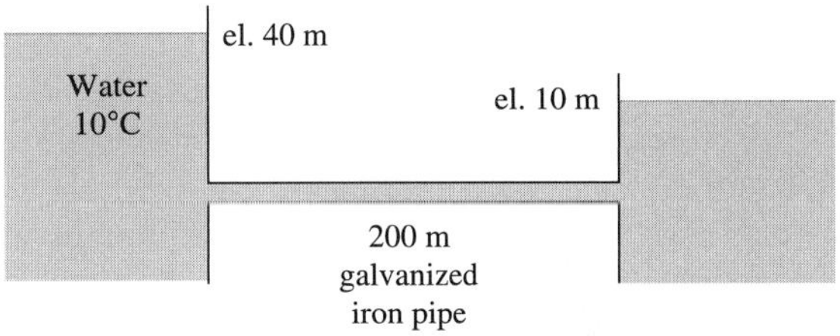

FIGURE P15.85

15.86 A pressure drop of 400 Pa is allowable in gas flow in a 400-m horizontal section of 12-cm-diameter wrought iron pipe. If the temperature and pressure are 40°C and 200 kPa absolute, find the maximum mass flux if the gas is:
(a) Air **(b)** Carbon dioxide **(c)** Hydrogen

15.87 A pressure drop of 30 psi is not to be exceeded over a 600-ft length of horizontal 4-ft-diameter concrete pipe transporting water at 60°F. What flow rate can be accommodated? Use:
(a) The Moody diagram
(b) Equation 15.85

15.88 Estimate the size of plastic tubing that should be selected if 0.002 m^3/s of fluid is to be transported such that the pressure drop does not exceed 200 kPa in a 100-m horizontal section. The fluid is:
(a) Water at 20°C **(b)** Glycerine at 60°C
(c) Kerosene at 20°C **(d)** SAE-10W oil at 40°C

15.89 Select a concrete pipe size that will transport $5 \text{ m}^3/\text{s}$ of 20°C water so that the head loss does not exceed 20 m in a 300-m horizontal pipe section. Use:
(a) The Moody diagram
(b) Equation 15.86

15.90 A farmer wishes to siphon 10°C water from a lake 1200 m from a field positioned a distance of 3 m below the surface of the lake. What size of drawn tubing should be selected if 400 L of water is desired each minute? Use:
(a) The Moody diagram
(b) Equation 15.86
Neglect all losses except that due to wall friction. Is the exiting kinetic energy also negligible?

15.91 Atmospheric air at 30°C is to be transported through a square sheet metal (smooth) conduit at the rate of $4 \text{ m}^3/\text{s}$. What should be the conduit dimensions so that the head loss does not exceed 10 m over a horizontal length of 200 m?

15.92 Water at 20°C is transported through a 2 cm × 4 cm smooth conduit and experiences a pressure drop of 80 Pa over a 2-m horizontal length. What is the flow rate?

15.93 A plastic conduit 4 cm × 10 cm transports water at 20°C. If a pressure drop of 100 Pa is measured by gages spaced 5 m apart on a horizontal section, find the flow rate.

15.94 An open rectangular 1.2-m-wide concrete (use $e = 1.5$ mm) channel transports 20°C water from a reservoir to a location 10 000 m away. Using the Moody diagram, estimate the flow rate if the channel is on a slope of 0.0015 and the water depth is:
(a) 0.3 m **(b)** 0.6 m **(c)** 0.9 m

Minor Losses

15.95 If the loss coefficient for a sudden expansion is based on the exit velocity V_2, determined the loss coefficient in terms of A_1 and A_2.

15.96 Explain, with reference to Eq. 12.41, why the high- and low-pressure regions of Fig. 15.14 exist. Also, sketch a velocity profile from the inside corner of the bend to the outside of the bend along a line from B to C as sketched in Fig. P15.96. Explain why a secondary flow results after the bend.

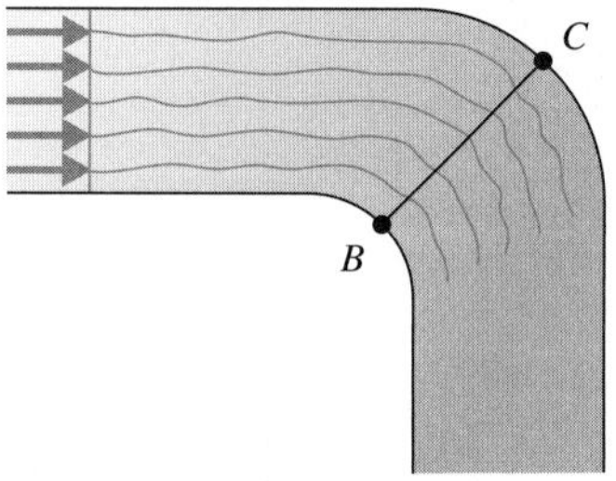

FIGURE P15.96

15.97 For each system shown in Fig. P15.97, find p_2 if $Q_f = 0.02 \text{ m}^3/\text{s}$ of air at 20°C and $p_1 = 50$ kPa.

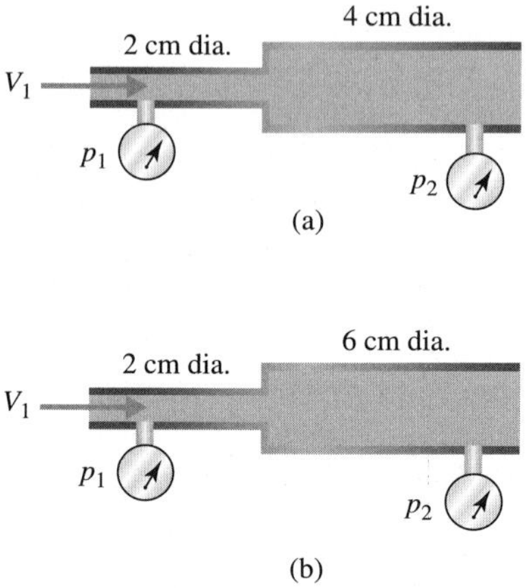

FIGURE P15.97

15.98 Replace the sudden enlargement of Problem 15.97 with a 20° expansion angle and rework the problem.

15.99 For each system shown in Fig. P15.99, estimate the loss coefficient based on V_2 using the data of Fig. 15.15.

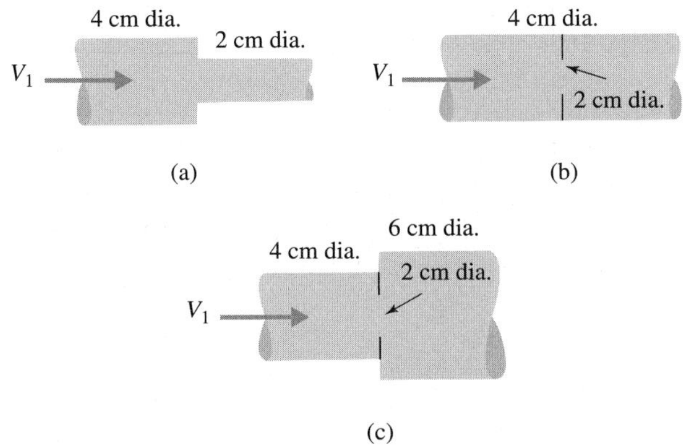

FIGURE P15.99

15.100 The flow rate is measured to be 0.12 ft^3/sec from the pipe shown in Fig. P15.100. Find the loss coefficient of the valve. Neglect wall friction.

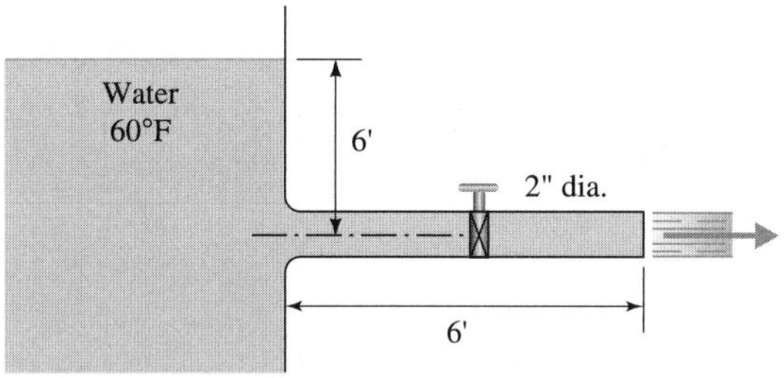

FIGURE P15.100

15.101 The flow rate is measured to be 6 L/s in the pipe shown in Fig. P15.101. Find the loss coefficient of the valve if H is:
(a) 4 cm **(b)** 8 cm

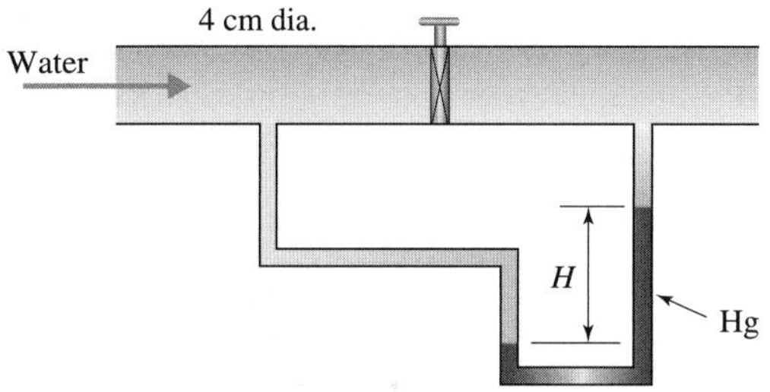

FIGURE P15.101

Simple Piping Systems

15.102 Find the flow rate from the pipe shown in Fig. P15.102. Sketch the EGL and the HGL.

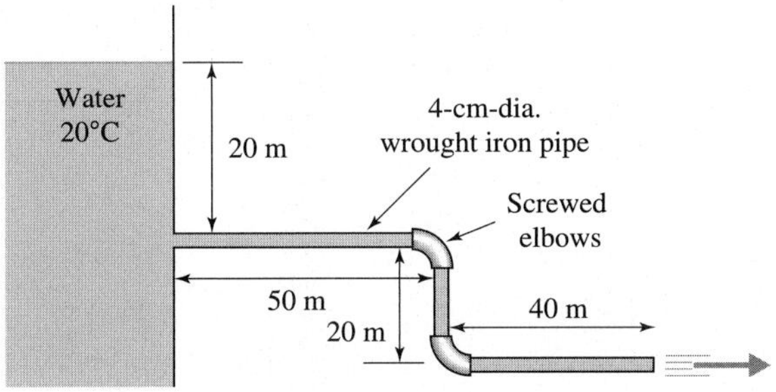

FIGURE P15.102

15.103 Water at 70°F flows from a 4-in.-diameter 1200-ft length of cast iron horizontal pipe that is attached to a reservoir with a square-edged entrance. A screwed globe valve that controls the flow is half open. Find the flow rate if the reservoir elevation above the pipe exit is:
(a) 15 ft **(b)** 30 ft **(c)** 60 ft

15.104 Estimate the flow rate to be expected through the plastic siphon shown in Fig. P15.104 if the diameter is:
(a) 4 cm **(b)** 8 cm **(c)** 12 cm

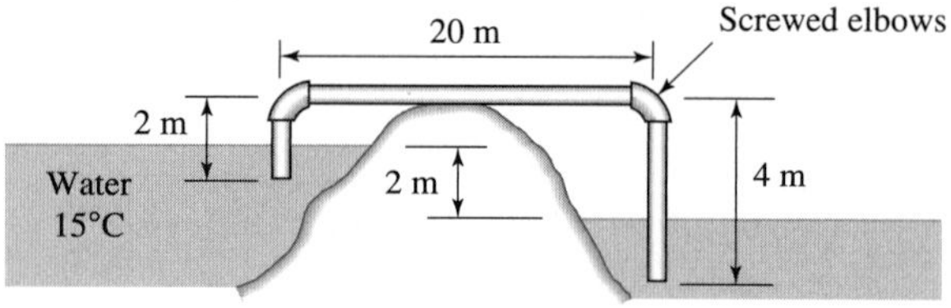

FIGURE P15.104

15.105 For the cast-iron piping shown in Fig. P15.105, calculate the flow rate and minimum pressure and sketch the HGL and the EGL if:
(a) $H = 10$ m **(b)** $H = 20$ m
(c) $H = 30$ m

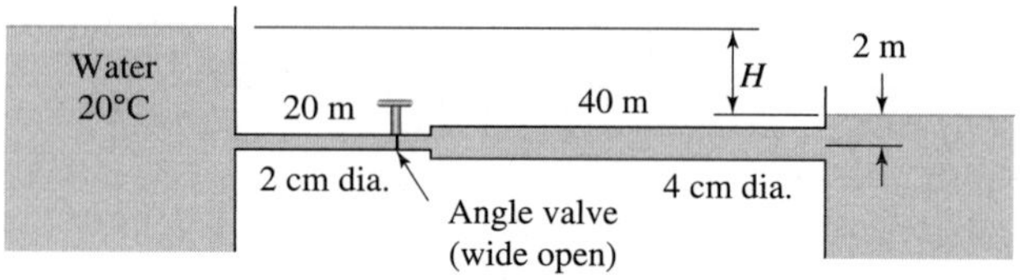

FIGURE P15.105

15.106 Water at 35°C flows from a bathtub through a 2.5-cm-diameter plastic pipe to a large air-filled drain pipe. There are two screwed elbows in the 10 m of pipe. If the drain pipe is 0.8 m below the water level in the tub, estimate how long it will take to collect 10 L of water.

15.107 A 3-cm-diameter plastic tube with screwed elbows is used to siphon the water as shown in Fig. P15.107. Estimate the maximum height H for which the siphon will operate.

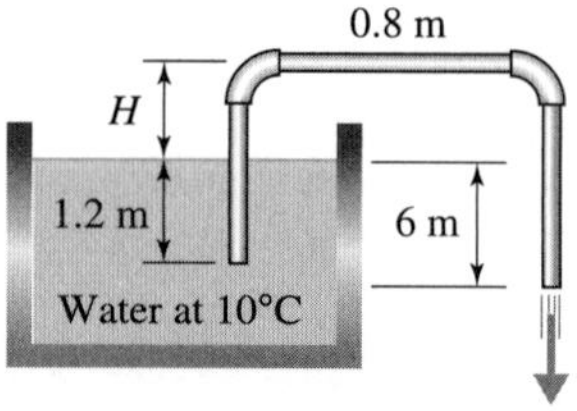

FIGURE P15.107

15.108 A 12-L lawn sprayer is filled with 8 L of 20°C water. It is 1.2 m high and has an 8-mm-diameter copper tube ($e \simeq 0$) that reaches the bottom (it is a little short). A 1.2-m-long 5-mm-diameter smooth hose is attached to the copper tube. The hose ends with a 2-mm-diameter nozzle. If the sprayer is pressurized to 100 kPa, estimate the initial velocity exiting the nozzle.

15.109 What is the maximum flow rate through the pipe shown in Fig. P15.109 if the elevation difference of the reservoir surfaces is:
(a) 80 m?
(b) 150 m?
(c) 200 m?

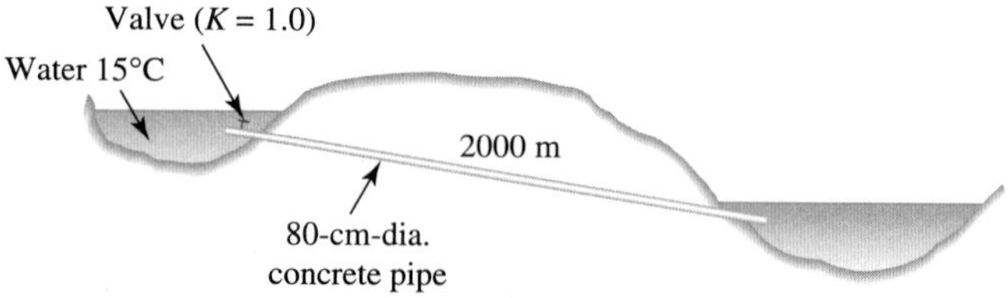

FIGURE P15.109

15.110 A 9.4-mm-diameter 60-m-long plastic pipe transports 20°C water from a spring to a pond 3 m below, similar to the situation shown in Problem 15.109. It is observed that the water alternates between a relatively fast moving stream to a relatively slow moving stream. Explain this phenomenon with supporting calculations.

15.111 Water at 60°F is to be pumped through 900 ft of cast iron pipe from a reservoir to a device that is 30 ft above the reservoir surface. It is to enter the device at 30 psi. Screwed components include two elbows, a square-edged entrance, and an angle valve. If the flow rate is to be 0.6 ft^3/sec, what pump power is needed (assume 80% efficiency) if the pipe diameter is:
(a) 1.5 in.?
(b) 3 in.?
(c) 4.5 in.?

15.112 What pump power (85% efficient) is needed for a flow rate of 0.01 m^3/s in the pipe shown in Fig. P15.112? What is the greatest distance from the left reservoir that the pump can be located?

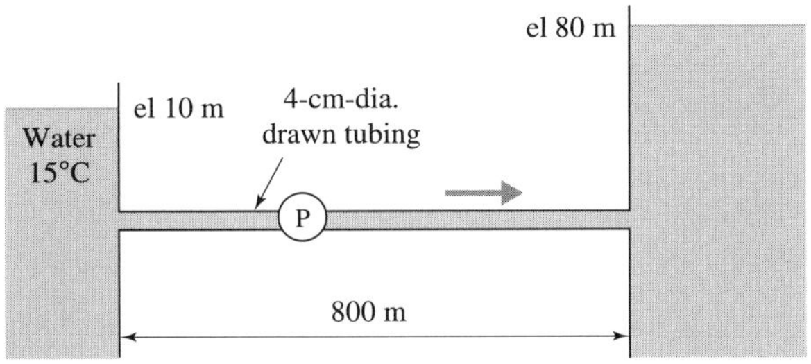

FIGURE P15.112

15.113 A flow rate of 2 m^3/s exists in the pipe shown in Fig. P15.113. What is the expected power output of the turbine (85% efficient) if the elevation difference of the reservoir surfaces is:
(a) 20 m? **(b)** 60 m? **(c)** 100 m?

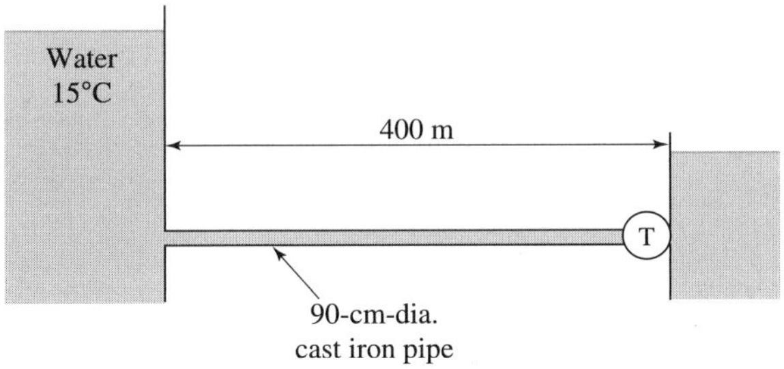

FIGURE P15.113

15.114 What pump power (75% efficient) is needed in the piping system shown in Fig. P15.114? What is the greatest distance from the reservoir that the pump can be located?

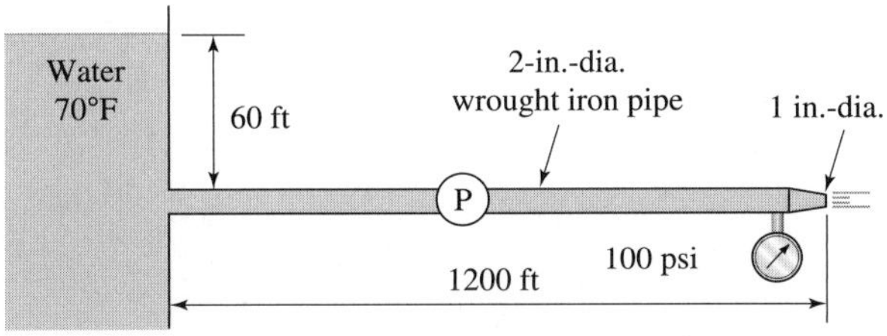

FIGURE P15.114

15.115 The pump shown in Fig. P15.115 has the characteristic curves shown in Example 15.18.
(a) Estimate the flow rate and the power required by the pump.
(b) Sketch the EGL and the HGL.
(c) If cavitation is possible, determine the maximum distance from the reservoir to locate the pump.

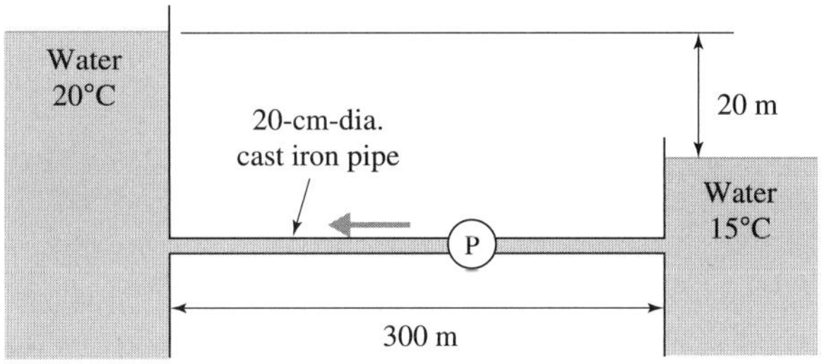

FIGURE P15.115

15.116 Reverse the flow direction in Problem 15.115 and redo the problem.

15.117 A turbine replaces the pump of Problem 15.115. Estimate the power output if $\eta_T = 0.88$. The turbine's characteristic curve is $H_T = 0.8Q_f$, where H_T is measured in meters and Q_f in L/s.

15.118 The pump shown in Fig. P15.118 has characteristic curves shown in Example 15.18. Estimate the flow rate and:
(a) Calculate the pump power requirement.
(b) Calculate the pressure at the pump inlet.
(c) Calculate the pressure at the pump outlet.
(d) Sketch the EGL and the HGL.

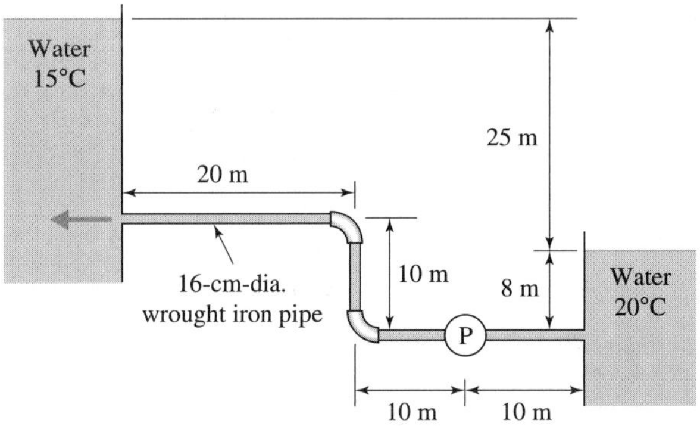

FIGURE P15.118

15.119 Reverse the flow direction in Problem 15.118 and redo the problem.

15.120 A turbine with a characteristic curve shown in Fig. P15.120 is inserted in the pipeline. Calculate the turbine power output. Assume that $\eta_T = 0.90$.

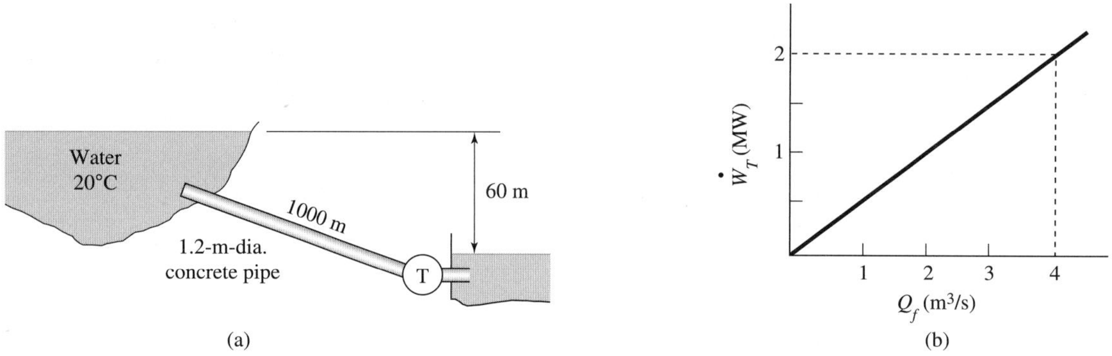

FIGURE P15.120

Open-Channel Flow

15.121 Using a control volume surrounding a finite length of the liquid in a channel flowing at constant depth, find the average shear stress on the walls if water flows in a 10-ft-wide rectangular channel at a depth of 6 ft on a 0.001 slope.

15.122 Using the approach mentioned in Problem 15.121, determine the average shear stress on the portion of the wall of a 40-cm-diameter circular conduit in contact with the water, which is flowing at a constant depth of 10 cm. The slope is 0.0016.

15.123 Calculate the flow rate in a 2-m-wide rectangular, planed wood channel on a slope of 0.001 if the depth is 60 cm. Use:
(a) The Chezy–Manning equation
(b) The Darcy–Weisbach equation

15.124 Water flows in a 6-ft-diameter finished concrete conduit on a 0.0012 slope. Predict the flow rate if the depth of flow is:
(a) 6 ft minus a little **(b)** 5.7 ft
(c) 3 ft **(d)** 1.5 ft
(e) 0.5 ft

15.125 For the channel shown in Fig. P15.125, find the flow rate and average velocity if $S = 0.001$ using:
(a) The Chezy–Manning equation
(b) The Darcy–Weisbach equation

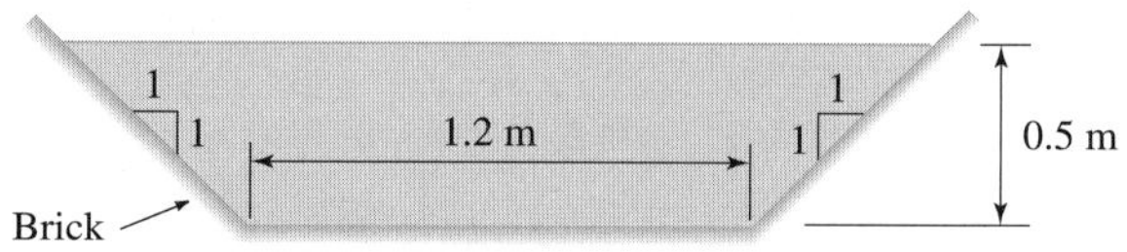

FIGURE P15.125

15.126 At what depth will 5 m^3/s of water flow in a 2-m-wide rectangular brick channel with $S = 0.001$? Use:
(a) The Chezy–Manning equation
(b) The Darcy–Weisbach equation

15.127 The cross section of a straight river is approximated as shown in Fig. P15.127. At what depth will 100 m^3/s of water flow? The slope is 0.001.

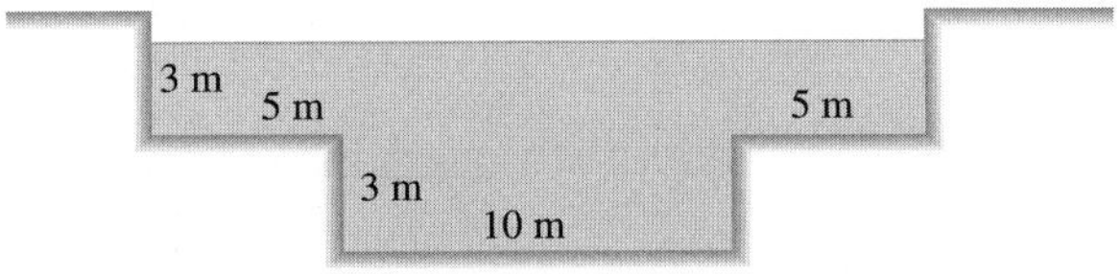

FIGURE P15.127

15.128 For the channel shown in Fig. P15.128, if $S = 0.0016$, find the flow depth if $Q_f = 10\ m^3/s$ and the channel is constructed with:
(a) Planed wood **(b)** Brick

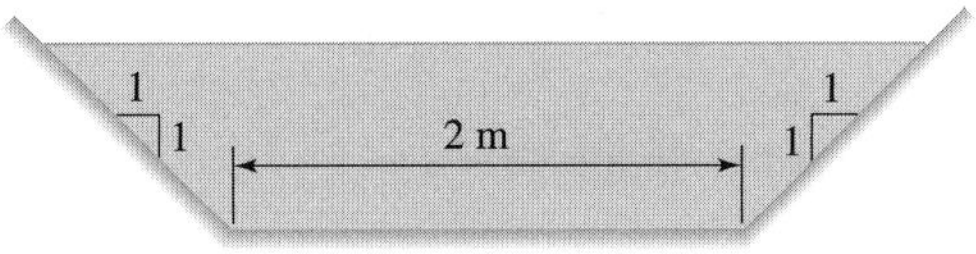

FIGURE P15.128

15.129 Water flows in a 4-ft-diameter sewer pipe at a flow rate of 24 ft^3/sec. Estimate the depth if the slope is 0.001.

15.130 Water flows in a 80-cm-diameter sewer pipe at a flow rate of 0.2 m^3/s. Determine the depth if the slope is 0.001.

A blimp is a lighter-than-air-vessel that is streamlined to reduce the drag forces encountered during its forward motion. Large blimps, perhaps 1000 m long, could be used as cruise ships to tour the globe; sea sickness would be avoided! (Courtesy of Goodyear Tires, Inc.)

16

External Flows

Outline

Chapter Objectives

The objectives of this chapter are to:

- ▲ Discuss separated and attached flows.
- ▲ Introduce the lift and drag coefficients.
- ▲ Determine the drag on a variety of bodies.
- ▲ Study the influence of vortex shedding and streamlining.
- ▲ Determine when cavitation is present.
- ▲ Calculate the lift and drag on airfoils.

- ▲ Superimpose several simple potential flows to construct a flow of interest.
- ▲ Analyze both laminar and turbulent boundary layers on a flat plate.
- ▲ Calculate convection heat transfer from boundary layer profiles.
- ▲ Provide numerous examples and problems demonstrating how the various quantities of interest are determined for the many exernal flows discussed in this chapter.

Nomenclature

The following variables are introduced in this chapter:

c	chord length
C_D	drag coefficient
c_f	local skin friction coefficient
C_f	skin friction coefficient
C_L	lift coefficient
f	frequency
h_c	convective heat transfer coefficient
k	added mass coefficient
k_f	fluid thermal conductivity
m_a	added mass
m_f	mass of displaced fluid
$\dot{mom}$	momentum flux
Nu	Nusselt number
Pr	Prandtl number
p_v	vapor pressure
q_f	flow rate per unit depth
q''_x	heat flux (W/m^2)
r_c	a cylinder radius
Re_x	Reynolds number based on a distance x
T_f	average boundary layer temperature
T_s	surface temperature
T^+	a dimensionless temperature
T_∞	bulk fluid temperature
u_τ	shear velocity
U_∞	free-stream velocity away from a body
$U(x)$	free-stream velocity at body surface
x^+, y^+	dimensionless distances
α	angle of attack, thermal diffusivity
δ	boundary layer thickness
δ_d	displacement thickness
δ_t	temperature boundary layer thickness
δ_v	viscous wall layer thickness
ϕ	velocity potential
Γ	vortex strength
μ	doublet strength
θ	momentum thickness
σ	cavitation number
τ_0	wall shear stress
ψ	stream function

16.1 INTRODUCTION

The study of external flows is of particular importance to the aeronautical engineer in the analysis of airflow around the various components of an aircraft. In fact, much of the present knowledge of external flows has been obtained from studies motivated by such aerodynamic problems. There is, however, substantial interest by other engineers in external flows; the flow of fluid around turbine blades, automobiles, buildings, athletic stadiums, smokestacks, spray droplets, bridge abutments, submarine pipelines, river sediment, and red blood cells suggest

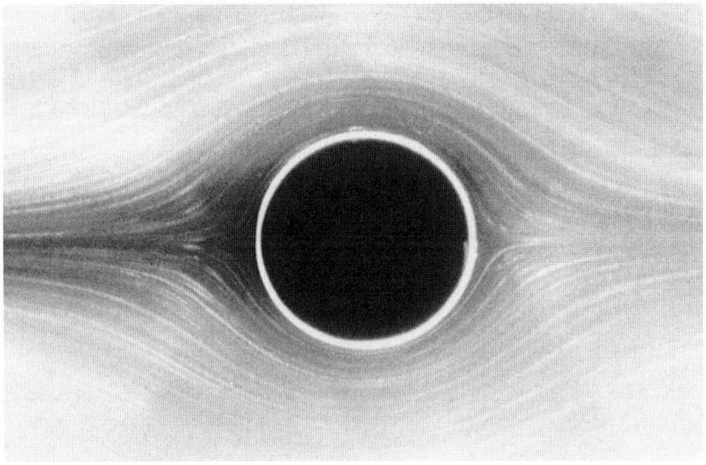

FIGURE 16.1 Flow past a circular cylinder at Re = 0.16. The flow is from left to right. It resembles superficially the pattern of potential flow. The flow of water is shown by aluminum dust. (Courtesy of The Parabolic Press, Stanford, California. Reprinted with permission.)

a variety of phenomena that can be understood only from the perspective of external flows.

It is a difficult task to determine the flow field external to a body and the pressure distribution on the surface of a body, even for the simplest geometry. To discuss this subject, consider low-Reynolds-number flows (Re < 5, or so) and high-Reynolds-number flows (Re > 1000). Low-Reynolds-number flows, called *creeping flows* or *Stokes flows*, seldom occur in engineering applications (flow around spray droplets, lubrication in small gaps, and flow in porous media would be exceptions) and are not presented in this book; they are left to the specialist. We will direct our attention to high-Reynolds-number flows only. We do, however, show a Stokes flow in Fig. 16.1.

Key Concept: *Low-Reynolds-number flows seldom occur in engineering applications.*

High-Reynolds-number flows can be subdivided into three major categories: (1) incompressible immersed flows involving such objects as automobiles, helicopters, submarines, low-speed aircraft, take-off and landing of commercial aircraft, buildings, and turbine blades; (2) flows of liquids that involve a free surface as experienced by a ship or a bridge abutment; and (3) compressible flows involving high-speed objects ($V > 100$ m/s) such as aircraft, missiles, and bullets. We will focus our attention on the first category of flows in this chapter and consider cases in which the object is far from a solid boundary or other objects. The flow becomes significantly influenced by the presence of a boundary or another object, as shown in Fig. 16.2; in part (d) the slender object must be at least five body lengths below the free surface before free surface effects can be neglected. Flows such as those shown in Fig. 16.2 are not included in an introductory presentation.

High-Reynolds-number incompressible immersed flows are divided into two categories: flows around blunt bodies and flows around streamlined bodies, as displayed in Fig. 16.3. The boundary layer (see Section 12.3.2) near the stagnation point is a laminar boundary layer, and for a sufficiently large Reynolds number,

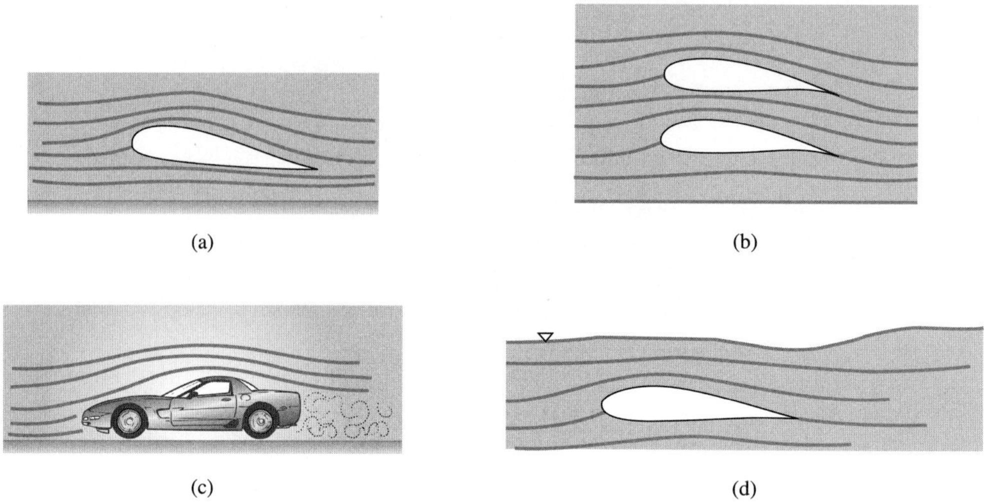

FIGURE 16.2 Examples of complicated immersed flows: (a) flow near a solid boundary; (b) flow between two turbine blades; (c) flow around an automobile; (d) flow near a free surface.

Separated region: *A region of recirculating flow.*

Wake: *A region of velocity defect that grows due to diffusion.*

undergoes transition downstream to a turbulent boundary layer, as shown; the flow may separate from the body and form a **separated region**, a region of recirculating flow, as shown for the blunt body, or it simply leaves the streamlined body at the trailing edge (there may be a small separated region here). The **wake**, which is characterized by a *velocity defect* is a growing (diffusion) region, and trails the body, as shown. The boundaries of the wake, the separated region, and the turbulent boundary layer are quite time dependent; in the sketch the time-average location of the wake is shown by the dashed lines. Shear stresses due to viscosity are concentrated in the thin boundary layer, the separated region, and the wake; outside these regions the flow is approximated by an inviscid flow.

From the figure it could be assumed that the separated region does not exchange mass with the free stream since mass does not cross a streamline. When viewed instantaneously, however, the separation streamline is highly time dependent, and due to this unsteady character, the separated region is able to exchange mass slowly with the free stream.

Key Concept: *The wake diffuses into the main flow and eventually disappears.*

Several comments should be made regarding the separated region and the wake. The separated region eventually closes; the wake keeps diffusing into the main flow and eventually disappears as its area becomes exceedingly large (the fluid regains the free-stream velocity). Time-average streamlines do not enter a separated region; they do enter a wake. The separated region is submerged within the wake.

Drag: *The force the flow exerts in the direction of the flow.*

Lift: *The force the flow exerts normal to the direction of flow.*

Flow around a blunt object is usually treated empirically, as was done for a turbulent flow in a conduit. We follow that procedure here. We are interested primarily in the **drag**, the force the flow exerts on the body in the direction of the flow. The **lift**, which acts normal to the direction of flow, will be of interest for airfoil shapes, as presented in Section 16.4. The actual details of the flow field are

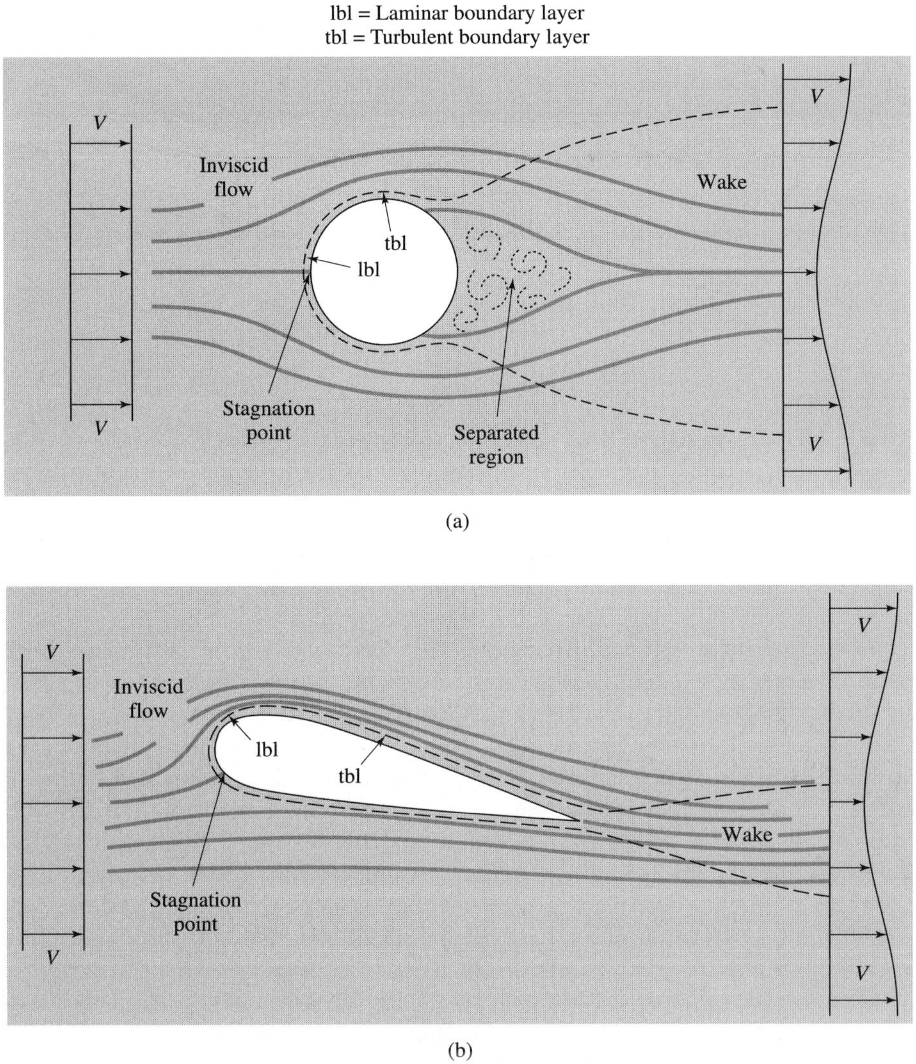

FIGURE 16.3 Flow around a blunt body and a streamlined body.

seldom of interest and are not presented in this introductory text. We present the drag F_D and lift F_L as dimensionless coefficients: the *drag coefficient* and *lift coefficient*, defined as

$$C_D = \frac{F_D}{\frac{1}{2}\rho V^2 A} \qquad C_L = \frac{F_L}{\frac{1}{2}\rho V^2 A} \tag{16.1}$$

where A is most often the projected area (projected on a plane normal to the direction of the flow); for airfoil shapes, the area is based upon the chord (see Fig. 16.4). Drag coefficients for several common shapes are presented in Section 16.3.1.

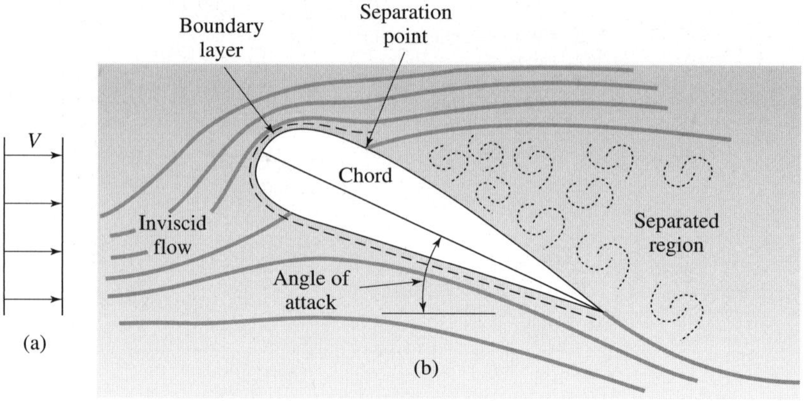

FIGURE 16.4 Streamlined body that is stalled.

Key Concept: *The drag on a blunt object is dominated by the flow in the separated region.*

Since the drag on a blunt object is dominated by the flow in the separated region, there is little interest in studying the boundary layer growth on the front part of a blunt body and the associated viscous shear at the wall. Hence interest is focused on the empirical data that provide the drag coefficient.

Flow around a streamlined body—the separated region is insignificantly small or nonexistent—provides the motivation for detailed study of the laminar and turbulent boundary layers. A boundary layer that develops on a plane streamlined surface, such as an airfoil, is usually sufficiently thin that the curvature of the surface can be ignored and the problem can be treated as a boundary layer developing on a flat plate with a nonzero pressure gradient. We will provide a detailed study of flow on a flat plate with zero pressure gradient; once that problem is understood, the influence of a pressure gradient can be discussed. If the flow in the boundary layer on a streamlined body can be determined, the drag can be calculated, since the drag is a result of the shear stress and pressure force acting on the body surface.

Key Concept: *The inviscid, free-stream flow exists outside the boundary layer.*

Outside the boundary layer there exists an inviscid, *free-stream flow*, as shown in Figs. 16.3 and 16.4. Initially, we will assume that the free-stream flow is known. Before the velocity profile in the boundary layer can be determined, it is necessary that the inviscid flow solution be known. It is found by completely ignoring the boundary layer, since it is so thin, and solving the appropriate inviscid equations. The inviscid flow solution is then used to provide the lift on the body, and the two quantities used in the boundary layer flow solution: the pressure gradient and the velocity at the boundary. With the inviscid flow known and the boundary layer flow determined, quantities of interest can be obtained in the flow about a streamlined body.

16.2 SEPARATION

Before we present the empirical data associated with the flow around blunt bodies, the general nature of separation will be discussed. Separation occurs when the main stream leaves the body, resulting in a separated region, as sketched in Fig.

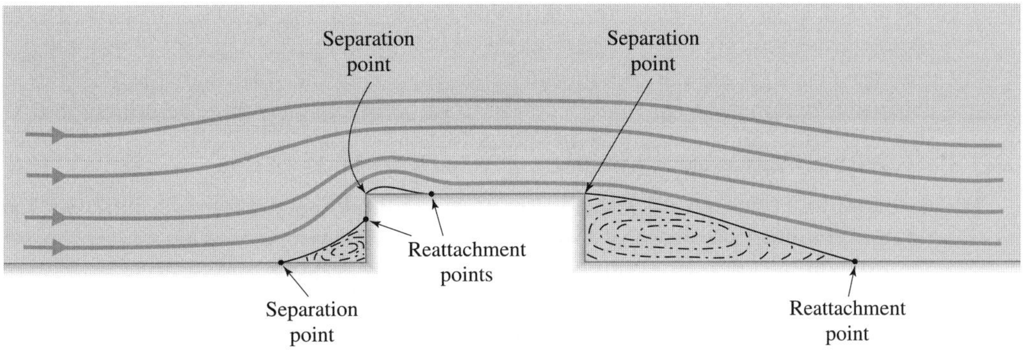

FIGURE 16.5 Separation due to abrupt geometry changes.

16.3a. When separation occurs on a streamlined body near the forward portion of an airfoil, as it will with a sufficiently large angle of attack (the angle the oncoming flow makes with the **chord**, a line connecting the trailing edge with the nose), the flow situation is referred to as **stall**, as shown in Fig. 16.4. Stall is highly undesirable on aircraft at cruise conditions and leads to inefficiencies when it occurs on turbine blades. It is used, however, to provide the high drag needed when landing an aircraft, or in certain maneuvers by stunt planes. For blunt bodies, however, separation is unavoidable at high Reynolds numbers and its effect must be understood.

Chord: *A line connecting the trailing edge with the nose.*

Stall: *The flow condition where separation occurs on a streamlined body near the forward portion.*

Key Concept: *For blunt bodies, separation is unavoidable at high Reynolds numbers.*

The location of the separation point is dependent primarily on the geometry of the body; if the body has an abrupt change in geometry, such as that shown in Fig. 16.5, separation will occur at, or near, the abrupt change; however, it will also occur upstream on the flat surface as shown. In addition, reattachment will occur at some location, as shown. Let us establish the criterion used in predicting the location of the separation point on a surface with no abrupt geometry change.

Consider the flow on the flat surface just before the step of Fig. 16.5. The region near the forward separation point is enlarged and shown in Fig. 16.6; the y-coordinate is normal to the wall and the x-coordinate is measured along the wall.

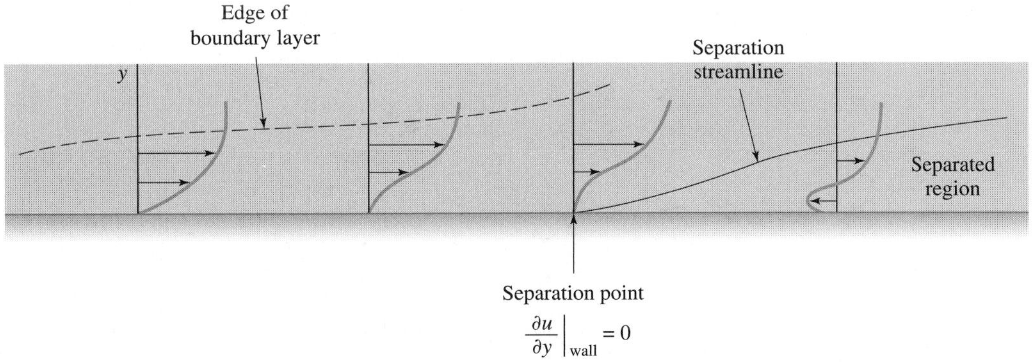

FIGURE 16.6 Flow separation on a flat surface due to an adverse pressure gradient.

Key Concept: *The separation point is defined as that point where $(\partial u/\partial y)_{wall} = 0$.*

Key Concept: *As the boundary layer before separation becomes turbulent, the separation point moves to the rear.*

Downstream of the separation point the x-component velocity near the wall is in the negative x-direction and thus at the wall $\partial u/\partial y$ must be negative. Upstream of the separation point the x-component velocity near the wall is in the positive x-direction, demanding that $\partial u/\partial y$ at the wall be positive. Hence we conclude that the separation point is defined as that point where $(\partial u/\partial y)_{\text{wall}} = 0$.

Observe that separation on the flat surface occurs as the flow is approaching a stagnation region where the velocity is low and the pressure is high. As the flow approaches the stagnation region the pressure increases, that is, $\partial p/\partial x > 0$; the pressure gradient is positive. Since separation is often undesirable, a positive pressure gradient is called an *adverse pressure gradient*; a negative gradient is a *favorable pressure gradient*. In general, the effect of an adverse pressure gradient results in decreasing velocities in the streamwise direction; if an adverse pressure gradient acts on a surface over a sufficient distance, separation may result. This is true even if the surface is a flat plate, such as the wall of a diffuser. See Section 16.6.7 for further discussion.

In addition to geometry and pressure gradient, several other parameters influence separation. These include the Reynolds number as a very important parameter, with the wall roughness, the *free-stream fluctuation intensity*[1] (the intensity of the disturbances that exist away from the boundary), and the wall temperature having less but occasionally significant influence.[2] Visualize, for example, flow around a sphere; at sufficiently low Reynolds numbers no separation will occur. As the Reynolds number increases to a particular value, separation will occur over a small area at the rear; this area will become larger and larger as the Reynolds number increases until at a sufficiently large Reynolds number no additional increase in separation area will be observed. The boundary layer before separation will still be laminar. An interesting phenomenon takes place as the boundary layer before separation becomes turbulent; there is a sudden movement of the separation point to the rear of the sphere, resulting in a substantial reduction in the separation area and thus the drag. This phenomenon is explained by comparing the velocity profile of a laminar boundary layer with that of a turbulent boundary layer, as displayed in Fig. 16.7. As was true in a pipe

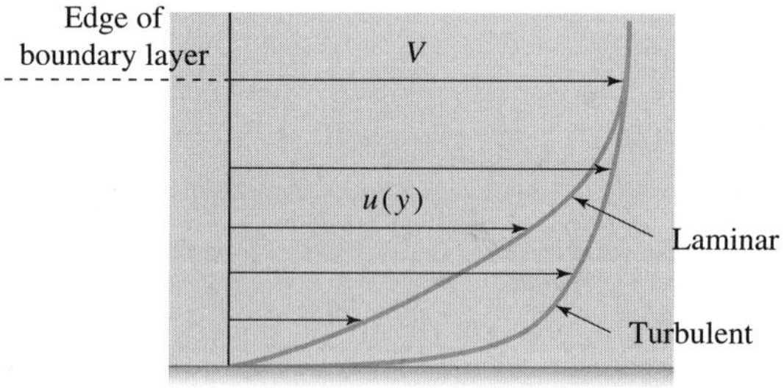

FIGURE 16.7 Comparison of laminar and turbulent velocity profiles.

[1]Free-stream fluctuation intensity is defined as $\sqrt{\overline{u'^2}}/V$, where u' is the fluctuation. A value of 0.001 is quite low and 0.1 is quite high. This is also called *free-stream turbulence*; the free stream is not turbulent, however.

[2]If a fluid is flowing past a body that is rigidly supported, the vibration level of the support system will also influence the separation phenomenon. External sound waves can also be significant.

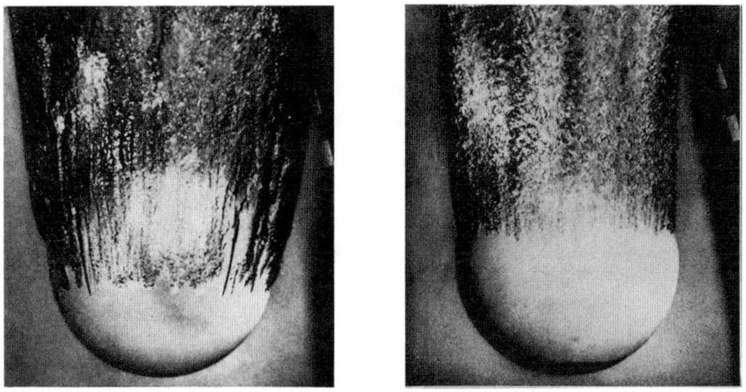

(a) (b)

FIGURE 16.8 Effect of boundary layer transition on separation: (a) laminar boundary layer before separation; (b) turbulent boundary layer before separation. (U.S. Navy photographs.)

flow, the turbulent profile has a much larger gradient near the wall (much larger wall shear stress) and thus the momentum of the fluid near the wall is substantially larger in the turbulent boundary layer. For a given geometry a greater distance is required to reduce the velocity near the wall to zero, resulting in the movement of the separation point to the rear, as is quite obvious in Fig. 16.8, where both spheres are moving with the same velocity (the sphere in (b) has sandpaper attached in the nose region). In Fig. 16.8a it is observed that separation occurs on the front half of the sphere, in a region of favorable pressure gradient. This separation is due to the centrifugal effects as the fluid moves around the sphere. This phenomenon of drag reduction is observed in the drop in the drag coefficient curves for a sphere and a cylinder, presented in the following section.

16.3 FLOW AROUND IMMERSED BODIES

16.3.1 Drag Coefficients

From our study of dimensional analysis we know that for a steady, incompressible flow in which gravity, thermal, and surface tension effects are neglected, the primary flow parameter that influences the flow is the Reynolds number; other occasionally important parameters include the relative wall roughness and the free-stream velocity fluctuation intensity.

The drag coefficient curves for two bodies that do not exhibit sudden geometric changes will be presented; the drag coefficients for the smooth sphere and the long smooth cylinder are shown in Fig. 16.9 over a large range of Reynolds numbers. At Re < 1 creeping flow with no separation results. For the sphere, this creeping flow problem has been solved, with the result that

$$C_D = \frac{24}{\text{Re}} \qquad \text{Re} < 1 \tag{16.2}$$

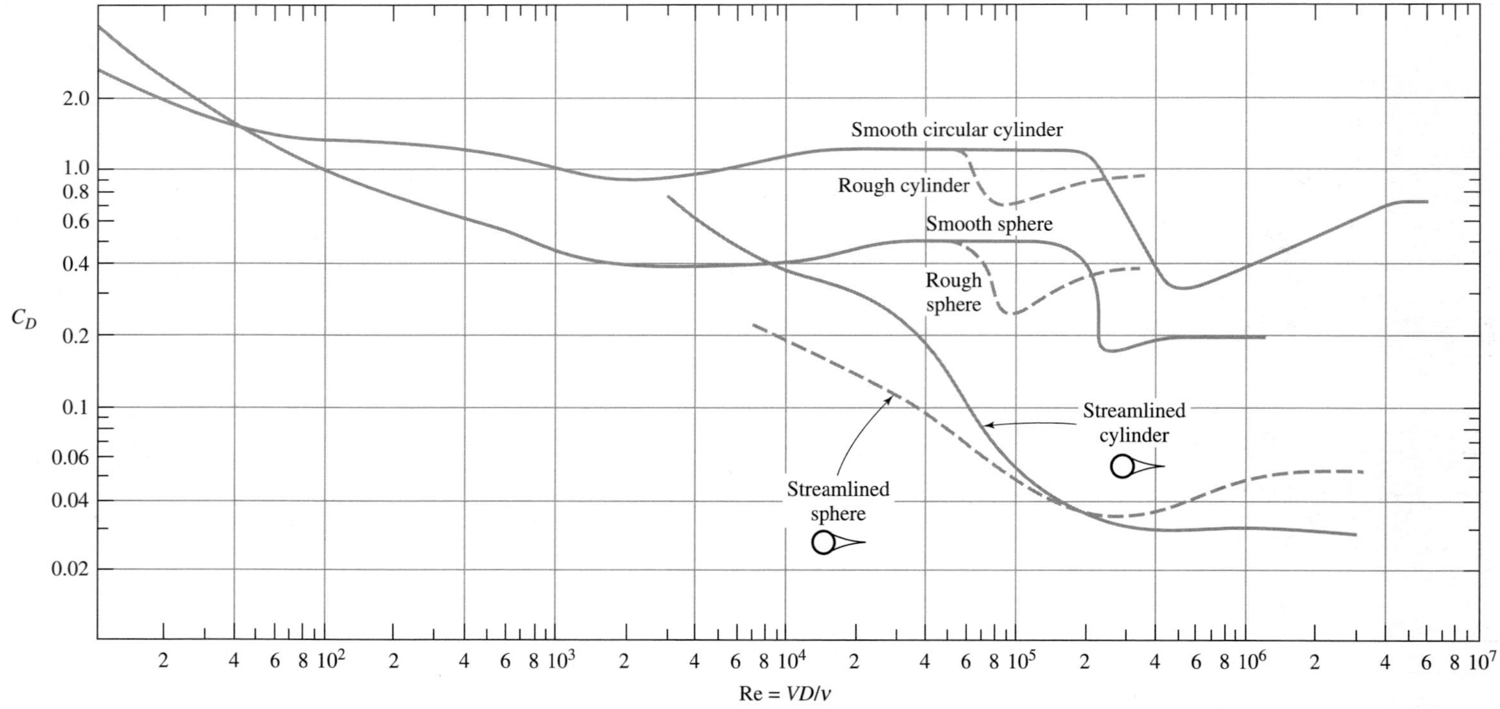

FIGURE 16.9 Drag coefficients for flow around a long cylinder and a sphere. (See E. Achenbach, *J. Fluid Mech.*, Vol. 46, 1971, and Vol. 54, 1972.)

Separation is observed at Re $\simeq$ 10 over a very small area on the rear of the body. The separated area increases as the Reynolds number increases until Re $\simeq$ 1000, where the separated region ceases to enlarge; during this growth of the separated region the drag coefficient decreases. At Re = 1000, 95% of the drag is due to form drag (the drag force due to the pressure acting on the body) and 5% is due to frictional drag (the drag force due to the shear stresses acting on the body).

The drag coefficient curve is relatively flat for smooth bodies over the range $10^3 < \text{Re} < 2 \times 10^5$. The boundary layer before the point of separation is laminar and the separated region is as shown in Fig. 16.8a. At Re $\simeq 2 \times 10^5$, for a smooth surface and with low free-stream fluctuation intensity, the boundary layer before separation undergoes transition to a turbulent state and the increased momentum in the boundary layer "pushes" the separation back, as shown in Fig. 16.8b, with a substantial decrease (a 60 to 80% drop) in the drag. If the surface is rough (dimples on a golf ball) or the free stream has high free-stream fluctuation intensity, the drop in the C_D curve may occur at Re $\simeq 8 \times 10^4$. Since a lower drag is usually desirable, surface roughness is often added; the dimples on the golf ball may increase the flight distance by 50 to 100%.

Key Concept: *Dimples on the golf ball may increase the flight distance by 50 to 100%.*

After the sudden drop in drag, the C_D curve is observed to again increase with increased Reynolds number. Experimental data are not readily available for Re > 10^6 for a sphere and Re > 6×10^7 for a cylinder; however, a value of $C_D = 0.2$ for a sphere at large Reynolds number appears acceptable. Some engineers use $C_D = 0.4$ for cylinders at large Reynolds numbers; however, the data presented here suggests that that is too low. Additional experimental data is needed.

For cylinders of finite length and for elliptic cylinders, the drag coefficients are presented in Table 16.1. The finite-length cylinders are assumed to have two free ends. If one end is fixed to a solid surface, its length must be doubled when using Table 16.1. Blunt objects with sudden geometry changes have separated regions that are relatively insensitive to the Reynolds number; the drag coefficients for some common shapes are given in Table 16.2.

TABLE 16.1 Drag Coefficients of Finite-Length Circular Cylinders[a] with Free Ends[b] and of Infinite-Length Elliptic Cylinders

Circular cylinder		Elliptic cylinder[c]		
$\frac{\text{Length}}{\text{Diameter}}$	$\frac{C_D}{C_{D\infty}}$	$\frac{\text{Major axis}}{\text{Minor axis}}$	Re	C_D
∞	1	2	4×10^4	0.6
40	0.82	4	10^5	0.46
20	0.76	4	2.5×10^4 to 10^5	0.32
10	0.68	8	2.5×10^4	0.29
5	0.62	8	2×10^5	0.20
3	0.62			
2	0.57			
1	0.53			

[a] $C_{D\infty}$ is the drag coefficient for the infinite-length circular cyulinder obtained in Fig. 16.9.
[b] If one end is fixed to a solid surface, double the length of the cylinder.
[c] Flow is in the direction of the major axis.

TABLE 16.2 Drag Coefficients for Various Blunt Objects

Object		Re	C_D
Square cylinder → □ w (w)	$L/w = \infty$	$>10^4$	2.0
	$L/w = 1$ (cube)	$>10^4$	1.1
rounded corners ($r = 0.2w$)	$L/w = \infty$	$=10^5$	1.2
Rectangular plates →| w	$L/w = \infty$	$>10^3$	2.0
	20	$>10^3$	1.5
	5	$>10^3$	1.2
	1	$>10^3$	1.1
Circular cylinder → □ D (L)	$L/D = 0.1$ (disk)	$>10^3$	1.1
	4	$>10^3$	0.9
	7	$>10^3$	1.0
Semicircular cylinder → D		$>10^4$	2.2
→ ᗡ		$>10^4$	1.2
Semicircular shell →)		$=2 \times 10^4$	2.3
→ (		$=2 \times 10^4$	1.1
Equilateral cylinders → ◁	2.0	$>10^4$	2.0
→ ▷	1	$>10^4$	1.4
Cone → ◁ (α)	$\alpha = 30°$	$>10^4$	0.6
	60°	$>10^4$	0.8
	90°	$>10^4$	1.2
Solid hemisphere → ◗		$>10^4$	1.2
→ ◖		$>10^4$	0.4
Hollow hemisphere → D		$>10^4$	1.4
→ ᗡ		$>10^4$	0.4
Parachute		$>10^7$	1.4
Automobile			
1920	—	$>10^5$	0.80
Modern, with square corners	—	$>10^5$	0.30
Modern, with rounded corners	—	$>10^5$	0.29
Van		$>10^5$	0.42
Bicycle, upright rider			1.1
racing, bent over			0.9
racing, drafting			0.5
Semitruck, standard			0.96
with streamlined deflector			0.76
with deflector and gap seal			0.70

Example 16.1

A square sign, 10 ft × 10 ft, is attached to the top of a 60-ft-high pole which is 12 in. in diameter (Fig. E16.1). Approximate the maximum moment that must be resisted by the base for a wind speed of 100 ft/sec.

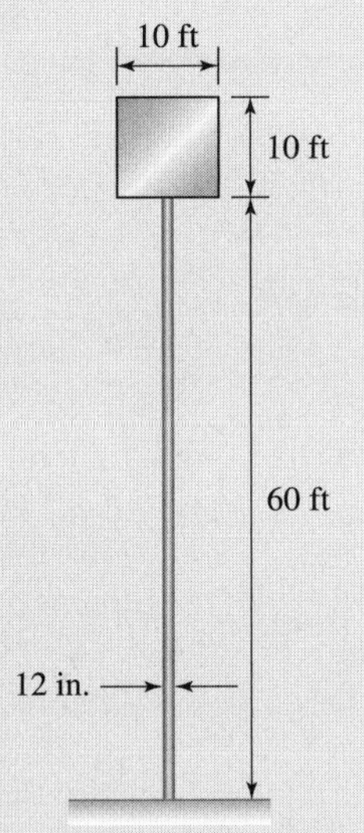

FIGURE E16.1

Solution

The maximum force F_1 acting on the sign occurs when the wind is normal to the sign; it is

$$F_1 = C_D \times \tfrac{1}{2}\rho V^2 A$$
$$= 1.1 \times \tfrac{1}{2} \times 0.0024 \times 100^2 \times 10^2 = 1320 \text{ lb}$$

where C_D is found in Table 16.2 and we use the standard value $\rho = 0.0024$ slug/ft^3 since it was not given. The force F_2 acting on the cylindrical pole is (using the projected area as $A = 60 \times 1$ ft^2)

$$F_2 = C_D \times \tfrac{1}{2}\rho V^2 A$$
$$= 0.8 \times \tfrac{1}{2} \times 0.0024 \times 100^2 \times 60 = 576 \text{ lb}$$

where C_D is found from Fig. 16.9 with $\text{Re} = 100 \times 1/1.6 \times 10^{-4} = 6.2 \times 10^5$, assuming a high-intensity fluctuation level (i.e., a rough cylinder); since neither end is free, we do not use the multiplication factor of Table 16.1.

The resisting moment that must be supplied by the supporting base is

$$M = d_1 F_1 + d_2 F_2$$
$$= 65 \times 1320 + 30 \times 576 = 103{,}000 \text{ ft-lb}$$

assuming that the forces act at the centers of their respective areas.

Example 16.2

Determine the terminal velocity of a 30-cm-diameter smooth sphere ($S = 1.02$) if it is dropped in (a) air at 20°C and (b) water at 20°C.

Solution

(a) When terminal velocity is reached by a falling object, the weight of the object is balanced by the drag force acting on the object. Using $\sum F = 0$ and Eq. 16.1, we have

$$W = F_D$$

$$\therefore \gamma_{\text{sphere}} \times \text{volume} = C_D \times \tfrac{1}{2}\rho V^2 A$$

The volume of a sphere is $\frac{4}{3}\pi R^3$ and $\gamma_{\text{sphere}} = S\gamma_{\text{water}}$; the projected area is $A = \pi R^2$. Thus

$$S\gamma_{\text{water}} \times \tfrac{4}{3}\pi R^3 = C_D \times \tfrac{1}{2}\rho V^2 \pi R^2$$

The velocity can now be expressed as

$$V = \left(\frac{8RS\gamma_{\text{water}}}{3\rho C_D}\right)^{1/2} = \left(\frac{8 \times 0.15 \times 1.02 \times 9800}{3 \times 1.20 \times C_D}\right)^{1/2} = \frac{57.7}{\sqrt{C_D}}$$

Let us assume that the Reynolds number will be quite large and use C_D from Fig. 16.9 as $C_D = 0.2$. Then

$$V = \frac{57.7}{\sqrt{0.2}} = 129 \text{ m/s}$$

We must check the Reynolds number to verify the C_D value assumed. It is

$$\text{Re} = \frac{VD}{\nu} = \frac{129 \times 0.3}{1.6 \times 10^{-5}} = 2.42 \times 10^6$$

This is beyond the end of the curve where data are unavailable; we will assume that the drag coefficient is unchanged at 0.2, so the terminal velocity is 129 m/s.

(b) For the sphere falling in water, we must include the buoyancy force B acting in the same direction as the drag force F_D. Hence the summation of forces yields

$$W = F_D + B$$

$$\therefore \gamma_{\text{sphere}} \times \text{volume} = C_D \times \tfrac{1}{2}\rho V^2 A + \gamma_{\text{water}} \times \text{volume}$$

This gives

$$(S-1)\gamma_{\text{water}} \times \tfrac{4}{3}\pi R^3 = C_D \times \tfrac{1}{2}\rho V^2 \pi R^2$$

Using $\rho = 1000 \text{ kg/m}^3$, there results

$$V = \left(\frac{8R(S-1)\gamma_{\text{water}}}{3\rho C_D}\right)^{1/2} = \left(\frac{8 \times 0.15 \times 0.02 \times 9800}{3 \times 1000 \times C_D}\right)^{1/2} = \frac{0.28}{\sqrt{C_D}}$$

We anticipate the Reynolds number being lower than in part (a), so let's assume that it is in the range $2 \times 10^4 < \text{Re} < 2 \times 10^5$. Then $C_D = 0.5$ and there results

$$V = 0.40 \text{ m/s}$$

This gives a Reynolds number of

$$\text{Re} = \frac{VD}{\nu} = \frac{0.40 \times 0.3}{10^{-6}} = 1.2 \times 10^5$$

This is in the required range, so the terminal velocity is expected to be 0.40 m/s. Of course, if the sphere were roughened (sand glued to the surface), the C_D value would be less and the velocity would be greater.

16.3.2 Vortex Shedding

Long blunt objects, such as circular cylinders, exhibit a particularly interesting phenomenon when placed normal to a fluid flow; vortices or eddies (regions of circulating fluid) are shed from the object, regularly and alternately from opposite sides, as shown in Fig. 16.10. The resulting flow downstream is often referred to as a *Kármán vortex street*. The vortices are shed in the Reynolds number range $40 < \text{Re} < 10\,000$, and are accompanied by turbulence above $\text{Re} = 300$. Photographs of low- and high-Reynolds-number vortex shedding are presented in Fig. 16.11.

Key Concept: *Vortices are shed regularly and alternately from opposite sides of circular cylinders.*

Dimensional analysis may be applied to find an expression for the shedding frequency. For high-Reynolds-number flows, that is, flows with insignificant viscous forces, the shedding frequency f, in hertz, depends only on the velocity

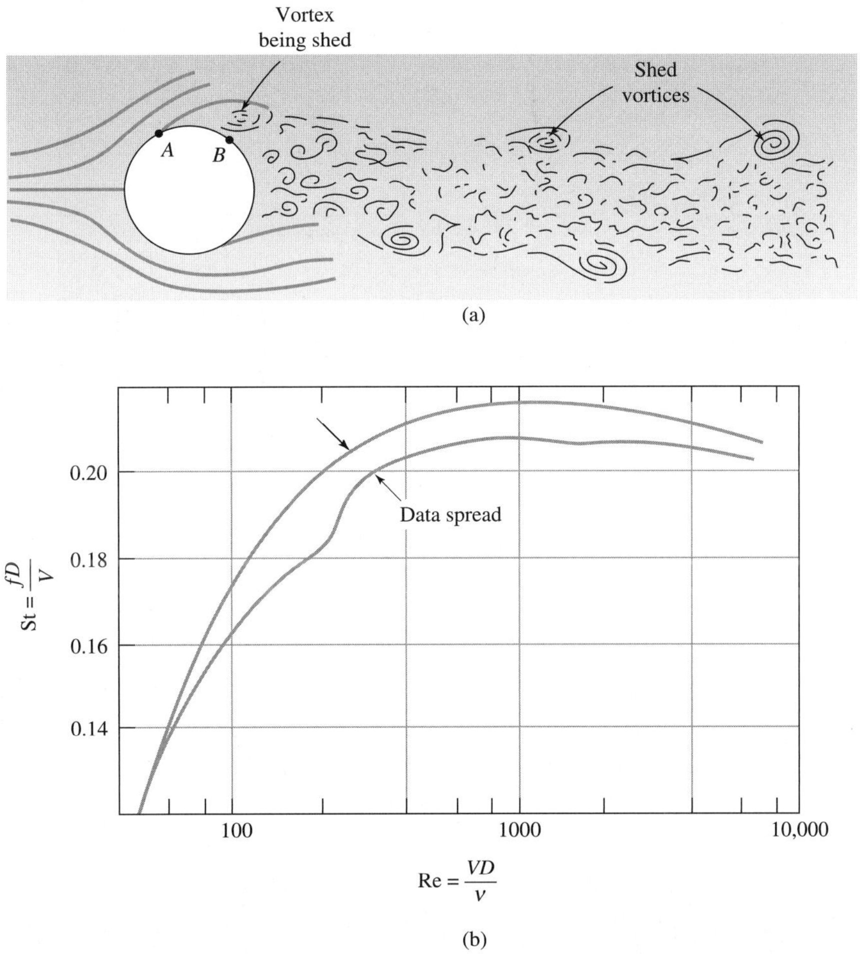

FIGURE 16.10 Vortex shedding from a cylinder: (a) vortex shedding; (b) Strouhal number versus Reynolds number. (From *NACA Rep. 1191*, by A. Roshko, 1954.)

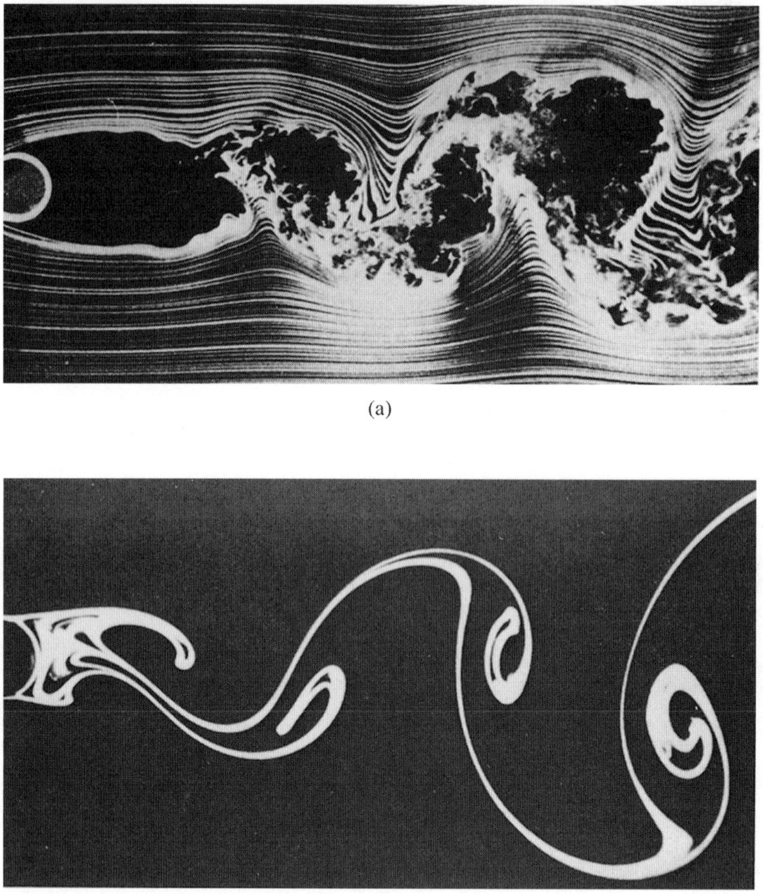

(a)

(b)

FIGURE 16.11 Vortex shedding at high and low Reynolds numbers: (a) Re = 10 000 (photograph by Thomas Corke and Hassan Nagib); (b) Re = 140 (photograph by Sadatoshi Taneda).

and diameter. Thus $f = f(V, D)$. Using dimensional analysis we can show that $fD/V = \text{const}$. The shedding frequency, expressed as a dimensionless quantity, is expressed as the Strouhal number,

$$\text{St} = \frac{fD}{V} \tag{16.3}$$

From the experimental results of Fig. 16.10, we observe that the Strouhal number is essentially constant (0.21) over the range $300 < \text{Re} < 10\,000$; hence, the frequency is directly proportional to the velocity over this relatively large Reynolds number range.

Example 16.3

The velocity of a slow-moving, 30°C air stream is to be measured using a cylinder and a pressure tap located between points A and B on the cylinder in Fig. 16.10. The velocity range is expected to be $0.1 < V < 1$ m/s. What size cylinder should be selected and what frequency would be observed by the pressure-measuring device for $V = 1$ m/s?

Solution

The Reynolds number should be in the vortex shedding range, say 4000. For the maximum velocity the diameter would be found as follows:

$$4000 = \frac{VD}{\nu}$$

$$= \frac{1.0 \times D}{1.6 \times 10^{-5}}$$

$$\therefore D = 0.064 \text{ m} \qquad \text{Select } D = 6 \text{ cm}$$

At $V = 0.1$ m/s the Reynolds number is $0.1 \times 0.06/1.6 \times 10^{-5} = 375$. Vortex shedding would occur, so this is acceptable. The expected vortex shedding frequency at $V = 1.0$ m/s is found using a Strouhal number from Fig. 16.10 of 0.21. Hence

$$0.21 = \frac{fD}{V}$$

$$= \frac{f \times 0.06}{1.0}$$

$$\therefore f = 3.5 \text{ hertz}$$

The engineer or architect must be very careful when designing structures such as towers and bridges that shed vortices. When a vortex is shed, a small force is applied to the structure; if the frequency of shedding is close to the natural frequency[3] (or one of the harmonics) of the structure, the phenomenon of resonance may occur in which the response to the applied force is multiplied by a large factor. For example, when resonance occurs on a television tower, the deflection of the tower due to the applied force may become so large that the supporting cables fail, leading to collapse of the structure. This has occurred many times, leading to severe damage and numerous deaths and injuries. The collapse of the Tacoma Narrows suspension bridge is undoubtedly the most spectacular failure due to vortex shedding. "Galloping" power lines, in which a power line alternates between the usual catenary and an inverted catenary, is another example that may lead to significant damage; this may occur when a power line has iced up, providing a much larger cross-sectional area for the wind.

Key Concept: *If the phenomenon of resonance occurs, the response to an applied force is multiplied by a large factor.*

[3]The natural frequency is that frequency with which a structure vibrates when given a "knock."

16.3.3 Streamlining

Streamlining: *Reducing the high pressure at the rear of the object allowing slow-moving surface flow to move rearward.*

Key Concept: *The angle at the trailing edge must not be greater than about 20° for streamlining to be effective.*

If the flow is to remain attached to the surface of a blunt object, such as a cylinder or a sphere, it must move into regions of higher and higher pressure as it progresses to the rear stagnation point. At sufficiently high Reynolds numbers (Re > 10) the slow-moving boundary layer flow near the surface is unable to make its way into the high-pressure region near the rear stagnation point, so it separates from the object. **Streamlining** reduces the high pressure at the rear of the object so that the slow-moving flow near the surface is able to negotiate its way into a slightly higher-pressure region. The fluid may not be able to make it all the way to the trailing edge of the streamlined object, but the separation region will be reduced to only a small percentage of the initial separated region on the blunt object. The included angle at the trailing edge must not be greater than about 20° or the separation region will be too large and the effect of streamlining will be negated. Drag coefficients for streamlined cylinders and spheres are displayed in Fig. 16.9.

When a body is streamlined, the surface area is increased substantially. This eliminates the majority of the pressure drag but increases the shear drag on the surface. To minimize drag, the idea is to minimize the sum of the pressure drag and the shear drag. Consequently, the streamlined body cannot be so long that the shear drag is larger than the pressure drag plus the shear drag for a shorter body. An optimization procedure is required. Such a procedure would lead to a thickness-to-chord length ratio of about 0.25 for a strut.

Obviously, for a low-Reynolds-number flow (Re < 10) the drag is due primarily to shear drag and thus streamlining is unnecessary; it would undoubtedly lead to increased drag since the surface area would be increased.

Finally, it should be pointed out that another advantage of streamlining is that the periodic shedding of vortices is usually eliminated. The vibrations produced by vortex shedding are often undesirable, so streamlining not only decreases drag but can eliminate the vibrations.

Example 16.4

A strut on a stunt plane traveling at 60 m/s is 4 cm in diameter and 24 cm long. Calculate the drag force acting on the strut as a circular cylinder and as a streamlined strut as shown in Fig. E16.4. Would you expect vortex shedding from the circular cylinder?

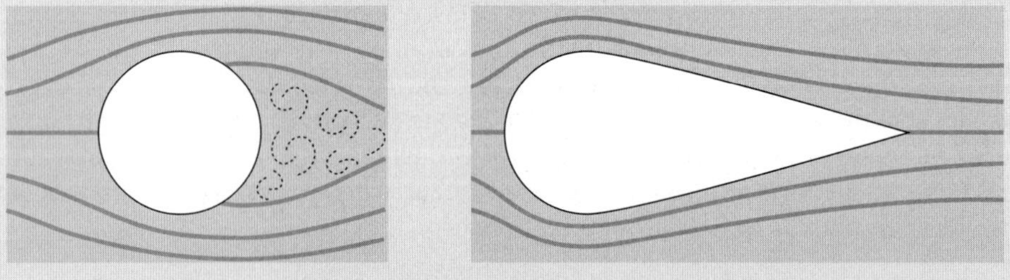

(a) (b)

FIGURE E16.4

(continued)

Solution

The Reynolds number associated with the cylinder and the streamlined strut is, assuming that $T = 20°C$,

$$\text{Re} = \frac{VD}{\nu}$$

$$= \frac{60 \times 0.04}{1.5 \times 10^{-5}} = 1.6 \times 10^5$$

Assuming a smooth surface, the drag coefficient is $C_D = 1.2$ from Fig. 16.9. The drag force on the circular cylinder is then

$$F_D = C_D \times \tfrac{1}{2}\rho V^2 A$$

$$= 1.2 \times \tfrac{1}{2} \times 1.20 \times 60^2 \times (0.24 \times 0.04) = 24.9 \text{ N}$$

For the streamlined strut of (b), Fig. 16.9 yields $C_D = 0.04$. The drag force is

$$F_D = C_D \times \tfrac{1}{2}\rho V^2 A$$

$$= 0.04 \times \tfrac{1}{2} \times 1.20 \times 60^2 \times (0.24 \times 0.04) = 0.82 \text{ N}$$

This is a reduction of 97% in the drag, a rather substantial percentage.

Vortex shedding is not to be expected on the circular cylinder; the Reynolds number is too high.

16.3.4 Cavitation

Cavitation: *A change of phase from liquid to vapor which occurs whenever the local pressure is less than the vapor pressure.*

Cavitation is a very rapid change of phase from liquid to vapor which may occur in a liquid whenever the local pressure is equal to or less than the vapor pressure. The first appearance of cavitation is at the position of lowest pressure in the field of flow. Four types of cavitation have been identified:

1. *Traveling cavitation*, which exists when vapor bubbles or cavities are formed, are swept downstream, and collapse.
2. *Fixed cavitation*, which exists when a fixed cavity of vapor exists as a separated region. The separated region may reattach to the body or the separated region may enclose the rear of the body and be closed by the main flow, in which case it is referred to as *supercavitation*.
3. *Vortex cavitation*, which is found in the high-velocity, and thus low-pressure, core of a vortex, often observed in the tip vortex leaving a propeller.
4. *Vibratory cavitation*, which may exist when a pressure wave moves in a liquid. A pressure wave consists of a pressure pulse, which has a high pressure followed by a low pressure. The low-pressure part of the wave (or vibration) can result in cavitation.

Key Concept: *The high instantaneous pressures may cause damage to stainless steel components.*

The first type of cavitation, in which vapor bubbles are formed and collapse, is associated with potential damage. The instantaneous pressures resulting from the collapse can be extremely high (perhaps 1400 MPa) and may cause damage to stainless steel components as happens on the propellers of ships.

TABLE 16.3 Drag Coefficients for Zero Cavitation Number for Blunt Objects

Two-dimensional body				Axisymmetric body		
Geometry		θ	$C_D(0)$	Geometry	θ	$C_D(0)$
Flat plate	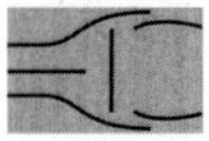	—	0.88	Disk	—	0.8
Circular cylinder		—	0.50	Sphere	—	0.30
Wedge	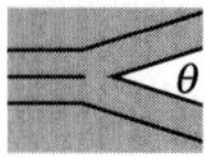	120	0.74	Cone	120	0.64
		90	0.64		90	0.52
		60	0.49	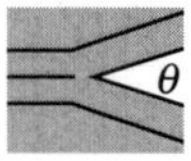	60	0.38
		30	0.28		30	0.20

Cavitation occurs whenever the *cavitation number* σ, defined by

$$\sigma = \frac{p_\infty - p_v}{\frac{1}{2}\rho V^2} \tag{16.4}$$

is less than the critical cavitation number σ_{crit}, which depends on the geometry of the body and the Reynolds number. Here p_∞ is the absolute pressure in the undisturbed free stream and p_v is the vapor pressure. As σ decreases below σ_{crit}, the cavitation increases in intensity, moving from traveling cavitation to fixed cavitation to supercavitation.

The drag coefficient of a body is dependent on the cavitation number and for small cavitation numbers is given by

$$C_D(\sigma) = C_D(0)(1 + \sigma) \tag{16.5}$$

where some values of $C_D(0)$ for common shapes are listed in Table 16.3 for $\text{Re} \simeq 10^5$.

The hydrofoil, an airfoil-type body that is used to lift a vessel out of the water, is a shape that is invariably associated with cavitation. Drag and lift coefficients and critical cavitation numbers are given in Table 16.4 for a typical hydrofoil with $10^5 < \text{Re} < 10^6$, where the Reynolds number is based on the chord length and the area used with C_D and C_L is the chord times the length.

TABLE 16.4 Drag and Lift Coefficients and Critical Cavitation Number for a Typical Hydrofoil

Angle (°)	Lift coefficient C_L	Drag coefficient C_D	Critical cavitation number σ_{crit}
−2	0.2	0.014	0.5
0	0.4	0.014	0.6
2	0.6	0.015	0.7
4	0.8	0.018	0.8
6	0.95	0.022	1.2
8	1.10	0.03	1.8
10	1.22	0.04	2.5

Example 16.5

A hydrofoil is to operate 20 in. below the surface of 60°F water at an angle of attack of 8° and travel at 45 ft/sec. If its chord length is 24 in. and it is 6 ft long, calculate its lift and drag. Is cavitation present?

Solution

The absolute pressure p_∞ is

$$p_\infty = \gamma h + p_{\text{atm}}$$

$$= 62.4 \times \frac{20}{12} + 2117 = 2221 \text{ psf absolute}$$

The vapor pressure is $p_v = 0.256$ psia, so

$$\sigma = \frac{p_\infty - p_v}{\frac{1}{2}\rho V^2}$$

$$= \frac{2221 - 0.256 \times 144}{\frac{1}{2} \times 1.94 \times 45^2} = 1.11$$

Answering the last question first, we see that this is less than 1.8, hence cavitation exists.

The lift force is, finding C_L in Table 16.4,

$$F_L = C_L \times \tfrac{1}{2}\rho V^2 A$$

$$= 1.1 \times \tfrac{1}{2} \times 1.94 \times 45^2 \times \left(\frac{24}{12} \times 6\right) = 25{,}900 \text{ lb}$$

The drag force is, finding C_D in Table 16.4,

$$F_D = C_D \times \tfrac{1}{2}\rho V^2 A$$

$$= 0.03 \times \tfrac{1}{2} \times 1.94 \times 45^2 \times (2 \times 6) = 707 \text{ lb}$$

16.3.5 Added Mass

The previous sections in this chapter have dealt with bodies moving at constant velocity. In this section we consider bodies accelerating from rest in a fluid. When a body accelerates, we say that an unbalanced force acts on the body; not only does the body accelerate but so does some of the fluid surrounding the body. The acceleration of the surrounding fluid requires an added force over and above the force required to accelerate only the body. A relatively simple way of accounting for the fluid mass being accelerated is to add a mass, called the *added mass* m_a, to the mass of the body. Summing forces in the direction of motion for a symmetrical body moving in the direction of its axis of symmetry we have, for horizontal motion,

Key Concept: *Not only does a body accelerate, but so does some of the fluid surrounding the body.*

$$F - F_D = (m + m_a)\frac{dV_B}{dt} \tag{16.6}$$

where V_B is the velocity of the body and F_D is the drag force. For an initial acceleration from rest F_D would be zero.

The added mass is related to the mass of the fluid m_f displaced by the body by the relationship

$$m_a = km_f \tag{16.7}$$

where k is an added mass coefficient. For a sphere $k = 0.5$; for an ellipsoid with major axis twice the minor axis and moving in the direction of the major axis, $k = 0.2$; for a long cylinder moving normal to its axis, $k = 1.0$. These values were calculated for inviscid flows and thus are applicable for motions starting from rest so that the viscous forces are negligible.

For dense bodies accelerating in the atmosphere the added mass is negligibly small and is typically ignored. Masses accelerating from rest in a liquid are more influenced by the added mass and it must usually be accounted for. Offshore structures that are subjected to oscillatory wave motions experience periodic forces the determination of which must include the added mass effect.

Example 16.6

A sphere with specific gravity 2.5 is released from rest in water. Calculate its initial acceleration. What is the percentage error if the added mass is ignored?

Solution

The summation of forces in the vertical direction, with zero drag, is

$$W - B = (m + m_a)\frac{dV_B}{dt}$$

where B is the buoyant force. Substituting in the appropriate quantities gives, letting $\forall$ = sphere volume,

$$S\gamma_{\text{water}}\forall - \gamma_{\text{water}}\forall = (\rho_{\text{water}}S\forall + 0.5\rho_{\text{water}}\forall)\frac{dV_B}{dt}$$

This gives

$$g(S - 1) = (S + 0.5)\frac{dV_B}{dt}$$

Hence

$$\frac{dV_B}{dt} = \frac{g(S-1)}{S+0.5} = \frac{9.8(2.5-1)}{2.5+0.5} = 4.90 \text{ m/s}^2$$

If we ignored the added mass, the acceleration would be

$$\frac{dV_B}{dt} = \frac{g(S-1)}{S} = \frac{9.8(2.5-1)}{2.5} = 5.88 \text{ m/s}^2$$

This is an error of 20%.

16.4 LIFT AND DRAG ON AIRFOILS

Separation occurs on a blunt body, such as a cylinder, due to the strong adverse pressure gradient in the boundary layer on the rear of the body. An airfoil is a streamlined body designed to reduce the adverse pressure gradient so that separation will not occur, usually with a small angle of attack, as shown in Fig. 16.12. Without separation the drag is due primarily to the wall shear stress, which results from viscous effects in the boundary layer.

The boundary layer on an airfoil is very thin, and thus it can be ignored when solving for the flow field (the streamline pattern and the pressure distribution) surrounding the airfoil. Since the boundary layer is so thin, the pressure on the wall is not significantly influenced by the boundary layer's existence. Hence the lift on an airfoil can be approximated by integrating the pressure distribution as given by the inviscid flow solution on the wall. In the next section we will demonstrate how this is done; in this section we simply give empirical results.

Key Concept: *The lift on an airfoil can be approximated by integrating the inviscid flow solution.*

The drag on an airfoil can be predicted by solving the boundary layer equations (simplified Navier–Stokes equations) for the shear stress on the wall, and performing the appropriate integration. The inviscid flow field must be known before the boundary layer equations can be solved since the pressure gradient on the wall and the inviscid flow velocity at the wall are needed as inputs in solving for the boundary layer flow. Boundary layer calculations will be presented in Section 16.6; in this section we present the empirical results for the drag.

The drag coefficient presented may seem quite low compared with the coefficients of the preceding section. For airfoils a much larger projected area is used, namely, the plan area, which is the chord c (see Fig. 16.12) times the length L of the airfoil. Thus the drag and lift coefficients are defined as

$$C_D = \frac{F_D}{\frac{1}{2}\rho V^2 c L} \tag{16.8}$$

$$C_L = \frac{F_L}{\frac{1}{2}\rho V^2 c L} \tag{16.9}$$

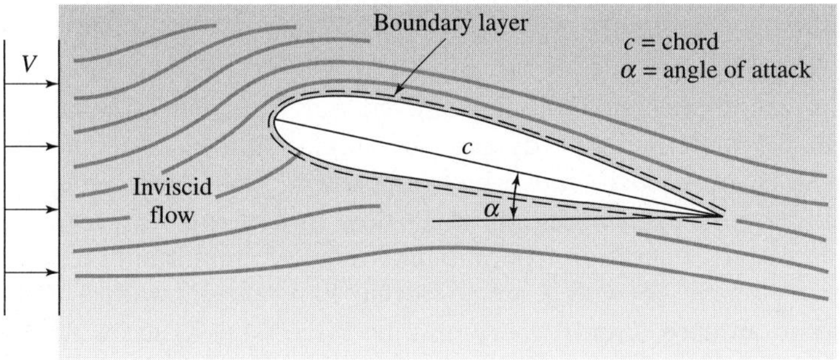

FIGURE 16.12 Flow around an airfoil at an angle of attack.

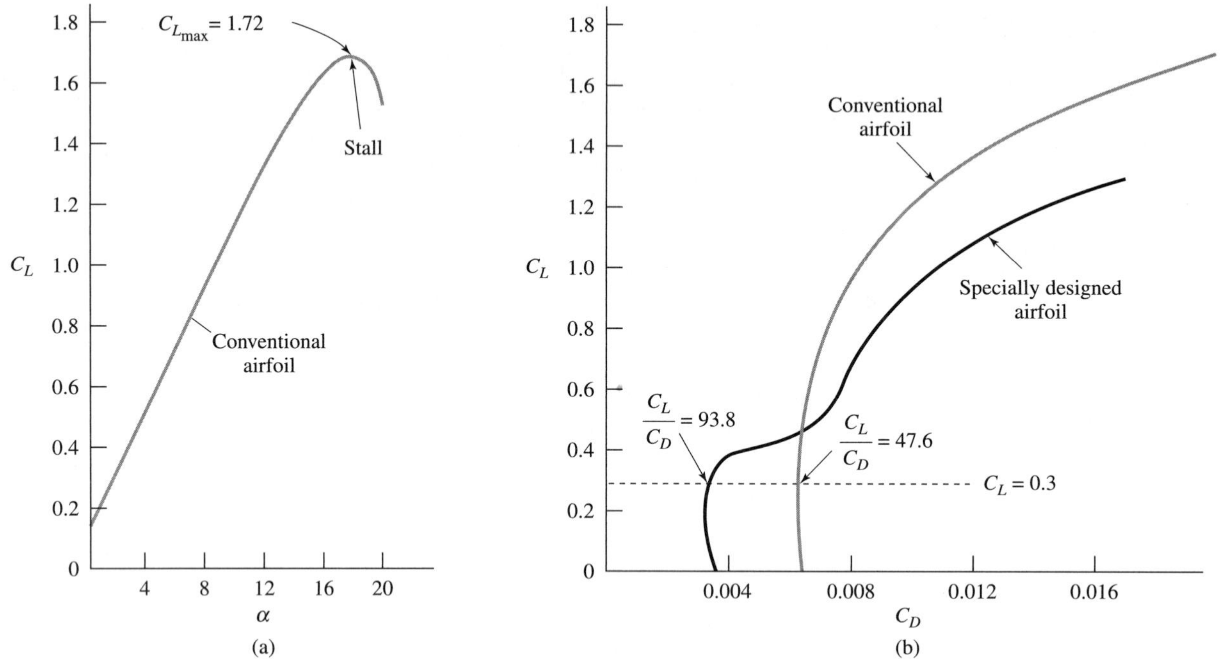

FIGURE 16.13 Lift and drag coefficients for airfoils with $\text{Re} = Vc/\nu \simeq 9 \times 10^6$.

Key Concept: *The design lift coefficient is near the minimum drag coefficient condition.*

For a typical airfoil the lift and drag coefficients are given in Fig. 16.13. For a specially designed airfoil the drag coefficient may be as low as 0.0035, but the maximum lift coefficient is about 1.5. The design lift coefficient (cruise condition) is about 0.3, which is near the minimum drag coefficient condition. This corresponds to an angle of attack of about 2°, far from the stall condition of about 16°.

Conventional airfoils are not symmetric; hence there is a positive lift coefficient at zero angle of attack. The lift is directly proportional to the angle of attack but deviates from the straight-line function just before stall. The drag coefficient also increases linearly up to an angle of attack of about five degrees for a conventional airfoil; then it increases in a nonlinear relation with angle of attack.

To take off and land at relatively low speeds, it is necessary to attain significantly higher lift coefficients than the maximum of 1.7 of Fig. 16.13. Or if a relatively low lift coefficient is to be accepted, the area $c \times L$ must be enlarged. Both are actually accomplished. Flaps are moved out from a section of each airfoil, resulting in an increased chord, and the angle of attack of the flap is also increased. Slots are used to move high-pressure air from the underside into the relatively low momentum boundary layer flow on the top side, as shown in Fig. 16.14; this prevents separation from the flap, thereby maintaining high lift. The lift coefficient can reach 2.5 with a single-slotted flap and 3.2 with a double-slotted flap. On some modern aircraft there may be three flaps in series with three slots along with a nose flap, to ensure that the boundary layer does not separate from the upper surface of the airfoil.

Key Concept: *Slots allow high-pressure air to energize the slow-moving air thereby preventing separation from the flap.*

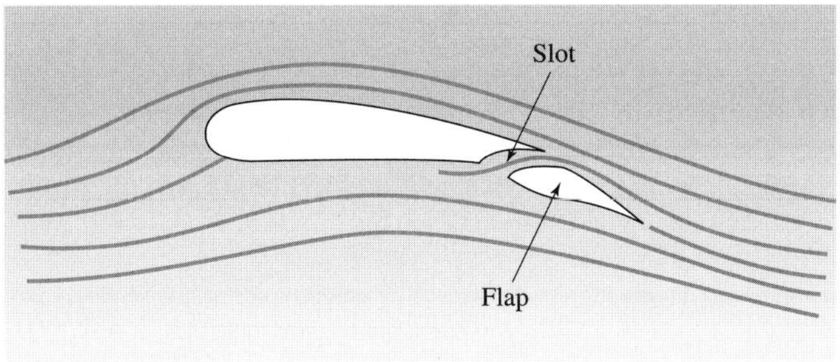

FIGURE 16.14 Flapped airfoil with slot for separation control.

The total lift on an aircraft is supplied primarily by the airfoil. The effective length of the airfoil when calculating the lift is taken to be the tip-to-tip distance, the **wingspan**, since the fuselage acts to produce the lift of the midsection of the airfoil. The drag calculation must include the shear acting on the airfoil, the fuselage, and the tail section.

Wingspan: *The effective length of an airfoil is the tip-to-tip distance.*

The drag coefficient is essentially constant on airfoils up to a Mach number of about 0.75. Then a sudden rise occurs until the Mach number reaches unity; see Fig. 16.15. The drag coefficient then slowly falls. Obviously, the condition of $M = 1$ is to be avoided. Thus aircraft either fly at $M < 0.75$ or $M > 1.5$ or so, to avoid the high drag coefficients near $M = 1$. Near $M = 1$ there are also regions of flow that oscillate from subsonic to supersonic. Such oscillations create forces that are best avoided.

It is useful to use swept-back airfoils since it is the component of velocity normal to the leading edge of the airfoil that must be used in calculating the Mach number in Fig. 16.15. Cruise speeds at $M = 0.8$ with swept-back wings are not uncommon. It should be pointed out, though, that fuel consumption depends on power required, and power is drag force times velocity; hence fuel consumption depends on the velocity cubed since the drag force depends on the velocity squared, assuming all other parameters are constant. A lower velocity results in a fuel savings even though the engines must operate longer when traveling a fixed distance.

Key Concept: *The component of velocity normal to the leading edge is used in calculating the Mach number.*

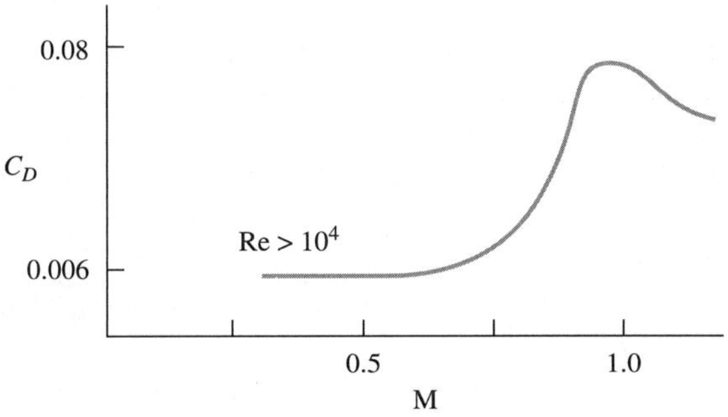

FIGURE 16.15 Drag coefficient as a function of Mach number (speed) for a typical unswept airfoil.

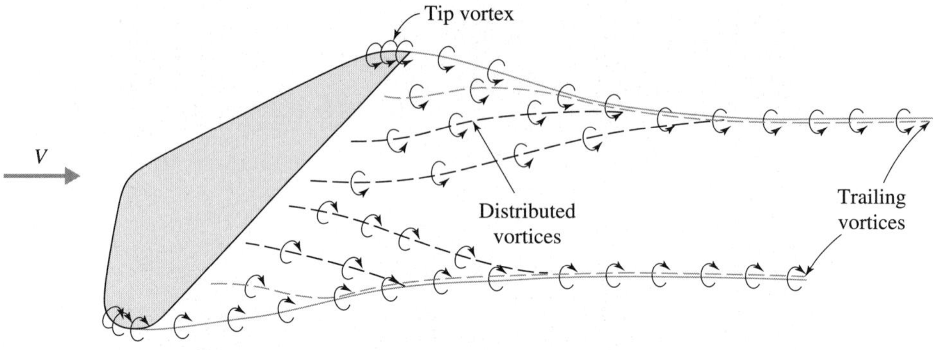

FIGURE 16.16 Trailing vortex.

A final comment on airfoils regards the influence of a finite airfoil. To understand flow around a finite airfoil we make reference to a vortex. Fluid particles rotate about the center of a vortex as they travel along in the flow field. There is a high pressure on the bottom and a low pressure on the top side of the airfoil sketched in Fig. 16.16 with a model airfoil shown in Fig. 16.17. This results in a movement of air from the bottom side to the top side around the ends of the airfoil, as shown, resulting in a strong tip vortex. Distributed vortices are also shed all along the airfoil and they all collect into two large trailing vortices. On a clear day, the two trailing vortices may show up as visible white streaks of water vapor

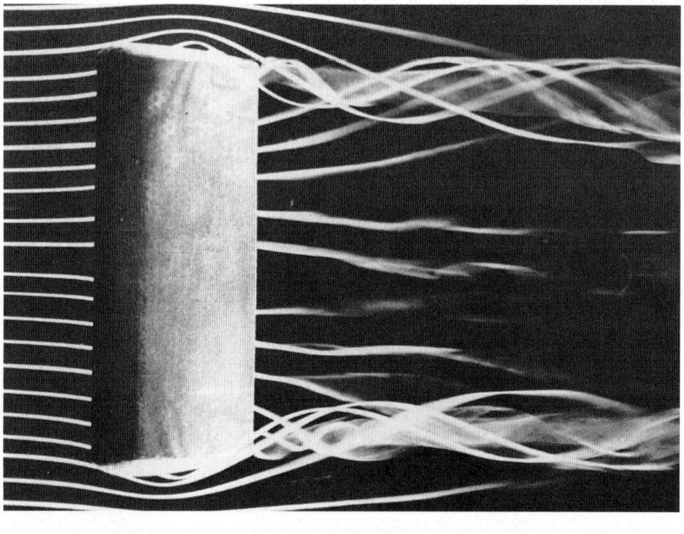

FIGURE 16.17 Trailing vortices from a rectangular wing. The flow remains attached over the entire wing surface. The centers of the vortex cores leave the trailing edge at the tips. The model is tested in a smoke tunnel at Reynolds number 100 000. (Courtesy of The Parabolic Press, Stanford, California. Reprinted with permission.)

behind a high-flying aircraft. The trailing vortices persist for a considerable distance (perhaps 15 km) behind a large aircraft, and their 90-m/s velocities can cause a small aircraft to flip over. Also, the trailing vortices induce a downwash, a downward velocity component, that must be accounted for in the design of the aircraft. The tail section is located up high to minimize the effect of this downwash.

Key Concept: *Trailing vortices behind a large aircraft can cause a small aircraft to lose control.*

Example 16.7

A light airplane weighs 10 000 N, its wingspan measures 12 m, its chord measures 1.8 m, and a payload of 2000 N is anticipated. Predict (a) the takeoff speed if an angle of attack of 8° is desired, (b) the stall speed of the conventional airfoil, and (c) the power required by the airfoil during cruise at 50 m/s.

Solution

(a) The lift on an airplane is equal to its weight. With the payload the total weight is 12 000 N; hence the lift-coefficient equation (16.9) gives the following:

$$V = \left(\frac{F_L}{\frac{1}{2}\rho C_L c L}\right)^{1/2}$$
$$= \left(\frac{12\,000}{\frac{1}{2} \times 1.20 \times 1.0 \times 1.8 \times 12}\right)^{1/2} = 30.4 \text{ m/s}$$

where we have used $C_L = 1.0$ at $\alpha = 8°$ from Fig. 16.13, and $\rho = 1.2$ kg/m^3 since light aircraft take off at ground level.

(b) The stall speed is found using a maximum lift coefficient of 1.72 from Fig. 16.13.

$$V_{\text{stall}} = \left(\frac{F_L}{\frac{1}{2}\rho C_L c L}\right)^{1/2}$$
$$= \left(\frac{12\,000}{\frac{1}{2} \times 1.20 \times 1.72 \times 1.8 \times 12}\right)^{1/2} = 23.2 \text{ m/s}$$

(c) The power demanded by the airfoil during cruise is equal to the drag force times the velocity. The design lift coefficient is assumed equal to 0.3 and thus from Fig. 16.13, assuming a conventional airfoil, $C_D = 0.0063$. This gives

$$F_D = \tfrac{1}{2}\rho V^2 c L C_D$$
$$= \tfrac{1}{2} \times 1.20 \times 50^2 \times 1.8 \times 12 \times 0.0063 = 204 \text{ N}$$

The power is then

$$\text{power} = F_D \times V$$
$$= 204 \times 50 = 10\,200 \text{ W} \quad \text{or} \quad 13.7 \text{ hp}$$

The total power would be significantly larger since the drag on the fuselage and tail section must be included.

16.5 POTENTIAL FLOW THEORY

16.5.1 Basic Flow Equations

Inviscid flow exists outside the boundary layer and the wake in high-Reynolds-number flow around bodies. For an airfoil the boundary layer is quite thin, and the inviscid flow provides a good approximation to the actual flow; it is used to predict the pressure distribution on the surface, thereby giving a good estimate of the lift. It will also provide us with the velocity $U(x)$ to be used as a boundary condition in the boundary layer solution of Section 16.6; from such a solution we can estimate the drag and predict possible points of separation. Consequently, the inviscid flow solution is very important in our study of external flows. Obviously, if we use the empirical results of previous sections, the details of the inviscid flow solution are unnecessary. If, on the other hand, we wish to predict quantities such as the lift and drag, and locate possible points of separation, by using the required differential equations, the inviscid flow solution is essential.

Key Concept: *The inviscid flow solution is very important in our study of external flows.*

Consider a velocity field that is given by the gradient of a scalar function ϕ, that is,

$$\mathbf{V} = \nabla\phi \tag{16.10}$$

in which ϕ is called a *velocity potential function*. Such a velocity field is called a **potential flow** (or an *irrotational flow*) and possesses the property that the vorticity $\boldsymbol{\omega}$, which is the curl of the velocity vector, is zero; this is expressed by

Potential flow: *A flow with zero vorticity.*

$$\boldsymbol{\omega} = \nabla \times \mathbf{V} = 0 \tag{16.11}$$

The fact that the vorticity is zero for a potential flow can easily be shown by letting $\mathbf{V} = \nabla\phi$ and expanding Eq. 16.10 in rectangular coordinates. A fluid particle that does not possess vorticity (i.e., is not rotating) cannot obtain vorticity without the action of viscosity; normal pressure forces and body forces that act through the mass center cannot impart rotation to a fluid particle.

Key Concept: *The vorticity is zero for a potential flow.*

Let us define another scalar function that will aid us in our study of plane fluid flows. We let

$$u = \frac{\partial\psi}{\partial y} \quad \text{and} \quad v = -\frac{\partial\psi}{\partial x} \tag{16.12}$$

and refer to the scalar function $\psi(x, y)$ as a **stream function**. By using the mathematical description of a streamline, $\mathbf{V} \times d\mathbf{r} = 0$, we see that, for a plane flow, $u\,dy - v\,dx = 0$. Substituting from Eqs. 16.12, this becomes

Stream function: *The stream function is constant along a streamline.*

$$\frac{\partial\psi}{\partial y}dy + \frac{\partial\psi}{\partial x}dx = 0 \tag{16.13}$$

This is, of course, $d\psi = 0$. Thus ψ is constant along a streamline. Example 16.9 will show that the difference $(\psi_2 - \psi_1)$ between any two streamlines is equal to the flow rate per unit depth between the two streamlines.

Before we present some simple functions, some additional observations concerning ϕ and ψ are made. Using Eqs. 16.10 and 16.12, we see that

$$u = \frac{\partial \phi}{\partial x} = \frac{\partial \psi}{\partial y} \qquad v = \frac{\partial \phi}{\partial y} = -\frac{\partial \psi}{\partial x} \tag{16.14}$$

These relationships between the derivatives of ϕ and ψ are the famous *Cauchy–Riemann equations* from the theory of complex variables. The functions ϕ and ψ form an *analytic function* $(\phi + i\psi)$ called the *complex velocity potential.* The theory of complex variables with all its powerful theorems is thus applicable to this restricted class of problems: namely, plane, incompressible, potential flows. An example will show that the streamlines and constant-potential lines intersect each other at right angles.

Key Concept: *The streamlines and constant-potential lines intersect each other at right angles.*

Example 16.8

A scalar potential function is given by $\phi = A \tan^{-1}(y/x)$. Find the stream function $\psi(x, y)$.

Solution

The relationship between ϕ and ψ is given by Eq. 16.14. We have

$$\frac{\partial \psi}{\partial y} = \frac{\partial \phi}{\partial x} = \frac{\partial}{\partial x}\left(A \tan^{-1}\frac{y}{x}\right) = -\frac{Ay}{x^2 + y^2}$$

This can be integrated as follows:

$$\int \frac{\partial \psi}{\partial y} dy = -A \int \frac{y}{x^2 + y^2} dy$$

$$\therefore \psi = -\frac{A}{2} \ln(x^2 + y^2) + f(x)$$

A function of x rather than a constant must be added since partial derivatives are being used. Now, let us differentiate this expression with respect to x. There results

$$\frac{\partial \psi}{\partial x} = -A \frac{x}{x^2 + y^2} + \frac{df}{dx}$$

This must equal $(-\partial\phi/\partial y)$ as demanded by Eq. 16.14; that is,

$$-\frac{Ax}{x^2 + y^2} + \frac{df}{dx} = -\frac{Ax}{x^2 + y^2}$$

Thus

$$\frac{df}{dx} = 0 \qquad \text{or} \qquad f = C$$

Since ϕ and ψ are used to find the velocity components by differentiation, the constant is of no concern; it is usually set equal to zero. Hence

$$\psi = -\frac{A}{2} \ln(x^2 + y^2)$$

Example 16.9

Show that the difference in the stream function between any two streamlines is equal to the flow rate per unit depth between the two streamlines. The flow rate per unit depth is denoted by q_f.

Solution

Consider the flow between two streamlines infinitesimally close, as shown in Fig. E16.9a. The flow rate per unit depth through the elemental area is, referring to Fig. E16.9b,

$$\begin{aligned} dq_f &= dq_{f1} + dq_{f2} \\ &= u\,dy - v\,dx \\ &= \frac{\partial \psi}{\partial y} dy + \frac{\partial \psi}{\partial x} dx = d\psi \end{aligned}$$

If this is integrated between two streamlines with $\psi = \psi_1$ and $\psi = \psi_2$, there results

$$q_f = \psi_2 - \psi_1$$

thereby proving the statement of the example.

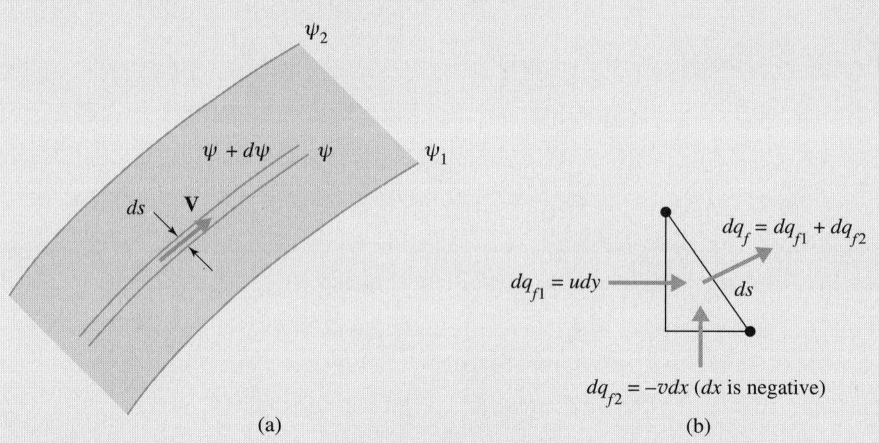

FIGURE E16.9

Example 16.10

Show that the streamlines and equipotential lines of a plane, incompressible, potential flow intersect one another at right angles.

(continued)

Solution

If, at a point, the slope of a streamline is the negative reciprocal of the slope of an equipotential line, the two lines are perpendicular to each other. The slope of a streamline (see Fig. E16.10a) is given by

$$\left.\frac{dy}{dx}\right|_{\psi=\text{const}} = \frac{v}{u}$$

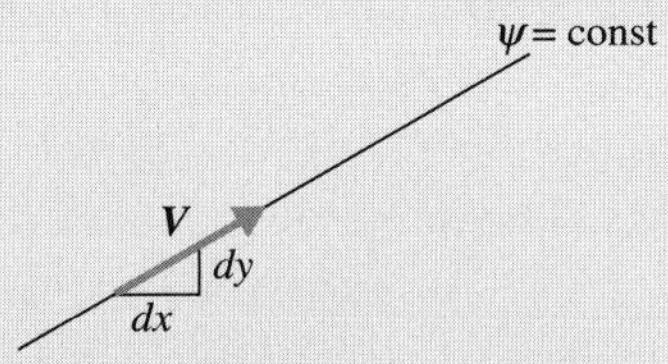

FIGURE E16.10a

The slope of an equipotential line is found from

$$d\phi = \frac{\partial \phi}{\partial x}dx + \frac{\partial \phi}{\partial y}dy = 0$$

since $\phi =$ constant along an equipotential line. This gives

$$\left.\frac{dy}{dx}\right|_{\phi=\text{const}} = -\frac{\partial \phi/\partial x}{\partial \phi/\partial y} = -\frac{u}{v}$$

Hence we see that the slope of the streamline is the negative reciprocal of the slope of the equipotential line, that is,

$$\left.\frac{dy}{dx}\right|_{\phi=\text{const}} = -\left(\left.\frac{dy}{dx}\right|_{\phi=\text{const}}\right)^{-1}$$

Thus, whenever the streamlines intersect the equipotential lines, they must do so at right angles. A sketch of the streamlines and equipotential lines (equally spaced at large distances from the body), known as a *flow net,* is shown in Fig. E16.10b for flow over a weir. Such a carefully constructed sketch can be used to approximate the velocities at points of interest in an inviscid flow. Pressures can then be found using Bernoulli's equation.

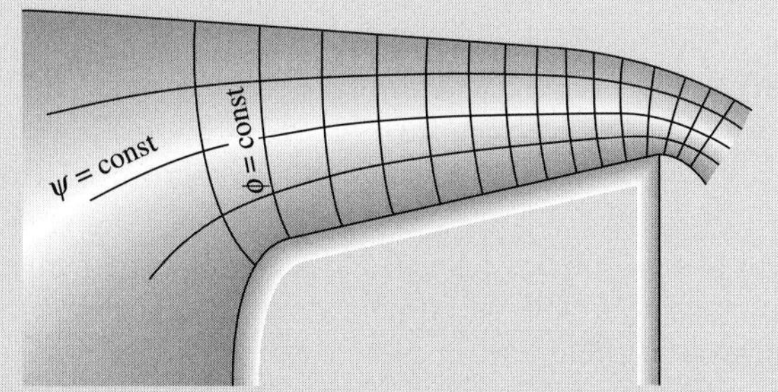

FIGURE E16.10b

16.5.2 Simple Solutions

Now, let us identify some relatively simple functions that are potential functions having zero vorticity. However, before we do, it is often more convenient to use polar coordinates. The velocity components take the following forms:

$$v_r = \frac{1}{r}\frac{\partial \psi}{\partial \theta} = \frac{\partial \phi}{\partial r} \qquad v_\theta = -\frac{\partial \psi}{\partial r} = \frac{1}{r}\frac{\partial \phi}{\partial \theta} \tag{16.15}$$

Key Concept: *The doublet can be visualized as a source and sink of equal strength separated by a very small distance.*

We now introduce the names of four simple flows, sketched in Fig. 16.18, and their corresponding functions. The names and functions are:

Uniform flow: $$\psi = U_\infty y \qquad \phi = U_\infty x \tag{16.16}$$

Line source: $$\psi = \frac{q_f}{2\pi}\theta \qquad \phi = \frac{q_f}{2\pi}\ln r \tag{16.17}$$

Irrotational vortex: $$\psi = \frac{\Gamma}{2\pi}\ln r \qquad \phi = \frac{\Gamma}{2\pi}\theta \tag{16.18}$$

Doublet: $$\psi = -\frac{\mu \sin\theta}{r} \qquad \phi = -\frac{\mu\cos\theta}{r} \tag{16.19}$$

The uniform flow velocity U_∞ is assumed to be in the x-direction; if a y-component is desired an appropriate term is simply added. The *source strength* q_f is the volume rate of flow per unit depth issuing from the source; a negative value would create a *sink*. The *vortex strength* Γ is the *circulation* about the origin, defined by

$$\Gamma = \oint_L \mathbf{V} \cdot \mathbf{ds} \tag{16.20}$$

where L must be a closed curve (a circle is usually used) around the origin and clockwise is positive. The *doublet strength* μ is for a doublet oriented in the negative x-direction; note the large arrow (in Fig. 16.18d) showing the direction of the doublet. Doublets oriented in other directions are seldom of interest and are not considered here.

Of the four flows presented above, the doublet is rather mysterious; it can be visualized as a source and a sink of equal strength separated by a very small distance. Its usefulness is in the creation of certain other flows of interest. The irrotational vortex is encountered when water swirls down a drain or into the turbine of a hydropower dam, or more spectacularly, in a tornado.

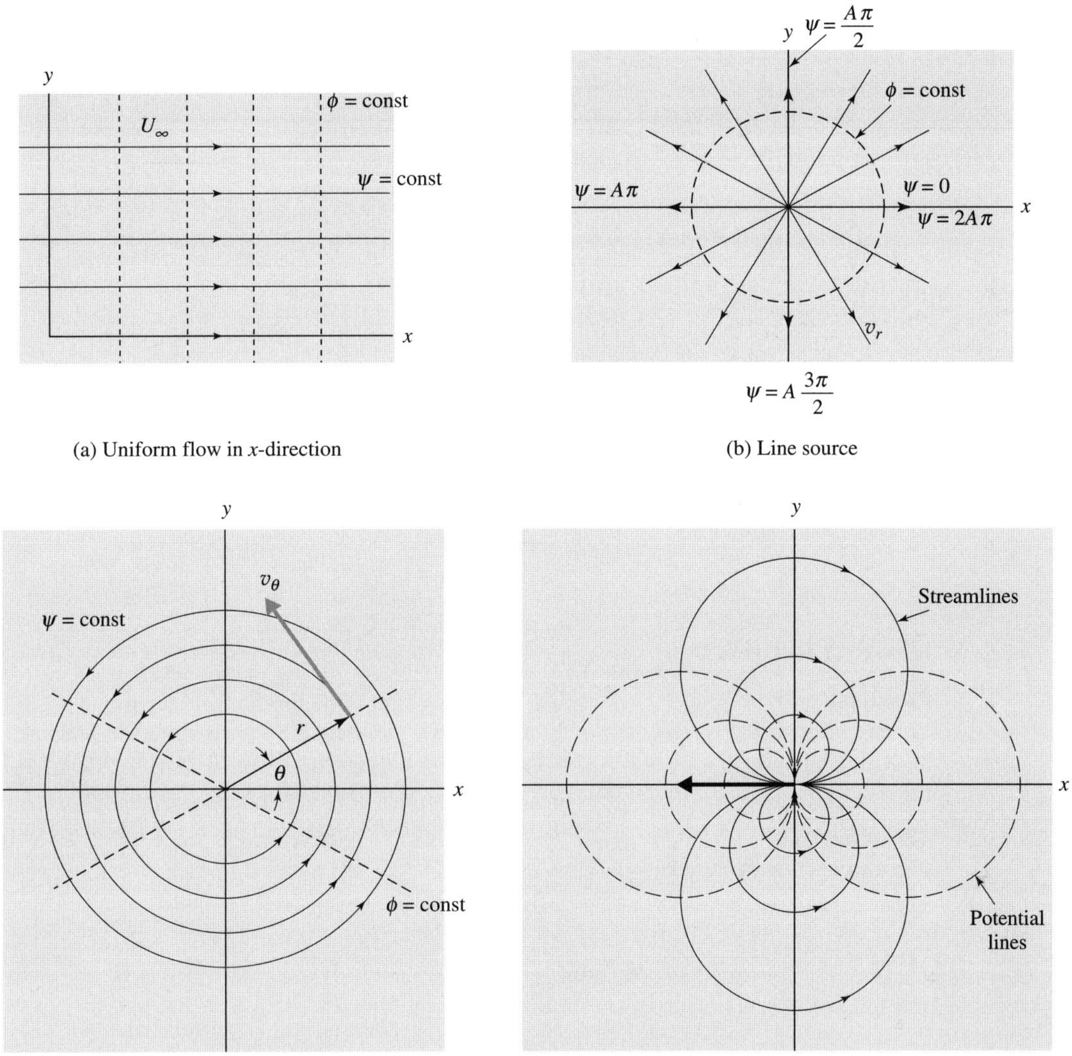

(a) Uniform flow in x-direction

(b) Line source

(c) Irrotational vortex

(d) Doublet

FIGURE 16.18 Four simple potential flows.

The velocity components for the four simple flows are easily shown, using Eqs. 16.14 and 16.15 for rectangular and polar coordinates, to be as follows:

$$\text{Uniform flow:} \quad u = U_\infty \qquad v = 0$$
$$v_r = U_\infty \cos\theta \qquad v_\theta = -U_\infty \sin\theta \tag{16.21}$$

$$\text{Line source:}\quad v_r = \frac{q}{2\pi r} \qquad v_\theta = 0$$
$$u = \frac{q_f}{2\pi}\frac{x}{x^2+y^2} \qquad v = \frac{q_f}{2\pi}\frac{y}{x^2+y^2} \tag{16.22}$$

$$\text{Irrotational vortex:}\quad v_r = 0 \qquad v_\theta = -\frac{\Gamma}{2\pi r}$$
$$u = -\frac{\Gamma}{2\pi}\frac{y}{x^2+y^2} \qquad v = \frac{\Gamma}{2\pi}\frac{x}{x^2+y^2} \tag{16.23}$$

$$\text{Doublet:}\quad v_r = -\frac{\mu\cos\theta}{r^2} \qquad v_\theta = -\frac{\mu\sin\theta}{r^2}$$
$$u = -\mu\frac{x^2-y^2}{(x^2+y^2)^2} \qquad v = -\mu\frac{2xy}{(x^2+y^2)^2} \tag{16.24}$$

Example 16.11

The pressure far from an irrotational vortex (a simplified tornado) in the atmosphere is zero gage. If the velocity at $r = 20$ m is 20 m/s, estimate the velocity and the pressure at $r = 2$ m. (The irrotational vortex ceases to be a good model for a tornado when r is small. In the "eye" of the tornado the motion is approximated by rigid-body motion.)

Solution

For an irrotational vortex, we know that

$$v_\theta = -\frac{\Gamma}{2\pi r}$$

Hence

$$\Gamma = -2\pi r v_\theta$$
$$= -2\pi \times 20 \times 20 = -800\pi \text{ m}^2/\text{s}$$

The velocity at $r = 2$ m is then

$$v_\theta = -\frac{-800\pi}{2\pi \times 2} = 200 \text{ m/s}$$

Bernoulli's equation for this incompressible, inviscid, steady flow then gives the pressure as follows assuming a stagnant atmosphere away from the tornado:

$$\overset{0}{\cancel{p_\infty}} + \frac{\overset{0}{\cancel{U_\infty^2}}}{2}\rho = p + \frac{v_\theta^2}{2}\rho$$
$$\therefore p = -\tfrac{1}{2}\rho v_\theta^2$$
$$= -\tfrac{1}{2} \times 1.20 \times 200^2 = -24\,000 \text{ Pa}$$

The negative sign denotes a vacuum. It is this vacuum that causes roofs of buildings to blow off during a tornado.

16.5.3 Superposition

The simple flows presented in Section 16.5.2 are of particular interest because they can be superimposed with each other to form more complicated irrotational flows of engineering importance. In fact, the most complicated incompressible, plane irrotational flow can be constructed using these simple flows. For example, suppose that the flow around an airfoil with a slotted flap is desired. We could divide the surface of the airfoil into a relatively large number (say, 200) of panels, locate a source or a sink (or alternatively, a doublet) at the center of each panel, add a uniform flow and an irrotational vortex, and then by adjusting[4] the panel source strengths, the desired inviscid flow could be created. The development of the model and the computer routine necessary to perform the calculations are considered beyond the scope of this book.

Key Concept: *The most complicated incompressible, plane irrotational flow can be constructed using these simple flows.*

In this section we demonstrate superposition by creating flow around a circular cylinder with and without circulation. First, superimpose a uniform flow and a doublet; there results

$$\psi = U_\infty y - \frac{\mu \sin \theta}{r} \tag{16.25}$$

The velocity component v_r is (let $y = r \sin \theta$)

$$\begin{aligned} v_r &= \frac{1}{r}\frac{\partial \psi}{\partial \theta} \\ &= U_\infty \cos \theta - \frac{\mu}{r^2} \cos \theta \end{aligned} \tag{16.26}$$

Let us ask the question: Is there a radius r_c for which $v_r = 0$? If we set $v_r = 0$ we find that

$$r_c = \sqrt{\frac{\mu}{U_\infty}} \tag{16.27}$$

At this radius v_r is identically zero for all θ and thus the circle $r = r_c$ must be a streamline.

The stagnation points are found by setting $v_\theta = 0$ on the circle $r = r_c$. There results

$$\begin{aligned} v_\theta &= -\frac{\partial \psi}{\partial r} \\ &= -U_\infty \sin \theta - \frac{\mu \sin \theta}{r_c^2} = -2U_\infty \sin \theta = 0 \end{aligned} \tag{16.28}$$

Thus we see that $v_\theta = 0$ at $\theta = 0°$ and $180°$. The flow is as sketched in Fig. 16.19a. We only have an interest in the flow external to the circular streamline $r = r_c$.

[4] The source strengths are adjusted so that the normal component of velocity at the center of each panel is zero. The vortex strength is adjusted so that the rear stagnation point occurs at the trailing edge.

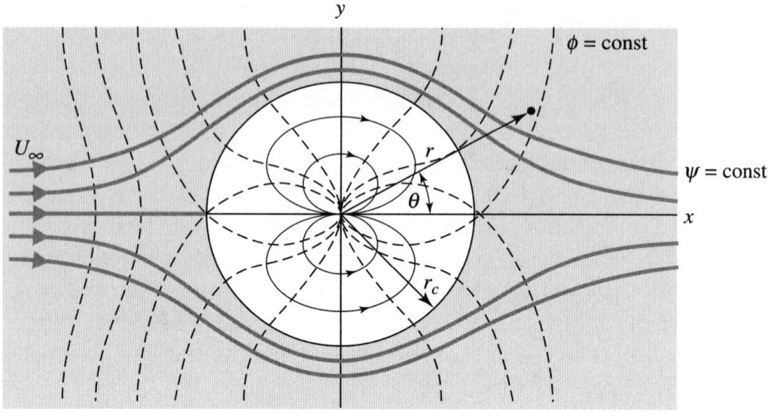

(a) Potential flow

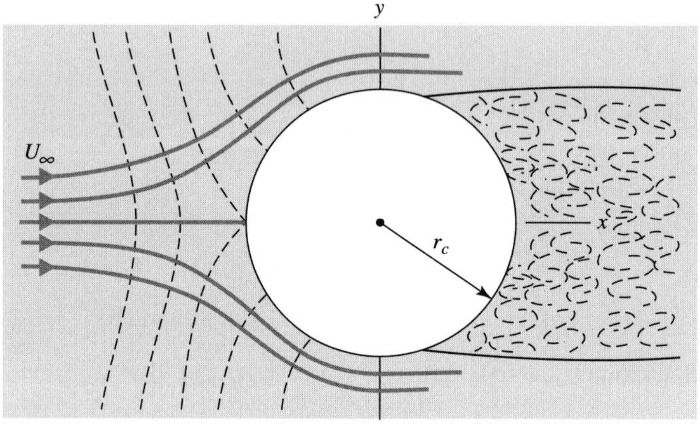

(b) Actual flow

FIGURE 16.19 Flow around a circular cylinder.

If the pressure distribution were desired on the cylinder, Bernoulli's equation could be used between the stagnation point where $V = 0$ and $p = p_0$ and some arbitrary point on the cylinder to yield

$$p_c = p_0 - \rho \frac{v_\theta^2}{2}$$

$$= p_0 - 2\rho U_\infty^2 \sin^2 \theta \tag{16.29}$$

Key Concept: *The pressure distribution up to the point of separation is very nearly the same as that predicted by potential flow.*

This provides a symmetric pressure distribution that yields zero drag and zero lift. The zero-lift prediction is acceptable for a real flow, but the zero-drag result is obviously unacceptable. This may suggest that we ignore the inviscid flow solution; however, compared with the actual flow situation of Fig. 16.19b, the measured pressure distribution on the cylinder up to the point of separation is very nearly the same as that predicted by the potential flow solution. Hence the potential flow solution is quite useful to us, even for blunt bodies that experience a sep-

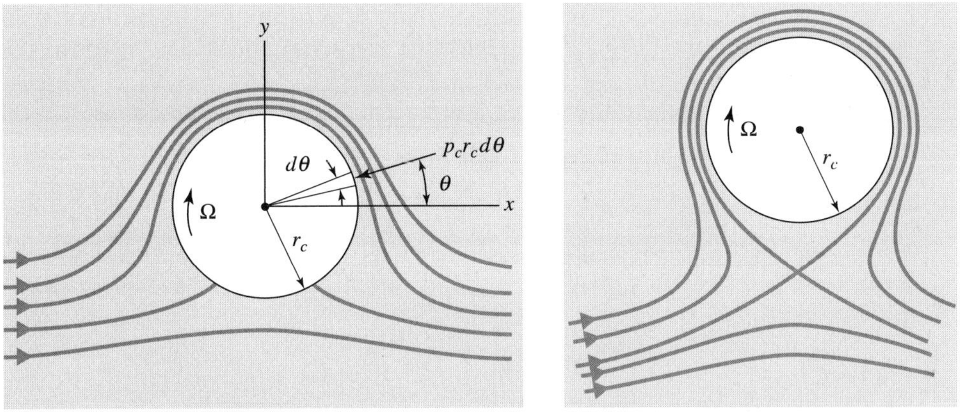

(a) $\Gamma < 4\pi U_\infty r_c$ (b) $\Gamma > 4\pi U_\infty r_c$

FIGURE 16.20 Flow around a circular cylinder with circulation.

arated flow. At low Reynolds numbers, viscous effects are not confined to a thin boundary layer, so potential flow theory is not useful.

Let us now consider flow around a rotating cylinder. This is accomplished by adding an irrotational vortex to the doublet and uniform flow so that

$$\psi = U_\infty y - \frac{\mu \sin \theta}{r} + \frac{\Gamma}{2\pi} \ln r \tag{16.30}$$

Since the vortex flow, consisting of circular streamlines, does not influence the velocity component v_r the cylinder $r = r_c$ remains unchanged. The stagnation points, however, are changed and are located by setting $v_\theta = 0$ on $r = r_c$; that is, letting $\mu = U_\infty r_c^2$,

$$\begin{aligned} v_\theta &= -\frac{\partial \psi}{\partial r} \\ &= -2U_\infty \sin \theta - \frac{\Gamma}{2\pi r_c} = 0 \end{aligned} \tag{16.31}$$

This gives the location of the stagnation points as shown in Fig. 16.20. In (a) the stagnation points are on the cylinder where $r = r_c$, but in (b) the circulation is sufficiently large that a single stagnation point is off the cylinder with $\theta = 270°$.

Bernoulli's equation gives the pressure distribution as

$$p_c = p_0 - \rho \frac{U_\infty^2}{2} \left(2 \sin \theta + \frac{\Gamma}{2\pi r_c U_\infty} \right)^2 \tag{16.32}$$

This can be integrated to give drag = 0 and the lift per unit length as

$$\begin{aligned} F_L &= -\int_\theta^{2v} p_c \sin \theta r_c \, d\theta \\ &= \rho U_\infty \Gamma \end{aligned} \tag{16.33}$$

Key Concept: *The expression $\rho U_\infty \Gamma$ provides an excellent approximation to the lift for all cylinders.*

This expression for the lift provides an excellent approximation to the lift for all cylinders, including the airfoil. Along with the zero-drag conclusion, it forms the *Kutta-Joukowsky theorem*.

Other superpositions of the simple flows are included in the problems.

Example 16.12

An 8-in.-diameter cylinder rotates clockwise at 1000 rpm in a 60°F-atmospheric airstream flowing at 15 ft/sec. Locate any stagnation points and find the minimum pressure on the cylinder.

Solution

The circulation is calculated (see Eq. 16.20) to be

$$\Gamma = \oint_L \mathbf{V} \cdot d\mathbf{s}$$

$$= 2\pi r_c^2 \Omega = 2\pi \times \left(\frac{4}{12}\right)^2 \times \frac{1000 \times 2\pi}{60} = 73.1 \text{ ft}^2/\text{sec}$$

This is greater than $4\pi U_\infty r_c = 4\pi \times 15 \times 4/12 = 62.8 \text{ ft}^2/\text{sec}$; hence the stagnation point is off the cylinder (see Fig. 16.20b) at

$$\theta = 270^\circ \qquad r_0 = -\frac{\Gamma}{4\pi U_\infty \sin 270^\circ} = -\frac{73.1}{4\pi \times 15 \times (-1)} = 0.388 \text{ ft}$$

Only one stagnation point exists.

The minimum pressure is located on the top of the cylinder where $\theta = 90^\circ$. Using Bernoulli's equation from the free stream to that point, we have, letting $p_\infty = 0$,

$$\overset{0}{\cancel{p_\infty}} + \rho \frac{U_\infty^2}{2} = p_{\min} + \frac{\rho}{2}(v_\theta)_{\max}^2$$

$$\therefore\ p_{\min} = \frac{\rho}{2}[U_\infty^2 - (v_\theta)_{\max}^2] = \frac{\rho}{2}\left[U_\infty^2 - \left(-2U_\infty \sin 90^\circ - \frac{\Gamma}{2\pi r_c}\right)^2\right]$$

$$= \frac{0.0024}{2}\left[15^2 - \left(2 \times 15 + \frac{73.1}{2\pi \times 4/12}\right)^2\right] = -4.78 \text{ psf}$$

16.6 BOUNDARY LAYER THEORY

16.6.1 General Background

In our study of high-Reynolds-number external flows we have observed that the viscous effects are confined to a thin layer of fluid, a boundary layer, next to the body and to the wake downstream of the body. For a streamlined body such as an airfoil, a good approximation to the drag can be obtained by integrating the viscous shear stress on the wall. To predict the wall shear, the velocity gradient at the wall must be known. This requires a complete solution of the flow field (i.e., a

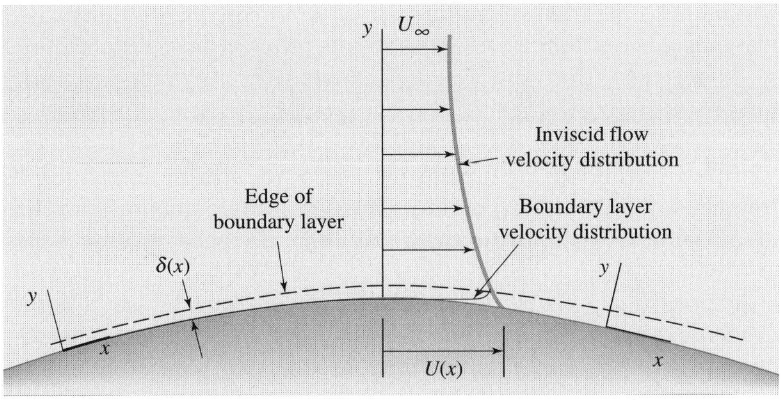

FIGURE 16.21 Boundary layer on a curved surface.

solution of the partial differential equations) inside the boundary layer. Such a solution also allows us to predict locations of possible separation. In this section we derive the integral equations and provide solution techniques for a boundary layer flow on a flat plate with a zero pressure gradient; this simplified flow has many applications. Nonzero-pressure-gradient flows on flat plates and curved surfaces are beyond the scope of this introductory presentation.

Let us now discuss some of the characteristics of a boundary layer. The edge of the boundary layer, with thickness designated by $\delta(x)$, cannot be observed in an actual flow; we arbitrarily define it to be that locus of points where the velocity is equal to 99% of the free-stream velocity [the free-stream velocity is the inviscid flow wall velocity $U(x)$], as shown in Fig. 16.21. Since the boundary layer is so thin, the pressure in the boundary layer is assumed to be the pressure $p(x)$ at the wall as predicted by the inviscid flow solution.

Key Concept: *The edge of the boundary layer cannot be observed in an actual flow.*

Key Concept: *The pressure in the boundary layer is the pressure at the wall of the inviscid flow solution.*

The boundary layer begins as a laminar flow with zero thickness at the leading edge of a flat plate, as shown in Fig. 16.22, or with some finite thickness at the stagnation point of a blunt object or an airfoil (see Fig. 16.3). After a distance x_T, which depends on the free-stream velocity, the viscosity, the pressure gradient,

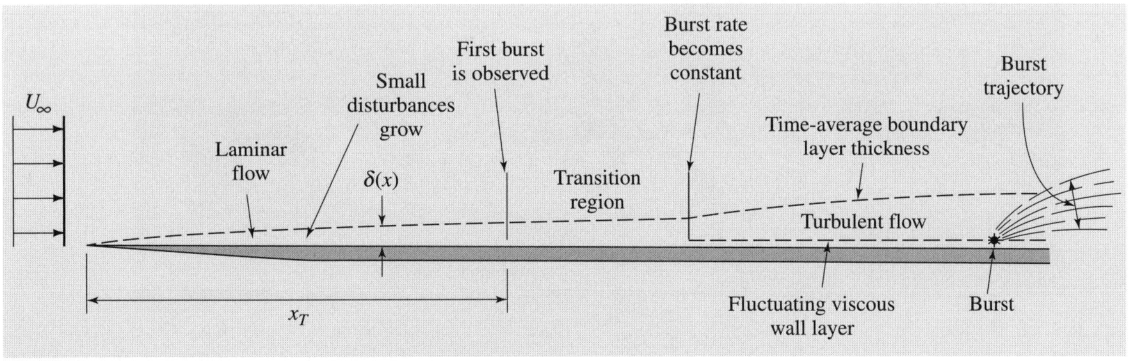

FIGURE 16.22 Boundary layer with transition.

the wall roughness, the free-stream fluctuation level, and wall rigidity, the laminar flow undergoes a transition process that results, after a short distance, in a turbulent flow, as sketched. For flow over a flat plate with zero pressure gradient, this transition process occurs when $U_\infty x_T/\nu = 3 \times 10^5$ for flow on rough plates or with high free-stream fluctuation intensity ($\sqrt{\overline{u'^2}}/U_\infty \simeq 0.1$), or $U_\infty x_T/\nu = 5 \times 10^5$ for flow on smooth rigid plates with low free-stream fluctuation intensity. For extremely low fluctuation levels in research labs, laminar flows have been observed on smooth rigid plates with carefully designed leading edges up to $U_\infty x_T/\nu = 10^6$.

The quantity $U_\infty x/\nu$ is the *local Reynolds number* and $U_\infty x_T/\nu$ is the *critical Reynolds number*. For a smooth rigid flat plate and a very low free-stream fluctuation level, a zero-pressure-gradient flow becomes unstable (i.e., small disturbances will grow) at a local Reynolds number of about 6×10^4. The small disturbances grow initially as a two-dimensional wave, then a three-dimensional wave,

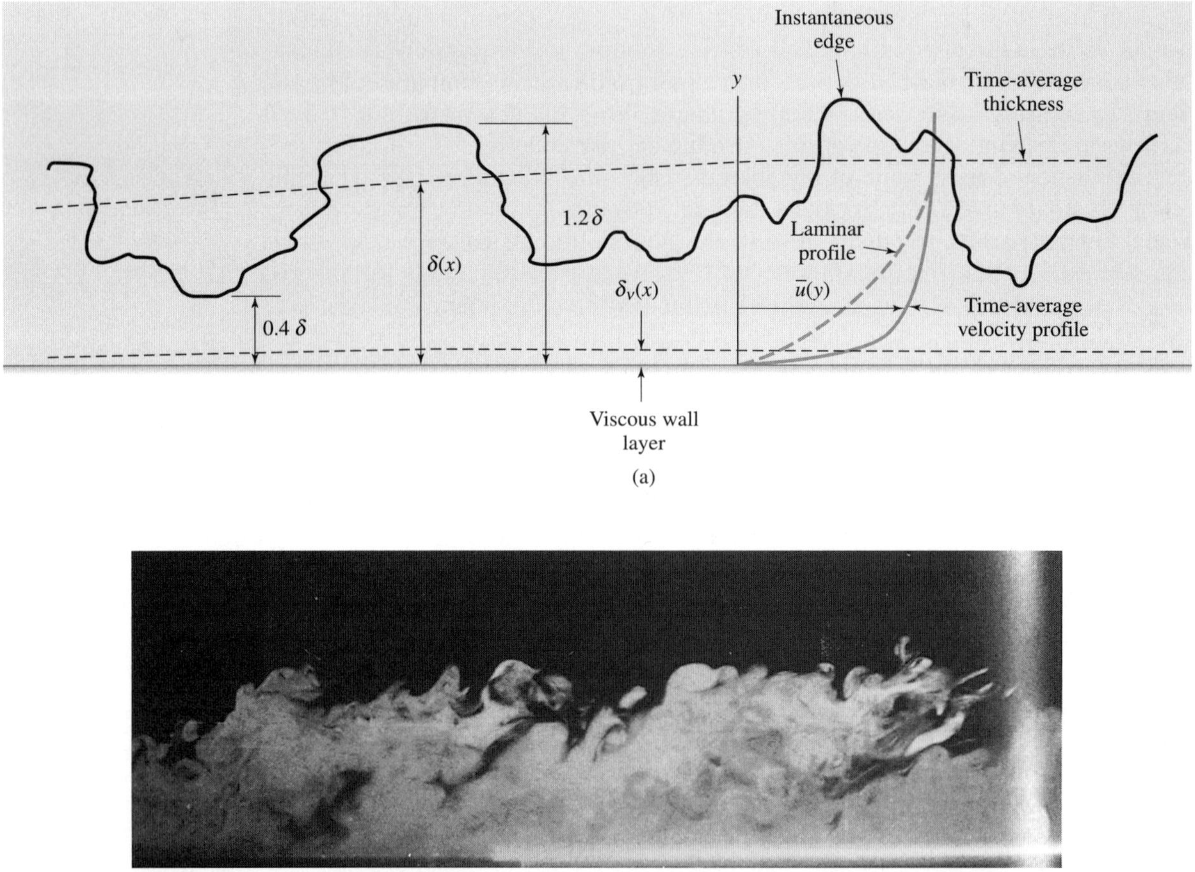

FIGURE 16.23 Turbulent boundary layer: (a) nomenclature sketch; (b) streamwise slice of the boundary layer. (Photograph by R. E. Falco.)

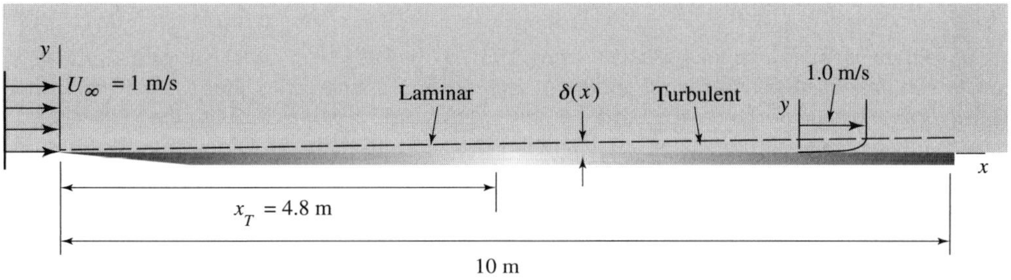

FIGURE 16.24 A relatively thick boundary layer in air with $\text{Re}_{\text{crit}} = 3 \times 10^5$.

and finally burst into a turbulent spot; the initial burst forms the beginning of the transition region. The transition region is relatively short and is usually ignored in calculations. The flow up to x_T is assumed to be laminar, and the flow after x_T is considered to be turbulent.

Key Concept: *The transition region is relatively short and is usually ignored in calculations.*

The turbulent boundary layer thickens much more rapidly than the laminar layer. It also has a substantially greater wall shear. A sketch of a turbulent boundary layer with its submerged viscous wall layer is shown in Fig. 16.23a, and an actual photograph in Fig. 16.23b. The time-average thickness is $\delta(x)$ and the time-average viscous wall layer thickness is $\delta_v(x)$. Both layers are actually quite time dependent. The instantaneous boundary layer thickness varies between 0.4δ and 1.2δ, as shown. The turbulent profile has a greater slope at the wall than does a laminar profile with the same boundary layer thickness, as can be observed in Fig. 16.23a.

Key Concept: *A turbulent profile has a greater slope at the wall than does a laminar profile.*

Finally, we should emphasize that the boundary layer is quite thin. A thick boundary layer shown by the dashed line is drawn to scale in Fig. 16.24. We have assumed a laminar flow up to x_T and a turbulent flow thereafter. For higher velocities the boundary layer thickness decreases. If we assumed that $U_\infty =$ 100 m/s, the boundary layer would hardly be noticed drawn to the same scale, yet all the viscous effects are confined in that thin layer; the velocity is brought to rest with very large gradients. Viscous dissipative effects in this thin layer are large enough to cause sufficiently high temperatures that satellites burn up on reentry.

Key Concept: *Viscous effects in the boundary layer cause satellites to burn up.*

16.6.2 Von Kármán Integral Equation

From the velocity profile in the boundary layer of Fig. 16.24 it is observed that the velocity goes from $u = 0.99U_\infty$ at $y = \delta$ to $u = 0$ at $y = 0$ over a very short distance (the thickness of the boundary layer). Hence it is not surprising that we can approximate the velocity profile, for both laminar and turbulent flow, with a good deal of accuracy. If the velocity profile can be assumed known, the continuity and momentum integral equations will enable us to predict the boundary layer thickness and the wall shear and thus the drag. So let us develop the integral equations for the boundary layer.

Key Concept: *We can approximate the velocity profile with a good deal of accuracy.*

Consider an infinitesimal control volume, shown in Fig. 16.25a. The integral continuity equation allows us to find $\dot{m}_{\text{top}}$ (see Fig. 16.25b). It is, assuming unit depth,

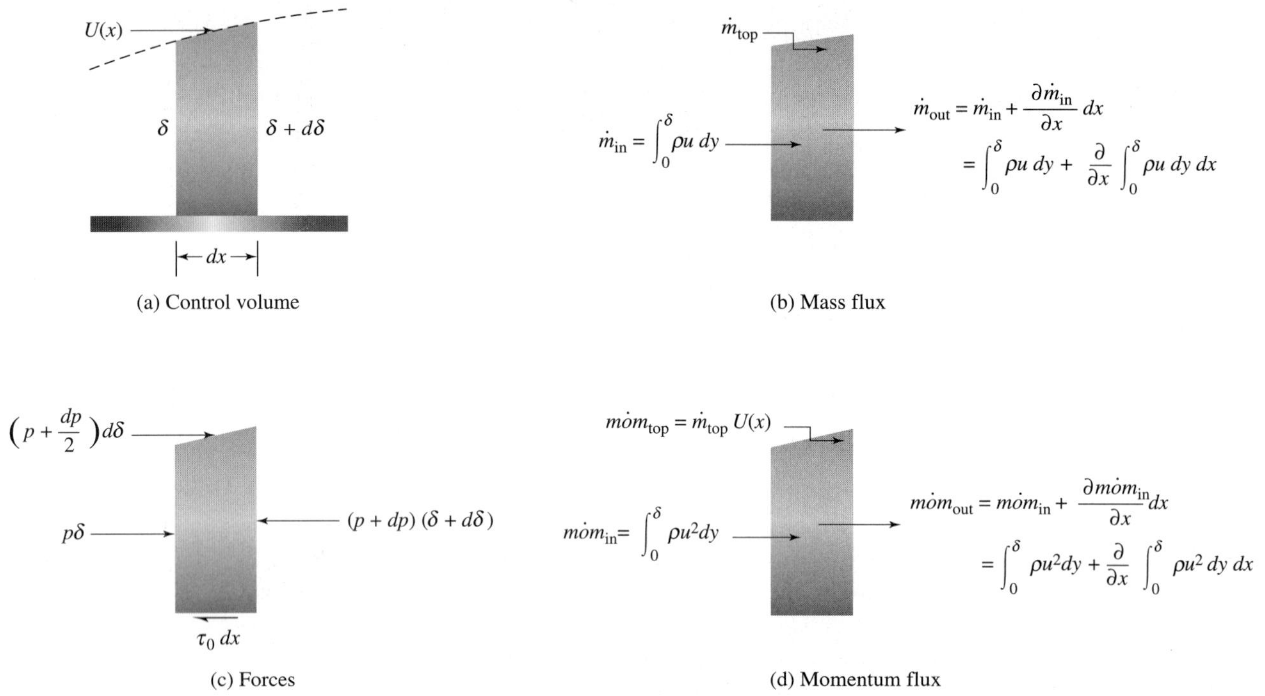

FIGURE 16.25 Control volume for a boundary layer with variable $U(x)$.

$$
\begin{aligned}
\dot{m}_{\text{top}} &= \dot{m}_{\text{out}} - \dot{m}_{\text{in}} \\
&= \frac{\partial}{\partial x}\int_0^\delta \rho u\, dy\, dx
\end{aligned}
\tag{16.34}
$$

The integral momentum equation takes the form

$$
\sum F_x = m\dot{o}m_{\text{out}} - m\dot{o}m_{\text{in}} - m\dot{o}m_{\text{top}} \tag{16.35}
$$

where $m\dot{o}m$ represents the momentum flux in the x-direction. Referring to Figs. 16.25c and 16.25d this becomes, neglecting high-order terms,

$$
-\delta\, dp - \tau_0\, dx = \frac{\partial}{\partial x}\int_0^\delta \rho u^2\, dy\, dx - \left(\frac{\partial}{\partial x}\int_0^\delta \rho u\, dy\, dx\right) U(x) \tag{16.36}
$$

where $\dot{m}_{\text{top}}$ is given in Eq. 16.34. We have not assumed that $U(x)$ is constant. Divide through by $-dx$ and obtain

$$
\tau_0 + \delta \frac{dp}{dx} = U(x)\frac{d}{dx}\int_0^\delta \rho u\, dy - \frac{d}{dx}\int_0^\delta \rho u^2\, dy \tag{16.37}
$$

where we have used ordinary derivatives since the integrals are only functions of x. This equation is often referred to as the *von Kármán integral equation*.

For flow over a flat plate with zero pressure gradient, so that $dp/dx = 0$ and $U(x) = U_\infty$, von Kármán's integral equation takes the simplified form

$$\tau_0 = \frac{d}{dx}\int_0^\delta \rho u U_\infty \, dy - \frac{d}{dx}\int_0^\delta \rho u^2 \, dy$$
$$= \frac{d}{dx}\int_0^\delta \rho u (U_\infty - u)\, dy \qquad (16.38)$$

If the velocity profile can be assumed, this equation along with $\tau_0(x) = \mu \partial u/\partial y|_{y=0}$ allows us to solve for both $\delta(x)$ and $\tau_0(x)$. This will be demonstrated in the following sections for both a laminar and a turbulent boundary layer.

Before we do this, however, there are two additional lengths often used in boundary layer theory. They are the *displacement thickness* δ_d and the *momentum thickness* θ, defined by

$$\delta_d = \frac{1}{U}\int_0^\delta (U - u)\, dy \qquad (16.39)$$

$$\theta = \frac{1}{U^2}\int_0^\delta u(U - u)\, dy \qquad (16.40)$$

The displacement thickness is the displacement of the streamlines in the free stream as a result of velocity deficits in the boundary layer, as can be shown by continuity considerations. The momentum layer thickness is the equivalent thickness of a fluid layer with velocity U with momentum equal to the momentum lost due to friction; the momentum thickness is often used as a characteristic length in turbulent boundary layer studies. The end-of-chapter problems will demonstrate the use of δ_d and θ. It should be noted, though, that von Kármán's integral equation (16.38) takes the form, assuming that $\rho =$ const.,

$$\tau_0 = \rho U_\infty^2 \frac{d\theta}{dx} \qquad (16.41)$$

16.6.3 Approximate Solution to the Laminar Boundary Layer

It is possible to use the von Kármán integral equation and obtain a fairly accurate approximation to the laminar boundary layer on a flat plate with zero pressure gradient. We have four conditions that a proposed velocity profile should satisfy:

$$\begin{aligned} u &= 0 && \text{at } y = 0 \\ u &= U_\infty && \text{at } y = \delta \\ \frac{\partial u}{\partial y} &= 0 && \text{at } y = \delta \\ \frac{\partial^2 u}{\partial y^2} &= 0 && \text{at } y = 0 \end{aligned} \qquad (16.42)$$

The first three of these conditions are obvious from a sketch of a velocity profile; the fourth condition comes from a partial differential equation (see Table K.3—the momentum equation in Cartesian coordinates) for the steady flow on the flat plate under consideration.

A cubic polynomial can satisfy the four conditions above; let us assume that

$$\frac{u}{U_\infty} = A + By + Cy^2 + Dy^3 \tag{16.43}$$

where A, B, C, and D can be functions of x. Using the four conditions, we see that

$$A = 0, \qquad B = \frac{3}{2\delta}, \qquad C = 0, \qquad D = -\frac{1}{2\delta^3} \tag{16.44}$$

Hence a good approximation for the velocity profile in a laminar flow is

$$\frac{u}{U_\infty} = \frac{3}{2}\frac{y}{\delta} + \frac{1}{2}\left(\frac{y}{\delta}\right)^3 \tag{16.45}$$

Let us now use this velocity profile to find $\delta(x)$ and $\tau_0(x)$. Von Kármán's integral equation (16.38) gives

$$\begin{aligned}\tau_0 &= \frac{d}{dx}\int_0^\delta \rho\left(\frac{3y}{2\delta} - \frac{y^3}{2\delta^3}\right)\left(1 - \frac{3y}{2\delta} - \frac{y^3}{2\delta^3}\right)U_\infty^2\,dy \\ &= 0.139\rho U_\infty^2 \frac{d\delta}{dx}\end{aligned} \tag{16.46}$$

At the wall we know that $\tau_0 = \mu \partial u/\partial y|_{y=0}$, or using the cubic profile (16.45),

$$\tau_0 = \mu\left(U_\infty \frac{3}{2\delta}\right) \tag{16.47}$$

Equating the foregoing expressions for $\tau_0(x)$, we find that

$$\delta\,d\delta = \frac{\frac{3}{2}\mu U_\infty}{0.139\rho U_\infty^2}dx = 10.8\frac{\nu}{U_\infty}dx \tag{16.48}$$

Using $\delta = 0$ at $x = 0$ (the leading edge), Eq. 16.48 is integrated to give

$$\delta = 4.65\sqrt{\frac{\nu x}{U_\infty}} = 4.65\frac{x}{\sqrt{\mathrm{Re}_x}} \tag{16.49}$$

where Re_x is the local Reynolds number. This is substituted back into Eq. 16.47, giving the wall shear as

$$\begin{aligned}\tau_0 &= 0.323\rho U_\infty^2\sqrt{\frac{\nu}{xU_\infty}} \\ &= \frac{0.323\rho U_\infty^2}{\sqrt{\mathrm{Re}_x}}\end{aligned} \tag{16.50}$$

The shearing stress is made dimensionless by dividing by $\frac{1}{2}\rho U_\infty^2$. The **local skin friction coefficient** c_f results; it is

Local skin friction coefficient: *A dimensionless wall shearing stress.*

$$c_f = \frac{\tau_0}{\frac{1}{2}\rho U_\infty^2}$$

$$= \frac{0.646}{\sqrt{U_\infty x/\nu}} = \frac{0.646}{\sqrt{\text{Re}_x}} \tag{16.51}$$

If the wall shear is integrated over the length L, there results, per unit width,

$$F_D = \int_0^L \tau_0\, dx = 0.646\rho U_\infty \sqrt{U_\infty L \nu}$$

$$= \frac{0.646\rho U_\infty^2 L}{\sqrt{\text{Re}_L}} \tag{16.52}$$

or in terms of the **skin friction coefficient** C_f,

Skin friction coefficient: *A dimensionless drag force.*

$$C_f = \frac{F_D}{\frac{1}{2}\rho U_\infty^2 L}$$

$$= \frac{1.29}{\sqrt{U_\infty L/\nu}} = \frac{1.29}{\sqrt{\text{Re}_L}} \tag{16.53}$$

where Re_L is the Reynolds number at the end of the flat plate.

Note that the shearing stress τ_0 becomes unbounded as $x \to 0$. Hence we would not expect $\tau_0(x)$ to be a very good approximation to the wall shear near the leading edge. The expression for the drag is acceptable, however.

The results above are quite good when compared with results from a solution of the partial differential equations. The boundary layer thickness is 7% too low; the constant in Eq. 16.49 should be 5 if an exact solution were obtained. The wall shear is less than 3% too low; the constant in Eq. 16.50 should be 0.332. The solutons to the partial differential equations would provide

$$\delta = 5\sqrt{\frac{\nu x}{U_\infty}} \tag{16.54}$$

$$c_f = \frac{0.646}{\sqrt{\text{Re}_x}} \tag{16.55}$$

$$C_f = \frac{1.29}{\sqrt{\text{Re}_L}} \tag{16.56}$$

Example 16.13

Assume a parabolic velocity profile and calculate the boundary layer thickness and the wall shear. Compare with the more accurate solutions.

Solution

The parabolic velocity profile is assumed to be

$$\frac{u}{U_\infty} = A + By + Cy^2$$

The fourth condition, which would be impossible to satisfy, of (16.42) is omitted; this leaves

$$0 = A$$

$$1 = A + B\delta + C\delta^2$$

$$0 = B + 2C\delta$$

A simultaneous solution provides

$$A = 0, \qquad B = \frac{2}{\delta}, \qquad C = -\frac{1}{\delta^2}$$

The velocity profile is then

$$\frac{u}{U_\infty} = 2\frac{y}{\delta} - \frac{y^2}{\delta^2}$$

This is substituted into von Kármán's integral equation (16.38) to obtain

$$\begin{aligned}\tau_0 &= \frac{d}{dx}\int_0^\delta \rho U_\infty^2 \left(2\frac{y}{\delta} - \frac{y^2}{\delta^2}\right)\left(1 - \frac{2y}{\delta} + \frac{y^2}{\delta^2}\right) dy \\ &= \frac{2}{15}\rho U_\infty^2 \frac{d\delta}{dx}\end{aligned}$$

We also use $\tau_0 = \mu \partial u/\partial y|_{y=0}$; that is,

$$\tau_0 = \mu U_\infty \frac{2}{\delta}$$

Equating the two expressions above, we obtain

$$\delta\, d\delta = 15 \frac{\nu}{U_\infty} dx$$

Using $\delta = 0$ at $x = 0$, this is integrated to

$$\delta = 5.48\sqrt{\frac{\nu x}{U_\infty}}$$

This is 18% higher than the value using the cubic but only 10% higher than the more accurate result of $5\sqrt{\nu x/U_\infty}$.

(continued)

The wall shear is found to be

$$\tau_0 = \frac{2\mu U_\infty}{\delta}$$

$$= 0.365\rho U_\infty^2 \sqrt{\frac{\nu}{xU_\infty}}$$

This is 13% higher than the value using the cubic and 10% higher than the more accurate value of $0.332\rho U_\infty^2 \sqrt{\nu/xU_\infty}$. Because the boundary layer is so thin, there is little difference between a cubic and a parabola or the actual profile; refer to the profile in Fig. 16.24.

16.6.4 Turbulent Boundary Layer: Power-Law Form

For turbulent boundary layer flow we have two methods for obtaining the information desired. Both methods utilize experimental data but the one that we present in this section is the simpler of the two. The second method, to be presented in the next section, provides us with more information than we usually desire for most applications; however, it is more accurate.

In the method to be presented first we fit the data for the velocity profile with a power-law equation. The power-law form is

$$\frac{\bar{u}}{U_\infty} = \left(\frac{y}{\delta}\right)^{1/n} \qquad n = \begin{cases} 7 & \mathrm{Re}_x < 10^7 \\ 8 & 10^7 < \mathrm{Re}_x < 10^8 \\ 9 & 10^8 < \mathrm{Re}_x < 10^9 \end{cases} \tag{16.57}$$

where

$$\mathrm{Re}_x = \frac{U_\infty x}{\nu} \tag{16.58}$$

Von Kármán's integral equation can now be applied following the steps used for laminar flow, except when the shear stress at the wall is evaluated. The power-law form (16.57) yields $(\partial \bar{u}/\partial y)_{y=0} = \infty$; hence the profile gives poor results near the wall, especially for the shear stress. So rather than using $\tau_0 = (\mu \partial \bar{u}/\partial y)_{y=0}$, we use an empirical relation; the *Blasius formula*, which relates the local skin friction coefficient to the boundary layer thickness, is

Key Concept: *The power-law form gives poor results at the wall.*

$$c_f = 0.046\left(\frac{\nu}{U_\infty \delta}\right)^{1/4} \tag{16.59}$$

or, relating τ_0 to c_f using the definition of c_f in Eq. 16.51, the shear stress relation is

$$\tau_0 = 0.023\rho U_\infty^2 \left(\frac{\nu}{U_\infty \delta}\right)^{1/4} \tag{16.60}$$

Von Kármán's integral equation provides us with a second expression for τ_0. Substitute the velocity profile of (16.57) with $\mathrm{Re}_x < 10^7$ into Eq. 16.38 and obtain

$$\tau_0 = \frac{d}{dx}\int_0^\delta \rho U_\infty^2 \left(\frac{y}{\delta}\right)^{1/7}\left[1-\left(\frac{y}{\delta}\right)^{1/7}\right] dy$$

$$= \frac{7}{72}\rho U_\infty^2 \frac{d\delta}{dx} \tag{16.61}$$

Combining the two expressions above for τ_0, we find that

$$\delta^{1/4}\, d\delta = 0.237\left(\frac{\nu}{U_\infty}\right)^{1/4} dx \tag{16.62}$$

Assuming a turbulent flow over the entire plate length L (the laminar portion is often quite short, i.e., $L \gg x_T$), there results

$$\delta = 0.38x\left(\frac{\nu}{U_\infty x}\right)^{1/5}$$

$$= 0.38x\,\mathrm{Re}_x^{-1/5} \qquad \mathrm{Re}_x < 10^7 \tag{16.63}$$

Substituting this expression for δ back into Eq. 16.59, we find that

$$c_f = 0.059\,\mathrm{Re}_x^{1/5} \qquad \mathrm{Re}_x < 10^7 \tag{16.64}$$

and, performing the required integration (see Eqs. 16.52 and 16.53), there results, with $n = 7$,

$$C_f = 0.073\,\mathrm{Re}_L^{-1/5} \qquad \mathrm{Re}_L < 10^7 \tag{16.65}$$

where

$$\mathrm{Re}_L = \frac{U_\infty L}{\nu}$$

The relations above can be stretched to $\mathrm{Re}_x \simeq 10^8$ without substantial error.

Key Concept: *There may be a significant laminar portion on the leading edge of a flat plate.*

If L is not much larger than x_T, say $L = 3x_T$, then there is a significant laminar portion on the leading part of a flat plate, and the skin friction coefficient can be modified as

$$C_f = 0.073\,\mathrm{Re}_L^{-1/5} - 1700\,\mathrm{Re}_L^{-1} \qquad \mathrm{Re}_L < 10^7 \tag{16.66}$$

This relationship is based on transition occurring at $\mathrm{Re}_{\mathrm{crit}} = 5 \times 10^5$. If $\mathrm{Re}_{\mathrm{crit}} = 3 \times 10^5$, the constant of 1700 is replaced by 1060; if $\mathrm{Re}_{\mathrm{crit}} = 6 \times 10^5$, it becomes 2080.

Finally, the displacement thickness and momentum thickness can be evaluated, using $n = 7$, to be

$$\begin{aligned} \delta_d &= 0.048x\,\mathrm{Re}_x^{-1/5} \\ \theta &= 0.037x\,\mathrm{Re}_x^{-1/5} \end{aligned} \tag{16.67}$$

Example 16.14

Estimate the boundary layer thickness at the end of a 4-m-long flat surface if the free-stream velocity is $U_\infty = 5$ m/s. Use atmospheric air at 30°C. Also, predict the drag force if the surface is 5 m wide. (a) Neglect the laminar portion of the flow and (b) account for the laminar portion using $\text{Re}_{\text{crit}} = 5 \times 10^5$.

Solution

(a) Let us first assume turbulent flow from the leading edge. The boundary layer thickness is given by Eq. 16.63. It is

$$\delta = 0.38x\,\text{Re}_x^{-1/5}$$
$$= 0.38 \times 4 \times \left(\frac{5 \times 4}{1.6 \times 10^{-5}}\right)^{-1/5} = 0.0917 \text{ m}$$

The drag force is, using Eq. 16.65,

$$F_D = C_f \times \tfrac{1}{2}\rho U_\infty^2 Lw$$
$$= 0.073\left(\frac{5 \times 4}{1.6 \times 10^{-5}}\right)^{-1/5} \times \frac{1}{2} \times 1.16 \times 5^2 \times 4 \times 5 = 1.28 \text{ N}$$

The predictions above assume that $\text{Re}_L < 10^7$. The Reynolds number is

$$\text{Re}_L = \frac{5 \times 4}{1.6 \times 10^{-5}} = 1.25 \times 10^6$$

Hence the calculations are acceptable.

(b) Now let us account for the laminar portion of the boundary layer. Referring to Fig. E16.14, the distance x_T is found as follows:

$$\text{Re}_{\text{crit}} = 5 \times 10^5 = \frac{U_\infty x_T}{\nu}$$

$$\therefore\ x_T = 5 \times 10^5 \times 1.6 \times \frac{10^{-5}}{5} = 1.6 \text{ m}$$

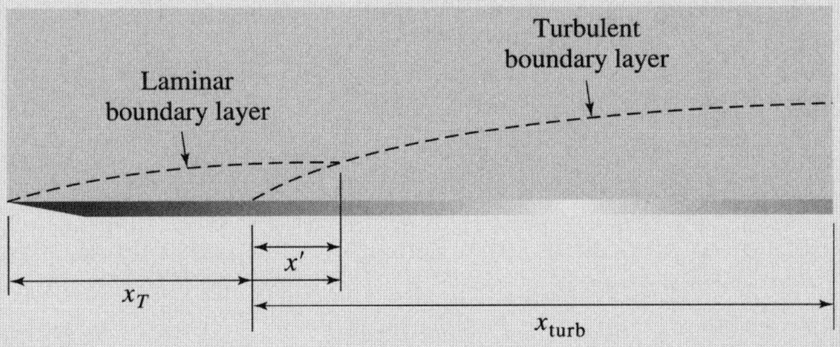

FIGURE E16.14

(continued)

The boundary layer thickness using Eq. 16.54 at x_T is,

$$\delta = 5\sqrt{\frac{x\nu}{U_\infty}}$$

$$= 5\sqrt{\frac{1.6 \times 1.6 \times 10^{-5}}{5}} = 0.0113 \text{ m}$$

The location of the fictitious origin of the turbulent flow (see Fig. E16.14) is found using Eq. 16.63 to be

$$x'^{4/5} = \frac{\delta}{0.38}\left(\frac{U_\infty}{\nu}\right)^{1/5}$$

$$\therefore\ x' = \left(\frac{0.0113}{0.38}\right)^{5/4}\left(\frac{5}{1.6 \times 10^{-5}}\right)^{1/4} = 0.292 \text{ m}$$

The distance x_{turb} is then $x_{\text{turb}} = 4 - 1.6 + 0.292 = 2.69$ m. Using Eq. 16.63, the thickness at the end of the surface is

$$\delta = 0.38x\left(\frac{\nu}{U_\infty x}\right)^{1/5}$$

$$= 0.38 \times 2.69 \times \left(\frac{1.6 \times 10^{-5}}{5 \times 2.69}\right)^{1/5} = 0.067 \text{ m}$$

The value of part (a) is 37% too high when compared with this more accurate value.

The more accurate drag force is found using Eq. 16.66 to be

$$F_D = C_f \times \tfrac{1}{2}\rho U_\infty^2 Lw$$

$$= [0.073\ \text{Re}_L^{-1/5} - 1700\ \text{Re}_L^{-1}] \times \frac{1}{2}\rho U_\infty^2 Lw$$

$$= \left[0.073\left(\frac{5 \times 4}{1.6 \times 10^{-5}}\right)^{-1/5} - 1700\left(\frac{5 \times 4}{1.6 \times 10^{-5}}\right)^{-1}\right] \times \frac{1}{2} \times 1.16 \times 5^2 \times 4 \times 5$$

$$= 0.88 \text{ N}$$

The prediction of part (a) is 45% too high. For relatively short surfaces it is obvious that significant errors result if the thinner laminar portion with its smaller shear stress is neglected.

16.6.5 Turbulent Boundary Layer: Empirical Form

The second method of predicting turbulent flow quantities on a flat plate with zero pressure gradient is based entirely on data. It is more accurate than the power-law form but also more complicated. The time-average turbulent velocity profile can be divided into two regions, the *inner region* and the *outer region*, as shown in Fig. 16.26. The inner region is characterized by the *self-similar* (the dimensionless

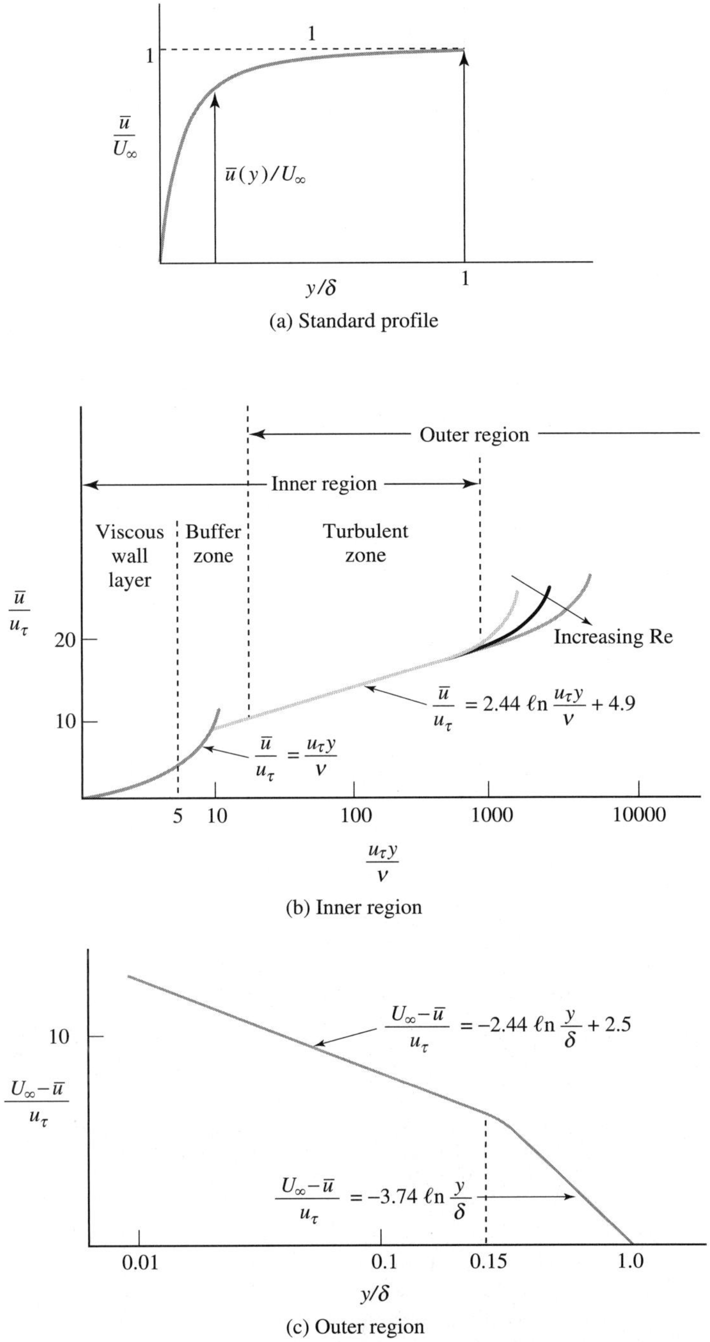

FIGURE 16.26 Velocity profile in a turbulent boundary layer.

dependent variable depends on only one dimensionless independent variable) relation,

$$\frac{\bar{u}}{u_\tau} = f\left(\frac{u_\tau y}{\nu}\right) \tag{16.68}$$

in which u_τ is the *shear velocity*, given by[5]

Key Concept: *u_τ is a ficticious velocity called the shear velocity.*

$$u_\tau = \sqrt{\frac{\tau_0}{\rho}} \tag{16.69}$$

The velocity profile in the outer region is given by the self-similar relation

$$\frac{U_\infty - \bar{u}}{u_\tau} = f\left(\frac{y}{\delta}\right) \tag{16.70}$$

where $(U_\infty - \bar{u})$ is called the *velocity defect.*

The inner region has three distinct zones; the viscous wall layer, the buffer zone, and the turbulent zone, as shown in Fig. 16.26b. The highly fluctuating *viscous wall layer* has a linear time-average profile given by

$$\frac{\bar{u}}{u_\tau} = \frac{u_\tau y}{\nu} \tag{16.71}$$

The quantity ν/u_τ is the characteristic length in the turbulent inner region; hence the dimensionless distance from the wall is denoted by

$$y^+ = \frac{u_\tau y}{\nu} \tag{16.72}$$

Key Concept: *The viscous wall layer is highly fluctuating.*

The viscous wall layer is very thin, extending up to $y^+ \simeq 5$. A logarithmic profile exists from $y^+ \simeq 50$ to $y/\delta \simeq 0.15$. In this self-similar *turbulent zone,*

$$\frac{\bar{u}}{u_\tau} = 2.44 \ln \frac{u_\tau y}{\nu} + 4.9 \qquad 50 < \frac{u_\tau y}{\nu} \qquad \frac{y}{\delta} < 0.15 \tag{16.73}$$

The location of the outer edge of the turbulent zone is strongly dependent on the Reynolds number. The value of $u_\tau y/\nu$ that locates the outer edge increases as the Reynolds number increases, as shown. A *buffer zone*, with no specified velocity profile, connects the two self-similar zones.

[5]The shear velocity is a fictitious velocity and is defined because the quantity $\sqrt{\tau_0/\rho}$ occurs often in empirical relations in turbulent boundary layer flows.

The outer region relates the velocity defect to y/δ. In the turbulent zone the velocity-defect profile is sketched in Fig. 16.26c and is

$$\frac{U_\infty - \bar{u}}{u_\tau} = -2.44 \ln \frac{y}{\delta} + 2.5 \qquad 50 < \frac{u_\tau y}{\nu} \qquad \frac{y}{\delta} < 0.15 \qquad (16.74)$$

Between $y/\delta = 0.15$ and $y/\delta = 1$ researchers fit the data with several relations; the one selected here is

$$\frac{U_\infty - \bar{u}}{u_\tau} = -3.74 \ln \frac{y}{\delta} \qquad \frac{y}{\delta} > 0.15 \qquad (16.75)$$

The equations above involve the shear velocity u_τ, which depends on the wall shear τ_0. The equation that is valid at the wall is Eq. 16.71, which, using $\tau_0 = \mu\, \partial u/\partial y|_{y=0}$, simply provides us with an identity. It does not allow us to calculate τ_0. Hence a relationship is necessary to provide us with τ_0 (or equally c_f). There are several such relationships used; one that gives excellent results is

$$c_f = \frac{0.455}{(\ln 0.06\, \text{Re}_x)^2} \qquad (16.76)$$

This local skin friction coefficient allows us to determine τ_0 and thus u_τ at any location of interest. Knowing u_τ, the velocity profiles can be used to calculate quantities of interest.

Key Concept: *The local skin friction coefficient allows us to determine τ_0 and thus u_τ.*

Assuming turbulent flow from the leading edge, the shear stress can be integrated to yield the drag. Then the skin friction coefficient becomes

$$C_f = \frac{0.523}{(\ln 0.06\, \text{Re}_L)^2} \qquad (16.77)$$

This relation is very good and can be used up to $\text{Re}_L = 10^9$ with an error of 2% or less. Even at $\text{Re}_L = 10^{10}$ the error is about 4%. To account for a laminar portion, the same term included in Eq. 16.66 can be subtracted from Eq. 16.77.

In concluding this section, a very useful relationship can be obtained by combining the two logarithmic profiles for the common turbulent zone. Substitute Eq. 16.73 into Eq. 16.74 and obtain

$$\frac{U_\infty}{u_\tau} = 2.44 \ln \frac{u_\tau \delta}{\nu} + 7.4 \qquad (16.78)$$

This equation allows an easy calculation of δ knowing u_τ.

Example 16.15

Estimate the thickness δ_v of the viscous wall layer and the boundary layer thickness at the end of a 15-ft-long flat plate if $U_\infty = 100$ ft/sec in 60°F-atmospheric air. Also, calculate the drag force on one side if the plate is 10 ft wide. Use the empirical data.

Solution

To find the viscous wall layer thickness we must know the shear velocity and hence the wall shear. The wall shear, using Eq. 16.76, and the shear velocity at $x = 15$ ft are

$$\begin{aligned}\tau_0 &= \tfrac{1}{2}\rho U_\infty^2 c_f \\ &= \frac{1}{2}\rho U_\infty^2 \frac{0.455}{(\ln 0.06\,\text{Re}_x)^2} \\ &= \frac{1}{2} \times 0.0024 \times 100^2 \frac{0.455}{\left(\ln 0.06 \dfrac{100 \times 15}{1.6 \times 10^{-4}}\right)^2} = 0.0311 \text{ psf}\end{aligned}$$

$$\therefore\ u_\tau = \sqrt{\frac{\tau_0}{\rho}} = \sqrt{\frac{0.0311}{0.0024}} = 3.6 \text{ ft/sec}$$

The viscous wall layer thickness is determined using Eq. 16.72 with $y^+ = 5$ as follows:

$$\frac{u_\tau \delta_v}{\nu} = 5$$

$$\therefore\ \delta_v = \frac{5\nu}{u_\tau} = \frac{5 \times 1.6 \times 10^{-4}}{3.6} = 2.22 \times 10^{-4} \text{ ft}$$

The boundary layer thickness at $x = 15$ ft is found using Eq. 16.78:

$$\frac{U_\infty}{u_\tau} = 2.44 \ln \frac{u_\tau \delta}{\nu} + 7.4$$

$$\frac{100}{3.6} = 2.44 \ln \frac{3.6 \times \delta}{1.6 \times 10^{-4}} + 7.4 \qquad \therefore\ \delta = 0.188 \text{ ft}$$

The drag force is calculated using Eq. 16.77 to be

$$\begin{aligned}F_D &= C_f \times \tfrac{1}{2}\rho U_\infty^2 L w \\ &= \frac{0.523}{(\ln 0.06\,\text{Re}_L)^2} \times \frac{1}{2}\rho U_\infty^2 L w \\ &= \frac{0.523}{\left(\ln 0.06 \dfrac{100 \times 15}{1.6 \times 10^{-4}}\right)^2} \times \frac{1}{2} \times 0.0024 \times 100^2 \times 15 \times 10 = 5.4 \text{ lb}\end{aligned}$$

The laminar portion of the boundary layer has been neglected.

Example 16.16

Estimate the maximum boundary layer thickness and the drag on the side of a ship that measures 40 m long with a submerged depth of 8 m assuming the side of the ship is approximated as a flat plate. The ship travels at 10 m/s. (a) Use the empirical methods and (b) compare with the results using the power law model.

Solution

(a) The boundary layer thickness is found from Eq. 16.78. First we must find τ_0 from Eq. 16.76 and then u_τ as follows:

$$\tau_0 = \frac{1}{2}\rho U_\infty^2 \frac{0.455}{(\ln 0.06\ \text{Re}_L)^2}$$

$$= \frac{1}{2} \times 1000 \times 10^2 \frac{0.455}{\left(\ln 0.06 \dfrac{10 \times 40}{10^{-6}}\right)^2} = 78.8\ \text{Pa}$$

$$\therefore u_\tau = \sqrt{\frac{\tau_0}{\rho}}$$

$$= \sqrt{\frac{78.8}{1000}} = 0.28\ \text{m/s}$$

The maximum boundary layer thickness is found using Eq. 16.78:

$$\frac{U_\infty}{u_\tau} = 2.44 \ln \frac{u_\tau \delta}{\nu} + 7.4$$

$$\frac{10}{0.28} = 2.44 \ln \frac{0.28\delta}{10^{-6}} + 7.4 \qquad \therefore \delta = 0.39\ \text{m}$$

The drag is

$$F_D = C_f \times \tfrac{1}{2}\rho U_\infty^2 Lw$$

$$= \frac{0.523}{\left(\ln 0.06 \dfrac{10 \times 40}{10^{-6}}\right)^2} \times \frac{1}{2} \times 1000 \times 10^2 \times 40 \times 8 = 29\,000\ \text{N}$$

(b) First, calculate the Reynolds number: $\text{Re} = 10 \times 40/10^{-6} = 4 \times 10^8$. We select $n = 9$. Equation (16.61) becomes

$$\tau_0 = \frac{d}{dx}\int_0^\delta \rho U_\infty^2 \left(\frac{y}{\delta}\right)^{1/9}\left[1 - \left(\frac{y}{\delta}\right)^{1/9}\right] dy$$

$$= \frac{9}{110}\rho U_\infty^2 \frac{d\delta}{dx}$$

Equating this to the τ_0 of Eq. 16.60, we find that

$$\delta^{1/4}\, d\delta = 0.281(\nu/U_\infty)^{1/4}\, dx$$

(continued)

Assume $\delta = 0$ at $x = 0$ and integrate. This provides

$$\delta = 0.433x \, \mathrm{Re}_x^{-1/5}$$

$$= 0.433(40)\left(\frac{10 \times 40}{10^{-6}}\right)^{-1/5} = 0.33 \text{ m}$$

This value is 15% too low.

The drag force is found to be

$$F_D = 0.071 \, \mathrm{Re}_L^{-1/5} \times \tfrac{1}{2}\rho U_\infty^2 L w$$

$$= 0.071\left(\frac{10 \times 40}{10^{-6}}\right)^{-1/5} \times \frac{1}{2} \times 1000 \times 10^2 \times 40 \times 8 = 21\,600 \text{ N}$$

This value is 25% too low. Obviously, the power-law equations are in significant error.

16.6.6 Convection Heat Transfer

In Section 3.6, we described convection heat transfer from a solid surface to a fluid in motion. We can now visualize this in terms of a temperature boundary layer at the solid surface illustrated in Fig. 16.27, similar to that described for the velocity boundary layer shown in Fig. 16.22. The thermal boundary layer thickness $\delta_t(x)$ can be greater than, less than, or equal to the velocity boundary layer thickness $\delta(x)$, according to properties of the fluid. Newton's law of cooling (Eq. 3.25), therefore, provides us with a means to describe the heat flux across this thermal boundary layer.

The convective heat transfer coefficient in Eq. 3.25 is dependent on many factors, including properties of the fluid, properties associated with the flow, and the geometry. Convection is a combination of energy transfer by conduction and advection. If we assume the velocity is zero at the wall, we can related convection heat transfer to conduction at the wall by

$$q_x'' = -k_f \left.\frac{\delta T}{\delta y}\right|_{y=0} = h_c(x)(T_s - T_\infty) \tag{16.79}$$

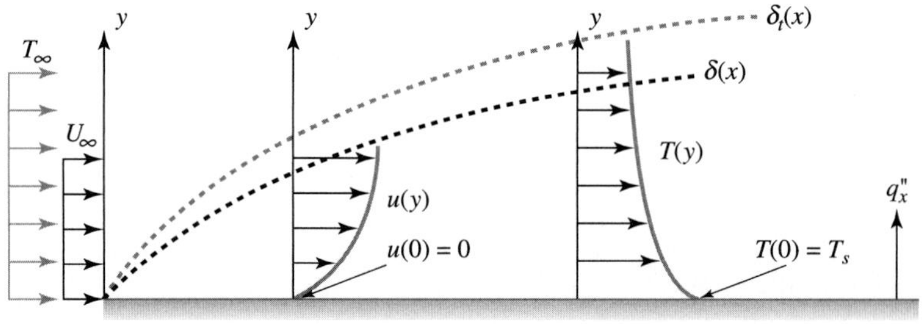

FIGURE 16.27 The velocity and thermal boundary layers for a fluid moving over a solid surface. The layer thicknesses are exaggerated to show the distributions.

where T_s is the surface temperature, T_∞ is the bulk fluid temperature, k_f is the thermal conductivity of the fluid, and the convective heat transfer coefficient h_c depends on x. The temperature T depends on both x and y, i.e., $T(x, y)$. Thus, the heat transfer coefficient can be expressed as

$$h_c(x) = \frac{-k_f \left.\dfrac{\partial T}{\partial y}\right|_{y=0}}{(T_s - T_\infty)} \tag{16.80}$$

We define T and y in dimensionless terms as

$$T^+ = \frac{T - T_s}{(T_s - T_\infty)} \qquad y^+ = \frac{y}{L} \tag{16.81}$$

where L is some characteristic length (such as the length of a plate). We can now express the heat transfer coefficient in dimensionless terms as the *Nusselt number* Nu_x, and the dimensionless temperature gradient at $y = 0$:

$$\mathrm{Nu}_x = \left.\frac{\partial T^+}{\partial y^+}\right|_{y^+=0} = \frac{L}{T_s - T_\infty} \left.\frac{\partial T}{\partial y}\right|_{y=0} \tag{16.82}$$

Using Eq. 16.80, this is expressed as

$$\mathrm{Nu}_x = \frac{L}{T_\infty - T_s} \frac{h_c(x)(T_s - T_\infty)}{k_f} = \frac{Lh_c(x)}{k_f} \tag{16.83}$$

Since the temperature gradient at $y = 0$ will be a function of flow conditions, the propertiers of the fluid, and the geometry, we can express the Nusselt number in the form

$$\mathrm{Nu}_x = f(x^+, \mathrm{Re}_x, \mathrm{Pr}) \tag{16.84}$$

where the Reynolds number and *Prandtl number* Pr are defined as

$$\mathrm{Re} = \frac{U_\infty x}{\nu} \qquad \mathrm{Pr} = \frac{\text{kinematic viscosity}}{\text{thermal diffusivity}} = \frac{\nu}{\alpha} = \frac{c_p \mu}{k_f} \tag{16.85}$$

The fluid properties are typically evaluated at the average boundary layer temperature T_f, defined as

$$T_f = \frac{T_s + T_\infty}{2} \tag{16.86}$$

Thus far, we have specified the heat transfer coefficient and the Nusselt number at a specific location x. In many cases, we would like to determine an average heat transfer coefficient over some characteristic length L:

$$\bar{h}_{cL} = \frac{1}{L} \oint_0^L h_c(x)\, dx \tag{16.87}$$

and

$$\overline{\mathrm{Nu}}_L = \frac{L\bar{h}_{cL}}{k_f} \tag{16.88}$$

As one might assume, the most difficult aspect of convection heat transfer is determining the convective heat transfer coefficient or the Nusselt number. Analytical and numerical solutions involve the solutions of both the flow field (such as described in Section 16.6.2) and the complete energy equation for temperature. As a result, analytical solutions can be obtained only for a few simplified cases such as that described in Section 16.6.2, and numerical solutions are often computationally intensive. Thus, in many cases, it is appropriate to take a simpler approach and use empirical correlations to determine the heat transfer coefficient. Based on the functional relationship indicated by Eq. (16.84), we write a generalized form for the average Nusselt number:

$$\overline{\mathrm{Nu}}_L = C \cdot \mathrm{Re}_L^m \cdot \mathrm{Pr}^n \tag{16.89}$$

where C, m, and n are constants. Table 16.5 shows correlations for Nu_L and Nu_x for a select number of cases. Note that all have been derived from empirical correlations with the exception of the ones for laminar flow over a flat plate, which can be determined analytically. For mixed flow conditions, we assume that the transition from laminar to turbulent flow occurs at a point (at x_{crit}, where $\mathrm{Re}_x = 5 \times 10^5$), and then we integrate the first from $x = 0$ to x_{crit} using the local relationship for laminar flow, and then from $x = x_{\mathrm{crit}}$ to L using the local relationship for turbulent flow. Finally, for spherical and cylindrical geometries, note that the Reynolds number is based on the diameter.

Therefore, to determine the heat flux due to convection from Eq. 3.25, we must first determine the convective heat transfer coefficient by calculating the Reynolds number and then find the appropriate correlation in Table 16.5 for the geometry and conditions given. These steps are illustrated in Example 16.17.

TABLE 16.5 Convective Heat Transfer Correlations for External Flow for Select Cases

Geometry	Re Range	Property Range	Property Temp	Correlation	Eq. No.
Flat Plate	$\mathrm{Re}_x < 5 \times 10^5$ (Laminar)	$0.6 < \mathrm{Pr} < 50$	T_f	$\mathrm{Nu}_x = 0.332\, \mathrm{Re}_x^{1/2}\, \mathrm{Pr}^{1/3}$	(16.90)
				$\overline{\mathrm{Nu}}_L = 2\, \mathrm{Nu}_x$	(16.91)
	$5 \times 10^5 < \mathrm{Re}_x < 10^8$ (Turbulent)	$0.6 < \mathrm{Pr} < 60$	T_f	$\mathrm{Nu}_x = 0.296\, \mathrm{Re}_x^{4/5}\, \mathrm{Pr}^{1/3}$	(16.92)
	$\mathrm{Re}_x < 10^8$ (Mixed)	$0.6 < \mathrm{Pr} < 60$	T_f	$\overline{\mathrm{Nu}}_L = (0.037\, \mathrm{Re}_L^{4/5} - 871)\, \mathrm{Pr}^{1/3}$	(16.93)
Cylinder	$0.4 < \mathrm{Re}_D < 4$	$\mathrm{Pr} \geq 0.7$	T_f	$\overline{\mathrm{Nu}}_D = 0.989\, \mathrm{Re}_D^{0.330}\, \mathrm{Pr}^{1/3}$	(16.94)
	$4 < \mathrm{Re}_D < 40$			$\overline{\mathrm{Nu}}_D = 0.911\, \mathrm{Re}_D^{0.385}\, \mathrm{Pr}^{1/3}$	(16.95)
	$40 < \mathrm{Re}_D < 4 \times 10^3$			$\overline{\mathrm{Nu}}_D = 0.683\, \mathrm{Re}_D^{0.466}\, \mathrm{Pr}^{1/3}$	(16.96)
	$4 \times 10^3 < \mathrm{Re}_D < 4 \times 10^4$			$\overline{\mathrm{Nu}}_D = 0.193\, \mathrm{Re}_D^{0.618}\, \mathrm{Pr}^{1/3}$	(16.97)
	$4 \times 10^4 < \mathrm{Re}_D < 7.6 \times 10^5$			$\overline{\mathrm{Nu}}_D = 0.027\, \mathrm{Re}_D^{0.805}\, \mathrm{Pr}^{1/3}$	(16.99)
Sphere	$3.5 \times \mathrm{Re}_D < 7.6 \times 10^4$	$0.71 < \mathrm{Pr} < 380$ $1.0 < \mu_\infty/\mu_s < 3.2$	T_∞	$\overline{\mathrm{Nu}}_D = 2 + (0.4\, \mathrm{Re}_D^{1/2} + 0.06\, \mathrm{Re}_D^{1/3}) \times \mathrm{Pr}^{0.4}(\mu_\infty/\mu_s)^{1/4}$	(16.99)

Example 16.17

Air at a pressure of 1 atm and a temperature of 300 K flows with a velocity of 10 m/s over a 3-m long flat plate that is maintained at a uniform temperature of 400 K. Using the properties of air at 350 K, determine the average heat flux from the plate to the air. At 350 K, $\nu = 2.092 \times 10^{-5}$ m^2/s, $k = 0.03$ W/m·K, and Pr = 0.70.

Solution

The heat flux can be determined from Eq. 16.79:

$$q_x'' = \bar{h}_{c_L}(T_s - T_\infty)$$

where $\bar{h}_{c_L}$ is the average heat transfer coefficient over the surface, which can be determined from the Nusselt number in Eq. 16.83:

$$h_{c_L} = \frac{\overline{\text{Nu}_L} \cdot k_f}{L}$$

The question now is to find the appropriate Nusselt number correlation. To do this, we must first determine the Reynolds number:

$$\text{Re}_L = \frac{U_\infty \cdot L}{\nu} = \frac{10 \cdot 3}{2.092 \times 10^{-5}} = 1.43 \times 10^6$$

Since $\text{Re}_L > 5 \times 10^5$, the flow at the end of the plate is turbulent. We can determine the critical length from the critical Reynolds number (5×10^5):

$$L_c = \frac{\text{Re}_{L_c} \cdot \nu}{U_\infty} = \frac{5 \times 10^5 \cdot 2.092 \times 10^{-5}}{10} = 1.05 \text{ m}$$

Thus, the flow is laminar over the first third of the plate and turbulent for the remainder (neglecting the transition region). From Table 16.5, the appropriate correlation is Eq. 16.93, which takes into account both laminar and turbulent regions:

$$\overline{\text{Nu}_{L_c}} = (0.037\,\text{Re}_L^{4/5} - 871)\,\text{Pr}^{1/3} = (0.037 \cdot (1.43 \times 10^6)^{4/5} - 871) \cdot 0.7^{1/3} = 1992.4$$

Therefore, the average heat transfer coefficient is

$$h_{c_L} = \frac{\overline{\text{Nu}_L} \cdot k_f}{L} = \frac{1992.4 \cdot 0.03}{3} = 19.92 \text{ W/m}^2\text{·K}$$

and the average heat flux over the surface is

$$q_x'' = \bar{h}_{c_L}(T_s - T_\infty) = 19.92 \cdot (400 - 300) = 1992 \text{ W/m}^2$$

16.6.7 Pressure Gradient Effects

In the foregoing sections we have concentrated our study of boundary layers on the flat plate with a zero pressure gradient. This is the simplest boundary layer flow and allows us to model many flows of engineering interest. The inclusion of a pressure gradient, even though relatively low, markedly alters the boundary layer flow. In fact, a strong negative pressure gradient (such as flow in a contraction)

Key Concept: *A strong negative pressure gradient may relaminarize a turbulent boundary layer.*

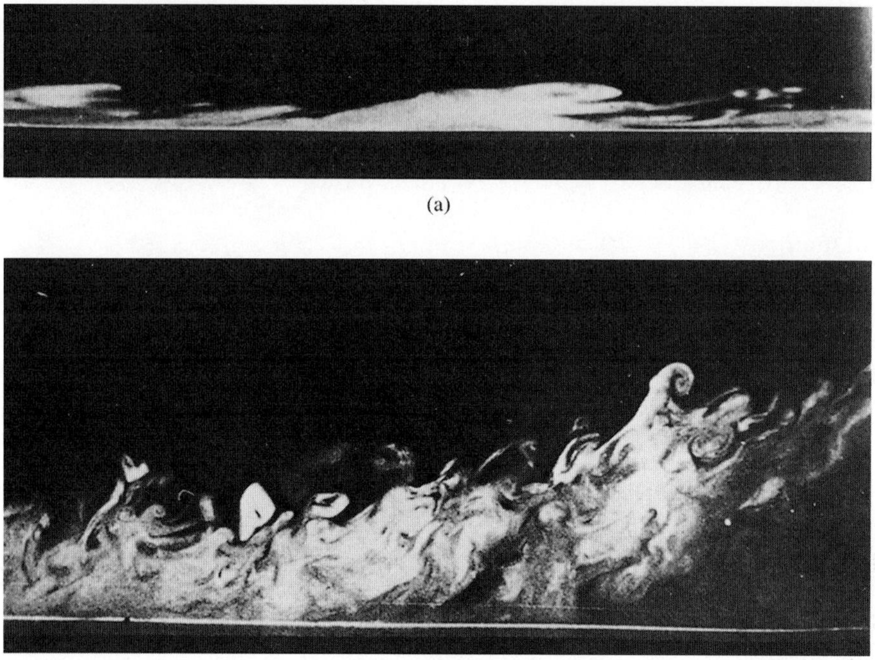

(a)

(b)

FIGURE 16.28 Influence of a strong pressure gradient on a turbulent flow: (a) a strong negative pressure gradient may relaminarize a flow; (b) a strong positive pressure gradient causes a strong boundary layer to thicken. (Photographs by R. E. Falco.)

may relaminarize a turbulent boundary layer; that is, the turbulence production in the viscous wall layer which sustains the turbulence ceases and a laminar boundary layer is reestablished. A positive pressure gradient quickly causes the boundary layer to thicken and eventually to separate. These two effects are shown in the photographs of Fig. 16.28.

The flow about any plane body with curvature, such as an airfoil, can be modeled as flow over a flat plate with a nonzero pressure gradient. The boundary layer thickness is so much smaller than the radius of curvature that the additional curvature terms drop out of the equations. The inviscid flow solution at the wall provides the pressure gradient dp/dx and the velocity $U(x)$ at the edge of the boundary layer. For axisymmetric flows, such as flow over the nose of an aircraft, boundary layer equations in cylindrical coordinates must be utilized.

The pressure gradient determines the value of the second derivative $\partial^2 u/\partial y^2$ at the wall. From the Navier–Stokes equations of Table K.3 it can be shown that (the details are not included here but with $u = v = w = 0$ and $\frac{\partial^2 u}{\partial x^2} = \frac{\partial^2 u}{\partial z^2} = 0$ at the wall, it is relatively obvious)

$$\frac{dp}{dx} = \mu \left.\frac{\partial^2 u}{\partial y^2}\right|_{y=0} \tag{16.100}$$

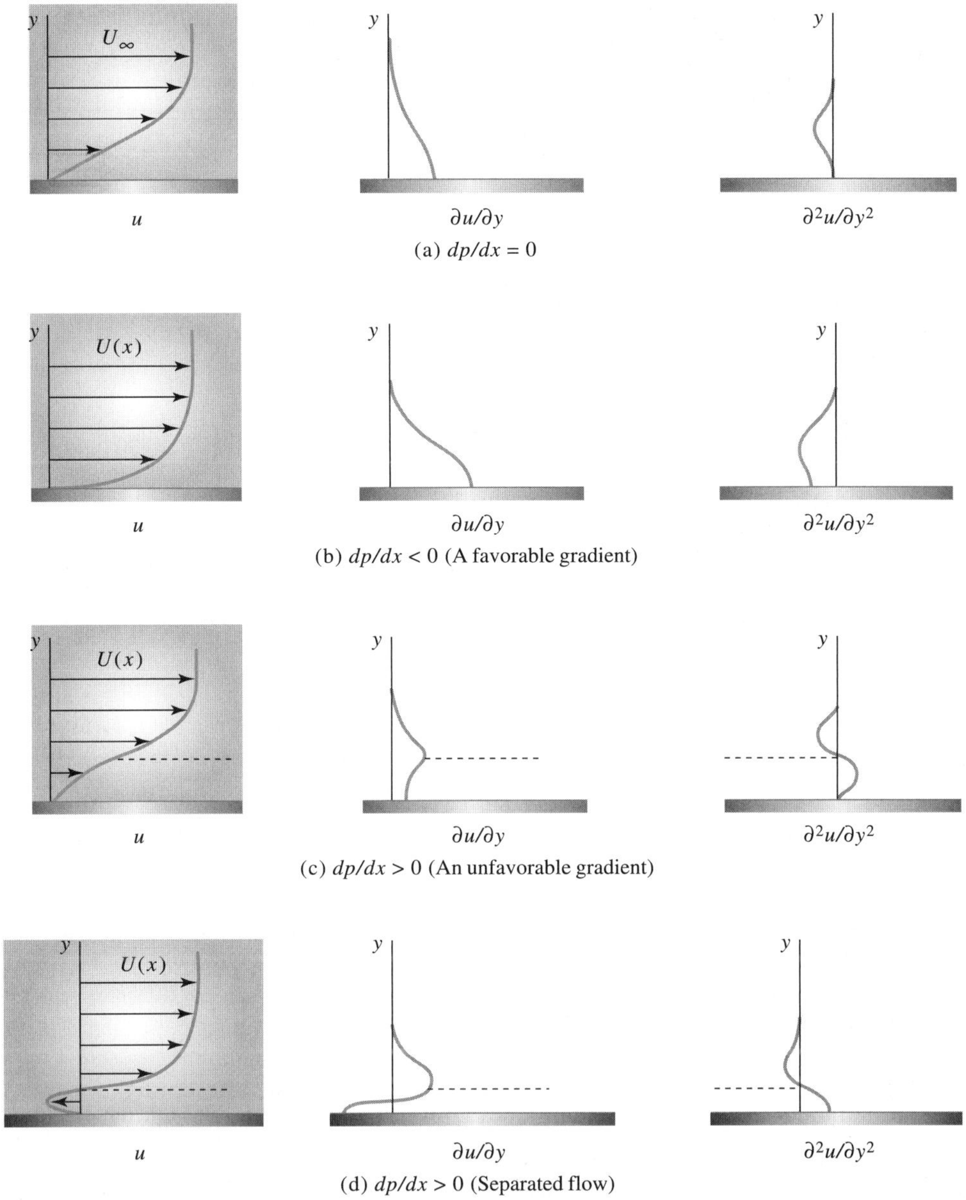

FIGURE 16.29 Influence of the pressure gradient.

for either a laminar or a turbulent boundary layer flow. For a zero pressure gradient the second derivative is zero at the wall; then since the first derivative is greatest at the wall and decreases as y increases, the second derivative must be negative for positive y. The profiles are sketched in Fig. 16.29a.

For a negative (favorable) pressure gradient the slope of the velocity profile near the wall is relatively large with a negative second derivative at the wall and

Key Concept: *For a negative pressure gradient, there is a reduced tendency for the flow to separate.*

throughout the layer. The momentum near the wall is larger than that of the zero pressure gradient flow, as shown in Fig. 16.29b, and thus there is a reduced tendency for the flow to separate. Turbulence production is discouraged and relaminarization may occur for a sufficiently large negative pressure gradient.

If a positive (unfavorable) pressure gradient is imposed on the flow the second derivative at the wall will be positive and the flow will be as sketched in part (c) or (d). If the unfavorable pressure gradient acts over a sufficient distance, part (d) will probably represent the flow situation with the flow separated from the surface. Near the wall the higher pressure downstream will drive the low momentum flow near the wall in the upstream direction, resulting in flow reversal, as shown. The point at which $\partial u/\partial y = 0$ at the wall locates the point of separation.

The problem of a laminar boundary layer with a pressure gradient can be solved using conventional numerical techniques with a computer. The procedure requires the solution of a simplified boundary layer differential equation. For a turbulent flow the apparent shear stress term (see Eq. 15.59) must be included. Experimental results are often necessary for turbulent flow problems, as was the situation for internal flows.

16.7 SUMMARY

The drag and lift coefficients are defined as

$$C_D = \frac{\text{Drag}}{\frac{1}{2}\rho V^2 A} \qquad C_L = \frac{\text{Lift}}{\frac{1}{2}\rho V^2 A} \tag{16.101}$$

where the area is the projected area for blunt objects, and the chord times the length for an airfoil.

Vortex shedding occurs from a cylinder whenever the Reynold's number is in the range $300 < \text{Re} < 10\,000$. The frequency of shedding is found from the Strouhal number

$$\text{St} = \frac{fD}{V} \tag{16.102}$$

where f is the frequency in hertz.

Plane potential flows are constructed by superimposing the following simple flows:

$$\begin{aligned} &\text{Uniform flow:} && \psi = U_\infty y \\ &\text{Line source:} && \psi = \frac{q_f}{2\pi}\theta \\ &\text{Irrotational vortex:} && \psi = \frac{\Gamma}{2\pi}\ln r \\ &\text{Doublet:} && \psi = -\frac{\mu}{r}\sin\theta \end{aligned} \tag{16.103}$$

The stream function for the rotating cylinder is given by

$$\psi_{\text{cylinder}} = U_\infty y - \frac{\mu}{r}\sin\theta + \frac{\Gamma}{2\pi}\ln r \tag{16.104}$$

where the cylinder radius is

$$r_c = \sqrt{\frac{\mu}{U_\infty}} \tag{16.105}$$

The velocity components are

$$\begin{aligned} u &= \frac{\partial\psi}{\partial y}, & v &= -\frac{\partial\psi}{\partial x} \\ v_r &= \frac{1}{r}\frac{\partial\psi}{\partial\theta}, & v_\theta &= -\frac{\partial\psi}{\partial r} \end{aligned} \tag{16.106}$$

For a laminar boundary layer on a flat plate with zero pressure gradient the exact solution provides

$$\delta = 5\sqrt{\frac{\nu x}{U_\infty}}, \qquad c_f = 0.664\sqrt{\frac{\nu}{xU_\infty}}, \qquad C_f = 1.33\sqrt{\frac{\nu}{LU_\infty}} \tag{16.107}$$

For a turbulent flow from the leading edge, the power-law profile with $\eta = 7$ gives

$$\delta = 0.38x\left(\frac{\nu}{xU_\infty}\right)^{1/5}, \qquad c_f = 0.059\left(\frac{xU_\infty}{\nu}\right)^{1/5}, \qquad C_f = 0.073\left(\frac{\nu}{LU_\infty}\right)^{1/5} \tag{16.108}$$

where the wall shear and drag force per unit width are, respectively,

$$\tau_0 = \frac{1}{2}c_f\rho U_\infty^2, \qquad F_D = \frac{1}{2}C_f\rho U_\infty^2 L \tag{16.109}$$

The convection heat transfer is calculated using the Nusselt number from Table 16.5 and

$$q_x'' = h_c(T_s - T_\infty) \tag{16.110}$$

where

$$\text{Nu}_x = \frac{Lh_c(x)}{k_f} \tag{16.111}$$

For a cylinder or a sphere, the average Nusselt number is used

$$\overline{\text{Nu}}_D = C\,\text{Re}_D^m\,\text{Pr}^n \tag{16.112}$$

where C, m, and n are found in Table 16.5.

REVIEW PROBLEMS FOR THE FE/EIT EXAMINATION

16.1FE The drag force on a streamlined shape is due primarily to:
A. The wake.
B. The component of the pressure force acting in the flow direction.
C. The shear stress.
D. The separated region near the trailing edge.

16.2FE A golf ball has dimples to increase its flight distance. Select the best reason that accounts for the longer flight distance of a dimpled ball compared to that of a smooth ball.
A. The shearing stress is smaller on the dimpled ball.
B. The dimpled ball has an effective smaller diameter.
C. The wake on the dimpled ball is smaller.
D. The pressure on the front of the smooth ball is larger.

16.3FE Flood waters at 10°C flow over an 8-mm diameter fence wire with a velocity of 0.8 m/s. Which of the following is true?
A. It is a Stokes flow with no separation.
B. The separated region covers most of the rear of the wire.
C. The drag is due to the relatively low pressure in the separated region.
D. The separated region covers only a small area at the rear of the wire.

16.4FE The drag on a 10-m-diameter spherical water storage tank in an 80 km/hr wind is approximately:
A. 6300 N **B.** 4700 N
C. 3200 N **D.** 2300 N

16.5FE A 4-m-long smooth cylinder experiences a drag of 60 N when subjected to an atmospheric air speed of 40 m/s. Estimate the diameter of the cylinder.
A. 127 mm **B.** 63 mm
C. 26 mm **D.** 4.1 mm

16.6FE Vortices are shed from a 2-cm-diameter cylinder due to a 4 m/s air stream. How far apart would you expect the vortices to be downstream of the cylinder?
A. 44 cm **B.** 23 cm
C. 9 cm **D.** 4 cm

16.7FE Streamlining reduces drag primarily by:
A. Reducing the wall shear.
B. Reducing the pressure in the stagnation region.
C. Reducing the separated flow area.
D. Eliminating the wake.

16.8FE Estimate the takeoff speed needed for a 1200-kg aircraft (including payload) if the angle of attack at takeoff is to be 10°. The effective wing area (chord times length) is 16 m^2.
A. 22 m/s **B.** 33 m/s
C. 44 m/s **D.** 55 m/s

16.9FE Consider air with a pressure of 1 atm and a temperature of 300 K flowing with a velocity of 5 m/s over a 2.0-m long flat plate that is maintained at a uniform temperature of 500 K. Using the properties of air at 400 K, determine the average heat transfer coefficient over the plate. At 400 K, $\nu = 2.641 \times 10^{-5}$ m^2/s, $k = 0.0338$ W/m·K, and Pr = 0.69.
A. 1.0 $W/m^2 \cdot K$ **B.** 3.0 $W/m^2 \cdot K$
C. 12.8 $W/m^2 \cdot K$ **D.** 6.1 $W/m^2 \cdot K$

PROBLEMS

Separated Flows

16.1 Sketch the flow over an airfoil at a large angle of attack for both attached flow and separated flow. Also, sketch the expected pressure distributions on the top and bottom surfaces for both flows. Identify regions of favorable and unfavorable pressure gradients.

16.2 A spherical particle is moving in 20°C atmospheric air at a velocity of 20 m/s. What must its diameter be for Re = 5 and Re = 10^5? Sketch the expected flow field at these Reynolds numbers. Identify all regions of the flow.

16.3 Sketch the flow that is expected over a semitruck (a tractor and trailer) where the trailer is substantially higher than the tractor with and without an air deflector attached to the roof of the tractor. Sketch a side view indicating any separated regions, boundary layers, and the wake.

16.4 Air is blowing by a long, rectangular building with the wind blowing parallel to the long sides. Sketch the top view showing the separated flow regions, the inviscid flow region, the boundary layers, and the wake.

16.5 A 0.8-in.-diameter sphere is to move with Re = 5. At what velocity does it travel if it is submerged in:
(a) Water at 60°F?
(b) Water at 180°F?
(c) Standard air at 60°F?

16.6 Air at 20°C flows around a cylindrical body at a velocity of 20 m/s. Calculate the Reynolds number if the body is:
(a) A 6-m-diameter smokestack.
(b) A 6-cm-diameter flagpole.
(c) A 6-mm-diameter wire.
Use $\text{Re} = VD/\nu$. Would a separated flow be expected?

16.7 The pressure distribution on the front of a 2-m-diameter disk (Fig. P16.7) is approximated by $p(r) = p_0(1 - r^2)$. If $V = 20$ m/s in this 20°C–atmospheric airflow, estimate the drag force and the drag coefficient for this disk. Assume that the pressure on the back side is zero.

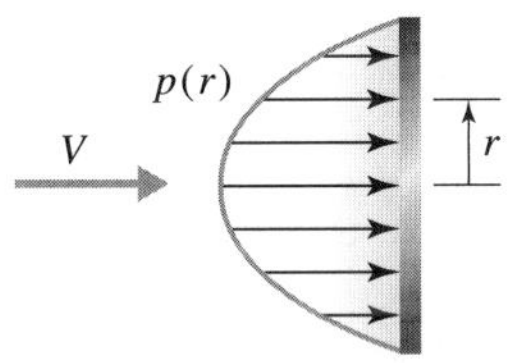

FIGURE P16.7

16.8 A flat plate 30 cm × 30 cm acts as a hydrofoil. If it is oriented at an angle of attack of 10°, estimate the lift and the drag if the pressure on the bottom side is 20 kPa and on the top side there is a vacuum of 10 kPa; neglect the effect of shear stress. Also, estimate the lift and drag coefficients if the hydrofoil velocity is 5 m/s. Use the surface area of the plate in the definition of the lift and drag coefficients.

16.9 The symmetric airfoil shown in Fig. P16.9 flies at an altitude of 12 000 m with an angle of attack of 5°. If $p_l = 26$ kPa and $p_u = 8$ kPa, estimate the lift and drag coefficients neglecting shear stresses.

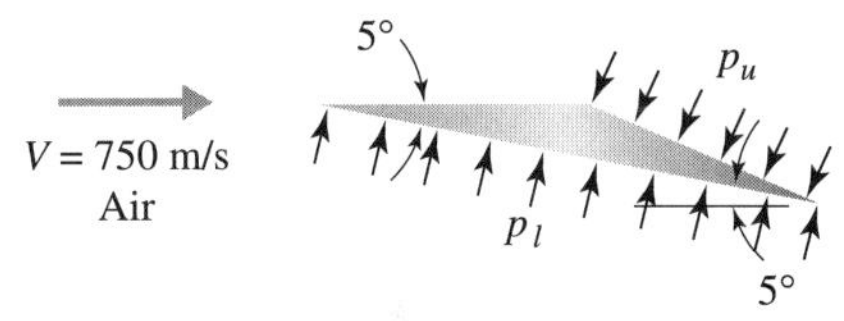

FIGURE P16.9

16.10 If the drag coefficient for a 10-cm-diameter sphere is given by $C_D = 1.0$, calculate the drag if the sphere is falling in the atmosphere:
(a) At sea level. **(b)** At 30 000 m.
(c) In 10°C water.

16.11 Calculate the drag on a smooth 50-cm-diameter sphere when subjected to a 20°C–atmospheric airflow of:
(a) 6 m/s **(b)** 15 m/s

16.12 A 4.45-cm-diameter golf ball is roughened to reduce the drag during its flight. If the Reynolds number at which the sudden drop occurs is reduced from 3×10^5 to 6×10^4 by the roughening (dimples), would you expect this to lengthen the flight of a golf ball significantly? Justify your reasoning with appropriate calculations.

16.13 A 4-in.-diameter smooth sphere experiences a drag of 0.5 lb when placed in standard 60°F air.
(a) What is the velocity of the airstream?
(b) At what increased speed will the sphere experience the same drag?

16.14 A smooth 20-cm-diameter sphere experiences a drag of 4.2 N when placed in a 20°C water channel. Calculate the drag coefficient and the Reynolds number.

16.15 A 2-m-diameter smokestack stands 60 m high. It is designed to resist a 40-m/s wind. At this speed, what total force would be expected, and what moment would the base be required to resist? Assume atmospheric–20°C air.

16.16 A flagpole is composed of three sections: a 5-cm-diameter top section that is 10 m long, a 7.5-cm-diameter middle section 15 m long, and a 10-cm-diameter bottom section 20 m long. Calculate the total force acting on the flagpole and the resisting moment provided by the base when subjected to a 25-m/s wind speed. Make the calculations for:
(a) A winter day at −30°C.
(b) A summer day at 35°C.

16.17 A drag force of 10 lb is desired at $Re = 10^5$ on a 6-ft-long cylinder in a 60°F–atmospheric airflow. What velocity should be selected, and what should be the cylinder diameter?

16.18 A 20-m-high structure is 2 m in diameter at the top and 8 m in diameter at the bottom as shown in Fig. P16.18. If the diameter varies linearly with height, estimate the total drag force due to a 30-m/s wind. Use atmospheric air at 20°C.

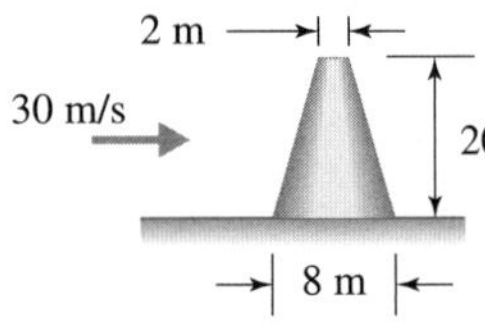

FIGURE P16.18

16.19 A steel sphere ($S = 7.82$) is dropped in water at 20°C. Calculate the terminal velocity if the diameter of the sphere is:
(a) 10 cm **(b)** 5 cm
(c) 1 cm **(d)** 2 mm

16.20 Estimate the terminal velocity of a 20-in.-diameter sphere as it falls in a 60°F atmosphere near the Earth if it has a specific gravity of:
(a) 0.005 **(b)** 0.02 **(c)** 1.0

16.21 Estimate the terminal velocity of a skydiver by making reasonable approximations of the arms, legs, head, and body. Assume air at 20°C.

16.22 Assuming that the drag on a modern automobile at high speeds is due primarily to form drag, estimate the power (horsepower) needed by an automobile with a 3.2-m^2 cross-sectional area to travel at:
(a) 80 km/h **(b)** 90 km/h **(c)** 100 km/h

16.23 The 2 m × 3 m sign shown in Fig. P16.23 weighs 400 N. What wind speed is required to blow over the sign? Neglect the weight of the thin legs.

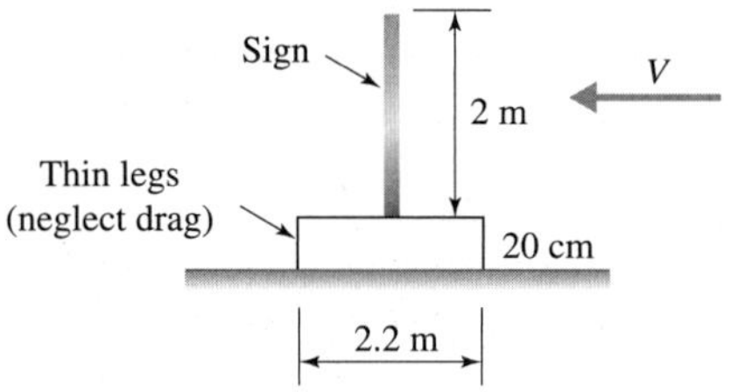

FIGURE P16.23

16.24 Calculate the drag force on a 60-cm-diameter 6-m-long cylinder if 20°C-air is blowing normal to its axis at 40 km/h and the cylinder:
(a) Is a section of a very long cylinder.
(b) Has both ends free.
(c) Is fixed to the ground with the top free.

16.25 An 80-kg paratrooper leaps from an elevation of 3000 m. Estimate the landing speed if the paratrooper:
(a) Curls up as tightly as possible.
(b) Uses an 8-m-diameter lightweight parachute.
(c) Uses a 2-m-diameter safety parachute.

16.26 A semitruck travels 200 000 km each year at an average speed of 90 km/h. Estimate the savings if a streamlined deflector is added to reduce the drag coefficient. Fuel costs $0.40 per liter and the truck without the deflector averages 1.2 km per liter of fuel.

16.27 A rectangular car carrier has a cross section 6 ft × 2 ft. Estimate the minimum added power (horsepower) required to travel at 60 mph because of the car carrier.

16.28 Assume that the velocity at the corners of an automobile where the rear-view mirror is located is 1.6 times the automobile's velocity. How much horsepower is required by the two 10-cm-diameter rear-view mirrors for an automobile speed of 100 km/h?

16.29 Atmospheric air at 25°C blows normal to a 4-m-long section of a cone, 30 cm in diameter at one end and 2 m in diameter at the other end, with a velocity of 20 m/s. Predict the drag on the object. Assume that $C_D = 0.4$ for each cylindrical element of the object.

16.30 An 80-cm-diameter balloon (Fig. P16.30) that weighs 0.5 N is filled with 20°C helium at 20 kPa. Neglecting the weight of the string, estimate V if α equals:

(a) 80° **(b)** 70°
(c) 60° **(d)** 50°

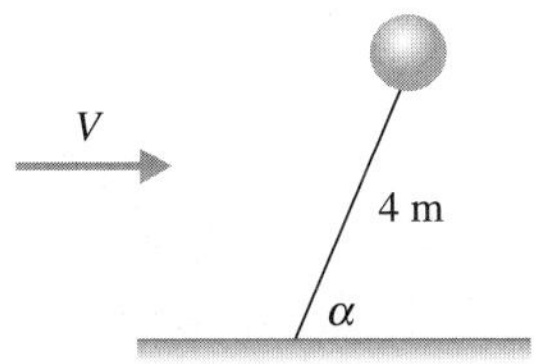

FIGURE P16.30

16.31 For the newly planted tree shown in Fig. P16.31, the soil–ball interface is capable of resisting a moment of 5000 N·m. Predict the minimum wind speed that might possibly tip the tree over. Assume that $C_D = 0.4$ for a cylinder in this air flow.

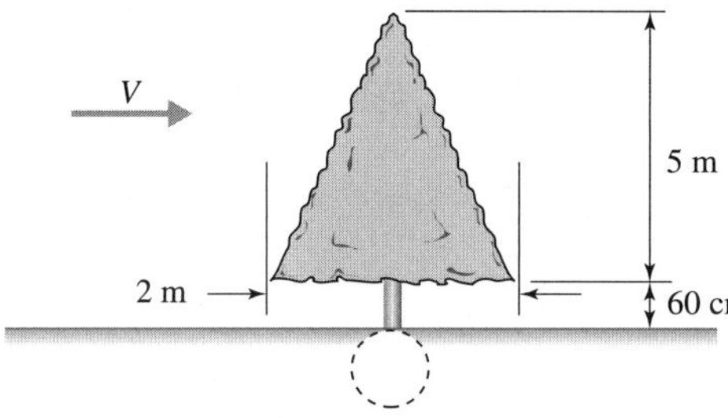

FIGURE P16.31

16.32 A sign 1.2 m × 0.6 m is attached to the top of a pizza delivery car. The car travels 10 hours a day, 6 days a week. Estimate the added cost in a year that the sign adds to the fuel used by the car. The average speed of the car is 40 km/h, fuel costs $0.30 per liter, the engine/drive train is 30% efficient, and the fuel contains 12 000 kJ/kg.

16.33 A bike rider is able to travel at an average speed of 25 mph while riding upright. It is determined that the rider's projected area is 0.56 m^2. If the rider assumes a racing position so that his projected area is 0.40 m^2, estimate his increased speed if his drag coefficient decreases 20%, assuming the same energy expenditure.

16.34 An automobile with a cross-sectional area of 3 m^2 is powered by a 40-hp engine. Estimate the maximum possible speed if the drive train is 90% efficient. (The engine is rated by the power produced before the transmission.)

Vortex Shedding

16.35 Over what range of velocities would you expect vortex shedding from a telephone wire 3 mm in diameter? Could you hear any of the vorticies being shed? (Human beings can hear frequencies between 20 and 20 000 Hz.)

16.36 A wire is being towed through 60°F water normal to its axis at a speed of 6 ft/sec. What diameter (both large and small) could the wire have so that vortex shedding would not occur?

16.37 It is quite difficult to measure low velocities. To determine the velocity in a low-speed airflow, the vorticies being shed from a 10-cm-diameter cylinder are observed to occur at 0.2 Hz. Estimate the airspeed if the temperature is 20°C.

16.38 Motion pictures show that vortices are shed from a 2-m-diameter cylinder at 0.002 Hz while it is moving in 20°C water. What is the cylinder velocity?

Streamlining

16.39 A 6-in.-diameter smokestack on a semitruck extends 6 ft straight up into the free stream. Estimate the horsepower needed because of the stack for a speed of 60 mph. If the stack were streamlined, estimate the reduced horsepower.

16.40 A wind speed of 3 m/s blows normal to an 8-cm-diameter smooth cylinder that is 2 m long. Calculate the drag force. The cylinder is now streamlined. What is the percentage reduction in the drag? Assume that $T = 20°C$.

16.41 Water flows by an 80-cm-diameter cylinder that protrudes 2 m up from the bottom of a river. For an average water speed of 2 m/s, estimate the drag on the cylinder. If the cylinder were streamlined, what would be the percentage reduction in drag?

16.42 Circular 2-cm-diameter tubes are used as supports on an ultralight airplane, designed to fly at 50 km/hr. If there are 20 linear meters of the tubes, estimate the horsepower needed by the tubes. If the tubes were streamlined, estimate the reduced horsepower required of the tubes.

16.43 A bike racer is able to ride 50 km/hr at top speed. Estimate the drag force just due to his head. If he wore a streamlined, tight-fitting helmet, estimate the reduced drag force.

Cavitation

16.44 The critical cavitation number for a streamlined strut is 0.7. Find the maximum velocity of the body to which the strut is attached if cavitation is to be avoided. The body is traveling 5 m beneath a water surface.

16.45 A lift force of 200 kN is desired at a speed of 12 m/s on the hydrofoil of Fig. P16.45, designed to operate at a depth of 40 cm. It has a 40-cm chord and is 10 m long. Calculate the angle of attack and the drag force. Is cavitation present under these conditions?

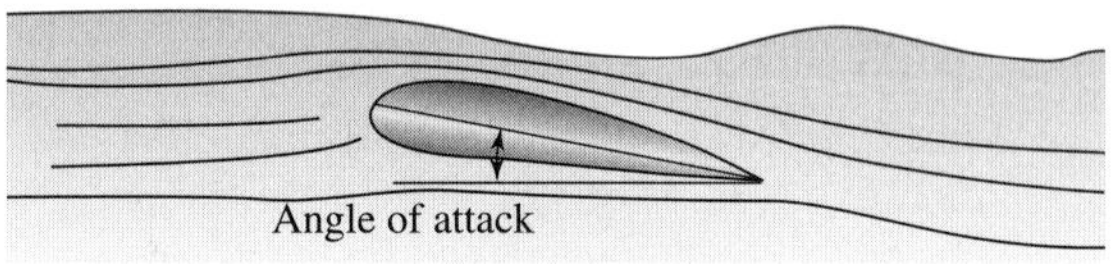

FIGURE P16.45

16.46 A hydrofoil, designed to operate at a depth of 16 in., has a 16-in. chord and is 30 ft long. A lift force of 50,000 lb is desired at a speed of 35 ft/sec. Calculate the angle of attack and the drag force. Is cavitation present under these conditions?

16.47 A body resembling a sphere has a diameter of approximately 0.8 m. It is towed at a speed of 20 m/s, 5 m below the water surface. Estimate the drag acting on the body.

16.48 A 2200-kg mine sweeper is designed to ride above the water with hydrofoils at the four corners providing the lift. If the hydrofoils have a 40-cm chord length, what total hydrofoil length is required if they are to operate 60 cm below the surface at a 6° angle of attack? The vehicle is to travel at 50 m/s.

Added Mass

16.49 A 40-cm-diameter sphere, which weighs 400 N, is released from rest while submerged in water. Calculate its initial acceleration:

(a) Neglecting the added mass.

(b) Including the added mass.

16.50 A submersible, whose length is twice its maximum diameter, resembles an ellipsoid. If its added mass is ignored, what is the percentage error in a calculation of its initial acceleration if its specific gravity is 1.2?

Lift and Drag on Airfoils

16.51 An aircraft with a mass, including payload, of 1000 kg is designed to cruise at a speed of 80 m/s at 10 km. The effective wing area is approximately 15 m^2. Determine the lift coefficient and the angle of attack. What power is required by the airfoil during cruise? Assume a conventional airfoil.

16.52 A 1500-kg aircraft is designed to carry a payload of 3000 N when cruising at 80 m/s at 10 km. The effective wing area is 20 m^2. Assuming a conventional airfoil, calculate:
(a) The takeoff speed if an angle of attack of 10° is desired.
(b) The stall speed when landing.
(c) The power required at cruise if 45% of the power is needed to move the airfoil.

16.53 In Problem 16.52 it was necessary to assume that takeoff was at sea level. What would be the takeoff speed in Wyoming, where the elevation is 2000 m?

16.54 The aircraft of Problem 16.52 is flown at 2 km rather than 10 km. Estimate the percentage increase or decrease in power required at cruise.

16.55 An additional load of 6000 N is added to the aircraft of Problem 16.52. Estimate the takeoff speed if the angle of attack remains at 10°.

16.56 Estimate the minimum landing speed for a 250 000-kg aircraft in an emergency situation if the angle of attack is selected to be near stall and:
(a) No slotted flaps are used.
(b) One slotted flap is used.
(c) Two slotted flaps are used.
Its wingspan is 60 m and the airfoil chord averages 8 m.

16.57 In Problem 16.56 it is assumed that the aircraft lands at standard sea-level conditions since neither elevation nor temperature is given. Calculate the percentage increase or decrease in the emergency landing speed if the aircraft is to land:
(a) In Denver where the elevation is 1600 m.
(b) At sea level when the temperature is very cold at −40°C.
(c) At sea level when the temperature is hot at 50°C.

16.58 A proposed aircraft is to resemble a huge airfoil, a flying wing (Fig. P16.58). Its wingspan will be 200 m and its chord will average 30 m. Estimate, assuming a conventional airfoil, the total mass of the aircraft, including payload, for a design speed of 800 km/h at an elevation of 8 km. Also, calculate the power requirement.

FIGURE P16.58

Vorticity, Velocity Potential, and Stream Function

16.59 Determine which of the following flows are irrotational and incompressible and find the velocity potential function, should one exist for each incompressible flow.
(a) $\mathbf{V} = 10x\hat{\mathbf{i}} + 20y\hat{\mathbf{j}}$
(b) $\mathbf{V} = 8y\hat{\mathbf{i}} + 8x\hat{\mathbf{j}} - 6z\hat{\mathbf{k}}$
(c) $\mathbf{V} = (x\hat{\mathbf{i}} + y\hat{\mathbf{j}})/\sqrt{x^2 + y^2}$
(d) $\mathbf{V} = (x\hat{\mathbf{i}} + y\hat{\mathbf{j}})/(x^2 + y^2)$

16.60 State the stream function and velocity potential corresponding to a uniform velocity of $100\hat{\mathbf{i}} + 50\hat{\mathbf{j}}$ using rectangular coordinates.

16.61 A flow is represented by the stream function $\psi = 40\tan^{-1}(y/x)$.
(a) Express the stream function in polar form.
(b) Is this an incompressible flow? Show why.
(c) Determine the velocity potential.
(d) Find the radius where the acceleration is -10 m/s².

16.62 The stream function for a flow is $\psi = 20\ln(x^2 + y^2)$ m²/s. Determine the velocity potential for this incompressible flow. If the pressure at a large distance from the origin is 20 kPa, what is the pressure at the point (0, 20 cm) if water is flowing?

16.63 A stream function is given by

$$\psi = 10y - \frac{10y}{x^2 + y^2}$$

(a) Find the velocity potential $\phi(x, y)$.
(b) Assuming water to be flowing, find the pressure along the x-axis if $p = 50$ kPa at $x = -\infty$.
(c) Locate any stagnation points.

16.64 The velocity potential for a flow is

$$\phi = 10x + 5 \ln(x^2 + y^2)$$

(a) Find the stream function $\psi(x, y)$.
(b) Assume that water is flowing and find the pressure along the x-axis if $p = 100$ kPa at $x = -\infty$.
(c) Locate any stagnation points.
(d) Find the acceleration at $x = -2$ m, $y = 0$.

16.65 The velocity profile in a wide 0.2-m-high channel is given by $u(y) = y - y^2/0.2$. Determine the stream function for this flow. Calculate the flow rate by integrating the velocity profile and by using $\Delta\psi$. Explain why a velocity potential does not exist by referring to Eq. 16.10.

Superposition of Simple Flows

16.66 The body formed by superimposing a source at the origin of strength 5π ft^2/sec and a uniform flow of 30 ft/sec is shown in Fig. P16.66.
(a) Locate any stagnation points.
(b) Find the y-intercept y_B of the body.
(c) Find the thickness of the body at $x = \infty$.
(d) Find u at $x = -12$ in., $y = 0$.

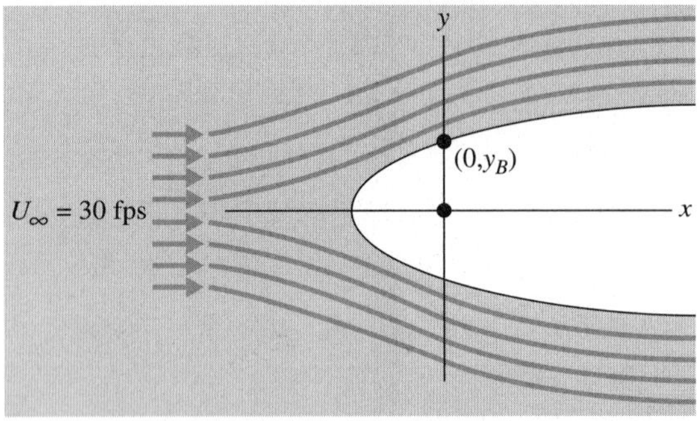

FIGURE P16.66

16.67 A source with strength π m^2/s and a sink of equal strength are located at $(-1$ m, $0)$ and $(1$ m, $0)$, respectively. They are combined with a uniform flow $U_\infty = 10$ m/s to form a *Rankine oval*. Calculate the length and maximum thickness of the oval. If $p = 10$ kPa at $x = -\infty$, find the minimum pressure if water is flowing.

16.68 An oval is formed from a source and sink of strengths 2π m^2/s located at $(-1, 0)$ and $(1, 0)$, respectively, combined with a uniform flow of 2 m/s. Locate any stagnation points, and find the velocity at $(-4, 0)$ and $(0, 4)$. Distances are in meters.

16.69 Two sources of strength 2π m^2/s are located at $(0, 1)$ and $(0, -1)$, respectively. Sketch the resulting flow and locate any stagnation points. Find the velocity at $(1, 1)$. Distances are in meters.

16.70 The two sources of Problem 16.69 are superimposed with a uniform flow. Sketch the flow, locate any stagnation points, and find the y-intercept of the body formed if:
(a) $U_\infty = 10$ m/s.
(b) $U_\infty = 1$ m/s.
(c) $U_\infty = 0.2$ m/s.

16.71 A doublet of strength 60 m^3/s is superimposed with a uniform flow of 8 m/s of water. Calculate:
(a) The radius of the resulting cylinder.
(b) The pressure increase from $x = -\infty$ to the stagnation point.
(c) The velocity $v_\theta(\theta)$ on the cylinder.
(d) The pressure decrease from the stagnation point to the point of minimum pressure on the cylinder.

16.72 A sink of strength 4π m^2/s is superimposed with a vortex of strength 20π m^2/s.
(a) Sketch a pathline of a particle initially occupying the point $(x = 0,\ y = 1$ m$)$. Use tangents at every 45° with straight-line extensions.
(b) Calculate the acceleration at $(0, 1)$.
(c) If $p(10, 10) = 20$ kPa, what is $p(0, 0.1)$ if atmospheric air is flowing?

16.73 The cylinder shown in Fig. P16.73 is formed by combining a doublet of strength 40 m^3/s with a uniform flow of 10 m/s.

(a) Sketch the velocity along the y-axis from the cylinder to $y = \infty$.

(b) Calculate the velocity at ($x = -4$ m, $y = 3$ m).

(c) Calculate the drag coefficient for the cylinder assuming potential flow over the front half and constant pressure over the back half equal to the value at $\theta = 90°$.

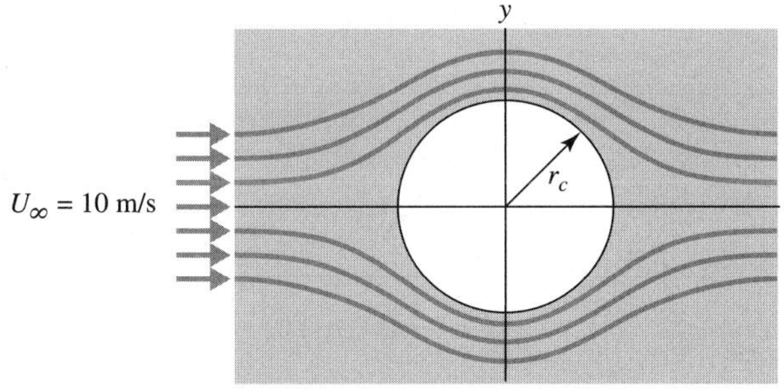

FIGURE P16.73

16.74 A 2-m-diameter cylinder is placed in a uniform water flow of 4 m/s.

(a) Sketch the velocity along the x-axis from the cylinder to $x = -\infty$.

(b) Find v_θ on the front half of the cylinder.

(c) Find $p(\theta)$ on the front half of the cylinder if $p = 50$ kPa at $x = -\infty$.

(d) Estimate the drag force on a 1-m length of the cylinder if the pressure over the rear half is constant and equal to the value at $\theta = 90°$.

16.75 Superimpose a free stream $U_\infty = 30$ ft/sec, a doublet $\mu = 400$ ft^3/sec, and a vortex $\Gamma = 1000$ ft^2/sec. Locate any stagnation points and predict the minimum and maximum pressure on the surface of the cylinder if $p = 0$ at $x = -\infty$ and atmospheric air is flowing.

16.76 A 0.8-m-diameter cylinder is placed in a 20-m/s atmospheric airflow. At what rotational speed should the cylinder rotate so that only one stagnation point exists on its surface? Calculate the minimum pressure acting on the cylinder if $p = 0$ at $x = -\infty$.

16.77 A 1.2-m-diameter cylinder rotates at 120 rpm in a 3-m/s atmospheric airstream. Locate any stagnation points and calculate the minimum and maximum pressures on the cylinder if $p = 0$ at $x = -\infty$.

16.78 The circulation around a 60-ft-long airfoil (measured tip to tip) is calculated to have a value of 15,000 ft^2/sec. Estimate the lift generated by the airfoil if the aircraft is flying at 30,000 ft with a speed of 350 ft/sec. Assume the flow to be incompressible.

16.79 The velocity field due to a source of strength 2π m^2/s per meter, located at (2 m, 2 m) in a 90° corner, is desired. Use the *method of images*, that is, add one or more sources at the appropriate locations, and determine the velocity field by finding $u(x, y)$ and $v(x, y)$.

Boundary Layers

16.80 How far from the leading edge can turbulence be expected on an airfoil traveling at 300 ft/sec if the elevation is:
(a) 0 ft? **(b)** 12 000 ft? **(c)** 30 000 ft?
Use $Re_{crit} = 6 \times 10^5$ and assume a flat plate with zero pressure gradient.

16.81 The boundary layer on a flat plate with zero pressure gradient is to be studied in a wind tunnel. How far from the leading edge would you expect to find turbulent flow if $U_\infty = 10$ m/s and:

(a) The plate is held rigid with high free-stream disturbance level?

(b) The plate is held rigid with low free-stream disturbance level?

(c) The plate is vibrated with low free-stream disturbance level?

(d) The plate is vibrated with high free-stream disturbance level?

(e) At what distance would you expect a small disturbance to grow for the flow of part (b)?

16.82 Repeat Problem 16.81 but place the flat plate in a water channel.

16.83 A laminar region is desired to be at least 2 m long on a smooth rigid flat plate. A wind tunnel and a water channel are available. What maximum speed can be selected for each? Assume low free-stream fluctuation intensity.

16.84 Determine the pressure $p(x)$ in the boundary layer and the velocity $U(x)$ at the edge of the boundary layer that would be present on the front of the cylinder of Problem 16.73. Let the pressure at the stagnation point be 20 kPa with water flowing. Let x be measured from the stagnation point; see Figure 16.19.

16.85 A boundary layer would develop as shown in Fig. P16.85 from the front stagnation point of Problem 16.74. Determine $U(x)$ and $p(x)$ that would be needed to solve for the boundary layer growth on the cylinder front. Measure x from the stagnation point.

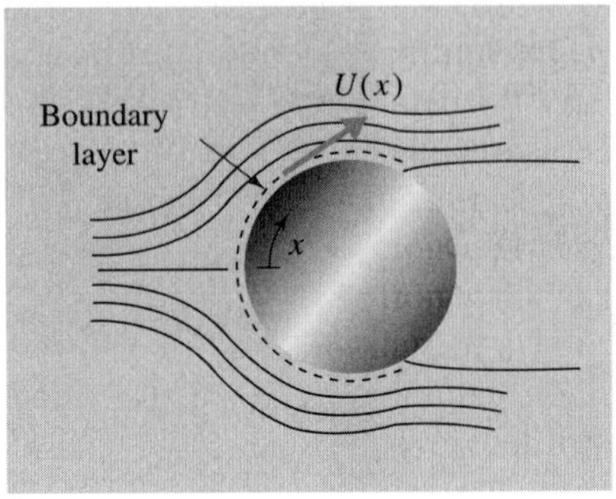

FIGURE P16.85

16.86 Assuming inviscid uniform flow of air through the contraction shown in Figure P16.86, estimate $U(x)$ and dp/dx, which are necessary to solve for the boundary layer growth on the flat plate. Assume a one-dimensional flow with $\rho = 1.0$ kg/m^3.

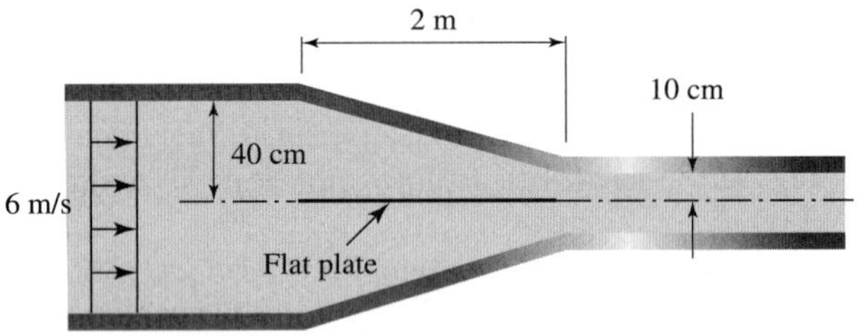

FIGURE P16.86

Von Kármán Integral Equation

16.87 Provide the detailed steps to the final form of Eq. 16.34 and Eq. 16.36. Refer to Figure 16.25.

16.88 Show that the von Kármán integral Equation 16.37 can be put in the form

$$\tau_0 = -\delta\frac{dp}{dx} + \rho\frac{d}{dx}\int_0^\delta u(U-u)\,dy - \rho\frac{dU}{dx}\int_0^\delta u\,dy$$

Note that the quantity $\int_0^\delta u\,dy$ is only a function of x.

16.89 Show that the von Kármán integral equation of Problem 16.88 can be written as

$$\tau_0 = \rho\frac{d}{dx}(\theta U^2) + \rho\delta_d U\frac{dU}{dx}$$

To accomplish this, we must show that Bernoulli's equation $p + \rho U^2/2 = \text{const.}$ can be differentiated to yield

$$\frac{dp}{dx} = -\rho U\frac{dU}{dx} = -\frac{\rho}{\delta}\frac{dU}{dx}\int_0^\delta U\,dy$$

16.90 Assume that $u = U_\infty \sin(\pi y/2\delta)$ in a zero pressure gradient boundary layer. Calculate:
(a) $\delta(x)$.
(b) $\tau_0(x)$.
(c) v at $y = \delta$ and $x = 3$ m.

16.91 Assume a linear velocity profile and find $\delta(x)$ and $\tau_0(x)$. Compute the percentage error when compared with the exact expressions for a laminar flow. Use $dp/dx = 0$.

16.92 A boundary layer profile is approximated with:

$$u = 3U_\infty \frac{y}{\delta} \qquad 0 < y \le \delta/6$$

$$u = U_\infty\left(\frac{y}{\delta} + \frac{1}{3}\right) \qquad \delta/6 < y \le \delta/2$$

$$u = U_\infty\left(\frac{y}{3\delta} + \frac{2}{3}\right) \qquad \delta/2 < y \le \delta$$

Determine $\delta(x)$ and $\tau_0(x)$; compute the percentage error when compared with the exact expressions for a laminar flow.

16.93 If the walls in a wind-tunnel test section are parallel, the velocity in the center portion of the tunnel will accelerate as shown in Fig. P16.93. To maintain a constant tunnel velocity so that $dp/dx = 0$, show that the walls should be displaced outward a distance $\delta_d(x)$. If a wind tunnel were square, how far should one wall be displaced outward for $dp/dx = 0$?

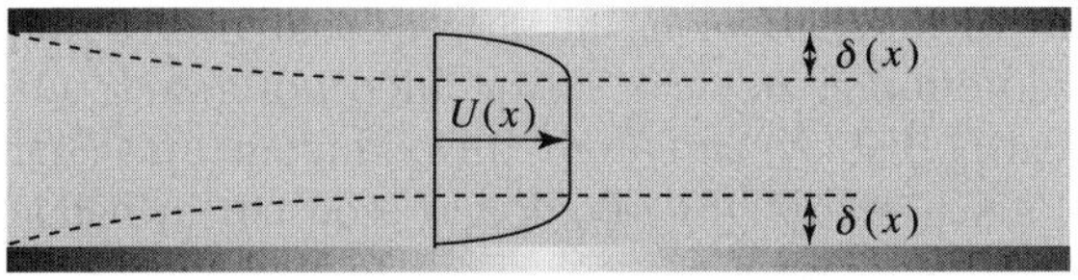

FIGURE P16.93

16.94 The velocity profile at a given x-location in the boundary layer (Fig. P16.94) is assumed to be

$$u(y) = 10\left(2\frac{y}{\delta} - \frac{y^2}{\delta^2}\right)$$

A streamline is 2 cm from the flat plate at the leading edge. How far is it from the plate when $x = 3$ m (i.e., what is h)? Also, calculate the displacement thickness at $x = 3$ m. Compare the displacement thickness to $(h - 2)$ cm.

16.95 It is desired that the test section in a wind tunnel experience a zero pressure gradient. If the test section has a square cross section, what should be the equation of the displacement of one of the walls (three walls will be straight and parallel or perpendicular) if the 30°C–air is pressurized to 160 kPa absolute? Assume a boundary layer profile of $u/U_\infty = 2y/\delta - y^2/\delta^2$. Assume that $y = 0$ at $x = 0$ of the displacement equation $y(x)$.

16.96 Find δ_d and θ for a laminar boundary layer assuming:
(a) A cubic profile.
(b) A parabolic profile.
(c) That $u = U_\infty \sin(\pi y/2\delta)$.
Compute percentage errors when compared with the exact values of $\delta_d = 1.72\sqrt{\nu x/U_\infty}$ and $\theta = 0.644\sqrt{\nu x/U_\infty}$.

16.97 A laminar flow is maintained in a boundary layer on a 20-ft-long 15-ft-wide flat plate with 60°F–atmospheric air flowing at 12 ft/sec. Assuming parabolic profile, calculate:
(a) δ at $x = 20$ ft.
(b) τ_0 at $x = 20$ ft.
(c) The drag force on one side.
(d) v at $y = \delta$ and $x = 10$ ft.

16.98 Work Problem 16.97, but assume a cubic profile.

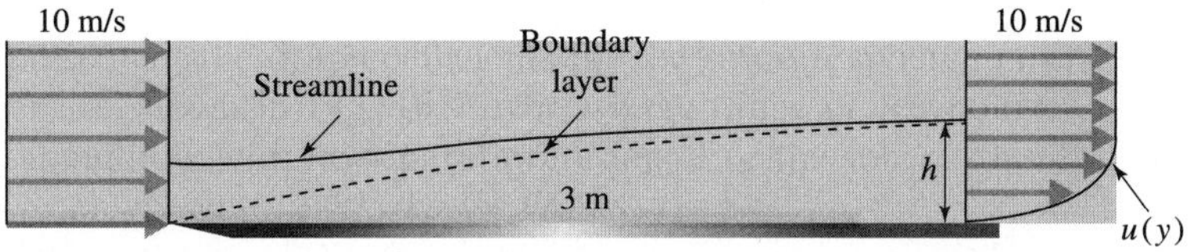

FIGURE P16.94

Laminar and Turbulent Boundary Layers

16.99 Atmospheric air at 20°C flows at 10 m/s over a 2-m-long 4-m-wide flat plate. Calculate the maximum boundary layer thickness and the drag force on one side assuming:
(a) Laminar flow over the entire length.
(b) Turbulent flow over the entire length.

16.100 Fluid flows over a flat plate at 20 m/s. Determine δ and τ_0 at $x = 6$ m if the fluid is:
(a) Atmospheric air at 20°C.
(b) Water at 20°C.
Neglect the laminar portion.

16.101 Assume a turbulent velocity profile $\bar{u} = U_\infty (y/\delta)^{1/7}$. Does this profile satisfy the conditions at $y = \delta$? Can it give the shear stress at the wall? Plot both a cubic laminar profile and the one-seventh power law profile on the same graph assuming the same boundary layer thickness.

16.102 Estimate the drag on one side of a 12-ft-long 51-ft-wide flat plate if atmospheric air at 60°F is flowing with a velocity of 20 ft/sec. Assume that:
(a) $\text{Re}_{\text{crit}} = 3 \times 10^5$ **(b)** $\text{Re}_{\text{crit}} = 5 \times 10^5$
(c) $\text{Re}_{\text{crit}} = 6 \times 10^5$

16.103 A 1-m-long flat plate with a sharp leading edge that is 2 m wide is towed parallel to itself in 20°C water at 1.2 m/s. Estimate the total drag if:
(a) $\text{Re}_{\text{crit}} = 3 \times 10^5$ **(b)** $\text{Re}_{\text{crit}} = 6 \times 10^5$
(c) $\text{Re}_{\text{crit}} = 9 \times 10^5$

16.104 Air moving at 60 km/h is considered to have a zero boundary layer thickness at a distance of 100 km from shore. At the beach, estimate the boundary layer thickness and the wall shear using:
(a) A one-seventh power law.
(b) Empirical data.
Use $T = 20$°C.

16.105 For the conditions of Problem 16.104, at the beach calculate:
(a) The thickness of the viscous wall layer.
(b) The displacement thickness.

16.106 Atmospheric air at 60°F flows over a flat plate at 300 ft/sec. At $x = 20$ ft, estimate:
(a) The local skin friction coefficient.
(b) The wall shear.
(c) The viscous wall layer thickness.
(d) The boundary layer thickness.

16.107 Water at 20°C flows over a flat plate at 10 m/s. At $x = 3$ m, estimate:
(a) The viscous wall layer thickness.
(b) The velocity at the edge of the viscous wall layer.
(c) The value of y at the outer edge of the turbulent zone.
(d) The boundary layer thickness.

16.108 Estimate the total shear drag on a ship traveling at 10 m/s if the sides are flat plates 10 m × 100 m with zero pressure gradients. What is the maximum boundary layer thickness?

16.109 The large 600-m-long, 100-m-diameter, cigar-shaped dirigible sketched in Fig. P16.109 is planned for cruises for retired professors. As a first estimate the drag is calculated assuming it to be a flat plate with zero pressure gradient, with the drag on the nose and rear areas neglected.
(a) Estimate the power required of each of four engines if it is to cruise at 15 m/s.
(b) Estimate the payload if one half of its volume is filled with helium, and its engines, equipment, and structure have a mass of 1.2×10^6 kg.

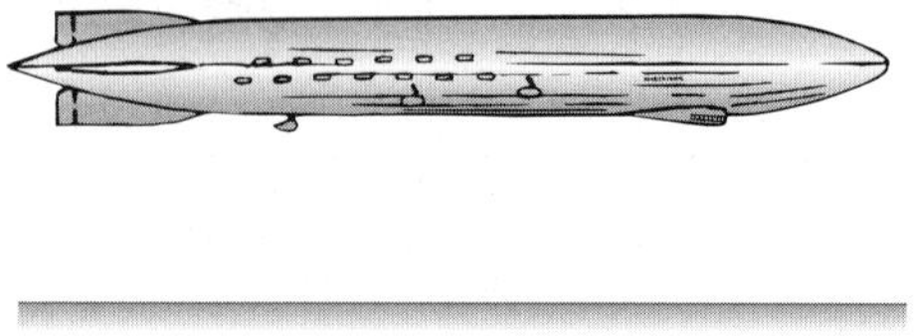

FIGURE P16.109

Convection Heat Transfer

16.110 Consider atmospheric air at 270 K in parallel flow at 15 m/s over a flat roof 20 m long that is maintained at a uniform temperature of 300 K. Using the properties of air at 285 K, determine the average heat flux from the roof to the air. (At 285 K, the properties of air are $\nu = 1.4555 \times 10^{-5}$ m^2/s, $k = 0.0251$ W/m·K, and Pr $= 0.711$.)

16.111 Water at 300 K flows over a flat heated surface, 2 m long, at 5 m/s. The heated surface is maintained at a uniform temperature of 360 K. Using the properties of water at 330 K, determine the average heat flux from the surface to the water. (At 330 K, the properties of water are $\nu = 4.968 \times 10^{-7}$ m^2/s, $k = 0.65$ W/m·K, and Pr $= 3.15$.)

16.112 Repeat Problem 16.111, assuming the fluid is air rather than water. Comment on the differences between your answers. (At 330 K, the properties of air are $\nu = 1.8878 \times 10^{-5}$ m^2/s, $k = 0.0285$ W/m·K, and Pr $= 0.703$.)

16.113 Consider atmospheric air at 270 K in parallel flow at 15 m/s over a flat roof 20 m long that is maintained at a uniform temperature of 300 K. Using the properties of air at 285 K, determine the average heat flux from the roof to the air. (At 285 K, the properties of air are $\nu = 1.4555 \times 10^{-5}$ m^2/s, $k = 0.0251$ W/m·K, and Pr $= 0.711$.)

16.114 Atmospheric air at 300 K flows at 20 m/s past a circular pipe with a outer diameter of 5 cm. The pipe is maintained at a uniform temperature of 400 K. Using the properties of air at 300 K, determine the average heat flux from the pipe to the air. (At 300 K, the properties of air are $\nu = 1.589 \times 10^{-5}$ m^2/s, $k = 0.0263$ W/m·K, and Pr $= 0.707$.) If the airflow is reduced by 50% to 10 m/s, what percent change is there in the average heat flux?

16.115 Air at 300 K flows past a long heated pipe with a diameter of 10 cm and maintained at 350 K. If the average heat flux from the pipe to the air is 10 000 W/m^2, what is the velocity of the air? (Use the properties of air at 300 K: $\nu = 1.589 \times 10^{-5}$ m^2/s, $k = 0.0263$ W/m·K, and Pr $= 0.707$.)

Pressure Gradient Effects

16.116 Sketch the velocity profiles near and normal to the cylinder's surface at each of the points indicated in Fig. P16.116. The flow separates at *C*.

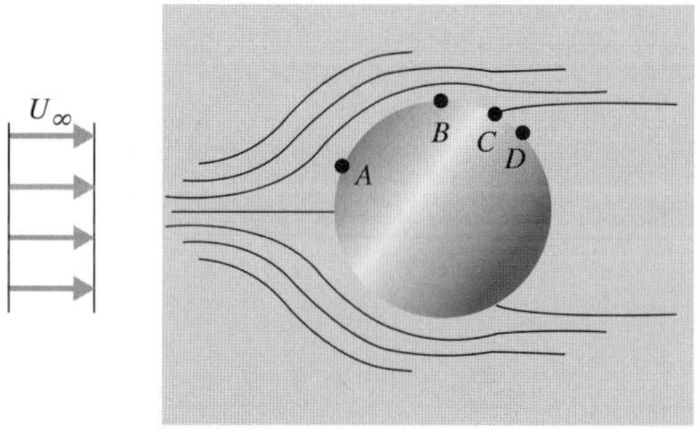

FIGURE P16.116

16.117 Sketch the expected boundary layer profiles at each of the points indicated in Fig. P16.117, showing relative thicknesses. The flow undergoes transition to turbulence just after point *A*. It separates at *D*. Show all profiles on the same plot. Indicate the sign of the pressure gradient at each point.

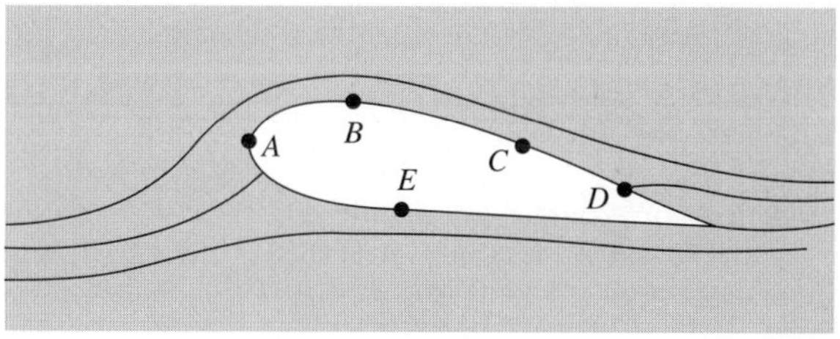

FIGURE P16.117

A new-generation space vehicle is rendered by an artist as it re-enters the earth's atmosphere. A shock wave is shown to be generated at the leading edge of the vessel, signifying that it is traveling faster than the speed of sound. (U.S. National Aeronautics and Space Administration)

17

Compressible Flow

Outline

Chapter Objectives

The objectives of this chapter are to:

- ▲ Present the various equations needed to solve the uniform-flow problems of a compressible gas flow.
- ▲ Apply the basic equations to isentropic flow through a nozzle.
- ▲ Introduce the normal shock wave.
- ▲ Solve for velocities and pressures when a shock wave exists in a converging–diverging nozzle.
- ▲ Study the oblique shock wave needed to turn a supersonic flow through an angle.
- ▲ Calculate the angle through which expansion waves can turn a flow around a convex corner.
- ▲ Provide numerous examples and problems that illustrate isentropic flow, flow through normal and oblique shock waves, and the turning of a supersonic flow around a convex corner.

Nomenclature

The following variables are introduced in this chapter:

A^*	area where M = 1	α	angle of the Mach cone
C_p	pressure recovery factor	β	angle of oblique shock wave
p^*	pressure where M = 1	η_N	nozzle efficiency
p_0	pressure where $V = 0$	μ	angle of Mach wave
p_r	receiver pressure	θ	wedge angle, Prandtl-Meyer function
s	specific entropy		
S	entropy		
T^*	temperature where M = 1		
T_0	temperature where $V = 0$		

17.1 INTRODUCTION

Compressible flow: *The flow of gas in which the density changes significantly between points on a streamline.*

In this chapter we consider flows of gases in which the density changes significantly between points on a streamline; such flows are called **compressible flows**. We will consider those problems that can be solved using the integral equations: the continuity equation, the energy equation, and for some problems, the momentum equation.

Key Concept: *Not all gas flows are compressible flows.*

Not all gas flows are compressible flows, neither are all compressible flows gas flows. At low speeds, less than a Mach number ($M = V/\sqrt{kRT}$) of about 0.3, gas flows may be treated as incompressible flows. This is justified because the density variations caused by the flow are negligible (less than about 3%). Incompressible gas flows occur in a large number of situations of engineering interest; many of these have been considered in earlier chapters. There are many flows, however, in which the density variations must be accounted for. Included among these are airflows around commercial and military aircraft, airflow through jet engines, and the flow of a gas in compressors and turbines.

There are examples of compressible effects important in liquid flows; water hammer and compression waves due to underwater blasts are examples of compressible liquid flows. Compressibility of rock accounts for the propagation through the earth's surface of longitudinal waves due to an earthquake. In this chapter we are concerned with compressibility effects in gas flows.

Let us introduce the effects of compressibility into the simplest of flow situations. The velocity at a given streamwise location in a conduit is assumed to be uniform and hence does not vary normal to the flow direction. For this simple uniform flow we recall that the continuity equation takes the form (see Eq. 13.27)

$$\dot{m} = \rho_1 A_1 V_1 = \rho_2 A_2 V_2 \tag{17.1}$$

The momentum equation for a uniform, compressible flow in the x-direction only takes the form (see Eq. 13.69)

$$\sum F = \dot{m}(V_2 - V_1) \tag{17.2}$$

The energy equation, neglecting potential energy changes, is written (see Eqs. 13.51 and 13.52)

$$\frac{\dot{Q}-\dot{W}_S}{\dot{m}}=\frac{V_2^2-V_1^2}{2}+h_2-h_1 \tag{17.3}$$

where we have used $h=\tilde{u}+p/\rho$. Assuming an ideal gas with constant specific heats, the energy equation takes the form

$$\frac{\dot{Q}-\dot{W}_S}{\dot{m}}=\frac{V_2^2-V_1^2}{2}+c_p(T_2-T_1) \tag{17.4}$$

or

$$\frac{\dot{Q}-\dot{W}_S}{\dot{m}}=\frac{V_2^2-V_1^2}{2}+\frac{k}{k-1}\left(\frac{p_2}{\rho_2}-\frac{p_1}{\rho_1}\right) \tag{17.5}$$

where we have used the thermodynamic relations

$$h_2-h_1=c_p(T_2-T_1) \tag{17.6a}$$

$$c_p=R+c_v \tag{17.6b}$$

$$k=\frac{c_p}{c_v} \tag{17.6c}$$

and the ideal-gas law

$$p=\rho RT \tag{17.7}$$

Should we have an interest in calculating the entropy change between two sections, we will use the definition of entropy as

$$\Delta S=\int\frac{\delta Q}{T}\bigg|_{\text{reversible}} \tag{17.8}$$

where δQ represents the differential heat transfer. Using the first law, this becomes, for an ideal gas with constant specific heat,

$$\Delta S=mc_p\ln\frac{T_2}{T_1}-mR\ln\frac{p_2}{p_1} \tag{17.9}$$

If a process is adiabatic ($Q=0$) and reversible (no losses) Eq. 17.8 shows that the entropy change is zero (i.e., the flow is *isentropic*). If the flow is isentropic, the relationship above can be used, along with the ideal-gas law and Eqs. 17.6b and c, to show that

$$\frac{T_2}{T_1} = \left(\frac{p_2}{p_1}\right)^{(k-1)/k} \qquad \frac{p_2}{p_1} = \left(\frac{\rho_2}{\rho_1}\right)^k \tag{17.10}$$

Note: Temperatures and pressures must be measured on absolute scales.

Let us now introduce a parameter, the Mach number, that will be of special interest throughout our study of compressible flow.

17.2 SPEED OF SOUND AND THE MACH NUMBER

Sound wave: *It travels with a velocity c relative to a stationary observer.*

The speed of sound is the speed at which a pressure disturbance of small amplitude travels through a fluid. It is analogous to the small ripple, a gravity wave, that travels radially outward when a pebble is dropped into a pond. To determine the speed of sound consider a small pressure disturbance, called a **sound wave**, to be passing through a pipe, as shown in Fig. 17.1. It travels with a velocity c relative to a stationary observer as shown in part (a); the pressure, density, and temperature will change by the small amounts Δp, $\Delta\rho$, and ΔT, respectively. There will also be an induced velocity ΔV in the fluid immediately behind the sound wave. To simplify the problem we will create a steady flow by having the observer travel at the speed of the wave so that the sound wave appears to be stationary. The flow will then approach the wave from the right, with the speed of sound c, as shown in Fig. 17.1b. The flow properties will all change across the wave and the velocity in the downstream flow will be expressed as $c + \Delta V$, where ΔV is the small change in velocity. If the flow speed is reduced, the wave would move to the right, and if the flow speed is increased it would move to the left. For the flow speed equal to the speed of sound, $V = c$, the sound wave would be stationary, as shown.

Let us apply the continuity equation and the energy equation to a small control volume enclosing the sound wave, shown in Fig. 17.1c. The continuity equation (17.1) takes the form

$$\rho A c = (\rho + \Delta\rho)A(c + \Delta V) \tag{17.11}$$

This can be rewritten as

$$\rho \Delta V = -c\Delta\rho \tag{17.12}$$

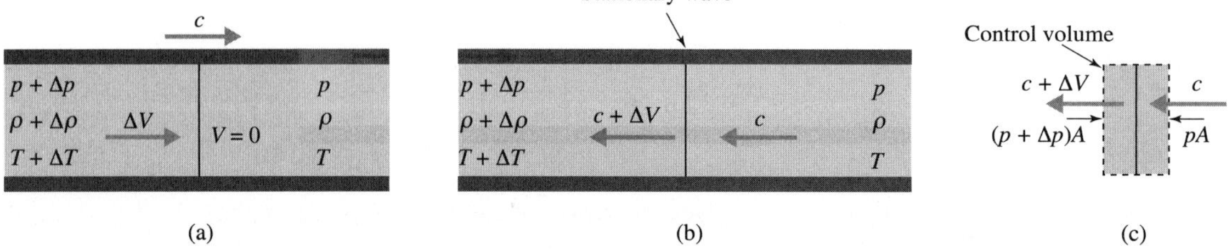

FIGURE 17.1 Sound wave: (a) stationary observer; (b) observer moving with the wave; (c) control volume enclosing the wave.

where we have neglected the higher-order term $\Delta\rho\,\Delta V$; that is, $\Delta\rho$ represents a small percentage change in ρ so that $\Delta\rho \ll \rho$.

The streamwise-component momentum equation, where only pressure forces exist on the small control volume, yields

$$pA - (p + \Delta p)A = \rho Ac(c + \Delta V - c) \tag{17.13}$$

which simplifies to, neglecting higher-order terms,

$$-\Delta p = \rho c\,\Delta V \tag{17.14}$$

Combining this with Eq. 17.12 results in

$$c = \sqrt{\frac{\Delta p}{\Delta\rho}} \tag{17.15}$$

Since the changes Δp and $\Delta\rho$ are quite small, we can write

$$\frac{\Delta p}{\Delta\rho} \simeq \frac{dp}{d\rho} \tag{17.16}$$

Small-amplitude, moderate-frequency waves (up to about 18 000 Hz) travel with no change in entropy (isentropically), so that

$$\frac{p}{\rho^k} = \text{const.} \tag{17.17}$$

This can be differentiated to give

$$\frac{dp}{d\rho} = k\frac{p}{\rho} \tag{17.18}$$

Using this in Eq. 17.15, the speed of sound c is given by

$$c = \sqrt{\frac{kp}{\rho}} \tag{17.19}$$

or, using the ideal-gas law,

$$c = \sqrt{kRT} \tag{17.20}$$

At high frequency, sound waves generate friction and the process ceases to be isentropic. It is better approximated by an isothermal process. For an ideal gas an isothermal approximation would lead to

$$c = \sqrt{RT} \tag{17.21}$$

For small waves traveling through liquids or solids the *bulk modulus* is used; it has dimensions of pressure and is equal to $\rho\,dp/d\rho$. For water it has a nominal value of 2110 MPa. It does vary slightly with temperature and pressure. Using Eq. 17.15, this leads to a speed of propagation of 1450 m/s for a small-amplitude pressure wave in water.

Key Concept: *A small-amplitude pressure wave in water travels 1450 m/s.*

An important quantity used in the study of compressible flow is the dimensionless velocity called the *Mach number*, introduced in Eq. 12.26 as

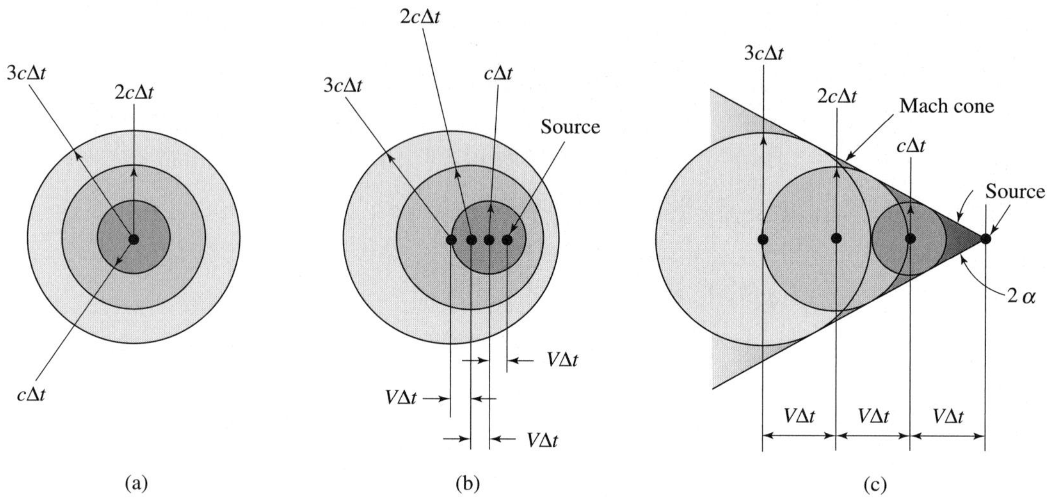

FIGURE 17.2 Sound waves propagating from a noise source: (a) stationary source; (b) moving source: $V < c$; (c) moving source, $V > c$.

$$\mathrm{M} = \frac{V}{c} \tag{17.22}$$

If $\mathrm{M} < 1$, the flow is a *subsonic* flow, and if $\mathrm{M} > 1$, it is a *supersonic* flow.

If a source of sound waves is at a fixed location, the waves travel radially away from the source with the speed of sound c. Figure 17.2a shows the position of the sound waves after a time increment Δt and after multiples of Δt. Part (b) shows a source moving with a velocity V which is less than the speed of sound. Note that the sound waves always propagate ahead of the source so that an airplane traveling at a speed less than the speed of sound will always "announce" its approach. This is not true, however, for an object traveling at a speed greater than the speed of sound, as shown in Fig. 17.2c. The region outside the cone is a *zone of silence*, so that an approaching object moving at a supersonic speed could not be heard until it passed overhead and the *Mach cone*, the cone shown, intercepted the observer. From the figure the angle α of the Mach cone is given by

Key Concept: *The approach of an object moving at $M > 1$ cannot be sensed.*

$$\alpha = \sin^{-1}\frac{c}{V} = \sin^{-1}\frac{1}{\mathrm{M}} \tag{17.23}$$

The discussion above is limited to small-amplitude sound waves, often called *Mach waves*. They are formed on the needle nose of an aircraft, or on the leading edge of an airfoil if the leading edge is sufficiently sharp. If the nose is blunt or if the leading edge is not sufficiently sharp, a supersonic aircraft will produce a large-amplitude wave called a **shock wave**. The shock wave will also form a zone of silence, but the initial angle at the source created by the shock wave will be larger than that of a Mach wave. Shock waves will be considered in subsequent sections.

Shock wave: *A large-amplitude wave which may be created by a blunt object.*

Example 17.1

A needle-nose projectile traveling at a speed with M = 3 passes 200 m above the observer of Fig. E17.1. Calculate the projectile's velocity and determine how far beyond the observer the projectile will first be heard.

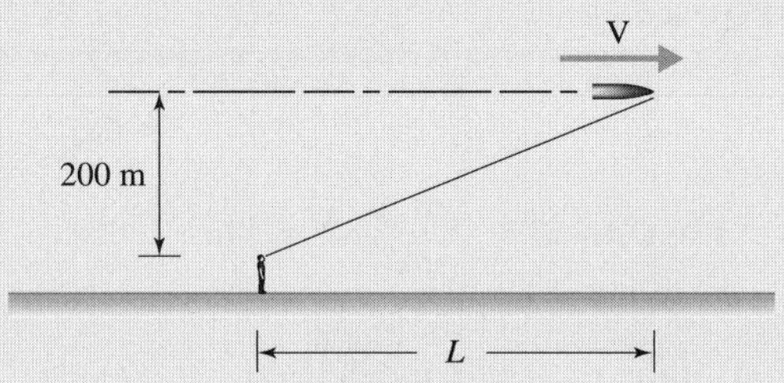

FIGURE E17.1

Solution
At a Mach number of 3 the velocity is

$$V = \mathrm{M}c = \mathrm{M}\sqrt{kRT}$$
$$= 3\sqrt{1.4 \times 287 \times 288} = 1021 \text{ m/s}$$

where a standard temperature of 15°C has been assumed since no temperature is given. Using h as the height and L as the distance beyond the observer (refer to Fig. 17.2c), we have

$$\sin \alpha = \frac{h}{\sqrt{L^2 + h^2}} = \frac{1}{\mathrm{M}}$$

With the information given,

$$\frac{200}{\sqrt{L^2 + 200^2}} = \frac{1}{3}$$

giving

$$L = 566 \text{ m}$$

Note: The units on kRT are $\frac{\text{N}\cdot\text{m}}{\text{kg}\cdot\text{K}} \times \text{K} = \frac{\text{N}\cdot\text{m}}{\text{N}\cdot\text{s}^2/\text{m}} = \frac{\text{m}^2}{\text{s}^2}$. The quantity k is dimensionless.

17.3 ISENTROPIC NOZZLE FLOW

There are many applications where gas flows through a section of tube or conduit that has a changing area in which a steady, uniform, isentropic flow is a good approximation to the actual flow situation. The diffuser near the front of a jet engine, exhaust gases passing through the blades of a turbine, the nozzles on a rocket engine, a broken natural gas line, and gas flow measuring devices are all

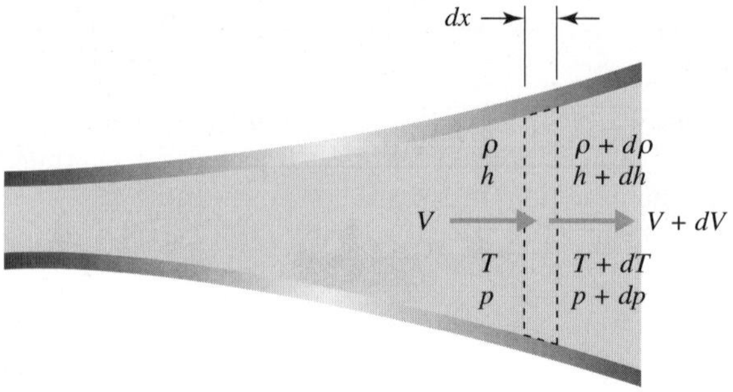

FIGURE 17.3 Uniform, isentropic flow.

Key Concept: *Steady, uniform, isentropic flow is a good approximation to many flow situations.*

examples of situations that can be modeled with a steady, uniform, isentropic flow. Consider the flow through the infinitesimal control volume shown in Fig. 17.3. With a changing area the continuity equation

$$\rho A V = \text{const.} \tag{17.24}$$

applied between two sections a distance dx apart takes the form

$$\rho A V = (\rho + d\rho)(A + dA)(V + dV) \tag{17.25}$$

Key Concept: *Retain only first-order terms in the differential quantities.*

Keeping only first-order terms in the differential quantities, Eq. 17.25 can be put in the form

$$\frac{dV}{V} + \frac{dA}{A} + \frac{d\rho}{\rho} = 0 \tag{17.26}$$

The energy equation can be written (see Eq. 17.5)

$$\frac{V^2}{2} + \frac{k}{k-1}\frac{p}{\rho} = \text{const.} \tag{17.27}$$

For the present application we have

$$\frac{V^2}{2} + \frac{k}{k-1}\frac{p}{\rho} = \frac{(V + dV)^2}{2} + \frac{k}{k-1}\frac{p + dp}{\rho + d\rho} \tag{17.28}$$

or again, retaining only first-order terms,

$$V\,dV + \frac{k}{k-1}\frac{\rho\,dp - p\,d\rho}{\rho^2} = 0 \tag{17.29}$$

For an isentropic process we use Eq. 17.18 and there results

$$V\,dV + k\frac{p}{\rho^2}d\rho = 0 \tag{17.30}$$

Substituting for $d\rho/\rho$ from Eq. 17.26, the above becomes

$$\frac{dV}{V}\left(\frac{\rho V^2}{kp} - 1\right) = \frac{dA}{A} \tag{17.31}$$

In terms of the speed of sound this is written as

$$\frac{dV}{V}\left(\frac{V^2}{c^2}-1\right)=\frac{dA}{A} \tag{17.32}$$

Introduce the Mach number and we have the very important relationship for an isentropic uniform flow in a changing area:

$$\frac{dV}{V}(\mathrm{M}^2-1)=\frac{dA}{A} \tag{17.33}$$

From Eq. 17.33 we can make the following observations:

1. If the area is increasing, $dA > 0$, and $\mathrm{M} < 1$, we see that dV must be negative, that is, $dV < 0$. The flow is decelerating for this subsonic flow.
2. If the area is increasing and $\mathrm{M} > 1$, we see that $dV > 0$; hence the flow is accelerating in the diverging section for this supersonic flow.
3. If the area is decreasing and $\mathrm{M} < 1$, then $dV > 0$, resulting in an accelerating flow.
4. If the area is decreasing and $\mathrm{M} > 1$, then $dV < 0$, indicating a decelerating flow.
5. At a throat where $dA = 0$, either $dV = 0$ or $\mathrm{M} = 1$, or possibly both.

If we define a **nozzle** as a device that accelerates the flow, we see that observations 2 and 3 describe a nozzle and observations 1 and 4 describe a **diffuser**, a device that decelerates the flow. The supersonic flow leads to rather surprising results: an accelerating flow in an enlarging area and a decelerating flow in a decreasing area. This is, in fact, the situation encountered in a traffic flow on a freeway; hence a supersonic flow might be used to model a traffic flow.

Nozzle: *A device that accelerates the flow.*

Diffuser: *A device that decelerates the flow.*

Note that the observations above prohibit a supersonic flow in a converging section attached to a reservoir. If a supersonic flow is to be generated by releasing a gas from a reservoir, there must be a converging section in which a subsonic flow accelerates to the throat where $\mathrm{M} = 1$ followed by a diverging section in which the flow continues to accelerate with $\mathrm{M} > 1$, as shown in Fig. 17.4. This is the type of nozzle observed on rockets used to place satellites into orbit.

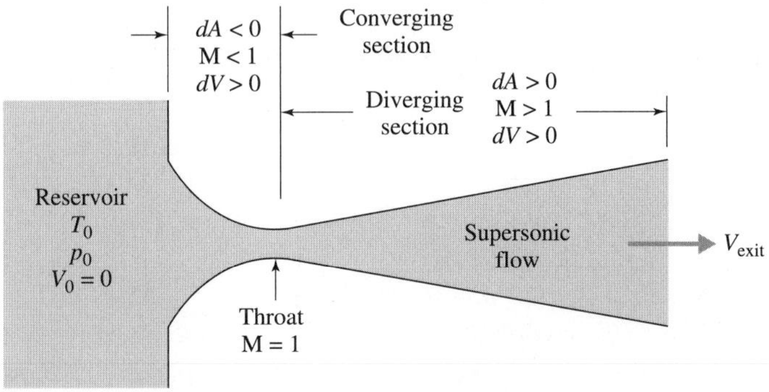

FIGURE 17.4 Supersonic nozzle.

The isentropic flow in the nozzle will now be considered in more detail. The energy equation between the reservoir where $V_0 = 0$ and any section can be written in the form

$$c_p T_0 = \frac{V^2}{2} + c_p T \tag{17.34}$$

Stagnation quantities: *Quantities with a zero subscript at a location where $V = 0$.*

Quantities with a zero subscript are often called **stagnation quantities** since they occur at a location where $V = 0$. Recognizing that $V = \mathrm{M}c$, $c_p/c_v = k$, $c_p = c_v + R$, and $c = \sqrt{kRT}$, we can write the energy equation as

$$\frac{T_0}{T} = 1 + \frac{k-1}{2}\mathrm{M}^2 \tag{17.35}$$

For our isentropic flow the pressure and density ratios are then (see Eqs. 17.10)

$$\begin{aligned} \frac{p_0}{p} &= \left(1 + \frac{k-1}{2}\mathrm{M}^2\right)^{k/(k-1)} \\ \frac{\rho_0}{\rho} &= \left(1 + \frac{k-1}{2}\mathrm{M}^2\right)^{1/(k-1)} \end{aligned} \tag{17.36}$$

If a supersonic flow occurs downstream of the throat, then $\mathrm{M} = 1$ at the throat and denoting this *critical area* with an asterisk (*) superscript, we have the following critical ratios:

$$\begin{aligned} \frac{T^*}{T_0} &= \frac{2}{k+1} \\ \frac{p^*}{p_0} &= \left(\frac{2}{k+1}\right)^{k/(k-1)} \\ \frac{\rho^*}{\rho_0} &= \left(\frac{2}{k+1}\right)^{1/(k-1)} \end{aligned} \tag{17.37}$$

Key Concept: *We reference the critical area even though an actual throat does not occur.*

We often make reference to the critical area even though an actual throat does not occur; we can imagine a throat occurring and call it the critical area. In fact, the isentropic flow table, Table I.1 in Appendix I, includes just such an area. For air with $k = 1.4$ the critical values are

$$p^* = 0.5283\, p_0 \qquad T^* = 0.8333\, T_0 \qquad \rho^* = 0.6340\, \rho_0 \tag{17.38}$$

We can determine an expression for the mass flux through the nozzle from the equation

$$\begin{aligned} \dot{m} &= \rho A V \\ &= \frac{p}{RT} A\mathrm{M}\sqrt{kRT} = \frac{p}{\sqrt{T}}\sqrt{\frac{k}{R}} A\mathrm{M} \end{aligned} \tag{17.39}$$

Using Eqs. 17.35 and 17.36, this can be expressed as

$$\dot{m} = \frac{p_0\left(1+\dfrac{k-1}{2}\mathrm{M}^2\right)^{k/(1-k)}}{\sqrt{T_0}\left(1+\dfrac{k-1}{2}\mathrm{M}^2\right)^{-1/2}}\sqrt{\frac{k}{R}}A\mathrm{M}$$

$$= p_0\sqrt{\frac{k}{RT_0}}\mathrm{M}A\left(1+\frac{k-1}{2}\mathrm{M}^2\right)^{(k+1)/2(1-k)} \tag{17.40}$$

If we choose the critical area where $\mathrm{M}^* = 1$, we see that

$$\dot{m} = p_0 A^*\sqrt{\frac{k}{RT_0}}\left(\frac{k+1}{2}\right)^{(k+1)/2(1-k)} \tag{17.41}$$

This shows that the mass flux in the nozzle is only dependent on the reservoir conditions and the critical area A^*. By combining Eqs. 17.40 and 17.41, the area ratio A/A^* can be written in terms of the Mach number as

$$\frac{A}{A^*} = \frac{1}{\mathrm{M}}\left[\frac{2+(k-1)\mathrm{M}^2}{k+1}\right]^{(k+1)/2(k-1)} \tag{17.42}$$

This ratio is included in Table I.1 for airflow.

As a final consideration in our study of isentropic nozzle flow, we will present the influence of reservoir pressure and receiver pressure on the mass flux. First, the converging nozzle will be presented; then the converging–diverging nozzle will follow. The converging nozzle is assumed to be attached to a reservoir, as shown in Fig. 17.5a, with fixed conditions; the pressure in the receiver can be lowered

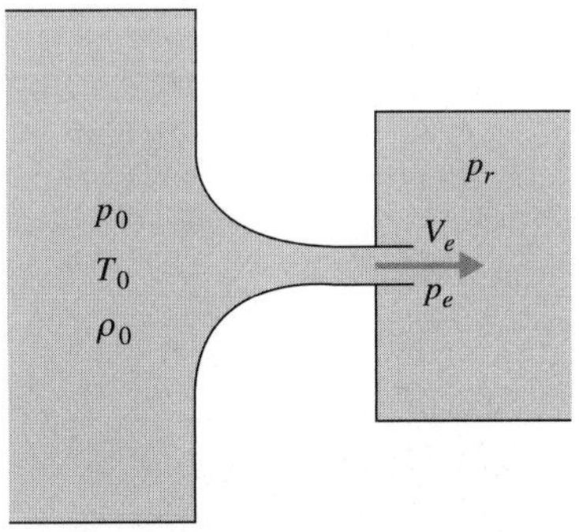

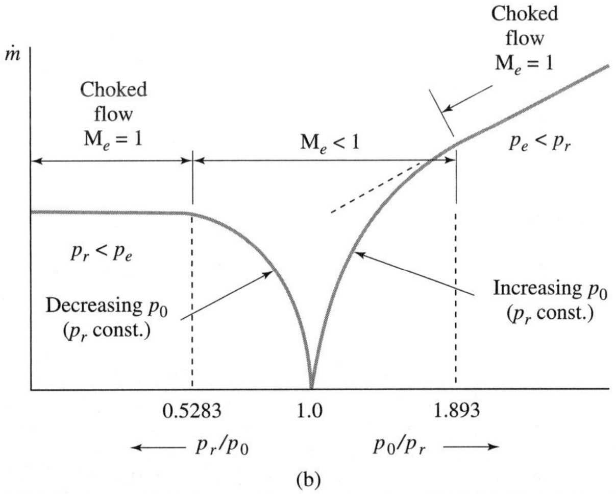

FIGURE 17.5 Converging nozzle.

to provide an increasing mass flux through the nozzle, as shown by the left curve in Fig. 17.5b. When the receiver pressure p_r reaches the critical pressure (for air $p_r = 0.5283\, p_0$) the Mach number M_e at the throat (the exit) is unity. As p_r is reduced below this critical value the mass flux will not increase, and the condition of *choked flow* occurs as shown by the left curve in Fig. 17.5b. If the throat is a critical area, that is, $M = 1$ at the throat, then $\dot{m}$ is only dependent on the throat area A^* and the reservoir conditions, as indicated by Eq. 17.41. Hence, reducing p_r below the critical pressure has no effect on the upstream flow. This is reasonable since disturbances travel at the speed of sound; if p_r is reduced the disturbances that would travel upstream thereby changing conditions cannot do so since the velocity of the stream at the exit is equal to the speed of sound thereby preventing any disturbances from propagating upstream.

Key Concept: *Reducing p_r below the critical pressure has no effect on the upstream flow.*

If, in the converging nozzle, we keep p_r constant and increase the reservoir pressure (keep T_0 constant also), a choked flow again occurs when $M_e = 1$; however, when p_0 is increased still further, we see from Eq. 17.41 that the mass flux will increase, as shown by the right curve of Fig. 17.5b. The pressure p_e is equal to p_r until the Mach number M_e is just equal to unity. This begins the condition of choked flow. The exit pressure p_e for the choked flow condition will be greater than the receiver pressure p_r, a condition that occurs when a gas line ruptures.

Key Concept: *If $p_e > p_r$ the flow exiting the nozzle is able to turn sharply.*

A note may be in order explaining how it is possible for the flow exit pressure p_e to exceed the receiver pressure p_r. If $p_e > p_r$, the flow exiting the nozzle is able to turn rather sharply, causing a flow pattern sketched in Fig. 17.6. This possible flow situation will be studied in Section 17.7.

Now, consider the converging–diverging nozzle, shown in Fig. 17.7, with reservoir and receiver as indicated. We will only present the condition of constant reservoir pressure and reduced receiver pressure. For this nozzle we sketch the

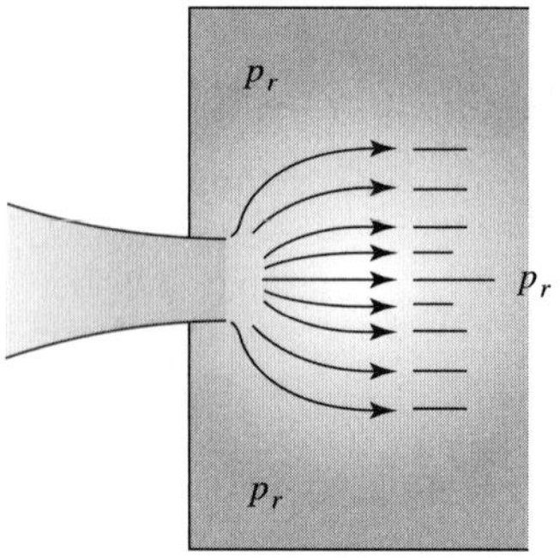

FIGURE 17.6 Nozzle exit flow for $p_e > p_r$.

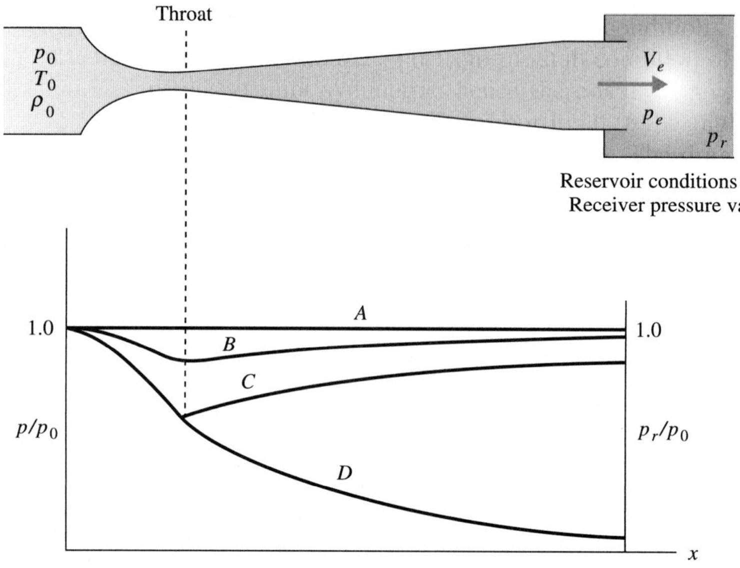

FIGURE 17.7 Converging–diverging nozzle.

pressure ratio p/p_0 as a function of location in the nozzle for various receiver pressure ratios p_r/p_0. If $p_r/p_0 = 1$, no flow occurs, corresponding to curve A. If p_r is reduced a small amount, curve B results and a subsonic flow exists throughout the nozzle with a minimum pressure occurring at the throat. As the pressure is reduced still further, a pressure is reached that will result in the Mach number at the throat just being unity, as sketched by curve C; the flow remains subsonic throughout, however. Another particular receiver pressure exists, considerably below that of curve C, that will also produce an isentropic flow throughout; it results in curve D. Any receiver pressure in between these two particular pressures will result in a nonisentropic flow in the nozzle; a shock wave, to be studied later, occurs which renders our assumption of isentropic flow invalid. If the receiver pressure is below that associated with curve D, we again find that the nozzle exit pressure p_e is greater than the receiver pressure p_r. The mass flux in the nozzle increases from that of curve A to that of curve C; but as the receiver pressure is reduced below that of curve C, no increase in mass flux can occur since the throat condition will remain unchanged. The mass flux is given by Eq. 17.41.

Final notes regard nozzle and diffuser effectiveness. The purpose of a nozzle is to convert enthalpy (which can be thought of as stored energy) into kinetic energy as described by Eq. 17.3 with $\dot{Q} = \dot{W}_S = 0$. The efficiency η_N of a nozzle is defined as

Key Concept: *The purpose of a nozzle is to convert stored energy into kinetic energy.*

$$\eta_N = \frac{(\Delta \mathrm{KE})_{\mathrm{actual}}}{(\Delta \mathrm{KE})_{\mathrm{isentropic}}} = \frac{h_0 - h_e}{h_0 - h_{es}} \qquad (17.43)$$

where h_e is the actual exit enthalpy and h_{es} is the isentropic exit enthalpy. Efficiencies are between 90 and 99% with larger nozzles having the higher percentages because the viscous wall effects that account for most of the losses are relatively small with the larger nozzles.

The purpose of a diffuser is to slow down the fluid and recover the pressure. For a diffuser we define the *pressure recovery factor* C_p to be

Key Concept: *The purpose of a diffuser is to recover the pressure.*

$$C_p = \frac{\Delta p_{\mathrm{actual}}}{\Delta p_{\mathrm{isentropic}}} \qquad (17.44)$$

Such factors vary between 40% when the flow actually separates from the wall (the included angle should be less than about 10° for the subsonic section to avoid this separation) to 85%. Vanes can be installed in wide-angled diffusers such that each passage between vanes expands at an angle of 10° or less. Viscous effects are significantly greater in the diffuser than in the nozzle because of the thicker viscous wall layers.

The reader may think that the flow in the supersonic diverging nozzle may also tend to separate from the wall; this is not the case. Expansion fans, a phenomenon provided by nature, to be studied in a subsequent section, allow the supersonic flow to turn rather sharp angles so that supersonic nozzles are constructed with large included angles, such as those attached to the rockets that propel satellites into orbit.

Key Concept: *Supersonic nozzles are constructed with large included angles.*

Example 17.2

Air exits from a reservoir maintained at 20°C and 500 kPa absolute into a receiver maintained at (a) 300 kPa absolute and (b) 200 kPa absolute. Estimate the mass flux if the exit area is 10 cm^2. Use the equations first and then the isentropic flow table, Table I.1. Refer to Fig. 17.5.

Solution

To estimate the mass flux we will assume isentropic flow. For air the receiver pressure that would result in $M_e = 1$ is

$$p_r = 0.5283 p_0 = 0.5283 \times 500 = 264.2 \text{ kPa}$$

For part (a) $M_e < 1$ since $p_r > 264.2$ kPa, and for part (b) choked flow occurs and $M_e = 1$ since $p_r < 264.2$ kPa.

(a) To find the exit Mach number Eq. 17.36 gives

$$1 + \frac{k-1}{2}M^2 = \left(\frac{p_0}{p}\right)^{(k-1)/k}$$

or

$$M_e^2 = \frac{2}{k-1}\left(\frac{p_0}{p}\right)^{(k-1)/k} - \frac{2}{k-1}$$

$$= \frac{2}{0.4}\left(\frac{500}{300}\right)^{0.2857} - \frac{2}{0.4} = 0.7857 \qquad \therefore M_e = 0.8864$$

The mass flux is given by Eq. 17.40 and is found to be

$$\dot{m} = p_0\sqrt{\frac{k}{RT_0}}MA\left(1 + \frac{k-1}{2}M^2\right)^{(k+1)/2(1-k)}$$

$$= 500\,000\sqrt{\frac{1.4}{287 \times 293}} \times 0.8864 \times 0.001\left(1 + \frac{0.4}{2} \times 0.8864^2\right)^{-2.4/0.8}$$

$$= 1.167 \text{ kg/s}$$

(b) Choked flow occurs and thus $M_e = 1$ and Eq. 17.41 yields

$$\dot{m} = p_0 A^*\sqrt{\frac{k}{RT_0}}\left(\frac{k+1}{2}\right)^{(k+1)/2(1-k)}$$

$$= 500\,000 \times 0.001\sqrt{\frac{1.4}{287 \times 293}}\left(\frac{2.4}{2}\right)^{-2.4/0.8} = 1.181 \text{ kg/s}$$

Now, let us use the isentropic flow table (Table I.1) and solve parts (a) and (b).

(a) For a pressure ratio of $p/p_0 = 300/500 = 0.6$, we interpolate to find

$$M_e = \frac{0.6041 - 0.6}{0.6041 - 0.5913} \times 0.02 + 0.88 = 0.886$$

$$\frac{T_e}{T_0} = \frac{0.6041 - 0.6}{0.6041 - 0.5913}(0.8606 - 0.8659) + 0.8659 = 0.864$$

$$T_e = 0.864 \times 293 = 253 \text{ K}$$

(continued)

The velocity and density are, respectively,

$$V = \text{M}c = 0.886\sqrt{1.4 \times 287 \times 253} = 282 \text{ m/s}$$

$$\rho = \frac{p}{RT} = \frac{300}{0.287 \times 253} = 4.13 \text{ kg/m}^3$$

The mass flux is then

$$\begin{aligned} \dot{m} &= \rho A V \\ &= 4.13 \times 0.001 \times 282 = 1.165 \text{ kg/s} \end{aligned}$$

(b) For choked flow we know that $\text{M}_e = 1$. The table gives

$$\frac{T_e}{T_0} = 0.8333 \qquad \text{and} \qquad \frac{p_e}{p_0} = 0.5283$$

Thus the temperature, velocity, and density are, respectively,

$$T = 0.8333 \times 293 = 244.2 \text{ K}$$

$$V = \text{M}c = 1 \times \sqrt{1.4 \times 287 \times 244.2} = 313.2 \text{ m/s}$$

$$\rho = \frac{p}{RT} = \frac{0.5283 \times 500}{0.287 \times 244.2} = 3.769 \text{ kg/m}^3$$

The mass flux is calculated to be

$$\begin{aligned} \dot{m} &= \rho A V \\ &= 3.769 \times 0.001 \times 313.2 = 1.180 \text{ kg/s} \end{aligned}$$

The results using the equations are essentially the same as those using the tables.

Example 17.3

The converging–diverging nozzle of Fig. E17.3, with an exit area of 40 cm^2 and a throat area of 10 cm^2, is attached to a reservoir with $T = 20°\text{C}$ and $p = 500$ kPa absolute. Determine the two exit pressures that result in $\text{M} = 1$ at the throat for an isentropic flow. Also, determine the associated exit temperatures and velocities.

Solution

The exit pressures we seek are associated with curves C and D of Fig. E17.3. The area ratio is

$$\frac{A}{A^*} = \frac{40}{10} = 4$$

We could solve Eq. 17.42 for M using a trial-and-error technique; however, it is simpler to use the isentropic flow table, Table I.1. There are two entries for $A/A^* = 4$. Interpolation gives

$$\left(\frac{p}{p_0}\right)_C = \frac{4.182 - 4.0}{4.182 - 3.673}(0.9823 - 0.9864) + 0.9864 = 0.9849$$

$$\left(\frac{p}{p_0}\right)_D = \frac{4.0 - 3.999}{4.076 - 3.999}(0.02891 - 0.02980) + 0.02980 = 0.02979$$

(continued)

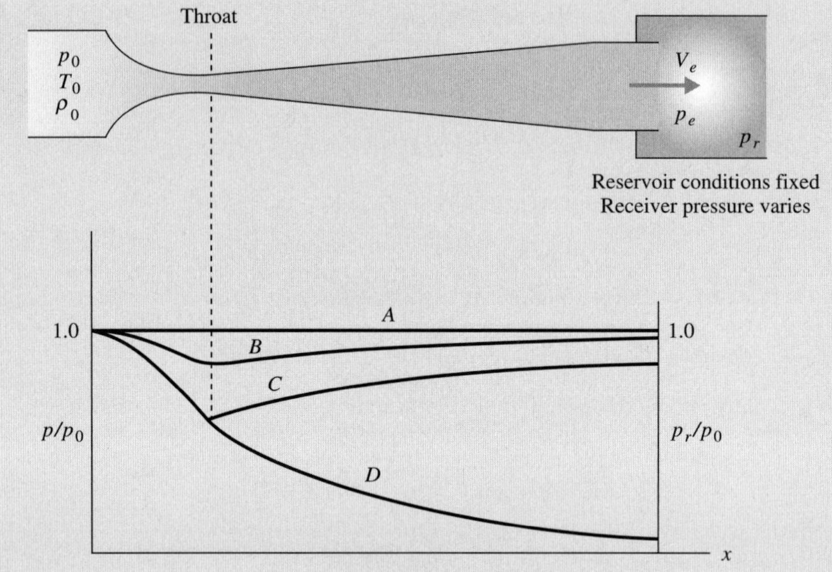

FIGURE E17.3

Hence the two exit pressures that will result in isentropic flow are

$$p_C = 492.4 \text{ kPa} \qquad \text{and} \qquad p_D = 14.9 \text{ kPa}$$

Note the very small pressure difference (7.6 kPa) between receiver and reservoir necessary to create the flow condition of curve C of Fig. E17.3.

The exit temperature ratios and Mach numbers are interpolated to be

$$\left(\frac{T}{T_0}\right)_C = 0.3576(0.9949 - 0.9961) + 0.9961 = 0.9957$$

$$\left(\frac{T}{T_0}\right)_D = 0.01299(0.3633 - 0.3665) + 0.3665 = 0.3665$$

$$\text{M}_C = 0.3576 \times 0.02 + 0.14 = 0.147$$

$$\text{M}_D = 0.01299 \times 0.02 + 2.94 = 2.94$$

The exit temperatures associated with curves C and D are thus

$$T_C = 0.9957 \times 293 = 291.7 \text{ K}$$

$$T_D = 0.3665 \times 293 = 107.4 \text{ K}$$

The exit velocities are found from $V = \text{M}c$ to be

$$V_C = 0.147\sqrt{1.4 \times 287 \times 291.7} = 50.3 \text{ m/s}$$

$$V_D = 2.94\sqrt{1.4 \times 287 \times 107.4} = 611 \text{ m/s}$$

Example 17.4

Gas flows are assumed to be incompressible flows at Mach numbers less than about 0.3. Determine the error involved in calculating the stagnation pressure for an airflow with $M = 0.3$.

Solution

For an incompressible airflow the energy equation (13.54) with no losses would give

$$p_0 = p_1 + \rho \frac{V_1^2}{2}$$

where $V_0 = 0$ at the stagnation point.

The isentropic flow equation (17.36), with $k = 1.4$, gives

$$p_0 = p_1(1 + 0.2M_1^2)^{3.5}$$

Use the binomial theorem $(1 + x)^n = 1 + nx + n(n-1)x^2/2! + \cdots$ and express this as, letting $x = 0.2M_1^2$,

$$p_0 = p_1(1 + 0.7M_1^2 + 0.175M_1^4 + \cdots)$$

This can be written as (see Eqs. 17.19 and 17.22)

$$\begin{aligned} p_0 - p_1 &= p_1 M_1^2(0.7 + 0.175M_1^2 + \cdots) \\ &= \frac{\rho_1 V_1^2}{1.4}(0.7 + 0.175M_1^2 + \cdots) \\ &= \rho_1 \frac{V_1^2}{2}(1 + 0.25M_1^2 + \cdots) \end{aligned}$$

Substituting $M_1 = 0.3$, we see that

$$p_0 - p_1 = \rho_1 \frac{V_1^2}{2}(1 + 0.0225 + \cdots)$$

Comparing this with the incompressible flow equation, we see that the error is only slightly greater than 2%. Hence it is reasonable to approximate a gas flow below $M = 0.3$ ($V \cong 100$ m/s for air at standard conditions) with an incompressible flow for many engineering applications.

17.4 NORMAL SHOCK WAVE

Small-amplitude disturbances travel at the speed of sound, as was established in a preceding section. In this section a large-amplitude disturbance will be studied. Its speed of propagation and its effect on other flow properties, such as pressure and temperature, will be considered. Large-amplitude disturbances occur in a number of situations. Examples include the flow in a gun barrel ahead of the projectile, the exit flow from a rocket or jet engine nozzle, the airflow around a supersonic aircraft, and the expanding front due to an explosion. Such large disturbances propagating through a gas are called *shock waves*. They can be oriented normal to the

Key Concept: *The changes that occur across a shock wave take place over an extremely short distance.*

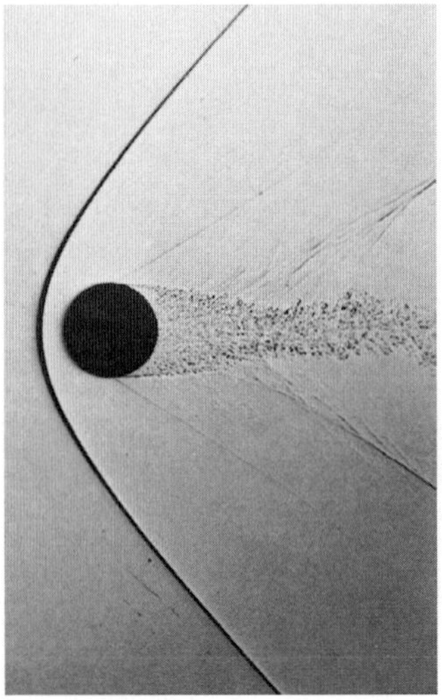

FIGURE 17.8 A shock wave is observed in front of a sphere at M = 1.53. (Photograph by A. C. Charters.)

Key Concept: *The three equations allow us to determine three unknowns.*

flow or at oblique angles. In this section we consider only the normal shock wave that occurs in a tube or directly in front of a blunt object; the photograph in Fig. 17.8 shows the shock wave in front of a sphere.

The property changes that occur across a shock wave take place over an extremely short distance. For usual conditions the distance is only several mean free paths of the molecules, on the order of 10^{-4} mm. Phenomena such as viscous dissipation and heat conduction that occur within the shock wave will not be studied here. We will treat the shock wave as a zero-thickness discontinuity in the flow and allow our integral control volume equations to relate the quantities of interest.

Consider a normal shock wave moving with velocity V_1. We can make it stationary by moving the flow in a tube at velocity V_1, as shown in Fig. 17.9. The continuity equation, recognizing that $A_1 = A_2$, is

$$\rho_1 V_1 = \rho_2 V_2 \tag{17.45}$$

The energy equation (17.5), with $\dot{Q} = \dot{W}_S = 0$, is

$$\frac{V_2^2 - V_1^2}{2} + \frac{k}{k-1}\left(\frac{p_2}{\rho_2} - \frac{p_1}{\rho_1}\right) = 0 \tag{17.46}$$

The momentum equation (17.2), with only pressure forces, becomes

$$p_1 - p_2 = \rho_1 V_1 (V_2 - V_1) \tag{17.47}$$

where the areas have divided out. The above three equations allow us to determine three unknown quantities; if ρ_1, V_1, and p_1 are known, we can find ρ_2, V_2, p_2, and subsequently T_2 and M_2, using the appropriate equations.

It is convenient, however, to express the equations in terms of the Mach numbers M_1 and M_2. This results in a set of equations that are simpler to solve than solving the three simultaneous equations listed above. To do this we write Eq. 17.47, using $\rho_1 V_1 = \rho_2 V_2$, in the form

$$p_1\left(1 + \frac{\rho_1 V_1^2}{p_1}\right) = p_2\left(1 + \frac{\rho_2 V_2^2}{p_2}\right) \tag{17.48}$$

Introducing $M^2 = V^2\rho/pk$ (see Eq. 17.20), the momentum equation becomes

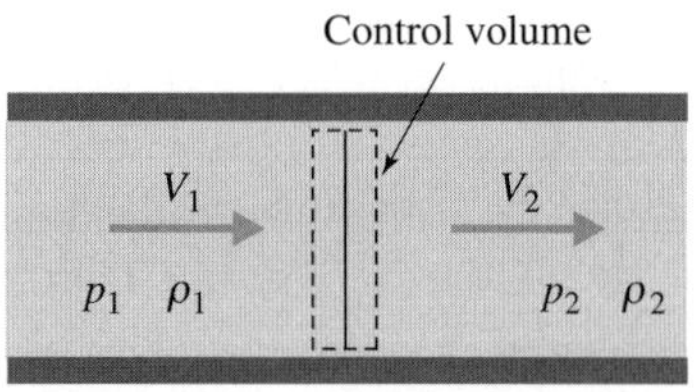

FIGURE 17.9 Stationary shock wave in a tube.

$$\frac{p_2}{p_1} = \frac{1 + k\mathrm{M}_1^2}{1 + k\mathrm{M}_2^2} \tag{17.49}$$

Similarly, the energy equation (17.46), with $p = \rho RT$, is written as

$$T_1\left(1 + \frac{k-1}{kRT_1}\frac{V_1^2}{2}\right) = T_2\left(1 + \frac{k-1}{kRT_2}\frac{V_2^2}{2}\right) \tag{17.50}$$

or, substituting $\mathrm{M}^2 = V^2/kRT$,

$$\frac{T_2}{T_1} = \frac{1 + \dfrac{k-1}{2}\mathrm{M}_1^2}{1 + \dfrac{k-1}{2}\mathrm{M}_2^2} \tag{17.51}$$

If we substitute $\rho = p/RT$ into the continuity equation (17.45), we have

$$\frac{p_1 V_1}{RT_1} = \frac{p_2 V_2}{RT_2} \tag{17.52}$$

which becomes, using $V = \mathrm{M}\sqrt{kRT}$,

$$\frac{p_2}{p_1}\frac{\mathrm{M}_2}{\mathrm{M}_1}\sqrt{\frac{T_1}{T_2}} = 1 \tag{17.53}$$

Substituting for the pressure and temperature ratios from Eqs. 17.49 and 17.51, the continuity equation takes the form

$$\frac{\mathrm{M}_1\left(1 + \dfrac{k-1}{2}\mathrm{M}_1^2\right)^{1/2}}{1 + k\mathrm{M}_1^2} = \frac{\mathrm{M}_2\left(1 + \dfrac{k-1}{2}\mathrm{M}_2^2\right)^{1/2}}{1 + k\mathrm{M}_2^2} \tag{17.54}$$

Hence, the downstream Mach number is related to the upstream Mach number by

$$\mathrm{M}_2^2 = \frac{\mathrm{M}_1^2 + \dfrac{2}{k-1}}{\dfrac{2k}{k-1}\mathrm{M}_1^2 - 1} \tag{17.55}$$

This allows us to express the pressure and temperature ratios in terms of M_1 only.

The momentum equation (17.49) then takes the form

$$\frac{p_2}{p_1} = \frac{2k}{k+1}\mathrm{M}_1^2 - \frac{k-1}{k+1} \tag{17.56}$$

and the energy equation becomes

$$\frac{T_2}{T_1} = \frac{\left(1 + \dfrac{k-1}{2}\mathrm{M}_1^2\right)\left(\dfrac{2k}{k-1}\mathrm{M}_1^2 - 1\right)}{\dfrac{(k+1)^2}{2(k-1)}\mathrm{M}_1^2} \tag{17.57}$$

For air, with $k = 1.4$, the preceding three equations reduce to

$$
\begin{aligned}
M_2^2 &= \frac{M_1^2 + 5}{7M_1^2 - 1} \\
\frac{p_2}{p_1} &= \frac{7M_1^2 - 1}{6} \\
\frac{T_2}{T_1} &= \frac{(M_1^2 + 5)(7M_1^2 - 1)}{36M_1^2}
\end{aligned}
\tag{17.58}
$$

From the first of these three equations, we observe:

Key Concept: *A finite wave that converts a subsonic flow into a supersonic flow is an impossibility.*

- If $M_1 = 1$, then $M_2 = 1$ and no shock wave exists.
- If $M_1 > 1$, then $M_2 < 1$ and the normal shock wave converts a supersonic flow into a subsonic flow.
- If $M_1 < 1$, then $M_2 > 1$ and a subsonic flow appears to be converted into a supersonic flow by the presence of a normal shock wave. This possibility is eliminated by the second law since it would demand a decrease in entropy by a process in an isolated system, an impossibility.

The impossibility stated above is observed by considering the entropy increase, given by

$$
\begin{aligned}
s_2 - s_1 &= c_p \ln \frac{T_2}{T_1} - R \ln \frac{p_2}{p_1} \\
&= c_p \ln \frac{2 + (k-1)M_1^2}{2 + (k-1)M_2^2} - R \ln \frac{1 + kM_1^2}{1 + kM_2^2}
\end{aligned}
\tag{17.59}
$$

For air, with $k = 1.4$, this is plotted in Fig. 17.10, relating M_2 to M_1, with Eq. 17.55. Note the impossible negative entropy change whenever $M_1 < 1$.

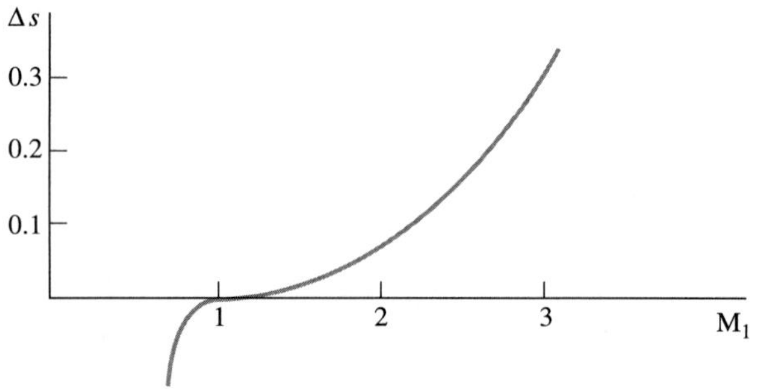

FIGURE 17.10 Entropy change for a normal shock in air.

The relationship among the thermodynamic properties, listed in the equations above, can be demonstrated graphically with reference to the T-s diagram in Fig. 17.11. The conditions upstream of the normal shock are designated by state 1, and downstream by state 2. Note the dashed line from state 1 to state 2 for the irreversible process that occurs inside the shock wave. The energy equation, with $\dot{Q} = \dot{W}_S = 0$, can be written as

$$\frac{V_1^2}{2c_p} + T_1 = \frac{V_2^2}{2c_p} + T_2 \tag{17.60}$$

The stagnation temperature is defined as the temperature that would exist if the flow is brought to rest isentropically. Thus the energy equation gives

$$T_{01} = T_{02} \tag{17.61}$$

as shown in the figure. The substantial decrease in stagnation pressure, $p_{02} < p_{01}$, is also observed in Fig. 17.11. If the entropy increases from state 1 to state 2, as it must, the stagnation pressure p_{02} must decrease, as shown, if we are to maintain $T_{01} = T_{02}$.

Gas tables are available that give the pressure ratio, the temperature ratio, the downstream Mach number, and the stagnation pressure ratio as a function of the upstream Mach number. Table I.2 is such a table for $k = 1.4$ and includes the ratios given by Eq. 17.58. Note that M_2 is always less than unity, p_2 is always greater than p_1, T_2 is always greater than T_1, and p_{02} is always less than p_{01}.

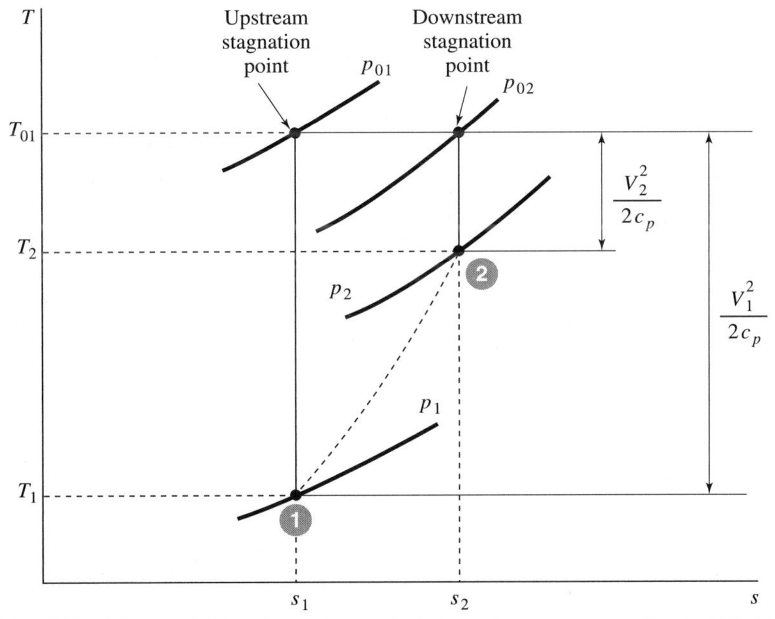

FIGURE 17.11 T–s diagram for a normal shock wave.

Example 17.5

The normal shock wave of Fig. E17.5 passes through stagnant air at 60°F and atmospheric pressure of 12 psi with a speed of 1500 ft/sec. Calculate the pressure and temperature downstream of the shock wave. Use (a) the equations and (b) the gas tables.

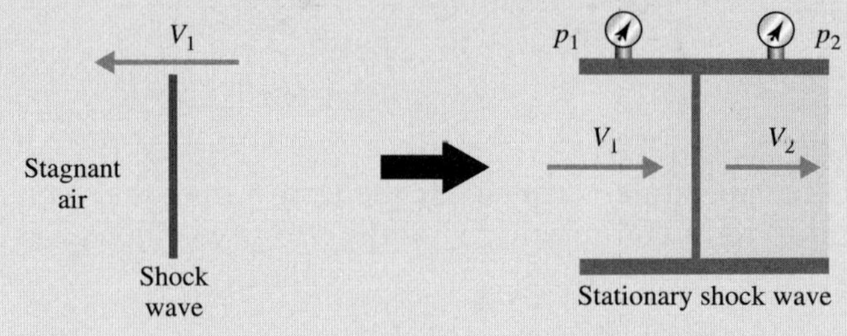

FIGURE E17.5

Solution

We consider the shock wave to be stationary with $V_1 = 1500$ ft/sec and $p_1 = 12$ psia.

(a) To use the simplified equations (17.58) we must know the upstream Mach number. It is

$$\mathrm{M}_1 = \frac{V_1}{c_1} = \frac{V_1}{\sqrt{kRT_1}}$$

$$= \frac{1500}{\sqrt{1.4 \times 1716 \times 520}} = 1.342$$

The pressure and temperature are then found to be

$$p_2 = \frac{p_1(7\mathrm{M}_1^2 - 1)}{6}$$

$$= \frac{12(7 \times 1.342^2 - 1)}{6} = 23.21 \text{ psia}$$

$$T_2 = \frac{T_1(\mathrm{M}_1^2 + 5)(7\mathrm{M}_1^2 - 1)}{36\mathrm{M}_1^2}$$

$$= \frac{520(1.342^2 + 5)(7 \times 1.342^2 - 1)}{36 \times 1.342^2} = 633.1°\text{R}$$

(b) From part (a), we use $\mathrm{M}_1 = 1.342$. Interpolation in Table I.2 yields

$$\frac{p_2}{p_1} = \frac{1.342 - 1.34}{1.36 - 1.34}(1.991 - 1.928) + 1.928 = 1.934$$

$$\frac{T_2}{T_1} = \frac{1.342 - 1.34}{1.36 - 1.34}(1.229 - 1.216) + 1.216 = 1.217$$

Using the information given, we have

$$p_2 = 12 \times 1.934 = 23.21 \text{ psia}$$

$$T_2 = 520 \times 1.217 = 632.8°\text{R}$$

Example 17.6

A normal shock wave propagates through otherwise stagnant air at standard conditions at a speed of 700 m/s. Determine the speed induced in the air immediately behind the shock wave as shown in Fig. E17.6.

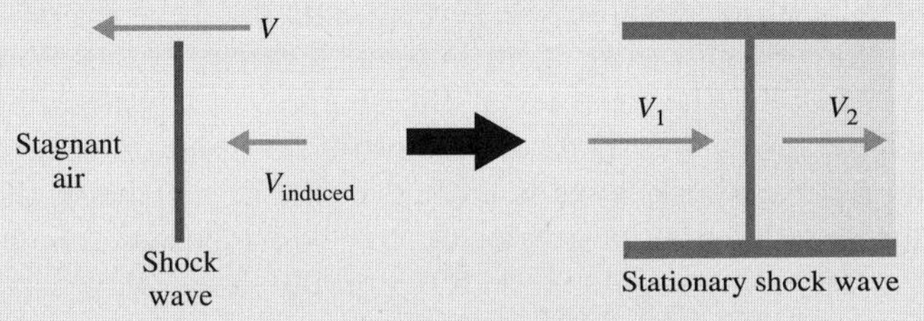

FIGURE E17.6

Solution

For standard conditions the temperature is 15°C. The upstream Mach number is thus

$$M_1 = \frac{V_1}{c_1} = \frac{V_1}{\sqrt{kRT_1}} = \frac{700}{\sqrt{1.4 \times 287 \times 288}} = 2.06$$

Using Eq. 17.58, we find that

$$M_2 = \left(\frac{M_1^2 + 5}{7M_1^2 - 1}\right)^{1/2} = \left(\frac{2.06^2 + 5}{7 \times 2.06^2 - 1}\right)^{1/2} = 0.567$$

$$T_2 = \frac{T_1(M_1^2 + 5)(7M_1^2 - 1)}{36M_1^2}$$

$$= \frac{288(2.06^2 + 5)(7 \times 2.06^2 - 1)}{36 \times 2.06^2} = 500.2 \text{ K}$$

This allows us to calculate

$$V_2 = M_2 c_2 = 0.567\sqrt{1.4 \times 287 \times 500.2} = 254.4 \text{ m/s}$$

This velocity assumes a flow with the shock wave stationary and the air approaching the shock wave at 700 m/s. If we superimpose a velocity of 700 m/s moving opposite to V_1, we find the induced velocity to be

$$V_{\text{induced}} = V_2 - V_1$$
$$= 254.4 - 700 = -446 \text{ m/s}$$

where the negative sign means that the induced velocity would be moving to the left if V_1 is to the right. The induced velocity would be in the same direction as the propagation of the shock wave. Such large induced velocities are responsible for much of the damage away from the bomb center caused by explosions of high-power bombs.

17.5 SHOCK WAVES IN CONVERGING–DIVERGING NOZZLES

Key Concept: *The location of the shock wave is usually prescribed in problems.*

The converging–diverging nozzle has already been presented for an isentropic flow; for receiver-to-reservoir pressure ratios between those of curves C and D of Figs. 17.7 and 17.12, shock waves exist in the flow either inside or outside the nozzle. If $p_r/p_0 = a$ (locate a on the vertical axis on the right of Fig. 17.12), a normal shock wave would exist at an internal location in the diverging portion of the nozzle. Usually, the location of the shock wave is prescribed in student problems since to locate the shock a trial-and-error solution is necessary. When $p_r/p_0 = b$ the normal shock wave is located in the exit plane of the nozzle. For a pressure ratio less than b but greater than e, two types of oblique shock wave patterns are observed, one with a central normal shock wave, as sketched for $p_r/p_0 = c$, and one with only oblique waves, as sketched for $p_r/p_0 = d$. These pressure ratios that result in oblique shock waves will not be considered here. As we move from d to e, the oblique shock waves become weaker and weaker until the isentropic flow is again realized at $p_r/p_0 = e$ with all shock waves absent. For pressure ratios below e a very complicated flow exists. The flow turns the corner at the nozzle exit rather abruptly due to expansion waves (isentropic waves to be considered in a subsequent section), then turns back due to the same expansion waves, resulting in a billowing out of the exhaust flow, as is visible from high-altitude satellite rocket engines. Let us work some examples now for the converging–diverging nozzle; no new equations are necessary.

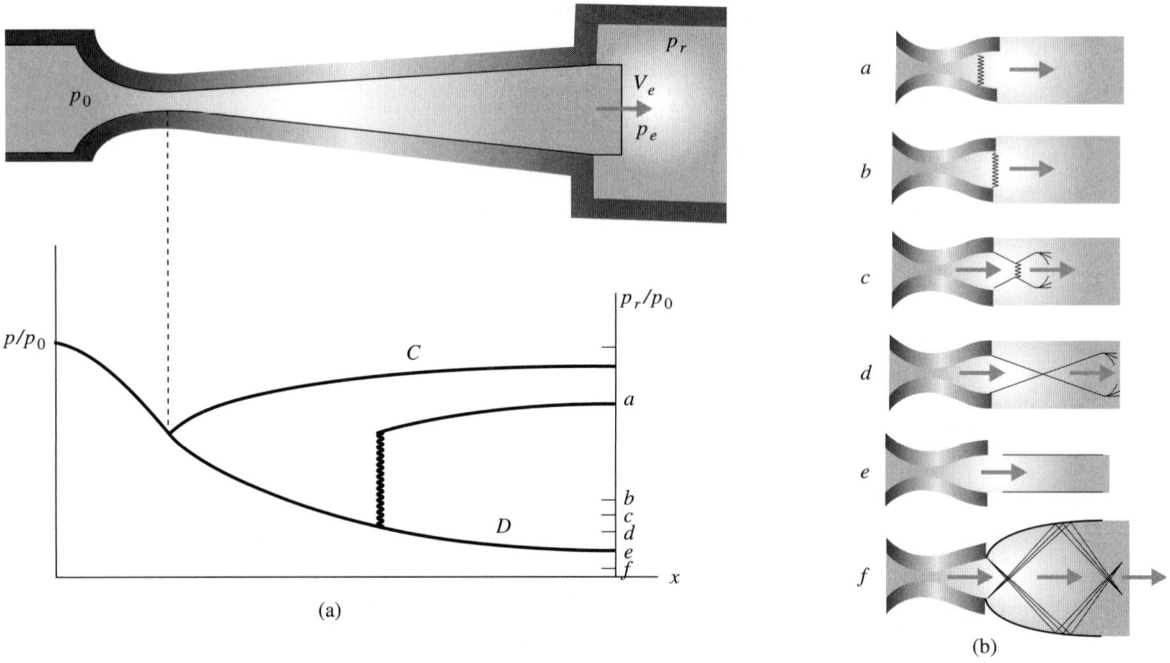

FIGURE 17.12 Converging–diverging nozzle.

Example 17.7

A converging–diverging nozzle has a throat diameter of 5 cm and an exit diameter of 10 cm. The reservoir is the laboratory, maintained at atmospheric conditions of 20°C and 90 kPa absolute. Air is constantly pumped from a receiver so that a normal shock wave stands across the exit plane of the nozzle. Determine the receiver pressure and the mass flux.

Solution

Isentropic flow occurs from the reservoir, to the throat, to the exit plane in front of the normal shock wave at state 1. Supersonic flow occurs downstream of the throat making the throat the critical area. Hence

$$\frac{A_1}{A^*} = \frac{10^2}{5^2} = 4$$

Interpolation in the isentropic flow table (Table I.1) gives

$$M_1 = 2.94 \qquad \frac{p_1}{p_0} = 0.0298$$

Hence the pressure in front of the normal shock is

$$\begin{aligned} p_1 &= p_0 \times 0.0298 \\ &= 90 \times 0.0298 = 2.68 \text{ kPa} \end{aligned}$$

From the normal shock table (Table I.2), using $M_1 = 2.94$, we find that

$$\frac{p_2}{p_1} = 9.918$$

$$\therefore\ p_2 = 9.918 \times 2.68 = 26.6 \text{ kPa}$$

This is the receiver pressure needed to orient the shock across the exit plane as shown for $p_r/p_0 = b$ in Fig. 17.12.

To find the mass flux through the nozzle, we need only consider the throat. Recognizing that $M_t = 1$, so that $V_t = c_t$, we can write

$$\dot{m} = \rho_t A_t V_t = \frac{p_t}{RT_t} A_t \sqrt{kRT_t} = p_t A_t \sqrt{\frac{k}{RT_t}}$$

The isentropic flow table yields

$$\frac{p_t}{p_0} = 0.5283 \qquad \frac{T_t}{T_0} = 0.8333$$

Thus the mass flux becomes

$$\begin{aligned} \dot{m} &= (0.5283 \times 90\,000) \times \frac{\pi \times 0.05^2}{4} \sqrt{\frac{1.4}{287 \times (0.8333 \times 293)}} \\ &= 0.417 \text{ kg/s} \end{aligned}$$

Remember, the pressure must be measured in pascals in the equation above.

Example 17.8

Air flows from a reservoir at 20°C and 200 kPa absolute through a 5-cm-diameter throat and exits from a 10-cm-diameter nozzle. Calculate the pressure needed to locate a normal shock wave at a position where the diameter is 7.5 cm.

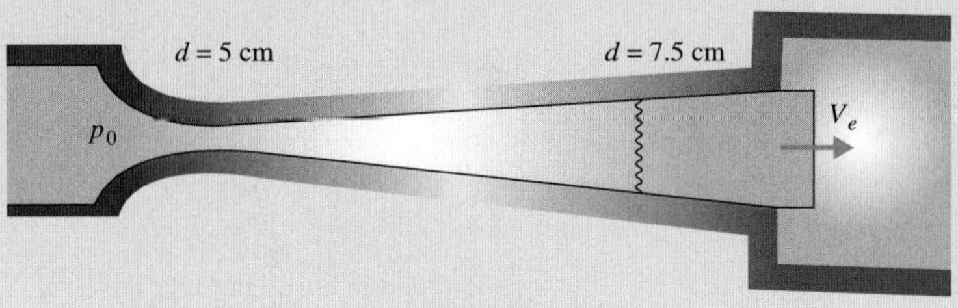

FIGURE E17.8

Solution

We will use the gas tables for this flow represented by the curve established by $p_r/p_0 = a$ in Fig. 17.12. The throat is a critical area since for a supersonic flow $M_t = 1$. The area ratio is

$$\frac{A_1}{A^*} = \frac{7.5^2}{5^2} = 2.25$$

For this area ratio we find, from Table I.1, that

$$M_1 = 2.33$$

Then from Table I.2, for this Mach number, we obtain

$$M_2 = 0.531 \qquad \frac{p_{02}}{p_{01}} = 0.570$$

The reservoir pressure $p_0 = p_{01}$. Thus

$$p_{02} = 0.570 \times 200 = 114 \text{ kPa}$$

Isentropic flow occurs from state 2 immediately after the normal shock wave to the exit. Hence, for $M_2 = 0.531$, we find from Table I.1 that

$$\frac{A_2}{A^*} = 1.285$$

so that, if A_e is the exit area,

$$\frac{A_e}{A^*} = \frac{A_2}{A^*} \times \frac{A_e}{A_2} = 1.285 \times \frac{10^2}{7.5^2} = 2.284$$

The Mach number and pressure ratio corresponding to this area ratio are

$$M_e = 0.265 \qquad \frac{p_e}{p_{0e}} = 0.952$$

For our isentropic flow between the shock and the exit, we know that $p_{02} = p_{0e}$; thus

$$p_e = p_{02} \times 0.952 = 114 \times 0.952 = 109 \text{ kPa}$$

Note the usefulness of the critical area ratio in obtaining the desired results.

Example 17.9

A pitot probe, the device used to measure the stagnation pressure in a flow, is inserted into an airstream and measures 300 kPa absolute, as shown in Fig. E17.9. The pressure in the flow is measured to be 75 kPa absolute. If the temperature at the stagnation point of the probe is measured as 150°C, determine the free-stream velocity V.

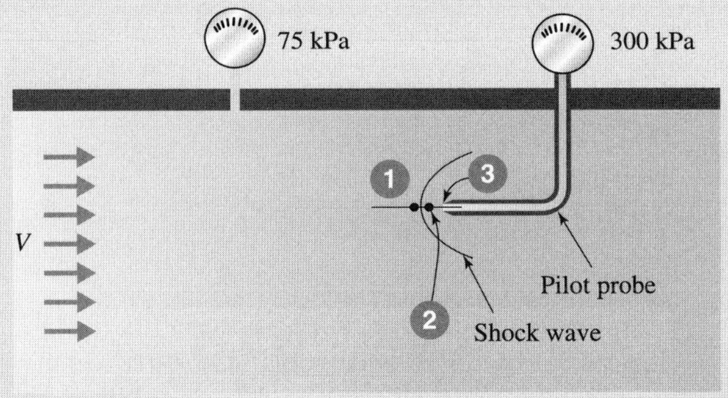

FIGURE E17.9

Solution

When a blunt object is placed in a supersonic flow a detached shock wave forms around the object, as it does around the front of the pitot probe shown. The flow that meets the front of the pitot probe at the stagnation point passes through a normal shock wave from state 1 to state 2; the subsonic flow at state 2 then decelerates isentropically to state 3, the stagnation point.

For the isentropic flow from state 2 to state 3 we can use Eq. 17.36,

$$\frac{p_3}{p_2} = \left(1 + \frac{k-1}{2}\mathrm{M}_2^2\right)^{k/(k-1)}$$

Across the normal shock we know that (see Eq. 17.56)

$$\frac{p_2}{p_1} = \frac{2k}{k+1}\mathrm{M}_1^2 - \frac{k-1}{k+1}$$

Also, the Mach numbers are related by Eq. 17.55,

$$\mathrm{M}_2^2 = \frac{(k-1)\mathrm{M}_1^2 + 2}{2k\mathrm{M}_1^2 - k + 1}$$

The three equations above can be combined, with some algebraic manipulation, to yield the *Rayleigh-pitot-tube formula* for supersonic flows, namely,

$$\frac{p_3}{p_1} = \frac{\left(\dfrac{k+1}{2}\mathrm{M}_1^2\right)^{k/(k-1)}}{\left(\dfrac{2k\mathrm{M}_1^2}{k+1} - \dfrac{k-1}{k+1}\right)^{1/(k-1)}}$$

(continued)

Substituting $k = 1.4$, $p_3 = 300$ kPa, and $p_1 = 75$ kPa. we have

$$\frac{300}{75} = \frac{(1.2\text{M}_1^2)^{3.5}}{(1.167\text{M}_1^2 - 0.1667)^{2.5}}$$

This can be solved by trial and error to give

$$\text{M}_1 = 1.65$$

The Mach number after the normal shock is interpolated from the shock table to be

$$\text{M}_2 = 0.654$$

Using the isentropic flow table with this Mach number, we interpolate the temperature at state 2 to be

$$\begin{aligned} T_2 &= T_3 \times 0.921 \\ &= 423 \times 0.921 = 389.6 \text{ K} \end{aligned}$$

The temperature in front of the normal shock is found by using the normal shock table:

$$\frac{T_2}{T_1} = 1.423$$

$$\therefore\ T_1 = \frac{T_2}{1.423} = 274 \text{ K}$$

Finally, the velocity before the normal shock is given by

$$\begin{aligned} V_1 &= \text{M}_1 c_1 = \text{M}_1\sqrt{kRT_1} \\ &= 1.65\sqrt{1.4 \times 287 \times 274} = 547 \text{ m/s} \end{aligned}$$

17.6 OBLIQUE SHOCK WAVES

In this section we investigate the oblique shock wave, a finite-amplitude wave that is not normal to the incoming flow. The flow approaching the oblique shock wave will be assumed to be in the x-direction. After the oblique shock the velocity vector will have a component normal to the flow direction. We will continue to assume that the flow before and after the oblique shock is uniform and steady.

Oblique shock waves form on the leading edge of a supersonic airfoil or in an abrupt corner, as sketched in Fig. 17.13. Oblique shock waves may also be found on axisymmetric bodies such as a nose cone or a bullet traveling at supersonic speeds. In this book, we consider only plane flows.

Key Concept: *The oblique shock wave turns the flow so that the velocity vector is parallel to the plane wall.*

The function of the oblique shock wave is to turn the flow so that the velocity vector $\mathbf{V}_2$ is parallel to the plane wall. The angle between the two velocity vectors introduces another variable into our analysis. The problem remains solvable, however, with the additional tangential momentum equation.

To analyze the oblique shock wave, consider a control volume enclosing a portion of the shock as shown in Fig. 17.14. The velocity vector upstream is assumed to be in the x-direction only; the oblique shock wave makes an angle β

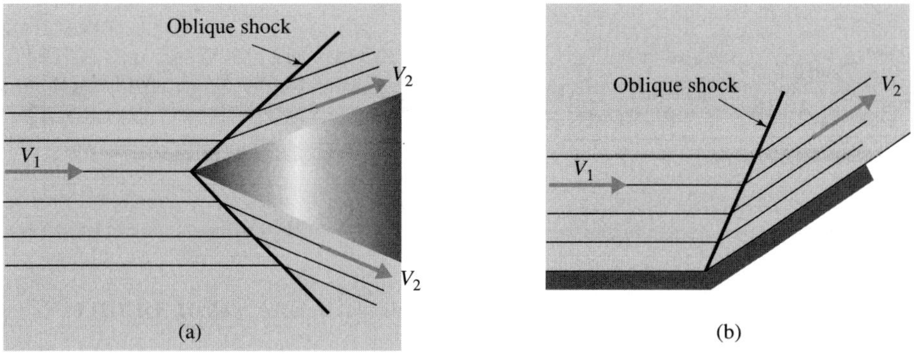

FIGURE 17.13 Oblique shock waves in a supersonic flow: (a) flow over a symmetrical wedge; (b) flow in a corner.

with the upstream velocity vector and turns the flow through the *deflection angle* or *wedge angle* θ so that V_2 is parallel to the wall. The components of the velocity vectors are shown normal and tangential to the oblique shock wave. The tangential components do not cause fluid to flow through the shock; hence the continuity equation, with $A_1 = A_2$, gives

$$\rho_1 V_{1n} = \rho_2 V_{2n} \tag{17.62}$$

The pressure forces act normal to the oblique shock and produce no tangential components. Thus the momentum equation expressed in the tangential direction requires that the tangential momentum into the control volume be equal to the tangential momentum leaving the control volume, that is,

$$\dot{m}_1 V_{1t} = \dot{m}_2 V_{2t} \tag{17.63}$$

or, using $\dot{m}_1 = \dot{m}_2$, we demand that

$$V_{1t} = V_{2t} \tag{17.64}$$

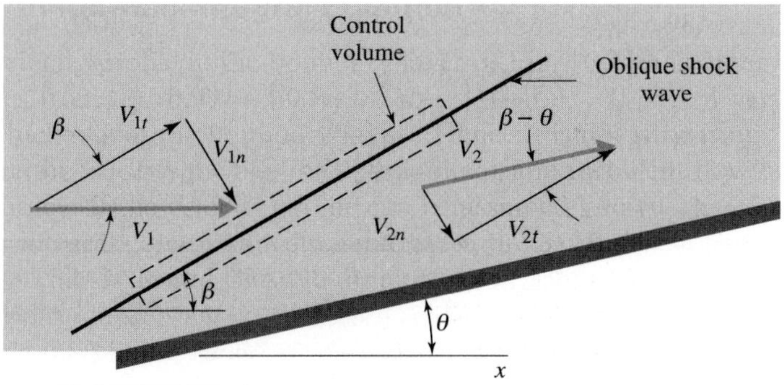

FIGURE 17.14 Control volume enclosing a small portion of an oblique shock wave.

The normal momentum equation takes the form

$$p_1 - p_2 = \rho_2 V_{2n}^2 - \rho_1 V_{1n}^2 \tag{17.65}$$

The energy equation, using $V^2 = V_n^2 + V_t^2$, may be written as

$$\frac{V_{1n}^2}{2} + \frac{k}{k-1}\frac{p_1}{\rho_1} = \frac{V_{2n}^2}{2} + \frac{k}{k-1}\frac{p_2}{\rho_2} \tag{17.66}$$

Key Concept: *The tangential components of the two velocity vectors do not enter the equations.*

where the tangential component terms have been canceled from both sides. Note that the tangential components of the two velocity vectors do not enter the continuity, normal momentum, or energy equations, the three equations used in the solution of the normal shock wave. Hence we may substitute V_{1n} and V_{2n} for V_1 and V_2, respectively, of the normal shock wave equations and obtain a solution. Either the normal shock wave equations or the normal shock wave table (Table I.2) may be used. Of course, we also replace M_1 and M_2 with M_{1n} and M_{2n}, respectively.

It is useful to relate the oblique shock angle β to the deflection angle θ. Using the continuity equation (17.62), with reference to Fig. 17.14, there results

$$\frac{\rho_2}{\rho_1} = \frac{V_{1n}}{V_{2n}} = \frac{V_{1t}\tan\beta}{V_{2t}\tan(\beta-\theta)} = \frac{\tan\beta}{\tan(\beta-\theta)} \tag{17.67}$$

From the normal shock wave equations (17.56) and (17.57) we can find the density ratio to be

$$\frac{\rho_2}{\rho_1} = \frac{p_2 T_1}{p_1 T_2} = \frac{(k+1)M_{1n}^2}{(k-1)M_{1n}^2 + 2} \tag{17.68}$$

Substituting this into Eq. 17.67 gives

$$\tan(\beta-\theta) = \frac{\tan\beta}{k+1}\left[k-1+\frac{2}{M_1^2\sin^2\beta}\right] \tag{17.69}$$

For a flow with a given M_1, this equation relates the oblique shock angle β to the wedge or corner angle θ. The three variables β, θ, and M_1 in the equation above are often plotted as in Fig. 17.15. We can observe several phenomena by studying the figure.

- For a specified upstream Mach number M_1 and a given wedge angle θ there are two possible oblique shock angles β, the large one corresponding to a "strong" shock and the smaller corresponding to a "weak" shock.
- For a given wedge angle θ there is a minimum Mach number for which there is only one oblique shock angle β.
- For a given wedge angle θ, if M_1 is less than the minimum for that particular curve, no oblique shock wave exists and the shock wave becomes detached, as shown in Fig. 17.16. Also, for a given M_1 there is a sufficiently large θ that will result in a detached shock wave.

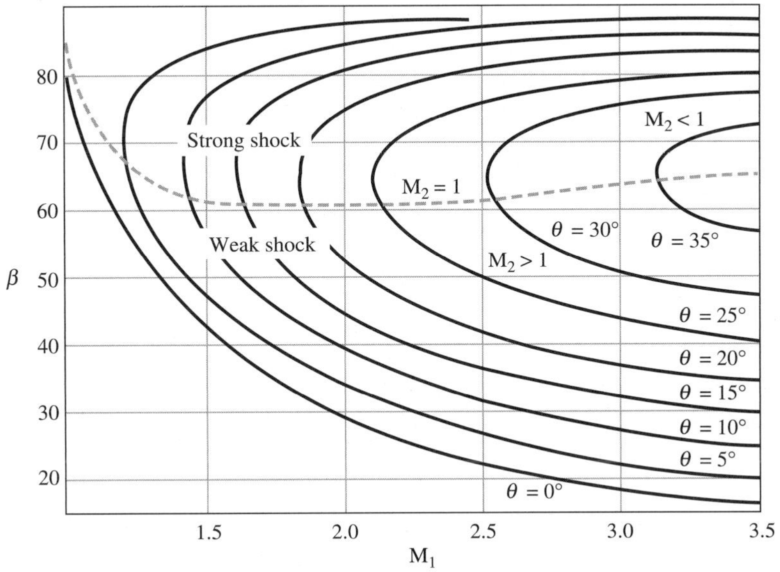

FIGURE 17.15 Oblique shock wave relationships for $k = 1.4$.

The pressure rise across the oblique shock wave determines whether a weak shock or a strong shock occurs. For a relatively small pressure rise a weak shock will occur with $M_2 > 1$. If the pressure rise is relatively large a strong shock occurs with $M_2 < 1$. Note that for the detached shocks around bodies a normal shock exists for the stagnation streamline; this is followed away from the stagnation point by the strong oblique shock, then the weak oblique shock, and eventually a Mach wave. For blunt bodies moving at supersonic speeds the shock wave is always detached.

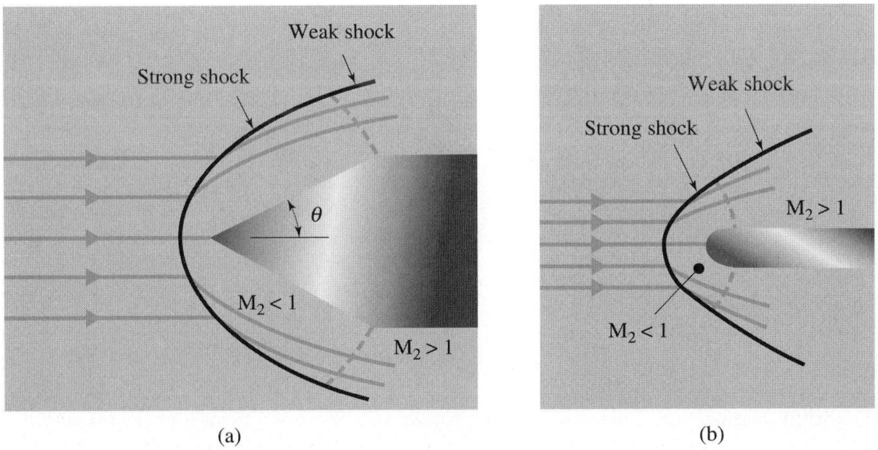

FIGURE 17.16 Detached shock waves: (a) flow around a wedge; (b) flow around a blunt object.

Example 17.10

Air flows over a wedge with $M_1 = 3$ as shown in Fig. E17.10. A weak shock reflects from the wall. Determine the values of M_3 and β_3 for the reflected wave.

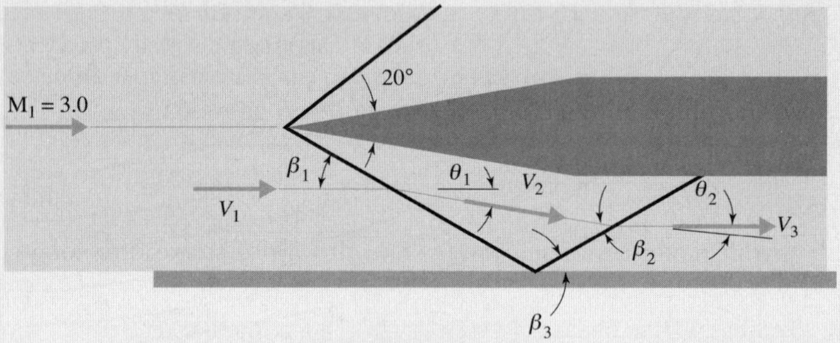

FIGURE E17.10

Solution

From Fig. 17.15 with $\theta_1 = 10°$ and $M_1 = 3.0$, we find for the weak shock that $\beta_1 = 27.5°$. This yields

$$M_{1n} = 3 \sin 27.5° = 1.39$$

From the shock table we interpolate to find

$$M_{2n} = 0.744 = M_2 \sin(27.5° - 10°)$$

$$\therefore M_2 = 2.48$$

The reflected shock must again turn the flow through an angle of 10°, that is, $\theta_2 = 10°$. For this wedge angle and $M_2 = 2.48$ from Fig. 17.15 for a weak shock, we see that $\beta_2 = 33°$. This results in

$$M_{2n} = 2.48 \sin 33° = 1.35$$

From the shock table

$$M_{3n} = 0.762 = M_3 \sin 23°$$

$$\therefore M_3 = 1.95$$

The desired angle is calculated to be

$$\beta_3 = \beta_2 - 10° = 23°$$

Note that Fig. 17.15 does not allow precise calculations. Equation 17.69 could be used, by trial and error, to improve the accuracy of the β's and hence the quantities that follow.

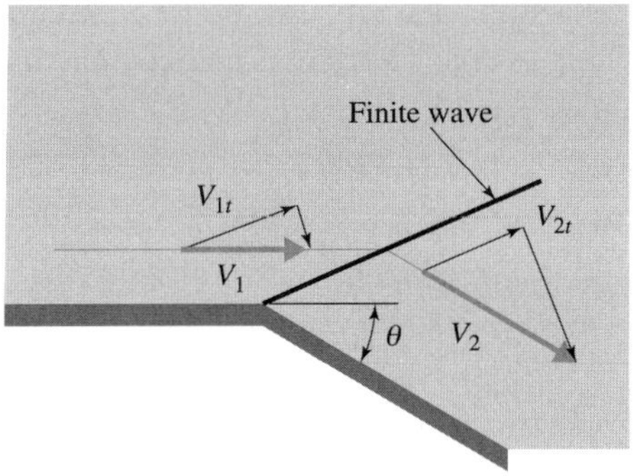

(a) Single finite wave

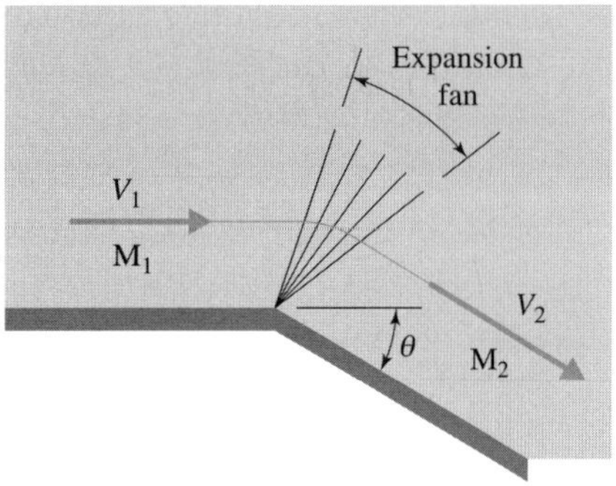

(b) Infinite number of Mach waves

FIGURE 17.17 Supersonic flow around a convex corner.

17.7 ISENTROPIC EXPANSION WAVES

In this section we consider the supersonic flow around a convex corner, as shown in Fig. 17.17. Let us first attempt to accomplish such a flow with the finite amplitude wave of Fig. 17.17a. The flow must turn the angle θ so that V_2 is parallel to the wall. The tangential component must be conserved because of momentum conservation. This would result in $V_2 > V_1$, as is obvious from the sketch. This would be the situation if a subsonic flow, $M_{1n} < 1$, could experience a finite increase to a supersonic flow, $M_{2n} > 1$. This, of course, is impossible because of the second law, as was noted in the discussion associated with Fig. 17.10. Consequently, we consider the turning of a finite wave around a convex corner an impossibility.

Consider a second possible mechanism that would allow the flow to turn the corner, a fan composed of an infinite number of Mach waves, emanating from the corner, as shown in Fig. 17.17b. The second law would not be violated with such a mechanism since each Mach wave is an isentropic wave. We will determine the effect of a single Mach wave on the flow and then integrate to obtain the total effect. Figure 17.18 shows the infinitesimal velocity change due to a single Mach wave. For the control volume enclosing the Mach wave, we know that the tan-

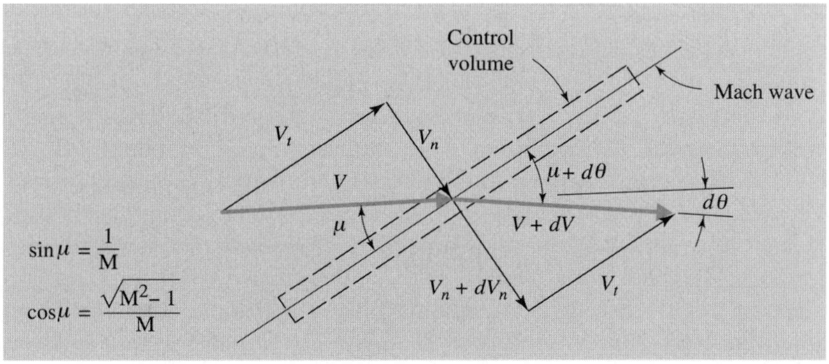

FIGURE 17.18 Mach wave.

gential momentum is conserved; thus the tangential velocity component remains unchanged as shown, as in the oblique shock wave. From the triangles in the figure we can write

$$V_t = V\cos\mu = (V + dV)\cos(\mu + d\theta) \tag{17.70}$$

Since $d\theta$ is small, this becomes,[1] using $\cos(\mu + d\theta) = \cos\mu - d\theta\sin\mu$,

$$V\sin\mu\, d\theta = \cos\mu\, dV \tag{17.71}$$

Substituting $\sin\mu = 1/\text{M}$ (see Eq. 17.23) and $\cos\mu = (\sqrt{\text{M}^2 - 1})/\text{M}$, we have

$$d\theta = \sqrt{\text{M}^2 - 1}\frac{dV}{V} \tag{17.72}$$

The relationship $V = \text{M}\sqrt{kRT}$ can be differentiated and rearranged to give

$$\frac{dV}{V} = \frac{d\text{M}}{\text{M}} + \frac{1}{2}\frac{dT}{T} \tag{17.73}$$

The energy equation, in the form $V^2/2 + kRT/(k-1) = \text{const.}$, may also be differentiated to yield

$$\frac{dV}{V} + \frac{1}{(k-1)\text{M}^2}\frac{dT}{T} = 0 \tag{17.74}$$

Eliminating dT/T by combining the two preceding equations, there results

$$\frac{dV}{V} = \frac{2}{2 + (k-1)\text{M}^2}\frac{d\text{M}}{\text{M}} \tag{17.75}$$

This can be substituted into Eq. 17.72, allowing a relationship between θ and M to be obtained. We have

$$d\theta = \frac{2\sqrt{\text{M}^2 - 1}}{2 + (k-1)\text{M}^2}\frac{d\text{M}}{\text{M}} \tag{17.76}$$

This can be integrated, using $\theta = 0$ at $\text{M} = 1$, to provide a relationship between the resulting Mach number (M_2 in Fig. 17.17) and the angle, provided that the incoming Mach number is unity; the relationship is

$$\theta = \left(\frac{k+1}{k-1}\right)^{1/2}\tan^{-1}\left[\frac{k-1}{k+1}(\text{M}^2 - 1)\right]^{1/2} - \tan^{-1}(\text{M}^2 - 1)^{1/2} \tag{17.77}$$

Prandtl–Meyer function: *The angle θ through which the supersonic flow turns.*

The angle θ, which is a function of M, is the **Prandtl–Meyer function**. It is tabulated for $k = 1.4$ in Table I.3 so that trial-and-error solutions to Eq. 17.77 for M are not necessary. Other changes that may be desired, such as the pressure or temperature changes, can be found from the isentropic-flow equations. The collection of Mach waves that turn the flow are referred to as *expansion waves.*

Key Concept: *The flow remains attached to the wall as it turns the corner, even for large angles.*

We will see, in working the examples and problems, that both the Mach number and the velocity increase as supersonic flow turns the convex corner. The flow remains attached to the wall as it turns the corner, even for large angles, a phenomenon not observed in a subsonic flow; a subsonic flow would separate from the abrupt corner, even for small angles. If we substitute $\text{M} = \infty$ in Eq.

[1] Recall the trigonometric identity, $\cos(\alpha + \beta) = \cos\alpha\cos\beta - \sin\alpha\sin\beta$. Then using $\cos d\theta = 1$ and $\sin d\theta = d\theta$, we have $\cos(\mu + d\theta) = \cos\mu - d\theta\sin\mu$.

17.57 Two oblique shocks intersect as shown in Fig. P17.57. Determine the angle of the reflected shocks if the airflow must leave parallel to its original direction. Also find M_3.

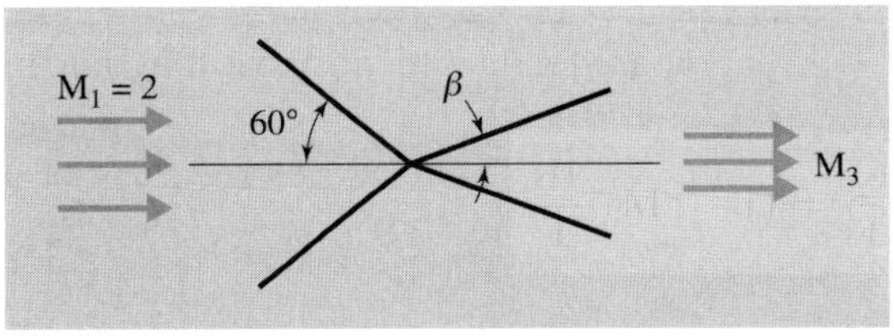

FIGURE P17.57

17.58 An oblique shock wave at a 35° angle is reflected from a plane wall. The upstream Mach number M_1 is 3.5 and $T_1 = 0°C$. Find V_3 after the reflected oblique shock wave for the airflow.

17.59 An oblique shock wave at a 35° angle is reflected from a plane wall. The upstream Mach number M_1 is 3.5 and $T_1 = 30°F$. Find V_3 after the reflected oblique shock wave for the airflow.

17.60 A supersonic inlet can be designed to have a normal shock wave oriented at the inlet, or a wedge can be used to provide a weak oblique shock wave, as shown in Fig. P17.60. Compare the pressure p_3 from the flow shown to the pressure that would exist behind a normal shock wave with no oblique shock.

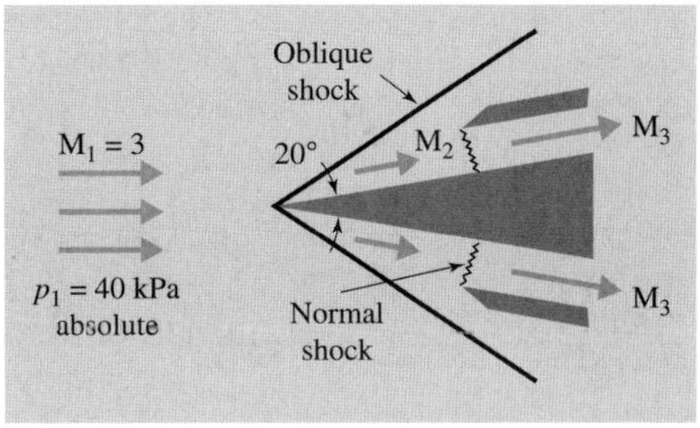

FIGURE P17.60

Expansion Waves

17.61 A supersonic airflow with $M_1 = 3$, $T_1 = -20°C$, and $p_1 = 20$ kPa absolute turns a convex corner of 25°. Calculate M_2, p_2, T_2, and V_2 after the expansion fan. Also calculate the included angle of the fan.

17.62 A supersonic airflow with $M_1 = 2$, $T_1 = 0°C$, and $p_1 = 20$ kPa absolute turns a convex corner. If $M_2 = 4$, what angle θ should the corner have? Also calculate T_2 and V_2.

17.63 A supersonic airflow with $M_1 = 2$, $T_1 = 30°F$, and $p_1 = 5$ psia turns a convex corner. If $M_2 = 4$, what angle θ should the corner have? Also calculate T_2 and V_2.

17.64 The flat plate shown in Fig. P17.64 is used as an airfoil at an angle of attack of 5°. Oblique shock waves and expansion fans allow the air to remain attached to the plate with the flow behind the airfoil parallel to the original direction. Calculate:

(a) The pressures on the upper and lower sides of the plate.

(b) The downstream Mach numbers M_{2u} and M_{2l}.

(c) The lift coefficient defined by $C_L = \text{lift}/(\frac{1}{2}\rho_1 V_1^2 A)$. Note that $\rho_1 V_1^2 = k M_1^2 p_1$.

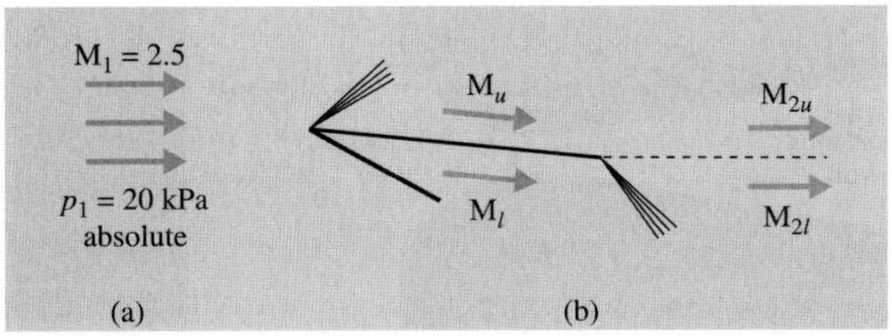

FIGURE P17.64

17.65 The supersonic airfoil shown in Fig. P17.65 is to fly at zero angle of attack. Calculate the drag coefficient $C_D = \text{drag}/(\frac{1}{2}\rho_1 V_1^2 A)$. Note that $\rho_1 V_1^2 = k M_1^2 p_1$.

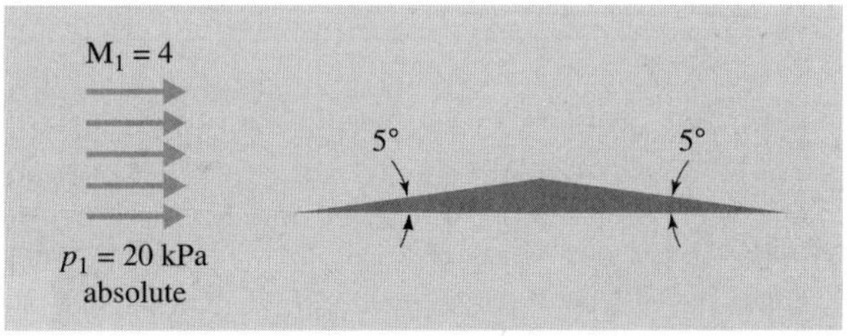

FIGURE P17.65

17.66 The airfoil in Problem 17.65 flies at an angle of attack of 5°. Determine the lift and drag coefficients, C_L and C_D. See Problem 17.64 and 17.65 for definitions of C_L and C_D.

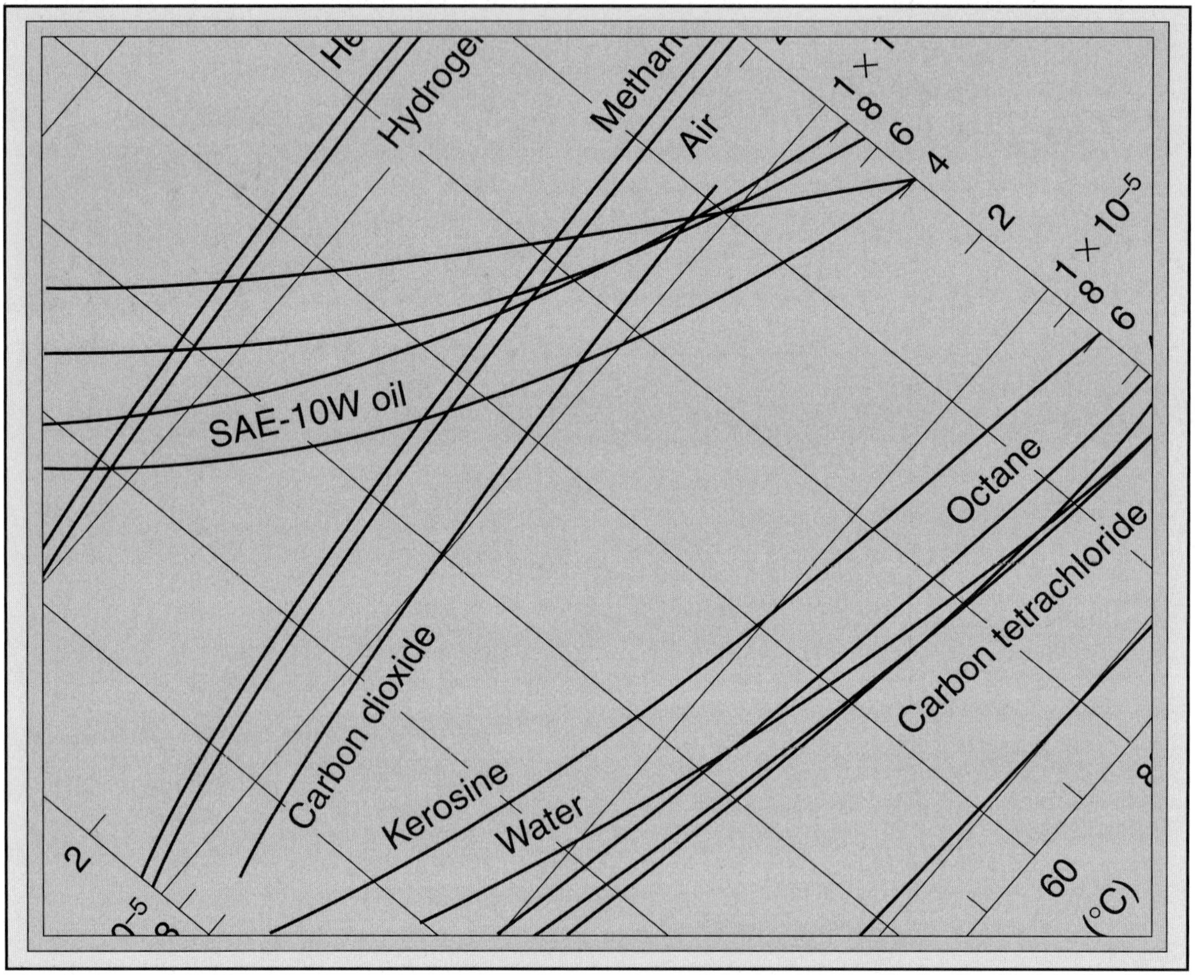

Air
SAE-10W oil
Carbon dioxide
Kerosine
Water
Octane
Carbon tetrachloride
60
(°C)

Appendix

A. UNITS AND CONVERSIONS

TABLE A.1 U.S. Engineering Units, SI Units, and Their Conversion Factors

Quantity	English units	International system[a] SI	Conversion factor
Length	inch	millimeter	1 in. = 25.4 mm
	foot	meter	1 ft = 0.3048 m
	mile	kilometer	1 mile = 1.609 km
Area	square inch	square centimeter	1 in^2 = 6.452 cm^2
	square foot	square meter	1 ft^2 = 0.09290 m^2
Volume	cubic inch	cubic centimeter	1 in^3 = 16.39 cm^3
	cubic foot	cubic meter	1 ft^3 = 0.02832 m^3
	gallon		1 gal = 0.003789 m^3
Mass	pound-mass	kilogram	1 lb_m = 0.4536 kg
	slug		1 slug = 14.59 kg
Density	slug/cubic foot	kilogram/cubic meter	1 slug/ft^3 = 515.4 kg/m^3
Force	pound-force	Newton	1 lb = 4.448 N
Work or torque	foot-pound	Newton-meter	1 ft-lb = 1.356 N·m
Pressure	pound/square inch	Newton/square meter (Pascal)	1 psi = 6895 Pa
	pound/square foot		1 psf = 47.88 Pa
Temperature	degree Fahrenheit	degree Celsius	___°F = $\frac{9}{5}$___°C + 32
	degree Rankine	kelvin	___°R = $\frac{9}{5}$___K
Energy	British thermal unit	joule	1 Btu = 1055 J
	calorie		1 cal = 4.186 J
	foot-pound		1 ft-lb = 1.356 J
Power	horsepower	Watt	1 hp = 745.7 W
	foot-pound/second		1 ft-lb/sec = 1.356 W
Velocity	foot/second	meter/second	1 ft/sec = 0.3048 m/s
Acceleration	foot/second squared	meter/second squared	1 ft/sec^2 = 0.3048 m/s^2
Frequency	cycle/second	Hertz	1 cps = 1.000 Hz
Viscosity	pound-second/square foot	Newton-second/square meter	1 lb-sec/ft^2 = 47.88 N·s/m^2

[a]The reversed initials in this abbreviation come from the French form of the name: Systéme International.

TABLE A.2 Conversion of Units

Length	Force	Mass	Velocity
1 cm = 0.3937 in.	1 lb = 0.4536 kg	1 oz = 28.35 g	1 mph = 1.467 ft/sec
1 m = 3.281 ft	1 lb = 0.4448 × 10^6 dyn	1 lb = 0.4536 kg	1 mph = 0.8684 knot
1 km = 0.6214 mile	1 lb = 32.17 pdl	1 slug = 32.17 lb	1 ft/sec = 0.3048 m/s
1 in. = 2.54 cm	1 kg = 2.205 lb	1 slug = 14.59 kg	1 m/s = 3.281 ft/sec
1 ft = 0.3048 m	1 pdl = 0.03108 lb	1 kg = 2.205 lb	1 km/h = 0.278 m/s
1 mile = 1.609 km	1 dyn = 2.248 × 10^{-6} lb	1 kg = 0.06852 slug	
1 mile = 5280 ft	1 lb = 4.448 N		
1 mile = 1760 yd	1 N = 0.2248 lb		

Work, energy, and power	Pressure	Volume	Flow rate
1 Btu = 778.2 ft-lb	1 psi = 2.036 in. Hg	1 ft^3 = 28.32 L	1 ft^3/min = 4.719 × 10^{-4} m^3/s
1 J = 10^7 ergs 1 J = 0.7376 ft-lb	1 psi = 27.7 in. H_2O	1 ft^3 = 7.481 gal (U.S.)	1 ft^3/sec = 0.02832 m^3/s
1 cal = 3.088 ft-lb	14.7 psi = 22.92 in. Hg	1 gal (U.S.) = 231 in^3	1 m^3/s = 35.31 ft^3/sec
1 cal = 0.003968 Btu	14.7 psi = 33.93 ft H_2O	1 gal (Brit.) = 1.2 gal (U.S.)	1 gal/min = 0.002228 ft^3/sec
1 kWh = 3413 Btu	14.7 psi = 1.0332 kg/cm^2	1 m^3 = 1000 L	1 ft^3/sec = 448.9 gal/min
1 Btu = 1.055 kJ	14.7 psi = 1.0133 bar	1 ft^3 = 0.02832 m^3	**Viscosity**
1 ft-lb = 1.356 J	1 kg/cm^2 = 14.22 psi	1 m^3 = 35.31 ft^3	1 stoke = 10^{-4} m^2/s
1 hp = 550 ft-lb/sec	1 in. Hg = 0.4912 psi		1 poise = 0.1 N·s/m^2
1 hp = 0.7067 Btu/sec	1 ft H_2O = 0.4331 psi		1 lb-sec/ft^2 = 47.88 N·s/m^2
1 hp = 0.7455 kW	1 psi = 6895 Pa		1 ft^2/sec = 0.0929 m^2/s
1 W = 1 J/s	1 psf = 47.88 Pa		
1 W = 1.0 × 10^7 dyn-cm/s	10^5 Pa = 1 bar		
1 erg = 10^{-7} J	1 kPa = 0.145 psi		
1 quad = 10^{15} Btu			
1 therm = 10^5 Btu			

B. MATERIAL PROPERTIES

TABLE B.1 Properties of the Standard Atmosphere

Altitude (m)	Temperature (K)	Pressure (kPa)	Density (kg/m^3)	Velocity of sound (m/s)
0	15.2	101.3	1.225	340
500	11.9	95.43	1.167	338
1000	8.7	89.85	1.112	336
2000	2.2	79.48	1.007	333
4000	−10.8	61.64	0.8194	325
6000	−23.8	47.21	0.6602	316
8000	−36.8	35.65	0.5258	308
10 000	−49.7	26.49	0.4136	300
12 000	−56.3	19.40	0.3119	295
14 000	−56.3	14.17	0.2278	295
16 000	−56.3	10.35	0.1665	295
18 000	−56.3	7.563	0.1216	295
20 000	−56.3	5.528	0.0889	295
30 000	−46.5	1.196	0.0184	302
40 000	−26.6	0.287	4.00×10^{-3}	317
50 000	−2.3	0.0798	1.03×10^{-3}	330
60 000	−17.2	0.0225	3.06×10^{-4}	321
70 000	−53.3	0.00551	8.75×10^{-5}	297
80 000	−92.3	0.00103	2.00×10^{-5}	269

TABLE B.1E Properties of the Standard Atmosphere

Altitude (ft)	Temperature (°F)	Pressure (lbf/ft^2)	Density ($slugs/ft^3$)	Velocity of sound (ft/sec)
0	59.0	2116	0.00237	1117
1000	55.4	2014	0.00231	1113
2000	51.9	1968	0.00224	1109
5000	41.2	1760	0.00205	1098
10 000	23.4	1455	0.00176	1078
15 000	5.54	1194	0.00150	1058
20 000	−12.3	973	0.00127	1037
25 000	−30.1	785	0.00107	1016
30 000	−48.0	628	0.000890	995
35 000	−65.8	498	0.000737	973
36 000	−67.6	475	0.000709	971
40 000	−67.6	392	0.000586	971
50 000	−67.6	242	0.000362	971
100 000	−51.4	23.2	3.31×10^{-5}	971

Note: Slug × 32.2 = lbm

TABLE B.2 Properties of Ideal Gases

Gas	Chemical Formula	Molar Mass	R		c_p		c_v		k
			kJ/kg·K	ft-lbf/lbm-°R	kJ/kg·K	Btu/lbm-°R	kJ/kg·K	Btu/lbm-°R	
Air	—	28.97	0.2870	53.34	1.003	0.240	0.717	0.171	1.400
Argon	Ar	39.95	0.2081	38.68	0.520	0.1253	0.312	0.0756	1.667
Butane	C_4H_{10}	58.12	0.1430	26.58	1.716	0.415	1.573	0.381	1.091
Carbon Dioxide	CO_2	44.01	0.1889	35.10	0.842	0.203	0.653	0.158	1.289
Carbon Monoxide	CO	28.01	0.2968	55.16	1.041	0.249	0.744	0.178	1.400
Ethane	C_2H_6	30.07	0.2765	51.38	1.766	0.427	1.490	0.361	1.186
Ethylene	C_2H_4	28.05	0.2964	55.07	1.548	0.411	1.252	0.340	1.237
Helium	He	4.00	2.0770	386.0	5.198	1.25	3.116	0.753	1.667
Hydrogen	H_2	2.02	4.1242	766.4	14.209	3.43	10.085	2.44	1.409
Methane	CH_4	16.04	0.5184	96.35	2.254	0.532	1.735	0.403	1.299
Neon	Ne	20.18	0.4120	76.55	1.020	0.246	0.618	0.1477	1.667
Nitrogen	N_2	28.01	0.2968	55.15	1.042	0.248	0.745	0.177	1.400
Octane	C_8H_{18}	114.23	0.0728	13.53	1.711	0.409	1.638	0.392	1.044
Oxygen	O_2	32.00	0.2598	48.28	0.922	0.219	0.662	0.157	1.393
Propane	C_3H_8	44.10	0.1886	35.04	1.679	0.407	1.491	0.362	1.126
Steam	H_2O	18.02	0.4615	85.76	1.872	0.445	1.411	0.335	1.327

Note: c_p, c_v, and k are at 300 K.
Source: G. J. Van Wylen and R. E. Sonntag, *Fundamentals of Classical Thermodynamics*, Wiley, New York, 1976.

TABLE B.3 Critical Constants

Substance	Formula	Molar Mass	Temperature		Pressure		Volume		Z_c
			K	°R	MPa	psia	ft³/lbmol	m³/kmol	
Air		28.97	133	239	3.76	547	0.0829	1.33	0.32
Ammonia	NH_3	17.03	405.5	729.8	11.28	1636	0.0724	1.16	0.243
Argon	Ar	39.94	151	272	4.86	705	0.0749	1.20	0.291
Benzene	C_6H_6	78.11	562	1012	4.92	714	0.2603	4.17	0.274
Butane	C_4H_{10}	58.12	425.2	765.2	3.80	551	0.2547	4.08	0.274
Carbon dioxide	CO_2	44.01	304.2	547.5	7.39	1070	0.0943	1.51	0.275
Carbon monoxide	CO	28.01	133	240	3.50	507	0.0930	1.49	0.294
Carbon tetrachloride	CCl_4	153.84	556.4	1001.5	4.56	661	0.2759	4.42	0.272
Dichlorodifluoromethane	CCl_2F_2	120.92	384.7	692.4	39.6	582		3.49	
Ethane	C_2H_6	30.07	305.5	549.8	4.88	708	0.1480	2.37	0.285
Ethylene	C_2H_4	28.05	282.4	508.3	5.12	742	0.1242	1.99	0.270
Helium	He	4.00	5.3	9.5	0.23	33.2	0.0578	0.93	0.308
Hydrogen	H_2	2.02	33.3	59.9	1.30	188	0.0649	1.04	0.304
Methane	CH_4	16.04	191.1	343.9	4.64	673	0.0993	1.59	0.290
Neon	Ne	20.18	44.5	80.1	2.73	395	0.0417	0.668	0.308
Nitrogen	N_2	28.02	126.2	227.1	3.39	492	0.0899	1.44	0.291
Oxygen	O_2	32.00	154.8	278.6	5.08	736	0.0780	1.25	0.292
Propane	C_3H_8	44.09	370.0	665.9	4.26	617	0.1998	3.20	0.277
Propene	C_3H_6	42.08	365.0	656.9	4.62	670	0.1810	2.90	0.274
Sulfur dioxide	SO_2	64.06	430.7	775.2	7.88	1143	0.1217	1.95	0.269
Water	H_2O	18.02	647.4	1165.3	22.1	3204	0.0568	0.90	0.230

Source: K. A. Kobe and R. E. Lynn, Jr., *Chem. Rev.*, **52**: 117–236 (1953).

TABLE B.4 Specific Heats of Liquids and Solids

c_p, kJ/kg·°C

Liquids

Substance	State	c_p	Substance	State	c_p
Water	1 atm, 25°C	4.177	Glycerin	1 atm, 10°C	2.32
Ammonia	sat., −20°C	4.52	Bismuth	1 atm, 425°C	0.144
	sat., 50°C	5.10	Mercury	1 atm, 10°C	0.138
Freon 12	sat., −20°C	0.908	Sodium	1 atm, 95°C	1.38
	sat., 50°C	1.02	Propane	1 atm, 0°C	2.41
Benzene	1 atm, 15°C	1.80	Ethyl Alcohol	1 atm, 25°C	2.43

Solids

Substance	T, °C	c_p	Substance	T, °C	c_p
Ice	−11	2.033	Lead	−100	0.118
	−2.2	2.10		0	0.124
Aluminum	−100	0.699		100	0.134
	0	0.870	Copper	−100	0.328
	100	0.941		0	0.381
Iron	20	0.448		100	0.393
Silver	20	0.233			

Source: Kenneth Wark, *Thermodynamics*, 3d ed., McGraw-Hill, New York, 1981.

TABLE B.4E Specific Heats of Liquids and Solids

c_p, Btu/lbm-°F

Liquids

Substance	State	c_p	Substance	State	c_p
Water	1 atm, 77°F	1.00	Glycerin	1 atm, 50°F	0.555
Ammonia	sat., −4°F	1.08	Bismuth	1 atm, 800°F	0.0344
	sat., 120°F	1.22	Mercury	1 atm, 50°F	0.0330
Freon 12	sat., −4°F	0.217	Sodium	1 atm, 200°F	0.330
	sat., 120°F	0.244	Propane	1 atm, 32°F	0.577
Benzene	1 atm, 60°F	0.431	Ethyl Alcohol	1 atm, 77°F	0.581

Solids

Substance	T, °F	c_p	Substance	T, °F	c_p
Ice	−76	0.392	Silver	−4	0.0557
	12	0.486	Lead	−150	0.0282
	28	0.402		30	0.0297
Aluminum	−150	0.167		210	0.0321
	30	0.208	Copper	−150	0.0785
	210	0.225		30	0.0911
Iron	−4	0.107		210	0.0940

Source: Kenneth Wark, *Thermodynamics*, 3d ed., McGraw-Hill, New York, 1981.

TABLE B.5 Constant–Pressure Specific Heats of Various Ideal Gases

$\theta \equiv T(\text{Kelvin})/100$

Gas	$\bar{c}_p$ kJ/kmol·K	Range K	Max. Error %
N_2	$39.060 - 512.79\theta^{-1.5} + 1072.7\theta^{-2} - 820.40\theta^{-3}$	300–3500	0.43
O_2	$37.432 + 0.020102\theta^{1.5} - 178.57\theta^{-1.5} + 236.88\theta^{-2}$	300–3500	0.30
H_2	$56.505 - 702.74\theta^{-0.75} + 1165.0\theta^{-1} - 560.70\theta^{-1.5}$	300–3500	0.60
CO	$69.145 - 0.70463\theta^{0.75} - 200.77\theta^{-0.5} + 176.76\theta^{-0.75}$	300–3500	0.42
OH	$81.546 - 59.350\theta^{0.25} + 17.329\theta^{0.75} - 4.2660\theta$	300–3500	0.43
NO	$59.283 - 1.7096\theta^{0.5} - 70.613\theta^{-0.5} + 74.889\theta^{-1.5}$	300–3500	0.34
H_2O	$143.05 - 183.54\theta^{0.25} + 82.751\theta^{0.5} - 3.6989\theta$	300–3500	0.43
CO_2	$-3.7357 + 30.529\theta^{0.5} - 4.1034\theta + 0.024198\theta^{2}$	300–3500	0.19
NO_2	$46.045 + 216.10\theta^{-0.5} - 363.66\theta^{-0.75} + 232.550\theta^{-2}$	300–3500	0.26
CH_4	$-672.87 + 439.74\theta^{0.25} - 24.875\theta^{0.75} + 323.88\theta^{-0.5}$	300–2000	0.15
C_2H_4	$-95.395 + 123.15\theta^{0.5} - 35.641\theta^{0.75} + 182.77\theta^{-3}$	300–2000	0.07

Source: G. J. Van Wylen and R. E. Sonntag, *Fundamentals of Classical Thermodynamics*, Wiley, New York, 1976.

TABLE B.5E Constant–Pressure Specific Heats of Various Ideal Gases

$\theta \equiv T(\text{Rankine})/180$

Gas	$\bar{c}_p$ Btu/lbmol-°R	Range °R	Max. Error %
N_2	$9.3355 - 122.56\theta^{-1.5} + 256.38\theta^{-2} - 196.08\theta^{-3}$	540–6300	0.43
O_2	$8.9465 + 4.8044 \times 10^{-3}\theta^{1.5} - 42.679\theta^{-1.5} + 56.615\theta^{-2}$	540–6300	0.30
H_2	$13.505 - 167.96\theta^{-0.75} + 278.44\theta^{-1} - 134.01\theta^{-1.5}$	540–6300	0.60
CO	$16.526 - 0.16841\theta^{0.75} - 47.985\theta^{-0.5} + 42.246\theta^{-0.75}$	540–6300	0.42
OH	$19.490 - 14.185\theta^{0.25} + 4.1418\theta^{0.75} - 1.0196\theta$	540–6300	0.43
NO	$14.169 - 0.40861\theta^{0.5} - 16.877\theta^{-0.5} + 17.899\theta^{-1.5}$	540–6300	0.34
H_2O	$34.190 - 43.868\theta^{0.25} + 19.778\theta^{0.5} - 0.88407\theta$	540–6300	0.43
CO_2	$-0.89286 + 7.2967\theta^{0.5} - 0.98074\theta + 5.7835 \times 10^{-3}\theta^{2}$	540–6300	0.19
NO_2	$11.005 + 51.650\theta^{-0.5} - 86.916\theta^{-0.75} + 55.580\theta^{-2}$	540–6300	0.26
CH_4	$-160.82 + 105.10\theta^{0.25} - 5.9452\theta^{0.75} + 77.408\theta^{-0.5}$	540–3600	0.15
C_2H_4	$-22.800 + 29.433\theta^{0.5} - 8.5185\theta^{0.75} + 43.683\theta^{-3}$	540–3600	0.07

Source: G. J. Van Wylen and R. E. Sonntag, *Fundamentals of Classical Thermodynamics*, Wiley, New York, 1976.

TABLE B.6 Enthalpy of Formation and Enthalpy of Vaporization

25°C (77°F), 1 atm					
Substance	Formula	$\bar{h}_f^\circ$ kJ/kmol	$\bar{h}_{fg}$ kJ/kmol	$\bar{h}_f^\circ$ Btu/lbmol	$\bar{h}_{fg}$ Btu/lbmol
Carbon	$C(s)$	0		0	
Hydrogen	$H_2(g)$	0		0	
Nitrogen	$N_2(g)$	0		0	
Oxygen	$O_2(g)$	0		0	
Carbon monoxide	$CO(g)$	−110 530		−47 540	
Carbon dioxide	$CO_2(g)$	−393 520		−169 300	
Water	$H_2O(g)$	−241 820		−104 040	
Water	$H_2O(l)$	−285 830	44 010	−122 970	
Hydrogen peroxide	$H_2O_2(g)$	−136 310	61 090	−58 640	26 260
Ammonia	$NH_3(g)$	−46 190		−19 750	
Oxygen	$O(g)$	249 170		+107 210	
Hydrogen	$H(g)$	218 000		+93 780	
Nitrogen	$N(g)$	472 680		+203 340	
Hydroxyl	$OH(g)$	39 040		+16 790	
Methane	$CH_4(g)$	−74 850		−32 210	
Acetylene (Ethyne)	$C_2H_2(g)$	226 730		+97 540	
Ethylene (Ethene)	$C_2H_4(g)$	52 280		+22 490	
Ethane	$C_2H_6(g)$	−84 680		−36 420	
Propylene (Propene)	$C_3H_6(g)$	20 410		+8790	
Propane	$C_3H_8(g)$	−103 850	15 060	−44 680	6480
n-Butane	$C_4H_{10}(g)$	−126 150	21 060	−54 270	9060
n-Pentane	$C_5H_{12}(g)$	−146 440	31 410		
n-Octane	$C_8H_{18}(g)$	−208 450	41 460	−89 680	17 835
Benzene	$C_6H_6(g)$	82 930	33 830	+35 680	14 550
Methyl alcohol	$CH_3OH(g)$	−200 890	37 900	−86 540	16 090
Ethyl alcohol	$C_2H_5OH(g)$	−235 310	42 340	−101 230	18 220

Sources: JANAF Thermochemical Tables, NSRDS-NBS-37, 1971; *Selected Values of Chemical Thermodynamic Properties*, NBS Technical Note 270-3, 1968; and API Res. Project 44, Carnegie Press, Carnegie Institute of Technology, Pittsburgh, 1953.

TABLE B.7 Enthalpy of Combustion and Enthalpy of Vaporization

		25°C (77°F), 1 atm			
Substance	Formula	−HHV kJ/kmol	$\bar{h}_{fg}$ kJ/kmol	−HHV Btu/lbmol	$\bar{h}_{fg}$ Btu/lbmol
Hydrogen	$H_2(g)$	−285 840		−122 970	
Carbon	$C(s)$	−393 520		−169 290	
Carbon monoxide	$CO(g)$	−282 990		−121 750	
Methane	$CH_4(g)$	−890 360		−383 040	
Acetylene	$C_2H_2(g)$	−1 299 600		−559 120	
Ethylene	$C_2H_4(g)$	−1 410 970		−607 010	
Ethane	$C_2H_6(g)$	−1 559 900		−671 080	
Propylene	$C_3H_6(g)$	−2 058 500		−885 580	
Propane	$C_3H_8(g)$	−2 220 000	15 060	−955 070	6480
n-Butane	$C_4H_{10}(g)$	−2 877 100	21 060	−1 237 800	9060
n-Pentane	$C_5H_{12}(g)$	−3 536 100	26 410	−1 521 300	11 360
n-Hexane	$C_6H_{14}(g)$	−4 194 800	31 530	−1 804 600	13 560
n-Heptane	$C_7H_{16}(g)$	−4 853 500	36 520	−2 088 000	15 710
n-Octane	$C_8H_{18}(g)$	−5 512 200	41 460	−2 371 400	17 835
Benzene	$C_6H_6(g)$	−3 301 500	33 830	−1 420 300	14 550
Toluene	$C_7H_8(g)$	−3 947 900	39 920	−1 698 400	17 180
Methyl alcohol	$CH_3OH(g)$	−764 540	37 900	−328 700	16 090
Ethyl alcohol	$C_2H_5OH(g)$	−1 409 300	42 340	−606 280	18 220

Note: Water appears as a liquid in the products of combustion.
Source: Kenneth Wark, *Thermodynamics,* 3d ed., McGraw-Hill, New York, 1981, pp. 834–835, Table A-23M.

TABLE B.8 Constants for the van der Waals and the Redlich-Kwong Equations of State

van der Waals equation

	a, kPa·m^6/kg^2	b, m^3/kg	a, lbf-ft^4/lbm^2	b, ft^3/lbm
Air	0.1630	0.00127	870	0.0202
Ammonia	1.468	0.00220	7850	0.0351
Carbon Dioxide	0.1883	0.000972	1010	0.0156
Carbon Monoxide	0.1880	0.00141	1010	0.0227
Freon 12	0.0718	0.000803	394	0.0132
Helium	0.214	0.00587	1190	0.0959
Hydrogen	6.083	0.0132	32 800	0.212
Methane	0.888	0.00266	4780	0.0427
Nitrogen	0.1747	0.00138	934	0.0221
Oxygen	0.1344	0.000993	720	0.0159
Propane	0.481	0.00204	2580	0.0328
Water	1.703	0.00169	9130	0.0271

Redlich-Kwong Equation

	a, kPa·m^6·K$^{1/2}$/kg^2	b, m^3/kg	a, lbf-ft^4-°R$^{1/2}$/lbm^2	b, ft^3/lbm
Air	1.905	0.000878	13 600	0.014
Ammonia	30.0	0.00152	215 000	0.0243
Carbon Dioxide	3.33	0.000674	24 000	0.0108
Carbon Monoxide	2.20	0.000978	15 900	0.0157
Freon 12	1.43	0.000557	10 500	0.00916
Helium	0.495	0.00407	3710	0.0665
Hydrogen	35.5	0.00916	257 000	0.147
Methane	12.43	0.00184	89 700	0.0296
Nitrogen	1.99	0.000957	14 300	0.0153
Oxygen	1.69	0.000689	12 200	0.0110
Propane	9.37	0.00141	67 600	0.0228
Water	43.9	0.00117	316 000	0.0188

TABLE B.9 Properties of Water

Temperature T (°C)	Density ρ (kg/m^3)	Viscosity μ (N·s/m^2)	Kinematic viscosity ν (m^2/s)	Surface tension σ (N/m)	Vapor pressure p_v (kPa)	Bulk modulus B (Pa)
0	999.9	1.792×10^{-3}	1.792×10^{-6}	0.0762	0.610	204×10^7
5	1000.0	1.519	1.519	0.0754	0.872	206
10	999.7	1.308	1.308	0.0748	1.13	211
15	999.1	1.140	1.141	0.0741	1.60	214
20	998.2	1.005	1.007	0.0736	2.34	220
30	995.7	0.801	0.804	0.0718	4.24	223
40	992.2	0.656	0.661	0.0701	3.38	227
50	988.1	0.549	0.556	0.0682	12.3	230
60	983.2	0.469	0.477	0.0668	19.9	228
70	977.8	0.406	0.415	0.0650	31.2	225
80	971.8	0.357	0.367	0.0630	47.3	221
90	965.3	0.317	0.328	0.0612	70.1	216
100	958.4	0.284×10^{-3}	0.296×10^{-6}	0.0594	101.3	207×10^7

TABLE B.9E Properties of Water

Temperature (°F)	Density (slug/ft^3)	Viscosity (lb-sec/ft^2)	Kinematic viscosity (ft^2/sec)	Surface tension (lb/ft)	Vapor pressure (psi)	Bulk modulus (psi)
32	1.94	3.75×10^{-5}	1.93×10^{-5}	0.518×10^{-2}	0.089	293 000
40	1.94	3.23	1.66	0.514	0.122	294 000
50	1.94	2.74	1.41	0.509	0.178	305 000
60	1.94	2.36	1.22	0.504	0.256	311 000
70	1.94	2.05	1.06	0.500	0.340	320 000
80	1.93	1.80	0.93	0.492	0.507	322 000
90	1.93	1.60	0.83	0.486	0.698	323 000
100	1.93	1.42	0.74	0.480	0.949	327 000
120	1.92	1.17	0.61	0.465	1.69	333 000
140	1.91	0.98	0.51	0.454	2.89	330 000
160	1.90	0.84	0.44	0.441	4.74	326 000
180	1.88	0.73	0.39	0.426	7.51	318 000
200	1.87	0.64	0.34	0.412	11.53	308 000
212	1.86	0.59×10^{-5}	0.32×10^{-5}	0.404×10^{-2}	14.7	300 000

TABLE B.10 Properties of Common Liquids at Atmospheric Pressure and Approximately 60 to 70°F (16 to 21°C)

Liquid	Specific weight γ		Density ρ		Surface tension[a] σ		Vapor pressure p_v	
	lbf/ft³	N/m³	slugs/ft³	kg/m³	lbf/ft	N/m	psia	kPa
Alcohol, ethyl	49.3	7744	1.53	789	0.0015	0.022	—	—
Benzene	56.2	8828	1.75	902	0.0020	0.029	1.50	10.3
Carbon tetrachloride	99.5	15 629	3.09	1593	0.0018	0.026	12.50	86.2
Gasoline	42.4	6660	1.32	680	—	—	—	—
Glycerin	78.6	12 346	2.44	1258	0.0043	0.063	2×10^{-6}	1.4×10^{-5}
Kerosene	50.5	7933	1.57	809	0.0017	0.025	—	—
Mercury	845.5	132 800	26.29	13 550	0.032	0.467	2.31×10^{-5}	1.59×10
SAE 10 oil	57.4	9016	1.78	917	0.0025	0.036	—	—
SAE 30 oil	57.4	9016	1.78	917	0.0024	0.035	—	—
Turpentine	54.3	8529	1.69	871	0.0018	0.026	7.7×10^{-3}	5.31×10^{-2}
Water	62.4	9810	1.94	1000	0.0050	0.073	0.34	2.34

[a]In contact with air.

TABLE B.11 Properties of Air at Atmospheric Pressure

Temperature T (°C)	Density ρ (kg/m³)	Viscosity μ (N·s/m²)	Kinematic viscosity ν (m²/s)	Velocity of sound c (m/s)
−50	1.582	1.46×10^{-5}	0.921×10^{-5}	299
−30	1.452	1.56	1.08×10^{-5}	312
−20	1.394	1.61	1.16	319
−10	1.342	1.67	1.24	325
0	1.292	1.72	1.33	331
10	1.247	1.76	1.42	337
20	1.204	1.81	1.51	343
30	1.164	1.86	1.60	349
40	1.127	1.91	1.69	355
50	1.092	1.95	1.79	360
60	1.060	2.00	1.89	366
70	1.030	2.05	1.99	371
80	1.000	2.09	2.09	377
90	0.973	2.13	2.19	382
100	0.946	2.17	2.30	387
200	0.746	2.57	3.45	436
300	0.616	2.93×10^{-5}	4.75×10^{-5}	480

TABLE B.11E Properties of Air at Atmospheric Pressure

Temperature (°F)	Density ($slug/ft^3$)	Viscosity ($lb\text{-}sec/ft^2$)	Kinematic viscosity (ft^2/sec)	Velocity of sound (ft/sec)
−20	0.00280	3.34×10^{-7}	11.9×10^{-5}	1028
0	0.00268	3.38	12.6	1051
20	0.00257	3.50	13.6	1074
40	0.00247	3.62	14.6	1096
60	0.00237	3.74	15.8	1117
68	0.00233	3.81	16.0	1125
80	0.00228	3.85	16.9	1138
100	0.00220	3.96	18.0	1159
120	0.00213	4.07	18.9	1180
160	0.00199	4.23	21.3	1220
200	0.00187	4.50	24.1	1258
300	0.00162	4.98	30.7	1348
400	0.00144	5.26	36.7	1431
1000	0.000844	7.87×10^{-7}	93.2×10^{-5}	1839

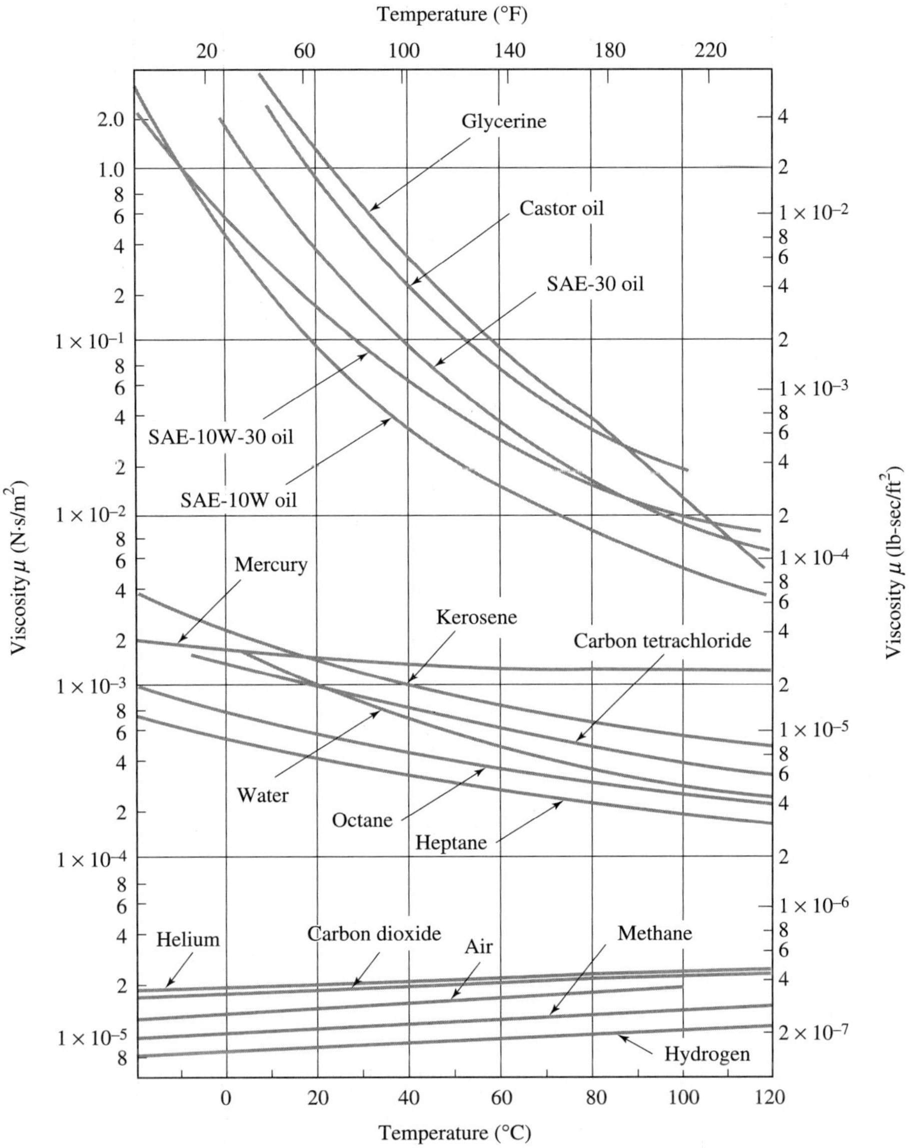

FIGURE B.1 Viscosity as a function of temperature. (From R. W. Fox and T. A. McDonald, *Introduction to Fluid Mechanics*, 2nd ed., John Wiley & Sons, Inc., New York, 1978.)

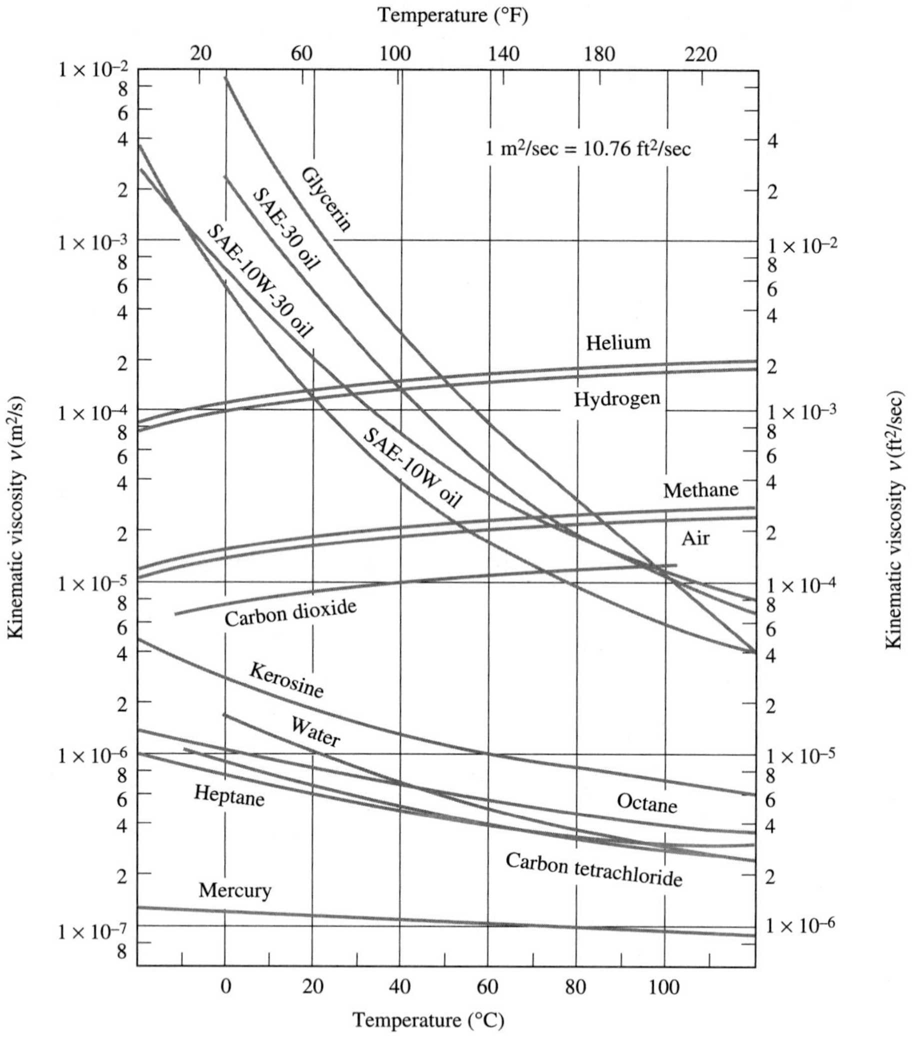

FIGURE B.2 Kinematic viscosity (at atmospheric pressure) as a function of temperature. (From R. W. Fox and A. T. McDonald, *Introduction to Fluid Mechanics*, 2nd ed., John Wiley & Sons, Inc., New York, 1978.)

C. THERMODYNAMIC PROPERTIES OF WATER (STEAM TABLES)

TABLE C.1 Properties of Saturated H_2O—Temperature Table

T, °C	p, MPa	Volume, m^3/kg		Energy, kJ/kg		Enthalpy, kJ/kg			Entropy, kJ/kg·K		
		v_f	v_g	u_f	u_g	h_f	h_{fg}	h_g	s_f	s_{fg}	s_g
0.010	0.0006113	0.001000	206.1	0.0	2375.3	0.0	2501.3	2501.3	0.0000	9.1571	9.1571
2	0.0007056	0.001000	179.9	8.4	2378.1	8.4	2496.6	2505.0	0.0305	9.0738	9.1043
5	0.0008721	0.001000	147.1	21.0	2382.2	21.0	2489.5	2510.5	0.0761	8.9505	9.0266
10	0.001228	0.001000	106.4	42.0	2389.2	42.0	2477.7	2519.7	0.1510	8.7506	8.9016
15	0.001705	0.001001	77.93	63.0	2396.0	63.0	2465.9	2528.9	0.2244	8.5578	8.7822
20	0.002338	0.001002	57.79	83.9	2402.9	83.9	2454.2	2538.1	0.2965	8.3715	8.6680
25	0.003169	0.001003	43.36	104.9	2409.8	104.9	2442.3	2547.2	0.3672	8.1916	8.5588
30	0.004246	0.001004	32.90	125.8	2416.6	125.8	2430.4	2556.2	0.4367	8.0174	8.4541
35	0.005628	0.001006	25.22	146.7	2423.4	146.7	2418.6	2565.3	0.5051	7.8488	8.3539
40	0.007383	0.001008	19.52	167.5	2430.1	167.5	2406.8	2574.3	0.5723	7.6855	8.2578
45	0.009593	0.001010	15.26	188.4	2436.8	188.4	2394.8	2583.2	0.6385	7.5271	8.1656
50	0.01235	0.001012	12.03	209.3	2443.5	209.3	2382.8	2592.1	0.7036	7.3735	8.0771
55	0.01576	0.001015	9.569	230.2	2450.1	230.2	2370.7	2600.9	0.7678	7.2243	7.9921
60	0.01994	0.001017	7.671	251.1	2456.6	251.1	2358.5	2609.6	0.8310	7.0794	7.9104
65	0.02503	0.001020	6.197	272.0	2463.1	272.0	2346.2	2618.2	0.8934	6.9384	7.8318
70	0.03119	0.001023	5.042	292.9	2469.5	293.0	2333.8	2626.8	0.9549	6.8012	7.7561
75	0.03858	0.001026	4.131	313.9	2475.9	313.9	2321.4	2635.3	1.0155	6.6678	7.6833
80	0.04739	0.001029	3.407	334.8	2482.2	334.9	2308.8	2643.7	1.0754	6.5376	7.6130
85	0.05783	0.001032	2.828	355.8	2488.4	355.9	2296.0	2651.9	1.1344	6.4109	7.5453
90	0.07013	0.001036	2.361	376.8	2494.5	376.9	2283.2	2660.1	1.1927	6.2872	7.4799
95	0.08455	0.001040	1.982	397.9	2500.6	397.9	2270.2	2668.1	1.2503	6.1664	7.4167
100	0.1013	0.001044	1.673	418.9	2506.5	419.0	2257.0	2676.0	1.3071	6.0486	7.3557
110	0.1433	0.001052	1.210	461.1	2518.1	461.3	2230.2	2691.5	1.4188	5.8207	7.2395
120	0.1985	0.001060	0.8919	503.5	2529.2	503.7	2202.6	2706.3	1.5280	5.6024	7.1304
130	0.2701	0.001070	0.6685	546.0	2539.9	546.3	2174.2	2720.5	1.6348	5.3929	7.0277
140	0.3613	0.001080	0.5089	588.7	2550.0	589.1	2144.8	2733.9	1.7395	5.1912	6.9307
150	0.4758	0.001090	0.3928	631.7	2559.5	632.2	2114.2	2746.4	1.8422	4.9965	6.8387
160	0.6178	0.001102	0.3071	674.9	2568.4	675.5	2082.6	2758.1	1.9431	4.8079	6.7510
170	0.7916	0.001114	0.2428	718.3	2576.5	719.2	2049.5	2768.7	2.0423	4.6249	6.6672
180	1.002	0.001127	0.1941	762.1	2583.7	763.2	2015.0	2778.2	2.1400	4.4466	6.5866
190	1.254	0.001141	0.1565	806.2	2590.0	807.5	1978.8	2786.4	2.2363	4.2724	6.5087
200	1.554	0.001156	0.1274	850.6	2595.3	852.4	1940.8	2793.2	2.3313	4.1018	6.4331
210	1.906	0.001173	0.1044	895.5	2599.4	897.7	1900.8	2798.5	2.4253	3.9340	6.3593
220	2.318	0.001190	0.08620	940.9	2602.4	943.6	1858.5	2802.1	2.5183	3.7686	6.2869
230	2.795	0.001209	0.07159	986.7	2603.9	990.1	1813.9	2804.0	2.6105	3.6050	6.2155
240	3.344	0.001229	0.05977	1033.2	2604.0	1037.3	1766.5	2803.8	2.7021	3.4425	6.1446
250	3.973	0.001251	0.05013	1080.4	2602.4	1085.3	1716.2	2801.5	2.7933	3.2805	6.0738
260	4.688	0.001276	0.04221	1128.4	2599.0	1134.4	1662.5	2796.9	2.8844	3.1184	6.0028
270	5.498	0.001302	0.03565	1177.3	2593.7	1184.5	1605.2	2789.7	2.9757	2.9553	5.9310
280	6.411	0.001332	0.03017	1227.4	2586.1	1236.0	1543.6	2779.6	3.0674	2.7905	5.8579
290	7.436	0.001366	0.02557	1278.9	2576.0	1289.0	1477.2	2766.2	3.1600	2.6230	5.7830
300	8.580	0.001404	0.02168	1332.0	2563.0	1344.0	1405.0	2749.0	3.2540	2.4513	5.7053
310	9.856	0.001447	0.01835	1387.0	2546.4	1401.3	1326.0	2727.3	3.3500	2.2739	5.6239
320	11.27	0.001499	0.01549	1444.6	2525.5	1461.4	1238.7	2700.1	3.4487	2.0883	5.5370
330	12.84	0.001561	0.01300	1505.2	2499.0	1525.3	1140.6	2665.9	3.5514	1.8911	5.4425
340	14.59	0.001638	0.01080	1570.3	2464.6	1594.2	1027.9	2622.1	3.6601	1.6765	5.3366
350	16.51	0.001740	0.008815	1641.8	2418.5	1670.6	893.4	2564.0	3.7784	1.4338	5.2122
360	18.65	0.001892	0.006947	1725.2	2351.6	1760.5	720.7	2481.2	3.9154	1.1382	5.0536
370	21.03	0.002213	0.004931	1844.0	2229.0	1890.5	442.2	2332.7	4.1114	0.6876	4.7990
374.136	22.088	0.003155	0.003155	2029.6	2029.6	2099.3	0.0	2099.3	4.4305	0.0000	4.4305

Sources: Keenan, Keyes, Hill, and Moore, *Steam Tables,* Wiley, New York, 1969; G. J. Van Wylen and R. E. Sonntag, *Fundamentals of Classical Thermodynamics,* Wiley, New York, 1973.

TABLE C.2 Properties of Saturated H_2O—Pressure Table

p, MPa	T, °C	Volume, m^3/kg		Energy, kJ/kg		Enthalpy, kJ/kg			Entropy, kJ/kg·K		
		v_f	v_g	u_f	u_g	h_f	h_{fg}	h_g	s_f	s_{fg}	s_g
0.000611	0.01	0.001000	206.1	0.0	2375.3	0.0	2501.3	2501.3	0.0000	9.1571	9.1571
0.0008	3.8	0.001000	159.7	15.8	2380.5	15.8	2492.5	2508.3	0.0575	9.0007	9.0582
0.001	7.0	0.001000	129.2	29.3	2385.0	29.3	2484.9	2514.2	0.1059	8.8706	8.9765
0.0012	9.7	0.001000	108.7	40.6	2388.7	40.6	2478.5	2519.1	0.1460	8.7639	8.9099
0.0014	12.0	0.001001	93.92	50.3	2391.9	50.3	2473.1	2523.4	0.1802	8.6736	8.8538
0.0016	14.0	0.001001	82.76	58.9	2394.7	58.9	2468.2	2527.1	0.2101	8.5952	8.8053
0.0018	15.8	0.001001	74.03	66.5	2397.2	66.5	2464.0	2530.5	0.2367	8.5259	8.7626
0.002	17.5	0.001001	67.00	73.5	2399.5	73.5	2460.0	2533.5	0.2606	8.4639	8.7245
0.003	24.1	0.001003	45.67	101.0	2408.5	101.0	2444.5	2545.5	0.3544	8.2240	8.5784
0.004	29.0	0.001004	34.80	121.4	2415.2	121.4	2433.0	2554.4	0.4225	8.0529	8.4754
0.006	36.2	0.001006	23.74	151.5	2424.9	151.5	2415.9	2567.4	0.5208	7.8104	8.3312
0.008	41.5	0.001008	18.10	173.9	2432.1	173.9	2403.1	2577.0	0.5924	7.6371	8.2295
0.01	45.8	0.001010	14.67	191.8	2437.9	191.8	2392.8	2584.6	0.6491	7.5019	8.1510
0.012	49.4	0.001012	12.36	206.9	2442.7	206.9	2384.1	2591.0	0.6961	7.3910	8.0871
0.014	52.6	0.001013	10.69	220.0	2446.9	220.0	2376.6	2596.6	0.7365	7.2968	8.0333
0.016	55.3	0.001015	9.433	231.5	2450.5	231.5	2369.9	2601.4	0.7719	7.2149	7.9868
0.018	57.8	0.001016	8.445	241.9	2453.8	241.9	2363.9	2605.8	0.8034	7.1425	7.9459
0.02	60.1	0.001017	7.649	251.4	2456.7	251.4	2358.3	2609.7	0.8319	7.0774	7.9093
0.03	69.1	0.001022	5.229	289.2	2468.4	289.2	2336.1	2625.3	0.9439	6.8256	7.7695
0.04	75.9	0.001026	3.993	317.5	2477.0	317.6	2319.1	2636.7	1.0260	6.6449	7.6709
0.06	85.9	0.001033	2.732	359.8	2489.6	359.8	2293.7	2653.5	1.1455	6.3873	7.5328
0.08	93.5	0.001039	2.087	391.6	2498.8	391.6	2274.1	2665.7	1.2331	6.2023	7.4354
0.1	99.6	0.001043	1.694	417.3	2506.1	417.4	2258.1	2675.5	1.3029	6.0573	7.3602
0.12	104.8	0.001047	1.428	439.2	2512.1	439.3	2244.2	2683.5	1.3611	5.9378	7.2980
0.14	109.3	0.001051	1.237	458.2	2517.3	458.4	2232.0	2690.4	1.4112	5.8360	7.2472
0.16	113.3	0.001054	1.091	475.2	2521.8	475.3	2221.2	2696.5	1.4553	5.7472	7.2025
0.18	116.9	0.001058	0.9775	490.5	2525.9	490.7	2211.1	2701.8	1.4948	5.6683	7.1631
0.2	120.2	0.001061	0.8857	504.5	2529.5	504.7	2201.9	2706.6	1.5305	5.5975	7.1280
0.3	133.5	0.001073	0.6058	561.1	2543.6	561.5	2163.8	2725.3	1.6722	5.3205	6.9927
0.4	143.6	0.001084	0.4625	604.3	2553.6	604.7	2133.8	2738.5	1.7770	5.1197	6.8967
0.6	158.9	0.001101	0.3157	669.9	2567.4	670.6	2086.2	2756.8	1.9316	4.8293	6.7609
0.8	170.4	0.001115	0.2404	720.2	2576.8	721.1	2048.0	2769.1	2.0466	4.6170	6.6636
1	179.9	0.001127	0.1944	761.7	2583.6	762.8	2015.3	2778.1	2.1391	4.4482	6.5873
1.2	188.0	0.001139	0.1633	797.3	2588.8	798.6	1986.2	2784.8	2.2170	4.3072	6.5242
1.4	195.1	0.001149	0.1408	828.7	2592.8	830.3	1959.7	2790.0	2.2847	4.1854	6.4701
1.6	201.4	0.001159	0.1238	856.9	2596.0	858.8	1935.2	2794.0	2.3446	4.0780	6.4226
1.8	207.2	0.001168	0.1104	882.7	2598.4	884.8	1912.3	2797.1	2.3986	3.9816	6.3802
2	212.4	0.001177	0.09963	906.4	2600.3	908.8	1890.7	2799.5	2.4478	3.8939	6.3417
3	233.9	0.001216	0.06668	1004.8	2604.1	1008.4	1795.7	2804.1	2.6462	3.5416	6.1878
4	250.4	0.001252	0.04978	1082.3	2602.3	1087.3	1714.1	2801.4	2.7970	3.2739	6.0709
6	275.6	0.001319	0.03244	1205.4	2589.7	1213.3	1571.0	2784.3	3.0273	2.8627	5.8900
8	295.1	0.001384	0.02352	1305.6	2569.8	1316.6	1441.4	2758.0	3.2075	2.5365	5.7440
9	303.4	0.001418	0.02048	1350.5	2557.8	1363.3	1378.8	2742.1	3.2865	2.3916	5.6781
10	311.1	0.001452	0.01803	1393.0	2544.4	1407.6	1317.1	2724.7	3.3603	2.2546	5.6149
12	324.8	0.001527	0.01426	1472.9	2513.7	1491.3	1193.6	2684.9	3.4970	1.9963	5.4933
14	336.8	0.001611	0.01149	1548.6	2476.8	1571.1	1066.5	2637.6	3.6240	1.7486	5.3726
16	347.4	0.001711	0.009307	1622.7	2431.8	1650.0	930.7	2580.7	3.7468	1.4996	5.2464
18	357.1	0.001840	0.007491	1698.9	2374.4	1732.0	777.2	2509.2	3.8722	1.2332	5.1054
20	365.8	0.002036	0.005836	1785.6	2293.2	1826.3	583.7	2410.0	4.0146	0.9135	4.9281
22.088	374.136	0.003155	0.003155	2029.6	2029.6	2099.3	0.0	2099.3	4.4305	0.0000	4.4305

Sources: Keenan, Keyes, Hill, and Moore, *Steam Tables*, Wiley, New York, 1969; G. J. Van Wylen and R. E. Sonntag, *Fundamentals of Classical Thermodynamics*, Wiley, New York, 1973.

TABLE C.3 Superheated Steam (T in °C, v in m^3/kg, u and h in kJ/kg, s in kJ/kg·K)

Superheated Steam												
T	v	u	h	s	v	u	h	s	v	u	h	s
	p = 0.010 MPa (45.81)				p = 0.050 MPa (81.33)				p = 0.10 MPa (99.63)			
Sat.	14.674	2437.9	2584.7	8.1502	3.240	2483.9	2645.9	7.5939	1.6940	2506.1	2675.5	7.3594
50	14.869	2443.9	2592.6	8.1749								
100	17.196	2515.5	2687.5	8.4479	3.418	2511.6	2682.5	7.6947	1.6958	2506.7	2676.2	7.3614
150	19.512	2587.9	2783.0	8.6882	3.889	2585.6	2780.1	7.9401	1.9364	2582.8	2776.4	7.6134
200	21.825	2661.3	2879.5	8.9038	4.356	2659.9	2877.7	8.1580	2.172	2658.1	2875.3	7.8343
250	24.136	2736.0	2977.3	9.1002	4.820	2735.0	2976.0	8.3556	2.406	2733.7	2974.3	8.0333
300	26.445	2812.1	3076.5	9.2813	5.284	2811.3	3075.5	8.5373	2.639	2810.4	3074.3	8.2158
400	31.063	2968.9	3279.6	9.6077	6.209	2968.5	3278.9	8.8642	3.103	2967.9	3278.2	8.5435
500	35.679	3132.3	3489.1	9.8978	7.134	3132.0	3488.7	9.1546	3.565	3131.6	3488.1	8.8342
600	40.295	3302.5	3705.4	10.1608	8.057	3302.2	3705.1	9.4178	4.028	3301.9	3704.7	9.0976
700	44.911	3479.6	3928.7	10.4028	8.981	3479.4	3928.5	9.6599	4.490	3479.2	3928.2	9.3398
800	49.526	3663.8	4159.0	10.6281	9.904	3663.6	4158.9	9.8852	4.952	3663.5	4158.6	9.5652
900	54.141	3855.0	4396.4	10.8396	10.828	3854.9	4396.3	10.0967	5.414	3854.8	4396.1	9.7767
1000	58.757	4053.0	4640.6	11.0393	11.751	4052.9	4640.5	10.2964	5.875	4052.8	4640.3	9.9764
1100	63.372	4257.5	4891.2	11.2287	12.674	4257.4	4891.1	10.4859	6.337	4257.3	4891.0	10.1659
1200	67.987	4467.9	5147.8	11.4091	13.597	4467.8	5147.7	10.6662	6.799	4467.7	5147.6	10.3463
1300	72.602	4683.7	5409.7	11.5811	14.521	4683.6	5409.6	10.8382	7.260	4683.5	5409.5	10.5183
	p = 0.20 MPa (120.23)								p = 0.40 MPa (143.63)			
Sat.	.8857	2529.5	2706.7	7.1272					.4625	2553.6	2738.6	6.8959
150	.9596	2576.9	2768.8	7.2795					.4708	2564.5	2752.8	6.9299
200	1.0803	2654.4	2870.5	7.5066					.5342	2646.8	2860.5	7.1706
250	1.1988	2731.2	2971.0	7.7086					.5951	2726.1	2964.2	7.3789
300	1.3162	2808.6	3071.8	7.8926					.6548	2804.8	3066.8	7.5662
400	1.5493	2966.7	3276.6	8.2218					.7726	2964.4	3273.4	7.8985
500	1.7814	3130.8	3487.1	8.5133					.8893	3129.2	3484.9	8.1913
600	2.013	3301.4	3704.0	8.7770					1.0055	3300.2	3702.4	8.4558
700	2.244	3478.8	3927.6	9.0194					1.1215	3477.9	3926.5	8.6987
800	2.475	3663.1	4158.2	9.2449					1.2372	3662.4	4157.3	8.9244
900	2.706	3854.5	4395.8	9.4566					1.3529	3853.9	4395.1	9.1362
1000	2.937	4052.5	4640.0	9.6563					1.4685	4052.0	4639.4	9.3360
1100	3.168	4257.0	4890.7	9.8458					1.5840	4256.5	4890.2	9.5256
1200	3.399	4467.5	5147.3	10.0262					1.6996	4467.0	5146.8	9.7060
1300	3.630	4683.2	5409.3	10.1982					1.8151	4682.8	5408.8	9.8780
					p = 0.60 MPa (158.85)				p = 0.80 MPa (170.43)			
Sat.					.3157	2567.4	2756.8	6.7600	.2404	2576.8	2769.1	6.6628
200					.3520	2638.9	2850.1	6.9665	.2608	2630.6	2839.3	6.8158
250					.3938	2720.9	2957.2	7.1816	.2931	2715.5	2950.0	7.0384
300					.4344	2801.0	3061.6	7.3724	.3241	2797.2	3056.5	7.2328
350					.4742	2881.2	3165.7	7.5464	.3544	2878.2	3161.7	7.4089
400					.5137	2962.1	3270.3	7.7079	.3843	2959.7	3267.1	7.5716
500					.5920	3127.6	3482.8	8.0021	.4433	3126.0	3480.6	7.8673
600					.6697	3299.1	3700.9	8.2674	.5018	3297.9	3699.4	8.1333
700					.7472	3477.0	3925.3	8.5107	.5601	3476.2	3924.2	8.3770
800					.8245	3661.8	4156.5	8.7367	.6181	3661.1	4155.6	8.6033
900					.9017	3853.4	4394.4	8.9486	.6761	3852.8	4393.7	8.8153
1000					.9788	4051.5	4638.8	9.1485	.7340	4051.0	4638.2	9.0153
1100					1.0559	4256.1	4889.6	9.3381	.7919	4255.6	4889.1	9.2050
1200					1.1330	4466.5	5146.3	9.5185	.8497	4466.1	5145.9	9.3855
1300					1.2101	4682.3	5408.3	9.6906	.9076	4681.8	5407.9	9.5575

TABLE C.3 (*continued*)

Superheated Steam

T	v	u	h	s	v	u	h	s	v	u	h	s
	p = 1.00 MPa (179.91)				p = 1.20 MPa (187.99)				p = 1.40 MPa (195.07)			
Sat.	.194 44	2583.6	2778.1	6.5865	.163 33	2588.8	2784.8	6.5233	.140 84	2592.8	2790.0	6.4693
200	.2060	2621.9	2827.9	6.6940	.169 30	2612.8	2815.9	6.5898	.143 02	2603.1	2803.3	6.4975
250	.2327	2709.9	2942.6	6.9247	.192 34	2704.2	2935.0	6.8294	.163 50	2698.3	2927.2	6.7467
300	.2579	2793.2	3051.2	7.1229	.2138	2789.2	3045.8	7.0317	.182 28	2785.2	3040.4	6.9534
350	.2825	2875.2	3157.7	7.3011	.2345	2872.2	3153.6	7.2121	.2003	2869.2	3149.5	7.1360
400	.3066	2957.3	3263.9	7.4651	.2548	2954.9	3260.7	7.3774	.2178	2952.5	3257.5	7.3026
500	.3541	3124.4	3478.5	7.7622	.2946	3122.8	3476.3	7.6759	.2521	3121.1	3474.1	7.6027
600	.4011	3296.8	3697.9	8.0290	.3339	3295.6	3696.3	7.9435	.2860	3294.4	3694.8	7.8710
700	.4478	3475.3	3923.1	8.2731	.3729	3474.4	3922.0	8.1881	.3195	3473.6	3920.8	8.1160
800	.4943	3660.4	4154.7	8.4996	.4118	3659.7	4153.8	8.4148	.3528	3659.0	4153.0	8.3431
900	.5407	3852.2	4392.9	8.7118	.4505	3851.6	4392.2	8.6272	.3861	3851.1	4391.5	8.5556
1000	.5871	4050.5	4637.6	8.9119	.4892	4050.0	4637.0	8.8274	.4192	4049.5	4636.4	8.7559
1100	.6335	4255.1	4888.6	9.1017	.5278	4254.6	4888.0	9.0172	.4524	4254.1	4887.5	8.9457
1200	.6798	4465.6	5145.4	9.2822	.5665	4465.1	5144.9	9.1977	.4855	4464.7	5144.4	9.1262
1300	.7261	4681.3	5407.4	9.4543	.6051	4680.9	5407.0	9.3698	.5186	4680.4	5406.5	9.2984
	p = 1.60 MPa (201.41)				p = 1.80 MPa (207.15)				p = 2.00 MPa (212.42)			
Sat.	.123 80	2596.0	2794.0	6.4218	.110 42	2598.4	2797.1	6.3794	.099 63	2600.3	2799.5	6.3409
225	.132 87	2644.7	2857.3	6.5518	.116 73	2636.6	2846.7	6.4808	.103 77	2628.3	2835.8	6.4147
250	.141 84	2692.3	2919.2	6.6732	.124 97	2686.0	2911.0	6.6066	.111 44	2679.6	2902.5	6.5453
300	.158 62	2781.1	3034.8	6.8844	.140 21	2776.9	3029.2	6.8226	.125 47	2772.6	3023.5	6.7664
350	.174 56	2866.1	3145.4	7.0694	.154 57	2863.0	3141.2	7.0100	.138 57	2859.8	3137.0	6.9563
400	.190 05	2950.1	3254.2	7.2374	.168 47	2947.7	3250.9	7.1794	.151 20	2945.2	3247.6	7.1271
500	.2203	3119.5	3472.0	7.5390	.195 50	3117.9	3469.8	7.4825	.175 68	3116.2	3467.6	7.4317
600	.2500	3293.3	3693.2	7.8080	.2220	3292.1	3691.7	7.7523	.199 60	3290.9	3690.1	7.7024
700	.2794	3472.7	3919.7	8.0535	.2482	3471.8	3918.5	7.9983	.2232	3470.9	3917.4	7.9487
800	.3086	3658.3	4152.1	8.2808	.2742	3657.6	4151.2	8.2258	.2467	3657.0	4150.3	8.1765
900	.3377	3850.5	4390.8	8.4935	.3001	3849.9	4390.1	8.4386	.2700	3849.3	4389.4	8.3895
1000	.3668	4049.0	4635.8	8.6938	.3260	4048.5	4635.2	8.6391	.2933	4048.0	4634.6	8.5901
1100	.3958	4253.7	4887.0	8.8837	.3518	4253.2	4886.4	8.8290	.3166	4252.7	4885.9	8.7800
1200	.4248	4464.2	5143.9	9.0643	.3776	4463.7	5143.4	9.0096	.3398	4463.3	5142.9	8.9607
1300	.4538	4679.9	5406.0	9.2364	.4034	4679.5	5405.6	9.1818	.3631	4679.0	5405.1	9.1329
	p = 2.50 MPa (223.99)				p = 3.00 MPa (233.90)				p = 3.50 MPa (242.60)			
Sat.	.079 98	2603.1	2803.1	6.2575	.066 68	2604.1	2804.2	6.1869	.057 07	2603.7	2803.4	6.1253
225	.080 27	2605.6	2806.3	6.2639								
250	.087 00	2662.6	2880.1	6.4085	.070 58	2644.0	2855.8	6.2872	.058 72	2623.7	2829.2	6.1749
300	.098 90	2761.6	3008.8	6.6438	.081 14	2750.1	2993.5	6.5390	.068 42	2738.0	2977.5	6.4461
350	.109 76	2851.9	3126.3	6.8403	.090 53	2843.7	3115.3	6.7428	.076 78	2835.3	3104.0	6.6579
400	.120 10	2939.1	3239.3	7.0148	.099 36	2932.8	3230.9	6.9212	.084 53	2926.4	3222.3	6.8405
450	.130 14	3025.5	3350.8	7.1746	.107 87	3020.4	3344.0	7.0834	.091 96	3015.3	3337.2	7.0052
500	.139 98	3112.1	3462.1	7.3234	.116 19	3108.0	3456.5	7.2338	.099 18	3103.0	3450.9	7.1572
600	.159 30	3288.0	3686.3	7.5960	.132 43	3285.0	3682.3	7.5085	.113 24	3282.1	3678.4	7.4339
700	.178 32	3468.7	3914.5	7.8435	.148 38	3466.5	3911.7	7.7571	.126 99	3464.3	3908.8	7.6837
800	.197 16	3655.3	4148.2	8.0720	.164 14	3653.5	4145.9	7.9862	.140 56	3651.8	4143.7	7.9134
900	.215 90	3847.9	4387.6	8.2853	.179 80	3846.5	4385.9	8.1999	.154 02	3845.0	4384.1	8.1276
1000	.2346	4046.7	4633.1	8.4861	.195 41	4045.4	4631.6	8.4009	.167 43	4044.1	4630.1	8.3288
1100	.2532	4251.5	4884.6	8.6762	.210 98	4250.3	4883.3	8.5912	.180 80	4249.2	4881.9	8.5192
1200	.2718	4462.1	5141.7	8.8569	.226 52	4460.9	5140.5	8.7720	.194 15	4459.8	5139.3	8.7000
1300	.2905	4677.8	5404.0	9.0291	.242 06	4676.6	5402.8	8.9442	.207 49	4675.5	5401.7	8.8723

TABLE C.3 (*continued*)

Superheated Steam

T	v	u	h	s	v	u	h	s	v	u	h	s
	p = 4.0 MPa (250.40)								p = 5.0 MPa (263.99)			
Sat.	.049 78	2602.3	2801.4	6.0701					.039 44	2597.1	2794.3	5.9734
275	.054 57	2667.9	2886.2	6.2285					.041 41	2631.3	2838.3	6.0544
300	.058 84	2725.3	2960.7	6.3615					.045 32	2698.0	2924.5	6.2084
350	.066 45	2826.7	3092.5	6.5821					.051 94	2808.7	3068.4	6.4493
400	.073 41	2919.9	3213.6	6.7690					.057 81	2906.6	3195.7	6.6459
450	.080 02	3010.2	3330.3	6.9363					.063 30	2999.7	3316.2	6.8186
500	.086 43	3099.5	3445.3	7.0901					.068 57	3091.0	3433.8	6.9759
600	.098 85	3279.1	3674.4	7.3688					.078 69	3273.0	3666.5	7.2589
700	.110 95	3462.1	3905.9	7.6198					.088 49	3457.6	3900.1	7.5122
800	.122 87	3650.0	4141.5	7.8502					.098 11	3646.6	4137.1	7.7440
900	.134 69	3843.6	4382.3	8.0647					.107 62	3840.7	4378.8	7.9593
1000	.146 45	4042.9	4628.7	8.2662					.117 07	4040.4	4625.7	8.1612
1100	.158 17	4248.0	4880.6	8.4567					.126 48	4245.6	4878.0	8.3520
1200	.169 87	4458.6	5138.1	8.6376					.135 87	4456.3	5135.7	8.5331
1300	.181 56	4674.3	5400.5	8.8100					.145 26	4672.0	5398.2	8.7055
	p = 6.0 MPa (275.64)				p = 7.0 MPa (285.88)				p = 8.0 MPa (295.06)			
Sat.	.032 44	2589.7	2784.3	5.8892	.027 37	2580.5	2772.1	5.8133	.023 52	2569.8	2758.0	5.7432
300	.036 16	2667.2	2884.2	6.0674	.029 47	2632.2	2838.4	5.9305	.024 26	2590.9	2785.0	5.7906
350	.042 23	2789.6	3043.0	6.3335	.035 24	2769.4	3016.0	6.2283	.029 95	2747.7	2987.3	6.1301
400	.047 39	2892.9	3177.2	6.5408	.039 93	2878.6	3158.1	6.4478	.034 32	2863.8	3138.3	6.3634
450	.052 14	2988.9	3301.8	6.7193	.044 16	2978.0	3287.1	6.6327	.038 17	2966.7	3272.0	6.5551
500	.056 65	3082.2	3422.2	6.8803	.048 14	3073.4	3410.3	6.7975	.041 75	3064.3	3398.3	6.7240
550	.061 01	3174.6	3540.6	7.0288	.051 95	3167.2	3530.9	6.9486	.045 16	3159.8	3521.0	6.8778
600	.065 25	3266.9	3658.4	7.1677	.055 65	3260.7	3650.3	7.0894	.048 45	3254.4	3642.0	7.0206
700	.073 52	3453.1	3894.2	7.4234	.062 83	3448.5	3888.3	7.3476	.054 81	3443.9	3882.4	7.2812
800	.081 60	3643.1	4132.7	7.6566	.069 81	3639.5	4128.2	7.5822	.060 97	3636.0	4123.8	7.5173
900	.089 58	3837.8	4375.3	7.8727	.076 69	3835.0	4371.8	7.7991	.067 02	3832.1	4368.3	7.7351
1000	.097 49	4037.8	4622.7	8.0751	.083 50	4035.3	4619.8	8.0020	.073 01	4032.8	4616.9	7.9384
1100	.105 36	4243.3	4875.4	8.2661	.090 27	4240.9	4872.8	8.1933	.078 96	4238.6	4870.3	8.1300
1200	.113 21	4454.0	5133.3	8.4474	.097 03	4451.7	5130.9	8.3747	.084 89	4449.5	5128.5	8.3115
1300	.121 06	4669.6	5396.0	8.6199	.103 77	4667.3	5393.7	8.5473	.090 80	4665.0	5391.5	8.4842
	p = 9.0 MPa (303.40)				p = 10.0 MPa (311.06)				p = 12.5 MPa (327.89)			
Sat.	.020 48	2557.8	2742.1	5.6772	.018 026	2544.4	2724.7	5.6141				
					.013 495	2505.1	2673.8	5.4624				
325	.023 27	2646.6	2856.0	5.8712	.019 861	2610.4	2809.1	5.7568				
350	.025 80	2724.4	2956.6	6.0361	.022 42	2699.2	2923.4	5.9443				
					.016 126	2624.6	2826.2	5.7118				
400	.029 93	2848.4	3117.8	6.2854	.026 41	2832.4	3096.5	6.2120	.020 00	2789.3	3039.3	6.0417
450	.033 50	2955.2	3256.6	6.4844	.029 75	2943.4	3240.9	6.4190	.022 99	2912.5	3199.8	6.2719
500	.036 77	3055.2	3386.1	6.6576	.032 79	3045.8	3373.7	6.5966	.025 60	3021.7	3341.8	6.4618
550	.039 87	3152.2	3511.0	6.8142	.035 64	3144.6	3500.9	6.7561	.028 01	3125.0	3475.2	6.6290
600	.042 85	3248.1	3633.7	6.9589	.038 37	3241.7	3625.3	6.9029	.030 29	3225.4	3604.0	6.7810
650	.045 74	3343.6	3755.3	7.0943	.041 01	3338.2	3748.2	7.0398	.032 48	3324.4	3730.4	6.9218
700	.048 57	3439.3	3876.5	7.2221	.043 58	3434.7	3870.5	7.1687	.034 60	3422.9	3855.3	7.0536
800	.054 09	3632.5	4119.3	7.4596	.048 59	3628.9	4114.8	7.4077	.038 69	3620.0	4103.6	7.2965
900	.059 50	3829.2	4364.8	7.6783	.053 49	3826.3	4361.2	7.6272	.042 67	3819.1	4352.5	7.5182
1000	.064 85	4030.3	4614.0	7.8821	.058 32	4027.8	4611.0	7.8315	.046 58	4021.6	4603.8	7.7237
1100	.070 16	4236.3	4867.7	8.0740	.063 12	4234.0	4865.1	8.0237	.050 45	4228.2	4858.8	7.9165
1200	.075 44	4447.2	5126.2	8.2556	.067 89	4444.9	5123.8	8.2055	.054 30	4439.3	5118.0	8.0987
1300	.080 72	4662.7	5389.2	8.4284	.072 65	4460.5	5387.0	8.3783	.058 13	4654.8	5381.4	8.2717

TABLE C.3 *(continued)*

Superheated Steam

T	v	u	h	s	v	u	h	s	v	u	h	s
	p = 15.0 MPa (342.24)				p = 17.5 MPa (354.75)				p = 20.0 MPa (365.81)			
Sat.	.010 337	2455.5	2610.5	5.3098	.007 920	2390.2	2528.8	5.1419	.005 834	2293.0	2409.7	4.9269
350	.011 470	2520.4	2692.4	5.4421								
400	.015 649	2740.7	2975.5	5.8811	.012 447	2685.0	2902.9	5.7213	.009 942	2619.3	2818.1	5.5540
450	.018 445	2879.5	3156.2	6.1404	.015 174	2844.2	3109.7	6.0184	.012 695	2806.2	3060.1	5.9017
500	.020 80	2996.6	3308.6	6.3443	.017 358	2970.3	3274.1	6.2383	.014 768	2942.9	3238.2	6.1401
550	.022 93	3104.7	3448.6	6.5199	.019 288	3083.9	3421.4	6.4230	.016 555	3062.4	3393.5	6.3348
600	.024 91	3208.6	3582.3	6.6776	.021 06	3191.5	3560.1	6.5866	.018 178	3174.0	3537.6	6.5048
650	.026 80	3310.3	3712.3	6.8224	.022 74	3296.0	3693.9	6.7357	.019 693	3281.4	3675.3	6.6582
700	.028 61	3410.9	3840.1	6.9572	.024 34	3398.7	3824.6	6.8736	.021 13	3386.4	3809.0	6.7993
800	.032 10	3610.9	4092.4	7.2040	.027 38	3601.8	4081.1	7.1244	.023 85	3592.7	4069.7	7.0544
900	.035 46	3811.9	4343.8	7.4279	.030 31	3804.7	4335.1	7.3507	.026 45	3797.5	4326.4	7.2830
1000	.038 75	4015.4	4596.6	7.6348	.033 16	4009.3	4589.5	7.5589	.028 97	4003.1	4582.5	7.4925
1100	.042 00	4222.6	4852.6	7.8283	.035 97	4216.9	4846.4	7.7531	.031 45	4211.3	4840.2	7.6874
1200	.045 23	4433.8	5112.3	8.0108	.038 76	4428.3	5106.6	7.9360	.033 91	4422.8	5101.0	7.8707
1300	.048 45	4649.1	5376.0	8.1840	.041 54	4643.5	5370.5	8.1093	.036 36	4638.0	5365.1	8.0442
	p = 25.0 MPa				p = 30.0 MPa				p = 40.0 MPa			
375	.001 973 1	1798.7	1848.0	4.0320	.001 789 2	1737.8	1791.5	3.9305	.001 640 7	1677.1	1742.8	3.8290
400	.006 004	2430.1	2580.2	5.1418	.002 790	2067.4	2151.1	4.4728	.001 907 7	1854.6	1930.9	4.1135
425	.007 881	2609.2	2806.3	5.4723	.005 303	2455.1	2614.2	5.1504	.002 532	2096.9	2198.1	4.5029
450	.009 162	2720.7	2949.7	5.6744	.006 735	2619.3	2821.4	5.4424	.003 693	2365.1	2512.8	4.9459
500	.011 123	2884.3	3162.4	5.9592	.008 678	2820.7	3081.1	5.7905	.005 622	2678.4	2903.3	5.4700
550	.012 724	3017.5	3335.6	6.1765	.010 168	2970.3	3275.4	6.0342	.006 984	2869.7	3149.1	5.7785
600	.014 137	3137.9	3491.4	6.3602	.011 446	3100.5	3443.9	6.2331	.008 094	3022.6	3346.4	6.0114
650	.015 433	3251.6	3637.4	6.5229	.012 596	3221.0	3598.9	6.4058	.009 063	3158.0	3520.6	6.2054
700	.016 646	3361.3	3777.5	6.6707	.013 661	3335.8	3745.6	6.5606	.009 941	3283.6	3681.2	6.3750
800	.018 912	3574.3	4047.1	6.9345	.015 623	3555.5	4024.2	6.8332	.011 523	3517.8	3978.7	6.6662
900	.021 045	3783.0	4309.1	7.1680	.017 448	3768.5	4291.9	7.0718	.012 962	3739.4	4257.9	6.9150
1000	.023 10	3990.9	4568.5	7.3802	.019 196	3978.8	4554.7	7.2867	.014 324	3954.6	4527.6	7.1356
1100	.025 12	4200.2	4828.2	7.5765	.020 903	4189.2	4816.3	7.4845	.015 642	4167.4	4793.1	7.3364
1200	.027 11	4412.0	5089.9	7.7605	.022 589	4401.3	5079.0	7.6692	.016 940	4380.1	5057.7	7.5224
1300	.029 10	4626.9	5354.4	7.9342	.024 266	4616.0	5344.0	7.8432	.018 229	4594.3	5323.5	7.6969

Sources: Kennan, Keyes, Hill, and Moore, *Steam Tables*, Wiley, New York, 1969; G. J. Van Wylen and R. E. Sonntag, *Fundamentals of Classical Thermodynamics*, Wiley, New York, 1973.

TABLE C.4 Compressed Liquid

T	$p = 5$ MPa (263.99)				$p = 10$ MPa (311.06)				$p = 15$ MPa (342.42)			
	v	u	h	s	v	u	h	s	v	u	h	s
0	0.000 997 7	0.04	5.04	0.0001	0.000 995 2	0.09	10.04	0.0002	0.000 992 8	0.15	15.05	0.0004
20	0.000 999 5	83.65	88.65	0.2956	0.000 997 2	83.36	93.33	0.2945	0.000 995 0	83.06	97.99	0.2934
40	0.001 005 6	166.95	171.97	0.5705	0.001 003 4	166.35	176.38	0.5686	0.001 001 3	165.76	180.78	0.5666
60	0.001 014 9	250.23	255.30	0.8285	0.001 012 7	249.36	259.49	0.8258	0.001 010 5	248.51	263.67	0.8232
80	0.001 026 8	333.72	338.85	1.0720	0.001 024 5	332.59	342.83	1.0688	0.001 022 2	331.48	346.81	1.0656
100	0.001 041 0	417.52	422.72	1.3030	0.001 038 5	416.12	426.50	1.2992	0.001 036 1	414.74	430.28	1.2955
120	0.001 057 6	501.80	507.09	1.5233	0.001 054 9	500.08	510.64	1.5189	0.001 052 2	498.40	514.19	1.5145
140	0.001 076 8	586.76	592.15	1.7343	0.001 073 7	584.68	595.42	1.7292	0.001 070 7	582.66	598.72	1.7242
160	0.001 098 8	672.62	678.12	1.9375	0.001 095 3	670.13	681.08	1.9317	0.001 091 8	667.71	684.09	1.9260
180	0.001 124 0	759.63	765.25	2.1341	0.001 119 9	756.65	767.84	2.1275	0.001 115 9	753.76	770.50	2.1210
200	0.001 153 0	848.1	853.9	2.3255	0.001 148 0	844.5	856.0	2.3178	0.001 143 3	841.0	858.2	2.3104
220	0.001 186 6	938.4	944.4	2.5128	0.001 180 5	934.1	945.9	2.5039	0.001 174 8	929.9	947.5	2.4953
240	0.001 226 4	1031.4	1037.5	2.6979	0.001 218 7	1026.0	1038.1	2.6872	0.001 211 4	1020.8	1039.0	2.6771
260	0.001 274 9	1127.9	1134.3	2.8830	0.001 264 5	1121.1	1133.7	2.8699	0.001 255 0	1114.6	1133.4	2.8576

T	$p = 20$ MPa (365.81)				$p = 30$ MPa				$p = 50$ MPa			
	v	u	h	s	v	u	h	s	v	u	h	s
0	0.000 990 4	0.19	20.01	0.0004	0.000 985 6	0.25	29.82	0.0001	0.000 976 6	0.20	49.03	0.0014
20	0.000 992 8	82.77	102.62	0.2923	0.000 988 6	82.17	111.84	0.2899	0.000 980 4	81.00	130.02	0.2848
40	0.000 999 2	165.17	185.16	0.5646	0.000 995 1	164.04	193.89	0.5607	0.000 987 2	161.86	211.21	0.5527
60	0.001 008 4	247.68	267.85	0.8206	0.001 004 2	246.06	276.19	0.8154	0.000 996 2	242.98	292.79	0.8052
80	0.001 019 9	330.40	350.80	1.0624	0.001 015 6	328.30	358.77	1.0561	0.001 007 3	324.34	374.70	1.0440
100	0.001 033 7	413.39	434.06	1.2917	0.001 029 0	410.78	441.66	1.2844	0.001 020 1	405.88	456.89	1.2703
120	0.001 049 6	496.76	517.76	1.5102	0.001 044 5	493.59	524.93	1.5018	0.001 034 8	487.65	539.39	1.4857
140	0.001 067 8	580.69	602.04	1.7193	0.001 062 1	576.88	608.75	1.7098	0.001 051 5	569.77	622.35	1.6915
160	0.001 088 5	665.35	687.12	1.9204	0.001 082 1	660.82	693.28	1.9096	0.001 070 3	652.41	705.92	1.8891
180	0.001 112 0	750.95	773.20	2.1147	0.001 104 7	745.59	778.73	2.1024	0.001 091 2	735.69	790.25	2.0794
200	0.001 138 8	837.7	860.5	2.3031	0.001 130 2	831.4	865.3	2.2893	0.001 114 6	819.7	875.5	2.2634
240	0.001 204 6	1016.0	1040.0	2.6674	0.001 192 0	1006.9	1042.6	2.6490	0.001 170 2	990.7	1049.2	2.6158
280	0.001 296 5	1204.7	1230.6	3.0248	0.001 275 5	1190.7	1229.0	2.9986	0.001 241 5	1167.2	1229.3	2.9537
320	0.001 443 7	1415.7	1444.6	3.3979	0.001 399 7	1390.7	1432.7	3.3539	0.001 338 8	1353.3	1420.2	3.2868
360	0.001 822 6	1702.8	1739.3	3.8772	0.001 626 5	1626.6	1675.4	3.7494	0.001 483 8	1556.0	1630.2	3.6291

Sources: Kennan, Keyes, Hill, and Moore, *Steam Tables*, Wiley, New York, 1969; G. J. Van Wylen and R. E. Sonntag, *Fundamentals of Classical Thermodynamics*, Wiley, New York, 1973.

TABLE C.5 Saturated Solid—Vapor

T, °C	Specific Volume			Internal Energy			Enthalpy			Entropy		
	p, kPa	Sat. Solid $v_i \times 10^3$	Sat. Vapor v_g	Sat. Solid u_i	Subl. u_{ig}	Sat. Vapor u_g	Sat. Solid h_i	Subl. h_{ig}	Sat. Vapor h_g	Sat. Solid s_i	Subl. s_{ig}	Sat. Vapor s_g
0.01	0.6113	1.0908	206.1	−333.40	2708.7	2375.3	−333.40	2834.8	2501.4	−1.221	10.378	9.156
0	0.6108	1.0908	206.3	−333.43	2708.8	2375.3	−333.43	2834.8	2501.3	−1.221	10.378	9.157
−2	0.5176	1.0904	241.7	−337.62	2710.2	2372.6	−337.62	2835.3	2497.7	−1.237	10.456	9.219
−4	0.4375	1.0901	283.8	−341.78	2711.6	2369.8	−341.78	2835.7	2494.0	−1.253	10.536	9.283
−6	0.3689	1.0898	334.2	−345.91	2712.9	2367.0	−345.91	2836.2	2490.3	−1.268	10.616	9.348
−8	0.3102	1.0894	394.4	−350.02	2714.2	2364.2	−350.02	2836.6	2486.6	−1.284	10.698	9.414
−10	0.2602	1.0891	466.7	−354.09	2715.5	2361.4	−354.09	3837.0	2482.9	−1.299	10.781	9.481
−12	0.2176	1.0888	553.7	−358.14	2716.8	2358.7	−358.14	2837.3	2479.2	−1.315	10.865	9.550
−14	0.1815	1.0884	658.8	−362.15	2718.0	2355.9	−362.15	2837.6	2475.5	−1.331	10.950	9.619
−16	0.1510	1.0881	786.0	−366.14	2719.2	2353.1	−366.14	2837.9	2471.8	−1.346	11.036	9.690
−20	0.1035	1.0874	1128.6	−374.03	2721.6	2347.5	−374.03	2838.4	2464.3	−1.377	11.212	9.835
−24	0.0701	1.0868	1640.1	−381.80	2723.7	2342.0	−381.80	2838.7	2456.9	−1.408	11.394	9.985
−28	0.0469	1.0861	2413.7	−389.45	2725.8	2336.4	−389.45	2839.0	2449.5	−1.439	11.580	10.141
−32	0.0309	1.0854	3600	−396.98	2727.8	2330.8	−396.98	2839.1	2442.1	−1.471	11.773	10.303
−36	0.0201	1.0848	5444	−404.40	2729.6	2325.2	−404.40	2839.1	2434.7	−1.501	11.972	10.470
−40	0.0129	1.0841	8354	−411.70	2731.3	2319.6	−411.70	2838.9	2427.2	−1.532	12.176	10.644

Sources: Kennan, Keyes, Hill, and Moore, *Steam Tables*, Wiley, New York, 1969; G. J. Van Wylen and R. E. Sonntag, *Fundamentals of Classical Thermodynamics*, Wiley, New York, 1973.

TABLE C.1E Properties of Saturated H_2O—Temperature Table

Temp, T, °F	Press. p, psia	Specific Volume: Sat. Liquid v_f	Specific Volume: Sat. Vapor v_g	Internal Energy: Sat. Liquid u_f	Internal Energy: Evap. u_{fg}	Internal Energy: Sat. Vapor u_g	Enthalpy: Sat. Liquid h_f	Enthalpy: Evap. h_{fg}	Enthalpy: Sat. Vapor h_g	Entropy: Sat. Liquid s_f	Entropy: Evap. s_{fg}	Entropy: Sat. Vapor s_g
32.018	0.08866	0.016022	3302	0.00	1021.2	1021.2	0.01	1075.4	1075.4	0.00000	2.1869	2.1869
35	0.09992	0.016021	2948	2.99	1019.2	1022.2	3.00	1073.7	1076.7	0.00607	2.1704	2.1764
40	0.12166	0.016020	2445	8.02	1015.8	1023.9	8.02	1070.9	1078.9	0.01617	2.1430	2.1592
45	0.14748	0.016021	2037	13.04	1012.5	1025.5	13.04	1068.1	1081.1	0.02618	2.1162	2.1423
50	0.17803	0.016024	1704.2	18.06	1009.1	1027.2	18.06	1065.2	1083.3	0.03607	2.0899	2.1259
60	0.2563	0.016035	1206.9	28.08	1002.4	1030.4	28.08	1059.6	1087.7	0.05555	2.0388	2.0943
70	0.3632	0.016051	867.7	38.09	995.6	1033.7	38.09	1054.0	1092.0	0.07463	1.9896	2.0642
80	0.5073	0.016073	632.8	48.08	988.9	1037.0	48.09	1048.3	1096.4	0.09332	1.9423	2.0356
90	0.6988	0.016099	467.7	58.07	982.2	1040.2	58.07	1042.7	1100.7	0.11165	1.8966	2.0083
100	0.9503	0.016130	350.0	68.04	975.4	1043.5	68.05	1037.0	1105.0	0.12963	1.8526	1.9822
110	1.2763	0.016166	265.1	78.02	968.7	1046.7	78.02	1031.3	1109.3	0.14730	1.8101	1.9574
120	1.6945	0.016205	203.0	87.99	961.9	1049.9	88.00	1025.5	1113.5	0.16465	1.7690	1.9336
130	2.225	0.016247	157.17	97.97	955.1	1053.0	97.98	1019.8	1117.8	0.18172	1.7292	1.9109
140	2.892	0.016293	122.88	107.95	948.2	1056.2	107.96	1014.0	1121.9	0.19851	1.6907	1.8892
150	3.722	0.016343	96.99	117.95	941.3	1059.3	117.96	1008.1	1126.1	0.21503	1.6533	1.8684
160	4.745	0.016395	77.23	127.94	934.4	1062.3	127.96	1002.2	1130.1	0.23130	1.6171	1.8484
170	5.996	0.016450	62.02	137.95	927.4	1065.4	137.97	996.2	1134.2	0.24732	1.5819	1.8293
180	7.515	0.016509	50.20	147.97	920.4	1068.3	147.99	990.2	1138.2	0.26311	1.5478	1.8109
190	9.343	0.016570	40.95	158.00	913.3	1071.3	158.03	984.1	1142.1	0.27866	1.5146	1.7932
200	11.529	0.016634	33.63	168.04	906.2	1074.2	168.07	977.9	1145.9	0.29400	1.4822	1.7762
210	14.125	0.016702	27.82	178.10	898.9	1077.0	178.14	971.6	1149.7	0.30913	1.4508	1.7599
212	14.698	0.016716	26.80	180.11	897.5	1077.6	180.16	970.3	1150.5	0.31213	1.4446	1.7567
220	17.188	0.016772	23.15	188.17	891.7	1079.8	188.22	965.3	1153.5	0.32406	1.4201	1.7441
230	20.78	0.016845	19.386	198.26	884.3	1082.6	198.32	958.8	1157.1	0.33880	1.3091	1.7289
240	24.97	0.016922	16.327	208.36	876.9	1085.3	208.44	952.3	1160.7	0.35335	1.3609	1.7143
250	29.82	0.017001	13.826	218.49	869.4	1087.9	218.59	945.6	1164.2	0.36772	1.3324	1.7001
260	35.42	0.017084	11.768	228.64	861.8	1090.5	228.76	938.8	1167.6	0.38193	1.3044	1.6864
270	41.85	0.017170	10.066	238.82	854.1	1093.0	238.95	932.0	1170.9	0.39567	1.2771	1.6731
280	49.18	0.017259	8.650	249.02	846.3	1095.4	249.18	924.9	1174.1	0.40986	1.2504	1.6602
290	57.53	0.017352	7.467	259.25	838.5	1097.7	259.44	917.8	1177.2	0.42360	1.2241	1.6477
300	66.98	0.017448	6.472	269.52	830.5	1100.0	269.73	910.4	1180.2	0.43720	1.1984	1.6356
320	89.60	0.017652	4.919	290.14	814.1	1104.2	290.43	895.3	1185.8	0.46400	1.1483	1.6123
340	117.93	0.017872	3.792	310.91	797.1	1108.0	311.30	879.5	1190.8	0.49031	1.0997	1.5901
360	152.92	0.018108	2.961	331.84	779.6	1111.4	332.35	862.9	1195.2	0.51617	1.0526	1.5688
380	195.60	0.018363	2.339	352.95	761.4	1114.3	353.62	845.4	1199.0	0.54163	1.0067	1.5483
400	247.1	0.018638	1.8661	374.27	742.4	1116.6	375.12	826.8	1202.0	0.56672	0.9617	1.5284
420	308.5	0.018936	1.5024	395.81	722.5	1118.3	396.89	807.2	1204.1	0.59152	0.9175	1.5091
440	381.2	0.019260	1.2192	417.62	701.7	1119.3	418.98	786.3	1205.3	0.61605	0.8740	1.4900
460	466.3	0.019614	0.9961	439.7	679.8	1119.6	441.4	764.1	1205.5	0.6404	0.8308	1.4712
480	565.5	0.020002	0.8187	462.2	656.7	1118.9	464.3	740.3	1204.6	0.6646	0.7878	1.4524
500	680.0	0.02043	0.6761	485.1	632.3	1117.4	487.7	714.8	1202.5	0.6888	0.7448	1.4335
520	811.4	0.02091	0.5605	508.5	606.2	1114.8	511.7	687.3	1198.9	0.7130	0.7015	1.4145
540	961.5	0.02145	0.4658	532.6	578.4	1111.0	536.4	657.5	1193.8	0.7374	0.6576	1.3950
560	1131.8	0.02207	0.3877	557.4	548.4	1105.8	562.0	625.0	1187.0	0.7620	0.6129	1.3749
580	1324.3	0.02278	0.3225	583.1	515.9	1098.9	588.6	589.3	1178.0	0.7872	0.5668	1.3540
600	1541.0	0.02363	0.2677	609.9	480.1	1090.0	616.7	549.7	1166.4	0.8130	0.5187	1.3317
620	1784.4	0.02465	0.2209	638.3	440.2	1078.5	646.4	505.0	1151.4	0.8398	0.4677	1.3075
640	2057.1	0.02593	0.1805	668.7	394.5	1063.2	678.6	453.4	1131.9	0.8681	0.4122	1.2803
660	2362	0.02767	0.14459	702.3	340.0	1042.3	714.4	391.1	1105.5	0.8990	0.3493	1.2483
680	2705	0.03032	0.11127	741.7	269.3	1011.0	756.9	309.8	1066.7	0.9350	0.2718	1.2068
700	3090	0.03666	0.07438	801.7	145.9	947.7	822.7	167.5	990.2	0.9902	0.1444	1.1346
705.44	3204	0.05053	0.05053	872.6	0	872.6	902.5	0	902.5	1.0580	0	1.0580

Source: Kennan, Keyes, Hill, and Moore, *Steam Tables*, Wiley, New York, 1969.

TABLE C.2E Properties of Saturated H_2O—Pressure Table

Press. p, psia	T, °F	Specific Volume: Sat. Liquid v_f	Specific Volume: Sat. Vapor v_g	Internal Energy: Sat. Liquid u_f	Internal Energy: Evap. u_{fg}	Internal Energy: Sat. Vapor u_g	Enthalpy: Sat. Liquid h_f	Enthalpy: Evap. h_{fg}	Enthalpy: Sat. Vapor h_g	Entropy: Sat. Liquid s_f	Entropy: Evap. s_{fg}	Entropy: Sat. Vapor s_g
1.0	101.70	0.016136	333.6	69.74	974.3	1044.0	69.74	1036.0	1105.8	0.13266	1.8453	1.9779
2.0	126.04	0.016230	173.75	94.02	957.8	1051.8	94.02	1022.1	1116.1	0.17499	1.7448	1.9198
3.0	141.43	0.016300	118.72	109.38	947.2	1056.6	109.39	1013.1	1122.5	0.20089	1.6852	1.8861
4.0	152.93	0.016358	90.64	120.88	939.3	1060.2	120.89	1006.4	1127.3	0.21983	1.6426	1.8624
5.0	162.21	0.016407	73.53	130.15	932.9	1063.0	130.17	1000.9	1131.0	0.23486	1.6093	1.8441
6.0	170.03	0.016451	61.98	137.98	927.4	1065.4	138.00	996.2	1134.2	0.24736	1.5819	1.8292
8.0	182.84	0.016526	47.35	150.81	918.4	1069.2	150.84	988.4	1139.3	0.26754	1.5383	1.8058
10	193.19	0.016590	38.42	161.20	911.0	1072.2	161.23	982.1	1143.3	0.28358	1.5041	1.7877
14.696	211.99	0.016715	26.80	180.10	897.5	1077.6	180.15	970.4	1150.5	0.31212	1.4446	1.7567
15	213.03	0.016723	26.29	181.14	896.8	1077.9	181.19	969.7	1150.9	0.31367	1.4414	1.7551
20	227.96	0.016830	20.09	196.19	885.8	1082.0	196.26	960.1	1156.4	0.33580	1.3962	1.7320
25	240.08	0.016922	16.306	208.44	876.9	1085.3	208.52	952.2	1160.7	0.35345	1.3607	1.7142
30	250.34	0.017004	13.748	218.84	869.2	1088.0	218.93	945.4	1164.3	0.36821	1.3314	1.6996
35	259.30	0.017073	11.900	227.93	862.4	1090.3	228.04	939.3	1167.4	0.38093	1.3064	1.6873
40	267.26	0.017146	10.501	236.03	856.2	1092.3	236.16	933.8	1170.0	0.39214	1.2845	1.6767
45	274.46	0.017209	9.403	243.37	850.7	1094.0	243.51	928.8	1172.3	0.40218	1.2651	1.6673
50	281.03	0.017269	8.518	250.08	845.5	1095.6	250.24	924.2	1174.4	0.41129	1.2476	1.6589
55	287.10	0.017325	7.789	256.28	840.8	1097.0	256.46	919.9	1176.3	0.41963	1.2317	1.6513
60	292.73	0.017378	7.177	262.06	836.3	1098.3	262.25	915.8	1178.0	0.42733	1.2170	1.6444
65	298.00	0.017429	6.657	267.46	832.1	1099.5	267.67	911.9	1179.6	0.43450	1.2035	1.6380
70	302.96	0.017478	6.209	272.56	828.1	1100.6	272.79	908.3	1181.0	0.44120	1.1909	1.6321
75	307.63	0.017524	5.818	277.37	824.3	1101.6	277.61	904.8	1182.4	0.44749	1.1790	1.6265
80	312.07	0.017570	5.474	281.95	820.6	1102.6	282.21	901.4	1183.6	0.45344	1.1679	1.6214
85	316.29	0.017613	5.170	286.30	817.1	1103.5	286.58	898.2	1184.8	0.45907	1.1574	1.6165
90	320.31	0.017655	4.898	290.46	813.8	1104.3	290.76	895.1	1185.9	0.46442	1.1475	1.6119
95	324.16	0.017696	4.654	294.45	810.6	1105.0	294.76	892.1	1186.9	0.46952	1.1380	1.6076
100	327.86	0.017736	4.434	298.28	807.5	1105.8	298.61	889.2	1187.8	0.47439	1.1290	1.6034
110	334.82	0.017813	4.051	305.52	801.6	1107.1	305.88	883.7	1189.6	0.48355	1.1122	1.5957
120	341.30	0.017886	3.730	312.27	796.0	1108.3	312.67	878.5	1191.1	0.49201	1.0966	1.5886
130	347.37	0.017957	3.457	318.61	790.7	1109.4	319.04	873.5	1192.5	0.49989	1.0822	1.5821
140	353.08	0.018024	3.221	324.58	785.7	1110.3	325.05	868.7	1193.8	0.50727	1.0688	1.5761
150	358.48	0.018089	3.016	330.24	781.0	1111.2	330.75	864.2	1194.9	0.51422	1.0562	1.5704
160	363.60	0.018152	2.836	335.63	776.4	1112.0	336.16	859.8	1196.0	0.52078	1.0443	1.5651
170	368.47	0.018214	2.676	340.76	772.0	1112.7	341.33	855.6	1196.9	0.52700	1.0330	1.5600
180	373.13	0.018273	2.533	345.68	767.7	1113.4	346.29	851.5	1197.8	0.53292	1.0223	1.5553
190	337.59	0.018331	2.405	350.39	763.6	1114.0	351.04	847.5	1198.6	0.53857	1.0122	1.5507
200	381.86	0.018387	2.289	354.9	759.6	1114.6	355.6	843.7	1199.3	0.5440	1.0025	1.5664
300	417.43	0.018896	1.5442	393.0	725.1	1118.2	394.1	809.8	1203.9	0.5883	0.9232	1.5115
400	444.70	0.019340	1.1620	422.8	696.7	1119.5	424.2	781.2	1205.5	0.6218	0.8638	1.4856
500	467.13	0.019748	0.9283	447.7	671.7	1119.4	449.5	755.8	1205.3	0.6490	0.8154	1.4645
600	486.33	0.02013	0.7702	469.4	649.1	1118.6	471.7	732.4	1204.1	0.6723	0.7742	1.4464
700	503.23	0.02051	0.6558	488.9	628.2	1117.0	491.5	710.5	1202.0	0.6927	0.7378	1.4305
800	518.36	0.02087	0.5691	506.6	608.4	1115.0	509.7	689.6	1199.3	0.7110	0.7050	1.4160
900	532.12	0.02123	0.5009	523.0	589.6	1112.6	526.6	669.5	1196.0	0.7277	0.6750	1.4027
1000	544.75	0.02159	0.4459	538.4	571.5	1109.9	542.4	650.0	1192.4	0.7432	0.6471	1.3903
1200	567.37	0.02232	0.3623	566.7	536.8	1103.5	571.7	612.3	1183.9	0.7712	0.5961	1.3673
1400	587.25	0.02307	0.3016	592.7	503.3	1096.0	598.6	575.5	1174.1	0.7964	0.5497	1.3461
1600	605.06	0.02386	0.2552	616.9	470.5	1087.4	624.0	538.9	1162.9	0.8196	0.5062	1.3258
1800	621.21	0.02472	0.2183	640.0	437.6	1077.7	648.3	502.1	1150.4	0.8414	0.4645	1.3060
2000	636.00	0.02565	0.18813	662.4	404.2	1066.6	671.9	464.4	1136.3	0.8623	0.4238	1.2861
2500	668.31	0.02860	0.13059	717.7	313.4	1031.0	730.9	360.5	1091.4	0.9131	0.3196	1.2327
3000	695.52	0.03431	0.08404	783.4	185.4	968.8	802.5	213.0	1015.5	0.9732	0.1843	1.1575
3203.6	705.44	0.05053	0.05053	872.6	0	872.6	902.5	0	902.5	1.0580	0	1.0580

Source: Keenan, Keyes, Hill, and Moore, *Steam Tables*, Wiley, New York, 1969.

TABLE C.3E Properties of Superheated Steam

°F	v	u	h	s	v	u	h	s	v	u	h	s
	$p = 1.0$ (101.70)				$p = 5.0$ (162.21)				$p = 10.0$ (193.19)			
Sat	333.6	1044.0	1105.8	1.9779	73.53	1063.0	1131.0	1.8441	38.42	1072.2	1143.3	1.7877
200	392.5	1077.5	1150.1	2.0508	78.15	1076.3	1148.6	1.8715	38.85	1074.7	1146.6	1.7927
240	416.4	1091.2	1168.3	2.0775	83.00	1090.3	1167.1	1.8987	41.32	1089.0	1165.5	1.8205
280	440.3	1105.0	1186.5	2.1028	87.83	1104.3	1185.5	1.9244	43.77	1103.3	1184.3	1.8467
320	464.2	1118.9	1204.8	2.1269	92.64	1118.3	1204.0	1.9487	46.20	1117.6	1203.1	1.8714
360	488.1	1132.9	1223.2	2.1500	97.45	1132.4	1222.6	1.9719	48.62	1131.8	1221.8	1.8948
400	511.9	1147.0	1241.8	2.1720	102.24	1146.6	1241.2	1.9941	51.03	1146.1	1240.5	1.9171
500	571.5	1182.8	1288.5	2.2235	114.20	1182.5	1288.2	2.0458	57.04	1182.2	1287.7	1.9690
600	631.1	1219.3	1336.1	2.2706	126.15	1219.1	1335.8	2.0930	63.03	1218.9	1335.5	2.0164
700	690.7	1256.7	1384.5	2.3142	138.08	1256.5	1384.3	2.1367	69.01	1256.3	1384.0	2.0601
800	750.3	1294.9	1433.7	2.3550	150.01	1294.7	1433.5	2.1775	74.98	1294.6	1433.3	2.1009
1000	869.5	1373.9	1534.8	2.4294	173.86	1373.9	1534.7	2.2520	86.91	1373.8	1534.6	2.1755
	$p = 14.696$ (211.99)				$p = 20$ (227.96)				$p = 40$ (267.26)			
Sat	26.80	1077.6	1150.5	1.7567	20.09	1082.0	1156.4	1.7320	10.501	1092.3	1170.0	1.6767
240	28.00	1087.9	1164.0	1.7764	20.47	1086.5	1162.3	1.7405				
280	29.69	1102.4	1183.1	1.8030	21.73	1101.4	1181.8	1.7676	10.711	1097.3	1176.6	1.6857
320	31.36	1116.8	1202.1	1.8280	22.98	1116.0	1201.0	1.7930	11.360	1112.8	1196.9	1.7124
360	33.02	1131.2	1221.0	1.8516	24.21	1130.6	1220.1	1.8168	11.996	1128.0	1216.8	1.7373
400	34.67	1145.6	1239.9	1.8741	25.43	1145.1	1239.2	1.8395	12.623	1143.0	1236.4	1.7606
500	38.77	1181.8	1287.3	1.9263	28.46	1181.5	1286.8	1.8919	14.164	1180.1	1284.9	1.8140
600	42.86	1218.6	1335.2	1.9737	31.47	1218.4	1334.8	1.9395	15.685	1217.3	1333.4	1.8621
700	46.93	1256.1	1383.8	2.0175	34.47	1255.9	1383.5	1.9834	17.196	1255.1	1382.4	1.9063
800	51.00	1294.4	1433.1	2.0584	37.46	1294.3	1432.9	2.0243	18.701	1293.7	1432.1	1.9474
1000	59.13	1373.7	1534.5	2.1330	43.44	1373.5	1534.3	2.0989	21.70	1373.1	1533.8	2.0223
1200	67.25	1465	1639.3	2.2003	49.41	1456.4	1639.2	2.1663	24.69	1456.1	1638.9	2.0897
	$p = 60$ (292.73)				$p = 80$ (312.07)				$p = 100$ (327.86)			
Sat	7.177	1098.3	1178.0	1.6444	5.474	1102.6	1183.6	1.6214	4.434	1105.8	1187.8	1.6034
320	7.485	1109.5	1192.6	1.6634	5.544	1106.0	1188.0	1.6271				
360	7.924	1125.3	1213.3	1.6893	5.886	1122.5	1209.7	1.6541	4.662	1119.7	1205.9	1.6259
400	8.353	1140.8	1233.5	1.7134	6.217	1138.5	1230.6	1.6790	4.934	1136.2	1227.5	1.6517
500	9.399	1178.6	1283.0	1.7678	7.017	1177.2	1281.1	1.7346	5.587	1175.7	1279.1	1.7085
600	10.425	1216.3	1332.1	1.8165	7.794	1215.3	1330.7	1.7838	6.216	1214.2	1329.3	1.7582
700	11.440	1254.4	1381.4	1.8609	8.561	1253.6	1380.3	1.8285	6.834	1252.8	1379.2	1.8033
800	12.448	1293.0	1431.2	1.9022	9.321	1292.4	1430.4	1.8700	7.445	1291.8	1429.6	1.8449
1000	14.454	1372.7	1533.2	1.9773	10.831	1372.3	1532.6	1.9453	8.657	1371.9	1532.1	1.9204
1200	16.452	1455.8	1638.5	2.0448	12.333	1455.5	1638.1	2.0130	9.861	1455.2	1637.7	1.9882
1400	18.445	1542.5	1747.3	2.1067	13.830	1542.3	1747.0	2.0749	11.060	1542.0	1746.7	2.0502
1600	20.44	1632.8	1859.7	2.1641	15.324	1632.6	1859.5	2.1323	12.257	1632.4	1859.3	2.1076
	$p = 120$ (341.30)				$p = 140$ (353.08)				$p = 160$ (363.60)			
Sat	3.730	1108.3	1191.1	1.5886	3.221	1110.3	1193.8	1.5761	2.836	1112.0	1196.0	1.5651
360	3.844	1116.7	1202.0	1.6021	3.259	1113.5	1198.0	1.5812				
400	4.079	1133.8	1224.4	1.6288	3.466	1131.4	1221.2	1.6088	3.007	1128.8	1217.8	1.5911
450	4.360	1154.3	1251.2	1.6590	3.713	1152.4	1248.6	1.6399	3.228	1150.5	1246.1	1.6230
500	4.633	1174.2	1277.1	1.6868	3.952	1172.7	1275.1	1.6682	3.440	1171.2	1273.0	1.6518
600	5.164	1213.2	1327.8	1.7371	4.412	1212.1	1326.4	1.7191	3.848	1211.1	1325.0	1.7034
700	5.682	1252.0	1378.2	1.7825	4.860	1251.2	1377.1	1.7648	4.243	1250.4	1376.0	1.7494
800	6.195	1291.2	1428.7	1.8243	5.301	1290.5	1427.9	1.8068	4.631	1289.9	1427.0	1.7916
1000	7.208	1371.5	1531.5	1.9000	6.173	1371.0	1531.0	1.8827	5.397	1370.6	1530.4	1.8677
1200	8.213	1454.9	1637.3	1.9679	7.036	1454.6	1636.9	1.9507	6.154	1454.3	1636.5	1.9358
1400	9.214	1541.8	1746.4	2.0300	7.895	1541.6	1746.1	2.0129	6.906	1541.4	1745.9	1.9980
1600	10.212	1632.3	1859.0	2.0875	8.752	1632.1	1858.8	2.0704	7.656	1631.9	1858.6	2.0556

TABLE C.3E (*continued*)

°F	v	u	h	s	v	u	h	s	v	u	h	s
	p = 180 (373.13)				p = 200 (381.86)				p = 300 (417.43)			
Sat	2.533	1113.4	1197.8	1.5553	2.289	1114.6	1199.3	1.5464	1.5442	1118.2	1203.9	1.5115
400	2.648	1126.2	1214.4	1.5749	2.361	1123.5	1210.8	1.5600				
450	2.850	1148.5	1243.4	1.6078	2.548	1146.4	1240.7	1.5938	1.6361	1135.4	1226.2	1.5365
500	3.042	1169.6	1270.9	1.6372	2.724	1168.0	1268.8	1.6239	1.7662	1159.5	1257.5	1.5701
600	3.409	1210.0	1323.5	1.6893	3.058	1208.9	1322.1	1.6767	2.004	1203.2	1314.5	1.6266
700	3.763	1249.6	1374.9	1.7357	3.379	1248.8	1373.8	1.7234	2.227	1244.6	1368.3	1.6751
800	4.110	1289.3	1426.2	1.7781	3.693	1288.6	1425.3	1.7660	2.442	1285.4	1421.0	1.7187
900	4.453	1329.4	1477.7	1.8175	4.003	1328.9	1477.1	1.8055	2.653	1326.3	1473.6	1.7589
1000	4.793	1370.2	1529.8	1.8545	4.310	1369.8	1529.3	1.8425	2.860	1367.7	1526.5	1.7964
1200	5.467	1454.0	1636.1	1.9227	4.918	1453.7	1635.7	1.9109	3.270	1452.2	1633.8	1.8653
1400	6.137	1541.2	1745.6	1.9849	5.521	1540.9	1745.3	1.9732	3.675	1539.8	1743.8	1.9279
1600	6.804	1631.7	1858.4	2.0425	6.123	1631.6	1858.2	2.0308	4.078	1630.7	1857.0	1.9857
	p = 400 (444.70)				p = 500 (467.13)				p = 600 (486.33)			
Sat	1.1620	1119.5	1205.5	1.4856	0.9283	1119.4	1205.3	1.4645	0.7702	1118.6	1204.1	1.4464
500	1.2843	1150.1	1245.2	1.5282	0.9924	1139.7	1231.5	1.4923	0.7947	1128.0	1216.2	1.4592
550	1.3833	1174.6	1277.0	1.5605	1.0792	1166.7	1266.6	1.5279	0.8749	1158.2	1255.4	1.4990
600	1.4760	1197.3	1306.6	1.5892	1.1583	1191.1	1298.3	1.5585	0.9456	1184.5	1289.5	1.5320
700	1.6503	1240.4	1362.5	1.6397	1.3040	1236.0	1356.7	1.6112	1.0727	1231.5	1350.6	1.5872
800	1.8163	1282.1	1416.6	1.6844	1.4407	1278.8	1412.1	1.6571	1.1900	1275.4	1407.6	1.6343
900	1.9776	1323.7	1470.1	1.7252	1.5723	1321.0	1466.5	1.6987	1.3021	1318.4	1462.9	1.6766
1000	2.136	1365.5	1523.6	1.7632	1.7008	1363.3	1520.7	1.7371	1.4108	1361.2	1517.8	1.7155
1100					1.8271	1406.0	1575.1	1.7731	1.5173	1404.2	1572.7	1.7519
1200	2.446	1450.7	1631.8	1.8327	1.9518	1449.2	1629.8	1.8072	1.6222	1447.7	1627.8	1.7861
1400	2.752	1538.7	1742.4	1.8956	2.198	1537.6	1741.0	1.8704	1.8289	1536.5	1739.5	1.8497
1600	3.055	1629.8	1855.9	1.9535	2.442	1628.9	1854.8	1.9285	2.033	1628.0	1853.7	1.9080
	p = 800 (518.36)				p = 1000 (544.75)				p = 2000 (636.00)			
550	0.6154	1138.8	1229.9	1.4469	0.4534	1114.8	1198.7	1.3966				
600	0.6776	1170.1	1270.4	1.4861	0.5140	1153.7	1248.8	1.4450				
650	0.7324	1197.2	1305.6	1.5186	0.5637	1184.7	1289.1	1.4822	0.2057	1091.1	1167.2	1.3141
700	0.7829	1222.1	1338.0	1.5471	0.6080	1212.0	1324.6	1.5135	0.2487	1147.7	1239.8	1.3782
800	0.8764	1268.5	1398.2	1.5969	0.6878	1261.2	1388.5	1.5664	0.3071	1220.1	1333.8	1.4562
900	0.9640	1312.9	1455.6	1.6408	0.7610	1307.3	1488.1	1.6120	0.3534	1276.8	1407.6	1.5126
1000	1.0482	1356.7	1511.9	1.6807	0.8305	1352.2	1505.9	1.6530	0.3945	1328.1	1474.1	1.5598
1100	1.1300	1400.5	1567.8	1.7178	0.8976	1396.8	1562.9	1.6908	0.4325	1377.2	1537.2	1.6017
1200	1.2102	1444.6	1623.8	1.7526	0.9630	1441.5	1619.7	1.7261	0.4685	1425.2	1598.6	1.6398
1400	1.3674	1534.2	1736.6	1.8167	1.0905	1531.9	1733.7	1.7909	0.5368	1520.2	1718.8	1.7082
1600	1.5218	1626.2	1851.5	1.8754	1.2152	1624.4	1849.3	1.8499	0.6020	1615.4	1838.2	1.7692
	p = 3000 (695.52)				p = 4000				p = 5000			
650					0.02447	657.7	675.8	0.8574	0.02377	648.0	670.0	0.8482
700	0.09771	1003.9	1058.1	1.1944	0.02867	742.1	763.4	0.9345	0.02676	721.8	746.6	0.9156
750	0.14831	1114.7	1197.1	1.3122	0.06331	960.7	1007.5	1.1395	0.03364	821.4	852.6	1.0049
800	0.17572	1167.6	1265.2	1.3675	0.10522	1095.0	1172.9	1.2740	0.05932	987.2	1042.1	1.1583
850	0.19731	1207.7	1317.2	1.4080	0.12833	1156.5	1251.5	1.3352	0.08556	1092.7	1171.9	1.2956
900	0.2160	1241.8	1361.7	1.4414	0.14622	1201.5	1309.7	1.3789	0.10385	1155.1	1251.1	1.3190
1000	0.2485	1301.7	1439.6	1.4967	0.17520	1272.9	1402.6	1.4449	0.13120	1242.0	1363.4	1.3988
1100	0.2772	1356.2	1510.1	1.5434	0.19954	1333.9	1481.6	1.4973	0.15302	1310.6	1452.2	1.4577
1200	0.3036	1408.0	1576.6	1.5848	0.2213	1390.1	1553.9	1.5423	0.17199	1371.6	1530.8	1.5066
1300					0.2414	1443.7	1622.4	1.5823	0.18918	1428.6	1603.7	1.5493
1400	0.3524	1508.1	1703.7	1.6571	0.2603	1495.7	1688.4	1.6188	0.20517	1483.2	1673.0	1.5876
1600	0.3978	1606.3	1827.1	1.7201	0.2959	1597.1	1816.1	1.6841	0.2348	1587.9	1805.2	1.6551

Source: Keenan, Keyes, Hill, and Moore, *Steam Tables*, Wiley, New York, 1969.

TABLE C.4E Compressed Liquid

°F	*v*	*u*	*h*	*s*	*v*	*u*	*h*	*s*	*v*	*u*	*h*	*s*
	$p = 500$ (467.13)				$p = 1000$ (544.75)				$p = 1500$ (596.39)			
Sat	0.019748	447.70	449.53	0.64904	0.021591	538.39	542.38	0.74320	0.023461	604.97	611.48	0.80824
32	0.015994	0.00	1.49	0.00000	0.015967	0.03	2.99	0.00005	0.015939	0.05	4.47	0.00007
50	0.015998	18.02	19.50	0.03599	0.015972	17.99	20.94	0.03592	0.015946	17.95	22.38	0.03584
100	0.016106	67.87	69.36	0.12932	0.016082	67.70	70.68	0.12901	0.016058	67.53	71.99	0.12870
150	0.016318	117.66	119.17	0.21457	0.016293	117.38	120.40	0.21410	0.016268	117.10	121.62	0.21364
200	0.016608	167.65	169.19	0.29341	0.016580	167.26	170.32	0.29281	0.016554	166.87	171.46	0.29221
250	0.016972	217.99	219.56	0.36702	0.016941	217.47	220.61	0.36628	0.016910	216.96	221.65	0.36554
300	0.017416	268.92	270.53	0.43641	0.017379	268.24	271.46	0.43552	0.017343	267.58	272.39	0.43463
350	0.017954	320.71	322.37	0.50249	0.017909	319.83	323.15	0.50140	0.017865	318.98	323.94	0.50034
400	0.018608	373.68	375.40	0.56604	0.018550	372.55	375.98	0.56472	0.018493	371.45	376.59	0.56343
450	0.019420	428.40	430.19	0.62798	0.019340	426.89	430.47	0.62632	0.019264	425.44	430.79	0.62470
500					0.02036	483.8	487.5	0.6874	0.02024	481.8	487.4	0.6853
550									0.02158	542.1	548.1	0.7469
	$p = 2000$ (636.00)				$p = 3000$ (695.52)				$p = 5000$			
Sat	0.025649	662.40	671.89	0.86227	0.034310	783.45	802.50	0.97320				
32	0.015912	0.06	5.95	0.00008	0.015859	0.09	8.90	0.00009	0.015755	0.11	14.70	−0.00001
50	0.015920	17.91	23.81	0.03575	0.015870	17.84	26.65	0.03555	0.015773	17.67	32.26	0.03508
100	0.016034	67.37	73.30	0.12839	0.015987	67.04	75.91	0.12777	0.015897	66.40	81.11	0.12651
200	0.016527	166.49	172.60	0.29162	0.016476	165.74	174.89	0.29046	0.016376	164.32	179.47	0.28818
300	0.017308	266.93	273.33	0.43376	0.017240	265.66	275.23	0.43205	0.017110	263.25	279.08	0.42875
400	0.018439	370.38	377.21	0.56216	0.018334	368.32	378.50	0.55970	0.018141	364.47	381.25	0.55506
450	0.019191	424.04	431.14	0.62313	0.019053	421.36	431.93	0.62011	0.018803	416.44	433.84	0.61451
500	0.02014	479.8	487.3	0.6832	0.019944	476.2	487.3	0.6794	0.019603	469.8	487.9	0.6724
560	0.02172	551.8	559.8	0.7565	0.021382	546.2	558.0	0.7508	0.020835	536.7	556.0	0.7411
600	0.02330	605.4	614.0	0.8086	0.02274	597.0	609.6	0.8004	0.02191	584.0	604.2	0.7876
640					0.02475	654.3	668.0	0.8545	0.02334	634.6	656.2	0.8357
680					0.02879	728.4	744.3	0.9226	0.02535	690.6	714.1	0.8873
700									0.02676	721.8	746.6	0.9156

Source: Keenan, Keyes, Hill, and Moore, *Steam Tables*, Wiley, New York, 1969.

TABLE C.5E Saturated Solid—Vapor

T, °F	Specific Volume			Internal Energy			Enthalpy			Entropy		
	p, psia	Sat Solid v_i	Sat Vapor $v_g \times 10^{-3}$	Sat Solid u_i	Subl. u_{ig}	Sat Vapor u_g	Sat Solid h_i	Subl. h_{ig}	Sat Vapor h_g	Sat Solid s_i	Subl. s_{ig}	Sat Vapor s_g
32.018	0.0887	0.01747	3.302	−143.34	1164.6	1021.2	−143.34	1218.7	1075.4	−0.292	2.479	2.187
32	0.0886	0.01747	3.305	−143.35	1164.6	1021.2	−143.35	1218.7	1075.4	−0.292	2.479	2.187
30	0.0808	0.01747	3.607	−144.35	1164.9	1020.5	−144.35	1218.9	1074.5	−0.294	2.489	2.195
25	0.0641	0.01746	4.506	−146.84	1165.7	1018.9	−146.84	1219.1	1072.3	−0.299	2.515	2.216
20	0.0505	0.01745	5.655	−149.31	1166.5	1017.2	−149.31	1219.4	1070.1	−0.304	2.542	2.238
15	0.0396	0.01745	7.13	−151.75	1167.3	1015.5	−151.75	1219.7	1067.9	−0.309	2.569	2.260
10	0.0309	0.01744	9.04	−154.17	1168.1	1013.9	−154.17	1219.9	1065.7	−0.314	2.597	2.283
5	0.0240	0.01743	11.52	−156.56	1168.8	1012.2	−156.56	1220.1	1063.5	−0.320	2.626	2.306
0	0.0185	0.01743	14.77	−158.93	1169.5	1010.6	−158.93	1220.2	1061.2	−0.325	2.655	2.330
−5	0.0142	0.01742	19.03	−161.27	1170.2	1008.9	−161.27	1220.3	1059.0	−0.330	2.684	2.354
−10	0.0109	0.01741	24.66	−163.59	1170.9	1007.3	−163.59	1220.4	1056.8	−0.335	2.714	2.379
−15	0.0082	0.01740	32.2	−165.89	1171.5	1005.6	−165.89	1220.5	1054.6	−0.340	2.745	2.405
−20	0.0062	0.01740	42.2	−168.16	1172.1	1003.9	−168.16	1220.6	1052.4	−0.345	2.776	2.431
−25	0.0046	0.01739	55.7	−170.40	1172.7	1002.3	−170.40	1220.6	1050.2	−0.351	2.808	2.457
−30	0.0035	0.01738	74.1	−172.63	1173.2	1000.6	−172.63	1220.6	1048.0	−0.356	2.841	2.485
−35	0.0026	0.01737	99.2	−174.82	1173.8	998.9	−174.82	1220.6	1045.8	−0.361	2.874	2.513
−40	0.0019	0.01737	133.8	−177.00	1174.3	997.3	−177.00	1220.6	1043.6	−0.366	2.908	2.542

Source: Keenan, Keyes, Hill, and Moore, *Steam Tables*, Wiley, New York, 1969.

D. THERMODYNAMIC PROPERTIES OF FREON 12

TABLE D.1 Saturated Refrigerant 134a: Temperature Table

Temp. °C	Pressure kPa	Specific Volume m^3/kg		Internal Energy kJ/kg		Enthalpy kJ/kg			Entropy kJ/kg·K	
		Sat. Liquid $v_f \times 10^3$	Sat. Vapor v_g	Sat. Liquid u_f	Sat. Vapor u_g	Sat. Liquid h_f	Evap. h_{fg}	Sat. Vapor h_g	Sat. Liquid s_f	Sat. Vapor s_g
−40	51.64	0.7055	0.3569	−0.04	204.45	0.00	222.88	222.88	0.0000	0.9560
−36	63.32	0.7113	0.2947	4.68	206.73	4.73	220.67	225.40	0.0201	0.9506
−32	77.04	0.7172	0.2451	9.47	209.01	9.52	218.37	227.90	0.0401	0.9456
−28	93.05	0.7233	0.2052	14.31	211.29	14.37	216.01	230.38	0.0600	0.9411
−26	101.99	0.7265	0.1882	16.75	212.43	16.82	214.80	231.62	0.0699	0.9390
−24	111.60	0.7296	0.1728	19.21	213.57	19.29	213.57	232.85	0.0798	0.9370
−22	121.92	0.7328	0.1590	21.68	214.70	21.77	212.32	234.08	0.0897	0.9351
−20	132.99	0.7361	0.1464	24.17	215.84	24.26	211.05	235.31	0.0996	0.9332
−18	144.83	0.7395	0.1350	26.67	216.97	26.77	209.76	236.53	0.1094	0.9315
−16	157.48	0.7428	0.1247	29.18	218.10	29.30	208.45	237.74	0.1192	0.9298
−12	185.40	0.7498	0.1068	34.25	220.36	34.39	205.77	240.15	0.1388	0.9267
−8	217.04	0.7569	0.0919	39.38	222.60	39.54	203.00	242.54	0.1583	0.9239
−4	252.74	0.7644	0.0794	44.56	224.84	44.75	200.15	244.90	0.1777	0.9213
0	292.82	0.7721	0.0689	49.79	227.06	50.02	197.21	247.23	0.1970	0.9190
4	337.65	0.7801	0.0600	55.08	229.27	55.35	194.19	249.53	0.2162	0.9169
8	387.56	0.7884	0.0525	60.43	231.46	60.73	191.07	251.80	0.2354	0.9150
12	442.94	0.7971	0.0460	65.83	233.63	66.18	187.85	254.03	0.2545	0.9132
16	504.16	0.8062	0.0405	71.29	235.78	71.69	184.52	256.22	0.2735	0.9116
20	571.60	0.8157	0.0358	76.80	237.91	77.26	181.09	258.36	0.2924	0.9102
24	645.66	0.8257	0.0317	82.37	240.01	82.90	177.55	260.45	0.3113	0.9089
26	685.30	0.8309	0.0298	85.18	241.05	85.75	175.73	261.48	0.3208	0.9082
28	726.75	0.8362	0.0281	88.00	242.08	88.61	173.89	262.50	0.3302	0.9076
30	770.06	0.8417	0.0265	90.84	243.10	91.49	172.00	263.50	0.3396	0.9070
32	815.28	0.8473	0.0250	93.70	244.12	94.39	170.09	264.48	0.3490	0.9064
34	862.47	0.8530	0.0236	96.58	245.12	97.31	168.14	265.45	0.3584	0.9058
36	911.68	0.8590	0.0223	99.47	246.11	100.25	166.15	266.40	0.3678	0.9053
38	962.98	0.8651	0.0210	102.38	247.09	103.21	164.12	267.33	0.3772	0.9047
40	1016.4	0.8714	0.0199	105.30	248.06	106.19	162.05	268.24	0.3866	0.9041
42	1072.0	0.8780	0.0188	108.25	249.02	109.19	159.94	269.14	0.3960	0.9035
44	1129.9	0.8847	0.0177	111.22	249.96	112.22	157.79	270.01	0.4054	0.9030
48	1252.6	0.8989	0.0159	117.22	251.79	118.35	153.33	271.68	0.4243	0.9017
52	1385.1	0.9142	0.0142	123.31	253.55	124.58	148.66	273.24	0.4432	0.9004
56	1527.8	0.9308	0.0127	129.51	255.23	130.93	143.75	274.68	0.4622	0.8990
60	1681.3	0.9488	0.0114	135.82	256.81	137.42	138.57	275.99	0.4814	0.8973
70	2116.2	1.0027	0.0086	152.22	260.15	154.34	124.08	278.43	0.5302	0.8918
80	2632.4	1.0766	0.0064	169.88	262.14	172.71	106.41	279.12	0.5814	0.8827
90	3243.5	1.1949	0.0046	189.82	261.34	193.69	82.63	276.32	0.6380	0.8655
100	3974.2	1.5443	0.0027	218.60	248.49	224.74	34.40	259.13	0.7196	0.8117

Source: Tables D.1 through D.3 are based on equations from D. P. Wilson and R. S. Basu, "Thermodynamic Properties of a New Stratospherically Safe Working Fluid—Refrigerant 134a," *ASHRAE Trans.*, Vol. 94, Pt. 2, 1988, pp. 2095–2118.

TABLE D.2 Saturated Refrigerant 134a: Pressure Table

Pressure kPa	Temp. °C	Specific Volume m³/kg		Internal Energy kJ/kg		Enthalpy kJ/kg			Entropy kJ/kg·K	
		Sat. Liquid $v_f \times 10^3$	Sat. Vapor v_g	Sat. Liquid u_f	Sat. Vapor u_g	Sat. Liquid h_f	Evap. h_{fg}	Sat. Vapor h_g	Sat. Liquid s_f	Sat. Vapor s_g
60	−37.07	0.7097	0.3100	3.14	206.12	3.46	221.27	224.72	0.0147	0.9520
80	−31.21	0.7184	0.2366	10.41	209.46	10.47	217.92	228.39	0.0440	0.9447
100	−26.43	0.7258	0.1917	16.22	212.18	16.29	215.06	231.35	0.0678	0.9395
120	−22.36	0.7323	0.1614	21.23	214.50	21.32	212.54	233.86	0.0879	0.9354
140	−18.80	0.7381	0.1395	25.66	216.52	25.77	210.27	236.04	0.1055	0.9322
160	−15.62	0.7435	0.1229	29.66	218.32	29.78	208.19	237.97	0.1211	0.9295
180	−12.73	0.7485	0.1098	33.31	219.94	33.45	206.26	239.71	0.1352	0.9273
200	−10.09	0.7532	0.0993	36.69	221.43	36.84	204.46	241.30	0.1481	0.9253
240	−5.37	0.7618	0.0834	42.77	224.07	42.95	201.14	244.09	0.1710	0.9222
280	−1.23	0.7697	0.0719	48.18	226.38	48.39	198.13	246.52	0.1911	0.9197
320	2.48	0.7770	0.0632	53.06	228.43	53.31	195.35	248.66	0.2089	0.9177
360	5.84	0.7839	0.0564	57.54	230.28	57.82	192.76	250.58	0.2251	0.9160
400	8.93	0.7904	0.0509	61.69	231.97	62.00	190.32	252.32	0.2399	0.9145
500	15.74	0.8056	0.0409	70.93	235.64	71.33	184.74	256.07	0.2723	0.9117
600	21.58	0.8196	0.0341	78.99	238.74	79.48	179.71	259.19	0.2999	0.9097
700	26.72	0.8328	0.0292	86.19	241.42	86.78	175.07	261.85	0.3242	0.9080
800	31.33	0.8454	0.0255	92.75	243.78	93.42	170.73	264.15	0.3459	0.9066
900	35.53	0.8576	0.0226	98.79	245.88	99.56	166.62	266.18	0.3656	0.9054
1000	39.39	0.8695	0.0202	104.42	247.77	105.29	162.68	267.97	0.3838	0.9043
1200	46.32	0.8928	0.0166	114.69	251.03	115.76	155.23	270.99	0.4164	0.9023
1400	52.43	0.9159	0.0140	123.98	253.74	125.26	148.14	273.40	0.4453	0.9003
1600	57.92	0.9392	0.0121	132.52	256.00	134.02	141.31	275.33	0.4714	0.8982
1800	62.91	0.9631	0.0105	140.49	257.88	142.22	134.60	276.83	0.4954	0.8959
2000	67.49	0.9878	0.0093	148.02	259.41	149.99	127.95	277.94	0.5178	0.8934
2500	77.59	1.0562	0.0069	165.48	261.84	168.12	111.06	279.17	0.5687	0.8854
3000	86.22	1.1416	0.0053	181.88	262.16	185.30	92.71	278.01	0.6156	0.8735

TABLE D.3 Superheated Refrigerant 134a

T °C	v m^3/kg	u kJ/kg	h kJ/kg	s kJ/kg·K	v m^3/kg	u kJ/kg	h kJ/kg	s kJ/kg·K
	$p = 0.06$ MPa (−37.07°C)				$p = 0.10$ MPa (−26.43°C)			
Sat.	0.31003	206.12	224.72	0.9520	0.19170	212.18	231.35	0.9395
−20	0.33536	217.86	237.98	1.0062	0.19770	216.77	236.54	0.9602
−10	0.34992	224.97	245.96	1.0371	0.20686	224.01	244.70	0.9918
0	0.36433	232.24	254.10	1.0675	0.21587	231.41	252.99	1.0227
10	0.37861	239.69	262.41	1.0973	0.22473	238.96	261.43	1.0531
20	0.39279	247.32	270.89	1.1267	0.23349	246.67	270.02	1.0829
30	0.40688	255.12	279.53	1.1557	0.24216	254.54	278.76	1.1122
40	0.42091	263.10	288.35	1.1844	0.25076	262.58	287.66	1.1411
50	0.43487	271.25	297.34	1.2126	0.25930	270.79	296.72	1.1696
60	0.44879	279.58	306.51	1.2405	0.26779	279.16	305.94	1.1977
70	0.46266	288.08	315.84	1.2681	0.27623	287.70	315.32	1.2254
80	0.47650	296.75	325.34	1.2954	0.28464	296.40	324.87	1.2528
90	0.49031	305.58	335.00	1.3224	0.29302	305.27	334.57	1.2799
	$p = 0.14$ MPa (−18.80°C)				$p = 0.18$ MPa (−12.73°C)			
Sat.	0.13945	216.52	236.04	0.9322	0.10983	219.94	239.71	0.9273
−10	0.14519	223.03	243.40	0.9606	0.11135	222.02	242.06	0.9362
0	0.15219	230.55	251.86	0.9922	0.11678	229.67	250.69	0.9684
10	0.15875	238.21	260.43	1.0230	0.12207	237.44	259.41	0.9998
20	0.16520	246.01	269.13	1.0532	0.12723	245.33	268.23	1.0304
30	0.17155	253.96	277.97	1.0828	0.13230	253.36	277.17	1.0604
40	0.17783	262.06	286.96	1.1120	0.13730	261.53	286.24	1.0898
50	0.18404	270.32	296.09	1.1407	0.14222	269.85	295.45	1.1187
60	0.19020	278.74	305.37	1.1690	0.14710	278.31	304.79	1.1472
70	0.19633	287.32	314.80	1.1969	0.15193	286.93	314.28	1.1753
80	0.20241	296.06	324.39	1.2244	0.15672	295.71	323.92	1.2030
90	0.20846	304.95	334.14	1.2516	0.16148	304.63	333.70	1.2303
100	0.21449	314.01	344.04	1.2785	0.16622	313.72	343.63	1.2573
	$p = 0.20$ MPa (−10.09°C)				$p = 0.24$ MPa (−5.37°C)			
Sat.	0.09933	221.43	241.30	0.9253	0.08343	224.07	244.09	0.9222
−10	0.09938	221.50	241.38	0.9256				
0	0.10438	229.23	250.10	0.9582	0.08574	228.31	248.89	0.9399
10	0.10922	237.05	258.89	0.9898	0.08993	236.26	257.84	0.9721
20	0.11394	244.99	267.78	1.0206	0.09399	244.30	266.85	1.0034
30	0.11856	253.06	276.77	1.0508	0.09794	252.45	275.95	1.0339
40	0.12311	261.26	285.88	1.0804	0.10181	260.72	285.16	1.0637
50	0.12758	269.61	295.12	1.1094	0.10562	269.12	294.47	1.0930
60	0.13201	278.10	304.50	1.1380	0.10937	277.67	303.91	1.1218
70	0.13639	286.74	314.02	1.1661	0.11307	286.35	313.49	1.1501
80	0.14073	295.53	323.68	1.1939	0.11674	295.18	323.19	1.1780
90	0.14504	304.47	333.48	1.2212	0.12037	304.15	333.04	1.2055
100	0.14932	313.57	343.43	1.2483	0.12398	313.27	343.03	1.2326

TABLE D.3 (*continued*)

T °C	v m³/kg	u kJ/kg	h kJ/kg	s kJ/kg·K	v m³/kg	u kJ/kg	h kJ/kg	s kJ/kg·K
	$p = 0.28$ MPa (−1.23°C)				$p = 0.32$ MPa (2.48°C)			
Sat.	0.07193	226.38	246.52	0.9197	0.06322	228.43	248.66	0.9177
0	0.07240	227.37	247.64	0.9238				
10	0.07613	235.44	256.76	0.9566	0.06576	234.61	255.65	0.9427
20	0.07972	243.59	265.91	0.9883	0.06901	242.87	264.95	0.9749
30	0.08320	251.83	275.12	1.0192	0.07214	251.19	274.28	1.0062
40	0.08660	260.17	284.42	1.0494	0.07518	259.61	283.67	1.0367
50	0.08992	268.64	293.81	1.0789	0.07815	268.14	293.15	1.0665
60	0.09319	277.23	303.32	1.1079	0.08106	276.79	302.72	1.0957
70	0.09641	285.96	312.95	1.1364	0.08392	285.56	312.41	1.1243
80	0.09960	294.82	322.71	1.1644	0.08674	294.46	322.22	1.1525
90	0.10275	303.83	332.60	1.1920	0.08953	303.50	332.15	1.1802
100	0.10587	312.98	342.62	1.2193	0.09229	312.68	342.21	1.2076
110	0.10897	322.27	352.78	1.2461	0.09503	322.00	352.40	1.2345
120	0.11205	331.71	363.08	1.2727	0.09774	331.45	362.73	1.2611
	$p = 0.40$ MPa (8.93°C)				$p = 0.50$ MPa (15.74°C)			
Sat.	0.05089	231.97	252.32	0.9145	0.04086	235.64	256.07	0.9117
10	0.05119	232.87	253.35	0.9182				
20	0.05397	241.37	262.96	0.9515	0.04188	239.40	260.34	0.9264
30	0.05662	249.89	272.54	0.9837	0.04416	248.20	270.28	0.9597
40	0.05917	258.47	282.14	1.0148	0.04633	256.99	280.16	0.9918
50	0.06164	267.13	291.79	1.0452	0.04842	265.83	290.04	1.0229
60	0.06405	275.89	301.51	1.0748	0.05043	274.73	299.95	1.0531
70	0.06641	284.75	311.32	1.1038	0.05240	283.72	309.92	1.0825
80	0.06873	293.73	321.23	1.1322	0.05432	292.80	319.96	1.1114
90	0.07102	302.84	331.25	1.1602	0.05620	302.00	330.10	1.1397
100	0.07327	312.07	341.38	1.1878	0.05805	311.31	340.33	1.1675
110	0.07550	321.44	351.64	1.2149	0.05988	320.74	350.68	1.1949
120	0.07771	330.94	362.03	1.2417	0.06168	330.30	361.14	1.2218
130	0.07991	340.58	372.54	1.2681	0.06347	339.98	371.72	1.2484
140	0.08208	350.35	383.18	1.2941	0.06524	349.79	382.42	1.2746

TABLE D.3 *(continued)*

T °C	v m³/kg	u kJ/kg	h kJ/kg	s kJ/kg·K	v m³/kg	u kJ/kg	h kJ/kg	s kJ/kg·K
	$p = 0.60$ MPa (21.58°C)				$p = 0.70$ MPa (26.72°C)			
Sat.	0.03408	238.74	259.19	0.9097	0.02918	241.42	261.85	0.9080
30	0.03581	246.41	267.89	0.9388	0.02979	244.51	265.37	0.9197
40	0.03774	255.45	278.09	0.9719	0.03157	253.83	275.93	0.9539
50	0.03958	264.48	288.23	1.0037	0.03324	263.08	286.35	0.9867
60	0.04134	273.54	298.35	1.0346	0.03482	272.31	296.69	1.0182
70	0.04304	282.66	308.48	1.0645	0.03634	281.57	307.01	1.0487
80	0.04469	291.86	318.67	1.0938	0.03781	290.88	317.35	1.0784
90	0.04631	301.14	328.93	1.1225	0.03924	300.27	327.74	1.1074
100	0.04790	310.53	339.27	1.1505	0.04064	309.74	338.19	1.1358
110	0.04946	320.03	349.70	1.1781	0.04201	319.31	348.71	1.1637
120	0.05099	329.64	360.24	1.2053	0.04335	328.98	359.33	1.1910
130	0.05251	339.38	370.88	1.2320	0.04468	338.76	370.04	1.2179
140	0.05402	349.23	381.64	1.2584	0.04599	348.66	380.86	1.2444
150	0.05550	359.21	392.52	1.2844	0.04729	358.68	391.79	1.2706
160	0.05698	369.32	403.51	1.3100	0.04857	368.82	402.82	1.2963
	$p = 0.80$ MPa (31.33°C)				$p = 0.90$ MPa (35.53°C)			
Sat.	0.02547	243.78	264.15	0.9066	0.02255	245.88	266.18	0.9054
40	0.02691	252.13	273.66	0.9374	0.02325	250.32	271.25	0.9217
50	0.02846	261.62	284.39	0.9711	0.02472	260.09	282.34	0.9566
60	0.02992	271.04	294.98	1.0034	0.02609	269.72	293.21	0.9897
70	0.03131	280.45	305.50	1.0345	0.02738	279.30	303.94	1.0214
80	0.03264	289.89	316.00	1.0647	0.02861	288.87	314.62	1.0521
90	0.03393	299.37	326.52	1.0940	0.02980	298.46	325.28	1.0819
100	0.03519	308.93	337.08	1.1227	0.03095	308.11	335.96	1.1109
110	0.03642	318.57	347.71	1.1508	0.03207	317.82	346.68	1.1392
120	0.03762	328.31	358.40	1.1784	0.03316	327.62	357.47	1.1670
130	0.03881	338.14	369.19	1.2055	0.03423	337.52	368.33	1.1943
140	0.03997	348.09	380.07	1.2321	0.03529	347.51	379.27	1.2211
150	0.04113	358.15	391.05	1.2584	0.03633	357.61	390.31	1.2475
160	0.04227	368.32	402.14	1.2843	0.03736	367.82	401.44	1.2735
170	0.04340	378.61	413.33	1.3098	0.03838	378.14	412.68	1.2992
180	0.04452	389.02	424.63	1.3351	0.03939	388.57	424.02	1.3245

TABLE D.3 *(continued)*

T °C	v m³/kg	u kJ/kg	h kJ/kg	s kJ/kg·K	v m³/kg	u kJ/kg	h kJ/kg	s kJ/kg·K
	$p = 1.00$ MPa (39.39°C)				$p = 1.20$ MPa (46.32°C)			
Sat.	0.02020	247.77	267.97	0.9043	0.01663	251.03	270.99	0.9023
40	0.02029	248.39	268.68	0.9066				
50	0.02171	258.48	280.19	0.9428	0.01712	254.98	275.52	0.9164
60	0.02301	268.35	291.36	0.9768	0.01835	265.42	287.44	0.9527
70	0.02423	278.11	302.34	1.0093	0.01947	275.59	298.96	0.9868
80	0.02538	287.82	313.20	1.0405	0.02051	285.62	310.24	1.0192
90	0.02649	297.53	324.01	1.0707	0.02150	295.59	321.39	1.0503
100	0.02755	307.27	334.82	1.1000	0.02244	305.54	332.47	1.0804
110	0.02858	317.06	345.65	1.1286	0.02335	315.50	343.52	1.1096
120	0.02959	326.93	356.52	1.1567	0.02423	325.51	354.58	1.1381
130	0.03058	336.88	367.46	1.1841	0.02508	335.58	365.68	1.1660
140	0.03154	346.92	378.46	1.2111	0.02592	345.73	376.83	1.1933
150	0.03250	357.06	389.56	1.2376	0.02674	355.95	388.04	1.2201
160	0.03344	367.31	400.74	1.2638	0.02754	366.27	399.33	1.2465
170	0.03436	377.66	412.02	1.2895	0.02834	376.69	410.70	1.2724
180	0.03528	388.12	423.40	1.3149	0.02912	387.21	422.16	1.2980
	$p = 1.40$ MPa (52.43°C)				$p = 1.60$ MPa (57.92°C)			
Sat.	0.01405	253.74	273.40	0.9003	0.01208	256.00	275.33	0.8982
60	0.01495	262.17	283.10	0.9297	0.01233	258.48	278.20	0.9069
70	0.01603	272.87	295.31	0.9658	0.01340	269.89	291.33	0.9457
80	0.01701	283.29	307.10	0.9997	0.01435	280.78	303.74	0.9813
90	0.01792	293.55	318.63	1.0319	0.01521	291.39	315.72	1.0148
100	0.01878	303.73	330.02	1.0628	0.01601	301.84	327.46	1.0467
110	0.01960	313.88	341.32	1.0927	0.01677	312.20	339.04	1.0773
120	0.02039	324.05	352.59	1.1218	0.01750	322.53	350.53	1.1069
130	0.02115	334.25	363.86	1.1501	0.01820	332.87	361.99	1.1357
140	0.02189	344.50	375.15	1.1777	0.01887	343.24	373.44	1.1638
150	0.02262	354.82	386.49	1.2048	0.01953	353.66	384.91	1.1912
160	0.02333	365.22	397.89	1.2315	0.02017	364.15	396.43	1.2181
170	0.02403	375.71	409.36	1.2576	0.02080	374.71	407.99	1.2445
180	0.02472	386.29	420.90	1.2834	0.02142	385.35	419.62	1.2704
190	0.02541	396.96	432.53	1.3088	0.02203	396.08	431.33	1.2960
200	0.02608	407.73	444.24	1.3338	0.02263	406.90	443.11	1.3212

TABLE D.1E Properties of Saturated Refrigerant 134a: Temperature Table

Temp. °F	Pressure psia	Specific Volume ft³/lbm		Internal Energy Btu/lbm		Enthalpy Btu/lbm			Entropy Btu/lbm·°R	
		Sat. Liquid v_f	Sat. Vapor v_g	Sat. Liquid u_f	Sat. Vapor u_g	Sat. Liquid h_f	Evap. h_{fg}	Sat. Vapor h_g	Sat. Liquid s_f	Sat. Vapor s_g
−40	7.490	0.01130	5.7173	−0.02	87.90	0.00	95.82	95.82	0.0000	0.2283
−30	9.920	0.01143	4.3911	2.81	89.26	2.83	94.49	97.32	0.0067	0.2266
−20	12.949	0.01156	3.4173	5.69	90.62	5.71	93.10	98.81	0.0133	0.2250
−15	14.718	0.01163	3.0286	7.14	91.30	7.17	92.38	99.55	0.0166	0.2243
−10	16.674	0.01170	2.6918	8.61	91.98	8.65	91.64	100.29	0.0199	0.2236
−5	18.831	0.01178	2.3992	10.09	92.66	10.13	90.89	101.02	0.0231	0.2230
0	21.203	0.01185	2.1440	11.58	93.33	11.63	90.12	101.75	0.0264	0.2224
5	23.805	0.01193	1.9208	13.09	94.01	13.14	89.33	102.47	0.0296	0.2219
10	26.651	0.01200	1.7251	14.60	94.68	14.66	88.53	103.19	0.0329	0.2214
15	29.756	0.01208	1.5529	16.13	95.35	16.20	87.71	103.90	0.0361	0.2209
20	33.137	0.01216	1.4009	17.67	96.02	17.74	86.87	104.61	0.0393	0.2205
25	36.809	0.01225	1.2666	19.22	96.69	19.30	86.02	105.32	0.0426	0.2200
30	40.788	0.01233	1.1474	20.78	97.35	20.87	85.14	106.01	0.0458	0.2196
40	49.738	0.01251	0.9470	23.94	98.67	24.05	83.34	107.39	0.0522	0.2189
50	60.125	0.01270	0.7871	27.14	99.98	27.28	81.46	108.74	0.0585	0.2183
60	72.092	0.01290	0.6584	30.39	101.27	30.56	79.49	110.05	0.0648	0.2178
70	85.788	0.01311	0.5538	33.68	102.54	33.89	77.44	111.33	0.0711	0.2173
80	101.37	0.01334	0.4682	37.02	103.78	37.27	75.29	112.56	0.0774	0.2169
85	109.92	0.01346	0.4312	38.72	104.39	38.99	74.17	113.16	0.0805	0.2167
90	118.99	0.01358	0.3975	40.42	105.00	40.72	73.03	113.75	0.0836	0.2165
95	128.62	0.01371	0.3668	42.14	105.60	42.47	71.86	114.33	0.0867	0.2163
100	138.83	0.01385	0.3388	43.87	106.18	44.23	70.66	114.89	0.0898	0.2161
105	149.63	0.01399	0.3131	45.62	106.76	46.01	69.42	115.43	0.0930	0.2159
110	161.04	0.01414	0.2896	47.39	107.33	47.81	68.15	115.96	0.0961	0.2157
115	173.10	0.01429	0.2680	49.17	107.88	49.63	66.84	116.47	0.0992	0.2155
120	185.82	0.01445	0.2481	50.97	108.42	51.47	65.48	116.95	0.1023	0.2153
140	243.86	0.01520	0.1827	58.39	110.41	59.08	59.57	118.65	0.1150	0.2143
160	314.63	0.01617	0.1341	66.26	111.97	67.20	52.58	119.78	0.1280	0.2128
180	400.22	0.01758	0.0964	74.83	112.77	76.13	43.78	119.91	0.1417	0.2101
200	503.52	0.02014	0.0647	84.90	111.66	86.77	30.92	117.69	0.1575	0.2044
210	563.51	0.02329	0.0476	91.84	108.48	94.27	19.18	113.45	0.1684	0.1971

Source: Tables D.1E through D.3E are based on equations from D. P. Wilson and R. S. Basu, "Thermodynamic Properties of a New Stratospherically Safe Working Fluid—Refrigerant 134a," *ASHRAE Trans.*, Vol. 94, Pt. 2, 1988, pp. 2095–2118.

TABLE D.2E Properties of Saturated Refrigerant 134a: Pressure Table

Pressure psia	Temp. °F	Specific Volume ft³/lbm		Internal Energy Btu/lbm		Enthalpy Btu/lbm			Entropy Btu/lbm·°R	
		Sat. Liquid v_f	Sat. Vapor v_g	Sat. Liquid u_f	Sat. Vapor u_g	Sat. Liquid h_f	Evap. h_{fg}	Sat. Vapor h_g	Sat. Liquid s_f	Sat. Vapor s_g
5	−53.48	0.01113	8.3508	−3.74	86.07	−3.73	97.53	93.79	−.0090	0.2311
10	−29.71	0.01143	4.3581	2.89	89.30	2.91	94.45	97.37	0.0068	0.2265
15	−14.25	0.01164	2.9747	7.36	91.40	7.40	92.27	99.66	0.0171	0.2242
20	−2.48	0.01181	2.2661	10.84	93.00	10.89	90.50	101.39	0.0248	0.2227
30	15.38	0.01209	1.5408	16.24	95.40	16.31	87.65	103.96	0.0364	0.2209
40	29.04	0.01232	1.1692	20.48	97.23	20.57	85.31	105.88	0.0452	0.2197
50	40.27	0.01252	0.9422	24.02	98.71	24.14	83.29	107.43	0.0523	0.2189
60	49.89	0.01270	0.7887	27.10	99.96	27.24	81.48	108.72	0.0584	0.2183
70	58.35	0.01286	0.6778	29.85	101.05	30.01	79.82	109.83	0.0638	0.2179
80	65.93	0.01302	0.5938	32.33	102.02	32.53	78.28	110.81	0.0686	0.2175
90	72.83	0.01317	0.5278	34.62	102.89	34.84	76.84	111.68	0.0729	0.2172
100	79.17	0.01332	0.4747	36.75	103.68	36.99	75.47	112.46	0.0768	0.2169
120	90.54	0.01360	0.3941	40.61	105.06	40.91	72.91	113.82	0.0839	0.2165
140	100.56	0.01386	0.3358	44.07	106.25	44.43	70.52	114.95	0.0902	0.2161
160	109.56	0.01412	0.2916	47.23	107.28	47.65	68.26	115.91	0.0958	0.2157
180	117.74	0.01438	0.2569	50.16	108.18	50.64	66.10	116.74	0.1009	0.2154
200	125.28	0.01463	0.2288	52.90	108.98	53.44	64.01	117.44	0.1057	0.2151
220	132.27	0.01489	0.2056	55.48	109.68	56.09	61.96	118.05	0.1101	0.2147
240	138.79	0.01515	0.1861	57.93	110.30	58.61	59.96	118.56	0.1142	0.2144
260	144.92	0.01541	0.1695	60.28	110.84	61.02	57.97	118.99	0.1181	0.2140
280	150.70	0.01568	0.1550	62.53	111.31	63.34	56.00	119.35	0.1219	0.2136
300	156.17	0.01596	0.1424	64.71	111.72	65.59	54.03	119.62	0.1254	0.2132
350	168.72	0.01671	0.1166	69.88	112.45	70.97	49.03	120.00	0.1338	0.2118
400	179.95	0.01758	0.0965	74.81	112.77	76.11	43.80	119.91	0.1417	0.2102
450	190.12	0.01863	0.0800	79.63	112.60	81.18	38.08	119.26	0.1493	0.2079
500	199.38	0.02002	0.0657	84.54	111.76	86.39	31.44	117.83	0.1570	0.2047

TABLE D.3E Superheated Refrigerant 134a Vapor

T °F	v ft³/lbm	u Btu/lbm	h Btu/lbm	s Btu/lbm·°R	v ft³/lbm	u Btu/lbm	h Btu/lbm	s Btu/lbm·°R
	$p = 10$ psia (−29.71°F)				$p = 15$ psia (−14.25°F)			
Sat.	4.3581	89.30	97.37	0.2265	2.9747	91.40	99.66	0.2242
−20	4.4718	90.89	99.17	0.2307				
0	4.7026	94.24	102.94	0.2391	3.0893	93.84	102.42	0.2303
20	4.9297	97.67	106.79	0.2472	3.2468	97.33	106.34	0.2386
40	5.1539	101.19	110.72	0.2553	3.4012	100.89	110.33	0.2468
60	5.3758	104.80	114.74	0.2632	3.5533	104.54	114.40	0.2548
80	5.5959	108.50	118.85	0.2709	3.7034	108.28	118.56	0.2626
100	5.8145	112.29	123.05	0.2786	3.8520	112.10	122.79	0.2703
120	6.0318	116.18	127.34	0.2861	3.9993	116.01	127.11	0.2779
140	6.2482	120.16	131.72	0.2935	4.1456	120.00	131.51	0.2854
160	6.4638	124.23	136.19	0.3009	4.2911	124.09	136.00	0.2927
180	6.6786	128.38	140.74	0.3081	4.4359	128.26	140.57	0.3000
200	6.8929	132.63	145.39	0.3152	4.5801	132.52	145.23	0.3072
	$p = 20$ psia (−2.48°F)				$p = 30$ psia (15.38°F)			
Sat.	2.2661	93.00	101.39	0.2227	1.5408	95.40	103.96	0.2209
0	2.2816	93.43	101.88	0.2238				
20	2.4046	96.98	105.88	0.2323	1.5611	96.26	104.92	0.2229
40	2.5244	100.59	109.94	0.2406	1.6465	99.98	109.12	0.2315
60	2.6416	104.28	114.06	0.2487	1.7293	103.75	113.35	0.2398
80	2.7569	108.05	118.25	0.2566	1.8098	107.59	117.63	0.2478
100	2.8705	111.90	122.52	0.2644	1.8887	111.49	121.98	0.2558
120	2.9829	115.83	126.87	0.2720	1.9662	115.47	126.39	0.2635
140	3.0942	119.85	131.30	0.2795	2.0426	119.53	130.87	0.2711
160	3.2047	123.95	135.81	0.2869	2.1181	123.66	135.42	0.2786
180	3.3144	128.13	140.40	0.2922	2.1929	127.88	140.05	0.2859
200	3.4236	132.40	145.07	0.3014	2.2671	132.17	144.76	0.2932
220	3.5323	136.76	149.83	0.3085	2.3407	136.55	149.54	0.3003
	$p = 40$ psia (29.04°F)				$p = 50$ psia (40.27°F)			
Sat.	1.1692	97.23	105.88	0.2197	0.9422	98.71	107.43	0.2189
40	1.2065	99.33	108.26	0.2245				
60	1.2723	103.20	112.62	0.2331	0.9974	102.62	111.85	0.2276
80	1.3357	107.11	117.00	0.2414	1.0508	106.62	116.34	0.2361
100	1.3973	111.08	121.42	0.2494	1.1022	110.65	120.85	0.2443
120	1.4575	115.11	125.90	0.2573	1.1520	114.74	125.39	0.2523
140	1.5165	119.21	130.43	0.2650	1.2007	118.88	129.99	0.2601
160	1.5746	123.38	135.03	0.2725	1.2484	123.08	134.64	0.2677
180	1.6319	127.62	139.70	0.2799	1.2953	127.36	139.34	0.2752
200	1.6887	131.94	144.44	0.2872	1.3415	131.71	144.12	0.2825
220	1.7449	136.34	149.25	0.2944	1.3873	136.12	148.96	0.2897
240	1.8006	140.81	154.14	0.3015	1.4326	140.61	153.87	0.2969
260	1.8561	145.36	159.10	0.3085	1.4775	145.18	158.85	0.3039
280	1.9112	149.98	164.13	0.3154	1.5221	149.82	163.90	0.3108

TABLE D.3E (*continued*)

T °F	*v* ft³/lbm	*u* Btu/lbm	*h* Btu/lbm	*s* Btu/lbm·°R	*v* ft³/lbm	*u* Btu/lbm	*h* Btu/lbm	*s* Btu/lbm·°R
	p = 60 psia (49.89°F)				p = 70 psia (58.35°F)			
Sat.	0.7887	99.96	108.72	0.2183	0.6778	101.05	109.83	0.2179
60	0.8135	102.03	111.06	0.2229	0.6814	101.40	110.23	0.2186
80	0.8604	106.11	115.66	0.2316	0.7239	105.58	114.96	0.2276
100	0.9051	110.21	120.26	0.2399	0.7640	109.76	119.66	0.2361
120	0.9482	114.35	124.88	0.2480	0.8023	113.96	124.36	0.2444
140	0.9900	118.54	129.53	0.2559	0.8393	118.20	129.07	0.2524
160	1.0308	122.79	134.23	0.2636	0.8752	122.49	133.82	0.2601
180	1.0707	127.10	138.98	0.2712	0.9103	126.83	138.62	0.2678
200	1.1100	131.47	143.79	0.2786	0.9446	131.23	143.46	0.2752
220	1.1488	135.91	148.66	0.2859	0.9784	135.69	148.36	0.2825
240	1.1871	140.42	153.60	0.2930	1.0118	140.22	153.33	0.2897
260	1.2251	145.00	158.60	0.3001	1.0448	144.82	158.35	0.2968
280	1.2627	149.65	163.67	0.3070	1.0774	149.48	163.44	0.3038
300	1.3001	154.38	168.81	0.3139	1.1098	154.22	168.60	0.3107
	p = 80 psia (65.93°F)				p = 90 psia (72.83°F)			
Sat.	0.5938	102.02	110.81	0.2175	0.5278	102.89	111.68	0.2172
80	0.6211	105.03	114.23	0.2239	0.5408	104.46	113.47	0.2205
100	0.6579	109.30	119.04	0.2327	0.5751	108.82	118.39	0.2295
120	0.6927	113.56	123.82	0.2411	0.6073	113.15	123.27	0.2380
140	0.7261	117.85	128.60	0.2492	0.6380	117.50	128.12	0.2463
160	0.7584	122.18	133.41	0.2570	0.6675	121.87	132.98	0.2542
180	0.7898	126.55	138.25	0.2647	0.6961	126.28	137.87	0.2620
200	0.8205	130.98	143.13	0.2722	0.7239	130.73	142.79	0.2696
220	0.8506	135.47	148.06	0.2796	0.7512	135.25	147.76	0.2770
240	0.8803	140.02	153.05	0.2868	0.7779	139.82	152.77	0.2843
260	0.9095	144.63	158.10	0.2940	0.8043	144.45	157.84	0.2914
280	0.9384	149.32	163.21	0.3010	0.8303	149.15	162.97	0.2984
300	0.9671	154.06	168.38	0.3079	0.8561	153.91	168.16	0.3054
320	0.9955	158.88	173.62	0.3147	0.8816	158.73	173.42	0.3122
	p = 100 psia (79.17°F)				p = 120 psia (90.54°F)			
Sat.	0.4747	103.68	112.46	0.2169	0.3941	105.06	113.82	0.2165
80	0.4761	103.87	112.68	0.2173				
100	0.5086	108.32	117.73	0.2265	0.4080	107.26	116.32	0.2210
120	0.5388	112.73	122.70	0.2352	0.4355	111.84	121.52	0.2301
140	0.5674	117.13	127.63	0.2436	0.4610	116.37	126.61	0.2387
160	0.5947	121.55	132.55	0.2517	0.4852	120.89	131.66	0.2470
180	0.6210	125.99	137.49	0.2595	0.5082	125.42	136.70	0.2550
200	0.6466	130.48	142.45	0.2671	0.5305	129.97	141.75	0.2628
220	0.6716	135.02	147.45	0.2746	0.5520	134.56	146.82	0.2704
240	0.6960	139.61	152.49	0.2819	0.5731	139.20	151.92	0.2778
260	0.7201	144.26	157.59	0.2891	0.5937	143.89	157.07	0.2850
280	0.7438	148.98	162.74	0.2962	0.6140	148.63	162.26	0.2921
300	0.7672	153.75	167.95	0.3031	0.6339	153.43	167.51	0.2991
320	0.7904	158.59	173.21	0.3099	0.6537	158.29	172.81	0.3060

TABLE D.3E (*continued*)

T °F	*v* ft³/lbm	*u* Btu/lbm	*h* Btu/lbm	*s* Btu/lbm·°R	*v* ft³/lbm	*u* Btu/lbm	*h* Btu/lbm	*s* Btu/lbm·°R
	p = 140 psia (100.6°F)				p = 160 psia (109.6°F)			
Sat.	0.3358	106.25	114.95	0.2161	0.2916	107.28	115.91	0.2157
120	0.3610	110.90	120.25	0.2254	0.3044	109.88	118.89	0.2209
140	0.3846	115.58	125.54	0.2344	0.3269	114.73	124.41	0.2303
160	0.4066	120.21	130.74	0.2429	0.3474	119.49	129.78	0.2391
180	0.4274	124.82	135.89	0.2511	0.3666	124.20	135.06	0.2475
200	0.4474	129.44	141.03	0.2590	0.3849	128.90	140.29	0.2555
220	0.4666	134.09	146.18	0.2667	0.4023	133.61	145.52	0.2633
240	0.4852	138.77	151.34	0.2742	0.4192	138.34	150.75	0.2709
260	0.5034	143.50	156.54	0.2815	0.4356	143.11	156.00	0.2783
280	0.5212	148.28	161.78	0.2887	0.4516	147.92	161.29	0.2856
300	0.5387	153.11	167.06	0.2957	0.4672	152.78	166.61	0.2927
320	0.5559	157.99	172.39	0.3026	0.4826	157.69	171.98	0.2996
340	0.5730	162.93	177.78	0.3094	0.4978	162.65	177.39	0.3065
360	0.5898	167.93	183.21	0.3162	0.5128	167.67	182.85	0.3132
	p = 180 psia (117.7°F)				p = 200 psia (125.3°F)			
Sat.	0.2569	108.18	116.74	0.2154	0.2288	108.98	117.44	0.2151
120	0.2595	108.77	117.41	0.2166				
140	0.2814	113.83	123.21	0.2264	0.2446	112.87	121.92	0.2226
160	0.3011	118.74	128.77	0.2355	0.2636	117.94	127.70	0.2321
180	0.3191	123.56	134.19	0.2441	0.2809	122.88	133.28	0.2410
200	0.3361	128.34	139.53	0.2524	0.2970	127.76	138.75	0.2494
220	0.3523	133.11	144.84	0.2603	0.3121	132.60	144.15	0.2575
240	0.3678	137.90	150.15	0.2680	0.3266	137.44	149.53	0.2653
260	0.3828	142.71	155.46	0.2755	0.3405	142.30	154.90	0.2728
280	0.3974	147.55	160.79	0.2828	0.3540	147.18	160.28	0.2802
300	0.4116	152.44	166.15	0.2899	0.3671	152.10	165.69	0.2874
320	0.4256	157.38	171.55	0.2969	0.3799	157.07	171.13	0.2945
340	0.4393	162.36	177.00	0.3038	0.3926	162.07	176.60	0.3014
360	0.4529	167.40	182.49	0.3106	0.4050	167.13	182.12	0.3082
	p = 300 psia (156.2°F)				p = 400 psia (179.9°F)			
Sat.	0.1424	111.72	119.62	0.2132	0.0965	112.77	119.91	0.2102
160	0.1462	112.95	121.07	0.2155				
180	0.1633	118.93	128.00	0.2265	0.0965	112.79	119.93	0.2102
200	0.1777	124.47	134.34	0.2363	0.1143	120.14	128.60	0.2235
220	0.1905	129.79	140.36	0.2453	0.1275	126.35	135.79	0.2343
240	0.2021	134.99	146.21	0.2537	0.1386	132.12	142.38	0.2438
260	0.2130	140.12	151.95	0.2618	0.1484	137.65	148.64	0.2527
280	0.2234	145.23	157.63	0.2696	0.1575	143.06	154.72	0.2610
300	0.2333	150.33	163.28	0.2772	0.1660	148.39	160.67	0.2689
320	0.2428	155.44	168.92	0.2845	0.1740	153.69	166.57	0.2766
340	0.2521	160.57	174.56	0.2916	0.1816	158.97	172.42	0.2840
360	0.2611	165.74	180.23	0.2986	0.1890	164.26	178.26	0.2912
380	0.2699	170.94	185.92	0.3055	0.1962	169.57	184.09	0.2983
400	0.2786	176.18	191.64	0.3122	0.2032	174.90	189.94	0.3051

E. THERMODYNAMIC PROPERTIES OF AMMONIA

TABLE E.1 Saturated Ammonia

T, °C	p, kPa	Specific Volume m^3/kg			Enthalpy kJ/kg			Entropy kJ/kg·K		
		Sat. Liquid v_f	Evap. v_{fg}	Sat. Vapor v_g	Sat. Liquid h_f	Evap. h_{fg}	Sat. Vapor h_g	Sat. Liquid s_f	Evap. s_{fg}	Sat. Vapor s_g
−50	40.88	0.001 424	2.6239	2.6254	−44.3	1416.7	1372.4	−0.1942	6.3502	6.1561
−46	51.55	0.001 434	2.1126	2.1140	−26.6	1405.8	1379.2	−0.1156	6.1902	6.0746
−44	57.69	0.001 439	1.9018	1.9032	−17.8	1400.3	1382.5	−0.0768	6.1120	6.0352
−42	64.42	0.001 444	1.7155	1.7170	−8.9	1394.7	1385.8	−0.0382	6.0349	5.9967
−40	71.77	0.001 449	1.5506	1.5521	0.0	1389.0	1389.0	0.0000	5.9589	5.9589
−38	79.80	0.001 454	1.4043	1.4085	8.9	1383.3	1392.2	0.0380	5.8840	5.9220
−36	88.54	0.001 460	1.2742	1.2757	17.8	1377.6	1395.4	0.0757	5.8101	5.8858
−34	98.05	0.001 465	1.1582	1.1597	26.8	1371.8	1398.5	0.1132	5.7372	5.8504
−32	108.37	0.001 470	1.0547	1.0562	35.7	1365.9	1401.6	0.1504	5.6652	5.8156
−30	119.55	0.001 476	0.9621	0.9635	44.7	1360.0	1404.6	0.1873	5.5942	5.7815
−28	131.64	0.001 481	0.8790	0.8805	53.6	1354.0	1407.6	0.2240	5.5241	5.7481
−26	144.70	0.001 487	0.8044	0.8059	62.6	1347.9	1410.5	0.2605	5.4548	5.7153
−24	158.78	0.001 492	0.7373	0.7388	71.6	1341.8	1413.4	0.2967	5.3864	5.6831
−22	173.93	0.001 498	0.6768	0.6783	80.7	1335.6	1416.2	0.3327	5.3188	5.6515
−20	190.22	0.001 504	0.6222	0.6237	89.7	1329.3	1419.0	0.3684	5.2520	5.6205
−18	207.71	0.001 510	0.5728	0.5743	98.8	1322.9	1421.7	0.4040	5.1860	5.5900
−16	226.45	0.001 515	0.5280	0.5296	107.8	1316.5	1424.4	0.4393	5.1207	5.5600
−14	246.51	0.001 521	0.4874	0.4889	116.9	1310.0	1427.0	0.4744	5.0561	5.5305
−12	267.95	0.001 528	0.4505	0.4520	126.0	1303.5	1429.5	0.5093	4.9922	5.5015
−10	290.85	0.001 534	0.4169	0.4185	135.2	1296.8	1432.0	0.5440	4.9290	5.4730
−8	315.25	0.001 540	0.3863	0.3878	144.3	1290.1	1434.4	0.5785	4.8664	5.4449
−6	341.25	0.001 546	0.3583	0.3599	153.5	1283.3	1436.8	0.6128	4.8045	5.4173
−4	368.90	0.001 553	0.3328	0.3343	162.7	1276.4	1439.1	0.6469	4.7432	5.3901
−2	398.27	0.001 559	0.3094	0.3109	171.9	1269.4	1441.3	0.6808	4.6825	5.3633
0	429.44	0.001 566	0.2879	0.2895	181.1	1262.4	1443.5	0.7145	4.6223	5.3369
2	462.49	0.001 573	0.2683	0.2698	190.4	1255.2	1445.6	0.7481	4.5627	5.3108
4	497.49	0.001 580	0.2502	0.2517	199.6	1248.0	1447.6	0.7815	4.5037	5.2852
6	534.51	0.001 587	0.2335	0.2351	208.9	1240.6	1449.6	0.8148	4.4451	5.2599
8	573.64	0.001 594	0.2182	0.2198	218.3	1233.2	1451.5	0.8479	4.3871	5.2350
10	614.95	0.001 601	0.2040	0.2056	227.6	1225.7	1453.3	0.8808	4.3295	5.2104
12	658.52	0.001 608	0.1910	0.1926	237.0	1218.1	1455.1	0.9136	4.2725	5.1861
14	704.44	0.001 616	0.1789	0.1805	246.4	1210.4	1456.8	0.9463	4.2159	5.1621
16	752.79	0.001 623	0.1677	0.1693	255.9	1202.6	1458.5	0.9788	4.1597	5.1385
18	803.66	0.001 631	0.1574	0.1590	265.4	1194.7	1460.0	1.0112	4.1039	5.1151
20	857.12	0.001 639	0.1477	0.1494	274.9	1186.7	1461.5	1.0434	4.0486	5.0920
22	913.27	0.001 647	0.1388	0.1405	284.4	1178.5	1462.9	1.0755	3.9937	5.0692
24	972.19	0.001 655	0.1305	0.1322	294.0	1170.3	1464.3	1.1075	3.9392	5.0467
26	1033.97	0.001 663	0.1228	0.1245	303.6	1162.0	1465.6	1.1394	3.8850	5.0244
28	1098.71	0.001 671	0.1156	0.1173	313.2	1153.6	1466.8	1.1711	3.8312	5.0023
30	1166.49	0.001 680	0.1089	0.1106	322.9	1145.0	1467.9	1.2028	3.7777	4.9805
32	1237.41	0.001 689	0.1027	0.1044	332.6	1136.4	1469.0	1.2343	3.7246	4.9589
34	1311.55	0.001 698	0.0969	0.0986	342.3	1127.6	1469.9	1.2656	3.6718	4.9374
36	1389.03	0.001 707	0.0914	0.0931	352.1	1118.7	1470.8	1.2969	3.6192	4.9161
38	1469.92	0.001 716	0.0863	0.0880	361.9	1109.7	1471.5	1.3281	3.5669	4.8950
40	1554.33	0.001 726	0.0815	0.0833	371.7	1100.5	1472.2	1.3591	3.5148	4.8740
42	1642.35	0.001 735	0.0771	0.0788	381.6	1091.2	1472.8	1.3901	3.4630	4.8530
44	1734.09	0.001 745	0.0728	0.0746	391.5	1081.7	1473.2	1.4209	3.4112	4.8322
46	1829.65	0.001 756	0.0689	0.0707	401.5	1072.0	1473.5	1.4518	3.3595	4.8113
48	1929.13	0.001 766	0.0652	0.0669	411.5	1062.2	1473.7	1.4826	3.3079	4.7905
50	2032.62	0.001 777	0.0617	0.0635	421.7	1052.0	1473.7	1.5135	3.2561	4.7696

Source: National Bureau of Standards Circular No. 142, *Tables of Thermodynamic Properties of Ammonia.*

TABLE E.2 Superheated Ammonia

p, kPa (T_{sat}, °C)		Temperature, °C											
		−20	−10	0	10	20	30	40	50	60	70	80	100
	v	2.4474	2.5481	2.6482	2.7479	2.8473	2.9464	3.0453	3.1441	3.2427	3.3413	3.4397	
50	h	1435.8	1457.0	1478.1	1499.2	1520.4	1541.7	1563.0	1584.5	1606.1	1627.8	1649.7	
(−46.54)	s	6.3256	6.4077	6.4865	6.5625	6.6360	6.7073	6.7766	6.8441	6.9099	6.9743	7.0372	
	v	1.6233	1.6915	1.7591	1.8263	1.8932	1.9597	2.0261	2.0923	2.1584	2.2244	2.2903	
75	h	1433.0	1454.7	1476.1	1497.5	1518.9	1540.3	1561.8	1583.4	1605.1	1626.9	1648.9	
(−39.18)	s	6.1190	6.2028	6.2828	−6.3597	6.4339	6.5058	6.5756	6.6434	6.7096	6.7742	6.8373	
	v	1.2110	1.2631	1.3145	1.3654	1.4160	1.4664	1.5165	1.5664	1.6163	1.6659	1.7155	1.8145
100	h	1430.1	1452.2	1474.1	1495.7	1517.3	1538.9	1560.5	1582.2	1604.1	1626.0	1648.0	1692.6
(−33.61)	s	5.9695	6.0552	6.1366	6.2144	6.2894	6.3618	6.4321	6.5003	6.5668	6.6316	6.6950	6.8177
	v	0.9635	1.0059	1.0476	1.0889	1.1297	1.1703	1.2107	1.2509	1.2909	1.3309	1.3707	1.4501
125	h	1427.2	1449.8	1472.0	1493.9	1515.7	1537.5	1559.3	1581.1	1603.0	1625.0	1647.2	1691.8
(−29.08)	s	5.8512	5.9389	6.0217	6.1006	6.1763	6.2494	6.3201	6.3887	6.4555	6.5206	6.5842	6.7072
	v	0.7984	0.8344	0.8697	0.9045	0.9388	0.9729	1.0068	1.0405	1.0740	1.1074	1.1408	1.2072
150	h	1424.1	1447.3	1469.8	1492.1	1514.1	1536.1	1558.0	1580.0	1602.0	1624.1	1646.3	1691.1
(−25.23)	s	5.7526	5.8424	5.9266	6.0066	6.0831	6.1568	6.2280	6.2970	6.3641	6.4295	6.4933	6.6167
	v		0.6199	0.6471	0.6738	0.7001	0.7261	0.7519	0.7774	0.8029	0.8282	0.8533	0.9035
200	h		1442.0	1465.5	1488.4	1510.9	1533.2	1555.5	1577.7	1599.9	1622.2	1644.6	1689.6
(−18.86)	s		5.6863	5.7737	5.8559	5.9342	6.0091	6.0813	6.1512	6.2189	6.2849	6.3491	6.4732
	v		0.4910	0.5135	0.5354	0.5568	0.5780	0.5989	0.6196	0.6401	0.6605	0.6809	0.7212
250	h		1436.6	1461.0	1484.5	1507.6	1530.3	1552.9	1575.4	1597.8	1620.3	1642.8	1688.2
(−13.67)	s		5.5609	5.6517	5.7365	5.8165	5.8928	5.9661	6.0368	6.1052	6.1717	6.2365	6.3613
	v			0.4243	0.4430	0.4613	0.4792	0.4968	0.5143	0.5316	0.5488	0.5658	0.5997
300	h			1456.3	1480.6	1504.2	1527.4	1550.3	1573.0	1595.7	1618.4	1641.1	1686.7
(−9.23)	s			5.5493	5.6366	5.7186	5.7963	5.8707	5.9423	6.0114	6.0785	6.1437	6.2693
	v			0.3605	0.3770	0.3929	0.4086	0.4239	0.4391	0.4541	0.4689	0.4837	0.5129
350	h			1451.5	1476.5	1500.7	1524.4	1547.6	1570.7	1593.6	1616.5	1639.3	1685.2
(−5.35)	s			5.4600	5.5502	5.6342	5.7135	5.7890	5.8615	5.9314	5.9990	6.0647	6.1910
	v			0.3125	0.3274	0.3417	0.3556	0.3692	0.3826	0.3959	0.4090	0.4220	0.4478
400	h			1446.5	1472.4	1497.2	1521.3	1544.9	1568.3	1591.5	1614.5	1637.6	1683.7
(−1.89)	s			5.3803	5.4735	5.5597	5.6405	5.7173	5.7907	5.8613	5.9296	5.9957	6.1228
	v			0.2752	0.2887	0.3017	0.3143	0.3266	0.3387	0.3506	0.3624	0.3740	0.3971
450	h			1441.3	1468.1	1493.6	1518.2	1542.2	1565.9	1589.3	1612.6	1635.8	1682.2
(1.26)	s			5.3078	5.4042	5.4926	5.5752	5.6532	5.7275	5.7989	5.8678	5.9345	6.0623

TABLE E.2 (*continued*)

p, kPa (T_{sat}, °C)		Temperature, °C											
		20	30	40	50	60	70	80	100	120	140	160	180
	v	0.2698	0.2813	0.2926	0.3036	0.3144	0.3251	0.3357	0.3565	0.3771	0.3975		
500	h	1489.9	1515.0	1539.5	1563.4	1587.1	1610.6	1634.0	1680.7	1727.5	1774.7		
(4.14)	s	5.4314	5.5157	5.5950	5.6704	5.7425	5.8120	5.8793	6.0079	6.1301	6.2472		
	v	0.2217	0.2317	0.2414	0.2508	0.2600	0.2691	0.2781	0.2957	0.3130	0.3302		
600	h	1482.4	1508.6	1533.8	1558.5	1582.7	1606.6	1630.4	1677.7	1724.9	1772.4		
(9.29)	s	5.3222	5.4102	5.4923	5.5697	5.6436	5.7144	5.7826	5.9129	6.0363	6.1541		
	v	0.1874	0.1963	0.2048	0.2131	0.2212	0.2291	0.2369	0.2522	0.2672	0.2821		
700	h	1474.5	1501.9	1528.1	1553.4	1578.2	1602.6	1626.8	1674.6	1722.4	1770.2		
(13.81)	s	5.2259	5.3179	5.4029	5.4826	5.5582	5.6303	5.6997	5.8316	5.9562	6.0749		
	v	0.1615	0.1696	0.1773	0.1848	0.1920	0.1991	0.2060	0.2196	0.2329	0.2459	0.2589	
800	h	1466.3	1495.0	1522.2	1548.3	1573.7	1598.6	1623.1	1671.6	1719.8	1768.0	1816.4	
(17.86)	s	5.1387	5.2351	5.3232	5.4053	5.4827	5.5562	5.6268	5.7603	5.8861	6.0057	6.1202	
	v		0.1488	0.1559	0.1627	0.1693	0.1757	0.1820	0.1942	0.2061	0.2178	0.2294	
900	h		1488.0	1516.2	1543.0	1569.1	1594.4	1619.4	1668.5	1717.1	1765.7	1814.4	
(21.54)	s		5.1593	5.2508	5.3354	5.4147	5.4897	5.5614	5.6968	5.8237	5.9442	6.0594	
	v		0.1321	0.1388	0.1450	0.1511	0.1570	0.1627	0.1739	0.1847	0.1954	0.2058	0.2162
1000	h		1480.6	1510.0	1537.7	1564.4	1590.3	1615.6	1665.4	1714.5	1763.4	1812.4	1861.7
(24.91)	s		5.0889	5.1840	5.2713	5.3525	5.4292	5.5021	5.6392	5.7674	5.8888	6.0047	6.1159
	v			0.1129	0.1185	0.1238	0.1289	0.1338	0.1434	0.1526	0.1616	0.1705	0.1792
1200	h			1497.1	1526.6	1554.7	1581.7	1608.0	1659.2	1709.2	1758.9	1808.5	1858.2
(30.96)	s			5.0629	5.1560	5.2416	5.3215	5.3970	5.5379	5.6687	5.7919	5.9091	6.0214
	v			0.0944	0.0995	0.1042	0.1088	0.1132	0.1216	0.1297	0.1376	0.1452	0.1528
1400	h			1483.4	1515.1	1544.7	1573.0	1600.2	1652.8	1703.9	1754.3	1804.5	1854.7
(36.28)	s			4.9534	5.0530	5.1434	5.2270	5.3053	5.4501	5.5836	5.7087	5.8273	5.9406
	v				0.0851	0.0895	0.0937	0.0977	0.1053	0.1125	0.1195	0.1263	0.1330
1600	h				1502.9	1534.4	1564.0	1592.3	1646.4	1698.5	1749.7	1800.5	1851.2
(41.05)	s				4.9584	5.0543	5.1419	5.2232	5.3722	5.5084	5.6355	5.7555	5.8699
	v				0.0739	0.0781	0.0820	0.0856	0.0926	0.0992	0.1055	0.1116	0.1177
1800	h				1490.0	1523.5	1554.6	1584.1	1639.8	1693.1	1745.1	1796.5	1847.7
(45.39)	s				4.8693	4.9715	5.0635	5.1482	5.3018	5.4409	5.5699	5.6914	5.8069
	v				0.0648	0.0688	0.0725	0.0760	0.0824	0.0885	0.0943	0.0999	0.1054
2000	h				1476.1	1512.0	1544.9	1575.6	1633.2	1687.6	1740.4	1792.4	1844.1
(49.38)	s				4.7834	4.8930	4.9902	5.0786	5.2371	5.3793	5.5104	5.6333	5.7499

Source: National Bureau of Standards Circular No. 142, *Tables of Thermodynamic Properties of Ammonia.*

TABLE E.1E Saturated Ammonia

T, °F	p, psia	Specific Volume, ft³/lbm			Enthalpy, Btu/lbm			Entropy, Btu/lbm-°R		
		Sat. Liquid v_f	Evap. v_{fg}	Sat. Vapor v_g	Sat. Liquid h_f	Evap. h_{fg}	Sat. Vapor h_g	Sat. Liquid s_f	Evap. s_{fg}	Sat. Vapor s_g
−60	5.55	0.0228	44.707	44.73	−21.2	610.8	589.6	−0.0517	1.5286	1.4769
−55	6.54	0.0229	38.375	38.38	−15.9	607.5	591.6	−0.0386	1.5017	1.4631
−50	7.67	0.0230	33.057	33.08	−10.6	604.3	593.7	−0.0256	1.4753	1.4497
−45	8.95	0.0231	28.597	28.62	−5.3	600.9	595.6	−0.0127	1.4495	1.4368
−40	10.41	0.02322	24.837	24.86	0	597.6	597.6	0.000	1.4242	1.4242
−35	12.05	0.02333	21.657	21.68	5.3	594.2	599.5	0.0126	1.3994	1.4120
−30	13.90	0.0235	18.947	18.97	10.7	590.7	601.4	0.0250	1.3751	1.4001
−25	15.98	0.0236	16.636	16.66	16.0	587.2	603.2	0.0374	1.3512	1.3886
−20	18.30	0.0237	14.656	14.68	21.4	583.6	605.0	0.0497	1.3277	1.3774
−15	20.88	0.02381	12.946	12.97	26.7	580.0	606.7	0.0618	1.3044	1.3664
−10	23.74	0.02393	11.476	11.50	32.1	576.4	608.5	0.0738	1.2820	1.3558
−5	26.92	0.02406	10.206	10.23	37.5	572.6	610.1	0.0857	1.2597	1.3454
0	30.42	0.02419	9.092	9.116	42.9	568.9	611.8	0.0975	1.2377	1.3352
5	34.27	0.02432	8.1257	8.150	48.3	565.0	613.3	0.1092	1.2161	1.3253
10	38.51	0.02446	7.2795	7.304	53.8	561.1	614.9	0.1208	1.1949	1.3157
15	43.14	0.02460	6.5374	6.562	59.2	557.1	616.3	0.1323	1.1739	1.3062
20	48.21	0.02474	5.8853	5.910	64.7	553.1	617.8	0.1437	1.1532	1.2969
25	53.73	0.02488	5.3091	5.334	70.2	548.9	619.1	0.1551	1.1328	1.2879
30	59.74	0.02503	4.8000	4.825	75.7	544.8	620.5	0.1663	1.1127	1.2790
35	66.26	0.02518	4.3478	4.373	81.2	540.5	621.7	0.1775	1.0929	1.2704
40	73.32	0.02533	3.9457	3.971	86.8	536.2	623.0	0.1885	1.0733	1.2618
45	80.96	0.02548	3.5885	3.614	92.3	531.8	624.1	0.1996	1.0539	1.2535
50	89.19	0.02564	3.2684	3.294	97.9	527.3	625.2	0.2105	1.0348	1.2453
55	98.06	0.02581	2.9822	3.008	103.5	522.8	626.3	0.2214	1.0159	1.2373
60	107.6	0.02597	2.7250	2.751	109.2	518.1	627.3	0.2322	0.9972	1.2294
65	117.8	0.02614	2.4939	2.520	114.8	513.4	628.2	0.2430	0.9786	1.2216
70	128.8	0.02632	2.2857	2.312	120.5	508.6	629.1	0.2537	0.9603	1.2140
75	140.5	0.02650	2.0985	2.125	126.2	503.7	629.9	0.2643	0.9422	1.2065
80	153.0	0.02668	1.9283	1.955	132.0	498.7	630.7	0.2749	0.9242	1.1991
85	166.4	0.02687	1.7741	1.801	137.8	493.6	631.4	0.2854	0.9064	1.1918
90	180.6	0.02707	1.6339	1.661	143.5	488.5	632.0	0.2958	0.8888	1.1846
95	195.8	0.02727	1.5067	1.534	149.4	483.2	632.6	0.3062	0.8713	1.1775
100	211.9	0.02747	1.3915	1.419	155.2	477.8	633.0	0.3166	0.8539	1.1705
105	228.9	0.02769	1.2853	1.313	161.1	472.3	633.4	0.3269	0.8366	1.1635
110	247.0	0.02790	1.1891	1.217	167.0	466.7	633.7	0.3372	0.8194	1.1566
115	266.2	0.02813	1.0999	1.128	173.0	460.9	633.9	0.3474	0.8023	1.1497
120	286.4	0.02836	1.0186	1.047	179.0	455.0	634.0	0.3576	0.7851	1.1427
125	307.8	0.02860	0.9444	0.973	185.1	448.9	634.0	0.3679	0.7679	1.1358

Source: National Bureau of Standards Circular No. 142, *Tables of Thermodynamic Properties of Ammonia.*

TABLE E.2E Superheated Ammonia

p, psia (T_{sat}, °F)		Temperature, F											
		0	20	40	60	80	100	120	140	160	180	200	220
	v	28.58	29.90	31.20	32.49	33.78	35.07	36.35	37.62	38.90	40.17	41.45	
10	h	618.9	629.1	639.3	649.5	659.7	670.0	680.3	690.6	701.1	711.6	722.2	
(−41.34)	s	1.477	1.499	1.520	1.540	1.559	1.578	1.596	1.614	1.631	1.647	1.664	
	v	18.92	19.82	20.70	21.58	22.44	23.31	24.17	25.03	25.88	26.74	27.59	
15	h	617.2	627.8	638.2	648.5	658.9	669.2	679.6	690.0	700.5	711.1	721.7	
(−27.29)	s	1.427	1.450	1.471	1.491	1.511	1.529	1.548	1.566	1.583	1.599	1.616	
	v	14.09	14.78	15.45	16.12	16.78	17.43	18.08	18.73	19.37	20.02	20.66	21.3
20	h	615.5	626.4	637.0	647.5	658.0	668.5	678.9	689.4	700.0	710.6	721.2	732.0
(−16.64)	s	1.391	1.414	1.436	1.456	1.476	1.495	1.513	1.531	1.549	1.565	1.582	1.598
	v	11.19	11.75	12.30	12.84	13.37	13.90	14.43	14.95	15.47	15.99	16.50	17.02
25	h	613.8	625.0	635.8	646.5	657.1	667.7	678.2	688.8	699.4	710.1	720.8	731.6
(−7.96)	s	1.362	1.386	1.408	1.429	1.449	1.468	1.486	1.504	1.522	1.539	1.555	1.571
	v	9.25	9.731	10.20	10.65	11.10	11.55	11.99	12.43	12.87	13.30	13.73	14.16
30	h	611.9	623.5	634.6	645.5	656.2	666.9	677.5	688.2	698.8	709.6	720.3	731.1
(−.57)	s	1.337	1.362	1.385	1.406	1.426	1.446	1.464	1.482	1.500	1.517	1.533	1.550
	v		8.287	8.695	9.093	9.484	9.869	10.25	10.63	11.00	11.38	11.75	12.12
35	h		622.0	633.4	644.4	655.3	666.1	676.8	687.6	698.3	709.1	719.9	730.7
(5.89)	s		1.341	1.365	1.386	1.407	1.427	1.445	1.464	1.481	1.498	1.515	1.531
	v		7.203	7.568	7.922	8.268	8.609	8.945	9.278	9.609	9.938	10.27	10.59
40	h		620.4	632.1	643.4	654.4	665.3	676.1	686.9	697.7	708.5	719.4	730.3
(11.66)	s		1.323	1.347	1.369	1.390	1.410	1.429	1.447	1.465	1.482	1.499	1.515
	v		6.363	6.694	7.014	7.326	7.632	7.934	8.232	8.528	8.822	9.115	9.406
45	h		618.8	630.8	642.3	653.5	664.6	675.5	686.3	697.2	708.0	718.9	729.9
(16.87)	s		1.307	1.331	1.354	1.375	1.395	1.414	1.433	1.450	1.468	1.485	1.501
	v			5.988	6.280	6.564	6.843	7.117	7.387	7.655	7.921	8.185	8.448
50	h			629.5	641.2	652.6	663.7	674.7	685.7	696.6	707.5	718.5	729.4
(21.67)	s			1.317	1.340	1.361	1.382	1.401	1.420	1.437	1.455	1.472	1.488
	v			4.933	5.184	5.428	5.665	5.897	6.126	6.352	6.576	6.789	7.019
60	h			626.8	639.0	650.7	662.1	673.3	684.4	695.5	706.5	717.5	728.6
(30.21)	s			1.291	1.315	1.337	1.358	1.378	1.397	1.415	1.432	1.449	1.466

p, psia (T_{sat}, °F)		60	80	100	120	140	160	180	200	240	280	320	360
	v	4.401	4.615	4.822	5.025	5.224	5.420	5.615	5.807	6.187	6.563		
70	h	636.6	648.7	660.4	671.8	683.1	694.3	705.5	716.6	738.9	761.4		
(37.7)	s	1.294	1.317	1.338	1.358	1.377	1.395	1.413	1.430	1.463	1.494		
	v	3.812	4.005	4.190	4.371	4.548	4.722	4.893	5.063	5.398	5.73		
80	h	634.3	646.7	658.7	670.4	681.8	693.2	704.4	715.6	738.1	760.7		
(44.4)	s	1.275	1.298	1.320	1.340	1.360	1.378	1.396	1.414	1.447	1.478		
	v	3.353	3.529	3.698	3.862	4.021	4.178	4.332	4.484	4.785	5.081		
90	h	631.8	644.7	657.0	668.9	680.5	692.0	703.4	714.7	737.3	760.0		
(50.47)	s	1.257	1.281	1.304	1.325	1.344	1.363	1.381	1.400	1.432	1.464		
	v	2.985	3.149	3.304	3.454	3.600	3.743	3.883	4.021	4.294	4.562		
100	h	629.3	642.6	655.2	667.3	679.2	690.8	702.3	713.7	736.5	759.4		
(56.05)	s	1.241	1.266	1.289	1.310	1.331	1.349	1.368	1.385	1.419	1.451		
	v		2.166	2.288	2.404	2.515	2.622	2.727	2.830	3.030	3.227	3.420	
140	h		633.8	647.8	661.1	673.7	686.0	698.0	709.9	733.3	756.7	780.0	
(74.79)	s		1.214	1.240	1.263	1.284	1.305	1.324	1.342	1.376	1.409	1.440	
	v			1.720	1.818	1.910	1.999	2.084	2.167	2.328	2.484	2.637	
180	h			639.9	654.4	668.0	681.0	693.6	705.9	730.1	753.9	777.7	
(89.78)	s			1.199	1.225	1.248	1.269	1.289	1.308	1.344	1.377	1.408	
	v				1.443	1.525	1.601	1.675	1.745	1.881	2.012	2.140	2.265
220	h				647.3	662.0	675.8	689.1	701.9	726.8	751.1	775.3	799.5
(102.42)	s				1.192	1.217	1.239	1.260	1.280	1.317	1.351	1.383	1.413
	v				1.302	1.380	1.452	1.521	1.587	1.714	1.835	1.954	2.069
240	h				643.5	658.8	673.1	686.7	699.8	725.1	749.8	774.1	798.4
(108.09)	s				1.176	1.203	1.226	1.248	1.268	1.305	1.339	1.371	1.402

Source: National Bureau of Standards Circular No. 142, *Tables of Thermodynamic Properties of Ammonia.*

F. IDEAL-GAS TABLES

TABLE F.1 Properties of Air

T, K	h, kJ/kg	p_r	u, kJ/kg	v_r	ϕ, kJ/kg·K	T, K	h, kJ/kg	p_r	u, kJ/kg	v_r	ϕ, kJ/kg·K
200	199.97	0.3363	142.56	1707.	1.29559	780	800.03	43.35	576.12	51.64	2.69013
220	219.97	0.4690	156.82	1346.	1.39105	820	843.98	52.49	608.59	44.84	2.74504
240	240.02	0.6355	171.13	1084.	1.47824	860	888.27	63.09	641.40	39.12	2.79783
260	260.09	0.8405	185.45	887.8	1.55848	900	932.93	75.29	674.58	34.31	2.84856
280	280.13	1.0889	199.75	738.0	1.63279	940	977.92	89.28	708.08	30.22	2.89748
290	290.16	1.2311	206.91	676.1	1.66802	980	1023.25	105.2	741.98	26.73	2.94468
300	300.19	1.3860	214.07	621.2	1.70203	1020	1068.89	123.4	776.10	23.72	2.99034
310	310.24	1.5546	221.25	572.3	1.73498	1060	1114.86	143.9	810.62	21.14	3.03449
320	320.29	1.7375	228.43	528.6	1.76690	1100	1161.07	167.1	845.33	18.896	3.07732
340	340.42	2.149	242.82	454.1	1.82790	1140	1207.57	193.1	880.35	16.946	3.11883
360	360.58	2.626	257.24	393.4	1.88543	1180	1254.34	222.2	915.57	15.241	3.15916
380	380.77	3.176	271.69	343.4	1.94001	1220	1301.31	254.7	951.09	13.747	3.19834
400	400.98	3.806	286.16	301.6	1.99194	1260	1348.55	290.8	986.90	12.435	3.23638
420	421.26	4.522	300.69	266.6	2.04142	1300	1395.97	330.9	1022.82	11.275	3.27345
440	441.61	5.332	315.30	236.8	2.08870	1340	1443.60	375.3	1058.94	10.247	3.30959
460	462.02	6.245	329.97	211.4	2.13407	1380	1491.44	424.2	1095.26	9.337	3.34474
480	482.49	7.268	344.70	189.5	2.17760	1420	1539.44	478.0	1131.77	8.526	3.37901
500	503.02	8.411	359.49	170.6	2.21952	1460	1587.63	537.1	1168.49	7.801	3.41247
520	523.63	9.684	374.36	154.1	2.25997	1500	1635.97	601.9	1205.41	7.152	3.44516
540	544.35	11.10	389.34	139.7	2.29906	1540	1684.51	672.8	1242.43	6.569	3.47712
560	565.17	12.66	404.42	127.0	2.33685	1580	1733.17	750.0	1279.65	6.046	3.50829
580	586.04	14.38	419.55	115.7	2.37348	1620	1782.00	834.1	1316.96	5.574	3.53879
600	607.02	16.28	434.78	105.8	2.40902	1660	1830.96	925.6	1354.48	5.147	3.56867
620	628.07	18.36	450.09	96.92	2.44356	1700	1880.1	1025	1392.7	4.761	3.5979
640	649.22	20.65	465.50	88.99	2.47716	1800	2003.3	1310	1487.2	3.944	3.6684
660	670.47	23.13	481.01	81.89	2.50985	1900	2127.4	1655	1582.6	3.295	3.7354
680	691.82	25.85	496.62	75.50	2.54175	2000	2252.1	2068	1678.7	2.776	3.7994
700	713.27	28.80	512.33	69.76	2.57277	2100	2377.4	2559	1775.3	2.356	3.8605
720	734.82	32.02	528.14	64.53	2.60319	2200	2503.2	3138	1872.4	2.012	3.9191
740	756.44	35.50	544.02	59.82	2.63280						

Source: J. H. Keenan and J. Kaye, *Gas Tables*, Wiley, New York, 1945.

TABLE F.2 Molar Properties of Nitrogen, N_2

$\bar{h}_f^\circ = 0$ kJ/kmol

T, K	$\bar{h}$ kJ/kmol	$\bar{u}$ kJ/kmol	$\bar{\phi}$ kJ/kmol·K	T, K	$\bar{h}$ kJ/kmol	$\bar{u}$ kJ/kmol	$\bar{\phi}$ kJ/kmol·K
0	0	0	0	1000	30 129	21 815	228.057
220	6391	4562	182.639	1020	30 784	22 304	228.706
240	6975	4979	185.180	1040	31 442	22 795	229.344
260	7558	5396	187.514	1060	32 101	23 288	229.973
280	8141	5813	189.673	1080	32 762	23 782	230.591
298	8669	6190	191.502	1100	33 426	24 280	231.199
300	8723	6229	191.682	1120	34 092	24 780	231.799
320	9306	6645	193.562	1140	34 760	25 282	232.391
340	9888	7061	195.328	1160	35 430	25 786	232.973
360	10 471	7478	196.995	1180	36 104	26 291	233.549
380	11 055	7895	198.572	1200	36 777	26 799	234.115
400	11 640	8314	200.071	1240	38 129	27 819	235.223
420	12 225	8733	201.499	1260	38 807	28 331	235.766
440	12 811	9153	202.863	1280	39 488	28 845	236.302
460	13 399	9574	204.170	1300	40 170	29 361	236.831
480	13 988	9997	205.424	1320	40 853	29 878	237.353
500	14 581	10 423	206.630	1340	41 539	30 398	237.867
520	15 172	10 848	207.792	1360	42 227	30 919	238.376
540	15 766	11 277	208.914	1380	42 915	31 441	238.878
560	16 363	11 707	209.999	1400	43 605	31 964	239.375
580	16 962	12 139	211.049	1440	44 988	33 014	240.350
600	17 563	12 574	212.066	1480	46 377	34 071	241.301
620	18 166	13 011	213.055	1520	47 771	35 133	242.228
640	18 772	13 450	214.018	1560	49 168	36 197	243.137
660	19 380	13 892	214.954	1600	50 571	37 268	244.028
680	19 991	14 337	215.866	1700	54 099	39 965	246.166
700	20 604	14 784	216.756	1800	57 651	42 685	248.195
720	21 220	15 234	217.624	1900	61 220	45 423	250.128
740	21 839	15 686	218.472	2000	64 810	48 181	251.969
760	22 460	16 141	219.301	2100	68 417	50 957	253.726
780	23 085	16 599	220.113	2200	72 040	53 749	255.412
800	23 714	17 061	220.907	2300	75 676	56 553	257.02
820	24 342	17 524	221.684	2400	79 320	59 366	258.580
840	24 974	17 990	222.447	2500	82 981	62 195	260.073
860	25 610	18 459	223.194	2600	86 650	65 033	261.512
880	26 248	18 931	223.927	2700	90 328	67 880	262.902
900	26 890	19 407	224.647	2800	94 014	70 734	264.241
920	27 532	19 883	225.353	2900	97 705	73 593	265.538
940	28 178	20 362	226.047	3000	101 407	76 464	266.793
960	28 826	20 844	226.728	3100	105 115	79 341	268.007
980	29 476	21 328	227.398	3200	108 830	82 224	269.186

Source: JANAF Thermochemical Tables, NSRDS-NBS-37, 1971.

TABLE F.3 Molar Properties of Oxygen, O_2

$\bar{h}_f^\circ = 0$ kJ/kmol

T	$\bar{h}$	$\bar{u}$	$\bar{\phi}$	T	$\bar{h}$	$\bar{u}$	$\bar{\phi}$
0	0	0	0	1020	32 088	23 607	244.164
220	6404	4575	196.171	1040	32 789	24 142	244.844
240	6984	4989	198.696	1060	33 490	24 677	245.513
260	7566	5405	201.027	1080	34 194	25 214	246.171
280	8150	5822	203.191	1100	34 899	25 753	246.818
298	8682	6203	205.033	1120	35 606	26 294	247.454
300	8736	6242	205.213	1140	36 314	26 836	248.081
320	9325	6664	207.112	1160	37 023	27 379	248.698
340	9916	7090	208.904	1180	37 734	27 923	249.307
360	10 511	7518	210.604	1200	38 447	28 469	249.906
380	11 109	7949	212.222	1220	39 162	29 018	250.497
400	11 711	8384	213.765	1240	39 877	29 568	251.079
420	12 314	8822	215.241	1260	40 594	30 118	251.653
440	12 923	9264	216.656	1280	41 312	30 670	252.219
460	13 535	9710	218.016	1300	42 033	31 224	252.776
480	14 151	10 160	219.326	1320	42 753	31 778	253.325
500	14 770	10 614	220.589	1340	43 475	32 334	253.868
520	15 395	11 071	221.812	1360	44 198	32 891	254.404
540	16 022	11 533	222.997	1380	44 923	33 449	254.932
560	16 654	11 998	224.146	1400	45 648	34 008	255.454
580	17 290	12 467	225.262	1440	47 102	35 129	256.475
600	17 929	12 940	226.346	1480	48 561	36 256	257.474
620	18 572	13 417	227.400	1520	50 024	37 387	258.450
640	19 219	13 898	228.429	1540	50 756	37 952	258.928
660	19 870	14 383	229.430	1560	51 490	38 520	259.402
680	20 524	14 871	230.405	1600	52 961	39 658	260.333
700	21 184	15 364	231.358	1700	56 652	42 517	262.571
720	21 845	15 859	232.291	1800	60 371	45 405	264.701
740	22 510	16 357	233.201	1900	64 116	48 319	266.722
760	23 178	16 859	234.091	2000	67 881	51 253	268.655
780	23 850	17 364	234.960	2100	71 668	54 208	270.504
800	24 523	17 872	235.810	2200	75 484	57 192	272.278
820	25 199	18 382	236.644	2300	79 316	60 193	273.981
840	25 877	18 893	237.462	2400	83 174	63 219	275.625
860	26 559	19 408	238.264	2500	87 057	66 271	277.207
880	27 242	19 925	239.051	2600	90 956	69 339	278.738
900	27 928	20 445	239.823	2700	94 881	72 433	280.219
920	28 616	20 967	240.580	2800	98 826	75 546	281.654
940	29 306	21 491	241.323	2900	102 793	78 682	283.048
960	29 999	22 017	242.052	3000	106 780	81 837	284.399
980	30 692	22 544	242.768	3100	110 784	85 009	285.713
1000	31 389	23 075	243.471	3200	114 809	88 203	286.989

Source: JANAF Thermochemical Tables, NSRDS-NBS-37, 1971.

TABLE F.4 Molar Properties of Carbon Dioxide, CO_2

$\bar{h}_f^\circ = -393\,520$ kJ/kmol

T	$\bar{h}$	$\bar{u}$	$\bar{\phi}$	T	$\bar{h}$	$\bar{u}$	$\bar{\phi}$
0	0	0	0	1020	43 859	35 378	270.293
220	6601	4772	202.966	1040	44 953	36 306	271.354
240	7280	5285	205.920	1060	46 051	37 238	272.400
260	7979	5817	208.717	1080	47 153	38 174	273.430
280	8697	6369	211.376	1100	48 258	39 112	274.445
298	9364	6885	213.685	1120	49 369	40 057	275.444
300	9431	6939	213.915	1140	50 484	41 006	276.430
320	10 186	7526	216.351	1160	51 602	41 957	277.403
340	10 959	8131	218.694	1180	52 724	42 913	278.361
360	11 748	8752	220.948	1200	53 848	43 871	279.307
380	12 552	9392	223.122	1220	54 977	44 834	280.238
400	13 372	10 046	225.225	1240	56 108	45 799	281.158
420	14 206	10 714	227.258	1260	57 244	46 768	282.066
440	15 054	11 393	229.230	1280	58 381	47 739	282.962
460	15 916	12 091	231.144	1300	59 522	48 713	283.847
480	16 791	12 800	233.004	1320	60 666	49 691	284.722
500	17 678	13 521	234.814	1340	61 813	50 672	285.586
520	18 576	14 253	236.575	1360	62 963	51 656	286.439
540	19 485	14 996	238.292	1380	64 116	52 643	287.283
560	20 407	15 751	239.962	1400	65 271	53 631	288.106
580	21 337	16 515	241.602	1440	67 586	55 614	289.743
600	22 280	17 291	243.199	1480	69 911	57 606	291.333
620	23 231	18 076	244.758	1520	72 246	59 609	292.888
640	24 190	18 869	246.282	1560	74 590	61 620	294.411
660	25 160	19 672	247.773	1600	76 944	63 741	295.901
680	26 138	20 484	249.233	1700	82 856	68 721	299.482
700	27 125	21 305	250.663	1800	88 806	73 840	302.884
720	28 121	22 134	252.065	1900	94 793	78 996	306.122
740	29 124	22 972	253.439	2000	100 804	84 185	309.210
760	30 135	23 817	254.787	2100	106 864	89 404	312.160
780	31 154	24 669	256.110	2200	112 939	94 648	314.988
800	32 179	25 527	257.408	2300	119 035	99 912	317.695
820	33 212	26 394	258.682	2400	125 152	105 197	320.302
840	34 251	27 267	259.934	2500	131 290	110 504	322.308
860	35 296	28 125	261.164	2600	137 449	115 832	325.222
880	36 347	29 031	262.371	2700	143 620	121 172	327.549
900	37 405	29 922	263.559	2800	149 808	126 528	329.800
920	38 467	30 818	264.728	2900	156 009	131 898	331.975
940	39 535	31 719	265.877	3000	162 226	137 283	334.084
960	40 607	32 625	267.007	3100	168 456	142 681	336.126
980	41 685	33 537	268.119	3200	174 695	148 089	338.109
1000	42 769	34 455	269.215				

Source: JANAF Thermochemical Tables, NSRDS-NBS-37, 1971.

TABLE F.5 Molar Properties of Carbon Monoxide, CO

$\bar{h}_f^\circ = -110\,530$ kJ/kmol

T	$\bar{h}$	$\bar{u}$	$\bar{\phi}$	T	$\bar{h}$	$\bar{u}$	$\bar{\phi}$
0	0	0	0	1040	31 688	23 041	235.728
220	6391	4562	188.683	1060	32 357	23 544	236.364
240	6975	4979	191.221	1080	33 029	24 049	236.992
260	7558	5396	193.554	1100	33 702	24 557	237.609
280	8140	5812	195.713	1120	34 377	25 065	238.217
300	8723	6229	197.723	1140	35 054	25 575	238.817
320	9306	6645	199.603	1160	35 733	26 088	239.407
340	9889	7062	201.371	1180	36 406	26 602	239.989
360	10 473	7480	203.040	1200	37 095	27 118	240.663
380	11 058	7899	204.622	1220	37 780	27 637	241.128
400	11 644	8319	206.125	1240	38 466	28 426	241.686
420	12 232	8740	207.549	1260	39 154	28 678	242.236
440	12 821	9163	208.929	1280	39 844	29 201	242.780
460	13 412	9587	210.243	1300	40 534	29 725	243.316
480	14 005	10 014	211.504	1320	41 226	30 251	243.844
500	14 600	10 443	212.719	1340	41 919	30 778	244.366
520	15 197	10 874	213.890	1360	42 613	31 306	244.880
540	15 797	11 307	215.020	1380	43 309	31 836	245.388
560	16 399	11 743	216.115	1400	44 007	32 367	245.889
580	17 003	12 181	217.175	1440	45 408	33 434	246.876
600	17 611	12 622	218.204	1480	46 813	34 508	247.839
620	18 221	13 066	219.205	1520	48 222	35 584	248.778
640	18 833	13 512	220.179	1560	49 635	36 665	249.695
660	19 449	13 962	221.127	1600	51 053	37 750	250.592
680	20 068	14 414	222.052	1700	54 609	40 474	252.751
700	20 690	14 870	222.953	1800	58 191	43 225	254.797
720	21 315	15 328	223.833	1900	61 794	45 997	256.743
740	21 943	15 789	224.692	2000	65 408	48 780	258.600
760	22 573	16 255	225.533	2100	69 044	51 584	260.370
780	23 208	16 723	226.357	2200	72 688	54 396	262.065
800	23 844	17 193	227.162	2300	76 345	57 222	263.692
820	24 483	17 665	227.952	2400	80 015	60 060	265.253
840	25 124	18 140	228.724	2500	83 692	62 906	266.755
860	25 768	18 617	229.482	2600	87 383	65 766	268.202
880	26 415	19 099	230.227	2700	91 077	68 628	269.596
900	27 066	19 583	230.957	2800	94 784	71 504	270.943
920	27 719	20 070	231.674	2900	98 495	74 383	272.249
940	28 375	20 559	232.379	3000	102 210	77 267	273.508
960	29 033	21 051	233.072	3100	105 939	80 164	274.730
980	29 693	21 545	233.752	3150	107 802	81 612	275.326
1000	30 355	22 041	234.421	3200	109 667	83 061	275.914
1020	31 020	22 540	235.079				

Source: JANAF Thermochemical Tables, NSRDS-NBS-37, 1971.

TABLE F.6 Molar Properties of Water, H_2O

$\bar{h}_f^\circ = -241\,810$ kJ/kmol

T	$\bar{h}$	$\bar{u}$	$\bar{\phi}$	T	$\bar{h}$	$\bar{u}$	$\bar{\phi}$
0	0	0	0	1020	36 709	28 228	233.415
220	7295	5466	178.576	1040	37 542	28 895	234.223
240	7961	5965	181.471	1060	38 380	29 567	235.020
260	8627	6466	184.139	1080	39 223	30 243	235.806
280	9296	6968	186.616	1100	40 071	30 925	236.584
298	9904	7425	188.720	1120	40 923	31 611	237.352
300	9966	7472	188.928	1140	41 780	32 301	238.110
320	10 639	7978	191.098	1160	42 642	32 997	238.859
340	11 314	8487	193.144	1180	43 509	33 698	239.600
360	11 992	8998	195.081	1200	44 380	34 403	240.333
380	12 672	9513	196.920	1220	45 256	35 112	241.057
400	13 356	10 030	198.673	1240	46 137	35 827	241.773
420	14 043	10 551	200.350	1260	47 022	36 546	242.482
440	14 734	11 075	201.955	1280	47 912	37 270	243.183
460	15 428	11 603	203.497	1300	48 807	38 000	243.877
480	16 126	12 135	204.982	1320	49 707	38 732	244.564
500	16 828	12 671	206.413	1340	50 612	39 470	245.243
520	17 534	13 211	207.799	1360	51 521	40 213	245.915
540	18 245	13 755	209.139	1400	53 351	41 711	247.241
560	18 959	14 303	210.440	1440	55 198	43 226	248.543
580	19 678	14 856	211.702	1480	57 062	44 756	249.820
600	20 402	15 413	212.920	1520	58 942	46 304	251.074
620	21 130	15 975	214.122	1560	60 838	47 868	252.305
640	21 862	16 541	215.285	1600	62 748	49 445	253.513
660	22 600	17 112	216.419	1700	67 589	53 455	256.450
680	23 342	17 688	217.527	1800	72 513	57 547	259.262
700	24 088	18 268	218.610	1900	77 517	61 720	261.969
720	24 840	18 854	219.668	2000	82 593	65 965	264.571
740	25 597	19 444	220.707	2100	87 735	70 275	267.081
760	26 358	20 039	221.720	2200	92 940	74 649	269.500
780	27 125	20 639	222.717	2300	98 199	79 076	271.839
800	27 896	21 245	223.693	2400	103 508	83 553	274.098
820	28 672	21 855	224.651	2500	108 868	88 082	276.286
840	29 454	22 470	225.592	2600	114 273	92 656	278.407
860	30 240	23 090	226.517	2700	119 717	97 269	280.462
880	31 032	23 715	227.426	2800	125 198	101 917	282.453
900	31 828	24 345	228.321	2900	130 717	106 605	284.390
920	32 629	24 980	229.202	3000	136 264	111 321	286.273
940	33 436	25 621	230.070	3100	141 846	116 072	288.102
960	34 247	26 265	230.924	3150	144 648	118 458	288.9
980	35 061	26 913	231.767	3200	147 457	120 851	289.884
1000	35 882	27 568	232.597	3250	150 250	123 250	290.7

Source: JANAF Thermochemical Tables, NSRDS-NBS-37, 1971.

TABLE F.1E Properties of Air

T, °R	h Btu/lbm	p_r	u Btu/lbm	v_r	ϕ Btu/lbm-°R
400	95.53	0.4858	68.11	305.0	0.52890
440	105.11	0.6776	74.93	240.6	0.55172
480	114.69	0.9182	81.77	193.65	0.57255
520	124.27	1.2147	88.62	158.58	0.59173
537	128.10	1.3593	91.53	146.34	0.59945
540	129.06	1.3860	92.04	144.32	0.60078
560	133.86	1.5742	95.47	131.78	0.60950
580	138.66	1.7800	98.90	120.70	0.61793
600	143.47	2.005	102.34	110.88	0.62607
620	148.28	2.249	105.78	102.12	0.63395
640	153.09	2.514	109.21	94.30	0.64159
660	157.92	2.801	112.67	87.27	0.64902
680	162.73	3.111	116.12	80.96	0.65621
700	167.56	3.446	119.58	75.25	0.66321
720	172.39	3.806	123.04	70.07	0.67002
740	177.23	4.193	126.51	65.38	0.67665
760	182.08	4.607	129.99	61.10	0.68312
780	186.94	5.051	133.47	57.20	0.68942
800	191.81	5.526	136.97	53.63	0.69558
820	196.69	6.033	140.47	50.35	0.70160
840	201.56	6.573	143.98	47.34	0.70747
860	206.46	7.149	147.50	44.57	0.71323
880	211.35	7.761	151.02	42.01	0.71886
900	216.26	8.411	154.57	39.64	0.72438
920	221.18	9.102	158.12	37.44	0.72979
940	226.11	9.834	161.68	35.41	0.73509
960	231.06	10.610	165.26	33.52	0.74030
980	236.02	11.430	168.83	31.76	0.74540
1000	240.98	12.298	172.43	30.12	0.75042
1020	245.97	13.215	176.04	28.59	0.75536
1040	250.95	14.182	179.66	27.17	0.76019
1060	255.96	15.203	183.29	25.82	0.76496
1080	260.97	16.278	186.93	24.58	0.76964
1100	265.99	17.413	190.58	23.40	0.77426
1120	271.03	18.604	194.25	22.30	0.77880
1160	281.14	21.18	201.63	20.293	0.78767
1200	291.30	24.01	209.05	18.514	0.79628
1240	301.52	27.13	216.53	16.932	0.80466
1280	311.79	30.55	244.05	15.518	0.81280
1320	322.11	34.31	231.63	14.253	0.82075
1360	332.48	38.41	239.25	13.118	0.82848
1400	342.90	42.88	246.93	12.095	0.83604
1440	353.37	47.75	254.66	11.172	0.84341
1480	363.89	53.04	262.44	10.336	0.85062
1520	374.47	58.78	270.26	9.578	0.85767
1560	385.08	65.00	278.13	8.890	0.86456
1600	395.74	71.73	286.06	8.263	0.87130
1640	406.45	78.99	294.03	7.691	0.87791
1680	417.20	86.82	302.04	7.168	0.88439
1720	428.00	95.24	310.09	6.690	0.89074
1760	438.83	104.30	318.18	6.251	0.89697
1800	449.71	114.03	326.32	5.847	0.90308
1900	477.09	141.51	346.85	4.974	0.91788
2000	504.71	174.00	367.61	4.258	0.93205
2200	560.59	256.6	409.78	3.176	0.95868
2400	617.22	367.6	452.70	2.419	0.98331
2600	674.49	513.5	496.26	1.8756	1.00623
2800	732.33	702.0	540.40	1.4775	1.02767
3000	790.68	941.4	585.04	1.1803	1.04779

Source: J. H. Keenan and J. Kaye, *Gas Tables*, Wiley, New York, 1945.

TABLE F.2E Molar Properties of Nitrogen, N_2

$\bar{h}_f^\circ = 0$ Btu/lbmol

T, °R	$\bar{h}$, Btu/lbmol	$\bar{u}$, Btu/lbmol	$\bar{\phi}$, Btu/lbmol-°R	T, °R	$\bar{h}$, Btu/lbmol	$\bar{u}$, Btu/lbmol	$\bar{\phi}$, Btu/lbmol-°R
300	2082.0	1486.2	41.695	1100	7695.0	5510.5	50.783
320	2221.0	1585.5	42.143	1120	7839.3	5615.2	50.912
340	2360.0	1684.8	42.564	1160	8129.0	5825.4	51.167
400	2777.0	1982.6	43.694	1200	8420.0	6037.0	51.413
440	3055.1	2181.3	44.357	1240	8712.6	6250.1	51.653
480	3333.1	2379.9	44.962	1280	9006.4	6464.5	51.887
520	3611.3	2578.6	45.519	1320	9301.8	6680.4	52.114
537	3729.5	2663.1	45.743	1360	9598.6	6897.8	52.335
540	3750.3	2678.0	45.781	1400	9896.9	7116.7	52.551
560	3889.5	2777.4	46.034	1440	10196.6	7337.0	52.763
580	4028.7	2876.9	46.278	1480	10497.8	7558.7	52.969
600	4167.9	2976.4	46.514	1520	10800.4	7781.9	53.171
620	4307.1	3075.9	46.742	1560	11104.3	8006.4	53.369
640	4446.4	3175.5	46.964	1600	11409.7	8232.3	53.561
660	4585.8	3275.2	47.178	1640	11716.4	8459.6	53.751
680	4725.3	3374.9	47.386	1680	12024.3	8688.1	53.936
700	4864.9	3474.8	47.588	1720	12333.7	8918.0	54.118
720	5004.5	3574.7	47.785	1760	12644.3	9149.2	54.297
740	5144.3	3674.7	47.977	1800	12956.3	9381.7	54.472
760	5284.1	3774.9	48.164	1900	13741.6	9968.4	54.896
780	5424.2	3875.2	48.345	2000	14534.4	10562.6	55.303
800	5564.4	3975.7	48.522	2200	16139.8	11770.9	56.068
820	5704.7	4076.3	48.696	2400	17767.9	13001.8	56.777
840	5845.3	4177.1	48.865	2600	19415.8	14252.5	57.436
860	5985.9	4278.1	49.031	2800	21081.1	15520.6	58.053
880	6126.9	4379.4	49.193	3000	22761.5	16803.9	58.632
900	6268.1	4480.8	49.352	3100	23606.8	17450.6	58.910
920	6409.6	4582.6	49.507	3200	24455.0	18100.2	59.179
940	6551.2	4684.5	49.659	3300	25306.0	18752.7	59.442
960	6693.1	4786.7	49.808	3400	26159.7	19407.7	59.697
980	6835.4	4889.3	49.955	3600	27874.4	20725.3	60.186
1000	6977.9	4992.0	50.099	3700	28735.1	21387.4	60.422
1020	7120.7	5095.1	50.241	3800	29597.9	22051.6	60.562
1040	7263.8	5198.5	50.380	3900	30462.8	22717.9	60.877
1060	7407.2	5302.2	50.516	5300	42728.3	32203.2	63.563
1080	7551.0	5406.2	50.651	5380	43436.0	32752.1	63.695

Source: J. H. Keenan and J. Kaye, *Gas Tables*, Wiley, New York, 1945.

TABLE F.3E Molar Properties of Oxygen, O_2

$\bar{h}_f^\circ = 0$ Btu/lbmol

T, °R	h	$\bar{u}$	$\bar{\phi}$	T, °R	$\bar{h}$	$\bar{u}$	$\bar{\phi}$
300	2073.5	1477.8	44.927	1280	9254.6	6712.7	55.386
320	2212.6	1577.1	45.375	1320	9571.6	6950.2	55.630
340	2351.7	1676.5	45.797	1360	9890.2	7189.4	55.867
400	2769.1	1974.8	46.927	1400	10210.4	7430.1	56.099
420	2908.3	2074.3	47.267	1440	10532.0	7672.4	56.326
440	3047.5	2173.8	47.591	1480	10855.1	7916.0	56.547
480	3326.5	2373.3	48.198	1520	11179.6	8161.1	56.763
520	3606.1	2573.4	48.757	1560	11505.4	8407.4	56.975
537	3725.1	2658.7	48.982	1600	11832.5	8655.1	57.182
540	3746.2	2673.8	49.021	1640	12160.9	8904.1	57.385
560	3886.6	2774.5	49.276	1680	12490.4	9154.1	57.582
580	4027.3	2875.5	49.522	1720	12821.1	9405.4	57.777
600	4168.3	2976.8	49.762	1760	13153.0	9657.9	57.968
620	4309.7	3078.4	49.993	1800	13485.8	9911.2	58.155
640	4451.4	3180.4	50.218	1900	14322.1	10549.0	58.607
660	4593.5	3282.9	50.437	2000	15164.0	11192.3	59.039
680	4736.2	3385.8	50.650	2200	16862.6	12493.7	59.848
700	4879.3	3489.2	50.858	2400	18579.2	13813.1	60.594
720	5022.9	3593.1	51.059	2600	20311.4	15148.1	61.287
740	5167.0	3697.4	51.257	2800	22057.8	16497.4	61.934
760	5311.4	3802.2	51.450	3000	23817.7	17860.1	62.540
780	5456.4	3907.5	51.638	3100	24702.5	18546.3	62.831
800	5602.0	4013.3	51.821	3200	25590.5	19235.7	63.113
820	5748.1	4119.7	52.002	3300	26481.6	19928.2	63.386
840	5894.8	4226.6	52.179	3400	27375.9	20623.9	63.654
860	6041.9	4334.1	52.352	3600	29173.9	22024.8	64.168
880	6189.6	4442.0	52.522	3700	30077.5	22729.8	64.415
900	6337.9	4550.6	52.688	3800	30984.1	23437.8	64.657
920	6486.7	4659.7	52.852	3900	31893.6	24148.7	64.893
940	6636.1	4769.4	53.012	4100	33721.6	25579.5	65.350
960	6786.0	4879.5	53.170	4200	34639.9	26299.2	65.571
980	6936.4	4990.3	53.326	4300	35561.1	27021.9	65.788
1000	7087.5	5101.6	53.477	4400	36485.0	27747.2	66.000
1020	7238.9	5213.3	53.628	4600	38341.4	29206.4	66.413
1040	7391.0	5325.7	53.775	4700	39273.6	29940.0	66.613
1060	7543.6	5438.6	53.921	4800	40208.6	30676.4	66.809
1080	7697.8	5552.1	54.064	4900	41146.1	31415.3	67.003
1100	7850.4	5665.9	54.204	5100	43029.1	32901.2	67.380
1120	8004.5	5780.3	54.343	5200	43974.3	33647.9	67.562
1160	8314.2	6010.6	54.614	5300	44922.2	34397.1	67.743
1200	8625.8	6242.8	54.879	5380	45682.1	34998.1	67.885
1240	8939.4	6476.9	55.136				

Source: J. H. Keenan and J. Kaye, *Gas Tables*, Wiley, New York, 1945.

TABLE F.4E Molar Properties of Carbon Dioxide, CO_2

$\bar{h}_f^\circ = -169\,300$ Btu/lbmol

T, °R	$\bar{h}$	$\bar{u}$	$\bar{\phi}$	T, °R	$\bar{h}$	$\bar{u}$	$\bar{\phi}$
300	2108.2	1512.4	46.353	1320	12376.4	9755.0	60.412
340	2407.3	1732.1	47.289	1340	12617.0	9955.9	60.593
380	2716.4	1961.8	48.148	1360	12858.5	10157.7	60.772
420	3035.7	2201.7	48.947	1380	13101.0	10360.5	60.949
460	3365.7	2452.2	49.698	1400	13344.7	10564.5	61.124
480	3534.7	2581.5	50.058	1420	13589.1	10769.2	61.298
500	3706.2	2713.3	50.408	1440	13834.5	10974.8	61.469
520	3880.3	2847.7	50.750	1460	14080.8	11181.4	61.639
537	4030.2	2963.8	51.032	1480	14328.0	11388.9	61.808
540	4056.8	2984.4	51.082	1500	14576.0	11597.2	61.974
580	4417.2	3265.4	51.726	1520	14824.9	11806.4	62.138
600	4600.9	3409.4	52.038	1540	15074.7	12016.5	62.302
620	4786.8	3555.6	52.343	1560	15325.3	12227.3	62.464
640	4974.9	3704.0	52.641	1580	15576.7	12439.0	62.624
660	5165.2	3854.6	52.934	1600	15829.0	12651.6	62.783
680	5357.6	4007.2	53.225	1620	16081.9	12864.8	62.939
700	5552.0	4161.9	53.503	1640	16335.7	13078.9	63.095
720	5748.4	4318.6	53.780	1660	16590.2	13293.7	63.250
740	5946.8	4477.3	54.051	1700	17101.4	13725.4	63.555
760	6147.0	4637.9	54.319	1800	18391.5	14816.9	64.292
780	6349.1	4800.1	54.582	1900	19697.8	15924.7	64.999
800	6552.9	4964.2	54.839	2000	21018.7	17046.9	65.676
820	6758.3	5129.9	55.093	2100	22352.7	18182.4	66.327
840	6965.7	5297.6	55.343	2200	23699.0	19330.1	66.953
860	7174.7	5466.9	55.589	2300	25056.3	20488.8	67.557
880	7385.3	5637.7	55.831	2400	26424.0	21657.9	68.139
900	7597.6	5810.3	56.070	2500	27801.2	22836.5	68.702
920	7811.4	5984.4	56.305	2600	29187.1	24023.8	69.245
940	8026.8	6160.1	56.536	2700	30581.2	25219.4	69.771
960	8243.8	6337.4	56.765	2800	31982.8	26422.4	70.282
980	8462.2	6516.1	56.990	2900	33391.5	27632.5	70.776
1000	8682.1	6696.2	57.212	3000	34806.6	28849.0	71.255
1020	8903.4	6877.8	57.432	3100	36227.9	30071.7	71.722
1040	9126.2	7060.9	57.647	3200	37654.7	31299.9	72.175
1060	9350.3	7245.3	57.861	3300	39086.7	32533.3	72.616
1080	9575.8	7431.1	58.072	3400	40523.6	33771.6	73.045
1100	9802.6	7618.1	58.281	3500	41965.2	35014.7	73.462
1120	10030.6	7806.4	58.485	3600	43411.0	36261.9	73.870
1140	10260.1	7996.2	58.689	3700	44860.6	37512.9	74.267
1160	10490.6	8187.0	58.889	3800	46314.0	38767.7	74.655
1180	10722.3	8379.0	59.088	3900	47771.0	40026.1	75.033
1200	10955.3	8572.3	59.283	4000	49231.4	41287.9	75.404
1220	11189.4	8766.6	59.477	4200	52162.0	43821.4	76.119
1240	11424.6	8962.1	59.668	4400	55105.1	46367.3	76.803
1260	11661.0	9158.8	59.858	4600	58059.7	48924.7	77.460
1280	11898.4	9356.5	60.044	4800	61024.9	51492.7	78.091
1300	12136.9	9555.3	60.229	5000	64000.0	54070.6	78.698

Source: J. H. Keenan and J. Kaye, *Gas Tables*, Wiley, New York, 1945.

TABLE F.5E Molar Properties of Carbon Monoxide, CO

$\bar{h}_f^\circ = -47\,550$ Btu/lbmol

T, °R	h	$\bar{u}$	$\bar{\phi}$	T, °R	$\bar{h}$	$\bar{u}$	$\bar{\phi}$
300	2081.9	1486.1	43.223	1140	8009.2	5745.4	52.598
320	2220.9	1585.4	43.672	1160	8156.1	5852.5	52.726
340	2359.9	1684.7	44.093	1180	8303.3	5960.0	52.852
360	2498.8	1783.9	44.490	1200	8450.8	6067.8	52.976
380	2637.9	1883.3	44.866	1220	8598.8	6176.0	53.098
400	2776.9	1982.6	45.223	1240	8747.2	6284.7	53.218
420	2916.0	2081.9	45.563	1260	8896.0	6393.8	53.337
440	3055.0	2181.2	45.886	1280	9045.0	6503.1	53.455
460	3194.0	2280.5	46.194	1300	9194.6	6613.0	53.571
480	3333.0	2379.8	46.491	1320	9344.6	6723.2	53.685
500	3472.1	2479.2	46.775	1340	9494.8	6833.7	53.799
520	3611.2	2578.6	47.048	1360	9645.5	6944.7	53.910
537	3729.5	2663.1	47.272	1380	9796.6	7056.1	54.021
540	3750.3	2677.9	47.310	1400	9948.1	7167.9	54.129
560	3889.5	2777.4	47.563	1420	10100.0	7280.1	54.237
580	4028.7	2876.9	47.807	1460	10404.8	7505.4	54.448
600	4168.0	2976.5	48.044	1500	10711.1	7732.3	54.655
620	4307.4	3076.2	48.272	1520	10864.9	7846.4	54.757
640	4446.9	3175.9	48.494	1540	11019.0	7960.8	54.858
660	4586.5	3275.8	48.709	1560	11173.4	8075.4	54.958
680	4726.2	3375.8	48.917	1580	11328.2	8190.5	55.056
700	4866.0	3475.9	49.120	1600	11483.4	8306.0	55.154
720	5006.1	3576.3	49.317	1620	11638.9	8421.8	55.251
740	5146.4	3676.9	49.509	1640	11794.7	8537.9	55.347
760	5286.8	3777.5	49.697	1660	11950.9	8654.4	55.441
780	5427.4	3878.4	49.880	1680	12107.5	8771.2	55.535
800	5568.2	3979.5	50.058	1700	12264.3	8888.3	55.628
820	5709.4	4081.0	50.232	1800	13053.2	9478.6	56.078
840	5850.7	4182.6	50.402	1900	13849.8	10076.6	56.509
860	5992.3	4284.5	50.569	2000	14653.2	10681.5	56.922
880	6134.2	4386.6	50.732	2100	15463.3	11293.0	57.317
900	6276.4	4489.1	50.892	2200	16279.4	11910.5	57.696
920	6419.0	4592.0	51.048	2300	17101.0	12533.5	58.062
940	6561.7	4695.0	51.202	2400	17927.4	13161.3	58.414
960	6704.9	4798.5	51.353	2500	18758.8	13794.1	58.754
980	6848.4	4902.3	51.501	2600	19594.3	14431.0	59.081
1000	6992.2	5006.3	51.646	2700	20434.0	15072.2	59.398
1020	7136.4	5110.8	51.788	2800	21277.2	15716.8	59.705
1040	7281.0	5215.7	51.929	2900	22123.8	16364.8	60.002
1060	7425.9	5320.9	52.067	3000	22973.4	17015.8	60.290
1080	7571.1	5426.4	52.203	3100	23826.0	17669.8	60.569
1100	7716.8	5532.3	52.337	3200	24681.2	18326.4	60.841
1120	7862.9	5638.7	52.468	3300	25539.0	18985.6	61.105
3400	26399.3	19647.3	61.362	4300	34231.2	25692.0	63.405
3500	27261.8	20311.2	61.612	4400	35109.2	26371.4	63.607
3600	28126.6	20977.5	61.855	4600	36869.3	27734.3	63.998
3700	28993.5	21645.8	62.093	4700	37751.0	28417.4	64.188
3800	29862.3	22316.0	62.325	5000	40402.7	30473.4	64.735
3900	30732.9	22988.0	62.551	5100	41288.6	31160.7	64.910
4000	31605.2	23661.7	62.772	5200	42175.5	31849.0	65.082
4100	32479.1	24337.0	62.988	5300	43063.2	32538.1	65.252
4200	33354.4	25013.8	63.198	5380	43774.1	33090.1	65.385

Source: J. H. Keenan and J. Kaye, *Gas Tables*, Wiley, New York, 1945.

TABLE F.6E Molar Properties of Water Vapor, H_2O

$\bar{h}_f^\circ = -104\,040$ Btu/lbmol

T, °R	$\bar{h}$ Btu/lbmol	$\bar{u}$ Btu/lbmol	$\bar{\phi}$ Btu/lbmol-°R
300	2367.6	1771.8	40.439
340	2686.0	2010.8	41.435
380	3004.4	2249.8	42.320
420	3323.2	2489.1	43.117
460	3642.3	2728.8	43.841
500	3962.0	2969.1	44.508
537	4258.0	3191.9	45.079
540	4282.4	3210.0	45.124
580	4603.7	3451.9	45.696
620	4926.1	3694.9	46.235
660	5250.0	3939.3	46.741
700	5575.4	4185.3	47.219
740	5902.6	4433.1	47.673
780	6231.7	4682.7	48.106
820	6562.6	4934.2	48.520
860	6895.6	5187.8	48.916
900	7230.9	5443.6	49.298
940	7568.4	5701.7	49.665
980	7908.2	5962.0	50.019
1020	8250.4	6224.8	50.360
1060	8595.0	6490.0	50.693
1100	8942.0	6757.5	51.013
1140	9291.4	7027.5	51.325
1180	9643.4	7300.1	51.630
1220	9998.0	7575.2	51.925
1260	10 354.9	7852.7	52.212
1300	10 714.5	8132.9	52.494
1340	11 076.6	8415.5	52.768
1380	11 441.4	8700.9	53.037
1420	11 808.8	8988.9	53.299
1460	12 178.8	9279.4	53.556
1500	12 551.4	9572.7	53.808
1600	13 494.4	10 317.6	54.418
1700	14 455.4	11 079.4	54.999
1800	15 433.0	11 858.4	55.559
1900	16 428	12 654	56.097
2100	18 467	14 297	57.119
2300	20 571	16 003	58.077
2500	22 735	17 771	58.980
2700	24 957	19 595	59.837
2900	27 231	21 472	60.650
3100	29 553	23 397	61.426
3300	31 918	25 365	62.167
3500	34 324	27 373	62.876
3700	36 765	29 418	63.557
3900	39 240	31 495	64.210
4100	41 745	33 603	64.839
4300	44 278	35 739	65.444
4500	46 836	37 900	66.028
4700	49 417	40 083	66.591
4900	52 019	42 288	67.135
5000	53 327	43 398	67.401

Source: J. H. Keenan and J. Kaye, *Gas Tables*, Wiley, New York, 1945.

G. PSYCHROMETRIC CHARTS

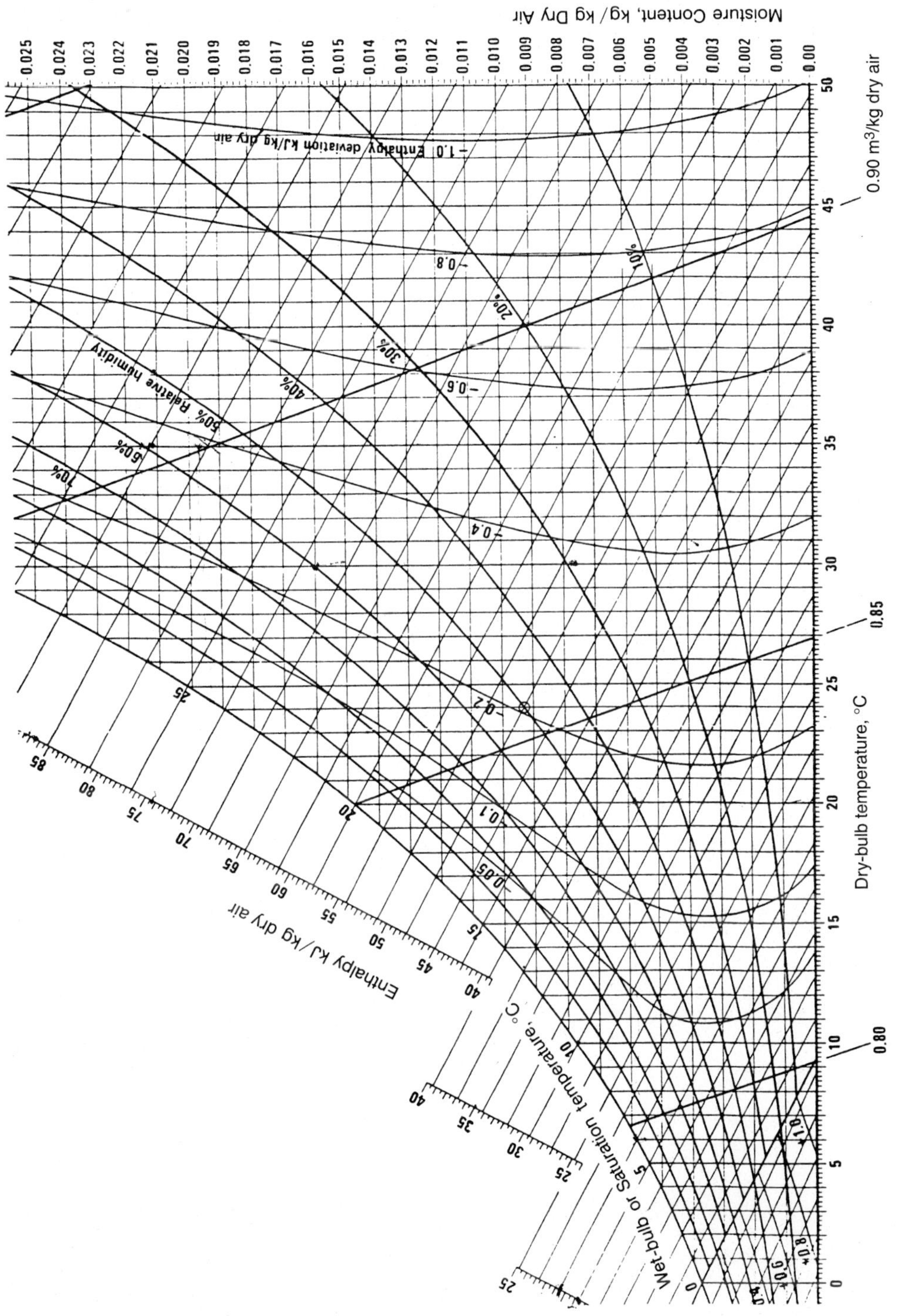

FIGURE G.1 Psychrometric chart, $p = 1$ atm. (*Carrier corporation.*)

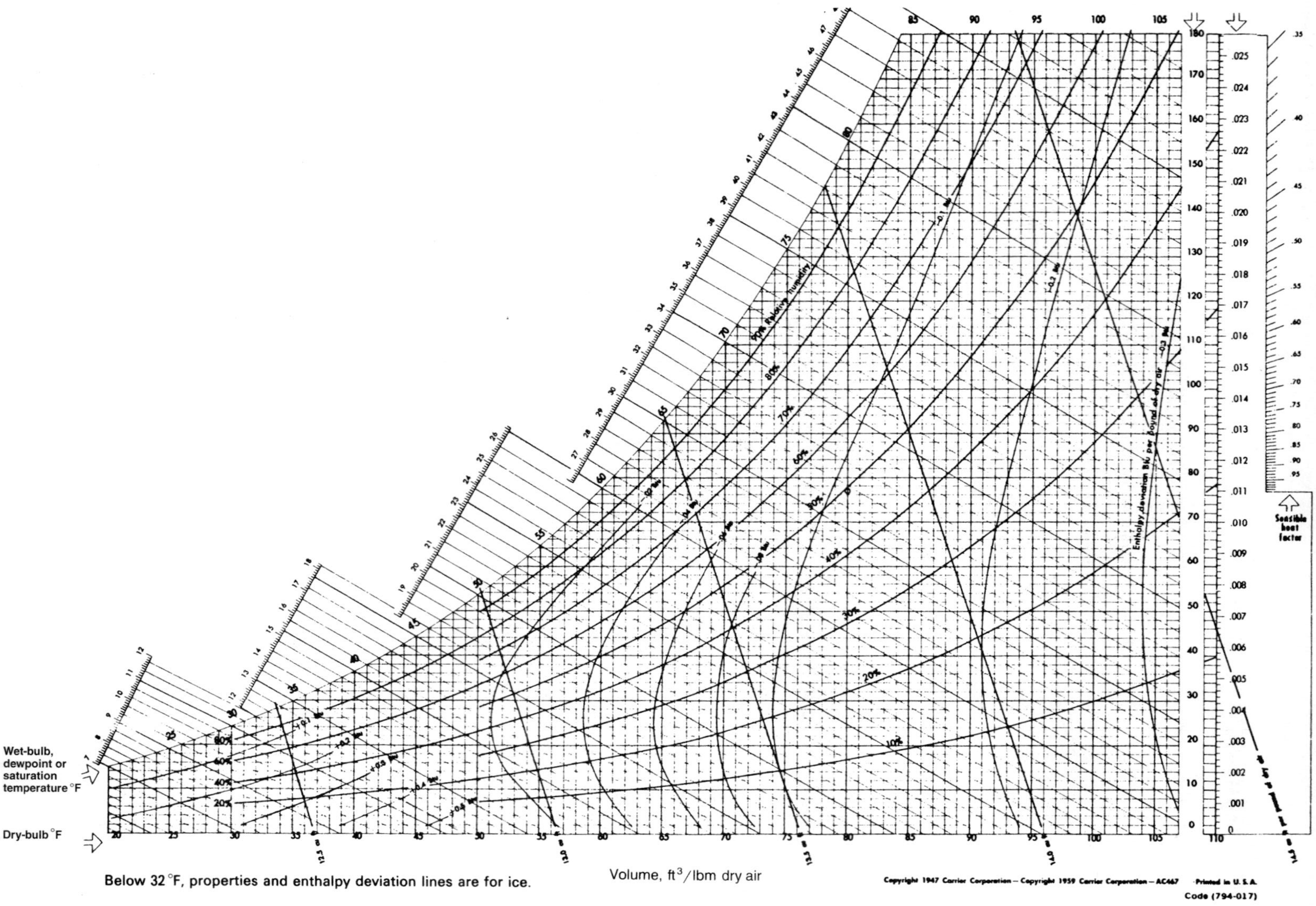

FIGURE G.1E Psychrometric chart, $p = 1$ atm. (*Carrier corporation.*)

H. COMPRESSIBILITY CHART

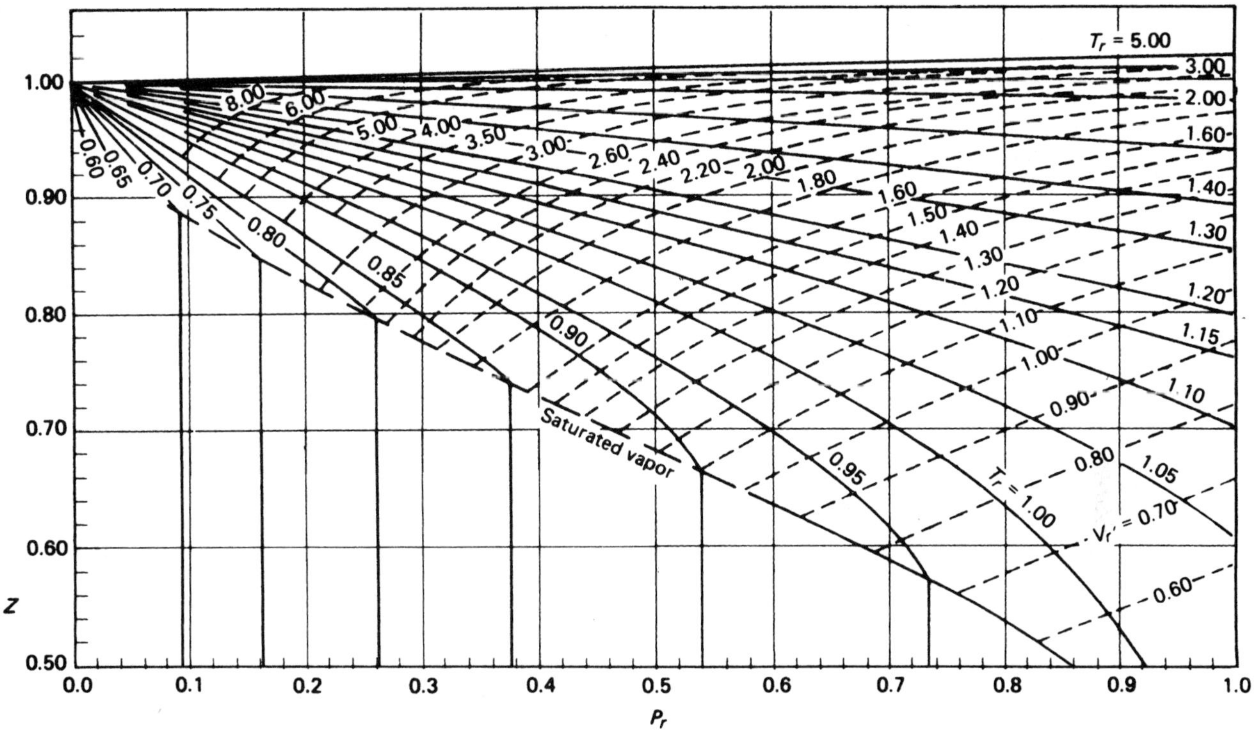

FIGURE H.1 Compressibility Chart.

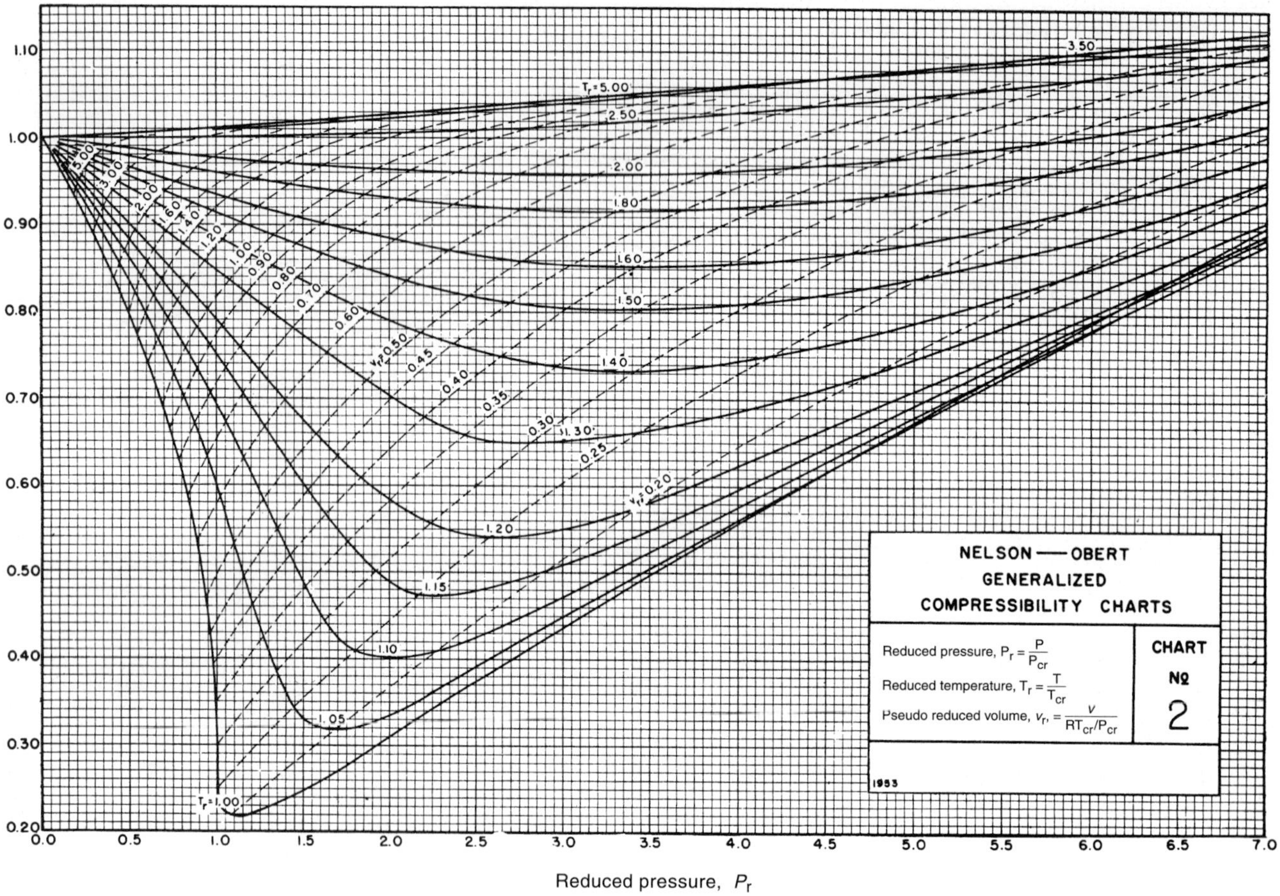

FIGURE H.1 (continued) [V. M. Faires, "Problems on Thermodynamics," Macmillan, New York, 1962. Data from L. C. Nelson and E. F. Obert, Generalized Compressibility Charts, Chem. Eng. **61**: 203 (1954).]

I. COMPRESSIBLE-FLOW TABLES FOR AIR

TABLE I.1 Isentropic Flow

M	p/p_0	T/T_0	A/A^*	M	p/p_0	T/T_0	A/A^*
0	1.0000	1.0000	∞	2.00	.1278	.5556	1.688
.04	.9989	.9997	14.4815	2.04	.1201	.5458	1.745
.08	.9955	.9987	7.2616	2.08	.1128	.5361	1.806
.12	.9900	.9971	4.8643	2.12	.1060	.5266	1.869
.16	.9823	.9949	3.6727	2.16	$.9956^{-1}$	.5173	1.935
.20	.9725	.9921	2.9635	2.20	$.9352^{-1}$	.5081	2.005
.24	.9607	.9886	2.4956	2.24	$.8785^{-1}$	.4991	2.078
.28	.9470	.9846	2.1656	2.28	$.8251^{-1}$	.4903	2.154
.32	.9315	.9799	1.9219	2.32	$.7751^{-1}$	.4816	2.233
.36	.9143	.9747	1.7358	2.36	$.7281^{-1}$	.4731	2.316
.40	.8956	.9690	1.5901	2.40	$.6840^{-1}$	.4647	2.403
.44	.8755	.9627	1.4740	2.44	$.6426^{-1}$	.4565	2.494
.48	.8541	.9560	1.3801	2.48	$.6038^{-1}$	.4484	2.588
.52	.8317	.9487	1.3034	2.52	$.5674^{-1}$	.4405	2.686
.56	.8082	.9410	1.2403	2.56	$.5332^{-1}$	.4328	2.789
.60	.7840	.9328	1.1882	2.60	$.5012^{-1}$	.4252	2.896
.64	.7591	.9243	1.1452	2.64	$.4711^{-1}$	.4177	3.007
.68	.7338	.9153	1.1097	2.68	$.4429^{-1}$	.4104	3.123
.72	.7080	.9061	1.0806	2.72	$.4165^{-1}$	.4033	3.244
.76	.6821	.8964	1.0570	2.76	$.3917^{-1}$	.3963	3.370
.80	.6560	.8865	1.0382	2.80	$.3685^{-1}$	.3894	3.500
.84	.6300	.8763	1.0237	2.84	$.3467^{-1}$	.3827	3.636
.88	.6041	.8659	1.0129	2.88	$.3263^{-1}$	.3761	3.777
.92	.5785	.8552	1.0056	2.92	$.3071^{-1}$	.3696	3.924
.96	.5532	.8444	1.0014	2.96	$.2891^{-1}$	.3633	4.076
1.00	.5283	.8333	1.000	3.00	$.2722^{-1}$	.3571	4.235
1.04	.5039	.8222	1.001	3.04	$.2564^{-1}$	.3511	4.399
1.08	.4800	.8108	1.005	3.08	$.2416^{-1}$	.3452	4.570
1.12	.4568	.7994	1.011	3.12	$.2276^{-1}$	.3393	4.747
1.16	.4343	.7879	1.020	3.16	$.2146^{-1}$	.3337	4.930
1.20	.4124	.7764	1.030	3.20	$.2023^{-1}$	.3281	5.121
1.24	.3912	.7648	1.043	3.24	$.1908^{-1}$	.3226	5.319
1.28	.3708	.7532	1.058	3.28	$.1799^{-1}$	.3173	5.523
1.32	.3512	.7416	1.075	3.32	$.1698^{-1}$	.3121	5.736
1.36	.3323	.7300	1.094	3.36	$.1602^{-1}$	.3069	5.956
1.40	.3142	.7184	1.115	3.40	$.1512^{-1}$	.3019	6.184
1.44	.2969	.7069	1.138	3.44	$.1428^{-1}$	.2970	6.420
1.48	.2804	.6954	1.163	3.48	$.1349^{-1}$	.2922	6.664
1.52	.2646	.6840	1.190	3.52	$.1274^{-1}$	.2875	6.917
1.56	.2496	.6726	1.219	3.56	$.1204^{-1}$	.2829	7.179
1.60	.2353	.6614	1.250	3.60	$.1138^{-1}$	.2784	7.450
1.64	.2217	.6502	1.284	3.64	$.1076^{-1}$	.2740	7.730
1.68	.2088	.6392	1.319	3.68	$.1018^{-1}$	.2697	8.020
1.72	.1966	.6283	1.357	3.72	$.9633^{-2}$	.2654	8.320
1.76	.1850	.6175	1.397	3.76	$.9116^{-2}$	.2613	8.630
1.80	.1740	.6068	1.439	3.80	$.8629^{-2}$	.2572	8.951
1.84	.1637	.5963	1.484	3.84	$.8171^{-2}$	.2532	9.282
1.88	.1539	.5859	1.531	3.88	$.7739^{-2}$	.2493	9.624
1.92	.1447	.5756	1.580	3.92	$.7332^{-2}$	.2455	9.977
1.96	.1360	.5655	1.633	3.96	$.6948^{-2}$	.2418	10.34

TABLE I.1 (*continued*)

M	p/p_0	T/T_0	A/A^*	M	p/p_0	T/T_0	A/A^*
4.00	$.6586^{-2}$	.2381	10.72	4.60	$.3053^{-2}$	.1911	18.02
4.04	$.6245^{-2}$	.2345	11.11	4.64	$.2906^{-2}$	.1885	18.63
4.08	$.5923^{-2}$	.2310	11.51	4.68	$.2768^{-2}$	.1859	19.26
4.12	$.5619^{-2}$	.2275	11.92	4.72	$.2637^{-2}$	.1833	19.91
4.16	$.5333^{-2}$	.2242	12.35	4.76	$.2512^{-2}$	.1808	20.58
4.20	$.5062^{-2}$	.2208	12.79	4.80	$.2394^{-2}$	.1783	21.26
4.24	$.4806^{-2}$	.2176	13.25	4.84	$.2283^{-2}$	.1759	21.97
4.28	$.4565^{-2}$	.2144	13.72	4.88	$.2177^{-2}$	.1735	22.70
4.32	$.4337^{-2}$	.2113	14.20	4.92	$.2076^{-2}$	.1712	23.44
4.36	$.4121^{-2}$	.2083	14.70	4.96	$.1981^{-2}$	.1689	24.21
4.40	$.3918^{-2}$	.2053	15.21	5.00	$.1890^{-2}$	.1667	25.00
4.44	$.3725^{-2}$	.2023	15.74	7.00	$.0242^{-2}$	.0926	104.14
4.48	$.3543^{-2}$	.1994	16.28	9.00	$.0474^{-3}$	.0582	327.19
4.52	$.3370^{-2}$	.1966	16.84	∞	0	0	∞
4.56	$.3207^{-2}$	.1938	17.42				

TABLE I.2 Normal-Shock Flow

M_1	M_2	p_2/p_1	T_2/T_1	p_{02}/p_{01}	M_1	M_2	p_2/p_1	T_2/T_1	p_{02}/p_{01}
1.00	1.000	1.000	1.000	1.000	2.16	.5525	5.277	1.822	.6464
1.04	.9620	1.095	1.026	.9999	2.20	.5471	5.480	1.857	.6281
1.08	.9277	1.194	1.052	.9994	2.24	.5418	5.687	1.892	.6100
1.12	.8966	1.297	1.078	.9982	2.28	.5368	5.898	1.929	.5921
1.16	.8682	1.403	1.103	.9961	2.32	.5321	6.113	1.965	.5745
1.20	.8422	1.513	1.128	.9928	2.36	.5275	6.331	2.002	.5572
1.24	.8183	1.627	1.153	.9884	2.40	.5231	6.553	2.040	.5401
1.28	.7963	1.745	1.178	.9827	2.44	.5189	6.779	2.079	.5234
1.32	.7760	1.866	1.204	.9758	2.48	.5149	7.009	2.118	.5071
1.36	.7572	1.991	1.229	.9676	2.52	.5111	7.242	2.157	.4991
1.40	.7397	2.120	1.255	.9582	2.56	.5074	7.479	2.198	.4754
1.44	.7235	2.253	1.281	.9476	2.60	.5039	7.720	2.238	.4601
1.48	.7083	2.389	1.307	.9360	2.64	.5005	7.965	2.280	.4452
1.52	.6941	2.529	1.334	.9233	2.68	.4972	8.213	2.322	.4307
1.56	.6809	2.673	1.361	.9097	2.72	.4941	8.465	2.364	.4166
1.60	.6684	2.820	1.388	.8952	2.76	.4911	8.721	2.407	.4028
1.64	.6568	2.971	1.416	.8799	2.80	.4882	8.980	2.451	.3895
1.68	.6458	3.126	1.444	.8640	2.84	.4854	9.243	2.496	.3765
1.72	.6355	3.285	1.473	.8474	2.88	.4827	9.510	2.540	.3639
1.76	.6257	3.447	1.502	.8302	2.92	.4801	9.781	2.586	.3517
1.80	.6165	3.613	1.532	.8127	2.96	.4776	10.06	2.632	.3398
1.84	.6078	3.783	1.562	.7948	3.00	.4752	10.33	2.679	.3283
1.88	.5996	3.957	1.592	.7765	3.04	.4729	10.62	2.726	.3172
1.92	.5918	4.134	1.624	.7581	3.08	.4706	10.90	2.774	.3065
1.96	.5844	4.315	1.655	.7395	3.12	.4685	11.19	2.823	.2960
2.00	.5774	4.500	1.688	.7209	3.16	.4664	11.48	2.872	.2860
2.04	.5707	4.689	1.720	.7022	3.20	.4643	11.78	2.922	.2762
2.08	.5643	4.881	1.754	.6835	3.24	.4624	12.08	2.972	.2668
2.12	.5583	5.077	1.787	.6649	3.28	.4605	12.38	3.023	.2577
3.32	.4587	12.69	3.075	.2489	4.24	.4290	20.81	4.432	.1135
3.36	.4569	13.00	3.127	.2404	4.28	.4281	21.20	4.499	.1098
3.40	.4552	13.32	3.180	.2322	4.32	.4272	21.61	4.566	.1062
3.44	.4535	13.64	3.234	.2243	4.36	.4264	22.01	4.633	.1028
3.48	.4519	13.96	3.288	.2167	4.40	.4255	22.42	4.702	$.9948^{-1}$
3.52	.4504	14.29	3.343	.2093	4.44	.4247	22.83	4.771	$.9628^{-1}$
3.56	.4489	14.62	3.398	.2022	4.48	.4239	23.25	4.840	$.9320^{-1}$
3.60	.4474	14.95	3.454	.1953	4.52	.4232	23.67	4.910	$.9022^{-1}$
3.64	.4460	15.29	3.510	.1887	4.56	.4224	24.09	4.981	$.8735^{-1}$
3.68	.4446	15.63	3.568	.1823	4.60	.4217	24.52	5.052	$.8459^{-1}$
3.72	.4433	15.98	3.625	.1761	4.64	.4210	24.95	5.124	$.8192^{-1}$
3.76	.4420	16.33	3.684	.1702	4.68	.4203	25.39	5.197	$.7934^{-1}$
3.80	.4407	16.68	3.743	.1645	4.72	.4196	25.82	5.270	$.7685^{-1}$
3.84	.4395	17.04	3.802	.1589	4.76	.4189	26.27	5.344	$.7445^{-1}$
3.88	.4383	17.40	3.863	.1536	4.80	.4183	26.71	5.418	$.7214^{-1}$
3.92	.4372	17.76	3.923	.1485	4.84	.4176	27.16	5.494	$.6991^{-1}$
3.96	.4360	18.13	3.985	.1435	4.88	.4170	27.62	5.569	$.6775^{-1}$
4.00	.4350	18.50	4.047	.1388	4.92	.4164	28.07	5.646	$.6567^{-1}$
4.04	.4339	18.88	4.110	.1342	4.96	.4158	28.54	5.723	$.6366^{-1}$
4.08	.4329	19.25	4.173	.1297	5.00	.4152	29.00	5.800	$.6172^{-1}$
4.12	.4319	19.64	4.237	.1254	7.00	.3974	57.00	10.469	$.1535^{-1}$
4.16	.4309	20.02	4.301	.1213	9.00	.3898	94.33	16.693	$.0496^{-1}$
4.20	.4299	20.41	4.367	.1173	∞	.3780	∞	∞	0

TABLE I.3 Prandtl-Meyer Function

M	θ	μ	M	θ	μ
1.00	0	90.00	1.96	25.271	30.68
1.04	.3510	74.06	2.00	26.380	30.00
1.08	.9680	67.81	2.04	27.476	29.35
1.12	1.735	63.23	2.08	28.560	28.74
1.16	2.607	59.55	2.12	29.631	28.14
1.20	3.558	56.44	2.16	30.689	27.58
1.24	4.569	53.75	2.20	31.732	27.04
1.28	5.627	51.38	2.24	32.763	26.51
1.32	6.721	49.25	2.28	33.780	26.01
1.36	7.844	47.33	2.32	34.783	25.53
1.40	8.987	45.58	2.36	35.771	25.07
1.44	10.146	43.98	2.40	36.746	24.62
1.48	11.317	42.51	2.44	37.708	24.19
1.52	12.495	41.14	2.48	38.655	23.78
1.56	13.677	39.87	2.52	39.589	23.38
1.60	14.861	38.68	2.56	40.509	22.99
1.64	16.043	37.57	2.60	41.415	22.62
1.68	17.222	36.53	2.64	42.307	22.26
1.72	18.397	35.55	2.68	43.187	21.91
1.76	19.565	34.62	2.72	44.053	21.57
1.80	20.725	33.75	2.76	44.906	21.24
1.84	21.877	32.92	2.80	45.746	20.92
1.88	23.019	32.13	2.84	46.573	20.62
1.92	24.151	31.39	2.88	47.388	20.32
2.92	48.190	20.03	4.00	65.785	14.48
2.96	48.980	19.75	4.04	66.309	14.33
3.00	49.757	19.47	4.08	66.826	14.19
3.04	50.523	19.20	4.12	67.336	14.05
3.08	51.277	18.95	4.16	67.838	13.91
3.12	52.020	18.69	4.20	68.333	13.77
3.16	52.751	18.45	4.24	68.821	13.64
3.20	53.470	18.21	4.28	69.302	13.51
3.24	54.179	17.98	4.32	69.777	13.38
3.28	54.877	17.75	4.36	70.245	13.26
3.32	55.564	17.53	4.40	70.706	13.14
3.36	56.241	17.31	4.44	71.161	13.02
3.40	56.907	17.10	4.48	71.610	12.90
3.44	57.564	16.90	4.52	72.052	12.78
3.48	58.210	16.70	4.56	72.489	12.67
3.52	58.847	16.51	4.60	72.919	12.56
3.56	59.474	16.31	4.64	73.344	12.45
3.60	60.091	16.13	4.68	73.763	12.34
3.64	60.700	15.95	4.72	74.176	12.23
3.68	61.299	15.77	4.76	74.584	12.13
3.72	61.899	15.59	4.80	74.986	12.03
3.76	62.471	15.42	4.84	75.383	11.92
3.80	63.044	15.26	4.88	75.775	11.83
3.84	63.608	15.10	4.92	76.162	11.73
3.88	64.164	14.94	4.96	76.544	11.63
3.92	64.713	14.78	5.00	76.920	11.54
3.96	65.253	14.63			

J. PROPERTIES OF AREAS AND VOLUMES

TABLE J.1 Areas

	Sketch	Area	Centroid	Second moment
Rectangle	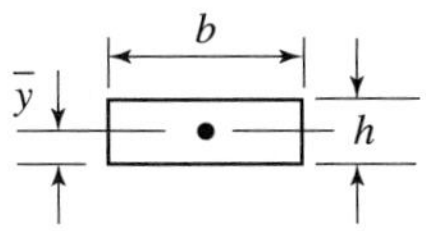	bh	$\bar{y} = h/2$	$\bar{I} = bh^3/12$ $\bar{I}_{xy} = 0$
Triangle	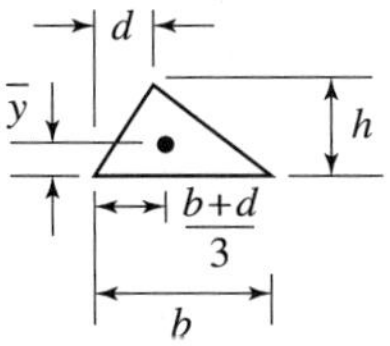	$bh/2$	$\bar{y} = h/3$	$\bar{I} = bh^3/36$ $\bar{I}_{xy} = (b - 2d)bh^3/72$
Circle	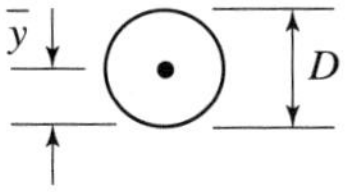	$\pi D^2/4$	$\bar{y} = r$	$\bar{I} = \pi D^4/64$
Semicircle	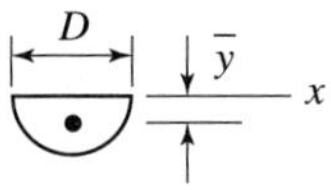	$\pi D^2/8$	$\bar{y} = 4r/3\pi$	$I_x = \pi D^4/128$
Ellipse	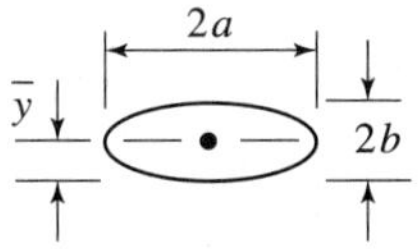	πab	$\bar{y} = b$	$\bar{I} = \pi ab^3/4$
Semiellipse	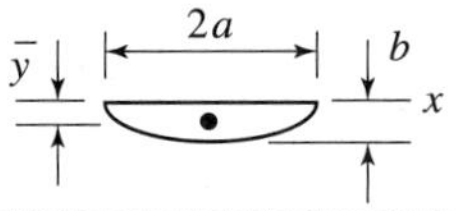	$\pi ab/2$	$\bar{y} = 4b/3\pi$	$I_x = \pi ab^3/8$

TABLE J.2 Volumes

	Sketch	Surface area	Volume	Centroid
Cylinder	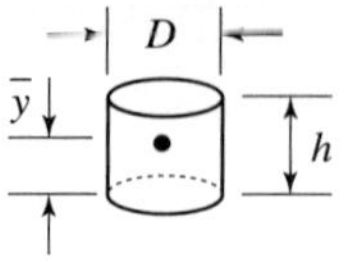	$\pi Dh + \pi D^2/2$	$\pi D^2 h/4$	$\bar{y} = h/2$
Sphere	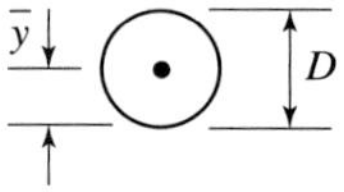	πD^2	$\pi D^3/6$	$\bar{y} = r$
Cone	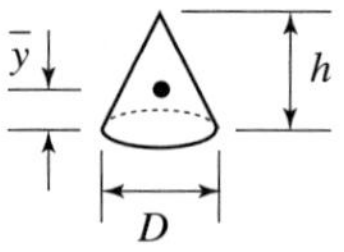	$\pi(r^2 + r\sqrt{r^2 + h^2})$	$\pi D^2 h/12$	$\bar{y} = h/4$
Hemisphere	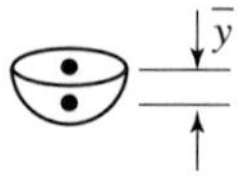	$3\pi D^2/4$	$\pi D^2/12$	$\bar{y} = 3r/8$

K. VECTOR RELATIONS

TABLE K.1 Vector Relationships

$\mathbf{A}\cdot\mathbf{B} = A_xB_x + A_yB_y + A_zB_z$

$\mathbf{A}\times\mathbf{B} = (A_yB_z - A_zB_y)\hat{\mathbf{i}} + (A_zB_x - A_xB_z)\hat{\mathbf{j}} + (A_xB_y - A_yB_x)\hat{\mathbf{k}}$

If $\mathbf{A}\perp\mathbf{B}$, $\mathbf{A}\cdot\mathbf{B}=0$

If $\mathbf{A}\parallel\mathbf{B}$, $\mathbf{A}\times\mathbf{B}=0$

gradient operator $= \boldsymbol{\nabla} = \dfrac{\partial}{\partial x}\hat{\mathbf{i}} + \dfrac{\partial}{\partial y}\hat{\mathbf{j}} + \dfrac{\partial}{\partial z}\hat{\mathbf{k}}$

divergence of $\mathbf{V} = \boldsymbol{\nabla}\cdot\mathbf{V} = \dfrac{\partial u}{\partial x} + \dfrac{\partial v}{\partial y} + \dfrac{\partial w}{\partial z}$

curl of $\mathbf{V} = \boldsymbol{\nabla}\times\mathbf{V} = \left(\dfrac{\partial w}{\partial y} - \dfrac{\partial v}{\partial z}\right)\hat{\mathbf{i}} + \left(\dfrac{\partial u}{\partial z} - \dfrac{\partial w}{\partial x}\right)\hat{\mathbf{j}} + \left(\dfrac{\partial v}{\partial x} - \dfrac{\partial u}{\partial y}\right)\hat{\mathbf{k}}$

Laplace's equation $= \boldsymbol{\nabla}^2\phi = 0$

Irrotational (conservative) vector field: $\boldsymbol{\nabla}\times\mathbf{V} = 0$

Stokes' theorem: $\oint_C \mathbf{V}\cdot d\mathbf{l} = \iint_A (\boldsymbol{\nabla}\times\mathbf{V})\cdot\hat{\mathbf{n}}\,dA$

Gauss' Theorem: $\oiint_A \mathbf{V}\cdot\ddot{\mathbf{n}}\,dA = \iiint_{\mathcal{V}} \boldsymbol{\nabla}\cdot\mathbf{V}\,d\mathcal{V}$

TABLE K.2 Constitutive Equations

Cartesian

$$\sigma_{xx} = -p + 2\mu\frac{\partial u}{\partial x} \qquad \tau_{xy} = \mu\left(\frac{\partial u}{\partial y} + \frac{\partial v}{\partial x}\right)$$

$$\sigma_{yy} = -p + 2\mu\frac{\partial v}{\partial y} \qquad \tau_{yz} = \mu\left(\frac{\partial v}{\partial z} + \frac{\partial w}{\partial y}\right)$$

$$\sigma_{zz} = -p + 2\mu\frac{\partial w}{\partial z} \qquad \tau_{xz} = \mu\left(\frac{\partial u}{\partial z} + \frac{\partial \varphi}{\partial x}\right)$$

Cylindrical

$$\sigma_{rr} = -p + 2\mu\frac{\partial v_r}{\partial r} \qquad \tau_{r\theta} = \mu\left[r\frac{\partial}{\partial r}\left(\frac{v_\theta}{r}\right) + \frac{1}{r}\frac{\partial v_r}{\partial \theta}\right]$$

$$\sigma_{\theta\theta} = -p + 2\mu\left(\frac{1}{r}\frac{\partial v_\theta}{\partial \theta} + \frac{v_r}{r}\right) \qquad \tau_{\theta z} = \mu\left[\frac{\partial v_\theta}{\partial z} + \frac{1}{r}\frac{\partial v_z}{\partial \theta}\right]$$

$$\sigma_{zz} = -p + 2\mu\frac{\partial v_z}{\partial z} \qquad \tau_{rz} = \mu\left[\frac{\partial v_r}{\partial z} + \frac{\partial v_z}{\partial r}\right]$$

Spherical

$$\sigma_{rr} = -p + 2\mu\frac{\partial v_r}{\partial r}$$

$$\sigma_{\partial\theta} = -p + 2\mu\left(\frac{1}{r}\frac{\partial v_\theta}{\partial \theta} + \frac{v_r}{r}\right)$$

$$\sigma_{\phi\phi} = -p + 2\mu\left(\frac{1}{r\sin\theta}\frac{\partial v_\phi}{\partial \phi} + \frac{v_r}{r} + \frac{v_\theta\cos\theta}{r}\right)$$

$$\tau_{r\theta} = \mu\left[r\frac{\partial}{\partial r}\left(\frac{v_\theta}{r}\right) + \frac{1}{r}\frac{\partial v_r}{\partial \theta}\right]$$

$$\tau_{\theta\phi} = \mu\left[\frac{\sin\theta}{r}\frac{\partial}{\partial\theta}\left(\frac{v_\phi}{\sin\theta}\right) + \frac{1}{r\sin\theta}\frac{\partial v_\theta}{\partial \phi}\right]$$

$$\tau_{r\phi} = \mu\left[\frac{1}{r\sin\theta}\frac{\partial v_r}{\partial \phi} + r\frac{\partial}{\partial r}\left(\frac{v_\phi}{r}\right)\right]$$

TABLE K.3 Fundamental Laws for Incompressible Flows

Continuity

Cartesian

$$\frac{\partial u}{\partial x}+\frac{\partial v}{\partial y}+\frac{\partial w}{\partial z}=0$$

Cylindrical

$$\frac{1}{r}\frac{\partial}{\partial r}(rv_r)+\frac{1}{r}\frac{\partial v_\theta}{\partial \theta}+\frac{\partial v_z}{\partial z}=0$$

Spherical

$$\frac{1}{r^2}\frac{\partial}{\partial r}(r^2 v_r)+\frac{1}{r\sin\theta}\frac{\partial}{\partial\theta}(v_\theta\sin\theta)+\frac{1}{r\sin\theta}\frac{\partial v_\phi}{\partial\phi}=0$$

Momentum

Cartesian

$$\frac{Du}{Dt}=-\frac{1}{\rho}\frac{\partial p}{\partial x}+g_x+\nu\nabla^2 u$$

$$\frac{Dv}{Dt}=-\frac{1}{\rho}\frac{\partial p}{\partial y}+g_y+\nu\nabla^2 v$$

$$\frac{Dw}{Dt}=-\frac{1}{\rho}\frac{\partial p}{\partial z}+g_z+\nu\nabla^2 w$$

$$\frac{D}{Dt}=\frac{\partial}{\partial t}+u\frac{\partial}{\partial x}+v\frac{\partial}{\partial y}+w\frac{\partial}{\partial z}$$

$$\nabla^2=\frac{\partial^2}{\partial x^2}+\frac{\partial^2}{\partial y^2}+\frac{\partial^2}{\partial z^2}$$

Cylindrical

$$\frac{Dv_r}{Dt}-\frac{v_\theta^2}{r}=-\frac{1}{\rho}\frac{\partial p}{\partial r}+g_r+\nu\left(\nabla^2 v_t-\frac{v_t}{r^2}-\frac{2}{r^2}\frac{\partial v_\theta}{\partial\theta}\right)$$

$$\frac{Dv_\theta}{Dt}+\frac{v_j v_\theta}{r}=-\frac{1}{\rho^r}\frac{\partial p}{\partial\theta}+g_\theta+\nu\left(\nabla^2 v_\theta+\frac{2}{r^2}\frac{\partial v_r}{\partial\theta}-\frac{v_\theta}{r^2}\right)$$

$$\frac{Dv_z}{Dt}=-\frac{1}{\rho}\frac{\partial p}{\partial z}+g_z+\nu\nabla^2 v_z$$

$$\frac{D}{Dt}=\frac{\partial}{\partial t}+v_r\frac{\partial}{\partial r}+\frac{v_\theta}{r}\frac{\partial}{\partial\theta}+v_z\frac{\partial}{\partial z}$$

$$\nabla^2=\frac{\partial^2}{\partial r^2}+\frac{1}{r}\frac{\partial}{\partial r}+\frac{1}{r^2}\frac{\partial^2}{\partial\theta^2}+\frac{\partial^2}{\partial z^2}$$

Spherical

$$\frac{Dv_r}{Dt}-\frac{v_\sigma^2+v_\phi^2}{r}=-\frac{1}{\rho}\frac{\partial p}{\partial r}+g_\gamma+\nu\left(\nabla^2 v_r-\frac{2v_t}{r^2}-\frac{2}{r^2}\frac{\partial v_\rho}{\partial\theta}-\frac{2v_\theta\cos\theta}{r^2}-\frac{2}{r^2\sin\theta}\frac{\partial v_\phi}{\partial\phi}\right)$$

$$\frac{Dv_\theta}{Dt}+\frac{v_r v_\theta-v_\phi^2\cos\theta}{r}=-\frac{1}{\rho r}\frac{\partial p}{\partial\theta}+g_\theta+\nu\left(\nabla^2 v_\theta+\frac{2}{r^2}\frac{\partial v_r}{\partial\theta}-\frac{v_\theta}{r^2\sin^2\theta}-\frac{2\cos\theta}{r^2\sin^2\theta}\frac{\partial v_\phi}{\partial\phi}\right)$$

$$\frac{Dv_\phi}{Dt}+\frac{v_\phi v_r}{r}+\frac{v_\theta u_\phi\cos\theta}{r}=-\frac{1}{\rho r\sin\theta}\frac{\partial p}{\partial\phi}+g_\phi+\nu\left(\nabla^2 u_\phi-\frac{v_\phi}{r^2\sin^2\theta}+\frac{2}{r^2\sin^2\theta}\frac{\partial v_r}{\partial\phi}+\frac{2\cos\theta\partial v_\theta}{r^2\sin^2\theta\partial\phi}\right)$$

$$\frac{D}{Dt}=\frac{\partial}{\partial t}+v_r\frac{\partial}{\partial r}+\frac{v_\theta}{r}\frac{\partial}{\partial\theta}+\frac{v_\phi}{r\sin\theta}\frac{\partial}{\partial\phi}$$

$$\nabla^2=\frac{1}{r^2}\frac{\partial}{\partial r}\left(r^2\frac{\partial}{\partial r}\right)+\frac{1}{r^2\sin\theta}\frac{\partial}{\partial\theta}\left(\sin\theta\frac{\partial}{\partial\theta}\right)+\frac{1}{r^2\sin^2\theta}\frac{\partial^2}{\partial\phi^2}$$

Energy

Cartesian

$$\rho\frac{Dh}{Dt}=K\nabla^2 T+2\mu\left[\left(\frac{\partial u}{\partial x}\right)^2+\left(\frac{\partial v}{\partial y}\right)^2+\left(\frac{\partial w}{\partial z}\right)^2+\frac{1}{2}\left(\frac{\partial u}{\partial y}+\frac{\partial v}{\partial x}\right)^2+\frac{1}{2}\left(\frac{\partial v}{\partial z}+\frac{\partial w}{\partial y}\right)^2+\frac{1}{2}\left(\frac{\partial u}{\partial z}+\frac{\partial w}{\partial x}\right)^2\right]$$

Cylindrical

$$\rho\frac{Dh}{Dt}=K\nabla^2 T+2\mu\left[\left(\frac{\partial v_r}{\partial r}\right)^2+\left(\frac{1}{r}\frac{\partial v_\theta}{\partial\theta}+\frac{v_r}{r}\right)^2+\left(\frac{\partial v_2}{\partial z}\right)^2+\frac{1}{2}\left(\frac{1}{r}\frac{\partial v_z}{\partial\theta}+\frac{\partial v_\theta}{\partial z}\right)^2+\frac{1}{2}\left(\frac{\partial v_r}{\partial z}+\frac{\partial v_z}{\partial r}\right)^2+\frac{1}{2}\left(\frac{1}{r}\frac{\partial v_r}{\partial\theta}+\frac{\partial v_\theta}{\partial r}-\frac{v_\theta}{r}\right)^2\right]$$

Spherical

$$\rho\frac{Dh}{Dt}=K\nabla^2 T+2\mu\left[\left(\frac{\partial v_r}{\partial r}\right)^2+\left(\frac{1}{r}\frac{\partial v_\theta}{\partial\theta}+\frac{v_r}{r}\right)^2+\left(\frac{1}{r\sin\theta}\frac{\partial v_\phi}{\partial\phi}+\frac{v_r}{r}\frac{v_\theta\cos\theta}{r}\right)^2\right]$$
$$+\mu\left[\left(\frac{1}{r\sin\theta}\frac{\partial v_\theta}{\partial\phi}+\frac{\sin\theta}{r}\frac{\partial}{\partial\theta}\left(\frac{D_\theta}{\sin\theta}\right)\right)^2+\left(\frac{1}{r\sin\theta}\frac{\partial v_t}{\partial\phi}+r\frac{\partial}{\partial r}\left(\frac{v_\phi}{r}\right)\right)^2+\left(r\frac{\partial}{\partial r}\left(\frac{v_\theta}{r}\right)+\frac{1}{r}\frac{\partial v_r}{\partial\theta}\right)^2\right]$$

Answers to Selected Problems

Chapter 1

1.2FE	**C**	
1.4FE	**B**	
1.6FE	**B**	
1.8FE	**D**	
1.10FE	**A**	
1.12FE	**A**	
1.14FE	**B**	
1.2	**(a)** 97.7 N	
1.4	No	
1.7	**(c)** 29.18 kg	**(e)** 2712 W
1.8	97.7 N	
1.10	-5.886 m/s^2	
1.12	39.86 kN	
1.14	**(a)** 10 194 kg	
1.16	7.55 m/s^2	
1.18	**(b)** 5.97×10^{24} kg	
1.20	**(a)** -73.33 kPa, 0.68 m, 0.5 kg, 4.905 N	
1.22	105 kPa, 787 mm, 10.7 m	
1.24	400 kPa	
1.26	396 kPa, 276 kPa, 180 kPa	
1.28	8005 Pa, 4584 N	
1.30	78.8°F, 538.8°R	
1.32	10^7 N/m	
1.34	521 kJ	

Chapter 2

2.2FE	**B**
2.4FE	**C**
2.6FE	**D**
2.8FE	**B**
2.10FE	**A**
2.12FE	**B**
2.14FE	**C**
2.16FE	**A**
2.2	**(a)** 0.0498
2.4	**(a)** 0.7466 ft^3/lbm
2.6	**(a)** 0.1024, 7.383 kPa
2.8	**(a)** 3.86 m^3/kg, 11.59 m^3
2.10	0.002 m^3, 2.76 m^3
2.12	**(a)** 307.2 ft^3
2.14	35.98 kg, 19.51 kg
2.16	0.0466 m^3
2.18	0.01728
2.20	**(a)** 2.506 m^3
2.22	**(b)** 0.0115 m^3
2.23	**(b)** 1.925 ft^3/lbm
2.24	1375 lbm, 1610 lbm
2.26	1620°C
2.28	**(a)** 26.3 kPa
2.30	**(a)** 0.1867 kg
2.32	**(b)** 19 300 kPa
2.33	**(c)** 267 psia

Chapter 3

3.2FE	**C**
3.4FE	**C**
3.6FE	**D**
3.8FE	**B**
3.10FE	**A**
3.12FE	**D**
3.2	153.8 kJ
3.4	**(e)** 28.3 kJ
3.6	**(b)** 248 kJ
3.8	−959 ft-lbf
3.10	2.654 kJ
3.11	**(b)** 0.804 kJ
3.14	2.54×10^8 ft-lbf
3.16	0.132 kJ
3.18	−809 kJ
3.20	−2360 ft-lbf
3.22	36.2 J
3.24	−24,000 ft-lbf
3.28	6250 N/m
3.30	2280 N·m
3.32	1320 W

Chapter 4

4.2FE	**A**
4.4FE	**A**
4.6FE	**C**
4.8FE	**D**
4.10FE	**D**
4.12FE	**D**
4.14FE	**C**
4.16FE	**D**
4.18FE	**C**
4.20FE	**B**
4.22FE	**C**
4.24FE	**C**
4.26FE	**A**
4.28FE	**B**
4.30FE	**D**
4.32FE	**B**
4.34FE	**D**
4.36FE	**B**
4.38FE	**C**
4.40FE	**B**
4.2	675 kJ
4.4	−0.49 Btu
4.5	**(c)** 25 kJ, 15 kJ
4.6	−200 kJ
4.8	−49 J
4.10	377.6 kJ
4.12	−3260 Btu
4.14	1505 kJ
4.16	**(a)** 147°F
4.18	785°C
4.20	**(a)** 787°C
4.22	**(b)** 1531 kJ
4.24	0.617 MPa, 890 kJ
4.26	**(b)** 0.388 Btu/lbm-°R
4.28	**(a)** 402.4 kJ/kg
4.30	418 kJ, 186 kJ
4.32	**(a)** 102°C
4.34	**(a)** 16.1°C
4.36	860 kJ
4.38	1200 kJ, 860 kJ
4.42	1800 kJ
4.44	**(a)** 49°C
4.46	−729 kJ/kg
4.48	753 kJ
4.50	**(a)** 314°C, 1.14 MPa
4.52	**(a)** 1465 kJ **(c)** −291 kJ
4.54	**(a)** 9850 kJ **(c)** 9550 kJ
4.56	−2.22 kJ
4.58	−116 kJ
4.60	174.7°C, 240 kPa
4.62	**(a)** 373.1 kJ
4.64	77.1 kJ
4.66	4.014 kJ
4.68	240 s
4.70	3.65 kg/s, 195 m/s
4.72	6.64 kg/s, 255 m/s
4.74	4.95 kg/s, 109 m/s
4.76	**(a)** 0.01348 kg/s·m^3
4.78	16 m/s
4.80	**(a)** 85.8 kJ/kg

4.82	569°C, 22.3, 600°C
4.84	1110 hp
4.86	558 kW
4.88	4.72 hp
4.90	6000 kW
4.92	21.2 kW
4.94	11,970 hp, 230 fps
4.96	539 kW
4.98	−70.5 kJ/s
4.100	1.55 m/s, 219°C
4.102	1.672 kg/s, −0.91°C
4.104	40,900 lbm/min
4.106	813°C, 32.15 kg
4.108	503 kJ
4.110	284°F, 4.36 lbm
4.112	**(b)** 624 kPa

Chapter 5

5.2FE	**D**
5.4FE	**C**
5.6FE	**B**
5.8FE	**C**
5.10FE	**C**
5.12FE	**D**
5.14FE	**A**
5.16FE	**D**
5.18FE	**C**
5.20FE	**A**
5.2	5.21
5.4	0.3
5.9	Impossible
5.10	12.8 kW
5.12	17.78 kW, 236.4°C
5.14	432.2°C
5.16	6.42%, 15.57
5.18	0.333
5.20	99°C
5.22	177 ft-lbf
5.24	6014 lbm/hr, 1.11 hp
5.26	72°C
5.28	1465 kJ
5.30	8 kJ/s
5.32	0.341 kJ/K
5.34	−15.72 Btu/°R
5.38	**(a)** 0.0445 m^3 **(c)** 0.0460 m^3
5.40	**(a)** 2.31 kJ/K **(c)** 2.40 kJ/K
5.42	**(a)** 0.349 Btu/°R **(c)** 0.352 Btu/°R
5.44	**(a)** 254 m^3 **(c)** 255 m^3
5.46	**(a)** −377 kJ/kg, −1.26 kJ/kg·K **(d)** −5420 kJ/kg, −18.1 kJ/kg·K
5.48	0.473 kJ/K
5.50	1974°R
5.52	**(a)** 812°F, 362 psia
5.54	**(a)** −53.1 kJ
5.56	**(a)** 2.91 kJ/kg·K, 829°C
5.58	**(a)** 3189 kJ/kg, 7.604 kJ/kg·K
5.60	**(a)** 39 kJ, 0.457 kJ/K
5.62	**(a)** 185 kJ
5.64	**(a)** 0°C, 0.064 kJ/K
5.66	0.563
5.68	433 kJ/kg, 24.9%
5.70	**(a)** 19.5 kW
5.76	**(a)** 0.0167 Btu/°R
5.78	**(c)** 0.104 Btu/°R
5.80	**(a)** 7.56 kJ/K
5.82	0.432 Btu/sec-°R
5.84	173.8 m/s
5.86	0.147 kg/s
5.88	**(a)** 2280 hp
5.90	71.3%
5.92	**(a)** 0.198 kg/s
5.94	**(b)** 6.09 lbm/sec
5.96	238 m/s

Chapter 6

6.2FE	**B**
6.4FE	**A**
6.6FE	**B**
6.8FE	**D**
6.10FE	**D**
6.12FE	**D**
6.14FE	**A**
6.16FE	**C**

6.2	28.3%
6.4	**(d)** 53.2 hp
6.6	**(a)** 37%
6.8	13.8 kJ/s, 666 kJ/s
6.10	**(c)** 9.2%
6.12	39.1%
6.14	34.2%
6.16	**(a)** 40.3%
6.18	42%
6.20	35.6%
6.22	45.8%
6.24	38.8%
6.26	12 600 kg/hr
6.28	45.7%
6.30	50.5%
6.32	33.5%
6.34	16 850 kg/hr
6.36	**(d)** 1972 kg/s
6.38	35.7%
6.40	31.5%
6.42	489,600 Btu/hr
6.44	**(b)** 20.8 hp
6.46	**(a)** 28.5 kW
6.48	2.59 kg/s
6.50	**(b)** 0.261 kg/s
6.52	**(c)** 14.84 kW **(e)** 0.940 kg/s
6.54	**(b)** 3.28 hp
6.56	**(c)** 406 hp

Chapter 7

7.2FE	B
7.4FE	A
7.6FE	C
7.8FE	D
7.10FE	B
7.12FE	D
7.14FE	D
7.16FE	B
7.2	522 kJ/kg, 22.2%
7.4	66.4%, 1872, 952 kPa
7.6	45.4%, 207.4 kW, 146.6 kPa
7.8	**(a)** 1.228 kg/s
7.10	104 kW, 214 kPa
7.12	3.376 kJ, 56.5%, 170 hp
7.14	2002 kJ/kg, 1298 kPa
7.16	57.5%, 383 kPa
7.18	**(a)** 254,000 ft-lbf/lbm, 144 psia
7.20	2.822, 63.5%, 144 hp
7.22	2.57, 1430 kPa
7.24	21.0, 5890 kJ/s
7.26	**(a)** 831 kJ/kg, 60%, 60%
7.28	3.29 kg/s
7.30	**(a)** 40.1%, 1419 hp, 0.486
7.32	208 kJ/kg, 734 kJ/kg, 0.221
7.34	**(a)** 0.559, 23.3%
7.36	59.8%
7.38	43.7%
7.40	**(a)** 38.1%, 0.346
7.42	**(a)** 0.518
7.44	**(a)** 51.1%, 0.489
7.46	56%
7.48	46.9%
7.50	**(a)** 1.737, 169 hp
7.52	125 kW, 0.282
7.54	163.5 K, 1.65, 0.354 kg/s
7.56	**(a)** 0.737

Chapter 8

8.2FE	B
8.4FE	C
8.6FE	A
8.8FE	D
8.10FE	A
8.2	63.6%, 75.5°F, 14.8 ft^3/lbm, 43.6 Btu/lbm
8.4	0.01001, 621 kg, 6.22 kg, 31 400 kJ
8.6	78.4%, 0.0461, 101.7°F, 15.63 ft^{3}lbm, 77.5 Btu/lbm
8.8	22.9%
8.10	34.1%
8.12	19.3°C
8.13	**(b)** 25%, 0.0036, −1°C, 29.5 kJ/kg
8.14	19.4°C
8.16	88%
8.18	**(a)** 32%, −7.2 Btu/sec

8.20	**(a)** 101°C, 1.20 kg, −6290 kJ
8.22	609 Btu/lbm, 0.514 lbm/min
8.24	19 kJ/kg, 98 kJ/kg
8.26	**(a)** 0.91 kg/s, 20.9 kJ/s
8.28	**(a)** 0.329 kg/min
8.30	**(a)** 94.2 kg/min
8.32	228 cfm, 367 Btu/min
8.34	**(a)** 73°F, 0.78 lbm/sec

Chapter 9

9.2	17.2, 15.1%, 59°C, 89.8%
9.4	15.7, 18.1%, 54.6°C, 87.0%
9.6	15.34, 18.8%, 53.4°C, 86.0%
9.8	29.0, 6.35%, 43.8°C
9.10	27.8, 6.87%, 41.8°C
9.12	44.1 ft^3/min
9.14	**(a)** 20.7, 50.5°C
9.16	65.4%, 7.78%
9.18	**(a)** 30.8, 76.0%, 40.3°C
9.20	65.9 m^3/min
9.22	11.3%
9.24	21.46
9.26	76.8%
9.28	−393 520 kJ/kmol
9.30	**(b)** −572,500 Btu/lbmol
9.32	**(a)** −968 400 kJ/kg
9.34	**(a)** 44%
9.36	−1 299 000 kJ/kmol
9.38	103.1 psia, −3730 Btu
9.40	−220,600 Btu, 74.5 psia
9.44	**(b)** 1833 K
9.46	1732 K
9.48	4740°R
9.49	2940 K

Chapter 10

10.2FE	**B**
10.4FE	**C**
10.6FE	**B**
10.8FE	**A**
10.10FE	**C**
10.12FE	**B**
10.4	**(a)** L/T^2
10.8	**(a)** 1.25×10^8 N **(e)** 5.2×10^{-2} m^2
10.10	**(a)** 5.555×10^{-5} m/s **(d)** 0.0472 m^3/s
10.12	10.06 lb
10.14	**(a)** 153.6 kPa **(d)** 78.8 kPa
10.16	62.8 kPa, 1.95%
10.18	−59°F, −50.6°C
10.20	2.4 N, 0.0095°
10.22	**(b)** 0.635 kg
10.24	0.014 psf
10.26	2.74, 1.04 hp
10.28	91×10^{-5} ft-lb
10.30	56.1 N·m
10.34	899 m
10.36	**(b)** 4940 fps
10.38	29.6 kPa, 59.3 kPa
10.40	0.13 m
10.42	$2\sigma \cos \beta / \rho g t$
10.44	No
10.48	141°F
10.50	1.226 kg/m^3, 12.03 N/m^3
10.52	0.1339 slug/ft^3, 2.01 slug
10.54	52.7 psig
10.56	**(a)** 15.27 m/s
10.58	69.2°C
10.60	−101 Btu
10.62	**(a)** 191 kJ
10.64	**(b)** 266.9 m/s
10.66	**(a)** 2654 m

Chapter 11

11.2FE	**D**
11.4FE	**A**
11.6FE	**A**
11.8FE	**A**
11.2	**(c)** 1334 Pa
11.4	2.31, 4.48 slug/ft^3
11.6	**(a)** 4.63 kPa
11.8	1.84 kPa

11.10	96.49 kPa, 96.44 kPa
11.12	**(a)** 28.9 kPa
11.14	232,700 ft
11.16	**(a)** 3,373 m
11.18	**(a)** 30.9 m/s **(d)** 148 fps
11.20	10.64 kPa
11.22	15.62 kPa
11.24	14 kPa
11.26	11.12 kPa
11.28	3.32 ft
11.30	6934 N
11.34	**(a)** 251 kN
11.36	**(a)** 740 kN, (0, −.167) m
11.38	523 kN
11.40	**(b)** 52.3 kN
11.42	**(b)** 0.667 m
11.44	**(a)** won't tip
11.46	616 kN
11.48	984 kN
11.50	120 N
11.52	**(b)** 1996 lb, 0.955
11.54	**(a)** 1.372 m
11.56	4.62 in.
11.58	8.8 kN
11.60	1.23×10^6
11.62	**(b)** 16.4 ft
11.64	**(a)** 17.15 psi
11.66	**(b)** 4.8 m/s^2
11.68	**(b)** 1163 kN
11.70	**(b)** 9435 Pa, 8859 Pa, 3924 Pa
11.72	**(a)** 7.62 rad/s **(c)** 7.41 rad/s
11.74	**(a)** 33.15 kPa **(c)** 42.4 kPa

Chapter 12

12.2FE	**C**
12.4FE	**C**
12.6FE	**C**
12.8FE	**A**
12.6	$udy - vdx$
12.8	**(a)** 2 m/s, 3.606 m/s
12.10	**(b)** $x^2y = -2$
12.12	**(a)** $-20\mathbf{k}$ **(c)** $6\mathbf{i} - 2\mathbf{k}$
12.16	**(a)** (8.2, 0, 0) m/s^2
12.18	$a_x = v_r \partial u/\partial r + u \partial u/\partial x$
12.20	−0.369 °C/s
12.22	−2500 kg/s·m^2
12.24	$\mathbf{V}\cdot\nabla + \partial/\partial t$
12.26	**(a)** $160\mathbf{j} - 600\mathbf{i}$ m/s^2
12.28	**(d)** 2D, (r, z)
12.30	(b)
12.32	(f) and (h)
12.34	(d) and (e)
12.36	Re = 39 000. Turb
12.38	Re = 14 100. Turb
12.40	**(a)** 3 cm
12.42	**(b)** M = 0.235. Incomp
12.44	$D\rho/Dt = 0$. Incomp
12.46	12.1 N
12.48	**(a)** $p_0 - 50x^2\rho$
12.50	**(b)** $\rho U_2/2$
12.52	6.32 m/s
12.54	**(b)** 4.97 m/s **(d)** 16.44 fps
12.56	12.76 m/s
12.58	**(b)** −6060 Pa
12.60	**(b)** 3.516 m/s, 193.6 kPa
12.62	**(b)** $\sqrt{[2g(H - h)]}$
12.64	51 m
12.66	−36.9 Pa
12.68	**(a)** Re $\approx 4 \times 10^4$. Separate
12.70	40 kPa
12.72	$>, <, >$

Chapter 13

13.2FE	**D**
13.4FE	**D**
13.6FE	**C**
13.8FE	**C**
13.10FE	**A**
13.12FE	**C**
13.14FE	**A**
13.12	$dm_v/dt = dm/dt - \rho Q_f$
13.14	3.968 slug/sec, 2.045 ft^3/sec
13.16	0.1509 m^3/s
13.18	319.8 m/s, 20.89 kg/s
13.20	**(a)** 3.33 m/s, 16.75 kg/s, 0.01675 m^3/s

13.22	**(a)** $12(1 - 576r^2)$ fps
	(c) $4(1 - 10\,000r^2)$ m/s
13.24	0.82 kg/s
13.26	4.528 slug/sec, 4.6 slug/sec
13.28	3.01×10^{-5} slug/sec-ft^3
13.30	2.57 kg/s
13.32	12.5 fps
13.34	0.008 m/s
13.36	**(a)** 1.11 mm/s
13.38	0.836°C
13.40	**(a)** 69.85 kW
13.42	7.93 ft, 1.76 ft
13.44	7.41 m/s
13.46	1.571 m
13.48	**(a)** 394.4 kPa
13.50	**(a)** 0.0374 m^3/s
13.52	**(a)** 0.0365 m^3/s **(c)** 1.318 cfs
13.54	**(a)** 30°C
13.56	**(a)** 0.131 m
13.58	23.5 hp
13.60	1.304 MW
13.62	**(a)** 299°C
13.64	5390 W
13.66	115 hp
13.68	0.815 m
13.70	**(b)** 1.056
13.72	**(a)** 0.32 m^3/s
13.74	10.18 kN
13.76	1072 lb
13.78	353 kN
13.80	618 kN
13.82	**(a)** 8.86 m/s
13.84	135 kPa
13.86	103 N, 759 N
13.88	**(a)** 55 fps **(c)** 25 fps
13.90	**(a)** 305 lb, 528 lb
	(b) 76.2 lb, 686 lb
13.92	**(b)** 679 N, 392 N
13.94	**(b)** 47°, 29.5°, 3.6 MW
13.96	**(a)** 11.08 kN, 240 kg/s, 80 kg/s
13.98	1.333 m/s
13.100	458 m
13.102	141 N
13.104	191 ft/sec per ft
13.106	3780 N
13.108	**(a)** 54.9 kW/m of width

Chapter 14

14.2FE	A
14.4FE	A
14.6FE	A
14.2	**(a)** FT/L **(f)** FL/T
14.4	$\rho Vl/\mu = C$
14.6	$V = \sqrt{gH/C}$
14.8	$F_D/\rho l^2 V^2 = f(d/l, \mu/\rho lV)$
14.10	$h/d = f(\sigma/\gamma d^2, \beta)$
14.12	$\sigma = CMy/I$
14.14	$V = C\sqrt{gH}$
14.16	$\Delta p/\rho V^2 = f(\nu/Vd, L/d, e/d)$
14.18	$Q_f/\sqrt{gR^5} = f(A/R^2, e/R, S)$
14.20	$F_D/\rho V^2 d^2 = f(\mu/\rho Vd, e/d, I)$
14.22	$F_D/\rho V^2 d^2 = f(\mu/\rho Vd, e/d, r/d, Cd^2)$
14.24	$F_L/\rho V^2 l^2 = f(c/V, t/l_c, \alpha)$
14.26	$F_D/\rho V^2 d^2 = f(\mu/\rho Vd, d/L, \rho/\rho_c, V/fd)$
14.28	$T/\rho\Omega^2 d^5 = f(f/\Omega, H/d, l/d, N, h/d)$
14.30	$d/D = f(V/V_j, \sigma/\rho V_j^2 D, \mu/\rho V_j D, \rho_a/\rho)$
14.32	$\mu/\rho VD = f(H/D, l/D, gD/V^2)$
14.34	$y_2/y_1 = f(gy_1/V_1^2)$
14.38	**(b)** 112 kg/s, 7390 kPa
14.40	1000 km/hr. Too high
14.42	33.1 m/s, 3.56
14.44	Wind tunnel
14.46	1.29 m/s, 2.16 MN
14.48	63.2 fps, 800 lb
14.50	278
14.52	$V_m > 37.5$ m/s
14.54	$V_m > 250$ m/s. A possible study
14.56	276 m/s, 262 m/s, 34.6 kPa abs
14.58	750 N·m, 12.5 rpm

Chapter 15

15.2FE	A
15.4FE	D
15.6FE	D
15.8FE	B
15.10FE	B
15.12FE	C
15.2	**(c)** 0.005 m/s
15.5	**(a)** 0.1 m/s
15.6	**(b)** 12.6 m
15.8	7.2 m, developed
15.10	3.7 m, 0.72 m
15.12	**(b)** 4.8 m, 0.4 m
15.16	268 ft, 0.04
15.18	**(b)** 510 Pa
15.20	0.091 ft, 3300 psf
15.22	0.704 mm
15.24	2210, 1.1 Pa
15.26	$R/\sqrt{2}$
15.28	**(a)** $0.707r_0$
15.30	0.0188 ft, 0.00163 psf
15.32	0.426 m/s, 0.0054 Pa
15.36	**(a)** 0.0274 $N{\cdot}s/m^2$
15.38	12.1 m^3/s, 60.4 m/s
15.40	0.028 m^3/s
15.42	0.0556 cfs
15.44	**(a)** 18.5 m/s
15.48	7.9 N·m
15.50	18.9 W
15.52	12.6 W
15.58	0.146 ft^2/sec, −0.118
15.60	**(b)** rough
15.62	**(a)** 0.0251 m/s
15.64	21.7 fps
15.66	−40.8 kPa/m
15.68	24.2 m/s
15.70	**(a)** 1.12 m/s, 0.0127 m^3/s
15.72	**(b)** 0.0146
15.74	**(a)** 0.064
15.76	**(b)** 59 ft
15.78	**(b)** 29.4 kPa
15.80	11,200 psf
15.82	147 kPa
15.84	**(a)** 0.0027 m^3/s
15.86	**(a)** 0.052 kg/s
15.88	**(a)** 0.032 m
15.90	**(a)** 0.121 m
15.92	0.000143 m^3/s
15.94	**(a)** 0.0351 m^3/s
15.98	**(a)** 52.6 kPa
15.100	11.7
15.102	0.0044 m^3/s
15.104	**(a)** 0.019 m^3/s
15.106	1.3 min
15.108	46.1 m/s
15.110	oscillatory flow
15.112	13.0 m, 117 kW
15.114	37.9 ft, 251 hp
15.116	0.23 m^3/s, 190 kW
15.118	**(a)** 195 kW
15.120	1.05 MW
15.122	0.921 Pa
15.124	**(b)** 171 cfs
15.126	**(a)** 1.71 m
15.128	**(b)** 1.38 m
15.130	0.389 m

Chapter 16

16.2FE	C
16.4FE	B
16.6FE	C
16.8FE	B
16.2	3.78×10^{-5} m
16.6	**(a)** 7.9×10^6
16.8	2659 N, 2.36
16.10	**(a)** 3.25×10^{-7} N
16.14	0.5, 1.46×10^5
16.16	**(a)** 1380 N, 25.7 kN·m
16.18	36 kN
16.20	**(a)** 27 fps
16.22	**(a)** 10 hp
16.24	**(a)** 93.6 N
16.26	$8800
16.28	1.24 hp

16.30	**(a)** 1.58 m/s
16.32	\$683
16.34	125 km/hr
16.36	0.24 in.
16.38	0.191 m/s
16.40	0.822 N, 70.8%
16.42	1.04 hp, 0.26 hp
16.44	20.6 m/s
16.46	1280 lb, No
16.48	0.69 m
16.50	16.7%
16.52	**(a)** 34.5 m/s
16.54	114%
16.56	69.8 m/s
16.58	2.38×10^6 kg, 657 hp
16.60	$100y - 50x$
16.62	$-40 \tan^{-1} y/x$
16.64	**(b)** $10y + 10 \tan^{-1} y/x$
16.66	**(a)** (−1 in., 0)
16.68	$(\sqrt{2}, 0), (-\sqrt{2}, 0)$
16.70	**(a)** 0.314 m
16.72	**(b)** (−104, 0) m^2/s
16.74	**(b)** $-8 \sin \theta$
16.76	100 rad/s, −3660 Pa
16.78	472,000 lb
16.80	**(a)** 3.74 in.
16.82	**(b)** 10 cm
16.84	$20 - 200 \sin^2 x/2$
16.86	$16/(2.67 - x)$ m/s
16.90	**(a)** $4.79\sqrt{\nu x/U_\infty}$
16.92	$0.451 \rho U_\infty^2 \mathrm{Re}_x^{-1/2}$
16.94	2.1 cm, 1 mm
16.96	**(a)** $1.74\sqrt{\nu x/U_\infty}$, 1.2%
16.98	**(a)** 0.0221 m
16.100	**(b)** 286 Pa
16.102	**(a)** 0.31 lb
16.104	**(a)** 235 m, 0.0618 Pa
16.106	**(b)** 0.229 psf
16.108	163 kN, 0.89 m

Chapter 17

17.2	$Rk/(k-1)$
17.6	$\sqrt{RT}$
17.8	1453 m/s
17.10	0.0069 sec
17.12	309 m
17.14	3.776 sec
17.16	−0.113 fps, 0.021°F
17.18	**(a)** 77.3 m/s
17.20	**(a)** 0.1473 kg/s
17.22	**(a)** 0.1423 kg/s
17.24	7.29 kg/s
17.26	0.202 slug/sec
17.28	97.4 kPa, 199.4 kPa abs
17.30	0.102 slug/sec
17.32	494 kPa, 4.29 kPa abs
17.34	0.319 ft, 684 fps
17.36	8.16 cm
17.38	22.1 cm^2, 772°C
17.40	1260 m/s
17.42	412 kN
17.44	**(a)** 3.77 kg/m^3, 808 kPa abs
17.48	908 m/s, 1600 kPa abs
17.50	391 kPa abs, 448°C
17.52	102 kPa abs, 0.471 kg/s
17.54	49.7 psia, 26.3 psia, 1020 fps
17.56	**(a)** 1.49, 620 m/s
17.58	780 m/s
17.60	555 kPa, 413 kPa abs
17.62	39.4°, 867 m/s
17.64	**(a)** 14.24 kPa, 26.4 kPa abs
17.66	0.0854, 0.010

Index